# Handbook of Diffusion MR Tractography
Imaging Methods, Biophysical Models, Algorithms and Applications

# Handbook of Diffusion MR Tractography

Imaging Methods, Biophysical Models, Algorithms and Applications

Edited by

**Flavio Dell'Acqua**
NATBRAINLAB, Department of Neuroimaging & Department of Forensics and Neurodevelopmental Sciences, Institute of Psychiatry, Psychology and Neuroscience, King's College London, London, United Kingdom

**Maxime Descoteaux**
Department of Computer Science, Faculty of Science, Université de Sherbrooke, Sherbrooke, QC, Canada; Sherbrooke Connectivity Imaging Laboratory, Sherbrooke, QC, Canada

**Alexander Leemans**
PROVIDI lab, Image Sciences Institute, University Medical Center Utrecht, Utrecht, The Netherlands

Academic Press is an imprint of Elsevier
125 London Wall, London EC2Y 5AS, United Kingdom
525 B Street, Suite 1650, San Diego, CA 92101, United States
50 Hampshire Street, 5th Floor, Cambridge, MA 02139, United States

Copyright © 2025 Elsevier Ltd. All rights are reserved, including those for text and data mining, AI training, and similar technologies.

Publisher's note: Elsevier takes a neutral position with respect to territorial disputes or jurisdictional claims in its published content, including in maps and institutional affiliations.

For accessibility purposes, images in this book are accompanied by alt text descriptions provided by Elsevier.

No part of this publication may be reproduced or transmitted in any form or by any means, electronic or mechanical, including photocopying, recording, or any information storage and retrieval system, without permission in writing from the publisher. Details on how to seek permission, further information about the Publisher's permissions policies and our arrangements with organizations such as the Copyright Clearance Center and the Copyright Licensing Agency, can be found at our website: www.elsevier.com/permissions.

This book and the individual contributions contained in it are protected under copyright by the Publisher (other than as may be noted herein).

**Notices**
Knowledge and best practice in this field are constantly changing. As new research and experience broaden our understanding, changes in research methods, professional practices, or medical treatment may become necessary.

Practitioners and researchers must always rely on their own experience and knowledge in evaluating and using any information, methods, compounds, or experiments described herein. In using such information or methods they should be mindful of their own safety and the safety of others, including parties for whom they have a professional responsibility.

To the fullest extent of the law, neither the Publisher nor the authors, contributors, or editors, assume any liability for any injury and/or damage to persons or property as a matter of products liability, negligence or otherwise, or from any use or operation of any methods, products, instructions, or ideas contained in the material herein.

ISBN 978-0-12-818894-1

For information on all Academic Press publications
visit our website at https://www.elsevier.com/books-and-journals

Publisher: Mara Conner
Acquisitions Editor: Tim Pitts
Editorial Project Manager: Emily Thomson
Production Project Manager: Kamesh R
Cover Designer: Greg Harris

Typeset by STRAIVE, India

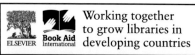

# Contents

Contributors    xv
Preface    xxi

## Part I
## From anatomy to tractography

### 1. The brain and its pathways

*Marco Catani*

| | |
|---|---|
| 1. Introduction | 3 |
| 2. Overview of the nervous system | 4 |
| 3. From molecules to neuronal circuits | 5 |
| 4. Cytoarchitectonic and neuronal connectivity | 7 |
| 5. White matter anatomy | 9 |
| 6. Imaging networks with diffusion MRI tractography | 12 |
| 7. Interhemispheric tract asymmetry during brain development and aging | 13 |
| 8. Conclusions | 15 |
| Acknowledgments | 15 |
| References | 15 |

### 2. Neurobiology of connections

*Giorgio M. Innocenti
and Kathleen S. Rockland*

| | |
|---|---|
| 1. Introduction | 19 |
| 2. The gray matter | 19 |
| 2.1 Cortical areas | 19 |
| 2.2 Cortical layers | 21 |
| 2.3 Vertical organization | 21 |
| 2.4 Gyrification | 22 |
| 2.5 Development of gray matter and of gyri | 22 |
| 2.6 Connections (outputs) | 22 |
| 2.7 Connections (inputs) | 23 |
| 2.8 Receptors | 24 |
| 3. The white matter | 25 |

| | |
|---|---|
| 4. The axon | 27 |
| 4.1 Mapping | 28 |
| 4.2 Differential amplification | 28 |
| 4.3 Development of synaptic connections | 29 |
| 4.4 Temporal transformations | 30 |
| 4.5 Axon diameter in development | 32 |
| 5. Conclusions | 33 |
| References | 34 |

### 3. Past and present of mapping brain connections

*Flavio Dell'Acqua, Mattia Veronese,
Ottavia Dipasquale, Daniel Martins,
Fedal Saini, and Marco Catani*

| | |
|---|---|
| 1. Early connectivity maps | 41 |
| 2. Cortical mapping and connectivity | 44 |
| 3. Associationist school and disconnection syndromes | 46 |
| 4. Modern connectivity | 47 |
| 5. Contemporary neuroimaging methods | 49 |
| 6. Conclusions | 51 |
| References | 52 |

### 4. From Brownian motion to the brain connectome: My perspective and historical view of diffusion MRI

*Denis Le Bihan*

| | |
|---|---|
| 1. Introduction | 55 |
| 2. The birth of diffusion MRI | 55 |
| 2.1 The concept | 55 |
| 2.2 The making | 56 |
| 2.3 First trials | 57 |
| 3. IVIM: Generalized diffusion MRI | 60 |
| 4. How IVIM and diffusion MRI entered the clinical field | 61 |

**vi** Contents

5. The emergence of DTI and tractography    63
    5.1 Other NIH "firsts"    66
6. Water, the forgotten biological molecule    67
7. IVIM and diffusion fMRI    69
    7.1 IVIM fMRI    69
    7.2 Diffusion fMRI    70
8. A global picture for the brain connectome    70
9. Conclusion    72
References    72

# Part II
# Diffusion MRI

## 5. Physics of diffusion imaging: Fundamentals

*Dmitry S. Novikov, Els Fieremans, and Hong-Hsi Lee*

1. Diffusion basics    77
    1.1 Langevin dynamics    77
    1.2 Central limit theorem    78
    1.3 Diffusion equation    80
    1.4 Diffusion in the presence of tissue microstructure    80
2. How to measure diffusion with NMR?    81
    2.1 Scales    81
    2.2 The mesoscopic Bloch-Torrey equation    82
    2.3 The diffusion-weighted signal    82
    2.4 Pulsed gradients    83
3. Diffusion as coarse-graining    86
    3.1 The double average    86
    3.2 The three regimes    88
4. Coarse-graining over an axon    89
    4.1 Axon caliber scale    89
    4.2 Undulation scale    90
    4.3 Axon as a "stick"    91
5. From microstructure to fiber orientations    92
    5.1 White matter dMRI signal as a convolution    92
    5.2 The Standard Model of diffusion in white matter    93
    5.3 Factorizing fiber ODF and fascicle response: Rotational invariants    94
    5.4 Specificity and relevance of SM parameters    96
6. Concluding remarks    96
References    97

## 6. Diffusion MRI acquisition for tractography: Diffusion encoding

*Jennifer S.W. Campbell, Steven H. Baete, Julien Cohen-Adad, J.-Donald Tournier, Filip Szczepankiewicz, Christian Beaulieu, Corey A. Baron, Merry Mani, Kawin Setsompop, Congyu Liao, Christine L. Tardif, Sjoerd B. Vos, Anastasia Yendiki, Ilana R. Leppert, Els Fieremans, and G. Bruce Pike*

1. Diffusion contrast    103
    1.1 Qualitative description of the origin of diffusion contrast    103
    1.2 Quantitative description of the origin of diffusion contrast    105
2. The diffusion encoding block    107
    2.1 Hardware for diffusion MRI    107
    2.2 Choice of $b$-value and sampling scheme for tractography    109
    2.3 Diffusion encoding related artifact reduction    112
    2.4 Diffusion encoding adaptations    116
3. Terminology guide    117
References    118

## 7. Diffusion MRI acquisition for tractography: Diffusion sequences

*Jennifer S.W. Campbell, Steven H. Baete, Julien Cohen-Adad, J.-Donald Tournier, Filip Szczepankiewicz, Christian Beaulieu, Corey A. Baron, Merry Mani, Kawin Setsompop, Congyu Liao, Christine L. Tardif, Sjoerd B. Vos, Anastasia Yendiki, Ilana R. Leppert, Els Fieremans, and G. Bruce Pike*

1. Introduction and a brief history    123
2. Single-shot diffusion sequences    124
3. Reduced sampling schemes    127
    3.1 Parallel imaging    127
    3.2 Partial Fourier acquisition    128
4. Segmented diffusion sequences    129
5. Simultaneous multislice and volumetric acquisitions    131
6. Imaging-related artifact reduction    132
    6.1 Ghosting    132
    6.2 Fat    134
    6.3 Cerebrospinal fluid effects    134
    6.4 Motion    135
    6.5 Other artifacts    135
7. Multicontrast sensitization of the diffusion sequence    135
8. The trade-off between spatial and angular resolution in tractography acquisition    137
9. Terminology guide    138
References    138

## 8. Diffusion MRI acquisition for tractography: Beyond the in vivo adult human brain

*Jennifer S.W. Campbell, Steven H. Baete, Julien Cohen-Adad, J.-Donald Tournier, Filip Szczepankiewicz, Christian Beaulieu, Corey A. Baron, Merry Mani, Kawin Setsompop, Congyu Liao, Christine L. Tardif, Sjoerd B. Vos, Anastasia Yendiki, Ilana R. Leppert, Els Fieremans, Alberto De Luca, Alexander Leemans, and G. Bruce Pike*

| | |
|---|---|
| 1. Ex vivo and preclinical diffusion imaging for brain tractography | 143 |
| 2. Pediatrics, neonates, and fetal diffusion imaging | 144 |
| 3. Diffusion imaging for tractography outside the brain | 145 |
| 4. Terminology guide | 147 |
| References | 148 |

## 9. Diffusion MRI processing and estimation

*Chantal M.W. Tax, Okan Irfanoglu, Jesper Andersson, and Alexander Leemans*

| | |
|---|---|
| 1. Introduction | 151 |
| 2. Correcting for artifacts | 151 |
|   2.1 Data quality assessment | 152 |
|   2.2 Signal drift | 153 |
|   2.3 Gibbs ringing | 154 |
|   2.4 Subject motion | 155 |
|   2.5 Geometric distortions | 157 |
|   2.6 Noise | 161 |
|   2.7 Outliers | 164 |
| 3. Estimation | 165 |
|   3.1 DTI and DKI | 165 |
|   3.2 Spherical deconvolution approaches | 166 |
|   3.3 Multicompartment models | 167 |
|   3.4 q-Space approaches | 168 |
| 4. Conclusion | 169 |
| References | 169 |

## 10. Single-shell diffusion models: From DTI to HARDI

*Flavio Dell'Acqua, Alexander Leemans, Matthew Dawson, and Maxime Descoteaux*

| | |
|---|---|
| 1. Introduction | 177 |
| 2. From diffusion anisotropy to diffusion tensor imaging | 177 |
|   2.1 Diffusion tensor imaging | 178 |
|   2.2 Computing the diffusion tensor | 179 |
|   2.3 Diffusion tensor metrics | 179 |
|   2.4 Diffusion tensor orientation | 181 |
|   2.5 Diffusion tensor limitations | 181 |
| 3. High angular resolution diffusion imaging (HARDI) | 182 |
|   3.1 Multitensor and multicompartmental approaches | 183 |
|   3.2 Nonparametric approaches | 184 |
|   3.3 Q-ball imaging | 185 |
|   3.4 Persistent angular structure and diffusion orientation transform | 187 |
| 4. Spherical deconvolution | 188 |
|   4.1 Solving the deconvolution problem | 189 |
|   4.2 Fiber response function | 191 |
|   4.3 Constrained spherical deconvolution | 193 |
|   4.4 Richardson-Lucy spherical deconvolution | 193 |
| 5. Optimal strategies for single-shell models | 195 |
| 6. Conclusion: Why single-shell? | 196 |
| References | 196 |

## 11. Multishell models

*Jan Morez, Alberto De Luca, Maximillian Pietsch, Daan Christiaens, Steven H. Baete, Marco Reisert, and Ben Jeurissen*

| | |
|---|---|
| 1. Introduction | 201 |
| 2. What is multishell data? | 202 |
| 3. Modeling of multishell data for fiber tracking | 203 |
|   3.1 Models of diffusion | 204 |
|   3.2 Models of fibrous tissue | 207 |
| 4. Beyond multishell acquisitions | 215 |
| 5. Summary and conclusion | 215 |
| References | 218 |

## 12. From diffusion models to fiber orientations

*Richard Stones, Maxime Descoteaux, and Flavio Dell'Acqua*

| | |
|---|---|
| 1. Introduction | 221 |
| 2. Discrete fiber populations | 221 |
| 3. Orientation distribution functions | 223 |
|   3.1 Diffusion orientation distribution function | 223 |
|   3.2 Fiber orientation distribution function | 225 |
| 4. ODF representation | 226 |
| 5. Extracting information from the ODF | 227 |

**viii** Contents

| | | |
|---|---|---|
| 5.1 | Local fiber orientation | 227 |
| 5.2 | Extracting fiber-specific information and fixels | 227 |
| 5.3 | Fiber dispersion and structural complexity | 228 |
| **6.** | **Additional orientation distribution functions** | **232** |
| 6.1 | Uncertainty ODF | 232 |
| 6.2 | Track orientation distribution function | 232 |
| **7.** | **Advanced processing of ODFs** | **232** |
| **8.** | **fODFs from nondiffusion methods** | **234** |
| **9.** | **Conclusion** | **234** |
| | **References** | **235** |

# Part III
# Tractography algorithms

## 13. Deterministic fiber tractography

*Alexander Leemans, Flavio Dell'Acqua, and Maxime Descoteaux*

| | | |
|---|---|---|
| **1.** | **Historical background** | **241** |
| **2.** | **Key ingredients of fiber tractography** | **241** |
| 2.1 | Overview | 241 |
| 2.2 | Mathematical framework of streamline FT | 243 |
| 2.3 | User-defined FT settings | 246 |
| 2.4 | Beyond streamlines | 248 |
| **3.** | **Methodological considerations** | **249** |
| **4.** | **Conclusion** | **251** |
| | **References** | **251** |

## 14. Probabilistic tractography

*Gabriel Girard, Dogu Baran Aydogan, Flavio Dell'Acqua, Alexander Leemans, Maxime Descoteaux, and Stamatios N. Sotiropoulos*

| | | |
|---|---|---|
| **1.** | **Introduction** | **257** |
| **2.** | **Orientation dispersion and uncertainty** | **259** |
| **3.** | **The main idea of probabilistic propagation and tracking** | **261** |
| **4.** | **Uncertainty-based probabilistic tractography** | **262** |
| 4.1 | Bootstrapping | 262 |
| 4.2 | Bayesian inference | 263 |
| 4.3 | Uncertainty of what? | 264 |
| **5.** | **Dispersion-based probabilistic tractography** | **264** |
| 5.1 | Streamline propagation | 264 |

| | | |
|---|---|---|
| **6.** | **Comparing probabilistic trajectories** | **266** |
| 6.1 | Synthetic data | 266 |
| 6.2 | In vivo data | 267 |
| **7.** | **Interpretation: What do these probabilistic estimates mean?** | **269** |
| 7.1 | Confidence intervals | 269 |
| 7.2 | Proportion of estimated connecting fiber | 270 |
| **8.** | **Conclusion** | **270** |
| | **References** | **271** |

## 15. Geodesic tractography

*Luc Florack, Rick Sengers, and Andrea Fuster*

| | | |
|---|---|---|
| **1.** | **Introduction** | **275** |
| **2.** | **The Riemann-DTI paradigm** | **276** |
| 2.1 | Heuristics | 276 |
| 2.2 | Mathematical formulation | 279 |
| 2.3 | Connecting geometry to DTI and DWI | 285 |
| 2.4 | The effect of data variability on geometry | 287 |
| **3.** | **Conclusion** | **289** |
| | **Appendix: Basic concepts from linear algebra and tensor calculus** | **289** |
| | **Acknowledgments** | **292** |
| | **References** | **292** |

## 16. Global tractography

*Alessandro Daducci, Simona Schiavi, Daan Christiaens, Robert Smith, and Daniel C. Alexander*

| | | |
|---|---|---|
| **1.** | **Introduction** | **297** |
| **2.** | **What does "global" really mean?** | **299** |
| 2.1 | Can we ever be truly global? | 299 |
| **3.** | **The global tractography paradigm** | **301** |
| 3.1 | "Generative" vs. "discriminative" approaches | 302 |
| **4.** | **Generative approaches** | **303** |
| 4.1 | Spin glass models | 304 |
| 4.2 | Gibbs tracker | 304 |
| 4.3 | Multitissue and multicompartment models | 305 |
| 4.4 | Streamline-wise approaches | 305 |
| **5.** | **Discriminative approaches** | **306** |
| 5.1 | Methods that filter streamlines | 306 |
| 5.2 | Methods that estimate streamline contributions | 308 |
| **6.** | **Extending the capabilities of global methods** | **309** |
| 6.1 | Adding advanced models of tissue microstructure | 309 |
| 6.2 | Adding multimodal data | 310 |

Contents **ix**

6.3 Adding anatomical information about white matter organization 311
7. Conclusion 312
Acknowledgments 312
References 313

## 17. Machine learning in tractography

*Peter Neher, Philippe Poulin, Daniel Jörgens, Marco Reisert, Itay Benou, and Klaus Maier-Hein*

1. Introduction 315
2. Training and validation data 317
   2.1 Hardware phantom, simulated, and ex vivo data 317
   2.2 In vivo human datasets 321
3. Current applications of machine learning in tractography 323
   3.1 Local modeling 324
   3.2 Sequence-based modeling 327
   3.3 Global modeling 331
   3.4 Streamline classification 333
4. What could be gained, what did we learn, and what challenges remain? 338
   4.1 General considerations on ML-based tractography 338
   4.2 Considerations on the state of the art in ML-based tractography validation 339
   4.3 Considerations on suitable datasets for ML-based tractography 341
5. Conclusion and take-home message 341
References 341

## 18. Improving tractography using anatomical priors and multimodal integration

*Etienne St-Onge, Gabriel Girard, Kurt G. Schilling, Alessandro Daducci, Samuel Deslauriers-Gauthier, Laurent Petit, and Maxime Descoteaux*

1. Introduction 347
2. Guiding tractography with tissue maps 347
   2.1 Binary tissue masks 348
   2.2 Probabilistic tissue maps 349
   2.3 Cortical surfaces 351
3. Anatomical constraints 352
   3.1 ROI-based tractography 352
   3.2 Bundle-specific tractography 353
4. Tractography with microstructure information 355
   4.1 Filtering tractograms with microstructure information 355
   4.2 Guiding trajectories using microstructure information 356
5. Tractography with functional maps 356
   5.1 Functional regions 356
   5.2 Functional connectivity 357
   5.3 Functional local orientation 357
6. Conclusion 358
References 359

## 19. Tractography in pathological anatomy: Some general considerations

*Guillaume Theaud, Manon Edde, Alexander Leemans, Flavio Dell'Acqua, Joseph Yuan-Mou Yang, and Maxime Descoteaux*

1. Introduction 363
2. Technical considerations for tractography within pathological brains 363
3. Practical considerations for tractography within pathological brains 365
   3.1 Real-time tractography as a visualization and data exploration tool 365
   3.2 The diffusion tensor imaging tractography pipeline 365
   3.3 Tractography pipelines robust to crossing fibers 368
4. Challenges of implementing tractography pipelines in pathological brains 370
   4.1 Semiautomatic and manual lesion filling 370
   4.2 Atlas-based WM masking 371
5. What can I do with my tractogram in practice? 371
   5.1 Interactive visualization 371
   5.2 Virtual white matter bundle dissection 371
   5.3 Connectomic analysis 372
6. Which tractography pipeline can I use in my application? 372
   6.1 Disorders consisting of white matter lesions 373
   6.2 Disorders consisting of cortical-based lesions 373

x Contents

6.3 Disorders with marked brain anatomy distortion — 375
6.4 Disorders with microscopic disease infiltrates and perilesional WM changes — 375
7. **Conclusion** — **376**
**Appendix** — **377**
**References** — **377**

## 20. Tractography visualization

*Maxime Chamberland, Charles Poirier,*
*Tom Hendriks, Dmitri Shastin,*
*Anna Vilanova, and Alexander Leemans*

1. **Introduction** — **381**
2. **Tractography rendering from a scientific visualization point-of-view** — **381**
   2.1 The many flavors of streamline rendering — 381
   2.2 Leveraging the graphical processing unit — 384
3. **Interacting with tractograms** — **384**
   3.1 Selecting streamlines with regions of interest — 385
   3.2 Interactively slicing tractograms via selection planes — 385
   3.3 Real-time tractography — 386
   3.4 Placing tractography in context — 387
4. **Advanced tractography visualization** — **388**
   4.1 Along-tract mapping and profiling — 388
   4.2 Web-based visualization — 388
   4.3 Uncertainty mapping — 388
   4.4 Immersive virtuality — 389
   4.5 Photorealistic rendering — 389
5. **Conclusion** — **390**
**References** — **390**

# Part IV
# From streamlines to tracts

## 21. Dissecting white matter pathways: A neuroanatomical approach

*Stephanie J. Forkel, Cesare Bortolami,*
*Lilit Dulyan, Rachel L.C. Barrett, and*
*Ahmad Beyh*

1. **Principles of anatomically guided manual dissections** — **398**
   1.1 Optimal diffusion maps for ROI delineation — 398

1.2 Tractography in clinical populations—Stroke and neurosurgical patients — 399
1.3 Anatomical placement of ROIs — 401
2. **Anatomical delineation in nonhuman primates** — **402**
3. **Extracting statistical indices** — **402**
4. **Atlas of neuroanatomical dissections** — **404**
   4.1 Superior longitudinal fasciculus — 404
   4.2 Cingulum — 404
   4.3 Uncinate fasciculus — 405
   4.4 Inferior longitudinal fasciculus — 405
   4.5 Inferior fronto-occipital fasciculus — 405
   4.6 Arcuate fasciculus — 406
   4.7 Frontal aslant tract — 406
   4.8 Fronto-insular tracts — 406
   4.9 Corticospinal tract — 408
   4.10 Anterior thalamic radiations — 408
   4.11 Frontostriatal projections — 408
   4.12 Pre-postcentral U-shaped fibers — 410
   4.13 Optic radiation (geniculocalcarine fasciculus) — 410
   4.14 Medial occipital longitudinal tract — 410
   4.15 Corpus callosum — 412
   4.16 Anterior commissure — 412
   4.17 Frontomarginal and fronto-orbito-polar tracts — 412
   4.18 Fornix — 414
   4.19 Vertical occipital fasciculus — 414
   4.20 Accumbofrontal pathway — 415
**Funding** — **416**
**References** — **416**

## 22. Dissecting white matter pathways: Automatic and semiautomatic approaches

*Eleftherios Garyfallidis,*
*Marc-Alexandre Côté, Francois Rheault,*
*and Emanuele Olivetti*

1. **Introduction** — **423**
2. **Why is it important?** — **424**
3. **What is a streamline?** — **424**
4. **Streamline distance functions** — **425**
   4.1 Comparing distances — 426
5. **Overview of state-of-the-art methods** — **427**
   5.1 Unsupervised and semisupervised methods — 427
   5.2 Supervised methods — 429

6. Shape similarities using bundle adjacency and fractal dimensions — 434
  6.1 The fractal dimension of a bundle mask — 435
  6.2 Streamline-based bundle atlases — 435
7. The impact of different pipeline choices — 435
8. Visualizing the virtual dissections — 435
9. Summary — 436
References — 436

## 23. Methods and statistics for diffusion MRI tractometry

*Maxime Chamberland, Samuel St-Jean, Derek K. Jones, Maxime Descoteaux, and Alexander Leemans*

1. Introduction — 439
2. Tractometry: Along-streamline analysis — 439
  2.1 Bundle segmentation — 440
  2.2 Streamline ordering — 440
  2.3 Core streamline definition — 441
  2.4 Measure assignment — 442
3. Statistical analysis — 442
  3.1 Choice of space for along-streamline analysis — 442
  3.2 Statistical tests for along-streamline analysis — 442
  3.3 Normative modeling and machine learning — 445
4. Comparison with other frameworks — 446
5. Conclusion — 447
Acknowledgments — 447
References — 447

## 24. Connectivity and connectomics

*Andrew Zalesky, Stamatios N. Sotiropoulos, Saad Jbabdi, and Alex Fornito*

1. Introduction — 451
2. Delineating nodes — 452
  2.1 What is a brain area? — 452
  2.2 Connectivity fingerprints — 454
  2.3 Brain parcellation—Why parcellate the brain? — 454
  2.4 Brain parcellation—The methods — 454
  2.5 Connectivity-based parcellation — 455
  2.6 Multimodal parcellation — 455
  2.7 Gradients and soft parcellations — 456
  2.8 Looking forward — 456
3. Mapping connectivity and connectomes — 457

  3.1 Deep white matter tracking — 457
  3.2 Path termination and superficial white matter tracking — 459
  3.3 Quantifying edges — 460
  3.4 Quantification biases — 461
  3.5 Summary — 462
4. Connectome accuracy and validation — 462
  4.1 In vivo validation — 463
  4.2 Connectome phantoms — 463
  4.3 Other evidence of connectome validity — 465
  4.4 Connectome thresholding — 466
  4.5 Connectome filters — 467
  4.6 Summary and future directions — 468
5. Graph analysis of the connectome — 469
  5.1 Building a graph model — 469
  5.2 Analyzing brain network connectivity — 470
  5.3 Analyzing brain network topology — 472
  5.4 The importance of null models — 475
  5.5 Summary — 475
References — 476

## 25. Tractography validation Part 1: Foundations, numerical simulations, and phantom models

*Tim B. Dyrby, Els Fieremans, Francois Rheault, Adam W. Anderson, Marco Palombo, Silvio Sarubbo, Peter Neher, and Kurt G. Schilling*

1. Anatomy, tractography, and validation — 485
  1.1 Anatomical length scales — 485
  1.2 Neuroimaging length scales — 487
  1.3 What needs to be validated? — 487
  1.4 Bridging anatomy and tractography — 487
2. Numerical simulations — 489
  2.1 Simulating the brain network — 489
  2.2 Simulating tissue microstructure along the brain network — 494
3. Physical phantoms — 497
  3.1 Isotropic liquids — 497
  3.2 Anisotropic phantoms for diffusion tractography — 498
  3.3 Which phantoms for which experiment? — 504
References — 504

**xii** Contents

## 26. Tractography validation Part 2: The use of anatomical model systems and measures for validation

*Tim B. Dyrby, Silvio Sarubbo, Francois Rheault, Els Fieremans, Adam W. Anderson, Marco Palombo, Peter Neher, Kathleen S. Rockland, and Kurt G. Schilling*

| | |
|---|---|
| 1. Anatomical model systems | 511 |
| 1.1 Species and validation considerations | 511 |
| 1.2 Anatomical model systems: Microdissection | 513 |
| 1.3 Anatomical model systems: Neuronal tracers | 518 |
| 1.4 Anatomical model systems: Validating fiber orientation | 525 |
| 2. Empirical validations | 530 |
| 2.1 What is empirical validation? | 530 |
| 2.2 What can be assessed? | 531 |
| 2.3 Empirical validation: Connectomes | 531 |
| 2.4 Empirical validation: Fiber bundles | 531 |
| 2.5 Empirical validation: Local reconstruction | 532 |
| 2.6 Considerations in empirical validation | 533 |
| 3. Measures for validation of tractography | 533 |
| 3.1 Measures for validating connections/connectomes | 533 |
| 3.2 Measures for validating bundles | 534 |
| 3.3 Measures for validation of local fiber orientation distribution | 534 |
| 3.4 Measures for validating microstructure | 535 |
| 3.5 Quantification considerations | 535 |
| References | 536 |

## 27. Tractography validation part 3: Lessons learned through validation studies

*Kurt G. Schilling, Francois Rheault, and Tim B. Dyrby*

| | |
|---|---|
| 1. Anatomy can be more complex than our models | 543 |
| 2. Reconstruction techniques capture fiber orientation distribution but are limited in extracting discrete peaks and orientation | 546 |

| | |
|---|---|
| 3. Tractography can reconstruct known WM anatomy—Valid path, shape, and position of WM bundles | 547 |
| 4. Tractography is a fair, but far from perfect, predictor of the presence of connections: There is an inherent sensitivity/specificity trade-off | 549 |
| 5. The strength of connections has useful predictive power | 550 |
| 6. Methods are reproducible, but there is significant variance across methods | 551 |
| 7. There exists no "optimal" combination of acquisition, reconstruction, or tractography parameters | 554 |
| 8. Limitations: Obstacles, biases, and challenges to overcome | 554 |
| 9. Bridging anatomy and tractography is challenging | 556 |
| 9.1 Bridging anatomy and tractography | 556 |
| 9.2 Validation as an iterative process | 557 |
| References | 557 |

## 28. Current challenges and opportunities for tractography

*Francois Rheault, Philippe Poulin, Alex Valcourt Caron, Etienne St-Onge, Kurt G. Schilling, Laurent Petit, Flavio Dell'Acqua, Alexander Leemans, and Maxime Descoteaux*

| | |
|---|---|
| 1. The rise of tractography in neuroscience | 565 |
| 2. Virtual reconstruction versus underlying anatomy: What is and what isn't? | 565 |
| 2.1 Local signal challenges | 566 |
| 2.2 Path generation challenge | 569 |
| 2.3 Origin and termination | 571 |
| 2.4 Tractography and spatial coverage | 571 |
| 3. Pathways and connectomes: How to interpret them? | 574 |
| 3.1 Bundle segmentation | 574 |
| 3.2 Effects on connectomics | 576 |
| 4. Challenges and opportunities for the dMRI tractography community | 576 |
| 4.1 Nomenclature and terminology | 576 |
| 4.2 Reaching consensus on anatomical definitions | 577 |
| 4.3 Limitations in anatomical validation | 577 |
| 5. Conclusion | 578 |
| References | 578 |

# Part V
# Tractography applications

## 29. Tractography: Applications to neurodevelopment, aging, and plasticity

*Catherine Lebel, David Salat, and Jason Yeatman*

| | |
|---|---|
| 1. Introduction | 583 |
| 1.1 Tractography measures | 583 |
| 1.2 Types of analysis | 584 |
| 1.3 Confounds | 586 |
| 1.4 Study cohorts and design | 586 |
| 2. Brain development | 586 |
| 2.1 Contributions of tractography | 587 |
| 2.2 In utero development of white matter tracts | 587 |
| 2.3 Understanding differential maturation in overlapping tracks | 589 |
| 2.4 Graph theory analysis | 590 |
| 3. Aging | 590 |
| 3.1 Contributions of tractography | 593 |
| 3.2 Cognitive associations | 593 |
| 4. Sex differences | 596 |
| 5. Atypical populations | 596 |
| 5.1 Developmental disorders | 596 |
| 5.2 Cerebrovascular and Alzheimer's disease | 597 |
| 5.3 Modifiable risk factors | 599 |
| 6. Plasticity | 600 |
| 6.1 Cognitive development is correlated with white matter properties | 601 |
| 6.2 Long-term impacts of childhood experience: White matter development in expert musicians | 601 |
| 6.3 The causal influence of environmental factors on white matter development: Intervention studies | 602 |
| 7. Conclusions and future directions | 602 |
| References | 604 |

## 30. Linking behavior with white matter networks

*Sanja Budisavljevic, Stephanie Ameis, Rok Berlot, Hanrietta Howells, and Marika Urbanski*

| | |
|---|---|
| 1. Introduction | 613 |
| 2. Socioemotional functions | 613 |
| 2.1 White matter tracts and socioemotional processing in a healthy population | 613 |

| | |
|---|---|
| 2.2 Frontal and limbic networks and impairments in social-emotional processing | 615 |
| 3. Cognitive functions | 616 |
| 3.1 Limbic white matter tracts and episodic memory in healthy populations | 616 |
| 3.2 Limbic white matter tracts and impairments in episodic memory | 617 |
| 3.3 White matter tracts and cognitive control in a healthy population | 617 |
| 3.4 White matter tracts and impairments in cognitive control | 618 |
| 4. Language functions | 618 |
| 4.1 Arcuate fasciculus and language functions in a healthy population | 618 |
| 4.2 Arcuate fasciculus and language impairments | 620 |
| 5. Motor functions | 620 |
| 6. Visuospatial functions | 622 |
| 6.1 Frontoparietal white matter tracts and visuospatial functions in a healthy population | 622 |
| 6.2 Frontoparietal white matter tracts in unilateral spatial neglect | 622 |
| 7. Conclusions | 623 |
| References | 624 |

## 31. Neurosurgical applications of clinical tractography

*Alberto Bizzi, Joseph Yuan-Mou Yang, Jahard Aliaga-Arias, Flavio Dell'Acqua, José Pedro Lavrador, and Francesco Vergani*

| | |
|---|---|
| 1. Introduction | 631 |
| 2. Neuro-oncology | 632 |
| 3. Intraoperative tractography | 636 |
| 4. Functional neurosurgery | 638 |
| 5. Neurovascular | 639 |
| 6. Skull base | 640 |
| 7. Pediatrics | 641 |
| 8. Traumatic brain injury | 643 |
| 9. Conclusions | 643 |
| References | 643 |

## 32. Preclinical and ex vivo tractography: Techniques and applications at high field

*Manisha Aggarwal*

| | |
|---|---|
| 1. Introduction | 653 |
| 2. Technical considerations | 654 |
| 2.1 Acquisition pulse sequences | 654 |
| 2.2 $q$-space sampling schemes | 655 |

**xiv** Contents

3. Applications 656
  3.1 Mesoscale connectivity mapping in the human brain 656
  3.2 Microimaging and preclinical applications 658
  3.3 Comparison of diffusion modeling and tractography approaches 659
4. Towards validation: Comparison with neuronal tracing, microscopy, and optical imaging modalities 661
5. Conclusion 663
References 663

## 33. Multicenter studies and harmonization: Problems, solutions, and open challenges

*Chantal M.W. Tax, Suheyla Cetin Karayumak, Kurt G. Schilling, Daniel Moyer, Bennett A. Landman, Neda Jahanshad, and Yogesh Rathi*

1. Introduction 669
2. The problem 670
  2.1 Quantifying variability 670
  2.2 Sources of variability 671
  2.3 Magnitude of variability 677
3. Proposed solutions 678
  3.1 Statistical approaches for harmonization of diffusion MRI measures 678
  3.2 Harmonization of DWI data 680
4. Open challenges 682
  4.1 Multicenter variability of microstructural estimates 682
  4.2 Fiber direction estimates and tractography 683
  4.3 Disease 683
  4.4 Longitudinal analysis 683
  4.5 Evaluation of harmonization 683
5. Conclusion 684
References 684

## Part VI
## Appendix

## Appendix A: Vectors and tensors 691

*Philippe Karan, Gabrielle Grenier, and Jon Haitz Legarreta*

A.1. Vectors 691
  A.1.1 Mathematics of the vector 691

A.2. Tensors 693
  A.2.1 Mathematics of the tensor 693
  A.2.2 Applications of tensors in dMRI tractography 695

## Appendix B: Numerical integration 697

*Jon Haitz Legarreta and Robert A. Dallyn*

B.1. Introduction 697
B.2. Methods 697
  B.2.1 The Euler method 698
  B.2.2 The Runge-Kutta method 699
  B.2.3 The Adams-Bashforth method 700
B.3. Aspects of numerical integration 701
B.4. Applications and relevance in tractography 702
References 703

## Appendix C: Interpolation, splines, and smoothing 705

*Matteo Battocchio and Simona Schiavi*

C.1. Data interpolation 705
  C.1.1 Diffusion image interpolation 705
C.2. Spline smoothing 706
References 709

## Appendix D: Spherical harmonics 711

*Gabrielle Grenier, Charles Poirier, and Jon Haitz Legarreta*

D.1. Definition and properties of spherical harmonics 711
D.2. Spherical harmonics as a basis 713
D.3. Applications and relevance in tractography 714
References 715

Index 717

# Contributors

*Numbers in parenthesis indicate the pages on which the authors' contributions begin.*

**Manisha Aggarwal** (653), Department of Radiology and Radiological Science, Johns Hopkins University School of Medicine, Baltimore, MD, United States

**Daniel C. Alexander** (297), Centre for Medical Image Computing, Department of Computer Science, University College London, London, United Kingdom

**Jahard Aliaga-Arias** (631), Department of Neurosurgery, King's College Hospital NHS Foundation Trust, London, United Kingdom

**Stephanie Ameis** (613), University of Toronto, Toronto, ON, Canada

**Adam W. Anderson** (485,511), Biomedical Engineering, Vanderbilt University, Nashville, TN, United States

**Jesper Andersson** (151), FMRIB Centre, Oxford, United Kingdom

**Dogu Baran Aydogan** (257), A.I. Virtanen Institute for Molecular Sciences, University of Eastern Finland, Kuopio; Department of Neuroscience and Biomedical Engineering, Aalto University School of Science, Espoo; Department of Psychiatry, Helsinki University Hospital, Helsinki, Finland

**Steven H. Baete** (103,123,143,201), Center for Advanced Imaging Innovation and Research (CAI2R), Department of Radiology, NYU Langone Health; New York University Grossman School of Medicine, New York, NY, United States

**Corey A. Baron** (103,123,143), Robarts Research Institute, Western University, London, ON, Canada

**Rachel L.C. Barrett** (397), Department of Forensics and Neurodevelopmental Sciences, Institute of Psychiatry, Psychology and Neuroscience, King's College London, London, United Kingdom

**Matteo Battocchio** (705), Department of Computer Science, University of Verona, Verona, Italy

**Christian Beaulieu** (103,123,143), University of Alberta, Edmonton, AB, Canada

**Itay Benou** (315), Department of Electrical and Computer Engineering, Ben-Gurion University of the Negev, Beer-Sheva, Israel

**Rok Berlot** (613), Department of Neurology, University Medical Centre Ljubljana; Faculty of Medicine, University of Ljubljana, Ljubljana, Slovenia

**Ahmad Beyh** (397), Department of Forensics and Neurodevelopmental Sciences, Institute of Psychiatry, Psychology and Neuroscience, King's College London; Laboratory of Neurobiology, Department of Cell and Developmental Biology, University College London, London, United Kingdom

**Alberto Bizzi** (631), Fondazione IRCCS Istituto Neurologico Carlo Besta, Milan, Italy

**Cesare Bortolami** (397), Brain Connectivity and Behaviour Laboratory, Sorbonne Universities, Paris, France

**Sanja Budisavljevic** (613), School of Medicine, University of St Andrews, St Andrews, United Kingdom

**Jennifer S.W. Campbell** (103,123,143), McConnell Brain Imaging Centre, Montreal Neurological Institute, McGill University, Montreal, QC, Canada

**Alex Valcourt Caron** (565), Sherbrooke Connectivity Imaging Laboratory (SCIL), University of Sherbrooke, Sherbrooke, QC, Canada

**Marco Catani** (3,41), Department of Neuroscience, Imaging, and Clinical Sciences (DNISC) and Istituto di Tecnologie Avanzate Biomediche (ITAB), University "G. D'Annunzio" of Chieti-Pescara, Chieti; IRCCS SYNLAB, SDN Research Institute, Naples, Italy

**Maxime Chamberland** (381,439), Eindhoven University of Technology, Eindhoven, The Netherlands

**Daan Christiaens** (201,297), Centre for the Developing Brain, School of Biomedical Engineering and Imaging Sciences, King's College London, London, United Kingdom; Department of Electrical Engineering (ESAT/PSI), KU Leuven, Leuven, Belgium

**Julien Cohen-Adad** (103,123,143), Polytechnique Montreal, Montreal, QC, Canada

**Marc-Alexandre Côté** (423), Microsoft Research, Montreal, QC, Canada

**Alessandro Daducci** (297,347), Department of Computer Science, University of Verona, Verona, Italy

**Matthew Dawson** (177), NATBRAINLAB, Department of Neuroimaging, Institute of Psychiatry, Psychology and Neuroscience, King's College London, London, United Kingdom

**Robert A. Dallyn** (697), NatBrainLab, Institute of Psychiatry, Psychology and Neuroscience, King's College London, London, United Kingdom

**Alberto De Luca** (143,201), Image Sciences Institute, Division Imaging & Oncology; Neurology Department, UMC Utrecht Brain Center, University Medical Center Utrecht, Utrecht, The Netherlands

**Flavio Dell'Acqua** (41,177,221,241,257,363,565,631), NATBRAINLAB, Department of Neuroimaging; Department of Forensics and Neurodevelopmental Sciences, Institute of Psychiatry, Psychology and Neuroscience, King's College London, London, United Kingdom

**Maxime Descoteaux** (177,221,241,257,347, 363,439,565), Sherbrooke Connectivity Imaging Laboratory (SCIL), Department of Computer Science, University of Sherbrooke, Sherbrooke, QC, Canada

**Samuel Deslauriers-Gauthier** (347), CRONOS, Inria Centre at Université Côte d'Azur, France

**Ottavia Dipasquale** (41), NATBRAINLAB, Department of Neuroimaging, Institute of Psychiatry, Psychology and Neuroscience, King's College London, London, United Kingdom

**Lilit Dulyan** (397), Donders Institute for Brain Cognition Behaviour, Radboud University, Nijmegen, The Netherlands

**Tim B. Dyrby** (485,511,543), Danish Research Centre for Magnetic Resonance, Centre for Functional and Diagnostic Imaging and Research, Copenhagen University Hospital—Amager and Hvidovre, Copenhagen; Department of Applied Mathematics and Computer Science, Technical University of Denmark (DTU), Kongens Lyngby, Denmark

**Manon Edde** (363), Sherbrooke Connectivity Imaging Laboratory (SCIL), University of Sherbrooke, Sherbrooke, QC, Canada

**Els Fieremans** (77,103,123,143,485,511), Bernard and Irene Schwartz Center for Biomedical Imaging, Department of Radiology, New York University Grossman School of Medicine, New York, NY, United States

**Luc Florack** (275), Department of Mathematics and Computer Science, Eindhoven University of Technology, The Netherlands

**Stephanie J. Forkel** (397), Donders Institute for Brain Cognition Behaviour, Radboud University, Nijmegen, The Netherlands; Brain Connectivity and Behaviour Laboratory, Sorbonne Universities, Paris, France; Centre for Neuroimaging Sciences, Department of Neuroimaging, Institute of Psychiatry, Psychology and Neuroscience, King's College London, London, United Kingdom

**Alex Fornito** (451), Turner Institute for Brain and Mental Health, School of Psychological Sciences, and Monash Biomedical Imaging, Monash University, Melbourne, VIC, Australia

**Andrea Fuster** (275), Department of Mathematics and Computer Science, Eindhoven University of Technology, The Netherlands

**Eleftherios Garyfallidis** (423), Department of Intelligent Systems Engineering, Luddy School of Informatics, Computing and Engineering, Indiana University Bloomington, Bloomington, IN, United States

**Gabriel Girard** (257,347), Department of Computer Science, University of Sherbrooke, Sherbrooke, QC, Canada; Signal Processing Laboratory (LTS5), École Polytechnique Fédérale de Lausanne (EPFL), Lausanne, Switzerland

**Gabrielle Grenier** (691,711), Department of Radiology, Brigham and Women's Hospital, Mass General Brigham/Harvard Medical School, Boston, MA, United States; Department of Computer Sciences, Université de Sherbrooke, Sherbrooke, QC, Canada

**Tom Hendriks** (381), Eindhoven University of Technology, Eindhoven, The Netherlands

**Hanrietta Howells** (613), Department of Medical Biotechnology and Translational Medicine, University of Milan, Milan, Italy

**Giorgio M. Innocenti** (19), Karolinska Institutet, Stockholm, Sweden; Ecole Polytechnique Federale Lausanne, Lausanne, Switzerland

**Okan Irfanoglu** (151), Quantitative Medical Imaging Section, National Institute of Biomedical Imaging and Bioengineering, National Institutes of Health, Bethesda, MD, United States

**Neda Jahanshad** (669), Imaging Genetics Center, Mark and Mary Stevens Neuroimaging and Informatics Institute, University of Southern California, Marina del Rey, CA, United States

**Saad Jbabdi** (451), Wellcome Centre for Integrative Neuroimaging (WIN), Oxford Centre for Functional Magnetic Resonance Imaging of the Brain (FMRIB), University of Oxford, Oxford, United Kingdom

**Ben Jeurissen** (201), imec-Vision Lab, Department of Physics, University of Antwerp, Antwerp, Belgium

**Derek K. Jones** (439), Cardiff University Brain Research Imaging Centre (CUBRIC), Cardiff University, Cardiff, United Kingdom

**Daniel Jörgens** (315), Krembil Research Institute, University Health Network, Toronto, Canada

**Philippe Karan** (691), Sherbrooke Connectivity Imaging Laboratory (SCIL), Universite de Sherbrooke, Sherbrooke, QC, Canada

**Suheyla Cetin Karayumak** (669), Brigham and Women's Hospital, Harvard Medical School, Boston, MA, United States

**Bennett A. Landman** (669), Department of Radiology and Radiological Sciences; Vanderbilt University Institute of Imaging Science, Vanderbilt University Medical Center; Department of Electrical and Computer Engineering, Vanderbilt University, Nashville, TN, United States

**José Pedro Lavrador** (631), Department of Neurosurgery, King's College Hospital NHS Foundation Trust, London, United Kingdom

**Denis Le Bihan** (55), NeuroSpin, Joliot Institute, Bât 145, CEA-Saclay Center, Paris-Saclay University, Gif-sur-Yvette, France

**Catherine Lebel** (583), Department of Radiology, University of Calgary; Alberta Children's Hospital Research Institute and Hotchkiss Brain Institute, Calgary, AB, Canada

**Hong-Hsi Lee** (77), Bernard and Irene Schwartz Center for Biomedical Imaging, Department of Radiology, New York University Grossman School of Medicine, New York, NY; Athinoula A. Martinos Center for Biomedical Imaging, Department of Radiology, Massachusetts General Hospital, Charlestown, MA, United States

**Alexander Leemans** (143,151,177,241,257,363,381, 439,565), PROVIDI lab, Image Sciences Institute, University Medical Center Utrecht, Utrecht, The Netherlands

**Jon Haitz Legarreta** (691,697,711), Department of Radiology, Brigham and Women's Hospital, Mass General Brigham/Harvard Medical School, Boston, MA, United States

**Ilana R. Leppert** (103,123,143), McConnell Brain Imaging Centre, Montreal Neurological Institute, McGill University, Montreal, QC, Canada

**Congyu Liao** (103,123,143), Stanford University, Stanford, CA, United States

**Klaus Maier-Hein** (315), German Cancer Research Center (DKFZ) Heidelberg, Division of Medical Image Computing, Heidelberg, Germany

**Merry Mani** (103,123,143), University of Iowa, Iowa City, IA, United States

**Daniel Martins** (41), NATBRAINLAB, Department of Neuroimaging, Institute of Psychiatry, Psychology and Neuroscience, King's College London, London, United Kingdom

**Jan Morez** (201), imec-Vision Lab, Department of Physics, University of Antwerp, Antwerp, Belgium

**Daniel Moyer** (669), CSAIL, Massachusetts Institute of Technology, Cambridge, MA, United States

**Peter Neher** (315,485,511), German Cancer Research Center (DKFZ) Heidelberg, Division of Medical Image Computing, Heidelberg, Germany

**Dmitry S. Novikov** (77), Bernard and Irene Schwartz Center for Biomedical Imaging, Department of Radiology, New York University Grossman School of Medicine, New York, NY, United States

**Emanuele Olivetti** (423), Fondazione Bruno Kessler, Trento, Italy

**Marco Palombo** (485,511), Cardiff University Brain Research Imaging Centre (CUBRIC), School of Psychology; School of Computer Science and Informatics, Cardiff University, Cardiff, United Kingdom

**Laurent Petit** (347,565), Groupe d'Imagerie Neurofonctionnelle, Institut des Maladies Neurodégénératives (GIN-IMN), UMR5293, CNRS, CEA, Université Bordeaux, Bordeaux, France

**Maximillian Pietsch** (201), Centre for the Developing Brain, School of Biomedical Engineering and Imaging Sciences, King's College London, London, United Kingdom

**G. Bruce Pike** (103,123,143), Hotchkiss Brain Institute, University of Calgary, Calgary, AB, Canada

**Charles Poirier** (381,711), Department of Computer Sciences, Universite de Sherbrooke; Sherbrooke Connectivity Imaging Lab, University of Sherbrooke, Sherbrooke, Canada

**Philippe Poulin** (315,565), Sherbrooke Connectivity Imaging Laboratory (SCIL), University of Sherbrooke, Sherbrooke, QC, Canada

**Yogesh Rathi** (669), Brigham and Women's Hospital, Harvard Medical School, Boston, MA, United States

**Marco Reisert** (201,315), Department of Radiology, University Medical Center Freiburg, Freiburg; Department of Stereotactic and Functional

**xviii** Contributors

Neurosurgery; Department of Diagnostic and Interventional Radiology, Medical Physics, Medical Center of the University of Freiburg, Medical Faculty of the University of Freiburg, Freiburg im Breisgau, Germany

**Francois Rheault** (423,485,511,543,565), Department of Electrical and Computer Engineering, Vanderbilt University; Vanderbilt University Institute of Imaging Science, Vanderbilt University Medical Center, Nashville, TN, United States; Department of Computer Science; Sherbrooke Connectivity Imaging Laboratory (SCIL), University of Sherbrooke, Sherbrooke, QC, Canada

**Kathleen S. Rockland** (19,511), Department of Anatomy & Neurobiology, Boston University; Boston University Chobanian and Avedisian School of Medicine, Boston, MA, United States

**Fedal Saini** (41), Department of Forensics and Neurodevelopmental Sciences, Institute of Psychiatry, Psychology and Neuroscience, King's College London, London, United Kingdom

**David Salat** (583), Department of Radiology, Harvard Medical School; Massachusetts General Hospital, Boston, MA, United States

**Silvio Sarubbo** (485,511), Department of Neurosurgery, "S. Chiara" Hospital, Trento, Italy

**Simona Schiavi** (297,705), Department of Computer Science, University of Verona, Verona; Department of Neuroscience, Rehabilitation, Ophthalmology, Genetics, Maternal and Child Health (DINOGMI), University of Genoa, Genoa, Italy

**Kurt G. Schilling** (347,485,511,543,565,669), Department of Radiology and Radiological Sciences; Vanderbilt University Institute of Imaging Science, Vanderbilt University Medical Center, Nashville, TN, United States

**Rick Sengers** (275), Department of Mathematics and Computer Science, Eindhoven University of Technology, The Netherlands

**Kawin Setsompop** (103,123,143), Stanford University, Stanford, CA, United States

**Dmitri Shastin** (381), Cardiff University Brain Research Imaging Centre (CUBRIC), School of Psychology, Cardiff University, Cardiff, United Kingdom

**Robert Smith** (297), The Florey Institute of Neuroscience and Mental Health, Heidelberg; The University of Melbourne, Melbourne, VIC, Australia

**Stamatios N. Sotiropoulos** (257,451), Wellcome Centre for Integrative Neuroimaging (WIN), Oxford Centre for Functional Magnetic Resonance Imaging of the Brain (FMRIB), University of Oxford, Oxford; Sir Peter Mansfield Imaging Centre, School of Medicine, University of Nottingham; NIHR Biomedical Research Centre, University of Nottingham, Nottingham University Hospitals NHS Trust, Nottingham, United Kingdom

**Samuel St-Jean** (439), Department of Biomedical Engineering, Faculty of Medicine and Dentistry, University of Alberta, Edmonton, AB, Canada; Clinical Sciences Lund, Lund University, Lund, Sweden

**Richard Stones** (221), Department of Forensics and Neurodevelopmental Sciences; NATBRAINLAB, Institute of Psychiatry, Psychology and Neuroscience, King's College London, London, United Kingdom

**Etienne St-Onge** (347,565), Department of Computer Science and Engineering, Université du Québec en Outaouais, Saint-Jérôme, Québec, Canada

**Filip Szczepankiewicz** (103,123,143), Lund University, Lund, Sweden; Brigham and Women's Hospital, Harvard Medical School, Boston, MA, United States

**Christine L. Tardif** (103,123,143), McConnell Brain Imaging Centre, Montreal Neurological Institute, McGill University, Montreal, QC, Canada

**Chantal M.W. Tax** (151,669), CUBRIC, Cardiff University, Cardiff, United Kingdom; UMC Utrecht, Utrecht University, Utrecht, The Netherlands

**Guillaume Theaud** (363), Sherbrooke Connectivity Imaging Laboratory (SCIL), University of Sherbrooke, Sherbrooke, QC, Canada

**J.-Donald Tournier** (103,123,143), King's College London, London, United Kingdom

**Marika Urbanski** (613), Department of Neurology, CHU Poitiers; Labcom I3M, Centre Hospitalier Universitaire de Poitiers, plateforme Ultra-Haut champ 3T-7T, Poitiers University Hospital, Laboratory of Applied Mathematics, CNRS UMR 7348, Poitiers, France

**Francesco Vergani** (631), Department of Neurosurgery, King's College Hospital NHS Foundation Trust, London, United Kingdom

**Mattia Veronese** (41), NATBRAINLAB, Department of Neuroimaging, Institute of Psychiatry, Psychology and Neuroscience, King's College London, London, United Kingdom; Department of Information Engineering, University of Padua, Padua, Italy

**Anna Vilanova** (381), Eindhoven University of Technology, Eindhoven; Delft University of Technology, Delft, The Netherlands

**Sjoerd B. Vos** (103,123,143), Centre for Microscopy, Characterisation, and Analysis, The University of Western Australia, Nedlands, WA, Australia

**Joseph Yuan-Mou Yang** (363,631), Neuroscience Advanced Clinical Imaging Service (NACIS), Department of Neurosurgery, The Royal Children's Hospital, Melbourne, VIC, Australia

**Jason Yeatman** (583), Department of Pediatrics, Stanford University, Stanford; Maternal & Child Health Research Institute, Palo Alto, CA, United States

**Anastasia Yendiki** (103,123,143), Massachusetts General Hospital, Harvard Medical School, Boston, MA, United States

**Andrew Zalesky** (451), Department of Biomedical Engineering; Melbourne Neuropsychiatry Centre, Department of Psychiatry, The University of Melbourne, Melbourne, VIC, Australia

# Preface

Over the past few decades, diffusion MRI tractography has evolved into an indispensable tool for exploring the intricate architecture of the human brain. This powerful technique has not only revolutionized neuroscience research but has also become a crucial asset in clinical practice, particularly in neurosurgery. By enabling the visualization and mapping of complex neural pathways, tractography has significantly advanced our understanding of brain connectivity and function, opening new horizons in neuroscience and providing invaluable insights for preoperative planning.

The journey of diffusion MRI tractography, from a novel imaging technique to an essential component of modern neuroscience and clinical practice, has been remarkable. As the field has grown, so has its complexity. Today, tractography stands at the intersection of multiple disciplines, including physics, mathematics, computer science, neuroscience, and medicine. This multidisciplinary nature has been both a strength and a challenge, fostering innovation while also creating potential barriers to communication and understanding across different areas of expertise.

Recognizing these challenges, and the need to consolidate knowledge from various disciplines into a single, accessible volume, we conceived the idea for this handbook. Its origins can be traced back to a series of educational courses and workshops conducted at the annual meetings of the Organization for Human Brain Mapping (OHBM) and the International Society for Magnetic Resonance in Medicine (ISMRM) between 2014 and 2020. These events consistently drew large audiences of researchers and clinicians eager to expand their understanding of diffusion tractography.

In response to this enthusiasm, we brought together leading experts from diverse backgrounds to contribute their perspectives and expertise. Our aim was to create a comprehensive resource that organizes and integrates the vast knowledge of the field, bridging gaps between different areas of specialization. This handbook covers a wide range of topics, from the fundamental principles of diffusion MRI physics to advanced tractography algorithms, and from the intricacies of data acquisition and processing to the latest clinical applications. We have included chapters on the historical development of the field, the current state-of-the-art techniques, and emerging trends that promise to shape the future of diffusion tractography.

We have crafted this book for a broad audience, offering multiple pathways through its content. For newcomers to the field, we provide clear, foundational knowledge to build upon. For seasoned researchers and clinicians, we offer in-depth discussions of advanced topics and cutting-edge developments. Technical experts will find detailed explanations of imaging techniques, while clinical practitioners will find useful suggestions that can be applied in clinical practice. By bringing together different disciplines within these pages, we hope to stimulate a deeper understanding across specialties and inspire collaboration that will advance both research and applications. Our goal is not only to present the current state of knowledge but also to stimulate new ideas and approaches that will push the boundaries of what is possible with tractography tomorrow.

As editors, we are deeply grateful to all the contributors who have shared their knowledge and insights, making this handbook possible. Their expertise and dedication reflect the collaborative spirit that has driven the field forward. We also extend our thanks to the countless researchers, clinicians, and patients whose work and experiences have shaped our understanding of brain connectivity and paved the way for the advancements described in this volume.

We invite you, the reader, to explore the pages that follow with curiosity and an open mind. Whether you are a researcher seeking to understand the latest methods, a clinician looking to apply tractography in your practice, or a student embarking on your journey in neuroscience, we hope this handbook will serve as a valuable guide and inspiration for your work in exploring the complexities of the human brain and beyond.

The Editors
**Flavio Dell'Acqua**
**Maxime Descoteaux**
**Alexander Leemans**

Part I

# From anatomy to tractography

# Chapter 1

# The brain and its pathways

## Marco Catani

*Department of Neuroscience, Imaging, and Clinical Sciences (DNISC) and Istituto di Tecnologie Avanzate Biomediche (ITAB), University "G. D'Annunzio" of Chieti-Pescara, Chieti, Italy*

## 1  Introduction

The human nervous system is by far the most complex product of evolution. Unexpectedly, what sets our brain apart from other animals is not the genetic divergence, estimated at little more than 1% compared to chimpanzees, nor the very short list of specialized neurons that arguably may exist only in humans. The complexity of our nervous system is generally explained by the remarkable number of neurons, estimated to be around 87 billion, and their ability to establish a staggering number of connections (Herculano-Houzel, 2012). Surprisingly, the brains of primates have fewer synapses per neuron compared to other species such as rodents (Wildenberg et al., 2021), but their brains contain a greater total number of neurons. The ability of single neurons to establish connections may therefore not represent a key factor to explain higher cognitive functions. Among primates, humans have the largest brains, and this explains why our brain with a total of a thousand trillion connections is the most interconnected information-processing device on the planet. Hence, to understand how the brain-mind works we need methods to describe the specific anatomy of brain connections characteristic of our species.

In the last two decades the development of methods to study brain connectivity has polarized the interest of major research groups, funders, and manufacturers. Traditional microscopy methods and novel approaches based on MRI are joining forces to study the neuronal connectivity at different levels of spatial resolution (Fig. 1). The ultimate goal for those using these methods is to describe the entire map of the human brain networks, in one word, the connectome (Sporns et al., 2005).

The study of brain connections has a long history dating back to postmortem blunt dissection performed by the pioneer anatomists of the 18th and 19th century (Catani et al., 2012, 2013a). This method, although rudimentary in our eyes, allowed us to delineate the trajectories of the major white matter bundles of the human brain. The development of animal tracing studies in the 20th century has further improved our knowledge of connections, adding more detailed descriptions of brain connectivity in the monkey brain and other species (Schmahmann and Pandya, 2006). The use of genetic manipulation combined with axonal tracing and three-dimensional imaging permits us to label individual groups of neurons and identify the intricate connectivity of small cortical regions (Dawson et al., 2023). This approach is suitable only for non-human animals and their findings cannot always be directly transposed to the human brain (Barrett et al., 2020).

In 1994, a real breakthrough came with the development of diffusion tensor imaging (DTI) to study the organization of white matter in the living human brain. With this method it was possible, for the first time, to measure and extract in vivo and noninvasively the organization and the microstructure of white matter fibers by quantifying the movement of water molecules inside the tissue (Basser et al., 1994). A few years later, tractography algorithms were proposed as a tool to mathematically reconstruct three-dimensional trajectories of the major white matter pathways (Mori et al., 1999; Conturo et al., 1999; Jones et al., 1999). This technique has rapidly become the most important tool for investigating the connectional anatomy of the healthy and pathological human brain (Fig. 1, right). In parallel, a new breed of diffusion MRI methods was developed with the aim of overcoming two of the major limitations of DTI tractography: the inability to resolve multiple fiber orientations inside the same voxel (i.e., the fiber crossing problem), and the lack of specificity of DTI indices (Dell'Acqua and Catani, 2012; Dell'Acqua and Tournier, 2019). Images of white matter tracts reconstructed with tractography are gradually replacing those obtained from postmortem blunt dissection in some of the most popular textbooks used for teaching anatomy, including *Gray's Anatomy* (Donkelaar ten, 2020; Catani and Zilles, 2020). In 2019 the International Federation of Associations of Anatomists (IFAA) approved the new *Terminologia Neuroanatomica (TNA)*, which represents the official international neuroanatomical terminology (FIPAT, 2017); all of the new white matter tracts contained in the revised TNA were first described using tractography.

---

**Handbook of Diffusion MR Tractography. https://doi.org/10.1016/B978-0-12-818894-1.00034-3**
Copyright © 2025 Elsevier Ltd. All rights are reserved, including those for text and data mining, AI training, and similar technologies.

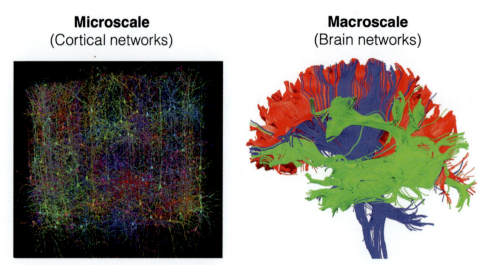

**FIG. 1** Visualization of structural brain networks. (Left) Advanced electron microscopy of cortical layers permits the reconstruction of intricate cortical networks formed by individual neurons, here displayed in different colors. (Right) Diffusion tensor imaging tractography uses signals acquired with magnetic resonance imaging (MRI) to recreate the trajectory of long white matter connections forming large-scale networks. *((Left) Dataset and visualization from www.microns-explorer.org (MICrONS Consortium).)*

Considering that diffusion MRI tractography is a unique method to probe the nervous tissue microstructure and to map the course of large-scale connections in the brain (Le Bihan, 2003), this chapter provides a summary of the rules governing neuroanatomical organization at different spatial resolution scales.

## 2 Overview of the nervous system

The nervous system is composed of an equal number of neuronal and nonneuronal cells (von Bartheld et al., 2016). Neurons have a central part termed the body (or soma) and propagate signals according to the polarization law: from their receiving appendices (dendrites) to their output projections (axons). To provide the energy required by constant neuronal activity and the synaptic formation, other nonneuronal cells work to supply nutrients and oxygen, stabilize the formation of new synapses while removing the old ones, insulate axonal membranes, and maintain a high metabolic rate and blood supply. A blood-brain barrier controls the neuronal environment and imposes severe restrictions on the types of substances that can pass from the bloodstream into nervous tissue.

The description of the neuron as the fundamental building block of the nervous system is a relatively recent discovery compared to other anatomical terms that have circulated for much longer in the anatomical jargon. This can explain some of the common misconceptions that are quickly learned by new students of neuroanatomy. For example, the traditional division of the cerebral tissue into *gray and white matter* based on the differences in color at the naked eye of the postmortem preparations can often be misleading if considered indicative of a neat partition between the neuronal body confined in the cortex (or subcortical nuclei) and its axon in the white matter, or cortical computational processes vs white matter conductive functions (Fig. 2A). Indeed, gray matter contains not only the bodies of neurons but also dendrites and axons. The degree of axonal myelination depends on the distance traveled. Axons confined to the local gray matter are either unmyelinated or lightly myelinated. Axons traveling long distances are heavily myelinated and their course is almost entirely within the white matter except for a relatively short initial and terminal segment in the gray matter. Hence, the gray matter of the cerebral cortex, basal ganglia, thalamus and hypothalamus, nuclei of the brainstem and cerebellum, and horns of the spinal cord is a tissue in which both neuronal bodies and axonal connections abound (Fig. 2B and C). Similarly, white matter lodges sparse neuronal bodies (Fig. 2B).

Finally, recent experimental studies showing dynamic exchanges between gray and white matter compartments for nonneuronal cell populations speak against a neat divide between gray and white matter. Oligodendrocytes, the cells responsible for ensheathing axons with myelin, migrate from the gray matter, where they lie in a quiescent form, to the white matter. As they do so, they transform from immature precursors into mature oligodendrocytes. This process is continuous, from perinatal life to old age and adaptively dynamic. The most important triggers for the migration process include the activation of cortical neurons and the release of neurotrophins and neurotransmitters (Mount and Monje, 2017).

Another traditional anatomical dichotomy separates the *central nervous system (CNS)* contained within the skull and the spine from the *peripheral nervous system* located outside *(PNS)*. The CNS consists of the brain and spinal cord, and contains the majority of neuronal cell bodies. The PNS includes all nervous tissue outside the skull and the spine and consists of

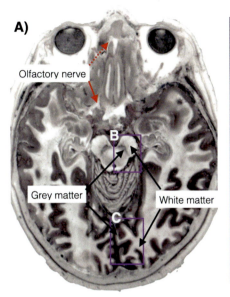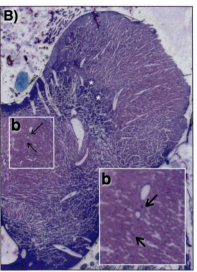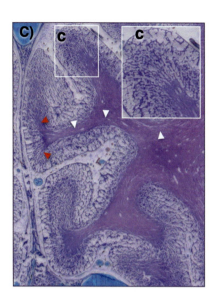

**FIG. 2** Gray and white matter of the human brain, and central vs peripheral division of the nervous system. (A) Gray and white matter regions in a postmortem axial section are indicated at the level of the midbrain and occipital cortex. Also, the intracranial *(red line)* and extracranial *(dashed red line)* segments of the olfactory nerve are visible in this slice. (B) Magnification of the midbrain region where groups of neurons are distributed along the white matter fibers of the mesencephalon as substantia nigra, pars reticularis *(white asterisks)*. Isolated neuronal bodies of the red nucleus *(black arrows)* are visible in an area (b) occupied almost entirely by the myelinated axons of the superior cerebellar peduncle. (C) Magnification of the axial section through the occipital gyri surrounding the calcarine sulcus. The myelinated axons of the optic radiations run through the white matter of the occipital lobe *(white arrowheads)* before entering the cortex *(red arrowheads)*. The cortex contains an intricate system of radial and perpendicular myelinated fibers (c). *(Images courtesy of Dr Peter Ratiu.)*

the cranial and spinal nerves, the peripheral autonomic nervous system (ANS), and the special sense organs (taste, olfaction, vision, hearing, and balance). This subdivision in CNS and PNS is both anatomical and functionally arbitrary as the cranial and spinal nerves originate or terminate within the central nervous system (Fig. 2A). The spinal cord and brain communicate with the rest of the body via the spinal and cranial nerves, respectively. These nerves contain afferent fibers that bring information into the CNS from sensory receptors, and efferent fibers that convey instructions from the CNS to the peripheral effector organs. The CNS controls many aspects of bodily function by reflex action mediated via interconnections of varying complexity between the afferent and efferent components of the spinal and cranial nerves. Numerous descending connections from the brain modulates this activity. Hence the functions of the PNS and CNS are coordinated and participate in most of the functions of the nervous system.

## 3 From molecules to neuronal circuits

During evolution, neurons became the most efficient signal-processing system of organic life by acquiring two unique properties. Like cells of the endocrine or immune system, neurons produce molecules that they release into the extracellular space to communicate among themselves. But unlike hormones or antibodies that use fluid transportation to reach their relatively distant targets, neurotransmitters travel a very short distance before they attach to receptors on the membrane of the receiving neuron. Both membranes of the releasing and receiving neurons bud out to form a synapse (from Greek sun- "together" + hapsis "joining") (Foster and Sherrington, 1897), whose function relies on several molecules including neurotransmitters and receptors, adhesion proteins, microtubules, and ion channels (Fig. 3A). Surprisingly, the majority of the genetic elements coding for the synaptic molecules were present before the synapse itself in aneural organisms, where they participated in other cellular processes such as cell division and intracellular transport (Viscardi et al., 2021).

One neurotransmitter can bind to multiple receptor types, and the final modulatory effect can be excitatory or inhibitory, depending on the receptor they interact with. Two types of receptors have been identified: ionotropic receptors are transmembrane ion-channel proteins that directly gate the ion inflow; metabotropic receptors activate a metabolic cascade that ultimately induces the opening of an ion-channel protein. The fast ionotropic neurotransmission is used mainly by the glutamatergic and GABAergic neurons, while other neurotransmitters interact mainly with slow metabotropic receptors (Catani and Zilles, 2020). The genes for ionotropic receptors emerged later in the evolution of neural organisms compared to metabotropic receptors, and this may have contributed to further advancement in synaptic signaling speed (Viscardi et al., 2021).

**6 PART | I From anatomy to tractography**

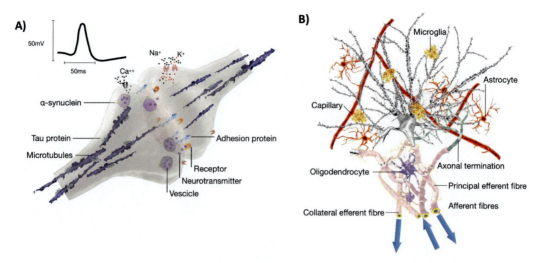

**FIG. 3** (A) The synapse is the point of contact between neurons where a number of proteins work to facilitate the adhesion between pre- and postsynaptic membranes (e.g., neurexin and neuroligin) or the release of vesicles containing neurotransmitters (e.g., α-synuclein). The microtubules that form the intraneuronal scaffolding are stabilized by the protein tau. Protein tau and the α-synuclein can accumulate and cause synaptic degeneration. The attachment of the neurotransmitter to receptors triggers a transient transmembrane calcium ion movement that can result in an inhibitory or an excitatory effect. (B) A single neuron can form several hundreds of synapses, a process that requires regulation of local blood flow, metabolic processes carried out by astrocytes, and the assistance of microglia cells. Neurons that are particularly active induce an increase in myelination of their axonal fibers by stimulating maturation and migration of oligodendrocyte precursors.

The receptor and synaptic turnover are continuous from the first few weeks of fetal life to death and are governed by the combined effect of genes, along with shared and specific environment. The brain potential for perception, behavior, and cognition is indeed embedded in the development of specific synaptic patterns and so are our memories and experiences of the world. Neurons do not do all the synaptic work by themselves; they need help from the astrocytes that regulate local blood flow and metabolism, and the microglia, the resident immune cells of the brain that contribute to consolidating synapses or their "pruning" (Fig. 3B).

The second unique property of neurons is a direct consequence of the "door-drop" synaptic communication they developed, which required an extraordinary elongation of their morphology and a faster way of transferring information from the dendritic to the axonal terminals. Considering that a neuron can stretch from less than a millimeter to more than a meter and the intracellular flow is extremely slow, about half a meter in 1 day, the neuron had to find an alternative communication vehicle. The solution was the transformation of the neuronal membrane into a semipermeable barrier capable of propagating a fast electrochemical signal, the action potential. This required new voltage-gated ion channel proteins whose genes are rooted very early in the evolution of neuronal organisms (>600 million years ago) (Viscardi et al., 2021). At rest, the neuron actively maintains an electrochemical gradient of K+ and Na+ across its membrane, resulting in a 70 mV potential between the internal (negative) and external (positive) surface. When neurotransmitters combine to receptors, the synaptic membrane modifies its permeability by opening its K+ and Na+ channels so that there is a sudden inversion of the resting state potential. The depolarization of the synaptic membrane causes the opening of the neighboring channels and like a ripple the action potential propagates along a direction parallel to the neuronal membrane.

The depolarization of a single synapse is not sufficient to generate an action potential as the process requires the summation of several inputs within the same dendritic branch and between branches (Fig. 3B). The summation process follows both linear and nonlinear dynamics and depends on the stimulus formats (single stimuli versus trains), conditions of spatial integration (within versus between dendritic branches), ratio between excitatory and inhibitory synapses, and receptive properties of the depolarizing membrane (Poirazi et al., 2003). All these variables are directly and indirectly influenced by the pattern of local intracortical and long-range connectivity (Figs. 3B and 4). Collateral fibers, for example, branch off the main axon and feed back on their own neuronal bodies or inhibitory interneurons to self-modulate their own firing. Feedforward afferents from upstream neurons and feedback projections from downstream neurons may increase the excitability of some neurons while decreasing the activity of others. Projections from the ascending reticular system provide both tonic and phasic activation (or inhibition) of cortical neurons and play a major role in the shift from non-REM to REM sleep and wakefulness. In turn, the frontal lobe neurons control the reticular system, thalamus, and the brainstem, whose modulating activity on the cortex varies according to the nature and complexity of the task.

## 4 Cytoarchitectonic and neuronal connectivity

There are about 87 billion neurons in the human brain and three main neuronal populations can be distinguished (Fig. 4) (Catani and Zilles, 2020). *Pyramidal cells* have the largest diameter and are the most prevalent neuronal type in the cortex. The dimensions of their neuronal body is directly related to the diameter and length of their axons. Hence, cortical regions with a high density of the largest pyramidal cells (also called Betz cells) are the primary cortical motor areas that send projections to the distant neurons of the spinal cord. Pyramidal cells are excitatory and use glutamate as neurotransmitter.

*Spinous (or spiny) stellate* cells are the second most numerous neuronal types in the cortex and like the pyramidal cells they use glutamate. They are named spinous for the high density of dendritic spines that cover their radiating dendrites. The spines are axonal protrusions that usually receive excitatory inputs and therefore their presence in large quantities indicates a receptive function of the spinous stellate cells. A high density of spinous stellate cells and a wide-spreading distribution across the superficial cortical layers is typical of primary auditory, somatosensory, and visual cortical areas representing the major target of thalamo-cortical inputs. Spinous stellate neurons are small in diameter and therefore when present in a high density they confer a granular aspect to the tissue under the microscope.

The third neuronal population, the *aspiny or sparsely spinous stellate cells*, are GABAergic (gamma-aminobutyric acid) inhibitory interneurons. Their axons are confined to gray matter and terminate on pyramidal cells and other interneurons. The morphology of these interneurons is variable, and several subtypes have been identified.

Approximately 95% of the cortical neuronal activity is mediated by fast excitatory (glutamate, 80%) and fast inhibitory (GABA, 15%) neurons. The remaining 5% is represented by modulatory neurons (dopamine, serotonin, noradrenaline, acetylcholine, etc.) located in small subcortical nuclei. The axons of these neurons gather into ascending bundles projecting to other subcortical nuclei or more diffusely to large cortical areas. The diffuse distribution of these monoaminergic projections and their influence on a large number of cortical and subcortical regions explain why they represent the main target of most pharmacological treatments for psychiatric and neurological disorders.

The cytoarchitectonic arrangement of the three populations of neurons described here is highly variable in the gray matter, from a inhomogeneous scattered pattern in some regions of the brainstem and spinal cord to a multilayering of the cerebral cortex (Fig. 5, left figures). In the cerebral cortex, fibers are arranged parallel to the cortical surface or perpendicularly to it (Fig. 5, right figures). Most of these fibers remain within the gray matter and travel a short distance before establishing contact with another neuron. The short fibers form local networks that in the cerebral cortex have a characteristic columnar arrangement. Each column spans across the six layered cortex and receives afferents from local neurons or from distant neurons (Mountcastle, 1957).

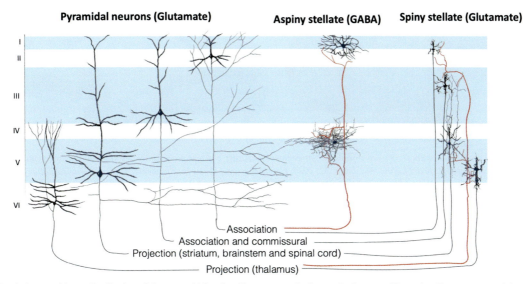

**FIG. 4** Morphology and layer distribution of the pyramidal and stellate neurons in the cerebral cortex. The main efferent targets of the pyramidal and spiny stellate cells (colored in *black*) and the main afferent to the aspiny and spiny stellate neurons (colored in *red*) are also indicated. *(Modified from Catani, M., Zilles, K., 2020. Cerebral hemispheres. In: Standring, S. (Ed.), Gray's Anatomy, the Anatomical Basis of Clinical Practice, forty-second ed. Elsevier, Amsterdam.)*

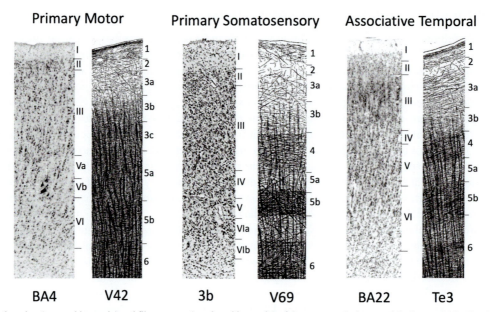

**FIG. 5** Cellular layering (cytoarchitectonic) and fiber pattern (myeloarchitectonic) of three neocortical areas of the human brain. Brodmann areas 4 and 3b are typical examples of heterotypical cortex whose layering is characterized by a high number of pyramidal cells and a low density of stellate cells (primary motor cortex) or the opposite pattern (primary somatosensory). Brodmann area 22 associative temporal area is an example of eulaminate cortex with an alternating distribution of pyramidal and stellate cells. There is some correspondence between cyto- and myeloarchitectonic layering. The prevalence of vertical fibers in area 4 is indicative of the presence of a high density of long projection fibers characterized by a large diameter and fewer collateral branches. In area 3b the lower number of vertical fibers and relatively higher density of horizontal fibers indicate the presence of numerous short intracortical connections and collateral fibers. This is typical of all primary sensory areas that connect mainly to neighboring areas through short association fibers. Finally, area 22 shows a myeloarchitectonic pattern intermediate between the primary motor and sensory areas. *(Modified from Catani, M., Zilles, K., 2020. Cerebral hemispheres. In: Standring, S. (Ed.), Gray's Anatomy, the Anatomical Basis of Clinical Practice, forty-second ed. Elsevier, Amsterdam.)*

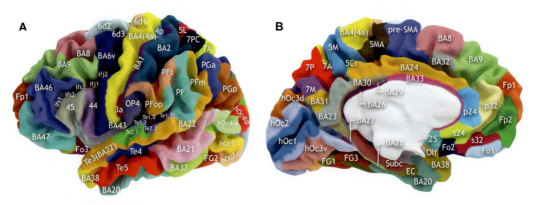

**FIG. 6** Parcellation of the human cerebral cortex according to cytoarchitectonic criteria. Areas are a combination of old Brodmann areas (BA) and new parcellation from the Julich group. *(From Catani, M., Zilles, K., 2020. Cerebral hemispheres. In: Standring, S. (Ed.), Gray's Anatomy, the Anatomical Basis of Clinical Practice, forty-second ed. Elsevier, Amsterdam. Images were created by Daniele Cancemi.)*

Both neuronal and fiber layering have been used to separate the cerebral cortex into cortical areas (or fields) (Fig. 6) (ffytche and Catani, 2005; Catani and Zilles, 2020). The majority of the *isocortex*, also named neocortex because it developed late during human brain evolution, is characterized by a well-defined six layers (eulaminate), where the four central layers show a distinct alternate distribution of pyramidal and spinous stellate cells, with a higher prevalence of pyramidal cells in layer III (external pyramidal layer) and V (internal pyramidal layer) and higher density of stellate cells in layers II (external granular) and IV (internal granular). The external layer I (plexiform) is scarcely populated by neuronal cells, while layer VI is called multiform or polymorphous for the presence of all three types of neurons. The eulaminate isocortex is typical of associative areas.

In addition to the eulaminate cortex, the koniocortex and the agranular/dysgranular cortices represent the opposite extremes of the *heterotypical* cortex. The granular koniocortex (konio means "dust" in Greek) is characterized by the low number of pyramidal cells and the highest density of spinous stellate neurons that also invade layers III and V. The koniocortex is typical of primary sensory areas.

Conversely the agranular/dysgranular cortex has a high density of pyramidal cells that from layers III and V invade layer IV, which is interrupted (dysgranular) or not visible as a distinct layer (agranular). The motor and premotor areas of the frontal and the anterior areas of the cingulate gyrus and insula are typically agranular/dysgranular zones. The remaining cortex is termed allocortex, which is typically found in the olfactory and limbic regions dedicated to olfactory and gustatory perception, memory, emotion regulation, reward, and social behavior. The allocortex is often continuous with other gray matter structures that developed early during brain evolution and its heterogeneous lamination pattern often includes fewer layers compared to the isocortex, with the exception of the entorhinal cortex that has more than six layers.

The anatomical variability of the isocortex reflects functional differences of individual zones that are strictly linked to their pattern of connectivity. However, while sharp borders between cytoarchitectonic areas have been defined, the passage from one area to another is often gradual and mediated by zones of transitions. This is an important observation also for functional considerations. For example, BA3a that lies between the agranular/dysgranular primary motor cortex (BA4) and the koniocortex of the somatosensory BA3b, displays cytoarchitectonic characteristics common to both its neighboring areas, such as low spinous stellate neurons and few larger pyramidal cells with an incipient layer IV. Nevertheless, even within functionally related areas, differences in cytoarchitectonic and connectivity can be quite striking. For example, in the somatosensory cortex layer V is thinner than layer IV in BA3b whereas the opposite is progressively found in areas BA1 and BA2. This indicates a gradual change in the ratio between the afferent thalamic (targeting layer IV) and efferent thalamic and associative (originating from layer V) fibers within the somatosensory areas.

## 5  White matter anatomy

The local networks interact with peripheral receptive organs, muscles, and other distant networks, through long myelinated axons traveling in the white matter of the spinal cord, brain stem and telencephalon. While the course of these long axons is already determined at birth, their morphology (cone diameter, terminal synapses) and degree of myelination changes constantly through life, especially in response to the stimulation of peripheral receptors or the cortical activity. These changes permit circuit specialization and learning. The emergence of cognizant perception, thought, and social behavior is therefore to be understood in terms of continuous remodeling of local and extended networks.

The axons that leave the cortex and travel within the white matter are grouped together according to common anatomical and functional properties. On average a single oligodendrocyte cell may ensheath up to 50 axons and by doing so groups them together into small bundles. The degree of myelin ensheathment of axons is a dynamic process directly linked to their functional activity. Cortical neurons engaged in a specific task trigger changes not only in the cortex, through synaptic pruning for example, but they also modify the anatomy of their axons. It has been demonstrated that high neuronal activity triggers the maturation and migration of oligodendrocyte precursors that leave the cortex and move along the long axons originating from the most active neurons (Mount and Monje, 2017). Increased myelination of these axons guarantees a higher speed of conduction. An action potential can travel at a speed of 0.5 m/s in unmyelinated axons or up to 150 m/s in highly myelinated axons. The variability of action potential speed is an important factor when considering the computational aspects of large brain networks and the efficient control of cognitive processes, movement and behavior. The conduction velocity is regulated adaptively and its variations, mediated by myelin, have profound effects on neuronal network function in terms of spike-time arrival, oscillation frequency, oscillator coupling, and propagation of brain waves (Pajevic et al., 2014).

In turn, myelinated axonal bundles gather into larger white matter fasciculi or tracts. The fasciculi of the telencephalon can be classified into projection, association, and commissural tracts (Fig. 7) (Catani and Thiebaut de Schotten, 2012).

*Projection tracts* connect cortical areas to subcortical nuclei and according to their axonal trajectory they can be further subdivided into ascending (from subcortical to cortical) and descending (from cortical to subcortical) projections. The major ascending projection tracts originate from the thalamus and metathalamus (i.e., medial and lateral geniculate nuclei) and travel within the internal capsule as thalamic, auditory, and optic radiations. The same fasciculi contain a large contingent of descending corticothalamic fibers that terminate either in the same thalamic nuclei that send them sensory information (feedback control) or different nuclei (feedforward processing). A comparatively smaller but functionally relevant group of ascending fibers, the mesolimbic and mesocortical bundles, originate from the mesencephalic nuclei and project to the frontal and limbic cortical areas. This extracapsular contingent contributes to the regulation of affect, motivation, and attention.

**10 PART | I From anatomy to tractography**

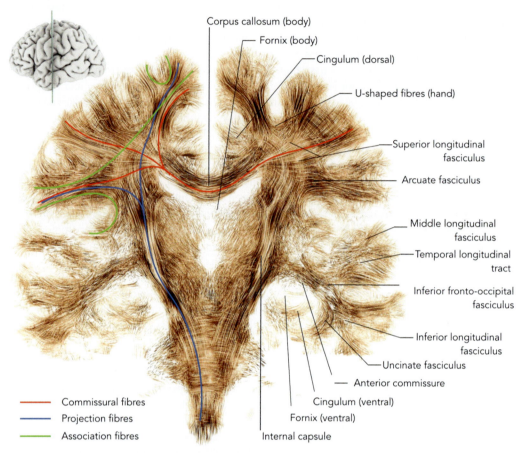

**FIG. 7** Tractography visualization of the tracts passing through a coronal slice at the level of the central sulcus. On the right-hand side, the location of the major projection, association, and commissural tracts are indicated. *From the latest edition of the Gray's Anatomy (Elsevier)—Chapter 32. Courtesy of Flavio Dell'Acqua and Marco Catani—Natbrainlab.*

The major nonthalamic descending pathways are represented by axons originating from pyramidal neurons of layer V and terminating in the basal ganglia (corticostriatal), tectum, cranial nerves (corticobulbar tract), pontine nuclei, and spinal cord (corticospinal system). Finally, the fornix is a projection tract that connects the allocortex of the hippocampal system and isocortex of the parahippocampal gyrus with the mammillary bodies, hypothalamus, and anterior thalamic nuclei. Projection fibers are well preserved among vertebrates for their adaptive function. They convey and elaborate sensory information for the control of movement and behavior that permits the organism greater chances of survival through automatic reflexes, exploration of the environment for foraging, and the modification of the environment.

*Association tracts* are cortico-cortical connections traveling within the same hemisphere. According to their length, association tracts can be separated into short U-shaped fibers connecting adjacent gyri and intermediate and long association fibers connecting distant gyri. The majority of the U-shaped fibers connect gyri within the same lobe (intralobar) while most of the long association fibers connect gyri from different lobes (interlobar). Association tracts have emerged late in brain evolution in parallel with the increase of the cortical surface of the hemispheres. Their development has facilitated the emergence of fine motor skills for toolmaking, language, more flexible cognitive functions, and complex social behavior. The major association tracts of the human brain are the arcuate fasciculus (for language and praxis), the superior longitudinal fasciculus I and II for visuospatial processing and eye and limb motor coordination, the inferior fronto-occipital fasciculus and inferior longitudinal fasciculus for visual processing, and the cingulum and uncinate fasciculus for memory, affective regulation, and social behavior.

*Commissural tracts* cross the cerebral midline to connect cortical or subcortical regions of the two hemispheres. Homotopic fibers interconnect corresponding regions in the two hemispheres and represent the most prevalent type of commissural connections. Heterotopic commissural fibers connect noncorresponding gray matter regions of the two hemispheres. The largest commissural tracts of the human brain are the corpus callosum and the anterior commissure. Commissural tracts are important for the development of lateralized cognitive functions and functions that require communication between the two hemispheres, such as bimanual coordination (Box 1).

### BOX 1 Meynert's rule and white matter concentric zones.

Theodor Meynert was the first to observe a direct relation between the length of the fibers and how deep they travel in the white matter (Fig. 8, left) (Meynert, 1885). Among the association fibers, for example, the short U-shaped fibers are always confined in the white matter just beneath the cortex, whereas longer association tracts occupy deeper white matter regions. This general organization of the association fibers can be easily visualized with diffusion tractography and has important implications for correlating symptoms with lesion location (Catani and ffytche, 2005). For example, the deeper the lesion, the higher the probability that symptoms are related to disconnection of distant cortical areas.

Another general principle of white matter organization is the subdivision of the white matter into a core zone and an external zone. This division is particularly evident on coronal images (Fig. 8, right). The core zone is occupied solely by commissural and projection pathways forming the walls of the lateral ventricles. The external zone contains all association fibers and the terminal portions of all projection and commissural tracts. The concentric zonal distribution of the white matter pathways is the result of serial modifications occurring during evolution and early fetal life. The early period of human brain development (up to 16 weeks) is characterized by the formation of commissural and projection pathways of the core zone, followed by the progressive expansion of the external zone driven by the exponential growth of association and U-shaped fibers.

Depending on the location, lesions to white matter have a significant impact on cognition (Lamar et al., 2008; Catani et al., 2012). The use of tractography in individual patients or tractography-derived atlases is improving localization of white matter damage and our clinical-anatomical inferences. Disorders that typically affect white matter of the inner core zone, for example, disrupt conduction of the action potential along large commissural and projection pathways and therefore often manifest with extrapyramidal motor deficits, pseudobulbar symptoms, sensory-motor coordination problems, and general slowness of cognitive functions (e.g., normotensive hydrocephalus, Binswanger's disease). More rarely, small lesions of the core zone can manifest with isolated motor or somatosensory deficits (internal capsule), or visual (optic radiations), auditory (auditory projections), or memory deficits (fornix).

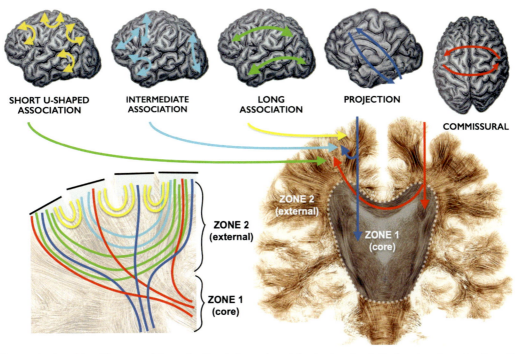

**FIG. 8** The organization of the white matter of the cerebral hemispheres. Association tracts are located in the external zone together with projection and commissural fibers. According to Meynert's rule the shortest U-shaped fibers are more superficial than intermediate and long association tracts. The core zone which develops earlier than the external zone is occupied only by the long projection and commissural fibers.

# 6  Imaging networks with diffusion MRI tractography

Before diffusion MRI tractography, many assumptions about human connections were taken for granted. Traditionally, network models based on axonal tracing studies in animals have been blindly transposed to the human brain (Schmahmann and Pandya, 2006; Petrides, 2013). While this approach is arguably justified for functions shared between humans and other primates, its validity is questionable for those lateralized abilities that are unique to our species. Indeed, accumulating findings suggest striking simian-to-human differences in a number of connections (Rilling et al., 2008; Thiebaut de Schotten et al., 2012; Barrett et al., 2020). Despite its limitations, tractography remains the only viable method to study the network anatomy of the living human brain (Catani et al., 2002; Jbabdi et al., 2015). Current uses of diffusion MRI tractography can be divided into methods for studying *structural connectivity* and methods for *white matter segmentation*. While the two approaches share similar methodologies, they differ in the nature of the questions they are capable of answering (Dell'Acqua and Catani, 2012).

Structural connectivity is the property of distant neurons located in the gray matter and directly linked by their axonal processes. Assessment of anatomical connectivity requires a precise quantification of the course of the axonal projections reaching their target regions. While this information can be reliably obtained with classical axonal tracing methods, diffusion MRI tractography has been considered to be limited in its ability to generate faithful representations of the true anatomy of fibers, especially in the proximity of the gray matter (Reveley et al., 2015). Among these limiting factors are the partial volume effect (loss of signal in a voxel due to presence of nonwhite matter tissue) and highly complex patterns of fiber architecture as an invariable source of low anisotropy diffusion signal in voxels close to the cortex (Vos et al., 2011). Hence, estimates of fiber orientation in these regions are highly uncertain. This uncertainty limits the ability to obtain reliable reconstructions of the course and termination of streamlines that propagate closely to cortical gray matter areas and subcortical nuclei. The risk of generating artifactual reconstructions is therefore very high when approaching the cortex and a diffusion tractography technique to delineate the exact axonal connectivity pattern between two distant neuronal populations remains elusive ( Jones, 2003; Thomas et al., 2014). See Chapters 24 and 28 for more on this.

The use of diffusion MRI tractography for white matter segmentation attempts to classify white matter voxels according to the likelihood of tract occupancy. As with the connectivity approach, the process requires selecting streamlines from regions of interest, although in this case both cortical and deep white matter regions can be used to isolate individual tracts (Catani and Thiebaut de Schotten, 2008). The core of each tract is then reconstructed, and voxels labeled according to the tract (or tracts) that most likely traverses them. This approach tries to identify the most likely location of tracts in each white matter voxel rather than the probability of two gray matter voxels being reciprocally connected. The advantage of this approach is that it does not require streamlines reaching target cortical or subcortical regions. White matter segmentation, which can be highly reliable for deep voxels located at the core of a white matter bundle where fibers run cohesively in parallel and interindividual variability is low (Thiebaut de Schotten et al., 2011a), has been used to describe general principles of white matter organization (Box 1). In more peripheral regions segmentation may be more difficult and lead to underestimation of the real anatomical extension of the tract. Furthermore, when tractography is performed using advanced methods for resolving fiber crossing, multiple tracts can be mapped within a single voxel (Dell'Acqua and Tournier, 2019), increasing the complexity of the final reconstructions. For an in-depth discussion on the most recent diffusion and tractography methods, see later chapters of this book (see Chapters 9–20).

Connectivity and segmentation approaches have been used to study network anatomy using indirect measures of tract volume and microstructural properties of fibers, such as axonal morphology, myelin content, and fiber organization (Beaulieu, 2002; Concha, 2014). The first measures of tract volume were the overall number of streamlines that compose a single tract or the total volume of voxels intersected by those streamlines. However, these approaches have limitations, as the number of streamlines may be affected by complex methodology factors that need to be accounted for in addition to biological explanations (seeding criteria, stopping thresholds, multiple regions of interest, etc.) (Catani et al., 2011; Jones et al., 2013). Histological properties that are likely to determine larger tract volumes are increased axonal diameter and myelination, high axonal density and fiber dispersion, presence of fiber crossing and branching (Fig. 9). For more, see also Chapters 10–12.

In addition to tract volume, for each voxel intersected by streamlines, other diffusion indexes can be extracted, and the total average calculated. Examples include fractional anisotropy, mean diffusivity, and parallel and radial diffusivity, which can provide important information on the microstructural properties of fibers and their organization (Fig. 9) (Le Bihan, 2003). Fractional anisotropy values, for example, can be modulated by multiple factors, such as axonal anatomy (intra-axonal composition, axon diameter, and membrane permeability), fiber myelination (myelin density, internodal distance, and myelin distribution), or fiber arrangement and morphology (axonal dispersion, axonal crossing, and axonal branching) (Song et al., 2003; Drobyshevsky et al., 2005; Beaulieu, 2002). See Chapters 23 and 29 for more. Other diffusion

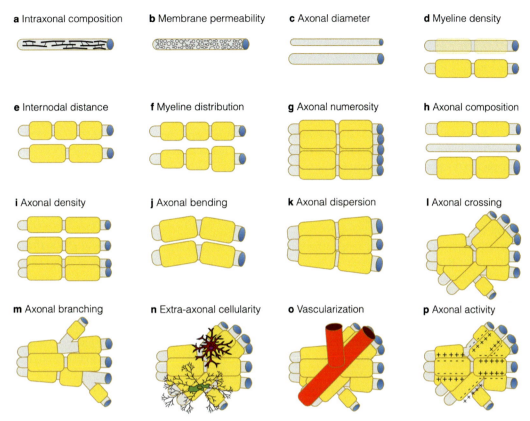

**FIG. 9** Biological factors influencing neuronal functioning and possibly diffusion measurements. (A) Intra-axonal components such as microtubules and filaments of the cytoskeleton are important for regulating axonal trafficking from the neuronal body to the synapses. (B) Changes in axonal membrane permeability are typical of the developing brain and frequently occur in many physiological and pathological conditions. (C) Larger axons conduct faster action potentials compared to smaller axons. (D) Higher myelin density facilitates faster conduction. (E and F) Internodal distance and myelin distribution vary along the axon with direct effect on speed of conduction. (G–I) Axonal count, composition, and density vary from tract to tract and during development and aging. (J–L) Axonal bending, dispersion, and crossing are geometrical aspects that greatly impact diffusion parameters, without necessarily affecting fiber function. (M) Axonal branching is an important physiological mechanism for signal propagation and filtering. (N and O) Glial cells (e.g., astrocytes, microglia) and vascularization have a fundamental role for the metabolism of white matter fibers but their impact on diffusion properties is poorly understood. (P) Axonal activity could also have an effect on diffusion properties of white matter fibers, although this remains yet to be demonstrated.

measurements may reveal more specific fiber properties. Changes in axial diffusivity measurements, for example, could be related to intraxonal composition, while radial diffusivity may be more sensitive to changes in membrane permeability and myelin density (Beaulieu, 2002; Concha, 2014; Paus et al., 2014). Other measurements based on nontensorial models have also been proposed to map microstructure along specific populations of fibers (e.g., apparent fiber density, AFD; hindrance modulated orientational anisotropy, HMOA; neurite orientation density and dispersion imaging, NODDI; etc.) (Dell'Acqua and Tournier, 2019). Further details on this topic are presented in Chapter 12. These in vivo diffusion-based measurements allowed network anatomy to be studied at different scales during development, adulthood, and aging. The application of diffusion imaging to the study of tract asymmetry between hemispheres and its correlation with performances in cognitive tests has provided insights into the human brain organization that were not available before the advent of tractography.

## 7 Interhemispheric tract asymmetry during brain development and aging

At first glance, structural symmetry appears as the basic plan of the vertebrate nervous system. Yet the closer we look, an uneven anatomy between the two sides becomes clearer (Geschwind and Galaburda, 1987; Frasnelli et al., 2012). Admittedly, some of these differences can be subtle, visible only under the microscope (Concha et al., 2012) or after advanced imaging analysis (Toga and Thompson, 2003). Nevertheless, the breaking of symmetry seems a constant finding, especially

**14 PART | I** From anatomy to tractography

when the observed anatomy is that of complex neuronal networks (Concha et al., 2012; Catani et al., 2007; Thiebaut de Schotten et al., 2011a,b; Chechlacz et al., 2015). A scientific interest in the structural asymmetry of the human networks is justified for its potential link to lateralization of cognitive functions that characterize our species (Geschwind and Galaburda, 1987; Toga and Thompson, 2003; Bishop, 2013).

The development of white matter tracts begins before birth and continues through life. Tractography has been used to visualize tract asymmetry in human fetuses (Takahashi et al., 2012; Mitter et al., 2015), preterm borns (Yoo et al., 2005), infants (Dubois et al., 2009; Ratnarajah et al., 2013), children and adolescents (Eluvathingal et al., 2007; Budisavljevic et al., 2015a,b; Lebel et al., 2012), adults (Budisavljevic et al., 2015a), and the elderly (Lebel et al., 2012; Shu et al., 2015). Taken together, these studies suggest that the development of network asymmetry is a temporally dynamic process that varies from tract to tract in the same brain (some tracts develop asymmetry earlier than others) and differs from subject to subject in the general population. In addition, preliminary evidence suggests that familial effects (genetic and shared environment) play a greater role in controlling the asymmetry of those tracts that mature earlier, whereas specific environmental factors are more important for tracts that continue to lateralize later in life (Budisavljevic et al., 2015a,b). See also Chapters 29 and 30.

Fetal brain tractography identifies long range fibers as early as 15 weeks of gestation (Song et al., 2015) with a clearer tract delineation in older fetuses (Mitter et al., 2015). In general, before birth fibers are unmyelinated and anisotropy signal originates primarily from an increase in axonal density and microarchitectural axonal complexity (Takahashi et al., 2012). Most studies have reported the presence of bilateral tracts, although in a significant number of fetal brains tracts can be seen only in the left or in the right hemisphere (Mitter et al., 2015; Song et al., 2015). The relative low resolution of diffusion images, the fetal movements, and the variations of the signal-to-noise ratio (SNR) between hemispheres in relation to the position of fetal head are factors that might affect the ability to identify tracts in one of the hemispheres (Mitter et al., 2015). It is, therefore, most likely that reported fetal interhemispheric differences are currently related to technological limitations rather than true underlying tract asymmetries.

Just after birth myelin is deposited along axons by maturing oligodendrocytes and this process leads to further changes in the diffusion signal (Drobyshevsky et al., 2005). Progressive myelination leads to a significant increase in fractional anisotropy, allowing for a more reliable identification and quantification of tract anatomy with tractography. In infants an asymmetry of fractional anisotropy is reported for some associative (arcuate fasciculus) and projection (corticospinal tract) pathways (O'Muircheartaigh et al., 2014; Dubois et al., 2009), a finding which confirms an intense postnatal myelination in the left hemisphere reported by studies using other imaging modalities (O'Muircheartaigh et al., 2013). This microstructural asymmetry is not accompanied by a clear volume asymmetry (Dubois et al., 2015; Song et al., 2015), which is usually detected later in life (Lebel et al., 2012; Budisavljevic et al., 2015a,b). At this early stage, interhemispheric asymmetry in volume may be more subtle and involve only some global measures of connectivity derived from graph analysis (Ratnarajah et al., 2013).

In older children and adolescents, the possibility of longer scanning sessions permits acquiring diffusion data of higher quality and more sensitive estimates of volume asymmetry in larger cohorts. One advantage is the possibility to separate large tracts into smaller bundles and identify more complex patterns of asymmetry within the same tract. The arcuate fasciculus, for example, can be further divided into three segments connecting different regions of the perisylvian language cortex (Catani et al., 2005, 2007). This analysis reveals different developmental trajectories and asymmetries for the three segments (Budisavljevic et al., 2015a,b). Before adolescence the long segment shows a significant leftward volume asymmetry, while the anterior segment displays an opposite pattern of asymmetry. Unlike these two segments, which remain lateralized through adolescence and early adulthood, the more posterior segment connecting temporoparietal regions shows a rightward asymmetry before adolescence but becomes progressively more symmetrical in adolescence. The loss of asymmetry is due to the reduction in volume of the connections in the right hemisphere. Twin studies suggest different factors drive these modifications: genetic and shared environmental factors for tracts that lateralize early, specific environmental factors for networks that continue to change later in life (Budisavljevic et al., 2015a,b, 2016).

Once established, further changes in the asymmetry pattern are unlikely to occur in the adult brain. This is not to say that in the adult population asymmetry is similar across subjects or between males and females (Catani et al., 2007). Our current understanding of network asymmetry suggests that in adult brains complex cognitive functions such as language, visuospatial attention, and emotion and memory processing rely on the coordinated activity of several tracts whose asymmetry is highly heterogeneous (Fig. 10). Within the same tract, measures of volume and microstructure (e.g., fractional anisotropy) are usually concordant, although in some tracts asymmetry is more evident only for one measure (e.g., long segment of the arcuate and the inferior longitudinal fasciculus in the language domain). Mapping tract interhemispheric differences has helped to gain important information on language recovery after left hemisphere stroke (Forkel et al., 2014).

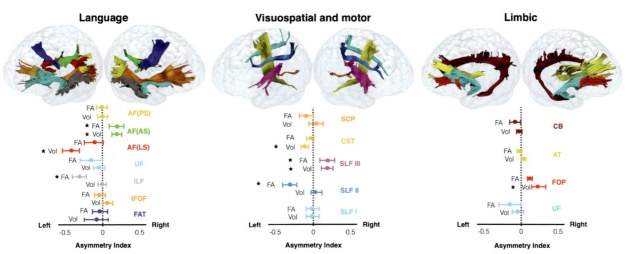

**FIG. 10** The complexity of white matter asymmetry for large-scale system networks dedicated to language, visuospatial attention and motor action, and emotion and memory (limbic). In the healthy adult population, most of the language tracts are bilateral or left lateralized except the anterior segment of the arcuate fasciculus, whose rightwards asymmetry could be related to prosody, musical cognition, or visuospatial attention. Asterisks indicate statistically significant asymmetries. AF(LS), arcuate fasciculus (long segment); AF(PS), arcuate fasciculus (posterior segment); AF(AS), arcuate fasciculus (anterior segment). *AT*, anterior thalamic projections; *CB*, cingulum bundle; *CST*, corticospinal tract; *FAT*, frontal aslant tract; *FOP*, frontal-orbitopolar tract; *IFOF*, inferior fronto-occipital fasciculus; *ILF*, inferior longitudinal fasciculus; *SCP*, superior cerebellar peduncle; *SLF I*, superior longitudinal fasciculus, superior branch; *SLF II*, superior longitudinal fasciculus, middle branch; *UF*, uncinate fascicles. *(Based on Catani, M., Allin, P.G.A., Husain, M., Pugliese, L., Mesulam, M.M., Murray, R.M., Jones, D.K., 2007. Symmetries in human brain pathways predict verbal recall. Proc. Natl. Acad. Sci. USA 104, 17163–17168, Thiebaut de Schotten, M., ffytche, D.H., Bizzi, A., Dell'Acqua, F., Allin, M., Walshe, M., Murray, R., Williams, S.C., Murphy, D.G., Catani, M., 2011. Atlasing location, asymmetry and inter-subject variability of white matter tracts in the human brain with MR Atlasing location, asymmetry and inter-subject variability of white matter tracts in the human brain with MR diffusion tractography. NeuroImage 54, 49–59, and Thiebaut de Schotten, M., Dell'Acqua, F., Forkel, S., Simmons, A., Murphy, D.G., Catani, M., 2011. A lateralized brain network for visuospatial attention. Nat. Neurosci. 14, 1245–1246.)*

In older age, changes in the diffusion parameters (e.g., decreased fractional anisotropy, increased "neurite" dispersion) are usually bilateral and related to progressive gliosis, myelin loss, and tract atrophy or disarray, which characterize white matter tissue in the elderly brain (Lundgaard et al., 2014; Phillips et al., 2013). It is also true that brain aging is highly heterogeneous and asymmetric changes could affect particular subgroups, such as elderly people at risk of dementia (ApoE carriers with greater right hemisphere differences in the dorsal cingulum) (Brown et al., 2011) or subjects with primary progressive aphasia with predominant left hemisphere damage (Catani et al., 2013a,b; D'Anna et al., 2016; Forkel et al., 2020).

## 8 Conclusions

In the last 20 years diffusion tractography has contributed to describing the connectional anatomy of the human brain more than any other method available. The improved characterization of major brain pathways, the quantification of tract asymmetry between hemispheres, and the description of tract variability among individuals has required an update of the international nomenclature and classical textbooks. There is still room for major developments in the field and admittedly for most brain disorders tractography remains only a research tool. But the utility of tractography for teaching anatomy is unquestionable and its clinical application is steadily spreading among neurosurgeons and neuroradiologists for the benefits of surgical patients.

## Acknowledgments

I would like to thank Samuel Craig for his helpful comments on the chapter, Peter Ratiu for providing histological images used for Fig. 1, and Daniele Cancemi for creating images of Fig. 6.

## References

Barrett, R.L.C., Dawson, M., Dyrby, T.B., Krug, K., Ptito, M., D'Arceuil, H., Croxson, P.L., Johnson, P.J., Howells, H., Forkel, S.J., Dell'Acqua, F., Catani, M., 2020. Differences in frontal network anatomy across primate species. J. Neurosci. 40 (10), 2094–2107.

Basser, P.J., Mattiello, J., LeBihan, D., 1994. Estimation of the effective self-diffusion tensor from the NMR spin echo. J Magn Reson B 103, 247–254.

Beaulieu, C., 2002. The basis of anisotropic water diffusion in the nervous system—a technical review. NMR Biomed. 15, 435–455.

Bishop, D.V.M., 2013. Cerebral asymmetry and language development: cause, correlate or consequence? Science 340, 1230531.

Brown, J.A., Terashima, K.H., Burggren, A.C., Ercoli, L.M., Miller, K.J., Small, G.W., Bookheimer, S.Y., 2011. Brain network local interconnectivity loss in aging APOE-4 allele carriers. Proc. Natl. Acad. Sci. USA 108, 20760–20765.

Budisavljevic, S., Dell'Acqua, F., Rijsdijk, F.V., Kane, F., Picchioni, M., McGuire, P., Toulopoulou, T., Georgiades, A., Kalidindi, S., Kravariti, E., Murray, R.M., Murphy, D.G., Craig, M.C., Catani, M., 2015a. Age-related differences and heritability of the perisylvian language networks. J. Neurosci. 35, 12625–12634.

Budisavljevic, S., Kawadler, J.M., Dell'Acqua, F., Rijsdijk, F.V., Kane, F., Picchioni, M., McGuire, P., Toulopoulou, T., Georgiades, A., Kalidindi, S., Kravariti, E., Murray, R.M., Murphy, D.G., Craig, M.C., Catani, M., 2015b. Heritability of the limbic networks. Soc. Cogn. Affect. Neurosci.

Budisavljevic, S., Dell'Acqua, F., Zanatto, D., Begliomini, C., Miotto, D., Motta, R., Castiello, U., 2016. Asymmetry and structure of the fronto-parietal networks underlie visuomotor processing in humans. Cereb. Cortex 11 (5), 746–757.

Catani, M., ffytche, D.H., 2005. The rises and falls of disconnection syndromes. Brain 128 (10), 2224–2239.

Catani, M., Thiebaut de Schotten, M., 2008. A diffusion tensor imaging tractography atlas for in vivo dissections. Cortex 44, 1105–1132.

Catani, M., Thiebaut de Schotten, M., 2012. Atlas of Human Brain Connections, Oxford University Press.

Catani, M., Zilles, K., 2020. Cerebral hemispheres. In: Standring, S. (Ed.), Gray's Anatomy, the Anatomical Basis of Clinical Practice, forty-second ed. Elsevier, Amsterdam.

Catani, M., Howard, R.J., Pajevic, S., Jones, D.K., 2002. Virtual in vivo interactive dissection of white matter fasciculi in the human brain. NeuroImage 17, 77–94.

Catani, M., Jones, D., ffytche, D., 2005. Perisylvian language networks of the human brain. Ann. Neurol. 57 (1), 8–16.

Catani, M., Allin, P.G.A., Husain, M., Pugliese, L., Mesulam, M.M., Murray, R.M., Jones, D.K., 2007. Symmetries in human brain pathways predict verbal recall. Proc. Natl. Acad. Sci. USA 104, 17163–17168.

Catani, M., Craig, M.C., Forkel, S.J., Kanaan, R., Picchioni, M., Toulopoulou, T., Shergill, S., Williams, S., Murphy, D.G., McGuire, P., 2011. Altered integrity of perisylvian language pathways in schizophrenia: relationship to auditory hallucinations. Biol. Psychiatry 70 (12), 1143–1150.

Catani, M., Dell'Acqua, F., Bizzi, A., Forkel, S.J., Williams, S.C., Simmons, A., Murphy, D.G., Thiebaut de Schotten, M., 2012. Beyond cortical localization in clinico-anatomical correlation. Cortex 48 (10), 1262–1287.

Catani, M., Thiebaut de Schotten, M., Slater, D., Dell'Acqua, F., 2013a. Connectomic approaches before the connectome. NeuroImage 80, 2–13.

Catani, M., Mesulam, M.M., Jakobsen, E., Malik, F., Martersteck, A., Wieneke, C., Thompson, C.K., Thiebaut De Schotten, M., Dell'Acqua, F., Weintraub, S., Rogalski, E., 2013b. A novel frontal pathway underlies verbal fluency in primary progressive aphasia. Brain 136 (8), 2619–2628.

Chechlacz, M., Gillebert, C.R., Vangkilde, S.A., Petersen, A., Humphreys, G.W., 2015. Structural variability within frontoparietal networks and individual differences in attentional functions: an approach using the theory of visual attention. J. Neurosci. 35, 10647–10658.

Concha, L., 2014. A macroscopic view of microstructure: using diffusion-weighted images to infer damage, repair, and plasticity of white matter. Neuroscience 276, 14–28.

Concha, M.L., Bianco, I.H., Wilson, S.W., 2012. Encoding asymmetry within neural circuits. Nat. Rev. Neurosci. 13, 832–843.

Conturo, T.E., Lori, N.F., Cull, T.S., Akbudak, E., Snyder, A.Z., Shimony, J.S., McKinstry, R.C., Burton, H., Raichle, M.E., 1999. Tracking neuronal fiber pathways in the living human brain. Proc. Natl. Acad. Sci. USA 96, 10422–10427.

D'Anna, L., Mesulam, M.M., Thiebaut De Schotten, M., Dell'Acqua, F., Murphy, D., Wieneke, C., Martersteck, A., Cobia, D., Rogalski, E., Catani, M., 2016. Frontotemporal networks and behavioral symptoms in primary progressive aphasia. Neurology 86 (15), 1393–1399.

Dawson, M., Gordon-Fleet, K., Yan, L., Tardos, V., He, H., Mui, K., Nawani, S., Asgarian, Z., Catani, M., Fernandes, C., Drescher, U., 2023. Sexual dimorphism in the social behaviour of Cntnap2-null mice correlates with disrupted synaptic connectivity and increased microglial activity in the anterior cingulate cortex. Commun. Biol. 6 (1), 846. https://doi.org/10.1038/s42003-023-05215-0. In press.

Dell'Acqua, F., Catani, M., 2012. Structural human brain networks: hot topics in diffusion tractography. Curr. Opin. Neurol. 25 (4), 375–383.

Dell'Acqua, Tournier, 2019. Modelling white matter with spherical deconvolution: how and why? NMR Biomed. 32, e3945.

Donkelaar ten, H.J., 2020. Clinical Neuroanatomy: Brain Circuitry and its Disorders. Springer.

Drobyshevsky, A., Song, S.K., Gamkrelidze, G., Wyrwicz, A.M., Derrick, M., Meng, F., Li, L., Ji, X., Trommer, B., Beardsley, D.J., Luo, N.L., Back, S.A., Tan, S., 2005. Developmental changes in diffusion anisotropy coincide with immature oligodendrocyte progression and maturation of compound action potential. J. Neurosci. 25, 5988–5997.

Dubois, J., Hertz-Pannier, L., Cachia, A., Mangin, J.F., Le Bihan, D., Dehaene-Lambertz, G., 2009. Structural asymmetries in the infant language and sensori-motor networks. Cereb. Cortex 19, 414–423.

Dubois, J., Poupon, C., Thirion, B., Simonnet, H., Kulikova, S., Leroy, F., Hertz-Pannier, L., Dehaene-Lambertz, G., 2015. Exploring the early organization and maturation of linguistic pathways in the human infant brain. Cereb. Cortex (Epub ahead of print).

Eluvathingal, T.J., Hasan, K.M., Kramer, L., Fletcher, J.M., Ewing-Cobbs, L., 2007. Quantitative diffusion tensor tractography of association and projection fibers in normally developing children and adolescents. Cereb. Cortex 17, 2760–2768.

ffytche, D.H., Catani, M., 2005. Beyond localization: from hodology to function. Philos. Trans. R. Soc. Lond. B Biol. Sci. 360 (1456), 767–779.

FIPAT, 2017. Terminologia Neuroanatomica. FIPAT.library.dal.ca. Federative International Programme for Anatomical Terminology,.

Forkel, S., Thiebaut de Schotten, M., Dell'Acqya, F., Kalra, L., Murphy, D., Williams, S., Catani, M., 2014. Anatomical predictors of aphasia recovery: A tractography study of bilateral perisylvian language networks. Brain 137 (7), 2027–2039. In press.

Forkel, S.J., Rogalski, E., Sancho, N.D., D'Anna, L., Laguna, P.L., Sridhar, J., Dell'acqua, F., Weintraub, S., Thompson, C., Mesulam, M.-M., Catani, M., 2020. Anatomical evidence of an indirect pathway for word repetition. Neurology 94 (6), E594–E606.

Foster, M., Sherrington, C.S., 1897. A Textbook of Physiology, Part Three: The Central Nervous System, seventh ed. MacMillan & Co Ltd, London.

Frasnelli, E., Vallortigara, G., Rogers, L.J., 2012. Left-right asymmetries of behaviour and nervous system in invertebrates. Neurosci. Biobehav. Rev. 36, 1273–1291.

Geschwind, N., Galaburda, A.M., 1987. Cerebral Lateralization: Biological Mechanisms, Associations and Pathology. MIT, Cambridge MA.

Herculano-Houzel, S., 2012. The remarkable, yet not extraordinary, human brain as a scaled-up primate brain and its associated cost. PNAS 109 (supplement_1), 10661–10668.

Jbabdi, S., Sotiropoulos, S.N., Haber, S.N., Van Essen, D.C., Behrens, T.E., 2015. Measuring macroscopic brain connections in vivo. Nat. Neurosci. 18, 1546–1555.

Jones, D.K., 2003. Determining and visualizing uncertainty in estimates of fiber orientation from diffusion tensor MRI. Magn. Reson. Med. 49, 7–12.

Jones, D., Simmons, A., Williams, S., Horsfield, M., 1999. Non-invasive assessment of axonal fiber connectivity in the human brain via diffusion tensor MRI. Magn. Reson. Med. 42 (1), 37–41. In press.

Jones, D.K., Knösche, T.R., Turner, R., 2013. White matter integrity, fiber count, and other fallacies: the do's and don'ts of diffusion MRI. NeuroImage 73, 239–254.

Lamar, M., Catani, M., Price, C.C., Heilman, K.M., Libon, D.J., 2008. The impact of region-specific leukoaraiosis on working memory deficits in dementia. Neuropsychologia 46 (10), 2597–2601.

Le Bihan, D., 2003. Looking into the functional architecture of the brain with diffusion MRI. Nat. Rev. Neurosci. 4, 469–480.

Lebel, C., Gee, M., Camicioli, R., Wieler, M., Martin, W., Beaulieu, C., 2012. Diffusion tensor imaging of white matter tract evolution over the lifespan. NeuroImage 60, 340–352.

Lundgaard, I., Osorio, M.J., Kress, B.T., Sanggaard, S., Nedergaard, M., 2014. White matter astrocytes in health and disease. Neuroscience 276, 161–173.

Meynert, T., 1885. Clinical Treatise on Diseases of the Fore-brain Based Upon a Study of Its Structure, Functions, and Nutrition. Putnam, London.

Mitter, C., Prayer, D., Brugger, P.C., Weber, M., Kasprian, G., 2015. In vivo tractography of fetal association fibers. PLoS One 10. eCollection.

Mori, S., Crain, B.J., Chacko, V.P., van Zijl, P.C., 1999. Three-dimensional tracking of axonal projections in the brain by magnetic resonance imaging. Ann. Neurol. 45, 265–269.

Mount, C.W., Monje, M., 2017. Wrapped to adapt: experience-dependent myelination. Neuron 95 (4), 743–756.

Mountcastle, V.B., 1957. Modality and topographic properties of single neurons of cat's somatic sensory cortex. J. Neurophysiol. 20 (4), 408–434.

O'Muircheartaigh, J., Dean, D.C., Dirks, H., Waskiewicz, N., Lehman, K., Jerskey, B.A., Deoni, S.C., 2013. Interactions between white matter asymmetry and language during neurodevelopment. J. Neurosci. 33 (41), 16170–16177.

O'Muircheartaigh, J., Dean III, D.C., Ginestet, C.E., Walker, L., Waskiewicz, N., Lehman, K., Dirks, H., Piryatinsky, I., Deoni, S.C., 2014. White matter development and early cognition in babies and toddlers. Hum. Brain Mapp. 35, 4475–4487.

Pajevic, S., Basser, P.J., Fields, R.D., 2014. Role of myelin plasticity in oscillations and synchrony of neuronal activity. Neuroscience 276, 135–147.

Paus, T., Pesaresi, M., French, L., 2014. White matter as a transport system. Neuroscience 276, 117–225.

Petrides, M., 2013. Neuroanatomy of Language Regions of the Human Brain. Academic Press.

Phillips, O.R., Clark, K.A., Luders, E., Azhir, R., Joshi, S.H., Woods, R.P., Mazziotta, J.C., Toga, A.W., Narr, K.L., 2013. Superficial white matter: effects of age, sex, and hemisphere. Brain Connect. 3, 146–159.

Poirazi, P., et al., 2003. Pyramidal neuron as a two-layer neural network. Neuron 37, 989–999.

Ratnarajah, N., Rifkin-Graboi, A., Fortier, M.V., Chong, Y.S., Kwek, K., Saw, S.-M., Godfrey, K.M., Gluckman, P.D., Meaney, M.J., Qiu, A., 2013. Structural connectivity asymmetry in the neonatal brain. NeuroImage 75, 187–194.

Reveley, C., Seth, A.K., Pierpaoli, C., Silva, A.C., Yu, D., Saunders, R.C., Leopold, D.A., Ye, F.Q., 2015. Superficial white matter fiber systems impede detection of long-range cortical connections in diffusion MR tractography. Proc. Natl. Acad. Sci. USA 112, 2820–2828.

Rilling, J.K., Glasser, M.F., Preuss, T.M., Ma, X., Zhao, T., Hu, X., Behrens, T.E., 2008. The evolution of the arcuate fasciculus revealed with comparative DTI. Nat. Neurosci. 11, 426–428.

Schmahmann, J.D., Pandya, D.N., 2006. Fiber Pathways of the Brain. Oxford University Press, Oxford.

Shu, N., Li, X., Ma, C., Zhang, J., Chen, K., Liang, Y., Chen, Y., Zhang, Z., 2015. Effects of APOE promoter polymorphism on the topological organization of brain structural connectome in nondemented elderly. Hum. Brain Mapp.

Song, S.K., Sun, S.W., Ju, W.K., Lin, S.J., Cross, A.H., Neufeld, A.H., 2003. Diffusion tensor imaging detects and differentiates axon and myelin degeneration in mouse optic nerve after retinal ischemia. NeuroImage 20, 1714–1722.

Song, J.W., Mitchell, P.D., Kolasinski, J., Grant, P.E., Galaburda, A.M., Takahashi, E., 2015. Asymmetry of white matter pathways in developing human brains. Cereb. Cortex.

Sporns, O., Tononi, G., Kötter, R., 2005. The human connectome: a structural description of the human brain. PLoS Comput. Biol. 1 (4), e42.

Takahashi, E., Folkerth, R.D., Galaburda, A.M., Grant, P.E., 2012. Emerging cerebral connectivity in the human fetal brain: an MR tractography study. Cereb. Cortex 22, 455–464.

Thiebaut de Schotten, M., ffytche, D.H., Bizzi, A., Dell'Acqua, F., Allin, M., Walshe, M., Murray, R., Williams, S.C., Murphy, D.G., Catani, M., 2011a. Atlasing location, asymmetry and inter-subject variability of white matter tracts in the human brain with MR diffusion tractography. NeuroImage 54, 49–59.

Thiebaut de Schotten, M., Dell'Acqua, F., Forkel, S., Simmons, A., Murphy, D.G., Catani, M., 2011b. A lateralized brain network for visuospatial attention. Nat. Neurosci. 14, 1245–1246.

Thiebaut de Schotten, M., Dell'Acqua, F., Valabregue, R., Catani, M., 2012. Monkey to human comparative anatomy of the frontal lobe association tracts. Cortex 48, 82–96.

Thomas, C., Ye, F.Q., Irfanoglu, M.O., Modi, P., Saleem, K.S., Leopold, D.A., Pierpaoli, C., 2014. Anatomical accuracy of brain connections derived from diffusion MRI tractography is inherently limited. Proc. Natl. Acad. Sci. USA 111, 16574–16579.

Toga, A.W., Thompson, P.M., 2003. Mapping brain asymmetry. Nat. Rev. Neurosci. 4, 37–48.

Viscardi, L.H., Imparato, D.O., Bortiolini, M.C., Dalmolin, R.J.S., 2021. Ionotropic receptors as a driving force behind human synapse establishment. Mol. Biol. Evol. 38, 735–744.

von Bartheld, C.S., Bahney, J., Herculano-Houzel, S., 2016. The search for true numbers of neurons and glial cells in the human brain: a review of 150 years of cell counting. J. Comp. Neurol. 524 (18), 3865–3895.

Vos, S.B., Jones, D.K., Viergever, M.A., Leemans, A., 2011. Partial volume effect as a hidden covariate in DTI analyses. NeuroImage 55, 1566–1576.

Wildenberg, G., Rosen, M., Lundell, J., Paukner, D., Freedman, D., Kasthuri, N., et al., 2021. Primate neuronal connections are sparse in cortex as compared to mouse. Cell Rep. 36 (11), 109709.

Yoo, S.S., Park, H.J., Soul, J.S., Mamata, H., Park, H., Westin, C.F., Bassan, H., Du Plessis, A.J., Robertson Jr., R.L., Maier, S.E., Ringer, S.A., Volpe, J.J., Zientara, G.P., 2005. In vivo visualization of white matter fiber tracts of preterm- and term-infant brains with diffusion tensor magnetic resonance imaging. Investig. Radiol. 40, 110–115.

# Chapter 2

# Neurobiology of connections

**Giorgio M. Innocenti[a,b,†] and Kathleen S. Rockland[c]**

[a]*Karolinska Institutet, Stockholm, Sweden,* [b]*Ecole Polytechnique Federale Lausanne, Lausanne, Switzerland,* [c]*Boston University Chobanian and Avedisian School of Medicine, Boston, MA, United States*

## 1 Introduction

The central nervous system consists of two main compartments, the gray matter and the white matter. The first relates to accumulations of neuronal cell bodies, which are assembled in nuclei and in laminated structures. The second consists of the accumulations of axons, interconnecting the neuronal cell bodies. This is but an approximate subdivision. Some neurons are sparsely distributed in the white matter where they might subserve axonal guidance in development and axon maintenance in the adult (e.g., Riederer et al., 2004; Mortazavi et al., 2016); and myelinated axons, entering and exiting, intermingle in gray matter, even as was long appreciated in studies of cortical myeloarchitectonics (cf. Amunts and Zilles, 2015). Furthermore, nonneuronal glia are numerous both in the gray and white matter. The glia consist of oligodendrocytes, astrocytes, and microglia, which will be mentioned only occasionally in this chapter.

Orderly spatial organization is characteristic of both gray and white matter. However, it is more evident and has been more extensively investigated in laminated structures such as the cerebral and cerebellar cortices. In this review we will concentrate on the cerebral cortex, more specifically on the neocortex, also called isocortex, from an anatomical mesoscopic resolution perspective.

## 2 The gray matter

### 2.1 Cortical areas

From the time of the great pioneer cartographers, evidence has supported the parcellation of cerebral cortex into multiple areas (Fig. 1). A generally accepted standard for area identification calls for agreement across multiple correlative criteria, most commonly, cytoarchitecture, myeloarchitecture, afferent and efferent connections, and neuronal activation properties (for a recent review, see Amunts and Zilles, 2015). This standard is achieved to a strong degree for primary sensory areas and entorhinal cortex; but for vast stretches of frontal, parietal, and temporal association cortices, the "pencil-sharp" borders of primary sensory areas are replaced by "ill-defined transitions" (Rosa and Tweedale, 2005; Palomero-Gallagher and Zilles, 2019). As stated for one representative example: "Would inferotemporal cortex best be seen as a very large area which changes in character gradually, or as a large number of smaller areas, each with indistinct boundaries (Rosa and Tweedale, 2005)?"

Areas are commonly not uniform but have subarea heterogeneity (Amunts and Zilles, 2015). (1) A conspicuous example is the heterogeneity of the primary and secondary cortical visual areas in primates (areas V1 and V2), as visualized by the periodic pattern of the metabolic enzyme cytochrome oxidase (CO). In these two areas, regions of high CO activity correspond in area V1 to thalamic input from the lateral geniculate nucleus or, in area V2, from the inferior pulvinar nucleus, and have also been related to segregated streams of V1 to V2 cortical connections (Sincich and Horton, 2005; and, for recent discussion, Garg et al., 2019; Vanni et al., 2020). (2) There is subarea interconnectivity by callosal connections; for V1 and V2, in several species, these are restricted to the representation of the vertical meridian at the border of the two areas, and extend only partially into V2 (reviewed in Innocenti et al., 1986). (3) A recent study of receptor density patterns in the macaque parietal lobe confirmed the location and extent of six areas previously defined by cytoarchitectonic, connectivity, or functional criteria, and provided new evidence for further subdivisions within a dorsal parietal area PE (Impieri et al., 2019).

There are topographically based subdivisions within what are usually considered single areas. In primary motor cortex, the layer-5 corticospinal Betz cells are larger in the most medial part, corresponding to the foot and leg representation

---

[†] Deceased, January 12, 2021.

*Handbook of Diffusion MR Tractography.* https://doi.org/10.1016/B978-0-12-818894-1.00033-1
Copyright © 2025 Elsevier Ltd. All rights are reserved, including those for text and data mining, AI training, and similar technologies.

**20 PART | I** From anatomy to tractography

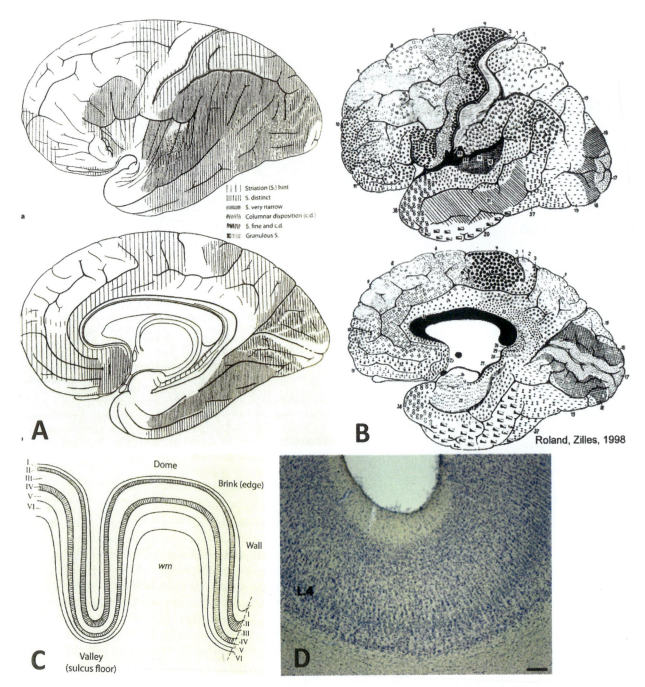

**FIG. 1** (A and B) Demarcation of cortical regions in humans according to (A) cellular verticality (Von Economo, 1927/2009) and (B) cellular architecture (Brodmann in Roland and Zilles, 1998), with distinguishable areas denoted by symbols and numbers. Lateral hemisphere view (at top) and medial surface view (at bottom). (C) Schematic (Von Economo, 1927/2009) to show laminar deformation at the gyral dome ("crown") and sulcal depth. The deeper layers are expanded or constricted, respectively, at the gyral crown and sulcal depth. (D) Nissl stain for cell bodies shows the constriction of the deeper layers and expansion of layer 1 at the sulcal depth. *L4*, layer 4. Scale bar = 200 μm. *(Panel (B) reproduced with permission from Roland, P.E., Zilles, K., 1998. Structural divisions and functional fields in the human cerebral cortex. Brain Res. Brain Res. Rev. 26, 87–105.)*

(Rivara et al., 2003). In area V1, representations of peripheral vision, in the anterior calcarine fissure, selectively receive projections from some parietal (Borra and Rockland, 2011) and auditory areas (Rockland and Ojima, 2003); and representations of central vision in V1, with some individual variability, send projections to extrastriate area V4 (Ungerleider et al., 2008). Extrastriate area MT/V5 receives connections from retrosplenial cortex (area 23v and prostriata), but only in the representation of the far visual field periphery, a zone which also has strong inputs from area V1 but little or no input from area V2 and ventral stream extrastriate areas (for marmoset: Palmer and Rosa, 2006). Subareal variations in input density (i.e., by quantification of synapses) would seem generally likely (DeFelipe et al., 2002), but have only scantily been investigated in humans or nonhuman primates (NHPs).

Thus the delineation and organization of discrete areas are not straightforward and remain to some extent problematic, both technically and conceptually. Functionally, in an early formulation of distributed processing, one reads in Brodmann that "'supposed elementary functional loci' [i.e., areas] are active in different numbers, in different degrees, and in differing combinations" (in Zilles, 2018; and see also, among others, Behrmann and Plaut, 2013, "Distributed circuits, not circumscribed centers..."). In a similar vein, focused on how areas functionally interact in a supraareal organization, evidence from recent imaging studies of multiple individuals reveals complex spatial topographies not evident in group-averaged studies (Braga et al., 2019: "...models should be open to the possibility that architectonic boundaries may not align with long-range connectivity in the adult brain in any simple manner for all zones of cortex").

## 2.2 Cortical layers

The early identification of cortical layers was based on differences in cell density and cell size (e.g., discussion in Zilles, 2018). From this perspective, six layers have conveniently been designated, from the cell-sparse uppermost layer 1 (zone of distal apical dendrites, along with scattered interneurons and glia) to the deep layer 6, bordering the white matter. This general framework distinguishes a "granular" layer 4 (where "granular" = stellate cells or, more commonly, small pyramids), approximately midway in the cortical depth; supragranular layers 1–3; and infragranular layers 5 and 6. Species- and area-specific variations are common, the most conspicuous being the elaborate sublaminar stratification of layer 4 in primate area V1, the radically diminished layer 4 in motor cortex, and the three laminar specializations of entorhinal cortex (elaborate layer 2 with cell islands, cell sparse layer 4 ("lamina dissecans"), and onionskin layer 6). (For myeloarchitectonics, see Amunts and Zilles, 2015.)

As a generalization, about 80% of cortical neurons are excitatory pyramidal neurons, so called for the shape of their cell body and outlined dendritic tree, and 20% inhibitory interneurons. Interneurons are further subdivided on the basis of dendritic morphology, postsynaptic targets, and neurochemical and genetic characteristics (Kubota et al., 2011; DeFelipe et al., 2013). With only a few exceptions, cortical interneurons are intrinsic to a given area and do not give rise to long-distance connections (but see Tomioka and Rockland, 2007). Subtype distribution is species- and area-dependent. For example, a mapping of "complex" basket cell terminations in the postmortem human brain shows a decreasing density across cortical areas 4, 3b, 13, and 18, approximately complementary to an increasing density for chandelier type terminations (respectively targeting cell somas or axon initial segments; Blazquel-Lorca et al., 2010). Differential distributions of three neurochemically distinct subpopulations of interneurons have also been reported in visual areas of the occipital and temporal lobes (DeFelipe et al., 1999), and medial prefrontal cortex (Gabbott and Bacon, 1999), both in the macaque monkey.

Until recently, excitatory pyramidal neurons appeared to be less heterogeneous than interneurons, although subtypes were distinguishable by dendritic morphology and long-distance connectivity, often in combination (Jones and Wise, 1977; Katz, 1987; Nhan and Callaway, 2012). Transcriptomic analyses, however, have rapidly provided evidence for a high degree of heterogeneity within pyramidal neuron populations (for rodent hippocampus, see Cembrowski and Spruston, 2019; Sugino et al., 2019).

## 2.3 Vertical organization

Early architectonic maps also observed cellular verticality, more pronounced in some areas than others (Fig. 1A, from figure 81 in Von Economo, 1927/2009). Later anatomical studies, using antibodies against cytoskeletal proteins (MAP2, microtubule associated protein), identified layer-5 pyramidal cell apical dendrites as key components of the cortical verticality, and further determined quantitative aspects of the microarchitecture (reviewed in Peters and Sethares, 1996; Peters et al., 1997; Rockland and Ichinohe, 2004; Innocenti and Vercelli, 2010). A detailed morphometric analysis of BA 25, 32, and 32' in humans, for example, reported per bundle a mean of 54 dendritic profiles varying in diameter from <1.0 to 4.0 µm (mean 2.3 µm; Gabbott, 2003; for macaque visual areas: Peters and Sethares, 1991; Peters et al., 1997). To some extent, the dendritic bundles ("minicolumns") correlate with alternating high or low density domains of thalamic terminations (~100 µm center-to-center), especially obvious at the border of layers 1 and 2 (for rodent and monkey visual areas: Ichinohe et al., 2003; Ji et al., 2015). The output organization has been technically more difficult to investigate, but one study using a suite of double retrograde tracers provides evidence for an orderly arrangement of different subpopulations of pyramidal cell dendritic bundles in relation to distinct combinations of output targets (Innocenti and Vercelli, 2010).

## 2.4 Gyrification

A basic feature of cerebral cortex in the human brain and that of many other mammals, including NHPs, is the presence of gyrification (Fig. 1C and D). It is often assumed that gyrification is an evolutionary strategy to increase cortical surface while minimizing cortical volume. It might also decrease the length of cortico-cortical connections (Zilles et al., 2013; Llinares-Benadero and Borrell, 2019). Gyrification imposes cellular heterogeneity of crown, wall, and sulcus, with pronounced laminar deformation at the gyral crowns and sulcal depths (respectively, exaggeration or compression of the infragranular layers). This deformation affects the alignment of supragranular pyramidal cell apical dendrites, which either fan out in relation to cell bodies (at gyral crowns), or crowd together and converge (at sulcal depths) (figure 13 in Rockland, 2019). Sulcal patterns are species specific but not fixed. They change over the lifespan; namely, sulcal width is reported to increase at a rate of about 0.7 mm/decade, while sulcal depth decreases at a rate of about 0.4 mm/decade (Kochunov et al., 2005). Age- and function-related changes in cortical folds have been related to changes in performance over the default mode network in older adults (Jockwitz et al., 2017). Quantitative and qualitative effects on connectivity are likely but remain to be investigated.

## 2.5 Development of gray matter and of gyri

The developing brain is characterized by proliferative zones where immature neurons are generated, and whence these subsequently migrate to their final location and acquire adult morphology. Two such zones give rise to the cortical gray matter in NHP: the periventricular (subventricular) zone and the median eminence at the base of the hypothalamus. The cerebral cortex develops by the appearance of a deep layer of ontogenetically older neurons, followed by addition of later generated neurons to the overlying, more superficial layers (Rakic, 1974). The appearance of gyri and sulci is concomitant with the development of the gray matter but what precisely causes the appearance of gyral folds has been long debated (Zilles et al., 2013; Garcia et al., 2018). The most likely hypothesis is that folding is caused by the differential development of superficial and deep layers and possibly by the mechanical imbalances that this causes (reviewed in Lui et al., 2011; Borrell, 2018; Garcia et al., 2018). This is linked to the enhanced proliferation in the ventricular zone, and more specifically, as mentioned previously, in the outer subventricular zone (Dehay et al., 2015). Consistent with this view, the naturally occurring or induced elimination of the deep cortical layers in development leads to narrowed gyri resembling the outcome in congenital microgyria (Innocenti and Berbel, 1991). In mice, changes in the intercellular adhesion of migrating neurons result in the development of macroscopic sulci with preserved lamination and radial glia morphology (Del Toro et al., 2017).

## 2.6 Connections (outputs)

Cortical output connections can be visualized by retrograde tracer injections in the target structures, which then reveal labeled somata and their laminar distribution in the source, originating area. On this basis, a rather clear laminar distribution has been elucidated; namely, cortical layers 1 and 4 have only few output neurons, if any; the supragranular layers (those above layer 4) are home to primarily cortico-cortical projections, and the infragranular layers (below layer 4) contain an approximately stratified mix of cortico- and cortico-subcortically projecting neurons (Shipp, 2007; Rockland, 2019). Some neurons in layer 4 give rise to callosal projections transitorially in development (De Leon Reyes et al., 2019). Among those that maintain the callosal projection in layer 4 (Innocenti and Fiore, 1976) are pyramidal neurons that lose their apical dendrite in development (Vercelli et al., 1992).

Pyramidal neurons in both supra- and infragranular layers usually have intrinsic axon collaterals (local, within the home area) as well as extrinsic projections (extending longer distances, beyond the home area of the originating neuron). Some supra- and infragranular pyramidal neurons do not have extrinsic, but only intrinsic projections (rat: Kita and Kita, 2012; Thomson, 2010; and monkey: Gilbert and Wiesel, 1983; Katz, 1987; Briggs and Callaway, 2001; Nassi and Callaway, 2009). It is likely that some of these neurons have eliminated a transient axon in development (Innocenti, 1986).

Details are species- and area-specific, with notable differences between primary and early sensory areas and association areas. For V2 and other early extrastriate visual areas in NHP (for current reviews: Shipp, 2007; Rockland, 2019; Vanni et al., 2020), layers 2, 3A, and 6 contain neurons with cortical "feedback" projections (i.e., terminating largely in layer 1 of the target area); layer 3 has neurons with cortical feedforward and/or callosal projections, terminating mainly in layer 4; layer 5 has neurons projecting subcortically to the superior colliculus and pulvinar, but also some cortico-cortical neurons; layer 6 has neurons that project to the thalamus, to the claustrum, or to cortical feedback targets, but also those that have only intrinsic collaterals (Thomson, 2010).

In higher-order association areas, some neurons in layer 3 project to the striatum, in addition to the more numerous corticostriatal projections from layer 5 (Borra et al., 2021; Haber, 2016; Shipp, 2017). Neurons in layer 5 contribute to cortico-cortical connections; but separate layer 5 subpopulations project to thalamic nuclei, the striatum, or amygdala. Neurons in layer 6 project cortically or to the thalamus or claustrum (and area-specific, striatum: Borra et al., 2021). Another gauge of area specialization is that the proportion of neurons projecting to various targets is not uniform, but varies across the areas. Area V1, for example, does not project to the striatum and V2 does so only weakly (Shipp, 2017); corticothalamic neurons in layer 6 outnumber those in layer 5, but the proportions are not uniform across areas (Erickson and Lewis, 2004; Xiao et al., 2009). Projections to the amygdala originate from supra- and infragranular layers, with area-specific differential density (Stefanacci and Amaral, 2000; Hoistad and Barbas, 2008; Cho et al., 2013). In summary, there are some general rules for laminar specific organization of connections, but these vary widely according to species, modality, and area.

## 2.7 Connections (inputs)

Cortical inputs originate from local intrinsic and more distant (a.k.a., extrinsic) sources. Of these, the greatest number of excitatory synapses is of local origin, from collaterals of other pyramidal neurons in the same area (Anderson et al., 1998; da Costa and Martin, 2011; Martin et al., 2017). Extrinsic excitatory inputs are from other cortical areas, ipsi- and contralateral; thalamic nuclei; claustrum; and for many areas, amygdala. Inhibitory inputs derive from local interneurons; but the subthalamic nucleus zona incerta sends widespread inhibitory input to layer 1 (Mitrofanis, 2005). Cholinergic, noradrenergic, dopaminergic, and serotonergic neuromodulatory inputs vary in distribution and density according to areas and species.

From extracellular injections of anterograde tracers (typically about or greater than 1.0 mm in diameter) in experimental animals, terminations are known to target specific cortical layers. In general, in NHP, layer 4 is a major input layer for thalamocortical terminations, although in association areas, the terminal zone is lower layer 3 and upper layer 4 (pulvino-cortical: Levitt et al., 1995; mediodorsal: Erickson and Lewis, 2004); and layer 1 is targeted by other thalamo-cortical subpopulations. Amygdalocortical terminations target lower layer 1 in area V1 but are more widely dispersed, except for layer 4, in temporal cortical areas (Freese and Amaral, 2005).

Cortico-cortical connections have been further divided into at least two broad subtypes (Fig. 2, and Box 1), often called "feedforward" and "feedback" (or "forward" and "back," Rockland and Pandya, 1979; Ungerleider and Desimone, 1986; reviewed in Angelucci et al., 2017; Rockland, 2019; Vanni et al., 2020). This is supported by multiple criteria; namely, (1) distinct layers of origin (respectively, layer 3 vs layers 2, 3A, and 6), (2) distinct layers of termination (respectively, layers 3 and/or 4 vs layers 1, 2, and variably layer 6), and (3) distinct spatial configuration of the terminal arbors (one to four spatially separate, compact arbors in layer 4 vs a highly divergent terminal segment in layers 1, 2, sometimes bearing delimited terminal clusters at irregular intervals). Fourth, applicable to early visual pathways, is a quasifunctional progression from primary to higher cortical areas ("feedforward") and the reciprocating "feedback" projections, from higher areas. A useful additional category is of "intermediate" or "lateral" connections ("intermediate" in Ungerleider and Desimone, 1986; Maunsell and Van Essen, 1983; D'Souza and Burkhalter, 2017). By this is understood areas in a "lateral" relationship (i.e., not "upper" or "lower" level of a hierarchy). The laminar signature of lateral connections is of terminations across multiple layers ("columnar") with neurons of origin also in multiple layers (particularly, layers 3 and 5). This category is widely applicable for describing connections among the association areas.

Excitatory intrinsic (a.k.a., "local" or "home area") connections, representing the horizontally extended collaterals ("daisy") of pyramidal neurons within any cortical area, are concentrated in layers 3 and 5 (reviewed in Rockland, 2019). Curiously, there is heterogeneity in the set of collaterals from a single pyramidal neuron, in that these can be myelinated or not. For four intracellularly filled and reconstructed neurons in cat area V1, two of four linear segments were myelinated (Koestinger et al., 2017). In area V1 of primates, the densely myelinated stria of Gennari in layer 4B consists of local collaterals of excitatory neurons and not thalamo-cortical axons. The number of collaterals, range and/or average, is not known, and to determine this would require a large sample of intracellularly filled neurons with complete axonal visualization.

An important comment on laminar specific terminations is the fact that basal, oblique, and apical dendrites extend beyond the home layer of the originating neuron soma. Thus data on the laminar dispersion of actual postsynaptic targets are still relatively lacking. Input to layer 1, for example, can be assumed to terminate on distal apical dendrites, but do these derive from neurons in layers 2, 3, or 5 (most layer-6 neurons do not extend dendrites to layer 1)? Electron microscopic studies, of which there are comparatively few, provide reliable data about whether a synapse is onto a pyramidal or interneuron dendrite; but further information about the location and identity of the parent neuron requires another approach, such as complete reconstruction from dendrite to soma. In a related issue, axons commonly have terminal arbors in more

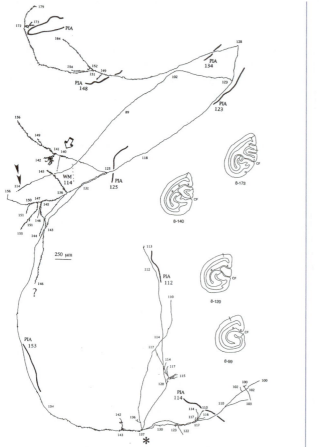
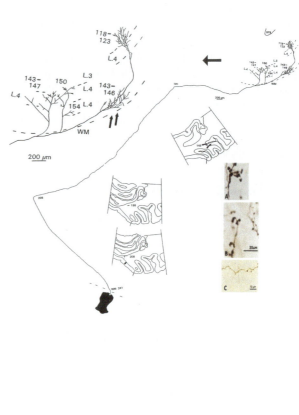

FIG. 2 Cortically projecting axons are phenotypically distinct in terms of layers of termination and spatial organization. In the sensory cortical areas, feedforward axons (at right) have 1–3 spatially delimited arbors (~250 μm in diameter), which typically target 4. An exception, shown here, are the projections from V1 to MT/V5 in the macaque. These have multiple arbors, but these distribute in layer 6 *(double arrows)* as well as layer 4. The V1 to MT/V5 projections are also unusual in that they terminate with relatively large boutons, ~3.0 μm in diameter at the light microscopic level (A and B at right). Most cortico-cortical terminations are more typically small (~1.0 μm in diameter, C at right). Feedback axons (at left) differ in several features. As shown here for an axon projecting from area MT to V1 in squirrel monkey, feedback axons are spatially divergent and distribute in layer 1. *Conventions:* Section outlines (coronal at left and horizontal at right) indicate anterior-posterior (at left) and dorsoventral (at right) axon extent. Numbers are used to denote the Z, depth axis, where 20 section numbers = 1.0 mm. For the feedback axon (at left) *double arrowheads* point to the main axon, as this enters the gray matter to ascend to layer 1 and bifurcates ("pia," section 131). At right, the *solid arrow* points to a higher magnification view of the distalmost portion of the V1 to MT/V5 axon in the gray matter. *Solid black* indicates injection site (Inj.). *(Reproduced with permissions from Rockland, K.S., Knutson, T., 2000. Feedback connections from area MT of the squirrel monkey to areas V1 and V2. J. Comp. Neurol. 425, 345–368 and Rockland, K.S., 1995. Morphology of individual axons projecting from area V2 to MT in the macaque. J. Comp. Neurol. 355, 15–26.)*

than one layer, but, without complete reconstruction of single axons, the multiple arbors are likely to be overlooked: for example, the smaller axon collaterals in layer 6 of some geniculocortical axons (Blasdel and Lund, 1983) and of V1 projections to area MT/V5 (Rockland, 1989).

## 2.8 Receptors

Receptors are molecular complexes that establish the connection between a neuron and its environment. They can be visualized by ligands, some of which, labeled by short-lived radioactivity, can be used in the human brain. The majority of receptors are specific to neurotransmitters used by axons afferent to the cortex or intrinsic to the cortex. A long series of papers from Karl Zilles and collaborators now exists on receptor density across areas ("receptor fingerprints") in the postmortem human and macaque monkey brains (e.g., Palomero-Gallagher and Zilles, 2019, and cited references). To briefly summarize, highest densities of most receptors are found in the supragranular layers, presumably corresponding to maximal synaptic density. The laminar profiles of receptor densities, however, "rarely coincide with the histologically defined borders of layers" (Palomero-Gallagher and Zilles, 2019); that is, receptor densities are best described as changing

Box 1 **Nomenclature of connections and axons.**

**Soma, or cell body (commonly 20 μm in diameter, but ranging both smaller and larger):** where the axon starts.

**Parent axon:** the portion of the axon between soma and first branching point (a.k.a., "proximal").

**Initial segment (∼20–60 μm long):** the beginning of the axon near the soma and the site of action potential initiation. Frequently, the target of inhibitory terminations from local parvalbumin-positive chandelier cells.

**Axonal arbor:** the delimited, concentrated cluster of terminations (a.k.a., synaptic boutons). Most cortical pyramidal neurons establish a proximal (intrinsic) arbor near the cell body and one or more distal arbors at distant target locations. Arbor diameter is highly variable but typical is ∼200 μm.

**Growth cone:** highly motile structure at the distal, "steering end" of growing axons.

**Branches:** axonal segments, calculated by convention in terms of the distal progression from cell soma to target structures. See "collateral."

**Axon collateral:** branches of one axon. These can be "intrinsic" (adjacent to the soma and in the same cortical area) or "extrinsic" (originating more distally from the axon, as this traverses the white matter) and directed to distant postsynaptic target structures. Intrinsic collaterals vary in number and not all neurons have intrinsic collaterals. Extrinsic collaterals are also variable. As is becoming increasingly clear, a given projection (e.g., V1 to extrastriate areas) can consist of some neurons that project to one target, but also others that project to variable numbers of multiple targets (e.g., Han et al., 2018).

**Horton-Strahler ordering:** Each axonal branch is denoted by a number, with the distalmost branches being n 1 (e.g., Tettoni et al., 1998; Anderson et al., 2002).

**Boutons:** Swellings along the axon ("en passant" bouton) or originating from a short stem ("boutons terminaux"). They usually correspond to synaptic boutons, although some passing boutons might be "varicosities" (e.g., packets of mitochondria) traveling between soma and distal terminals. One bouton may correspond to several synapses (Anderson et al., 1998).

**Conduction and transmission compartments:** axonal segments with or without synaptic boutons.

**Feedforward and feedback connections:** Axons interconnecting two areas, but originating from neurons in different layers and terminating in different layers. The terms were initially used in the visual system, where areas V1, V2, and V4 were considered the initiation of sequential "feedforward" pathways, with reciprocating "feedback" connections (see Rockland and Pandya, 1979; Maunsell and Van Essen, 1983 for an early usage). Association area connections might better be called "lateral."

**Convergence:** many to one. Can refer to a neuron (receiving hundreds or thousands of connections) or area (receiving multiple input connections).

**Divergence:** one to many. Can refer to a neuron (projecting to multiple targets, and hundreds or thousands of other neurons in total) or area (with neurons projecting to multiple targets).

**Drivers (Class 1) and modulators (Class 2):** axon types characterized by different morphological and physiological properties (Sherman and Guillery, 2011) referring to their supposed efficacy in activating the postsynaptic targets (see text).

**Bundle, fascicle, band, stria, tract:** terms designating Identifiable assemblies of tightly spaced axons; sometimes used as synonymous although traditionally ascribed specifically to given topographical entities (e.g., corticospinal tract, arcuate fasciculus, diagonal band of Broca, acoustic striae, etc.)

**Axonal dispersion:** should strictly refer to axons as these acquire loose spacing; synonymous with "defasciculation." Not to be confused with:

**Axonal misalignment:** axons with crossing or wavy trajectories within a tract, bundle, etc.

**Decussation:** crossing of axonal assemblies to noncorresponding (heterotopic) levels of the contralateral brain hemisphere (e.g., chiasm, pyramidal tract).

**Commissure:** crossing of axonal assemblies to corresponding (homotopic) levels of the contralateral brain hemisphere. Some commissures, e.g., corpus callosum, contain decussating (heterotopic) axons (e.g., corticostriatal; callosal projections; Innocenti et al., 2017).

**Streamline:** Computational construct obtained by an algorithmic procedure iteratively following the main orientations of water diffusion in the white matter, presumably representing several tightly spaced axons, not necessarily the same axons all along. Some of the definitions above can be extended to streamlines.

in a gradual manner, probably due to the fact that the majority of receptors are located not on cell bodies but on dendrites, which are spatially extended through the layers.

# 3  The white matter

The organization of white matter was first investigated by gross dissection (Klingler's technique). This consists of manual dissection of axonal bundles in fixed, and repeatedly frozen and defrozen brains. By this process, the gross orientation of major fiber bundles can be delineated. The technique does not provide precise information on origin and termination of axons but is still useful for comparison with the main white matter bundles identifiable with dMRI (De Benedictis et al.,

2016; Zemmoura et al., 2016; Sarubbo et al., 2019). Later, anterograde labeling after tracer injections was introduced in experimental animals (cf. Schmahmann and Pandya, 2006). A relatively simple scheme distinguishes at least six major fiber bundles. These are ipsilateral cortico-cortical association fibers, corticostriatal fibers, contralateral cortico-cortical fibers, corticopontine fibers, and the thalamocortical and corticothalamic projections (reviewed in Schmahmann et al., 2008). These all have further complex subdivisions, and data are still sparse on the finer composition in terms of fiber diameters, the relative fiber density contributed by different sources, or their spatial organization within the major bundle. In more recent developments, fiber diameter and orientation can be accessed in postmortem tissue by diattenuation imaging (Menzel et al., 2019), a modification of polarized light imaging (PLI), offering significantly improved resolution; and a new tissue preparation method (MAGIC) that allows high-resolution, label-free fluorescence imaging of myelinated fibers (Costantini et al., 2021).

In all species, as best we know, the considerable number of corticofugal axons (exiting the cortex) travel in a compact fascicle (Fig. 3) until they reach a decision point such as the entrance to the internal capsule or the corpus callosum (CC), where they defasciculate (disperse). Such tight axonal bundles are not revealed by diffusion tractography, probably because of the bidirectional nature of streamline visualization. Indeed, by contrast, corticopetal axons (toward a cortical target) reach the cortex scattered through the white matter. In addition, some corticofugal axons reach nearby locations coursing in the gray matter; and in gyrencephalic brains, axons usually described as "U fibers" reach nearby locations coursing under the gray matter at the fundus of sulci. Up to now, there are relatively few high-resolution studies of bundle trajectories, all necessarily in NHP (cf. Lehman et al., 2011; Morecraft et al., 2017).

The system of U-fibers is frequently described as short association fibers interlinking adjacent gyri (discussed in Schmahmann and Pandya, 2006). While apparent U-fibers can readily be identified in myelin stains and PLI (Takemura et al., 2020), and after injections of anterograde tracers in experimental animals, their source is still under investigation. That is, only some gyri interconnect via U-fibers (for tractography, see Oishi et al., 2011); and some seeming U-fibers consist of axons passing close to a sulcal depth but in passage to a further target (as: axons from the inferior parietal lobule passing below the intraparietal sulcus, but directed to cingulate cortex on the medial surface (personal observations)).

**FIG. 3** Right: Collage of sequential sections with axons anterogradely labeled by an injection in area 9 of the macaque monkey. The majority of axons initially course in a tight bundle until they separate (lower inset) to course to the internal capsule (IC) and the corpus callosum (CC). Other axons labeled by the same injection travel within or subjacent to the gray matter to nearby cortex (inset at left). Dorsal is at top, medial is to the left. Left: Histological reconstruction of axons coursing to area 9/46v from the injection shown at right. (The image is rotated counterclockwise to accommodate formatting.) Axonal segments (microtome sectioned) are charted in three sections spaced 240 μm apart and then color coded (*blue*, intermediary; *cyan*, most superficial section; *yellow*, deepest). Axons are superimposed on the myelin counterstained middle section. Insets are histological photographs of labeled axons. *(Data from G. Innocenti.)*

# 4 The axon

Cortical neurons exhibit enormous diversity. Diversity is the combined result of evolution and development, and it is presumably the key to the relation between brain structure and function. Neuronal diversity is expressed by somatodendritic morphological features, including the size of the cell body, the length and distribution of the dendrites, presence and typology of dendritic spines, and variety of receptors and ionic channels (Markram et al., 2004; Kanari et al., 2019 and above). Neurons also differ in the morphology of their axons in terms of diameter, length, distribution of branches (Winnubst et al., 2019, among others), and type of neurotransmitter released.

We will focus on the morphology of long axons in particular, as these interconnect cortical sites or cortical and subcortical structures. The trajectory of homogeneous groups of axons can be mapped with diffusion MRI with progressively improving resolution. We have summarized the most commonly used axonal nomenclature in Box 2. The diversity of axonal morphology can, to some extent, be related to connectional type: thalamocortical axons, cortical feedforward or feedback connections, as well as the different types of cortico-descending axons (Shipp, 2007; Rockland, 2020). Excitatory axons of different categories and species seem to obey similar principles of organization (Tettoni et al., 1998; Rouiller and Welker, 2000).

One of the principles is that axons have morphologically and functionally distinguishable proximal-distal parts (i.e., adjacent to and distant from the soma, respectively). Closely adjacent to the soma, most excitatory axons have a sheaf of inhibitory terminations embracing the specialized axon initial segment. Beyond this is the conduction compartment, by which the action potential is transported to the terminations, without establishing synaptic contacts. The distal, transmission compartment establishes chemical synaptic contacts, most of which use glutamate or aspartate as a transmitter, although other neurotransmitters—GABA, catecholamines, acetylcholine—can be used in different systems (Tettoni et al., 1998; Anderson et al., 2002). Some axons do not observe this separation, but establish synaptic contacts all along their trajectory; for example, the parallel fibers of the cerebellum or the corticostriatal axons (Parent and Parent, 2006; Innocenti et al., 2017), and some cortico-cortical axons (feedback: Rockland and Drash, 1996). These axons exhibit in fact a very broadly distributed transmission compartment, although this can still be discerned as "distal" and in fact is commonly restricted to the gray matter.

The fact that axons are specialized for the transmission of information in the form of axon potentials should not obscure the fact that they also implement molecular transport back and forth between the cell bodies and terminals and also with the environment. Furthermore, it is possible that molecular exchanges among neurons take place outside axons, via the extracellular spaces (volume conduction; Marcoli et al., 2015).

The view of the axon as an electric cable, akin to what is found in man-made devices, has influenced much of the work in the "classical" era of the neurosciences and up to about two decades ago when new anatomical techniques allowed to visualize and reconstruct single axons in a detail that could reveal their enormous complexity (e.g., Winnubst et al., 2019, in mouse). Indeed the classical view is profoundly misleading. The axon performs at least three computational operations at the core of brain function. The first operation is **mapping**, i.e., the axon maps the position of its soma of origin onto the

---

**Box 2 Nomenclature of axonal ultrastructure.**

**Axoplasm**: cytoplasm of the axon, containing cytoskeleton, water, organelles.

  **Cytoskeleton**: includes microtubules, neurofilaments, microfilaments, actin specializations (Leterrier et al., 2017).

  **Microfilaments:** actin, organized in rings.

  **Microtubules:** tubulin dimers that mediate anterograde and retrograde transport via kinesin and dynein transporters.

  **Neurofilaments:** light, medium, and high (LMH) interwoven proteins together with alpha-internexin, which assure the radial organization of the axon.

  **Myelin:** cytoplasmic lamellae ($\sim$40% water) of an oligodendrocyte wrapped around a segment of axon. The dry mass consists of 70%–85% lipids and 15%–30% proteins (Quarles, 2002).

  **g-ratio:** the ratio between inner and outer axonal diameter (inclusive of myelin).

  **Ranvier nodes:** periodic interruptions of the myelin sheath, and site of ionic exchanges involved in action potential conduction.

  **Chemical synapse:** site of transmission of the nerve impulse. The synaptic complex consists of the presynaptic, axonal element, containing synaptic vesicles and, frequently, mitochondria; a postsynaptic dendritic, somatic, or (for chandelier cells) axonal element; and 200–400 Å wide synaptic cleft. Excitatory and inhibitory synapses are associated with differences in the synaptic vesicles, chemical neurotransmitter (glutamate vs GABA), and synaptic cleft.

  **Electrical synapse** (a.k.a., **gap junction**): a close membrane apposition (with a gap of $\sim$20 A), between two pre- and postsynaptic neuronal elements (respectively, axonal and dendritic). Similar close astrocyte-astrocyte appositions are reported.

distribution of its synaptic boutons. This may provide a representation and replication of the parent neuron state in its multiple target brain region. The second operation is **differential amplification**, i.e., the axon can be viewed as the device that multiplies the action potential generated at the cell body into the neurotransmitter release at its synaptic boutons. In this way an axon can have stronger or weaker impact on the target depending on the number, distribution, and size of synaptic boutons. The third operation is **temporal transformations**, i.e., the axon can conduct action potentials at different speed and over different lengths, therefore generating delays between the activation of the cell body and that of the axon terminals. All these operations are initially under the control of developmental mechanisms that we will shortly mention. To fully understand the operations performed by the single axon, the properties of the postsynaptic location where the axon terminates should also be understood, which is, unfortunately, usually not the case. Here concepts of decremental vs linearized dendritic transmission become important, as well as dendritic hot-spots and other issues (Gidon et al., 2020).

## 4.1 Mapping

Axons engaged in long-range projections vary in the distribution of their branches even within individual pathways (e.g., in mouse; Han et al., 2018; Morita et al., 2019). Axons engaged in callosal connections between visual areas of the cat, in general terminate in one or two closely spaced terminal arbors, but a few callosal axons distribute more broadly, with up to seven terminal clusters over a territory of several hundred micrometers (Houzel et al., 1994).

In the occipital lobe of the monkey some extreme cases of axons diverging to broadly separate targets have been reported (Rockland and Drash, 1996). These were all "feedback" axons from temporal areas to lower-order areas, suggesting that divergence might be greater in these axons than in the "feed-forward" axons.

The axons with the broadest distribution of terminal arbors, calculated as total length, might be the cholinergic axons to cortex, which in the mouse reach up to 50 cm (Wu et al., 2014), although other axonal systems in rodents might reach comparable values. Even larger axonal arbors might be expected in primates, including humans.

One crucial question is: what is achieved by the divergence of axonal projections? One would expect some systematic relationship between the activation properties (receptive fields or others) of a parent neuron and those of the neurons targeted by its axon. In the primary visual cortex a systematic relationship has been reported between the orientation specificity of a parent neuron and that of its targets within V1, contacted via collaterals (Gilbert and Wiesel, 1989) although such an absolute relationship was later questioned (Kisvarday and Eysel, 1992; Martin et al., 2014). Another factor is receptive field size, which tends to increase from lower-order (closer to thalamus) to higher-order (further away from thalamus) areas. The information carried by the highly divergent feedback axons might mediate differences in receptive field size. Consistent with this possibility, the inactivation of feedback connections from V2 to V1 decreases the inhibitory responses of the receptive field periphery of V1 neurons (Nassi et al., 2013; Nurminen et al., 2018).

Developmental data suggest that functional matching between an axon and its targets is required for the stabilization of juvenile connections. The formation of axonal connections involves a complex series of operations. Axons grow in the white matter guided by attractive and repulsive cues. These cues include the response to diffusible substances as well as interactions with white matter structures, neurons, glial elements, and other axons. Molecules driving axonal growth have been identified (reviewed in Raper and Mason, 2010; Kolodkin and Tessier-Lavigne, 2011). After reaching the targets, some axons "wait," form ephemeral branches, or establish transient interactions with elements of the target. In some of the best studied cases it appears that axons are endowed with markers that signal their origin (positional cues) and predispose the axons to find complementary cues in the target. A large number of axons appear unable to find such cues and are eliminated (Innocenti and Price, 2005; Innocenti, 2020). The final validation of the connection, i.e., of the geometry of the terminal arbor, is activity dependent.

## 4.2 Differential amplification

The strength or efficacy of connections is important but currently hard to evaluate. It is often equated with the number of retrogradely filled neurons in a tracer experiment; but the data from anterograde experiments may be the more telling. "Strength" might depend on the number of connections reaching a target as well as on the distribution and type of synaptic boutons (best visualized by anterograde tracers) carried by the individual axons. In other words, the strength of connections critically depends on the number, density, and distribution of synapses, and their size. It also depends on the dendritic location of the synapses and convergence with other inputs, and may change depending on the state of the neuron, i.e., whether it was partially depolarized by subthreshold activations. In neocortex, the typically powerful excitatory drive of thalamocortical axons is associated with a distal axonal section with long terminal branches, richly endowed with synaptic boutons and occasionally giant synapses (Tettoni et al., 1998; Groh et al., 2008; Florence and Casagrande, 1987). The majority of boutons are found on the most peripheral branches (Tettoni et al., 1998; Anderson et al., 2002), which can equally indicate random or cell specific distribution of contacts (Hellwig et al., 1994; Anderson et al., 2002).

An increasing number of electron microscopic studies provide data on how synaptic structure relates to the physiological transmission strength; for example, by identifying functionally relevant human-specific specializations (Yakoubi et al., 2019), or by demonstrating multivesicular release (Holler et al., 2021).

In the corticothalamic projections, two classes of axonal terminations have been described across systems and species (Rouiller and Welker, 2000). One class is characterized by large synaptic boutons (2–5 µm in diameter), the other by smaller synaptic boutons, usually <1 µm in diameter. These differences gave rise to the robust classification of corticothalamic axons as, respectively, "drivers" (originating from a few layer-5 neurons per area) and "modulators" (more numerous, and from layer 6) (Sherman and Guillery, 2006, 2011; but see Bickford, 2016, for discussion). The two classes have other distinguishing properties, in addition to synaptic size; namely, number of terminations (respectively, few and many), activation by postsynaptic ionotropic or metabotropic receptors, and size of EPSPs, among others (Table 7.1 in Sherman and Guillery, 2006). This classification ("driving" vs "modulatory") has been extended to cortico-cortical axons, but with fewer supporting criteria.

Innocenti and Caminiti (2017) sampled at light microscopic level (LM) 249 terminal boutons and 520 passing boutons distributed on thick (0.64 µm on average) branches of callosal axons to the primary motor cortex of the monkey. The average diameter of the first (boutons terminaux) was 1.05 µm (sd 0.34) and for the second (boutons en passant) 0.97 µm (sd 0.34). Corresponding boutons on thin (0.21 µm, on average) branches were, respectively, 0.66 (n =94; sd 0.17) and 0.38 µm (n = 418; sd 012). Since the callosal axons to motor cortex range among the largest CC axons (Tomasi et al., 2012), these findings suggest, taking bouton size alone as a major criterion, that CC axons might fall in the modulatory class. However, as noted earlier, continuing work on synaptic structure and function highlights the complexity of these issues and the intricate interactions of bouton size and other parameters, such as neurotransmitter release sites (Yakoubi et al., 2019; Holler et al., 2021).

In a series of LM/EM studies in visual areas of macaque, Anderson and Martin (2002, 2006, 2009) investigated bouton size as a measure of specialization and efficacy of connections between V1 and MT, V2 and MT, V1 and V2, and V2 and V4. At the LM level, these were reported as relatively uniform and small (average diameter = 0.5 µm; Anderson and Martin, 2009), with the notable exception of some terminations from V1 to MT (diameter = 3.0 µm; Anderson et al., 1998) and from V2 to V1 ($d$ = 1.0 µm; Anderson and Martin, 2009). Taking postsynaptic density size (EM) as an estimate for synaptic efficacy, these authors report an average across connections of 0.11 µm$^2$ with relatively little average variation, although synapses in all connections exhibited a wide range of size (0.032–0.17 µm$^2$; Anderson and Martin, 2009). The apparently large size in LM of the V1 boutons in area MT was ascribed to a complex ultrastructural topography including multisynaptic contacts, postsynaptic perforations, and frequent engulfment of postsynaptic spines (Anderson et al., 1998; Anderson and Martin, 2002).

The interpretation that cortico-cortical axons might be modulatory is to some extent supported by the sparseness of their axonal terminal arbors, when compared with the thalamocortical axons (Tettoni et al., 1998; Florence and Casagrande, 1987). Important to note, however, is the distinction between intrinsic and extrinsic connectivity for the same neuron, where the number of intrinsic synapses is numerically the greater. Moreover, synaptic populations of the same axon terminating in different sites can differ significantly. This statement is supported by (in mice) EM analysis of individual identified thalamocortical axons targeting both motor and somatosensory cortices, which reports 1.5–2.0 larger synapses in the motor cortex (1.13–1.42 µm$^2$ in motor; 0.74–0.96 µm$^2$ in somatosensory). The conclusion is that an axon can impact its multiple targets with different synaptic strength (Rodriguez-Moreno et al., 2020), which is another aspect of the differential amplification implemented by the axon.

## 4.3 Development of synaptic connections

What determines the features of the distal, transmission part of an axon? In development, axons undergo a phase of exuberant production of synapses (Huttenlocher et al., 1982; Rakic et al., 1986; Aggoun-Zouaoui et al., 1996; Bressoud and Innocenti, 1999; Innocenti and Price, 2005). This is followed by a phase of synaptic reduction, which in several cases appears to be activity dependent (Zufferey et al., 1999; Riccomagno and Kolodkin, 2015). Experiments aimed at determining if the size of synaptic boutons depends on the axon itself or on the target have been performed by switching retinal projections to the somatosensory or auditory thalamus in development (Campbell and Frost, 1988). The results stress the role of the target since the retinal axons established synaptic boutons recognizably similar to those established by somatosensory axons rather than typical retinal boutons.

## 4.4 Temporal transformations

The finite speed of action potential conduction causes the axon to behave as a delay line. In other words, the activation of the molecular machinery that leads to neurotransmitter release is delayed relative to the action potential generated in the cell body. As a first approximation, the delay depends on the speed of action potential propagation and on the length of the axon. The speed of action potential propagation (conduction velocity; CV) depends on: (1) the diameter of the axon, (2) the thickness of the myelin sheath (for myelinated axons), (3) the distance between Ranvier nodes, since the conduction is "saltatory," i.e., jumps from node to node, although a submyelin conduction compartment was recently described (Cohen et al., 2020). The structural factors determining the speed of action potential conduction have been known since the work of Hursh (1939), Rushton (1951), and Waxman and Bennet (1972), and have been revisited recently by the elegant work of Drakesmith et al. (2019).

This simple theoretical relationship is complicated by several factors. First, the diameter varies locally along the axon (Gao et al., 2019; Lee et al., 2019). One can therefore expect locally uneven conduction speed although, to our knowledge, this has never been demonstrated physiologically. However, over long distances, the diameter of axonal bundles appears to remain on average constant (Innocenti et al., 2019) which facilitates a computation of average conduction delays of axonal pathways. Second, the "g-ratio" between inner and outer axon diameter, the latter including myelin thickness, tends, as predicted (Rushton, 1951; Smith and Koles, 1970), to stabilize around a value of 0.7 for large axons (above 2 µm; Fig. 3 in Innocenti, 2017). It is smaller for smaller axons, which, therefore, appear to have a thicker myelin sheath. Furthermore, there appears to be considerable scatter in the g-ratio, although it is unclear if this is the consequence of chance measurements along transversely sectioned axons. Third, the internodal distance is proportional to axon diameter, but with some degree of scatter among individual axons (Hess and Young, 1952; Ibrahim et al., 1995). Finally, the speed of action potential conduction appears to depend on the previous history of the axon (Swadlow et al., 1980; Swadlow, 1985). A further complication derives from theoretical considerations, suggesting failure of action potential propagation at branching point (Segev and Schneidmann, 1999) which, however, was not confirmed for all cases (Popovic et al., 2011). Further attention should be focused on the possibility that axons of different diameters may conduct different spike frequencies (Perge et al., 2012).

The speed of action potential conduction impacts temporal transformations along individual axons as well as across different axons. Transformation within single axons can be guessed on the grounds of axonal morphology. Clearly there are two extreme axonal morphologies. In one, axons deliver synaptic boutons and terminal arbors over long and sometimes tortuous trajectories. Examples are the parallel fibers in the cerebellum, where single axons can synapse along their trajectories with multiple Purkinje cells. Corticostriatal axons present a similar, divergent morphology (Parent and Parent, 2006; Innocenti et al., 2017; Morita et al., 2019). Similar albeit less extreme morphologies were found among the visual callosal axons of the cat (Houzel et al., 1994) and, as mentioned earlier, many cortical feedback axons to layer 1 have a widely divergent geometry (Suzuki et al., 2000; Rockland, 2020). The second type is axons that branch far from their termination, and then take nearly parallel courses to nearby but spatially separate terminal arbors. In the latter case, simulation of action potential propagation suggested that these axonal morphologies, which clearly violate Cajal's principle of conservation of "protoplasme nerveux transmetteur" (Innocenti et al., 1994), are instrumental in conducting action potential synchronously to spatially separate targets. A "hybrid" morphology might be distinguished where an axon is mainly divergent, with boutons distributed over long segments of several millimeters, but where small clusters (~50–100 µm across) of boutons also occur intermittently (Rockland and Drash, 1996; Rockland and Knutson, 2000; Suzuki et al., 2000; Anderson and Martin, 2009).

The renewed interest in temporal transformations across axons was triggered by the finding that axons of different diameter occur in different pathways, and sometimes within the same tract (as the CC, Caminiti et al., 2009, or the corticospinal tract, Innocenti et al., 2019). In the CC of the monkey, an anteroposterior differential trend of axon diameters was described (LaMantia and Rakic, 1990). This was found to correspond to pathways linking different cortical areas (Caminiti et al., 2009; Tomasi et al., 2012). The thickest axons were in the midbody of the CC corresponding to connections between primary motor and somatosensory areas. Thick CC axons also interconnect areas V1 and V2, coursing in the splenium. Thinner axons were found elsewhere in the CC, particularly in the genu, connecting prefrontal areas (Fig. 4).

From axon diameter, CV can be computed and with pathway length, the conduction delay between areas of the two hemispheres (Caminiti et al., 2009, 2013; Tomasi et al., 2012). The shortest conduction delays were found for the motor and primary somatosensory areas, and the longest delays for the temporal areas. From the delays calculated in rhesus monkey, the delays were extrapolated to humans, taking into account the measured differences in axon diameters in the two species and the volumetric differences of the brains. The overall message of these studies was, in confirmation of Ringo et al. (1994), that the human brain had become "slower" compared to the monkey. Interestingly, the spread

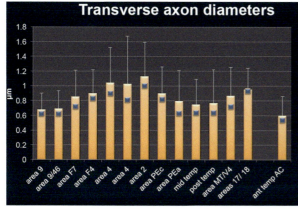
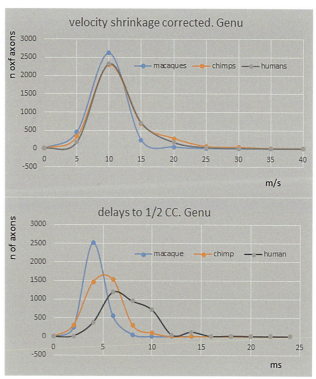
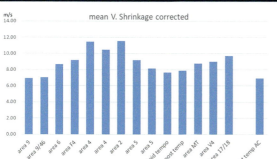

FIG. 4 Top Left: Means (with standard deviations) and medians *(squares)* of measurements of commissural axon diameters in transverse sections. Bottom Left: Mean computed conduction velocities for the same axons. Velocities are calculated from axon diameters assuming a 30% shrinkage due to fixation. Top Right: Distribution of mean conduction velocity (V) computed for axons crossing the genu of the corpus callosum in macaques, chimpanzee, and humans, corrected for histological shrinkage. Notice that a comparable small fraction of axons with $V \geq 15$ m/s is added to chimpanzee and humans. Bottom Right: Computed conduction delays to the mid genu in the three species. Notice that the increased axon diameters do not compensate for the increased length of the axons in chimpanzee and humans, which therefore acquire longer conduction times. *(Top Left and Bottom Left: Reproduced with permission from Tomasi, S., Caminiti, R., Innocenti, G.M., 2012. Areal differences in diameter and length of corticofugal projections. Cereb. Cortex 22, 1463–1472. Top Right: Data from Caminiti, R., Ghaziri, H., Galuske, R., Hof, P.R., Innocenti, G.M., 2009. Evolution amplified processing with temporally dispersed slow neuronal connectivity in primates. Proc. Natl. Acad. Sci. U. S. A. 106, 19551–19556.)*

of conduction delays increased. Apparently, evolution compensated for the volumetric increase of the brain only up to 400–500 cc, roughly the volume of the brain of *Australopithecus* but not beyond (Caminiti et al., 2009).

For callosal axons, the conduction delays calculated from histology match the electrophysiological measurements (Tomasi et al., 2012). For the cortico-descending pathways the thickest axons were found to originate from the primary motor cortex and from the supplementary motor cortex (Fig. 5). Thinner axons originated from the premotor and parietal areas (Innocenti et al., 2019). For the cortico-descending projections, the conduction velocities calculated from histology were slower and the conduction delays longer than the electrophysiological measurements (Firmin et al., 2014; Innocenti et al., 2019). The difference might be due to an electrode bias favoring the antidromic activation of large cell bodies (Kraskov et al., 2019), which also give rise to thicker axons (Humphrey and Corrie, 1978), although other explanations might be possible (Kraskov et al., 2019).

Axonal CV and conduction delays are fundamental players in cortical function. Previous work has focused on the role of conduction delays in the generation of cortical rhythms (Caminiti et al., 2009; partially reviewed in Innocenti et al., 2019). However, their role may be more fundamental in "information flow" among cortical areas. If the overall role of cortico-cortical axons is modulatory, information transfer along cortico-cortical pathways might become efficacious only in situations of temporal and spatial convergence as early proposed by the synfire chains hypothesis (Abeles, 1982; Gewaltig et al., 2001). The convergence might, in general terms, involve cortico-cortical axons with identical targets and/or reentrant cortico-thalamo-cortical chains.

The importance of axon diameters and their causal association in determining CV/conduction delays in cortical circuits has not escaped the dMRI community, which has, to some extent, been able to replicate the anatomical findings. Unfortunately, the estimates seem to be biased to axons above 2.5 μm in diameters (Romascano et al., 2020), which are the

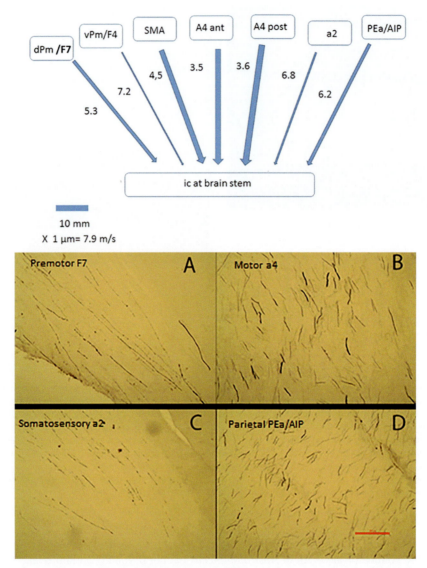

**FIG. 5** Cortical descending axons. Upper panel (schematic representation): *Thickness of arrows* is proportional to median axon thickness; *length of arrows* corresponds to that of streamlines from cortical ROIs to distal internal capsule (IC); *numbers* are the conduction delays in milliseconds over the same distance. The scale bar relates to length of the arrows, thickness of the axons, and conduction velocity for a segment of the corresponding length and thickness. Bottom: Photomicrographs of axons anterogradely labeled by BDA injections in (A) premotor cortex area F7; (B) motor cortex area 4; (C) somatosensory cortex area 2; (D) parietal cortex PEa/AIP. Calibration bar is 200 μm. *(Reproduced with permission from Innocenti, G.M., Caminiti, R., Rouiller, E.M., Knott, G., Dyrby, T.B., Descoteaux, M., Thiran, J.P., 2019. Diversity of cortico-descending projections: histological and diffusion MRI characterization in the monkey. Cereb. Cortex 29, 788–801.)*

minority in the central nervous system. In the future this limit might be overcome by the combined usage of scanners capable of stronger magnetic gradients, innovative pulse sequences, and modeling based on the anatomical ground truth.

## 4.5 Axon diameter in development

Axon diameter and myelination are key factors in the temporal transformations carried out by the axons. The developmental trajectory of axon diameters was described in the CC of the cat by Berbel and Innocenti (1988) and in that of the monkey by LaMantia and Rakic (1990). In the cat the distribution of axon diameters ranges up to 1 μm at embryonic age E58, and the fraction of axons between 0.5 and 1 μm progressively increases up to postnatal day P150, with a progressive increase in myelinated axons beginning at P26-39. Similarly, in the monkey the histogram of axon diameter shows 0.4–0.6 μm at E85, and the fraction of axons measuring between 0.4 and 0.6 μm progressively increases up to P60, with a concomitant addition of myelinated axons beginning at P30-P60.

A rich literature exists on the developmental trajectory of myelination, which would be impossible to summarize here (see also Nave, 2010; Tomassy et al., 2016; Stassart et al., 2018; Sango et al., 2019; Cohen et al., 2020). A few key elements are: (1) Radial axonal growth and myelination are interrelated. In CNS development, myelination starts on axons that have reached roughly $0.2\,\mu m$ in diameter. The radial growth of the axons depends on cytoskeletal proteins, in particular the neurofilaments (Box 2). (2) Myelination is required for the axon to grow and be maintained. (3) Oligodendrocytes have an intrinsic tendency to produce myelin. In vitro they have been found to myelinate nano-fibers above $0.4\,\mu m$ in diameter (reviewed in Hayashi and Suzuki, 2019). For myelination of small axons a number of molecular players is involved. A single oligodendrocyte usually myelinates several axons. (4) Myelination is activity dependent. This may not be surprising given the important role of myelin in tuning the CV of axons and the importance of axonal CV in generating conduction delays, discussed before. Nevertheless, it is surprising, and unexplained, how oligodendrocytes following an initial phase of exuberant myelination seem to be able to preferentially select active axons for myelination (Hines et al., 2015). (5) There may be different types of oligodendrocytes as suggested by Del Rio Hortega (Hayashi and Suzuki, 2019).

# 5 Conclusions

Brain morphology is a strong predictor of brain function and will probably provide the foundation for large scale continued investigations of brain function. Along this line, two issues should be considered.

An important issue is the strength of cortico-cortical connections. The view that the strength might be derived from the number of neurons engaged in the connections (i.e., via retrograde tracer injections) is misleading. As described earlier, several other parameters must be considered. These include (1) the diameter of the cortical axons, i.e., the velocity of information transfer between neurons; (2) the density of synaptic terminations at the target; (3) the size of synaptic boutons and the synaptic efficacy; (4) the nature of the targets, as inhibitory or excitatory neurons; (5) the degree of convergence in the projections. To these must be added the properties of the dendrites postsynaptic to the axonal terminations, and interactions of the immediate postsynaptic assemblies.

A second issue is the currently popular view that cerebral cortex is an ensemble of areas interconnected in a hierarchical fashion. In this view, the delays implemented by axonal conduction must play a crucial role. Nevertheless, the hierarchical concept is far from being supported unequivocally. First, the traditional view tends to underplay the role of cortico-thalamo-cortical connections. Second, the timing of activation of the various areas only in part supports the hierarchical concept (Nowak and Bullier, 1997; de Lafuente and Romo, 2006; Pessoa and Adolphs, 2010; Kauvar et al., 2020). Rather, the timing of neuronal activation in the different areas overlaps so that neurons "higher" in the hierarchy might activate earlier than neurons low in the hierarchy. Third, lesions or inactivation of cortical areas, aimed at testing the hierarchy concept, are bound to eliminate simultaneously both cortico-cortical connections and cortico-thalamo-cortical loops.

One set of cortico-cortical connections that is less directly vulnerable to the confound of the cortico-thalamo-cortical loops are the callosal connections. Available evidence from studies in the visual cortex in cats shows that inactivation of the primary visual areas does not eliminate visual responses in the contralateral area, but rather modulates them (Schmidt et al., 2010). The modulation is implemented by excitatory and inhibitory inputs, which can alter the response properties of the contralateral neurons to visual stimuli (Payne et al., 1991; Makarov et al., 2008).

Several attempts to link resting state fMRI to cortical connectivity have been put forth in the past. On the whole these studies have so far failed to provide the simple link expected between fMRI activity and connections. Although cutting the CC in epileptic patients alters interhemispheric functional connectivity (Roland et al., 2017), sectioning the CC in the monkey has not modified interhemispheric functional connectivity as expected, unless the anterior commissure is sectioned as well (O'Reilly et al., 2013). The authors state: "We conclude that functional connectivity is likely driven by cortico-cortical white matter connections but with complex network interactions such that a near-normal pattern of functional connectivity can be maintained by just a few indirect structural connections."

There may be other ways of testing the function of cortico-cortical connections in humans based on knowledge derived from animal studies. For example, changes in fMRI signal and EEG coherence (Knyazeva et al., 1999, 2006; Carmeli et al., 2005) were obtained with isooriented visual stimuli in the two hemifields, which from the animal literature can be interpreted as interconnecting visual neurons in the two hemispheres.

Methodologies for the morphological, in vivo study of the human brain are constantly improving, together with improved techniques for "whole brain" analyses. Clearly, these will enable a rapidly expanded knowledge base and predictions about the normal and/or exceptional human brain.

# References

Abeles, M., 1982. Local Cortical Circuits. Springer-Verlag, Berlin.

Aggoun-Zouaoui, D., Kiper, D.C., Innocenti, G.M., 1996. Growth of callosal terminal arbors in primary visual areas of the cat. Eur. J. Neurosci. 8, 1132–1148.

Amunts, K., Zilles, K., 2015. Architectonic mapping of the human brain beyond Brodmann. Neuron 88, 1086–1107.

Anderson, J.C., Martin, K.A., 2002. Connection from cortical area V2 to MT in macaque monkey. J. Comp. Neurol. 443, 56–70.

Anderson, J.C., Martin, K.A.C., 2006. Synaptic connection from cortical area V4 to V2 in macaque monkey. J. Comp. Neurol. 495, 709–721.

Anderson, J.C., Martin, K.A., 2009. The synaptic connections between cortical areas V1 and V2 in macaque monkey. J. Neurosci. 29, 11283–11293.

Anderson, J.C., et al., 1998. The connection from cortical area V1 to V5: a light and electron microscopic study. J. Neurosci. 18, 10525–10540.

Anderson, J.C., et al., 2002. Chance or design? Some specific considerations concerning synaptic boutons in cat visual cortex. J. Neurocytol. 31, 211–229.

Angelucci, A., et al., 2017. Circuits and mechanisms for surround modulation in visual cortex. Annu. Rev. Neurosci. 40, 425–451.

Behrmann, M., Plaut, D.C., 2013. Distributed circuits, not circumscribed centers, mediate visual recognition. Trends Cogn. Neurosci. 17, 210–219.

Berbel, P., Innocenti, G.M., 1988. The development of the corpus callosum in cats: a light- and electron-microscopic study. J. Comp. Neurol. 276, 132–156.

Bickford, M.E., 2016. Thalamic circuit diversity: modulation of the driver/modulator framework. Front. Neural Circuits 9, 86. 1–8.

Blasdel, G.G., Lund, J.S., 1983. Termination of afferent axons in macaque striate cortex. J. Neurosci. 3, 1389–1413.

Blazquel-Lorca, L., et al., 2010. GABAergic complex basket formations in the human neocortex. J. Comp. Neurol. 518, 4917–4937.

Borra, E., Rockland, K.S., 2011. Projections to early visual areas v1 and v2 in the calcarine fissure from parietal association areas in the macaque. Front. Neuroanat. 5, 35. 1–12.

Borra, E., Rizzo, M., Gerbella, M., Rozzi, S., Luppino, G., 2021. Laminar origin of corticostriatal projections to the motor putamen in the macaque brain. J. Neurosci. 41, 1455–1469.

Borrell, V., 2018. How cells fold the cerebral cortex. J. Neurosci. 38, 776–783.

Braga, R.M., van Dijk, K.R.A., Polimeni, J., Eldaief, M.C., Buckner, R.C., 2019. Parallel distributed networks resolved at high resolution reveal close juxtaposition of distinct regions. J. Neurophysiol. 121, 1513–1534.

Bressoud, R., Innocenti, G.M., 1999. Typology, early differentiation, and exuberant growth of a set of cortical axons. J. Comp. Neurol. 406, 87–108.

Briggs, F., Callaway, E.M., 2001. Layer-specific input to distinct cell types in layer 6 of monkey primary visual cortex. J. Neurosci. 21, 3600–3608.

Caminiti, R., Ghaziri, H., Galuske, R., Hof, P.R., Innocenti, G.M., 2009. Evolution amplified processing with temporally dispersed slow neuronal connectivity in primates. Proc. Natl. Acad. Sci. U. S. A. 106, 19551–19556.

Caminiti, R., Carducci, F., Piervincenzi, C., Battaglia-Mayer, A., Confalone, G., Visco-Comandini, F., Pantano, P., Innocenti, G.M., 2013. Diameter, length, speed, and conduction delay of callosal axons in macaque monkeys and humans: comparing data from histology and magnetic resonance imaging diffusion tractography. J. Neurosci. 33, 14501–14511.

Campbell, G., Frost, D.O., 1988. Synaptic organization of anomalous retinal projections to the somatosensory and auditory thalamus: target-controlled morphogenesis of axon terminals and synaptic glomeruli. J. Comp. Neurol. 272, 383–408.

Carmeli, C., Knyazeva, M.G., Innocenti, G.M., De Feo, O., 2005. Assessment of EEG synchronization based on state-space analysis. NeuroImage 25, 339–354.

Cembrowski, M.S., Spruston, N., 2019. Heterogeneity within classical cell types is the rule: lessons from hippocampal pyramidal neurons. Nat. Rev. Neurosci. 20, 193–204.

Cho, Y.T., et al., 2013. Cortico-amygdala-striatal circuits are organized as hierarchical subsystems through the primate amygdala. J. Neurosci. 33, 14017–14030.

Cohen, C.C.H., Popovic, M.A., Klooster, J., Weil, M.-T., Mobius, W., Nave, K.-A., Kole, M.H.P., 2020. Saltatory conduction along myelinated axons involves a periaxonal nanocircuit. Cell 180, 311–322.

D'Souza, R., Burkhalter, A., 2017. A laminar organization for selective cortico-cortical communication. Front. Neuroanat. 11, 17. 1–13.

Costantini, I., Baria, E., Sorelli, M., Matusche, F., Giardini, F., Menzel, M., et al., 2021. Autofluorescence enhancement for label-free imaging of myelinated fibers in mammalian brains. Sci. Rep. 11, 8038.

da Costa, N.M., Martin, K.A.C., 2011. How thalamus connects to spiny stellate cells in the cat's visual cortex. J. Neurosci. 31, 2925–2937.

De Benedictis, A., Petit, L., Descoteaux, M., Marras, C.E., Barbareschi, M., Corsini, F., Dallabona, M., Chioffi, F., Saubbo, S., 2016. New insights in the homotopic and heterotopic connectivity of the frontal portion of the human corpus callosum revealed by microdissection and diffusion tractography. Hum. Brain Mapp. 37, 4718–4735.

de Lafuente, V., Romo, R., 2006. Neural correlate of subjective sensory experience gradually builds up across cortical areas. Proc. Natl. Acad. Sci. U. S. A. 103, 14266–14271.

De Leon Reyes, N.S., Mederos, S., Varela, I., Weiss, L.A., Perea, G., Galazo, M.J., Nieto, M., 2019. Transient callosal projections of L4 neurons are eliminated for the acquisition of local connectivity. Nat. Commun. 10, 4549.

DeFelipe, J., et al., 1999. Distribution and patterns of connectivity of interneurons containing calbindin, calretinin, and parvalbumin in visual areas of the occipital and temporal lobes of the macaque monkey. J. Comp. Neurol. 412, 515–526.

DeFelipe, J., et al., 2002. Microstructure of the neocortex: comparative aspects. J. Neurocytol. 31, 299–316.

DeFelipe, J., et al., 2013. New insights into the classification and nomenclature of cortical GABAergic interneurons. Nat. Rev. Neurosci. 14, 202–216.

Dehay, C., Kennedy, H., Kosik, K.S., 2015. The outer subventricular zone and primate-specific cortical complexification. Neuron 85, 683–694.

Del Toro, D., Ruff, T., Cederfjall, E., Villalba, A., Seyit-Bremer, G., Borrell, V., Klein, R., 2017. Regulation of cerebral cortex folding by controlling neuronal migration via FLRT adhesion molecules. Cell 169, 621–635.

Drakesmith, M., Harms, R., Rudrapatna, S.U., Parker, G.D., Evans, C.J., Jones, D.K., 2019. Estimating axon conduction velocity in vivo from microstructural MRI. NeuroImage 203, 116186.

Erickson, S.L., Lewis, D.A., 2004. Cortical connections of the lateral mediodorsal thalamus in cynomolgus monkeys. J. Comp. Neurol. 473, 107–127.

Firmin, L., Field, P., Maier, M.A., Kraskov, A., Kirkwood, P.A., Nakajima, K., Lemon, R.N., Glickstein, M., 2014. Axon diameters and conduction velocities in the macaque pyramidal tract. J. Neurophysiol. 112, 1229–1240.

Florence, S.L., Casagrande, V.A., 1987. Organization of individual afferent axons in layer IV of striate cortex in a primate. J. Neurosci. 7, 3850–3868.

Freese, J.L., Amaral, D.G., 2005. The organization of projections from the amygdala to visual cortical areas TE and V1 in the macaque monkey. J. Comp. Neurol. 486, 295–317.

Gabbott, P.L., 2003. Radial organisation of neurons and dendrites in human cortical areas 25, 32, and 32′. Brain Res. 992, 298–304.

Gabbott, P.L., Bacon, S.J., 1999. Local circuit neurons in the medial prefrontal cortex (areas 24a,b,c, 25 and 32) in the monkey: I. Cell morphology and morphometrics. J. Comp. Neurol. 364, 567–608.

Gao, R., Asano, S.M., Upadhyayula, S., Pisarev, I., et al., 2019. Cortical column and whole-brain imaging with molecular contrast and nanoscale resolution. Science 363, eaau8302.

Garcia, K.E., Kroenke, C.D., Bayly, P.V., 2018. Mechanics of cortical folding: stress, growth and stability. Philos. Trans. R. Soc. B Biol. Sci. 373 (1759), 1–10.

Garg, A.K., et al., 2019. Color and orientation are jointly coded and spatially organized in primate primary visual cortex. Science 364, 1275–1279.

Gewaltig, M.O., Diesmann, M., Aertsen, A., 2001. Propagation of cortical synfire activity: survival probability in single trials and stability in the mean. Neural Netw. 14, 657–673.

Gidon, A., et al., 2020. Dendritic action potentials and computation in human layer 2/3 cortical neurons. Science 367, 83–87.

Gilbert, C.D., Wiesel, T.N., 1983. Clustered intrinsic connections in cat visual cortex. J. Neurosci. 3, 1116–1133.

Gilbert, C.D., Wiesel, T.N., 1989. Columnar specificity of intrinsic horizontal and corticocortical connections in cat visual cortex. J. Neurosci. 9, 2432–2442.

Groh, A., de Kock, C.P., Wimmer, V.C., Sakmann, B., Kuner, T., 2008. Driver or coincidence detector: modal switch of a corticothalamic giant synapse controlled by spontaneous activity and short-term depression. J. Neurosci. 28, 9652–9663.

Haber, S.N., 2016. Corticostriatal circuitry. Dialogues Clin. Neurosci. 18, 7–21.

Han, Y., Kebschull, J.M., Campbell, R.A.A., Cowan, D., Imhof, F., Zador, A.M., Mrsic-Flogel, T.D., 2018. The logic of single-cell projections from visual cortex. Nature 556, 51–56.

Hayashi, C., Suzuki, N., 2019. Heterogeneity of oligodendrocytes and their precursor cells. Adv. Exp. Med. Biol. 1190, 53–62.

Hellwig, B., Schuz, A., Aertsen, A., 1994. Synapses on axon collaterals of pyramidal cells are spaced at random intervals: a Golgi study in the mouse cerebral cortex. Biol. Cybern. 71, 1–12.

Hess, A., Young, J.Z., 1952. The nodes of Ranvier. Proc. R. Soc. Lond. B Biol. Sci. 140, 301–320.

Hines, J.H., Ravanelli, A.M., Schwindt, R., Scott, E.K., Appel, B., 2015. Neuronal activity biases axon selection for myelination in vivo. Nat. Neurosci. 18, 683–689.

Hoistad, M., Barbas, H., 2008. Sequence of information processing for emotions through pathways linking temporal and insular cortices with the amygdala. NeuroImage 40, 1016–1033.

Holler, S., Koestinger, G., Martin, K.A.C., Schuhknecht, G.F.P., Stratford, K.J., 2021. Structure and function of a neocortical synapse. Nature 591, 111–116.

Houzel, J.C., Milleret, C., Innocenti, G., 1994. Morphology of callosal axons interconnecting areas 17 and 18 of the cat. Eur. J. Neurosci. 6, 898–917.

Humphrey, D.R., Corrie, W.S., 1978. Properties of pyramidal tract neuron system within a functionally defined subregion of primate motor cortex. J. Neurophysiol. 41, 216–243.

Hursh, J.B., 1939. Conduction velocity and diameter of nerve fibers. Am. J. Phys. 27, 131–139.

Huttenlocher, P.R., de Courten, C., Garey, L.J., Van der Loos, H., 1982. Synaptogenesis in human visual cortex—evidence for synapse elimination during normal development. Neurosci. Lett. 33, 247–252.

Ibrahim, M., Butt, A.M., Berry, M., 1995. Relationship between myelin sheath diameter and internodal length in axons of the anterior medullary velum of the adult rat. J. Neurol. Sci. 133, 119–127.

Ichinohe, N., Fujiyama, F., Kaneko, T., Rockland, K.S., 2003. Honeycomb-like mosaic at the border of layers 1 and 2 in the cerebral cortex. J. Neurosci. 23, 1372–1382.

Impieri, D., Zilles, K., Niu, M., Rapan, L., Schubert, N., Galletti, C., Palomero-Gallagher, N., 2019. Receptor density pattern confirms and enhances the anatomic-functional features of the macaque superior parietal lobule areas. Brain Struct. Funct. 224, 2733–2756.

Innocenti, G.M., 1986. Postnatal development of corticocortical connections. Ital. J. Neurol. Sci. (Suppl. 5), 25–28.

Innocenti, G.M., 2017. Network causality, axonal computations, and Poffenberger. Exp. Brain Res. 235, 2349–2357.

Innocenti, G.M., 2020. The target of exuberant projections in development. Cereb. Cortex 30, 3820–3826.

Innocenti, G.M., Berbel, P., 1991. Analysis of an experimental cortical network: i) architectonics of visual areas 17 and 18 after neonatal injections of ibotenic acid; similarities with human microgyria. J. Neural Transplant. Plast. 2, 1–28.

Innocenti, G.M., Caminiti, R., 2017. Axon diameter relates to synaptic bouton size: structural properties define computationally different types of cortical connections in primates. Brain Struct. Funct. 222, 1169–1177.

Innocenti, G.M., Fiore, L., 1976. Morphological correlates of visual field transformation in the corpus callosum. Neurosci. Lett. 2, 245–252.

Innocenti, G.M., Price, D.J., 2005. Exuberance in the development of cortical networks. Nat. Rev. Neurosci. 6, 955–965.

Innocenti, G.M., Vercelli, A., 2010. Dendritic bundles, minicolumns, columns, and cortical output units. Front. Neuroanat. 4, 11. 1–7.

Innocenti, G.M., Clarke, S., Kraftsik, R., 1986. Interchange of callosal and association projections in the developing visual cortex. J. Neurosci. 6, 1384–1409.

Innocenti, G.M., Lehmann, P., Houzel, J.C., 1994. Computational structure of visual callosal axons. Eur. J. Neurosci. 6, 918–935.

Innocenti, G.M., Dyrby, T.B., Andersen, K.W., Rouiller, E.M., Caminiti, R., 2017. The crossed projection to the striatum in two species of monkey and in humans: behavioral and evolutionary significance. Cereb. Cortex 6, 3217–3230.

Innocenti, G.M., Caminiti, R., Rouiller, E.M., Knott, G., Dyrby, T.B., Descoteaux, M., Thiran, J.P., 2019. Diversity of cortico-descending projections: histological and diffusion MRI characterization in the monkey. Cereb. Cortex 29, 788–801.

Ji, W., et al., 2015. Modularity in the organization of mouse primary visual cortex. Neuron 87, 632–643.

Jockwitz, C., Caspers, S., Lux, S., Jutten, K., Schleicher, A., Eickhoff, S.B., Amunts, K., Zilles, K., 2017. Influence of age and cognitive performance on resting-state brain networks of older adults in a population-based cohort. Cortex 89, 28–44.

Jones, E.G., Wise, S.P., 1977. Somatotopic and columnar organization in the corticotectal projection of the rat somatic sensory cortex. Brain Res. 133, 223–235.

Kanari, L., Ramaswamy, S., Shi, Y., Morand, S., Meystre, J., Perin, R., Abdellah, M., Wang, Y., Hess, K., Markram, H., 2019. Objective morphological classification of neocortical pyramidal cells. Cereb. Cortex 29, 1719–1735.

Katz, L.C., 1987. Local circuitry of identified projection neurons in cat visual cortex brain slices. J. Neurosci. 7, 1223–1249.

Kauvar, I.V., Machado, T.A., Yuen, E., Kochalka, J., Choi, M., Allen, W.E., et al., 2020. Cortical observation by synchronous multifocal optical sampling reveals widespread population encoding of actions. Neuron 107, 351–367.

Kisvarday, Z.K., Eysel, U.T., 1992. Cellular organization of reciprocal patchy networks in layer III of cat visual cortex (area 17). Neuroscience 46, 275–286.

Kita, T., Kita, H., 2012. The subthalamic nucleus is one of multiple innervation sites for long-range corticofugal axons: a single-axon tracing study in the rat. J. Neurosci. 32, 5990–5999.

Knyazeva, M.G., Kiper, D.C., Vildavski, V.Y., Despland, P.A., Maeder-Ingvar, M., Innocenti, G.M., 1999. Visual stimulus-dependent changes in inter-hemispheric EEG coherence in ferrets. J. Neurophysiol. 82, 3082–3094.

Knyazeva, M.G., Fornari, E., Meuli, R., Innocenti, G., Maeder, P., 2006. Imaging of a synchronous neuronal assembly in the human visual brain. Neuro-Image 29, 593–604.

Kochunov, P., et al., 2005. Age-related morphology trends of cortical sulci. Hum. Brain Mapp. 26, 210–220.

Koestinger, G., et al., 2017. Synaptic connections formed by patchy projections of pyramidal cells in the superficial layers of cat visual cortex. Brain Struct. Funct. 222, 3025–3042.

Kolodkin, A.L., Tessier-Lavigne, M., 2011. Mechanisms and molecules of neuronal wiring: a primer. Cold Spring Harb. Perspect. Biol. 3, a001727.

Kraskov, A., Baker, S., Soteropoulos, D., Kirkwood, P., Lemon, R., 2019. The corticospinal discrepancy: where are all the slow pyramidal tract neurons? Cereb. Cortex 29, 3977–3981.

Kubota, Y., et al., 2011. Selective coexpression of multiple chemical markers defines discrete populations of neocortical GABAergic neurons. Cereb. Cortex 21, 1803–1817.

LaMantia, A.S., Rakic, P., 1990. Axon overproduction and elimination in the corpus callosum of the developing rhesus monkey. J. Neurosci. 10, 2156–2175.

Lee, H.H., Yaros, K., Veraart, J., Pathan, J.L., Liang, F.X., Kim, S.G., Novikov, D.S., Fieremans, E., 2019. Along-axon diameter variation and axonal orientation dispersion revealed with 3D electron microscopy: implications for quantifying brain white matter microstructure with histology and diffusion MRI. Brain Struct. Funct. 224, 1469–1488.

Lehman, J.F., Greenberg, B.D., McIntyre, C.C., Rasmussen, S.A., Haber, S.N., 2011. Rules ventral prefrontal cortical axons use to reach their targets: implications for diffusion tensor imaging tractography and deep brain stimulation for psychiatric illness. Neuroscience 31, 10392–10402.

Leterrier, C., Dubey, P., Roy, S., 2017. The nano-architecture of the axonal cytoskeleton. Nat. Rev. Neurosci. 18, 713–726.

Levitt, J.B., et al., 1995. Connections between the pulvinar complex and cytochrome oxidase-defined compartments in visual area V2 of macaque monkey. Exp. Brain Res. 104, 419–430.

Llinares-Benadero, C., Borrell, V., 2019. Deconstructing cortical folding: genetic, cellular and mechanical determinants. Nat. Rev. Neurosci. 20, 161–176.

Lui, J.H., Hansen, D.V., Kriegstein, A.R., 2011. Development and evolution of the human neocortex. Cell 146, 18–36.

Makarov, V.A., Schmidt, K.E., Castellanos, N.P., Lopez-Aguado, L., Innocenti, G.M., 2008. Stimulus-dependent interaction between the visual areas 17 and 18 of the 2 hemispheres of the ferret (*Mustela putorius*). Cereb. Cortex 28, 1951–1960. https://doi.org/10.1093/cercor/bhm222.

Marcoli, M., Agnati, L.F., Benedetti, F., Genedani, S., Guidolin, D., Ferraro, L., Maura, G., Fuxe, K., 2015. On the role of the extracellular space on the holistic behavior of the brain. Rev. Neurosci. 26, 489–506.

Markram, H., Toledo-Rodriguez, M., Wang, Y., Gupta, A., Silberberg, G., Wu, C., 2004. Interneurons of the neocortical inhibitory system. Nat. Rev. Neurosci. 5, 793–807.

Martin, K.A., et al., 2014. Superficial layer pyramidal cells communicate heterogeneously between multiple functional domains of cat primary visual cortex. Nat. Commun. 5, 1–13.

Martin, K.A.C., et al., 2017. A biological blueprint for the axons of superficial layer pyramidal cells in cat primary visual cortex. Brain Struct. Funct. 222, 3407–3430.

Maunsell, J.H., Van Essen, D.C., 1983. The connections of the middle temporal visual area (MT) and their relationship to a cortical hierarchy in the macaque monkey. J. Neurosci. 3, 2563–2586.

Menzel, M., et al., 2019. Diattenuation Imaging reveals different brain tissue properties. Sci. Rep. 9, 1–12.

Mitrofanis, J., 2005. Some certainty for the "zone of uncertainty"? Exploring the function of the zona incerta. Neuroscience 130, 1–15.

Morecraft, R.J., Binneboese, A., Stilwell-Morecraft, K.S., Ge, J., 2017. Localization of orofacial representation in the corona radiata, internal capsule and cerebral peduncle in *Macaca mulatta*. J. Comp. Neurol. 525, 3429–3457.

Morita, K., Im, S., Kawaguchi, Y., 2019. Differential striatal axonal arborizations of the intratelencephalic and pyramidal-tract neurons: analysis of the data in the MouseLight database. Front. Neural Circuits 13, 71. 1–16 https://doi.org/10.3389/fncir.2019.00071.

Mortazavi, F., Wang, X., Rosene, D.L., Rockland, K.S., 2016. White matter neurons in young adult and aged rhesus monkey. Front. Neuroanat. 10, 15. 1–10.

Nassi, J.J., Callaway, E.M., 2009. Parallel processing strategies of the primate visual system. Nat. Rev. Neurosci. 10, 360–372.

Nassi, J.J., et al., 2013. Corticocortical feedback contributes to surround suppression in V1 of the alert primate. J. Neurosci. 33, 8504–8517.

Nave, K.A., 2010. Myelination and the trophic support of long axons. Nat. Rev. Neurosci. 11, 275–283.

Nhan, H.L., Callaway, E.M., 2012. Morphology of superior colliculus- and middle temporal area-projecting neurons in primate primary visual cortex. J. Comp. Neurol. 520, 52–80.

Nowak, L.C., Bullier, J., 1997. The timing of information transfer in the visual system. In: Rockland, K.S., Kaas, J.H., Peters, A. (Eds.), Cerebral Cortex. vol. 12. Plenum Press, New York, pp. 205–241 (Chapter 5).

Nurminen, L., et al., 2018. Top-down feedback controls spatial summation and response amplitude in primate visual cortex. Nat. Commun., 2281. 1–13.

Oishi, K., Huang, H., Yoshioka, T., et al., 2011. Superficially located white matter structures commonly seen in the human and the macaque brain with diffusion tensor imaging. Brain Connect. 1, 37–47.

O'Reilly, J.X., Croxson, P.L., Jbabdi, S., Sallet, J., Noonan, M.P., Mars, R.B., Browning, P.G., Wilson, C.R., Mitchell, A.S., Miller, K.L., Rushworth, M.F., Baxter, M.G., 2013. A causal effect of disconnection lesions on interhemispheric functional connectivity in rhesus monkeys. Proc. Natl. Acad. Sci. 110, 13982–13987.

Palmer, S.M., Rosa, M.G., 2006. A distinct anatomical network of cortical areas for analysis of motion in far peripheral vision. Eur. J. Neurosci. 24, 2389–2405.

Palomero-Gallagher, N., Zilles, K., 2019. Cortical layers: cyto-, myelo-, receptor- and synaptic architecture in human cortical areas. NeuroImage 197, 716–741.

Parent, M., Parent, A., 2006. Single-axon tracing study of corticostriatal projections arising from primary motor cortex in primates. J. Comp. Neurol. 496, 202–213.

Payne, B.R., Siwek, D.F., Lomber, S.G., 1991. Complex transcallosal interactions in visual cortex. Vis. Neurosci. 6, 283–289.

Perge, J.A., Niven, J.E., Mugnaini, E., Balasubramanian, V., Sterling, P., 2012. Why do axons differ in caliber? J. Neurosci. 32, 626–638.

Pessoa, L., Adolphs, R., 2010. Emotion processing and the amygdala: from a 'low road' to 'many roads' of evaluating biological significance. Nat. Rev. Neurosci. 11, 773–783.

Peters, A., Sethares, C., 1991. Organization of pyramidal neurons in area 17 of monkey visual cortex. J. Comp. Neurol. 306, 1–23.

Peters, A., Sethares, C., 1996. Myelinated axons and the pyramidal cell modules in monkey primary visual cortex. J. Comp. Neurol. 365, 232–255.

Peters, A., et al., 1997. The organization of pyramidal cells in area 18 of the rhesus monkey. Cereb. Cortex 7, 405–421.

Popovic, M.A., Foust, A.J., McCormick, D.A., Zecevic, D., 2011. The spatio-temporal characteristics of action potential initiation in layer 5 pyramidal neurons: a voltage imaging study. J. Physiol. 589, 4167–4187.

Quarles, R.H., 2002. Myelin sheaths: glycoproteins involved in their formation, maintenance and degeneration. Cell. Mol. Life Sci. 59, 1851–1871.

Rakic, P., 1974. Neurons in rhesus monkey visual cortex: systematic relation between time of origin and eventual disposition. Science 183, 425–427.

Rakic, P., Bourgeois, J.P., Eckenhoff, M.F., Zecevic, N., Goldman-Rakic, P.S., 1986. Concurrent overproduction of synapses in diverse regions of the primate cerebral cortex. Science 232, 232–235.

Raper, J., Mason, C., 2010. Cellular strategies of axonal pathfinding. Cold Spring Harb. Perspect. Biol. 2, 1–21.

Riccomagno, M.M., Kolodkin, A.L., 2015. Sculpting neural circuits by axon and dendrite pruning. Annu. Rev. Cell Dev. Biol. 31, 779–805.

Riederer, B.M., Berbel, P., Innocenti, G.I., 2004. Neurons in the corpus callosum of the cat during postnatal development. Eur. J. Neurosci. 19, 2039–2046.

Ringo, J.L., Doty, R.W., Demeter, S., Simard, P.Y., 1994. Time is of the essence: a conjecture that hemispheric specialization arises from interhemispheric conduction delay. Cereb. Cortex 4, 331–343.

Rivara, C.B., Sherwood, C.C., Bouras, C., Hof, P.R., 2003. Stereologic characterization and spatial distribution patterns of Betz cells in the human primary motor cortex. Anat. Rec. A Discov. Mol. Cell. Evol. Biol. 270, 137–151.

Rockland, K.S., 1989. Bistratified distribution of terminal arbors of individual axons projecting from area V1 to middle temporal area (MT) in the macaque monkey. Vis. Neurosci. 3, 155–170.

Rockland, K.S., 2019. What do we know about laminar connectivity? NeuroImage 197, 772–784.

Rockland, K.S., 2020. What we can learn from the complex architecture of single axons. Brain Struct. Funct. 225, 1327–1347.

Rockland, K.S., Drash, G.W., 1996. Collateralized divergent feedback connections that target multiple cortical areas. J. Comp. Neurol. 373, 529–548.

Rockland, K.S., Ichinohe, N., 2004. Some thoughts on cortical minicolumns. Exp. Brain Res. 158, 265–277.

Rockland, K.S., Knutson, T., 2000. Feedback connections from area MT of the squirrel monkey to areas V1 and V2. J. Comp. Neurol. 425, 345–368.

Rockland, K.S., Ojima, H., 2003. Multisensory convergence in calcarine visual areas in macaque monkey. Int. J. Psychophysiol. 50, 19–26.

Rockland, K.S., Pandya, D.N., 1979. Laminar origins and terminations of cortical connections of the occipital lobe in the rhesus monkey. Brain Res. 179, 3–20.

Rodriguez-Moreno, J., et al., 2020. Area-specific synapse structure in branched posterior nucleus axons reveals a new level of complexity in thalamo-cortical networks. J. Neurosci. 40, 2663–2679.

Roland, P.E., Zilles, K., 1998. Structural divisions and functional fields in the human cerebral cortex. Brain Res. Brain Res. Rev. 26, 87–105.

Roland, J.L., Snyder, A.Z., Hacker, C.D., Mitra, A., Shimony, J.S., Limbrick, D.D., Raichle, M.E., Smyth, M.D., Leuthardt, E.C., 2017. On the role of the corpus callosum in interhemispheric functional connectivity in humans. Proc. Natl. Acad. Sci. U. S. A. 114, 13278–13283.

Romascano, D., Barakovic, M., Rafael-Patino, J., Dyrby, T.B., Thiran, J.-P., Daducci, A., 2020. ActiveAxadd: toward non-parametric and orientationally invariant axon diameter distribution mapping using PGSE. Magn. Reson. Med. 83 (6), 2322–2330. https://doi.org/10.1002/mrm.28053.

Rosa, M.G., Tweedale, R., 2005. Brain maps, great and small: lessons from comparative studies of primate visual cortical organization. Philos. Trans. R. Soc. Lond. Ser. B Biol. Sci. 360, 665–691.

Rouiller, E.M., Welker, E., 2000. A comparative analysis of the morphology of corticothalamic projections in mammals. Brain Res. Bull. 53, 727–741.

Rushton, W.A.H., 1951. A theory of the effects of fibre size in medullated nerve. J. Physiol. 115, 101–122.

Sango, K., Yamauchi, J., Ogata, T., Susuki, K., 2019. Myelin: Basic and Clinical Advances, first ed. Springer Nature, Singapore. 378.

Sarubbo, S., Petit, L., De Benedictis, A., Chioffi, F., Ptito, M., Dyrby, T.B., 2019. Uncovering the inferior fronto-occipital fascicle and its topological organization in non-human primates: the missing connection for language evolution. Brain Struct. Funct. 224, 1553–1567.

Schmahmann, J.D., Pandya, D.N., 2006. Fiber Pathways of the Brain. Oxford University Press, New York.

Schmahmann, J.D., Smith, E.E., Eichler, F.S., Filley, C.M., 2008. Cerebral white matter: neuroanatomy, clinical neurology, and neurobehavioral correlates. Ann. N. Y. Acad. Sci. 1142, 266–309.

Schmidt, K.E., Lomber, S.G., Innocenti, G.M., 2010. Specificity of neuronal responses in primary visual cortex is modulated by interhemispheric corticocortical input. Cereb. Cortex 20, 2776–2786.

Segev, I., Schneidmann, E., 1999. Axons as computing devices: basic insights gained from models. J. Physiol. Paris 93, 263–270.

Sherman, S.M., Guillery, R.W., 2006. Exploring the Thalamus and Its Role in Cortical Function. The MIT Press, Cambridge, MA.

Sherman, S.M., Guillery, R.W., 2011. Distinct functions for direct and transthalamic corticocortical connections. J. Neurophysiol. 106, 1068–1077.

Shipp, S., 2007. Structure and function of the cerebral cortex. Curr. Biol. 17, R443–R449.

Shipp, S., 2017. The functional logic of corticostriatal connections. Brain Struct. Funct. 222, 669–706.

Sincich, L.C., Horton, J.C., 2005. Input to V2 thin stripes arises from V1 cytochrome oxidase patches. J. Neurosci. 25, 10087–10093.

Smith, R., Koles, Z., 1970. Myelinated nerve fibers: computed effect of myelin thickness on conduction velocity. Am. J. Phys. 219, 1256–1258.

Stassart, R.M., Möbius, W., Nave, K.A., Edgar, J.M., 2018. The axon-myelin unit in development and degenerative disease. Front. Neurosci. 12, 467. 1–22.

Stefanacci, L., Amaral, D.G., 2000. Topographic organization of cortical inputs to the lateral nucleus of the macaque monkey amygdala: a retrograde tracing study. J. Comp. Neurol. 421, 52–79.

Sugino, K., et al., 2019. Mapping the transcriptional diversity of genetically and anatomically defined cell populations in the mouse brain. eLife 8, e38619. 1–29.

Suzuki, W., et al., 2000. Divergent backward projections from the anterior part of the inferotemporal cortex (area TE) in the macaque. J. Comp. Neurol. 422, 206–428.

Swadlow, H.A., 1985. Physiological properties of individual cerebral axons studied in vivo for as long as one year. J. Neurophysiol. 54, 1346–1362.

Swadlow, H.A., Kocsis, J.D., Waxman, S.G., 1980. Modulation of impulse conduction along the axonal tree. Annu. Rev. Biophys. Bioeng. 9, 143–179.

Takemura, H., Palomero-Gallagher, N., Axer, M., Grassel, D., Jorgensen, M.J., Woods, R., Zilles, K., 2020. Anatomy of nerve fiber bundles at micrometer-resolution in the vervet monkey visual system. Elife, e55444. https://doi.org/10.7554/eLife.55444.

Tettoni, L., Gheorghita-Baechler, F., Bressoud, R., Welker, E., Innocenti, G.M., 1998. Constant and variable aspects of axonal phenotype in cerebral cortex. Cereb. Cortex 8, 543–552.

Thomson, A., 2010. Neocortical layer 6, a review. Front. Neuroanat. 4, 13. 1–14.

Tomasi, S., Caminiti, R., Innocenti, G.M., 2012. Areal differences in diameter and length of corticofugal projections. Cereb. Cortex 22, 1463–1472.

Tomassy, G.S., Dershowitz, L.B., Arlotta, P., 2016. Diversity matters: a revised guide to myelination. Trends Cell Biol. 26, 135–147.

Tomioka, R., Rockland, K.S., 2007. Long-distance corticocortical GABAergic neurons in the adult monkey white and gray matter. J. Comp. Neurol. 505, 526–538.

Ungerleider, L., Desimone, R., 1986. Cortical connections of visual area MT in the macaque. J. Comp. Neurol. 248, 190–222.

Ungerleider, L.G., et al., 2008. Cortical connections of area V4 in the macaque. Cereb. Cortex 18, 477–499.

Vanni, S., Hokkanen, H., Werner, F., Angelucchi, A., 2020. Anatomy and physiology of macaque visual cortical areas V1, V2, and V5/MT: bases for biologically realistic models. Cereb. Cortex 30, 3483–3517.

Vercelli, A., Assal, F., Innocenti, G.M., 1992. Emergence of callosally projecting neurons with stellate morphology in the visual cortex of the kitten. Exp. Brain Res. 90, 346–358.

Von Economo, C., 1927/2009. Cellular Structure of the Human Cerebral Cortex (translated and edited: L.C. Triarhou). Karger, Basel.

Waxman, S.G., Bennet, M.V.L., 1972. Relative conduction velocities of small myelinated and non-myelinated fibres in the central nervous system. Nat. New Biol. 238, 217–219.

Winnubst, J., Bas, E., Ferreira, T.A., Wu, Z., Economo, M.N., Edson, P., Arthur, B.J., Bruns, C., Rokicki, K., Schauder, D., Olbris, D.J., Murphy, S.D., et al., 2019. Reconstruction of 1,000 projection neurons reveals new cell types and organization of long-range connectivity in the mouse brain. Cell 179, 268–281.

Wu, H., Williams, J., Nathans, J., 2014. Complete morphologies of basal forebrain cholinergic neurons in the mouse. eLife, 02444.

Xiao, D., et al., 2009. Laminar and modular organization of prefrontal projections to multiple thalamic nuclei. Neuroscience 161, 1067–1081.

Yakoubi, R., Rollenhagen, A., von LeHe, M., Shao, Y., Satzler, K., Lubke, J.H.R., 2019. Quantitative three-dimensional reconstructions of excitatory synaptic boutons in layer 5 of the adult human temporal lobe neocortex: a fine-scale electron microscopic analysis. Cereb. Cortex 29, 2797–2814.

Zemmoura, I., Blanchard, E., Raynal, P.-I., Rousselot-Denis, C., Destrieux, C., Velut, S., 2016. How Klingler's dissection permits exploration of brain structural connectivity? An electron microscopy study of human white matter. Brain Struct. Funct. 221, 2477–2486.

Zilles, K., 2018. Brodmann: a pioneer of human brain mapping—his impact on concepts of cortical organization. Brain 141, 3262–3278.

Zilles, K., et al., 2013. Development of cortical folding during evolution and ontogeny. Trends Neurosci. 36, 275–284.

Zufferey, P.D., Jin, F., Nakamura, H., Tettoni, L., Innocenti, G.M., 1999. The role of pattern vision in the development of cortico-cortical connections. Eur. J. Neurosci. 11, 2669–2688.

# Chapter 3

# Past and present of mapping brain connections

Flavio Dell'Acqua[a,b], Mattia Veronese[a,c], Ottavia Dipasquale[a], Daniel Martins[a], Fedal Saini[b], and Marco Catani[d,e]

[a]NATBRAINLAB, Department of Neuroimaging, Institute of Psychiatry, Psychology and Neuroscience, King's College London, London, United Kingdom, [b]Department of Forensics and Neurodevelopmental Sciences, Institute of Psychiatry, Psychology and Neuroscience, King's College London, London, United Kingdom, [c]Department of Information Engineering, University of Padua, Padua, Italy, [d]IRCCS SYNLAB, SDN Research Institute, Naples, Italy, [e]Department of Neuroscience, Imaging, and Clinical Sciences (DNISC) and Istituto di Tecnologie Avanzate Biomediche (ITAB), University "G. D'Annunzio" of Chieti-Pescara, Chieti, Italy

The quest to map the intricate network of the brain's connections is not a recent idea, but rather it has deep roots in history. This concept has captured the imagination of early scientists and researchers who saw the brain as the center of the human mental faculties. Since the early days of neuroscience, numerous techniques and models have been developed, each one reflecting the main theories and level of technology that has characterized its respective era. In this chapter, we review the intricate journey that, starting from early neuroanatomy and physiological theories, has led to the development of the modern concept of brain connectivity. This idea has shaped and guided the field of neuroscience for many years, and it now forms the basis for the use of tractography as an essential tool to investigate in vivo the complex structural organization of the human brain.

## 1 Early connectivity maps

For many centuries it was believed that human mental faculties such as sensations, imagination, and memory originate from a hydraulic system of ventricles or "cells" in the brain. Fluid movements within the ventricles were thought to influence a person's mental states and behaviors, while mental health problems were believed to originate from an imbalance in the flow of such fluids. The idea, commonly known as *ventricular theory*, was based on brain dissections originally described by Herophilus of Alexandria in the 4th century BC and two centuries later by Galen (Clarke and O'Malley, 1996). The brain representation that derived from this theory consisted of a diagram depicting a variable number of intercommunicating ventricles (Fig. 1, Left). These ventricular reservoirs or cells can be thought of as nodes, each with specialized functions, that were reciprocally interconnected through a system of interventricular foramina. These early representations of the brain already contain fundamental elements of contemporary brain description. They introduced the concept of function localization within the brain. Implicit to this are the modern concepts of anatomical differentiation and functional specialization. The brain is thus conceptualized as a collection of different functional areas interconnected through a *network* system. The ventricular cell organization also closely resembles the modern concept of connectome where brain functions are not just localized but emerge from the interaction of different nodes within the network.

The ventricular model persisted throughout the Middle Ages, maintaining its influence even into the Renaissance, despite some initial experimental evidence already started to challenge its validity. Around 1508, Leonardo da Vinci presented evidence contradicting the traditional structural depiction of the ventricular system by using a wax cast of the ventricles from a dissected brain (Pevsner, 2002). A few years later, Vesalius demonstrated that the ventricular anatomy described by Galen was questionable, as it did not align with the evidence gathered from brain dissections (Vesalius, 1543). A new scientific era was also dawning, embracing the experimental method (Galilei, 1638), and the strong belief in the close relationship between anatomy and function (Catani, 2007). In this transformative era, where postmortem dissections became the predominant method for studying the nervous system, important anatomical discoveries were made, such as the differentiation between the "cerebrum" (i.e., gray matter or cortex) and the "medulla" (i.e., white matter), accomplished by Arcangelo Piccolomini in 1586. This basic distinction between gray and white matter is still very relevant

Handbook of Diffusion MR Tractography. https://doi.org/10.1016/B978-0-12-818894-1.00003-3
Copyright © 2025 Elsevier Ltd. All rights are reserved, including those for text and data mining, AI training, and similar technologies.

**FIG. 1** *Left*, Drawing of the ventricular system from a manuscript dating back to around 1300. Five distinct cells or ventricles are named according to their functions, with individual foramina connecting each cell in a network-like fashion. *Middle*, Representation of the intricate organization of white matter as a complex hydraulic system, as described by Descartes. *Right*, Postmortem dissections of white matter pathways, showing anatomically consistent trajectories from the centrum ovale, including projection fibers and commissural fibers (partially disconnected). *(Left, from Unnamed Miniaturist, c.1300, University Library, Cambridge.* Middle, *from: Descartes, R., 1664. L'Homme de René Descartes et un Traité de la formation du fœtus. Charles Angot, Paris.* Right, *from: Vieussens, R., 1685. Neurographia universalis. Jean Certe, Lyon.)*

today. In many neuroimaging studies, brain segmentation is indeed one of the very first steps in the analysis pipeline for structural MRI data. Functional and structural connectivity analyses are also particularly dependent on the accurate segmentation of white and gray matter regions.

In the 17th century, there was a further increase in interest surrounding the study of white matter anatomy. Many scientists at the time realized that white matter was not merely a uniform tissue required only for mechanical support. Instead, it was composed of fibers whose paths or trajectories could be traced and described. White matter was an intricate medium consisting of tubular filaments where fluid could flow between central "cells" and peripheral nerves (Descartes, 1662). Around the same time, it was recognized that white matter fibers originate from the cortex (Malpighi, 1666) and could be responsible for motor and sensory functions (Willis, 1664). Scholars understood that white matter fibers were responsible for coordinating a broad spectrum of behaviors, ranging from simple reflex responses all the way to complex cognitive functions (Descartes, 1664). Similarly, Steno in 1669 described white matter *fibers* as highly and precisely organized within the brain, having an important role since "*...all the diversity of our sensations and movements depend upon them*" (Steno, 1669).

Despite these advancements, inaccuracies in anatomical details were still evident in maps from this period. For instance, Descartes' depiction of the brain (Fig. 1, Middle) aimed to illustrate the complexity of white matter connections through a more conceptual approach, rather than faithfully portraying actual anatomical details (Descartes, 1664). However, other maps also focused on presenting genuine anatomical findings (Fig. 1, Right; Vieussens, 1685). From all these studies a new view of the white matter emerged, and, from this moment, the concept of brain connectivity became associated with the idea of fibers and nerves.

For many years, the study of white matter remained linked with a "hydraulic" understanding of the brain physiology. According to the original ventricular theory, cognitive functions were believed to emerge from the passage of fluid through ventricles and hollow fibers. Therefore, revealing or mapping the exact anatomy of these fibers would have provided new insights into brain functions. However, as new anatomical discoveries were made, the ventricular theory also changed and adapted. Brain functions were no longer localized within the ventricles but instead were compartmentalized within individual gyri containing distinct "animal spirits" (Willis, 1664). It was now the constant movement of these fluids constrained within white matter tracts giving origin to the different cognitive abilities. Similarly, the complex organization of cortical gyri in the human brain was then explained by the need to accommodate more complex faculties compared to other animals.

It was only at the beginning of the 19th century that brain maps of white matter connections started to conform with more contemporary standards. White matter was now described with callosal, projection, and association pathways for the first time. During this time, neuroanatomists started to identify and provide detailed anatomical descriptions of individual association tracts (Burdach, 1822; Reil, 1809, 1812). At the same time, the increased interest in cortical localization of brain functions contributed to the formulation of the organology theory by Gall and Spurzheim, which describes the brain as an organ made up of multiple "organs," represented by anatomical convolutions that receive fiber from different regions (Gall and Spurzheim, 1810).

Despite incorporating some modern elements, Gall and Spurzheim theory soon faced a decline in reputation. One of the main reasons was the failure of organology to incorporate new concepts regarding the physiology of the nervous system, which had emerged from the earlier experimental work of Galvani and Aldini (see below). Moreover, Gall and Spurzheim later proposed that more developed functions were localized in correspondingly well-developed cortical organs, leading them to conclude that larger organs would leave more pronounced impressions on the skull. In this way, the famous phrenological theory saw the light, with the aim of ascertaining the *"moral dispositions of man and animal, by the configuration of their heads"* (Gall and Spurzheim, 1810). Despite facing substantial discreditation as a scientific theory already by 1840, the ideas of phrenology still managed to rapidly disseminate and gain significant consideration worldwide.

Meanwhile in Italy, Luigi Galvani observed that the muscles of frog legs would contract when exposed to electrical stimuli. This marked the dawn of a new scientific paradigm known as *animal electricity* (Galvani, 1791) (Fig. 2, Left). Building upon this work, Galvani's nephew and assistant, Giovanni Aldini, significantly advanced and popularized the concepts of animal electricity and galvanism. Going beyond his uncle's experiments, Aldini applied electrical currents to larger animals and even the limbs of executed criminals, triggering movements in their bodies (Aldini, 1804) (Fig. 2, Right). Not surprisingly, just a few years later in 1817, Mary Shelley published *Frankenstein*.

These events sparked intense debates across Europe, setting the stage for a series of new experiments that marked the beginning of modern electrophysiology. The introduction of electricity attracted eminent figures from various fields, fostering an exchange of ideas between the fields of anatomy and physics. This interdisciplinary approach led to remarkable innovations and insights. For instance, during this period the notion emerged that the cortex functioned as an "electrical generator" within the human brain (Baillarger, 1840). This idea was likely influenced from the striking similarities between the contemporary understanding of cortex layering and the lamellar organization of the "voltaic pile" (Volta, 1800). Additionally, the introduction of new recording methods (Matteucci, 1830) led to the discovery and description of the action potential in muscles (DuBois-Reymond, 1848) and peripheral nerves (Bernstein, 1868; von Helmholtz, 1850). It was also during this period that the concept of reflex was expanded beyond the spinal cord to also include the brain (Griesinger, 1843). This idea would later revolutionize the field of physiological psychology (Pavlov, 1903).

During the second half of the 19th century, even more advancements in cellular biology and microscopy methods contributed to further expanding our understanding of the brain and its organization. In 1839, the German scientist Theodor Schwann proposed the cell theory, which postulated that cells are the fundamental building blocks of all living organisms (Schwann, 1839). This powerful concept prompted numerous microscopists to explore and describe new cellular features of the brain itself. For instance, Kölliker classified the cortex into myeloarchitectonic and cytoarchitectonic, based on the distribution patterns of fibers and cells, respectively (Kölliker, 1859), while Meynert observed that different brain regions exhibited different type of cortical layering (Meynert, 1869).

Histological staining methods were also advancing rapidly during this period. By applying chemical dyes, staining was used to enhance visibility and contrast in biological tissue, enabling researchers to observe and analyze previously invisible

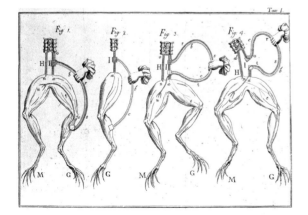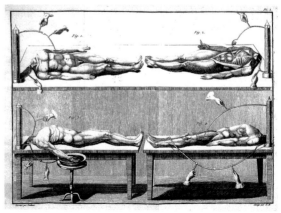

**FIG. 2** *Left*, A representation of Galvani's frog experiment, capturing the moment of muscle contraction when the frog's legs were stimulated with electricity. *Right*, A depiction of Aldini's experiments on bodies of criminals, where electrical currents were applied to elicit movement in various body parts. *(Left, from: Galvani, L., 1797. Memorie sulla elettricità animale di Luigi Galvani [...] al celebre abate Lazzaro Spallanzani. Bologna. Right, from: Aldini, G., 1804. Essai théorique et expérimental sur le galvanisme. De l'Imprimerie de Fournier fils, Paris.)*

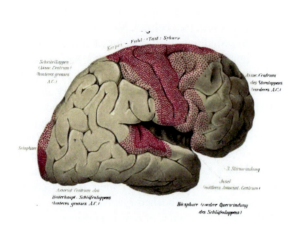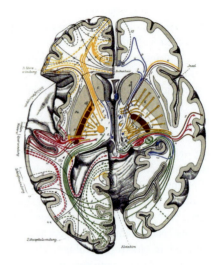

FIG. 3 *Left*, Cortical myelogenetic maps as identified by Flechsig according to their degree of myelination at birth. More myelinated areas at birth, such as primary sensory and motor areas, are marked with *denser dots*, followed by regions with intermediate myelination *(sparse dots)*. Areas that myelinated only after birth, including large associative regions in the frontal, parietal, temporal, and occipital lobes are indicated *without dots*. *Right*, Reconstruction depicting the trajectory of projection fibers based on myelogenetic maps. *(*Left and Right*, from: Flechsig, P., 1896. Gehirn und seele. Verlag von Veit, Leipzig.)*

structures. During this period Carl Weigert and Vittorio Marchi introduced a new staining method targeting myelin, accompanied by more precise microtomes for studying sequential tissue sections (Bentivoglio and Mazzarello, 2009). By applying myelin staining in brains of fetuses or newborns, Paul Emile Flechsig successfully mapped the chronological maturation sequence of white matter pathways in the human brain. Given the fact that projection tracts undergo early myelination, the researcher was able to trace the source and path of the corticospinal tract (Fig. 3) (Flechsig, 1876). In his pioneering work, Flechsig also unveiled the existence of asymmetry during its crossing, or decussation, at the medulla level.

Around the same period, Bernhard von Gudden perfected a particular type of experimental ablation technique, later named the Gudden Method, to study the consequences of sensory deprivation on brain structures in animal models. By surgically removing sensory organs or cutting the neural pathways that relay sensory information to the brain, he successfully induced secondary degeneration and atrophy of the corresponding sensory areas (Gudden, 1870). Expanding on the research conducted by his mentor Gudden, Constantin von Monakow furthered the investigations into thalamic and motor projection fibers. By selectively removing specific regions of the cortex in animals, Monakow successfully traced the regressive degeneration of neural tracts back to the thalamic nuclei and other subcortical nuclei (Monakow, 1897). Monakow was also the first to recognize the impact of a brain lesion on remote areas of the brain, a phenomenon he termed "diaschisis" (Monakow, 1914). All these new discoveries led many researchers to start to recognize that a complete understanding of the nervous system could only be achieved through an accurate depiction of the brain's intricate connections (Dejerine and Dejerine-Klumpke, 1895, 1901; von Bechterew, 1900).

The insights generated during this period did not remain confined to university labs. Medical professionals began utilizing their understanding of neuroanatomy to inform their clinical approaches, marking the rise of the *clinical-anatomical correlation* method. This approach replaced phrenology and gained popularity in identifying potential functional connections within cortical areas (Bouillaud, 1825; Broca, 1861). Central to this approach was the notion that the brain should be comprehended as a cohesive entity, a system comprising interconnected and integrated regions. Disorders of the nervous system were now attributed to either cortical damage or disconnection syndromes (Fig. 4; Dejerine and Dejerine-Klumpke, 1895, 1901). However, the anatomical understanding of the human brain was still far from being accurate, often requiring a good amount of approximation. Complex anatomical details were often simplified into diagrams, frequently leading to important oversimplifications (Catani and Ffytche, 2005; Catani and Mesulam, 2008).

## 2 Cortical mapping and connectivity

By the end of the 19th century two significant developments took place in the study of the human brain. For the first time, with the use of novel staining techniques and microscopy, neuroscientists managed to visualize and describe the

 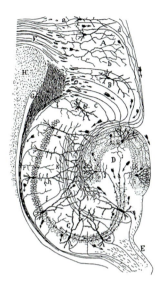

**FIG. 4** *Left*, Golgi's illustration of hippocampal histology, depicted using his pioneering staining method known as the black reaction. *Right*, Santiago Ramon y Cajal's rendition of hippocampal histology. Cajal, employing the black reaction, not only provided detailed visualizations but also introduced groundbreaking computational concepts inspired by his observations, including likely directionality of impulse propagation, as indicated by the *arrows*. *(Left, from: Golgi, C., 1903. Opera omnia, Ulrico Hoepli. Right, from: Cajal, S., 1911. Histologie du Systeme Nerveux de l'Homme et des Vertebres. Maloine, Paris.)*

organization of individual neural circuits (Cajal, 1893; Golgi, 1873). Ramon y Cajal introduced the neuron doctrine, presenting a set of fundamental biological principles governing neuronal organization (Cajal, 1911). He proposed that axon and dendrite shapes are constrained by biophysical factors such as cytoplasmic volume, space, and conduction time. From here, Cajal further presented the law of dynamic polarization, which allowed him to also determine the directionality of information between neurons. This law identified dendrites and the cell body as the primary structures responsible for receiving action potentials as input, while axons serve as the main output.

The second major development was the introduction of whole-brain maps looking at differences in the histological organization of the cortex. By using new staining methods, scientists segmented the human cortex based on differences in the cyto- and myeloarchitecture across different regions. As we saw in the previous section, it was during this time that primary and associative cortical areas as well as trajectories of maturation of long white matter projection tracts were described (Flechsig, 1896). Detailed histological examination of brain tissue later led to the development of increasingly comprehensive brain atlases (von Economo and Koskinas, 1925) and to the localization of various cerebral functions and clinical symptoms (Campbell, 1905).

However, during this time, diverging opinions also emerged among neuroscientists regarding the real purpose of cortical maps. Specifically, there was the question as to whether these maps could be used to directly associate function with specific regions or they were primarily anatomical division without direct relevance to functions. On one side, proponents of the localizationist school argued that by studying the brain's structure and organization, one could uncover the functions associated with specific regions. The idea was that cortical parcellations based on cytoarchitectonic and myeloarchitectonic held the potential to unveil important functional divisions within the brain. The goal here was to establish robust scientific foundations for understanding mental illness and to create an anatomically based classification system for psychiatric disorders (Baillarger, 1840; Meynert, 1869; Wernicke, 1906). In England, Alfred Campbell focused his research on the localization of cerebral functions by studying patients with brain injuries and neurological conditions. He examined cytoarchitecture and myeloarchitecture to investigate the impact of brain damage on various cognitive and sensory abilities. Campbell's work culminated in the creation of a detailed map of the human and other primate brains, with 17 distinct cortical fields, some of which were labeled according to specific functions, such as "visuo-sensory," "visuo-psychic," "audito-sensory," and "audito-psychic" (Fig. 5, left) (Campbell, 1905). Campbell's contributions extended beyond the mapping of the brain cytoarchitecture. His work was one of the first attempts to integrate clinical, anatomical, and physiological data to provide a comprehensive understanding of brain function (Catani and Ffytche, 2005).

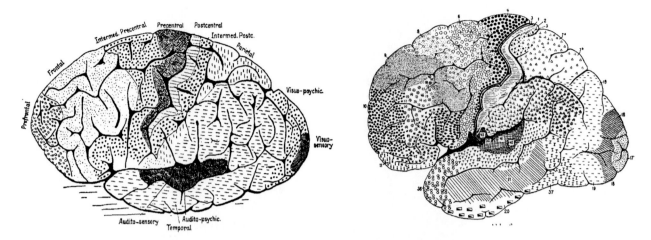

FIG. 5  *Left*, Campbell's map delineating the organization of the cortex into 17 distinct fields based on differences in cortical cyto- and myeloarchitecture and the assigned functional roles, such as "visuo-sensory" and "visuo-psychic." *Right*, Brodmann's cortical subdivision strictly guided by cytoarchitectonic criteria. Brodmann strongly rejected any association between individual areas and functional localization. (Left, modified from: *Campbell, A.W., 1905. Histological Studies on the Localisation of Cerebral Function. Cambridge University Press, Cambridge.* Right, modified from: *Brodmann, K., 1909. Vergleichende Lokalisationslehre der Grosshirnrinde in ihren Prinzipien dargestellt auf Grund des Zellenbaues. Barth, Leipzig.*)

In contrast to localizationism, other researchers contended that cognitive processes and behaviors emerge from the coordinated activity of multiple brain areas working in concert, rather than being localized in isolated regions. Consequently, cortical maps were regarded purely as anatomical representations, without any direct implication on function. Korbinian Brodmann was a proponent of this school of thought himself. His well-known cytoarchitectural parcellation of the brain identified 44 distinct areas in humans, with his complete numbering system including up to 52 areas across the primate brain. Each area was labeled with a simple number (Fig. 5, right). In his work, Brodmann made it clear that the brain must be considered as a *"whole"* with *"only one psychic center"* (Brodmann, 1909). In his view, higher functions were widespread over different cortical regions in a way that made it impossible to differentiate distinct functional centers. It is a bit ironic that today the Brodmann parcellation has become one of the reference maps for the cortical localization of multiple brain functions, where the same numbers are often directly associated with precise cortical functions.

Cortical mapping reached its peak with the contributions of Cécile and Oskar Vogt, and Constantin von Economo and George Koskinas. The Vogts conducted a comprehensive analysis of the myeloarchitectonic variations in a wide range of specimens, including both healthy and diseased brains. Their effort led to the meticulous mapping of over 200 anatomical areas and the detailed characterization of the cellular architecture within the cerebral cortex (Vogt and Vogt, 1926). However, despite their ambitious endeavor, which reached colossal proportions, the mapping of the temporal and occipital cortex remained incomplete. In 1925 Constantin von Economo and George Koskinas published a monumental atlas that presented a comprehensive depiction of the human cerebral cortex divided into 107 areas. Each area was accompanied by a detailed cytoarchitectonic examination including quantitative assessments of various characteristics such as cortical thickness and volume, shape, size, cell count, cell density, and the organization of cells into distinct patterns and layers. Despite being hailed by some as the ultimate reference on cortical cartography, Economo and Koskinas' atlas did not receive widespread acceptance within the scientific community. This can be attributed, in part, to the impractical size of this atlas and intrinsic limitations, like poorly defined boundaries between some of the smallest areas. The first half of the 20th century also saw a general decline in interest in macroscopic parcellations and localizationism. This was likely a consequence of the increasing popularity of whole brain approaches to understanding brain functions.

## 3 Associationist school and disconnection syndromes

In parallel to these debates, the second half of the 19th century also saw the emergence of the associationist school, initiated by the work of Theodor Meynert and Karl Wernicke. Meynert's initial research focused on the anatomical divisions of the

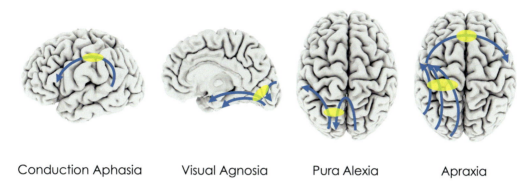

**FIG. 6** Schematic representation of the classical disconnection syndromes: The *blue arrows* in the figure represent the pathways involved in the syndromes, while the lesion is shown in *yellow*. From left to right, Conduction Aphasia caused by a lesion disconnecting Broca's and Wernicke's areas; Visual Agnosia, caused by lesions that disrupt visual information transfer from the visual cortex to other brain area; Pura Alexia, where a lesion disconnects a visual verbal center from the visual areas in both hemispheres; Apraxia, where the left-hand motor area is disconnected from other brain regions.

brain, looking into the intricacies of its cyto- and myeloarchitectural organization. However, in contrast to the strict localizationism, Meynert's work emphasized the importance of connections between different brain regions. Meynert was the first to recognize the importance of *association* fibers connecting regions within the same hemisphere. He organized white matter into three main groups of fibers. Together with association fibers, he also classified *projection fibers* as ascending or descending fibers to or from the cortex and *commissural fibers* that cross the midline and connect areas of the cortex in both hemispheres. Meynert further subdivided the associative fibers into *long association* and *short* or *U-shaped association* fibers. For more detailed information on Meynert's classification, see also Chapter 1. Extending on Meynert theories, Karl Wernicke continued the dissemination and expanded the associationist idea by introducing the concept of *disconnection syndromes*. This was a new clinical approach suggesting that certain mental and neurological disorders could arise also from disrupted communication between brain regions, rather than damage to specific locations (Wernicke, 1885). Wernicke's associationist idea was that, with the exception of sensory and motor regions, higher cognitive functions did not reside within a single cortical location. Instead, they were thought to be the product of complex interactions among several different areas of the brain. This school of thought expanded and evolved over time, thanks also to the works of Ludwig Lichtheim, Joseph Jules Dejerine, and Hugo Liepmann. Disconnection syndromes provided explanations for several neurological disorders including conduction aphasia (Wernicke, 1874), a disorder characterized by impaired repetition and paraphasic speech despite normal comprehension and fluency; visual agnosia (Lissauer, 1890), the inability to recognize visual objects while still having intact vision; apraxia (Liepmann, 1900), a disorder of motor planning and execution; and pure alexia (Dejerine, 1892), a condition resulting in reading impairment (Fig. 6).

Unfortunately, while successful at explaining important clinical symptoms, the associationist approach also lost its appeal and found itself affected by the strong influence of the contemporary localizationist approach. Over time, clinical and research interest started to shift more in favor of cortical specialization and cortical maps. This and the subsequent descriptions of cortical regions responsible also for higher functions was a problem for the original associationist approach (Campbell, 1905). Nevertheless, as previously mentioned, the early 20th century saw a general decline in both the localizationist and associationist approach in favor of a more holistic perspective (Brodmann, 1909). By the mid-20th century, both cortical localization and disconnection syndromes had lost most of their prominence (Head, 1926; Lashley, 1950). The interest was now focused on resolving the fine complexity of cortical architecture and its electrophysiology.

## 4 Modern connectivity

During the mid-20th century, significant advances in histological and electrophysiological methods contributed to expanding our understanding of cortical architecture. This shift was significant also because it moved from human ex-vivo anatomy to in-vivo animal anatomy and neurophysiology. Implanted electrodes, for the first time, allowed recordings of the action potential from multiple groups of axons and neurons. Combined with axonal tracing and histological methods, this allowed the investigation of precise cortical networks and their functions. Notable examples include the work of Mountcastle and Powell (1959) on the somatosensory cortex of monkeys. They provided the first descriptions of the "columnar organization" and proposed it as the basic unit of cortical organization. A few years later, Hubel and Wiesel (1962), looking at the visual cortex of cats, described the concept of "oriented receptor field." They observed that

**48 PART | I** From anatomy to tractography

different groups of neurons responded to lines of light projected at different angles. This was the first time that researchers were able to map and understand how visual information was encoded in the visual cortex.

In parallel, new tracing techniques also started to emerge during the late 1960s. They relied on the active transport of proteins and enzymes along axons. Tracers injected into specific brain regions would enter neurons and be transported either in the direction of the neuron's terminations (anterograde direction) or in the opposite direction (retrograde direction; Morecraft et al., 2014). This development facilitated the identification and tracing of individual axons, allowing for the precise definition of their cortical termination (Nauta and Gygax, 1951; Fink and Heimer, 1967; Mesulam, 1982; Petrides and Pandya, 1984). Further progress was also made when different types of tracers (including viral tracers) were combined with multiple injections allowing the simultaneous mapping not only of individual connections but more extended cortical and subcortical networks (Morecraft et al., 2014). This approach allowed for more precise and detailed description of cortico-cortical and cortico-subcortical pathways, including feedforward and feedback connections as well as hierarchical and parallel connections. This way of mapping brain connectivity was fundamental for the development of new computational models of different brain functions (Fig. 7) (Felleman and Van Essen, 1991; Mesulam, 1998).

While tracing methods provided significant advancements in the understanding of neural pathways, they also presented a major limitation: the assumption that findings from animal studies can also be directly translated to human anatomy. This assumption was true for general anatomical principles and the main sensory and motor functions, but for higher cognitive functions, such as language, it was more difficult or impossible to develop animal models replicating human abilities.

Nevertheless, during the whole 20th century, the application of disconnection procedures in monkeys, along with the integration of behavioral investigations contributed to revitalizing a network approach to brain functions (Mishkin, 1966). This also inspired and guided new medical interventions in humans. Specifically, cortical-subcortical connections were of particular interest, opening new avenues for studying neurological and psychological disorders in humans. This led to investigating previously unexplored fields of behavioral research, such as emotions and social behavior. In 1937, James

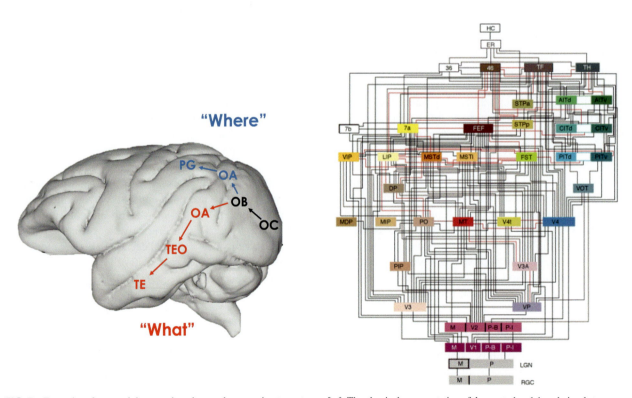

**FIG. 7** Examples of connectivity maps based on nonhuman primate anatomy. *Left,* The classical representation of the ventral and dorsal visual streams, as proposed by Mishkin et al. (1983). This figure highlights the dual-stream nature of visual processing with the ventral "what" and dorsal "where" pathways. *Right,* Felleman and Van Essen's (1991) famous representation of the visual system illustrates the hierarchical organization of the connectivity of visual areas. This detailed graph constitutes a major contribution to our understanding of visual processing networks. *(Right, Reproduced with permission from Oxford University Press: Felleman, D.J., Van Essen, D.C., 1991. Distributed hierarchical processing in the primate cerebral cortex. Cereb. Cortex 1 (1), 1–47.)*

Papez described the limbic circuit, a complex network of connections that found the basis for emotions and social interactions (Papez, 1937). Around the same time, neurosurgical procedures also started to be viewed as a promising treatment for mental disorders. Early experiments on chimpanzees showed reduced aggressive behavior after surgical disconnection of the thalamic projections to the frontal lobe (i.e., lobotomy) without noticeable effects on other brain functions according to these reports (Fulton, 1951; Jacobsen et al., 1935). Following these results, Antonio Egas Moniz introduced leucotomy (i.e., surgical cutting of white matter fibers) by disconnecting fronto-thalamic connections in patients with severe psychoses (Moniz, 1937). These procedures were based on the assumption that severing certain neural connections could alleviate mental distress. Walter Freeman and James Watts further refined the procedure and applied frontal lobotomy in more than 3000 chronic psychiatric patients (Freeman, 1957; Freeman and Watts, 1945). However, these procedures often also showed significant and detrimental effects on these patients. Thanks also to the introduction of effective antipsychotic drugs, the practice of performing surgery on severe mental health disorders ended by the late 1960s. Neurosurgery also began to be used as a treatment for epilepsy by performing callosotomy, a procedure aimed to prevent the spread of epileptic activity between hemispheres (Dandy, 1936). Initial studies showed that "split-brains" or complete callosotomy had only minor negative side effects on cognition. However, Roger Sperry studies later showed that, when extensively tested, complete callosotomy did indeed induce significant cognitive effects. Sperry's work was also instrumental in providing conclusive evidence of the functional specialization of the cerebral hemispheres (Sperry, 1974). For his pioneering research, Sperry would later share the Nobel Prize in 1981 together with Hubel and Wiesel for their work in the field of brain specialization and visual information processing, respectively.

Fueled by the increasing number of clinical and research findings, the second half of the 20th century saw a revival of interest in studying the effect of white matter disconnection. This culminated with Norman Geschwind and the beginning of the neoassociationist school (Geschwind, 1965a,b). Geschwind not only reintroduced classical associationist ideas to modern neurology, but also further expanded upon them. He integrated these with new concepts, linking them with contemporary findings from neuroscience, and emphasized the importance of white matter connectivity. This approach significantly advanced the understanding of brain connectivity and its relation to function and behavior, attracting the interest of many clinicians and researchers. This set the stage for more advanced network models of brain function. For instance, in the 1980s, Marsel Mesulam built on Geschwind's ideas and proposed network models incorporating both cortical hubs and heteromodal association areas (Mesulam, 1990). Meanwhile, Damasio also put forth models of convergence zones, suggesting that certain brain regions integrate information from lower-order sensory regions, in a hierarchical manner (Damasio, 1989).

While theoretical network models of brain connectivity advanced significantly in the 1980s–1990s, only invasive animal tracing studies provided detailed empirical wiring diagrams of brain connectivity. Detailed maps of anatomical connections across the whole human brain remained elusive (Mesulam, 2012). Models of large-scale connectivity in the human brain were primarily conceptual and lacked ground-truth anatomical validation. Recognizing this limitation, in a 1993 commentary in *Nature*, Francis Crick and Edward Jones explicitly stated that new noninvasive methods to map the living human brain were critically needed to move beyond the limitations of animal models (Crick and Jones, 1993).

In 1994, Peter Basser, James Mattiello, and Denis Le Bihan published the first paper on diffusion tensor imaging (DTI; Basser et al., 1994), showing that this technique could indeed noninvasively extract microstructural information from living biological tissues and infer the spatial organization and orientation of fiber bundles. Within a few years, DTI enabled mathematical reconstruction of major white matter pathways (Conturo et al., 1999; Mori et al., 1999) across the entire living human brain, marking the start of the new field of diffusion MR tractography. Tractography techniques rapidly advanced in the 2000s. With new models, algorithms, and improved MRI data, more detailed mapping of structural brain networks became possible. Today, there are still many limitations and caveats regarding the use and interpretation of tractography reconstructions. Nevertheless, in just over 20 years since its inception, diffusion tractography has established itself as an essential tool for clinical and neuroscience applications. The theory, methods, challenges, applications, and impact of tractography will be described in more detail in the next chapters of this book. This chapter concludes with a brief overview of imaging modalities that complement diffusion imaging and tractography by probing the functional and metabolic aspects of brain connectivity.

# 5 Contemporary neuroimaging methods

In contemporary neuroscience, neuroimaging methods have inaugurated a new era in the study of functional and anatomical connectivity in the living human brain. Many of these methods are still evolving at a rapid pace and some have found use for clinical applications Electroencephalography (EEG), developed in the 1920s, stood for many decades as the only noninvasive way to directly monitor brain activity, with applications like mapping the alpha rhythm (Berger, 1929). EEG gained more traction as a tool for connectivity studies in the 1990s with the introduction of multichannel recordings and more advanced analysis methods. Techniques like coherence analysis assessed functional connections by measuring correlation of electrical potentials over time (Nunez et al., 1997). Similarly, magnetoencephalography (MEG), introduced in the 1960s, increased localization of neuromagnetic signals. Important developments included the demonstration of the tonotopic organization of the auditory cortex (Romani et al., 1982) and source localization in the somatosensory cortex (Wood et al., 1985). EEG and MEG methods are still widely used and new analysis techniques are developed today thanks to their high temporal resolution. However, the intrinsic limited spatial resolution hinders accurate anatomical mapping of smaller functional areas.

This limitation led to the development and widespread adoption of functional MRI (fMRI), a new imaging technique introduced in 1990 and based on the blood-oxygen-level-dependent (BOLD) MR contrast mechanism (Ogawa et al., 1990). Although with decreased temporal resolution, fMRI allows the spatial localization of brain activity to individual cortical and subcortical regions. Building on BOLD fMRI, in 1995, Biswal et al. were the first to demonstrate that during rest the left and right hemispheric regions of the primary motor network are not silent, but show intrinsically correlated low-frequency BOLD fluctuations, indicating a functionally connected network (Biswal et al., 1995). Since this pioneering work, several other studies not only replicated those findings, but also showed that similar patterns can be found between regions not only known to be coactive during specific tasks, like the primary visual network and auditory network, but also for higher order cognitive networks and the default-mode network (Smith et al., 2009; Van Den Heuvel and Pol, 2010; Yeo et al., 2011). Collectively, these studies have demonstrated the existence of several low-frequency, spontaneous fluctuations that are highly correlated between multiple brain regions. These findings are the foundation of the rapidly expanding field known as *resting state functional connectivity* analysis.

During the 1990s, researchers also began utilizing graph theory and network analysis to computationally model the brain's connectivity architecture. Key contributions came from Karl Friston on functional integration and segregation (Friston, 1994) and Giulio Tononi on neural complexity (Tononi et al., 1994), while Olaf Sporns pioneered the application of graph theory analysis to study the brain's topological network organization (Sporns et al., 2000). Initially, these computational models were more theoretical without direct empirical anatomical models. More recently, the application of graph-theory modeling has also included structural information from tractography methods to account for anatomical or structural connectivity. Overall, this has contributed to expand our view of the organization of the functional and structural connectome (Sporns, 2013a; Wang et al., 2010). These studies suggest that the brain can be described as a complex system that topologically organizes in a nontrivial manner (e.g., small-world architecture and modular structure) to support efficient information processing (Bassett and Bullmore, 2017; Turkheimer et al., 2022). Graph theory-based approaches model the brain as a collection of nodes and edges. Nodes indicate brain regions, and edges represent the relationships between them (e.g., functional or structural connectivity). Several graph metrics can be used to investigate the organizational mechanisms underlying the relevant networks, such as indices of node degree, clustering, efficiency, and centrality (Farahani et al., 2019). Functional brain networks have been shown to organize intrinsically as highly modular small-world architectures capable of efficiently transferring information at a low wiring cost (Fig. 8, left). Hence, graph theory modeling offers an opportunity to study and combine both structural and functional connectivity and investigate how they relate to each other in healthy individuals and across a range of brain pathologies (Bassett and Bullmore, 2017). Chapter 24 will cover in detail the methods and applications behind structural connectivity and connectomics analysis.

The study of brain functional connectivity is not restricted to fMRI. Even a molecular neuroimaging modality like positron emission tomography (PET), which measures the radioactivity of a positron-emitting tracer injected in a living body, can contribute to the investigation of brain connectivity in many forms. In fact, the default mode network was first discovered in 2001 not by using fMRI but by Raichle et al. using PET $^{15}$O-labeled radiotracers (Raichle et al., 2001). In this study, the authors measured oxygen extraction fraction (OEF)—the ratio of oxygen metabolism to blood flow. This quantity was known to decrease during brain activation but to increase at rest. As expected, Raichle et al. found OEF increased in the visual cortex at rest with eyes closed. However, they also observed other areas with decreased OEF compared to the brain average, suggesting these regions were active in the baseline "default" state (Raichle and Snyder, 2007). While the methodologies for detecting decreased OEF with PET and increased BOLD with fMRI are similar, PET is associated with superior accuracy for the quantification of the cerebral blood flow and metabolic rate of oxygen. However, one disadvantage of PET in functional connectivity studies is the very low temporal resolution due to the long scan time of each acquisition (on the order of minutes) when compared with fMRI, which instead can exhibit temporal resolutions on the order of seconds or hundreds of milliseconds (Watabe and Hatazawa, 2019).

The application of molecular neuroimaging in functional connectivity studies has also given rise to the concept of "metabolic" functional connectivity (Watabe and Hatazawa, 2019). Glucose is the fundamental metabolic substrate in

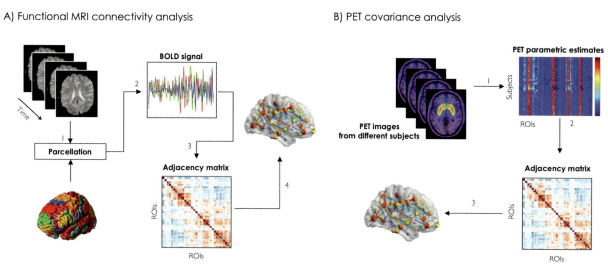

**FIG. 8** Schematic visualization of functional connectivity analysis with different modalities. *Left*, Functional MRI connectivity analysis: (1) Brain regions are segmentation with anatomical atlas; (2) The temporal signal is extracted, preprocessed, and filtered; (3) Cross-regional correlations are performed across all regions; (4) Thresholds are applied to identify adjacency matrix and networks. *Right*, PET covariance analysis: (1) Regional PET parametric estimates are derived from a population of individuals. (2) Pairwise interregional correlations are performed across individuals. (3) Computation of adjacency matrix and networks at population level.

the brain (Paschen et al., 1985) and brain glucose consumption can be accurately measured with $^{18}$F-FDG PET imaging (Schöll et al., 2014). Because glucose metabolism reflects the local brain activity, this measure can be used to study functional connectivity (Wehrl et al., 2013). This approach has been validated in recent studies using simultaneous or combined PET/fMRI for functional connectivity analysis (Hellyer et al., 2017; Savio et al., 2017). Analysis of brain metabolic connectivity has found important applications in dementia research, as reduced metabolic connections may anticipate the neurodegenerative processes typical of this disorder (Morbelli et al., 2012) (Fig. 8, right).

Finally, in recent years, the contribution of PET to functional connectivity studies has moved beyond the application of covariance statistics to raw neuroimaging measures (Sala et al., 2023; Veronese et al., 2019). PET and similar techniques offer the unique opportunity to map molecular targets in the brain, which cannot be investigated by fMRI due to the lack of specificity to different neurochemical systems and neurotransmitters (Kringelbach et al., 2020; Richiardi et al., 2015). To overcome this limitation, multimodal neuroimaging methods have been introduced to investigate the complexity of dynamical neuronal systems (Sporns, 2013b). Following this idea, recent works have introduced methods like the receptor-enriched analysis of functional connectivity by targets (REACT), a multimodal approach that allows the estimation of the whole-brain functional connectivity associated with the distribution of distinct molecular targets and neurotransmitters provided by PET (Dipasquale et al., 2019). This is just a brief overview of some neuroimaging methods developed to study functional brain connectivity. The field is very dynamic, with new techniques being introduced regularly. Readers are encouraged to explore the latest literature for ongoing and recent developments in this area.

## 6 Conclusions

The intricacies of brain connectivity have captivated neuroscientists for centuries, from early ventricular theories to today's advanced network models. This chapter has traced the evolution of brain connectivity research across different periods, highlighting the tools and the theories that defined each era. Early theories relied on postmortem dissections and animal methods, and debates between different schools of thought. Modern neuroimaging has allowed us to move closer to the living human brain, allowing examination of healthy anatomy and pathological conditions. Diffusion MRI and tractography have unlocked new ways of exploring the intricate anatomical wiring of the human brain, while techniques like fMRI, MEG, and PET have revealed important functional and metabolic dynamics of the brain.

As the field continues to evolve, new tools and fields of research will continue to emerge, further deepening our insights into the intricate networks and complexities of the brain. Yet, as extensive as our knowledge has become, it is clear that we are not near the end of this voyage. The vast landscape of brain connectivity remains a frontier with much more to explore.

# References

Aldini, G., 1804. Essai théorique et expérimental sur le galvanisme. De l'Imprimerie de Fournier fils, Paris.

Baillarger, J.G.F., 1840. Recherches sur la structure de la couche corticale des circonvolutions du cerveau. Chez J.-B. Baillière.

Basser, P.J., Mattiello, J., LeBihan, D., 1994. MR diffusion tensor spectroscopy and imaging. Biophys. J. 66 (1), 259–267.

Bassett, D.S., Bullmore, E.T., 2017. Small-world brain networks revisited. Neuroscientist 23 (5), 499–516.

Bentivoglio, M., Mazzarello, P., 2009. The anatomical foundations of clinical neurology. In: Handbook of Clinical Neurology. vol. 95. Elsevier, pp. 149–168.

Berger, H., 1929. Über das Elektrenkephalogramm des Menschen. Arch. Psychiatr. Nervenkr. 87 (1), 527–570. https://doi.org/10.1007/BF01797193.

Bernstein, J., 1868. Ueber den zeitlichen Verlauf der negativen Schwankung des Nervenstroms. Arch. Gesamte Physiol. Menschen Tiere 1 (1), 173–207.

Biswal, B., Zerrin Yetkin, F., Haughton, V.M., Hyde, J.S., 1995. Functional connectivity in the motor cortex of resting human brain using echo-planar MRI. Magn. Reson. Med. 34 (4), 537–541.

Bouillaud, J., 1825. Traité clinique et physiologique de l'encéphalite: ou inflammation du cerveau,.. et de ses suites. Chez JB Baillière.

Broca, P., 1861. Nouvelle observation d'aphémie produite par une lésion de la moitié postérieure des deuxième et troisième circonvolutions frontales. Bull. Soc. Anat. Paris 36, 398–407.

Brodmann, K., 1909. Vergleichende Lokalisationslehre der Grosshirnrinde in ihren Prinzipien dargestellt auf Grund des Zellenbaues. Barth.

Burdach, K.F., 1822. Vom Baue und Leben des Gehirns. vol. 2 Dyk'schen Buchhandlung, Leipzig.

Cajal, S.R., 1893. Manual de histología normal y de técnica micrográfica. Librería de Pascual Aguilar.

Cajal, S., 1911. Histologie du Systeme Nerveux de l'Homme et des Vertebres. Maloine, Paris.

Campbell, A.W., 1905. Histological Studies on the Localisation of Cerebral Function. University Press.

Catani, M., 2007. From hodology to function. Brain 130 (3), 602–605.

Catani, M., Ffytche, D.H., 2005. The rises and falls of disconnection syndromes. Brain 128 (10), 2224–2239.

Catani, M., Mesulam, M., 2008. The arcuate fasciculus and the disconnection theme in language and aphasia: history and current state. Cortex 44 (8), 953–961.

Clarke, E., O'Malley, C.D., 1996. The Human Brain and Spinal Cord: A Historical Study Illustrated by Writings From Antiquity to the Twentieth Century. Norman Publishing.

Conturo, T.E., Lori, N.F., Cull, T.S., Akbudak, E., Snyder, A.Z., Shimony, J.S., McKinstry, R.C., Burton, H., Raichle, M.E., 1999. Tracking neuronal fiber pathways in the living human brain. Proc. Natl. Acad. Sci. 96 (18), 10422–10427.

Crick, F., Jones, E., 1993. Backwardness of human neuroanatomy. Nature 361 (6408), 109–110.

Damasio, A.R., 1989. Time-locked multiregional retroactivation: a systems-level proposal for the neural substrates of recall and recognition. Cognition 33 (1–2), 25–62.

Dandy, W.E., 1936. The Brain. J.B. Lippincott Company.

Dejerine, J., 1892. Contribution à l'étude anatomopathologique et clinique des différents variétés de cécité verbale. Mém. Soc. Biol. 4, 61–90.

Dejerine, J., Dejerine-Klumpke, A., 1895. Anatomie des centres nerveux. vol. 1 Rueff.

Dejerine, J., Dejerine-Klumpke, A., 1901. Anatomie des centres nerveux. vol. 2 Rueff et Cie, Paris.

Descartes, R., 1662. De homine. Apud Franciscum Moyardum & Petrum Leffen.

Descartes, R., 1664. L'Homme de René Descartes et un Traité de la formation du fœtus. Charles Angot, Paris.

Dipasquale, O., Selvaggi, P., Veronese, M., Gabay, A.S., Turkheimer, F., Mehta, M.A., 2019. Receptor-Enriched Analysis of functional connectivity by targets (REACT): a novel, multimodal analytical approach informed by PET to study the pharmacodynamic response of the brain under MDMA. NeuroImage 195, 252–260.

DuBois-Reymond, E., 1848. Untersuchungen über thierische Elektricität. 1. G. Reimer, Berlin.

Farahani, F.V., Karwowski, W., Lighthall, N.R., 2019. Application of graph theory for identifying connectivity patterns in human brain networks: a systematic review. Front. Neurosci. 13, 585.

Felleman, D.J., Van Essen, D.C., 1991. Distributed hierarchical processing in the primate cerebral cortex. Cereb. Cortex 1 (1), 1–47.

Fink, R.P., Heimer, L., 1967. Two methods for selective silver impregnation of degenerating axons and their synaptic endings in the central nervous system. Brain Res. 4 (4), 369–374.

Flechsig, P., 1876. Die Leitungsbahnen im Gehirn und Rückenmark des Menschen: auf Grund Entwickelungsgeschichtlicher Untersuchungen. W. Engelmann.

Flechsig, P., 1896. Gehirn und seele. Verlag von Veit, Leipzig.

Freeman, W., 1957. Frontal lobotomy 1936-56. A follow-up study of 3000 patients from one to twenty years. Am. J. Psychiatry 113, 877–886.

Freeman, W., Watts, J.W., 1945. Prefrontal lobotomy: the problem of schizophrenia. Am. J. Psychiatry 101, 739–748.

Friston, K.J., 1994. Functional and effective connectivity in neuroimaging: a synthesis. Hum. Brain Mapp. 2 (1–2), 56–78.

Fulton, J.F., 1951. Frontal Lobotomy and Affective Behavior: A Neurophysiological Analysis. W. W. Norton & Co Inc.

Galilei, G., 1638. Discorsi e dimostrazioni matematiche: intorno à due nuoue scienze, attenenti alla mecanica & i movimenti locali.. Con una appendice del centro di grauità d'alcuni solidi. Appresso gli Elseuirii.

Gall, F., Spurzheim, G., 1810. Anatomie et physiologie du système nerveux en général, et du cerveau en particulier. F. Schoell, Paris.

Galvani, L., 1791. De viribus electricitatis in motu musculari: Commentarius. Tip. Istituto delle Scienze, Bologna.

Geschwind, N., 1965a. Disconnexion syndromes in animals and man. I. Brain 88 (2), 237–294.

Geschwind, N., 1965b. Disconnexion syndromes in animals and man. II. Brain 88 (3), 585–644.

Golgi, C., 1873. Sulla struttura della sostanza grigia del cervelo. Gazzetta Medica Italiana. Lombardia 33, 244.

Griesinger, W., 1843. Ueber psychische Reflexactionen. Arch. Physiol. Heilkd. 2, 76–113.

Gudden, 1870. Experimentaluntersuchungen über das peripherische und centrale Nervensystem. Arch. Psychiatr. Nervenkr. 2 (3), 693–723.

Head, H., 1926. Aphasia and Kindred Disorders of Speech. The University Press.

Hellyer, P.J., Barry, E.F., Pellizzon, A., Veronese, M., Rizzo, G., Tonietto, M., Schütze, M., Brammer, M., Romano-Silva, M.A., Bertoldo, A., 2017. Protein synthesis is associated with high-speed dynamics and broad-band stability of functional hubs in the brain. NeuroImage 155, 209–216.

Hubel, D.H., Wiesel, T.N., 1962. Receptive fields, binocular interaction and functional architecture in the cat's visual cortex. J. Physiol. 160 (1), 106–154.

Jacobsen, C.F., Wolfe, J.B., Jackson, T.A., 1935. An experimental analysis of the functions of the frontal association areas in primates. J. Nerv. Ment. Dis. 82, 1–14.

Kölliker, A., 1859. Handbuch der Gewebelehre des Menschen. vol. 560 Wilhelm Engelmann, Leipzig.

Kringelbach, M.L., Cruzat, J., Cabral, J., Knudsen, G.M., Carhart-Harris, R., Whybrow, P.C., Logothetis, N.K., Deco, G., 2020. Dynamic coupling of whole-brain neuronal and neurotransmitter systems. Proc. Natl. Acad. Sci. 117 (17), 9566–9576.

Lashley, K.S., 1950. In Search of the Engram.

Liepmann, H., 1900. Das krankheitsbild der apraxie (motorische asymbolie): Auf grund eines falles von einseitiger apraxie. Monatsschr. Psychiatr. Neurol. 8, 15–44.

Lissauer, H., 1890. Ein Fall von Seelenblindheit nebst einem Beitrage zur Theorie derselben. Arch. Psychiatr. Nervenkr. 21 (2), 222–270.

Malpighi, M., 1666. De cerebri cortice. Montius, Bologna.

Matteucci, C., 1830. Sulla contrazione provata dagli animali all'aprirsi del circolo elettrico in che trovansi. Casali, Forlì.

Mesulam, M., 1982. Tracing Neural Connections With Horseradish Peroxidase. Wiley, New York.

Mesulam, M.M., 1990. Large-scale neurocognitive networks and distributed processing for attention, language, and memory. Ann. Neurol. 28 (5), 597–613.

Mesulam, M.-M., 1998. From sensation to cognition. Brain J. Neurol. 121 (6), 1013–1052.

Mesulam, M., 2012. The evolving landscape of human cortical connectivity: facts and inferences. NeuroImage 62 (4), 2182–2189.

Meynert, T., 1869. Neue Untersuchungen über den Bau der Grosshirnrinde und seine örtlichen Verschiedenheiten: Vortrag, gehalten in der Sitzung der k.k. Gesellschaft der Ärzte am 20. November 1868. Medizinische Jahrbücher, p. 17.

Mishkin, M., 1966. Visual mechanisms beyond the striate cortex. In: Russel, R. (Ed.), Frontiers in Physiological Psychology. Academic Press, New York.

Mishkin, M., Ungerleider, L.G., Macko, K.A., 1983. Object vision and spatial vision: two cortical pathways. Trends Neurosci. 6, 414–417.

Monakow, C., 1897. Gehirnpathologie. Alfred Hölder, Vienna.

Monakow, C., 1914. Die Lokalisation im Grosshirn und der Abbau der Funktion durch kortikale Herde. JF Bergmann.

Moniz, E., 1937. Prefrontal leucotomy in the treatment of mental disorders. Am. J. Psychiatry 93, 1379–1385.

Morbelli, S., Drzezga, A., Perneczky, R., Frisoni, G.B., Caroli, A., van Berckel, B.N., Ossenkoppele, R., Guedj, E., Didic, M., Brugnolo, A., 2012. Resting metabolic connectivity in prodromal Alzheimer's disease. A European Alzheimer Disease Consortium (EADC) project. Neurobiol. Aging 33 (11), 2533–2550.

Morecraft, R.J., Ugolini, G., Lanciego, J.L., Wouterlood, F.G., Pandya, D.N., 2014. Classic and contemporary neural tract-tracing techniques. In: Diffusion MRI. Elsevier, pp. 359–399.

Mori, S., Crain, B.J., Chacko, V.P., Van Zijl, P.C., 1999. Three-dimensional tracking of axonal projections in the brain by magnetic resonance imaging. Ann. Neurol. 45 (2), 265–269.

Mountcastle, V.B., Powell, T.P., 1959. Neural mechanisms subserving cutaneous sensibility, with special reference to the role of afferent inhibition in sensory perception and discrimination. Bull. Johns Hopkins Hosp. 105, 201–232.

Nauta, W.J., Gygax, P., 1951. Silver impregnation of degenerating axon terminals in the central nervous system: (1) technic. (2) Chemical notes. Stain Technol. 26 (1), 5–11.

Nunez, P.L., Srinivasan, R., Westdorp, A.F., Wijesinghe, R.S., Tucker, D.M., Silberstein, R.B., Cadusch, P.J., 1997. EEG coherency: I: statistics, reference electrode, volume conduction, Laplacians, cortical imaging, and interpretation at multiple scales. Electroencephalogr. Clin. Neurophysiol. 103 (5), 499–515.

Ogawa, S., Lee, T.M., Kay, A.R., Tank, D.W., 1990. Brain magnetic resonance imaging with contrast dependent on blood oxygenation. Proc. Natl. Acad. Sci. U. S. A. 87 (24), 9868–9872.

Papez, J.W., 1937. A proposed mechanism of emotion. Arch. Neurol. Psychiatry 38 (4), 725–743.

Paschen, W., Mies, G., Bodsch, W., Yamori, Y., Hossmann, K., 1985. Regional cerebral blood flow, glucose metabolism, protein synthesis, serum protein extravasation, and content of biochemical substrates in stroke-prone spontaneously hypertensive rats. Stroke 16 (5), 841–845.

Pavlov, I.D., 1903. The experimental psychology and psychopathology of animals. In: Psychopathology and Psychiatry. Routledge, pp. 13–30.

Petrides, M., Pandya, D.N., 1984. Projections to the frontal cortex from the posterior parietal region in the rhesus monkey. J. Comp. Neurol. 228 (1), 105–116.

Pevsner, J., 2002. Leonardo da Vinci's contributions to neuroscience. Trends Neurosci. 25 (4), 217–220.

Raichle, M.E., Snyder, A.Z., 2007. A default mode of brain function: a brief history of an evolving idea. NeuroImage 37 (4), 1083–1090.

Raichle, M.E., MacLeod, A.M., Snyder, A.Z., Powers, W.J., Gusnard, D.A., Shulman, G.L., 2001. A default mode of brain function. Proc. Natl. Acad. Sci. 98 (2), 676–682.

Reil, J.C., 1809. Die sylvische grube. Arch. Physiol. 9. 195–208, 911.

Reil, J.C., 1812. Die vördere Commissur im großen Gehirn. Arch. Physiol. 11 (1), 89–100.

Richiardi, J., Altmann, A., Milazzo, A.-C., Chang, C., Chakravarty, M.M., Banaschewski, T., Barker, G.J., Bokde, A.L., Bromberg, U., Büchel, C., 2015. Correlated gene expression supports synchronous activity in brain networks. Science 348 (6240), 1241–1244.

Romani, G.L., Williamson, S.J., Kaufman, L., 1982. Tonotopic organization of the human auditory cortex. Science 216 (4552), 1339–1340.

Sala, A., et al., 2023. Brain connectomics: time for a molecular imaging perspective? Trends Cogn. Sci. 27 (4), 353–366.

Savio, A., Fünger, S., Tahmasian, M., Rachakonda, S., Manoliu, A., Sorg, C., Grimmer, T., Calhoun, V., Drzezga, A., Riedl, V., 2017. Resting-state networks as simultaneously measured with functional MRI and PET. J. Nucl. Med. 58 (8), 1314–1317.

Schöll, M., Damián, A., Engler, H., 2014. Fluorodeoxyglucose PET in neurology and psychiatry. PET Clin. 9 (4), 371–390.

Schwann, T., 1839. Mikroskopische Untersuchungen Über Die Übereinstimmung in Der Struktur und Dem Wachstum Der Tiere und Pflanzen. G. Reimer, Berlin.

Smith, S.M., Fox, P.T., Miller, K.L., Glahn, D.C., Fox, P.M., Mackay, C.E., Filippini, N., Watkins, K.E., Toro, R., Laird, A.R., 2009. Correspondence of the brain's functional architecture during activation and rest. Proc. Natl. Acad. Sci. 106 (31), 13040–13045.

Sperry, R.W., 1974. Lateral specialization in the surgically separated hemispheres. In: Schmitt, F., Worden, F. (Eds.), Neurosciences Third Study Program, Chapter I. Neurosciences Third Study Program, vol. 3. MIT Press, Cambridge, pp. 5–19.

Sporns, O., 2013a. The human connectome: origins and challenges. NeuroImage 80, 53–61.

Sporns, O., 2013b. Network attributes for segregation and integration in the human brain. Curr. Opin. Neurobiol. 23 (2), 162–171.

Sporns, O., Tononi, G., Edelman, G.M., 2000. Theoretical neuroanatomy: relating anatomical and functional connectivity in graphs and cortical connection matrices. Cereb. Cortex 10 (2), 127–141.

Steno, N., 1669. Discours de Monsieur Stenon, sur l'anatomie du cerveau à Messieurs de l'Assemblée, qui se fait chez Monsieur Theuenot. Chez Robert de Ninuille.

Tononi, G., Sporns, O., Edelman, G.M., 1994. A measure for brain complexity: relating functional segregation and integration in the nervous system. Proc. Natl. Acad. Sci. 91 (11), 5033–5037.

Turkheimer, F.E., Rosas, F.E., Dipasquale, O., Martins, D., Fagerholm, E.D., Expert, P., Váša, F., Lord, L.-D., Leech, R., 2022. A complex systems perspective on neuroimaging studies of behavior and its disorders. Neuroscientist 28 (4), 382–399.

Van Den Heuvel, M.P., Pol, H.E.H., 2010. Exploring the brain network: a review on resting-state fMRI functional connectivity. Eur. Neuropsychopharmacol. 20 (8), 519–534.

Veronese, M., Moro, L., Arcolin, M., Dipasquale, O., Rizzo, G., Expert, P., Khan, W., Fisher, P.M., Svarer, C., Bertoldo, A., 2019. Covariance statistics and network analysis of brain PET imaging studies. Sci. Rep. 9 (1), 2496.

Vesalius, A., 1543. De Humani Corporis Fabrica. Oporini, Basel.

Vieussens, R., 1685. Neurographia universalis. Jean Certe, Lyon.

Vogt, C., Vogt, O., 1926. Die vergleichend-architektonische und die vergleichend-reizphysiologische Felderung der Großhirnrinde unter besonderer Berücksichtigung der menschlichen. Naturwissenschaften 14 (50–51), 1190–1194.

Volta, A., 1800. On the electricity excited by the mere contact of conducting substances of different kinds. Philos. Trans. R. Soc. Lond. 90, 403–431.

von Bechterew, W., 1900. Les voies de conduction du cerveau et de la moelle. Storck and Doin.

von Economo, C.F., Koskinas, G.N., 1925. Die cytoarchitektonik der hirnrinde des erwachsenen menschen. J. Springer.

von Helmholtz, H., 1850. Messungen über den zeitlichen Verlauf der Zuckung animalischer Muskeln und die Fortpflanzungsgeschwindigkeit der Reizung in den Nerven. In: Archiv für Anatomie, Physiologie und wissenschaftliche Medicin. Veit & Comp., Berlin, pp. 276–364.

Wang, J., Zuo, X., He, Y., 2010. Graph-based network analysis of resting-state functional MRI. Front. Syst. Neurosci. 4, 1419.

Watabe, T., Hatazawa, J., 2019. Evaluation of functional connectivity in the brain using positron emission tomography: a mini-review. Front. Neurosci. 13, 775.

Wehrl, H.F., Hossain, M., Lankes, K., Liu, C.-C., Bezrukov, I., Martirosian, P., Schick, F., Reischl, G., Pichler, B.J., 2013. Simultaneous PET-MRI reveals brain function in activated and resting state on metabolic, hemodynamic and multiple temporal scales. Nat. Med. 19 (9), 1184–1189.

Wernicke, K., 1874. Der aphasische Symptomencomplex: eine psychologische Studie auf anatomischer Basis. Cohn & Weigert.

Wernicke, K., 1885. Some new studies on aphasia. Fortschr. Med., 824–830.

Wernicke, C., 1906. Grundrisse der psychiatrie (Foundations of Psychiatry). Theirne, Leipzig.

Willis, T., 1664. Cerebri anatome: cui accessit nervorum descriptio et usus. typis Ja. Flescher, impensis Jo. Martyn & Ja. Allestry apud insigne Campanae in Coemeterio D. Pauli.

Wood, C.C., Cohen, D., Cuffin, B.N., Yarita, M., Allison, T., 1985. Electrical sources in human somatosensory cortex: identification by combined magnetic and potential recordings. Science 227 (4690), 1051–1053.

Yeo, B.T., Krienen, F.M., Sepulcre, J., Sabuncu, M.R., Lashkari, D., Hollinshead, M., Roffman, J.L., Smoller, J.W., Zöllei, L., Polimeni, J.R., 2011. The organization of the human cerebral cortex estimated by intrinsic functional connectivity. J. Neurophysiol. 106 (3), 1125–1165.

Chapter 4

# From Brownian motion to the brain connectome: My perspective and historical view of diffusion MRI

Denis Le Bihan

*NeuroSpin, Joliot Institute, Bât 145, CEA-Saclay Center, Paris-Saclay University, Gif-sur-Yvette, France*

## 1 Introduction

Diffusion magnetic resonance imaging (dMRI) came into existence in the mid-1980s and has been incredibly successful during the last 35 years (more than 1,400,000 entries in Google Scholar for "dMRI" at time of writing). Clinical applications started in neurology, especially for the management of acute stroke patients. It soon also became a standard to investigate brain white matter diseases with diffusion tensor imaging (DTI, 400,000 entries in Google Scholar), which was shown to reveal abnormalities in white matter fiber integrity in neurological or even psychiatric disorders, and which allowed for the first time to obtain, completely noninvasively, stunning maps in color and three dimensions of brain connections (the "Human Brain Connectome"), as beautifully exemplified in this book. More recently, dMRI has also been used as a means to produce maps of neuronal activation, an alternative to BOLD (blood oxygenation level dependent) functional magnetic resonance imaging (fMRI), which only indirectly relies on brain hemodynamics. Over time, clinical diffusion MRI has been extended, especially in oncology for the diagnosis and monitoring of cancer lesions not only in the brain but now in almost all organs: dMRI is becoming a reference imaging modality for prostate and breast cancer. dMRI can be transformed into virtual MR elastography to evaluate tissue stiffness, for instance in the liver to stage fibrosis. Indeed, dMRI started in my hands (back in 1984, when I was a radiology resident) in the liver, to address a challenge that one of my radiologist mentors gave me: would it be possible with magnetic resonance imaging (MRI) to differentiate liver tumors from angiomas? There were no MRI contrast media clinically available at that time. I had the fuzzy intuition that, perhaps, a molecular diffusion measurement of water would result in low values in solid tumors because of molecular movement restriction, while "diffusion" would be somewhat enhanced in flowing blood. This view started everything: implementation of dMRI and demonstration of how dMRI could be used both to characterize tissues and to evaluate blood microcirculation (or perfusion) in tissues. This chapter provides a recollection of the early times and milestones of dMRI and gives a glimpse into some of the outstanding questions that the concept might allow us to address.

## 2 The birth of diffusion MRI

### 2.1 The concept

From my early teenage years, I have been a fervent admirer of Einstein's work. Unexpectedly, in his PhD thesis (as well as in an article published in 1905, "annum mirabilis," together with his famous articles on the photoelectric effect and the special relativity theory) Einstein linked diffusion (a visible phenomenon) to Brownian motion (a microscopic phenomenon) in the context of the theory of heat to prove the existence of (invisible) atoms and molecules (unbelievably, almost half of the scientists at that time denied their existence) (Einstein, 1905). The botanist Robert Brown observed in 1827 through a microscope that pollen grains suspended in water move spontaneously in a random manner, but he was not able to determine the mechanism that caused this motion (he even considered it as the secret of life!). Independently, the macroscopic phenomenon of "diffusion," referring to the net movement of a substance from a region of high concentration to a region of low concentration, was described mathematically with a set of equations by Adolf Fick in 1855. Einstein conjectured that Brownian motion resulted from particles being moved by individual molecules and how their

---

*Handbook of Diffusion MR Tractography.* https://doi.org/10.1016/B978-0-12-818894-1.00018-5
Copyright © 2025 Elsevier Ltd. All rights reserved, including those for text and data mining, AI training, and similar technologies.

displacement was linked to the diffusion coefficient of Fick's laws, thus bridging for the first time the concepts of diffusion and Brownian motion, and, hence, the microscopic and the macroscopic scales. Based on this explanation of Brownian motion, Jean Perrin was soon able to experimentally determine the existence and the size of the water molecule from diffusion measurements in 1908, for which he got the Nobel Prize in 1926. Amazingly, many years later it became possible to get images representing self-diffusion of water molecules in body tissues, completely noninvasively, thanks to dMRI. The master concept that drives dMRI and that explains its huge success is that during their diffusion-driven displacements, water molecules probe tissue structure at a *micrometer* scale well beyond the usual *millimetric* image resolution, a kind of virtual biopsy.

More specifically, in a free medium, during a given time interval, molecular displacements obey a three-dimensional Gaussian distribution: molecules travel randomly in space over a distance that is statistically well described by a "diffusion coefficient," $D$. This coefficient depends only on the size (mass) of the molecules, the temperature, and the nature (viscosity) of the medium. On a statistical basis, the mean-squared distance traveled by diffusing molecules *along one spatial direction* in a given interval of time $T_d$ (diffusion time) is:

$$\langle X^2 \rangle = 2DT_d \tag{1}$$

where $\langle X^2 \rangle$ is the average mean-squared diffusion distance along this direction (in 3D one has to replace 2 by 6). In the case of "free" water molecules diffusing in water at 37°C, the diffusion coefficient is $3.10^{-9}$ m$^2$ s$^{-1}$. This means that about 32% of the molecules have reached at least 17 μm during 50 ms (typical diffusion time for dMRI), while only 5% of them have traveled over distances greater than 34 μm.

In practice, however, the actual diffusion distance is reduced in biological tissues compared to free water, and the displacement distribution is no longer Gaussian, as water molecules move in tissues bouncing, crossing, contouring, or interacting with many tissue components, such as cell membranes, fibers, or macromolecules. In other words, while over very short times diffusion mainly reflects the local intrinsic viscosity, at the longer diffusion times used with dMRI the effects of the obstacles become predominant. The noninvasive observation of the water diffusion-driven displacement distributions in vivo thus provides us uniquely with clues to the fine structural features and geometric organization of tissues, and to the variations in those features according to changes in physiological or pathological states.

## 2.2 The making

In the 1950s, soon after the introduction of NMR (nuclear magnetic resonance) by Bloch and Purcell (Nobel Prize 1942), it was realized that, due to the poor field homogeneity of magnets, diffusion had a deleterious effect on signals. Pioneers such as Hahn, Carr, Torrey, and others worked out theories to explain how diffusion in magnetic field gradients would impact NMR signals and show how diffusion effects (then a nuisance) could be avoided or removed by using ad-hoc NMR sequences. It was not until the 1960s that Stejskal and Tanner introduced their gradient pulse scheme to purposely measure diffusion in samples, including biological tissues (Stejskal and Tanner, 1965). But with NMR their measurements could not be localized. (I had the privilege of inviting Dr. Stejskal, now deceased, to the first International Society of Magnetic Resonance in Medicine (ISMRM) workshop on dMRI that I organized in Bethesda in 1990, and of meeting Dr. Tanner at the ISMRM meeting in Singapore in 2016.)

Although Sir Peter Mansfield, one of the fathers of MRI who received the 2003 Nobel Prize with Paul Lauterbur, suggested that the combination of diffusion magnetic pulse gradient experiments and MRI was theoretically feasible, he did not actually implement it. Some preliminary work was done by Wesbey et al. (1984) to determine whether diffusion had a detectable effect in conventional MRI scans and whether modifying the acquisition parameters (such as the amplitude of the slice selection gradient through changes in slice thickness) could possibly increase the effect. However, it was clear to me that specific, localized measurements of diffusion would, instead, require special treatment to get accurate measurements in tissues. I thought that dedicated magnetic gradient pulses, such as the Stejskal-Tanner scheme, should be used for diffusion encoding, but the problem was to mix such pulses with those used in the MRI sequence for spatial encoding and quantity the resulting effect. It was not trivial (many at that time thought it was not even possible to get images of diffusion in vivo), and some technical issues had to be worked out, some still outstanding today (such as geometric distortion resulting from uncompensated eddy currents) and the object of an abundant literature. The novelty was to *localize* the diffusion measurements, that is, to obtain *maps* of the diffusion coefficients in tissues, which had never been done before, especially in vivo, with any technique. I was very excited and, in a matter of weeks, "dMRI," as we know it and still use it today, was born, implemented, and patented (Fig. 1).

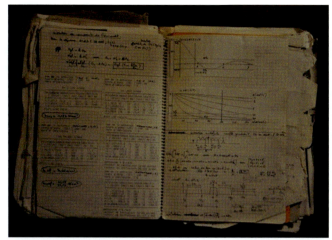
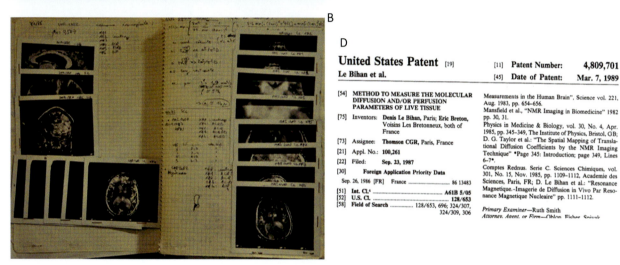

FIG. 1 Early debuts of diffusion MRI. (A) Photo of my notebook showing the gradient pulse scheme used for diffusion encoding and imaging, calculation of the $b$ factor, simulations, and issues we had at the time with the incompletely filtered out 50 Hz currents on the gradient system; (B) Examples of some early diffusion MR images of the brain, the first pulses (this page is dated April 4, 1986) showing strong motion artifacts and initial (and brave!) attempts to estimate diffusion distances; (C) Copy of Eq. (3) in Le Bihan and Breton (1985) where the "$b$ factor" is introduced and defined for the first time (terms in the equation can be found in Le Bihan and Breton, 1985); (D) Front page of the diffusion patent (this one extending the first application from 1985).

My first dMRI paper appeared in 1985 in French in the journal of the Academy of Sciences. In this paper (I was so proud that it was contributed by the famous Anatole Abragam!) (Le Bihan and Breton, 1985), I introduced the basic dMRI equation (Fig. 1C):

$$A = \exp(-bD) \qquad (2)$$

where $A$ is the diffusion-driven signal attenuation, $D$ the diffusion coefficient, and $b$ is the "$b$" factor (I can say now that I took it from "Bihan"), which I thought would advantageously replace a complex double time integral equation quantifying the degree of diffusion weighting of any MRI pulse sequence. A key point is that the Stejskal-Tanner equation could no longer be used due to the presence of many cross-terms between diffusion and imaging gradient pulses. However, this paper did not get much attention at the time, clearly because it was written in French (up to now it has been cited about 500 times).

## 2.3 First trials

Some of my first diffusion measurements were made using a homemade 2 T whole body system designed for spectroscopy in Orsay (Atomic Energy Commission, CEA). Fig. 2 is a photo of the setup at the time; I am sitting at the Bruker console.

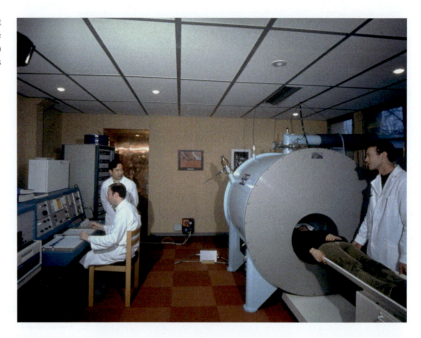

FIG. 2 Setup of our NMR scanner. The system at CEA Frederic Joliot laboratory consisted of a homemade 2T magnet and a Bruker console (no gradient system) (1984). I am sitting at the console. My other colleagues are Tran-Dinh Son and Philippe Jehenson.

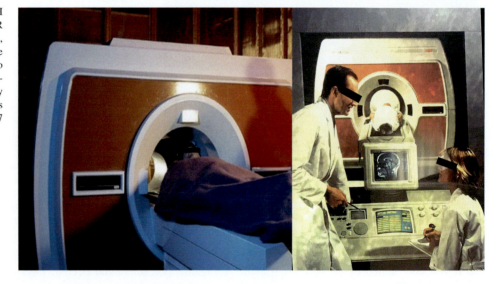

FIG. 3 Clinical 0.35T MRI scanner from CGR company. CGR built the first French MRI scanners, starting at 0.15T with a resistive magnet and later extending to 0.35T and 0.5T with superconducting magnets. The company installed a few of them in hospitals before being bought in June 1987 by GEMS.

We can even see the printout of the echo signals decaying under diffusion. To my astonishment, I discovered this photo many years later inside an Air France inflight magazine stored behind the seat in front of me! This early system had no gradient hardware and I was just installing my phantoms slightly off center, so as to get some field "gradient" along the $z$-axis. It worked pretty well, and I gained some confidence.

The first *images* were obtained on an almost "homemade" 0.5T scanner called Magniscan from the then Compagnie Générale de Radiologie (CGR), a French company located in Buc near Versailles in France (now GEMS European Headquarters) (Fig. 3).

The first trials in the liver were very disappointing, mainly because of huge motion artifacts from respiration (we had to wait until 1999 for the landmark paper by Yamada et al., 1999, which demonstrated that my idea to separate angiomas from liver tumors was correct). First, the MRI scanner I used operated at 0.35T and the signal-to-noise ratio was very low. Second, gradient hardware barely allowed strengths beyond 8 or 10 mT/m to be reached, resulting in $b$ values no larger than 100 or 200 mm/s$^2$ and gradient coil designs did not compensate for eddy current, leading to severe artifacts (Fig. 1). Third, there was no echo-planar imaging (EPI), just plain 2 dimensional Fourier transform (2DFT) spin-echo sequences. Acquisition times necessary for diffusion encoding were very long (close to 10 min for each $b$ value) and,

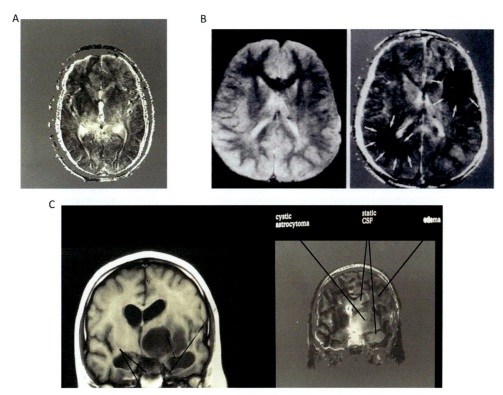

FIG. 4 Early diffusion MRI images of the brain. (A) Normal volunteer (probably myself); (B) Brain ischemia in a child (left: T2w image, right: ADC image) showing large areas with decreased ADC; (C) Cystic tumor surrounded by edema (left: T2w image, right: ADC image). The visualization of CSF flow patterns (IVIM effect) in the dilated ventricles was particularly impressive at the time (those images are reprinted from a scientific exhibit presented at the 1986 Radiological Society of North American (RSNA) meeting in Chicago).

as respiratory gating was not available either, motion artifacts were atrocious in the body. Today, dMRI in the liver is performed routinely and many studies have shown its usefulness to assess liver diseases.

So, I gave up and quickly switched to the brain, as it was after all my background (I started my residency as a neurosurgeon before switching to neuroradiology) and a much less mobile organ. I scanned my own brain and those of some of my colleagues (a common practice at the time!) before scanning patients (Fig. 4). It worked beautifully, and that move was a great achievement: dMRI with its neurological potential was established, and the rest is history.

The world's first diffusion images of the brain were made public in August 1985 at the SMRM meeting in London (Fig. 5). I became nervous when, installing my slides in the speaker preparation room, I found seated right next to me another scientist assembling slides—on dMRI. This scientist was Dieter Merboldt, who introduced a diffusion-sensitized stimulated-echo sequence with images of the hand (Merboldt et al., 1985), without quantification of diffusion. That year there were only three abstracts on diffusion imaging at the Society for Magnetic Resonance in Medicine (SMRM) meeting (the third one was a poster by Taylor and Bushel showing diffusion measurements in a chicken egg using steady gradient pulses for diffusion and ignoring imaging pulses (Taylor and Bushell, 1985)).

Still, dMRI was very sensitive to motion artifacts (it is still!), to the point that some colleagues (including prominent ones) yelled at me during talks at SMRM meetings, saying it was just not possible to measure diffusion in the brain, despite my efforts to explain that incoherent molecular motion and coherent macroscopic head motion could be sorted out. It was very discouraging, but, unfortunately for them and fortunately for dMRI, I am stubborn (a feature often attributed to people from Brittany, where my ancestors are from). I kept going and it paid off, as dMRI progressively gained momentum (I had two SMRM abstracts in 1986, three in 1987). Ironically, some of my early detractors worked full time on dMRI later. Finally, my next (and first) paper in *Radiology* (Le Bihan et al., 1986), where the very first brain diffusion MR images obtained in patients were presented, was much better received. This seminal paper has been cited to date nearly 4000 times and is the most cited article for neuroradiology in the *Radiology* journal published since its inception in 1923. Nowadays, there are several full oral and poster sessions dedicated to dMRI methodology and applications at meetings organized by ISMRM, European Society for Magnetic Resonance in Medicine and Biology (ESMRMB), European Society of Radiology (ECR), and Radiological Society of North American (RSNA), and dMRI has become one of the largest communities at the ISMRM.

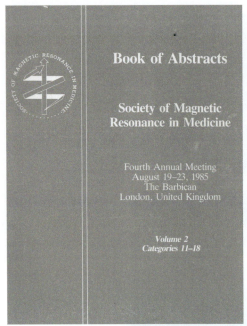
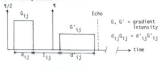

FIG. 5 First presentations of diffusion MRI at the (I)SMRM meetings. Top: The first dMRI images of the brain (my brain) were presented during the 1985 SMRM meeting in London (presentation #1238). Bottom: At the Montreal meeting of 1986 I had two presentations (the abstract of the one shown here was assigned #1 of all meeting abstracts, which I thought was a good presage).

## 3 IVIM: Generalized diffusion MRI

Meanwhile, during those years, I was investigating the idea that dMRI could also provide information on perfusion. I came up with the view that the movement of the blood in the microvasculature could be modeled as a pseudo diffusion process at the imaging macroscopic scale. With the true (molecular) diffusion process, because of their own thermal energy, molecules move and collide with each other. Each collision results in a change in the motion direction of each molecule, and the overall process is well described by a random walk, as described by Einstein. Borrowing (we would rather say today copying/pasting) his view, I considered that, at a completely different scale and in addition to molecular diffusion, the pattern of flowing water molecules in blood, changing direction many times between each successive capillary segment, could be described as a pseudo diffusion process if those segments are disposed in space in a pseudo-random manner. The overall movement mimics a random walk and Einstein's mathematical model of diffusion should work as well, although blood microcirculation (and, indeed, molecular motion) is deterministic. Amazingly, while the difference in spatial scale between the processes of diffusion (nanometers) and pseudo diffusion (tens of micrometers) extends across five orders of magnitude, respectively associated diffusion and pseudo diffusion coefficients only differ by roughly one order of magnitude ($D$, the molecular diffusion coefficient of water in tissues, is about $1 \times 10^{-3}$ mm$^2$/s, while $D^*$, the pseudo diffusion coefficient associated with blood flow, is about $10 \times 10^{-3}$ mm$^2$/s in the brain). This is because those coefficients combine effects of elementary particle velocity (time) and distance: molecular diffusion is a very fast process considering molecular distances, while blood flow pseudo diffusion is comparatively much slower, but involves distances of tens of micrometers. This proximity in values for $D$ and $D^*$ means that diffusion MR images are prone to contamination by blood microcirculation effects, especially when using very low $b$ values.

After some brainstorming I came up, with my mentor Maurice Guéron (at the French Ecole Polytechnique, where I was simultaneously completing my PhD in physics), with the concept of IntraVoxel Incoherent Motion, or IVIM, to cover the overall molecular displacements to which "diffusion" MRI could be sensitive. It was very clear to me that the early results of dMRI measurements I obtained in the brain would include perfusion effects (especially with the very low $b$ values used at the time), among other things, and not only true diffusion, which led me to introduce the term *apparent diffusion coefficient* or *ADC* in my 1986 *Radiology* article. I found later that Tanner had also mentioned "apparent diffusion coefficients" earlier, but not in the context of a generalized framework to describe incoherent microscopic motion. By "apparent diffusion" Tanner was only referring to the "relative diffusion" observed in tissues compared to free diffusion, because of restriction effects. The theoretical framework for IVIM and the demonstration of the validity of the concepts in phantoms and in vivo were exposed in a Scientific Exhibit at RSNA (which was awarded a Magna Cum Laude!) and in my second seminal *Radiology* paper (Le Bihan et al., 1988) (which is the second most cited article for neuroradiology in the *Radiology* journal published since its inception in 1923), accompanied by a terrific editorial by Thomas Dixon (Dixon, 1988) (Fig. 6).

Because of my IVIM article, dMRI has been associated for many years with perfusion imaging, hence the many "diffusion/perfusion" sessions at meetings and workshops, or even book chapters, or journal keywords, although they refer to completely different phenomena, both physically and biologically. This unexpected association has been a little bit puzzling for some of my colleagues, who at some point teased me with such aphorisms as "diffusion, perfusion, … confusion," printed on T-shirts worn by some of them at the first SMRM-sponsored workshop on diffusion/perfusion MRI, which I organized in Bethesda in 1990 (Fig. 7).

It is fair to say that perfusion-driven IVIM MRI was technically challenging and a subject of controversy for many years. Separation of perfusion from diffusion requires good signal-to-noise ratios and stable gradient hardware, which were difficult to attain with low-field MRI systems, although other groups published encouraging results in the brain or other organs. "Perfusion," as seen with IVIM MRI, had to be redefined to accommodate the physiological viewpoint (blood flow to deliver nutriments to tissues) and the radiological viewpoint (flow of blood in the vascular compartment) (Le Bihan and Turner, 1992). Indeed, the exact nature of what is measured with IVIM MRI became the object of PhD theses, articles, and even ISMRM workshops. IVIM is today enjoying a spectacular revival, especially in oncology, as quantitative images of perfusion can be obtained, even in the fetus and the placenta, without the need for any tracer or contrast agent, to the extent that, with some colleagues from Asia, the United States, and the EU, we just coedited a 500-page textbook on IVIM MRI (Le Bihan et al., 2018).

## 4 How IVIM and diffusion MRI entered the clinical field

dMRI received a lot of attention at that time following an unexpected breakthrough from Michael Moseley and his team at the University of California in San Francisco (UCSF). The team was using dMRI (in fact, unknowingly, IVIM) to study *perfusion* in an acute cat brain ischemia model based on the occlusion of the middle cerebral artery. However, their major finding was a puzzling decrease of the water *diffusion* coefficient occurring during the early phase of acute brain ischemia, while standard (T1, T2) MRI remained completely normal (Moseley et al., 1990a). This finding gave a tremendous boost to dMRI, which was still a pure research lab tool, attracting clinicians and convincing manufacturers to make diffusion imaging work reliably on their systems for clinical applications.

To accomplish this, I had developed "fast" acquisition methods with gradient-echo techniques that had just been introduced, such as steady-state free precession (SSFP) sequences, but they were far from perfect. I had joined the Clinical Center at NIH at that time, where I immensely benefited from a collaboration with Robert Turner, who moved to NIH soon after me. With his unique expertise in gradient hardware and EPI (Robert came from the notorious Nottingham laboratory of Peter Mansfield, the father of EPI), we could obtain the first IVIM-EPI images (Turner et al., 1990). Because EPI

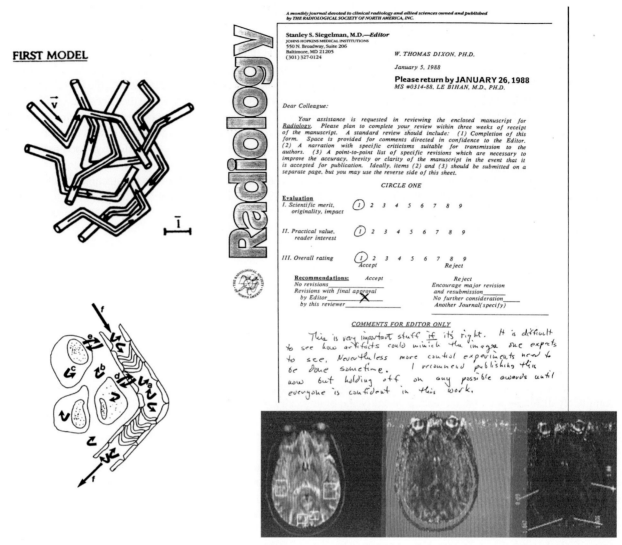

**FIG. 6** First IVIM article published in *Radiology* (1988). Left: Schemes of the 1988 Radiology manuscript explaining the differences between genuine diffusion (in tissues and blood) and pseudo diffusion from microcirculation. Top right: Laudatory review of the manuscript by Thomas Dixon. Bottom right: First attempt of IVIM fMRI with a visual stimulation (left: T2w slice, middle: IVIM slice, right: subtraction of IVIM images obtained with and without the visual stimulation showing an increase in ADC in visual cortex). *(Top right: Courtesy Thomas Dixon.)*

is a "single-shot" acquisition technique, motion artifacts were virtually eliminated. With EPI, IVIM and diffusion images could be obtained in a matter of seconds. Indeed, it was not until the availability of EPI on clinical MRI scanners that diffusion and IVIM MRI could really take off. This endeavor was not trivial, as we had to get gradient coils and power supplies with outstanding performances for the time (Fig. 8). With the help of our colleagues Joe Maier, Bob Vavrek, and James MacFall from General Electric Medical Systems (Milwaukee), Robert could install his gradient coil prototype. Young generations of radiologists should realize how lucky they are to benefit from tremendous advances in MRI technology made over the last 40 years, combining EPI with parallel imaging using multiple channels, reducing echo times, making acquisitions less vulnerable to motion with respiratory triggering, and enjoying state-of-the art gradient hardware above 40 mT/m.

dMRI changed forever the destiny of acute brain stroke patients. Until the 1990s there was no definite diagnosis for patients undergoing an acute brain stroke (caused by a clot blocking an artery in the brain). Differential diagnosis between a clot and a hemorrhage was more a guess, and there was no treatment available at all. Nearly half of the patients would die and the other would stay with permanent handicaps, such as being hemiplegic or aphasic for life. The social burden, the economic cost, everything was—and still is, in fact—considerable. At that time, some colleagues from Harvard could demonstrate for the first time that, using EPI dMRI, they could reproduce the results obtained at UCSF in the cat model (Warach et al., 1992; Chien et al., 1992): within the first hours after the acute stroke a bright area corresponding to the area where

**FIG. 7** Layout of the flyer of the first SMRM workshop on diffusion and microcirculation MRI. The workshop was held close to NIH. Five years after the introduction of diffusion MRI the list of scientists interested or active in the field was already impressive. Some are still active in the field, while others have retired or passed away, notably Edward Stejskal.

neurons were suffering and dying was visible on the diffusion-weighted MR images while standard (T1, T2) MR images remained completely normal. The possibility of establishing an accurate diagnosis made it possible for pharmaceutical companies to launch clinical trials on thrombolytic agents to dissolve the clots in vessels. Overnight, it became possible to treat and save patients, providing that thrombolysis treatment was administered soon enough (within the first hours after the onset of the stroke): some hemiplegic or aphasic patients could recover instantaneously, escaping the terrible destiny that was lying ahead of them by remaining handicapped for the rest of their life. dMRI was the tool that started it all and permitted this revolution in the management of those patients.

This breakthrough pushed MRI manufacturers to implement EPI and dMRI on their systems. Later on, other technical improvements came, such as "parallel" acquisition techniques to allow signals to be collected simultaneously using an array of several radiofrequency coils, further reducing residual artifacts, such as geometric distortion due to magnetic inhomogeneities. Diffusion EPI-MRI, which has become a pillar of medical imaging, is now installed on virtually all clinical MRI scanners available on the planet.

## 5 The emergence of DTI and tractography

It is important to notice that with MRI only the molecular displacement component along the magnetic field gradient direction is detectable. However, diffusion is truly a three-dimensional process, and water molecular mobility in tissues is not necessarily the same in all directions. Diffusion *anisotropy* may result from the presence of obstacles that limit

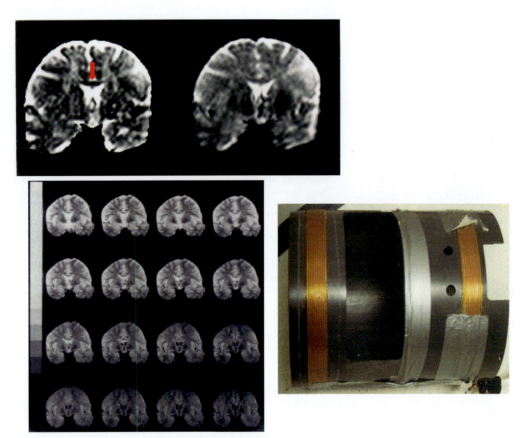

**FIG. 8** First diffusion/IVIM EPI images (obtained at the National Institutes of Health (NIH) in 1990). Bottom left: Set of coronal images of a normal subject obtained at $b$ values from 0 to $1000 s/mm^2$ within 1 min. The differential signal attenuation in CSF and gray and white matter with diffusion weighting is clearly visible, especially as the images are free from motion artifacts. Top left: ADC images showing diffusion anisotropy effects in white matter. The corpus callosum *(arrow)*, especially, appears dark because diffusion weighting could be applied only along the body axis direction with the Helmholtz gradient coil that Robert Turner had built, fitting the head (right).

molecular movement in some directions more than others. Slight anisotropic diffusion effects were observed in excised biological tissues in the 1970s in tissues with strongly oriented components, such as rat skeletal muscles (Hansen, 1971; Cleveland et al., 1976).

However, the discovery in vivo that water diffusion was anisotropic in the central nervous system white matter (spinal cord, then the brain) came as an original dMRI finding (Moseley et al., 1990b; Chenevert et al., 1990). Diffusion anisotropy in white matter grossly originates from its specific organization in bundles of more or less myelinated axonal fibers running in parallel: diffusion measured with MRI in the direction of the fibers (whatever the species or the fiber type) is about three to six times faster than in the perpendicular direction. Immediately after this discovery I thought about reverting this observation to determine and map the orientation of white matter fibers in the brain: assuming the direction of the fibers was parallel to the direction with the highest diffusion, we could infer their orientation simply by sampling the diffusion coefficient along multiple directions, setting up the founding principle that later underlined tractography. Our first attempt with Philippe Douek, a French medical resident working with me at the NIH, was very crude, using only two main directions and a simple color display (Fig. 9A) as a proof of concept (Douek et al., 1991): with this work white matter acquired colors based on the orientation of their fiber bundles at each location, the ancestor of tractography.

From those basic images to the gorgeous fiber tract 3D displays that now make the cover of journals or textbooks (including *Gray's Anatomy*! (Standring and Tunstall, 2015)), there was another big step, which was made possible through my encounter with Peter Basser. We met by chance at an NIH Research Festival in October 1990, where I gave a talk on dMRI. Peter was working on ionic fluxes in tissues and had a poster. Peter appropriately commented that the correct way to deal with anisotropic diffusion was to switch to a *tensor* formalism, of which I was aware, but the problem was that the way dMRI measurements were analyzed could not directly give access individually to the six (instead of nine for symmetry reasons) components of the diffusion tensor. So far, our approach had been oversimplistic and limited in scope: diffusion

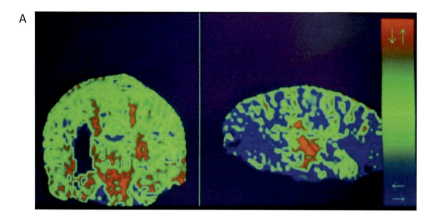

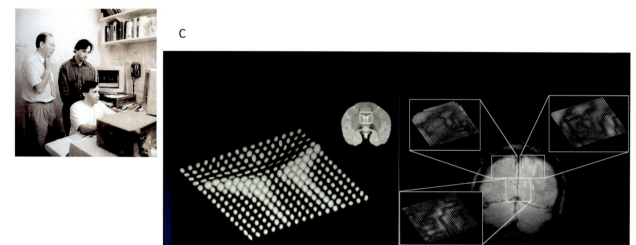

**FIG. 9** First color-encoded white matter fiber orientation maps and DTI debuts. (A) First demonstration in the brain of a normal subject that one may obtain information on the orientation of white matter fibers based on diffusion anisotropy (left: coronal slice, right: sagittal slice). Only two directions were acquired (longitudinal and vertical to the brain). *Blue* corresponds to voxels with an ADC larger in the anterior-posterior direction, *red* to the vertical direction, and *green* to the absence of preferred direction (either because fibers are oriented in a perpendicular direction or because there is no anisotropy). (B) Polaroid photo of D. Le Bihan (left) and P. Basser (right) looking at work presented by J. Mattiello. (C) Left: First diffusion ellipsoid maps obtained in a macaque brain. Anisotropy is clearly seen in the corpus callosum, while fast, isotropic diffusion is displayed in ventricles. Right: First images in a normal subject.

coefficients were considered as scalar quantities that varied according to the measurement directions. However, except in particular cases where the main diffusion directions coincided with those of the MRI scanner gradient, this approach did not allow us to properly determine the direction of the highest diffusivity because of diffusion cross-terms between perpendicular directions. After some brainstorming, Peter and I came up in 1991 with what is known today as DTI: dMRI signals acquired along any particular direction (obtained by applying simultaneous gradient pulses along the $X$, $Y$, and $Z$ axes) are weighted by a linear combination of the diffusion tensor components. dMRI signals acquired along six noncolinear directions represent a set of six equations that can be solved to retrieve the six individual diffusion tensor components in the MRI gradient frame at each location. After ad-hoc transformation (diagonalization) the eigenvectors and eigenvalues ($\lambda$) of the diffusion tensor could be obtained in the frame of the tissue at each location, giving its orientation in space and diffusivity along and perpendicularly to this direction. Meanwhile, the (scalar) *b-factor* got promoted to the rank of a *b-matrix* (Mattiello et al., 1994), thanks to the hard math work of our postdoc James Mattiello, to handle the multiple cross-terms between diffusion encoding and imaging gradient pulses along each axis, but also between each axis, a daunting task with EPI (Mattiello et al., 1997) (with age the *b-matrix* even turned recently to a fat *b-tensor* (Westin et al., 2014)Westin et al., 2014). We were initially not allowed to conduct experiments in human subjects, so we used vegetables and pork meat (and later acquired data in the monkey brain before the first attempts on the human brain) to prove the concept was working, and published our results at the 1992 SMRM meeting and in another seminal article in the JMR journal (Basser et al., 1994a), which has now been cited 4500 times. As it is difficult to display tensor data, we introduced in the *Biophysical Journal*

(Basser et al., 1994b) (this article has been cited even more, 7200 times) the concept of "diffusion ellipsoids" representing in 3D the diffusion distance covered in space by molecules in a given diffusion time (Fig. 9).

After my return to France, Peter Basser worked with Carlo Pierpaoli, who I had also met at NIH, to acquire DTI images of the human brain (Pierpaoli et al., 1996). DTI data can provide exquisite information on tissue microstructure and architecture (Basser, 1995; Le Bihan, 1995a) but we realized that the DTI concept was not straightforward and that its full potential was not easy to grasp at first sight. Subsequently, with my French CEA colleagues we wrote a thorough review article in *JMRI* in 2001 on the concepts and applications of DTI (Le Bihan et al., 2001). This article has been a kind of blockbuster, with more than 4400 citations so far. With DTI, three types of information are obtained at once (Le Bihan et al., 2001):

- . The mean diffusivity MD (similar to the isotropic ADC), which characterizes the overall mean-squared displacement of molecules (average ellipsoid size) and the overall presence of obstacles to diffusion;
- . The degree of anisotropy, which describes how much molecular displacements vary in space (ellipsoid eccentricity) and is related to the presence and coherence of oriented structures. A popular index is the fractional anisotropy (FA), which must be preferred to other nonnormalized quantities such as the eigenvalues difference $\lambda 1 - \lambda 3$, which is extremely sensitive to noise and depends on MD values (Iima et al., 2020);
- . The main direction of diffusivities (main ellipsoid axes), which is linked to the orientation in space of the structures.

However, DTI (at an image voxel level) cannot "see" anisotropy present at a microscopic level (because of the presence of oriented structures, such as dendrites in brain cortex), due to the averaging effect over the many different directions present in the voxel. It later became possible to unravel the presence of such microscopic anisotropic structures using appropriate dMRI sequences encoding the coupling between multiple diffusion directions (Ozarslan and Basser, 2008; Westin et al., 2014) .

At the beginning, DTI provided overall orientation of white matter bundles in each voxel (3D vector field maps), based on the assumption that only one direction is predominant in each voxel, and that diffusivity is the highest along this direction. Still, "connecting" subsequent voxels on the basis of their respective fiber orientation to infer some continuity in the fibers required another layer of algorithmic approaches, which came a few years later independently from different groups (Poupon et al., 1998; Conturo et al., 1999; Mori et al., 1999; Jones et al., 2000), allowing for the first time the reconstruction in 3D and in color of some of the major fiber bundles of the living human brain. Since then, enormous progress has been made to resolve some of the limitations of the time (such as the presence in some voxels of low anisotropy, fiber crossing, or fanning) and explore other avenues (such as probabilistic approaches (Behrens et al., 2003; Hagmann et al., 2003; Parker et al., 2003)). Accurate tractography, whatever the approach, also benefited a lot from technical improvements that allowed directions to be sampled with a high angular resolution (beyond the minimum six required for basic DTI), so-called HARDI (Frank, 2002), for instance with the Q-ball (Tournier et al., 2004) and DSI methods (Wedeen et al., 2005), and the use of high $b$ values to increase the contribution of intra-axonal water. All those issues will be covered in much more detail in the next chapters of this book.

Interestingly, although DTI is an established method to investigate brain white matter, the origin of this anisotropy is not completely clear. Obviously, the orientational restriction of water diffusion within axons (anisotropically restricted diffusion) contributes a lot to the overall anisotropy. Myelin certainly plays an important role, modulating the degree of anisotropy, as observed in demyelinating diseases, but anisotropy is also observed in unmyelinated fibers (indeed, FA increases during brain maturation when fibers get myelinated). The packed, parallel arrangement of the fibers in bundles may be sufficient to explain the presence of anisotropy even in the absence of myelination, especially in very dense and compact bundles. Extracellular water also contributes to the anisotropy effect, especially at low diffusion weighting: perpendicularly to the fibers, water molecules must diffuse along tortuous pathways around fibers, which "slow" them down. Neurofilaments within the axons might also play some role. Overall, experimental evidence points toward the importance of the spatial organization of the membranes and intra-axonal content, together with the presence of myelin, but all is not clear yet, even more so in gray matter.

## 5.1 Other NIH "firsts"

Diffusion of other molecular moieties of interest, such as some metabolites or neurotransmitters, can also be studied with MR spectroscopy, as we showed for the first time in the human brain with Stefan Posse (Posse et al., 1993). With José Delannoy and Ronald Levin we also implemented temperature imaging using dMRI (Le Bihan et al., 1989) (we were quoted as having made the most expensive thermometer in the world) and developed a device providing hyperthermia inside the MRI scanner for cancer treatment. For this work we received the Sylvia Sorkin Greenfield Award in Medical Physics (Delannoy et al., 1990).

From Brownian motion to the brain connectome: Historical view of diffusion MRI Chapter | 4  67

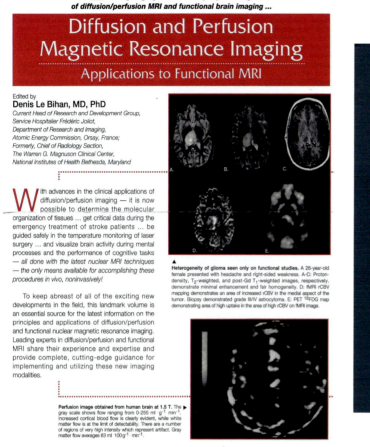
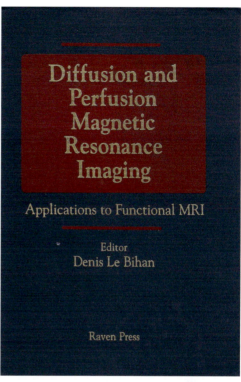

**FIG. 10** First textbook on diffusion and functional MRI (1995). Left: Pamphlet published by Raven Press to advertise the book. Right: Book cover. Interestingly, about half the book is dedicated to fMRI. Hence, this is the first textbook on both diffusion MRI (including DTI) and fMRI.

In 1994, I was called back to France, after 7 incredible years spent at the NIH. This 1987–1994 period was probably the golden age of brain MRI. Important concepts were introduced at that time, such as fMRI and dMRI (with DTI), and it was certainly a time of great excitement. In this context, an important achievement for me was the publication in 1995 of the first textbook on dMRI and fMRI following the workshop on dMRI and fMRI that I coorganized in Bethesda in 1993 (Le Bihan, 1995b) (Fig. 10). It was a tremendous success, but the book quickly went out of print, as Raven Press was merged with another publisher that was not interested in printing other editions. Much later I could find a few copies still available on Amazon, for an astronomical cost!

## 6 Water, the forgotten biological molecule

My return to France did not mean I was going to retire. Indeed, one can spend an entire (professional) life trying to understand the detailed mechanisms of the diffusion process in biological tissues underlying what we see on diffusion MR images. To help me with this intimidating task, I tried first to gather most of the diffusion thinkers of the time at a 2002 ISMRM-sponsored workshop on this topic in Saint-Malo, as I thought that a seaside meeting would be inspiring to scientists motivated by water diffusion. The workshop was a great success, socially and scientifically, with lots of brainstorming and interactions. But the conclusion was that we could not really explain, for instance, why the ADC decreases so much in brain acute ischemia, how cell swelling could lead to decreased diffusion, or precisely why diffusion anisotropy occurs in white matter. Sure, some interesting models were around, but there was always a piece of experimental evidence to disprove those models. For instance, it was thought that diffusion would be faster in the extracellular space than in the intracellular space, explaining how diffusion could decrease with cell swelling. However, it was pointed out that the slow diffusion component (if this would represent a cellular compartment) represents at best only a third of the overall water, while the intracellular-space volume in the brain is more than 80% water. Moreover, some studies had reported that fast diffusion was also found within cells.

Tissues are complex, organized structures in which diffusion largely differs from the Brownian motion modeled by Einstein. Such non-Gaussian diffusion effects become clearly visible when using high $b$ values, which is now possible thanks to progress made in gradient hardware. While achievable $b$ values at the onset of dMRI were in the range of $100 \, s/mm^2$, they climbed to around $1000 \, s/mm^2$ in the 1990s, benefiting from the EPI gradient hardware, and easily reach $3000 \, s/mm^2$ today (even above $20,000 \, s/mm^2$ in some prototype gradient systems made available for the Human Connectome Project or in preclinical systems where gradient amplitudes of $1000 \, mT/m$ are not uncommon, compared to $30–80 \, mT$ in clinical systems). The ADC concept (also often referred to as the monoexponential model) reaches here its limitation. While by its definition it intrinsically encompasses both Gaussian and non-Gaussian diffusion, the reason of its outstanding and lasting clinical success, it cannot give a proper account of the signal attenuation non-Gaussian behavior which becomes apparent at high $b$ values. Yet, extremely valuable information on tissue structure can be specifically found in non-Gaussian diffusion. Several fitting models have been introduced to *empirically* handle this non-Gaussian behavior, such as the polynomial or kurtosis model (also called diffusion kurtosis imaging, DKI), the biexponential model, the statistical model, the stretched exponential model, and others (see Yablonskiy and Sukstanskii, 2010 for a review and other articles in the special edition of *NMR in Biomedicine*). With such models new parameters have emerged beyond the ADC, which have shown great potential to characterize pathologic or physiologic conditions, although they only give "empirical" information on the degree of diffusion non-Gaussianity and nothing specific on tissue features. Recently, we introduced a model-free approach, the so-called signature index (Iima and Le Bihan, 2016), to directly classify tissues based on their signal behavior, including altogether IVIM, Gaussian, and non-Gaussian diffusion effects, or even effects that have not been identified to date, without fitting signals and without estimating any model parameter. Tissues are identified by a direct comparison (calculation of distances) of their signal patterns at multiple $b$ values with those of a library of known (or simulated) signature tissues (Fig. 11A).

Other "explanatory" models have been designed to provide estimates of physical parameters of tissue features, such as the axon diameter in white matter (CHARMED (Assaf and Basser, 2005) and AxCaliber models (Assaf et al., 2008)) or the neurite distribution of orientations (NODDI model (Zhang et al., 2012)). Unfortunately, reverse engineering to retrieve distinct tissue features from the diffusion signal is generally ill-posed. Many different tissue patterns may result in the same signal behavior, so that, even if the basic diffusion mechanisms are known, some a priori knowledge of the tissue structure and strong assumptions (for instance on cell geometry, membrane permeability, mean and distribution of cell size) must be made to make precise inferences about tissue features. Overall, cell membranes play a central role as a *barrier* that hinders water diffusion. However, a common drawback of most models is that effects of membrane permeability (in terms of water exchange or cell residence times) have been largely ignored, for the simple reason that it is generally unknown, while such effects could have a big impact on the observed diffusion signal behavior, not to say the estimated values of the model parameters. Hence, it may be illusory to simulate in vivo biological tissue features with a sufficient level of realistic details obtained from ex vivo samples.

Beside "mechanistic" or geometric factors that may modulate the random walk diffusion of water or other molecules in tissues, one should not forget that there are "functional" or molecular features that are specific to the water molecule and interfere with the diffusion process. Water is a small, highly polarized molecule that does not exist by itself but belongs to a dense network of water molecules that interact with other molecular structures, such as macromolecules or proteins, through their hydrogen bonds. What we actually see with dMRI is the movement of hydrogen nuclei continuously breaking and reforming those bonds to propagate within the network. I tried to propose a global diffusion model based on the physical structure of water in tissues to complement interpretation based only on geometrical compartments (Le Bihan, 2007). I then discovered that the presence (or, more exactly, the amount) of structured water in cells was in itself a subject of great controversy among physicists and biologists, so I organized a symposium on this topic in 2008 in Japan and coedited a book on the subject with the title *Water: The Forgotten Biological Molecule* (Le Bihan and Fukuyama, 2010) to gather our current knowledge on the status of water in biological tissues, with relevance to dMRI. We have largely underestimated the importance of water in biology, from protein and membrane dynamics to cell physiology. Besides protein-bonding through hydrogen bonds, water also forms molecular networks whose properties, including diffusion, may be altered in the vicinity of charged cellular or intracellular membranes. Given the large surface area/volume ratio of most cells, such a membrane-interacting water network probably constitutes an important fraction of tissue water. Hence, any fluctuation in the shape, size, or density of cells (or subcellular components) would induce large variations in the membrane surface area, which could impact the dMRI signal. This emerging research area, looking for some yet unclear roles of water in biological

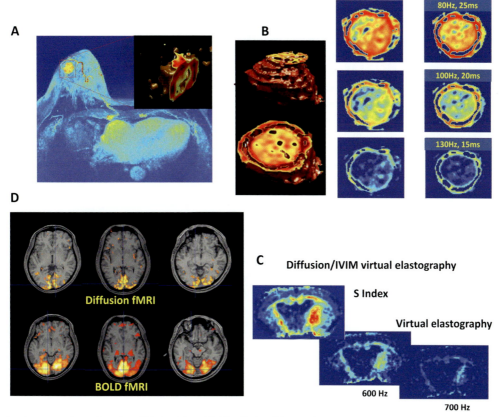

FIG. 11 New developments of non-Gaussian diffusion MRI. (A) The Sindex (Signature) approach allows a virtual biopsy of lesions to be obtained. Tissue types are identified based on the similarity of their diffusion MRI signal patterns at high $b$ values (here up to 1500 s/mm$^2$) with those of a library of reference (signature) tissues (here breast tissue with cancer lesion). (B) Virtual MR Elastography. Non-Gaussian diffusion MRI signals can also be used to quantitatively estimate tissue stiffness (in kPa). Elasticity maps can then be used to simulate traveling mechanical waves at any frequency, creating new contrasts, as shown here in the liver of a patient with a hepatocarcinoma. (C) In the brain, diffusion-based virtual elastography produces maps that provide new information. In this example of an implemented rat brain glioma model, the propagation of the tumor cells in a rat brain model along fiber tracts appears clearly visible at a frequency of 700 Hz. (D) Non-Gaussian diffusion MRI with high $b$ values (here $b = 2400$ s/mm$^2$) can be used to obtain maps of neural activity (here visual stimulation). The putative mechanism is a change in neural tissue microstructure (neural swelling and neuromechanical coupling) concomitant to neural activity. The activation pattern appears more accurate in space and time than with BOLD fMRI, which is only indirectly linked to neural activation through the neurovascular coupling principle.

processes, should both greatly benefit from dMRI and shed light on its mechanisms. Moreover, an unexpected development emerged very recently, virtual MR elastography, as I showed that dMRI could be used to obtain quantitative estimates of liver tissue stiffness (Fig. 11B and C), hypothesizing that tissue microscopic features (such as fibers and cell membranes) would affect both water diffusion and tissue elastic properties in a similar way.

In the context of this book, one has to say, however, that perhaps the exact mechanisms underlying water diffusion in tissues might not be essential for tractography per se. Indeed, most tractography algorithms just need, as inputs, the main directions of the fibers present in each voxel, derived from the directional dMRI signal patterns. This assumption is true to some extent, but tractography accuracy may gain from some knowledge of the underlying microstructure (Daducci et al., 2016) and may even fail without this knowledge when one is able to address low level connectivity, for instance within cortical areas.

## 7 IVIM and diffusion fMRI

### 7.1 IVIM fMRI

Back in the middle of the 1980s, one of my major goals for IVIM imaging was to produce maps of brain perfusion to investigate brain function, given the known coupling between neural activation, metabolism, and blood flow (so-called neurovascular coupling (Roy and Sherrington, 1890)). I recall scanning the daughter of my mentor back in 1986 (the visual

70   PART | I   From anatomy to tractography

stimulation was very crude, just waving a color print from a magazine). We obtained some encouraging results, but not strong enough to write an article (Fig. 6). IVIM-based fMRI did not survive competing methods that appeared at about the same time, first a stillborn approach based on contrast agents (Belliveau et al., 1991) taken up almost immediately thereafter with the blood oxygen level dependent (BOLD) concept (Ogawa et al., 1992). BOLD fMRI was clearly much easier to implement and much more sensitive, so there was no real room for the challenging IVIM method. Later work, however, has proven the validity of the IVIM concept, with an increase in the IVIM perfusion parameters in brain activated regions, and the potential of the approach to aid in our understanding of the different vascular contributions to the fMRI signal (Song et al., 1996, 2018). IVIM MRI has also been used in combination with BOLD fMRI in a kind of negative way, to increase BOLD fMRI spatial resolution by removing the contribution of flow in large arteries or veins feeding or draining large neuronal territories (Song et al., 1996).

## 7.2   Diffusion fMRI

However, another promising avenue that has emerged in the early 2000s (although this is something I wanted to try as soon as 1994, when I left NIH) is the possibility to use dMRI to detect brain activation, instead of BOLD (Darquie et al., 2001; Le Bihan et al., 2006). I am now referring not to IVIM and perfusion, but to genuine diffusion changes occurring in tissues during neuronal activation. To do so, I expatriated again, in 2005, this time to Kyoto. Britanny people also have the reputation of being skilled seafaring sailors (this is not my case): their genes make them restless, so they need to keep moving, as animated by Brownian motion.

Although BOLD fMRI has been extremely successful for the functional neuroimaging community, it presents well-known limitations. The degree and mechanism of the coupling between neuronal activation, metabolism, and blood flow are not fully understood and may even fail in some pathological conditions or in the presence of drugs, even though the brain apparently works normally. Also, the spatial functional resolution of vascular-based functional neuroimaging is ultimately limited, because vessels responsible for the increase in blood flow and blood volume feed or drain somewhat large territories that include clusters of neurons with potentially different functions. Similarly, the physiological delay necessary for the mechanisms triggering the vascular response to work intrinsically limits the temporal resolution of BOLD fMRI. In contrast to vascular-based approaches, dMRI has the potential to reveal changes in the intrinsic water physical properties during brain activation, which could be more intimately linked to the neuronal activation mechanisms (there is an abundant literature on the neuronal swelling that accompanies cell depolarization (Andrew and MacVicar, 1994; Cohen et al., 1968) especially in dendritic spines: there are about $8 \times 10^8$ synapses/mm$^3$ in the cortex, 20,000/neuron, with a dendritic length of 10mm/neuron or 400m/mm$^3$) and lead to an improved spatial and temporal resolution (Fig. 11D). Several preclinical studies relying on pharmacological challenges interfering with neurovascular coupling or cell swelling have confirmed that (1) the diffusion and BOLD fMRI responses could be decoupled, confirming their differential mechanisms; (2) the diffusion fMRI response is not dependent on neurovascular coupling, but, instead, sensitive to underlying neural swelling status; and (3) the diffusion fMRI response follows neural activity status closely and more accurately than BOLD fMRI, especially under anesthetic or vasoactive drug conditions (Abe et al., 2017a,b; Tsurugizawa et al., 2013, 2017). Indeed, those microstructural effects reflected in the diffusion behavior of water during activation might indeed be an *active* component of this process, as water homeostasis and water movement have without a doubt a central role in brain physiology (neurons do not express aquaporin channels, oppositely to astrocytes, and cannot regulate their volume). What is the contribution to brain tissue function of those rapid "mechanical" changes that have been noticed in tissue microstructure (such as the twitching of the dendrite spines (Crick, 1982))? Could we envision that, because laws of nature are most often reversible, neurons could be acting as piezoelectric sensors that get depolarized when "feeling" the movements of neighboring cells, a faster alternative to synaptic transmission for information fluxes within neural networks? Here also dMRI could help in both understanding and exploiting those mechanisms.

## 8   A global picture for the brain connectome

Diffusion-based tractography is an important approach to interpret fMRI data and establish the networks underlying cognitive processes. An important trend has been to match functional connectivity, as obtained with either task-based or resting fMRI, and anatomical connectivity inferred from dMRI (Koch et al., 2002) to derive estimates of the human "connectome," whole brain maps of cortico-cortical connectivity (Biswal et al., 2010; Gong et al., 2009). However, at this stage, one has to keep in mind that mainly the large white matter bundles (rather than smaller intracortical connections) dominate tractography estimates, and that there is no indication on the directional and functional status of the information flow along the tracts. Visualizing the activity flow along each tract noninvasively certainly remains a holy grail. A detailed knowledge of the anatomical connections (in terms of length and size of the fibers, as obtained from the diffusion tensor measurements) might also give in the future some information on the propagation times between activated foci (nodes) and, thus, indirectly provide clues on the timing of the activation of each node of the network. This kind of information is required to explore synchronizations between cortical regions.

Indeed, because the propagation speed of action potentials in the brain connectome has a finite limit, "present" and "simultaneity" are ill-defined in the brain, and concepts borrowed from the special and general relativity theories might prove relevant (Le Bihan, 2020). Especially, one may consider that time and space, as in the universe, might be tightly mingled in the brain, so that the brain functional and structural features could be unified through a combined brain "spacetime." This four-dimensional brain spacetime would obey a kind of relativistic principle and present a functional curvature generated by brain activity, in a way similar to how gravitational masses give our four-dimensional universe spacetime its curvature. This coherent physical and biological framework at global brain level allows the merging of gray matter "activity" and white matter "transmission" into a unique framework to describe how neural activity flows within the Human Brain Connectome. This spacetime is curved by the activity or energy ("mass") present in brain nodes made of gray matter clusters. The activity generated by those nodes flows along connections following "brainlines" that are geodesics (shortest path in four dimensions) in this brain curved spacetime. In other words, geodesic tracks representing activity flow within the brain follow specific connections that dynamically depend on the changing "gradients" among them derived from the activity potential in the continuously changing brain spacetime landscape (Fig. 12).

Such a view may already shed light on brain functional features and dysfunction phenotypes (clinical expression of diseases) observed in some neuropsychiatric and consciousness disorders (Le Bihan, 2020, 2023). Obviously, work remains to validate this new approach and explore its potential to "explain" or model the workings of the brain, in normal or diseased conditions. To do so anatomical connectivity (from diffusion-based tractography, which is today the only available approach) and functional data (from BOLD or diffusion fMRI, EEG, or MEG) will need to be collected at the resolution and level that the framework is addressing, until some kind of "functional tractography" becomes available (Deslauriers-Gauthier et al., 2019). There is hope that ultra-high field MRI with powerful magnets and strong gradient hardware (Le Bihan and Schild, 2017) could soon provide us with such data.

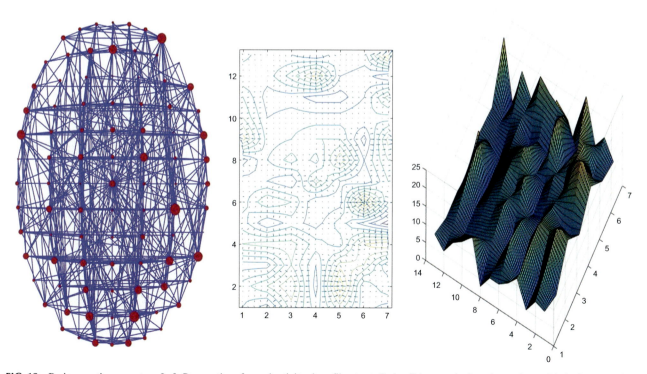

FIG. 12 Brain spacetime curvature. Left: Propagation of neural activity along fiber tracts ("edges") between brain nodes can be modeled using a pseudo-diffusion relativistic model in a 4D brain spacetime with a speed limit $c^*$. Each node is associated with a "mass" depending on its activity that curves the brain spacetime (right). Activity flow just follows the corresponding geodesics (minimizing distances in this four dimensions spacetime) (middle). Reciprocally, activity flow modifies the degree of activation of the nodes and, hence, its curvature, which changes dynamically.

# 9 Conclusion

Diffusion, DTI, and IVIM MRI are now well-established modalities within the medical imaging landscape. However, it took about 10 years after 1985 for dMRI to enter the routine clinical field in hospitals, more than 10 years from the first DTI abstracts to the generalized usage of tractography in the brain, and more than 25 years for IVIM to be used clinically to evaluate perfusion in the body. After all, diffusion is a slow process progressing, not linearly, but according to the square root of time. While T1 and T2 relaxation, magnetic susceptibility, magnetization contrast, etc. exist only because of the presence of the MRI scanner magnetic field, diffusion is not per se an "MRI parameter." MRI is just a sophisticated tool to measure diffusion in the body locally and noninvasively. Water is central to life and its molecular diffusion obviously occurs outside MRI magnets. Indeed, all biological processes require molecules to interact, whether for DNA or RNA replication, protein transcription, protein and enzyme activity, cross-membrane transport, neurotransmission, and so on. And to interact, molecules must first meet, at their own pace, so that diffusion appears as the universal process that Mother Nature (and evolution) has capitalized upon for this purpose. More than an imaging modality, dMRI should be considered as a powerful, genuine multidisciplinary concept at our fingertips to investigate cell physiology and life. So, let us keep our eyes open, as dMRI may well have more surprises ahead for us.

# References

Abe, Y., Tsurugizawa, T., Le Bihan, D., 2017a. Water diffusion closely reveals neural activity status in rat brain loci affected by anesthesia. PLoS Biol. 15, 1–24. https://doi.org/10.1371/journal.pbio.2001494.

Abe, Y., Van Nguyen, K., Tsurugizawa, T., Ciobanu, L., Le Bihan, D., 2017b. Modulation of water diffusion by activation-induced neural cell swelling in *Aplysia californica*. Sci. Rep. 7, 1–8. https://doi.org/10.1038/s41598-017-05586-5.

Andrew, R.D., MacVicar, B.A., 1994. Imaging cell volume changes and neuronal excitation in the hippocampal slice. Neuroscience 62, 371–383. https://doi.org/10.1016/0306-4522(94)90372-7.

Assaf, Y., Basser, P.J., 2005. Composite hindered and restricted model of diffusion (CHARMED) MR imaging of the human brain. NeuroImage 27 (1), 48–58.

Assaf, Y., Blumenfeld-Katzir, T., Yovel, Y., Basser, P.J., 2008. AxCaliber: a method for measuring axon diameter distribution from diffusion MRI. Magn. Reson. Med. 59 (6), 1347–1354.

Basser, P., 1995. Inferring microstructural features and the physiological state of tissues from diffusion-weighted images. NMR Biomed. 8 (7), 333–344.

Basser, P.J., Mattiello, J., Le Bihan, D., 1994a. Estimation of the effective self-diffusion tensor from the NMR spin echo. J. Magn. Reson. 103, 247–254.

Basser, P.J., Mattiello, J., Le Bihan, D., 1994b. MR diffusion *tensor* spectroscopy and imaging. Biophys. J. 66, 259–267.

Behrens, T.E.J., Woolrich, M.W., Jenkinson, M., JohansenBerg, H., Nunes, R.G., Clare, S., Matthews, P.M., Brady, J.M., Smith, S.M., 2003. Characterization and propagation of uncertainty in diffusion-weighted MR imaging. Magn. Reson. Med. 50, 1077–1088.

Belliveau, J.W., Kennedy, D.N., McKinstry, R.C., Burchbinder, B.R., Weisskoff, R.M., Cohen, M.S., Vevea, J.M., Brady, T.J., Rosen, B.R., Buchbinder, B.R., 1991. Functional mapping of the human visual cortex by magnetic resonance imaging. Science 254, 716–719.

Biswal, B.B., Mennes, M., Zuo, X.N., Gohel, S., Kelly, C., Smith, S.M., Beckmann, C.F., Adelstein, J.S., Buckner, R.L., Colcombe, S., Dogonowski, A.M., Ernst, M., Fair, D., Hampson, M., Hoptman, M.J., Hyde, J.S., Kiviniemi, V.J., Kotter, R., Li, S.J., Lin, C.P., Lowe, M.J., Mackay, C., Madden, D.J., Madsen, K.H., Margulies, D.S., Mayberg, H.S., McMahon, K., Monk, C.S., Mostofsky, S.H., Nagel, B.J., Pekar, J.J., Peltier, S.J., Petersen, S.E., Riedl, V., Rombouts, S.A., Rypma, B., Schlaggar, B.L., Schmidt, S., Seidler, R.D., Siegle, G.J., Sorg, C., Teng, G.J., Veijola, J., Villringer, A., Walter, M., Wang, L., Weng, X.C., Whitfield-Gabrieli, S., Williamson, P., Windischberger, C., Zang, Y.F., Zhang, H.Y., Castellanos, F.X., Milham, M.P., 2010. Toward discovery science of human brain function. Proc. Natl. Acad. Sci. U. S. A. 107, 4734–4739.

Chenevert, T.L., Brunberg, J.A., Pipe, J.G., 1990. Anisotropic diffusion within human white matter: demonstration with NMR techniques *in vivo*. Radiology 177, 401–405.

Chien, D., Kwong, K.K., Gress, D.R., Buonanno, F.S., Buxton, R.B., Rosen, B.R., 1992. MR diffusion imaging of cerebral infarction in humans. Am. J. Neuroradiol. 13, 1097–1102.

Cleveland, G.G., Chang, D.C., Hazelwood, C.F., Rorschach, H.E., 1976. Nuclear magnetic resonance measurement of skeletal muscle. Anisotropy of the diffusion coefficient of the intracellular water. Biophys. J. 16, 1043–1053.

Cohen, L.B., Keynes, R.D., Hille, B., 1968. Light scattering and birefringence changes during nerve activity. Nature 218, 438–441. https://doi.org/10.1038/218438a0.

Conturo, T.E., Lori, N.F., Cull, T.S., Akbudak, E., Snyder, A.Z., Shimony, J.S., McKinstry, R.C., Burton, H., Raichle, M.E., 1999. Tracking neuronal fiber pathways in the living human brain. PNAS 96, 10422–10427.

Crick, F., 1982. Do dendritic spines twitch? Trends Neurosci. 5, 44–46.

Daducci, A., Dal, P., Descoteaux, M., Thiran, J.P., 2016. Microstructure informed tractography: pitfalls and open challenges. Front. Neurosci. 10, 247. https://doi.org/10.3389/fnins.2016.00247.

Darquie, A., Poline, J.B., Poupon, C., Saint-Jalmes, H., Le Bihan, D., 2001. Transient decrease in water diffusion observed in human occipital cortex during visual stimulation. PNAS 98, 9391–9395.

Delannoy, J., Le Bihan, D., Hoult, D.I., Levin, R.L., 1990. Hyperthermia system combined with a magnetic resonance imaging unit. Med. Phys. 17 (5), 855–860.

Deslauriers-Gauthier, S., Lina, J.M., Butler, R., Whittingstall, K., Gilbert, G., Bernier, P.M., Deriche, R., Descoteaux, M., 2019. White matter information flow mapping from diffusion MRI and EEG. NeuroImage 201, 116017. https://doi.org/10.1016/j.neuroimage.2019.116017.

Dixon, W.T., 1988. Separation of diffusion and perfusion in intravoxel incoherent motion MR imaging: a modest proposal with tremendous potential. Radiology 168, 566–567.

Douek, P., Turner, R., Pekar, J., Patronas, N., Le Bihan, D., 1991. MR color mapping of myelin fiber orientation. J. Comput. Assist. Tomogr. 15 (6), 923–929.

Einstein, A., 1905. Über die von der molecularkinetischen Theorie der Wärme geforderte Bewegung von in ruhenden Flüssigkeiten suspendierten Teilchen. Ann. Phys. (Leipzig) 17, 549–569.

Frank, L.R., 2002. Anisotropy in high angular resolution diffusion-weighted MRI. Magn. Reson. Med. 45, 935–939.

Gong, G., He, Y., Concha, L., Lebel, C., Gross, D.W., Evans, A.C., Beaulieu, C., 2009. Mapping anatomical connectivity patterns of human cerebral cortex using in vivo diffusion tensor imaging tractography. Cereb. Cortex 19, 524–536.

Hagmann, P., Thiran, J.P., Jonasson, L., Vandergheynst, P., Clarke, S., Maeder, P., Meuli, R., 2003. DTI mapping of human brain connectivity: statistical fibre tracking and virtual dissection. NeuroImage 19, 545–554.

Hansen, J.R., 1971. Pulsed NMR study of water mobility in muscle and brain tissue. Biochim. Biophys. Acta 230, 482–486.

Iima, M., Le Bihan, D., 2016. Clinical intravoxel incoherent motion and diffusion MR imaging: past, present, and future. Radiology 278 (1), 13–32.

Iima, M., Partridge, S.C., Le Bihan, D., 2020. Six DWI questions you always wanted to know but were afraid to ask: clinical relevance for breast diffusion MRI. Eur. Radiol. https://doi.org/10.1007/s00330-019-06648-0.

Jones, D.K., Simmons, A., Williams, S.C., Horsfield, M.A., 2000. Non-invasive assessment of axonal fiber connectivity in the human brain via diffusion tensor MRI. Magn. Reson. Med. 42, 37–41.

Koch, M.A., Norris, D.G., Hund-Georgiadis, M., 2002. An investigation of functional and anatomical connectivity using magnetic resonance imaging. NeuroImage 16, 241–250.

Le Bihan, D., 1995a. Molecular diffusion, tissue microdynamics and microstructure. NMR Biomed. 8 (7), 375–386.

Le Bihan, D. (Ed.), 1995b. Diffusion and Perfusion Magnetic Resonance Imaging. Applications to Functional MRI. Raven Press, New York.

Le Bihan, D., 2007. The "wet mind": water and functional neuroimaging. Phys. Med. Biol. 52, R57–R90.

Le Bihan, D., 2020. On time and space in the brain: a relativistic pseudo-diffusion framework. Brain Multiphys. 1, 1000016.

Le Bihan, D., 2023. From black holes entropy to consciousness: the dimensions of the brain connectome. Entropy 25 (12), 1645. https://doi.org/10.3390/e25121645.

Le Bihan, D., Breton, E., 1985. Imagerie de diffusion in vivo par résonance magnétique nucléaire. C. R. Acad. Sci. Paris 301 (Série II), 1109–1112.

Le Bihan, D., Fukuyama, H. (Eds.), 2010. Water: The Forgotten Biological Molecule. Pan Stanford Publishing, Singapore.

Le Bihan, D., Schild, T., 2017. Human brain MRI at 500MHz, scientific perspectives and technological challenges. Supercond. Sci. Technol. 30, 033003 (20 pp.).

Le Bihan, D., Turner, R., 1992. The capillary network: a link between IVIM and classical perfusion. Magn. Reson. Med. 27, 171–178.

Le Bihan, D., Breton, E., Lallemand, D., Grenier, P., Cabanis, E., Laval-Jeantet, M., 1986. MR imaging of intravoxel incoherent motions: application to diffusion and perfusion in neurologic disorders. Radiology 161, 401–407.

Le Bihan, D., Breton, E., Lallemand, D., Aubin, M.L., Vignaud, J., Laval-Jeantet, M., 1988. Separation of diffusion and perfusion in intravoxel incoherent motion MR imaging. Radiology 168, 497–505.

Le Bihan, D., Delannoy, J., Levin, R.L., 1989. Temperature mapping with MR imaging of molecular diffusion: application to hyperthermia. Radiology 171 (3), 853–857.

Le Bihan, D., Mangin, J.F., Poupon, C., Clark, C.A., Pappata, S., Molko, N., Chabriat, H., 2001. Diffusion tensor imaging: concepts and applications. J. Magn. Reson. Imaging 13 (4), 534–546.

Le Bihan, D., Urayama, S., Aso, T., Hanakawa, T., Fukuyama, H., 2006. Direct and fast detection of neuronal activation in the human brain with diffusion MRI. PNAS 103, 8263–8268.

Le Bihan, D., Iima, M., Federau, C., Sigmund, E.S. (Eds.), 2018. Intravoxel Incoherent Motion (IVIM) MRI: Principles and Applications. Pan Stanford Publishing, Singapore.

Mattiello, J., Basser, P., Le Bihan, D., 1994. Analytical expressions for the b matrix in NMR diffusion imaging and spectroscopy. J. Magn. Reson. Ser. A 108, 131–141.

Mattiello, J., Basser, P., Le Bihan, D., 1997. The b matrix in diffusion tensor echo-planar imaging. Magn. Reson. Med. 37, 292–300.

Merboldt, K.D., Hanicke, W., Frahm, J., 1985. Self-diffusion NMR imaging using stimulated echoes. J. Magn. Reson. 64, 479–486.

Mori, S., Crain, B.J., Chacko, V.P., Van Zijl, P.C.M., 1999. Three-dimensional tracking of axonal projections in the brain by magnetic resonance imaging. Ann. Neurol. 45, 265–269.

Moseley, M., Cohen, Y., Mintorovitch, J., Chileuitt, L., Shimizu, H., Kucharczyk, J., et al., 1990a. Early detection of regional cerebral ischemia in cats: comparison of diffusion- and T2-weighted MRI and spectroscopy. Magn. Reson. Med. 14 (2), 330–346.

Moseley, M.E., Cohen, Y., Kucharczyk, J., 1990b. Diffusion-weighted MR imaging of anisotropic water diffusion in cat central nervous system. Radiology 176, 439–446.

Ogawa, S., Tank, D.W., Menon, R.S., Ellerman, J.M., Kim, S.G., Merkle, H., Ugurbil, K., 1992. Intrinsic signal changes accompanying sensory stimulation—functional brain mapping with magnetic resonance imaging. PNAS 89, 5951–5955.

Ozarslan, E., Basser, P., 2008. Microscopic anisotropy revealed by NMR double pulsed field gradient experiments with arbitrary timing parameters. J. Chem. Phys. 128, 154511.

Parker, G.J., Haroon, H.A., Wheeler-Kingshott, C.A., 2003. A framework for a streamline-based probabilistic index of connectivity (PICo) using a structural interpretation of MRI diffusion measurements. J. Magn. Reson. Imaging 18, 242–254.

Pierpaoli, C., Jezzard, P., Basser, P.J., Barnett, A., DiChiro, G., 1996. Diffusion tensor MR imaging of the human brain. Radiology 201, 637–648.

Posse, S., Cuenod, C.A., Le Bihan, D., 1993. Human brain: proton diffusion MR spectroscopy. Radiology 188 (3), 719–725.

Poupon, C., Mangin, J.F., Frouin, V., Regis, J., Poupon, F., Le Bihan, D., Bloch, I., 1998. Regularization of MR diffusion tensor maps for tracking brain white matter bundles. In: Medical Image Computing and Computer-Assisted Intervention—MICCAI, Oct. 98. LNCS-1496, MIT, Springer-Verlag, pp. 489–498.

Roy, C.W., Sherrington, C.S., 1890. On the regulation of the blood supply of the brain. J. Physiol. 11, 85–108.

Song, A.W., Wong, E.C., Tan, S.G., Hyde, J.S., 1996. Diffusion weighted fMRI at 1.5 T. Magn. Reson. Med. 35, 155–158.

Song, A.W., Bruce, I., Petty, C., Chen, N.K., 2018. IVIM fMRI: brain activation with a high spatial specificity and resolution. In: Le Bihan, D., Iima, M., Federau, C., Sigmund, E.S. (Eds.), Intravoxel Incoherent Motion (IVIM) MRI: Principles and Applications. Pan Stanford Publishing, Singapore.

Standring, S., Tunstall, R., 2015. Gray's Anatomy: The Anatomical Basis of Clinical Practice, forty-first ed. Elsevier.

Stejskal, E.O., Tanner, J.E., 1965. Spin diffusion measurements: spin echoes in the presence of a time-dependent field gradient. J. Chem. Phys. 42, 288–292.

Taylor, D.G., Bushell, M.C., 1985. The spatial mapping of translational diffusion coefficients by the NMR imaging technique. Phys. Med. Biol. 30, 345–349.

Tournier, J.D., Calamante, F., Gadian, D.G., Connelly, A., 2004. Direct estimation of the fiber orientation density function from diffusion-weighed MRI data using spherical deconvolution. NeuroImage 23, 1176–1185.

Tsurugizawa, T., Ciobanu, L., Le Bihan, D., 2013. Water diffusion in brain cortex closely tracks underlying neuronal activity. Proc. Natl. Acad. Sci. https://doi.org/10.1073/pnas.1303178110.

Tsurugizawa, T., Abe, Y., Le Bihan, D., 2017. Water apparent diffusion coefficient correlates with gamma oscillation of local field potentials in the rat brain nucleus accumbens following alcohol injection. J. Cereb. Blood Flow Metab. 37, 3193–3202. https://doi.org/10.1177/0271678X16685104.

Turner, R., Le Bihan, D., Maier, J., Vavrek, R., Hedges, L.K., Pekar, J., 1990. Echo-planar imaging of intravoxel incoherent motions. Radiology 177, 407–414.

Warach, S., Chien, D., Li, W., Ronthal, M., Edelman, R.R., 1992. Fast magnetic resonance diffusion-weighted imaging of acute human stroke. Neurology 42, 1717–1723.

Wedeen, V.J., Hagmann, P., Tseng, W.Y.I., Reese, T.G., Weisskoff, R.M., 2005. Mapping complex tissue architecture with diffusion spectrum magnetic resonance imaging. Magn. Reson. Med. 54, 1377–1386.

Wesbey, G.E., Moseley, M.E., Ehman, R.L., 1984. Translational molecular self-diffusion in magnetic resonance imaging. II. Measurement of the self-diffusion coefficient. Investig. Radiol. 19, 491–498.

Westin, C.-F., Szczepankiewicz, F., Pasternak, O., Ozarslan, E., Topgaard, D., Knutsson, H., Nilsson, M., 2014. Measurement tensors in diffusion MRI: generalizing the concept of diffusion encoding. Med. Image Comput. Assist. Interv. 17 (Pt3), 209–216. https://doi.org/10.1007/978-3-319-10443-0_27.

Yablonskiy, D.A., Sukstanskii, A.L., 2010. Theoretical models of the diffusion weighted MR signal. NMR Biomed. 23, 661–681.

Yamada, I., Aung, W., Himeno, Y., Nakagawa, T., Shibuya, H., 1999. Diffusion coefficients in abdominal organs and hepatic lesions: evaluation with intravoxel incoherent motion echo-planar MR imaging. Radiology 210, 617–623.

Zhang, H., Schneider, T., Wheeler-Kingshott, C.A., Alexander, D.C., 2012. NODDI: practical in vivo neurite orientation dispersion and density imaging of the human brain. NeuroImage 61 (4), 1000–1016.

# Part II

# Diffusion MRI

# Chapter 5

# Physics of diffusion imaging: Fundamentals

**Dmitry S. Novikov[a], Els Fieremans[a], and Hong-Hsi Lee[a,b]**

[a]*Bernard and Irene Schwartz Center for Biomedical Imaging, Department of Radiology, New York University Grossman School of Medicine, New York, NY, United States,* [b]*Athinoula A. Martinos Center for Biomedical Imaging, Department of Radiology, Massachusetts General Hospital, Charlestown, MA, United States*

## 1 Diffusion basics

Diffusion is a physical phenomenon of stochastic thermal motion of molecules. There are two equivalent approaches for describing the inherent randomness of this phenomenon: looking at the *stochastic dynamics* of a molecule or writing down the deterministic diffusion equation for the evolution of the *probability distribution* of molecular displacements. We begin with the former, and then connect with the latter. We first focus on free diffusion in uniform media, and then describe how diffusion becomes restricted or hindered in the presence of tissue microstructure.

### 1.1 Langevin dynamics

We begin with the equation of motion for a molecule, to determine its trajectory $x(t)$. Conceptually, as compared with the deterministic Newtonian evolution, the trajectory $x(t)$ of any particular molecule is random. This randomness arises due to the stochastic force $f(t)$ entering the equation of motion ("$ma = F$") for a random walker:

$$m\partial_t^2 x(t) = f(t) - \eta\partial_t x. \tag{1}$$

For thermal fluctuations in liquids, the inertial term $m\partial_t^2 x$ is negligible compared to the viscous force $-\eta\partial_t x$ proportional to molecular velocity, which makes the latter determined by $f$:

$$\partial_t x = f(t)/\eta \equiv v(t). \tag{2}$$

The physical meaning of this equation is that a molecule has no "free will" (inertia), and its velocity $\partial_t x$ is completely set by its environment exemplified by the random force $f(t)$ rescaled by the viscosity.[a]

While the order of the differential equation (1) is lowered from 2 to 1, it is still nontrivial, because the right-hand side of Eq. (2) is a *random function* $v(t)$. This is the stochastic, or Langevin, dynamics. Technically, one has to solve Eq. (2) for any $v(t)$, and then average the observable quantities over the distribution of functions $v(t)$ that reflects the molecular environment.

Much like for any random variable $\xi$, where one can parameterize the probability density function (PDF) $\mathcal{P}(\xi)$ via its moments $\langle \xi^n \rangle$, or the reduced moments (cumulants) (van Kampen, 1981) (with the lower-order ones generally being more important), the Langevin dynamics is defined by specifying the corresponding moments or cumulants in the space of random functions $v(t)$. The lowest-order one is the mean velocity (drift), $\langle v(t) \rangle$. In the absence of flow, or in the frame of reference comoving with it, the mean velocity is zero, $\langle v(t) \rangle \equiv 0$, which we will consider in what follows without the loss of generality.

The next-order cumulant is the velocity autocorrelation function $\langle v(t)v(t') \rangle$. In simple liquids, the velocity autocorrelation function is an extremely sharp peak (such that we can consider it to be infinitely sharp):

$$\langle v_i(t)v_j(t') \rangle \simeq 2D_0 \cdot \delta_{ij}\delta(t - t'), \tag{3}$$

---

a. Strictly speaking, viscosity itself emerges at the scale somewhat greater than molecular size, so the linear response coefficient $\eta$ is not exactly equal to the bulk liquid viscosity. This is not essential, as all we need here is that $\partial_t x$ is determined by some stochastic drive $v(t)$ applied to a molecule by its environment; subsequent argument is made in terms of statistical properties of $v(t)$.

---

**Handbook of Diffusion MR Tractography. https://doi.org/10.1016/B978-0-12-818894-1.00031-8**

Copyright © 2025 Elsevier Ltd. All rights reserved, including those for text and data mining, AI training, and similar technologies.

where $\mathbf{v}=\{v_i\}$ is the velocity vector with Cartesian components $i=\{1, 2, 3\}=\{x, y, z\}$, and $\delta_{ij}$ is Kronecker's symbol.[b] In reality, the width of this peak is $\Delta t \sim 1–10$ ps (Singer et al., 2016)—orders of magnitude below the nuclear magnetic resonance (NMR)-relevant times of milliseconds. The proportionality coefficient $D_0$, determining the strength of the molecular correlations of the stochastic force, is by definition the *diffusion constant*. Physically, Eq. (3) means that the motion along different coordinate axes is independent, and also that a molecule loses memory about its previous direction and velocity after just a few picoseconds.

In the absence of net flow, the mean molecular displacement along each coordinate $\langle x_i(t)\rangle \equiv 0$ (without the loss of generality, we set $x|_{t=0}=0$ due to the translation invariance in a uniform liquid). Integrating Eq. (3) twice,[c] we obtain the positive mean-squared displacement along coordinate $x$ (Einstein, 1905).

$$\langle x^2(t)\rangle \equiv \sum_{i=1}^{d}\langle x_i^2(t)\rangle = 2dD_0 \cdot t \tag{4}$$

in $d$ spatial dimensions growing with the diffusion time $t$.

Eq. (4) is one of the most fundamental relations in diffusion. It implies that the root-mean-squared displacement (the *diffusion length*)

$$L(t) \equiv \sqrt{\langle x^2(t)\rangle} = \sqrt{2dD_0 \cdot t} \tag{5}$$

that is, a typical distance a molecule would be displaced by (in any direction)—is growing as a square root of diffusion time $t$ (Einstein, 1905; Fig. 1). This time dependence is much *slower* (at sufficiently long $t$) than that for a ballistic motion $x \propto t$. This is the result of stochastic evolution in the absence of coherent flow.

## 1.2 Central limit theorem

The intuition behind Eq. (4), as well as the solution to Eq. (2), arises from making parallels with the central limit theorem (CLT) (van Kampen, 1981), which (roughly) states that the sum of a large number of independent random variables is asymptotically Gaussian-distributed.

Let us discretize the time and the displacement $x(t)$ along a given coordinate as a sum of $N=t/\Delta t \gg 1$ random displacements $\Delta x_n$. Choosing the discretization time step $\Delta t$ to be equal to or greater than the molecular correlation time of a few picoseconds, cf. Section 1.1, the sum of independent displacements

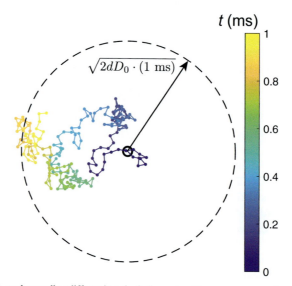

**FIG. 1** An example of free diffusion. A random walker diffuses in a $d=2$-dimensional homogeneous medium characterized by the intrinsic diffusivity $D_0$. The diffusion displacement grows as a square root of diffusion time $t$ in Eq. (5).

---

b. $\delta_{ij}=1$ when $i=j$, and 0 otherwise.

c. The double integration $\int_0^t d\tau d\tau' \, \langle v_i(\tau)v_j(\tau')\rangle = 2D_0\delta_{ij}\int_0^t d\tau \int du \, \delta(u)$ is over the square $[0, t] \times [0, t]$, where the integrand is nonzero only on the diagonal. The integration in the transverse-to-diagonal direction $u=\tau'-\tau$ removes the $\delta$-function, and the remaining integration yields $t$, the side of the square.

$$x(t) \equiv x_N = \Delta x_1 + \Delta x_2 + \cdots + \Delta x_N$$

falls within the realm of CLT: it is asymptotically distributed as a Gaussian random variable,

$$\mathcal{P}(x_N) \equiv \mathcal{P}(x|t) \simeq \frac{1}{\sqrt{2\pi\sigma^2}} e^{-x^2/2\sigma^2}, \tag{6}$$

where the variance

$$\sigma^2 = \sum_{n=1}^{N} \langle (\Delta x_n)^2 \rangle \simeq N \cdot \langle (\Delta x)^2 \rangle = \frac{\langle (\Delta x)^2 \rangle}{\Delta t} \cdot t \equiv 2D_0 t$$

along each spatial dimension grows with the number of steps, yielding Eq. (4). Hence, the CLT provides the meaning of the diffusion constant

$$D_0 = \frac{\langle (\Delta x)^2 \rangle}{2\Delta t} \tag{7}$$

as the mean-squared displacement $\langle (\Delta x)^2 \rangle$ along each dimension over an elementary time step $\Delta t$, if one discretizes Eq. (2) with the correlation function given by Eq. (3). This is, in fact, how Monte Carlo simulations of diffusion are set up (Hall and Alexander, 2009; Fieremans and Lee, 2018; Lee et al., 2021), for example, Fig. 1 or Fig. 2.

Generalizing, in $d$ spatial dimensions, the displacement PDF

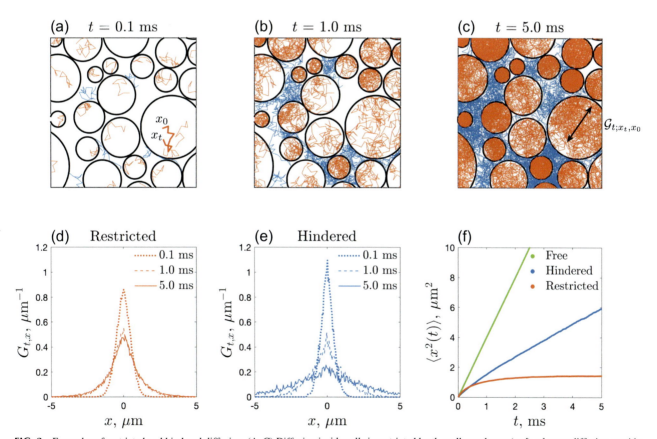

FIG. 2 Examples of restricted and hindered diffusion. (A–C) Diffusion inside cells is restricted by the cell membrane (*red*), whereas diffusion outside cells is hindered (*blue*). The diameter of cells ranges from 0.35 to 6.25 μm, based on the inner axonal diameter distribution in histology (Aboitiz et al., 1992). The box outline has a length of 6.57 μm. (D) The ensemble-average propagator (22) of restricted diffusion (inside cells) with respect to the diffusion displacement $x$. The average propagator is calculated by projecting the simulated particle density to the $x$-axis. At short $t \sim 0.1$ ms, the propagator is roughly Gaussian; at long $t \gtrsim 1$ ms, the propagator is the power spectrum of pore shape, cf. Eq. (27), and does not change with time. (E) The ensemble-average propagator (22) of hindered diffusion (outside cells) with respect to $x$. Similarly, at short $t \sim 0.1$ ms, the propagator is roughly Gaussian; at long $t \gtrsim 1$ ms, the propagator gradually extends its range with time. (F) The mean-squared displacement $\langle x^2(t) \rangle$ for free (*green*), hindered (*blue*), and restricted diffusion (*red*). For the free diffusion, $\langle x^2(t) \rangle \sim t$, Eq. (4). For hindered diffusion, $\langle x^2(t) \rangle$ grows slower than for the free diffusion. For restricted diffusion, $\langle x^2(t) \rangle$ grows at short $t$, and approaches a plateau at long $t$, Eq. (13).

$$\mathcal{P}(x|t) = \frac{1}{(4\pi D_0 t)^{d/2}} e^{-x^2/4D_0 t}, \quad t > 0, \tag{8}$$

where $x^2 = \sum_{i=1}^{d} x_i^2$, yielding Eq. (4).

## 1.3 Diffusion equation

Eq. (8) is the Gaussian *diffusion propagator*, which can be thought of as an envelope (or the density) of the cloud of random walkers emanating from the origin at $t = 0$. Notice that the CLT helped us make a transition from looking at the dynamics of individual random walkers, governed by the Langevin's equation (2) with a stochastic term $f(t)$, to the evolution of the *probability density* of random walkers, Eqs. (6) and (8). It turns out that this density is governed by a *deterministic* differential equation

$$\partial_t \mathcal{P} = D_0 \nabla^2 \mathcal{P} \tag{9}$$

with the initial condition $\mathcal{P}|_{t=0} = \delta(x)$, which can be straightforwardly verified.

The *diffusion equation* (9) provides an equivalent description of the diffusion process. The randomness that was explicit in the Langevin's formulation is implicit in the description of Eq. (9): this equation has deterministic parameters ($D_0$) but is written on the probability density for the random variable $x$, the displacement over the time $t$.

To dig deeper into the physics of the diffusion equation (9), one notices that formally it is a combination of two first-order differential equations

$$\partial_t \mathcal{P} = -\nabla \cdot j, \tag{10}$$

$$j = -D_0 \nabla \mathcal{P}, \tag{11}$$

each with a distinct meaning. This meaning is revealed by focusing on the vector field $j(t, x)$.

Eq. (10) is the conservation law in the differential form, with a meaning that the number $Q(t) = \int_V d^d x \, \mathcal{P}(x|t)$ of molecules in a given volume $V$ can only change by their inflow or outflow through the boundary $\partial V$, but not by the creation or destruction of molecules:

$$\partial_t Q = -\int_V d^d x \, \nabla \cdot j(t, x) = -\int_{\partial V} ds \cdot j(t, x).$$

We used the divergence (Gauss) theorem in Eq. (10), which reveals that the meaning of the vector $j(t, x)$ is the *current*, or the *diffusive flux*.

Eq. (11) is Fick's law (Fick, 1855), which relates the flux $j$ back to the (probability, or particle) density $\mathcal{P}$. This law reflects an obvious property: it is the gradient of the concentration that creates net current (there is no current in equilibrium); this current is directed against the density gradient, to return the system to equilibrium (hence the minus sign). Eq. (11) is a property of the medium, and so it is approximate: We characterize all the complexity at the molecular level with just a single parameter—which, miraculously, again happens to be the diffusion constant $D_0$. For simple liquids on the scales exceeding the molecular dimensions (nanometers) and temporal correlations (picoseconds), this law is very precise, and so is the diffusion equation (9).

By contrasting the equivalent microscopic (Langevin) and macroscopic (diffusion equation) approaches to describing the diffusion, we see that the same parameter, $D_0$, has two complementary meanings: the elementary mean-squared displacement (7) over the molecular correlation time, as well as the kinetic coefficient in Eq. (11) relating the diffusive flux to the concentration of random walkers.

The diffusion constant $D_0$ differs for different molecules and also depends on temperature (Holz et al., 2000). For water at 37°C, $D_0 \approx 3.0 \mu m^2/ms$. This value is a blessing: together with the relatively long NMR times limited by $T_2 \sim 100$ ms, and ultimately by $T_1 \sim 1000$ ms, water diffusion coefficient enables probing the range of displacements $L(t) \sim 1-100 \mu m$, Eq. (5), which is commensurate with the cellular architecture.

## 1.4 Diffusion in the presence of tissue microstructure

So far, we have considered free diffusion. In the case of tissue microstructure, Fig. 2 illustrates the strong effect of impermeable cell barriers on both intra- and extracellular diffusion. It is quite clear that the restrictions to diffusion change the functional form of the propagator (8). It is no longer Gaussian; technically, it acquires higher-order *cumulants* (Kiselev, 2010) that quantify the deviation from the Gaussian PDF.

The loss of Gaussianity is expected: the cell walls introduce negative correlations between subsequent steps $\Delta x_n$ (a step toward a wall is rejected if the previous step has brought the molecule sufficiently close to the wall). Hence, the displacements $\Delta x_n$ are not independent random variables, and there are no reasons for the CLT to apply. Likewise, the velocity autocorrelation function (3) acquires *memory* (a negative tail due to the presence of barriers, as the step toward the wall must be followed by the reflection from it, i.e., the velocity sign change).

All these considerations imply that the mean-squared displacement in general does not grow in proportion to $t$ (Fig. 2). Equivalently, the (cumulative) diffusion coefficient, which is defined as the coefficient of proportionality

$$D(t) \equiv \frac{\langle x^2(t) \rangle}{2d \cdot t}, \tag{12}$$

becomes $t$-dependent. It monotonically decreases from its free (intrinsic) value $D_0$ to the so-called *tortuosity limit* $D_\infty \equiv D(t)|_{t \to \infty} < D_0$.

For *hindered diffusion* (e.g., in the extracellular space), the limit $D_\infty > 0$, and the CLT is restored at sufficiently long times. Indeed, for $t \gg t_c$, where $t_c$ is the time to diffuse past the correlation length $\ell_c$ of the packing of cells, one can rediscretize the Brownian path into temporal steps longer than $t_c \sim \ell_c^2/D_\infty$ and spatial steps exceeding $\ell_c$. These displacements eventually become independent (as the correlation length is by definition the scale beyond which the memory about packing structure fades away), and the propagator becomes Gaussian, albeit with the variance growing with $t$ with a smaller rate than for the free diffusion. One says that the local diffusivity fluctuations are *self-averaging*, yielding an effective coarse-grained medium (Novikov et al., 2014, 2019), as discussed in Section 3.

For *restricted diffusion* (inside impermeable cells), the diffusion displacement is limited by the cell size $\rho$ at long times, that is, $\langle x^2(t) \rangle \sim \rho^2$ for $t \gg t_c$, with $t_c = \rho^2/D_0$ the time to diffuse across the cell. Based on Eq. (12), the intracellular cumulative diffusivity

$$D(t) \sim \frac{\rho^2}{t} \tag{13}$$

approaches $D_\infty = 0$ as $1/t$ at long times.

## 2 How to measure diffusion with NMR?

### 2.1 Scales

When talking about the Langevin dynamics, we had in mind a stochastic force $f(t)$ determined by all molecular degrees of freedom describing rotations, vibrations, and translations, relevant at the nanometer and picosecond level. At much longer $t \gtrsim 1$ ms relevant for NMR experiments, their effect averages out to produce effective parameters such as the relaxation rate constants $R_1$ and $R_2$, and the diffusion coefficient $D$. The parameters $R_1$ and $R_2$ emerge in the Bloembergen-Purcell-Pound theory and enter the Bloch equations describing the semiclassical evolution of the macroscopic magnetization (Bloembergen et al., 1948).

In biological tissues, such as brain, the parameters $R_1(x)$, $R_2(x)$, and $D(x)$ acquire *spatial dependence* at the scale $\sim 0.1-10\,\mu m$, set by the cellular architecture, much coarser than molecular dimensions (Fig. 3). These spatial variations

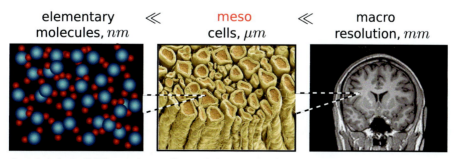

**FIG. 3** The mesoscopic scale in brain dMRI, as an intermediate scale between the elementary (molecular) and the macroscopic (resolution).

**82 PART | II** Diffusion MRI

become relevant at the corresponding $t \sim 1-1000$ ms time scales of diffusion magnetic resonance imaging (dMRI)—much slower than the picosecond time scales at which the local relaxation rate constants and the local diffusion coefficient emerge.

From a physics standpoint, the spatial variations of $R_1(x)$, $R_2(x)$, and $D(x)$ (with the latter including boundary conditions associated with cell membranes) occur at the *mesoscopic scale* (Fig. 3). The term "mesoscopic" originated in condensed matter physics some decades ago (Imry, 1997), signifying focusing on the intermediate scales ("meso"), in-between elementary (say, atomic or molecular) and macroscopic (associated with the sample size or the measurement resolution). By design, this term is relative, depending on which spatiotemporal scales are deemed small and large.

For dMRI, the mesoscopic scale corresponds to tissue heterogeneities at the scale defined by the MRI-controlled diffusion length (5), typically ranging on the order of microns to tens of microns. In the dMRI literature, it is commonly referred to as the *microstructure scale*. This scale is commensurate with immense structural complexity of tissue architecture, giving rise to possibility of tracing neuronal fibers by connecting the mesoscopic and macroscopic scales.

## 2.2 The mesoscopic Bloch-Torrey equation

The diffusion equation (9) governs the evolution of the probability density $\mathcal{P}(x|t)$ in uniform liquids. Up to a normalization, one can think of $\mathcal{P}(x|t) \propto c(x,t)$, the concentration of molecules that are somehow "labeled" relative to other molecules (an analogy is the spread of ink in water).

In nonuniform media, such as tissues, the evolution of the concentration is governed by the diffusion equation

$$\partial_t c(t,x) = \nabla \cdot (D(x)\nabla c(t,x)), \tag{14}$$

in which the spatially dependent mesoscopic diffusion coefficient $D(x)$ reflects the tissue microstructure. Eq. (14) is the combination of particle conservation (10)

$$\partial_t c(t,x) = -\nabla \cdot j(t,x),$$

where $j$ is the particle current (flux), and the local embodiment of Fick's law (11), which says that the concentration flux is linearly proportional to the concentration gradient, now with a *locally varying* $D(x)$:

$$j(t,x) = -D(x)\nabla c(t,x).$$

With NMR, we measure not the concentration of molecules but their transverse magnetization $m(t, x)$ (in the rotating frame). This magnetization is a vector in the plane transverse to the main field $B_0$; a two-dimensional vector can be represented as a complex number. Hence, we can write the corresponding diffusion-like equation for this complex field $m(t, x)$, which incorporates relaxation, susceptibility effects, and other MR contrast mechanisms. The effective theory of magnetization can be described by the mesoscopic Bloch-Torrey equation (Novikov et al., 2019):

$$\partial_t m(t,x) = \nabla \cdot (D(x)\nabla m(t,x)) - (R_2(x) + i\Omega(t,x))m(t,x), \tag{15}$$

where $R_2(x)$ is the locally varying transverse relaxation rate (that originates at the molecular level; Bloembergen et al., 1948), and

$$\Omega(t,x) = \Omega(x) + g(t) \cdot x \tag{16}$$

is the Larmor frequency offset. The first term $\Omega(x)$ originates due to static susceptibility-induced field heterogeneities. The second term is caused by the externally applied diffusion-encoding Larmor frequency gradient $g(t) = \gamma \nabla B_0(t)$ (where $\gamma$ is the proton gyromagnetic ratio).[d] The dMRI signal is then proportional to the total magnetization within a voxel, that is, $S \propto \int m(t,x)\mathrm{d}x$.

The externally applied time-dependent gradient $g(t)$ of the Larmor frequency will be key to measuring the diffusion with MR (effectively, by labeling spin positions), as we will now describe.

## 2.3 The diffusion-weighted signal

To measure diffusion with magnetic resonance, one generally applies diffusion-sensitizing gradients $g(t)$ of the Larmor frequency and measures the resulting voxel-wise magnetization. The gradients cause this so-called diffusion-weighted

---

d. Maxwell's equations forbid having a pure nonzero field gradient, since $\nabla B = 0$. Hence, strictly speaking, there will be nonlinear in $x$ corrections to Eq. (16) due to the so-called concomitant fields, neglected in what follows.

signal to undergo additional transverse relaxation; this signal suppression is generally stronger for those spins that manage to get displaced further along the direction of the gradients, such that their phases go further out of balance at the time of echo.

Consider a general time-dependent gradient $g(\tau)$ modifying the Larmor frequency in Eq. (16). The Larmor precession phase

$$\phi(t) \equiv \phi[g(\tau), x(\tau)] = \int_0^t [\Omega(x(\tau)) + g(\tau) \cdot x(\tau)] d\tau \tag{17}$$

is accumulated during the time interval $t$ of gradient application for a spin following the Brownian path $x(\tau)$.[e] The first term, taken on different paths $x(\tau)$, leads to the phase dispersion due to the static Larmor frequency offset $\Omega(x)$, for example, due to tissue iron and other susceptibility sources, and causes the usual transverse MR relaxation (Kiselev and Novikov, 2018). The second term is controlled by the measurement. In this way, the spin precession phase (17) is a *functional* of the applied gradient $g(\tau)$: it depends on the function $g(\tau)$ for $0 < \tau < t$. For a given spin, $\phi$ is also a functional of its Brownian path $x(\tau)$.

The signal (the voxel-wise magnetization in the transverse plane) is given by the sum of the phase factor $e^{-i\phi(t)}$ over all possible Brownian paths $x(\tau)$ starting at a given point $x_0$ (accounting for the magnetization of a "spin packet" emanating from $x_0$), followed by summing over all such spin packets with all possible positions $x_0$ inside a voxel:

$$S[g(\tau)] = S_0 \int n(x_0) \, dx_0 \sum_{\text{All paths } x(\tau) \text{ from } x_0} e^{-i\phi[g(\tau), x(\tau)]}. \tag{18}$$

In the simplest case of water everywhere inside a voxel, the density $n(x_0) = 1/V$, where $V$ is the voxel volume. In this way, Eq. (18) accounts for the contributions of all spins in a voxel.[f] Of course, the measured signal is a functional of the applied gradient $g(\tau)$, as is the spin precession phase. The overall normalization factor $S_0$ includes proton density and depends on hardware and scan parameters. It is not essential, as it is easily canceled by normalizing the measurement by the "unweighted" signal $S[g \equiv 0]$.

Eq. (18) is completely general (albeit not easily tractable). If we sample Brownian paths $x(\tau)$ emanating from every allowed point $x_0$ within the voxel with the density $n(x_0)$, for example, using the Monte Carlo method (implementing the Langevin dynamics), we would get the diffusion-weighted signal for any $g(\tau)$. Obviously, the signal "knows" (indirectly) about tissue microstructure, embodied in how the Brownian paths are sampled, including all permeable and impermeable restrictions (corresponding to the boundary conditions for Eq. 15) and variations in the local diffusion coefficient $D(x)$. For the Monte Carlo dynamics, such local sampling rules are summarized by Fieremans and Lee (2018) and Lee et al. (2021).

In what follows, we consider the workhorse of the dMRI, the pulsed-gradient method, based on the particular functional form of the gradient $g(\tau)$. This will allow us to make a more concrete connection between the signal and tissue microstructure. We will also neglect the effects of intrinsic $\Omega(x)$, considering pure diffusion without the relaxation.[g]

## 2.4 Pulsed gradients

The most commonly used sequence to measure dMRI signals is the pulsed-gradient spin-echo (Stejskal and Tanner, 1965) (PGSE, Fig. 4A). For PGSE, the gradient is a combination of two pulses. The refocusing rf pulse in-between effectively reverses the polarity of the first pulse, such that the gradient acting on spins reads

---

e. Here, the effective gradient waveform $g(\tau)$ is assumed, accounting for the gradient polarity reversal due to any refocusing pulses (Callaghan, 1991).

f. The sum over all paths is called Feynman's path integral. It involves constructing the measure $d[x(\tau)]$ in the space of random functions $x(\tau)$, the so-called Wiener measure. Such functional integration is necessary to keep in mind for the general $\phi[g(\tau), x(\tau)]$. This expression becomes much simpler for the case of pulse gradients considered in the following.

g. The molecular-level transverse relaxation due to finite water $R_2$ in Eq. (15) leads to the overall factor $e^{-R_2 t}$ renormalizing the signal amplitude $S_0$; it gets factored out when determining the diffusion properties, by normalizing the signal (18) by $S[g \equiv 0]$. The differences in the molecular-level $R_2$ values of each nonexchanging tissue "compartment" result in their $T_2$-weighted relative contributions in the overall diffusion signal. The nontrivial situation arises when the spatially varying $\Omega(x)$ at the scale of the diffusion length interferes with the applied gradient. This interference between the applied and intrinsic Larmor frequency variations causes the deviations of the "apparent" MR-measured diffusion characteristics from their genuine ones (Zhong et al., 1991; Does et al., 1999; Kiselev, 2004; Cho et al., 2009; Pampel et al., 2010; Álvarez et al., 2017; Novikov et al., 2018b). We will not consider such interference effects in what follows.

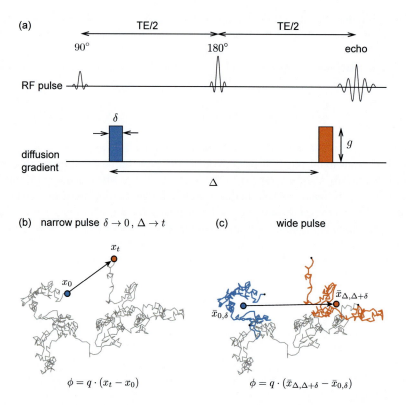

**FIG. 4** (A) Pulse-gradient spin echo (PGSE) with echo time (TE), time interval between gradient pairs (Δ), diffusion gradient pulse width (δ), and gradient strength (g). (B) For ideal narrow pulse PGSE, δ → 0, and Δ is redefined as diffusion time t. The diffusion phase φ is determined by the diffusion wave vector q = gδ and diffusion displacement of positions x(τ) at τ = 0 and τ = t. (C) For realistic wide pulse PGSE, the diffusion phase is determined by the diffusion displacement of the "center-of-mass" positions $\bar{x}_{t_1,t_2}$ averaged within each gradient pulse application. The *blue* (*red*) path is the diffusion trajectory of a random walker during the first (second) gradient pulse application, and the centers of mass are marked as big circles.

$$g(\tau) = \begin{cases} -g, & 0 < \tau < \delta; \\ g, & \Delta < \tau < \Delta + \delta. \end{cases}$$

In the *narrow pulse limit* (δ → 0, Δ → t) of PGSE, and for Ω ≡ 0, the spin phase encodes the molecular displacement (Fig. 4B):

$$\phi = q \cdot (x_t - x_0), \quad \text{with} \quad q = g\delta. \tag{19}$$

Because the spin phase now does not depend on the path x(τ) but only on the difference between the final and initial points, all paths x(τ) that end up in the same final point $x_t$ yield equal contributions to the sum in Eq. (18), as they carry the same phase factor $e^{-i\phi}$. The sum over all paths with a given displacement $x_t - x_0$ over time t samples the *local* diffusion propagator $\mathcal{G}_{t;x_t,x_0}$, that is, the PDF for a molecule to diffuse from position $x_0$ to $x_t$ during time t. It is the fundamental solution of the mesoscopic diffusion equation (or more generally, mesoscopic Bloch-Torrey equation), Eq. (15), with the initial condition $\mathcal{G}_{t;x_t,x_0}|_{t\to+0} = \delta(x_t - x_0)$. Therefore Eq. (18) becomes

$$S(t,q) = S_0 \int dx_t dx_0 \, n(x_0) \, \mathcal{G}_{t;x_t,x_0} \, e^{-iq(x_t - x_0)}, \tag{20}$$

where the integration over all final points $x_t$ weighted with the factor $\mathcal{G}_{t;x_t,x_0}$ realizes the sum over paths.

Changing the integration variable from $x_t$ to the displacement $x = x_t - x_0$, we can see that Eq. (20) becomes equivalent to the spatial Fourier transform

$$\frac{S(t,q)}{S_0} = G_{t,q} = \int dx \, e^{-iqx} G_{t,x} \tag{21}$$

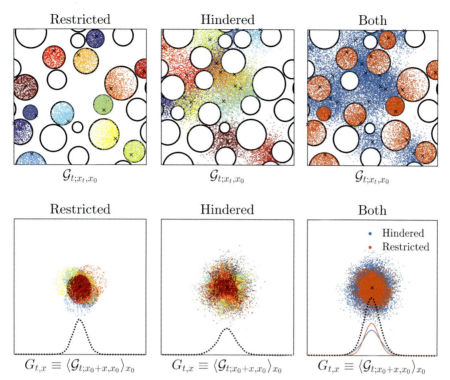

**FIG. 5** The double average in diffusion MR. For restricted and hindered diffusion, and both, the *top row* shows the local propagator $\mathcal{G}_{t;x_t,x_0}$ initiated from each position $x_0$ (cross marker), manifested by 1000 possible Brownian paths $x_t$ (colored points) for each $x_0$. Circular boundaries are impermeable. The *bottom row* shows the ensemble-average $G_{t,x} \equiv \langle \mathcal{G}_{t;x_0+x,x_0} \rangle_{x_0}$ over all initial positions $x_0$ within the MRI voxel, with its one-dimensional projection shown below (*dashed line*). The envelope over the Brownian paths $x_t$ initiated from each $x_0$ in the top row yields the local propagator $\mathcal{G}_{t;x_t,x_0}$, and the further ensemble-average over all initial positions $x_0$ yields the translation-invariant $G_{t,x}$ in the bottom row. The second averaging step (ii) washes out most of the local information at a cellular level within a voxel. Fortunately, the voxel-averaged propagator $G_{t,x}$ remains non-Gaussian, and thus potentially informative.

of the propagator

$$G_{t,x} = \langle \mathcal{G}_{t;x_0+x,x_0} \rangle_{x_0} \equiv \int dx_0\, n(x_0)\, \mathcal{G}_{t;x_0+x,x_0} \quad (22)$$

that is an *ensemble-average* of the local $\mathcal{G}_{t;x_t,x_0}$ over all positions $x_0$ within an NMR sample/MRI voxel, as illustrated in Figs. 2 and 5, and discussed in detail by Callaghan (1991), Grebenkov (2007), Kiselev (2017), and Novikov et al. (2019). The three-dimensional wave vector $q = g\delta$ is created by a Larmor frequency gradient $g$ applied over a short duration $\delta$.

Eq. (21) is one of the key relations in diffusion MR. It connects the (ideal) narrow-pulse measurement with the fundamental property of the medium, the ensemble-averaged diffusion propagator $G$. The latter contains information about tissue microstructure.

It is important to note that at the ensemble-averaging step (22), $\mathcal{G} \to G$, most of the valuable local information about different cellular environments within a voxel gets washed out. The good news is that the voxel-averaged propagator $G_{t,x}$ remains non-Gaussian, and thus potentially informative. Roughly speaking, the functional form of $G_{t,x}$ tells just how non-uniform and restrictive an *average* local environment of size $\sim L(t)$ is in a given voxel. Hence, intuitively, one can probe tissue microstructure by studying how distinct is the object $G_{t,q}$ from a simple Gaussian

$$G_{t,q}^{(0)}\Big|_{t>0} = e^{-bD_0}, \quad b = q^2 t, \quad (23)$$

which is precisely the Fourier transform of the free diffusion PDF (Eq. 8).

The propagator $G_{t,q}$ depends on $t$ and $q$ separately. However, the Gaussian propagator (23) depends on the so-called *b*-factor, the combination of $q$ and $t$. For the finite pulse duration,

$$b = q^2(\Delta - \delta/3).$$

**86 PART | II** Diffusion MRI

While the $b$-value can only characterize a dMRI measurement if a signal is a sum of contributions with Gaussian diffusion, it has historically remained a standard parameter since the days when basic diffusion NMR sequence (Stejskal and Tanner, 1965) was combined with imaging (Le Bihan et al., 1986). Today, one typically uses $b$ as a measure of $q$ at fixed $t$, such that one specifies either the pair $(q, t)$ or $(b, t)$ to characterize the experiment. For non-Gaussian diffusion, specifying the whole pulse shape becomes crucial (Callaghan, 1991; Mitra and Halperin, 1995; Grebenkov, 2007; Lasič et al., 2014; Shemesh et al., 2016; Topgaard, 2017; Novikov et al., 2019).

In realistic wide pulse PGSE (Fig. 4C), diffusion phase is related to the "center-of-mass" positions averaged during each gradient pulse application (Mitra and Halperin, 1995):

$$\phi = q \cdot (\bar{x}_{\varDelta, \varDelta + \delta} - \bar{x}_{0, \delta}),$$

where

$$\bar{x}_{t_1, t_2} = \frac{1}{t_2 - t_1} \int_{t_1}^{t_2} x(\tau) \mathrm{d}\tau.$$

We note that the involvement of center of mass within each gradient pulse width indicates the coarse-graining effect (cf. Section 3) over not just the diffusion time (as discussed there), but also over pulse duration (Neuman, 1974; Åslund and Topgaard, 2009; Burcaw et al., 2015; Lee et al., 2018b), offering an additional dimension of experiments characterized by the triple $(q, \varDelta, \delta)$ (Novikov, 2021).

PGSE signals are weighted by the transverse (spin-spin) relaxation via $\sim e^{-\mathrm{TE}/T_2}$ with transverse relaxation time $T_2$ and echo time TE. Since typical tissue $T_2 \sim 50-100$ ms, PGSE is preferred for measurements with diffusion time $t \lesssim 100$ ms.

For long diffusion times $t \gtrsim 100$ ms, PGSE signals are too low to maintain a reasonable signal-to-noise ratio (SNR) due to the $T_2$ relaxation, since $\mathrm{TE} \geq t$ and $e^{-\mathrm{TE}/T_2}$ becomes small. Instead, using a stimulated echo (STEAM) sequence, one can increase the diffusion time by increasing mixing time TM, leading to longitudinal (spin-lattice) relaxation weighting $\sim e^{-\mathrm{TM}/T_1}$ with the longitudinal relaxation time $T_1 > T_2$. This is due to the phase-cycling scheme of STEAM sequence, spoiling all coherence pathways but one of the stimulated echoes (Tal and Gonen, 2015). In the brain, the signal decay due to $T_1$ relaxation is much slower than that of $T_2$ relaxation, and thus STEAM signals have higher SNR than PGSE signals at long diffusion times. However, dMRI signals measured by STEAM have several confounding factors. For example, the crusher gradients in STEAM result in a nontrivial contribution to the diffusion-weighting ($b$-value) growing with the mixing time, even for a nondiffusion weighted signal (Kleinnijenhuis et al., 2018).

## 3 Diffusion as coarse-graining

### 3.1 The double average

Without the knowledge of detailed mechanisms, we can understand an observed physical phenomenon in a complex system via an *effective theory*, which leads to a representation described by only a limited number of relevant parameters. To achieve this "minimalism" in science, the basic principle is captured by *coarse-graining*—that is, averaging the system's dynamics over short temporal and spatial length scales in a statistical ensemble to derive the effective dynamics on coarser temporal and spatial scales (and in principle, perform this process iteratively); the system's number of degrees of freedom is therefore reduced by averaging out the irrelevant parameters (Wilson, 1983; Cardy, 1996).

In this way, coarse-graining over molecular interactions in liquids gives rise to effective theories represented by the Navier-Stokes and diffusion equations, governed by just a couple of effective parameters: viscosity and diffusion coefficient. These effective theories, describing the relaxation of momentum and density, correspondingly, become valid at the scales of a few nanometers. In a homogeneous liquid, that is the end of the story.

The effect of tissue microstructure on the measured diffusion signal can be qualitatively understood using coarse-graining over the increasing diffusion length (5). In fact, the very goal of modeling in our field is to adequately average the diffusive dynamics of spin-carrying (e.g., water) molecules over the microstructure layer of complexity, and derive the effective theory of the dMRI signal acquired over a macroscopic voxel. Such a theory will tell which microstructure-sensitive parameters are contained in the macroscopic signal and hence can be mapped. For tissue microstructure quantification with MR, the coarse-graining and effective medium way of thinking was introduced in recent review articles (Novikov et al., 2019; Novikov, 2021).

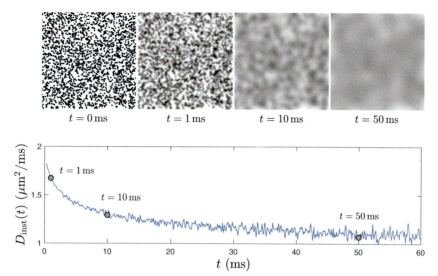

**FIG. 6** Diffusion as a coarse-graining effect. At long diffusion times $t$, fine details in microstructure are self-averaged due to diffusion over a coarse-graining length scale $L(t) \propto \sqrt{t}$ (*upper row*). The fluctuation of local diffusivity $\delta D(t;x)$ is gradually self-averaged over time, leading to a diffusivity time-dependence $\propto \langle (\delta D(t;x))^2 \rangle |_{L(t)}$. In this simulation, intrinsic diffusivity is $D_0 = 2 \mu m^2/ms$, and bulk diffusivity in $t \to \infty$ limit is $D_\infty \simeq 1.02 \mu m^2/ms$. The value of simulated $D_{\text{inst}}(t)$ (Eq. 26) has larger fluctuations at long times due to the numerical imprecision accumulated over MC steps.

Technically, averaging over the microstructure-level complexity involves performing a *double average* (Novikov, 2021), as illustrated in Figs. 5 and 6:

(i) Average over the Brownian paths initiated from each point $x_0$, yielding the local propagator $\mathcal{G}_{t;x_t,x_0}$ introduced in Section 2.4; its examples for a few $x_0$ are shown in Fig. 5 with different colors. This is the coarse-graining step. It qualitatively makes each local environment homogenized over a domain of the size of the diffusion length $L(t)$, with the envelope $\mathcal{G}_{t;x_t,x_0}$ playing a role of a low-pass filter over the microstructure (Fig. 6).

(ii) Ensemble average over all initial positions $x_0$ in a voxel, as in Eq. (22). This step, shown in the bottom row of Fig. 5, selects typical local contributions to the overall signal for a given voxel, and swipes rare atypical ones under the experimental noise floor.

The path-averaging step (i) is naturally performed by the diffusing water molecules (or by Monte Carlo simulated paths) in a given structural arrangement, as the "packet" of molecules spreads from the initial point $x_0$ according to its envelope $\mathcal{G}_{t;x_t,x_0}$.[h] This envelope is determined by the particular arrangement of cells around $x_0$; following Eq. (4), its variance defines the local time-dependent diffusion coefficient

$$D(t;x_0) \equiv \frac{\langle (x_t - x_0)^2 \rangle}{2d \cdot t}, \quad \langle (x_t - x_0)^2 \rangle = \int dx_t (x_t - x_0)^2 \mathcal{G}_{t;x_t,x_0} \quad (24)$$

at the point $x_0$. If we were to start anywhere else within the range of the diffusion length $L(t)$ from $x_0$, the diffusing molecules would sense roughly the same structure, and have a similarly behaving mean-squared displacement in Eq. (24). Thus we realize that the local diffusion coefficient $D(t;x)$, with $x$ in the domain of size $\sim L(t)$ around $x_0$, is roughly uniform in space. This is why the diffusion length $L(t)$ can be seen as a smoothing filter window over the structure (Fig. 6). Of course, the vicinity of another point $x_0'$ far away from $x_0$ would have its own structure, and its own $D(t;x_0')$, possibly quite different from $D(t;x_0)$. By the same token, the whole local propagator $\mathcal{G}_{t;x_t,x_0}$ does not strongly depend on $x_0$ if we move $x_0$ by less than $L(t)$, and can notably change if we explore a different local environment, moving $x_0 \to x_0'$ such that $|x_0' - x_0| \gg L(t)$. (This discussion implies that we do not move $x_0$ across an impermeable boundary, i.e., we stay within a given "compartment.")

The ensemble-averaging step (ii) makes the resulting propagator (22) *translation-invariant*, that is, dependent on the displacement $x = x_t - x_0$ rather than on the points $x_t$ and $x_0$ separately.

---

h. We imply that the diffusion is not fully restricted—for example, in the extracellular space, Fig. 5 (hindered, middle column) and Fig. 6, or inside axons that are much longer than any attainable $L(t)$, Fig. 7 discussed later. For the fully restricted diffusion, coarse-graining stops whenever the diffusion length $L(t)$ grows to match the compartment size (Fig. 5, left column).

**FIG. 7** Which geometric features of axons are we potentially sensitive to with dMRI? Starting from a realistic axon shape segmented from three-dimensional electron microscopy (Lee et al., 2019), coarse-graining over the increasing diffusion length $L(t)$ first highlights the caliber variations (beads or varicosities) at the $\rho \sim 1\mu m$ scale, followed by the undulations with wavelength on the order of $\lambda_u \sim 30\mu m$. The fully coarse-grained "stick" compartment of negligible radius is characterized by one-dimensional Gaussian diffusion.

## 3.2 The three regimes

The coarse-graining way of thinking tells us that, depending on the diffusion time, the tissue as effectively "seen" by the diffusing spins looks qualitatively different (Fig. 6). To classify these differences, let us introduce the *structural length and time scales*

$$\ell_s = \{\rho \text{ or } \ell_c\} \quad \text{and} \quad t_s = \frac{\ell_s^2}{D}, \tag{25}$$

where $\ell_s$ can either be the size $\rho$ of cells (for the restricted diffusion, e.g., axon or soma diameter) or the correlation length $\ell_c$ of their packing, which is usually on the order of the distance between their centers (Burcaw et al., 2015).

Depending on the relation between the diffusion length $L(t)$ (the coarse-graining window) and $\ell_s$, one considers the three regimes (we follow Novikov et al., 2019, Section 1.5):

(i) **No coarse-graining**, $L(t) \ll \ell_s$. Diffusion time is short, $t \ll t_s$, and each spin senses its immediate vicinity characterized by its own intrinsic diffusion coefficient $D(x_0)$—for example, cytoplasmic diffusivity, diffusivity inside cell nucleus, diffusivity of water in the extracellular space. An example is the $t=0$ panel of Fig. 6. Restrictions provide a small $\sim \sqrt{t}$ correction to the overall measured diffusion coefficient (Mitra et al., 1993).

(ii) **Coarse-graining over the structure**, $L(t) \gtrsim \ell_s$, when the diffusion time $t \gtrsim t_s$ is such that the diffusion length matches the characteristic scale of tissue microstructure. Here, the *transient* (time-dependent) effects lead to the possibility to quantify $\ell_s$. The coarse-graining in this regime is schematically shown in the remaining panels of the top row of Fig. 6.

For hindered diffusion, this regime is characterized by the asymptotic power-law decay (Novikov et al., 2014)

$$D_{\text{inst}}(t) \equiv \frac{\partial}{\partial t}\frac{\langle x^2(t) \rangle}{2d} \simeq D_\infty + \text{const} \cdot t^{-\vartheta}, \quad t \gg t_s \tag{26}$$

of the instantaneous diffusion coefficient $D_{\text{inst}}(t)$, which yields the corresponding power-law tails in the conventional pulse-gradient diffusion coefficient (12), $D(t) = \frac{1}{t}\int_0^t d\tau D_{\text{inst}}(\tau)$, and in the higher-order cumulants such as kurtosis (Burcaw et al., 2015; Dhital et al., 2019; Lee et al., 2020c). From this approach, the disorder correlation length $\ell_c$ of beads (Novikov et al., 2014; Fieremans et al., 2016; Jespersen et al., 2018; Lee et al., 2020b) along neurites (axons and dendrites), and packing correlation length of axonal fibers in the fiber cross-section (Burcaw et al., 2015; Fieremans et al., 2016; Lee et al., 2018b) were recently estimated to be $\ell_c \sim 1\mu m$, in accord with histology (Shepherd et al., 2002; Lee et al., 2019).

For restricted diffusion, the coarse-graining eventually stops, and the compartment effectively shrinks to a point when its size $\rho \ll L(t)$. The diffusivity scaling (13) enables mapping the compartment size $\rho$, for example, cancer cell size (Reynaud, 2017), as well as axonal diameter (cf. Section 4.1).

**(iii) Complete coarse-graining**, $L(t) \to \infty$: Gaussian compartment(s). At times $t \gg t_s$, spins have sampled large enough domains that their statistical properties have become similar—in other words, the local diffusivities (24) around different $x_0$ are practically the same. Physicists say that the diffusion process becomes *self-averaging*, that is, each domain well represents the whole macroscopic tissue compartment. When $t \to \infty$, the compartment effectively looks completely homogeneous, characterized by the long-time ("tortuosity") limit $D_\infty = D(t)|_{t\to\infty}$ of the diffusion coefficient, and the diffusion becomes Gaussian. Generally, the diffusion becomes anisotropic within a given tissue compartment, such that the diffusion tensor eigenvalues approach their (different) tortuosity limits. This is the picture of multiple Gaussian compartments (in general, anisotropic). This picture—for intra- and extraaxonal spaces and for the cerebrospinal fluid (CSF)—is behind many diffusion models (Jespersen et al., 2007, 2010; Fieremans et al., 2010, 2011; Zhang et al., 2012; Sotiropoulos et al., 2012; Jelescu et al., 2016a; Jensen et al., 2016; Reisert et al., 2017; Novikov et al., 2018c; Veraart et al., 2018; Lampinen et al., 2017), which fall under the overarching umbrella of the Standard Model (SM) for diffusion in white matter (WM) (Novikov et al., 2019, Section 3), described in Section 5.2.

We note that the role of the diffusion wave vector $q$ is in providing a snapshot of the coarse-graining process at the length scale $|x| \lesssim 1/q$—technically, these $x$ define the main contribution to the Fourier integral in Eq. (21), as the exponential $e^{-iqx}$ does not strongly oscillate. Applying large $q$, one can access the structure (size, shape) even at long times $t \gg t_s$—that is, from "under" the blurring window $L(t)$.[i] This is the *diffusion-diffraction regime* of Callaghan (1991), where, for instance, the signal

$$G_{t,q}|_{t\gg t_s} \propto \Gamma(q), \quad \Gamma(q) = |v(q)|^2 \tag{27}$$

from the intracellular space is given by the power-spectrum $\Gamma(q)$ of the *closed pore* shape $v(x)$ for $t \gg t_s$. Likewise, a power-spectrum $\Gamma(q)$ of the *connected pore* space (e.g., extracellular space) can be probed when the diffusion time exceeds the correlation length of the packing of the "grains" (e.g., impermeable cells) (Mitra et al., 1992). Employing both the strength and the duration of the diffusion gradient makes the phase diagram of the microstructure dMRI at least three-dimensional (Novikov, 2021; Kiselev, 2020).

# 4 Coarse-graining over an axon

To illustrate coarse-graining as an effective theory, we consider diffusion inside a myelinated axon, assuming no exchange of protons in the intraaxonal compartment with the myelin or extraaxonal compartment at the relevant time scales. Indeed, water exchange between intra- and extraaxonal spaces is deemed negligible, because the echo time for clinical and pre-clinical (50–150 ms) use is much shorter than the exchange time $\sim 1$ s (defined by the reciprocal of exchange rate) between the two compartments (Quirk et al., 2003; Fieremans et al., 2010; Nilsson et al., 2013; Veraart et al., 2019). In addition, exchange with the myelin water compartment can be neglected due to its short $T_2$ time (Mackay et al., 1994) as compared to clinical and preclinical TE.

As illustrated in Fig. 7, axons have complicated shapes (Abdollahzadeh et al., 2019; Lee et al., 2019). If we were to successively blur the structural details at an increasing diffusion length scale $L(t)$ in the spirit of Fig. 6, we can distinguish (at least) the following structural hierarchy: variable cross-sections with irregular shapes ($\rho \sim 1\,\mu m$—Section 4.1); undulations ($\lambda_u \sim 30\,\mu m$—Section 4.2); and finally a featureless one-dimensional channel ("stick"), if we were to look at an axon from afar (Section 4.3).

## 4.1 Axon caliber scale

At the scale $L(t)$ below the axonal caliber $\rho \sim 1\mu m$, intraaxonal diffusion is fully three-dimensional. Diffusion (along and transverse) is in principle sensitive to the axon's cross-sectional shape, its local caliber (a measure of its cross-sectional size), and caliber variations along its length, as we now outline.

*Sensitivity to axon caliber (transverse).* The contribution $G_{t,x}$ from the restricted compartment of Fig. 2 is sensitive to the size $\rho$ of the cells (circles). Ideally, to measure axon caliber, one should achieve the diffusion-diffraction regime, where the wave vector $q = g\delta$ matches the inverse axon caliber, $q \sim 1/\rho$, such that $b = q^2 t$ and $D(t)$ given by Eq. (13) together yield $-\ln S \sim (q\rho)^2$, to estimate $\rho$. However, this regime is very difficult to access even on animal systems, as it also implies gradient pulses shorter than the diffusion time across the axon, $\delta \ll t_s \lesssim 1$ms (for the notion of $q$ to be well-defined).

---

i. Using $q \gg 1/L(t)$ away from the diffusion dispersion relation $Dq^2 t \sim 1$ of Eq. (4) has its parallels with probing the atomic structure of a liquid with a beam of neutrons with momentum $\hbar q$ corresponding to $q \sim (0.1\,\text{nm})^{-1}$, that is, reaching for the structure "under" the scale where its coarse-grained continuous hydrodynamic description becomes valid.

**90 PART | II** Diffusion MRI

Practically, experiments fall into the opposite, *diffusion-narrowing regime* (Robertson, 1966; Murday and Cotts, 1968; Neuman, 1974), where $\delta \gg t_s$, and the signal attenuation

$$-\ln S \sim \frac{g^2 \rho^4 \delta}{D_0} \sim (g\delta \cdot \rho)^2 \cdot \frac{t_s}{\delta} \ll (q\rho)^2 \tag{28}$$

is parametrically *weaker* than in the diffusion-diffraction regime, by the factor of the small parameter $t_s/\delta \ll 1$.

The weak diffusion-narrowing attenuation (28) makes calibers $\rho \lesssim 1\mu m$ of typical axons in the human brain invisible even to scanners with Connectome gradients; on clinical systems, signal attenuation (28) for such axons is about $10^{-6}-10^{-5}$ (Burcaw et al., 2015; Nilsson et al., 2017).[j] Volume-weighting the $\rho^4$ dependence in Eq. (28) means that the technique, when applicable, measures the *effective axonal radius* (Burcaw et al., 2015)

$$r_{\rm eff} = \left( \langle \rho^6 \rangle / \langle \rho^2 \rangle \right)^{1/4} \tag{29}$$

dominated by the sixth moment of axonal radius distribution, and, hence, is heavily skewed by its *tail*, as was recently validated in rodents and in humans (Veraart et al., 2020).

The effective radius (29) can be viewed as a result of the double average (Section 3): (i) averaging over the Brownian paths within each axon yields the attenuation (28); (ii) the ensemble average over all axons for $-\ln S \ll 1$ collapses a complicated distribution of axonal radii onto the ratio (29) of its moments, which can be thought of as an effective theory parameter.

We also note that the hindered diffusion in the extraaxonal space coarse-grains the axonal packing geometry in such a way that the extraaxonal transverse diffusivity acquires the time dependence with $\vartheta = 1$ exponent in Eq. (26), yielding the $(\ln t)/t$ tail in the cumulative diffusion coefficient (12). This logarithmic enhancement dominates in the overall diffusivity time-dependence, making the measured $D(t)$ transverse to the fiber much more sensitive to the extraaxonal geometry, and specifically to the packing correlation length $\ell_c$, which is on the order of the *outer* axonal diameter (Burcaw et al., 2015).

*Sensitivity to axon caliber variations (along).* Axon caliber and cross-sectional shape vary along its length, giving rise to *caliber variations*, such as varicosities or beads (Shepherd et al., 2002; Budde and Frank, 2010). Lee et al. (2020b) and Abdollahzadeh et al. (2024) showed that the placements of axonal beads has a finite correlation length (short-range disorder). Such randomness results in the power-law tail (26) for the *along-axon* diffusivity and kurtosis with the exponent $\vartheta = 1/2$ predicted by Novikov et al. (2014), confirmed in simulations of artificial (Palombo et al., 2018) and electron microscopy-derived axonal geometry (Lee et al., 2020b), and observed in rat (Does et al., 2003; Novikov et al., 2014), human WM (Fieremans et al., 2016; Arbabi et al., 2020; Lee et al., 2020b; Tan et al., 2020), and fixed spinal cord (Jespersen et al., 2018).

The $\sim t^{-1/2}$ tail (26) is a result of coarse-graining the three-dimensional axon down to a one-dimensional effective medium, where the transverse-to-axon degrees of freedom map onto the along-axon structural disorder (Novikov et al., 2014). The amplitude of the $\sim t^{-1/2}$ tail (26) is related to the parameters of the structural disorder, such as the correlation length of the bead placements (Fieremans et al., 2016; Lee et al., 2020b; Abdollahzadeh et al. (2024)).

## 4.2 Undulation scale

When $L(t)$ reaches an order-of-magnitude larger scale $\lambda_u \sim 30\mu m$, the caliber can be effectively shrunk to a point, such that we can focus on the wavy shape of an axonal skeleton. This wave is referred to as an *undulation* (Nilsson et al., 2012) and contributes to the time-dependent diffusion along and transverse to the axon (Brabec et al., 2020; Lee et al., 2020a).

In particular, undulations can strongly contribute to *apparent* axonal caliber (even if we were to neglect the actual axon thickness $\rho$). Naively, one would imagine that coarse-graining the wavy skeleton of an amplitude $w_0$ and period $\lambda_u$ over $L(t) \gtrsim \lambda_u$ should make it look as a tube of thickness $\sim w_0$, such that the attenuation $-\ln S \sim (qw_0)^2$, yielding the effective caliber $\sim w_0$. This intuition is correct for narrow pulses, when $\delta$ is much shorter than the time $t_u \sim \lambda_u^2/D_0$ to diffuse along the undulation. For wide pulses, $\delta \gg t_u$, the intuition is similar to that of the diffusion-narrowing limit (Eq. 28). Specifically, the correlation time scale $t_s$, over which the successive random phase contributions become independent, becomes $t_u$. Signal attenuation corresponds to summing $N \sim \delta/t_u \gg 1$ independent spin phase variances,

---

j. This smallness becomes apparent when one converts a run-of-the-mill clinical scanner gradient of 40 mT/m to the Larmor frequency gradient $g = 0.0107(\mu m/ms)^{-1}$. Numerically, the attenuation (28) is most naturally expressed in terms of the *radius* $\rho$; typical axonal radius $\rho = 0.5\,\mu m$, $D_0 \approx 2-3\,\mu m^2/ms$, and $\delta \sim 10 ms$, multiplied by the small numerical prefactor 7/48 for the cylinder that was dropped in Eq. (28), together yield this result.

$$-\ln S \sim (gw_0 \cdot t_u)^2 \cdot N \sim \frac{g^2 w_0^2 \lambda_u^2 \delta}{D_0} \equiv \frac{g^2 r_{und}^4 \delta}{D_0} \qquad (30)$$

defining the effective axon radius $r_{und} \sim \sqrt{w_0 \lambda_u}$ due to the undulations (Lee et al., 2020a, Eq. 26).

Since the undulation wavelength greatly exceeds axon caliber, $\lambda_u \gg \rho$, whereas its amplitude $w_0 \sim 1\,\mu m$ is on the order of axon caliber (Lee et al., 2019; Abdollahzadeh et al., 2019), in realistic experimental settings we are more sensitive to undulations than to the actual axonal caliber, $r_{und} \gg r_{eff}$. Hence, undulations strongly bias the estimation of axon diameter at moderate diffusion weighting (Lee et al., 2020a). On the other hand, the undulation effect on the along-axon diffusion is less significant (Lee et al., 2020a, Appendix E).

## 4.3 Axon as a "stick"

Finally, for a longer diffusion length, $L(t) \gg \lambda_u$, and for diffusion weighting not enough to resolve $\rho$ and $w_0$, the caliber and undulation effects can be neglected, and an axon can be simplified down to a featureless narrow "stick" (zero-radius cylinder, with effective diffusion constant $D_a < D_0$; Fig. 7). The signal from such sticks was studied by Kroenke et al. (2004) for NAA diffusion in rat brain, followed by the water diffusion study of Jespersen et al. (2007). The dMRI signal from any collection of sticks, averaged over all gradient directions in a $b$-shell, becomes equivalent (Jespersen et al., 2013; Kaden et al., 2016) to the Callaghan's model (Callaghan et al., 1979) of *isotropically* distributed sticks, yielding the universal $b^{-1/2}$ scaling

$$\int_0^{\pi/2} d\theta \sin\theta e^{-bD_a \cos^2\theta} \simeq \sqrt{\frac{\pi}{4bD_a}}, \quad bD_a \gg 1. \qquad (31)$$

Here, $\theta$ is the angle between the stick and the applied gradient and $D_a$ the intraaxonal diffusivity. The nontrivial $b^{-1/2} \sim q^{-1}$ functional form, coming from the intrastick compartment, was observed only recently in vivo (McKinnon et al., 2017; Veraart et al., 2019, 2020), validating the picture of sticks for axons (and, perhaps, for glia processes) in WM (Fig. 8). Its counterpart for the planar tensor encoding is the $b^{-1} \sim q^{-2}$ decay (Afzali et al., 2020).

Going back to the way we obtained Eq. (31), all we really needed was to assume no exchange between intra- and extraaxonal space and that diffusion over the diffusion time $t$ was happening along a locally straight one-dimensional segment. Hence, the scaling (31) will also apply for *short t*, such that $L(t) \lesssim \lambda_u$, that is, the undulations are longer than the diffusion length, during which a large $b$-value is accumulated by applying strong gradients. In this case, the axon is effectively split

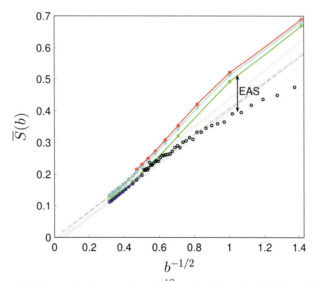

**FIG. 8** Intraaxonal signal from water and NAA showing the asymptotic $b^{-1/2}$ scaling (31) at high $b$. Water diffusion signal from four human subjects (from Veraart et al., 2019), isotropically averaged for each shell, is plotted as a function of $b^{-1/2}$, exhibiting an asymptotically linear scaling at $b \gtrsim 6\,\mu m^2/ms$. Black circles show the diffusion-weighted direction-averaged NAA spectroscopic signal from a rat brain (data from Kroenke et al., 2004), scaled by the $T_2$-weighted intraaxonal water fraction to match the human data, with the $b$-values rescaled to adjust for about a fivefold lower NAA diffusion coefficient. *(Reproduced with permission from Fig. 5C of Novikov, D.S., Kiselev, V.G., Jespersen, S.N., 2018. On modeling. Magn. Reson. Med. 79 (6), 3172–3193. https://doi.org/10.1002/mrm.27101, Wiley Inc.)*

into locally straight segments of size $\sim L(t)$; the contribution from each such segment, averaged over all gradient directions, yields Eq. (31)—albeit with $D_a$ corresponding to an effective diffusion coefficient coarse-grained at the scale $L(t)$. Therefore angular averaging over each $b$-shell gives a way to factor out not just the axonal orientation dispersion $\mathcal{P}(\hat{n})$ (as described in Section 5), but also the effect of the undulations (Section 4.2), which becomes instrumental in mapping axonal caliber.

The most remarkable property of the high-$b$ intraaxonal signal (31) is that it decays *much slower* than the extraaxonal signal, and hence it was found to persist for very large $6 \lesssim b \le 10 \, \mathrm{ms}/\mu\mathrm{m}^2$ on a clinical scanner (Veraart et al., 2019) (cf. Fig. 8) as well as for $b \le 25 \, \mathrm{ms}/\mu\mathrm{m}^2$ on the human Connectome scanner and for $b \le 100 \, \mathrm{ms}/\mu\mathrm{m}^2$ on an animal system (Veraart et al., 2020). In other words, the slow scaling (31) tells that practically all we see at such strong diffusion weightings in WM is the intraaxonal contribution! This is an example of a "spectroscopic" property of dMRI signal, where understanding distinct *functional forms* allows us to separate contributions from distinct tissue building blocks (Novikov et al., 2018a). The mismatch between the purely intraaxonal (NAA) signal and that from water (both from intra- and extraaxonal spaces) at $b \lesssim 6 \, \mu\mathrm{m}^2/\mathrm{ms}$ exemplifies the contribution of the extraaxonal space (EAS, Fig. 8), which is being exponentially suppressed by increasing the diffusion weighting due to nonzero diffusion coefficient in any direction—as contrasted with the power-law decaying stick signal (Eq. 31).

# 5 From microstructure to fiber orientations

## 5.1 White matter dMRI signal as a convolution

A common anatomical signature of any macroscopic WM voxel is its domain structure: locally, the fibers are well-aligned into domains, called *fascicles*; yet within a millimeter-scale voxel, many fascicles may have different properties and orientations. In other words, a dMRI signal is a sum of the contributions from distinct fascicles.

How well-defined are the fascicles from the point of diffusion? The intuition gained from Sections 3 and 4 tells us that any structure smaller than the diffusion length (5) gets coarse-grained by the diffusion, whereas the parts of a voxel separated by more than $L(t)$ are ensemble-averaged (not coarse-grained), that is, they add up weighted by their water fractions (and, generally, with the $e^{-\mathrm{TE}/T_2}$ weights that may differ due to the differences in the local $T_2$ NMR relaxation times). Given that $L(t) \sim 10 \, \mu\mathrm{m}$ in human dMRI measurements is rather short, a typical WM fascicle is at least as thick as $L(t)$, hence differently oriented fascicles are generally not "mixed" by the coarse-graining.

This basic picture can be formalized as follows. The signal from a voxel, measured in the unit direction $\hat{g}$

$$S_{\hat{g}}(b) = \int_{|\hat{n}|=1} \mathrm{d}\hat{n} \, \mathcal{P}(\hat{n}) \mathcal{K}(b, \hat{g}; \hat{n}) \tag{32}$$

is given by the sum over fascicles pointing in the directions $\hat{n}$, distributed according to their *orientation distribution function* (ODF) $\mathcal{P}(\hat{n})$. Each fascicle (labeled by its orientation $\hat{n}$) contributes an elementary signal $\mathcal{K}(b, \hat{g}; \hat{n})$.

Of course, at this point all the physics of diffusion is hidden in $\mathcal{K}(b, \hat{g}; \hat{n})$. Modeling this function will be key to relating the microstructure to the fiber orientations. For that, let us make two simplifying assumptions:

1. All fascicles in a voxel have the same microstructural properties, that is, they respond in the same way to a diffusion gradient, if they were to be identically aligned.[k]
2. Each fascicle is axially symmetric. While approximate, this assumption seems to be quite natural from visibly inspecting axonal segmentations.

These two assumptions together allow for a significant simplification: The fascicle response function $\mathcal{K}(b, \hat{g}; \hat{n}) = \mathcal{K}(b, \hat{g} \cdot \hat{n})$ becomes the function of the *relative angle* between $\hat{g}$ and $\hat{n}$. In other words, in the basis where any fascicle is aligned along the $z$-axis, its response $\mathcal{K}(b, \hat{g}; \hat{n}) \equiv \mathcal{K}(b, \cos\theta)$ is the function of the polar angle $\theta$.

The resulting translational invariance on a unit sphere $|\hat{n}| = 1$ renders the sum (32) into a *convolution* (Healy et al., 1998)

---

k. Strictly speaking, this assumption is incorrect—it implies that different fiber tracts crossing in a given voxel have the same microstructure. Assaf and Basser (2005) and De Santis et al. (2016) prompted assigning different (albeit constant throughout the brain) fiber responses to different tracts to deconvolve the ODF (Sherbondy et al., 2010; Tournier et al., 2011; Girard et al., 2017). A practically relevant question is whether it is possible to detect biases when parametrizing all fascicles in the same way. A recent study (Christiaens et al., 2020) showed that the corresponding fit residuals are sufficiently random, that is, differences between response functions for different fascicles may typically fall below the noise floor.

$$S_{\hat{g}}(b) = \int_{|\hat{n}|=1} d\hat{n} \mathcal{P}(\hat{n}) \mathcal{K}(b, \hat{g} \cdot \hat{n}) \tag{33}$$

between the fiber ODF $\mathcal{P}(\hat{n})$ and the response kernel $\mathcal{K}$ from a perfectly aligned fiber segment (fascicle) pointing in the direction $\hat{n}$. Here, the ODF is normalized to

$$\int d\hat{n} \mathcal{P}(\hat{n}) \equiv 1, \quad d\hat{n} \equiv \frac{\sin\theta d\theta\, d\phi}{4\pi}. \tag{34}$$

The measure $d\hat{n}$ is chosen so that $\int d\hat{n} \cdot 1 \equiv 1$, while $\mathcal{K}|_{b=0} = S_0 \equiv S|_{b=0}$.

The general representation (33) gave rise to a number of methods for deconvolving the fiber ODF from the dMRI signal for a given $|q|$ shell in $q$-space, using different empirical forms of the kernel (Tournier et al., 2004, 2007; Anderson, 2005; Dell'Acqua et al., 2007; Jian and Vemuri, 2007; Kaden et al., 2007; White and Dale, 2009), considered in Part III of this book.

## 5.2 The Standard Model of diffusion in white matter

The kernel $\mathcal{K}$ can still be quite complicated, for example, generally it can include exchange between intra- and extrastick spaces (e.g., potentially for unmyelinated axons and in gray matter; Lee et al., 2020c; Jelescu and Novikov, 2020; Jelescu et al., 2022; Olesen et al., 2022). However, having major WM tracts in mind, where dozens of myelin layers prevent the exchange from happening on the relevant diffusion time scales (Quirk et al., 2003; Fieremans et al., 2010; Nilsson et al., 2013; Veraart et al., 2019), we make an extra two assumptions in addition to those of Section 5.1:

3. No exchange between "sticks" and extraaxonal space.
4. Both intra- and extraaxonal diffusion fall into the long-time regime (iii) of Section 3.

Assumptions 1–4 together lead to the general picture of anisotropic Gaussian compartments (Fig. 9). This corresponds to the elementary fascicle response kernel (Reisert et al., 2017; Novikov et al., 2018c)

$$\mathcal{K}(b,\xi) = S_0 [f e^{-bD_a \xi^2} + (1-f-f_w) e^{-bD_e^\perp - b(D_e^\parallel - D_e^\perp)\xi^2} + f_w e^{-bD_w}], \quad \xi = \hat{g} \cdot \hat{n}, \tag{35}$$

being a sum of exponential (in $b$) contributions from the aligned intraaxonal and extraaxonal spaces, respectively modeled by a stick compartment (Section 4.3), by the axially symmetric Gaussian compartment with transverse and longitudinal diffusivities $D_e^\perp$ and $D_e^\parallel$ and the principal direction along the stick (Fig. 9). The last term describes the isotropic "free water" (CSF) compartment with fraction $f_w$ and diffusivity $D_w$ and is often ignored.

The myelin water compartment is typically neglected due to its short $T_2$ time (Mackay et al., 1994) as compared to the echo times TE employed in clinical dMRI. We emphasize that the fractions $f$ and $f_w$ are the relative *signal fractions* and not the absolute water volume fractions, due to neglecting myelin water, as well as due to generally different $T_2$ values for the intra- and extraneurite compartments (Dortch et al., 2013; Veraart et al., 2018), and for the free water.

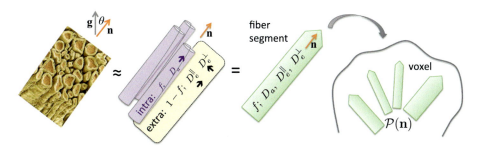

FIG. 9 Standard Model of diffusion in neuronal tissue (Novikov et al., 2019, Section 3). In the long time limit (iii), elementary fiber segments (fascicles), consisting of intra- and extraneurite compartments, are described by at least four independent parameters: $f, D_a, D_e^\parallel$, and $D_e^\perp$, with a possible addition of the isotropic free water compartment with fraction $f_w$. Within a macroscopic imaging voxel, such segments contribute to the directional dMRI signal according to their orientational dispersion $\mathcal{P}(\hat{n})$.

The isotropic water compartment has $D_w \approx 3\mu m^2/ms$, cf. Section 1.3. Because of the ODF normalization $\int d\hat{n}\mathcal{P}(\hat{n})\equiv 1$, the isotropic third term can be included in the kernel or added separately to signal (33); the choice to include it in the kernel (35) makes the formulation (33) more elegant.

The overarching model (33)–(35) includes nearly all previously used models (Jespersen et al., 2007, 2010; Fieremans et al., 2010, 2011; Zhang et al., 2012; Sotiropoulos et al., 2012; Jelescu et al., 2016a; Jensen et al., 2016; Reisert et al., 2017; Lampinen et al., 2017; Novikov et al., 2018c; Veraart et al., 2018) of diffusion in the WM as particular cases. Because of the overall popularity and inclusiveness of this picture, it was suggested to call the model (33)–(35) the standard model (SM) of diffusion in white matter (Novikov et al., 2019), as a tongue-in-cheek analogy with the SM in particle physics.

## 5.3 Factorizing fiber ODF and fascicle response: Rotational invariants

### 5.3.1 General formulation

We now outline the recently proposed family of approaches to SM parameter estimation, where the key point will be to rely on the convolution property (33) to split the estimation into two pieces: the kernel and the ODF. Much like convolutions become products in the Fourier domain, the convolution (33) between the individual fiber response $\mathcal{K}$ and the ODF $\mathcal{P}$ becomes a product in the "spherical Fourier" domain, that is, the spherical harmonics (SH) basis (Healy et al., 1998):

$$S_{lm}(b) = p_{lm}K_l(b), \tag{36}$$

where $p_{lm}$ are the SH coefficients

$$\mathcal{P}(\hat{n}) = 1 + \sum_{l=2,4,\dots} \sum_{m=-l}^{l} p_{lm}Y_{lm}(\hat{n}) \tag{37}$$

and $K_l(b)$ are the projections of the kernel $\mathcal{K}(b,\xi)$ onto the Legendre polynomials $(-1)^{l/2}P_l(\xi)$ with $l = 0, 2, \dots$ (Jespersen et al., 2007; Novikov et al., 2018c; Reisert et al., 2017).

The factorization (36) can be used to separate the estimation of $p_{lm}$ from that of the scalar parameters $x = \{f, D_a, D_e^{\parallel}, D_e^{\perp}, \dots\}$ of the kernel (Fig. 9). Since any rotation transforms SH components $S_{lm}$ and $p_{lm}$ according to a unitary transformation belonging to the $(2l + 1)$-dimensional irreducible representation of the SO(3) group labeled by "angular momentum" $l$, the 2-norms

$$\| p_{lm} \|^2 \equiv \sum_{m=-l}^{l} |p_{lm}|^2 \quad \text{and} \quad \| S_{lm} \|^2 \equiv \sum_{m=-l}^{l} |S_{lm}|^2$$

are invariant under rotations. It is thus convenient to introduce[1] *rotational invariants* $p_l \equiv \| p_{lm} \| /\mathcal{N}_l$ and $S_l \equiv \| S_{lm} \| /\mathcal{N}_l$, where normalization $\mathcal{N}_l = \sqrt{4\pi(2l + 1)}$ is chosen so that $0 \leq p_l \leq 1$. Hence, Eq. (36) for the $(l, m)$ SH components give rise to the corresponding equations for the rotational invariants (Novikov et al., 2018c; Reisert et al., 2017),

$$S_l(b|x) = p_l K_l(b|x), \quad l = 0, 2, \dots. \tag{38}$$

The invariant $p_0 \equiv 1$ is trivial (ODF normalization); the remaining ODF invariants $p_l$, one for each $l$, characterize its anisotropy regardless of the chosen basis.

The system of Eq. (38) provides a way to (almost) factorize the estimation of the kernel parameters $x$, and subsequently the ODF parameters $p_{lm}$ using Eq. (36). We used the word "almost" because the $l > 0$ ODF invariants couple $S_l$ and $K_l$. However, for each $l > 0$, there is only a single unknown $p_l$ added to the set of kernel parameters $x$, instead of $2l + 1$ parameters $p_{lm}$.

The $l = 0$ invariant in Eq. (38) has been independently introduced as "powder average" and "spherical mean" (Jespersen et al., 2013; Lasič et al., 2014; Kaden et al., 2016; Hansen et al., 2015; Szczepankiewicz et al., 2016). The ODF factorization in this case simply follows from swapping the order of integrations over $\hat{g}$ and $\hat{n}$:

$$S_{l=0} \propto \int d\hat{g} \int d\hat{n}\mathcal{P}(\hat{n})\mathcal{K}(b, \hat{g} \cdot \hat{n})$$

$$= \int d\hat{n}\mathcal{P}(\hat{n}) \int d\hat{g}\mathcal{K}(b, \hat{g} \cdot \hat{n}) \equiv \int_0^1 d\cos\theta \mathcal{K}(b, \cos\theta),$$

---

1. The idea to operate with a single "energy" $L_2$ norm per each "frequency" band $l$ of SH has been previously applied, for example, to the problem of shape matching in computer graphics (Kazhdan et al., 2003) and recently for dMRI data harmonization (Mirzaalian et al., 2016).

since $\int d\hat{g} \mathcal{K}(b, \hat{g} \cdot \hat{n})$ is independent of fiber direction $\hat{n}$ due to the translational invariance on a unit sphere, and the ODF is normalized to $\int d\hat{n} \mathcal{P}(\hat{n}) \equiv 1$. The above equation gives the projection of kernel (35) onto the $l = 0$ Legendre polynomial $P_0(\xi) \equiv 1$, where $\xi = \cos\theta$ in our case; for a stick compartment, this projection yields Eq. (31).

Eq. (38) formally yields an infinite family of rotational invariants $K_l(b)$ (Reisert et al., 2017; Novikov et al., 2018c), one for every $l = 2, 4, \ldots$. However, it turns out that by far the most useful is the next-order, $l = 2$ invariant, since the projections of $e^{-bD\xi^2}$ onto the Legendre polynomials with $l > 2$, giving the compartment contributions to $K_l(b, x)$, become progressively smaller numerically, and too slowly varying with $b$ (Jespersen et al., 2007), which makes adding equations for higher $l$ impractical. We also note that including the $l > 0$ invariants in system Eq. (38) is only possible for anisotropic ODFs, with $p_l > 0$. Physically, this is expected since the less symmetric the system, the more inequivalent ways there exist for probing it.

### 5.3.2 Parameter estimation degeneracies

SM parameter estimation amounts to inverting the nonlinear relations (38) with respect to model parameters $x$ and $p_l$. One can perform it using machine-learning (Reisert et al., 2017) or max-likelihood (Novikov et al., 2018c) approaches. Regardless of the method, the problem is degenerate, that is, multiple equivalent solutions to Eq. (38) exist, reflecting the relative lack of sensitivity of the measurement to the model parameters. In particular, the "fit landscape" (the profile of the objective function to be minimized during fitting), its discrete and continuous degeneracies (i.e., distinct minima and flat directions), and the associated issues of accuracy and precision were pointed out numerically (Jelescu et al., 2016a). The topology of this landscape was subsequently derived analytically via the low-$b$ expansion of Eq. (38) (Novikov et al., 2018c). Widely adopted model assumptions and constraints (Zhang et al., 2012; Kaden et al., 2016), which help stabilize the fit, were shown to fail (Novikov et al., 2018c; Lampinen et al., 2017).

The degeneracy exists due to the multicompartmental nature of the kernel (35) (Novikov et al., 2018c), much like for the biexponential model (Kiselev and Il'yasov, 2007). One can say that any standard (directional) dMRI measurement effectively undersamples the scalar part of the model (33), often not providing enough relations between the kernel parameters, and oversamples the ODF part. In other words, the system's true complexity lies within the kernel's parameters hidden in functions $K_l(b|x)$, Eq. (38), while the ODF is in some sense "on the surface." However, this is good news for fiber tractography, as the estimation of $p_{lm}$ (that can be fed into the fiber tracking routine) is largely insensitive to the degeneracies of the scalar sector of the problem.

### 5.3.3 Resolving degeneracies via complementary measurements

Parameter estimation degeneracies may be cured by acquiring sufficiently "orthogonal" (complementary) information, that helps increase the curvature of the fit landscape along the most uninformative (flat) directions.

The freedom of choosing the gradient waveform $g(t)$ in three dimensions (Section 2.3), and hence the shape of $q(t)$ and of the corresponding $b$-tensor (generalizing the $b$-value) (Westin et al., 2016; Topgaard, 2017), significantly helps in resolving fit degeneracies. For the SM, Skinner et al. (2017) used double diffusion encoding to suppress the extraaxonal compartment and then measure the diffusivity inside axons in rat spinal cord; Jensen and Helpern (2018) and Dhital et al. (2019) combined planar and linear encodings into different flavors of triple diffusion encoding, which together with orientational averaging enabled accessing the intraaxonal diffusivity. Coelho et al. (2019b) empirically found that the SM fit landscape becomes notably less degenerate by including the planar tensor encoding, whereas Reisert et al. (2019) proved this statement analytically using an expansion at low $b$, and Fieremans et al. (2018) showed that employing the spherical tensor encoding has in practice a similar effect, making use of the earlier observation (Szczepankiewicz et al., 2015) that spherical tensor encoding provides a sufficiently complementary contrast. For this problem, Coelho et al. (2019a, 2022), and Lampinen et al. (2020) and Coelho et al. (2024) optimized the experiment design based on the Cramér-Rao bound, by optimizing $b$-tensor shapes for all diffusion and covariance tensor parameters, and by optimizing the $q(t)$ waveforms for the SM parameters, correspondingly. Finally, Jespersen et al. (2018) and Lee et al. (2018a) used the diffusion time dependence to break the compartment degeneracy of a time-dependent extension of the SM, corresponding to the incomplete coarse-graining regime (ii) of Section 3.

Employing multiple NMR contrasts, beyond diffusion, helps achieve even greater complementarity. Multimodal examples in the brain include varying the inversion time (De Santis et al., 2016) to resolve fiber crossings; varying the echo time (Veraart et al., 2018; Lampinen et al., 2019; Coelho et al., 2024) to alleviate fit degeneracies for the SM, based on the distinct $T_2$ values in the intra- and extracellular spaces (Dortch et al., 2013); altering compartment relaxation properties by intracerebroventricular injection of contrast agent (Silva et al., 2002; Kunz et al., 2018) to measure compartment diffusivities independently, and thereby establish prior knowledge for breaking the SM degeneracy; and employing the $T_2^*$

**96  PART | II** Diffusion MRI

contrast together with strong diffusion weighting enabling the separation between intra- and extraaxonal compartments (Kleban et al., 2020).

Fiber tractography can also provide the complementary information by coupling the knowledge of the ODF into the problem (36); a global method of estimating all parameters (kernel and ODF) simultaneously (the concept of *mesoscopic fiber tracking*, or MesoFT) was suggested by Reisert et al. (2014).

## 5.4  Specificity and relevance of SM parameters

Here, we highlight how both the *scalar parameters* $f$, $D_a$, $D_e^{\parallel}$, $D_e^{\perp}$, and $f_w$, and the *spherical tensor parameters* (the SH coefficients of the ODF $\mathcal{P}(\hat{n})$), carry distinct biophysical significance of increased specificity and understanding of disease, as well as improved fiber tractography, the main subject of this book.

The ability to estimate scalar parameters of the kernel (35) would make dMRI measurements *specific*—rather than just sensitive—to micrometer-level manifestations of disease processes, such as demyelination (Fieremans et al., 2012; Novikov and Fieremans, 2012; Jelescu et al., 2016b) ($D_e^{\perp}$), axonal/dendritic loss (Jelescu et al., 2016b; Khan et al., 2016a,b; Vestergaard-Poulsen et al., 2011) ($f$), beading (Budde and Frank, 2010), inflammation and edema ($f_w$, as well as, potentially, mostly $D_a$ for cytotoxic and mostly $D_e^{\parallel}$, $D_e^{\perp}$ for vasogenic edema; Unterberg et al., 2004). Combining $f$ with the extraaxonal volume fraction, derived either from tortuosity modeling based on $D_e^{\perp}$ (Novikov and Fieremans, 2012; Fieremans et al., 2012) or from the myelin volume fraction via relaxometry, would ultimately allow one to determine axonal $g$-ratio (Stikov et al., 2015).

Since the precise nature and pathological changes in microarchitecture of restrictions leading to the scalar parameter values are unknown, ideally, to become specific to pathology, one needs to estimate $f$, $f_w$, $D_a$, $D_e^{\parallel}$, and $D_e^{\perp}$ separately. As discussed in Sections 5.3.2 and 5.3.3, this prompts for comprehensive diffusion protocols that include "orthogonal" measurements and validation methods (Coronado-Leija et al., 2024)to gain better understanding of the relevant tissue features of modeling.

In practice, the scalar kernel often needs to be estimated from noisy dMRI data obtained using clinical protocols that consist of few low-$b$ linear encoding diffusion shells, including large-scale studies such as the Human Connectome Project (Sotiropoulos et al., 2013), Adolescent Brain Cognitive Development study (Siemens protocol) (Casey et al., 2018), and UK Biobank (Alfaro-Almagro et al., 2018), which is not feasible without imposing additional model assumptions or constraints, as discussed in Section 5.3.2. Multiple models have been proposed for SM estimation (Fieremans et al., 2011; Zhang et al., 2012; Kaden et al., 2016) that are popular because of their robustness, a prerequisite for clinical translation, yet each superimpose additional constraints on both the scalar SM parameters and the fiber ODF (Novikov et al., 2019). As an illustration, Jelescu et al. (2015) compared the performance of SM parameters in normal human development, as derived using WMTI (Fieremans et al., 2011) and NODDI (Zhang et al., 2012), respectively, and showed quantitative estimates that are model-dependent, exhibiting biases and limitations related to the model assumptions. This illustration clearly calls for extreme caution when interpreting modeling studies based on limited clinical dMRI data, where accuracy is typically sacrificed in favor of precision. Efforts of improved SM estimation methods based on applying soft priors and machine learning approaches are underway (Coelho et al., 2024; Liao et al., 2024).

For fiber tractography, deconvolving the voxel-wise fiber ODF, instead of relying on the empirical directions obtained by Fourier-transforming the dMRI signal from the $q$ to $x$ space, provides a more adequate starting point. Luckily (for fiber tractography), the typical dMRI protocols largely oversample the directional information, and estimation of the ODF is largely insensitive to the degeneracies of the SM scalar kernel estimation.

The dMRI protocol can also be tailored to increase specificity and accuracy of fiber tractography based on the insights gained from biophysical modeling. In particular, the observed power-law scaling (31) in WM (Veraart et al., 2019; McKinnon et al., 2017) implies that the intraaxonal contribution dominates at sufficiently high $b$ (e.g., $b \geq$ $6\,\mathrm{ms}/\mu\mathrm{m}^2 = 6000\,\mathrm{s}/\mathrm{mm}^2$ at TE $= 105$ ms; Veraart et al., 2019). Hence, increasing the diffusion weighing not only improves angular resolution but also effectively suppresses contributions from extraaxonal signal and edema, leading to cleaner and more accurate fiber tracking (Fan et al., 2016; Farquharson et al., 2013).

## 6  Concluding remarks

While fiber tracking has been for many years an arena for developing algorithms of connecting the tracts based on empirically determined dMRI signal anisotropy, fundamentally the dMRI signal emerges on the scale of WM microstructure. Taking into account the biophysical foundations of the measured signal can lead to synergies between the microstructure and the structural connectivity communities. Microstructure-informed fiber tracking is poised to merge the microstructural and the macroscopic scales, and provide local scalar, orientational, and global connectivity metrics across the brain.

# References

Abdollahzadeh, A., Belevich, I., Jokitalo, E., Tohka, J., Sierra, A., 2019. Automated 3D axonal morphometry of white matter. Sci. Rep. 20452322. 9, 6084. https://doi.org/10.1038/s41598-019-42648-2.

Abdollahzadeh, A., Coronado-Leija, R., Lee, H.-H., Sierra, A., Fieremans, E., Novikov, D.S., 2024. Quantifying changes of axonal shape in traumatic brain injury with time-dependent diffusion. Proc. ISMRM, 304.

Aboitiz, F., Scheibel, A.B., Fisher, R.S., Zaidel, E., 1992. Fiber composition of the human corpus callosum. Brain Res. 598 (1–2), 143–153.

Afzali, M., Aja-Fernández, S., Jones, D.K., 2020. Direction-averaged diffusion-weighted MRI signal using different axisymmetric B-tensor encoding schemes. Magn. Reson. Med. 15222594. 84 (3), 1579–1591. https://doi.org/10.1002/mrm.28191.

Alfaro-Almagro, F., Jenkinson, M., Bangerter, N.K., Andersson, J.L.R., Griffanti, L., Douaud, G., Sotiropoulos, S.N., Jbabdi, S., Hernandez-Fernandez, M., Vallee, E., et al., 2018. Image processing and quality control for the first 10,000 brain imaging datasets from UK Biobank. NeuroImage 166, 400–424.

Álvarez, G.A., Shemesh, N., Frydman, L., 2017. Internal gradient distributions: a susceptibility-derived tensor delivering morphologies by magnetic resonance. Sci. Rep. 7, 3311. https://doi.org/10.1038/s41598-017-03277-9.

Anderson, A.W., 2005. Measurement of fiber orientation distributions using high angular resolution diffusion imaging. Magn. Reson. Med. 07403194. 54 (5), 1194–1206. https://doi.org/10.1002/mrm.20667.

Arbabi, A., Kai, J., Khan, A.R., Baron, C.A., 2020. Diffusion dispersion imaging: mapping oscillating gradient spin-echo frequency dependence in the human brain. Magn. Reson. Med. 0740-3194. 83 (6), 2197–2208. https://doi.org/10.1002/mrm.28083.

Åslund, I., Topgaard, D., 2009. Determination of the self-diffusion coefficient of intracellular water using PGSE NMR with variable gradient pulse length. J. Magn. Reson. 201 (2), 250–254.

Assaf, Y., Basser, P.J., 2005. Composite hindered and restricted model of diffusion (CHARMED) MR imaging of the human brain. NeuroImage 27 (1), 48–58.

Bloembergen, N., Purcell, E.M., Pound, R.V., 1948. Relaxation effects in nuclear magnetic resonance absorption. Phys. Rev. 73 (7), 679–712. https://doi.org/10.1103/PhysRev.73.679.

Brabec, J., Lasič, S., Nilsson, M., 2020. Time-dependent diffusion in undulating thin fibers: impact on axon diameter estimation. NMR Biomed. 0952-3480. 33 (3). https://doi.org/10.1002/nbm.4187.

Budde, M.D., Frank, J.A., 2010. Neurite beading is sufficient to decrease the apparent diffusion coefficient after ischemic stroke. Proc. Natl. Acad. Sci. USA 107 (32), 14472–14477. https://doi.org/10.1073/pnas.1004841107.

Burcaw, L.M., Fieremans, E., Novikov, D.S., 2015. Mesoscopic structure of neuronal tracts from time-dependent diffusion. NeuroImage 114, 18–37.

Callaghan, P.T., 1991. Principles of Nuclear Magnetic Resonance Microscopy. Clarendon Press.

Callaghan, P.T., Jolley, K.W., Lelievre, J., 1979. Diffusion of water in the endosperm tissue of wheat grains as studied by pulsed field gradient nuclear magnetic resonance. Biophys. J. 0006-3495. 28 (1), 133–141. https://doi.org/10.1016/S0006-3495(79)85164-4.

Cardy, J., 1996. Scaling and Renormalization in Statistical Physics. vol. 5 Cambridge University Press.

Casey, B.J., Cannonier, T., Conley, M.I., Cohen, A.O., Barch, D.M., Heitzeg, M.M., Soules, M.E., Teslovich, T., Dellarco, D.V., Garavan, H., et al., 2018. The Adolescent Brain Cognitive Development (ABCD) study: imaging acquisition across 21 sites. Dev. Cogn. Neurosci. 32, 43–54.

Cho, H., Ryu, S., Ackerman, J.L., Song, Y., 2009. Visualization of inhomogeneous local magnetic field gradient due to susceptibility contrast. J. Magn. Reson. 1090-7807. 198 (1), 88–93. https://doi.org/10.1016/j.jmr.2009.01.024.

Christiaens, D., Veraart, J., Cordero-Grande, L., Price, A.N., Hutter, J., Hajnal, J.V., Tournier, J.D., 2020. On the need for bundle-specific microstructure kernels in diffusion MRI. NeuroImage 10959572. 208 (July 2019), 116460. https://doi.org/10.1016/j.neuroimage.2019.116460.

Coelho, S., Pozo, J.M., Jespersen, S.N., Frangi, A.F., 2019a. Optimal experimental design for biophysical modelling in multidimensional diffusion MRI. In: Medical Image Computing and Computer Assisted Intervention - MICCAI 2019. MICCAI 2019. Lecture Notes in Computer Science, 11766. Springer, pp. 617–625.

Coelho, S., Pozo, J.M., Jespersen, S.N., Jones, D.K., Frangi, A.F., 2019b. Resolving degeneracy in diffusion MRI biophysical model parameter estimation using double diffusion encoding. Magn. Reson. Med. 0740-3194. 82 (1), 395–410. https://doi.org/10.1002/mrm.27714.

Coelho, S., Baete, S.H., Lemberskiy, G., Ades-Aron, B., Barrol, G., Veraart, J., Novikov, D.S., Fieremans, E., 2022. Reproducibility of the Standard Model of diffusion in white matter on clinical MRI systems. NeuroImage 257, 119290. https://doi.org/10.1016/j.neuroimage.2022.119290.

Coelho, S., Liao, Y., Szczepankiewicz, F., Veraart, J., Chung, S., Lui, Y.W., Novikov, D.S., Fieremans, E., 2024. Assessment of precision and accuracy of brain white matter microstructure using combined diffusion MRI and relaxometry. Hum. Brain Mapp. 45, e26725. https://doi.org/10.1002/hbm.26725.

Coronado-Leija, R., Abdollahzadeh, A., Lee, H.-H., Coelho, S., Ades-Aron, B., Liao, Y., Salo, R.A., Tohka, J., Sierra, A., Novikov, D.S., Fieremans, E., 2024. Volume electron microscopy in injured rat brain validates white matter microstructure metrics from diffusion MRI. Imaging Neurosci. https://doi.org/10.1162/imag_a_00212.

De Santis, S., Barazany, D., Jones, D.K., Assaf, Y., 2016. Resolving relaxometry and diffusion properties within the same voxel in the presence of crossing fibres by combining inversion recovery and diffusion-weighted acquisitions. Magn. Reson. Med. 15222594. 75 (1), 372–380. https://doi.org/10.1002/mrm.25644.

Dell'Acqua, F., Rizzo, G., Scifo, P., Clarke, R.A., Scotti, G., Fazio, F., 2007. A model-based deconvolution approach to solve fiber crossing in diffusion-weighted MR imaging. IEEE Trans. Biomed. Eng. 54 (3), 462–472. https://doi.org/10.1109/TBME.2006.888830.

Dhital, B., Reisert, M., Kellner, E., Kiselev, V.G., 2019. NeuroImage intra-axonal diffusivity in brain white matter. NeuroImage 1053-8119. 189 (January), 543–550. https://doi.org/10.1016/j.neuroimage.2019.01.015.

Does, M.D., Zhong, J., Gore, J.C., 1999. In vivo measurement of ADC change due to intravascular susceptibility variation. Magn. Reson. Med. 07403194. 41 (2), 236–240. https://doi.org/10.1002/(SICI)1522-2594(199902)41:2<236::AID-MRM4>3.0.CO;2-3.

Does, M.D., Parsons, E.C., Gore, J.C., 2003. Oscillating gradient measurements of water diffusion in normal and globally ischemic rat brain. Magn. Reson. Med. 49, 206–215.

Dortch, R.D., Harkins, K.D., Juttukonda, M.R., Gore, J.C., Does, M.D., 2013. Characterizing inter-compartmental water exchange in myelinated tissue using relaxation exchange spectroscopy. Magn. Reson. Med. 07403194. 70 (5), 1450–1459. https://doi.org/10.1002/mrm.24571.

Einstein, A., 1905. On the motion of small particles suspended in liquids at rest required by the molecular-kinetic theory of heat. Ann. Phys. 17, 549–560.

Fan, Q., Witzel, T., Nummenmaa, A., Van Dijk, K.R.A., Van Horn, J.D., Drews, M.K., Somerville, L.H., Sheridan, M.A., Santillana, R.M., Snyder, J., Hedden, T., Shaw, E.E., Hollinshead, M.O., Renvall, V., Zanzonico, R., Keil, B., Cauley, S., Polimeni, J.R., Tisdall, D., Buckner, R.L., Wedeen, V.J., Wald, L.L., Toga, A.W., Rosen, B.R., 2016. MGH-USC Human Connectome Project datasets with ultra-high b-value diffusion MRI. NeuroImage 124, 1108–1114.

Farquharson, S., Tournier, J.D., Calamante, F., Fabinyi, G., Schneider-Kolsky, M., Jackson, G.D., Connelly, A., 2013. White matter fiber tractography: why we need to move beyond DTI: clinical article. J. Neurosurg. 118 (6), 1367–1377.

Fick, A., 1855. Ueber diffusion. Ann. Phys. 170 (1), 59–86.

Fieremans, E., Lee, H.H., 2018. Physical and numerical phantoms for the validation of brain microstructural MRI: a cookbook. NeuroImage 182, 39–61.

Fieremans, E., Novikov, D.S., Jensen, J.H., Helpern, J.A., 2010. Monte Carlo study of a two-compartment exchange model of diffusion. NMR Biomed. 23, 711–724.

Fieremans, E., Jensen, J.H., Helpern, J.A., 2011. White matter characterization with diffusional kurtosis imaging. NeuroImage 58 (1), 177–188.

Fieremans, E., Jensen, J.H., Helpern, J.A., Kim, S., Grossman, R.I., Inglese, M., Novikov, D.S., 2012. Diffusion distinguishes between axonal loss and demyelination in brain white matter. Proc. Int. Soc. Magn. Reson. Med. 20, 714.

Fieremans, E., Burcaw, L.M., Lee, H.H., Lemberskiy, G., Veraart, J., Novikov, D.S., 2016. In vivo observation and biophysical interpretation of time-dependent diffusion in human white matter. NeuroImage 129, 414–427. https://doi.org/10.1016/j.neuroimage.2016.01.018.

Fieremans, E., Veraart, J., Ades-Aron, B., Szczepankiewicz, F., Nilsson, M., Novikov, D.S., 2018. Effect of combining linear with spherical tensor encoding on estimating brain microstructural parameters. In: Proceedings 26th Annual Meeting ISMRM, Paris, France, p. 254.

Girard, G., Daducci, A., Petit, L., Thiran, J.P., Whittingstall, K., Deriche, R., Wassermann, D., Descoteaux, M., 2017. AxTract: toward microstructure informed tractography. Hum. Brain Mapp. 10659471. 38 (11), 5485–5500. https://doi.org/10.1002/hbm.23741.

Grebenkov, D.S., 2007. NMR survey of reflected Brownian motion. Rev. Mod. Phys. 79 (3), 1077–1137. https://doi.org/10.1103/RevModPhys.79.1077.

Hall, M.G., Alexander, D.C., 2009. Convergence and parameter choice for Monte-Carlo simulations of diffusion MRI. IEEE Trans. Med. Imaging 28 (9), 1354–1364.

Hansen, B., Lund, T.E., Sangill, R., Jespersen, S.N., 2015. Neurite density from an isotropic diffusion model. Proc. Int. Soc. Magn. Reson. Med. 23, 3043.

Healy, D.M., Hendriks, H., Kim, P.T.P., 1998. Spherical deconvolution. J. Multivar. Anal. 0047259X. 67, 1–22. https://doi.org/10.1006/jmva.1998.1757.

Holz, M., Heil, S.R., Sacco, A., 2000. Temperature-dependent self-diffusion coefficients of water and six selected molecular liquids for calibration in accurate 1H NMR PFG measurements. Phys. Chem. Chem. Phys. 14639076. 2 (20), 4740–4742. https://doi.org/10.1039/b005319h.

Imry, Y., 1997. Introduction to Mesoscopic Physics. Oxford University Press, New York.

Jelescu, I.O., Novikov, D.S., 2020. Water exchange time between gray matter compartments in vivo. Proc. Int. Soc. Magn. Reson. Med. 28, 715.

Jelescu, I.O., Veraart, J., Adisetiyo, V., Milla, S.S., Novikov, D.S., Fieremans, E., 2015. One diffusion acquisition and different white matter models: how does microstructure change in human early development based on WMTI and NODDI? NeuroImage 107, 242–256.

Jelescu, I.O., Veraart, J., Fieremans, E., Novikov, D.S., 2016a. Degeneracy in model parameter estimation for multi-compartmental diffusion in neuronal tissue. NMR Biomed. 29 (1), 33–47.

Jelescu, I.O., Zurek, M., Winters, K.V., Veraart, J., Rajaratnam, A., Kim, N.S., Babb, J.S., Shepherd, T.M., Novikov, D.S., Kim, S.G., Fieremans, E., 2016b. In vivo quantification of demyelination and recovery using compartment-specific diffusion MRI metrics validated by electron microscopy. NeuroImage 132, 104–114. https://doi.org/10.1016/j.neuroimage.2016.02.004.

Jelescu, I.O., de Skowronski, A., Geffroy, F., Palombo, M., Novikov, D.S., 2022. Neurite exchange imaging (NEXI): a minimal model of diffusion in gray matter with inter-compartment water exchange. Neuroimage 256, 119277. https://doi.org/10.1016/j.neuroimage.2022.119277.

Jensen, J.H., Helpern, J.A., 2018. Characterizing intra-axonal water diffusion with direction-averaged triple diffusion encoding MRI. NMR Biomed. 31, e3930. https://doi.org/10.1002/nbm.3930.

Jensen, J.H., Russell Glenn, G., Helpern, J.A., 2016. Fiber ball imaging. NeuroImage 10959572. 124, 824–833. https://doi.org/10.1016/j.neuroimage.2015.09.049.

Jespersen, S.N., Kroenke, C.D., Ostergaard, L., Ackerman, J.J.H., Yablonskiy, D.A., 2007. Modeling dendrite density from magnetic resonance diffusion measurements. NeuroImage 34 (4), 1473–1486.

Jespersen, S.N., Bjarkam, C.R., Nyengaard, J.R., Chakravarty, M.M., Hansen, B., Vosegaard, T., Ostergaard, L., Yablonskiy, D., Nielsen, N.C., Vestergaard-Poulsen, P., 2010. Neurite density from magnetic resonance diffusion measurements at ultrahigh field: comparison with light microscopy and electron microscopy. NeuroImage 49 (1), 205–216. https://doi.org/10.1016/j.neuroimage.2009.08.053.

Jespersen, S.N., Lundell, H., Sønderby, C.K., Dyrby, T.B., 2013. Orientationally invariant metrics of apparent compartment eccentricity from double pulsed field gradient diffusion experiments. NMR Biomed. 09523480. 26 (12), 1647–1662. https://doi.org/10.1002/nbm.2999.

Jespersen, S.N., Olesen, J.L., Hansen, B., Shemesh, N., 2018. Diffusion time dependence of microstructural parameters in fixed spinal cord. NeuroImage 10959572. 182 (August 2017), 329–342. https://doi.org/10.1016/j.neuroimage.2017.08.039.

Jian, B., Vemuri, B.C., 2007. A unified computational framework for deconvolution to reconstruct multiple fibers from diffusion weighted MRI. IEEE Trans. Med. Imaging 26 (11), 1464–1471. https://doi.org/10.1109/TMI.2007.907552.

Kaden, E., Knösche, T.R., Anwander, A., 2007. Parametric spherical deconvolution: inferring anatomical connectivity using diffusion MR imaging. NeuroImage 10538119. 37 (2), 474–488. https://doi.org/10.1016/j.neuroimage.2007.05.012.

Kaden, E., Kruggel, F., Alexander, D.C., 2016. Quantitative mapping of the per-axon diffusion coefficients in brain white matter. Magn. Reson. Med. 75, 1752–1763.

Kazhdan, M., Funkhouser, T., Rusinkiewicz, S., 2003. Rotation invariant spherical harmonic representation of 3D shape descriptors. In: Symposium on Geometry Processing, vol. 6, pp. 156–164.

Khan, A.R., Chuhutin, A., Wiborg, O., Kroenke, C.D., Nyengaard, J.R., Hansen, B., Jespersen, S.N., 2016a. Biophysical modeling of high field diffusion MRI demonstrates micro-structural aberration in chronic mild stress rat brain. NeuroImage 1053-8119. https://doi.org/10.1016/j.neuroimage.2016.07.001.

Khan, A.R., Chuhutin, A., Wiborg, O., Kroenke, C.D., Nyengaard, J.R., Hansen, B., Jespersen, S.N., 2016b. Summary of high field diffusion MRI and microscopy data demonstrate microstructural aberration in chronic mild stress rat brain. Data Brief 2352-3409. 8, 934. https://doi.org/10.1016/j.dib.2016.06.061.

Kiselev, V.G., 2004. Effect of magnetic field gradients induced by microvasculature on NMR measurements of molecular self-diffusion in biological tissues. J. Magn. Reson. 10907807. 170 (2), 228–235. https://doi.org/10.1016/j.jmr.2004.07.004.

Kiselev, V.G., 2010. The cumulant expansion: an overarching mathematical framework for understanding diffusion NMR. In: Jones, D.K. (Ed.), Diffusion MRI: Theory, Methods, and Applications. Oxford University Press, pp. 152–168.

Kiselev, V.G., 2017. Fundamentals of diffusion MRI physics. NMR Biomed. 30 (3), e3602.

Kiselev, V.G., 2020. Microstructure with diffusion MRI: what scale we are sensitive to? J. Neurosci. Methods 347, 108910. https://doi.org/10.1016/j.jneumeth.2020.108910.

Kiselev, V.G., Il'yasov, K.A., 2007. Is the "biexponential diffusion" biexponential? Magn. Reson. Med. 57 (3), 464–469. https://doi.org/10.1002/mrm.21164.

Kiselev, V.G., Novikov, D.S., 2018. NeuroImage transverse NMR relaxation in biological tissues. NeuroImage 1053-8119. 182 (December 2017), 149–168. https://doi.org/10.1016/j.neuroimage.2018.06.002.

Kleban, E., Tax, C.M.W., Rudrapatna, U.S., Jones, D.K., Bowtell, R., 2020. Strong diffusion gradients allow the separation of intra- and extra-axonal gradient-echo signals in the human brain. NeuroImage 10959572. 217 (March), 116793. https://doi.org/10.1016/j.neuroimage.2020.116793.

Kleinnijenhuis, M., Mollink, J., Lam, W.W., Kinchesh, P., Khrapitchev, A.A., Smart, S.C., Jbabdi, S., Miller, K.L., 2018. Choice of reference measurements affects quantification of long diffusion time behaviour using stimulated echoes. Magn. Reson. Med. 79 (2), 952–959.

Kroenke, C.D., Ackerman, J.J.H., Yablonskiy, D.A., 2004. On the nature of the NAA diffusion attenuated MR signal in the central nervous system. Magn. Reson. Med. 52 (5), 1052–1059. https://doi.org/10.1002/mrm.20260.

Kunz, N., da Silva, A.R., Jelescu, I.O., 2018. Intra- and extra-axonal axial diffusivities in the white matter: which one is faster? NeuroImage 10959572. 181, 314–322. https://doi.org/10.1016/j.neuroimage.2018.07.020.

Lampinen, B., Szczepankiewicz, F., Mårtensson, J., van Westen, D., Sundgren, P.C., Nilsson, M., 2017. Neurite density imaging versus imaging of microscopic anisotropy in diffusion MRI: a model comparison using spherical tensor encoding. NeuroImage 10538119. 147, 517–531. https://doi.org/10.1016/j.neuroimage.2016.11.053.

Lampinen, B., Szczepankiewicz, F., Novén, M., Westen, D.V., Hansson, O., Englund, E., Mårtensson, J., Westin, C.F., Nilsson, M., 2019. Searching for the neurite density with diffusion MRI: challenges for biophysical modeling. Hum. Brain Mapp. 40, 2529–2545. https://doi.org/10.1002/hbm.24542.

Lampinen, B., Szczepankiewicz, F., Mårtensson, J., van Westen, D., Hansson, O., Westin, C.F., Nilsson, M., 2020. Towards unconstrained compartment modeling in white matter using diffusion-relaxation MRI with tensor-valued diffusion encoding. Magn. Reson. Med. 15222594. 84 (3), 1605–1623. https://doi.org/10.1002/mrm.28216.

Lasič, S., Szczepankiewicz, F., Eriksson, S., Nilsson, M., Topgaard, D., 2014. Microanisotropy imaging: quantification of microscopic diffusion anisotropy and orientational order parameter by diffusion MRI with magic-angle spinning of the q-vector. Front. Phys. 2, 11.

Le Bihan, D., Breton, E., Lallemand, D., Grenier, P., Cabanis, E., Laval-Jeantet, M., 1986. MR imaging of intravoxel incoherent motions: application to diffusion and perfusion in neurologic disorders. Radiology 0033-8419. 161 (2), 401–407. https://doi.org/10.1148/radiology.161.2.3763909.

Lee, H.H., Fieremans, E., Novikov, D.S., 2018a. LEMONADE(t): exact relation of time-dependent diffusion signal moments to neuronal microstructure. Proceedings of the International Society for Magnetic Resonance in Medicine, 884.

Lee, H.H., Fieremans, E., Novikov, D.S., 2018b. What dominates the time dependence of diffusion transverse to axons: intra- or extra-axonal water? NeuroImage 10959572. 182, 500–510. https://doi.org/10.1016/j.neuroimage.2017.12.038.

Lee, H.H., Yaros, K., Veraart, J., Pathan, J.L., Liang, F.X., Kim, S.G., Novikov, D.S., Fieremans, E., 2019. Along-axon diameter variation and axonal orientation dispersion revealed with 3D electron microscopy: implications for quantifying brain white matter microstructure with histology and diffusion MRI. Brain Struct. Funct. 1863-2661. 224 (4), 1469–1488. https://doi.org/10.1007/s00429-019-01844-6.

Lee, H.H., Jespersen, S.N., Fieremans, E., Novikov, D.S., 2020a. The impact of realistic axonal shape on axon diameter estimation using diffusion MRI. NeuroImage 10538119. https://doi.org/10.1016/j.neuroimage.2020.117228.

Lee, H.H., Papaioannou, A., Kim, S.L., Novikov, D.S., Fieremans, E., 2020b. A time-dependent diffusion MRI signature of axon caliber variations and beading. Commun. Biol. 23993642. 3, 354. https://doi.org/10.1038/s42003-020-1050-x.

Lee, H.H., Papaioannou, A., Novikov, D.S., Fieremans, E., 2020c. In vivo observation and biophysical interpretation of time-dependent diffusion in human cortical gray matter. NeuroImage 10538119. 222, 117054. https://doi.org/10.1016/j.neuroimage.2020.117054.

Lee, H.H., Fieremans, E., Novikov, D.S., 2021. Realistic microstructure simulator (RMS): Monte Carlo simulations of diffusion in three-dimensional cell segmentations of microscopy images. J. Neurosci. Methods 350, 109018.

Liao, Y., Coelho, S., Chen, J., Ades-Aron, B., Pang, M., Stepanov, V., Osorio, R., Shepherd, T., Lui, Y.W., Novikov, D.S., Fieremans, E., 2024. Mapping tissue microstructure of brain white matter in vivo in health and disease using diffusion MRI. Imaging Neurosci. 2, 1–17. https://doi.org/10.1162/imag_a_00102.

Mackay, A., Whittall, K., Adler, J., Li, D., Paty, D., Graeb, D., 1994. In vivo visualization of myelin water in brain by magnetic resonance. Magn. Reson. Med. 15222594. 31 (6), 673–677. https://doi.org/10.1002/mrm.1910310614.

McKinnon, E.T., Jensen, J.H., Glenn, G.R., Helpern, J.A., 2017. Dependence on b-value of the direction-averaged diffusion-weighted imaging signal in brain. Magn. Reson. Imaging 36, 121–127.

Mirzaalian, H., Ning, L., Savadjiev, P., Pasternak, O., Bouix, S., Michailovich, O., Grant, G., Marx, C.E., Morey, R.A., Flashman, L.A., George, M.S., Mcallister, T.W., Andaluz, N., Shutter, L., Coimbra, R., Zafonte, R.D., Coleman, M.J., Kubicki, M., Westin, C.F., Stein, M.B., Shenton, M.E., Rathi, Y., 2016. NeuroImage inter-site and inter-scanner diffusion MRI data harmonization. NeuroImage 1053-8119. 135, 311–323. https://doi.org/10.1016/j.neuroimage.2016.04.041.

Mitra, P.P., Halperin, B.I., 1995. Effects of finite gradient-pulse widths in pulsed-field-gradient diffusion measurements. J. Magn. Reson. Ser. A 10641858. 113, 94–101. https://doi.org/10.1006/jmra.1995.1060.

Mitra, P.P., Sen, P.N., Schwartz, L.M., Le Doussal, P., 1992. Diffusion propagator as a probe of the structure of porous media. Phys. Rev. Lett. 0031-9007. 68 (24), 3555–3558. https://doi.org/10.1103/PhysRevLett.68.3555.

Mitra, P.P., Sen, P.N., Schwartz, L.M., 1993. Short-time behavior of the diffusion coefficient as a geometrical probe of porous media. Phys. Rev. B Condens. Matter Mater. Phys. 0163-1829. 47 (14), 8565–8574. https://doi.org/10.1103/PhysRevB.47.8565.

Murday, J.S., Cotts, R.M., 1968. Self-diffusion coefficient of liquid lithium. J. Chem. Phys. 00219606. 48 (11), 4938. https://doi.org/10.1063/1.1668160.

Neuman, C.H., 1974. Spin echo of spins diffusing in a bounded medium. J. Chem. Phys. 60 (11), 4508–4511.

Nilsson, M., Lätt, J., Ståhlberg, F., van Westen, D., Hagslätt, H., 2012. The importance of axonal undulation in diffusion MR measurements: a Monte Carlo simulation study. NMR Biomed. 25 (5), 795–805.

Nilsson, M., Lätt, J., van Westen, D., Brockstedt, S., Lasič, S., Ståhlberg, F., Topgaard, D., 2013. Noninvasive mapping of water diffusional exchange in the human brain using filter-exchange imaging. Magn. Reson. Med. 69 (6), 1572–1580.

Nilsson, M., Lasič, S., Drobnjak, I., Topgaard, D., Westin, C.F., 2017. Resolution limit of cylinder diameter estimation by diffusion MRI: the impact of gradient waveform and orientation dispersion. NMR Biomed. 09523480. 30, e3711. https://doi.org/10.1002/nbm.3711.

Novikov, D.S., 2021. The present and the future of microstructure MRI: from a paradigm shift to normal science. J. Neurosci. Methods 351, 108947. https://doi.org/10.1016/j.jneumeth.2020.108947.

Novikov, D.S., Fieremans, E., 2012. Relating extracellular diffusivity to cell size distribution and packing density as applied to white matter. Proc. Int. Soc. Magn. Reson. Med. 20, 1829.

Novikov, D.S., Jensen, J.H., Helpern, J.A., Fieremans, E., 2014. Revealing mesoscopic structural universality with diffusion. Proc. Natl. Acad. Sci. USA 111 (14), 5088–5093.

Novikov, D.S., Kiselev, V.G., Jespersen, S.N., 2018a. On modeling. Magn. Reson. Med. 79 (6), 3172–3193. https://doi.org/10.1002/mrm.27101.

Novikov, D.S., Reisert, M., Kiselev, V.G., 2018b. Effects of mesoscopic susceptibility and transverse relaxation on diffusion NMR. J. Magn. Reson. 293, 134–144. https://doi.org/10.1016/j.jmr.2018.06.007.

Novikov, D.S., Veraart, J., Jelescu, I.O., Fieremans, E., 2018c. Rotationally-invariant mapping of scalar and orientational metrics of neuronal microstructure with diffusion MRI. NeuroImage 10538119. 174, 518–538. https://doi.org/10.1016/j.neuroimage.2018.03.006.

Novikov, D.S., Fieremans, E., Jespersen, S.N., Kiselev, V.G., 2019. Quantifying brain microstructure with diffusion MRI: theory and parameter estimation. NMR Biomed. 32, e3998.

Olesen, J.L., Ostergaard, L., Shemesh, N., Jespersen, S.N., 2022. Diffusion time dependence, power-law scaling, and exchange in gray matter. Neuroimage 251, 118976. https://doi.org/10.1016/j.neuroimage.2022.118976.

Palombo, M., Ligneul, C., Hernandez-Garzon, E., Valette, J., 2018. Can we detect the effect of spines and leaflets on the diffusion of brain intracellular metabolites? NeuroImage 10959572. 182 (May 2017), 283–293. https://doi.org/10.1016/j.neuroimage.2017.05.003.

Pampel, A., Jochimsen, T.H., Möller, H.E., 2010. BOLD background gradient contributions in diffusion-weighted fMRI - comparison of spin-echo and twice-refocused spin-echo sequences. NMR Biomed. 10991492. 23 (6), 610–618. https://doi.org/10.1002/nbm.1502.

Quirk, J.D., Bretthorst, G.L., Duong, T.Q., Snyder, A.Z., Springer Jr., C.S., Ackerman, J.J.H., Neil, J.J., 2003. Equilibrium water exchange between the intra- and extracellular spaces of mammalian brain. Magn. Reson. Med. 50 (3), 493–499. https://doi.org/10.1002/mrm.10565.

Reisert, M., Kiselev, V.G., Dihtal, B., Kellner, E., Novikov, D.S., 2014. MesoFT: unifying diffusion modelling and fiber tracking. In: Medical Image Computing and Computer-Assisted Intervention—MICCAI 2014, Springer, pp. 201–208.

Reisert, M., Kellner, E., Dhital, B., Hennig, J., Kiselev, V.G., 2017. Disentangling micro from mesostructure by diffusion MRI: a Bayesian approach. NeuroImage 10959572. 147, 964–975. https://doi.org/10.1016/j.neuroimage.2016.09.058.

Reisert, M., Kiselev, V.G., Dhital, B., 2019. A unique analytical solution of the white matter standard model using linear and planar encodings. Magn. Reson. Med. 81, 3819–3825. https://doi.org/10.1002/mrm.27685.

Reynaud, O., 2017. Time-dependent diffusion MRI in cancer: tissue modeling and applications. Front. Phys. 2296-424X. 5 (November), 1–16. https://doi.org/10.3389/fphy.2017.00058.

Robertson, B., 1966. Spin-echo decay of spins diffusing in a bounded region. Phys. Rev. 0031899X. (1), 273–277. https://doi.org/10.1103/PhysRev.151.273.

Shemesh, N., Jespersen, S.N., Alexander, D.C., Cohen, Y., Drobnjak, I., Dyrby, T.B., Finsterbusch, J., Koch, M.A., Kuder, T., Laun, F., Lawrenz, M., Lundell, H., Mitra, P.P., Nilsson, M., Özarslan, E., Topgaard, D., Westin, C.F., 2016. Conventions and nomenclature for double diffusion encoding NMR and MRI. Magn. Reson. Med. 15222594. 75 (1), 82–87. https://doi.org/10.1002/mrm.25901.

Shepherd, G.M.G., Raastad, M., Andersen, P., 2002. General and variable features of varicosity spacing along unmyelinated axons in the hippocampus and cerebellum. Proc. Natl. Acad. Sci. USA 0027-8424. 99 (9), 6340–6345. https://doi.org/10.1073/pnas.052151299.

Sherbondy, A.J., Rowe, M.C., Alexander, D.C., 2010. MicroTrack: an algorithm for concurrent projectome and microstructure estimation. Med. Image Comput. Comput. Assist. Interv. 03029743. 6361 LNCS (Part 1), 183–190. https://doi.org/10.1007/978-3-642-15705-9_23.

Silva, M.D., Omae, T., Helmer, K.G., Li, F., Fisher, M., Sotak, C.H., 2002. Separating changes in the intra- and extracellular water apparent diffusion coefficient following focal cerebral ischemia in the rat brain. Magn. Reson. Med. 07403194. 48 (5), 826–837. https://doi.org/10.1002/mrm.10296.

Singer, P., Asthagiri, D., Chapman, W., Hirasaki, G., 2016. Molecular dynamics simulations of NMR relaxation and diffusion of bulk hydrocarbons and water. J. Magn. Reson. 277. https://doi.org/10.1016/j.jmr.2017.02.001.

Skinner, N.P., Kurpad, S.N., Schmit, B.D., Muftuler, L.T., Budde, M.D., 2017. Rapid in vivo detection of rat spinal cord injury with double-diffusion-encoded magnetic resonance spectroscopy. Magn. Reson. Med. 77, 1639–1649. https://doi.org/10.1002/mrm.26243.

Sotiropoulos, S.N., Behrens, T.E.J., Jbabdi, S., 2012. Ball and rackets: inferring fiber fanning from diffusion-weighted MRI. NeuroImage 60 (2), 1412–1425.

Sotiropoulos, S.N., Jbabdi, S., Xu, J., Andersson, J.L., Moeller, S., Auerbach, E.J., Glasser, M.F., Hernandez, M., Sapiro, G., Jenkinson, M., et al., 2013. Advances in diffusion MRI acquisition and processing in the Human Connectome Project. NeuroImage 80, 125–143.

Stejskal, E.O., Tanner, J.E., 1965. Spin diffusion measurements: spin echoes in the presence of a time-dependent field gradient. J. Chem. Phys. 42 (1), 288–292.

Stikov, N., Campbell, J.S.W., Stroh, T., Lavelée, M., Frey, S., Novek, J., Nuara, S., Ho, M.K., Bedell, B.J., Dougherty, R.F., Leppert, I.R., Boudreau, M., Narayanan, S., Duval, T., Cohen-Adad, J., Picard, P.A., Gasecka, A., Côté, D., Pike, G.B., 2015. In vivo histology of the myelin g-ratio with magnetic resonance imaging. NeuroImage 118, 397–405. https://doi.org/10.1016/j.neuroimage.2015.05.023.

Szczepankiewicz, F., Lasič, S., van Westen, D., Sundgren, P.C., Englund, E., Westin, C.F., Ståhlberg, F., Lätt, J., Topgaard, D., Nilsson, M., 2015. Quantification of microscopic diffusion anisotropy disentangles effects of orientation dispersion from microstructure: applications in healthy volunteers and in brain tumors. NeuroImage 104, 241–252.

Szczepankiewicz, F., van Westen, D., Englund, E., Westin, C.F., Ståhlberg, F., Lätt, J., Sundgren, P.C., Nilsson, M., 2016. The link between diffusion MRI and tumor heterogeneity: mapping cell eccentricity and density by diffusional variance decomposition (DIVIDE). NeuroImage 1053-8119. 142, 522–532. https://doi.org/10.1016/j.neuroimage.2016.07.038.

Tal, A., Gonen, O., 2015. Spectroscopic localization by simultaneous acquisition of the double-spin and stimulated echoes. Magn. Reson. Med. 73 (1), 31–43.

Tan, E.T., et al., 2020. Oscillating diffusion-encoding with a high gradient-ampliude and high slew-rate head-only gradient for human brain imaging. Magn. Reson. Med. 84, 950. https://doi.org/10.1002/mrm.28180.

Topgaard, D., 2017. Multidimensional diffusion MRI. J. Magn. Reson. 10960856. 275, 98–113. https://doi.org/10.1016/j.jmr.2016.12.007.

Tournier, J.D., Calamante, F., Gadian, D.G., Connelly, A., 2004. Direct estimation of the fiber orientation density function from diffusion-weighted MRI data using spherical deconvolution. NeuroImage 23 (3), 1176–1185.

Tournier, J.D., Calamante, F., Connelly, A., 2007. Robust determination of the fibre orientation distribution in diffusion MRI: non-negativity constrained super-resolved spherical deconvolution. NeuroImage 35 (4), 1459–1472. https://doi.org/10.1016/j.neuroimage.2007.02.016.

Tournier, J.D., Mori, S., Leemans, A., 2011. Diffusion tensor imaging and beyond. Magn. Reson. Med. 1522-2594. 65 (6), 1532–1556. https://doi.org/10.1002/mrm.22924.

Unterberg, A.W., Stover, J., Kress, B., Kiening, K.L., 2004. Edema and brain trauma. Neuroscience 129 (4), 1019–1027.

van Kampen, N.G., 1981. Stochastic Processes in Physics and Chemistry, first ed. Elsevier, Oxford.

Veraart, J., Novikov, D.S., Fieremans, E., 2018. TE dependent Diffusion Imaging (TEdDI) distinguishes between compartmental T2 relaxation times. NeuroImage 10959572. 182 (September 2017), 360–369. https://doi.org/10.1016/j.neuroimage.2017.09.030.

Veraart, J., Fieremans, E., Novikov, D.S., 2019. On the scaling behavior of water diffusion in human brain white matter. NeuroImage 185, 379–387.

Veraart, J., Nunes, D., Rudrapatna, U., Fieremans, E., Jones, D.K., Novikov, D.S., Shemesh, N., 2020. Noninvasive quantification of axon radii using diffusion MRI. eLife 9, e49855. https://doi.org/10.7554/eLife.49855.

Vestergaard-Poulsen, P., Wegener, G., Hansen, B., Bjarkam, C.R., Blackband, S.J., Nielsen, N.C., Jespersen, S.N., 2011. Diffusion-weighted MRI and quantitative biophysical modeling of hippocampal neurite loss in chronic stress. PLoS One 1932-6203. 6 (7), e20653. https://doi.org/10.1371/journal.pone.0020653.

Westin, C.F., Knutsson, H., Pasternak, O., Szczepankiewicz, F., Özarslan, E., van Westen, D., Mattisson, C., Bogren, M., O'Donnell, L.J., Kubicki, M., Topgaard, D., Nilsson, M., 2016. Q-space trajectory imaging for multidimensional diffusion MRI of the human brain. NeuroImage 10959572. 135, 345–362. https://doi.org/10.1016/j.neuroimage.2016.02.039.

White, N.S., Dale, A.M., 2009. Optimal diffusion MRI acquisition for fiber orientation density estimation: an analytic approach. Hum. Brain Mapp. 10659471. 30 (11), 3696–3703. https://doi.org/10.1002/hbm.20799.

Wilson, K.G., 1983. The renormalization group and critical phenomena. Rev. Mod. Phys. 55 (3), 583.

Zhang, H., Schneider, T., Wheeler-Kingshott, C.A., Alexander, D.C., 2012. NODDI: practical in vivo neurite orientation dispersion and density imaging of the human brain. NeuroImage 61 (4), 1000–1016. https://doi.org/10.1016/j.neuroimage.2012.03.072.

Zhong, J., Kennan, R.P., Gore, J.C., 1991. Effects of susceptibility variations on NMR measurements of diffusion. J. Magn. Reson. 00222364. 95 (2), 267–280. https://doi.org/10.1016/0022-2364(91)90217-H.

# Chapter 6

# Diffusion MRI acquisition for tractography: Diffusion encoding

Jennifer S.W. Campbell[a], Steven H. Baete[b], Julien Cohen-Adad[c], J.-Donald Tournier[d], Filip Szczepankiewicz[e,f], Christian Beaulieu[g], Corey A. Baron[h], Merry Mani[i], Kawin Setsompop[j], Congyu Liao[j], Christine L. Tardif[a], Sjoerd B. Vos[k], Anastasia Yendiki[l], Ilana R. Leppert[a], Els Fieremans[m], and G. Bruce Pike[n]

[a]McConnell Brain Imaging Centre, Montreal Neurological Institute, McGill University, Montreal, QC, Canada, [b]New York University Grossman School of Medicine, New York, NY, United States, [c]Polytechnique Montreal, Montreal, QC, Canada, [d]King's College London, London, United Kingdom, [e]Lund University, Lund, Sweden, [f]Brigham and Women's Hospital, Harvard Medical School, Boston, MA, United States, [g]University of Alberta, Edmonton, AB, Canada, [h]Robarts Research Institute, Western University, London, ON, Canada, [i]University of Iowa, Iowa City, IA, United States, [j]Stanford University, Stanford, CA, United States, [k]Centre for Microscopy, Characterisation, and Analysis, The University of Western Australia, Nedlands, WA, Australia, [l]Massachusetts General Hospital, Harvard Medical School, Boston, MA, United States, [m]Bernard and Irene Schwartz Center for Biomedical Imaging, Department of Radiology, New York University Grossman School of Medicine, New York, NY, United States, [n]Hotchkiss Brain Institute, University of Calgary, Calgary, AB, Canada

## 1 Diffusion contrast

### 1.1 Qualitative description of the origin of diffusion contrast

The key element of a diffusion encoded magnetic resonance sequence is a strong magnetic field gradient. This gradient induces signal decay via spin dephasing, providing contrast in the signal magnitude proportional to the amount of random displacement that diffusing water molecules undergo. To understand how diffusion contrast is encoded in MR acquisition, we need to have a closer look at the behavior of the magnetic dipoles or *spins* of the nuclei. In MRI, the nuclei in question are typically hydrogen (a single proton). When placed in an external magnetic field, the spins of all protons sum to create a net nonzero magnetic momentum or spin magnetization vector precessing around the magnetic field. As a radiofrequency (RF) excitation pulse rotates the spin magnetization toward the transverse plane, the transverse signal from the aligned precessing spins will generate a strong MRI signal. As time progresses, spins will lose phase alignment due to slightly different precessional frequencies that depend on the total magnetic field experienced by each individual spin isochromat, causing the MRI signal amplitude to decay over time ($T_2^*$-decay).

Spin dephasing in MRI in general is both a source of contrast ($T_2^*$ decay, $T_2$ decay, susceptibility weighted imaging (Haacke et al., 1999)) and a nuisance, due to the implications for signal-to-noise ratio (SNR). However, spin dephasing can be deliberately induced to encode diffusion contrast using a pair of magnetic field gradients (Fig. 1A and C), as introduced by Stejskal and Tanner (1965) in nuclear magnetic resonance (NMR). This diffusion encoding technique, using two gradient lobes, is sometimes called the Stejskal-Tanner technique, or alternatively the pulsed field gradient (PFG) technique. All gradients create linear changes in the main magnetic field in their direction of application, and they are commonly used to manipulate the local precessional frequency for spatial image encoding. In diffusion MRI (dMRI), the first of the pair of magnetic field gradients encodes the average spin position as a phase variation in the direction of the diffusion gradient, while the second gradient reverses this process. If diffusion is absent, the effects of the gradient pair will cancel, the spin phases will align (they will be "in phase") and the full signal magnitude will recover (excluding other nonstatic dephasing effects, i.e., $T_2$ decay). In contrast, when the spins experience random motion such as diffusion during and/or between the two gradients, the spin phases no longer line up at the time of the readout, resulting in a net attenuation of the measured signal. The amount of spin dephasing caused by random spin displacement over the voxel therefore determines the diffusion encoding, or diffusion weighting, of the image.

Handbook of Diffusion MR Tractography. https://doi.org/10.1016/B978-0-12-818894-1.00001-X
Copyright © 2025 Elsevier Ltd. All rights reserved, including those for text and data mining, AI training, and similar technologies.

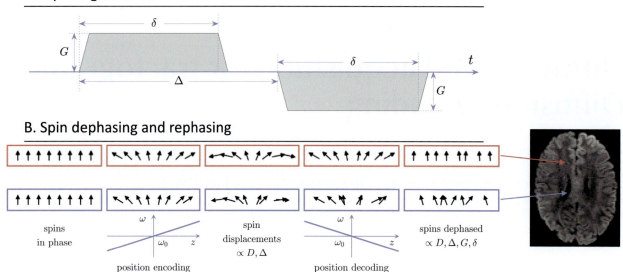

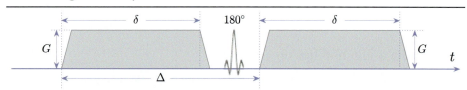

**FIG. 1** Spin dephasing and rephasing for diffusion encoding in one direction. Diffusion encoding can be achieved using a bipolar gradient pulse, as illustrated in A. However, typical diffusion weighting is achieved using a spin echo, as illustrated in C; thus the final image (B) is both $T_2$ weighted and diffusion weighted. The effect of diffusion itself is the same for both schemes A and C, and is illustrated in panel B. After excitation, spins precess in phase. The first diffusion gradient encodes the average spin position as a phase variation in the direction of the diffusion gradient, and the second gradient reverts this phase encoding. Diffusion-induced displacements along the gradient direction between and during encoding and decoding cause irreversible spin dephasing and signal loss. Spins in areas with higher diffusivity (e.g., cerebrospinal fluid, *blue pixel*) will experience large displacements and thus greater signal loss than spins in areas with lower diffusivity (e.g., white matter, *red pixel*). The diffusion encoding gradient is in the $z$ direction (through-plane in the image). *(Figure based on Pipe, J., 2014. Pulse sequences for diffusion-weighted MRI. In: Johansen-Berg, H., Behrens, T.E.J. (Eds.), Diffusion MRI: From Quantitative Measurement to In Vivo Neuroanatomy. Elsevier.)*

The diffusion contrast will scale with

(i) the amount of diffusive motion in the voxel, quantified to the first order by the diffusion coefficient, $D$, often quoted in mm$^2$/s or µm$^2$/ms (see Chapter 5).
(ii) the time between the gradient lobe onsets ($\Delta$, often referred to as the diffusion time; see Fig. 1).
(iii) the strength of the diffusion encoding, determined by the diffusion gradient area—the product of the gradient magnitude ($G$) and duration ($\delta$). The (small) diffusion weighting caused by all nondiffusion gradients present (imaging gradients, spoilers, etc.) should be included in any fitting that is applied to the data, particularly for diffusion sequences with long diffusion times (Mattiello et al., 1997).

Voxels with higher average diffusion coefficients will experience more spin dephasing and greater signal attenuation in the diffusion weighted (DW) image (Fig. 1B). To isolate the diffusion contrast in an image ($S$) from other effects such as $T_1$-recovery, $T_2$- and $T_2^*$-decay, and off-resonance effects, $S$ is compared to a second measurement ($S_0$) with the diffusion gradients set to zero (Fig. 2).

Signal loss due to static field (denoted $B_0$) inhomogeneities is commonly mitigated by placing a 180 degree refocusing RF pulse between the two diffusion gradients to form a pulsed gradient spin echo (PGSE) sequence (Stejskal and Tanner, 1965). The polarity of the second diffusion gradient is then inverted to account for the phase reversal induced by the extra

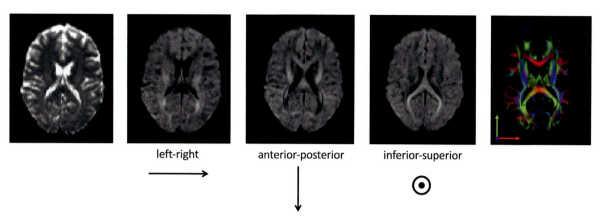

**FIG. 2** Diffusion weighted images. The diffusion tractography experiment will include at least one measurement with no diffusion weighting (left), and multiple measurements with different diffusion encoding directions (indicated by *arrows* below the images). The variation in direction of diffusion encoding changes the contrast when oriented fiber structure exists. The image on the right is a map of the fractional anisotropy (FA) from the diffusion tensor, color coded by the principal eigenvector of the diffusion tensor, which indicates the major fiber tract orientations. These images show the same slice of a healthy adult human brain. The image intensity window/level is different for the leftmost image in order to better show the image contrast. The acquisition parameters, which will be explained later in this chapter, were: magnetic field strength 3 T, 32 channel head coil, $b = 1000\,\text{s/mm}^2$, 2 mm isotropic voxels, BW = 1300 Hz/px, TE = 94 ms, TR = 8.6 s, phase encode direction AP, GRAPPA $R = 2$, and PPF 6/8, and two signal averages.

RF pulse (Fig. 1C). The long duration of the diffusion encoding gradient block, required to achieve high diffusion sensitivity, means that the echo time is long and thus the PGSE DW images and the accompanying $S_0$ images are heavily $T_2$ weighted.

The pair of diffusion gradients (Fig. 1C) sensitizes the MRI sequence only to diffusion displacements parallel (or antiparallel) to the direction of the magnetic field gradient. To form a complete picture of spin movements in an anisotropic tissue sample, we must repeat the experiment with diffusion gradient pairs applied in different angular directions (see Fig. 2). By combining these measurements, we can extract the tissue microstructure directionality for use in fiber tractography.

## 1.2 Quantitative description of the origin of diffusion contrast

Here, we will introduce the variables used in diffusion imaging acquisition and analyze the form of the signal attenuation caused by the diffusion encoding gradients mathematically. In dMRI, the vector $\boldsymbol{q}$ and the term "q-space" are used in a fashion analogous to the concept of k-space (see, e.g., Haacke et al., 1999) in the MRI spatial encoding. Like the vector $\boldsymbol{k}$ in the image readout, the vector $\boldsymbol{q}$ is the integral of all gradient waveforms $\boldsymbol{G}$ over time:

$$\boldsymbol{q}(t) = \gamma \int_0^t \boldsymbol{G}(\tau) d\tau, \quad (1)$$

where $\gamma$ is the gyromagnetic ratio. The magnitude of $\boldsymbol{q}$ (denoted $q$) evolves during the diffusion encoding block, and its maximal value is $q_{max}$. For the Stejskal-Tanner sequence (Fig. 1), $q_{max} = \gamma G \delta$. This sequence, in which the direction of the diffusion encoding gradient $\boldsymbol{G}$ is constant, is said to sample one point in what is called q-space.

The diffusion time, $t_d$, is the effective time over which the diffusion phenomenon is observed. For the PGSE sequence, it is given by the time between diffusion gradient pulses, $\Delta$, with the finite duration ($\delta$) of the pulses acting as a low-pass filter on the measured parameters (Novikov et al., 2019):

$$t_d \approx \Delta. \quad (2)$$

The diffusion length is the root-mean-squared distance diffusing molecules travel during the diffusion time, and its relation to the diffusion time is given by the Einstein relation (Chapter 5). Quantitatively, after the application of the diffusion encoding gradients, the resulting signal after spin dephasing will be the vector sum of the individual spin vectors. Hence, for an original signal $S_0$, the DW signal will be

$$S = S_0 \langle e^{i\phi} \rangle, \quad (3)$$

**106 PART | II** Diffusion MRI

where $\phi$ denotes the phase of a spin. Explicitly, the signal is then dependent on the probability $p(\phi)$ that a spin acquires phase $\phi$ during the diffusion encoding:

$$S = S_0 \int_{-\pi}^{\pi} e^{i\phi} p(\phi) \mathrm{d}\phi. \tag{4}$$

The phase $\phi$ accrued by any one spin depends on the magnetic fields it experiences along its random walk. If it displaces by vector $\mathrm{d}\mathbf{r}(\tau)$ at any time $\tau$, its phase at time $t$ will be

$$\phi(t) = -\gamma \int_0^t B(\tau) \mathrm{d}\tau, \tag{5}$$

where $B(\tau) = \mathbf{G}(\tau) \cdot \int_0^\tau \mathrm{d}\mathbf{r}(\tau') \mathrm{d}\tau'$, the instantaneous magnetic field seen by the spin, in the rotating frame (Haacke et al., 1999).

Invoking Einstein's relation for a Gaussian diffusion process, it can be shown (Haacke et al., 1999) that Eq. (4) becomes

$$S = S_0 e^{-\int_0^{TE} \mathbf{q}(t)^T \mathbf{D} \mathbf{q}(t) \mathrm{d}t}, \tag{6}$$

with $\mathbf{D}$ the diffusion tensor. We now define the $b$-matrix, sometimes called the $b$-tensor (Westin et al., 2016), as the outer product

$$\mathbf{b} = \int_0^{TE} \mathbf{q}(t) \mathbf{q}(t)^T \mathrm{d}t. \tag{7}$$

Hence, the signal can be expressed as a function of $\mathbf{b}$:

$$\begin{aligned} S &= S_0 e^{-\mathbf{b}:\mathbf{D}} \\ &= S_0 e^{-\sum_{i,j=1}^{3} b_{ij} D_{ij}}. \end{aligned} \tag{8}$$

For a sequence where the diffusion encoding gradient direction is constant (as is the case for the diffusion encoding sequences we have discussed thus far), the scalar quantity

$$b = \mathrm{trace}(\mathbf{b}) \tag{9}$$

is generally used to describe the magnitude of diffusion weighting. It is often quoted in s/mm$^2$ or ms/µm$^2$. Coupled with the diffusion encoding gradient direction $\hat{\mathbf{G}}$, the diffusion encoding is fully described. For the PGSE sequence, ignoring the contribution of imaging gradients to $b$,

$$\begin{aligned} b &= \gamma^2 G^2 \delta^2 \left[ \Delta - \frac{\delta}{3} \right] \\ &= q_{\mathrm{max}}^2 \left[ \Delta - \frac{\delta}{3} \right] \end{aligned} \tag{10}$$

The $b$-value is a function of both $q_{\mathrm{max}}$ and time. Both of these parameters are important in the diffusion encoding experiment: $q_{\mathrm{max}}$ determines how much signal attenuation occurs for a given amount of spin displacement, while the time over which this spin displacement occurs determines the total magnitude of displacement. The $S_0$ image, used in most dMRI computations to isolate the diffusion contrast, is called the $b=0$ or $b_0$ image ($S_0 = S(b=0)$). In any one diffusion encoding direction, we have

$$S = S_0 e^{-bD}, \tag{11}$$

where $D$ is the scalar diffusion coefficient in that orientation. Noting that the tensor or scalar diffusion coefficient $D$ describes, by definition, a Gaussian process, deviations from a Gaussian phase distribution will cause the signal decay as a function of $b$ to be nonexponential.

Diffusion in biological tissue is non-Gaussian, because the tissue microstructure restricts the motion of the diffusing molecules (Chapter 5). However, in the in vivo adult human brain tissue at $b$-values up to roughly $b = 1000 \, \text{s/mm}^2$, the signal difference between an idealized Gaussian and the true non-Gaussian diffusive process is typically ignored, and a Gaussian approximation for the diffusion displacement distribution is often used (see Chapter 10 on diffusion tensor imaging). At higher $b$-values, deviations from Gaussianity become more evident. More generally, there is a Fourier relationship between the DW signal and the diffusion displacement distribution or ensemble average diffusion propagator, $p(\boldsymbol{r}, t)$, which describes the probability that the spins displace by vector $\boldsymbol{r}$ in time $t$. Hence, for $t = t_d$, and assuming $\delta \to 0$ (Callaghan, 2010; Stejskal and Tanner, 1965),

$$S(\boldsymbol{q}_{\text{max}}, t_d) = S_0 \iiint_{\mathbb{R}^3} p(\boldsymbol{r}, t_d) e^{-i q_{\text{max}} \cdot \boldsymbol{r}} \mathrm{d}\boldsymbol{r}^3. \tag{12}$$

Measurement of the diffusion propagator $p(\boldsymbol{r}, t_d)$ using Eq. (12) would require the acquisition to probe a large extent of q-space. Another general expression for the signal is the cumulant expansion (Kiselev, 2010), which expresses the log of the signal as a function of powers of $b$. Diffusion tensor imaging uses the first-order (in $b$) cumulant, and above $b$ $\sim 1000 \, \text{s/mm}^2$, higher-order signal representations are more appropriate. The $b$-value determines at what point the expansion should be cut off.

The case where the diffusion encoding gradient direction is constant corresponds to a $b$-tensor with one nonzero eigenvalue. This is termed linear $b$-tensor encoding, and at the time of writing, acquisitions designed for tractography use, almost exclusively, linear encoding.

Note that any applied gradient will induce phase dispersion; therefore there are many ways other than the PGSE sequence to achieve diffusion encoding. The diffusion encoding waveforms need to return to the center of q-space, and there is normally a spin echo. An example is the twice-refocused spin echo (TRSE) (Reese et al., 2003), which reduces eddy current induced artifacts (see Section 2.3).

## 2 The diffusion encoding block

The ability to achieve diffusion encoding in an MRI sequence depends on the gradient capabilities of the system, as well as other hardware features. Intertwined with the choice of hardware, which may or may not be under the control of the investigator, is the specific choice of diffusion encoding parameters (e.g., the $b$-values and diffusion encoding directions). The diffusion encoding can also be tuned to reduce artifacts.

### 2.1 Hardware for diffusion MRI

This section will describe four key components of the MRI system and their roles in dMRI. These are the main magnet, gradient coils, shim coils, and RF coils.

### 2.1.1 Magnet

In an MRI system, the main field strength ($B_0$, not to be confused with the unweighted $b = 0$ diffusion scans) is generated by a magnet that is usually superconducting. Research centers and hospitals are typically equipped with 1.5 or 3 tesla (T) systems for scanning humans. Increasing numbers of 7T human systems are being installed, with over 150 units worldwide at the time of writing. Another avenue of recent investigation is the use of low field strength (0.55 T) for demanding imaging techniques like diffusion imaging (Campbell-Washburn et al., 2019). A higher field strength is both a blessing and a curse. While the SNR increases quasiproportionally with the field strength in tesla (Gallichan, 2018), other effects and artifacts also manifest more strongly (Ladd et al., 2018). Specific absorption rate (SAR) increases quadratically with field strength, and subjects might experience nausea and dizziness if they are moved through the field too rapidly at high field strength. RF transmit inhomogeneities (see Section 2.1.4) are more pronounced, causing incomplete refocusing in spin echo acquisitions such as those used in dMRI. Image distortions, already pronounced when using the echo planar imaging (EPI) readouts typically used in dMRI, increase proportionally with the field strength. These difficulties can however be mitigated with efficient hardware and pulse sequences (Andersson et al., 2003; Wu et al., 2018; Mani et al., 2020). Chapter 7 provides more detail on such techniques (see Fig. 7 of that chapter). Additionally, $T_2$ and $T_2^*$ relaxation times decrease with field strength, making the requirement for a short diffusion encoding block and fast readout even more critical than at lower field strengths.

**108 PART | II** Diffusion MRI

### 2.1.2 Gradient coils

As we have seen, the MRI scanner's magnetic field gradients play a key role in generating diffusion-based contrast. The gradients are located inside the MRI bore, between the main magnet and the RF transmit coil, and consist of three independent sets of coils that can create linear gradients along the scanner's $\hat{x}, \hat{y},$ and $\hat{z}$ directions. Gradients are typically characterized by their maximum amplitude and their slew rate, which is the maximum rate of change of the magnetic field gradient amplitude with time. Typical nominal maximum amplitude values for human systems are 40–80 mT/m, but this value can be much higher on specialized systems; the Connectom 1.0 and 2.0 scanners have maximum amplitudes of 300 mT/m and 500 mT/m, respectively (Huang et al., 2021; Setsompop et al., 2013). A notable benefit of strong gradients is the ability to achieve high $b$-values. Another benefit is the possibility to minimize the time required (and hence TE) to achieve a given $b$-value, thereby recovering a large amount of SNR compared with weaker gradient implementations (McNab et al., 2013). This SNR increase has been postulated to explain improved tractography at higher gradient strength (Chamberland et al., 2018).

The system's slew rate depends on its capability to switch rapidly between large currents. Nominal values in clinical systems are about 200 mT/m/ms, although this value is typically limited by the system, as the changing magnetic field can produce nerve stimulation (specialized head-only gradient systems (e.g., Foo et al., 2020) can minimize peripheral nerve stimulation and can therefore have higher slew rates, up to 900 mT/m/ms (Feinberg et al., 2023)). A high slew rate allows the desired diffusion and readout gradient magnitude to be reached faster, facilitating further reduction of the scanning time, and hence increasing SNR. It is also beneficial for diffusion experiments that require fast switching of the diffusion encoding gradient magnitude (see Section 2.4).

Systems with strong gradients have been associated with transient physiological effects, such as magnetophosphenes (Setsompop et al., 2013), which are perceived flashes of light resulting from the magnetic stimulation of the retina. These were seen during a stimulation study but not during normal use for brain imaging (Ladd et al., 2018; Setsompop et al., 2013).

### 2.1.3 Shim coils

MRI scanners may employ both active and passive shimming to minimize the static magnetic field inhomogeneity in the imaging region, thereby reducing image distortion and the refocusable contribution to $T_2^*$, which is called $T_2'$. The active shimming procedure is done using the shim coils to create field variations that counter the measured variation in the main magnet's field. The gradient coils are typically used for first-order shimming, with vendor-integrated special shim coils for second-order (typically at 3 T) up to third-order (typically at 7 T) shimming. The shimming process involves the estimation of a field map followed by the computation of a set of "shim coefficients" that describe the amount of current that should go into each gradient and shim coil to best correct the field nonuniformity. However, the default shim coils that are integrated in the MR system are not enough to fully compensate for the high degree of spatial variability in the magnetic field across the brain-spine axis, especially at 7 T and higher, which has motivated the development of custom technology (Topfer et al., 2016; Hetherington et al., 2006; Stockmann et al., 2016). Shim coils can also be used to compensate, in real time, for respiratory-induced fluctuations in the $B_0$ field, which can create ghosting and displacement artifacts in EPI (Verma and Cohen-Adad, 2014; Van Gelderen et al., 2007; Topfer et al., 2018). These advanced techniques are under active academic development; it is hoped that vendors will rapidly translate some of these innovations into widely available commercial products. Note that even with shimming of the $B_0$ field, the 180 degree refocusing RF pulse in the PGSE diffusion preparation also serves to refocus unwanted field inhomogeneities and improve image quality and SNR.

### 2.1.4 Radiofrequency coils

As we have seen, RF pulses play a role in dMRI, both in creating measurable signal through spin excitation and in refocusing the signal. RF coils are also used to receive the signal during readout. RF coils transmit and/or receive the magnetic component of the electromagnetic field (see, e.g., Gruber et al., 2018; Cohen-Adad and Wald, 2014 for more detail). The rule of thumb when designing and choosing the right coil for a given application is that RF transmission should be as homogeneous as possible (assessed by the so-called $B_1^+$ or flip angle map), while RF reception ($B_1^-$) should be as sensitive as possible (image SNR is directly related to the reception). At 3 T or lower, RF transmission is typically done by the so-called body coil, which is integrated into the MR system bore, and has excellent homogeneity. For RF reception, most manufacturers now provide array coils, which are composed of multiple loops (typically 20–64 elements on modern high-channel-count receiver arrays) laid out on the coil former so as to cover a specific body region. Beyond sensitivity, another key aspect of receive coils is their geometry (the number of coil elements and how they are laid out), which directly impacts the performance of parallel acquisition techniques (see Chapter 7), as well as simultaneous multislice (SMS, also called multiband (MB)) techniques (see Chapter 7). Most dMRI studies of the brain use head receive coils; however, flexible

Diffusion MRI acquisition for tractography: Diffusion encoding **Chapter | 6** **109**

surface coils can be used to achieve high-quality targeted diffusion imaging, for example of the occipital cortex (Frass-Kriegl et al., 2018; Movahedian Attar et al., 2020).

## 2.2 Choice of *b*-value and sampling scheme for tractography

A prime concern in the design of dMRI acquisition is the selection of the *b*-value and the corresponding distribution of DW gradient directions. For the purposes of tractography of the brain, the most commonly used protocols use the single-shell high angular resolution diffusion imaging (HARDI) approach (Tuch et al., 2002). High angular resolution is needed for many tractography applications, because it aids in the disentanglement of complex fiber configurations, such as fanning and crossing (Vos et al., 2016); a recent estimate of the occurrence of complex fiber configurations in brain white matter voxels is 60%–90% ( Jeurissen et al., 2013).

A single-shell HARDI acquisition scheme consists of a relatively large number ($N_g$) of uniformly distributed DW directions collected over a single *b*-value "shell," with one image volume for each DW direction, along with a small number of $b = 0$ volumes. This approach is widely used for tractography primarily because it focuses almost exclusively on characterizing the angular features of the signal, making it arguably the most efficient strategy for estimating fiber orientations (the most important information needed for tractography). The protocol is primarily characterized by the *b*-value and the corresponding number of directions used to sample the shell, among other considerations, discussed in following text.

### 2.2.1 Choice of b-value

As previously mentioned, the *b*-value dictates the amount of DW *contrast* imparted to the signal. Higher *b*-values provide more contrast at the expense of lower signal; the optimal *b*-value is therefore a compromise between the two. In order to resolve the different fiber orientations that might be present at any given location, the signals that correspond to these orientations need to be distinguishable from each other. At low *b*-values, each orientation will be associated with a high, but smoothly varying, signal (see Fig. 3). As the *b*-value is increased, this signal will be reduced, but its variation with angle will be greater. This makes it possible to resolve differently oriented signals more easily. But if the *b*-value is increased further, the signal becomes buried in the noise.

The specific issue of *b*-value selection for fiber orientation estimation has been investigated in a number of studies (White and Dale, 2009; Tournier et al., 2013; Alexander and Barker, 2005), with a consensus emerging around $b = 2500\text{--}3000 \, \text{s/mm}^2$ for multifiber orientation estimates in in vivo adult human white matter. This is higher than would typically be used for cases where only one fiber orientation is estimated per voxel (Jones et al., 1999), including clinical studies, where the *b*-value is usually $\sim 1000 \, \text{s/mm}^2$. The images produced therefore have a lower SNR for a given voxel size. It is important to note that what matters here is the *contrast-to-noise ratio* (CNR) of the data *across* DW directions, rather than the raw SNR of individual images (see Fig. 4).

The (maximum) *b*-value, once selected, dictates many of the other parameters in the sequence. The echo time is invariably set to the lowest achievable value in order to minimize artifacts and signal loss due to $T_2$ decay. This means using the maximum available gradient strength, with long DW gradient pulses (i.e., high values of $\delta$) and the shortest pulse separation ($\Delta$) that can be made to fit within the constraints of the PGSE sequence. Higher *b*-values will necessitate longer $\delta$ and longer TE. The time between subsequent acquisitions of a given slice, the repetition time (TR), will also be longer for higher *b*-values, because of the longer TE. The effective diffusion time $t_d$ is typically commensurate with the pulse duration itself, but this aspect is rarely (if ever) considered for tractography applications in the central nervous system (CNS). If the structures that restrict the motion of spins are larger, or the diffusion is slower, a deliberately longer diffusion time may be desired, but this is typically achieved by alteration of the sequence (see Section 2.4).

The use of high *b*-values also introduces other concerns. First among these is the increase in gradient heating which, if excessive, can cause scans to abort, and/or force the use of longer repetition times to reduce the overall duty cycle of the sequence (Hutter et al., 2018a). This will increase the overall scan time over that which would otherwise be achievable. Furthermore, the reduced SNR and increased contrast is problematic for some postprocessing correction techniques, required to deal with residual eddy-current distortion and/or subject motion (Graham et al., 2016). However, modern techniques are available now that handle these issues reliably (Andersson and Sotiropoulos, 2016; Andersson et al., 2017) (see Chapter 9). There are also concerns with eddy current generation itself and gradient-induced vibration due to the use of strong gradients (see Sections 2.3.1 and 2.3.6).

Although not directly relevant to tractography as such, an additional benefit of higher *b*-values is that the signal becomes more specific to the restricted water inside neurites (Novikov et al., 2019; Raffelt et al., 2012; Veraart et al., 2019; Genc et al., 2020). This is beneficial in microstructural modeling (Assaf and Basser, 2005), allowing for improved estimation of

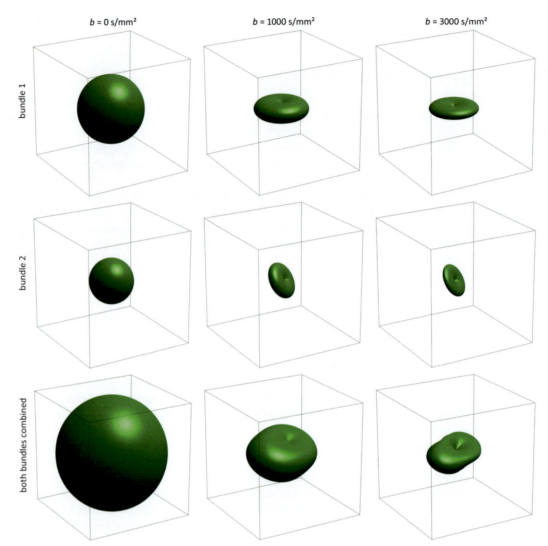

**FIG. 3** For a given $b$-value, the dMRI signal can be represented as a function of orientation as shown here. Each column shows the signal at a different $b$-value (0, 1000, and 3000 s/mm$^2$, respectively), displayed as a polar plot, or orientation distribution function (ODF). The bottom row shows the dMRI signal that would be measured for a voxel containing a 60/40 mix of two coherently oriented fiber bundles. The first bundle's fibers are aligned along the vertical axis, and its signal contribution is shown in the first row. The second bundle's fibers are aligned at an angle of 60 degrees to the vertical axis, and its signal contribution is shown in the second row. The signal in the bottom row is the sum of these signals. Note that the orientation contrast is stronger at higher $b$-values, making it easier to discern the contributions from the two distinct bundles. However, higher $b$-values also reduce the overall amount of signal and hence the SNR. The best parameters to resolve these orientations are therefore a compromise and should aim to maximize the overall contrast-to-noise ratio.

axon morphometry (Nilsson et al., 2017) (see Section 2.4). This is particularly relevant when coupled with multishell acquisition strategies (see the following), which allow the signal from different types of tissue to be separated (Jeurissen et al., 2014), further reducing the scope for contamination between different tissue types.

### 2.2.2 Distribution of DW directions

Having established the desired $b$-value, we also need to set the number and direction of the diffusion encoding directions. This is itself dependent on the $b$-value: higher $b$-values introduce more angular contrast and sharper angular features into the signal. To capture these sharper features, we need to sample the angular domain more densely, and hence need to use more sampling directions to cover the orientation domain fully (Tournier et al., 2013) (see Fig. 3). Therefore higher $b$-values of 2500–3000 s/mm$^2$ require on the order of 45–60 DW directions to ensure that any directional biases are negligible. Of note, single fiber estimates using lower $b$-values of $\sim$1000 s/mm$^2$ can be made with as few as six diffusion encoding directions, but there is still a benefit to acquiring data at more DW directions (e.g., 30 (Jones et al., 1999)), both

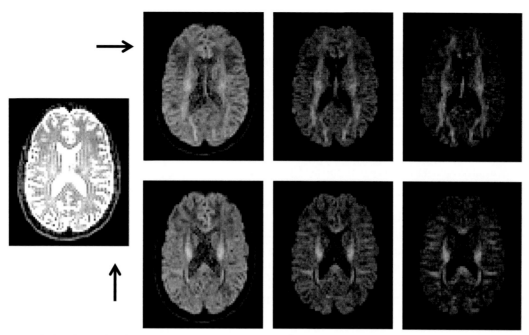

FIG. 4 Images acquired at (from left to right) $b = 0$, 1000, 2000, and 3000 s/mm$^2$. As the $b$-value increases, the SNR decreases, but the CNR increases. The CNR across directions is critical for estimation of the fiber orientations. The diffusion encoding is in the $\hat{x}$ direction for the images in the top row and $\hat{y}$ direction for the images in the bottom row *(arrows)*. The images were acquired at 3 T using a 64 channel head coil. Other scan parameters, to be explained later in this chapter, were 2.5 mm isotropic voxels, BW = 1795 Hz/px, TE = 110 ms, TR = 8.5 s, and phase encode direction AP. *(Data from BATMAN tutorial (https://osf.io/fkyht/), courtesy of Marlene Tahedl, Department for Biomedical Imaging, University of Regensburg.)*

for orientation estimation and SNR. Fifteen directions have been suggested to be the minimum to estimate multiple fiber orientations (Frank, 2002), but again this is suboptimal for performance.

The distribution of these diffusion gradient directions is also very important. In the brain, they need to cover the angular domain as uniformly as possible to avoid any biases. When the fiber orientation is known a priori, for instance in the spinal cord, this sampling can be reduced. There is, unfortunately, no closed-form mathematical solution for this uniform sampling, at least not for arbitrary numbers. However, the electrostatic repulsion model (Jones et al., 1999; Papadakis et al., 2000; Bak and Nielsen, 1997) performs well here, and is often used for this task.

It is important to note that in the context of dMRI, opposite directions are equivalent: they provide the same contrast because diffusion is a symmetric process. It is not uncommon for the set of directions to be entirely contained in one hemisphere, and yet still be optimal in terms of its dMRI encoding capability. However, opposite directions do have the opposite effect in terms of the eddy-current distortions they produce (see Section 2.3.1). It is therefore recommended to ensure the sampling directions *also* cover the full sphere relatively uniformly. This can be achieved by modifying an otherwise optimal set of directions by inverting some of the directions to point in the opposite direction (which has no impact on their performance for diffusion encoding), in order to produce a set of orientations that also has minimal overall bias in terms of the eddy-current distortions.

Ordering is also an important consideration due to subject motion, field drift, and gradient duty cycle. This is especially true in noncompliant cohorts, or with long scan times. Here, it is best to order the directions to minimize any bias that might be introduced by a change during the sequence. For instance, if all $b = 0$ volumes are acquired first, followed by all the DW volumes, the effect of subject motion early on in the scan might be problematic, since this would introduce a systematic misalignment between the $b = 0$ volumes and the rest of the scan (though this can largely be addressed with modern correction methods). A simple solution to this problem is to ensure all the $b$-values are acquired throughout the scan, in this case by interspersing the $b = 0$ volumes throughout the DW volumes; most scanner interfaces allow the user to define custom gradient tables to allow this to be done. In addition, noncompliant subjects are more prone to interruptions in the scan. To maximize the chances that the data acquired are nonetheless usable, it is important to ensure that similar directions are not acquired one after the other. A number of strategies have been proposed to achieve this (Hutter et al., 2018a; Dubois et al., 2006).

**112  PART | II** Diffusion MRI

## 2.2.3  Multishell acquisition schemes

Multishell HARDI refers to the acquisition of multiple "shells," each with a distinct $b$-value, within a single protocol. With the advent of SMS techniques (see Chapter 7), these approaches can now be used routinely within realistic acquisition times.

This has led to the development of improved analysis approaches to take advantage of the additional information provided by these types of acquisitions (see Chapter 11). For instance, these types of data have been shown to provide improved resolution of the diffusion orientation distribution function (ODF) (Aganj et al., 2010; Descoteaux et al., 2011), or the ability to remove unwanted contamination from the WM fiber ODF due to signal from other tissue types (such as gray matter and cerebrospinal fluid (CSF)) ( Jeurissen et al., 2014). Given these latest advances, the use of such multishell acquisitions is likely to increase considerably over the next few years.

With multishell sampling, the issue of optimal selection of $b$-values and their corresponding number of directions necessarily becomes more complex, given the many combinations possible. Similar considerations to those previously described apply, but the different contribution that each shell will make to the overall information content of the data collected needs to be taken into account (Tournier et al., 2020). Moreover, different techniques may impose different constraints (e.g., some may require the acquisition of the same directions per shell). While research on this optimization is ongoing, it is relatively common to acquire 2–3 DW shells (along with an adequate number of $b = 0$ volumes), with the number of DW directions increasing with $b$-value (both to ensure adequate sampling of the angular content, and to counter the inevitable reduction in SNR at higher $b$). Tools for evenly distributing the diffusion encoding directions over multiple shells are also available (e.g., Caruyer et al., 2013).

## 2.3  Diffusion encoding related artifact reduction

This section provides a brief overview of MRI artifacts that can be addressed by modifications to the diffusion encoding gradient waveform; artifact reduction related to the rest of the diffusion encoding sequence is discussed in Chapter 7. Corrections by postprocessing are discussed in Chapter 9.

### 2.3.1  Eddy currents

When the magnetic field gradient is modulated, eddy currents are induced in the conductive materials of the scanner hardware. These currents create an undesired gradient proportional to the slew rate of the applied gradient ($dG/dt$) and opposite to its direction. Assuming a single constant for the eddy current decay ($\lambda$), they can be approximated by (Van Vaals and Bergman, 1990; Bernstein et al., 2004)

$$G_{ec}(t) = -w\frac{dG}{dt} \otimes H(t)\exp(-t/\lambda), \tag{13}$$

where $w$ is a weight factor, $\otimes$ denotes convolution, and $H(t)$ is the Heaviside function. Eddy currents caused by diffusion encoding waveforms are especially prominent because these employ high slew rates for long durations, more so than those typically used for slice selection and readout. The eddy currents distort both the imaging gradients (see Chapter 7) and the diffusion encoding gradient waveform itself. Notably, the image distortion caused by eddy currents—observed as shearing, scaling, shifting, and ghosting—will depend on the direction and strength of the diffusion encoding ( Jezzard et al., 1998), making the artifacts heterogeneous across the diffusion acquisition. We have already noted that it is beneficial to apply diffusion gradients across both hemispheres of q-space in a typical tractography protocol to distribute the artifacts more homogeneously (Section 2.2). To reduce the artifacts, the desired gradient shape can be recovered by so-called "preemphasis," where the effect of eddy currents is predicted and accounted for during the execution of the gradient pulses (Bernstein et al., 2004), as shown in Fig. 5A. Clinical systems are calibrated to compensate for eddy currents by automatic preemphasis. Even so, the impact of eddy currents on image fidelity in dMRI must frequently be addressed in postprocessing (Chapter 9).

Eddy currents can also be considered explicitly in the design of the diffusion encoding waveform by using, for example, bipolar pulses (Alexander et al., 1997; Finsterbusch, 2009), adjusting the timing of asymmetric pulses in a double spin echo sequence, sometimes called the TRSE approach (Reese et al., 2003; Finsterbusch, 2010), and numerical eddy current minimization for arbitrary sequence timing (Aliotta et al., 2018). Eddy currents have also been accounted for in the optimization of nonlinear q-space trajectories (see Section 2.4) (Sjölund et al., 2015; Yang and McNab, 2019). We note that additional optimization constraints that minimize eddy currents also impact the encoding efficiency. Since the artifacts can be largely corrected posthoc, the trade-off between image quality and encoding efficiency must be carefully considered.

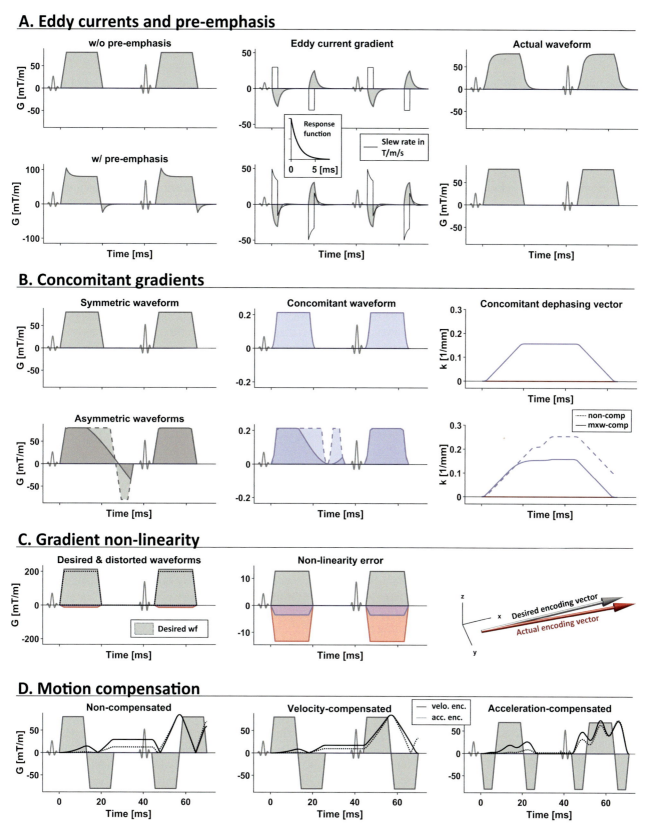

FIG. 5 (A) Examples of diffusion gradient waveforms affected by eddy currents. The first row is without preemphasis such that the actual waveform no longer has a trapezoidal shape. When preemphasis is used, the waveforms requested from the amplifier are nontrapezoidal, but designed to cancel the eddy currents such that the actual waveform is close to the desired trapezoids. The plots in the middle show the eddy current gradients and the slew rate (*solid black line*). Note that the eddy currents have been exaggerated for visual clarity. (B) Examples of symmetric and asymmetric gradient waveforms. The first example is a conventional symmetric design, and the effect of the concomitant gradient cancels ($k=0$ at the end of encoding). In the second row, asymmetric waveforms with (*solid*) and without (*broken*) Maxwell compensation are displayed; compensation ensures that the concomitant gradients are balanced, whereas noncompensated waveforms have residual dephasing vectors that may cause signal errors. (C) Gradient nonlinearity can have a relevant impact on the gradient waveform and the resulting diffusion encoding by changing its b-value and direction. The simulation assumes relatively high gradient amplitudes, such as those on a Siemens Connectom scanner, where gradient nonlinearity is more pronounced than on most clinical scanners. (D) Examples of waveforms that use variable levels of velocity (*solid*) and acceleration encoding (*broken*). Note that the moment curves are arbitrarily scaled for visibility, and that the acceleration-compensated waveform (right) is also compensated for flow.

**114 PART | II** Diffusion MRI

At the time of writing, both approaches are used for tractography. The PGSE sequence employing a pair of monopolar trapezoidal gradients (Fig. 1C) has more pronounced eddy current effects, but higher encoding efficiency. The TRSE is commonly used, at the expense of encoding efficiency.

### 2.3.2 Concomitant field gradients (Maxwell terms)

Linear field gradients are accompanied by undesired gradients that are known as concomitant gradients or Maxwell terms. They can be predicted from the main magnetic field strength ($B_0$) and the desired gradient waveform, such that

$$G_c(t, r) \approx \frac{1}{4B_0} \begin{bmatrix} G_z^2(t) & 0 & -2G_x(t)G_z(t) \\ 0 & G_z^2(t) & -2G_y(t)G_z(t) \\ -2G_x(t)G_z(t) & -2G_y(t)G_z(t) & 4G_x^2(t) + 4G_y^2(t) \end{bmatrix} r, \tag{14}$$

where $r$ is the position vector relative to the isocenter (Meier et al., 2008; Baron et al., 2012; Szczepankiewicz et al., 2019). These gradients are generally too weak to have a relevant effect on the $b$-matrix, but for diffusion encoding waveforms that are asymmetric around the 180 degree refocusing pulse, they may introduce a residual dephasing vector $k \neq 0$ where $k = \gamma \int G_c(t) \mathrm{d}t$ (Fig. 5B). This may cause severe signal loss that can be mistaken for diffusion (Baron et al., 2012), leading to gross errors in microstructural models and fiber orientation estimates (Szczepankiewicz et al., 2019). The effects of concomitant gradients can be mitigated by using symmetric gradient waveform designs, e.g., the PGSE configuration with identical trapezoids on either side of the refocusing pulse (Fig. 1C), online compensation (Meier et al., 2008), or waveforms that are asymmetric but natively compensated for concomitant gradient effects (Szczepankiewicz et al., 2019; Zhou et al., 1998).

### 2.3.3 Gradient nonlinearity

Gradient coils have a relatively small volume where the gradients are approximately linear; in regions of nonlinearity, the desired diffusion encoding is not achieved, leading to inaccurate model estimation and tractography (Mesri et al., 2020). This error is usually inflated if the gradient coils are designed to deliver ultrastrong gradients, where strength and slew rate are prioritized over linearity (Hidalgo-Tobon, 2010). Nonlinearity can be described by a transform ($L$) at position ($r$) such that the actual gradient waveform can be approximated by (Bammer et al., 2003)

$$G_{nl}(t, r) = L(r)G(t). \tag{15}$$

If $G_{nl}$ is known for each voxel, the actual $b$-matrix can be calculated from Eq. (7). By doing so, the spatially varying $b$-matrix—or any other parameter based on $G(t)$—can be used in the analysis to recover accuracy (Tan et al., 2013; Jovicich et al., 2006). In principle, the impact of gradient nonlinearity on the $b$-matrix can be avoided by minimizing the gradient amplitude and extending the encoding time ($b \propto |G|^2 t^3$); however, this approach is rarely feasible in practice. Fig. 5C shows the impact of gradient nonlinearity on the direction and $b$-value for linear $b$-tensor encoding. Gradient nonlinearity may also disrupt concomitant gradient compensation; however, design strategies exist that are robust to such effects (Szczepankiewicz et al., 2020a).

### 2.3.4 System drift

System drift can refer to any effect that causes an MRI system to exhibit a systematic temporal fluctuation. Although the sources can be many, an effect that is especially plausible in dMRI is resistive heating caused by the electrical current used to produce the magnetic field gradients. Assuming equal resistance ($R$) and coil sensitivity in all gradient axes (Hidalgo-Tobon, 2010), the power ($P$) deposited in the system is related to the sum of squared currents ($|I|^2$) and $b$-value according to

$$P = |I|^2 R \propto |G|^2 \propto b. \tag{16}$$

If there exists a correlation between the measured signal and the temperature of the gradient system, there is a risk of inducing confounding effects in the acquisition stage. For example, if experiments are executed in order, from low to high $b$, the scanner acquires low-$b$-data in a cold state, and high-$b$-data in a hot state, with maximal correlation between temperature and signal variation (Szczepankiewicz et al., 2020b). This effect was shown to cause signal error above 10% in $b = 0$ images dispersed throughout a 15-min multishell diffusion acquisition, and had a relevant impact on the resulting tractography (Vos et al., 2017). As in the case of motion during the acquisition, distributing the different $b$-values and encoding directions out over the acquisition can remove the correlation between the $b$-matrix and the signal bias (Vos et al., 2017); this strategy can also improve the overall performance of the system, especially for high-duty-cycle waveforms (Hutter et al., 2018b). Additionally, real-time monitoring of $B_0$ field drift can be used to reduce image shift and to improve fat suppression (Benner et al., 2006).

### 2.3.5 Incoherent motion

Diffusion weighting manifests as dephasing by random movement in a field gradient (Section 1), but similar dephasing can also be caused by other kinds of incoherent motion (termed intravoxel incoherent motion (IVIM)). For example, incoherent blood flow in the capillaries may be mistaken for diffusion at low $b$-values (Le Bihan et al., 1986). Other sources of motion include cardiac pulsation and pulmonary motion, which may influence diffusion measurements in brain (Skare and Andersson, 2001; Habib et al., 2010). In these cases, it is spatial incoherence instead of temporal incoherence that causes phase dispersion, hence, individual spins can be assumed to have constant velocity or acceleration over the duration of the diffusion encoding, making it possible to isolate the effect. Since the dephasing is proportional to the motion encoding of the gradient waveform (Ahlgren et al., 2016), it can be suppressed by using waveforms designed to have negligible higher moments, for example, by minimizing the length of the $n$th moment vector

$$m_n = \left\| \gamma \int_0^\tau \boldsymbol{G}(t) t^n \mathrm{d}t \right\|, \tag{17}$$

where $m_0 = 0$ to satisfy the spin echo condition, $n = 1$ encodes velocity, $n = 2$ encodes acceleration, etc. Motion compensated diffusion encoding waveform designs have recently been proposed (Aliotta et al., 2017; Lasič et al., 2020; Szczepankiewicz et al., 2020c), and are being explored in particular for diffusion imaging outside the brain; examples are shown in Fig. 5D. Pulsation artifacts can also be mitigated with gated imaging—at a cost to the total scan time (Nunes et al., 2005)—or in postprocessing (Zwiers, 2010).

### 2.3.6 Vibration

Driving a current in the gradient coils and cables delivering the current to them exerts Lorentz forces on the hardware. When the currents are switched rapidly—as is common for diffusion encoding waveforms, especially at their start and end—vibrations may be induced. These vibrations may cause acoustic noise, which may be unpleasant or even damaging to the subject (Rösch et al., 2016), as well as introduce artifacts into the images (Berl et al., 2015; Hiltunen et al., 2006; Gallichan et al., 2010). In a single-shell tractography acquisition, these forces will be identical regardless of $b$-value, since the gradients are typically used at full strength. What does change with $b$-value is how often the gradient switching occurs, and whether this matches a mechanical resonant frequency of the scanner. Hence, for example, the artifact can be more evident at $b = 1000 \, \mathrm{s/mm^2}$ than at $b = 3000 \, \mathrm{s/mm^2}$ in certain cases. Vibration artifacts are generally more pronounced for diffusion encoding in a particular orientation (Gallichan et al., 2010), and hence can cause the signal to artifactually resemble that from oriented fiber structure in that orientation. This is demonstrated in Fig. 6. Vibration-induced artifacts can be mitigated with mechanical adjustment, and with postprocessing (Gallichan et al., 2010).

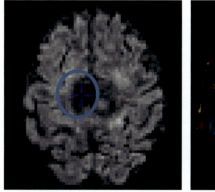
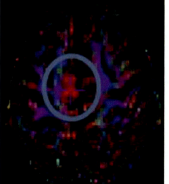

**FIG. 6** Vibration artifact. Tissue motion from vibration causes phase dispersion and hence signal attenuation in DW images (left), shown here to be more pronounced when the diffusion encoding direction is close to $\hat{x}$. This results in artifactually high anisotropy and bias in orientation estimates, as seen on the directionally encoded FA map at right.

## 2.4 Diffusion encoding adaptations

We have seen that the diffusion encoding of an MRI sequence is encapsulated in the $b$-matrix, which is a function of both $q$ and time (Section 1.2). We have discussed the range of values that is typically used for tractography (Section 2.2). The q-space sampling scheme and the diffusion time can be varied independently, expanding the scope of information obtainable from the dMRI experiment. Here, we will discuss variations on the previously described diffusion encoding strategies that can expand the information that guides tractography.

One variant on the time component of the diffusion encoding is the use of stimulated echoes in a stimulated echo acquisition mode (STEAM) sequence (Haase et al., 1986; Steidle and Schick, 2006), which is discussed in more detail in Chapter 8, Section 3. This technique allows the effective diffusion time $t_d$ to be longer than the effective time over which $T_2$ decay occurs. STEAM allows the diffusion time to go up to hundreds of milliseconds (Fieremans et al., 2016), and has been used to investigate the impact of diffusion time in microstructural modeling (Fieremans et al., 2016). Other diffusion encoding variants that affect the diffusion time include oscillating gradient spin echo (OGSE) encoding (Gross and Kosfeld, 1969), a technique for reducing the diffusion time to values on the order of a few milliseconds (Baron and Beaulieu, 2013; Does et al., 2003). OGSE is not currently available as a product sequence and its use on clinical scanners is the subject of research (Arbabi et al., 2020). Peripheral nerve stimulation and the challenge of achieving high $b$-values must be considered in such a sequence.

The major variants on the q-space extent and sampling scheme for tractography are single- and multi-$b$-shell protocols. The additional shells in a multishell protocol can be chosen for the purpose of microstructural modeling. There is interest in combining diffusion-based microstructural modeling with tractography to probe microstructure along specific tracts (Kamiya et al., 2017). Conversely, the microstructure can be used to guide tractography (Girard et al., 2015). Diffusion microstructural compartment modeling has been done with acquisitions with two nonzero $b$-shells and a maximal $b$-value comparable to that typically used for tractography (Zhang et al., 2012; Jelescu et al., 2016). However, multicompartment modeling has been shown to be improved by using many magnitudes of $b$ (Scherrer and Warfield, 2012), with values up to $10,000 \, s/mm^2$, for example, in order to incorporate a finite estimate of the axon diameter (Assaf and Basser, 2005). Compartment modeling suffers from degeneracy in the solution space (Jelescu et al., 2016), and improvements are an area of active research (Coelho et al., 2019; Reisert et al., 2019).

Another variant on the q-space sampling is the employment of a nonlinear q-space trajectory within one diffusion encoding scan, i.e., applying diffusion weighting sequentially in different directions in one excitation (Cory and Garroway, 1990; Mori and Van Zijl, 1995; Wong et al., 1995). The $b$-tensor for such an acquisition has more than one nonzero eigenvalue; in other words, it has rank greater than one. This technique provides additional information that can be used for microstructural inference; it allows us to distinguish tissue structures whose diffusion characteristics are not different when averaged over the scale of a voxel, but whose diffusion characteristics are different on the scale of the diffusion length (Szczepankiewicz et al., 2015; Lasič et al., 2014). Fig. 7 shows an example sequence with a waveform that yields a "spherical" $b$-tensor (Szczepankiewicz et al., 2020b), meaning that there are three equal nonzero eigenvalues (i.e., a rank 3 $b$-tensor). This diffusion encoding sensitizes the signal to the mean diffusivity of compartments on the scale of the diffusion length. Combining acquisitions with $b$-tensors with different shapes, typically including the

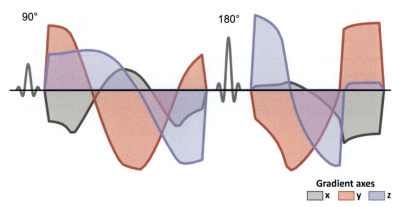

**FIG. 7** Spherical $b$-tensor encoding sequence diagram illustrating RF and gradient waveforms. This sequence sensitizes the signal to diffusion in all directions equally, meaning the signal is weighted by the compartmental mean diffusivity. Here, the gradients have been numerically optimized to yield efficient spherical $b$-tensor encoding that is also compensated for concomitant gradient effects (Szczepankiewicz et al., 2019).

linear *b*-tensors commonly used in tractography, has been shown to improve microstructural modeling because it can disentangle microstructure from orientation dispersion (Szczepankiewicz et al., 2015), improves estimation of fiber orientation functions (Cottaar et al., 2020), and allows for disambiguation of the diffusion parameters of distinct compartments (Coelho et al., 2019; Reisert et al., 2019; Lampinen et al., 2020).

There are many other variants on diffusion encoding, and a comprehensive coverage of these is beyond the scope of this book.

# 3 Terminology guide

| | |
|---|---|
| $B_0$ | The strength of the main magnetic field of an MRI scanner. It is aligned along the long axis of the bore of the magnet ($\hat{z}$ orientation). |
| *b*-matrix/*b*-tensor | A $3 \times 3$ matrix that quantifies the magnitude and direction(s) of diffusion weighting. |
| *b*-value | A parameter that quantifies the magnitude of diffusion weighting. When there is no diffusion weighting, $b = 0\,\text{s/mm}^2$, and the image is sometimes called the $b = 0$ or $b_0$ image. |
| DW, DWI | Diffusion weighting, diffusion weighted imaging, or a diffusion weighted image. |
| gradient | MRI scanners are equipped with gradient coils that linearly modulate the main field strength $B_0$ in the $\hat{x}, \hat{y},$ and $\hat{z}$ directions. |
| HARDI | High angular resolution diffusion imaging. A diffusion imaging protocol wherein multiple DWIs are acquired, each with linear *b*-tensor encoding along a different direction. Typically done with more than 30 diffusion encoding directions and $b > 1000\,\text{s/mm}^2$. |
| k-space | Term for the acquisition spatial frequency Fourier space in MRI. The vector $k$ is the integral of the position encoding gradient over time. There is a Fourier relationship between the k-space representation of an image and the image itself. |
| PFG | Pulsed field gradient. Term for a diffusion encoding scheme with two diffusion encoding gradient lobes. Typically performed with a 180 degree RF pulse to create a spin echo, and alternately called a Stejskal-Tanner or pulsed gradient spin echo (PGSE) encoding scheme. |
| PGSE | Pulsed gradient spin echo. Term for a diffusion encoding scheme with two monopolar unidirectional diffusion encoding gradients separated by a 180 degree RF pulse. Alternately called a Stejskal-Tanner or pulsed field gradient (PFG) encoding scheme. |
| q-space | Term for the diffusion encoding Fourier space in MRI. The vector $q$ is the integral of the diffusion encoding gradient over time. There is a Fourier relationship between the q-space representation of the signal and the probability the spins displace by vector $r$ over the course of the diffusion experiment. |
| $t_d$ | The diffusion time. The effective time over which the diffusion phenomenon is observed. |
| RF | Radio frequency. Applied RF radiation is used to create MRI signal and contrast, and the signal we measure is in the RF band. |
| shim | The process or hardware used to make the MRI magnetic field homogeneous. |
| slew rate | The rate of change of the magnetic field gradient in MRI. |
| SNR | Signal-to-noise ratio. |
| spin | Used as a noun, a term for an MR visible species or magnetic dipole thereof (informal). MR visible nuclear species have nonzero spin angular momentum, i.e., an odd number of protons and/or neutrons. |
| STEAM | Stimulated echo acquisition mode. In diffusion MRI, a way to store magnetization longitudinally in order to shorten TE and/or lengthen the diffusion time such that the diffusion time can be longer relative to TE. |
| $T_1$ | The longitudinal relaxation time. |
| $T_2$ | The transverse relaxation time. |
| TE | The echo time. |
| tractometry | The technique of assessing scalar quantities along reconstructed streamlines from tractography. |
| TRSE | Twice-refocused spin echo, a technique for reducing eddy current-induced distortion. |

# References

Aganj, I., Lenglet, C., Sapiro, G., Yacoub, E., Ugurbil, K., Harel, N., 2010. Reconstruction of the orientation distribution function in single- and multiple-shell q-ball imaging within constant solid angle. Magn. Reson. Med. 64 (2), 554–566.

Ahlgren, A., et al., 2016. Quantification of microcirculatory parameters by joint analysis of flow-compensated and non-flow-compensated intravoxel incoherent motion (IVIM) data. NMR Biomed. 29 (5), 640–649.

Alexander, D.C., Barker, G.J., 2005. Optimal imaging parameters for fiber-orientation estimation in diffusion MRI. NeuroImage 27 (2), 357–367.

Alexander, A.L., Tsuruda, J.S., Parker, D.L., 1997. Elimination of eddy current artifacts in diffusion-weighted echo-planar images: the use of bipolar gradients. Magn. Reson. Med. 38 (6), 1016–1021.

Aliotta, E., Wu, H.H., Ennis, D.B., 2017. Convex optimized diffusion encoding (CODE) gradient waveforms for minimum echo time and bulk motion–compensated diffusion-weighted MRI. Magn. Reson. Med. 77 (2), 717–729.

Aliotta, E., Moulin, K., Ennis, D.B., 2018. Eddy current-nulled convex optimized diffusion encoding (EN-CODE) for distortion-free diffusion tensor imaging with short echo times. Magn. Reson. Med. 79 (2), 663–672.

Andersson, J.L.R., Sotiropoulos, S.N., 2016. An integrated approach to correction for off-resonance effects and subject movement in diffusion MR imaging. NeuroImage 125, 1063–1078.

Andersson, J.L., Skare, S., Ashburner, J., 2003. How to correct susceptibility distortions in spin-echo echo-planar images: application to diffusion tensor imaging. NeuroImage 20 (2), 870–888.

Andersson, J.L.R., Graham, M.S., Drobnjak, I., Zhang, H., Filippini, N., Bastiani, M., 2017. Towards a comprehensive framework for movement and distortion correction of diffusion MR images: within volume movement. NeuroImage 152, 450–466.

Arbabi, A., Kai, J., Khan, A.R., Baron, C.A., 2020. Diffusion dispersion imaging: mapping oscillating gradient spin-echo frequency dependence in the human brain. Magn. Reson. Med. 83 (6), 2197–2208.

Assaf, Y., Basser, P.J., 2005. Composite hindered and restricted model of diffusion (CHARMED) MR imaging of the human brain. NeuroImage 27 (1), 48–58.

Bak, M., Nielsen, N.C., 1997. Repulsion, a novel approach to efficient powder averaging in solid-state NMR. J. Magn. Reson. 125 (1), 132–139.

Bammer, R., et al., 2003. Analysis and generalized correction of the effect of spatial gradient field distortions in diffusion-weighted imaging. Magn. Reson. Med. 50 (3), 560–569.

Baron, C.A., Beaulieu, C., 2013. Oscillating Gradient Spin-Echo (OGSE) diffusion tensor imaging of the human brain. Magn. Reson. Med. 72, 726–736.

Baron, C.A., Lebel, R.M., Wilman, A.H., Beaulieu, C., 2012. The effect of concomitant gradient fields on diffusion tensor imaging. Magn. Reson. Med. 68 (4), 1190–1201.

Benner, T., Van Der Kouwe, A.J.W., Kirsch, J.E., Sorensen, A.G., 2006. Real-time RF pulse adjustment for Bo drift correction. Magn. Reson. Med. 56 (1), 204–209.

Berl, M.M., et al., 2015. Investigation of vibration-induced artifact in clinical diffusion-weighted imaging of pediatric subjects. Hum. Brain Mapp. 36 (12), 4745–4757.

Bernstein, M.A., King, K.F., Zhou, X.J., 2004. Part III: gradients. In: Handbook of MRI Pulse Sequences. Elsevier.

Callaghan, P.T., 2010. Physics of diffusion. In: Jones, D.K. (Ed.), Diffusion MRI: Theory, Methods, and Applications. Oxford University Press.

Campbell-Washburn, A.E., et al., 2019. Opportunities in interventional and diagnostic imaging by using high-performance low-field-strength MRI. Radiology 293 (2), 384–393.

Caruyer, E., Lenglet, C., Sapiro, G., Deriche, R., 2013. Design of multishell sampling schemes with uniform coverage in diffusion MRI. Magn. Reson. Med. 69 (6), 1534–1540.

Chamberland, M., Tax, C.M.W., Jones, D.K., 2018. Meyer's loop tractography for image-guided surgery depends on imaging protocol and hardware. NeuroImage Clin. 20, 458–465.

Coelho, S., Pozo, J.M., Jespersen, S.N., Jones, D.K., Frangi, A.F., 2019. Resolving degeneracy in diffusion MRI biophysical model parameter estimation using double diffusion encoding. Magn. Reson. Med. 82 (1), 395–410.

Cohen-Adad, J., Wald, L.L., 2014. Array coils. In: Quantitative MRI of the Spinal Cord. Academic Press, pp. 59–67.

Cory, D.G., Garroway, A.N., 1990. Measurement of translational displacement probabilities by NMR: an indicator of compartmentation. Magn. Reson. Med. 14 (3), 435–444.

Cottaar, M., et al., 2020. Improved fibre dispersion estimation using b-tensor encoding. NeuroImage 215, 116832.

Descoteaux, M., Deriche, R., Le Bihan, D., Mangin, J.F., Poupon, C., 2011. Multiple q-shell diffusion propagator imaging. Med. Image Anal. 15 (4), 603–621.

Does, M.D., Parsons, E.C., Gore, J.C., 2003. Oscillating gradient measurements of water diffusion in normal and globally ischemic rat brain. Magn. Reson. Med. 49 (2), 206–215.

Dubois, J., Poupon, C., Lethimonnier, F., Le Bihan, D., 2006. Optimized diffusion gradient orientation schemes for corrupted clinical DTI data sets. Magn. Reson. Mater. Phys. Biol. Med. 19 (3), 134–143.

Feinberg, D.A., et al., 2023. Next-generation MRI scanner designed for ultra-high-resolution human brain imaging at 7 Tesla. Nat. Methods 20, 2048–2057.

Fieremans, E., Burcaw, L.M., Lee, H.-H., Lemberskiy, G., Veraart, J., Novikov, D.S., 2016. In vivo observation and biophysical interpretation of time-dependent diffusion in human white matter. NeuroImage 129, 414–427.

Finsterbusch, J., 2009. Eddy-current compensated diffusion weighting with a single refocusing RF pulse. Magn. Reson. Med. 61 (3), 748–754.

Finsterbusch, J., 2010. Double-spin-echo diffusion weighting with a modified eddy current adjustment. Magn. Reson. Imaging 28 (3), 434–440.

Foo, T.K.F., et al., 2020. Highly efficient head-only magnetic field insert gradient coil for achieving simultaneous high gradient amplitude and slew rate at 3.0T (MAGNUS) for brain microstructure imaging. Magn. Reson. Med. 83 (6), 2356–2369.

Frank, L.R., 2002. Characterization of anisotropy in high angular resolution diffusion-weighted MRI. Magn. Reson. Med. 47 (6), 1083–1099.

Frass-Kriegl, R., et al., 2018. Flexible 23-channel coil array for high-resolution magnetic resonance imaging at 3 Tesla. PLoS One 13 (11), e0206963.

Gallichan, D., 2018. Diffusion MRI of the human brain at ultra-high field (UHF): a review. NeuroImage 168, 172–180.

Gallichan, D., Scholz, J., Bartsch, A., Behrens, T.E., Robson, M.D., Miller, K.L., 2010. Addressing a systematic vibration artifact in diffusion-weighted MRI. Hum. Brain Mapp. 31 (2), 193–202.

Genc, S., Tax, C.M.W., Raven, E.P., Chamberland, M., Parker, G.D., Jones, D.K., 2020. Impact of b-value on estimates of apparent fibre density. Hum. Brain Mapp. 41 (10), 2583–2595.

Girard, G., Fick, R., Descoteaux, M., Deriche, R., Wassermann, D., 2015. AxTract: microstructure-driven tractography based on the ensemble average propagator. Inf. Process. Med. Imaging 24, 675–686.

Graham, M.S., Drobnjak, I., Zhang, H., 2016. Realistic simulation of artefacts in diffusion MRI for validating post-processing correction techniques. NeuroImage 125, 1079–1094.

Gross, B., Kosfeld, R., 1969. Anwendung der spin-echo-methode dermessung der selbstdiffusion. Messtechnik 77, 171–177.

Gruber, B., Froeling, M., Leiner, T., Klomp, D.W.J., 2018. RF coils: a practical guide for nonphysicists. J. Magn. Reson. Imaging 48 (3), 590–604.

Haacke, E.M., Brown, R.W., Thompson, M.R., Venkatesan, R., 1999. Magnetic Resonance Imaging. John Wiley and Sons Inc.

Haase, A., et al., 1986. MR imaging using stimulated echoes (STEAM). Radiology 160 (3), 787–790.

Habib, J., Auer, D.P., Morgan, P.S., 2010. A quantitative analysis of the benefits of cardiac gating in practical diffusion tensor imaging of the brain. Magn. Reson. Med. 63 (4), 1098–1103.

Hetherington, H.P., Chu, W.-J., Gonen, O., Pan, J.W., 2006. Robust fully automated shimming of the human brain for high-field 1H spectroscopic imaging. Magn. Reson. Med. 56 (1), 26–33.

Hidalgo-Tobon, S.S., 2010. Theory of gradient coil design methods for magnetic resonance imaging. Concepts Magn. Reson. A 36A (4), 223–242.

Hiltunen, J., Hari, R., Jousmäki, V., Müller, K., Sepponen, R., Joensuu, R., 2006. Quantification of mechanical vibration during diffusion tensor imaging at 3 T. NeuroImage 32 (1), 93–103.

Huang, S.Y., Witzel, T., Keil, B., Scholz, A., Davids, M., Dietz, P., Rummert, E., Ramb, R., Kirsch, J.E., Yendiki, A., Fan, Q., Tian, Q., Ramos-Llordén, G., Lee, H.H., Nummenmaa, A., Bilgic, B., Setsompop, K., Wang, F., Avram, A.V., Komlosh, M., Benjamini, D., Magdoom, K.N., Pathak, S., Schneider, W., Novikov, D.S., Fieremans, E., Tounekti, S., Mekkaoui, C., Augustinack, J., Berger, D., Shapson-Coe, A., Lichtman, J., Basser, P.J., Wald, L.L., Rosen, B.R., 2021. Connectome 2.0: developing the next-generation ultra-high gradient strength human MRI scanner for bridging studies of the micro-, meso- and macro-connectome. Neuroimage 243, 118530.

Hutter, J., et al., 2018a. Time-efficient and flexible design of optimized multishell HARDI diffusion. Magn. Reson. Med. 79 (3), 1276–1292.

Hutter, J., et al., 2018b. Highly efficient diffusion MRI by slice-interleaved free-waveform imaging (SIFI). In: Proceedings of the 26th International Society of Magnetic Resonance in Medicine.

Jelescu, I.O., Veraart, J., Fieremans, E., Novikov, D.S., 2016. Degeneracy in model parameter estimation for multi-compartmental diffusion in neuronal tissue. NMR Biomed. 29 (1), 33–47.

Jeurissen, B., Leemans, A., Tournier, J.D., Jones, D.K., Sijbers, J., 2013. Investigating the prevalence of complex fiber configurations in white matter tissue with diffusion magnetic resonance imaging. Hum. Brain Mapp. 34 (11), 2747–2766.

Jeurissen, B., Tournier, J.-D.D., Dhollander, T., Connelly, A., Sijbers, J., 2014. Multi-tissue constrained spherical deconvolution for improved analysis of multi-shell diffusion MRI data. NeuroImage 103, 411–426.

Jezzard, P., Barnett, A.S., Pierpaoli, C., 1998. Characterization of and correction for eddy current artifacts in echo planar diffusion imaging. Magn. Reson. Med. 39 (5), 801–812.

Jones, D.K., Horsfield, M.A., Simmons, A., 1999. Optimal strategies for measuring diffusion in anisotropic systems by magnetic resonance imaging. Magn. Reson. Med. 42 (3), 515–525.

Jovicich, J., et al., 2006. Reliability in multi-site structural MRI studies: effects of gradient non-linearity correction on phantom and human data. NeuroImage 30 (2), 436–443.

Kamiya, K., et al., 2017. Diffusion imaging of reversible and irreversible microstructural changes within the corticospinal tract in idiopathic normal pressure hydrocephalus. NeuroImage Clin. 14, 663–671.

Kiselev, V.G., 2010. The cumulant expansion: an overarching mathematical framework for understanding diffusion NMR. In: Jones, D.K. (Ed.), Diffusion MRI: Theory, Methods, and Applications. Oxford University Press.

Ladd, M.E., et al., 2018. Pros and cons of ultra-high-field MRI/MRS for human application. Prog. Nucl. Magn. Reson. Spectrosc. 109, 1–50. Elsevier B.V.

Lampinen, B., et al., 2020. Towards unconstrained compartment modeling in white matter using diffusion-relaxation MRI with tensor-valued diffusion encoding. Magn. Reson. Med. 84 (3), 1605–1623.

Lasič, S., Szczepankiewicz, F., Eriksson, S., Nilsson, M., Topgaard, D., 2014. Microanisotropy imaging: quantification of microscopic diffusion anisotropy and orientational order parameter by diffusion MRI with magic-angle spinning of the q-vector. Front. Phys. 2.

Lasič, S., et al., 2020. Motion-compensated b-tensor encoding for in vivo cardiac diffusion-weighted imaging. NMR Biomed. 33 (2).

Le Bihan, D., Breton, E., Lallemand, D., Grenier, P., Cabanis, E., Laval-Jeantet, M., 1986. MR imaging of intravoxel incoherent motions: application to diffusion and perfusion in neurologic disorders. Radiology 161 (2), 401–407.

Mani, M., Aggarwal, H.K., Magnotta, V., Jacob, M., 2020. Improved MUSSELS reconstruction for high-resolution multi-shot diffusion weighted imaging. Magn. Reson. Med. 83 (6), 2253–2263.

**120 PART | II** Diffusion MRI

Mattiello, J., Basser, P.J., Le Bihan, D., 1997. The b matrix in diffusion tensor echo-planar imaging. Magn. Reson. Med. 37 (2), 292–300.

McNab, J.A., et al., 2013. The Human Connectome Project and beyond: initial applications of 300 mT/m gradients. NeuroImage 80, 234–245.

Meier, C., Zwanger, M., Feiweier, T., Porter, D., 2008. Concomitant field terms for asymmetric gradient coils: consequences for diffusion, flow, and echo-planar imaging. Magn. Reson. Med. 60 (1), 128–134.

Mesri, H.Y., David, S., Viergever, M.A., Leemans, A., 2020. The adverse effect of gradient nonlinearities on diffusion MRI: from voxels to group studies. NeuroImage 205.

Mori, S., Van Zijl, P.C.M., 1995. Diffusion weighting by the trace of the diffusion tensor within a single scan. Magn. Reson. Med. 33 (1), 41–52.

Movahedian Attar, F., et al., 2020. Mapping short association fibers in the early cortical visual processing stream using in vivo diffusion tractography. Cereb. Cortex 30, 4496–4514.

Nilsson, M., Lasič, S., Drobnjak, I., Topgaard, D., Westin, C.-F., 2017. Resolution limit of cylinder diameter estimation by diffusion MRI: the impact of gradient waveform and orientation dispersion. NMR Biomed. 30 (7), e3711.

Novikov, D.S., Fieremans, E., Jespersen, S.N., Kiselev, V.G., 2019. Quantifying brain microstructure with diffusion MRI: theory and parameter estimation. NMR Biomed. 32 (4). John Wiley and Sons Ltd.

Nunes, R.G., Jezzard, P., Clare, S., 2005. Investigations on the efficiency of cardiac-gated methods for the acquisition of diffusion-weighted images. J. Magn. Reson. 177 (1), 102–110.

Papadakis, N.G., Murrills, C.D., Hall, L.D., Huang, C.L.H., Adrian Carpenter, T., 2000. Minimal gradient encoding for robust estimation of diffusion anisotropy. Magn. Reson. Imaging 18 (6), 671–679.

Raffelt, D., et al., 2012. Apparent Fibre Density: a novel measure for the analysis of diffusion-weighted magnetic resonance images. NeuroImage 59 (4), 3976–3994.

Reese, T.G., Heid, O., Weisskoff, R.M., Wedeen, V.J., 2003. Reduction of eddy-current-induced distortion in diffusion MRI using a twice-refocused spin echo. Magn. Reson. Med. 49 (1), 177–182.

Reisert, M., Kiselev, V.G., Dhital, B., 2019. A unique analytical solution of the white matter standard model using linear and planar encodings. Magn. Reson. Med. (September 2018), 1–7.

Rösch, J., et al., 2016. Quiet diffusion-weighted head scanning: initial clinical evaluation in ischemic stroke patients at 1.5T. J. Magn. Reson. Imaging 44 (5), 1238–1243.

Scherrer, B., Warfield, S.K., 2012. Parametric representation of multiple white matter fascicles from cube and sphere diffusion MRI. PLoS One 7 (11), e48232.

Setsompop, K., et al., 2013. Pushing the limits of in vivo diffusion MRI for the Human Connectome Project. NeuroImage 80, 220–233.

Sjölund, J., Szczepankiewicz, F., Nilsson, M., Topgaard, D., Westin, C.-F.F., Knutsson, H., 2015. Constrained optimization of gradient waveforms for generalized diffusion encoding. J. Magn. Reson. 261, 157–168.

Skare, S., Andersson, J.L.R., 2001. On the effects of gating in diffusion imaging of the brain using single shot EPI. Magn. Reson. Imaging 19 (8), 1125–1128.

Steidle, G., Schick, F., 2006. Echoplanar diffusion tensor imaging of the lower leg musculature using eddy current nulled stimulated echo preparation. Magn. Reson. Med. 55 (3), 541–548.

Stejskal, E.O., Tanner, J.E., 1965. Spin diffusion measurements: spin echoes in the presence of a time-dependent field gradient. J. Chem. Phys. 42 (1), 288–292.

Stockmann, J.P., et al., 2016. A 32-channel combined RF and B0 shim array for 3T brain imaging. Magn. Reson. Med. 75 (1), 441–451.

Szczepankiewicz, F., et al., 2015. Quantification of microscopic diffusion anisotropy disentangles effects of orientation dispersion from microstructure: applications in healthy volunteers and in brain tumors. NeuroImage 104, 241–252.

Szczepankiewicz, F., Westin, C., Nilsson, M., 2019. Maxwell-compensated design of asymmetric gradient waveforms for tensor-valued diffusion encoding. Magn. Reson. Med. 82 (4), 1424–1437.

Szczepankiewicz, F., Eichner, C., Anwander, A., Westin, C.-F., Paquette, M., 2020a. The impact of gradient non-linearity on Maxwell compensation when using asymmetric gradient waveforms for tensor-valued diffusion encoding. In: Proceedings of the 28th International Society of Magnetic Resonance in Medicine.

Szczepankiewicz, F., Westin, C.-F., Nilsson, M., 2020b. Gradient waveform design for tensor-valued encoding in diffusion MRI. J. Neurosci. Methods, 109007.

Szczepankiewicz, F., et al., 2020c. Motion-compensated gradient waveforms for tensor-valued diffusion encoding by constrained numerical optimization. Magn. Reson. Med. 85, 2117–2126.

Tan, E.T., Marinelli, L., Slavens, Z.W., King, K.F., Hardy, C.J., 2013. Improved correction for gradient nonlinearity effects in diffusion-weighted imaging. J. Magn. Reson. Imaging 38 (2), 448–453.

Topfer, R., et al., 2016. A 24-channel shim array for the human spinal cord: design, evaluation, and application. Magn. Reson. Med. 76 (5), 1604–1611.

Topfer, R., Foias, A., Stikov, N., Cohen-Adad, J., 2018. Real-time correction of respiration-induced distortions in the human spinal cord using a 24-channel shim array. Magn. Reson. Med. 80 (3), 935–946.

Tournier, J.D., Calamante, F., Connelly, A., 2013. Determination of the appropriate b value and number of gradient directions for high-angular-resolution diffusion-weighted imaging. NMR Biomed. 26 (12), 1775–1786.

Tournier, J., et al., 2020. A data-driven approach to optimising the encoding for multi-shell diffusion MRI with application to neonatal imaging. NMR Biomed. 33 (9).

Tuch, D.S., Reese, T.G., Wiegell, M.R., Makris, N., Belliveau, J.W., Van Wedeen, J., 2002. High angular resolution diffusion imaging reveals intravoxel white matter fiber heterogeneity. Magn. Reson. Med. 48 (4), 577–582.

Van Gelderen, P., De Zwart, J.A., Starewicz, P., Hinks, R.S., Duyn, J.H., 2007. Real-time shimming to compensate for respiration-induced B0 fluctuations. Magn. Reson. Med. 57 (2), 362–368.

Van Vaals, J.J., Bergman, A.H., 1990. Optimization of eddy-current compensation. J. Magn. Reson. 90 (1), 52–70.

Veraart, J., Fieremans, E., Novikov, D.S., 2019. On the scaling behavior of water diffusion in human brain white matter. NeuroImage 185, 379–387.

Verma, T., Cohen-Adad, J., 2014. Effect of respiration on the B0 field in the human spinal cord at 3T. Magn. Reson. Med. 72 (6), 1629–1636.

Vos, S.B., et al., 2016. Trade-off between angular and spatial resolutions in in vivo fiber tractography. NeuroImage 129, 117–132.

Vos, S.B., Tax, C.M.W., Luijten, P.R., Ourselin, S., Leemans, A., Froeling, M., 2017. The importance of correcting for signal drift in diffusion MRI. Magn. Reson. Med. 77 (1), 285–299.

Westin, C.-F., et al., 2016. Q-space trajectory imaging for multidimensional diffusion MRI of the human brain. NeuroImage 135, 345–362.

White, N.S., Dale, A.M., 2009. Optimal diffusion MRI acquisition for fiber orientation density estimation: an analytic approach. Hum. Brain Mapp. 30 (11), 3696–3703.

Wong, E.C., Cox, R.W., Song, A.W., 1995. Optimized isotropic diffusion weighting. Magn. Reson. Med. 34 (2), 139–143.

Wu, X., et al., 2018. High-resolution whole-brain diffusion MRI at 7T using radiofrequency parallel transmission. Magn. Reson. Med. 80 (5), 1857–1870.

Yang, G., McNab, J.A., 2019. Eddy current nulled constrained optimization of isotropic diffusion encoding gradient waveforms. Magn. Reson. Med. 81 (3), 1818–1832.

Zhang, H., Schneider, T., Wheeler-Kingshott, C.A., Alexander, D.C., 2012. NODDI: practical in vivo neurite orientation dispersion and density imaging of the human brain. NeuroImage 61 (4), 1000–1016.

Zhou, X.J., Tan, S.G., Bernstein, M.A., 1998. Artifacts induced by concomitant magnetic field in fast spin-echo imaging. Magn. Reson. Med. 40 (4), 582–591.

Zwiers, M.P., 2010. Patching cardiac and head motion artefacts in diffusion-weighted images. NeuroImage 53 (2), 565–575.

# Chapter 7

# Diffusion MRI acquisition for tractography: Diffusion sequences

Jennifer S.W. Campbell[a], Steven H. Baete[b], Julien Cohen-Adad[c], J.-Donald Tournier[d], Filip Szczepankiewicz[e,f], Christian Beaulieu[g], Corey A. Baron[h], Merry Mani[i], Kawin Setsompop[j], Congyu Liao[j], Christine L. Tardif[a], Sjoerd B. Vos[k], Anastasia Yendiki[l], Ilana R. Leppert[a], Els Fieremans[m], and G. Bruce Pike[n]

[a]McConnell Brain Imaging Centre, Montreal Neurological Institute, McGill University, Montreal, QC, Canada, [b]New York University Grossman School of Medicine, New York, NY, United States, [c]Polytechnique Montreal, Montreal, QC, Canada, [d]King's College London, London, United Kingdom, [e]Lund University, Lund, Sweden, [f]Brigham and Women's Hospital, Harvard Medical School, Boston, MA, United States, [g]University of Alberta, Edmonton, AB, Canada, [h]Robarts Research Institute, Western University, London, ON, Canada, [i]University of Iowa, Iowa City, IA, United States, [j]Stanford University, Stanford, CA, United States, [k]Centre for Microscopy, Characterisation, and Analysis, The University of Western Australia, Nedlands, WA, Australia, [l]Massachusetts General Hospital, Harvard Medical School, Boston, MA, United States, [m]Bernard and Irene Schwartz Center for Biomedical Imaging, Department of Radiology, New York University Grossman School of Medicine, New York, NY, United States, [n]Hotchkiss Brain Institute, University of Calgary, Calgary, AB, Canada

## 1 Introduction and a brief history

The first reports on the measurement of water diffusion properties with imaging in the human brain used "conventional" spin echo, spin-warp methods with monopolar Stejskal-Tanner diffusion-sensitizing gradients (Le Bihan et al., 1986; Thomsen et al., 1987). However, the acquisition times of these "multishot" methods were long, limiting clinical utility, and the image quality was poor because of severe motion artifacts, primarily due to intershot phase variations from the strong (for the time) and lengthy diffusion gradient pulses. Mansfield's previous development of echo planar imaging (EPI) (Mansfield, 1984, 1977), a.k.a. "snapshot" MRI, would be the key to reducing acquisition time and motion artifacts. The interest in EPI at the time was mainly in body applications with no diffusion weighting. The method was hampered by hardware limitations, e.g., gradient strengths (4 mT/m at that time).

In 1990, Turner and Le Bihan introduced the single-shot, spin echo diffusion weighted EPI (SS-SE-DW-EPI) sequence (Turner and Le Bihan, 1990), which has become the de facto standard for diffusion MRI to this day, due to its speed and insensitivity to motion. The brain was of primary interest, and single-shot EPI's first use in human brain showed good quality coronal images free of motion artifacts. It yielded reasonable quantitative measurements of diffusion coefficients, and demonstrated the known, but at the time not understood, property of anisotropic diffusion in human brain white matter (Turner et al., 1990), essential for tractography. Notably, single-shot EPI enabled the acquisition of multiple images at different $b$-values (here, $b = 0$ and 15 shells up to $b \sim 900\,s/mm^2$), demonstrating the signal attenuation curve, and multiple gradient-encoding directions (two orthogonal directions), demonstrating the anisotropy. The single-shot EPI pulse sequence is shown in Fig. 1 and is discussed in more detail in Section 2.

Thus diffusion weighted single-shot EPI appeared even early on to be quite promising for human brain studies (Stehling et al., 1991; Turner et al., 1991). The most important clinical demonstration of single-shot diffusion EPI was for the early detection of acute stroke in humans (5 years after the initial sequence!) (Warach et al., 1995), although the value of single-shot diffusion EPI for stroke had been demonstrated much earlier in animal models (Kucharczyk et al., 1991). Gradient performance, including deleterious eddy currents, was and still is a concern for EPI. It took until almost the turn of this millennium for the gradient hardware to be sufficiently robust for use on standard clinical scanners. Notably, two of the pioneering human brain tractography studies used single-shot diffusion EPI (Conturo et al., 1999; Jones et al., 1999); the third study used multishot diffusion EPI (Basser et al., 2000). Multishot techniques have continued to be used in diffusion studies where higher resolution, lower distortion, and shorter echo times are paramount, and modern implementations have increased the speed of such acquisitions. In this chapter, we will discuss the acquisition variants that can be used for tractography, including acquisition parameters for single-shot EPI, 2D multishot developments, and volumetric acquisitions. The choice of acquisition scheme is generally a trade-off between time, SNR, spatial resolution and coverage, and image quality.

Handbook of Diffusion MR Tractography. https://doi.org/10.1016/B978-0-12-818894-1.00019-7
Copyright © 2025 Elsevier Ltd. All rights are reserved, including those for text and data mining, AI training, and similar technologies.

A typical diffusion MRI acquisition for tractography will also include at minimum a higher resolution sequence (typically a standard $T_1$ weighted image, e.g., MP-RAGE) that shows basic anatomical structure, but not fiber orientation. Some studies also include other sequences that have complementary contrasts that can inform or enhance tractography (see Chapters 18 and 24). The tractographer should always ensure that an optimal, state-of-the-art acquisition is being used for their application, as research on this topic is constantly evolving.

## 2 Single-shot diffusion sequences

The single-shot EPI readout's speed and efficiency comes from the use of a train of bipolar gradient pulses that generate multiple gradient echoes, one for each line in a 2D slice of k-space, after one slice-selective RF excitation (see Fig. 1). In contrast, conventional spin-warp MRI methods acquire only one echo with a single readout line after each RF excitation. Thus EPI acquires all the data for a single slice very quickly, on the order of 100 ms (e.g., 120 ms from the first spoiler to the end of the readout in Fig. 1), which effectively eliminates intraslice motion-induced errors. In Fig. 1, the k-space traversal is achieved by navigating to the top of k-space first, acquiring one readout (RO) line ($k_x$ in this example), using a "blip" gradient in the phase encode (PE) direction ($k_y$ in this example) to move to the next phase encoding position, reading a second horizontal line in the opposite direction for the next echo, and so on in a back-and-forth pattern collinear with the Cartesian grid until the desired k-space is filled. The downside of single-shot imaging is that it limits the spatial resolution, given that the number of phase encode gradient oscillations one can do before the signal decays away due to $T_2^*$ is limited.

The acquisition is repeated for each slice for a given b-tensor (a linear b-tensor for tractography, i.e., a b-value and direction). The slices are usually interleaved to mitigate artifacts from imperfections in the RF pulse slice-selection profiles. In the time (TR) between acquisitions of a given slice, the signal recovers with $T_1$ recovery. The SNR-efficient regime for TR is the regime in which there is a good trade-off between the data acquisition duration for each slice and the wait time for sufficient signal recovery for subsequent acquisition of that slice (Engström et al., 2015); this corresponds to a TR of 1.5–3.5 s at 3 T. Given the number of slices required to cover the region of interest, and the time required to achieve diffusion encoding and read out the signal, the TR can easily be longer than this, with the spins in each slice spending the majority of their time sitting idle at full magnetization, rather than being imaged. Simultaneous multislice imaging has largely mitigated this problem (see Section 5) at standard slice thickness.

While preparing a diffusion imaging protocol, in addition to the number of directions and b shells, it is important to consider the quality of the image underlying each b/q-value point. The single-shot EPI-related options at one's discretion are typically: resolution, field-of-view, phase-encoding direction, slice orientation, interslice gap, readout bandwidth, and

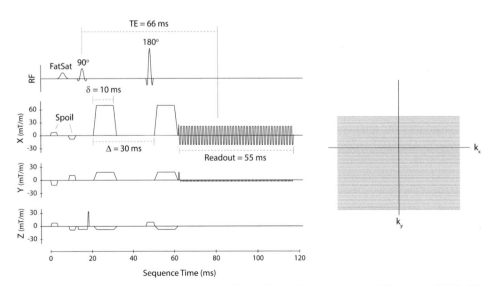

**FIG. 1** Single-shot, spin echo EPI diffusion pulse sequence. Sequence diagram for a typical sequence on a 3 T scanner with 80 mT/m gradients showing waveforms and timings for an FOV of 220 mm, 110 matrix, axial z-slice, $2 \times 2 \times 2$ mm$^3$ voxels, and $b = 1000$ s/mm$^2$. This sequence has a phase partial Fourier (PPF) single-shot k-space trajectory and no parallel imaging (PPF and parallel imaging are described in Section 3). The sequence (read from left to right) includes gradient spoilers for residual transverse magnetization, fat saturation, a 90 degree slice-selective excitation RF pulse, a first diffusion trapezoidal gradient pulse along three axes simultaneously ($x, y, z = 0.96, 0.25, -0.12$), a 180 degree slice-selective refocusing RF pulse, a second diffusion gradient identical to the first one yielding a diffusion time of ~30 ms, and a 6/8 phase partial Fourier blipped EPI readout train (82 trapezoids yielding 82 echoes with 110 points each, or 9020 points digitized over 55 ms). The k-space trajectory is shown at right. *(Figure credit: Robert Stobbe, University of Alberta.)*

acceleration methods to reduce the readout train length. This section will focus on all but the latter; nearly all single-shot EPI diffusion MRI experiments use some sort of $k$-space acceleration or reduction technique, and these will be described in the sections that follow this one. Other parameters, to be covered in more detail in Section 6, include the use of fat saturation (or selective water excitation) to minimize signal from skull fat.

Image resolution is one of the first decisions when setting up an imaging protocol, although it is always a balancing act between scan time (given the number of $b$-values and gradient directions desired), SNR (given the $b$-values desired), the anatomy to be studied, and the coverage needed (usually whole brain for tractography of the brain). Smaller voxels provide lower SNR proportional to volume. Many in vivo adult human brain studies have settled on 2 mm thick slices with $2 \times 2$ mm$^2$ nominal in-plane resolution, to yield "isotropic" voxels that minimize streamline bias in tractography. Clinical diffusion protocols often use thicker slices (3 or 5 mm) because they allow full brain coverage to be achieved in a shorter TR and hence shorter scan time. Alteration of the slice thickness has limited consequences on the pulse sequence timings other than total scan time for a given coverage. Recent technical advances have made the use of thinner slices (e.g., 1.5 mm, in which case full coverage of the cerebrum and cerebellum would take ~96 slices) more achievable in a shorter timeframe (see Section 5). Given a long scan time, such as that used by the Human Connectome Project, such techniques, namely simultaneous multislice (SMS, see Section 5), have enabled even higher resolution (1.25 mm) (Sotiropoulos et al., 2013a). Thin slices (1 mm) can also be achieved in a clinically relevant time by focusing on a specific region, e.g., the hippocampus (Treit et al., 2018). The in-plane resolution is determined by the field-of-view (FOV) and the matrix size; for example, an FOV of 220 mm with a 110 matrix would yield a nominal in-plane voxel size of 2 mm.

Fig. 2 demonstrates the effect of acquiring different voxel shapes and sizes. It shows the average DW images of the same individual for a thick slice of 5 mm with $1.5 \times 1.5$ mm$^2$ in-plane voxel size, and isotropic scans of 2 and 1.5 mm. These images were all acquired rapidly, and demonstrate that there is sufficient SNR at 3 T for a minimal diffusion tensor acquisition at $b = 1000$ s/mm$^2$. It is quite evident that the anatomical visualization is markedly improved on the images with thinner slices, and higher in-plane resolution coupled with thin slices is necessary to see certain details in three dimensions, for instance, the thin cortex and small structures. This is critical for making measurements in small brain regions or detecting focal injury, and for performing accurate tracking, particularly of smaller tracts, because there is less partial volume averaging with cerebrospinal fluid (CSF), gray matter, and neighboring tracts.

Many artifacts in single-shot EPI occur along the phase-encode direction, due to the slow traversal through $k$-space in that direction relative to the rapid readout direction. These include distortion from phase accrual due to $B_0$ field inhomogeneities (due to magnetic susceptibility, eddy currents, and Maxwell effects), and blurring due to $T_2^*$-induced signal amplitude variation during the readout. Most diffusion MRI brain studies acquire axial-oblique slices with phase-encoding along the anterior-posterior (AP) (or posterior-anterior (PA)) direction, because the brain has more left-right symmetry and left-right distortions can accidentally be interpreted as true effects (Fig. 3). It is common to acquire images with both the AP and PA phase encoding directions in order to correct, at least in part, for the susceptibility-induced distortions in postprocessing (Andersson et al., 2003) (see Chapter 9). Less commonly, the right to left and left to right directions can be acquired, enabling a shorter TE via a shorter echo train.

The length of the EPI readout train is determined by the speed of the acquisition of $k$-space points and time between $k$-space echoes. The length of the readout train is inversely proportional to the bandwidth (BW), provided a concomitant change in echo spacing occurs. Increasing the BW decreases the SNR: SNR $\propto 1/\sqrt{\text{BW}}$; however, the reduction in signal decay due to the shortened readout and earlier TE mitigates this SNR loss. Doubling the bandwidth halves the distortions and $T_2^*$ blurring, and common practice is to adjust the BW such that the echo spacing is minimal at field strengths of 3 T and higher. The difference between low and high BW is shown in Fig. 4. The higher BW reduces the echo train length by a factor of nearly two.

The pulse sequence timing must be adjusted such that the center of $k$-space is read at TE. The pulse sequence shown in Fig. 1 has a gap after the first diffusion gradient before the 180-degree refocusing pulse to ensure this occurs. The shorter the readout before the center of $k$-space, the smaller this gap will be, and the more efficient the sequence; $k$-space acceleration and hardware advances can help in this respect.

For more detail on single-shot EPI and its use in diffusion acquisition, please refer to the following EPI dedicated book (Schmitt et al., 1998) and diffusion MRI book chapters (Skare and Bammer, 2010; Pipe, 2014).

There are other, non-EPI, single-shot diffusion MRI approaches that may have efficiency and artifact advantages over EPI, but they are not widely available. Single-shot spiral-out trajectories can reduce distortion and provide higher SNR via a drastically reduced TE, because the acquisition begins at the center of $k$-space (Li et al., 1999). The primary disadvantage of this approach is that spiral trajectories are much more sensitive to eddy current-associated fields, whether they originate from the diffusion or imaging gradients. Recently, single-shot spiral diffusion image quality has been improved by accounting for off-resonance and field dynamics to correctly follow the actual $k$-space trajectory (Wilm et al., 2017),

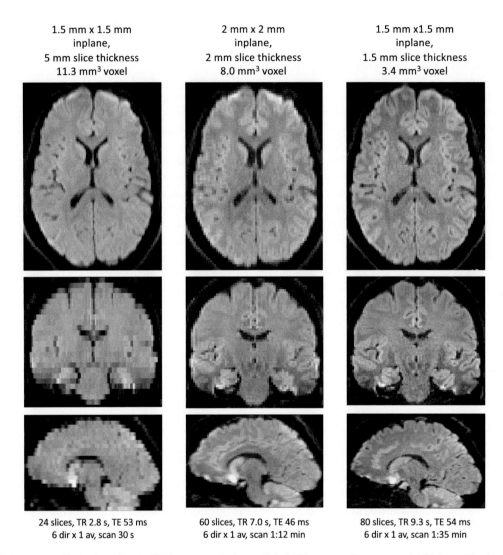

**FIG. 2** Resolution matters. Single-shot spin echo EPI images acquired on a clinical 3 T scanner with a commercial 64 channel head receiver array. For demonstration, data were acquired across six directions with $b = 1000\,\text{s/mm}^2$, one signal average, and different voxel sizes. Axial, coronal, and sagittal average DW images are shown. Other scan parameters (to be explained in following text and in Section 3) were: BW $\sim$1500 Hz/px, GRAPPA $R = 2$, and 6/8 PPF. The "prescan normalize" option has been selected to remove $B_1^-$ receiver field inhomogeneity.

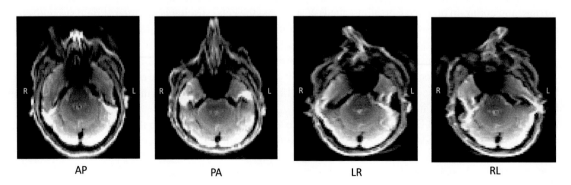

**FIG. 3** Susceptibility induced distortions based on phase-encoding direction. The slow traversal of $k$-space along the phase-encoding direction creates distortions that depend on the (signed) direction of the PE gradient. Shown here are the differences between the typical anterior-posterior or posterior-anterior and left-right/right-left phase encoding directions. The images were acquired at 3 T with a 64 channel head coil and 2.6 mm isotropic resolution, $b = 0\,\text{s/mm}^2$, TE $= 50$ ms, and TR $= 3.0$ s. Other scan parameters (explained in following text and in Sections 3 and 5) were: BW $= 2320$ Hz/px, GRAPPA $R = 2$, PPF $= 6/8$, and MB factor 3.

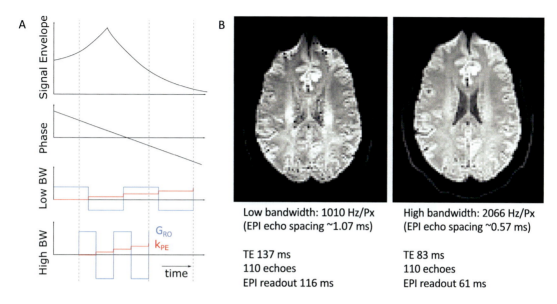

FIG. 4 Readout bandwidth. (A) An increased readout bandwidth is achieved by increasing the readout gradient amplitude ($G_{RO}$), and this shortens the readout duration. The shorter TE minimizes $T_2$ signal loss, which helps to offset the $1/\sqrt{BW}$ SNR signal loss. The shortened readout also reduces the $T_2^*$-induced signal variation throughout the readout, and the phase accrual due to $B_0$ inhomogeneity. (B) This results in decreased artifacts and increased SNR, shown on average DW images acquired at 3 T with a 64 channel head coil, $b = 1000\,\text{s/mm}^2$, and 2 mm isotropic voxels.

and by taking advantage of a strong, high slew rate head insert gradient system (Wilm et al., 2020), although in both cases the need for additional custom hardware limits widespread applicability. In addition to applications in the CNS, single-shot spiral imaging has been used for diffusion imaging of the heart (Gorodezky et al., 2018). GRASE (Liu et al., 1996), RARE/FSE (Alsop, 1997; Schick, 1997), and low flip angle GRE methods (Nolte et al., 2000; Merrem et al., 2017) have also been proposed as single-shot readout options for diffusion imaging.

## 3 Reduced sampling schemes

Contrary to many other MRI acquisitions, the primary motivation for reduced in-plane sampling schemes in dMRI is not to reduce the total scan duration. Rather, much of the required time is already dedicated to generating diffusion contrast, and the primary benefits of reduced sampling by parallel imaging, as with sampling faster by increasing the readout bandwidth, are reduced blurring and distortions (Section 3.1). In the special case of partial Fourier acquisitions, the primary benefit is increased SNR via decreased TE (Section 3.2). This section covers reduced sampling by collecting fewer k-space data points.

### 3.1 Parallel imaging

As with bandwidth increases, in-plane parallel imaging accelerates traversal through $k$-space. Here, the acceleration is enabled by skipping entire lines of $k$-space that are filled in during image reconstruction by exploiting redundancy between and spatially varying sensitivity of the multiple receiver channels that all record the MRI signal simultaneously (Fig. 5A). As the parallel imaging acceleration rate increases, both the TE and the level of distortions decrease (Fig. 5B). The details of specific algorithms are beyond the scope of this chapter but, nevertheless, the most widely used algorithms to implement parallel imaging reconstruction are based on data manipulations in the image-domain (SENSE, (Pruessmann et al., 1999)) and k-space (SMASH or GRAPPA, (Sodickson and Manning, 1997; Griswold et al., 2002)). Notably, both approaches have generally comparable performance. Similar to bandwidth increases, these distortion/blurring reductions are bought with the currency of SNR: SNR $\propto 1/\sqrt{R}\, g$, where $R$ is the reduction rate and $g$ is the so-called coil geometry "$g$-factor" (Pruessmann et al., 1999) (Fig. 5C). The $g$-factor SNR losses are mitigated with increasing numbers of receiver coils, and practically limit acceleration factors to a factor of ~2–3 along each dimension. SMS, to be described in Section 5, is also a parallel imaging technique, but it is parallel in the slice direction, and its primary purpose is to reduce scan time.

These EPI-based parallel imaging techniques are available as standard options on all modern clinical systems. However, there are emerging techniques that may offer the ability to accelerate image acquisition further. Parallel imaging

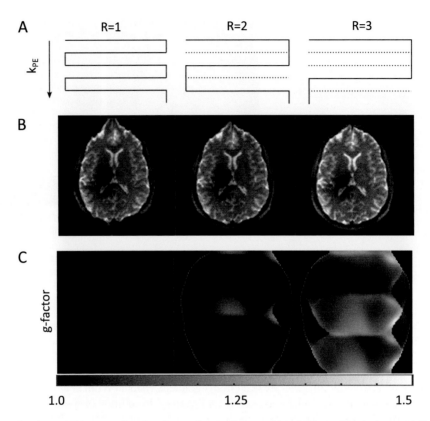

**FIG. 5** As the parallel acceleration rate $R$ increases from 1 to 3, more $k$-space lines are skipped (A), resulting in decreased distortions (B) with a concomitant spatially dependent amplification of noise that is quantified by the $g$-factor (C). Images acquired at 7 T with posterior-anterior phase encoding, $b = 0$ s/mm$^2$, 2 mm isotropic voxels, and maximal BW. GRAPPA reconstruction was employed.

undersampling in single-shot spiral trajectories occurs along multiple directions, which reduces g-factor losses compared to EPI undersampling along a single direction. As described in Section 2, recent advances have increased the promise of spiral readout trajectories.

### 3.2 Partial Fourier acquisition

Real-valued images have the property that their k-space is Hermitian symmetric (Fig. 6A). Accordingly, in this ideal situation, only half of the $k$-space would need to be acquired. However, in practice MRI images are complex valued due to phase accrual from $B_0$ inhomogeneity. Luckily, this phase generally varies slowly over the image, which means only the central region of $k$-space violates Hermitian symmetry. Thus it is possible to acquire a central symmetric region along with one side of k-space (Fig. 6B), and use Hermitian symmetry to fill in the opposing side of k-space. Naive application of Hermitian symmetry would leave an artifact-inducing sharp boundary between the acquired and filled-in data, but various algorithms have been introduced that mitigate this issue (e.g., homodyne filtering (Noll et al., 1991) and projection onto convex sets (Xu and Mark Haacke, 2001)). The trade-off for PPF imaging is that excessively aggressive undersampling leads to blurring and image ringing near sharp boundaries, and artifacts in regions with rapidly varying phase. Typically, partial Fourier fractions of lower than 3/4 (i.e., the first 1/4 of $k$-space is not acquired; usually denoted 6/8 by MRI scanner manufacturers) are not recommended for dMRI.

Unlike parallel imaging and increasing the readout bandwidth, partial Fourier imaging does not affect the rate of traversal through $k$-space, which causes it to have little impact on $B_0$ distortions. It does, however, significantly reduce the TE by twice the readout time reduction (Fig. 6C), which leads to a gain in SNR. The SNR gain is offset by some SNR loss because fewer data samples are acquired, but the TE reductions generally outweigh these losses (Fig. 6D). In Fig. 6E and the pulse sequence illustrated in Fig. 1, the $k$-space readout is a 6/8 PPF readout, where 82 of 110 lines of $k$-space are read, 27 before the center of $k$-space (which corresponds to TE), and 55 after. The gap after the first diffusion sensitizing gradient in Fig. 1 would have been even longer if this were not the case.

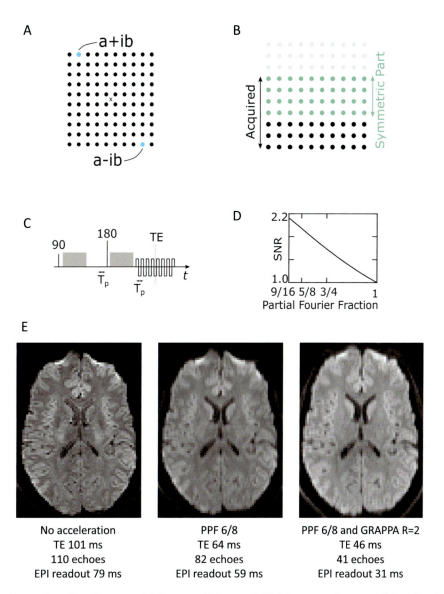

FIG. 6 (A) Real-valued images have Hermitian symmetric k-space, which means half of k-space can be generated from the other half by flipping the image and changing the sign on the imaginary part. (B) In practice, MRI images have slowly varying phase that can be captured from a symmetrically acquired portion of k-space, shown in *green*. (C) Partial Fourier encoding allows removal of lines of k-space that have total $T_p$ duration, which causes a $\sim 2T_p$ reduction in TE due to timing requirements of the 180-degree RF pulse. (D) Reductions in TE enabled by partial Fourier encoding result in greatly increased SNR (simulation for $T_2 = 50$ ms). (E) Average DW images for 6 diffusion encoding directions for (from left to right) full k-space coverage, 6/8 PPF, and 6/8 PPF combined with GRAPPA $R = 2$. With 6/8 PPF, SNR is enhanced; however, the PPF image is blurred in the PE direction. EPI/DTI processed on the scanner with no filtering except for prescan normalize to remove $B_1^-$ inhomogeneity. The images were acquired at 3 T with a 64 channel head coil, $b = 1000$ s/mm$^2$, 2 mm isotropic voxels, 1 signal average, and BW 1516 Hz/px.

PPF and parallel imaging are typically used together in dMRI, affording a very short TE (Fig. 6E). While the primary goal of the combination is enhanced image quality, the shortened TE allows for a shortened TR and hence overall scan time, meaning more diffusion encoding directions and b-shells can be acquired per unit time.

## 4 Segmented diffusion sequences

Segmented diffusion sequences provide a way to increase SNR and reduce distortion and blurring, at the expense of total scan time. In this approach, the typical long single-shot EPI readout is split into multiple segments, acquired with multiple excitations. As with in-plane parallel imaging, this reduces the TE and increases the speed of traversal of the phase encode direction of k-space. The reduction in k-space coverage per segment can be greater than with parallel imaging, because the missing data are acquired with another excitation to form a fully sampled k-space. Moreover, SNR is not lost by shortening the readout duration, because the total time for the readout of all segments is still long. The segmented strategy is especially important for in vivo high-resolution diffusion scans (Holdsworth et al., 2008) and in vivo diffusion studies at ultrahigh field strength (Heidemann et al., 2010), and is critical for high-resolution ex vivo diffusion studies (Miller et al., 2011) (Chapter 8). Segmented readouts can also be implemented in conjunction with partial Fourier imaging.

In the case of high-resolution studies, the $k$-space is sampled over a larger matrix size. Due to the increased number of phase encodings, a typical single-shot EPI readout will be longer, with a longer TE. This means that the SNR loss due to $T_2$ decay, geometric distortions, and $T_2^*$ induced blurring is even more severe than with standard resolution. Segmented EPI can reduce these effects by a factor proportional to the number of segments employed in such acquisitions. Similarly, in ultrahigh field diffusion MRI (e.g., 7 T MRI), the $T_2$ and $T_2^*$ relaxation times of the tissue are shortened and susceptibility-induced distortions are enhanced, meaning the distortion and blurring are exacerbated even at TEs typical of low-resolution diffusion MRI. Hence, a shorter TE than is typically achieved in 3 T MRI is required for good quality diffusion imaging at 7 T. Using segmented sequences, the higher SNR available at 7 T can be harnessed to achieve high-quality, high-resolution diffusion scans.

The EPI readout can be segmented in several ways to achieve the TE reduction. Two typical variations are: (i) those that segment the EPI trajectory along the phase encoding direction (referred to as interleaved EPI), and (ii) those that segment the EPI trajectory along the readout encoding direction (referred to as readout segmented EPI) (Holdsworth et al., 2019; Wu and Miller, 2017) (see Fig. 7). Interleaved EPI affords a higher reduction in TE for a given number of segments compared to readout segmented EPI (Wang et al., 2018a). Variations of these methods using rotated EPI segments (e.g., short-axis propeller EPI) or using non-Cartesian readouts (e.g., multishot spirals) have also been shown to be of great utility. Although segmented EPI reduces the TE per TR, such methods require multiple TRs to fully sample a given k-space (per diffusion weighting). Hence, these methods prolong scan time. Reduction in TR and total scan time can be achieved by combining these methods with advanced acquisition methods such as simultaneous multislice imaging (Mani et al., 2020a; Chang et al., 2015a; Herbst et al., 2017) (see Section 5).

Because of the multiple TRs required to fill in the $k$-space for a given diffusion weighting, in vivo segmented EPI-based diffusion imaging suffers from motion-related artifacts from various sources: (i) the head motion (rotations and translations) arising from bulk subject motion leads to k-space inconsistency, causing ghosting artifacts in the final image; and (ii) even in the absence of bulk subject motion, other involuntary motion such as CSF motion driven by respiration and cardiac pulsations are inevitable in in vivo imaging. Such involuntary motion is encoded by the large diffusion gradients as phase, meaning each segment will have a different phase. The inevitable intersegment phase inconsistency in segmented EPI scans prevents the $k$-space data from the different segments from being combined to form a single fully sampled $k$-space for standard MRI reconstruction. Specialized reconstructions that prevent the phase inconsistency from causing destructive phase interference are required to reconstruct the diffusion weighted images from segmented EPIs for in vivo studies.

Two approaches are commonly used to reconstruct DWIs from segmented acquisitions. The first approach employs a phase compensation method, where the phase of each segment is estimated and compensated in a second step. Traditionally, the phase estimation is performed using navigator echoes for both interleaved and readout segmented EPI (Porter and Heidemann, 2009). However, recent methods avoid the navigator echoes and estimate the phase from the $k$-space data of the individual segments themselves (Uecker et al., 2009; Chen et al., 2013). With phase estimation, a phase-compensated reconstruction can be formulated; composite sensitivity-based methods that combine the phase of each

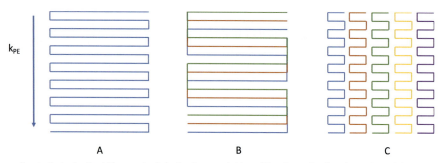

**FIG. 7** The $k$-space encoding trajectories for (A) a standard single-shot acquisition, (B) a three-shot interleaved acquisition, and (C) a five-shot readout segmented trajectory. The $k$-space trajectory of each shot is shown as a different color.

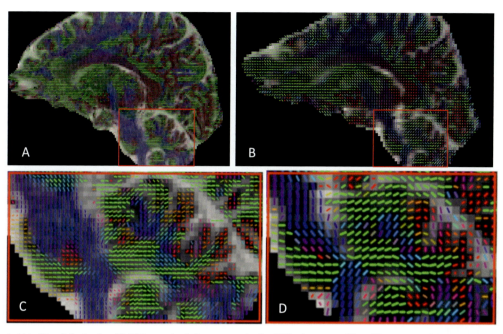

**FIG. 8** Segmented (A, C) and single-shot (B, D) human brain images acquired at 7 T in the same subject. The segmented dataset affords higher resolution (1.1 mm isotropic voxels) and less distortion than the single-shot dataset (2 mm isotropic voxels). Shown are the color-coded principal eigenvectors of the diffusion tensor for the entire brain as well as a small region indicated by the *red* box. The segmentation was done using four shots and parallel imaging-based reconstruction without phase compensation (Mani et al., 2020b). For both acquisitions, images were acquired using a 32-channel phased array coil, and monopolar PGSE encoding was performed with $b = 1000 \, \text{s/mm}^2$ and 60 diffusion directions. For the segmented sequence, TE = 53 ms, PPF = 62.5%, and the total acquisition time was 40 min for 100 slices. A dielectric pad was employed to improve the signal drop off toward the inferior brain regions.

segment with the coil sensitivities and treat those as encoding functions provide adequate reconstruction in such cases (Liu et al., 2005; Chu et al., 2015). However, as the number of segments becomes higher, it becomes harder to estimate the phase from the data. In such cases, a second approach using advanced parallel imaging methods that exploits phase relations between the segments in the form of low-rank constraints has been shown to provide adequate reconstructions (Mani et al., 2017, 2020b) (shown in Fig. 8).

## 5 Simultaneous multislice and volumetric acquisitions

As we have seen, classic diffusion MRI acquisitions with full brain coverage do not have a TR in the SNR efficient regime. Through the use of simultaneous multislice techniques, where multiple slices share the same diffusion preparation and data acquisition periods, TR can be shortened considerably, by a factor equal to that of the slice-acceleration or multiband factor (denoted MB or sometimes $R_{\text{slice}}$, not to be confused with the in-plane parallel imaging acceleration factor $R$, which we call $R_{\text{in-plane}}$ for clarity in this discussion). This shortens the overall scan time and increases SNR efficiency. The SMS approach to parallel imaging was first introduced by Larkman et al. (2001) and adapted for EPI by Nunes et al. (2006), with its first application to diffusion imaging by Feinberg et al. (2010). In SMS, the slice-selective RF pulses are selective for multiple slices at once. Unlike standard in-plane parallel imaging, parallel imaging in the slice direction does not shorten the acquisition period for each imaging slice and therefore does not incur the $\sqrt{R}$ SNR penalty. However, the g-factor noise penalty from the slice unaliasing reconstruction of SMS data is still problematic for typically SNR-starved diffusion acquisitions, limiting MB acceleration capability. The use of the controlled aliasing approach (controlled aliasing in parallel imaging (CAIPI) results in higher acceleration—CAIPIRINHA (Breuer et al., 2005)) via the blipped-CAIPI method for SMS-EPI (Setsompop et al., 2012a) has mitigated this issue, allowing an MB factor of 3–4 to be used with minimal noise penalty on 32–64 channel receiver arrays. It is important to note that the combined use of slice and in-plane parallel imaging accelerations will result in an increased g-factor noise penalty. This limits the total acceleration factor ($R_{\text{total}}$) to approximately sixfold, e.g., $R_{\text{in-plane}} \times \text{MB} = 2 \times 3$, which results in a 10%–25% noise penalty (Setsompop et al., 2012b). Fig. 9 shows an image acquired with MB = 3. Note that when it is otherwise possible to bring TR below the SNR-optimal range, the choice is normally to keep it longer than the minimum possible to allow for $T_1$ recovery and hence optimal SNR. Blipped-CAIPI SMS-EPI is now a matured technology and is available as an FDA-approved product on all major MRI vendors.

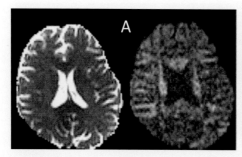 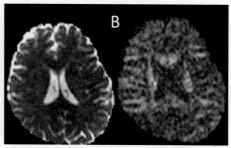

**FIG. 9** SMS allows comparable data to be acquired in a fraction of the non-SMS scan time. These maps were computed from data acquired with (A) no MB acceleration and (B) MB = 3 acceleration. Scanning parameters for both scans were 3 T field strength and TRSE diffusion encoding (see Chapter 6) with $b = 0$ (left) and $3000 \text{ s/mm}^2$ (right). The scans had 2.0 mm isotropic voxels, 63 slices, TE = 118 ms, BW 1658 Hz/px, PPF 6/8, and a total of 64 diffusion encoding directions, of which one is shown. For the conventional acquisition in (A), TR = 9.2 s and the total acquisition time was 10 min. For the SMS acquisition (B), TR = 3.1 s and the total scan time was 3.4 min.

With the push toward submillimeter isotropic resolution diffusion imaging for more detailed brain structural information, data acquisition across more imaging slices is needed to achieve whole-brain coverage. With more slices, the MB acceleration capability from SMS becomes insufficient to achieve a TR in the SNR-efficient regime. To achieve shorter TRs, 3D multislab acquisitions can be used, where data acquisition is performed across multiple slabs instead of slices of the brain. Here, an extra dimension of phase encoding ($k_z$) is performed across one or more slabs (Golay et al., 2002) in a multishot manner (Section 4), with shot-to-shot phase correction incorporated into the reconstruction. The use of 3D multislab imaging for dMRI is a recent development (Engström and Skare, 2013; Chang et al., 2015b; Wu et al., 2016a), and to achieve good reconstruction, postprocessing algorithms have also been developed to mitigate the slab boundary artifacts at short TRs (Van et al., 2015; Wu et al., 2016b; Liao et al., 2020). Moreover, to further improve the acquisition efficiency, simultaneous multislab EPI (SMSb) approaches have also been proposed (Frost et al., 2013; Bruce et al., 2017; Setsompop et al., 2018; Dai et al., 2019). When compared to multislab EPI, SMSb-EPI has the benefit of better parallel imaging performance along the slab direction, because it has a larger field of view in the $z$ direction across simultaneous multislabs. Furthermore, since the slabs are thinner in SMSb-EPI, it is easier to achieve a sharper slab transition with RF; this results in reduced slab boundary artifacts and a reduction in the need for slab oversampling. A novel slab-encoding SMSb-EPI method termed gSlider has enabled robust submillimeter diffusion imaging with high SNR-efficiency (Setsompop et al., 2018; Wang et al., 2018b; Liao et al., 2019). It uses RF slab encoding instead of a gradient in the slab ($z$) dimension. The gSlider method allows for self-navigation of shot-to-shot phase corruption and bulk subject motion. Fig. 10 shows 860 μm isotropic resolution brain images using gSlider-SMSb. These SNR efficient 3D acquisitions are still predominantly research-based.

## 6 Imaging-related artifact reduction

In Chapter 6, artifacts that can be impacted by the exact form of the diffusion encoding in a diffusion weighted sequence are discussed. The following discussion relates to artifacts that can be impacted by the choices made in the remainder of the acquisition.

### 6.1 Ghosting

Because EPI collects data by rastering back and forth across $k$-space, any delay between the start of data collection and the start of the EPI gradients results in a mismatch between the odd and even lines (Fig. 11A). To make matters worse, eddy current fields induced by readout gradient switching will also have opposite polarity, causing the phase to alternate between lines (Fig. 11B). The result of these issues is a "Nyquist N/2 ghost" (Fig. 11C). On clinical systems, these ghosts are corrected via the acquisition of additional data with the phase-encode gradients disabled, allowing the timing delay and alternating phase between $k$-space lines to be characterized and corrected for. However, residual ghosting may still be observed, and strategies to reduce it may be to reshim (the level of alternating phase is modulated by off-resonance), adjust slice orientation (certain oblique slices may produce more problematic eddy currents), or by servicing that includes recalibration of the eddy current precompensation factors.

Diffusion MRI acquisition for tractography: Diffusion sequences **Chapter | 7** 133

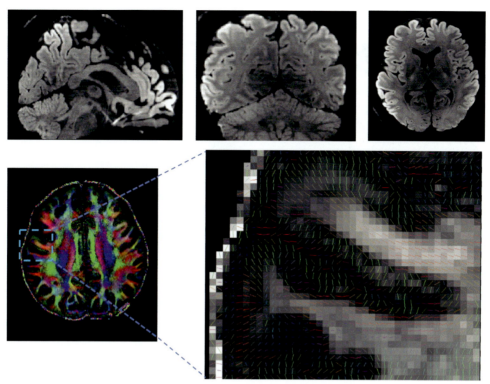

**FIG. 10** 860 μm isotropic resolution whole-brain diffusion data acquired using gSlider-SMSb on a clinical 3 T scanner with a commercial 64 channel head/neck receiver array. Data were acquired across 64 diffusion directions at $b = 1000$ s/mm$^2$. Three averages were acquired to boost the SNR, with a total acquisition time of ∼60 min. Shown are (top row) the average DW images and (bottom row) directionally encoded FA map, and a small region of the diffusion tensor principal eigenvector map overlaid on the scalar FA. Multiple voxels across gray matter can be observed at this resolution, along with the expected sharp dark band of FA at the gray-white boundaries where the fibers turn sharply as they enter the gray matter.

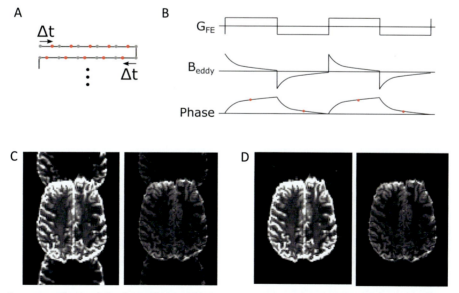

**FIG. 11** (A) A delay between gradients and signal recording results in k-space sampling positions being shifted relative to each other on alternating lines. (B) Gradient switching results in eddy current fields with polarity related to the direction of switching. These fields result in phase accrual (most impactful at the center of the k-space line, shown by the *red dots*) that differs between alternating lines of k-space. (C) Without correction, the Nyquist N/2 ghost can be observed on the $b = 0$ (left) and diffusion weighted images (right). (D) With correction, a residual Nyquist N/2 ghost can be seen on the $b = 0$ image (left), where the ghost of the bright CSF is significant relative to the tissue signal intensity. This is less significant in the diffusion weighted image (right).

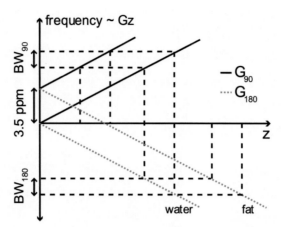

**FIG. 12** Fat signal removal using reversed gradients for the 90- and 180-degree pulses. Fat is shifted in frequency by 3.5 ppm compared to water, which results in a shift of the excited slice relative to the water protons. Reversing the gradient polarity for the 180-degree refocusing pulse causes the shifting to be in the opposite direction, and the fat signal is not refocused.

## 6.2 Fat

The primary problem caused by fat is that its large frequency (chemical) shift relative to water protons, combined with the slow $k$-space traversal speed through the phase encode direction, causes the fat to be shifted into the brain tissue. Further, the optimal Nyquist N/2 ghost correction is different between water and fat due to the frequency shift, which will impair ghost correction algorithms. The two most common methods to eliminate fat are chemically selective presaturation and 90–180-degree slice select gradient mismatches. The former method employs a module before the normal excitation pulse that excites only fat and then spoils the signal, which leaves negligible longitudinal magnetization from fat remaining for the 90-degree dMRI excitation. Alternatives are a water-selective excitation and adjustment of the RF pulse timing (Ivanov et al., 2010). The gradient mismatch and RF timing methods cause the refocusing slice to be offset from the excitation slice for fat, which leads to signal spoiling by the crushers surrounding the 180-degree pulse (Fig. 12). Notably, these methods can be combined for even stronger fat suppression (as seen in Fig. 1), at the expense of some SNR loss for slightly off-resonant locations.

## 6.3 Cerebrospinal fluid effects

The strong $T_2$ weighting of dMRI images means that CSF is bright in $b=0$ images relative to tissue, causing dMRI tissue metrics to be dominated by CSF in voxels that contain both tissue and CSF. While not an artifact per se, the fact that the metrics are weighted by signal fraction means that it is difficult to measure characteristics of the tissue when there is partial volume averaging with CSF. This issue is particularly problematic in white matter tracts near CSF (e.g., fornix) or in cortical gray matter. Acquisition-based methods to ameliorate this issue include: (i) CSF-nulled inversion recovery (Kwong et al., 1991); (ii) acquiring the data in multiple slabs, each with a short TR to preferentially suppress long $T_1$ CSF signal (Baron and Beaulieu, 2015); and (iii) adding at least some small degree of diffusion weighting on all acquisitions to preferentially suppress the highly diffusive CSF (Baron and Beaulieu, 2015). Notably, all these methods do not rely on any a priori assumptions, but they do require at least some degree of SNR and/or scan time trade-off in tissue. Postprocessing approaches include applying a multiple tissue model that aims to solve for the volume ratios of CSF and tissue in each voxel (Pasternak et al., 2009), at the expense of potential model inaccuracies (the problem is generally ill-posed and regularization is required). Generally, these models are better conditioned for multishelled acquisitions, including a lower $b$-value ($\sim$300–500 s/mm$^2$) to characterize the CSF compartment (Hoy et al., 2014).

Another issue related to the high signal of CSF is Gibbs ringing, which is strongest near the ventricles. To make matters worse, the ringing has opposing polarity between the $b=0$ and diffusion weighted acquisitions, which leads to ringing amplification in quantitative maps of diffusivity (Veraart et al., 2016a; Perrone et al., 2015). The CSF signal reduction techniques described previously also mitigate this issue. An effective postprocessing option is ringing removal via subvoxel shifting (Kellner et al., 2016; Lee et al., 2021) (see Chapter 9).

## 6.4 Motion

Postprocessing methods for dealing with rigid body head motion are described in Chapter 9. At the acquisition end, a powerful option for motion correction is real-time monitoring and prospective reorientation of acquired slices (Aksoy et al., 2011). Notably, this approach mitigates signal loss for nonsteady state effects that may otherwise occur when some regions get excited during different slice interleaves due to motion into an adjacent slice location. The trade-offs for this approach include a requirement for line-of-sight to a marker rigidly affixed to the head, patient comfort (the most accurate markers are held by the teeth), and the need for additional peripheral devices and specialized pulse sequences. The IVIM-induced signal dropout in diffusion weighted images (Chapter 6) from brain pulsation is particularly strong near the brainstem, and can be mitigated using cardiac gating or less aggressive partial Fourier acquisition (Chung et al., 2010), with a trade-off of decreased SNR efficiency.

## 6.5 Other artifacts

Distortions, blurring, and Gibbs ringing are a predominant artifact in dMRI, and we have discussed their source and mitigation in Sections 2–3.

At high field strengths, $B_1^+$ inhomogeneity can result in regional SNR loss. Diffusion MRI is particularly sensitive to this issue due to the use of 180 degree refocusing RF pulses. This issue can be mitigated by using single spin echo sequences instead of double spin echo sequences, and/or $B_1^+$ shimming on systems that support it.

A potential source of dMRI parameter bias is noise floor effects, which result from taking an absolute value of the complex image data. The smaller the voxel size, the closer to the noise floor the signal will be. Additionally, the higher the $b$-value, the lower the SNR of the individual images. This has prompted denoising to become an area of active research. The noise floor effects can be mitigated by using the adaptive combine multichannel coil combination technique (Walsh et al., 2000). This can be combined with accounting for Rician noise distributions in parameter fitting (Alexander, 2008) or estimating the noise floor using random matrix theory (Ades-Aron, 2018; Chen et al., 2024; Veraart et al., 2016b,c) with subsequent analytical correction for the finite noise floor (Koay and Basser, 2006) (see Chapter 9). Another approach is to save the complex image data and take the real part after removing the slowly varying image phase to eliminate the noise floor (and retain Gaussian noise characteristics) (Prah et al., 2010; Eichner et al., 2015). Notably, this topic is an active area of research and the optimal method to remove this phase is not yet clear.

Retrospective approaches for reducing these and other artifacts will be covered in Chapter 9.

## 7 Multicontrast sensitization of the diffusion sequence

The choice of diffusion sequence parameters has repercussions for the interpretation of diffusion weighted data that may affect tractography. Conversely, the parameters can be chosen to provide additional information beyond diffusivity within the context of a diffusion weighted sequence. The $T_1$ and $T_2$ relaxation times are sensitive to the local composition and microstructure of neuronal tissue. This means that distinct intravoxel microstructural compartments and fiber populations may be characterized by different relaxation times. For example, the water between the layers of the myelin sheath has a shorter $T_2$ than the intra- and extracellular water (MacKay et al., 1994). Similarly, fiber populations of the brain that have different microstructural features (e.g., axon caliber or myelin thickness) are also characterized by different relaxation times (Yeatman et al., 2014). $T_1$ and $T_2$ contrast are part of the diffusion-weighted imaging sequence, and can be harnessed to highlight or suppress certain anatomical structures, or to aid in the inference of microstructural parameters.

Typical diffusion-weighted spin echo images are $T_2$-weighted due to the long echo time. The $T_2$-weighting effectively acts as a filter, weighing the signal contribution of each compartment by its respective $T_2$. The $T_2$-weighted signal contribution can be modulated by varying the TE of the acquisition, subject to the limitations that timing and SNR impose on it. The choice of TE and hence the $T_2$-weighting of the diffusion sequence will impact the shape of the estimated fiber orientation distribution functions (fODFs, which are computed distributions of the signal contributions that come from fibers in each direction on the sphere—see Chapters 10 and 12). This impact on the fODFs thus affects the resulting tractography results. As shown in Fig. 13, the $T_2$-weighting can help enhance tracts with longer $T_2$ times that are otherwise barely visible. The diffusion acquisition protocol can also be modified to include multiple TEs to allow $T_2$ to be estimated quantitatively. The simultaneous estimation of the diffusion parameters and the $T_2$ of multiple compartments, an approach currently in its infancy, may aid in robustly estimating the compartment size and diffusivity (e.g., Veraart et al., 2018; Lampinen et al., 2019), and facilitate demodulation of the fODF (Ning et al., 2020).

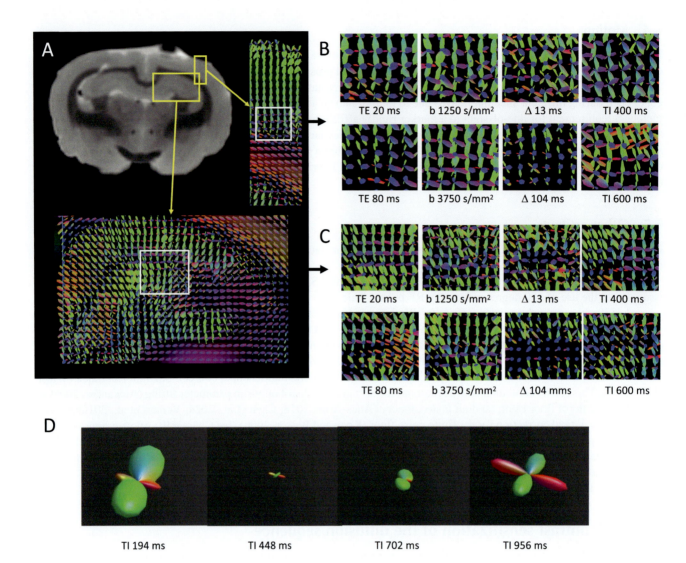

**FIG. 13** Impact of $T_1$ and $T_2$ weighting of diffusion MRI on the fiber orientation distribution function. Images (A–C) of a rabbit brain highlight different tracts within the gray matter (Assaf, 2019). Image (D) shows the impact of $T_1$ weighting on the white matter fODFs in human brain at the decussation of the cingulate bundle *(green)* and the genu of the corpus callosum *(red)*. The inversion prepulse used here decreases the signal magnitude from both tracts, but to a different extent, depending on the $T_1$. *((A–C) Reproduced from Assaf, Y., 2019. Imaging laminar structures in the gray matter with diffusion MRI. Neuro-Image 197, 677–688.)*

$T_1$ weighting can also be used to enhance or suppress tissue compartments by varying the repetition time or, more commonly, by adding an inversion-preparation pulse followed by a time delay before the spin echo acquisition. As we have seen (Section 6.3), TR and inversion recovery approaches are used to suppress the signal from CSF. The decrease in TR afforded by SMS imaging affects the relative signal contribution of long $T_1$ species, such as the gray matter and CSF. As shown in Fig. 13, the inversion time (TI) can be varied to highlight different white matter tracts as well. While measurements at multiple TEs are required for quantitative $T_2$ estimation, measurements at multiple inversion times enable quantitative $T_1$ estimation. Inversion recovery (IR) prepared diffusion data can be used to estimate and assign $T_1$ times to subvoxel compartments with distinct diffusion signatures, such as intravoxel white matter fascicle crossings (De Santis et al., 2016a,b; Leppert et al., 2021).

Other complementary contrast mechanisms, such as magnetization transfer (MT) (Ronen et al., 2006), can also be included in the diffusion encoding sequence by employing a preparatory block before the PGSE acquisition. Because IR and MT prepared diffusion images have minimal direct contribution from the myelin water compartment due to its short $T_2$, the resulting contrast will be biased toward the intra- and extracellular compartments, which have longer relaxation times (Labadie et al., 2014) and lower MT (Ronen et al., 2006). While their potential is still the subject of research, all

Diffusion MRI acquisition for tractography: Diffusion sequences **Chapter | 7 137**

of these complementary contrasts could possibly guide or restrict tractography algorithms in areas with complex intravoxel tract geometries (see Chapter 18).

## 8 The trade-off between spatial and angular resolution in tractography acquisition

The imaging and diffusion encoding blocks are not entirely unrelated, as there exists a trade-off between spatial and angular resolution in a diffusion acquisition. We have noted that complex subvoxel fiber geometry necessitates high angular resolution in diffusion MRI acquisition for human brain tractography. A higher spatial resolution, however, would allow one to separate neighboring bundles from each other independent of whether their orientations are distinct, and would be more spatially specific (Bruce et al., 2017; Vos et al., 2016). We have seen that high spatial resolution (Section 2) and high angular resolution (Chapter 6) typically come at the cost of scan time and reduced SNR. For higher spatial resolution, TR is lengthened because: (i) more slices are needed because they are thinner; and (ii) the lengthened TE required by a longer EPI readout train leads to a lengthened TR. For higher angular resolution, when it is achieved by an increase in the $b$-value, the TR is lengthened because of the need for a longer diffusion encoding block and hence longer TE. The need for more diffusion encoding directions also lengthens the scan time. For higher spatial resolution, the reduced SNR occurs because of smaller voxel volumes and longer TE. For higher angular resolution, again when it is achieved by an increase in the $b$-value, the reduced SNR occurs because of the longer TE. Increasing the number of diffusion encoding directions effectively increases the total SNR, however. Significant increases in spatial resolution typically need to be balanced with additional volumes to recover some of the lost SNR, leading to an extra increase in scan time. Defining a dMRI acquisition protocol therefore requires a trade-off between spatial and angular resolution, with both coming at the expense of time. The chosen resolutions greatly depend on the study application.

Most of the research on this trade-off has focused on comparing the optimal $b$-value, number of gradient orientations, and spatial resolution in single-shell acquisitions. In terms of voxel-wise fiber orientation estimation, the effective angular resolution, or angular resolution limit, is typically expressed as the minimal discernible angle between two fiber populations. It is a factor not just of the acquisition ($b_{max}$, $N_g$, and SNR), but also requires the method for estimation of the local fiber orientation to properly represent the possible resolution limit, i.e., an acquisition with $N_g = 100$ and $b_{max} = 3000\,s/mm^2$ can discern two fibers crossing at closer angles when using more parameters to represent the diffusion or fiber orientation distribution function (Descoteaux et al., 2007; Tournier et al., 2013; Schilling et al., 2017).

In an investigation of in vivo tractography, it was concluded that spatial resolution higher than that used in current clinical practice only appears to benefit tractography of crossing fibers if it can be achieved without reducing the angular resolution (Vos et al., 2016). In regions of crossing fibers, large bundles that dominate the signal can be tracked even at lower angular resolution, but the smaller bundles they cross cannot, even at the higher spatial resolutions employed in the study. The main bulk of the three major bundles investigated was reconstructed in all spatial resolutions and most angular resolutions, but specific cortical terminations were missed at low angular resolution for all but the highest spatial resolution. However, spatial resolution is key for dissociating distinct but proximal parallel fascicles (Liebrand et al., 2020). This trade-off may be different for investigations of relatively smaller structures, e.g., cranial nerve tractography (Danyluk et al., 2020) or tractography-based parcellation of deep gray matter (Patriat et al., 2018).

Multishell acquisitions that allow for estimation of different compartments can cope better with partial volume averaging between tissue types (see Chapter 6). The methodology and benefits of these methods will be discussed in Chapter 11, but their use requires additional gradient directions at intermediate $b$-values, further shifting the balance toward acquisition time spent on enhancing angular resolution.

The recent improvements in spatial resolution that have become available through 2D SMS imaging and time-efficient 3D dMRI acquisitions (Section 5) will allow us to further probe the benefits of increased spatial resolution in vivo. Using 3D EPI on a preclinical system ex vivo, high spatial resolution has been shown, by comparison to optical measurements, to have a greater impact than high angular resolution or high $b$-values on the accuracy of crossing-fiber models of axonal orientations in small human samples ( Jones et al., 2020).

Recent efforts have also focused on combining multiple acquisitions with different trade-offs to ensure sufficient SNR in each, e.g., acquiring high spatial resolution data with lower $b$-values and higher $b$-values at lower spatial resolution. This can then be combined to benefit from the high angular resolution in deep white matter and higher spatial resolution at tissue boundaries (Sotiropoulos et al., 2013b, 2016; Fan et al., 2017). Although conceptually attractive, acquisition and processing pipelines for such approaches are not readily available.

**138 PART | II** Diffusion MRI

# 9 Terminology guide

| | |
|---|---|
| $B_0$ | The strength of the main magnetic field of an MRI scanner. It is aligned along the long axis of the bore of the magnet ($\hat{z}$ orientation) |
| $b$-Matrix/$b$-tensor | A $3 \times 3$ matrix that quantifies the magnitude and direction(s) of diffusion weighting |
| $b$-Value | A parameter that quantifies the magnitude of diffusion weighting. When there is no diffusion weighting, $b = 0\,\text{s/mm}^2$, and the image is sometimes called the $b = 0$ or $b_0$ image |
| DW, DWI | Diffusion weighting, diffusion weighted imaging, or a diffusion weighted image |
| EPI | Echo planar imaging: $k$-space traversal by scanning multiple lines of $k$-space in one excitation, ordered in a raster |
| Gradient | MRI scanners are equipped with gradient coils that linearly modulate the main field strength $B_0$ in the $\hat{x}, \hat{y}$, and $\hat{z}$ directions |
| $k$-Space | Term for the acquisition spatial frequency Fourier space in MRI. The vector $\mathbf{k}$ is the integral of the position encoding gradient over time. There is a Fourier relationship between the $k$-space representation of an image and the image itself |
| MB | Multiband; the multiband factor. Also called simultaneous multislice (SMS) |
| RF | Radio frequency. Applied RF radiation is used to create MRI signal and contrast, and the signal we measure is in the RF band |
| Shim | The process or hardware used to make the MRI magnetic field homogeneous |
| Slew rate | The rate of change of the magnetic field gradient in MRI |
| SMS | Simultaneous multislice. Also called multiband (MB) |
| SNR | Signal-to-noise ratio |
| $T_1$ | The longitudinal relaxation time |
| $T_2$ | The transverse relaxation time |
| TE | The echo time |
| TR | The repetition time |
| Tractometry | The technique of assessing scalar quantities along reconstructed streamlines from tractography |

# References

Ades-Aron, B., et al., 2018. Evaluation of the accuracy and precision of the diffusion parameter estimation with Gibbs and NoisE removal pipeline. NeuroImage 183, 532–543.

Aksoy, M., et al., 2011. Real-time optical motion correction for diffusion tensor imaging. Magn. Reson. Med. 66 (2), 366–378.

Alexander, D.C., 2008. A general framework for experiment design in diffusion MRI and its application in measuring direct tissue-microstructure features. Magn. Reson. Med. 60 (2), 439–448.

Alsop, D.C., 1997. Phase insensitive preparation of single-shot RARE: application to diffusion imaging in humans. Magn. Reson. Med. 38 (4), 527–533.

Andersson, J.L., Skare, S., Ashburner, J., 2003. How to correct susceptibility distortions in spin-echo echo-planar images: application to diffusion tensor imaging. NeuroImage 20 (2), 870–888.

Assaf, Y., 2019. Imaging laminar structures in the gray matter with diffusion MRI. NeuroImage 197, 677–688.

Baron, C.A., Beaulieu, C., 2015. Acquisition strategy to reduce cerebrospinal fluid partial volume effects for improved DTI tractography. Magn. Reson. Med. 73 (3), 1075–1084.

Basser, P.J., Pajevic, S., Pierpaoli, C., Duda, J., Aldroubi, A., 2000. In vivo fiber tractography using DT-MRI data. Magn. Reson. Med. 44 (4), 625–632.

Breuer, F.A., Blaimer, M., Heidemann, R.M., Mueller, M.F., Griswold, M.A., Jakob, P.M., 2005. Controlled aliasing in parallel imaging results in higher acceleration (CAIPIRINHA) for multi-slice imaging. Magn. Reson. Med. 53 (3), 684–691.

Bruce, I.P., Chang, H.-C., Petty, C., Chen, N.-K., Song, A.W., 2017. 3D-MB-MUSE: a robust 3D multi-slab, multi-band and multi-shot reconstruction approach for ultrahigh resolution diffusion MRI. NeuroImage 159, 46–56.

Chang, H.-C., Guhaniyogi, S., Chen, N., 2015a. Interleaved diffusion-weighted improved by adaptive partial-Fourier and multiband multiplexed sensitivity-encoding reconstruction. Magn. Reson. Med. 73 (5), 1872–1884.

Chang, H.C., et al., 2015b. Human brain diffusion tensor imaging at submillimeter isotropic resolution on a 3Tesla clinical MRI scanner. NeuroImage 118, 667–675.

Chen, N.K., Guidon, A., Chang, H.C., Song, A.W., 2013. A robust multi-shot scan strategy for high-resolution diffusion weighted MRI enabled by multi-plexed sensitivity-encoding (MUSE). NeuroImage 72, 41–47.

Chen, J., Ades-Aron, B., Lee, H.-H., Mehrin, S., Pang, M., Novikov, D.S., Veraart, J., Fieremans, E., 2024. Optimization and validation of the DESIGNER preprocessing pipeline for clinical diffusion MRI in white matter aging. Imaging Neurosci. 2, 1–17.

Chu, M.-L., Chang, H.-C., Chung, H.-W., Truong, T.-K., Bashir, M.R., Chen, N., 2015. POCS-based reconstruction of multiplexed sensitivity encoded MRI (POCSMUSE): a general algorithm for reducing motion-related artifacts. Magn. Reson. Med. 74 (5), 1336–1348.

Chung, S.W., Courcot, B., Sdika, M., Moffat, K., Rae, C., Henry, R.G., 2010. Bootstrap quantification of cardiac pulsation artifact in DTI. NeuroImage 49 (1), 631–640.

Conturo, T.E., et al., 1999. Tracking neuronal fiber pathways in the living human brain. Proc. Natl. Acad. Sci. USA 96 (18), 10422–10427.

Dai, E., et al., 2019. A 3D k-space Fourier encoding and reconstruction framework for simultaneous multi-slab acquisition. Magn. Reson. Med. 82 (3), 1012–1024.

Danyluk, H., Sankar, T., Beaulieu, C., 2020. High spatial resolution nerve-specific DTI protocol outperforms whole-brain DTI protocol for imaging the trigeminal nerve in healthy individuals. NMR Biomed.

De Santis, S., Barazany, D., Jones, D.K., Assaf, Y., 2016a. Resolving relaxometry and diffusion properties within the same voxel in the presence of crossing fibres by combining inversion recovery and diffusion-weighted acquisitions. Magn. Reson. Med. 75 (1), 372–380.

De Santis, S., Assaf, Y., Jeurissen, B., Jones, D.K., Roebroeck, A., 2016b. T1 relaxometry of crossing fibres in the human brain. NeuroImage 141, 133–142.

Descoteaux, M., Angelino, E., Fitzgibbons, S., Deriche, R., 2007. Regularized, fast, and robust analytical Q-ball imaging. Magn. Reson. Med. 58 (3), 497–510.

Eichner, C., et al., 2015. Real diffusion-weighted MRI enabling true signal averaging and increased diffusion contrast. NeuroImage 122, 373–384.

Engström, M., Skare, S., 2013. Diffusion-weighted 3D multislab echo planar imaging for high signal-to-noise ratio efficiency and isotropic image resolution. Magn. Reson. Med. 70 (6), 1507–1514.

Engström, M., Mårtensson, M., Avventi, E., Skare, S., 2015. On the signal-to-noise ratio efficiency and slab-banding artifacts in three-dimensional multislab diffusion-weighted echo-planar imaging. Magn. Reson. Med. 73 (2), 718–725.

Fan, Q., et al., 2017. HIgh b-value and high resolution integrated diffusion (HIBRID) imaging. NeuroImage 150, 162–176.

Feinberg, D.A., et al., 2010. Multiplexed echo planar imaging for sub-second whole brain FMRI and fast diffusion imaging. PLoS One 5 (12), e15710.

Frost, R., Jezzard, P., Porter, D.A., Miller, K.L., 2013. Simultaneous multi-slab acquistiion in 3D multi-slab diffusion-weighted readout-segmented echo-planar imaging. In: *Proc 21st Intl Soc Mag Reson Med*, p. 3176.

Golay, X., Jiang, H., van Zijl, P.C.M., Mori, S., 2002. High-resolution isotropic 3D diffusion tensor imaging of the human brain. Magn. Reson. Med. 47 (5), 837–843.

Gorodezky, M., Scott, A.D., Ferreira, P.F., Nielles-Vallespin, S., Pennell, D.J., Firmin, D.N., 2018. Diffusion tensor cardiovascular magnetic resonance with a spiral trajectory: an in vivo comparison of echo planar and spiral stimulated echo sequences. Magn. Reson. Med. 80 (2), 648–654.

Griswold, M.A., et al., 2002. Generalized autocalibrating partially parallel acquisitions (GRAPPA). Magn. Reson. Med. 47 (6), 1202–1210.

Heidemann, R.M., et al., 2010. Diffusion imaging in humans at 7T using readout-segmented EPI and GRAPPA. Magn. Reson. Med. 64 (1), 9–14.

Herbst, M., Deng, W., Ernst, T., Stenger, V.A., 2017. Segmented simultaneous multi-slice diffusion weighted imaging with generalized trajectories. Magn. Reson. Med. 78 (4), 1476–1481.

Holdsworth, S.J., Skare, S., Newbould, R.D., Guzmann, R., Blevins, N.H., Bammer, R., 2008. Readout-segmented EPI for rapid high resolution diffusion imaging at 3T. Eur. J. Radiol. 65 (1), 36–46.

Holdsworth, S.J., O'Halloran, R., Setsompop, K., 2019. The quest for high spatial resolution diffusion-weighted imaging of the human brain in vivo. NMR Biomed. 32 (4), e4056.

Hoy, A.R., Koay, C.G.G., Kecskemeti, S.R., Alexander, A.L., 2014. Optimization of a free water elimination two-compartment model for diffusion tensor imaging. NeuroImage 103, 323–333.

Ivanov, D., Schäfer, A., Streicher, M.N., Heidemann, R.M., Trampel, R., Turner, R., 2010. A simple low-SAR technique for chemical-shift selection with high-field spin-echo imaging. Magn. Reson. Med. 64 (2), 319–326.

Jones, D.K., Simmons, A., Williams, S.C.R., Horsfield, M.A., 1999. Non-invasive assessment of axonal fiber connectivity in the human brain via diffusion tensor MRI. Magn. Reson. Med. 42 (1), 37–41.

Jones, R., et al., 2020. Insight into the fundamental trade-offs of diffusion MRI from polarization-sensitive optical coherence tomography in ex vivo human brain. NeuroImage 214, 116704.

Kellner, E., Dhital, B., Kiselev, V.G., Reisert, M., 2016. Gibbs-ringing artifact removal based on local subvoxel-shifts. Magn. Reson. Med. 76 (5), 1574–1581.

Koay, C.G., Basser, P.J., 2006. Analytically exact correction scheme for signal extraction from noisy magnitude MR signals. J. Magn. Reson. 179 (2), 317–322.

Kucharczyk, J., Mintorovitch, J., Asgari, H.S., Moseley, M., 1991. Diffusion/perfusion MR imaging of acute cerebral ischemia. Magn. Reson. Med. 19 (2), 311–315.

Kwong, K.K., McKinstry, R.C., Chien, D., Crawley, A.P., Pearlman, J.D., Rosen, B.R., 1991. CSF-suppressed quantitative single-shot diffusion imaging. Magn. Reson. Med. 21 (1), 157–163.

Labadie, C., et al., 2014. Myelin water mapping by spatially regularized longitudinal relaxographic imaging at high magnetic fields. Magn. Reson. Med. 71 (1), 375–387.

Lampinen, B., et al., 2019. Searching for the neurite density with diffusion MRI: challenges for biophysical modeling. Hum. Brain Mapp., 1–17. no. October 2018.

Larkman, D.J., Hajnal, J.V., Herlihy, A.H., Coutts, G.A., Young, I.R., Ehnholm, G.S., 2001. Use of Multicoil Arrays for Separation of Signal from Multiple Slices Simultaneously Excited.

Le Bihan, D., Breton, E., Lallemand, D., Grenier, P., Cabanis, E., Laval-Jeantet, M., 1986. MR imaging of intravoxel incoherent motions: application to diffusion and perfusion in neurologic disorders. Radiology 161 (2), 401–407.

Lee, H.-H., Novikov, D.S., Fieremans, E., 2021. Removal of partial Fourier-induced Gibbs (RPG) ringing artifacts in MRI. Magn. Reson. Med., 2733–2750.

Leppert, I.R., et al., 2021. Efficient whole-brain tract-specific $T_1$ mapping at 3T with slice-shuffled inversion-recovery diffusion-weighted imaging. Magn. Reson. Med. 86 (2), 738–753.

Li, T.Q., Takahashi, A.M., Hindmarsh, T., Moseley, M.E., 1999. ADC mapping by means of a single-shot spiral MRI technique with application in acute cerebral ischemia. Magn. Reson. Med. 41 (1), 143–147.

Liao, C., et al., 2019. Phase-matched virtual coil reconstruction for highly accelerated diffusion echo-planar imaging. NeuroImage 194, 291–302.

Liao, C., et al., 2020. High-fidelity, high-isotropic-resolution diffusion imaging through gSlider acquisition with and $T_1$ corrections and integrated $\Delta B_0/Rx$ shim array. Magn. Reson. Med. 83 (1), 56–67.

Liebrand, L.C., van Wingen, G.A., Vos, F.M., Denys, D., Caan, M.W.A., 2020. Spatial versus angular resolution for tractography-assisted planning of deep brain stimulation. NeuroImage Clin. 25.

Liu, G., Van Gelderen, P., Duyn, J., Moonen, C.T.W., 1996. Single-shot diffusion MRI of human brain on a conventional clinical instrument. Magn. Reson. Med. 35 (5), 671–677.

Liu, C., Moseley, M.E., Bammer, R., 2005. Simultaneous phase correction and SENSE reconstruction for navigated multi-shot DWI with non-cartesian k-space sampling. Magn. Reson. Med. 54 (6), 1412–1422.

MacKay, A., Whittall, K., Adler, J., Li, D., Paty, D., Graeb, D., 1994. In vivo visualization of myelin water in brain by magnetic resonance. Magn. Reson. Med. 31 (6), 673–677.

Mani, M., Jacob, M., Kelley, D., Magnotta, V., 2017. Multi-shot sensitivity-encoded diffusion data recovery using structured low-rank matrix completion (MUSSELS). Magn. Reson. Med. 78 (2), 494–507.

Mani, M., et al., 2020a. SMS MUSSELS: a navigator-free reconstruction for simultaneous multi-slice-accelerated multi-shot diffusion weighted imaging. Magn. Reson. Med. 83 (1), 154–169.

Mani, M., Aggarwal, H.K., Magnotta, V., Jacob, M., 2020b. Improved MUSSELS reconstruction for high-resolution multi-shot diffusion weighted imaging. Magn. Reson. Med. 83 (6), 2253–2263.

Mansfield, P., 1977. Multi-planar image formation using NMR spin echoes. J. Phys. C Solid State Phys. 10 (3), L55–L58.

Mansfield, P., 1984. Real-time echo-planar imaging by NMR. Br. Med. Bull. 40 (2), 187–190.

Merrem, A., et al., 2017. Rapid diffusion-weighted magnetic resonance imaging of the brain without susceptibility artifacts. Investig. Radiol. 52 (7), 428–433.

Miller, K.L., et al., 2011. Diffusion imaging of whole, post-mortem human brains on a clinical MRI scanner. NeuroImage 57 (1), 167–181.

Ning, L., Gagoski, B., Szczepankiewicz, F., Westin, C.F., Rathi, Y., 2020. Joint relaxation-diffusion imaging moments to probe neurite microstructure. IEEE Trans. Med. Imaging 39 (3), 668–677.

Noll, D.C., Nishimura, D.G., Macovski, A., 1991. Homodyne detection in magnetic resonance imaging. IEEE Trans. Med. Imaging 10 (2), 154–163.

Nolte, U.G., Finsterbusch, J., Frahm, J., 2000. Rapid isotropic diffusion mapping without susceptibility artifacts: whole brain studies using diffusion-weighted single-shot STEAM MR imaging. Magn. Reson. Med. 44 (5), 731–736.

Nunes, R.G., Hajnal, J.V., Golay, X., Larkman, D.J., 2006. Simultaneous slice excitation and reconstruction for single shot EPI. In: *Proc 14th Intl Soc Mag Reson Med*, p. 293.

Pasternak, O., Sochen, N., Gur, Y., Intrator, N., Assaf, Y., 2009. Free water elimination and mapping from diffusion MRI. Magn. Reson. Med. 62 (3), 717–730.

Patriat, R., et al., 2018. Individualized tractography-based parcellation of the globus pallidus pars interna using 7T MRI in movement disorder patients prior to DBS surgery. NeuroImage 178, 198–209.

Perrone, D., Aelterman, J., Pižurica, A., Jeurissen, B., Philips, W., Leemans, A., 2015. The effect of Gibbs ringing artifacts on measures derived from diffusion MRI. NeuroImage.

Pipe, J., 2014. Pulse sequences for diffusion-weighted MRI. In: Johansen-Berg, H., Behrens, T.E.J. (Eds.), Diffusion MRI: From Quantitative Measurement to In Vivo Neuroanatomy. Elsevier.

Porter, D.A., Heidemann, R.M., 2009. High resolution diffusion-weighted imaging using readout-segmented echo-planar imaging, parallel imaging and a two-dimensional navigator-based reacquisition. Magn. Reson. Med. 62 (2), 468–475.

Prah, D.E., Paulson, E.S., Nencka, A.S., Schmainda, K.M., 2010. A simple method for rectified noise floor suppression: phase-corrected real data reconstruction with application to diffusion-weighted imaging. Magn. Reson. Med. 64 (2), 418–429.

Pruessmann, K.P., Weiger, M., Scheidegger, M.B., Boesiger, P., 1999. SENSE: sensitivity encoding for fast MRI. Magn. Reson. Med. 42, 952–962.

Ronen, I., Moeller, S., Ugurbil, K., Kim, D.S., 2006. Investigation of multicomponent diffusion in cat brain using a combined MTC-DWI approach. Magn. Reson. Imaging 24 (4), 425–431.

Schick, F., 1997. SPLICE: sub-second diffusion-sensitive MR imaging using a modified fast spin-echo acquisition mode. Magn. Reson. Med. 38 (4), 638–644.

Schilling, K., Gao, Y., Janve, V., Stepniewska, I., Landman, B.A., Anderson, A.W., 2017. Can increased spatial resolution solve the crossing fiber problem for diffusion MRI? NMR Biomed. 30 (12), e3787.

Schmitt, F., Stehling, M.K., Turner, R., 1998. Echo-Planar Imaging. Springer, Berlin, Heidelberg.

Setsompop, K., Gagoski, B.A., Polimeni, J.R., Witzel, T., Wedeen, V.J., Wald, L.L., 2012a. Blipped-controlled aliasing in parallel imaging for simultaneous multislice echo planar imaging with reduced g-factor penalty. Magn. Reson. Med.

Setsompop, K., et al., 2012b. Improving diffusion MRI using simultaneous multi-slice echo planar imaging. NeuroImage 63 (1), 569–580.

Setsompop, K., et al., 2018. High-resolution in vivo diffusion imaging of the human brain with generalized slice dithered enhanced resolution: simultaneous multislice (gSlider-SMS). Magn. Reson. Med. 79 (1), 141–151.

Skare, S., Bammer, R., 2010. EPI-based pulse sequences for diffusion tensor MRI. In: Jones, D.K. (Ed.), Diffusion MRI: Theory, Methods, and Applications. Oxford University Press.

Sodickson, D.K., Manning, W.J., 1997. Simultaneous acquisition of spatial harmonics (SMASH): fast imaging with radiofrequency coil arrays. Magn. Reson. Med. 38 (4), 591–603.

Sotiropoulos, S.N., et al., 2013a. Advances in diffusion MRI acquisition and processing in the human connectome project. NeuroImage 80, 125–143.

Sotiropoulos, S.N., Jbabdi, S., Andersson, J.L., Woolrich, M.W., Ugurbil, K., Behrens, T.E.J., 2013b. RubiX: combining spatial resolutions for bayesian inference of crossing fibers in diffusion MRI. IEEE Trans. Med. Imaging 32 (6), 969–982.

Sotiropoulos, S.N., et al., 2016. Fusion in diffusion MRI for improved fibre orientation estimation: an application to the 3T and 7T data of the human connectome project. NeuroImage 134, 396–409.

Stehling, M.K., Turner, R., Mansfield, P., 1991. Echo-planar imaging: magnetic resonance imaging in a fraction of a second. Science 254 (5028), 43–50.

Thomsen, C., Henriksen, O., Ring, P., 1987. In vivo measurement of water self diffusion in the human brain by magnetic resonance imaging. Acta Radiol. 28 (3), 353–361.

Tournier, J.D., Calamante, F., Connelly, A., 2013. Determination of the appropriate b value and number of gradient directions for high-angular-resolution diffusion-weighted imaging. NMR Biomed. 26 (12), 1775–1786.

Treit, S., Steve, T., Gross, D.W., Beaulieu, C., 2018. High resolution in-vivo diffusion imaging of the human hippocampus. NeuroImage 182, 479–487.

Turner, R., Le Bihan, D., 1990. Single-shot diffusion imaging at 2.0 tesla. J. Magn. Reson. 86 (3), 445–452.

Turner, R., Le Bihan, D., Maier, J., Vavrek, R., Hedges, L.K., Pekar, J., 1990. Echo-planar imaging of intravoxel incoherent motion. Radiology 177 (2), 407–414.

Turner, R., Le Bihan, D., Scott Chesnicks, A., 1991. Echo-planar imaging of diffusion and perfusion. Magn. Reson. Med. 19 (2), 247–253.

Uecker, M., Karaus, A., Frahm, J., 2009. Inverse reconstruction method for segmented multishot diffusion-weighted MRI with multiple coils. Magn. Reson. Med. 62 (5), 1342–1348.

Van, A.T., Aksoy, M., Holdsworth, S.J., Kopeinigg, D., Vos, S.B., Bammer, R., 2015. Slab profile encoding (PEN) for minimizing slab boundary artifact in three-dimensional diffusion-weighted multislab acquisition. Magn. Reson. Med. 73 (2), 605–613.

Veraart, J., Fieremans, E., Jelescu, I.O., Knoll, F., Novikov, D.S., 2016a. Gibbs ringing in diffusion MRI. Magn. Reson. Med. 76 (1), 301–314.

Veraart, J., Fieremans, E., Novikov, D.S., 2016b. Diffusion MRI noise mapping using random matrix theory. Magn. Reson. Med. 76 (5), 1582–1593.

Veraart, J., Novikov, D.S., Christiaens, D., Ades-Aron, B., Sijbers, J., Fieremans, E., 2016c. Denoising of diffusion MRI using random matrix theory. NeuroImage 142, 394–406.

Veraart, J., Novikov, D.S., Fieremans, E., 2018. TE dependent diffusion imaging (TEdDI) distinguishes between compartmental T2 relaxation times. NeuroImage 182, 360–369.

Vos, S.B., et al., 2016. Trade-off between angular and spatial resolutions in in vivo fiber tractography. NeuroImage 129, 117–132.

Walsh, D.O., Gmitro, A.F., Marcellin, M.W., 2000. Adaptive reconstruction of phased array MR imagery. Magn. Reson. Med. 43 (5), 682–690.

Wang, Y., et al., 2018a. A comparison of readout segmented EPI and interleaved EPI in high-resolution diffusion weighted imaging. Magn. Reson. Imaging 47, 39–47.

Wang, F., et al., 2018b. Motion-robust sub-millimeter isotropic diffusion imaging through motion corrected generalized slice dithered enhanced resolution (MC-gSlider) acquisition. Magn. Reson. Med. 80.

Warach, S., Gaa, J., Siewert, B., Wielopolski, P., Edelman, R.R., 1995. Acute human stroke studied by whole brain echo planar diffusion-weighted magnetic resonance imaging. Ann. Neurol. 37 (2), 231–241.

Wilm, B.J., et al., 2017. Single-shot spiral imaging enabled by an expanded encoding model: demonstration in diffusion MRI. Magn. Reson. Med. 77 (1), 83–91.

Wilm, B.J., Hennel, F., Roesler, M.B., Weiger, M., Pruessmann, K.P., 2020. Minimizing the echo time in diffusion imaging using spiral readouts and a head gradient system. Magn. Reson. Med. mrm.28346.

Wu, W., Miller, K.L., 2017. Image Formation in Diffusion MRI: A Review of Recent Technical Developments behalf of International Society for Magnetic Resonance in Medicine. Authors J. Magn. Reson. Imaging,.

Wu, W., et al., 2016a. High-resolution diffusion MRI at 7T using a three-dimensional multi-slab acquisition. NeuroImage 143, 1–14.

Wu, W., Koopmans, P.J., Frost, R., Miller, K.L., 2016b. Reducing slab boundary artifacts in three-dimensional multislab diffusion MRI using nonlinear inversion for slab profile encoding (NPEN). Magn. Reson. Med. 76 (4), 1183–1195.

Xu, Y., Mark Haacke, E., 2001. Partial Fourier imaging in multi-dimensions: a means to save a full factor of two in time. J. Magn. Reson. Imaging 14 (5), 628–635.

Yeatman, J.D., Wandell, B.A., Mezer, A.A., 2014. Lifespan maturation and degeneration of human brain white matter. Nat. Commun. 5 (1), 1–12.

# Chapter 8

# Diffusion MRI acquisition for tractography: Beyond the in vivo adult human brain

Jennifer S.W. Campbell[a], Steven H. Baete[b], Julien Cohen-Adad[c], J.-Donald Tournier[d], Filip Szczepankiewicz[e,f], Christian Beaulieu[g], Corey A. Baron[h], Merry Mani[i], Kawin Setsompop[j], Congyu Liao[j], Christine L. Tardif[a], Sjoerd B. Vos[k], Anastasia Yendiki[l], Ilana R. Leppert[a], Els Fieremans[m], Alberto De Luca[n,o], Alexander Leemans[p], and G. Bruce Pike[q]

[a]McConnell Brain Imaging Centre, Montreal Neurological Institute, McGill University, Montreal, QC, Canada, [b]New York University Grossman School of Medicine, New York, NY, United States, [c]Polytechnique Montreal, Montreal, QC, Canada, [d]King's College London, London, United Kingdom, [e]Lund University, Lund, Sweden, [f]Brigham and Women's Hospital, Harvard Medical School, Boston, MA, United States, [g]University of Alberta, Edmonton, AB, Canada, [h]Robarts Research Institute, Western University, London, ON, Canada, [i]University of Iowa, Iowa City, IA, United States, [j]Stanford University, Stanford, CA, United States, [k]Centre for Microscopy, Characterisation, and Analysis, The University of Western Australia, Nedlands, WA, Australia, [l]Massachusetts General Hospital, Harvard Medical School, Boston, MA, United States, [m]Bernard and Irene Schwartz Center for Biomedical Imaging, Department of Radiology, New York University Grossman School of Medicine, New York, NY, United States, [n]Image Sciences Institute, Division Imaging & Oncology, University Medical Center Utrecht, Utrecht, The Netherlands, [o]Neurology Department, UMC Utrecht Brain Center, University Medical Center Utrecht, Utrecht, The Netherlands, [p]PROVIDI lab, Image Sciences Institute, University Medical Center Utrecht, Utrecht, The Netherlands, [q]Hotchkiss Brain Institute, University of Calgary, Calgary, AB, Canada

## 1 Ex vivo and preclinical diffusion imaging for brain tractography

Ex vivo and preclinical tractography have several applications, including validating methodology and assessing connectivity in interventional experiments. These will be discussed in detail in Chapter 32. The acquisition methods developed by the Human Connectome Project (Sotiropoulos et al., 2013), which include simultaneous multislice (SMS), are also applicable to the imaging of nonhuman primates (NHPs) in vivo (Autio et al., 2020). In vivo imaging of NHPs, which is typically performed using clinical scanners, offers the possibility of collecting diffusion and functional MRI data in the same animals, thus facilitating comparative studies of structural and functional organization of brain networks across species. However, most of the recent interest in diffusion imaging in animals has been in the context of postmortem validation. Scanning of small animals and/or ex vivo brain samples can be done on preclinical systems where the physical dimensions of the hardware can be reduced, which comes at the benefit of higher performance of the magnet and gradient systems and hence higher spatial and angular resolution, which we have seen benefit fiber reconstructions (Jones et al., 2020; Yendiki et al., 2022). This is particularly relevant for animal imaging, where the spatial resolution should be scaled according to the size of the brain of each species relative to the human brain (Herculano-Houzel, 2009), which creates challenges that are discussed below. For ex vivo diffusion imaging of whole human brains, the bore size of preclinical systems is often insufficient, meaning conventional clinical scanners are commonly used.

Preclinical systems are specialized MRI scanners that take advantage of the smaller sample size, as well as relaxed gradient and radio frequency (RF) safety limits. The bore size is smaller than that of clinical scanners, at 10–30 cm. It is possible to use a higher field strength, typically 7–11.7 T, and it is easier to produce a homogeneous $B_0$ field over the size of the sample because it is smaller. Preclinical systems also have higher gradient strength than clinical scanners, typically 400–1000 mT/m, and higher slew rates ranging from 600 to 6000 mT/m/ms, which are unachievable on clinical systems because of concerns regarding peripheral nerve stimulation. In addition to the preceding, recent work (Henriques et al., 2020) has shown that the signal-to-noise ratio (SNR) can be boosted on some preclinical systems by adopting a cryogenic cooling system of RF receive coils to reduce the impact of thermal noise.

It has long been known that diffusion anisotropy is preserved in postmortem brain tissue after fixation, even though diffusivity is reduced (Sun et al., 2005). The preservation of anisotropy allows tractography to be performed postmortem, with the benefit of higher resolution as discussed previously. However, the reduction of diffusivity implies that, to achieve

*Handbook of Diffusion MR Tractography. https://doi.org/10.1016/B978-0-12-818894-1.00026-4*
Copyright © 2025 Elsevier Ltd. All rights are reserved, including those for text and data mining, AI training, and similar technologies.

diffusion contrast equivalent to in vivo imaging, the $b$-values used ex vivo should be increased by the same factor by which diffusivity decreases. This is because it is the product $bD$ that determines the relative signal loss. This factor has been found empirically to be around 2–5 (Sun et al., 2005; D'Arceuil and de Crespigny, 2007; McNab et al., 2009; Dyrby et al., 2011). The need for such high $b$-values is a major challenge of ex vivo diffusion imaging, given the intrinsic loss of SNR due to high $b$-value and the longer the echo time (TE) required to achieve it. Preclinical scanners, which have much stronger gradients than large-bore systems, are advantageous in this regard.

A third parameter that needs to be considered for ex vivo imaging is the $T_2$ relaxation time, which is reduced by fixation in paraformaldehyde. To mitigate this problem, the sample is typically soaked in phosphate buffered saline (PBS) for a duration dependent on the sample size, up to multiple weeks, exchanging the PBS daily, to remove the formalin (Roebroeck et al., 2019). An alternative is in situ scanning before fixation (Jonkman et al., 2019). In some cases, the diffusion weighted stimulated echo acquisition mode (DW-STEAM) sequence (Section 3) is used in order to lengthen the diffusion time relative to TE (Fritz et al., 2019). Short TEs are crucial for ex vivo scans to achieve good SNR and to minimize artifacts. Hence, segmented echo planar imaging (EPI) is advantageous for such studies (Miller et al., 2011), especially when acquisition time is not the limiting factor.

Despite these challenges, ex vivo imaging can achieve much higher resolution and SNR compared to what is feasible in vivo, as it allows us to lengthen the acquisition time and place the sample much closer to the coil, while avoiding motion or respiratory artifacts. Early ex vivo animal studies used slow, spin-echo sequences for diffusion imaging, and thus allowed the collection of relatively few directions at low $b$-values, which were sufficient for diffusion tensor-based tractography (e.g., 30 directions at $b = 1200\,\text{s/mm}^2$ (Gao et al., 2013)). The adoption of 3D EPI has provided ex vivo animal imaging with much faster acquisition times and high SNR efficiency, thus achieving the higher $b$-values and angular resolutions that are necessary for more advanced modeling of crossing fibers for tractography (Schmahmann et al., 2007; Schilling et al., 2018; Safadi et al., 2018; Liu et al., 2020). The introduction of human scanners with ultrahigh gradients has ushered in new opportunities in this domain. While acquisition protocols for this purpose are still an open area of research, both multiecho 2D and single-echo 3D segmented EPI have shown promise for achieving submillimeter spatial resolution at very high $b$-values (McNab et al., 2013; Eichner et al., 2020). In addition to SE, steady-state free precession (SSFP) strategies have been used to achieve high-resolution ex vivo (Miller et al., 2012). SSFP acquisitions can be particularly advantageous ex vivo because they can achieve higher diffusion $b$-values than spin echo EPI with the same hardware, facilitating fiber tractography with high spatial resolution (up to 0.5 mm isotropic at 3 T (Benner et al., 2009)). A challenge in the use of SSFP is that the measured signal originates from a mixture of echoes and stimulated echoes, which makes exact quantitative computations from the signal difficult.

## 2 Pediatrics, neonates, and fetal diffusion imaging

There is increasing interest in mapping the evolution of connectivity throughout normal and abnormal human brain development, highlighted by large-scale studies such as the ABCD study (Casey et al., 2018), the developing Human Connectome Project (dHCP) (Bastiani et al., 2019), and the lifespan Human Connectome Project (Somerville et al., 2018). Diffusion imaging of the brain in early life poses a unique set of challenges, including dramatic changes in diffusion contrast and magnitude due to maturation processes, difficulty in preventing head motion, and a need for high spatial resolution due to small brain sizes. While all major fiber bundles develop in utero, their microstructural properties, for example myelination and axonal density, develop rapidly through the first 2 years of life (Dubois et al., 2014; Qiu et al., 2015; Batalle et al., 2017; Kimpton et al., 2020). Differences in microstructure lead to lower anisotropy and higher diffusivity compared to the adult brain. The higher diffusivity in the fetal and infant brain implies that diffusion-weighted image contrast comparable to that in the adult brain can be achieved with lower $b$-values. Furthermore, lower $b$-values are desirable due to their lower sensitivity to head motion, which is a major concern as imaging is usually performed on asleep but unsedated infants. However, at higher $b$-values, the signal is dominated by the intra-axonal compartment (see Chapters 5 and 6), and in this compartment the anisotropy discrepancy between the neonate and adult brain is less pronounced. Lower anisotropy (at lower $b$-values) makes the preferential orientation of diffusion, and hence the orientation of fiber bundles, which tractography relies on, harder to detect. The use of higher $b$-values for neonatal diffusion imaging has been increasing in recent years, including in the dHCP acquisition (Bastiani et al., 2019). After infancy, microstructural change becomes slower, although it continues to be observed through early adulthood for some fiber bundles (Lebel et al., 2019). As a result, the $b$-values used in childhood usually match those used in adult subjects.

As with adult populations, 2D EPI is the sequence of choice for in vivo diffusion imaging of the developing brain. In utero, where movement of the fetus is a particular concern, keeping the acquisition time for each volume as short as possible is key. As a result, typical scan protocols involve the acquisition of a few thick slices. Early demonstrations of the feasibility of deterministic tensor tractography in the fetal brain used axial 2D EPI acquisitions with 12–32 diffusion gradient directions, a single $b$-value in the range 500–1000 s/mm$^2$, in-plane resolution around 2 mm, and slice thickness in the range 2–4.5 mm

(Kasprian et al., 2008; Zanin et al., 2011; Jakab et al., 2015; Mitter et al., 2015; Marami et al., 2017). Alternatively, one can acquire three sets of such thick-slice scans in orthogonal orientations (axial, coronal, and sagittal), and use them to reconstruct isotropic-resolution, motion-corrected volumes (Fogtmann et al., 2014). Additional confounds in fetal imaging are maternal respiration (Jakab et al., 2017) and the presence of fat in the tissues surrounding the fetal head, which can result in chemical shift artifact if not saturated. To achieve a homogeneous $B_0$, it is optimal to shim over a small region; hence, shimming over a small region while still suppressing fat signal from the surrounding tissue can be beneficial (Gaspar et al., 2019).

For the infant brain, early acquisition protocols that enabled diffusion tractography used 2D EPI with low angular resolution (6–32 gradient directions), a single, relatively low $b$-value (600–1000 s/mm$^2$), and anisotropic voxels (1–2 mm in-plane resolution, 2–6 mm slice thickness) (Pannek et al., 2014). Despite their limitations, such acquisitions allowed the reconstruction of the major white matter bundles, not only with deterministic diffusion tensor tractography but also with some of the more advanced approaches that are typically used in the adult brain, including automated probabilistic tractography in the individual subject's data (Anblagan et al., 2015; Zöllei et al., 2019). More recently, studies have been performed in the infant brain using protocols more similar to the adult protocol used by the Human Connectome Project (Sotiropoulos et al., 2013). This includes high $b$-values (up to 3000 s/mm$^2$), high angular resolution scanning (60–280 gradient directions) (Pieterman et al., 2017; Salvan et al., 2017), multishell protocols (three or more shells with $b$-values from 400 to 3000 s/mm$^2$), high isotropic resolution (1.5 mm), and in some cases SMS (Hutter et al., 2018; Howell et al., 2019). For children and adolescents, the state-of-the-art protocols that have been developed for large-scale connectomics studies are similar to those of the adult Human Connectome Project. A key additional consideration is robustness of the protocol to aborted scans (Hutter et al., 2018).

## 3 Diffusion imaging for tractography outside the brain

Fiber tractography has also found relevant applications outside the brain, allowing for reconstruction of the architecture of the heart, and of the structure of several skeletal muscles, such as those in the lower and upper leg, forearm, tongue, and brachial plexus. Fiber tractography has also been successfully applied in the study of the organization of nerves, spinal nerves, prostate, and kidneys.

Key challenges in diffusion imaging outside the brain are motion, $B_0$ and $B_1$ inhomogeneity, and chemical shift artifact. Additionally, optimizing the $b$-value and TE for the diffusivity, anisotropy, and $T_2$ of the tissue of interest are important. As in the human brain in vivo, single-shot spin echo EPI is used for the majority of tractography studies outside the brain. In addition to spin echo EPI, other imaging sequence variants have been employed to address issues such as the shorter $T_2$ in muscle compared to the brain. For example, spiral instead of EPI readouts can help shorten the TE (Gorodezky et al., 2018) to increase the SNR, and STEAM (Merboldt et al., 1991; Haase et al., 1986) has been employed outside the brain (Steidle and Schick, 2006). This technique allows the effective diffusion time $t_d$ to be longer than the effective time over which $T_2$ decay occurs. This is beneficial, for example, in muscle, where $T_2$ is short and the size of the anisotropic muscle fiber cells is larger than that of neurons (10–100 μm diameter (Strijkers et al., 2010)), making it advantageous to have a shorter TE and longer diffusion time.

There are several variants on DW-STEAM, but it can be done by replacing the 180-degree RF pulse responsible for the SE in the pulsed gradient spin echo (PGSE) sequence (see Chapter 6) with two 90-degree RF pulses separated by a mixing time (TM) (see Fig. 1). The first of these 90-degree pulses stores the (dephased) magnetization along the scanner's $\hat{z}$ axis, where it no longer experiences $T_2$ decay. The magnetization is then flipped back into the transverse plane, where the second diffusion encoding gradient completes the diffusion encoding, and an echo is formed. In cardiac imaging, a TM of one cardiac cycle has been employed to reduce bulk motion effects (Gorodezky et al., 2018). The STEAM preparation causes a 50% signal loss compared to the standard PGSE sequence, so care must be taken in applications with low SNR. It is also particularly important to account for the contribution of imaging gradients to the $b$-value of the STEAM sequence, given its long diffusion time.

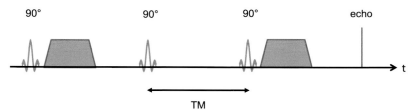

**FIG. 1** STEAM sequence diagram illustrating RF and gradient waveforms. In this variant of STEAM, the 180-degree pulse of the standard PGSE sequence (see Chapter 6 Fig. 1C and Chapter 7 Fig. 1) is replaced by two 90-degree pulses, thereby storing the magnetization along the $z$ axis. The time between the two 90-degree pulses is the mixing time (TM). A spoiling gradient (not shown) during TM dephases the spin echo signal.

Compared to brain applications, fiber tractography outside the brain is typically less demanding in terms of diffusion encoding. Given that crossing fibers are not expected in most body tissues and organs (although they have been demonstrated in the tongue muscle (Mazzoli et al., 2016)), diffusion tensor imaging (DTI)-based fiber tractography is the most popular approach outside the brain. Accordingly, acquiring data with a relatively low maximum $b$-value and less than 30 gradient directions is typically sufficient. The optimal $b$-value for diffusion imaging outside the brain depends on the diffusivity of the tissue, with higher $b$-values being optimal for tissues with lower mean diffusivity, as is the case ex vivo. For single-fiber estimates, the range 400–500 s/mm$^2$ is considered optimal for muscle (Strijkers et al., 2010), 500–700 s/mm$^2$ for kidney (Borrelli et al., 2019), 700–800 s/mm$^2$ for peripheral nerve (Khalil et al., 2010), and 1000–1100 s/mm$^2$ for spinal cord (Jones et al., 1999). Multishell protocols are also used outside the brain, often to account for potential confounders such as partial volume effects with blood pseudodiffusion (i.e., intravoxel incoherent motion (IVIM—see Chapters 4 and 6)) or free diffusing fluids (e.g., cerebrospinal fluid around the spine). Alternatively, preparation modules aimed at suppressing specific contributions by means of inversion recovery could be considered (e.g., fluid attenuated inversion recovery diffusion weighted imaging (FLAIR-DWI) (Baron et al., 2015)). Importantly, failing to account for such effects can lead to biases in tensor estimates in the skeletal muscle (De Luca et al., 2017) and likely other tissues, impacting fiber tractography performance.

Outside the brain, bulk motion and flow artifacts can be detrimental to diffusion imaging, and can be taken into account by employing first- or second-order compensated gradients (Stoeck et al., 2016; Aliotta et al., 2017; Nguyen et al., 2016; Wetscherek et al., 2015) (see Chapter 6), although often at the price of a longer echo time. Cardiac gating is used for imaging the heart, and also for the spine and kidney, because of pulsatile CSF and blood and urine flow, respectively. Respiratory gating or breath-hold techniques are used in the abdomen.

The abundance of fat in body imaging can challenge the quality of diffusion imaging outside the brain. Incomplete fat suppression typically manifests as a ghosting artifact shifted from the original location. Because water molecules trapped in fatty tissues have low mobility and abundance, the fat ghosting artifact can locally reduce the SNR and alter diffusion tensor estimates, ultimately impacting fiber tractography (Arrigoni et al., 2018; Hooijmans et al., 2015). Fat suppression is highly important for fiber tractography of skeletal muscles. Often, the best results are achieved by combining fat suppression with adiabatic inversion recovery, slice-select gradient reversal, and saturation for olefinic fat (Hernando et al., 2011) (see Chapter 6).

Fig. 2 shows two examples of the application of fiber tractography outside the brain. The left panel (A) reports a reconstruction of the lumbosacral nerves in a postmortem subject. In this study, the authors acquired both nonfixed postmortem

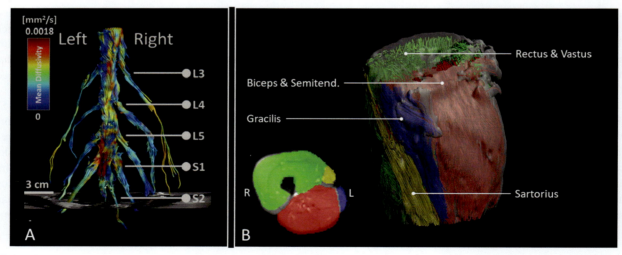

**FIG. 2** (A) A reconstruction of lumbosacral nerves obtained by performing DTI-based fiber tractography of a nonfixed postmortem sample. The streamlines are *color encoded* by the mean diffusivity. Lumbar vertebrae (L) and sacral vertebrae (S). The acquisition was performed at 1.5 T with 8 signal averages of one $b = 0$ s/mm$^2$ image and 15 directions at $b = 2000$ s/mm$^2$. The voxel size was 3 mm isotropic, and the acquisition parameters were TE = 82 ms, TR = 13.5 s, and SENSE factor 2. (B) Fiber reconstruction of the muscle groups of the human thigh using DTI-based fiber tractography. Each muscle group is color encoded according to the segmentation shown in the bottom left corner, which is shown according to the radiological convention (left (L) in the image = right (R) in space). The semitransparent surface shows the outer border of the leg muscles. The acquisition was performed at 3 T with 5 $b = 0$ s/mm$^2$ images and 16 directions each at $b = 250$ s/mm$^2$ and $b = 400$ s/mm$^2$. The voxel size was $2 \times 2 \times 6$ mm$^3$, and the acquisition consisted of 40 slices, with TE = 46 ms, TR = 6 s, SENSE factor 2, and fat suppression including gradient reversal. *((A) Reproduced with permission from Haakma, W., Pedersen, M., Froeling, M., Uhrenholt, L., Leemans, A., Boel, L.W.T., 2016. Diffusion tensor imaging of peripheral nerves in non-fixed post-mortem subjects. Forensic Sci. Int. 263, 139–146.)*

and in vivo data of the lumbosacral area, on 1.5 T and 3 T scanners, respectively, using single shell spin echo EPI diffusion acquisitions (Haakma et al., 2016). Tractography was performed by incorporating prior anatomical knowledge by accurate manual placement of tract seeding and exclusion regions, and of tract termination criteria based on the scalar DTI metrics fractional anisotropy (FA) and mean diffusivity (MD). Overall, tract reconstructions proved to be highly reproducible across multiple scans, both in vivo and postmortem. The right panel (B) of Fig. 2 shows an example application of fiber tractography to dissect four muscle groups of the human thigh. The diffusion dataset was acquired on a 3 T scanner and consisted of a 2 shell protocol with 16 directions at both $b = 250$ and $400 \, \text{s/mm}^2$, as w. In this case, fiber tractography was reconstructed by seeding a fiber tract in each voxel, then segmented using manually annotated masks (bottom left corner in Fig. 2B). Analysis of segmented muscles can be used to evaluate apparent fiber length (Mazzoli et al., 2016), or pennation angles (Oudeman et al., 2016), for example.

For more information about fiber tractography outside the central nervous system, the reader may refer to other texts that cover this application (e.g., Strijkers et al., 2010).

# 4 Terminology guide

| | |
|---|---|
| $B_0$ | The strength of the main magnetic field of an MRI scanner. It is aligned along the long axis of the bore of the magnet ($\hat{z}$ orientation). |
| $b$-matrix/$b$-tensor | A $3 \times 3$ matrix that quantifies the magnitude and direction(s) of diffusion weighting. |
| $b$-value | A parameter that quantifies the magnitude of diffusion weighting. When there is no diffusion weighting, $b = 0 \, \text{s/mm}^2$, and the image is sometimes called the $b = 0$ or $b_0$ image. |
| DW, DWI | Diffusion weighting, diffusion weighted imaging, or a diffusion weighted image. |
| EPI | Echo planar imaging; k-space traversal by scanning multiple lines of k-space in one excitation, ordered in a raster. |
| gradient | MRI scanners are equipped with gradient coils that linearly modulate the main field strength $B_0$ in the $\hat{x}, \hat{y}$, and $\hat{z}$ directions. |
| k-space | Term for the acquisition spatial frequency Fourier space in MRI. The vector $\mathbf{k}$ is the integral of the position encoding gradient over time. There is a Fourier relationship between the k-space representation of an image and the image itself. |
| MB | Multiband; the multiband factor. Also called simultaneous multislice (SMS). |
| PGSE | Pulsed gradient spin echo. Term for a diffusion encoding scheme with two monopolar unidirectional diffusion encoding gradients separated by a 180-degree RF pulse. Alternately called a Stejskal-Tanner or pulsed field gradient (PFG) encoding scheme. |
| $t_d$ | The diffusion time. The effective time over which the diffusion phenomenon is observed. |
| RF | Radio frequency. Applied RF radiation is used to create MRI signal and contrast, and the signal we measure is in the RF band. |
| shim | The process or hardware used to make the MRI magnetic field homogeneous. |
| slew rate | The rate of change of the magnetic field gradient in MRI. |
| SMS | Simultaneous multislice. Also called multiband (MB). |
| SNR | Signal-to-noise ratio. |
| STEAM | Stimulated echo acquisition mode. In diffusion MRI, a way to store magnetization longitudinally in order to shorten TE and/or lengthen the diffusion time such that the diffusion time can be longer relative to TE. |
| $T_1$ | The longitudinal relaxation time. |
| $T_2$ | The transverse relaxation time. |
| TE | The echo time. |
| TR | The repetition time. |

# References

Aliotta, E., Wu, H.H., Ennis, D.B., 2017. Convex optimized diffusion encoding (CODE) gradient waveforms for minimum echo time and bulk motion–compensated diffusion-weighted MRI. Magn. Reson. Med. 77 (2), 717–729.

Anblagan, D., et al., 2015. Tract shape modeling detects changes associated with preterm birth and neuroprotective treatment effects. NeuroImage Clin. 8, 51–58.

Arrigoni, F., et al., 2018. Multiparametric quantitative MRI assessment of thigh muscles in limb-girdle muscular dystrophy 2A and 2B. Muscle Nerve 58 (4), 550–558.

Autio, J.A., et al., 2020. Towards HCP-Style macaque connectomes: 24-Channel 3T multi-array coil, MRI sequences and preprocessing. NeuroImage 215, 116800.

Baron, C.A., et al., 2015. Reduction of diffusion-weighted imaging contrast of acute ischemic stroke at short diffusion times. Stroke 46 (8), 2136–2141.

Bastiani, M., et al., 2019. Automated processing pipeline for neonatal diffusion MRI in the developing Human Connectome Project. NeuroImage 185, 750–763.

Batalle, D., et al., 2017. Early development of structural networks and the impact of prematurity on brain connectivity. NeuroImage 149, 379–392.

Benner, T., Bakkour, A., Wang, R., Dickerson, B.C., 2009. High resolution ex-vivo diffusion imaging and fiber tracking. In: Proceedings of the 17th International Society for Magnetic Resonance in Medicine Annual Meeting, p. 3536.

Borrelli, P., Cavaliere, C., Basso, L., Soricelli, A., Salvatore, M., Aiello, M., 2019. Diffusion tensor imaging of the kidney: design and evaluation of a reliable processing pipeline. Sci. Rep. 9 (1), 12789.

Casey, B.J., et al., 2018. The Adolescent Brain Cognitive Development (ABCD) study: imaging acquisition across 21 sites. Dev. Cogn. Neurosci. 32, 43–54. Elsevier Ltd.

D'Arceuil, H., de Crespigny, A., 2007. The effects of brain tissue decomposition on diffusion tensor imaging and tractography. NeuroImage 36 (1), 64–68.

De Luca, A., Bertoldo, A., Froeling, M., 2017. Effects of perfusion on DTI and DKI estimates in the skeletal muscle. Magn. Reson. Med. 78 (1), 233–246.

Dubois, J., Dehaene-Lambertz, G., Kulikova, S., Poupon, C., Hüppi, P.S., Hertz-Pannier, L., 2014. The early development of brain white matter: a review of imaging studies in fetuses, newborns and infants. Neuroscience 276, 48–71. Elsevier Ltd.

Dyrby, T.B., Baaré, W.F.C., Alexander, D.C., Jelsing, J., Garde, E., Søgaard, L.V., 2011. An ex vivo imaging pipeline for producing high-quality and high-resolution diffusion-weighted imaging datasets. Hum. Brain Mapp. 32 (4), 544–563.

Eichner, C., et al., 2020. Increased sensitivity and signal-to-noise ratio in diffusion-weighted MRI using multi-echo acquisitions. NeuroImage 221.

Fogtmann, M., et al., 2014. A unified approach to diffusion direction sensitive slice registration and 3-D DTI reconstruction from moving fetal brain anatomy. IEEE Trans. Med. Imaging 33 (2), 272–289.

Fritz, F.J., Sengupta, S., Harms, R.L., Tse, D.H., Poser, B.A., Roebroeck, A., 2019. Ultra-high resolution and multi-shell diffusion MRI of intact ex vivo human brains using kT-dSTEAM at 9.4T. NeuroImage 202.

Gao, Y., Choe, A.S., Stepniewska, I., Li, X., Avison, M.J., Anderson, A.W., 2013. Validation of DTI tractography-based measures of primary motor area connectivity in the squirrel monkey brain. PLoS One 8 (10), e75065.

Gaspar, A.S., et al., 2019. Optimizing maternal fat suppression with constrained image-based shimming in fetal MR. Magn. Reson. Med. 81 (1), 477–485.

Gorodezky, M., Scott, A.D., Ferreira, P.F., Nielles-Vallespin, S., Pennell, D.J., Firmin, D.N., 2018. Diffusion tensor cardiovascular magnetic resonance with a spiral trajectory: an in vivo comparison of echo planar and spiral stimulated echo sequences. Magn. Reson. Med. 80 (2), 648–654.

Haakma, W., Pedersen, M., Froeling, M., Uhrenholt, L., Leemans, A., Boel, L.W.T., 2016. Diffusion tensor imaging of peripheral nerves in non-fixed post-mortem subjects. Forensic Sci. Int. 263, 139–146.

Haase, A., et al., 1986. MR imaging using stimulated echoes (STEAM). Radiology 160 (3), 787–790.

Henriques, R.N., Jespersen, S.N., Shemesh, N., 2020. Correlation tensor magnetic resonance imaging. NeuroImage 211, 116605.

Herculano-Houzel, S., 2009. The human brain in numbers: a linearly scaled-up primate brain. Front. Hum. Neurosci. 3 (NOV), 31.

Hernando, D., et al., 2011. Removal of olefinic fat chemical shift artifact in diffusion MRI. Magn. Reson. Med. 65 (3), 692–701.

Hooijmans, M.T., et al., 2015. Evaluation of skeletal muscle DTI in patients with Duchenne muscular dystrophy. NMR Biomed. 28 (11), 1589–1597.

Howell, B.R., et al., 2019. The UNC/UMN Baby Connectome Project (BCP): an overview of the study design and protocol development. NeuroImage 185, 891–905.

Hutter, J., et al., 2018. Time-efficient and flexible design of optimized multishell HARDI diffusion. Magn. Reson. Med. 79 (3), 1276–1292.

Jakab, A., et al., 2015. Disrupted developmental organization of the structural connectome in fetuses with corpus callosum agenesis. NeuroImage 111, 277–288.

Jakab, A., Tuura, R., Kellenberger, C., Scheer, I., 2017. In utero diffusion tensor imaging of the fetal brain: a reproducibility study. NeuroImage Clin. 15, 601–612.

Jones, D.K., Horsfield, M.A., Simmons, A., 1999. Optimal strategies for measuring diffusion in anisotropic systems by magnetic resonance imaging. Magn. Reson. Med. 42 (3), 515–525.

Jones, R., et al., 2020. Insight into the fundamental trade-offs of diffusion MRI from polarization-sensitive optical coherence tomography in ex vivo human brain. NeuroImage 214, 116704.

Jonkman, L.E., Kenkhuis, B., Geurts, J.J.G., van de Berg, W.D.J., 2019. Post-mortem MRI and histopathology in neurologic disease: a translational approach. Neurosci. Bull. 35 (2), 229–243. Springer.

Kasprian, G., et al., 2008. In utero tractography of fetal white matter development. NeuroImage 43 (2), 213–224.

Khalil, C., Budzik, J.F., Kermarrec, E., Balbi, V., Le Thuc, V., Cotten, A., 2010. Tractography of peripheral nerves and skeletal muscles. Eur. J. Radiol. 76 (3), 391–397.

Kimpton, J.A., et al., 2020. Diffusion magnetic resonance imaging assessment of regional white matter maturation in preterm neonates. Neuroradiology, 1–11.

Lebel, C., Treit, S., Beaulieu, C., 2019. A review of diffusion MRI of typical white matter development from early childhood to young adulthood. NMR Biomed. 32 (4), e3778.

Liu, C., et al., 2020. A resource for the detailed 3D mapping of white matter pathways in the marmoset brain. Nat. Neurosci. 23 (2), 271–280.

Marami, B., et al., 2017. Temporal slice registration and robust diffusion-tensor reconstruction for improved fetal brain structural connectivity analysis. NeuroImage 156, 475–488.

Mazzoli, V., et al., 2016. Assessment of passive muscle elongation using Diffusion Tensor MRI: correlation between fiber length and diffusion coefficients. NMR Biomed. 29 (12), 1813–1824.

McNab, J.A., Jbabdi, S., Deoni, S.C.L., Douaud, G., Behrens, T.E.J., Miller, K.L., 2009. High resolution diffusion-weighted imaging in fixed human brain using diffusion-weighted steady state free precession. NeuroImage 46 (3), 775–785.

McNab, J.A., et al., 2013. The Human Connectome Project and beyond: initial applications of 300 mT/m gradients. NeuroImage 80, 234–245.

Merboldt, K.-D., Hänicke, W., Frahm, J., 1991. Diffusion imaging using stimulated echoes. Magn. Reson. Med. 19 (2), 233–239.

Miller, K.L., et al., 2011. Diffusion imaging of whole, post-mortem human brains on a clinical MRI scanner. NeuroImage 57 (1), 167–181.

Miller, K.L., McNab, J.A., Jbabdi, S., Douaud, G., 2012. Diffusion tractography of post-mortem human brains: optimization and comparison of spin echo and steady-state free precession techniques. NeuroImage 59 (3), 2284–2297.

Mitter, C., Prayer, D., Brugger, P.C., Weber, M., Kasprian, G., 2015. In vivo tractography of fetal association fibers. PLoS One 10 (3), e0119536.

Nguyen, C., et al., 2016. In vivo diffusion-tensor MRI of the human heart on a 3 tesla clinical scanner: an optimized second order (M2) motion compensated diffusion-preparation approach. Magn. Reson. Med. 76 (5), 1354–1363.

Oudeman, J., Nederveen, A.J., Strijkers, G.J., Maas, M., Luijten, P.R., Froeling, M., 2016. Techniques and applications of skeletal muscle diffusion tensor imaging: a review. J. Magn. Reson. Imaging 43 (4), 773–788.

Pannek, K., Scheck, S.M., Colditz, P.B., Boyd, R.N., Rose, S.E., 2014. Magnetic resonance diffusion tractography of the preterm infant brain: a systematic review. Dev. Med. Child Neurol. 56 (2), 113–124.

Pieterman, K., et al., 2017. Cerebello-cerebral connectivity in the developing brain. Brain Struct. Funct. 222 (4), 1625–1634.

Qiu, A., Mori, S., Miller, M.I., 2015. Diffusion tensor imaging for understanding brain development in early life. Annu. Rev. Psychol. 66 (1), 853–876.

Roebroeck, A., Miller, K.L., Aggarwal, M., 2019. Ex vivo diffusion MRI of the human brain: technical challenges and recent advances. NMR Biomed. 32 (4), e3941.

Safadi, Z., et al., 2018. Functional segmentation of the anterior limb of the internal capsule: linking white matter abnormalities to specific connections. J. Neurosci. 38 (8), 2106–2117.

Salvan, P., et al., 2017. Language ability in preterm children is associated with arcuate fasciculi microstructure at term. Hum. Brain Mapp. 38 (8), 3836–3847.

Schilling, K.G., Janve, V., Gao, Y., Stepniewska, I., Landman, B.A., Anderson, A.W., 2018. Histological validation of diffusion MRI fiber orientation distributions and dispersion. NeuroImage 165.

Schmahmann, J.D., et al., 2007. Association fibre pathways of the brain: parallel observations from diffusion spectrum imaging and autoradiography. Brain 130 (3), 630–653.

Somerville, L.H., et al., 2018. The Lifespan Human Connectome Project in Development: a large-scale study of brain connectivity development in 5–21 year olds. NeuroImage 183, 456–468.

Sotiropoulos, S.N., et al., 2013. Advances in diffusion MRI acquisition and processing in the Human Connectome Project. NeuroImage 80, 125–143.

Steidle, G., Schick, F., 2006. Echoplanar diffusion tensor imaging of the lower leg musculature using eddy current nulled stimulated echo preparation. Magn. Reson. Med. 55 (3), 541–548.

Stoeck, C.T., Von Deuster, C., Genet, M., Atkinson, D., Kozerke, S., 2016. Second-order motion-compensated spin echo diffusion tensor imaging of the human heart. Magn. Reson. Med. 75 (4), 1669–1676.

Strijkers, G.J., Drost, M.R., Nicolay, K., 2010. Diffusion imaging in muscle. In: Jones, D.K. (Ed.), Diffusion MRI: Theory, Methods, and Applications. Oxford University Press, pp. 672–689.

Sun, S.W., et al., 2005. Formalin fixation alters water diffusion coefficient magnitude but not anisotropy in infarcted brain. Magn. Reson. Med. 53 (6), 1447–1451.

Wetscherek, A., Stieltjes, B., Laun, F.B., 2015. Flow-compensated intravoxel incoherent motion diffusion imaging. Magn. Reson. Med. 74 (2), 410–419.

Yendiki, A., Aggarwal, M., Axer, M., Howard, A.F.D., van Walsum, A.V.C., Haber, S.N., 2022. Post mortem mapping of connectional anatomy for the validation of diffusion MRI. NeuroImage 256, 119146.

Zanin, E., et al., 2011. White matter maturation of normal human fetal brain. An in vivo diffusion tensor tractography study. Brain Behav. 1 (2), 95–108.

Zöllei, L., Jaimes, C., Saliba, E., Grant, P.E., Yendiki, A., 2019. TRActs constrained by UnderLying INfant anatomy (TRACULInA): an automated probabilistic tractography tool with anatomical priors for use in the newborn brain. NeuroImage 199, 1–17.

# Chapter 9

# Diffusion MRI processing and estimation

Chantal M.W. Tax[a,b], Okan Irfanoglu[c], Jesper Andersson[d], and Alexander Leemans[e]

[a]CUBRIC, Cardiff University, Cardiff, United Kingdom, [b]UMC Utrecht, Utrecht University, Utrecht, The Netherlands, [c]Quantitative Medical Imaging Section, National Institute of Biomedical Imaging and Bioengineering, National Institutes of Health, Bethesda, MD, United States, [d]FMRIB Centre, Oxford, United Kingdom, [e]PROVIDI lab, Image Sciences Institute, University Medical Center Utrecht, Utrecht, The Netherlands

## 1 Introduction

Due to the ever-growing interest in diffusion magnetic resonance imaging (dMRI), novel approaches to optimal processing of dMRI data are being continuously developed. In this chapter, we describe the necessary processing steps to "prepare" the dMRI data for subsequent analysis and give an overview of commonly used and state-of-the-art processing methods. We consider the most commonly acquired type of acquisition: a (single-shot) echo-planar imaging (EPI) dMRI dataset acquired with multiple gradient directions and potentially multiple $b$-values, along with one or several anatomical (e.g., T1- and/or T2-weighted) MR images that are often acquired as part of a more general protocol.

In Section 2, we will discuss common artifacts in dMRI that are often still present in the data despite regular scanner quality assurance (e.g., by imaging phantoms). Some of these artifacts are well known and many software packages provide tools to correct for them; these corrections are more or less "standard" in a dMRI pipeline. Other artifacts have largely gone unnoticed but have gained renewed attention recently. Section 3 focuses on how to reliably estimate diffusion parameters once a specific model is chosen.

Note that there is no general consensus about the order in which data processing steps are applied and that these steps are often interwoven. Some artifacts, for example, can be corrected for during parameter estimation, or dMRI models can be used to spot and simultaneously correct for artifacts. In the following, we will often display image maps derived from DTI, such as the color-coded fractional anisotropy (FA) map, to evaluate the performance of the processing tools. As most researchers are familiar with these maps, it will facilitate the interpretation of the results.

## 2 Correcting for artifacts

One of the first steps in dMRI processing after the data has been acquired is to check image quality and correct for potential artifacts (Andersson and Skare, 2010; Jones and Cercignani, 2010; Jones and Leemans, 2011; Pierpaoli, 2010; Tax et al., 2016, 2022; Tournier et al., 2011). Artifacts can be considered as errors in the measurements that can lead to misinterpretation of the data. In the dMRI processing pipeline, measurements from multiple acquisitions will be combined, for example, from different diffusion images or from a T1-weighted image and a dMRI dataset. Consequently, both artifacts that occur in individual images or that systematically occur across multiple images can have significant consequences on the reliability of subsequent analysis. For example, artifacts can cause the estimated FA values to deviate from their true values, affecting a quantitative comparison in group studies. In addition, errors in estimates of the diffusion tensor (DT) first eigenvector can negatively bias diffusion tensor imaging (DTI)-based tractography and might change tract geometry or cause pathways to terminate in different brain areas. This in turn affects connectivity studies or hampers correct interpretation in neurosurgical applications.

A detailed inspection of dMRI data quality is often omitted since it is generally time-consuming. Given its importance, however, we will discuss in Section 2.1 a few straightforward and efficient tools for dMRI data quality checks that can be performed prior to and after artifact correction. Sections 2.2–2.7 will consider various types of artifacts and their correction strategies. Note that while there are many specialized data acquisition techniques for reducing artifacts (Andersson and Skare, 2010; Peterson and Bammer, 2016; Pierpaoli, 2010; Tax et al., 2016), we will focus here on the processing techniques. Sections 2.2 and 2.3 will discuss two artifacts that have been largely ignored in the diffusion community but that have gained attention recently: signal drift and Gibbs ringing. Whereas these artifacts are mainly acquisition related, other

---

Handbook of Diffusion MR Tractography. https://doi.org/10.1016/B978-0-12-818894-1.00006-9
Copyright © 2025 Elsevier Ltd. All rights are reserved, including those for text and data mining, AI training, and similar technologies.

artifacts are more subject related: Section 2.4 discusses motion-related artifacts such as bulk motion, cardiac pulsation, and signal instability between slices. Section 2.5 considers sources of artifacts resulting in geometric image distortions: eddy currents, susceptibility differences in tissues, and gradient nonlinearities. Finally, Sections 2.6 and 2.7 will discuss noise and signal outliers, respectively.

## 2.1 Data quality assessment

Visual inspection of the individual diffusion images can reveal some artifacts (e.g., instabilities between slices, Fig. 1A; slice dropouts, Fig. 1B; and signal decay due to vibration, Fig. 1C), but others are hard to spot in diffusion images. An efficient strategy for highlighting artifacts while avoiding a time-consuming and detailed slice-by-slice inspection of each dMRI dataset is to combine the display of diffusion images in various ways (Heemskerk et al., 2013; Tax et al., 2016; Tournier et al., 2011). For instance, quickly toggling between or looping through the same slice of different diffusion images can already reveal large signal dropouts, the presence of subject motion, and the magnitude of geometric distortions.

An image of the standard deviation of the nondiffusion-weighted images (non-DWIs) can highlight motion resulting from cardiac pulsation (Fig. 1D), while an image of the standard deviation of the DWIs can reveal image misalignment due to subject motion and eddy-current distortions as indicated by the bright rim in Fig. 1E. With further inspection of the color-coded FA map one can also recognize other artifacts. Examples include bulk subject motion leading to misalignment

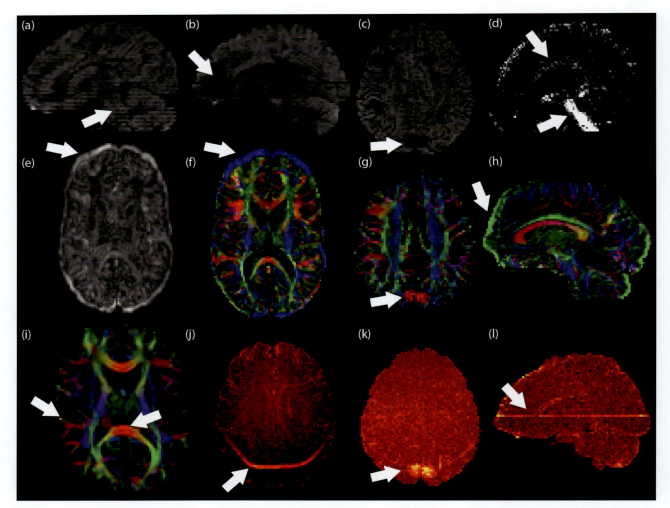

**FIG. 1** Examples of common dMRI artifacts. Some artifacts can be recognized in the diffusion images, such as instabilities between slices (A), slice dropout (B), and signal decay due to vibrations (C). Other artifacts can be more easily spotted with "standard-deviation" maps, such as cardiac pulsation around the brainstem and ventricles (D), and motion and eddy-current distortions (E); FA color-coded maps ((F) motion, (G) vibration, (H) eddy-current distortions, (I) artificially high/low FA values, in the corpus callosum likely due to Gibbs ringing); and residual maps ((J) chemical shift, (K) vibration, (L) slice dropout).

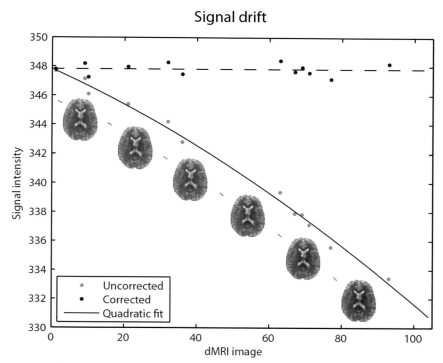

FIG. 2 Global signal decrease as a function of dMRI scan number. The *gray dots* represent the average signal of the non-DWIs. The *solid line* is a quadratic fit through these signal intensities, and the *black dots* represent the corrected average signal intensities for the non-DWIs (the DWIs are corrected in a similar way). In this example, there was an estimated 4.9% signal loss from the first to the last image.

between images (Fig. 1F), vibration (Fig. 1G, which is the same slice as Fig. 1C), eddy-current distortions (Fig. 1H), and abnormally high and low FA values (Fig. 1I).

Residuals, which represent the difference between a measurement and the prediction of the signal according to an estimated dMRI model or representation, can highlight artifacts that are not always clearly visible on individual DWIs or DTI-derived maps. An image of the average residual per voxel across DWIs should ideally be uniform. Large mean residuals either indicate that the chosen model is not optimal, or that artifacts are present (e.g., chemical shift artifacts, Fig. 1J; vibration artifacts, Fig. 1K, which is the same slice as Fig. 1C and G), or signal dropouts (Fig. 1L, which is the same slice as Fig. 1B). Plotting the average residual per DWI across voxels can highlight problems in data along specific diffusion-weighted (DW) gradients. Finally, identifying and quantifying outliers, which are measurements that are classified as being too distant from expected observations, can identify data quality issues (Chang et al., 2005, 2012; Heemskerk et al., 2013; Pannek et al., 2012; Tax et al., 2015; see also Section 2.7).

In addition to manual inspection strategies, some automated pipelines for evaluating data quality have been developed. Such pipelines can be used to facilitate quality checking in large cohort studies. Liu et al. (2010) and Oguz et al. (2014) developed an open-source software tool that aims to detect and reject volumes affected by interslice instabilities, residual motion artifacts, and vibration artifacts according to the method in Farzinfar et al. (2013). Roalf et al. (2016) compute quality metrics such as mean and maximum amount of intensity-based outliers, mean relative subject motion, and the signal-to-noise ratio (SNR) over time to classify data as being of poor, good, or excellent quality. A poor dataset is fully excluded from subsequent analysis. Although automated pipelines aim to improve the data fidelity, they are not always sensitive to artifacts and should therefore be extensively validated against visual quality assessment (Berl et al., 2015). Despite being labor intensive, keeping a strong (visual) connection to the data is still of paramount importance to warrant the reliability of subsequent analyses.

## 2.2 Signal drift

The combination of the EPI readout and the rapid switching of diffusion gradients puts a high load on the magnetic resonance imaging (MRI) system, resulting in heating and temporal instability of the scanner. Consequently, a change in global signal intensity can often be observed over time (Vos et al., 2017); see Fig. 2. This signal drift varies considerably in magnitude and temporal pattern when comparing scanners of different vendors. However, the exact underlying mechanism of signal drift is not always clear. For example, heating may cause signal drift through drift in the main magnetic field (B0), or through an altered transmission energy and flip angle over time. Signal drift can cause significant nontrivial effects on diffusion metrics and fiber tractography. Especially when the diffusion images are acquired in an ordered fashion, for example, from high to low $b$-value, a systematic artificial signal decrease at the end of the acquisition can be interpreted as overestimated magnitude of diffusion. Alternatively, when all gradient directions at the end of the acquisition protocol are primarily in one particular orientation, a systematic bias can occur along that orientation. Moreover, signal drift can also cause biases in acquisitions in which the diffusion directions and weighting are randomized.

During acquisition, it is possible to specifically correct for B0 drift (e.g., Benner et al., 2006), but B0 drift is potentially only a partial cause of the total signal drift. However, dynamically updating the center frequency might not be an option on several clinical scanners and does not fully correct for signal drift. Vos et al. (2017) therefore proposed a complementary and straightforward method to correct for signal drift during processing. They suggest to acquire the non-DWI volumes interleaved (i.e., every $n$th image volume is a non-DWI volume). The global signal decrease can then be estimated from these non-DWIs over time within a brain mask (created for each non-DWI to take into account subject motion). From a quadratic fit on these average signal intensities, they obtain a global rescaling factor as a function of image number that can be applied to each individual image for correction of the artifactual signal decrease (Fig. 2). More recently, this temporal modeling of signal drift has been extended by Hansen et al. (2019) to also incorporate any spatially dependent signal drifts.

## 2.3 Gibbs ringing

MRIs are typically reconstructed by applying an inverse Fourier transform to a set of sampled points in finite k-space. Since k-space holds the spatial frequencies of the MRI, leaving out the outer peripheral regions of k-space thus means that high-frequency information is lost. Such a truncation in k-space (or multiplication with a window function) is equivalent to the convolution with a sinc function in image space. The oscillating lobes of the sinc function, in turn, result in signal oscillations or "under- and overshoots" around sharp intensity edges in the image (Fig. 3A). Although this Gibbs ringing is a well-known MRI artifact, it has gone largely unnoticed in dMRI processing. After the qualitative notion that Gibbs ringing has an effect on DTI derived measures (Parker et al., 2001), its effect on dMRI has only relatively recently been quantitatively assessed (Veraart et al., 2016a; Perrone et al., 2015; Kellner et al., 2016). It was found that Gibbs ringing artifacts are more pronounced in the non-DWI because of the high intensity "jumps," which can, for example, be observed at the interface of cerebrospinal fluid (CSF) and the corpus callosum (CC, Fig. 3A). The Gibbs oscillations can lead to physically implausible signals (PIS), that is, a situation in which signals with a higher $b$-value are higher than signals with a lower $b$-value (whereas we expect that the signal decays with $b$-value). Therefore a PIS map can potentially identify Gibbs ringing artifacts (Fig. 3B; Tournier et al., 2011; Perrone et al., 2015). The ringing effect is often amplified in maps of dMRI metrics (e.g., Fig. 1I) and can, for example, result in highly abnormal DTI features (Veraart et al., 2016a; Perrone et al., 2015; Kellner et al., 2016).

There are several ways to correct for Gibbs ringing artifacts in MRI, including filtering to smooth out oscillations (Bakir and Reeves, 2000), piecewise reconstruction of smooth regions while preserving edges (Archibald and Gelb, 2002), extrapolating k-space data constrained by total variation (TV) regularization (Sarra, 2006), and performing a local subvoxel-shift to specifically avoid sampling of the ringing extrema (Kellner et al., 2016). In the context of dMRI, TV regularization approaches have shown to be able to alleviate the issue of Gibbs ringing (Veraart et al., 2016a; Perrone et al., 2015). Here, regularization refers to the introduction of additional information to better be able to solve such a complex problem. TV methods are based on the assumption that images with spurious detail have a high TV: the sum of the absolute gradient of the image is high. In practice, TV correction comes down to finding an image that is much similar to the original image but

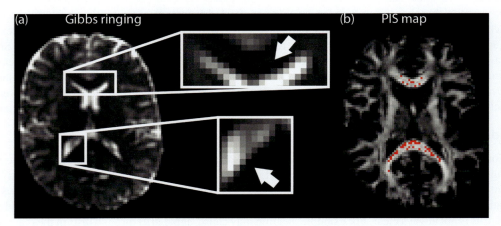

**FIG. 3** (A) Signal oscillations (or "under- and overshoots") around sharp intensity edges in the non-DWI at the interface of CSF and the CC. (B) PIS map indicating physically implausible signals where a DWI signal is larger than the non-DWI signal, overlaid on an FA image.

has less total intensity difference in neighboring voxels (i.e., the functional that is minimized includes a term based on the residuals when comparing the original and predicted image, and a TV term multiplied by a regularization parameter). This approach can be extended to capture higher-order variations of the image: in this case it is not expected that the image is piecewise constant, and, for example, linear variations in intensity are "allowed." This can avoid the "staircase effect" that might occur in corrected images (Veraart et al., 2016a). A drawback of TV methods is that the regularization parameter has to be tuned for each application individually: the optimal value is dependent on the noise level of the image and the degree of Gibbs ringing. Generally, it was found that its value should be lower for larger ringing effects, and is close to the noise standard deviation (Veraart et al., 2016a; Perrone et al., 2015). Kellner et al. (2016) reported that their approach, which is based on the resampling of the image such that the sinc function is sampled at its zero crossings, has advantages over TV in that it is more robust to parameter choices and independent of the noise ratio.

## 2.4 Subject motion

### 2.4.1 The effects of subject motion

Subject motion is potentially the most destructive, and hardest to fully correct, of all the artifacts that mar diffusion imaging. The most obvious effect of motion is the gross displacement of the object (brain) in one volume relative to another (Fig. 4A). If not corrected this introduces volume-to-volume changes in any given voxel, which can have profound effects on the modeling. Because of the strong contrast in both the $b = 0$ and the DWIs even a displacement by a subvoxel distance can cause signal changes in excess of 50%. With regards to this gross movement, the problem is very similar to that encountered in other multivolume EPI acquisitions such as fMRI or arterial spin labelling (ASL).

However, there is another artifact caused by subject movement that is particular to diffusion imaging. If the subject movement coincides temporally with the diffusion encoding of the sequence, it will lead to signal loss (Wedeen et al., 1994). This is hardly surprising given that diffusion images are specifically tailored to have a signal loss dependent on the amount of movement of water molecules. It is perhaps more surprising that not all types of movement cause signal loss; only rotations do (Storey et al., 2007).

In addition, there are other sources of unwanted and artifactual variance introduced by subject movement. These are all of smaller magnitude than the two outlined previously, but can be important to consider in cases or populations with larger than usual subject movement. These include:

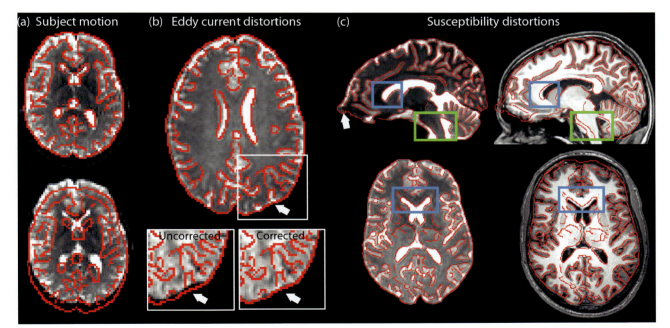

**FIG. 4** (A) Subject motion: a rigid bulk motion of the head (translations and rotations) results in misalignment of images that are acquired at two different time points in the sequence. *Top*: first non-DWI of the acquisition, with edges indicated in *red*. *Bottom*: a non-DWI later in the acquisition; the edges of the top-image are overlaid to indicate the misalignment. (B) Geometric distortions resulting from eddy currents. *Top*: non-DWI that is not affected by eddy-current distortions; edges are outlined in *red*. *Bottom left*: edges of non-DWI overlaid on a DWI that is affected by eddy-current distortions. *Bottom right*: improved alignment of the non-DWI and DWI after eddy-current distortion correction using registration. (C) Geometric distortions resulting from susceptibility gradients. *Left*: non-DWI with edges indicated in *red*. The *white arrow* shows an example of signal "smearing." *Right*: edges of non-DWI overlaid on an anatomical T1 image. *Green and blue squares* indicate misalignments.

**Intravolume movement** This refers to movement that occurs during the acquisition of the slices that jointly constitute a volume. One effect of this is the gross effect of the slices no longer lining up, distorting the volume that results from stacking together the slices. In that respect it is similar to the gross effect of movement described earlier, but because it is not corrected by most motion correction algorithms it is included here.

**Spin-history effects** This effect is related to the intravolume movement, in that it is most pronounced in the presence of such movement. The signal that is available on excitation of a slice depends on how long ago that same slice was previously excited. In the absence of subject movement, each slice is excited once every repetition time, and the same amount of signal is available each time. But consider the case where a slice is excited and read, after which the subject suddenly moves by exactly one slice, such that the same spins are excited again. Because of the very short time between the excitations, it will have much less signal available to it. This was a quite artificial and simplistic example, and spin history can cause a complex and heterogeneous mixture of hypo- and hyperintensities (see, e.g., Bhagalia and Kim, 2008; Yancey et al., 2011).

**Susceptibility-by-movement interaction** The susceptibility-induced off-resonance field is caused by the object itself, and as a first approximation follows the object. That is, if an object in one location causes a given field, then that object translated to a new location will yield a correspondingly translated version of the field. But any rotation of the object around an axis other than the direction of the magnetic flux of the main field will result in a new field that is not a rotated version of the original field. Hence, in the presence of subject movement with a rotational component, each volume will be distorted (slightly) differently.

**Movement in a nonhomogeneous receive field** Modern coil arrays have a "bowl-shaped" receive bias field, that is, signal originating from the center of the field of view (FOV) is detected with much poorer sensitivity than that originating from parts closer to the edges (closer to the coils; see, e.g., Kim et al., 2011). This means that between volumes a given location in the brain can move from a part of the FOV where the signal is detected with high sensitivity, to a part where it is located with poorer sensitivity, or vice versa. Thus, even after the volumes have been aligned geometrically, there will still be uncorrected signal modulation.

It should be stressed again that most datasets are not seriously affected by the secondary effects of movement listed earlier. The first two are an issue mainly in the presence of very sudden movement, and the latter two only in the presence of large movements.

### 2.4.2 Correcting subject motion

In this section, we will mainly discuss how to correct for the gross effects of movement and movement-induced signal loss. As described previously, the other, secondary, effects of movement tend not to be a big problem for most datasets.

To understand the particular problems associated with motion correction of diffusion images, it is useful to have a basic understanding of general image registration. There are really only three things that are needed for an image registration method.

**A transformation model** A model for transforming the coordinates and for the interpolation of coordinates between grid points, such that given some parameters one can calculate a transformed image. In the case of motion correction, the transform is the rigid body model (a subset of an affine transform) and the parameters are three translations and three rotations. DWIs carry orientation information; the images were acquired with diffusion sensitization in certain directions. Therefore, if rotations are necessary to align the image, the orientational information (captured in the B-matrix) should be rotated accordingly (Leemans and Jones, 2009).

**A cost-function** A function of a stationary (reference) image, a moving image, and the parameters of the transformation model, which has a minimum when the two images are aligned.

**An optimization method** A method for efficiently finding the previously mentioned minimum.

The main difficulty specific to diffusion imaging is how to devise a cost-function. DWIs are by design different from each other even when in the same location. A second difficulty is that all the different DWIs are distorted in different ways due to being affected by different eddy-current-induced off-resonance fields. Hence, it is not clear what to use for a reference, that is, what to align to.

Various attempts have been made to circumvent these problems. One suggestion has been to register all DWIs to a $b = 0$ image, thereby avoiding the problem of the reference image potentially being distorted (Rohde et al., 2004; Mohammadi et al., 2010). The large differences in image contrast between the images, and the ensuing cost-function difficulty, were addressed by using mutual information (MI) (Rohde et al., 2004; Mohammadi et al., 2010). However, the MI cost-function (Maes et al., 1997) was designed for dissimilar images where the "information" was largely the same, but where the contrast was different. T1- and T2-weighted images are a good example of this, where the information, in the nonpathological case, is essentially one of tissue type and very similar for the two. But the contrasts are very different, which prevents the use of "simpler" cost-functions. However, DWIs do by design have different information, which means that MI works increasingly poorly with increasing $b$-values.

A different approach has been to recognize that there is information pertinent to an image acquired with diffusion gradient $g_i$, in the other DWIs. This idea has been pursued in a model-based way (Andersson and Skare, 2002), by registering only to volumes acquired with a similar direction (Zhuang et al., 2013) or by registering to predictions from a Gaussian process (Andersson and Sotiropoulos, 2016). In particular, the latter has proven to be successful, and has been shown to perform better than an implementation of MI registration within the software package FSL (Smith et al., 2004).

### Correcting jointly with eddy-current-induced distortions

There are important reasons why motion parameters and eddy-current (see Section 2.5.1) parameters should be jointly estimated. A very significant part of the eddy-current-induced off-resonance fields consist of a linear combination of fields that vary linearly with location along the three main axes of the scanner. A linearly varying field along one of the axes orthogonal to the phase-encoding direction will cause a shear in the plane spanned by that axis and the phase-encode direction (Jezzard and Balaban, 1995). If, for example, the phase-encoding is along the $y$-axis, then a field that varies linearly along the $z$-axis will cause a $yz$-shear.

A rotation around any axis can be exactly composed as three shears in the plane orthogonal to the rotation axis (Eddy et al., 1996). For example, a rotation around the $x$-axis (the rotation most commonly seen in subject movement) can be composed as a $zy$-shear, followed by a $yz$-shear, and finished with another $zy$-shear. What this means is that a shear "looks like" a partial rotation, and this has important considerations for the estimation of both rotations (movement) and shears (eddy-current-induced distortions). Consider, for example, the case where a rotation around the $x$-axis is present in the data and where one attempts to estimate eddy-current-induced distortions and subject movement sequentially, starting with the former. The rotation would then be "interpreted" as a $yz$-shear, and after correction the rotated image volume would be sheared so as to, imperfectly, match the reference. When subsequently estimating movement, the rotation will have been largely accounted for by the shear, and the true rotation is likely to be vastly underestimated. And because, in this scenario, the movement model has no way of modeling a shear, it cannot undo that and instead find the "true" rotation. The scenarios may differ, but the only way to avoid this is to jointly estimate subject movement and eddy-current-induced distortions.

## 2.5 Geometric distortions

Some artifacts become more pronounced because of the EPI readout and manifest predominantly in the phase-encoding direction. This is caused by the way EPI uses the gradients to spatially encode the signal, that is, the single long "zig-zag-like" readout: the traversal in k-space is very slow in the phase-encoding direction compared to the frequency encoding direction. In other words, the "effective gradient strength" in hertz per millimeter in the phase-encoding direction—obtained by adding all the "blips" and distributing this over the acquisition period—is much lower than in the frequency encoding direction, almost on the order of magnitude of field inhomogeneities (Andersson and Skare, 2010). Hence, when the true applied field is not the field that we think we apply (off-resonance field), this has an amplified effect in EPI typically along the phase-encoding axis: the signal is reconstructed at the wrong location resulting in a geometrically distorted image.

Two important sources of off-resonance fields are eddy currents (Section 2.5.1), susceptibility differences in tissues (Section 2.5.2), and gradient nonuniformities (Section 2.5.3). Other off-resonance sources, which can usually be reduced during acquisition and therefore are not detailed here, are concomitant fields (higher-order terms in the gradients for spatial encoding; Baron et al., 2012; Du et al., 2002), incorrect shim, and local inhomogeneities in the field provided by the magnet (see, e.g., Peterson and Bammer, 2016). Residual effects of concomitant fields can often be "captured" by distortion correction during processing. Other common types of dMRI data artifacts that can be linked to or amplified by the EPI readout are Nyquist ghosting (Buonocore and Gao, 1997; Zhang and Wehrli, 2004) and chemical shift artifacts (Sarlls et al., 2011). However, these are not discussed here, as these artifacts are typically corrected at the data acquisition stage.

### 2.5.1 Eddy currents

Eddy currents are important sources of off-resonance fields that particularly affect DW acquisitions because of the additional diffusion gradients. To make a DWI the gradients have to be switched on rapidly, remain on for a reasonable time to achieve the desired diffusion weighting, and switched off rapidly. Such rapid changes of a magnetic field cause so-called eddy currents in conductors of the MRI, which in turn induce additional magnetic gradient fields. This effect is more problematic in dMRI than in most other acquisitions because the gradients for diffusion encoding are on for a longer time than typical gradients for spatial encoding. In the latter case, the eddy currents from switching on and off may compensate each other (Jones and Cercignani, 2010). The off-resonance fields resulting from eddy currents cause geometric distortions in the phase-encoding direction when they are constant over time during the EPI readout (Fig. 4B). However, they cause image blurring when they are time-varying, but this appears to be a minor issue (Andersson and Sotiropoulos, 2016). Finally, eddy currents can change the diffusion weighting in a nontrivial way, which affects subsequent modeling and is hard to correct for in a clinical setting. Acquisition techniques have been developed that reduce eddy currents (e.g., Reese et al., 2003) or that infer information on the eddy-current off-resonance field at the cost of an increased scan time (e.g., Jezzard et al., 1998; see also Andersson and Skare, 2010, for an overview). Instead, here we focus on processing techniques for the correction of the geometric distortions.

Registration-based techniques can be used to correct for eddy-current induced distortions during processing. Similar to registration-based motion correction, an appropriate transformation model is required that can describe the deformations resulting from eddy currents. Importantly, the eddy-current-induced distortions vary with the diffusion-encoding direction and $b$-value, where high $b$-value DWIs are affected more and non-DWIs remain virtually unaffected since no diffusion gradients are applied (Fig. 4B).

Proposed approaches vary in the degrees of freedom in their transformation model and whether they correct whole-brain or slice-wise. For example, if the field resulting from eddy currents is spatially linear, the distortions are affine: in-plane shearing (from eddy currents along the read direction), in-plane scaling (from eddy currents along the phase-encoding direction), and in-plane translation (from eddy currents along the slice selection direction) (Jezzard et al., 1998). Some software packages therefore use a whole-brain affine transformation per image to correct for both subject motion and eddy-current distortions. This results in the estimation of 8–12 parameters per image, where the "simplest" case only takes into account the distortions mentioned before (e.g., Mohammadi et al., 2010).

However, the distortions can vary dependent on the slice position, for example, when residual eddy-current fields from a previous slice add up to those of the current slice. This cannot be captured by a whole-brain affine transformation (Jones and Cercignani, 2010). Slice-by-slice correction for the in-plane residual distortions improves the accuracy of the registration (Mohammadi et al., 2010). Moreover, the assumption that eddy-current fields are spatially linear is likely to be incorrect.

Higher-order terms play an important role since it cannot be expected that eddy currents only reside in the gradient coils and that gradient systems are perfectly linear (Andersson and Sotiropoulos, 2016; Rohde et al., 2004). Therefore a quadratic model with 8 or 10 parameters (Andersson and Sotiropoulos, 2016; Rohde et al., 2004) or a cubic model with 16 or more parameters (Andersson and Sotiropoulos, 2016) can be adopted to correct for eddy currents. Simultaneous correction for subject motion (6 parameters) and eddy currents assuming a quadratic model (8 parameters) thus requires 14 parameters to be estimated per image. It is important to note that the volume of a voxel might change as a result of the registration, for example, in the case of scaling or shearing. This has the consequence that the intensity of the voxel should be modulated proportional to the volumetric change (e.g., by multiplying by the Jacobian determinant). Ignoring this step can lead to a wrong interpretation of the rate of diffusion in a given voxel and direction (Jones and Cercignani, 2010).

Several registration-based methods have been proposed to correct for rigid subject motion and/or eddy-current-induced distortions. To be able to use the simple SSD similarity metric, however, it remains challenging to find a good reference image that is not heavily affected by eddy-current distortions and has a similar contrast to DWIs. The non-DWI is free of eddy-current distortions but is very different in contrast compared to the DWIs. Therefore some studies proposed to acquire additional images with a more similar contrast, such as low $b$-value DWIs, for example, $b = 300$ s/mm$^2$ (Haselgrove and Moore, 1996) or a CSF suppressed non-DWI (Bastin, 2001). As the acquisition of additional images is not always possible in a clinical setting, we focus on registration-based methods that mostly rely on the acquired set of diffusion images. These methods can be broadly divided into conventional registration methods and prediction- or extrapolation-based registration methods (Nilsson et al., 2015).

Conventional registration methods register the acquired images themselves. An approach that overcomes the problem of registering images with different contrast is the method of Rohde et al. (2004), which uses normalized MI as a similarity metric (Studholme et al., 1999). Registration approaches based on MI have shown to accurately align diffusion images where all the images are generally registered to the first non-DWI, and are integrated in several diffusion MRI software packages (e.g., Leemans et al., 2009; Pierpaoli et al., 2010). Extrapolation-based methods register a predicted image from a model fit to the acquired set of diffusion images (Andersson and Skare, 2002; Bai and Alexander, 2008; Ben-Amitay et al.,

2012; Nilsson et al., 2015). These methods are therefore able to use similarity metrics such as SSD. Andersson and Skare (2002) developed "eddy_correct," which is included in FSL (Smith et al., 2004). They first fit a DT to the raw distorted data and use the tensor fit to predict DWIs with the same gradient directions as those acquired. The acquired images are then registered to the corresponding simulated images using the SSD as similarity measure, from which a new tensor fit is obtained. This process is repeated.

Registering high $b$-value images may require a more targeted approach due to the more significant contrast differences, considerable signal attenuation, and undefined tissue interfaces compared with low $b$-value images. The use of MI does not entirely solve these issues, and predictions of DWIs from the DT are inaccurate at high $b$-values. Mohammadi et al. (2010) register the DWIs to a median DWI per $b$-value shell, and Zhuang et al. (2013) only registered high $b$-value DWIs that are close in orientation.

In addition, several extrapolation-based methods based on other models than the DT were proposed. Ben-Amitay et al. (2012) register the high $b$-value DWIs in a multishell dMRI dataset by first rigidly registering the low $b$-value images to the non-DWI using MI. Subsequently, they simulate DWIs at high $b$-value using the composite hindered and restricted model of diffusion (CHARMED; Assaf and Basser, 2005), with the CHARMED parameters predicted from DT-derived indices obtained from the low $b$-value data. The acquired high $b$-value DWIs are then registered to the corresponding simulated DWIs using MI and a global affine transformation. Nilsson et al. (2015) stated that this DT-based extrapolation may negatively impact the registration in regions of partial volume with CSF, and proposed to correct for this by separating the DT into a tissue and CSF component. In addition, they adjust for artificial diffusion anisotropy in gray matter (Nilsson et al., 2015). They used a stretched-exponential model (Bennett et al., 2003) to extrapolate the signal and found visible improvement of the registration over the CHARMED-based method.

Andersson and Sotiropoulos (2016) estimate the eddy-current-induced field and predict estimates of undistorted DWIs based on Gaussian processes. Subsequently, they transform this prediction back to the distorted acquisition space to compare it with the acquired DWIs. They thereby avoid transformation and interpolation of the original data at this stage. Based on the difference of the original and predicted image, the motion and eddy-current distortion parameters are updated, and the whole process is repeated. This approach also has the option to simultaneously correct for susceptibility gradients (Andersson et al., 2003; see also Section 2.5.2), and is used in the Human Connectome Project (HCP) pipeline (Glasser et al., 2013; Sotiropoulos et al., 2013). Whereas most clinical acquisitions sample one-half of q-space, the authors suggest that the performance of the method can be improved by acquiring measurements with opposing gradient sign (Bodammer et al., 2004), where the angle between the gradient axes is small. This will "negate" the distortions and thus provide more information to drive the process. Alternatively, acquisition of gradient directions with opposite phase-encoding directions holds valuable information on the distortions. This method ("eddy") is implemented in FSL and the authors report improved results over "eddy_correct."

### 2.5.2 Susceptibility gradients

The magnetic susceptibility of a material refers to the degree of magnetization of an object when placing it in a magnetic field. Inside tissue, the magnetic field is slightly lower than the applied magnetic field due to its diamagnetic nature. This results in off-resonance fields (or macroscopic inhomogeneities) at the scale of the voxel size. These inhomogeneities are dependent on the composition and shape of the imaged body part and have a complex spatial distribution, particularly at air/tissue interfaces. The distortions resulting from such a susceptibility-induced off-resonance field (also known as susceptibility or EPI distortions) can be recognized as regions of signal "pile-up"—where the signal of several voxels is compressed into one voxel—or signal "smearing"—where the signal from one voxel is stretched over several voxels—see Fig. 4C, white arrow.

Since the susceptibility distortions only depend on the properties of the subject, all diffusion images are distorted in the same manner if one neglects subject motion. Susceptibility distortions would therefore not cause problems for local modeling per se, but they do affect tractography analyses (Irfanoglu et al., 2012; Jones and Cercignani, 2010), and introduce an extra source of intersubject variability. In addition, they cause a mismatch with anatomical images (Fig. 4C), which hampers a more complete picture of the structural organization of the brain. Therefore even though EPI distortion correction is not standard in many dMRI software packages, it is advised to be included in a typical dMRI processing pipeline.

A few things to keep in mind when correcting for susceptibility distortions: (1) they are dominant along the phase-encoding direction as discussed before, (2) resampling and interpolation of the images required for susceptibility distortion correction should ideally be combined with eddy-current distortion, motion, and other corrections; (3) the intensity in a voxel could be modulated when volumetric changes occur during correction (Jezzard and Balaban, 1995), but this is not strictly necessary because the modulation is the same for each DW volume; and (4) this step should not be accompanied by a rotation of the B-matrix because orientation information on a voxel level is not affected by this artifact (Irfanoglu et al., 2015).

Several approaches exist to reduce susceptibility distortions, generally requiring the acquisition of additional data. One method is to map the field inhomogeneity at every location. This can be achieved by considering the phase of two acquisitions (commonly gradient echo non-EPI, but other choices are also possible) with different echo times: one would expect that the accumulated phase is the same for the two acquisitions if no field inhomogeneities are present. Any phase difference between the acquisitions can thus be attributed to field inhomogeneities, and from the phase difference and the echo time difference, one can compute the field inhomogeneity (Jezzard and Balaban, 1995).

One way to correct for susceptibility distortions is to first convert such a field map to a voxel displacement map along the phase-encoding direction (e.g., Reber et al., 1998; Pintjens et al., 2008), that can be used to "unwarp" the distorted image onto an undistorted one. In practice, however, this approach has its drawbacks: the signal pile-up and smearing problem cannot be solved since there is not sufficient information as to how to exactly redistribute the signal back to its correct positions; it is a many-to-one mapping problem (Jones and Cercignani, 2010). It also faces other challenges (Holland et al., 2010), such as the risk of subject motion during the acquisition leading to errors in the field map, and the nontrivial process of unwrapping of the phase maps (i.e., correcting for "phase jumps" by adding multiples of $2\pi$ to the phase).

Another correction strategy, which can potentially deal with the signal redistribution problem and was shown to perform better than field mapping, involves the acquisition of pairs of images with reversed and/or varying phase-encoding directions (Bowtell et al., 1994; Chang and Fitzpatrick, 1992). In the case of reversed phase-encoding direction (e.g., bottom-up vs. top-down k-space traversal), the distortion direction is also reversed: regions of signal pile-up in one image correspond to regions of signal smearing in the corresponding image. The pair of such images provides sufficient information on the geometrical distortions and the redistribution of signals to create an anatomically faithful undistorted image. Such approaches, often referred to as "blip-up blip-down" methods, vary in the way they employ the images with different phase-encoding directions for the estimation of the distortion fields.

Morgan et al. (2004) and Bowtell et al. (1994) correct the distortions along every image line in phase-encoding direction, assuming that corresponding points in two reversed phase-encoding images are equidistant to the "true" anatomical location. This 1D correction per line, however, assumes that the subject is in the same exact position with the phase-encoding orientation still aligned with the image axis during the reversed scan, thereby ignoring subject motion. This can result in discontinuous displacement fields, and is sensitive to noise. Andersson et al. (2003) developed the approach TOPUP, which operates in 3D and is integrated in FSL (Smith et al., 2004) and used for the HCP data (Sotiropoulos et al., 2013). The authors propose to estimate a field map from the reversed phase-encoding images by modeling the image formation process of spin-echo EPI. Generally, they only use pairs of non-DWIs (either one pair at the beginning of the sequence, but ideally multiple pairs throughout the acquisition) to save acquisition time. However, the entire data (DWI and non-DWI) can be used as if it is acquired in a reversed phase-encoding fashion. Importantly, this approach takes into account subject motion between two images in a pair when determining the deformation field. From the estimated displacement field and the blip-up blip-down images, they are able to reconstruct the undistorted images through either a least-squares estimation process or Jacobian-determinant manipulation of the signal, a step that is integrated with the eddy-current correction tools in FSL (Andersson and Skare, 2002; Andersson and Sotiropoulos, 2016). However, the displacement fields in this approach are modeled with a finite set of basis functions, which limits the spatial variation (i.e., the highest variation that can be described is limited by the highest frequency in the used cosine basis; Holland et al., 2010; Ruthotto et al., 2012).

Holland et al. (2010) propose an efficient and fast variational approach by smoothing the reversed phase-encoding non-DWIs so that they look rather similar, register them, and update the total deformation field according to the computed refinement. In the next iterations, the width of the smoothing kernel is decreased until there is no smoothing. This approach allows for more flexible (nonparametric) transformations. Ruthotto et al. (2012, 2013) extend this approach by introducing an additional nonlinear regularization term that guarantees diffeomorphic transformations (i.e., the mapping is one-to-one and the map and its inverse can be infinitely differentiated), which generates anatomically more plausible fields.

The correction method from Hédouin et al. (2017) also aims to generate a diffeomorphic field by adopting the symmetric block-matching principles to the EPI distortion correction problem and the authors show that the quality of their correction is comparable to TOPUP with computation times being significantly faster. A critical note is that when only non-DWIs are used for the estimation of the correction fields, care must be taken that the correction is adequate in regions where the non-DWI has uniform contrast. In such regions, there might not be sufficient local contrast information to "drive" the correction process (Irfanoglu et al., 2015).

In a clinical routine, however, one often does not have access to more advanced acquisitions such as field maps and opposite phase-encoding direction acquisitions, or there is not enough time to acquire the additional scans. In this case, registration of the dMRI data to a structural "undistorted" image can be used to correct for susceptibility distortions, as such images are often acquired as part of the protocol. Registration methods have shown to perform similarly (or better/worse depending on the region) to field mapping (Tao et al., 2009; Wu et al., 2008). The transformation model for susceptibility correction typically includes nonrigid deformations only along the phase-encoding orientation. Registration of dMRI data to anatomical scans is challenging because images often do not necessarily have the same spatial resolution and coverage (i.e., in T1 images, for instance, typically a larger field of view is taken compared to DWIs).

In addition, the choices of similarity measure (e.g., SSD, MI), target image (e.g., T1, T2), and source image (e.g., FA, non-DWI, mean of the DWIs) may all affect the registration performance. Different methods have been developed for registration of dMRI data to anatomical scans. Kybic et al. (2000) register the non-DWI to a T2 image using the mean-squared difference as similarity metric and a B-spline modeled deformation field. Tao et al. (2009) developed a variational approach to register the non-DWI to the T2, where the term that forces similarity between the images depends on the derivative of the squared difference between the structural image and the transformed and Jacobian-determinant modulated non-DWI. They find that modeling the displacement field with higher-order B-spline deformations (i.e., a dense displacement field) better accounts for the distortions present in EPI. In contrast to relying on the difference between images, Wu et al. (2008) use MI as similarity metric combined with B-splines to register the non-DWI to the T2. The MI metric allows for registration of different dMRI quantities with different modalities, such as the FA, mean of the DWIs, and "anisotropic power maps" (Dell'Acqua et al., 2014) to the T1, and has been implemented in various software tools (Leemans et al., 2009; Pierpaoli et al., 2010). Glodeck et al. (2016) use MI to estimate a dense displacement field for the registration of the non-DWI to the T1, and introduce additional anisotropic regularization to make the registration process more robust.

Finally, hybrid approaches have been proposed that combine previously discussed methods. For example, Gholipour et al. (2011) use a field map to guide the registration of DWIs to a T2 image, where they used MI as similarity measure. This resulted in a better match between pairs of reversed phase-encoding images and between EPI and T1 images, compared with correction based on either registration or on field maps alone. Schilling et al. (2019) propose to employ deep-learning-based machine-learning techniques to estimate an undistorted non-DWI from a distorted non-DWI and a T1 image. They provide this distortion-free non-DWI to TOPUP with an infinite phase-encoding bandwidth to estimate a deformation field and take advantage of other features of TOPUP and EDDY.

Irfanoglu et al. (2015) propose an approach called DR-BUDDI in which they combine information from blip-up blip-down non-DWIs, DWIs (DTs in subsequent versions), and an undistorted T2 image to guide the registration. They use a deformation model capable of dealing with very large distortions in EPI (the popular symmetric diffeomorphic time-varying velocity-based model in the ANTS software package; Avants et al., 2008), and use an anisotropic regularization term with a cross-correlation similarity metric. In addition, they allow for incorporation of eddy-current and motion correction transformations for the blip-up and blip-down dataset independently, and therefore they account for the possibility that the phase-encoding direction in "reversed" images is inconsistent. The authors show that DR-BUDDI outperforms other methods (e.g., Andersson et al., 2003; Holland et al., 2010) in the presence of motion and large distortions and the quality of correction of white matter fiber bundles in regions where the non-DWI has uniform contrast, such as the pons, is superior due to the use of DWIs.

### 2.5.3 Gradient nonuniformities

Nonuniformities of the magnetic field gradients caused by hardware imperfections are known to affect the DWI gradients (Bammer et al., 2003). Violating the assumption of gradient linearity over the FOV results in errors in the dMRI measures, complicating data interpretation. In this context, several approaches have been proposed to model and correct gradient nonlinearities in dMRI, ranging from polynomial fits to more sophisticated methods based on signal modeling and optimization (e.g., Janke et al., 2004; Mesri et al., 2020; Doran et al., 2005). These methods have improved the accuracy and reproducibility of dMRI measurements, and they are expected to have a significant impact on clinical and research applications of dMRI in neuroscience, neurology, and oncology.

## 2.6 Noise

One important source of data quality deterioration affects all acquisitions: noise. Noise introduces a random scatter in the data causing the measured value to deviate from the true value. This can cause parameter estimates to be biased (i.e., there is a systematic deviation) and/or physically implausible. Here, we will focus on thermal noise, which arises from random fluctuations in the receive coil electronics and the subject to be imaged. The noise in the complex image domain is Gaussian for each receiving coil under certain assumptions (Henkelman, 1985). If we furthermore assume that a single coil fully samples k-space, the complex signal in image space is the sum of the true signal (i.e., the signal not contaminated by noise) and a complex additive white Gaussian noise term with zero mean and variance $\sigma^2$ ($N(0, \sigma^2)$). Hence, the measured signal has a mean that is equal to the true signal, and a variance $\sigma^2$ (St-Jean et al., 2016, 2020).

The magnitude image derived from the data in the complex image domain has a probability distribution corresponding to a Rician distribution with scale parameter $\sigma$. This distribution is more complicated than a Gaussian distribution: it has a mean that is not equal to the true signal value but also depends on $\sigma$, and a variance that in turn depends on the true signal value. Assuming a Gaussian distribution in the estimation process will thus bias the parameter estimates. Fortunately, in the case of high SNR (i.e., when the true signal is much larger than $\sigma$), the distribution of the derived magnitude signal converges to a Gaussian distribution. The mean of this Rician distribution depends on the true signal value and an "offset" related to $\sigma$ (Gudbjartsson and Patz, 1995; Fig. 5A, right), and is sometimes referred to as an offset-Gaussian function. In the image background, which consists of air and where the true signal is thus zero, the Rician PDF reduces to that of a Rayleigh distribution (Fig. 5A, left). For nonzero $\sigma$ the mean of this distribution is also nonzero, which gives rise to a minimum signal measurable even if the true signal is zero (also called the rectified noise floor; Jones and Basser, 2004). At low SNR this significantly affects DWI measurements and has a deleterious effect in Gaussian estimation procedures.

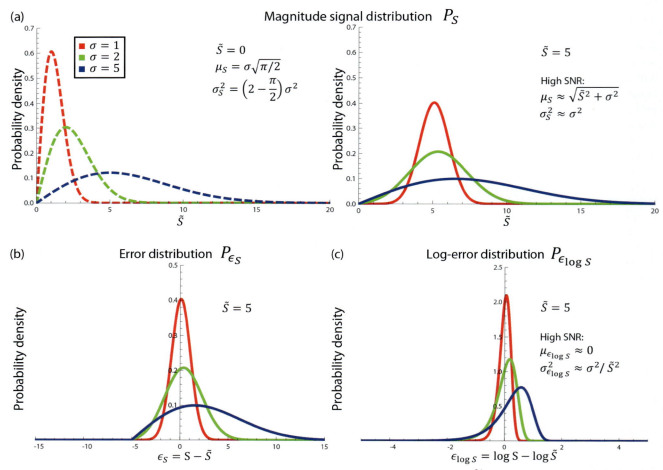

**FIG. 5** (A) The measured magnitude signal (here denoted by $S$) follows a Rician distribution with parameters $\tilde{S}$ indicating the true signal and $\sigma$ indicating the standard deviation of the normally distributed complex signal. When the true signal is zero, the distribution reduces to a Rayleigh distribution (*left*). In the case of high SNR ($\tilde{S} \gg \sigma$), the distribution starts to approximate a Gaussian distribution. (B) The error distribution for a signal with $\tilde{S} = 5$ for different $\sigma$; for higher SNR, this starts to look like a Gaussian. (C) The error distribution for a log-transformed signal with $\tilde{S} = 5$ for different $\sigma$; for SNR > 2 this starts to approximate a Gaussian.

The Rician distribution is valid in the case of a single coil. However, the signal is typically acquired with multiple coils, and the composite magnitude signal is reconstructed by combining the signals from each coil. If the composite signal is calculated by taking the sum-of-squares of the signals, there is no correlation between the coils, and the variance of the noise is the same in each coil, the composite magnitude signal follows a noncentral chi distribution (Constantinides et al., 1997). This distribution is a more general case of the Rician distribution, and reduces to the Rician distribution for a single coil. If the acquired signals are correlated between coils, the number of "effective coils" decreases and the noncentral chi distribution is an approximation of the real noise distribution. While several approaches can take the Rician distribution into account, fewer are generalized to the noncentral chi distribution.

The Rician and noncentral chi distribution assumptions are valid when k-space is fully sampled. Parallel imaging approaches are frequently used to shorten the acquisition time by acquiring only a subsampled version of k-space and combining the information from different coils. Such techniques can change the distribution of the noise dependent on the reconstruction process of the image (Aja-Fernández and Vegas-Sánchez-Ferrero, 2016). Examples of parallel imaging techniques include generalized autocalibrated partially parallel acquisitions (GRAPPA; Griswold et al., 2002) and sensitivity encoding (SENSE; Pruessmann et al., 1999). In the case of GRAPPA, the noncentral-chi model does not hold but can be used as an approximation for the magnitude-signal distribution. In the case of SENSE, the magnitude-signal is Rician distributed, just as in single-coil acquisitions (St-Jean et al., 2020).

In addition to acquisition with multiple coils and parallel imaging approaches, other acquisition and processing factors can influence the noise distribution. Importantly, some preprocessing steps that we discussed in Section 2.2 to correct for artifacts can also alter the noise distribution in a nontrivial way. For example, motion, eddy current, and susceptibility distortion correction require the interpolation of the noisy data. The combined effect of various artifact correction techniques on the noise distribution has not been extensively studied thus far.

Having knowledge of the noise variance is useful in parameter estimation, denoising techniques, outlier detection, and tuning regularization parameters, and there are several methods that try to estimate it from the data (Aja-Fernández et al., 2009). For example, the noise can be estimated from background areas of the image (e.g., Chang et al., 2005; Henkelman, 1985; Koay et al., 2009b; Sijbers et al., 2007) or from the image object (e.g., Chang et al., 2012; Coupé et al., 2010; Sijbers et al., 1998; Veraart et al., 2013a). The correlation between coils and the use of parallel imaging may result in signal distributions that are spatially varying. In this case, there is thus not a single SNR or noise standard deviation that can be estimated for the image, but a standard deviation for every location. The spatial dependence of the noise distribution can be visualized in so-called noise maps. Several methods have been proposed to estimate spatially varying noise maps from the data (e.g., Aja-Fernández et al., 2015; Glenn et al., 2015; Landman et al., 2009; Tabelow et al., 2015; Veraart et al., 2013a, 2016b). Alternatively, noise maps can be obtained as part of the dMRI acquisition by switching of the radiofrequency (RF) power and gradients but keeping other acquisition parameters the same (e.g., Froeling et al., 2017).

### 2.6.1 Dealing with noise during the estimation process

Once we have agreed on a dMRI biophysical or phenomenological model, we are left with the important challenge of how to obtain precise and accurate estimates of the parameters from the acquired dMRIs. Here, precision refers to estimates being consistent when measurements are repeated many times, and accuracy to estimates having no systematic error. Solving such an inverse problem is, unfortunately, challenging because (1) a model is an approximation and never exactly fits the measurements; (2) measurements always contain measurement noise; (3) if an exact solution does exist, it is not necessarily unique, that is, multiple (physically plausible) solutions can potentially result in the same predicted signals; and (4) the process of finding an inverse solution can be unstable: a small perturbation in the measurement can in some problems generate large changes in the estimates; this is called an ill-conditioned problem. Computations in our application domain are often ill-posed. Regularization, which refers to the introduction of additional information to solve an ill-posed problem or to prevent overfitting of the noise, can help to obtain more reliable and stable results.

Ideally, more measurements are performed than parameters to be estimated so that the system of equations is overdetermined. During parameter estimation we can then take into account the presence of noise in the measured data. Parameter estimation in dMRI is an active area of research: many approaches exist that vary in their underlying assumptions, incorporated prior information, and means of optimization (i.e., the method used to find a "best guess") (Koay et al., 2006). For example, some approaches explicitly take into account the Rician noise model and the noise floor associated with it. In addition, estimation approaches are mostly targeted toward a particular reconstruction approach. As such, it is difficult to unify the subject of estimation in dMRI, and we will concisely review common estimation techniques for some dMRI approaches in Section 3.

## 2.6.2 Dealing with noise prior to or after the estimation process

The effect of noise can also be reduced prior to parameter estimation by noise bias correction and denoising. The goal is to estimate or reconstruct the "true signal intensity image" given the noisy image. Many approaches have been developed for denoising of MRIs in general (e.g., Aja-Fernández et al., 2008; Manjón et al., 2010), or specifically for denoising in dMRI (e.g., Bao et al., 2013; Becker et al., 2014; Brion et al., 2013; Gramfort et al., 2014; Lam et al., 2014; Manjón et al., 2013; St-Jean et al., 2016; Tristán-Vega and Aja-Fernández, 2010). Some approaches estimate the true signal intensity and the noise variance simultaneously (e.g., Koay and Basser, 2006; Sijbers and den Dekker, 2004), while other approaches use noise estimates as input. Gudbjartsson and Patz (1995) propose to correct for the Rician noise bias by using the relationship between the mean magnitude signal and the true signal for high SNR (Fig. 5A, right), and apply this also to low SNR data. Koay and Basser (2006) derive the analytically exact correction and show that the correction in Gudbjartsson and Patz (1995) is a special case for high SNR. However, the corrected data are not Gaussian distributed after application of these methods. Koay et al. (2009a) aim to solve the noise-induced bias by transforming noncentral chi-distributed signals to Gaussian-distributed signals. Nonlocal means filters (Buades et al., 2005) restore the intensity in a voxel by taking a weighted average of voxels in the image. These voxels do not necessarily have to be close in position, hence the term non-local. The weights of the voxels depend on how similar these voxels are to the target voxel. Manjón et al. (2010) extended this approach to deal with spatially varying Rician noise. Such approaches may increase the reliability of dMRI parameter estimates (e.g., Zhou et al., 2015).

Denoising approaches specific to dMRI typically take advantage of the multiple acquired images in a dMRI dataset: they depict the same structure with different diffusion-weighting settings. Becker et al. (2014) proposed an approach that combines information from multiple shells to denoise dMRI data by using weighted means of neighboring points in imaging and q-space. By combining information from the "position-orientation" space, that is, the voxel position and the diffusion sensitizing orientation, the algorithm benefits from the high-SNR and low-orientation contrast in the position space and low-SNR high-orientation contrast in the orientation space. St-Jean et al. (2016) decompose the position-orientation space into patches to capture the local spatial and angular structure of the signal. In this way, they can account for spatially varying noise. To denoise the image, they find a local sparse representation by learning a dictionary instead of predefining it. Denoising based on principal component analysis (PCA) has become popular recently, based on the assumption that only a few principal components are necessary to describe the signal. The thermal noise principal components can be identified by their characteristic Marchenko Pastur distribution in the PCA eigenspectrum (Veraart et al., 2016b).

Alternatively, contextual processing can be performed as a postprocessing step after estimation to reduce the noise and enhance the structures present in dMRI data (e.g., Coulon et al., 2004; Westin et al., 2006). For example, diffusion Orientation Distribution Functions (dODFs) can be sharpened by using the orientational context at each location (Descoteaux et al., 2009; Florack, 2008). Other approaches work on the joint space of positions and orientations to enhance coherent ODF lobes and suppress incoherent structures (Barmpoutis et al., 2008; Dela Haije et al., 2014; Duits and Franken, 2011; Portegies et al., 2015; Reisert and Skibbe, 2012). These techniques have shown to be promising in clinical applications (Prčkovska et al., 2015; Tax et al., 2014a).

## 2.7 Outliers

In many practical situations, the data is not affected by thermal noise alone. Measurement artifacts that are not or insufficiently corrected for in the preprocessing stage can result in signal values that are very unlikely or even physically impossible, also called outliers. Examples include outliers resulting from vibration, cardiac pulsation, and system-related artifacts such as temporal scanner instabilities, spike noise, and signal dropouts. Alternatively, outliers can result from insufficient correction for chemical shift, breathing and head motion, and eddy-current/susceptibility distortions. Note that thermal noise can also result in outliers, and that it is difficult to disentangle these from outliers resulting from measurement artifacts. Some artifacts such as physiological noise are likely to result in signal dropouts (Chang et al., 2012), while other artifacts may lead to a signal increase. Outliers can significantly bias the estimates, and it is therefore important to minimize the effect of outliers before subsequent analysis.

Robust methods aim at reducing the effect of outliers during model estimation. Various robust estimation procedures have been developed specifically for dMRI (Chang et al., 2005, 2012; Collier et al., 2015; Mangin et al., 2002; Pannek et al., 2012; Parker et al., 2012; Berl et al., 2015; Tax et al., 2015; Tobisch et al., 2016; Zhou et al., 2011; Zwiers, 2010). The Robust Estimation of Tensors by Outlier Rejection (RESTORE) algorithm has been widely used in DTI and reduces the weight of outliers in an iterative procedure (Chang et al., 2005). An adapted version of this algorithm, called informed

RESTORE or iRESTORE, mostly targets signal dropouts and constrains the number of data points that can be excluded as outliers (Chang et al., 2012). Both RESTORE and iRESTORE are based on a nonlinear least square (NLS) in each iteration and are thus computationally expensive. Other approaches have used iterative reweighted linear least squares (IRLLS) fitting to speed up the process, without significant reduction of accuracy and precision of the estimates (Collier et al., 2015; Tax et al., 2015). The Robust Extraction of Kurtosis INDices with Linear Estimation (REKINDLE) approach is designed for robust estimation in DTI, diffusion kurtosis imaging (DKI), and other models that can be linearized, and takes into account the heteroscedasticity of the data after linearization. Other methods use higher-order models or representations to reduce the effect of outliers (Pannek et al., 2012; Tobisch et al., 2016). In general, care should be taken that rejecting outliers does not result in an ill-defined problem and that the fit is not biased by rejections of points clustered along certain directions (Chen et al., 2015). In such cases, results should be interpreted with extreme caution.

## 3 Estimation

### 3.1 DTI and DKI

Parameter estimation in DTI has been intensively investigated since its proposal, but even for this relatively "straightforward" case estimation remains challenging. For example, noise can cause the eigenvalues to become negative, which is physically implausible (note that the DT should be positive definite). The estimation approaches in DTI can be subdivided into (1) least squares (LS), (2) maximum likelihood (ML), and (3) Bayesian methods.

LS approaches are particularly appealing because they have been extensively studied and can be fast (Koay et al., 2006; Tristán-Vega and Aja-Fernández, 2010). They try to find a "best guess" by minimizing an objective function based on the sum of squared residuals, where a residual represents the difference between the measurement and the model prediction. Here, we will discuss nonlinear LS, linear least squares (LLS), and weighted linear least squares (WLLS). Since the DT has a nonlinear (exponential) relationship with the signal, the estimation process can be written as a nonlinear regression function, and NLS can be employed to estimate the DT. The nonlinear regression problem can be solved with off-the-shelf optimization algorithms such as the Levenberg-Marquardt optimization strategy. However, NLS estimation faces some challenges. First, NLS has to find an optimum in a nonlinear parameter landscape by gradually moving around and thus takes a long time. In addition, a risk exists of getting stuck in local optima; we therefore might not be entirely sure if the estimated parameters are also the "best guess" globally. Finally, when writing the problem as a nonlinear regression function including an error term, it is assumed that this error term has zero mean and a given variance. It can be shown that in the case of Rician distributed measurements this is generally not true (Fig. 5B), and as a result the parameter estimates will be biased (i.e., they will have a systematic error). This problem becomes more severe for lower SNR. It can be ameliorated (but not entirely solved), for example, by applying NLS where the model prediction is replaced by its approximate expectation under a Rician PDF (Jones and Basser, 2004). More specifically, the objective function is not directly the sum of squared residuals, but an offset is added to the model prediction before calculating the residuals. This is sometimes referred to as the offset-Gaussian objective function. Omitting this bias correction can result, for example, in an underestimation of the apparent diffusion coefficient (ADC).

In addition to NLS, LLS can be adopted by linearizing (i.e., taking the log-transform of the measured data), which makes it much easier to implement and faster than NLS. Another advantage is that the error terms in the linear regression equation turn out to be normally distributed for each measurement when the SNR of the DWIs is at least 2 (Fig. 5C; Koay et al., 2006; Salvador et al., 2005). However, the variance is not equal for all error terms in this case: as a result of the log-transform it now depends on the magnitude of the true underlying signal. This phenomenon is called heteroscedasticity (homoscedasticity thus refers to equal variance across measurements). This means that LLS is not optimal in that it assigns an equal weight to each residual; we should instead assign a higher weight to those measurements with a lower variance. This can be achieved with a WLLS fit, where the weight of each measurement is equal to the square of the true underlying signal. WLLS estimation theoretically yields the best linear unbiased estimator with the highest precision within its class. However, here lies another problem: we only have the measured signals that are affected by noise and thus not optimal to determine the weight. Instead, an iterative weighted linear least squares (IWLLS) can be used: predictions of the true signals are generated from an initial (W)LLS fit, and these are in turn used as weights for the next iteration. In practice, IWLLS shows a high performance in terms of accuracy and precision and may even be preferred over NLS (Veraart et al., 2013b).

Constrained versions of NLS, LLS, and WLLS exist to prevent the DT eigenvalues becoming negative. In DTI, this can be achieved quite elegantly: instead of the representation of the DT in its elements, we can simply write it differently. The Cholesky representation, for example, implicitly constrains the eigenvalues to be positive (in fact, the well-known eigenvalue decomposition is yet another way to represent the DT). Instead of estimating the tensor elements directly, we can then estimate elements of this other representation and compute the tensor elements from them (Koay et al., 2006).

ML estimation maximizes the probability of measuring the observed data given a set of parameters and an assumed noise model. Here, we can thus explicitly incorporate the Rician distribution (ML reduces to LS when the noise model is Gaussian). Also in this framework the DT can be constrained to be positive definite by representing it in a different way (e.g., as a Log-Euclidean metric; Fillard et al., 2007). Alternatively, when representing the DT using eigenvalue decomposition, the likelihood function can be multiplied by a prior on the DT eigenvalues and the non-DWI signal to ensure positivity (Andersson, 2008). In the latter case, we have entered the area of Bayesian statistics: maximum a posteriori (MAP) estimation maximizes the probability of the parameters given the data according to Bayes rule (which includes the term for ML estimation, a prior term, and a "data likelihood" term that is usually constant). Including a Rician noise model using this framework was found to result in less biased estimates, but in a poorer precision of the estimates compared to a Gaussian noise model for very low SNR (Andersson, 2008). The authors suggest that in group studies the high precision may be more important, and Gaussian noise models could remain to be used.

In addition to introducing priors to limit the possible range of the estimates, spatial regularization in the context of DTI can be performed. The purpose of spatial regularization is to suppress the influence of noise by, for instance, using neighborhood information, and to obtain a "smoother" appearance of the tensor field and derived features. Spatial priors can be introduced during the estimation process itself, which turns this into MAP estimation. For example, Fillard et al. (2007) include an anisotropic prior that preserves edges while smoothing homogeneous regions in an ML framework, and Liu et al. (2013) combine smoothness constraints with an NLS term. Both works enforce positive definiteness of the DT and show ability to effectively deal with noise.

Parameter estimation in DKI bears a resemblance to estimation in DTI since it can be written as a similar linear or nonlinear regression equation. For example, LS approaches have also been adopted for DKI estimation (e.g., Lu et al., 2006; Veraart et al., 2013b), where IWLLS shows high performance characteristics. In the context of DKI, however, the occurrence of physically implausible parameter estimates as a result of noise is an even bigger problem than in DTI. Specifically, the constraints that the kurtosis along any direction is likely positive in biological tissue and has a theoretical upper bound are often violated. Tabesh et al. (2011) proposed an LLS approach with linear constraints only along the discrete imaging gradient directions, which can be solved by a computationally heavy algorithm or a faster heuristic algorithm that approximates the optimal solution. Ghosh et al. (2014) evaluated a weighted least squares (WLS) approach in which the DT and kurtosis tensor (KT) were reparameterized (Cholesky and ternary quartic, respectively) to fulfill the positivity constraints along all directions and thus not only the imaging directions. The maximum kurtosis constraint was applied explicitly along the imaging directions by using efficient sequential quadratic programming. Glenn et al. (2015) used WLLS with noise bias correction to account for bias in DKI estimates derived from measurements with low SNR. In addition to LS approaches, constrained ML approaches have been developed to properly account for the Rician noise, thereby avoiding overestimation of the kurtosis values (Ghosh et al., 2014; Veraart et al., 2011).

## 3.2 Spherical deconvolution approaches

Spherical deconvolution (SD) approaches rely on a response function as input. Choosing an incorrect response function can result, for example, in spurious lobes and angular deviations (Parker et al., 2013). Even though being an oversimplification, it is generally assumed that there is a single response function throughout the whole brain. Several methods have been proposed to determine a suitable white matter RF. For example, it can be modeled by a tensor with predefined axial and radial diffusivities (Dell'Acqua et al., 2007; Tournier et al., 2004). However, at the $b$-values typically used for SD, the monoexponential assumption underlying the DT is not valid anymore. Such a DT-predicted single-fiber response thus does not include the restriction effects that typically can be measured at this high $b$-value. Instead, Tournier et al. (2007) proposed to estimate the response function from the data by selecting the 300 highest FA voxels, or only voxels with an FA exceeding a predefined FA threshold and averaging their signals. This method is still based on a DT fit to the data, and it is not always straightforward which FA threshold to use, for example, in ex vivo, spinal cord, or muscle dMRI data. Alternatively, a region of interest can be drawn in single-fiber regions such as the CC to estimate the single-fiber response function. However, this method introduces a certain level of user-dependence.

Tax et al. (2014b) proposed to recursively calibrate the response function from the data by iteratively removing voxels that exhibit more than one fODF peak with SD. This method aims to find single-fiber-population voxels using a more intuitive criterion than FA: a voxel is removed from calibration when the second peak exceeds a predefined peak ratio threshold compared to the first peak. The white matter response function is suboptimal in regions of partial volume effect with CSF and gray matter, which may result in an increase in the detection of false peaks and a decreased precision of the detected fiber orientations (Roine et al., 2014). Therefore the RF can be modified, for example, based on tissue fractions estimated from anatomical data (Jeurissen et al., 2014; Roine et al., 2015).

There are multiple implementations of the SD approach of which the majority can be cast in a unified deconvolution framework (Jian and Vemuri, 2007). In these approaches, the fODF is generally expressed as a linear combination of basis functions, where each basis function is multiplied by a weight. This allows the convolution problem to be written as a linear problem: the signal vector is a matrix multiplication of a coefficient vector with a design matrix that can analytically or numerically be computed from the response function and the basis functions. Estimation of the fODF then comes down to estimating the basis function coefficients. Regarding the choice of basis functions, SD methods can be classified into two groups (Cheng et al., 2014): (1) SD based on a continuous representation such as the spherical harmonics (SH) basis (Tournier et al., 2004, 2007) and (2) SD based on a discrete representation—a mixture of rotated versions of the response in discrete directions (Dell'Acqua et al., 2007, 2010; Ramirez-Manzanares et al., 2007). The latter representation holds a close connection to multicompartment models in which each compartment has a different orientation. In this case, the coefficients are the weights for each compartment. The drawback of this representation is that the angular resolution is limited by the number of directions on the sphere along which the response function is rotated. Other examples of basis functions include higher-order tensors (Feng et al., 2015; Weldeselassie et al., 2012) and Wisharts (Jian et al., 2007).

Solving the linear SD problem faces some challenges. For example, if a weight has to be estimated for every compartment, then the number of parameters easily exceeds the number of measurements (the system is under-determined). In this case, there exists no unique LS solution and the problem is ill-posed. In addition, the problem is often ill-conditioned: noise has a tremendous influence on the stability of SD approaches and can cause spurious lobes and physically implausible negative weights. Regularization and constraints are therefore commonly employed based on the assumptions that (1) only a few weights are nonzero, a property called sparsity, and (2) the fODF is a nonnegative function.

A popular SD approach is constrained spherical deconvolution (CSD), in which the fODF is expressed in SH basis functions (Tournier et al., 2007). This method iteratively obtains an improved fODF estimate employing Tikhonov regularization, which has an analytical solution and is in fact a regularized LLS estimation. More specifically, the objective function consists of the sum of squared residuals and a sum of the squared negative and small weights below a certain threshold. The regularization employed in CSD essentially drives negative and small values to zero while promoting large values, but it does not guarantee positivity or sparsity. This approach can be further extended with a correction for the Rician bias and spatial regularization (e.g., Tournier et al., 2013).

Another commonly used SD approach is Richardson-Lucy (RL) SD in which the solution can be constrained to be positive (Dell'Acqua et al., 2007). This method uses a set of DTs with predefined diffusivities oriented along many directions on the sphere as basis. The weight of each tensor is estimated in an iterative procedure. A damped version of this algorithm (dRL) can reduce the influence of isotropic partial volume effects from CSF or gray matter without adapting the white matter response function (Dell'Acqua et al., 2010); dRL is less sensitive to the choice of response function and causes less spurious peaks than CSD, but has a lower ability to resolve crossing fibers that have a low anisotropy (Parker et al., 2013). The ability of solving fiber crossings was shown to be improved by including spatial regularization and Rician/noncentral-chi noise models (Canales-Rodríguez et al., 2015; Liu et al., 2015).

Reconstructing a sparse fODF from a limited number of measurements can also be obtained by using concepts from the theory of compressed sensing (Donoho, 2006) and convex optimization. Minimizing the $l_0$-norm—equaling the number of nonzero entries—theoretically results in the sparsest solution. In practice, however, $l_0$-minimization problems are difficult to solve and it is often proposed to minimize the $l_1$-norm instead—which is the sum of the absolute coefficients—or a weighted version of the $l_1$-norm, or a combination of the $l_1$-norm and $l_2$-norm (Daducci et al., 2014; Feng et al., 2015; Jian and Vemuri, 2007; Landman et al., 2012; Ramirez-Manzanares et al., 2007). These sparse regularizations can be accompanied by spatial regularization to promote spatial coherence of, for instance, fiber directions or weights (Auría et al., 2015; Ramirez-Manzanares et al., 2007; Ye et al., 2016).

Nonnegativity constraints are not always enforced in the previously discussed works. To this end, Jian and Vemuri (2007) propose to solve a nonnegative LS problem using quadratic programming. Alternatively, Cheng et al. (2014) guarantee nonnegativity over the whole sphere instead of discretized points on the sphere by representing the fODF differently: the square root of the fODF is written as a linear combination of SH basis functions (see also Chapter 10).

## 3.3 Multicompartment models

Accurate and precise estimation with such routines is highly challenged by the large number of parameters to be estimated. This can even lead to degeneracies, with multiple equally plausible sets of parameters resulting in the same signal (Jelescu et al., 2016). Therefore simplifying assumptions are sometimes made to reduce the number of parameters and increase the precision of the estimates, for example, equal diffusivities in different compartments (e.g., Zhang et al., 2012). However, such assumptions can lead to significant bias in other parameters (Lampinen et al., 2017), and recent works rely instead on more extensive acquisitions for parameter estimation without fixing parameters (Coelho et al., 2022; Jelescu et al., 2016). Nonlinear estimation techniques are commonly adopted for parameter estimation of multicompartment models. Ferizi et al. (2014, 2015), Jones and Basser (2004), and Panagiotaki et al. (2012) use a Levenberg-Marquardt algorithm for NLS fitting in single-fiber population voxels with an offset-Gaussian objective function. Zhang et al. (2012), Alexander (2008), Alexander et al. (2010), and Tariq et al. (2016) use ML estimation with a Gauss-Newton optimization technique, sometimes followed by a Markov chain Monte Carlo procedure (Alexander et al., 2010). These estimation methods are computationally expensive and time-consuming, particularly in large cohort studies.

To address the computational issue in multicompartment model fitting, Daducci et al. (2015) propose to decouple the estimation of the number and orientation of distinct fiber populations, and the assessment of microstructural properties per fiber population. When an estimate of the underlying fiber population direction(s) is obtained (e.g., from DTI or SD), they can express the multicompartment model for each population as a linear system. The signal vector is written as a matrix multiplication of a coefficient vector with a design matrix or dictionary. In this case, the dictionary can be built from the expected signal attenuations at the measured q-space points for a range of parameter settings for each compartment. For example, in the case of NODDI, the dictionary is constructed from the signal attenuations corresponding to a range of intracellular volume fractions and orientation dispersions. This framework drastically accelerates the fitting and has shown to produce estimates that are in agreement with estimates from conventional fitting procedures.

Some works aim to improve or facilitate the estimation of microstructural parameters by first fitting a more general representation. For example, Fieremans et al. (2011) and Hui et al. (2015) derive biophysical parameters by first fitting the DKI model. The purpose is twofold: it facilitates a clearer interpretation of the KT in terms of microstructure, and the multicompartment model estimation can benefit from advanced estimation methods that are already available for DKI. However, such approaches can only capture the information that is available in the DT and KT, and is not sensitive to microstructural features that affect higher-order properties of the signal. Fieremans et al. (2011) derive analytical relationships between KT features and parameters of a physical model for a single-fiber direction within a voxel. Hui et al. (2015) developed a framework that can be used with more general models (e.g., crossing fibers, gray matter), but that requires nonlinear optimization of the parameters. Alternatively, Fick et al. first fit a regularized version of the MAP method (Özarslan et al., 2013) and then extrapolate the signal as a preprocessing step for the fitting of multicompartment models (Fick et al., 2016). They show a reduced variance of the estimates of AxCaliber (Assaf et al., 2008) and NODDI (Zhang et al., 2012) after this preprocessing step.

The fODF can be factored out (thereby reducing the complexity of estimation) by taking the powder average or spherical mean (Kaden et al., 2016) or computing rotationally invariant SH (Novikov et al., 2018).

## 3.4  q-Space approaches

In Q-ball imaging (Tuch, 2004), integration over a circle in q-space can be done by using interpolation or analytically by using SH (Descoteaux et al., 2007; Hess et al., 2006). In diffusion spectrum imaging (DSI), the truncation of q-space beyond a certain value can lead to Gibbs ringing artifacts in the ensemble average propagator (EAP) when taking the discrete Fourier transform. Therefore the q-space data is typically first multiplied by a filter (e.g., Hanning filter) to enforce a smooth attenuation of the signal with q-value (Wedeen et al., 2005). The signal is assumed to be zero beyond a certain q-value and is "zero-padded" before taking the Fourier transform. Only the real and nonnegative parts of the resulting EAP are maintained. To compute the dODF, the EAP is typically radially integrated from zero up to a certain maximum radius. In practice, however, there is no consensus as to what the interval of integration should be to avoid propagation of ringing and other artifacts into the dODF. Paquette et al. (2014) propose to integrate from a nonzero radius to a maximum radius dependent on the SNR. Tian et al. (2016) propose to use the unfiltered q-space data and truncate the EAP beyond its "extent," which was approximated by the mean displacement distance (as upper bound the ADC parallel to the CC was taken) (see also Chapters 10 and 12).

Methods that are based on basis functions to reconstruct the EAP require a reliable estimation of the coefficients. The fitting of such bases is often sensitive to noise, and several regularization methods have been proposed. These methods generally enforce the reconstructed signal to be smooth (angularly and/or radially). Examples of regularizations include Laplace-Beltrami regularization (solid harmonic basis) (Descoteaux et al., 2011) combined with a radial low-pass filter (SPF basis) (Assemlal et al., 2009), and Laplacian regularization (SPF basis) (Caruyer and Deriche, 2012), SHORE and MAP-MRI bases (Fick et al., 2016). Fick et al. use Laplacian regularization in an LS framework to fit the MAP basis, and show that their method outperforms a previously proposed method for MAP MRI fitting (Özarslan et al., 2013) and Laplacian regularization of the SPF basis (Caruyer and Deriche, 2012) in terms of signal fitting and reconstruction of the EAP and dODF (Fick et al., 2016).

# 4 Conclusion

Fiber tractography results are known to depend on the diffusion MRI approach used to derive the fiber orientations and the specific user-defined and algorithm-specific parameter settings (Jeurissen et al., 2019). To minimize further variability in reconstructing fiber pathways, proper preprocessing of diffusion MRI data and model estimation strategies are required (e.g., Leemans and Jones, 2009; Gallichan et al., 2010). This chapter provided an overview of the most common artifacts in diffusion MRI data and how to adequately correct for them, and of some strategies to reliably estimate diffusion MRI parameter maps and orientation distributions. Machine learning approaches to artifact correction and estimation are being developed and show promising results, and after thorough evaluation have the potential to address traditionally intractable problems.

# References

Aja-Fernández, S., Vegas-Sánchez-Ferrero, G., 2016. Statistical Analysis of Noise in MRI: Modeling, Filtering and Estimation. Springer, Cham.

Aja-Fernández, S., Alberola-Lopez, C., Westin, C.F., 2008. Noise and signal estimation in magnitude MRI and RICIAN distributed images: a LMMSE approach. IEEE Trans. Image Process. 17 (8), 1383–1398. https://doi.org/10.1109/TIP.2008.925382.

Aja-Fernández, S., Tristán-Vega, A., Alberola-López, C., 2009. Noise estimation in single- and multiple-coil magnetic resonance data based on statistical models. Magn. Reson. Imaging 27 (10), 1397–1409. https://doi.org/10.1016/j.mri.2009.05.025.

Aja-Fernández, S., Pieciak, T., Vegas-Sánchez-Ferrero, G., 2015. Spatially variant noise estimation in MRI: a homomorphic approach. Med. Image Anal. 20 (1), 184–197. https://doi.org/10.1016/j.media.2014.11.005.

Alexander, D.C., 2008. A general framework for experiment design in diffusion MRI and its application in measuring direct tissue-microstructure features. Magn. Reson. Med. 60 (2), 439–448. https://doi.org/10.1002/mrm.21646.

Alexander, D.C., Hubbard, P.L., Hall, M.G., Moore, E.A., Ptito, M., Parker, G.J., Dyrby, T.B., 2010. Orientationally invariant indices of axon diameter and density from diffusion MRI. NeuroImage 52 (4), 1374–1389. https://doi.org/10.1016/j.neuroimage.2010.05.043. https://www.sciencedirect.com/science/article/pii/S1053811910007755.

Andersson, J.L.R., 2008. Maximum a posteriori estimation of diffusion tensor parameters using a Rician noise model: why, how and but. NeuroImage 42 (4), 1340–1356. https://doi.org/10.1016/j.neuroimage.2008.05.053. https://www.sciencedirect.com/science/article/pii/S1053811908007131.

Andersson, J.L.R., Skare, S., 2002. A model-based method for retrospective correction of geometric distortions in diffusion-weighted EPI. NeuroImage 16 (1), 177–199. https://doi.org/10.1006/nimg.2001.1039.

Andersson, J., Skare, S., 2010. Image distortion and its correction in diffusion MRI. In: Diffusion MRI: Theory, Methods, and Applications, pp. 285–302.

Andersson, J.L., Sotiropoulos, S.N., 2016. An integrated approach to correction for off-resonance effects and subject movement in diffusion MR imaging. NeuroImage 125, 1063–1078.

Andersson, J.L., Skare, S., Ashburner, J., 2003. How to correct susceptibility distortions in spin-echo echo-planar images: application to diffusion tensor imaging. NeuroImage 20 (2), 870–888.

Archibald, R., Gelb, A., 2002. A method to reduce the Gibbs ringing artifact in MRI scans while keeping tissue boundary integrity. IEEE Trans. Med. Imaging 21, 305–319. https://doi.org/10.1109/TMI.2002.1000255.

Assaf, Y., Basser, P.J., 2005. Composite hindered and restricted model of diffusion (CHARMED) MR imaging of the human brain. NeuroImage 27 (1), 48–58. https://doi.org/10.1016/j.neuroimage.2005.03.042. https://www.sciencedirect.com/science/article/pii/S1053811905002259.

Assaf, Y., Blumenfeld-Katzir, T., Yovel, Y., Basser, P.J., 2008. Axcaliber: a method for measuring axon diameter distribution from diffusion MRI. Magn. Reson. Med. 59 (6), 1347–1354. https://doi.org/10.1002/mrm.21577. https://onlinelibrary.wiley.com/doi/abs/10.1002/mrm.21577.

Assemlal, H.E., Tschumperlé, D., Brun, L., 2009. Efficient and robust computation of PDF features from diffusion MR signal. Med. Image Anal. 13 (5), 715–729. https://doi.org/10.1016/j.media.2009.06.004. https://www.sciencedirect.com/science/article/pii/S1361841509000486.

Auría, A., Daducci, A., Thiran, J.P., Wiaux, Y., 2015. Structured sparsity for spatially coherent fibre orientation estimation in diffusion MRI. NeuroImage 115, 245–255. https://doi.org/10.1016/j.neuroimage.2015.04.049. https://www.sciencedirect.com/science/article/pii/S1053811915003493.

Avants, B.B., Epstein, C.L., Grossman, M., Gee, J.C., 2008. Symmetric diffeomorphic image registration with cross-correlation: evaluating automated labeling of elderly and neurodegenerative brain. Med. Image Anal. 12 (1), 26–41.

Bai, Y., Alexander, D.C., 2008. Model-based registration to correct for motion between acquisitions in diffusion MR imaging. In: 2008 5th IEEE International Symposium on Biomedical Imaging: From Nano to Macro, pp. 947–950.

Bakir, T., Reeves, S., 2000. A filter design method for minimizing ringing in a region of interest in MR spectroscopic images. IEEE Trans. Med. Imaging 19, 585–600. https://doi.org/10.1109/42.870664.

Bammer, R., Markl, M., Barnett, A., Acar, B., Alley, M.T., Pelc, N.J., Glover, G.H., Moseley, M.E., 2003. Analysis and generalized correction of the effect of spatial gradient field distortions in diffusion-weighted imaging. Magn. Reson. Med. 50 (3), 560–569.

Bao, L., Robini, M., Liu, W., Zhu, Y., 2013. Structure-adaptive sparse denoising for diffusion-tensor MRI. Med. Image Anal. 17 (4), 442–457. https://doi.org/10.1016/j.media.2013.01.006. https://www.sciencedirect.com/science/article/pii/S1361841513000078.

Barmpoutis, A., Vemuri, B.C., Howland, D., Forder, J.R., 2008. Extracting tractosemas from a displacement probability field for tractography in DW-MRI. In: Metaxas, D., Axel, L., Fichtinger, G., Székely, G. (Eds.), Medical Image Computing and Computer-Assisted Intervention - MICCAI 2008, Springer, Berlin, Heidelberg, pp. 9–16.

Baron, C.A., Lebel, R.M., Wilman, A.H., Beaulieu, C., 2012. The effect of concomitant gradient fields on diffusion tensor imaging. Magn. Reson. Med. 68 (4), 1190–1201. https://doi.org/10.1002/mrm.24120. https://onlinelibrary.wiley.com/doi/abs/10.1002/mrm.24120.

Bastin, M.E., 2001. On the use of the FLAIR technique to improve the correction of eddy current induced artefacts in MR diffusion tensor imaging. Magn. Reson. Imaging 19 (7), 937–950. https://doi.org/10.1016/S0730-725X(01)00427-1. https://www.sciencedirect.com/science/article/pii/S0730725X01004271.

Becker, S.M.A., Tabelow, K., Mohammadi, S., Weiskopf, N., Polzehl, J., 2014. Adaptive smoothing of multi-shell diffusion weighted magnetic resonance data by msPOAS. NeuroImage 95, 90–105. https://doi.org/10.1016/j.neuroimage.2014.03.053. https://www.sciencedirect.com/science/article/pii/S1053811914002146.

Ben-Amitay, S., Jones, D.K., Assaf, Y., 2012. Motion correction and registration of high b-value diffusion weighted images. Magn. Reson. Med. 67 (6), 1694–1702. https://doi.org/10.1002/mrm.23186. https://onlinelibrary.wiley.com/doi/abs/10.1002/mrm.23186.

Benner, T., van der Kouwe, A.J.W., Kirsch, J.E., Sorensen, A.G., 2006. Real-time RF pulse adjustment for B0 drift correction. Magn. Reson. Med. 56 (1), 204–209. https://doi.org/10.1002/mrm.20936. https://onlinelibrary.wiley.com/doi/abs/10.1002/mrm.20936.

Bennett, K.M., Schmainda, K.M., Bennett (Tong), R., Rowe, D.B., Lu, H., Hyde, J.S., 2003. Characterization of continuously distributed cortical water diffusion rates with a stretched-exponential model. Magn. Reson. Med. 50 (4), 727–734. https://doi.org/10.1002/mrm.10581. https://onlinelibrary.wiley.com/doi/abs/10.1002/mrm.10581.

Berl, M.M., Walker, L., Modi, P., Irfanoglu, M.O., Sarlls, J.E., Nayak, A., Pierpaoli, C., 2015. Investigation of vibration-induced artifact in clinical diffusion-weighted imaging of pediatric subjects. Hum. Brain Mapp. 36 (12), 4745–4757. https://doi.org/10.1002/hbm.22846. https://onlinelibrary.wiley.com/doi/abs/10.1002/hbm.22846.

Bhagalia, R., Kim, B., 2008. Spin saturation artifact correction using slice-to-volume registration motion estimates for fMRI time series. Med. Phys. 35 (2), 424–434.

Bodammer, N., Kaufmann, J., Kanowski, M., Tempelmann, C., 2004. Eddy current correction in diffusion-weighted imaging using pairs of images acquired with opposite diffusion gradient polarity. Magn. Reson. Med. 51 (1), 188–193. https://doi.org/10.1002/mrm.10690. https://onlinelibrary.wiley.com/doi/abs/10.1002/mrm.10690.

Bowtell, R.W., McIntyre, D.J.O., Commandre, M.J., Glover, P.M., Mansfield, P., 1994. Correction of geometric distortion in echo planar images. In: Proceedings of 2nd Annual Meeting of the SMR, p. 411. San Francisco.

Brion, V., Poupon, C., Riff, O., Aja-Fernández, S., Tristán-Vega, A., Mangin, J.F., Le Bihan, D., Poupon, F., 2013. Noise correction for HARDI and HYDI data obtained with multi-channel coils and sum of squares reconstruction: an anisotropic extension of the LMMSE. Magn. Reson. Imaging 31 (8), 1360–1371. https://doi.org/10.1016/j.mri.2013.04.002. https://www.sciencedirect.com/science/article/pii/S0730725X13001367.

Buades, A., Coll, B., Morel, J.M., 2005. A review of image denoising algorithms, with a new one. Multiscale Model. Simul. 4 (2), 490–530. https://doi.org/10.1137/040616024.

Buonocore, M.H., Gao, L., 1997. Ghost artifact reduction for echo planar imaging using image phase correction. Magn. Reson. Med. 38 (1), 89–100. https://doi.org/10.1002/mrm.1910380114. https://onlinelibrary.wiley.com/doi/abs/10.1002/mrm.1910380114.

Canales-Rodríguez, E.J., Daducci, A., Sotiropoulos, S.N., Caruyer, E., Aja-Fernández, S., Radua, J., Mendizabal, J.M.Y., Iturria-Medina, Y., Melie-García, L., Alemán-Gómez, Y., Thiran, J.P., Sarró, S., Pomarol-Clotet, E., Salvador, R., 2015. Spherical deconvolution of multichannel diffusion MRI data with non-Gaussian noise models and spatial regularization. PLoS ONE 10 (10), 1–29. https://doi.org/10.1371/journal.pone.0138910.

Caruyer, E., Deriche, R., 2012. Diffusion MRI signal reconstruction with continuity constraint and optimal regularization. Med. Image Anal. 16 (6), 1113–1120. https://doi.org/10.1016/j.media.2012.06.011. https://www.sciencedirect.com/science/article/pii/S136184151200093X.

Chang, H., Fitzpatrick, J.M., 1992. A technique for accurate magnetic resonance imaging in the presence of field inhomogeneities. IEEE Trans. Med. Imaging 11 (3), 319–329.

Chang, L.C., Jones, D.K., Pierpaoli, C., 2005. RESTORE: robust estimation of tensors by outlier rejection. Magn. Reson. Med. 53 (5), 1088–1095. https://doi.org/10.1002/mrm.20426. https://onlinelibrary.wiley.com/doi/abs/10.1002/mrm.20426.

Chang, L.C., Walker, L., Pierpaoli, C., 2012. Informed RESTORE: a method for robust estimation of diffusion tensor from low redundancy datasets in the presence of physiological noise artifacts. Magn. Reson. Med. 68 (5), 1654–1663. https://doi.org/10.1002/mrm.24173. https://onlinelibrary.wiley.com/doi/abs/10.1002/mrm.24173.

Chen, Y., Tymofiyeva, O., Hess, C.P., Xu, D., 2015. Effects of rejecting diffusion directions on tensor-derived parameters. NeuroImage 109, 160–170. https://doi.org/10.1016/j.neuroimage.2015.01.010. https://www.sciencedirect.com/science/article/pii/S1053811915000166.

Cheng, J., Deriche, R., Jiang, T., Shen, D., Yap, P.T., 2014. Non-Negative Spherical Deconvolution (NNSD) for estimation of fiber Orientation Distribution Function in single-/multi-shell diffusion MRI. NeuroImage 101, 750–764. https://doi.org/10.1016/j.neuroimage.2014.07.062. https://www.sciencedirect.com/science/article/pii/S1053811914006454.

Coelho, S., Baete, S.H., Lemberskiy, G., Ades-Aron, B., Barrol, G., Veraart, J., Novikov, D.S., Fieremans, E., 2022. Reproducibility of the standard model of diffusion in white matter on clinical MRI systems. NeuroImage 257, 119290.

Collier, Q., Veraart, J., Jeurissen, B., den Dekker, A.J., Sijbers, J., 2015. Iterative reweighted linear least squares for accurate, fast, and robust estimation of diffusion magnetic resonance parameters. Magn. Reson. Med. 73 (6), 2174–2184. https://doi.org/10.1002/mrm.25351. https://onlinelibrary.wiley.com/doi/abs/10.1002/mrm.25351.

Constantinides, C.D., Atalar, E., McVeigh, E.R., 1997. Signal-to-noise measurements in magnitude images from NMR phased arrays. Magn. Reson. Med. 38 (5), 852–857. https://doi.org/10.1002/mrm.1910380524. https://onlinelibrary.wiley.com/doi/abs/10.1002/mrm.1910380524.

Coulon, O., Alexander, D.C., Arridge, S., 2004. Diffusion tensor magnetic resonance image regularization. Med. Image Anal. 8 (1), 47–67. https://doi.org/10.1016/j.media.2003.06.002. https://www.sciencedirect.com/science/article/pii/S1361841503000720.

Coupé, P., Manjón, J.V., Gedamu, E., Arnold, D., Robles, M., Collins, D.L., 2010. Robust Rician noise estimation for MR images. Med. Image Anal. 14 (4), 483–493. https://doi.org/10.1016/j.media.2010.03.001. https://www.sciencedirect.com/science/article/pii/S1361841510000241.

Daducci, A., Van De Ville, D., Thiran, J.P., Wiaux, Y., 2014. Sparse regularization for fiber ODF reconstruction: from the suboptimality of $\ell$2 and $\ell$1 priors to $\ell$0. Med. Image Anal. 18 (6), 820–833. https://doi.org/10.1016/j.media.2014.01.011. https://www.sciencedirect.com/science/article/pii/S1361841514000243.

Daducci, A., Canales-Rodríguez, E.J., Zhang, H., Dyrby, T.B., Alexander, D.C., Thiran, J.P., 2015. Accelerated microstructure imaging via convex optimization (AMICO) from diffusion MRI data. NeuroImage 105, 32–44. https://doi.org/10.1016/j.neuroimage.2014.10.026. https://www.sciencedirect.com/science/article/pii/S1053811914008519.

Dela Haije, T.C.J., Duits, R., Tax, C.M.W., Westin, C.F., Vilanova, A., Burgeth, B., 2014. Sharpening fibers in diffusion weighted MRI via erosion. Visualization and Processing of Tensors and Higher Order Descriptors for Multi-Valued Data. Mathematics and Visualization. Springer, Berlin, Heidelberg, pp. 97–126, https://doi.org/10.1007/978-3-642-54301-2_5.

Dell'Acqua, F., Rizzo, G., Scifo, P., Clarke, R.A., Scotti, G., Fazio, F., 2007. A model-based deconvolution approach to solve fiber crossing in diffusion-weighted MR imaging. IEEE Trans. Biomed. Eng. 54 (3), 462–472. https://doi.org/10.1109/TBME.2006.888830.

Dell'Acqua, F., Scifo, P., Rizzo, G., Catani, M., Simmons, A., Scotti, G., Fazio, F., 2010. A modified damped Richardson-Lucy algorithm to reduce isotropic background effects in spherical deconvolution. NeuroImage 49 (2), 1446–1458. https://doi.org/10.1016/j.neuroimage.2009.09.033. https://www.sciencedirect.com/science/article/pii/S105381190901012X.

Dell'Acqua, F., Lacerda, L., Catani, M., Simmons, A., 2014. Anisotropic Power Maps: a diffusion contrast to reveal low anisotropy tissues from HARDI data. In: Proceedings of the International Society for Magnetic Resonance in Medicine, vol. 22, p. 730.

Descoteaux, M., Angelino, E., Fitzgibbons, S., Deriche, R., 2007. Regularized, fast, and robust analytical Q-ball imaging. Magn. Reson. Med. 58 (3), 497–510. https://doi.org/10.1002/mrm.21277. https://onlinelibrary.wiley.com/doi/abs/10.1002/mrm.21277.

Descoteaux, M., Deriche, R., Knosche, T.R., Anwander, A., 2009. Deterministic and probabilistic tractography based on complex fibre orientation distributions. IEEE Trans. Med. Imaging 28 (2), 269–286. https://doi.org/10.1109/TMI.2008.2004424.

Descoteaux, M., Deriche, R., Le Bihan, D., Mangin, J.F., Poupon, C., 2011. Multiple q-shell diffusion propagator imaging. Med. Image Anal. 15 (4), 603–621. https://doi.org/10.1016/j.media.2010.07.001. https://www.sciencedirect.com/science/article/pii/S1361841510000939.

Donoho, D.L., 2006. Compressed sensing. IEEE Trans. Inf. Theory 52 (4), 1289–1306. https://doi.org/10.1109/TIT.2006.871582.

Doran, S.J., Charles-Edwards, L., Reinsberg, S.A., Leach, M.O., 2005. A complete distortion correction for MR images: I. Gradient warp correction. Phys. Med. Biol. 50 (7), 1343. https://doi.org/10.1088/0031-9155/50/7/001.

Du, Y.P., Joe Zhou, X., Bernstein, M.A., 2002. Correction of concomitant magnetic field-induced image artifacts in nonaxial echo-planar imaging. Magn. Reson. Med. 48 (3), 509–515. https://doi.org/10.1002/mrm.10249. https://onlinelibrary.wiley.com/doi/abs/10.1002/mrm.10249.

Duits, R., Franken, E., 2011. Left-invariant diffusions on the space of positions and orientations and their application to crossing-preserving smoothing of HARDI images. Int. J. Comput. Vis. 92 (3), 231–264. https://doi.org/10.1007/s11263-010-0332-z.

Eddy, W.F., Fitzgerald, M., Noll, D.C., 1996. Improved image registration by using Fourier interpolation. Magn. Reson. Med. 36, 923–931.

Farzinfar, M., Oguz, I., Smith, R.G., Verde, A.R., Dietrich, C., Gupta, A., Escolar, M.L., Piven, J., Pujol, S., Vachet, C., Gouttard, S., Gerig, G., Dager, S., McKinstry, R.C., Paterson, S., Evans, A.C., Styner, M.A., 2013. Diffusion imaging quality control via entropy of principal direction distribution. NeuroImage 82, 1–12. https://doi.org/10.1016/j.neuroimage.2013.05.022. https://www.sciencedirect.com/science/article/pii/S1053811913005168.

Feng, Y., Wu, Y., Rathi, Y., Westin, C.F., 2015. Sparse deconvolution of higher order tensor for fiber orientation distribution estimation. Artif. Intell. Med. 65 (3), 229–238. https://doi.org/10.1016/j.artmed.2015.09.004. https://www.sciencedirect.com/science/article/pii/S0933365715001219.

Ferizi, U., Schneider, T., Panagiotaki, E., Nedjati-Gilani, G., Zhang, H., Wheeler-Kingshott, C.A.M., Alexander, D.C., 2014. A ranking of diffusion MRI compartment models with in vivo human brain data. Magn. Reson. Med. 72 (6), 1785–1792. https://doi.org/10.1002/mrm.25080. https://onlinelibrary.wiley.com/doi/abs/10.1002/mrm.25080.

Ferizi, U., Schneider, T., Witzel, T., Wald, L.L., Zhang, H., Wheeler-Kingshott, C.A.M., Alexander, D.C., 2015. White matter compartment models for in vivo diffusion MRI at 300 mT/m. NeuroImage 118, 468–483. https://doi.org/10.1016/j.neuroimage.2015.06.027. https://www.sciencedirect.com/science/article/pii/S1053811915005303.

Fick, R.H.J., Wassermann, D., Caruyer, E., Deriche, R., 2016. MAPL: tissue microstructure estimation using Laplacian-regularized MAP-MRI and its application to HCP data. NeuroImage 134, 365–385. https://doi.org/10.1016/j.neuroimage.2016.03.046. https://www.sciencedirect.com/science/article/pii/S1053811916002512.

Fieremans, E., Jensen, J.H., Helpern, J.A., 2011. White matter characterization with diffusional kurtosis imaging. NeuroImage 58 (1), 177–188. https://doi.org/10.1016/j.neuroimage.2011.06.006. https://www.sciencedirect.com/science/article/pii/S1053811911006148.

Fillard, P., Pennec, X., Arsigny, V., Ayache, N., 2007. Clinical DT-MRI estimation, smoothing, and fiber tracking with Log-Euclidean metrics. IEEE Trans. Med. Imaging 26 (11), 1472–1482. https://doi.org/10.1109/TMI.2007.899173.

Florack, L., 2008. Codomain scale space and regularization for high angular resolution diffusion imaging. In: 2008 IEEE Computer Society Conference on Computer Vision and Pattern Recognition Workshops, pp. 1–6, https://doi.org/10.1109/CVPRW.2008.4562967.

Froeling, M., Tax, C.M.W., Vos, S.B., Luijten, P.R., Leemans, A., 2017. "MASSIVE" brain dataset: multiple acquisitions for standardization of structural imaging validation and evaluation. Magn. Reson. Med. 77 (5), 1797–1809. https://doi.org/10.1002/mrm.26259. https://onlinelibrary.wiley.com/doi/abs/10.1002/mrm.26259.

Gallichan, D., Scholz, J., Bartsch, A., Behrens, T.E., Robson, M.D., Miller, K.L., 2010. Addressing a systematic vibration artifact in diffusion-weighted MRI. Hum. Brain Mapp. 31 (2), 193–202. https://doi.org/10.1002/hbm.20856. https://onlinelibrary.wiley.com/doi/abs/10.1002/hbm.20856.

Gholipour, A., Kehtarnavaz, N., Scherrer, B., Warfield, S.K., 2011. On the accuracy of unwarping techniques for the correction of susceptibility-induced geometric distortion in magnetic resonance Echo-planar images. Proc. Annu. Int. Conf. IEEE Eng. Med. Biol. Soc. 2011, 6997–7000. https://doi.org/10.1109/IEMBS.2011.6091769.

Ghosh, A., Milne, T., Deriche, R., 2014. Constrained diffusion kurtosis imaging using ternary quartics & MLE. Magn. Reson. Med. 71 (4), 1581–1591. https://doi.org/10.1002/mrm.24781. https://onlinelibrary.wiley.com/doi/abs/10.1002/mrm.24781.

Glasser, M.F., Sotiropoulos, S.N., Wilson, J.A., Coalson, T.S., Fischl, B., Andersson, J.L., Xu, J., Jbabdi, S., Webster, M., Polimeni, J.R., Van Essen, D.C., Jenkinson, M., 2013. The minimal preprocessing pipelines for the Human Connectome Project. NeuroImage 80, 105–124. https://doi.org/10.1016/j.neuroimage.2013.04.127. https://www.sciencedirect.com/science/article/pii/S1053811913005053.

**172 PART | II** Diffusion MRI

Glenn, G.R., Tabesh, A., Jensen, J.H., 2015. A simple noise correction scheme for diffusional kurtosis imaging. Magn. Reson. Imaging 33 (1), 124–133. https://doi.org/10.1016/j.mri.2014.08.028. https://www.sciencedirect.com/science/article/pii/S0730725X14002598.

Glodeck, D., Hesser, J., Zheng, L., 2016. Distortion correction of EPI data using multimodal nonrigid registration with an anisotropic regularization. Magn. Reson. Imaging 34 (2), 127–136. https://doi.org/10.1016/j.mri.2015.10.032.

Gramfort, A., Poupon, C., Descoteaux, M., 2014. Denoising and fast diffusion imaging with physically constrained sparse dictionary learning. Med. Image Anal. 18 (1), 36–49. https://doi.org/10.1016/j.media.2013.08.006. https://www.sciencedirect.com/science/article/pii/S1361841513001205.

Griswold, M.A., Jakob, P.M., Heidemann, R.M., Nittka, M., Jellus, V., Wang, J., Kiefer, B., Haase, A., 2002. Generalized autocalibrating partially parallel acquisitions (GRAPPA). Magn. Reson. Med. 47 (6), 1202–1210. https://doi.org/10.1002/mrm.10171. https://onlinelibrary.wiley.com/doi/abs/10.1002/mrm.10171.

Gudbjartsson, H., Patz, S., 1995. The Rician distribution of noisy MRI data. Magn. Reson. Med. 34 (6), 910–914. https://doi.org/10.1002/mrm.1910340618. https://onlinelibrary.wiley.com/doi/abs/10.1002/mrm.1910340618.

Hansen, C.B., Nath, V., Hainline, A.E., Schilling, K.G., Parvathaneni, P., Bayrak, R.G., Blaber, J.A., Irfanoglu, O., Pierpaoli, C., Anderson, A.W., Rogers, B.P., Landman, B.A., 2019. Characterization and correlation of signal drift in diffusion weighted MRI. Magn. Reson. Imaging 57, 133–142.

Haselgrove, J.C., Moore, J.R., 1996. Correction for distortion of echo-planar images used to calculate the apparent diffusion coefficient. Magn. Reson. Med. 36 (6), 960–964. https://doi.org/10.1002/mrm.1910360620. https://onlinelibrary.wiley.com/doi/abs/10.1002/mrm.1910360620.

Hédouin, R., Commowick, O., Bannier, E., Scherrer, B., Taquet, M., Warfield, S.K., Barillot, C., 2017. Block-matching distortion correction of echo-planar images with opposite phase encoding directions. IEEE Trans. Med. Imaging 36 (5), 1106–1115. https://doi.org/10.1109/TMI.2016.2646920.

Heemskerk, A., Leemans, A., Plaisier, A., Pieterman, K., Lequin, M., Dudink, J., 2013. Acquisition guidelines and quality assessment tools for analyzing neonatal diffusion tensor MRI data. Am. J. Neuroradiol. 34. https://doi.org/10.3174/ajnr.A3465.

Henkelman, R.M., 1985. Measurement of signal intensities in the presence of noise in MR images. Med. Phys. 12 (2), 232–233. https://doi.org/10.1118/1.595711. https://aapm.onlinelibrary.wiley.com/doi/abs/10.1118/1.595711.

Hess, C.P., Mukherjee, P., Han, E.T., Xu, D., Vigneron, D.B., 2006. Q-ball reconstruction of multimodal fiber orientations using the spherical harmonic basis. Magn. Reson. Med. 56 (1), 104–117. https://doi.org/10.1002/mrm.20931. https://onlinelibrary.wiley.com/doi/abs/10.1002/mrm.20931.

Holland, D., Kuperman, J.M., Dale, A.M., 2010. Efficient correction of inhomogeneous static magnetic field-induced distortion in Echo Planar Imaging. NeuroImage 50, 175–183.

Hui, E.S., Russell Glenn, G., Helpern, J.A., Jensen, J.H., 2015. Kurtosis analysis of neural diffusion organization. NeuroImage 106, 391–403. https://doi.org/10.1016/j.neuroimage.2014.11.015. https://www.sciencedirect.com/science/article/pii/S105381191400932X.

Irfanoglu, M.O., Walker, L., Sarlls, J., Marenco, S., Pierpaoli, C., 2012. Effects of image distortions originating from susceptibility variations and concomitant fields on diffusion MRI tractography results. NeuroImage 15 (61), 275–288.

Irfanoglu, M.O., Modi, P., Nayak, A., Hutchinson, E.B., Sarlls, J., Pierpaoli, C., 2015. DR-BUDDI: (Diffeomorphic Registration for Blip-Up blip-Down Diffusion Imaging) method for correcting echo planar imaging distortions. NeuroImage 106, 284–289.

Janke, A., Zhao, H., Cowin, G.J., Galloway, G.J., Doddrell, D.M., 2004. Use of spherical harmonic deconvolution methods to compensate for nonlinear gradient effects on MRI images. Magn. Reson. Med. 52 (1), 115–122. https://doi.org/10.1002/mrm.20122. https://onlinelibrary.wiley.com/doi/abs/10.1002/mrm.20122.

Jelescu, I.O., Veraart, J., Fieremans, E., Novikov, D.S., 2016. Degeneracy in model parameter estimation for multi-compartmental diffusion in neuronal tissue. NMR Biomed 29 (1), 33–47.

Jeurissen, B., Tournier, J.D., Dhollander, T., Connelly, A., Sijbers, J., 2014. Multi-tissue constrained spherical deconvolution for improved analysis of multi-shell diffusion MRI data. NeuroImage 103, 411–426. https://doi.org/10.1016/j.neuroimage.2014.07.061. https://www.sciencedirect.com/science/article/pii/S1053811914006442.

Jeurissen, B., Descoteaux, M., Mori, S., Leemans, A., 2019. Diffusion MRI fiber tractography of the brain. NMR Biomed. 32 (4), e3785. https://doi.org/10.1002/nbm.3785. https://analyticalsciencejournals.onlinelibrary.wiley.com/doi/abs/10.1002/nbm.3785.

Jezzard, P., Balaban, R.S., 1995. Correction for geometric distortion in echo planar images from B0 field variations. Magn. Reson. Med. 34, 65–73.

Jezzard, P., Barnett, A.S., Pierpaoli, C., 1998. Characterization of and correction for eddy current artifacts in echo planar diffusion imaging. Magn. Reson. Med. 39 (5), 801–812. https://doi.org/10.1002/mrm.1910390518. https://onlinelibrary.wiley.com/doi/abs/10.1002/mrm.1910390518.

Jian, B., Vemuri, B.C., 2007. A unified computational framework for deconvolution to reconstruct multiple fibers from diffusion weighted MRI. IEEE Trans. Med. Imaging 26 (11), 1464–1471. https://doi.org/10.1109/TMI.2007.907552.

Jian, B., Vemuri, B.C., Özarslan, E., Carney, P.R., Mareci, T.H., 2007. A novel tensor distribution model for the diffusion-weighted MR signal. NeuroImage 37 (1), 164–176. https://doi.org/10.1016/j.neuroimage.2007.03.074. https://www.sciencedirect.com/science/article/pii/S105381190700273X.

Jones, D.K., Basser, P.J., 2004. "Squashing peanuts and smashing pumpkins": how noise distorts diffusion-weighted MR data. Magn. Reson. Med. 52 (5), 979–993. https://doi.org/10.1002/mrm.20283. https://onlinelibrary.wiley.com/doi/abs/10.1002/mrm.20283.

Jones, D.K., Cercignani, M., 2010. Twenty-five pitfalls in the analysis of diffusion MRI data. NMR Biomed. 23 (7), 803–820. https://doi.org/10.1002/nbm.1543.

Jones, D.K., Leemans, A., 2011. Diffusion Tensor Imaging. Humana Press, Totowa, NJ, pp. 127–144, https://doi.org/10.1007/978-1-61737-992-5_6.

Kaden, E., Kelm, N.D., Carson, R.P., Does, M.D., Alexander, D.C., 2016. Multi-compartment microscopic diffusion imaging. NeuroImage 139, 346–359.

Kellner, E., Dhital, B., Kiselev, V.G., Reisert, M., 2016. Gibbs-ringing artifact removal based on local subvoxel-shifts. Magn. Reson. Med. 76 (5), 1574–1581. https://doi.org/10.1002/mrm.26054. https://onlinelibrary.wiley.com/doi/abs/10.1002/mrm.26054.

Kim, K., Habas, P.A., Rajagopalan, V., Scott, J.A., Corbett-Detig, J.M., Rousseau, F., Barkovich, A.J., Glenn, O.A., Studholme, C., 2011. Bias field inconsistency correction of motion-scattered multislice MRI for improved 3D image reconstruction. IEEE Trans. Med. Imaging 30 (9), 1704–1712.

Koay, C.G., Basser, P.J., 2006. Analytically exact correction scheme for signal extraction from noisy magnitude MR signals. J. Magn. Reson. 179 (2), 317–322. https://doi.org/10.1016/j.jmr.2006.01.016. https://www.sciencedirect.com/science/article/pii/S109078070600019X.

Koay, C.G., Chang, L.C., Carew, J.D., Pierpaoli, C., Basser, P.J., 2006. A unifying theoretical and algorithmic framework for least squares methods of estimation in diffusion tensor imaging. J. Magn. Reson. 182 (1), 115–125. https://doi.org/10.1016/j.jmr.2006.06.020. https://www.sciencedirect.com/science/article/pii/S1090780706001790.

Koay, C.G., Özarslan, E., Basser, P., 2009a. A signal-transformational framework for breaking the noise floor and its applications in MRI. J. Magn. Reson. 197, 108–119. https://doi.org/10.1016/j.jmr.2008.11.015.

Koay, C.G., Özarslan, E., Pierpaoli, C., 2009b. Probabilistic Identification and Estimation of Noise (PIESNO): a self-consistent approach and its applications in MRI. J. Magn. Reson. 199 (1), 94–103. https://doi.org/10.1016/j.jmr.2009.03.005. https://www.sciencedirect.com/science/article/pii/S1090780709000767.

Kybic, J., Thevenaz, P., Nirkko, A., Unser, M., 2000. Unwarping of unidirectionally distorted EPI images. IEEE Trans. Med. Imaging 19 (2), 80–93.

Lam, F., Babacan, S.D., Haldar, J.P., Weiner, M.W., Schuff, N., Liang, Z.P., 2014. Denoising diffusion-weighted magnitude MR images using rank and edge constraints. Magn. Reson. Med. 71 (3), 1272–1284. https://doi.org/10.1002/mrm.24728. https://onlinelibrary.wiley.com/doi/abs/10.1002/mrm.24728.

Lampinen, B., Szczepankiewicz, F., Mårtensson, J., van Westen, D., Sundgren, P.C., Nilsson, M., 2017. Neurite density imaging versus imaging of microscopic anisotropy in diffusion MRI: a model comparison using spherical tensor encoding. NeuroImage 147, 517–531.

Landman, B.A., Bazin, P.L., Prince, J.L., 2009. Estimation and application of spatially variable noise fields in diffusion tensor imaging. Magn. Reson. Imaging 27 (6), 741–751. https://doi.org/10.1016/j.mri.2009.01.001. https://www.sciencedirect.com/science/article/pii/S0730725X09000058.

Landman, B.A., Bogovic, J.A., Wan, H., ElShahaby, F.E.Z., Bazin, P.L., Prince, J.L., 2012. Resolution of crossing fibers with constrained compressed sensing using diffusion tensor MRI. NeuroImage 59 (3), 2175–2186. https://doi.org/10.1016/j.neuroimage.2011.10.011. https://www.sciencedirect.com/science/article/pii/S1053811911011724.

Leemans, A., Jones, D., 2009. The $B$-matrix must be rotated when correcting for subject motion in DTI data. Magn. Reson. Med. 61, 1336–1349. https://doi.org/10.1002/mrm.21890.

Leemans, A., Jeurissen, B., Sijbers, J., Jones, D.K., 2009. ExploreDTI: a graphical toolbox for processing, analyzing, and visualizing diffusion MR data. Proc. Int. Soc. Magn. Reson. Med. 17, 3537.

Liu, Z., Wang, Y., Gerig, G., Gouttard, S., Tao, R., Fletcher, T., Styner, M., 2010. Quality control of diffusion weighted images. In: Brent, J.L., William, W.B. (Eds.), Medical Imaging 2010: Advanced PACS-based Imaging Informatics and Therapeutic Applications, vol. 7628. SPIE, p. 76280J, https://doi.org/10.1117/12.844748.

Liu, M., Vemuri, B.C., Deriche, R., 2013. A robust variational approach for simultaneous smoothing and estimation of DTI. NeuroImage 67, 33–41. https://doi.org/10.1016/j.neuroimage.2012.11.012. https://www.sciencedirect.com/science/article/pii/S1053811912011081.

Liu, X., Yuan, Z., Guo, Z., Xu, D., 2015. A localized Richardson-Lucy algorithm for fiber orientation estimation in high angular resolution diffusion imaging. Med. Phys. 42 (5), 2524–2539. https://doi.org/10.1118/1.4917082. https://aapm.onlinelibrary.wiley.com/doi/abs/10.1118/1.4917082.

Lu, H., Jensen, J.H., Ramani, A., Helpern, J.A., 2006. Three-dimensional characterization of non-Gaussian water diffusion in humans using diffusion kurtosis imaging. NMR Biomed. 19 (2), 236–247. https://doi.org/10.1002/nbm.1020. https://analyticalsciencejournals.onlinelibrary.wiley.com/doi/abs/10.1002/nbm.1020.

Maes, F., Collignon, A., Vandermeulen, D., Marchal, G., Seutens, P., 1997. Multimodality image registration by maximization of mutual information. IEEE Trans. Med. Imaging 16, 187–198.

Mangin, J.F., Poupon, C., Clark, C., Le Bihan, D., Bloch, I., 2002. Distortion correction and robust tensor estimation for MR diffusion imaging. Med. Image Anal. 6 (3), 191–198. https://doi.org/10.1016/S1361-8415(02)00079-8. https://www.sciencedirect.com/science/article/pii/S1361841502000798.

Manjón, J.V., Coupé, P., Martí-Bonmatí, L., Collins, D.L., Robles, M., 2010. Adaptive non-local means denoising of MR images with spatially varying noise levels. J. Magn. Reson. Imaging 31 (1), 192–203. https://doi.org/10.1002/jmri.22003. https://onlinelibrary.wiley.com/doi/abs/10.1002/jmri.22003.

Manjón, J.V., Coupé, P., Concha, L., Buades, A., Collins, D.L., Robles, M., 2013. Diffusion weighted image denoising using overcomplete local PCA. PLoS ONE 8 (9), 1–12. https://doi.org/10.1371/journal.pone.0073021.

Mesri, H.Y., David, S., Viergever, M.A., Leemans, A., 2020. The adverse effect of gradient nonlinearities on diffusion MRI: from voxels to group studies. NeuroImage 205, 116127. https://doi.org/10.1016/j.neuroimage.2019.116127.

Mohammadi, S., Möller, H.E., Kugel, H., Müller, D.K., Deppe, M., 2010. Correcting eddy current and motion effects by affine whole-brain registrations: evaluation of three-dimensional distortions and comparison with slicewise correction. Magn. Reson. Med. 64 (4), 1047–1056.

Morgan, P.S., Bowtell, R.W., McIntyre, D.J.O., Worthington, B.S., 2004. Correction of spatial distortion in EPI due to inhomogeneous static magnetic fields using reversed gradient method. J. Magn. Reson. Imaging 19, 499–507.

Nilsson, M., Szczepankiewicz, F., van Westen, D., Hansson, O., 2015. Extrapolation-based references improve motion and eddy-current correction of high b-value DWI data: application in Parkinson's disease dementia. PLoS ONE 10 (11), 1–22. https://doi.org/10.1371/journal.pone.0141825.

Novikov, D.S., Veraart, J., Jelescu, I.O., Fieremans, E., 2018. Rotationally-invariant mapping of scalar and orientational metrics of neuronal microstructure with diffusion MRI. NeuroImage 174, 518–538.

Oguz, I., Farzinfar, M., Matsui, J., Budin, F., Liu, Z., Gerig, G., Johnson, H., Styner, M., 2014. DTIPrep: quality control of diffusion-weighted images. Front. Neuroinf. 8, 4. https://doi.org/10.3389/fninf.2014.00004.

Özarslan, E., Koay, C.G., Shepherd, T.M., Komlosh, M.E., İrfanoğlu, M.O., Pierpaoli, C., Basser, P.J., 2013. Mean apparent propagator (MAP) MRI: a novel diffusion imaging method for mapping tissue microstructure. NeuroImage 78, 16–32. https://doi.org/10.1016/j.neuroimage.2013.04.016. https://www.sciencedirect.com/science/article/pii/S1053811913003431.

Panagiotaki, E., Schneider, T., Siow, B., Hall, M.G., Lythgoe, M.F., Alexander, D.C., 2012. Compartment models of the diffusion MR signal in brain white matter: a taxonomy and comparison. NeuroImage 59 (3), 2241–2254. https://doi.org/10.1016/j.neuroimage.2011.09.081. https://www.sciencedirect.com/science/article/pii/S1053811911011566.

**174 PART | II** Diffusion MRI

Pannek, K., Raffelt, D., Bell, C., Mathias, J.L., Rose, S.E., 2012. HOMOR: higher order model outlier rejection for high b-value MR diffusion data. NeuroImage 63 (2), 835–842. https://doi.org/10.1016/j.neuroimage.2012.07.022. https://www.sciencedirect.com/science/article/pii/S10538119 12007331.

Paquette, M., Merlet, S., Deriche, R., Descoteaux, M., 2014. DSI 101: better ODFs for free. In: Proceedings of the 22nd Annual Meeting of the ISMRM, p. 7018.

Parker, G.J.M., Barker, G.J., Wheeler-Kingshot, C.A., 2001. Gibbs ringing and negative ADC values. In: Proceedings of the ISMRM Ninth Meeting.

Parker, G.D., Marshall, D., Rosin, P.L., Drage, N., Richmond, S., Jones, D.K., 2012. RESDORE: robust estimation in spherical deconvolution by outlier rejection. https://api.semanticscholar.org/CorpusID:5013426.

Parker, G.D., Marshall, D., Rosin, P.L., Drage, N., Richmond, S., Jones, D.K., 2013. A pitfall in the reconstruction of fibre ODFs using spherical deconvolution of diffusion MRI data. NeuroImage 65, 433–448. https://doi.org/10.1016/j.neuroimage.2012.10.022. https://www.sciencedirect.com/science/article/pii/S1053811912010257.

Perrone, D., Aelterman, J., Pižurica, A., Jeurissen, B., Philips, W., Leemans, A., 2015. The effect of Gibbs ringing artifacts on measures derived from diffusion MRI. NeuroImage 120, 441–455. https://doi.org/10.1016/j.neuroimage.2015.06.068. https://www.sciencedirect.com/science/article/pii/S105381191500573X.

Peterson, E., Bammer, R., 2016. Survivor's guide to DTI acquisition. In: Van Hecke, W., Emsell, L., Sunaert, S. (Eds.), Diffusion Tensor Imaging. Springer, New York, NY, pp. 89–126, https://doi.org/10.1007/978-1-4939-3118-7_6.

Pierpaoli, C., 2010. Artifacts in diffusion MRI. In: Diffusion MRI: Theory, Methods, and Applications, pp. 303–318.

Pierpaoli, C., Walker, L., Irfanoglu, M.O., Barnett, A., Basser, P., Chang, L.-C., Koay, C., Pajevic, S., Rohde, G., Sarlls, J., Wu, M., 2010. TORTOISE: an integrated software package for processing of diffusion MRI data. In: ISMRM 18th Annual Meeting, Stockholm, Sweden (abstract #1597).

Pintjens, W., Poot, D.H.J., Verhoye, M., Linden, A.V.D., Sijbers, J., 2008. Susceptibility correction for improved tractography using high field DT-EPI. In: Proceedings of SPIE Medical Imaging, vol. 6914.

Portegies, J.M., Fick, R.H.J., Sanguinetti, G.R., Meesters, S.P.L., Girard, G., Duits, R., 2015. Improving fiber alignment in HARDI by combining contextual PDE flow with constrained spherical deconvolution. PLoS ONE 10 (10), 1–33. https://doi.org/10.1371/journal.pone.0138122.

Prčkovska, V., Andorrà, M., Villoslada, P., Martinez-Heras, E., Duits, R., Fortin, D., Rodrigues, P., Descoteaux, M., 2015. Contextual diffusion image post-processing aids clinical applications. In: Hotz, I., Schultz, T. (Eds.), Visualization and Processing of Higher Order Descriptors for Multi-Valued Data, Springer International Publishing, Cham, pp. 353–377.

Pruessmann, K.P., Weiger, M., Scheidegger, M.B., Boesiger, P., 1999. SENSE: sensitivity encoding for fast MRI. Magn. Reson. Med. 42 (5), 952–962. https://doi.org/10.1002/(SICI)1522-2594(199911)42:5<952::AID-MRM16>3.0.CO;2-S. https://onlinelibrary.wiley.com/doi/abs/10.1002/%28SICI %291522-2594%28199911%2942%3A5%3C952%3A%3AAID-MRM16%3E3.0.CO%3B2-S.

Ramirez-Manzanares, A., Rivera, M., Vemuri, B.C., Carney, P., Mareci, T., 2007. Diffusion basis functions decomposition for estimating white matter intravoxel fiber geometry. IEEE Trans. Med. Imaging 26 (8), 1091–1102. https://doi.org/10.1109/TMI.2007.900461.

Reber, P.J., Wong, E.C., Buxton, R.B., Frank, L.R., 1998. Correction of off resonance-related distortion in echo-planar imaging using EPI-based field maps. Magn. Reson. Med. 39 (2), 328–330.

Reese, T.G., Heid, O., Weisskoff, R.M., Wedeen, V.J., 2003. Reduction of eddy-current-induced distortion in diffusion MRI using a twice-refocused spin echo. Magn. Reson. Med. 49 (1), 177–182. https://doi.org/10.1002/mrm.10308. https://onlinelibrary.wiley.com/doi/abs/10.1002/mrm.10308.

Reisert, M., Skibbe, H., 2012. Left-invariant diffusion on the motion group in terms of the irreducible representations of SO(3). ArXiv abs/1202.5414. https://api.semanticscholar.org/CorpusID:10414092.

Roalf, D.R., Quarmley, M., Elliott, M.A., Satterthwaite, T.D., Vandekar, S.N., Ruparel, K., Gennatas, E.D., Calkins, M.E., Moore, T.M., Hopson, R., Prabhakaran, K., Jackson, C.T., Verma, R., Hakonarson, H., Gur, R.C., Gur, R.E., 2016. The impact of quality assurance assessment on diffusion tensor imaging outcomes in a large-scale population-based cohort. NeuroImage 125, 903–919. https://doi.org/10.1016/j.neuroimage.2015.10.068. https://www.sciencedirect.com/science/article/pii/S1053811915009854.

Rohde, G.K., Barnett, A.S., Basser, P.J., Marenco, S., Pierpaoli, C., 2004. Comprehensive approach for correction of motion and distortion in diffusion-weighted MRI. Magn. Reson. Med. 51, 103–114.

Roine, T., Jeurissen, B., Perrone, D., Aelterman, J., Leemans, A., Philips, W., Sijbers, J., 2014. Isotropic non-white matter partial volume effects in constrained spherical deconvolution. Front. Neuroinf. 8. https://doi.org/10.3389/fninf.2014.00028. https://www.frontiersin.org/articles/10.3389/fninf. 2014.00028.

Roine, T., Jeurissen, B., Perrone, D., Aelterman, J., Philips, W., Leemans, A., Sijbers, J., 2015. Informed constrained spherical deconvolution (iCSD). Med. Image Anal. 24 (1), 269–281. https://doi.org/10.1016/j.media.2015.01.001. https://www.sciencedirect.com/science/article/pii/S1361841515000080.

Ruthotto, L., Kugel, H., Olesch, J., Fischer, B., Modersitzki, J., Burger, M., Wolters, C.H., 2012. Diffeomorphic susceptibility artifact correction of diffusion-weighted magnetic resonance images. Phys. Med. Biol. 57 (18), 5715.

Ruthotto, L., Mohammadi, S., Heck, C., Modersitzki, J., Weiskopf, N., 2013. Hyperelastic susceptibility artifact correction of DTI in SPM. In: Meinzer, H.-P., Deserno, T.M., Handels, H., Tolxdorff, T. (Eds.), Bildverarbeitung Für die Medizin 2013, Springer, Berlin, Heidelberg, pp. 344–349.

Salvador, R., Peña, A., Menon, D.K., Carpenter, T.A., Pickard, J.D., Bullmore, E.T., 2005. Formal characterization and extension of the linearized diffusion tensor model. Hum. Brain Mapp. 24 (2), 144–155. https://doi.org/10.1002/hbm.20076. https://onlinelibrary.wiley.com/doi/abs/10.1002/hbm. 20076.

Sarlls, J.E., Pierpaoli, C., Talagala, S.L., Luh, W.M., 2011. Robust fat suppression at 3T in high-resolution diffusion-weighted single-shot echo-planar imaging of human brain. Magn. Reson. Med. 66 (6), 1658–1665. https://doi.org/10.1002/mrm.22940. https://onlinelibrary.wiley.com/doi/abs/10. 1002/mrm.22940.

Sarra, S., 2006. Digital total variation filtering as postprocessing for Chebyshev pseudospectral methods for conservation laws. Numer. Algorithms 41, 17–33. https://doi.org/10.1007/s11075-005-9003-5.

Schilling, K.G., Blaber, J., Huo, Y., Newton, A., Hansen, C., Nath, V., Shafer, A.T., Williams, O., Resnick, S.M., Rogers, B., Anderson, A.W., Landman, B.A., 2019. Synthesized b0 for diffusion distortion correction (Synb0-DisCo). Magn. Reson. Imaging 64, 62–70.

Sijbers, J., den Dekker, A.J., 2004. Maximum likelihood estimation of signal amplitude and noise variance from MR data. Magn. Reson. Med. 51 (3), 586–594. https://doi.org/10.1002/mrm.10728. https://onlinelibrary.wiley.com/doi/abs/10.1002/mrm.10728.

Sijbers, J., den Dekker, A.J., Van Audekerke, J., Verhoye, M., Van Dyck, D., 1998. Estimation of the noise in magnitude MR images. Magn. Reson. Imaging 16 (1), 87–90. https://doi.org/10.1016/S0730-725X(97)00199-9. https://www.sciencedirect.com/science/article/pii/S0730725X97001999.

Sijbers, J., Poot, D., den Dekker, A.J., Pintjens, W., 2007. Automatic estimation of the noise variance from the histogram of a magnetic resonance image. Phys. Med. Biol. 52 (5), 1335. https://doi.org/10.1088/0031-9155/52/5/009.

Smith, S.M., Jenkinson, M., Woolrich, M.W., Beckmann, C.F., Behrens, T.E., Johansen-Berg, H., Bannister, P.R., Luca, M.D., Drobnjak, I., Flitney, D.E., Niazy, R.K., Saunders, J., Vickers, J., Zhang, Y., Stefano, N.D., Brady, J.M., Matthews, P.M., 2004. Advances in functional and structural MR image analysis and implementation as FSL. NeuroImage 23 (1), 208–219.

Sotiropoulos, S.N., Jbabdi, S., Xu, J., Andersson, J.L., Moeller, S., Auerbach, E.J., Glasser, M.F., Hernandez, M., Sapiro, G., Jenkinson, M., Feinberg, D.A., Yacoub, E., Lenglet, C., Essen, D.C.V., Ugurbil, K., Behrens, T.E.J., 2013. Advances in diffusion MRI acquisition and processing in the Human Connectome Project. NeuroImage 80, 125–143. https://doi.org/10.1016/j.neuroimage.2013.05.057.

St-Jean, S., Coupé, P., Descoteaux, M., 2016. Non Local Spatial and Angular Matching: enabling higher spatial resolution diffusion MRI datasets through adaptive denoising. Med. Image Anal. 32, 115–130. https://doi.org/10.1016/j.media.2016.02.010. https://www.sciencedirect.com/science/article/pii/S1361841516000335.

St-Jean, S., De Luca, A., Tax, C.M.W., Viergever, M.A., Leemans, A., 2020. Automated characterization of noise distributions in diffusion MRI data. Med. Image Anal. 65, 101758. https://doi.org/10.1016/j.media.2020.101758. https://www.sciencedirect.com/science/article/pii/S1361841520301225.

Storey, P., Frigo, F.J., Hinks, R.S., Mock, B.J., Collick, B.D., Baker, N., Marmurek, J., Graham, S.J., 2007. Partial k-space reconstruction in single-shot diffusion-weighted echo-planar imaging. Magn. Reson. Med. 57, 614–619.

Studholme, C., Hill, D.L.G., Hawkes, D.J., 1999. An overlap invariant entropy measure of 3D medical image alignment. Pattern Recogn. 32 (1), 71–86. https://doi.org/10.1016/S0031-3203(98)00091-0. https://www.sciencedirect.com/science/article/pii/S0031320398000910.

Tabelow, K., Voss, H.U., Polzehl, J., 2015. Local estimation of the noise level in MRI using structural adaptation. Med. Image Anal. 20 (1), 76–86. https://doi.org/10.1016/j.media.2014.10.008. https://www.sciencedirect.com/science/article/pii/S1361841514001546.

Tabesh, A., Jensen, J.H., Ardekani, B.A., Helpern, J.A., 2011. Estimation of tensors and tensor-derived measures in diffusional kurtosis imaging. Magn. Reson. Med. 65 (3), 823–836. https://doi.org/10.1002/mrm.22655. https://onlinelibrary.wiley.com/doi/abs/10.1002/mrm.22655.

Tao, R., Fletcher, P.T., Gerber, S., Whitaker, R.T., 2009. A variational image-based approach to the correction of susceptibility artifacts in the alignment of diffusion weighted and structural MRI. Inf. Process. Med. Imaging 21, 651–663.

Tariq, M., Schneider, T., Alexander, D.C., Gandini Wheeler-Kingshott, C.A., Zhang, H., 2016. Bingham-NODDI: mapping anisotropic orientation dispersion of neurites using diffusion MRI. NeuroImage 133, 207–223. https://doi.org/10.1016/j.neuroimage.2016.01.046. https://www.sciencedirect.com/science/article/pii/S1053811916000616.

Tax, C.M.W., Duits, R., Vilanova, A., ter Haar Romeny, B.M., Hofman, P., Wagner, L., Leemans, A., Ossenblok, P., 2014a. Evaluating contextual processing in diffusion MRI: application to optic radiation reconstruction for epilepsy surgery. PLoS ONE 9 (7), 1–19. https://doi.org/10.1371/journal.pone.0101524.

Tax, C.M.W., Jeurissen, B., Vos, S.B., Viergever, M.A., Leemans, A., 2014b. Recursive calibration of the fiber response function for spherical deconvolution of diffusion MRI data. NeuroImage 86, 67–80. https://doi.org/10.1016/j.neuroimage.2013.07.067. https://www.sciencedirect.com/science/article/pii/S1053811913008367.

Tax, C., Otte, W., Viergever, M., Dijkhuizen, R., Leemans, A., 2015. REKINDLE: robust extraction of kurtosis INDices with linear estimation. Magn. Reson. Med. 73. https://doi.org/10.1002/mrm.25165.

Tax, C., Vos, S., Leemans, A., 2016. Checking and correcting DTI data. In: Diffusion Tensor Imaging: A Practical Handbook, pp. 127–150.

Tax, C.M.W., Bastiani, M., Veraart, J., Garyfallidis, E., Irfanoglu, M.O., 2022. What's new and what's next in diffusion MRI preprocessing. NeuroImage 249, 118830.

Tian, Q., Rokem, A., Folkerth, R.D., Nummenmaa, A., Fan, Q., Edlow, B.L., McNab, J.A., 2016. Q-space truncation and sampling in diffusion spectrum imaging. Magn. Reson. Med. 76 (6), 1750–1763. https://doi.org/10.1002/mrm.26071. https://onlinelibrary.wiley.com/doi/abs/10.1002/mrm.26071.

Tobisch, A., Stöcker, T., Groeschel, S., Schultz, T., 2016. Iteratively reweighted L1-fitting for model-independent outlier removal and regularization in diffusion MRI. In: 2016 IEEE 13th International Symposium on Biomedical Imaging (ISBI), pp. 911–914. https://api.semanticscholar.org/CorpusID:18597690.

Tournier, J.D., Calamante, F., Gadian, D.G., Connelly, A., 2004. Direct estimation of the fiber orientation density function from diffusion-weighted MRI data using spherical deconvolution. NeuroImage 23 (3), 1176–1185. https://doi.org/10.1016/j.neuroimage.2004.07.037. https://www.sciencedirect.com/science/article/pii/S1053811904004100.

Tournier, J.D., Calamante, F., Connelly, A., 2007. Robust determination of the fibre orientation distribution in diffusion MRI: non-negativity constrained super-resolved spherical deconvolution. NeuroImage 35 (4), 1459–1472. https://doi.org/10.1016/j.neuroimage.2007.02.016. https://www.sciencedirect.com/science/article/pii/S1053811907001243.

Tournier, J.D., Mori, S., Leemans, A., 2011. Diffusion tensor imaging and beyond. Magn. Reson. Med. 65 (6), 1532–1556. https://doi.org/10.1002/mrm.22924. https://onlinelibrary.wiley.com/doi/abs/10.1002/mrm.22924.

Tournier, J.D., Calamante, F., Connelly, A., 2013. A robust spherical deconvolution method for the analysis of low SNR or low angular resolution diffusion data. In: Proceedings of the 21st Annual Meeting of the ISMRM, p. 772.

Tristán-Vega, A., Aja-Fernández, S., 2010. DWI filtering using joint information for DTI and HARDI. Med. Image Anal. 14 (2), 205–218. https://doi.org/10.1016/j.media.2009.11.001. https://www.sciencedirect.com/science/article/pii/S1361841509001388.

Tuch, D.S., 2004. Q-ball imaging. Magn. Reson. Med. 52 (6), 1358–1372. https://doi.org/10.1002/mrm.20279. https://onlinelibrary.wiley.com/doi/abs/10.1002/mrm.20279.

Veraart, J., Van Hecke, W., Sijbers, J., 2011. Constrained maximum likelihood estimation of the diffusion kurtosis tensor using a Rician noise model. Magn. Reson. Med. 66 (3), 678–686. https://doi.org/10.1002/mrm.22835. https://onlinelibrary.wiley.com/doi/abs/10.1002/mrm.22835.

Veraart, J., Rajan, J., Peeters, R.R., Leemans, A., Sunaert, S., Sijbers, J., 2013a. Comprehensive framework for accurate diffusion MRI parameter estimation. Magn. Reson. Med. 70 (4), 972–984. https://doi.org/10.1002/mrm.24529. https://onlinelibrary.wiley.com/doi/abs/10.1002/mrm.24529.

Veraart, J., Sijbers, J., Sunaert, S., Leemans, A., Jeurissen, B., 2013b. Weighted linear least squares estimation of diffusion MRI parameters: strengths, limitations, and pitfalls. NeuroImage 81, 335–346. https://doi.org/10.1016/j.neuroimage.2013.05.028. https://www.sciencedirect.com/science/article/pii/S1053811913005223.

Veraart, J., Fieremans, E., Jelescu, I.O., Knoll, F., Novikov, D.S., 2016a. Gibbs ringing in diffusion MRI. Magn. Reson. Med. 76 (1), 301–314. https://doi.org/10.1002/mrm.25866. https://onlinelibrary.wiley.com/doi/abs/10.1002/mrm.25866.

Veraart, J., Fieremans, E., Novikov, D.S., 2016b. Diffusion MRI noise mapping using random matrix theory. Magn. Reson. Med. 76 (5), 1582–1593. https://doi.org/10.1002/mrm.26059. https://onlinelibrary.wiley.com/doi/abs/10.1002/mrm.26059.

Vos, S.B., Tax, C.M.W., Luijten, P.R., Ourselin, S., Leemans, A., Froeling, M., 2017. The importance of correcting for signal drift in diffusion MRI. Magn. Reson. Med. 77 (1), 285–299. https://doi.org/10.1002/mrm.26124. https://onlinelibrary.wiley.com/doi/abs/10.1002/mrm.26124.

Wedeen, V.J., Weisskoff, R.M., Poncelet, B.P., 1994. MRI signal void due to in-plane motion is all-or-none. Magn. Reson. Med. 32 (1), 116–120.

Wedeen, V.J., Hagmann, P., Tseng, W.Y.I., Reese, T.G., Weisskoff, R.M., 2005. Mapping complex tissue architecture with diffusion spectrum magnetic resonance imaging. Magn. Reson. Med. 54 (6), 1377–1386. https://doi.org/10.1002/mrm.20642. https://onlinelibrary.wiley.com/doi/abs/10.1002/mrm.20642.

Weldeselassie, Y.T., Barmpoutis, A., Stella Atkins, M., 2012. Symmetric positive semi-definite Cartesian Tensor fiber orientation distributions (CT-FOD). Med. Image Anal. 16 (6), 1121–1129. https://doi.org/10.1016/j.media.2012.07.002. https://www.sciencedirect.com/science/article/pii/S1361841512000941.

Westin, C.F., Martin-Fernandez, M., Alberola-Lopez, C., Ruiz-Alzola, J., Knutsson, H., 2006. Tensor Field Regularization Using Normalized Convolution and Markov Random Fields in a Bayesian Framework. Springer, Berlin, Heidelberg, pp. 381–398, https://doi.org/10.1007/3-540-31272-2_24.

Wu, M., Chang, L.C., Walker, L., Lemaitre, H., Barnett, A.S., Marenco, S., Pierpaoli, C., 2008. Comparison of EPI distortion correction methods in diffusion tensor MRI using a novel framework. Proc. MICCAI 11, 321–329.

Yancey, S.E., Rotenberg, D.J., Tam, F., Chiew, M., Ranieri, S., Biswas, L., Anderson, K.J.T., Baker, S.N., Wright, G.A., Graham, S.J., 2011. Spin-history artifact during functional MRI: potential for adaptive correction. Med. Phys. 38 (8), 4634–4646.

Ye, C., Zhuo, J., Gullapalli, R.P., Prince, J.L., 2016. Estimation of fiber orientations using neighborhood information. Med. Image Anal. 32, 243–256. https://doi.org/10.1016/j.media.2016.05.008. https://www.sciencedirect.com/science/article/pii/S1361841516300378.

Zhang, Y., Wehrli, F.W., 2004. Reference-scan-free method for automated correction of Nyquist ghost artifacts in echoplanar brain images. Magn. Reson. Med. 51 (3), 621–624. https://doi.org/10.1002/mrm.10724. https://onlinelibrary.wiley.com/doi/abs/10.1002/mrm.10724.

Zhang, H., Schneider, T., Wheeler-Kingshott, C.A., Alexander, D.C., 2012. NODDI: practical in vivo neurite orientation dispersion and density imaging of the human brain. NeuroImage 61 (4), 1000–1016. https://doi.org/10.1016/j.neuroimage.2012.03.072. https://www.sciencedirect.com/science/article/pii/S1053811912003539.

Zhou, Z., Liu, W., Cui, J., Wang, X., Arias, D., Wen, Y., Bansal, R., Hao, X., Wang, Z., Peterson, B.S., Xu, D., 2011. Automated artifact detection and removal for improved tensor estimation in motion-corrupted DTI data sets using the combination of local binary patterns and 2D partial least squares. Magn. Reson. Imaging 29 (2), 230–242. https://doi.org/10.1016/j.mri.2010.06.022. https://www.sciencedirect.com/science/article/pii/S0730725X10002109.

Zhou, M.X., Yan, X., Xie, H.B., Zheng, H., Xu, D., Yang, G., 2015. Evaluation of non-local means based denoising filters for diffusion kurtosis imaging using a new phantom. PLoS ONE 10 (2), 1–15. https://doi.org/10.1371/journal.pone.0116986.

Zhuang, J., Lu, Z.L., Vidal, C.B., Damasio, H., 2013. Correction of eddy current distortions in high angular resolution diffusion imaging. J. Magn. Reson. Imaging 37 (6), 1460–1467.

Zwiers, M.P., 2010. Patching cardiac and head motion artefacts in diffusion-weighted images. NeuroImage 53 (2), 565–575. https://doi.org/10.1016/j.neuroimage.2010.06.014. https://www.sciencedirect.com/science/article/pii/S105381191000858X.

# Chapter 10

# Single-shell diffusion models: From DTI to HARDI

**Flavio Dell'Acqua[a], Alexander Leemans[b], Matthew Dawson[a], and Maxime Descoteaux[c]**

[a]*NATBRAINLAB, Department of Neuroimaging, Institute of Psychiatry, Psychology and Neuroscience, King's College London, London, United Kingdom,* [b]*PROVIDI lab, Image Sciences Institute, University Medical Center Utrecht, Utrecht, The Netherlands,* [c]*Sherbrooke Connectivity Imaging Laboratory (SCIL), Department of Computer Science, University of Sherbrooke, Sherbrooke, QC, Canada*

## 1 Introduction

In the previous chapters, we have seen how the diffusion signal can be encoded and how different acquisition strategies can be applied to collect data for tractography applications. In this chapter, we present an overview of the major diffusion techniques that have played a significant role in the development of tractography methods and continue to be widely used today in research and clinical applications using single-shell dMRI acquisitions. This chapter aims to be accessible to a broad readership, using an easy-to-understand formalism. The goal is to provide a quick reference and practical advice on how to get the best out of these approaches. The chapter begins by presenting the diffusion tensor and discussing how this model has been a fundamental step in the history of diffusion imaging and tractography. The high angular resolution diffusion imaging (HARDI) framework is then introduced, showing its role in enabling the creation of new generations of diffusion models capable of extracting multiple fiber orientations within individual voxels. We review multitensor methods and their ability to extract discrete fiber orientations, as well as models like q-ball imaging (QBI) and PASMRI, which allow the extraction of complete orientation distribution functions for each voxel. The final part of the chapter is dedicated to spherical deconvolution methods, their development, and the refinements that, over the years, have made these methods the go-to solution for many contemporary tractography applications.

## 2 From diffusion anisotropy to diffusion tensor imaging

In the early 1990s, when Moseley and colleagues first demonstrated that acute ischemia resulted in a lower apparent diffusion coefficient (ADC) in affected brain regions, they also noted that the measured ADC value appeared to be directionally dependent in white matter (Moseley et al., 1990a,b). While in gray matter, diffusion appeared uniform in all directions, or isotropic, in white matter areas like the corpus callosum, signal intensity varied significantly with the diffusion gradient's direction, suggesting that diffusion was highly anisotropic in these regions. Something about the organization of axons meant the ADC value was dependent on the relative angle between WM fibers and the direction of the applied diffusion-encoding gradients. While previous studies from Hansen (Hansen, 1971) and Cleveland (Cleveland et al., 1976) already had shown anisotropy in ex vivo tissues, Moseley's experiments proved, for the first time, that with diffusion MRI it was possible to probe the tissue microstructural and spatial organization in vivo and noninvasively.

While anisotropic diffusion in white matter was initially attributed to myelin sheaths acting as barriers to water diffusion (Moseley et al., 1990b; Chenevert et al., 1990), later studies showed that the origin of anisotropy cannot be associated with a single specific biological feature (Beaulieu and Allen, 1994a,b). Instead, the movement of water molecules is restricted or hindered by the general structural organization of the tissue itself. Anisotropy is especially pronounced in white matter because the structured alignment of axons allows water molecules to diffuse faster along the direction of the fibers. In contrast, perpendicular to their orientation, the diffusion is significantly reduced. This reduction is caused by the hindering effect of the densely packed myelin and membrane structures on water molecules, as well as restricted diffusion inside individual axons. The packing, orientational coherence, and density of cellular membranes and myelin sheets, along with intra- and extra-axonal diffusivity, all contribute to and modulate the final anisotropy (Pierpaoli and Basser, 1996; Beaulieu, 2002). The discovery of anisotropic diffusion in living biological tissue is a crucial achievement. When properly

*Handbook of Diffusion MR Tractography.* https://doi.org/10.1016/B978-0-12-818894-1.00010-0
Copyright © 2025 Elsevier Ltd. All rights are reserved, including those for text and data mining, AI training, and similar technologies.

modeled, anisotropy becomes the key to quantifying and understanding the complexity of white matter organization. Ultimately, anisotropy is the fundamental diffusion property required by all tractography algorithms.

## 2.1 Diffusion tensor imaging

One of the first efforts to model diffusion anisotropy in biological tissues was the diffusion tensor (DT), introduced by Peter Basser, James Mattiello, and Denis Le Bihan in 1994 (Basser et al., 1994). The diffusion tensor is a $3 \times 3$ symmetric matrix whose values describe the three-dimensional diffusion displacement of water molecules. While the ADC is a unidimensional measure of the apparent diffusivity along a particular direction under the assumption of Gaussian diffusion, the diffusion tensor (DT) can be seen as a natural extension to three dimensions, modeled as a trivariate Gaussian distribution along three orthogonal axes. In its more general form, the tensor can be expressed as:

$$D = \begin{bmatrix} D_{xx} & D_{xy} & D_{xz} \\ D_{yx} & D_{yy} & D_{yz} \\ D_{zx} & D_{zy} & D_{zz} \end{bmatrix} \qquad (1)$$

where, due to its symmetry ($D_{ij} = D_{ji}$), the tensor is fully defined by just six independent parameters. Using a more intuitive representation, each tensor can be decomposed into its eigenvalues and eigenvectors through the operation of diagonalization as:

$$D = V \Lambda V^T = [v_1 \; v_2 \; v_3] \begin{bmatrix} \lambda_1 & 0 & 0 \\ 0 & \lambda_2 & 0 \\ 0 & 0 & \lambda_3 \end{bmatrix} [v_1 \; v_2 \; v_3]^T \qquad (2)$$

This operation transforms the tensor into a matrix, $\Lambda$, with its three eigenvalues, ($\lambda_1 \geq \lambda_2 \geq \lambda_3$), which capture the apparent diffusivity along three orthogonal axes, and a matrix $V$, where its columns, ($v_1, v_2, v_3$), are three eigenvectors defining these three orthogonal axes or principal directions, in relation to the main reference system. In practice, this expression simplifies the representation of the tensor where the three eigenvectors describe how the tensor is oriented in space, and each eigenvalue quantifies the diffusivity along its respective eigenvector (Fig. 1A). A graphical and mathematically equivalent

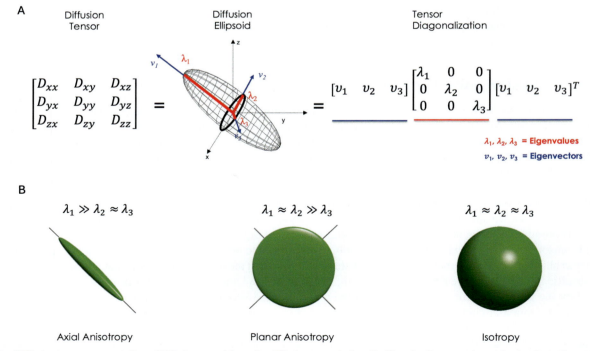

FIG. 1 Diffusion tensor representations. (A) In its general form, the diffusion tensor is described by a $3 \times 3$ symmetric matrix. An alternative and equivalent representation is the diffusion ellipsoid, which can be used to visualize the three eigenvalues ($\lambda_1, \lambda_2, \lambda_3$) and three eigenvectors ($v_1, v_2, v_3$) obtained from tensor diagonalization. (B) Typical diffusion profiles described by the tensor model: an elongated ellipsoid indicates axial anisotropy with predominant diffusion in one direction, a flattened ellipsoid suggests planar anisotropy with significant diffusion within a plane, and a spherical shape indicates isotropy with equal diffusion in all directions.

representation of the diffusion tensor is given by the *diffusion ellipsoid*. This visualization offers a complete description of all tensor properties, where both the *shape* and the *size* are controlled by the eigenvalues and its spatial *orientation* by the eigenvectors. Using only three eigenvalues, it is possible to describe and quantify different types of diffusion profiles: e.g., an elongated ellipsoid indicates highly directional anisotropic diffusion, characteristic of well-aligned fiber bundles, while a spherical ellipsoid suggests isotropic diffusion, typical in less structurally organized tissues, like gray matter (Fig. 1B).

## 2.2 Computing the diffusion tensor

To estimate the diffusion tensor, in addition to a nondiffusion weighted ($b = 0$) image, a minimum of six diffusion-weighted (DW) images must be acquired along noncollinear and noncoplanar gradient directions. However, in practice, to improve noise stability and reduce biases introduced by the choice of a limited set of gradient directions, a larger number of directions are often advised, with 20–30 directions considered a good compromise between acquisition time and the final quality of the tensor estimation (Skare et al., 2000; Jones, 2004). Different approaches have been proposed to estimate the DT from diffusion-weighted MRI data. The most common strategies are based on the use of linear least squares (LLS), weighted linear least squares (WLLS), and nonlinear least squares (NLLS) approaches.

The LLS approach involves linearizing the diffusion signal equation by taking the logarithm of the measured signal intensities. The diffusion tensor elements can then be directly estimated by computing the solution of the ordinary least squares problem. While computationally fast, this approach assumes that the noise in the log-transformed data is Gaussian and homoscedastic (i.e., has constant variance), which is not true due to the nonlinear nature of the log transformation. To account for the heteroscedastic noise in the log-transformed data, the WLLS approach introduces a weighting scheme based on an estimate of the noise variance for each measurement. By assigning lower weights to noisier measurements, with minimal impact on computational efficiency, the WLLS approach provides a more robust estimate of the diffusion tensor compared to LLS. Finally, the NLLS approach attempts to fit the diffusion tensor model directly to the raw signal intensities, without the need for log transformation. This approach uses iterative optimization algorithms (e.g., Levenberg–Marquardt) to find the tensor elements that best explain the observed data. Depending on the minimization function adopted, additional constraints can also be added to account for Rician noise statistics or to enforce nonnegativity on the tensor eigenvalues. Although more computationally intensive than the linear methods, the NLLS approach provides more accurate estimates of the diffusion tensor, particularly at low signal-to-noise ratio (SNR) (Koay et al., 2006; Jones, 2009). More advanced techniques have been proposed to also address outlier rejection. The RESTORE approach (Chang et al., 2005) introduces an iterative reweighting scheme to identify and reject outliers in the data, further enhancing the robustness of the tensor estimation. Non-Euclidean frameworks, such as the Log-Euclidean metrics, have also been developed to further improve the interpolation and regularization of diffusion tensors, providing more physically consistent results and increased numerical stability (Arsigny et al., 2006) (see also Chapter 9).

## 2.3 Diffusion tensor metrics

### 2.3.1 Trace (TR) and mean diffusivity (MD)

The trace and mean diffusivity are very closely related metrics that provide a rotationally invariant measure of the overall mobility, or diffusivity, of water molecules within each voxel. The *trace* is defined as the sum of the three diagonal elements of the diffusion tensor ($D_{xx}, D_{yy}, D_{zz}$), which can be proven to be equivalent to the sum of the three tensors' eigenvalues. Dividing the trace by 3 gives the *mean diffusivity*, which is a measure of the orientation-averaged mean diffusivity:

$$MD = \frac{Tr(D)}{3} = \frac{\lambda_1 + \lambda_2 + \lambda_3}{3} \tag{3}$$

At moderate to low $b$-values (e.g., $b \leq 1500 \, s/mm^2$), both metrics are known to be mostly uniform across the whole brain parenchyma (i.e., $MD \sim 0.7 \times 10^{-3} \, mm^2/s$) (Pierpaoli et al., 1996) (Fig. 2A). Although these metrics do not provide a significant contrast between gray and white matter, they are particularly useful in detecting small microstructural changes, making them very sensitive diffusion maps (Luque Laguna et al., 2020).

### 2.3.2 Fractional anisotropy (FA)

Fractional anisotropy (FA) is a widely used diffusion index that quantifies the degree of diffusion anisotropy in biological tissues. It is a normalized measure of variance among the three eigenvalues of the diffusion tensor, with values ranging from 0 to 1. While the mean diffusivity (MD) represents the average mobility of water molecules, FA provides complementary information about the directionality and coherence of the diffusion process. FA is mathematically defined as:

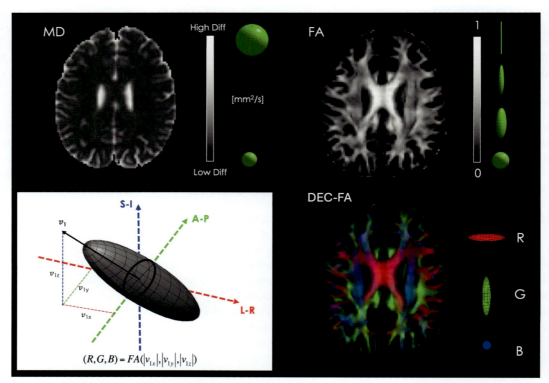

**FIG. 2** Diffusion tensor imaging maps. Clockwise from top left: Mean Diffusivity (MD), Fractional Anisotropy (FA), and Directional Encoded Color FA (DEC-FA) maps. Lower left: schematic representation of the color-encoding applied in DEC-FA maps, where red (R), green (G), and blue (B) represent left–right, anterior–posterior, and superior–inferior orientations, respectively.

$$FA = \frac{\sqrt{3((\lambda_1 - \langle\lambda\rangle)^2 + ((\lambda_2 - \langle\lambda\rangle)^2 + ((\lambda_3 - \langle\lambda\rangle)^2}}{\sqrt{2(\lambda_1^2 + \lambda_2^2 + \lambda_3^2)}} \tag{4}$$

Higher FA values indicate that one or more eigenvalues deviate significantly from their mean, suggesting a strong orientation-dependent diffusivity. Conversely, low FA values correspond to all eigenvalues being close to, or equal to, their mean diffusivity, suggesting a lack of orientational dependence and a more uniform diffusivity. In the healthy human brain, FA ranges from near 0 in gray matter to 0.8–0.9 or higher values in highly compact and coherently organized white matter, such as the body of the corpus callosum (Basser and Pierpaoli, 1996; Pierpaoli and Basser, 1996; Pierpaoli et al., 1996) (Fig. 2). Reductions in FA have been observed in various pathological conditions involving demyelination and neurodegeneration, making FA an important diffusion metric to detect and investigate biological changes in white matter. However, without strong clinical hypothesis, today it is not recommended to refer to FA only as a general index of white matter integrity. FA is sensitive to multiple and different aspects of the tissue microstructure and its organization, including fiber coherence, intrinsic anatomical variability, crossing fibers, and partial volume effects, which may not necessarily relate to actual white matter damage (Alexander et al., 2000; Jones et al., 2013).

### 2.3.3 Other metrics

In addition to FA and MD, several other metrics can be derived from the diffusion tensor. Individual eigenvalue maps or combinations of eigenvalues are often used to describe *axial* (parallel) diffusivity, $D_\parallel = \lambda_1$, and *radial* (perpendicular) diffusivity, $D_\perp = (\lambda_2 + \lambda_3)/2$, in relation to the main tensor orientation (Wheeler-Kingshott and Cercignani, 2009). Additional maps have been developed to better characterize the shape of the diffusion tensor. Examples include the *mode* of the tensor (Ennis and Kindlmann, 2006) and the *Westin* metrics used to describe the tensor shape with indices of "linearity," "sphericity," and "planarity" (Westin et al., 1997).

## 2.4 Diffusion tensor orientation

As we have seen in the previous sections, the spatial orientation of the diffusion tensor is encoded by its three eigenvectors. In the presence of anisotropy, the eigenvector associated with the largest eigenvalue, known as the principal eigenvector, indicates the orientation of maximum diffusivity within a voxel. For coherently organized white matter, the principal eigenvector aligns with the longitudinal axis of the fibers, providing a direct correspondence between the diffusion tensor orientation and the underlying white matter architecture. While early diffusion maps had already attempted to estimate the orientation of white matter using individual values or ratios of ADC maps (Beaulieu, 2002), it is only with the introduction of the diffusion tensor that it becomes possible to obtain a first quantitative, vectorial, and rotation-invariant representation of the microstructural organization in living biological tissue (Basser et al., 1994). Like the discovery of anisotropy, the introduction of tensor formalism is a crucial moment for diffusion imaging and the future development of tractography methods (Mori et al., 1999; Conturo et al., 1999; Jones et al., 1999b; Basser et al., 2000; Poupon et al., 2000; Parker, 2000).

One of the earliest applications of the DT orientational information was the introduction of fiber orientation maps (Pierpaoli, 1997; Jones et al., 1997). Among these, the most widely recognized are the directional encoded color (DEC) maps, also known as color-coded FA maps (Pajevic and Pierpaoli, 1999). In the DEC maps, the orientation of the principal eigenvector is represented by a combination of three colors: red, green, and blue (RGB). Each color is assigned to one of the three orthogonal axes of the reference frame: red for left–right, green for anterior–posterior, and blue for superior–inferior. The intensity of each color is then determined by the absolute value of the corresponding RGB component of the principal eigenvector. This approach allows for an immediate visual representation of the white matter's orientation: fibers oriented in the left–right direction are red, anterior–posterior in green, and superior–inferior in blue. In general, the principal eigenvector will have nonzero components along multiple axes, resulting in a mixture of red, green, and blue colors. To incorporate anisotropy information in the map, the color intensities are also modulated by the fractional anisotropy (FA) value, such that highly anisotropic regions appear bright and vivid, while less anisotropic regions appear darker and less saturated. This modulation helps to highlight areas with strong, coherent fiber orientation and to suppress noise in regions with low anisotropy, such as gray matter or cerebrospinal fluid (CSF) (Fig. 2).

The use of color-coded FA maps not only facilitates the assessment of white matter orientation but also simplifies the differentiation of adjacent white matter bundles running along different orientations. Applications have been particularly significant in the field of radiology and neurosurgical planning. The color coding helps surgeons to discriminate between various white matter tracts, enabling them to better see how anatomy has been altered by lesions or tumors. Even before performing tractography, this visual differentiation is useful in recognizing critical tracts during surgery, minimizing the risk of postoperative neurological deficits. In some circumstances, the combination of FA values and color coding is also used for qualitatively assessing white matter degeneration or tumor infiltration (Assaf and Pasternak, 2008; Sundgren et al., 2004; Potgieser et al., 2014).

However, the main use of tensor orientation information is in tractography reconstructions. Most of the chapters in this book discuss in detail how tractography reconstructions are obtained using different algorithms, diffusion models, and their applications. Here, we want to highlight how, thanks to the introduction of diffusion tensor tractography, it has been possible, for the first time, to reconstruct in vivo most of the major white matter pathways, which were previously described only using postmortem dissection methods. Because of this, DT tractography has rapidly emerged as an essential tool in neuroscience, mental health research, and clinical settings, allowing both the segmentation of white matter tracts and quantification of along-tract properties. Unfortunately, these early reconstructions have also revealed some important limitations of DT tractography that have affected its general applicability and the interpretation of some of its results. The next section provides an overview of these limitations.

## 2.5 Diffusion tensor limitations

The use of only three eigenvalues is a very efficient and compact way to describe the size and shape of the three-dimensional profile of the diffusion ellipsoid under the assumption of Gaussian diffusion. However, this representation is inadequate when the complexity of the biological environment requires a deeper characterization. In essence, what is not captured by the three eigenvalues is effectively lost and not included in the tensor model. A typical example of this is when the diffusion signal originates from multiple fiber orientations within a voxel, such as *crossing fibers*. In this case, the diffusion tensor can only provide an average representation of the underlying diffusion. Here, each eigenvalue represents a mixture of all contributions within a voxel, typically producing a smoother or rounder ellipsoidal profile, with decreased anisotropy. This issue is critical not only in the interpretation of FA maps, where crossing fibers can lead to FA drops in perfectly healthy white matter, but also in tractography where the decrease in anisotropy may cause tractography algorithms to stop in regions with no clear microstructural organization (Jones, 2008; Dell'Acqua and Catani, 2012).

**182 PART | II** Diffusion MRI

Similarly, the three eigenvectors and the overall spatial orientation of the tensor are also the result of all microstructural contributions within a voxel. The main tensor orientation is adequate in the presence of a single dominant fiber population. However, in the presence of multiple fiber populations, the tensor model is not able to describe multiple orientations separately; instead, it will always provide an average ellipsoid profile whose primary orientation may or may not reflect the real microstructural organization. In such cases, the principal eigenvector may not align with any real white matter orientation, leading to artifacts like spurious reconstructions or tractography stopping inside white matter regions.

For many years, the crossing-fiber problem was a very active field of research. While it was initially assumed that crossing was only a problem limited to a few white matter regions, multiple studies have now shown that, in reality, crossing configurations are one of the most prevalent configurations in the brain, with estimates suggesting that between 70% and 90% of white matter voxels of the human brain have two or more crossing fiber populations (Behrens et al., 2007; Descoteaux et al., 2009; Dell'Acqua et al., 2013; Jeurissen et al., 2013). While the term "crossing fiber" has been very popular in the diffusion and tractography literature, it is important to realize that a whole range of fiber configurations are also not modeled by the tensor model. These include configurations such as kissing fibers, where fibers touch and separate without crossing, or fanning fibers, where fibers simply fan out from a compact bundle. Additionally, even certain single fiber orientations cannot be completely modeled by the tensor when, within a voxel, fibers bend, twist, or branch. As we will see in the rest of this chapter, crossing fibers and other configurations have been successfully solved by later models. However, some configurations remain unsolved today, still limiting the accuracy of tractography algorithms. In particular, among all unsolved configurations, most can be associated with the intrinsic antipodal symmetry of the diffusion signal, which prevents modeling asymmetric fiber configurations within a voxel (Reisert et al., 2012; Bastiani et al., 2017; Poirier and Descoteaux, 2023). Chapters 12 and 28 discuss in more detail recent results on asymmetric fiber orientations and the current open challenges of diffusion modeling and tractography methods.

## 3  High angular resolution diffusion imaging (HARDI)

As we have seen in the previous sections, the diffusion tensor only provides a simplified three-dimensional description of the diffusion propagator. Attempts to obtain its complete characterization were made as far back as 1988, in the field of NMR, when Paul Callaghan suggested 6-D q-space imaging experiments to map the full diffusion propagator (Callaghan et al., 1988). Unfortunately, q-space imaging and later diffusion spectrum imaging (DSI) (Wedeen et al., 2005) have always required a large number of encoding directions and multiple diffusion weightings, reaching very high $b$-values (e.g., $b \geq 8000 \, \text{s/mm}^2$). These requirements have been a major obstacle and have limited the broad application of propagator-based imaging techniques.

Given these practical challenges, researchers interested in improving tractography and better mapping white matter structural organization have focused instead on capturing the angular characteristics of the diffusion signal. This approach has led to the introduction of high angular resolution diffusion imaging (HARDI) (Frank, 2001; Tuch et al., 2002), an acquisition strategy where the DW data is collected along multiple directions using a uniform angular sampling while keeping a constant $b$-value. Compared to previous approaches, data is not collected specifically to fit a predefined model or a finite number of parameters (i.e., the six independent elements of the diffusion tensor). Within the HARDI framework, the focus is shifted to the data itself and on how to best sample the diffusion signal to encode the available angular information (Fig. 3).

To effectively represent and analyze the angular information captured in HARDI acquisitions, spherical harmonics (SH) have emerged as an essential tool for many diffusion models and applications. Equivalent to the Fourier transform for spherical data, SH offers a convenient way to encode and manipulate the diffusion signal and diffusion profiles (Frank, 2001, 2002; Descoteaux et al., 2006). Particularly, due to the real and symmetric nature of the diffusion signal, only real, even-order SH coefficients are necessary for its accurate and complete representation. This not only allows a compact representation of the HARDI signal using a limited set of SH coefficients, but also helps in noise reduction by removing harmonics or "frequencies" that are not physically plausible (Frank, 2002). Moreover, as we will later see in more detail, the use of SH also allows a precise quantification of the minimum number of diffusion gradient directions given a chosen $b$-value (Tournier et al., 2013). For more details and a full mathematical description of SH, see Appendix D.

Traditionally, HARDI has typically been implemented as single-shell acquisitions where diffusion gradients are applied at a single $b$-value or the equivalent q-value in the case of q-space imaging approaches. These acquisitions remain relatively simple to implement and have scan times only marginally longer than DTI sequences, thus making them easily accessible for clinical and clinical-research settings. However, the concepts behind HARDI have also been extended to multishell acquisitions. For the remainder of this chapter, we will focus on single-shell HARDI models. Detailed discussions of multishell HARDI methods, including advantages and implementation challenges, are presented in the next chapter.

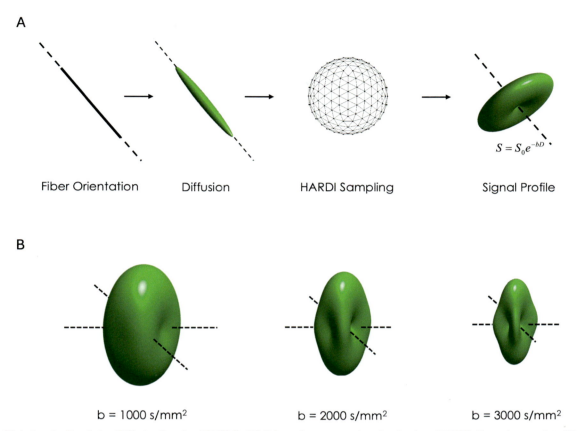

**FIG. 3** High Angular Resolution Diffusion Imaging (HARDI). (A) Schematic representation showing how HARDI allows the complete 3D characterization of the diffusion signal profile using uniform angular sampling. (B) Example of a signal profile for two crossing fibers with increasing $b$-value. Note the increase in angular contrast at higher $b$-values and the proportional decrease in signal amplitude, making high $b$-values better for resolving crossing fibers but also more sensitive to noise contamination.

### 3.1 Multitensor and multicompartmental approaches

Because the diffusion tensor has proven to be effective in modeling a single fiber orientation, the use of multiple tensors, or multitensor approaches, has been the logical next step to model multiple fiber orientations. In their more general form, these models assume that each fiber population is an independent compartment described by its own diffusion tensor with no molecular exchange between compartments. The total voxel signal is expressed as the sum of individual compartments:

$$S(b, \boldsymbol{u}) = S_0 \sum_{i=1}^{N} f_i e^{-b \boldsymbol{u} \boldsymbol{D}_i \boldsymbol{u}^T} \tag{5}$$

where $S_0$ is the nondiffusion weighted signal, $f_i$ and $\boldsymbol{D}_i$ are the relative volume fraction and diffusion tensor of the $i$-th compartment, respectively; $\boldsymbol{u}$ is the direction of the applied diffusion gradients, and $b$ is the $b$-value. Unfortunately, it has been evident since the early methods adopting this framework that pure multitensor models are ill-posed due to the large number of free parameters (i.e., seven parameters per fiber orientation) (Tuch, 2002). Without introducing any additional constraint, the same diffusion signal can be explained by multiple fiber configurations, leading to unstable results. To account for this, multitensor models have been limited to extracting usually no more than two distinct fiber populations, and additional constraints have been applied on the shape or the type of tensor to ensure convergence to a single solution. Examples include enforcing diffusion tensors with axial symmetry ($\lambda_2 = \lambda_3$) (Alexander and Barker, 2005; Kreher et al., 2005) or allowing each tensor eigenvalue to assume values only within certain ranges or even fixing it to predefined values (Tuch, 2002).

The inclusion of stronger constraints effectively enabled the transition from multitensor approaches to more general multicompartmental models, where specific compartment features (e.g., diffusivities, volume fractions, etc.) are described by a finite set of parameters. An example is the *ball-and-stick* model (Behrens et al., 2003), which deviates from classical multitensor approaches by introducing a simpler, yet effective, model. In this approach, the HARDI signal is described by two or more distinct compartments: the "ball," a single parameter, fully isotropic Gaussian diffusivity compartment, and

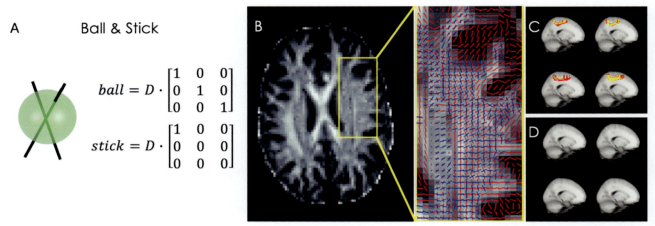

**FIG. 4** The ball-and-stick model. (A) Schematic representation of the ball-and-stick model. Here, all anisotropic information is captured by one or multiple sticks, and the voxel diffusivity by the ball compartment. (B) Examples of multiple fiber orientations resolved by the ball-and-stick model. (C and D) Probabilistic tractography of the medial SLF obtained using two sticks (C) vs using the one-stick model (D). *Panels (B and C) adapted from Behrens TEJ, Johansen Berg H, Jbabdi S, Rushworth MFS, Woolrich MW (2007) Probabilistic diffusion tractography with multiple fibre orientations: what can we gain? NeuroImage 34:144–155.*

one or more "stick" compartments, each one modeling a fully restricted single fiber orientation described by three parameters (i.e., two angles to describe its spatial orientation and one volume fraction). Each stick is identical and can be considered a totally anisotropic diffusion tensor with diffusion occurring only in the fiber orientation and none perpendicularly ($\lambda_2 = \lambda_3 = 0$). Because of its simplicity and the reduced number of parameters compared to a complete multitensor approach, this approach can fit multiple orientations, with stable results up to two, and sometimes even three, fiber compartments per voxel (Hosey et al., 2005; Behrens et al., 2007) (Fig. 4). This model has proven to be effective with single-shell and clinically feasible protocols and it is still actively used in tractography applications today.

It is interesting to observe that this model has also served as a prototype for different families of diffusion models. On one hand, this model was one of the first to introduce compartments with different types of diffusion (i.e., Gaussian and restricted diffusion). This idea has been further expanded by microstructure imaging models where different analytical diffusion models are used to separate and quantify different compartments (Assaf et al., 2004; Assaf and Basser, 2005; Zhang et al., 2012; Panagiotaki et al., 2012; Alexander et al., 2017; Novikov et al., 2018). On the other hand, by assuming the same signal profile for each fiber orientation, the ball-and-stick model also introduces, for the first time, a convolution operation between the distribution of fiber orientations and a kernel, or fiber response, modeling the signal for a single fiber orientation (Behrens et al., 2003). This idea is at the foundation of all spherical deconvolution approaches, as we will see later.

The ball-and-stick model has been further extended by replacing each "stick" population with distributions of sticks to capture more complex configurations, including measures of coherence or anatomical dispersion within each fiber population. Examples of these advancements include the parametric spherical deconvolution approach as proposed by Kaden et al. (2007) and the ball-and-rackets model (Sotiropoulos et al., 2012), where Bingham distributions are also used to capture, along with the main orientation of each fiber population, measures of fiber dispersion. For more details, see also Chapter 12. Other distributions such as matrix-variate gamma distributions (Scherrer et al., 2013) or more complex combinations of compartments (Jespersen et al., 2007) have also been explored in the literature. However, it is important to point out that in more complex and recent approaches, multiple *b*-value acquisitions are always required. As we will see in the next chapter, multiple *b*-values are an essential requirement to separate distinct signal attenuation profiles and extract individual properties and diffusivities from each compartment. This is one of the main reasons why, historically, microstructure imaging models have moved away from single-shell HARDI acquisitions, often compromising also on angular sampling and the ability to extract multiple fiber orientations (Dell'Acqua and Tournier, 2019). Only recently, with the advent of multiband acquisitions, has this issue been resolved, allowing for both adequate angular sampling and multiple diffusion weightings to support more complex modeling. For more details see Chapter 11.

An important limitation common to all multitensor and multicompartmental approaches is the challenge of determining a priori how many fiber orientations should be fitted for each voxel. Model selection is a critical step in these approaches because any error made in assigning the correct number of fiber compartments will inevitably affect the estimation of the final fiber orientations and their consistency with neighboring voxels. This is obviously a crucial point for tractography applications, but it also impacts microstructure imaging, where incomplete fiber orientation modeling will inevitably also affect the rest of the model metrics. Different model selection and statistical approaches have been proposed in the literature, ranging from simple correlation between the predicted and original signal (Tuch, 2002), to F-test and classification strategies based on spherical harmonics decomposition of the signal (Frank, 2002; Alexander et al., 2002; Parker and Alexander, 2003), as well as more sophisticated automatic relevance determination techniques based on a Bayesian framework (Behrens et al., 2003). It is important to acknowledge that model selection methods also frequently exhibit a bias toward simpler models, often discarding complex configurations in the presence of noisy data (Behrens et al., 2007).

## 3.2 Nonparametric approaches

In contrast to multitensor and multicompartmental models, another family of techniques, often defined as "nonparametric," has gained significant traction and popularity for its ability to extract the orientation of multiple fiber populations without the need for explicit model selection or assumptions about the number of fiber populations. More precisely, while the methods described previously can be considered "parametric," as they estimate a finite set of parameters for each compartment, nonparametric approaches estimate a continuous spherical function, commonly known as the orientation distribution function (ODF). This function captures the angular complexity and structural organization of the diffusion signal within a voxel. The main benefit of nonparametric approaches is that by computing the ODF, one can automatically extract the number of intra-voxel fiber orientations, which correspond to the local maxima of the spherical function, without the need for model selection. The orientation of each fiber component is determined by the direction in which the maxima point, and the relative or absolute contribution of each fiber component, can be inferred from the amplitude of the ODF at those maxima. The focus for the rest of the chapter will be on introducing the main methods used to estimate ODFs or equivalent functions and highlighting the strengths and limitations of each approach. A detailed discussion of the common properties and operations that can then be applied to ODFs and extracted fiber orientations is presented in Chapter 12.

## 3.3 Q-ball imaging

QBI is a popular nonparametric, model-independent approach developed by David S. Tuch (Tuch, 2002; Tuch et al., 2003; Tuch, 2004) to resolve multiple intravoxel fiber orientations. As such, QBI can extract the ODF directly from single-shell HARDI data without requiring model selection on the number of fiber populations or even making a priori assumptions about the diffusion process. QBI is based on the Funk-Radon transform (FRT), also known as the spherical Radon transform. This operation transforms a function defined on the sphere into a new function on the same sphere. For each point, the value of the FRT is computed by integrating the original function along the great circle perpendicular to the line connecting the center of the sphere to the chosen point. Tuch demonstrated that, when applied to diffusion data, the FRT could be used to directly extract a close approximation of the orientation distribution of the real spin-density without the need to first estimate the diffusion propagator as required by other q-space imaging-based methods. As we will see in detail in the next chapter, diffusion spectrum imaging and q-space imaging approaches require either complex q-space Cartesian sampling or large multishell acquisitions to accurately reconstruct the diffusion propagator (Wedeen et al., 2005). Once computed, the ODF can be explicitly extracted as a radial projection of the diffusion propagator, $P(r\boldsymbol{u})$, over the sphere as[a]:

$$ODF(\boldsymbol{u}) = \int_0^\infty P(r\boldsymbol{u})r^2 dr \tag{6}$$

where $\boldsymbol{u}$ is a unit vector, and $r$ is the radial coordinate in the propagator space. While this expression returns the angular spin density of the diffusing water molecules, it also reveals that the ODF does not preserve all information present in the diffusion propagator. It discards radial information, maintaining only its angular contrast. By focusing only on the angular information and directly computing the ODF from single-shell data, QBI effectively eliminates the need for time-consuming Cartesian sampling, enabling a more efficient and straightforward estimation of the ODF. Using the FRT, given a direction $\boldsymbol{u}$, the corresponding ODF value is immediately computed by integrating the diffusion signal along the equator, or the great circle, perpendicular to the chosen direction as:

$$ODF(\boldsymbol{q}, \boldsymbol{u}) \approx \int_{\boldsymbol{u} \perp \boldsymbol{q}} E(\boldsymbol{q}) d\boldsymbol{q} \tag{7}$$

Here, $E(\boldsymbol{q})$ represents the HARDI signal, normalized to the nondiffusion weighted image, acquired at constant q-value uniformly sampled along multiple $\boldsymbol{q}$-vectors; $\boldsymbol{u} \perp \boldsymbol{q}$ indicates that the integral is performed over the equator perpendicular to the direction $\boldsymbol{u}$. For more details about q-vectors and q-space formalism, see Chapters 5 and 6.

As anticipated, Tuch demonstrated that integrating the signal over the equator is a close approximation to the radial integration of the propagator (Tuch, 2004). In practice, the true relationship between the ODF obtained by the FRT

---

a. This expression already includes the $r^2$ term to account for a constant solid angle as presented in the DSI paper (Wedeen et al., 2005) and only later introduced in QBI with QBI-CSA (Aganj et al., 2010) as explained later in the text.

and the radial projection of the propagator is described by a more complex expression, where instead of an infinitely thin radial projection, the QBI computation introduces a finite "beam" around the orientation $u$, leading to a blurring in the resulting ODF (Tuch, 2004). The width of this beam depends also on the q-value applied during data acquisition. Lower q-values will lead to wider beams and increased blurring, but also more stable results, because more data is weighted by the beam. On the contrary, higher q-values will result in decreased blurring but potentially also noisier ODFs.

The original QBI formulation has undergone several improvements and refinements since its introduction. Hess et al. (2006) reformulated the reconstruction of the ODF and derived an analytic solution to compute the integration over the great circle using spherical harmonics. This representation allows the SH coefficients of the final ODF to be derived directly from the SH coefficients of the diffusion signal through a simple matrix multiplication. The use of SH allows for a more compact and efficient representation of the ODF, reducing data storage requirements and facilitating further processing and analysis. Descoteaux et al. (2007) further improved QBI by introducing an effective regularization strategy in the ODF estimation also using spherical harmonics. This regularization, based on the Laplace-Beltrami operator, allows more stable ODF reconstructions, reduces noise artifacts, and enables practical use of higher SH orders (i.e., $l \geq 4$) at the expense of only a minor angular resolution penalty in the final ODF. Descoteaux et al. (2009) later introduced the concept of ODF sharpening as a postprocessing step for ODF. This sharpening operation applies spherical deconvolution directly in the ODF space to remove the intrinsic angular blurring of the diffusion process present on the original ODF, and recovers a fiber-ODF, or fODF. As we will see later in the chapter, this function more accurately represents the underlying organization of white matter as it directly attempts to map the actual angular distribution of fiber orientations. The reconstructed fODF is significantly sharper and can resolve smaller crossing angles compared to the original ODF, with significant improvements also in tractography reconstructions (Descoteaux et al., 2009). Another significant improvement with this approach was the introduction of QBI with constant solid angle (QBI-CSA) from Aganj et al. (2010), and similarly from Tristán-Vega et al. (2009). QBI-CSA addresses an important limitation of the original QBI formulation, where the radial projection of the propagator was assumed to be linear, not taking into account the effect of quadratic growth of the volume element on the sphere with respect to the distance. By not considering a constant solid angle, regions farther from the origin of the propagator, which typically contain the highest angular contrast, are penalized compared to nearer regions with lower angular contrast. By accounting for a constant solid angle, this modification not only enhances the accuracy of the ODF by removing the need for additional corrections and normalization terms but also substantially increases the sharpness of the reconstructed ODFs, providing a more accurate representation of the underlying white matter architecture. Examples of different QBI implementations are shown in (Fig. 5). Additional improvements have also been proposed over the years to extend QBI to multishell acquisition to further improve accuracy and angular resolution. Examples include (Khachaturian et al., 2007; Wu and Alexander, 2007; Aganj et al., 2010; Canales-Rodríguez et al., 2009; Yeh et al., 2010). A more recent evolution of QBI is fiber ball imaging, which uses instead an inverse Funk transform to estimate fODF. Compared to QBI, fiber ball imaging has shown promising results with sharper fODF profiles (Jensen et al., 2016).

Overall QBI is an effective method to resolve multiple intravoxel fiber orientations, and thanks to its relatively simple numerical implementation, it has been widely adopted in multiple tractography applications. QBI has been successfully combined with probabilistic and bootstrapping tractography methods and has found application in clinical-research settings, including neurosurgical planning (Berman et al., 2008; Descoteaux et al., 2009; Sotiropoulos et al., 2010; Bucci et al., 2013; Mandelli et al., 2014). However, one of the main limitations of QBI is the need for high $b$-values (or q-values) to achieve good angular resolution, which typically requires $b$-values in the range of, or above, 4000 or $6000 \, \text{s/mm}^2$. While modern MRI scanners can achieve these values, they still impose significant trade-offs between the final choice of $b$-values, signal-to-noise ratio, and spatial resolution. It is important to note also that QBI, DSI, and most q-space based imaging methods require high $b$-values because of the nature of their recovered ODF. In model-independent approaches, the ODF is closely related to the angular characteristics of the diffusion propagator, and high $b$-values are needed to maximize its angular contrast. Even in the presence of perfectly parallel fibers, an ODF will always display a rounded profile because of the way the diffusion information is encoded in the measured diffusion signal (Dell'Acqua and Tournier, 2019). As we will see later in the chapter, this is an important difference compared to fODFs recovered by spherical deconvolution methods and sharpening techniques. Methods computing fODF explicitly model out the angular blurring introduced by the diffusion process, allowing the fODF to provide significantly sharper profiles, while also requiring lower $b$-values. To avoid confusion and to clearly separate methods recovering fODF from methods based on diffusion propagator characteristics, the recent literature has adopted the term diffusion-ODF (dODF) to specifically refer to all diffusion-based ODFs as obtained by QBI, DSI, and other methods associated with q-space imaging. For more details about the difference between dODF and fODF, please see also Chapter 12.

Despite these limitations, QBI and model-independent approaches remain particularly useful when typical model assumptions are not valid or when a clear model of the tissue microstructure under investigation is unavailable. In these circumstances, model-independent approaches offer a practical and robust method to investigate the tissue organization without hypotheses. This can be particularly useful in ex vivo fixed tissues, where the biophysics characteristics are largely

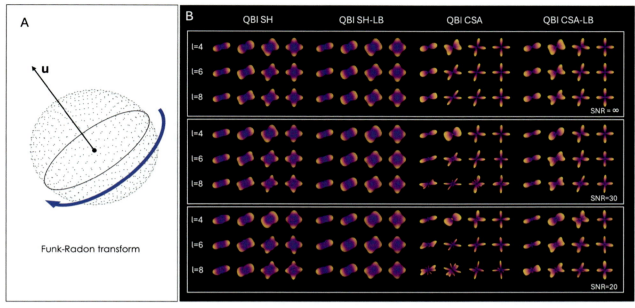

**FIG. 5** Q-ball imaging: (A) Schematic representation of the Funk-Radon transform used by QBI to estimate the ODF profile. (B) Comparison of ODF profiles reconstructed by different QBI implementations: using spherical harmonics (QBI-SH), using spherical harmonics and Laplace-Beltrami regularization, $\lambda = 0.004$ (QBI_SH_LB), using constant solid angle (QBI-CSA), using constant solid angle and Laplace-Beltrami regularization, $\lambda = 0.006$ (QBI CSA-LB). Simulated crossing angles are 30, 50, 70, 90 degrees, $b$-value $= 4000 \, \text{s/mm}^2$, 90 DWI directions.

different compared to the corresponding in vivo tissue, or in pathological regions, like stroke, where it is difficult to define a general biophysical model for the whole diffusion signal. In such cases, a model-independent approach may be helpful to explore and obtain a first unbiased representation of the underlying microstructure organization.

### 3.4 Persistent angular structure and diffusion orientation transform

Beside QBI, additional methods using spherical functions have been developed over the years to extract multiple fiber orientations and describe complex white matter organization. These include the persistent angular structure-MRI (PASMRI) by Jansons and Alexander (2003) and the diffusion orientation transform (DOT) by Ozarslan et al. (2006). Like QBI, both methods aim to quantify a spherical function that is closely related to the angular features of the propagator without the need for its complete estimation and by using only single-shell acquisitions.

At its core, PASMRI solves an inverse problem where the goal is to estimate the (radially) *persistent angular structure* (PAS), defined as a spherical function, $\widetilde{p}(u)$. This function when analyzed through its Fourier transform in three-dimensional space at a fixed radius $r$ provides the best fit to the measured HARDI signal. More precisely, since single-shell data encodes only angular but not radial information from the propagator, the full propagator is assumed here to be projected onto a single spherical function at fixed radius—the PAS function. This function effectively preserves the information about the angular structure of the propagator but discards its radial information. With this approach, the Fourier integral of the propagator reduces to an integral over the sphere, directly linking the measured HARDI signal and the PAS function (Jansons and Alexander, 2003). To practically estimate the PAS profile, PASMRI employs a maximum entropy approach, which involves minimizing the relative information of the PAS profile against a uniform spherical distribution. This ensures that any deviation from a spherical profile is supported by the data. Minimizing the relative information promotes solutions that contain the least information and leads to smoother representation of the PAS, while reducing noise instabilities. The maximum-entropy parametrization for the PAS can be expressed as:

$$\widetilde{p}(\boldsymbol{u}) = exp\left(\lambda_0 + \sum_{i=1}^{N} \lambda_i \cos\left(r\boldsymbol{q}_i \cdot \boldsymbol{u}\right)\right) \qquad (8)$$

where $N$ is the number of measurements, $\boldsymbol{q}_i$ is the i-th q-vector from the HARDI acquisition, and $\lambda_i$ are Lagrange multipliers that define the profile of the PAS and are estimated during fitting. Thanks to the exponential function always being positive for any real value, this parametrization inherently ensures the nonnegativity of the estimated PAS profile. The parameter $r$ defines the radius where the PAS is estimated. This radius sets the scale at which angular features are analyzed and, in practice, can be used as a regularization term to control between the sharpness of the recovered PAS profile and robustness to noise. Small values of $r$ will generate only smooth profiles; higher values will generate sharper profiles but will also increase sensitivity to noise and instabilities. A nonlinear optimization strategy based on the Levenberg-Marquart algorithm is used to fit the PAS parameters by minimizing the least squares fit between the measured diffusion signal and the Fourier transform of the PAS profile.

Compared to other methods, PASMRI provides clear and sharp fiber orientation estimates using relatively low $b$-values (i.e., 1500–2000 s/mm$^2$). PASMRI has been successfully applied in tractography and validation studies, showing good agreement with chemical tracing techniques (Dyrby et al., 2007, 2011) and expected brain anatomy (Parker and Alexander, 2005). While still being related to the propagator, the PAS profile is usually sharper than a traditional dODF and resembles more an fODF profile. Indeed, direct links between the PAS and the spherical deconvolution framework have been found by Seunarine and Alexander (2006), showing that the HARDI signal can also be expressed as a convolution between PAS and a kernel defined as $R(\boldsymbol{q}, \boldsymbol{u}) = r^{-2} cos(r\boldsymbol{q} \cdot \boldsymbol{u})$. However, PAS profiles obtained by directly deconvolving the signal have shown poorer results and more instabilities compared to the original maximum entropy approach as adopted in PASMRI (Seunarine and Alexander, 2006). Interestingly, a maximum entropy spherical deconvolution (MESD) algorithm has also been proposed by Alexander (2005a). Derived directly from the PASMRI framework, this algorithm produces extremely sharp and stable fODF profiles, thanks to the nonnegativity constraint introduced by the maximum entropy formalism. As we will see later, spherical deconvolution methods heavily rely on regularization strategies to obtain stable solutions, and nonnegativity is one of the most effective strategies for different methods.

Unfortunately, the main limitation of both PASMRI and MESD lies in their very high computational cost. PASMRI does not guarantee global convergence, requiring multiple restarts per voxel to ensure convergence to a stable solution (Alexander, 2005b). Additionally, each iteration of the nonlinear optimization involves computing numerical integrals for each voxel. This limitation has significantly restricted the adoption of these methods in practical applications on whole brain data due to the significant computation time required.

The diffusion orientation transform (DOT) is another approach, originally introduced by Ozarslan et al. (2006), that can be used to map complex tissue organization and extract multiple fiber orientations. The DOT transforms the diffusion profile computed from a single-shell HARDI acquisition into a probability profile that approximates the propagator at a fixed radius. This transformation is achieved by computing the Fourier transform in spherical coordinates by analytically evaluating the radial part of the diffusion signal under the assumption of monoexponential attenuation. The resulting probability values are calculated on a fixed radius, with smaller radii providing smoother and more robust-to-noise estimates, and larger radii showing sharper profiles with increased angular contrast but also increased sensitivity to noise instabilities. As for the previous methods, the output of the DOT is also a function defined on the sphere, but it differs from a dODF because the probability is computed here only at a fixed radius and not as a radial projection of the whole propagator. Nevertheless, results obtained from ex vivo data show that DOT profiles closely resemble conventional dODFs, making DOT another possible alternative to QBI and similar approaches. The assumption of multiexponential radial decay of the signal has also been investigated in the original DOT paper as a further extension of the method. The results have shown better sensitivity in resolving crossing fibers with significantly sharper profiles. However, while the monoexponential decay assumption requires only a single-shell acquisition, introducing multiexponential decays also requires longer multishell acquisitions.

# 4 Spherical deconvolution

As we have seen in previous sections, the idea of modeling the diffusion signal as a spherical convolution has been present since the early days of multicompartmental models and has reappeared multiple times in various diffusion techniques by different authors (Behrens et al., 2003; Alexander, 2005a; Descoteaux et al., 2009). Today, in the literature, spherical deconvolution (SD) refers to methods that directly deconvolve the measured diffusion signal using a kernel, or fiber response function, to estimate the underlying distribution of fiber orientations, or the fODF (Anderson and Ding, 2002; Tournier et al., 2004). In the simplest form, SD approaches assume that the total signal from a brain voxel, $S_T(\boldsymbol{u})$, is the sum of signal contributions only from fiber compartments, and that each compartment can be described by a common signal profile or single fiber response function, $R(\boldsymbol{v}, \boldsymbol{u})$ (Fig. 6). Under these assumptions the forward biophysical model can be expressed as:

$$S_T(\boldsymbol{v}) = \sum_{i=1}^{N} f_i S_i(\boldsymbol{v}) = fODF(\boldsymbol{u}) \otimes R(\boldsymbol{v}, \boldsymbol{u}) \tag{9}$$

where $f_i$ and $S_i(\boldsymbol{v})$ are the volume fraction and signal originating from the $i$-$th$ fiber population, $\boldsymbol{v}$ is a sampling direction of the HARDI signal, $\boldsymbol{u}$ is a vector defining a direction along the fODF, and $\otimes$ is the convolution operation. The fODF is then recovered by inverting this relation as:

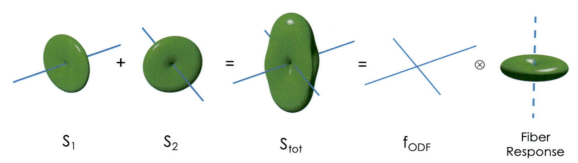

**FIG. 6** Spherical deconvolution: A graphical representation of the spherical convolution model: multiple fiber populations contribute with additive signals, S$_1$ and S$_2$, to the total voxel signal, S$_{tot}$. Under the assumption of a common fiber signal profile, this is equivalent to a convolution over the sphere of a fiber orientation distribution with a chosen fiber response function.

$$fODF(\boldsymbol{u}) = S_T(\boldsymbol{v}) \otimes {}^{-1} R(\boldsymbol{v}, \boldsymbol{u}) \qquad (10)$$

It is important to note that, on real data, signal contributions from other tissues, such as gray matter and cerebrospinal fluid (CSF), are likely to be present and contaminate the overall voxel signal (Behrens et al., 2003; Kaden et al., 2007; Dell'Acqua et al., 2007). The practical impact of these additional signal contributions on SD models are discussed later in this chapter. In this section, we focus on the properties and benefits of this initial formulation.

Besides the simplicity of this model, the main advantage of the SD framework is that by directly modeling the blurring effect of both the diffusion process and the MRI acquisition into the fiber response function, it is possible to recover a clean and sharp fODF profile that closely represents the real biological organization of white matter. This is crucial for tractography because fiber-tracking algorithms need a precise estimate of fiber orientations. Moreover, because of the continuous nature of the fODF function, it is possible to recover not only discrete fiber orientations but also more complex information, such as dispersion and coherence of each individual fiber populations, can be captured. Another benefit of the deconvolution operation is that SD methods do not require extremely high $b$-values. Optimal results have been reported in the literature to be around $b = 2500–3000 \, s/mm^2$, which is substantially less than the requirements for model-independent approaches or q-space based models. Finally, thanks to the linearity of the convolution operation, different numerical algorithms and approaches can be implemented to efficiently perform the deconvolution operation and obtain fast and reliable results (Dell'Acqua and Tournier, 2019). Thanks to these properties, SD methods are gaining widespread adoption in tractography applications and have established themselves as reliable methods across various fields, including neuroscience and psychiatric research, as well as clinical applications like neurosurgical planning. See also Chapters 29–31. In the next sections, we explain the challenges behind solving the deconvolution operation on real data. We then discuss the importance and implications of choosing different fiber responses and, finally, we present two popular spherical deconvolution approaches that have been widely applied in various tractography applications.

### 4.1 Solving the deconvolution problem

While the convolution model is simple and intuitive, solving the associated inverse problem on real brain data is a significantly more complex task. In early algorithms (Anderson and Ding, 2002; Tournier et al., 2004; Anderson, 2005) the use of spherical harmonics offered an elegant solution to describe both convolution and deconvolution operations in the spherical domain. By representing the diffusion signal, fiber response function, and fODF as linear combinations of SH basis functions, the convolution operation becomes a simple matrix multiplication of the spherical harmonic coefficients, simplifying computations (Dennis et al., 1998). Unfortunately, these early methods also quickly showed that, without strong regularization strategies, the solutions were unstable. Mathematically, the deconvolution operation is a well-defined inverse problem with a unique solution, but it is also an ill-posed problem because it doesn't satisfy Hadamard's condition of stability, where small changes in the input should also lead to small changes in the solution (Bertero and Boccacci, 1998). This makes the numerical implementation of spherical deconvolution an ill-conditioned problem, where even small changes in the signal (i.e., noise) or errors in the fiber response can lead to large changes or instabilities in the estimated fODF. One approach to controlling instabilities is to regularize the solution by either reducing the contribution of high-order harmonics through low-pass filtering or directly limiting the solution to a restricted set of spherical harmonics (e.g., $l \leq 6$) (Tournier et al., 2004; Anderson, 2005). While reducing the order of spherical harmonics does improve results, the fODF becomes significantly smoother, sacrificing angular resolution in favor of stability of the solution. Harmonic truncation also renders the fODF more vulnerable to Gibbs ringing artifacts, which can further introduce spurious fODF lobes (Anderson, 2005; Tournier et al., 2007).

Early approaches were also affected by the presence of fODF lobes with unphysical negative amplitudes, which limited their practical use in tractography applications. To address this issue, the next major improvement in SD algorithms was the introduction of nonnegative constraints. These constraints reduce or remove negative fODF lobes, making the fODF not only physically meaningful but also substantially more stable against noise and imprecision in the fiber response (Alexander, 2005a; Tournier et al., 2007; Dell'Acqua et al., 2007; Jian and Vemuri, 2007). In addition to or in combination with nonnegativity, other regularization strategies have been investigated over the years. These include maximum entropy frameworks (Alexander, 2005a), smoothness constraints (Sakaie and Lowe, 2007; Patel et al., 2010) and sparsity in the recovered fODF (Jian and Vemuri, 2007). Depending on how strongly enforced, sparsity can further help to improve solutions by limiting the number of discrete fODF lobes (Canales-Rodríguez et al., 2019). However, while nonnegativity constraints are generally safe and also effectively promote sparsity, applying sparsity constraints too strongly may introduce biases, especially in cases of significant dispersion or curvature. In such cases, sparsity constraints may collapse broad distributions into single peaks and suppress small but important secondary fiber orientations (Dell'Acqua and Tournier, 2019).

As mentioned earlier in previous sections, parametric SD approaches have also been explored in the literature as part of other frameworks. Unlike conventional SD techniques, these methods do not directly deconvolve the signal or estimate a continuous fODF. Instead, they compute a discrete parametrization of fiber orientations (Hosey et al., 2005; Behrens et al., 2003, 2007) or represent the distribution of fiber orientations using a set of discrete Bingham distributions (Kaden et al., 2007; Sotiropoulos et al., 2012). For more details on fiber orientation representations and Bingham distributions, see Chapter 12. In general, the main objective of all deconvolution strategies is to adopt strong constraints that increase the stability of the solution while maximizing the angular resolution of the recovered fODF. The introduction of nonnegative constraints marks an important point in the history of diffusion models because it effectively enables SD methods to be successfully applied in tractography applications. Among different algorithms, the constrained spherical deconvolution (CSD) algorithm has become a widely adopted nonnegative constrained SD algorithm (Tournier et al., 2007). More details about this algorithm are given in the next sections (Fig. 7).

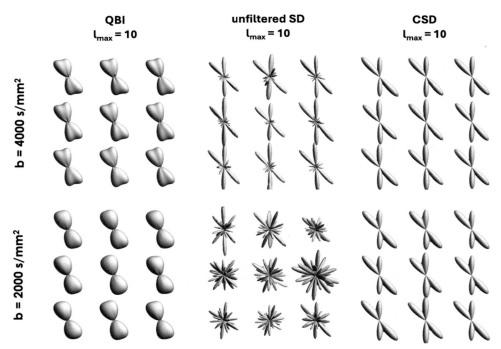

**FIG. 7** Comparison of QBI, unfiltered SD, and CSD: dODF and fODF reconstructions obtained from a phantom made of capillaries crossing at 45 degrees and increasing *b*-values. While SD can provide increased angular resolution, it is with the introduction of nonnegative constrained algorithms, like CSD, that crossing fibers can be resolved reliably and with high angular resolution. *Modified from Tournier, J. D., Yeh, C. H., Calamante, F., Cho, K. H., Connelly, A., & Lin, C. P. (2008). Resolving crossing fibres using constrained spherical deconvolution: Validation using diffusion-weighted imaging phantom data.* NeuroImage, *42(2), 617–625.*

As previously anticipated, the original convolution model assumes that the signal originates only from fiber populations, ignoring contributions from other tissues. While this assumption is reasonable in healthy white matter, extra signal contributions from partial volume effect at the interface between tissues (e.g., CSF, gray matter) or in the presence of pathological white matter (e.g., edema, lesions), can "contaminate" the convolution model (Dell'Acqua et al., 2010; Roine et al., 2014). It is important to note that, in practice, at $b$-values optimal for SD ($b = 2500–3000 \, \text{s/mm}^2$), the diffusion weighting already suppresses most of these extra contributions. In fact, these contributions generally correspond to components with moderate to high diffusivity components and relatively low or no anisotropy. The overall effect of these signal contributions can be seen as a uniform baseline or signal offset added to the total voxel signal and, indirectly, to the fODF itself. While not directly affecting the fiber orientation estimate, this offset partially deactivates or reduces the efficacy of various nonnegative constraints, allowing the solution to oscillate below the baseline, leading to more instabilities and spurious fODF lobes (Dell'Acqua et al., 2010; Roine et al., 2014).

To account for this, new regularization strategies have been introduced. Among different approaches, a successful approach is the damped Richardson-Lucy spherical deconvolution (dRL-SD) algorithm (Dell'Acqua et al., 2010). This algorithm has been shown to effectively reduce the negative effects of partial volume contamination from other tissues while still providing reliable fiber orientation estimates using single-shell data. More details about the dRL-SD algorithm are given in the next sections. Other approaches, based on parametric spherical deconvolution implementations, model out the isotropic contributions (Behrens et al., 2007; Kaden et al., 2007; Sotiropoulos et al., 2012). Alternatively, tissue segmentations from structural T1 weighted data have been used to modify the fiber response to account for non-WM tissue contributions (Roine et al., 2015).

As we will see in the next chapter, the use of multishell data enables the direct estimation of multiple tissue compartments and the separation of signal contributions from different tissues, allowing cleaner and more accurate fODF profiles also in regions affected by partial volume. The multishell, multitissue CSD (MSMT-CSD) (Jeurissen et al., 2014) and the generalized Richardson-Lucy (gRL) (Guo et al., 2020) algorithms are two examples of effective multishell spherical deconvolution implementations. For more details, see Chapter 11. Interestingly, a modified version of the MSMT-CSD algorithm has also been developed to work with single-shell data. This method exploits the nondiffusion weighted signal of the dataset as additional "second" shell, allowing an iterative fitting on multiple tissue types (Dhollander and Connelly, 2016). While the benefits of multishell data are obvious in regions with partial volume between different tissue types, it is worth mentioning that in deep white matter, fODFs obtained from single-shell data are very similar, if not identical, to those obtained from multishell data. This can be considered an effective demonstration of how already close and fairly accurate the original convolution model is for pure white matter.

## 4.2 Fiber response function

The choice of the fiber response plays an important role in all SD methods because it effectively captures the diffusion model that is used to guide the deconvolution operation. There are multiple ways of estimating the fiber response function and they can be broadly organized in two categories: *model based* and *empirically measured* fiber responses. Model based fiber responses assume a user-defined fiber response associated with a specific diffusion model like an axial symmetric tensor, a stick, or more complex diffusion models (Behrens et al., 2007; Dell'Acqua et al., 2007; Hosey et al., 2005; Jian and Vemuri, 2007). The key benefit of this approach is the easy implementation, leaving the choice of the model and its parameters completely under control of the user. This is particularly useful when spherical deconvolution (SD) algorithms are robust to imprecision in the fiber response, and the choice of fiber response can be used as an additional regularization parameter to modulate between angular resolution and noise stability. This further relaxes the need for precise quantification or knowledge of the exact fiber response (Dell'Acqua et al., 2013; Dell'Acqua and Tournier, 2019). Model-based fiber responses can also be easily applied to a large number of subjects or different cohorts, allowing results to be directly compared. Among different models, fiber responses based on a tensor model are easy to implement and have been shown to produce good results, offering a very close approximation to more complex models, such as those based on restricted diffusion. In practice, it has been demonstrated that for single-shell data with a fixed $b$-value, the shape of the fiber response profile can be fully characterized by a single parameter $\alpha = D_{\parallel} - D_{\perp}$ (Dell'Acqua et al., 2013; Dell'Acqua and Tournier, 2019). However, a possible disadvantage associated with model-based fiber responses is that with particular datasets (i.e., postmortem data, fixed tissues, or nonbrain data such as muscles), conventional "healthy" in vivo white matter fiber response functions need to be replaced with fiber responses that more closely reflect the correct biophysical system, and these models are not always readily available for all tissues.

Alternatively, with empirically measured fiber response functions, the signal profile of the fiber response is directly sampled from the data by selecting regions with the highest anisotropy and with a single coherent fiber orientation (Tournier et al., 2007). While not completely model independent, this approach offers a more agnostic view of the real biophysical system, since no assumptions are made about the diffusion characteristics of the tissue, and it can be applied to any type of data. Moreover, when sampling the fiber response, it is possible to account for microscopic axonal dispersion and oscillations along the main fiber orientation, which are expected even in the most coherent fiber orientations (Schilling et al., 2016; Ghosh et al., 2016). Unfortunately, sampled fiber response functions are also not free from limitations. Sampling fiber responses for different datasets may preclude or make comparison more difficult when different fiber response functions are used across subjects. In these circumstances, an average fiber response function must be chosen to guarantee that the same deconvolution is performed across all subjects or different populations.

Depending on the deconvolution algorithm chosen, errors or imprecisions in the fiber response may have a different effect on the final fODF. For example, methods based on spherical harmonics, while efficiently exploiting the properties of the SH basis and being able to resolve small crossing angles, tend to be sensitive to harmonics truncation and imprecision in the fiber response. This often introduces spurious fiber orientations that are the spherical equivalent of Gibbs ringing (Parker et al., 2013). To account for this, iterative calibration methods have been proposed to optimize fiber response functions and minimize instabilities (Tax et al., 2014; Tournier et al., 2013).

A common criticism of SD approaches is that the fiber response function may vary across brain regions. This is because different axonal diameters and tissue properties can lead to different signals between fiber populations. Multiple studies have investigated this issue and have shown that, in practice, assuming a common fiber response is a safe and reasonable approximation of the underlying white matter organization (Dell'Acqua and Tournier, 2019). A strong point in support of this assumption is the fact that, when using clinical scanners with conventional hardware (i.e., gradient strength $\leq 80\,mT/m$), the intra-axonal signal for all axons up to 10 μm in diameter is indistinguishable from the signal obtained by an ideal stick model (Raffelt et al., 2012; Dell'Acqua et al., 2013; Drobnjak et al., 2016; Veraart et al., 2019). This is a crucial point since most axons in the human brain are well below this size (Aboitiz et al., 1999; Caminiti et al., 2009). By using high $b$-values, as we have seen previously, it is possible to significantly reduce signal contributions from CSF, gray matter, and extra-axonal water, leaving the intra-axonal signal as the main contribution to the overall voxel signal. Under these conditions, each fiber population is indeed described by the same signal profile (Veraart et al., 2019). Moreover, recent studies employing multidimension diffusion acquisitions indicate that the microstructural anisotropy in white matter is also very uniform throughout the brain, further reinforcing the assumption of similar characteristics across white matter and a constant fiber response (Lampinen et al., 2017). It is important to point out that, even in the presence of relatively large changes in axial and radial diffusivity in the fiber signal, most of the changes affect mainly the *scale* of the overall signal profile more than its *shape*. Due to the linearity of the deconvolution operation, scale changes in the fiber signal directly translate to changes in the amplitude of the corresponding fODF lobe, preserving the correct fiber estimates (Dell'Acqua et al., 2013; Dell'Acqua and Tournier, 2019). As we will see in Chapter 12, exploiting the mismatch between a "healthy" fiber response and the signal from pathological fiber populations forms the basis for implementing fiber-specific indices like apparent fiber density (AFD), hindrance modulated orientational anisotropy (HMOA), and fixel-based analyses (Raffelt et al., 2012; Dell'Acqua et al., 2013; Raffelt et al., 2017a) (Fig. 8).

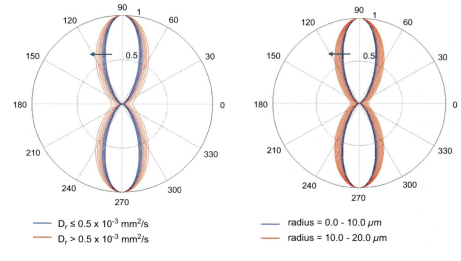

**FIG. 8** Fiber response function: Two-dimensional normalized signal profiles describing fiber response functions using Gaussian diffusion (left) and restricted diffusion inside cylinder (right). Significant changes in shape become evident only for large changes in radial diffusivity or for axonal diameters greater than 10 μm. *Modified from Dell'Acqua, F., Simmons, A., Williams, S. C. R., & Catani, M. (2013). Can spherical deconvolution provide more information than fiber orientations? Hindrance modulated orientational anisotropy, a true-tract specific index to characterize white matter diffusion. Hum. Brain Mapp., 34(10), 2464–2483.*

## 4.3 Constrained spherical deconvolution

Constrained spherical deconvolution (CSD) improves upon the original spherical deconvolution methods based on spherical harmonics by introducing an effective nonnegativity constraint in the deconvolution algorithm to overcome the intrinsic instabilities of the deconvolution operation (Tournier et al., 2007). Unlike other approaches, this regularization is imposed as a soft regularizer, meaning that while negative lobes are penalized, a strictly nonnegative fODF solution is not guaranteed. This choice offers two main advantages: first, it makes the computation significantly more efficient; second, it maximizes angular resolution at the expense of allowing some degree of Gibbs ringing, which is inevitable due to the truncation of higher harmonic orders in the SH representation (Tournier et al., 2007; Dell'Acqua and Tournier, 2019). The CSD algorithm solves the Tikhonov regularization problem:

$$\widehat{f} = \min_{f} \; \|Hf - m\|^2 + \lambda \|Af\|_-^2 \tag{11}$$

where $f$ represents the SH coefficients of the fiber ODF that minimize the least-squares difference between the measure data $m$ and predicted signal $Hf$. The second term in the expression is the regularization term, which ensures the nonnegativity by minimizing the sum-of-squares of all negative amplitudes in the fODF. Multiplying $f$ with a matrix $H = QR$ implements the forward convolution operation using spherical harmonics, where $R$ is a diagonal matrix representing the coefficients of the response function, and $Q$ is a matrix that maps the SH coefficients to the same set of directions as the measured DW signal (Dennis et al., 1998). To implement the regularization, the matrix $A$ maps the SH coefficients of the fODF to a set of dense uniformly distributed directions on a sphere (e.g., 300 directions) while the term $\|x\|_-^2$ represents the squared norm, selecting only the negative components of the vector $x$. The strength of this regularization is controlled by the value of $\lambda$ (Tournier et al., 2007). The algorithm is initialized with an initial estimate based on the original filtered SH deconvolution (Tournier et al., 2004). Although parameters may vary due to data quality and specific research questions, CSD is typically applied using SH orders up to $l = 8$ and $b$-values around 2000–3000 s/mm$^2$. The algorithm requires only 5–10 iterations per voxel to converge, resulting in computation times of around 10 s for a whole-brain dataset on a modern desktop computer (Tournier et al., 2007; Dell'Acqua and Tournier, 2019). CSD specifically uses a response function that is determined empirically from the data, rather than one derived from a specific model. Additionally, the deconvolution is performed directly on the raw DW signal without prior normalization to the nondiffusion weighted signal. The choice to use raw DW data instead of normalized data may seem at first to compromise precise quantification. However, the use of a fiber response directly sampled from the data helps to account for and model out any scaling factor introduced by the MRI scanner. More importantly, by using only raw data, CSD maintains the inherent linearity between the fODF and the diffusion signal, particularly in regions with significant partial volume effects from tissues with large T2 differences (e.g., CSF). In such regions, normalized DWI signals may be excessively penalized by the large signal contribution of the CSF fraction in the nondiffusion weighted image (Dell'Acqua and Tournier, 2019). However, while this may be beneficial in regions with large T2 differences, the use of subject-specific fiber response, as previously mentioned, may make comparison among subjects more complicated. In this case, additional corrections and intensity normalization need to be applied to harmonize subjects to a common fiber response (Raffelt et al., 2017b). CSD has become one of the most widely used SD methods in tractography, particularly when working with single-shell data. Its applications span both clinical and research settings, with its implementation available also in clinical software. An extension to multishell data accounting for the presence of multiple tissues has been introduced by Jeurissen et al. (2014) and will be discussed in detail in the next chapter.

## 4.4 Richardson-Lucy spherical deconvolution

The original Richardson-Lucy (RL) algorithm (Richardson, 1972; Lucy, 1974) was first developed in the field of astronomical image restoration and inverse problems. The RL algorithm is a particular case of the expectation maximization algorithm (Dempster et al., 1977; Shepp and Vardi, 1982), where a Bayesian framework is applied to restore images corrupted by optical distortions and Poisson noise (Molina et al., 2001). This algorithm has evolved over the years with subsequent versions incorporating Gaussian (Daube-Witherspoon and Muehllehner, 1986) and Rician noise (Canales-Rodríguez et al., 2015) and integrating numerous regularization strategies (Bratsolis and Sigelle, 2001; Molina et al., 2001; White, 1994). RL algorithms are particularly suited for the diffusion problem. First, the algorithm's intrinsic robustness to noise ensures well-conditioned solutions. Second, its resilience to errors in the fiber response function allows users to use the response function also as a regularization parameter, further balancing between angular resolution and noise stability, and removing the need to know the exact fiber response. Finally, the algorithm's multiplicative form does not require matrix inversion and implicitly enforces nonnegative solutions. The first implementation of the RL algorithm for spherical deconvolution assuming Gaussian noise is (Dell'Acqua et al., 2007):

$$\left[\boldsymbol{f}^{(k+1)}\right]_i = \left[\boldsymbol{f}^{(k)}\right]_i \frac{\left[\boldsymbol{H}^T \boldsymbol{s}\right]_i}{\left[\boldsymbol{H}^T \boldsymbol{H} \boldsymbol{f}^{(k)}\right]_i} \tag{12}$$

where, for the $k$-th algorithm iteration, the $[f]_i$ is the $i$-th element of an $n{\times}1$ vector $f$ representing the fODF along a uniform set of n-directions (e.g., 752) and $s$ is the $m \times 1$ vector containing the values of the HARDI signal acquired along m-directions (e.g., 60). $\boldsymbol{H}$ is an $m{\times}n$ matrix, where each column describes the fiber response function oriented along one of the n-directions. In practice, this algorithm only requires forward convolutions, in the form of fast matrix–vector multiplications, and element-wise divisions. Since $\boldsymbol{H}$ and $s$ are nonnegative by design, given an initial condition $[\boldsymbol{f}^{(0)}]_i > 0$, nonnegativity of the solution is always guaranteed. This version of the RL algorithm has been shown to be globally convergent by De Pierro (1993). Moreover, RL algorithms exhibit semiconvergence properties (Bertero and Boccacci, 1998, 2000), allowing the number of algorithm iterations to serve as a practical regularization parameter controlling the sharpness of the recovered fODF. While initial calibration is required to set up the fiber response function and determine the optimal number of iterations, these settings are not subject-specific and remain fairly constant for a broad range of datasets. However, they can be fine-tuned to maximize results for specific acquisition protocols or varying levels of data quality (Dell'Acqua et al., 2007).

A second and more widely applied version of the spherical deconvolution algorithm is the damped Richardson-Lucy algorithm (dRL-SD), which has been specifically designed to account for signal contamination in the presence of partial volume effects due to pathological or isotropic tissue contributions (Dell'Acqua et al., 2010). The dRL algorithm follows a regularization strategy based on the original work of (White, 1994) and leverages the linearity of the convolution model, where larger signal contributions correspond to larger fODF lobe amplitudes. Starting from an initial spherical uniform condition, the dRL algorithm introduces a damping function to the original RL algorithm that adaptively adjusts the strength of the regularization based on the fODF lobe amplitudes. As the iterations progress, the damping function progressively reduces regularization for lobes with growing amplitudes, which are supported by large signal fractions. Conversely, for decreasing amplitudes and smaller fODF lobes, the regularization penalizes and delays the emergence of spurious components that are not supported by significant signal contributions and are more likely to originate from noise or partial volume effects. In practice, damping is active for very small amplitudes, which correspond to the amplitude of isotropic tissues and background noise. For larger fODF amplitudes, the damping smoothly deactivates, and the dRL algorithm numerically converges to the standard RL algorithm (Dell'Acqua et al., 2010). This approach has been shown to significantly mitigate the effects of partial volume contamination, preserving angular resolution and enhancing the reliability of the estimated fiber orientations. Tractography studies have successfully applied the algorithm to investigate complex white matter anatomy in the healthy and the diseased brain (Catani et al., 2012, 2017; De Schotten et al., 2011; Vergani et al., 2016). Like CSD, the dRL algorithm has found applications in multiple research fields, from neuroscience to psychiatric research, and the algorithm has also been implemented in clinical software (Fig. 9).

Further improvements of the RL algorithm include a version designed to account for Rician and noncentral Chi noise distribution, and to incorporate spatial regularization based on neighboring voxels and the total-variation approach (Canales-Rodríguez et al., 2015).

It is important to note that, while these RL algorithms have been primarily used for single-shell diffusion MRI data, their underlying framework, which defines the main matrix $\boldsymbol{H}$ as a dictionary of fiber responses, already allows for the deconvolution of both single-shell and multishell data by matching $\boldsymbol{H}$ with the corresponding signal. A recent study has introduced a comprehensive multishell extension called generalized Richardson-Lucy (GRL) (Guo et al., 2020). This framework not only enables the extraction of different tissue types like individual WM, GM, and CSF fractions, but also demonstrates promising results in perfusion mapping. By incorporating the intravoxel incoherent motion (IVIM) model and utilizing data from extremely low $b$-values, GRL can estimate the contribution of blood pseudodiffusion to the diffusion MRI signal, opening up new possibilities for multicompartment tissue analysis.

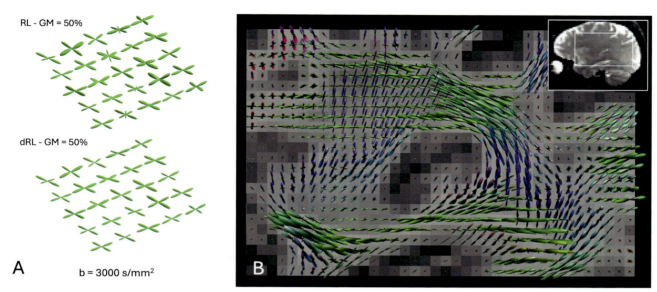

FIG. 9 Damped Richardson-Lucy: (A) Numerical\ simulation showing how the introduction of specific regularization to account for partial volume contamination allows a further reduction in spurious fiber orientation and instabilities in spherical deconvolution. (B) A sagittal slice from a healthy brain in a region corresponding to the arcuate fasciculus showing sharp and coherent fODF with minimal presence of spurious orientations.

## 5 Optimal strategies for single-shell models

As discussed in the previous sections, each model comes with specific acquisition recommendations to obtain the best results. Therefore, while it is not possible to identify a single-shell acquisition that can be optimal across all models, it is possible to identify general guidelines for the different types of models. For example, for DTI and its derived metrics, given an average ADC value for the tissue under examination, $b$-values close to 1/ADC are generally considered to be optimal (Jones et al., 1999a,b; Alexander and Barker, 2005). In practice, $b = 1000\,s/mm^2$ has become a very common choice for DTI studies looking at in vivo human brain data. However, to resolve multiple fiber orientations, higher $b$-values are needed to sufficiently increase the angular contrast in the measured diffusion signal. Unfortunately, increasing $b$-values also decreases the signal-to-noise ratio, and a balance must be found between the final $b$-value, the spatial resolution achievable, and the noise stability of the model. As we have seen, dODF models require significantly higher $b$-values to achieve optimal results (i.e., $b = 4000$–$6000\,s/mm^2$ or more). On the other hand, methods extracting fODF or discrete fiber orientations have consistently shown good results from $b > 1500\,s/mm^2$ and optimal results with $b$-values in the range of 2500–3000 s/mm$^2$, making these values a natural choice for many tractography studies (Alexander and Barker, 2005; Tournier et al., 2013; White and Dale, 2009; Dell'Acqua and Tournier, 2019). Once the $b$-value has been identified, the next important choice is to find the optimal number of gradient directions. Tournier et al. (2013) has proposed an elegant approach to quantify the minimum number of gradient directions required for HARDI acquisitions. By decomposing the HARDI signal into spherical harmonics, it can be demonstrated that the signal from a single, coherently oriented fiber population has the highest angular frequency content among all possible fiber configurations. Due to the linearity of the SH basis, frequencies not present within individual fiber populations cannot emerge in more complex fiber configurations. Hence, by identifying the maximum SH order required to encode the signal from a single fiber, we also identify the maximum order for any possible fiber configuration. Given a chosen $b$-value, the minimum number of directions is simply equal to the number of SH coefficients required to describe the signal up to the identified SH order (Fig. 10).

For instance, at $b = 1000\,s/mm^2$, the maximum SH order is typically $l = 6$, which corresponds to 28 coefficients. At $b = 3000\,s/mm^2$, the maximum SH order is typically $l = 8$, corresponding to 45 coefficients and requiring at least 45 gradient directions. In practice, to maximize SNR, more directions than the theoretical minimum are always acquired. This is because multiple factors may influence the overall signal quality, such as field strength, and the use of partial Fourier, parallel imaging, and simultaneous multislice acquisitions to accelerate scan times. As such, at least 60–90 directions

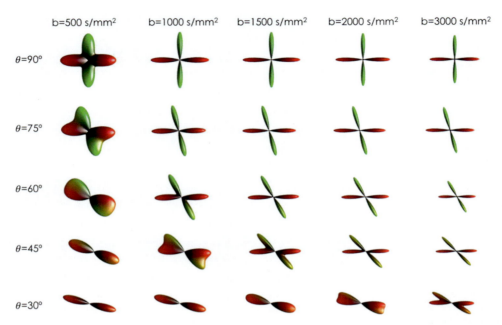

**FIG. 10** Effect of different $b$-values on fODF angular resolution. From left to right, higher $b$-values give higher angular resolutions and increased ability to detect multiple fiber orientations at smaller angles.

for $b$-values up to $3000 \, s/mm^2$ are commonly reported in the literature and used in clinical applications on 3T systems with 2 mm isotropic resolution and scan times between 5 and 10 min (see also Chapter 6).

## 6 Conclusion: Why single-shell?

Single-shell diffusion methods still have an important role in modern tractography. They are historically the first techniques to be developed for tractography applications and are the foundation upon which later multishell methods have been built. Despite the substantial advancement of multishell techniques, the relative simplicity of single-shell diffusion imaging remains a significant and appealing reason for some tractography applications. Single-shell approaches provide a straightforward conceptual framework for understanding and interpreting diffusion data, which is invaluable for educational purposes and for those new to the field of tractography. Also, single-shell data, when acquired at relatively high $b$-values, already suppress most of the fast diffusion components that may contaminate multicompartmental or spherical deconvolution models. The simpler acquisitions and the smaller amount of data required to produce optimal results are important factors to consider when designing complex experimental setups or in some clinical settings where scan time remains limited. In these cases, selecting simple approaches can be a more pragmatic choice in situations where the additional information provided by multiple $b$-values is not required or when, in the case of tractography, an adequate angular sampling must be prioritized over multiple shells. Lastly, the reduced complexity of single-shell data often results in reduced data management and processing demands. Nevertheless, it is important to stress that the benefits of multishell data are numerous and cannot be ignored. As such, there is no doubt that more and more acquired data will be multishell, and more tractography applications will naturally move to it as improvements in MR hardware and computational pipelines continue. The next chapter will explore in detail the diffusion models used in multishell acquisitions and their benefits.

## References

Aboitiz, F., Scheibel, A.B., Fisher, R.S., Zaidel, E., 1999. Fiber composition of the human corpus callosum. Brain Res. 598 (143–153), 122.
Aganj, I., Lenglet, C., Sapiro, G., Yacoub, E., Ugurbil, K., Harel, N., 2010. Reconstruction of the orientation distribution function in single- and multiple-shell q-ball imaging within constant solid angle. Magn. Reson. Med. 64 (2), 554–566.
Alexander, D.C., 2005a. Maximum entropy spherical deconvolution for diffusion MRI. In: Proc. Information Processing in Medical Imaging, pp. 76–87.
Alexander, D.C., 2005b. Multiple-fibre reconstruction algorithms for diffusion MRI. Ann. N. Y. Acad. Sci. 1046, 113–133.
Alexander, D.C., Barker, G.J., 2005. Optimal imaging parameters for fiber-orientation estimation in diffusion MRI. Neuroimage 27 (2), 357–367.
Alexander, A.L., Hasan, K., Kindlmann, G., Parker, D.L., Tsuruda, J.S., 2000. A geometric analysis of diffusion tensor measurements of the human brain. Magn. Reson. Med. 4 (4), 283–291.

Alexander, D.C., Barker, G.J., Arridge, S.R., 2002. Detection and modeling of non-Gaussian apparent diffusion coefficient profiles in human brain data. Magn. Reson. Med. 48 (2), 331–340.

Alexander, D.C., Dyrby, T.B., Nilsson, M., Zhang, H., 2017. Imaging brain microstructure with diffusion MRI: practicality and applications. NMR Biomed., e3841.

Anderson, A.W., 2005. Measurement of fiber orientation distributions using high angular resolution diffusion imaging. Magn. Reson. Med. 54 (5), 1194–1206.

Anderson, A., Ding, Z., 2002. Sub-voxel measurement of fiber orientation using high angular resolution diffusion tensor imaging. In: Proc. 10th Annual Meeting of the ISMRM Honolulu, 2002. Berkeley, USA: ISMRM, p. 440.

Arsigny, V., Fillard, P., Pennec, X., Ayache, N., 2006. Log-Euclidean metrics for fast and simple calculus on diffusion tensors. Magn. Reson. Med. 56 (2), 411–421.

Assaf, Y., Basser, P.J., 2005. Composite hindered and restricted model of diffusion (CHARMED) MR imaging of the human brain. Neuroimage 27 (1), 48–58.

Assaf, Y., Pasternak, O., 2008. Diffusion tensor imaging (DTI)-based white matter mapping in brain research: a review. J. Mol. Neurosci. 34 (1), 51–61.

Assaf, Y., Freidlin, R.Z., Rohde, G.K., Basser, P.J., 2004. New modeling and experimental framework to characterize hindered and restricted water diffusion in brain white matter. Magn. Reson. Med. 52 (5), 965–978.

Basser, P.J., Pierpaoli, C., 1996. Microstructural and physiological features of tissue elucidated by quantitative-diffusion tensor MRI. J. Magn. Reson. B 111, 209–219.

Basser, P.J., Mattiello, J., Le Bihan, D., 1994. Estimation of the effective self-diffusion tensor from the NMR spin echo. J. Magn. Reson. B 103, 247–254.

Basser, P.J., Pajevic, S., Pierpaoli, C., Duda, J., Aldroubi, A., 2000. In vivo fiber tractography using DT-MRI data. Magn. Reson. Med. 44 (4), 625–632.

Bastiani, M., Cottaar, M., Dikranian, K., Ghosh, A., Zhang, H., Alexander, D.C., Behrens, T.E., Jbabdi, S., Sotiropoulos, S.N., 2017. Improved tractography using asymmetric fibre orientation distributions. Neuroimage 158, 205–218.

Beaulieu, C., 2002. The basis of anisotropic water diffusion in the nervous system—a technical review. NMR Biomed. 15, 435–455.

Beaulieu, C., Allen, P.S., 1994a. Determinants of anisotropic water diffusion in nerves. Magn. Reson. Med. 31, 394–400.

Beaulieu, C., Allen, P.S., 1994b. Water diffusion in the giant axon of the squid: implications for diffusion-weighted MRI of the nervous system. Magn. Reson. Med. 32, 579–583.

Behrens, T.E.J., Woolrich, M.W., Jenkinson, M., Johansen-Berg, H., Nunes, R.G., Clare, S., Matthews, P.M., Brady, J.M., Smith, S.M., 2003. Characterization and propagation of uncertainty in diffusion-weighted MR imaging. Magn. Reson. Med. 50, 1077–1088.

Behrens, T.E.J., Johansen Berg, H., Jbabdi, S., Rushworth, M.F.S., Woolrich, M.W., 2007. Probabilistic diffusion tractography with multiple fibre orientations: what can we gain? Neuroimage 34, 144–155.

Berman, J.I., Chung, S., Mukherjee, P., Hess, C.P., Han, E.T., Henry, R.G., 2008. Probabilistic streamline q-ball tractography using the residual bootstrap. Neuroimage 39 (1), 215–222.

Bertero, M., Boccacci, P., 1998. Introduction to Inverse Problems in Imaging, first ed. CRC Press.

Bertero, M., Boccacci, P., 2000. Image restoration methods for the Large Binocular Telescope (LBT). Astron. Astrophys. Suppl. Ser. 147 (323–333), 112.

Bratsolis, E., Sigelle, M., 2001. A spatial regularization method preserving local photometry for Richardson-Lucy restoration. Astron. Astrophys. 375, 1120–1128.

Bucci, M., Mandelli, M.L., Berman, J.I., Amirbekian, B., Nguyen, C., Berger, M.S., Henry, R.G., 2013. Quantifying diffusion MRI tractography of the corticospinal tract in brain tumors with deterministic and probabilistic methods. Neuroimage Clin. 3, 361–368.

Callaghan, P.T., Eccles, C.D., Xia, Y., 1988. NMR microscopy of dynamic displacements: k-space and q-space imaging. J. Phys. E 21, 820.

Caminiti, R., Ghaziri, H., Galuske, R., Hof, P.R., Innocenti, G.M., 2009. Evolution amplified processing with temporally dispersed slow neuronal connectivity in primates. Proc. Natl. Acad. Sci. U. S. A. 106, 19551–19556.

Canales-Rodríguez, E.J., Melie-García, L., Iturria-Medina, Y., 2009. Mathematical description of q-space in spherical coordinates: exact q-ball imaging. Magn. Reson. Med. 61 (6), 1350–1367.

Canales-Rodríguez, E.J., Daducci, A., Sotiropoulos, S.N., Caruyer, E., Aja-Fernández, S., Radua, J., Mendizabal, J.M.Y., Iturria-Medina, Y., Melie-García, L., Alemán-Gómez, Y., Thiran, J.P., Sarró, S., Pomarol-Clotet, E., Salvador, R., 2015. Spherical deconvolution of multichannel diffusion MRI data with non-Gaussian noise models and spatial regularization. PLoS One 10 (10), 1–29.

Canales-Rodríguez, E.J., Legarreta, J.H., Pizzolato, M., Rensonnet, G., Girard, G., Patino, J.R., Barakovic, M., Romascano, D., Alemán-Gómez, Y., Radua, J., Pomarol-Clotet, E., Salvador, R., Thiran, J.P., Daducci, A., 2019. Sparse wars: a survey and comparative study of spherical deconvolution algorithms for diffusion MRI. Neuroimage 184, 140–160.

Catani, M., Dell'Acqua, F., Vergani, F., Malik, F., Hodge, H., Roy, P., Valabregue, R., Thiebaut de Schotten, M., 2012. Short frontal lobe connections of the human brain. Cortex 48 (2), 273–291.

Catani, M., Robertsson, N., Beyh, A., Huynh, V., de Santiago Requejo, F., Howells, H., Barrett, R.L.C., Aiello, M., Cavaliere, C., Dyrby, T.B., Krug, K., Ptito, M., D'Arceuil, H., Forkel, S.J., Dell'Acqua, F., 2017. Short parietal lobe connections of the human and monkey brain. Cortex 97, 339–357.

Chang, L.-C., Jones, D.K., Pierpaoli, C., 2005. RESTORE: robust estimation of tensors by outlier rejection. Magn. Reson. Med. 53, 1088–1095.

Chenevert, T.L., Brunberg, J.A., Pipe, J.G., 1990. Anisotropic diffusion within human white matter: demonstration with NMR techniques in vivo. Radiology 177, 401–405.

Cleveland, G.G., Chang, D.C., Hazelwood, C.F., Rorschach, H.E., 1976. Nuclear magnetic resonance measurement of skeletal muscle. Anisotropy of the diffusion coefficient of the intracellular water. Biophys. J. 16, 1043–1053.

Conturo, T.E., Lori, N.F., Cull, T.S., Akbudak, E., Snyder, A.Z., Shimony, J.S., McKinstry, R.C., Burton, M., Raichle, M.E., 1999. Tracking neuronal fiber pathways in the living human brain. Proc. Natl. Acad. Sci. U. S. A. 96, 10422–10427.

Daube-Witherspoon, M.E., Muehllehner, G., 1986. An iterative image space reconstruction algorithm suitable for volume ECT. IEEE Trans. Med. Imaging 5, 61–66.

De Pierro, A.R., 1993. On the relation between the ISRA and the EM algorithm for positron emission tomography. IEEE Trans. Med. Imaging 12, 328–333.

De Schotten, M.T., Dell'Acqua, F., Forkel, S.J., Simmons, A., Vergani, F., Murphy, D.G.M., Catani, M., Dell'Acqua, F., Forkel, S.J., Simmons, A., Vergani, F., Murphy, D.G.M., Catani, M., 2011. A lateralized brain network for visuospatial attention. Nat. Neurosci. 14 (10), 1245–1246.

Dell'Acqua, F., Catani, M., 2012. Structural human brain networks. Curr. Opin. Neurol. 25 (4), 375–383.

Dell'Acqua, F., Tournier, J.D., 2019. Modelling white matter with spherical deconvolution: how and why? NMR Biomed. 32 (4), 1–18.

Dell'Acqua, F., Scifo, P., Rizzo, G., Catani, M., Simmons, A., Scotti, G., Fazio, F., 2010. A modified damped Richardson-Lucy algorithm to reduce isotropic background effects in spherical deconvolution. Neuroimage 49 (2), 1446–1458.

Dell'Acqua, F., Simmons, A., Williams, S.C.R., Catani, M., 2013. Can spherical deconvolution provide more information than fiber orientations? Hindrance modulated orientational anisotropy, a true-tract specific index to characterize white matter diffusion. Hum. Brain Mapp. 34 (10), 2464–2483.

Dell'Acqua, F., Rizzo, G., Scifo, P., Clarke, R.A., Scotti, G., Fazio, F., 2007. A model-based deconvolution approach to solve fiber crossing in diffusion-weighted MR imaging. IEEE Trans. Biomed. Eng. 54, 462–472.

Dempster, A.P., Laird, N.M., Rubin, D.B., 1977. Maximum likelihood from incomplete data via the EM algorithm. J. R. Stat. Soc. B. 39, 1–38.

Dennis, M., Healy, J., Hendriks, H., Kim, P.T., 1998. Spherical deconvolution. J. Multivar. Anal. 67, 1–22.

Descoteaux, M., Angelino, E., Fitzgibbons, S., Deriche, R., 2006. Apparent diffusion coefficients from high angular resolution diffusion imaging: estimation and applications. Magn. Reson. Med. 56 (2), 395–410.

Descoteaux, M., Angelino, E., Fitzgibbons, S., Deriche, R., 2007. Regularized, fast, and robust analytical Q-ball imaging. Magn. Reson. Med. 58, 497–510.

Descoteaux, M., Deriche, R., Knösche, T., Anwander, A., 2009. Deterministic and probabilistic tractography based on complex fiber orientation distributions. IEEE Trans. Med. Imaging 28 (2), 269–286.

Dhollander, T., Connelly, A., 2016. A novel iterative approach to reap the benefits of multi-tissue CSD from just single-shell (+b=0) diffusion MRI data. In: Proceedings of the 24th Annual Meeting of ISMRM, Singapore, p. 3010.

Drobnjak, I., Zhang, H., Ianuş, A., Kaden, E., Alexander, D.C., 2016. PGSE, OGSE, and sensitivity to axon diameter in diffusion MRI: insight from a simulation study. Magn. Reson. Med. 75, 688–700.

Dyrby, T.B., Søgaard, L.V., Parker, G.J., Alexander, D.C., Lind, N.M., Baaré, W.F.C., Hay-Schmidt, A., Eriksen, N., Pakkenberg, B., Paulson, O.B., Jelsing, J., 2007. Validation of in vitro probabilistic tractography. Neuroimage 37 (4), 1267–1277.

Dyrby, T.B., Baaré, W.F.C., Alexander, D.C., Jelsing, J., Garde, E., Søgaard, L.V., 2011. An ex vivo imaging pipeline for producing high-quality and high-resolution diffusion-weighted imaging datasets. Hum. Brain Mapp. 32 (4), 544–563.

Ennis, D.B., Kindlmann, G., 2006. Orthogonal tensor invariants and the analysis of diffusion tensor magnetic resonance images. Magn. Reson. Med. 55 (1), 136–146.

Frank, L.R., 2001. Anisotropy in high angular resolution diffusion-weighted MRI. Magn. Reson. Med. 45, 935–939.

Frank, L.R., 2002. Characterization of anisotropy in high angular resolution diffusion-weighted MRI. Magn. Reson. Med. 47 (6), 1083–1099.

Ghosh, A., Alexander, D., Zhang, H., 2016. Crossing versus fanning: Model comparison using HCP data. In: Computational Diffusion MRI. Springer, Cham, Switzerland, pp. 159–169.

Guo, F., Leemans, A., Viergever, M.A., Dell'Acqua, F., De Luca, A., 2020. Generalized Richardson-Lucy (GRL) algorithm for analyzing multi-shell diffusion MRI data. Neuroimage 218, 116948.

Hansen, J.R., 1971. Pulsed NMR study of water mobility in muscle and brain tissue. Biochim. Biophys. Acta 230, 482–486.

Hess, C.P., Mukherjee, P., Han, E.T., Xu, D., Vigneron, D.B., 2006. Q-ball reconstruction of multimodal fiber orientations using the spherical harmonic basis. Magn. Reson. Med. 56, 104–117.

Hosey, T., Williams, G., Ansorge, R., 2005. Inference of multiple fiber orientations in high angular resolution diffusion imaging. Magn. Reson. Med. 54, 1480–1489.

Jansons, K.M., Alexander, D.C., 2003. Persistent angular structure: new insights from diffusion magnetic resonance imaging data. Inverse Probl. 19 (5), 1031–1046.

Jensen, J.H., Russell Glenn, G., Helpern, J.A., 2016. Fiber ball imaging. Neuroimage 124, 824–833.

Jespersen, S.N., Kroenke, C.D., Østergaard, L., Ackerman, J.J.H., Yablonskiy, D.A., 2007. Modeling dendrite density from magnetic resonance diffusion measurements. Neuroimage 34 (4), 1473–1486.

Jeurissen, B., Leemans, A., Tournier, J.D., Jones, D.K., Sijbers, J., 2013. Investigating the prevalence of complex fiber configurations in white matter tissue with diffusion magnetic resonance imaging. Hum. Brain Mapp. 34 (11), 2747–2766.

Jeurissen, B., Tournier, J.-D., Dhollander, T., Connelly, A., Sijbers, J., 2014. Multi-tissue constrained spherical deconvolution for improved analysis of multi-shell diffusion MRI data. Neuroimage 103, 411–426.

Jian, B., Vemuri, B.C., 2007. A unified computational framework for deconvolution to reconstruct multiple fibers from diffusion weighted MRI. IEEE Trans. Med. Imaging 26, 1464–1471.

Jones, D.K., 2004. The effect of gradient sampling schemes on measures derived from diffusion tensor MRI: a Monte Carlo study. Magn. Reson. Med. 51, 807–815.

Jones, D.K., 2008. Studying connections in the living human brain with diffusion MRI. Cortex 44 (8), 936–952.

Jones, D.K., 2009. Gaussian modeling of the diffusion signal. In: Diffusion MRI: From Quantitative Measurement to In-Vivo Neuroanatomy. Elsevier, p. 54. Chapter 3, 37.

Jones, D.K., Williams, S., Horsfield, M.A., 1997. Full representation of white-matter fibre direction on one map via diffusion tensor analysis. In: Proceedings of the 5th Fifth Annual Meeting of the International Society of Magnetic Resonance in Medicine. Berkeley, CA: ISMRM 1741.

Jones, D.K., Horsfield, M.A., Simmons, A., 1999a. Optimal strategies for measuring diffusion in anisotropic systems by magnetic resonance imaging. Magn. Reson. Med. 42, 515–525.

Jones, D.K., Simmons, A., Williams, S.C.R., Horsfield, M.A., 1999b. Non-invasive assessment of axonal fiber connectivity in the human brain via diffusion tensor MRI. Magn. Reson. Med. 42, 37–41.

Jones, D.K., Knösche, T.R., Turner, R., 2013. White matter integrity, fiber count, and other fallacies: the do's and don'ts of diffusion MRI. Neuroimage 73, 239–254.

Kaden, E., Knösche, T.R., Anwander, A., 2007. Parametric spherical deconvolution: inferring anatomical connectivity using diffusion MR imaging. Neuroimage 37 (2), 474–488.

Khachaturian, M.H., Wisco, J.J., Tuch, D.S., 2007. Boosting the sampling efficiency of q-ball imaging using multiple wavevector fusion. Magn. Reson. Med. 57 (2), 289–296.

Koay, C.G., Chang, L.C., Carew, J.D., Pierpaoli, C., Basser, P.J., 2006. A unifying theoretical and algorithmic framework for least squares methods of estimation in diffusion tensor imaging. J. Magn. Reson. 182, 115–125.

Kreher, B.W., Schneider, J.F., Mader, I., Martin, E., Hennig, J., Il'Yasov, K. A., 2005. Multitensor approach for analysis and tracking of complex fiber configurations. Magn. Reson. Med. 54 (5), 1216–1225.

Lampinen, B., Szczepankiewicz, F., Mårtensson, J., van Westen, D., Sundgren, P.C., Nilsson, M., 2017. Neurite density imaging versus imaging of microscopic anisotropy in diffusion MRI: a model comparison using spherical tensor encoding. Neuroimage 147, 517–531.

Lucy, L.B., 1974. An iterative technique for the rectification of observed distributions. Astron. J. 79, 745.

Luque Laguna, P.A., Combes, A.J.E., Streffer, J., Einstein, S., Timmers, M., Williams, S.C.R., Dell'Acqua, F., 2020. Reproducibility, reliability and variability of FA and MD in the older healthy population: a test-retest multiparametric analysis. NeuroImage Clin. 26, 102168.

Mandelli, M.L., Caverzasi, E., Binney, R.J., Henry, M.L., Lobach, I., Block, N., Amirbekian, B., Dronkers, N., Miller, B.L., Henry, R.G., Gorno-Tempini, M.L., 2014. Frontal white matter tracts sustaining speech production in primary progressive aphasia. J. Neurosci. 34 (29), 9754–9767.

Molina, R., Nunez, J., Cortijo, F.J., Mateos, J., 2001. Image restoration in astronomy: a Bayesian perspective. IEEE Signal Process. Mag. 18, 11–29.

Mori, S., Crain, B.J., Chacko, V.P., van Zijl, P.C.M., 1999. Three-dimensional tracking of axonal projections in the brain by magnetic resonance imaging. Ann. Neurol. 45 (2), 265–269.

Moseley, M.E., Cohen, Y., Mintorovitch, J., Chileuitt, L., Shimizu, H., Kucharczyk, J., Wendland, M.F., Weinstein, P.R., 1990a. Early detection of regional brain ischemia in cats: comparison of diffusion- and T2-weighted MRI and spectroscopy. Magn. Reson. Med. 14, 330–346.

Moseley, M.E., Cohen, Y., Kucharczyk, J., 1990b. Diffusion weighted MR imaging of anisotropic water diffusion in cat central nervous system. Radiology 187, 439–446.

Novikov, D.S., Fieremans, E., Jespersen, S.N., Kiselev, G., 2018. Quantifying brain microstructure with diffusion MRI: theory and parameter estimation. NMR Biomed., e3998.

Ozarslan, E., Shepherd, T.M., Vemuri, B.C., Blackband, S.J., Mareci, T.H., 2006. Resolution of complex tissue microarchitecture using the diffusion orientation transform (DOT). Neuroimage 31 (3), 1086–1103.

Pajevic, S., Pierpaoli, C., 1999. Color schemes to represent the orientation of anisotropic tissues from diffusion tensor data: application to white matter fiber tract mapping in the human brain. Magn. Reson. Med. 43, 526–540.

Panagiotaki, E., Schneider, T., Siow, B., Hall, M.G., Lythgoe, M.F., Alexander, D.C., 2012. Compartment models of the diffusion MR signal in brain white matter: a taxonomy and comparison. Neuroimage 59, 2241–2254.

Parker, G.J.M., 2000. Tracing fiber tracts using fast marching. In: Proceedings of the 8th Annual Meeting of the International Society for Magnetic Resonance in Medicine, ISMRM, Berkeley, CA, p. 85.

Parker, G.J.M., Alexander, D.C., 2003. Probabilistic Monte Carlo based mapping of cerebral connections utilising whole-brain crossing fibre information. Inf. Process. Med. Imaging 18, 684–695.

Parker, G.J.M., Alexander, D.C., 2005. Probabilistic anatomical connectivity derived from the microscopic persistent angular structure of cerebral tissue. Philos. Trans. R. Soc. Lond. B Biol. Sci. 360 (1457), 893–902.

Parker, G.D., Marshall, D., Rosin, P.L., Drage, N., Richmond, S., Jones, D.K., 2013. A pitfall in the reconstruction of fibre ODFs using spherical deconvolution of diffusion MRI data. Neuroimage 65, 433–448.

Patel, V., Shi, Y., Thompson, P.M., Toga, A.W., 2010. Mesh-based spherical deconvolution: a flexible approach to reconstruction of non-negative fiber orientation distributions. Neuroimage 51 (3), 1071–1081.

Pierpaoli, C., 1997. Oh no! One more method for colour mapping of fiber tract direction using diffusion MR imaging data. In: Proceedings of the 5th Fifth Annual Meeting of the International Society of Magnetic Resonance in Medicine. Berkeley, CA: ISMRM, p. 1743.

Pierpaoli, C., Basser, P.J., 1996. Toward a quantitative assessment of diffusion anisotropy. Magn. Reson. Med. 36, 893–906.

Pierpaoli, C., Jezzard, P., Basser, P.J., Barnett, A.S., 1996. Diffusion tensor MR imaging of the human brain. Radiology 201, 637–648.

Poirier, C., Descoteaux, M., 2023. A unified filtering method for estimating asymmetric orientation distribution functions: where and how asymmetry occurs in the brain. BioRxiv (2022.12.18.520881).

Potgieser, A.R., Wagemakers, M., van Hulzen, A.L., de Jong, B.M., Hoving, E.W., Groen, R.J., 2014. The role of diffusion tensor imaging in brain tumor surgery: a review of the literature. Clin. Neurol. Neurosurg. 124, 51–58.

Poupon, C., Clark, C.A., Frouin, V., Regis, J., Bloch, I., Le Bihan, D., Mangin, J., 2000. Regularization of diffusion-based direction maps for the tracking of brain white matter fasciculi. Neuroimage 12, 184–195.

Raffelt, D., Tournier, J.D., Rose, S., Ridgway, G.R., Henderson, R., Crozier, S., Salvado, O., Connelly, A., 2012. Apparent fibre density: a novel measure for the analysis of diffusion-weighted magnetic resonance images. Neuroimage 59 (4), 3976–3994.

Raffelt, D.A., Tournier, J.D., Smith, R.E., Vaughan, D.N., Jackson, G., Ridgway, G.R., Connelly, A., 2017a. Investigating white matter fibre density and morphology using fixel-based analysis. Neuroimage 144, 58–73.

Raffelt, D., Dhollander, T., Tournier, J.D., Tabbara, R., Smith, R.E., Pierre, E., Connelly, A., 2017b. Bias field correction and intensity normalisation for quantitative analysis of apparent fibre density. In: Proceedings of the 25th Annual Meeting of ISMRM, Honolulu, US, p. 3541.

Reisert, M., Kellner, E., Kiselev, V.G., 2012. About the geometry of asymmetric fiber orientation distributions. IEEE Trans. Med. Imaging 31 (6), 1240–1249.

Richardson, W.H., 1972. Bayesian-based iterative method of image restoration. J. Opt. Soc. Am. 62, 55–59.

Roine, T., Jeurissen, B., Perrone, D., Aelterman, J., Leemans, A., Philips, W., Sijbers, J., 2014. Isotropic non-white matter partial volume effects in constrained spherical deconvolution. Front. Neuroinform. 8, 1–9.

Roine, T., Jeurissen, B., Perrone, D., Aelterman, J., Philips, W., Leemans, A., Sijbers, J., 2015. Informed constrained spherical deconvolution (iCSD). Med. Image Anal. 24 (1), 269–281.

Sakaie, K.E., Lowe, M.J., 2007. An objective method for regularization of fiber orientation distributions derived from diffusion-weighted MRI. Neuroimage 34 (1), 169–176.

Scherrer, B., Schwartzman, A., Taquet, M., Prabhu, S.P., Sahin, M., Akhondi-Asl, A., Warfield, S.K., 2013. Characterizing the distribution of anisotropic micro-structural environments with diffusion-weighted imaging (DIAMOND). Med. Image Comput. Comput. Assist. Interv. 16 (Pt 3), 518–526.

Schilling, K., Janve, V., Gao, Y., Stepniewska, I., Landman, B.A., Anderson, A.W., 2016. Comparison of 3D orientation distribution functions measured with confocal microscopy and diffusion MRI. Neuroimage 2016 (129), 185–197.

Seunarine, K.K., Alexander, D.C., 2006. Linear persistent angular structure MRI and nonlinear spherical deconvolution for diffusion MRI. In: Proc. 14th Annual Meeting of the ISMRM, Seattle, p. 2726.

Shepp, L.A., Vardi, Y., 1982. Maximum likelihood reconstruction for emission tomography. IEEE Trans. Med. Imaging 1, 113–122.

Skare, S., Hedehus, M., Moseley, M.E., Li, T.Q., 2000. Condition number as a measure of noise performance of diffusion tensor data acquisition schemes with MRI. J. Magn. Reson. 147, 340–352.

Sotiropoulos, S.N., Bai, L., Morgan, P.S., Constantinescu, C.S., Tench, C.R., 2010. Brain tractography using Q-ball imaging and graph theory: improved connectivities through fibre crossings via a model-based approach. Neuroimage 49 (3), 2444–2456.

Sotiropoulos, S.N., Behrens, T.E.J., Jbabdi, S., 2012. Ball and rackets: inferring fiber fanning from diffusion-weighted MRI. Neuroimage 60 (2), 1412–1425.

Sundgren, P.C., Dong, Q., Gómez-Hassan, D., Mukherji, S.K., Maly, P., Welsh, R., 2004. Diffusion tensor imaging of the brain: review of clinical applications. Neuroradiology 46 (5), 339–350.

Tax, C.M.W., Jeurissen, B., Vos, S.B., Viergever, M.A., Leemans, A., 2014. Recursive calibration of the fiber response function for spherical deconvolution of diffusion MRI data. Neuroimage 86, 67–80.

Tournier, J.D., Calamante, F., Gadian, D.G., Connelly, A., 2004. Direct estimation of the fiber orientation density function from diffusion-weighted MRI data using spherical deconvolution. Neuroimage 23 (3), 1176–1185.

Tournier, J.-.D., Calamante, F., Connelly, A., 2007. Robust determination of the fibre orientation distribution in diffusion MRI: non-negativity constrained super- resolved spherical deconvolution. Neuroimage 35, 1459–1472.

Tournier, J.-.D., Calamante, F., Connelly, A., 2013. Determination of the appropriate b value and number of gradient directions for high-angular-resolution diffusion-weighted imaging. NMR Biomed. 26, 1775–1786.

Tristán-Vega, A., Westin, C.F., Aja-Fernández, S., 2009. Estimation of fiber orientation probability density functions in high angular resolution diffusion imaging. Neuroimage 47 (2), 638–650.

Tuch, D.S., 2002. Diffusion MRI of Complex Tissue Structure. (PhD thesis, Doctor of Philosophy in Biomedical Imaging at the Massachusetts Institute of Technology, Boston, US).

Tuch, D.S., 2004. Q-ball imaging. Magn. Reson. Med. 52 (6), 1358–1372.

Tuch, D.S., Reese, T.G., Wiegell, M.R., Makris, N., Belliveau, J.W., Wedeen, V.J., 2002. High angular resolution diffusion imaging reveals intravoxel white matter fiber heterogeneity. Magn. Reson. Med. 48, 577–582.

Tuch, D.S., Reese, T.G., Wiegell, M.R., Wedeen, V.J., 2003. Diffusion MRI of complex neural architecture. Neuron 40 (5), 885–895.

Veraart, J., Fieremans, E., Novikov, D.S., 2019. On the scaling behavior of water diffusion in human brain white matter. Neuroimage 185, 379–387.

Vergani, F., Martino, J., Morris, C., Attems, J., Ashkan, K., Dell'Acqua, F., 2016. Anatomic connections of the subgenual cingulate region. Neurosurgery 79 (3), 465–472.

Wedeen, V.J., Hagmann, P., Tseng, W.Y.I., Reese, T.G., Weisskoff, R.M., 2005. Mapping complex tissue architecture with diffusion spectrum magnetic resonance imaging. Magn. Reson. Med. 54 (6), 1377–1386.

Westin, C.F., Peled, S., Gudbjartsson, H., Kikinis, R., Jolesz, F.A., 1997. Geometrical diffusion measures for MRI from tensor basis analysis. In: Proceedings of the 5th Annual Meeting of ISMRM, Vancouver, Canada, p. 1742.

Wheeler-Kingshott, C.A.M., Cercignani, M., 2009. About 'axial' and 'radial' diffusivities. Magn. Reson. Med. 61, 1255–1260.

White, R.L., 1994. Image restoration using the damped Richardson-Lucy method. In: Proc. SPIE 2198, Instrumentation in Astronomy VIII. https://doi.org/10.1117/12.176819.

White, N.S., Dale, A.M., 2009. Optimal diffusion MRI acquisition for fiber orientation density estimation: an analytic approach. Hum. Brain Mapp. 30, 3696–3703.

Wu, Y.-C., Alexander, A.L., 2007. Hybrid diffusion imaging. Neuroimage 36 (3), 617–629.

Yeh, F.C., Wedeen, V.J., Tseng, W.Y.I., 2010. Generalized q-sampling imaging. IEEE Trans. Med. Imaging 29 (9), 1626–1635.

Zhang, H., Schneider, T., Wheeler-Kingshott, C.A., Alexander, D.C., 2012. NODDI: practical in vivo neurite orientation dispersion and density imaging of the human brain. Neuroimage 61, 1000–1016.

# Chapter 11

# Multishell models

**Jan Morez[a], Alberto De Luca[b,c], Maximillian Pietsch[d], Daan Christiaens[d,e], Steven H. Baete[f], Marco Reisert[g,h], and Ben Jeurissen[a]**

[a]imec-Vision Lab, Department of Physics, University of Antwerp, Antwerp, Belgium, [b]Image Sciences Institute, Division Imaging & Oncology, University Medical Center Utrecht, Utrecht, The Netherlands, [c]Neurology Department, UMC Utrecht Brain Center, University Medical Center Utrecht, Utrecht, The Netherlands, [d]Centre for the Developing Brain, School of Biomedical Engineering and Imaging Sciences, King's College London, London, United Kingdom, [e]Department of Electrical Engineering (ESAT/PSI), KU Leuven, Leuven, Belgium, [f]Center for Advanced Imaging Innovation and Research (CAI2R), Department of Radiology, NYU Langone Health, New York, NY, United States, [g]Department of Stereotactic and Functional Neurosurgery, Medical Center of the University of Freiburg, Medical Faculty of the University of Freiburg, Freiburg im Breisgau, Germany, [h]Department of Diagnostic and Interventional Radiology, Medical Physics, Medical Center of the University of Freiburg, Medical Faculty of the University of Freiburg, Freiburg im Breisgau, Germany

## 1 Introduction

At the core of every fiber-tracking algorithm lies the local model that relates the raw diffusion-weighted (DW) signal to the underlying fiber orientations. While there are many models that can extract fiber orientations, they all rely on the same assumption: when a number of fibers are aligned along a common orientation, the diffusion of water molecules will be hindered to a greater extent perpendicularly to this orientation than along it (Behrens et al., 2014). By measuring the diffusion-weighted MRI (DW-MRI) signal for each imaging voxel along a number of noncollinear orientations, the local orientation of the fibers can be assessed throughout the tissue of interest. These local fiber orientations can then be pieced together to infer long-range pathways connecting distant anatomical regions, a process that is called fiber tracking or fiber tractography (Jeurissen et al., 2019).

As seen in the previous chapter, one of the most widely used models for characterizing fiber orientation in terms of measured diffusion signal is the diffusion tensor (Basser et al., 1994). Diffusion tensor imaging (DTI) requires only a few DW-MRI images to be acquired (see Fig. 1A) and only modest computing resources for the model fit. Since its inception in 1999 (Basser et al., 2000; Conturo et al., 1999; Mori et al., 1999), DTI tractography has been successfully used to delineate the major white matter (WM) pathways and has been used in a wide range of neuroscientific studies (Catani et al., 2002; Catani and Thiebaut de Schotten, 2008; Wakana et al., 2004). An important limitation of the diffusion tensor, however, is that it is only capable of resolving a single fiber population per voxel. In voxels of complex fiber architecture (e.g., crossing fiber populations or partial volume effects between adjacent fiber populations), the diffusion tensor is often a poor representation of the underlying fiber orientations (Alexander, 2005). This can cause false-negative connections, in which tracking can terminate prematurely, or false-positive connections, in which tracks can veer off to an unconnected adjacent tract (Behrens et al., 2007; Jeurissen et al., 2011).

Several methods have been proposed to extract multiple fiber orientations from DW-MRI, many of them relying on a high angular resolution diffusion imaging (HARDI) protocol (Tuch et al., 2002). HARDI measures the DW signal using a larger number of directions than required for DTI, capturing the higher angular frequencies that are not adequately modeled by DTI (see Fig. 1B). Typically, HARDI acquisitions use a constant, strong diffusion weighting strength, or $b$-value, for all directions, resulting in a spherical, single-shell, acquisition scheme in $q$-space. While initially the adoption of multifiber tracking was hampered by unacceptably long scan times and limited software availability, nowadays it has been integrated into a plethora of software packages and can be readily used on clinically feasible datasets. These single-shell HARDI models have been covered at length in the previous chapter.

Recent developments in MRI hardware and reconstruction methods now permit the acquisition of large amounts of data within relatively short scan times (Chapter 7). This makes it possible to acquire so-called multishell data, with diffusion sensitization applied along many directions over multiple $b$-value *shells* (see Fig. 1C). While, initially, the use of multishell imaging was restricted to capturing nonmonoexponential decay (e.g., using diffusion kurtosis imaging (DKI) (Jensen et al., 2005) or multiexponential models (Assaf and Basser, 2005)) or to fit advanced microstructure models (Alexander et al.,

---

Handbook of Diffusion MR Tractography. https://doi.org/10.1016/B978-0-12-818894-1.00008-2
Copyright © 2025 Elsevier Ltd. All rights are reserved, including those for text and data mining, AI training, and similar technologies.

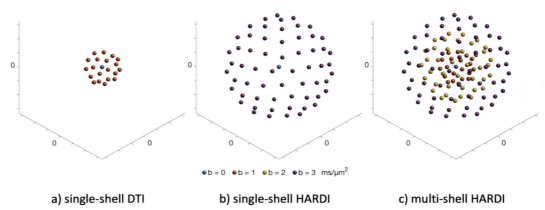

**FIG. 1** Diffusion sampling strategies: Single-shell DTI (A) vs single-shell HARDI (B) vs multishell HARDI (C) in $q$-space.

2019; Novikov et al., 2018; Panagiotaki et al., 2012), multishell imaging is now starting to be recommended as a requirement for most research diffusion studies. In this chapter, we will review multishell models, with a particular focus on what they can add for fiber tracking.

## 2 What is multishell data?

Before embarking on a journey along the different multishell models available, it is important to understand the concepts of $q$-space and $b$-value, also covered in Chapter 6.

Like any other gradient in the magnetic field, a diffusion gradient introduces a position-dependent phase shift of the transverse magnetization of the spins that is determined by the so-called $q$-vector:

$$q = \gamma g \delta \qquad (1)$$

with $g$ the diffusion encoding gradient vector and $\delta$ the diffusion encoding gradient duration. Note that the diffusion encoding gradient vector $g$ embodies both its strength $\|g\|$ as well as its orientation $\hat{g} = g/\|g\|$. In turn, the $q$-vector embodies both the extent of the position-dependent phase-shift $\|q\|$ as well as the orientation along which this phase-shift is introduced $\hat{q} = q/\|q\|$. The so-called $q$-space is the space of all possible $q$-vectors and each one of a series of DW images is characterized by a specific $q$-vector. This is why a diffusion experiment is often said to be "sampling the $q$-space." What this means is that a series of images is acquired with distinct diffusion-encoding gradient directions and strengths. At the origin of $q$-space, the position-dependent phase shift is $\|q\| = 0$, which means that no phase shift is introduced. As one moves outwards in $q$-space, the position-dependent phase shift introduced by the diffusion gradient will gradually increase.

The actual diffusion-weighting or $b$-value is then given by:

$$b = \|q\|^2 \Delta \qquad (2)$$

with $\Delta$ the so-called diffusion time, i.e., the time given to the molecules to diffuse before they are being rephased. Images for which $b = 0$ are images without diffusion weighting. By increasing the $b$-value one can sensitize the images to diffusion along the gradient direction $\hat{g}$, as Brownian motion along this axis leads to dephasing and a net signal loss. In this chapter, we will assume the diffusion time is fixed, so "increasing the $b$-value" should always be interpreted as increasing $\|q\|$.

With these definitions, we can understand the difference between multishell and single-shell diffusion acquisitions. Single-shell acquisitions only sample the angular domain of $q$-space: the diffusion gradient directions $\hat{g}$ are varied, while the diffusion-weighting or $b$-value is kept constant (see Fig. 2, left). Multishell acquisitions, on the other hand, sample both the angular and radial domains by varying both the direction as well as the amount of diffusion weighting. By allowing only a limited number of $b$-values to be sampled, the $q$-space sampling pattern consists of multiple concentric spheres or *shells* (see Fig. 2, right). As a result, such schemes are characterized by the number of shells acquired and the specific $b$-value and number of directions for each shell.

Multishell acquisitions typically require longer scanning times, so it is fair to ask: what is their potential added value? On close inspection of Fig. 2, one can observe that, as the $b$-value increases, the increased diffusion-weighting will result in more diffusion-induced signal loss and the rate at which this happens can change dramatically, depending on the structure

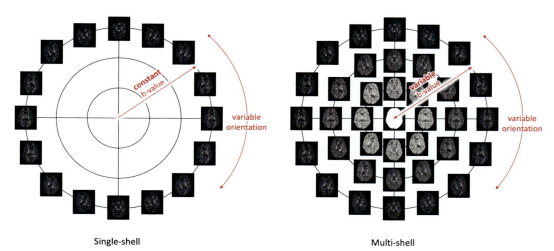

**FIG. 2** Schematic representation of a single- vs multishell acquisition scheme in $q$-space. For simplicity, the $q$-space sampling is restricted here to a 2D circle, rather than a 3D sphere. Note that a single-shell acquisition only samples the $q$-space in the angular domain (diffusion gradient directions), whereas a multishell acquisition also samples the radial domain (diffusion weighting strength or $b$-value). The image contrast at each point in $q$-space can be appreciated from a representative axial slice.

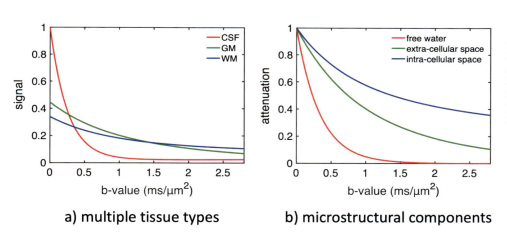

**FIG. 3** Spherical average of the diffusion-weighted signal as a function of $b$-value. The unique $b$-value-dependent signal for different tissue types (A) or microscopic substrates (B) opens up the possibility of multicompartment modeling.

of interest. For example, diffusion is free in cerebrospinal fluid (CSF) in the ventricles, causing a rapid decrease of the signal with increased $b$-value. In the parenchyma, on the other hand, diffusion is hindered/restricted by hydrophobic cell membranes and myelin sheaths, causing a much more gradual decrease of the signal (see Fig. 3A). As such, one can tell apart cortical gray matter (GM) from WM tissue, just by looking at their characteristic $b$-value-dependent curves (Jeurissen et al., 2014). Within the parenchyma, it is also possible to distinguish between the intra- and extracellular diffusion (see Fig. 3B). As the water molecules moving *between* the cells are merely *hindered* by the cell membranes, they can still move relatively freely, resulting in a faster decay curve. On the other hand, the water molecules *inside* the neurons are much more *restricted* in their random walk, resulting in a slower decay curve (Assaf and Basser, 2005). These tissue- or microstructure-dependent decay curves allow us to tease apart their contributions to the signal and open the door to compartment-specific modeling. While, at first sight, multicompartment modeling appears to only be of interest to the microstructural modeling community, it can be demonstrated that it can also offer tremendous advantages in the context of fiber tracking.

## 3 Modeling of multishell data for fiber tracking

In this section, we will provide an overview of the existing models for multishell data, in particular those that have been applied in the context of fiber tracking.

## 3.1 Models of diffusion

The models in this section merely describe the diffusion phenomenon within an MRI voxel without attributing it to a particular biological substrate. They range from model-free approaches that obtain the full diffusion probability density function (PDF) (Wedeen et al., 2005) to the more widely used diffusion tensor (Basser et al., 1994) and diffusion kurtosis model (Jensen et al., 2005).

### 3.1.1 Diffusion spectrum imaging

The diffusion PDF or spin propagator $p(\mathbf{r})$ quantifies the fraction of particles that will have been displaced by $\mathbf{r}$ within the fixed diffusion time $\Delta$, or, equivalently, the likelihood that a single particle will undergo that displacement. Stejskal and Tanner (1965) demonstrated that the signal attenuation in $\mathbf{q}$-space $S(\mathbf{q})/S(\mathbf{0})$ can be expressed as the 3D Fourier transform of this PDF:

$$S(\mathbf{q})/S(\mathbf{0}) = \int_{\mathbb{R}^3} p(\mathbf{r}) e^{i\mathbf{q}^T\mathbf{r}} d\mathbf{r} \qquad (3)$$

Consequently, one can obtain the diffusion PDF $p(\mathbf{r})$ directly from the signal attenuation in $\mathbf{q}$-space $S(\mathbf{q})/S(\mathbf{0})$ by applying the inverse Fourier transform:

$$p(\mathbf{r}) = \int_{\mathbb{R}^3} (S(\mathbf{q})/S(\mathbf{0})) e^{-i\mathbf{q}^T\mathbf{r}} d\mathbf{q} \qquad (4)$$

This approach is called $\mathbf{q}$-space imaging (Callaghan et al., 1988). A common approach to $\mathbf{q}$-space imaging is diffusion spectrum imaging (DSI) (Fig. 4) (Wedeen et al., 2005). In practice, DSI measures $S(\mathbf{q})$ for each of a Cartesian grid of $\mathbf{q}$-vectors and then reconstructs a discrete version of $p(\mathbf{r})$ by applying the 3D inverse fast Fourier transform (IFFT). Note that a Cartesian sampling scheme is not a multishell acquisition in the strict sense (see Section 2). However, recent developments have generalized DSI to enable multishell sampling on radial lines (Radial DSI (Baete et al., 2016)) or any sufficiently dense sampling of $\mathbf{q}$-space (Fig. 5) (Generalized $\mathbf{q}$-Sampling Imaging, GQI (Yeh et al., 2010)).

As DSI obtains the diffusion PDF directly from the measurements by means of the inverse Fourier transform, it makes no assumptions about the underlying microstructure or functional form of the PDF. As a result, it is able to represent the

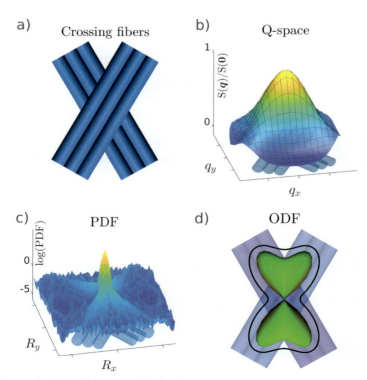

**FIG. 4** Illustration of the signal attenuation $S(\mathbf{q})/S(\mathbf{0})$ in $\mathbf{q}$-space (B), the spin propagator $p(\mathbf{r})$ (PDF, C), and the resulting orientation distribution function (ODF, D) of a simplified model of two crossing fiber bundles (A). The orientation of the crossing bundles is added to (B), (C), and (D) for clarity.

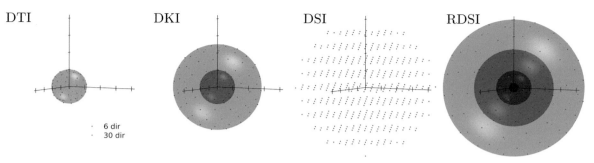

FIG. 5 Single- (DTI) and multishell (DKI, DSI, RDSI) diffusion sampling schemes.

diffusion patterns that result from multiple compartments and even multiple fiber orientations (Wedeen et al., 2005, 2008, 2012). The discrete representation of $p(r)$ that one obtains from the IFFT is not directly useful for estimating the fiber orientations, since it is a function of 3D space. In practice, one often calculates the radial integral of $p(r)$, which is called the diffusion orientation density function (dODF) $\psi(r)$:

$$\psi(\hat{r}) = \int_0^\infty p(\alpha\hat{r})\alpha^2 d\alpha \qquad (5)$$

where $\hat{r} = r/\|r\|$ is a unit vector and $\alpha$ the displacement magnitude (Aganj et al., 2009; Tuch, 2004). The factor $\alpha^2$ accounts for the size of the infinitesimal solid angle element of the PDF. Note that GQI does not include this $\alpha^2$ factor (Yeh et al., 2010).

At first glance, the dODF derived with DSI appears to be the ideal propagator to drive fiber tracking as it is able to resolve crossing fibers (Wedeen et al., 2008) and the technique has even been dubbed "high-definition fiber tracking" (Fernandez-Miranda et al., 2012). Indeed, it does intuitively make sense to propagate fiber trajectories along the orientation of least hindrance or maximal diffusion. This implies that the amount of diffusion in a particular direction is directly proportional to the underlying proportion of WM fibers along that direction. However, due to the nature of diffusion, the dODF does not necessarily reflect the underlying fiber ODF (Tuch et al., 2003). Although water molecules are most likely to diffuse along the fiber orientation, diffusion along other, even perpendicular, orientations is still common. As a consequence, diffusion from closely aligned fiber orientations will be blurred together, implying that only a single fiber population is present (Barnett, 2009; Jensen and Helpern, 2016; Kuo et al., 2008). Additionally, the overlapping of diffusion from different fiber populations is known to introduce a bias in the orientations of least hindrance (Tournier et al., 2008; Zhan and Yang, 2006). These issues can only be addressed properly by the introduction of a suitable model for diffusion in WM (Descoteaux et al., 2009; Tuch et al., 2003) (see Section 4).

Another important limitation of DSI is the large amount of data required to perform the inverse Fourier transform and the correspondingly long acquisition time.

### 3.1.2 From diffusion tensor to diffusion kurtosis imaging

The number of $q$-space samples required to invert Eq. (3) can be reduced dramatically by assuming a specific functional form for the diffusion PDF. This is exactly the idea behind one of the most popular modeling approaches, namely DTI (Basser et al., 1994). Indeed, by assuming that the diffusion PDF takes the form of a 3D anisotropic Gaussian distribution, Eq. (3) simplifies to:

$$S(q) = S_0 e^{-q^T D q \Delta} \qquad (6)$$

or

$$S(b, \hat{g}) = S_0 e^{-b \hat{g}^T D \hat{g}} = S_0 e^{-b \sum_{i=1}^{3} \sum_{j=1}^{3} \hat{g}_i \hat{g}_j D_{ij}} \qquad (7)$$

with $S_0$ the signal without diffusion weighting and $D$ the $3 \times 3$ covariance matrix of molecular displacement, better known as the apparent diffusion tensor. Eq. (7) allows one to estimate the apparent diffusion tensor from as little as six DW images $S(b, \hat{g})$ with a constant modest diffusion weighting strength $b$, along with a reference image without diffusion weighting. This model is described at great length in the previous chapter. A typical DTI experiment uses a single-shell acquisition;

however, DTI is also compatible with data acquired at multiple shells. These modest acquisition requirements make DTI one of the most widely used models to extract microstructural information, including metrics of fractional anisotropy (FA) and the WM fiber orientation that can be used for tractography (Basser et al., 2000; Catani et al., 2002; Catani and Thiebaut de Schotten, 2008).

The biggest drawback of the DTI model is that it provides only a composite view, treating each imaging voxel as a homogeneous compartment. In voxels containing complex fiber architecture (e.g., multiple crossing or "kissing" fiber populations), this assumption breaks down and fiber orientations extracted with DTI poorly reflect the underlying anatomy (Tuch et al., 2002) (see previous chapter for details on this). Moreover, while DTI is compatible with multishell data, at high b-values the DTI model is known to provide a poor fit due to its inability to represent nonmonoexponential signal decay that results from multiple underlying diffusivities originating from multiple compartments.

A more appropriate model for multishell data is the diffusion kurtosis model (Jensen et al., 2005):

$$S(b,\hat{g}) = S_0 e^{-b\sum_{i=1}^{3}\sum_{j=1}^{3}\hat{g}_i\hat{g}_j D_{ij} + \frac{1}{6}b^2 \overline{D}^2 \sum_{i=1}^{3}\sum_{j=1}^{3}\sum_{k=1}^{3}\sum_{l=1}^{3}\hat{g}_i\hat{g}_j\hat{g}_k\hat{g}_l W_{ijkl}} \tag{8}$$

with $W$ the 3 x 3 x 3 x 3 apparent kurtosis tensor and $\overline{D} = \text{Trace}(D)/3$ the mean diffusivity. DKI is a popular choice for multishell data, as its high-order nature allows it to capture nonmonoexponential signal decay that is prevalent in living tissue due to diffusivity heterogeneity (see Fig. 6). This heterogeneity can be readily visualized, e.g., in mean kurtosis (MK) maps, where tissue with homogeneous diffusivities (such as CSF) will have an MK close to 0, whereas tissues with more heterogeneous diffusivities (such as in GM and WM) exhibit values >0. Note that a DKI approximation of the signal allows for an analytical reconstruction of the underlying dODF, which is capable of resolving crossing fibers (Lazar et al., 2008). However, compared to DSI, which is based on the IFFT, the DKI approximation to the dODF provides only a simplified model.

### 3.1.3 Pros and cons of models of diffusion

All of the approaches described in this section are modeling the *diffusion phenomenon* within an MRI voxel without attributing the MRI signal to a particular microscopic substrate. This relieves the user from choosing a particular microstructural model. However, it does make these approaches highly *unspecific* to the biological tissue properties of interest. DTI, DKI, and DSI are all providing estimates or approximations of the underlying *diffusion* ODF; however, as explained in Section 3.1.1, the diffusion ODF is not necessarily the ideal propagator for fiber tracking. Indeed, diffusivity is high along WM axons but also in the ventricles, yet diffusion models make no attempt at separating these phenomena. As a result, while these models might accurately describe the diffusion process within a voxel, fiber tracking requires an accurate description of the tissue microstructure underpinning the local diffusion process. Therefore the diffusion PDF and its simpler cousin, the diffusion ODF, are fundamentally ill-equipped for fiber tracking.

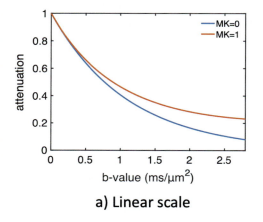
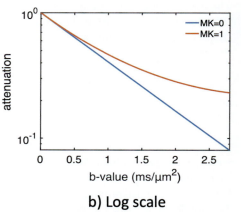

**FIG. 6** Monoexponential (mean kurtosis MK=0)—vs nonmonoexponential (MK=1) signal decay on a linear scale (A) and a log scale (B). DTI can only capture monoexponential decay *(blue curve)*. To capture nonmonoexponential decay *(red curve)* a higher-order model like DKI is required. Note that using a log scale, monoexponential decay is represented by a straight line and nonmonoexponential decay can be more readily observed.

a) Linear scale

b) Log scale

## 3.2 Models of fibrous tissue

The models in this section break down the DW-MRI signal in each imaging voxel into a sum of contributions from several compartments, where each compartment is assumed to correspond to a certain tissue type or a certain cellular component.

### 3.2.1 Data-driven approaches

Data-driven approaches assume no strict functional form to represent the DW-MRI signal of each of the compartments but obtain compartment-specific decay curves directly from the data. The most popular incarnation of such a data-driven approach, in particular in the context of fiber tracking, is that of spherical deconvolution (SD) (Dell'Acqua and Tournier, 2019; Tournier et al., 2004).

**Spherical (de)convolution: Principles**

SD is originally a single-shell HARDI method that allows estimating the full fiber orientation distribution function (fODF) in each brain voxel, regardless of the number of underlying fiber orientations (Tournier et al., 2004). It assumes that the DW signal in WM voxels, originating from various fiber populations, is given by the *spherical convolution* of the DW signal profile for a typical fiber population (*response function*) with the fODF, the fraction of the fibers as a function of orientation. The fODF can then be estimated from the DW signal by performing the inverse operation, i.e., *spherical deconvolution*. A visualization of the spherical convolution operation in a voxel containing two perpendicular fiber populations is shown in Fig. 7.

More formally, $S(\hat{g})$, measured along diffusion encoding direction $\hat{g}$ at a constant $b$-value, is the result of the spherical convolution of a response function $R(\hat{g}; \hat{n})$ and an fODF $F(\hat{n})$, such that

$$S(\hat{g}) = \int_{S^2} R(\hat{g}; \hat{n}) F(\hat{n}) d\hat{n} = R(\hat{g}; \hat{n}) \otimes F(\hat{n}) \tag{9}$$

where $R(\hat{g}; \hat{n})$ represents the signal that would be measured along $\hat{g}$ in the case of a single fiber population oriented along $\hat{n}$ (with $\|\hat{n}\| = 1$) and $F(\hat{n})$ represents the density of fibers oriented along $\hat{n}$. If $R(\hat{g}; \hat{n})$ is known, $F(\hat{n})$ can be obtained through spherical deconvolution (Tournier et al., 2004):

$$F(\hat{n}) = R(\hat{g}; \hat{n}) \otimes^{-1} S(\hat{g}) \tag{10}$$

Because deconvolution is typically ill-conditioned, a nonnegativity constraint is imposed to prevent unphysical negative amplitudes in the estimated fODF, a technique usually referred to as constrained spherical deconvolution (CSD (Tournier et al., 2007)):

$$F(\hat{n}) = R(\hat{g}; \hat{n}) \otimes^{-1} S(\hat{g}) \quad \text{s.t.} \quad F(\hat{n}) \geq 0 \tag{11}$$

Most deconvolution algorithms will use linear basis functions, such as spherical harmonics (SH) (Anderson, 2005; Tournier et al., 2004), to represent a continuous fODF, but some rely on a dense discretization of the fODF on the sphere (Dell'Acqua et al., 2007). In either case, the previous equation gives rise to a constrained linear least-squares problem of the form (Jeurissen et al., 2014):

$$\hat{x} = \underset{x}{\mathrm{argmin}} \|Ax - y\|_2^2 \quad \text{subject to} \quad Cx \geq 0 \tag{12}$$

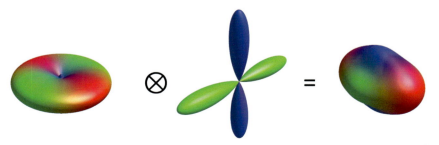

**FIG. 7** Spherical convolution: The single-shell DW signal (right) is assumed to be the spherical convolution of the response function (left) and the fODF (middle).

where $x$ is the unknown vector of coefficients (or discrete amplitudes) of the ODF, $y$ is the vector of DW signal intensities measured on a single shell in $q$-space, $A$ is the matrix relating the fODF coefficients (or discrete amplitudes) to the measured signal through spherical convolution, and $C$ is the matrix relating the fODF coefficients to the fODF amplitudes in a dense set of directions, effectively enforcing nonnegativity of the ODF. Note that when $x$ contains discrete fODF amplitudes, $C$ simplifies to the identity matrix.

Solving the constrained problem, it becomes possible to perform the spherical deconvolution operation with drastically reduced noise sensitivity, allowing reliable fODF estimates on clinically feasible DW-MRI data (Jeurissen et al., 2011, 2013; Tournier et al., 2007).

## Multicompartment spherical deconvolution

While CSD has been very successful at estimating the fODF in pure WM voxels, in voxels containing a mixture with other tissue types such as GM and CSF, the WM response function alone is no longer appropriate and CSD overestimates WM density and produces unreliable, noisy fODF estimates.

It has been shown that Eq. (12) can be extended to both support data acquired at *unique b-value shells* and to accommodate multiple compartments, a technique usually referred to as multitissue or multicompartment CSD (MT-CSD) (Jeurissen et al., 2014). This gives rise to a constrained linear least-squares problem of the form:

$$\begin{bmatrix} \hat{x}_1 \\ \vdots \\ \hat{x}_n \end{bmatrix} = \underset{\begin{bmatrix} x_1 \\ \vdots \\ x_n \end{bmatrix}}{\operatorname{argmin}} \left\| \begin{bmatrix} A_{1,1} & \cdots & A_{1,n} \\ \vdots & \ddots & \vdots \\ A_{m,1} & \cdots & A_{m,n} \end{bmatrix} \begin{bmatrix} x_1 \\ \vdots \\ x_n \end{bmatrix} - \begin{bmatrix} y_1 \\ \vdots \\ y_m \end{bmatrix} \right\|_2^2 \quad \text{subject to} \quad \begin{bmatrix} C_1 & 0 & 0 \\ 0 & \ddots & 0 \\ 0 & 0 & C_n \end{bmatrix} \begin{bmatrix} x_1 \\ \vdots \\ x_n \end{bmatrix} \geq 0 \quad (13)$$

where $x_j$ is the unknown vector of coefficients (or discrete amplitudes) of the ODF of the $j$th compartment, $y_i$ is the vector of DW signal intensities measured at the $i$th $b$-value shell, $A_{i,j}$ is the $b$-value- and compartment-specific convolution matrix relating the ODF coefficients (or discrete amplitudes) of the $j$th compartment to the DW signal intensities measured at the $i$th $b$-value shell, and $C_j$ is the compartment-specific matrix relating the ODF coefficients to the corresponding amplitudes, effectively enforcing nonnegativity of the ODF of each compartment. As a result, data acquired with different $b$-values all contribute to resolving different compartments as well as fiber orientations, all within the same large system of equations (see Fig. 8 for a graphical representation of this system of equations). Note that Eq. (13) can be generalized to arbitrary $q$-space samplings that do not sample the $q$-space in a shell-wise manner, enabling the use of this technique also in the case of Cartesian sampling (see Fig. 5, DSI) or in the case of unintentional deviations from perfect shell-wise acquisitions (e.g., in the case of gradient nonlinearities) (Morez et al., 2021).

Typically, MT-CSD takes into account the presence of the three major tissue types (CSF, GM, and WM) by leveraging the unique $b$-value dependency of each tissue type (see Fig. 3). As such, MT-CSD can obtain tissue density maps directly from the DW data and produces more accurate and precise fODFs in voxels containing GM and/or CSF (see Fig. 9).

These effects directly benefit fiber tracking (see Fig. 10). First, the increased precision and reduced number of spurious peaks result in less noisy tractograms near the WM-CSF and WM-GM interfaces, with better-defined fine structures. Second, as MT-CSD ensures that the fODF amplitudes are directly proportional to the density of the WM fibers, the fODF amplitudes derived with MT-CSD make a much more reliable stopping criterion for fiber tracking. With single-shell CSD,

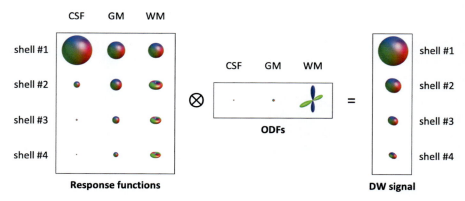

**FIG. 8** Multitissue spherical convolution with multiple $b$-values. In this particular example, there are four unique $b$-value shells available in the data and the three modeled compartments, CSF, GM, and WM, are present in this voxel. Note that the spherical means of the different response functions are shown in the corresponding tissue-specific decay curves in Fig. 3.

Multishell models **Chapter | 11 209**

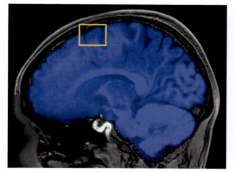
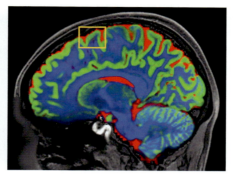

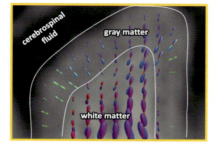

**FIG. 9** Apparent tissue densities (top) and WM fODFs (bottom) obtained with single-tissue CSD of single-shell data (left) vs multitissue CSD of multishell data (right). *(Modified from Jeurissen, B., Tournier, J.-D., Dhollander, T., Connelly, A., Sijbers, J., 2014. Multi-tissue constrained spherical deconvolution for improved analysis of multi-shell diffusion MRI data. NeuroImage 103, 411–426.)*

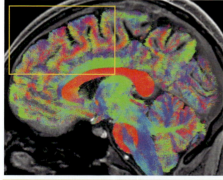
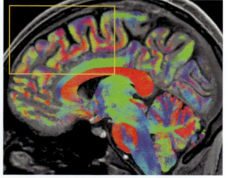
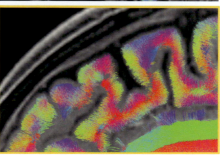

**FIG. 10** Slab visualization of a tractogram obtained with single-tissue fODFs of single-shell data (left) vs multi-tissue fODFs of multishell data (right). *(Modified from Jeurissen, B., Tournier, J.-D., Dhollander, T., Connelly, A., Sijbers, J., 2014. Multi-tissue constrained spherical deconvolution for improved analysis of multi-shell diffusion MRI data. NeuroImage, 103 411–426.)*

relatively large fODF amplitude thresholds are required to avoid tracking inside CSF and isotropic GM. MT-CSD, on the other hand, enables fiber tracking with significantly lower fODF thresholds compared to single-shell CSD, without resulting in spurious tracks near the tissue interfaces. This improves the ability to detect small WM structures that would otherwise be left out.

In addition to the improvements in fiber tracking, the additional apparent densities of GM and CSF provided by MT-CSD have also been used to study brain tissue composition (Giraldo et al., 2022; Jillings et al., 2020; Mito et al., 2020).

### Multiple anisotropic compartments

An interesting and challenging generalization of multitissue or multicompartment spherical deconvolution is to allow for multiple anisotropic compartments (De Luca et al., 2020; Pietsch et al., 2019). Multitissue SD with more than one anisotropic and optionally additional isotropic response functions allows delineating and modeling the angular dependency of more than a single tissue compartment.

This becomes relevant when a single anisotropic response function does not adequately capture all anisotropic compartments of interest in a given dataset. Notable examples thereof can be found in deep GM and in the WM-GM interface (De Luca et al., 2020), and in differentially maturating brain parenchyma in newborns (Pietsch et al., 2019).

Achieving appropriate fODF reconstructions in the GM is a key step to improving the study of brain connectivity by characterizing the fiber trajectories up to their (sub)cortical start- or endpoints. Diffusion properties of the fibers in the cortical and deep GM are remarkably distinct from those in WM: GM exhibits on average lower diffusion anisotropy and lower restriction effects at high $b$-value. For this reason, conventional spherical deconvolution approaches using a single anisotropic response function (specific to WM) are prone to reconstruct suboptimal fODFs in GM, including spurious peaks, or small lobes that are difficult to distinguish from the noise floor.

To address these issues De Luca et al. (2020) modeled the angular properties of fibers in both WM and GM. When applied to multishell datasets with a good signal-to-noise ratio, such as those from the Human Connectome Project, the framework quantifies two separate fODFs, one specific to WM and one specific to GM, as exemplified in Fig. 11B and C. An unsolved challenge when quantifying multiple fODFs is how to use their information content to perform fiber tractography. In the previously mentioned work, the authors have suggested computing a new fODF by linearly combining the individual FODs weighted by their corresponding signal fraction. The resulting fODF (Fig. 11D) can then be used in combination with conventional tractography strategies to reconstruct fiber tracts up to their cortical start/endpoint (Fig. 11E and F). Although straightforward, the linear combination of multiple fODFs is unlikely to be the most optimal to leverage the information content and tissue specificity of multiple fODFs, and future research in this direction remains necessary.

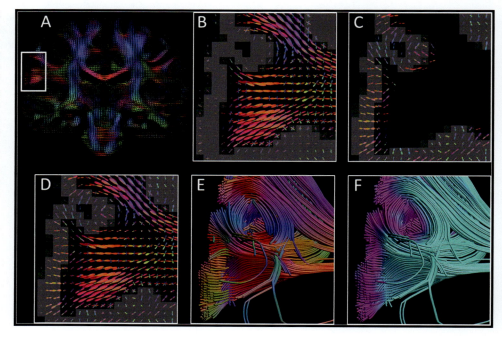

**FIG. 11** Multitissue CSD with multiple anisotropic compartments: (A) Visualization of the fODFs reconstructed in a coronal slice; (B) The WM-specific fODFs; (C) The GM-specific fODFs; (D) A linear combination of (B) and (C) showing contiguous fiber orientations between WM and GM; (E) Fiber tractogram obtained from the fODFs shown in (D) by seeding in the cortical GM; (F) Same as E, but coloring the trajectories according to the closest tissue type, as derived from a corresponding $T_1$-weighted image. *Turquoise* = WM, *Purple* = GM.

In the developing human brain in the weeks and months around birth, brain size and composition change markedly. In this time period, developmental processes such as neuronal proliferation, maturation and retraction, axonal growth, glial proliferation and myelination, developing cortical cytoarchitecture, and synaptogenesis fundamentally change the microstructural properties of the parenchyma. Radial glial cells span the cerebrum from the ventricles to the cortex and provide a "scaffolding" for migrating neurons.

Diffusion signal properties in cortical and WM areas change with age and are heterogeneous across the brain (Pietsch et al., 2019). In contrast to adult data, voxels in cortical GM and WM can have very similar properties while intratissue variability is high in developing WM. In Pietsch et al. (2019), a set of three tissue response functions was used to model tissue maturation across the age range of a neonatal population from the developing Human Connectome Project (http://www.developingconnectome.org/). The approach uses MT-CSD with a CSF response function to model free water and two anisotropic response functions, sampled from WM in the age-extremes of the cohort, to model the transition of WM tissue properties. The set of resulting ODFs is not *specific* to tissue classes such as CSF, GM, or WM but aims at capturing the changing signal properties throughout the brain.

Fig. 12 shows the resulting decomposition of two scans from a single prematurely born subject imaged at 32 and at 43 weeks postmenstrual age (wPMA). The two fODFs are combined by summation and used for whole-brain tracking. The background image shows the relatively high density of free fluid-like signal throughout the parenchyma at 32 wPMA that overall decreases with age with the notable exception of the periventricular crossroads (circle in A). This area is challenging for tracking as both tissue fODFs exhibit very low density. The ganglionic eminence or germinal zone is encircled in column B. It is a tightly packed transient structure that features prominently in fetal and prematurely born neonates and disappears by the first year of age. At 32 wPMA, the cortex is densely packed and exhibits a strong radial organization with signal properties much closer to mature WM than to mature cortical GM (column B).

## Response function estimation and factorization

All spherical deconvolution methods require an estimate of the compartment-specific response functions. Ideally, these would correspond to the DW signal that would be acquired for a unit volume of that specific compartment. E.g., in the

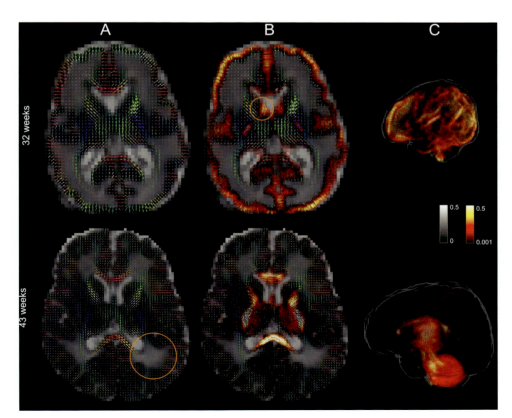

FIG. 12 Tissue decomposition and tracking in a single neonatal subject. The fluid density associated with the CSF response function is shown as the background image in columns A and B. The glyphs show the sum of the two tissue fODFs. Columns B and C show streamlines obtained via whole-brain tracking on the sum of the anisotropic fODFs, colored by the (orientationally resolved) amplitude in the fODF associated with the most mature WM tissue signal. In column C, brain mask outlines together with the streamlines are shown to scale and in 3D. Circles mark transient areas in the fluid-rich periventricular crossroads (A) and in the ganglionic eminence (B).

case of traditional single-shell deconvolution, the WM response function would correspond to the DW signal that would be acquired for a unit volume of WM fibers all coherently aligned along a single axis.

Two distinct approaches have been proposed in the past: either imposing a strict model for the tissue response, e.g., an axially symmetric tensor model to represent the WM response (Anderson, 2005; Behrens et al., 2007; Dell'Acqua et al., 2007; Hosey et al., 2005; Jian and Vemuri, 2007; Morez et al., 2021) or through direct empirical measurement from representative voxels identified by either explicit thresholding of brain regions with high FA for the WM response (Tournier et al., 2004, 2007), anatomical segmentation and thresholding from T1-weighted scans for multitissue responses (Jeurissen et al., 2014), or through iterative calibration of the fiber response (Dhollander et al., 2016; Tax et al., 2014; Tournier et al., 2013).

A particularly exciting approach to multicompartment response function estimation is to cast the MT-CSD model as a *factorization* problem, in which the ODF and the signal response of each tissue are both treated as unknowns in the fit (Christiaens et al., 2017). Instead of assuming the response functions as known when solving Eq. (13) for the tissue ODFs, the generalized fitting problem

$$R_t, F_t = \underset{R_t, F_t}{\mathrm{argmin}} \left\| S - \sum_t R_t \otimes F_t \right\|, \quad F_t > 0 \tag{14}$$

is defined with the local ODFs $F_t$ per voxel and the global responses $R_t$ of all tissue types as unknowns. As explained earlier, when the responses and ODFs are represented in the SH basis, the spherical convolution can be rewritten as a multiplication of the corresponding SH coefficients. In the SH basis and on a full image level, Eq. (14) hence becomes a seminonnegative tensor factorization problem (Christiaens et al., 2017). An important issue to consider is that the factorization in this form has infinite solutions; for example, the scaling of each of the responses is undefined. It is therefore necessary to impose additional constraints, for instance imposing sparsity of the ODFs (Reisert et al., 2014) or constraining the response functions to be a convex combination of the input data (Christiaens et al., 2017). The key advantage of these factorization approaches is that they are able to estimate response functions for any number of (an)isotropic signal components, without assumptions on what tissues these components represent and without heuristic estimates of their response functions. In fact, a three-component factorization in adult brain data produces components associated with WM, GM, and CSF, thus confirming that these are the "natural" tissue classes to use with MT-CSD. In brain development, in pathology, in other organs, or ex vivo, multicomponent factorization can identify responses and ODFs associated with other tissues, as shown, for example, in Fig. 13. As such, a factorization perspective offers a fully unsupervised means to generalize MT-CSD beyond healthy adult brain data without prior assumptions about the tissue.

### Pros and cons of data-driven approaches

A common criticism of spherical deconvolution approaches is that they assume that the response functions are constant throughout the brain, an assumption that might not be valid. For instance, WM bundles can differ in their underlying axonal diameters, which would violate this assumption. Gauging axonal diameters requires very high $b$-values only achievable on MRI scanners with strong gradients. At the moderate $b$-values achievable on a clinical scanner, axons effectively present as "sticks," supporting the fiber response assumption (Veraart et al., 2020). Moreover, recent studies using tensor-valued diffusion encoding suggest that, in WM, the microstructural anisotropy is remarkably uniform across the brain, which supports the constant response function assumption (Lampinen et al., 2017).

Notwithstanding these limitations, assuming a constant response function for each compartment is highly beneficial from a practical and computational standpoint, as this results in a linear least-squares problem that can be solved robustly and very efficiently. Compared to the models in Section 3.2.2, which require slow nonlinear solvers that often require overnight computation and are fraught with local minima, multicompartment spherical deconvolution approaches can process a full dataset in 1–5 minutes on a modern computer (Jeurissen et al., 2014; Jeurissen and Szczepankiewicz, 2021).

A tremendous advantage of spherical deconvolution is that it also directly provides the most natural input for fiber tracking, namely the fiber ODF, whereas the models in Section 3.1 provide estimates of the diffusion PDF and its derived diffusion ODF. The fODF not only provides the necessary orientation estimates of the fiber populations within each voxel (Jeurissen et al., 2011), but it also provides the necessary density information to determine when to stop fiber tracking (Jeurissen et al., 2014) and allows for quantitative probabilistic fiber tracking that reflects the underlying fiber densities (Smith et al., 2013, 2015). This unique feature, together with its fast and robust computation, currently makes (multitissue) spherical deconvolution one of the most effective models to drive fiber tracking (see Fig. 14).

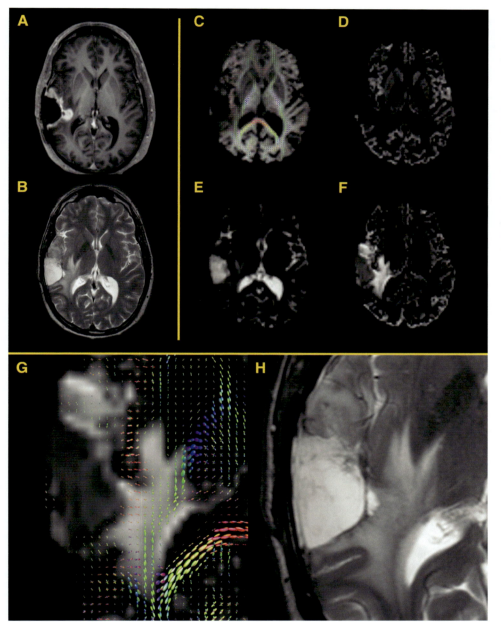

**FIG. 13** (A and B) Postsurgical T1w and T2w images of a patient with a brain tumor. (C–F) A four-component factorization of the DW-MRI recovers the white matter fiber orientation, gray matter, and CSF, as well as an additional component that segments the edema surrounding the resected tumor. (G and H) A close-up of the WM ODFs in this region shows the neural fibers traversing the edema tissue. *(Modified from Christiaens, D., Sunaert, S., Suetens, P., Maes, F., 2017. Convexity-constrained and nonnegativity-constrained spherical factorization in diffusion-weighted imaging. NeuroImage 146, 507–517.)*

### 3.2.2 Model-driven approaches

Model-driven approaches assume a strict functional form to represent the DW-MRI signal of each compartment. Moreover, rather than fixing the intrinsic diffusion parameters of each of these compartments and focusing on estimating the densities of these compartments, as is the case for the models in Section 3.2.1, these model-driven approaches attempt to capture both the intrinsic diffusion parameters of each of these compartments as well as their relative contributions.

**The standard model of microstructure**

Microstructure modeling has grown to become a very broad field of research that merits a review of its own. In this chapter, we will only discuss the overarching concepts in the context of multishell modeling. For a more in-depth look at microstructure modeling, please consult Chapter 5.

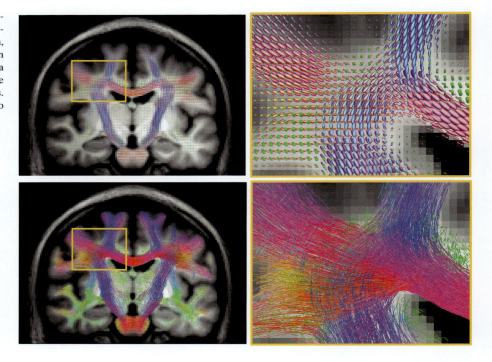

FIG. 14 Fiber ODFs obtained with multishell constrained spherical deconvolution (top). At each voxel in the brain, the continuous WM fODF provides an estimate of the density of WM as a function of orientation, regardless of the number of underlying fiber populations. This makes the fODF the ideal input to drive fiber tracking (bottom).

Rather than modeling a canonical signal response for the tissue, microstructure imaging aims to model the contributions of the various cellular compartments that constitute WM. Usually, these comprise at least restricted diffusion in the intra-axonal space and hindered diffusion in the extracellular space (Alexander et al., 2019; Jelescu and Budde, 2017; Novikov et al., 2018; Zhang et al., 2012). While many different functional forms have been proposed for each type of compartment (Panagiotaki et al., 2012), the prevailing minimal model of diffusion in WM is one in which intra-axonal space is modeled by a "stick" with 0 radial diffusivity and the extracellular space is modeled as axially symmetric Gaussian diffusion (Novikov et al., 2018). The intra-axonal axial diffusivity, the extracellular axial and radial diffusivity, and the relative signal fraction of both compartments are the model parameters. Nonlinear fitting to the local DW-MRI signal can thus map microstructural variation across WM and detect potential changes in pathology.

An important issue in microstructural modeling is the need to model the complex local fiber architecture and its effect on the signal. Many models have been proposed, e.g., a Watson distribution of the local fiber dispersion (Zhang et al., 2012), but the most general approach is to use spherical convolution with a fiber ODF (Jespersen et al., 2007; Novikov et al., 2018; Reisert et al., 2017), akin to multishell CSD. With this approach, modeling the tissue microstructure boils down to modeling the local signal response function, assumed to be rotation-invariant at least at the voxel level (Christiaens et al., 2020). With these approaches, microstructure modeling can inform tractography by providing locally tuned fiber ODFs (Daducci et al., 2016; Girard et al., 2017).

In fact, the spherical convolution of the multicompartment model of microstructure in terms of intra- and extra-axonal contributions with general fiber orientation function leads to inherent ambiguities when trying to invert the model (Coelho et al., 2019; Reisert et al., 2019). The first solution involves an fODF (fiber orientation distribution function) with minimal fiber dispersion, resulting in a very sharp fODF, coupled with low intra-axonal diffusivities (as seen in Fieremans et al., 2013). The second solution entails fODFs with smoother characteristics and higher intra-axonal diffusion (as employed in Zhang et al., 2012). Recent dedicated measurements (Dhital et al., 2019; Kunz et al., 2018) have provided a resolution to this problem, confirming the latter solution as the correct representation in normal-appearing WM.

### Pros and cons of model-driven approaches

The main advantage of model-based approaches is that they can potentially uncover more specific information about the tissue microstructure and the evolution of pathology. The quantitative metrics estimated in microstructure imaging have well-defined units on an interpretable scale, and are therefore well suited—provided good calibration—across subjects and across imaging sites. However, cross-scanner calibration is far from easy, so currently additional normalization remains necessary for multicenter studies (Ning et al., 2020; Pinto et al., 2020).

One drawback of model-based approaches, especially in comparison to the data-driven approaches discussed earlier, is the increased computational complexity due to the need for ill-conditioned multiexponential fitting. Depending on the fitting method, there are also specific requirements for the data acquisition protocol, and recent work has shown that stable estimates of the standard model parameters require sampling protocols beyond standard multishell HARDI schemes (Lampinen et al., 2020). However, the drawback can be partly diminished by Bayesian, machine learning-driven approaches (Reisert et al., 2017), which are trained on exhaustive signal simulations. In this way, the demands of protocols and computational time can be drastically reduced, while well-chosen prior distributions can resolve the ambiguities of the model inversion. Computation times are in a range of a few minutes and protocols are even suitable in acute settings with less than 2 minutes of acquisition time.

## 4 Beyond multishell acquisitions

Recent advancements in DW imaging have facilitated the use of nonconventional gradient waveforms that enable diffusion encoding in more than one direction per shot. These novel techniques, known as multidimensional DW-MRI or "tensor-valued" diffusion encoding (Morez et al., 2023; Szczepankiewicz et al., 2021; Topgaard, 2017; Westin et al., 2016) use a tensor to describe the diffusion encoding, as it can no longer be captured by the conventional $b$-value and encoding direction. The benefit of using $b$-tensors is that their shape is linked to how diffusion anisotropy manifests in the measured signal (Lasič et al., 2014). For example, by employing linear and spherical $b$-tensors, the sensitivity of the signal to diffusional anisotropy can be maximized or minimized, respectively (Eriksson et al., 2013; Mori and van Zijl, 1995; Szczepankiewicz et al., 2015; Wong et al., 1995). This contrast mechanism can therefore be used to gain additional information about the WM fibers. Previously, tensor-valued diffusion encoding techniques were primarily employed for quantifying scalar tissue parameters, often discarding the information that is specific to WM fiber orientation. However, recent work has leveraged mixtures of $b$-tensor shapes to discriminate multiple tissue types and infer WM fiber orientations (Jeurissen and Szczepankiewicz, 2021; Karan et al., 2022). In particular, nonlinear weighting schemes can resolve the aforementioned ambiguity of the standard WM model (Coelho et al., 2019; Reisert et al., 2019). In fact, additional planar shaped $b$-value schemes are enough to select the correct high-dispersive solution. As a consequence, priors on the model parameters during model estimation can be avoided. Spherical encodings are actually not enough to resolve the ambiguity (Coelho et al., 2019; Reisert et al., 2019).

The multicompartment spherical deconvolution framework (Section 3.2.1, Eq. 13) extends naturally to data acquired with multiple $b$-tensor shapes. Whereas in the original multishell CSD framework, the index $i$ enumerates the unique $b$-values, or shells within the dataset, in the extended framework, $i$ indexes the unique combinations of $b$-values and $b$-tensor shapes within the dataset. As a result, data acquired with different $b$-values and $b$-tensor shapes can contribute to resolving different compartments as well as fiber orientations, all within the same system of equations (see Fig. 15 for a graphical representation of this system of equations).

When carefully designed, such a combination between $b$-values and $b$-tensor shapes can improve the estimation of the apparent tissue densities and the WM fiber ODF, as well as downstream tractography and connectomics, without requiring additional scan time (Jeurissen and Szczepankiewicz, 2021).

## 5 Summary and conclusion

Compared to single-shell acquisitions, multishell acquisitions can accommodate a much wider range of models (see Fig. 16). Broadly speaking, these models can be categorized into models of *diffusion* and models of *fibrous tissue*. A diffusion model merely describes the diffusion phenomenon within an MRI voxel without attributing the MRI signal to a particular substrate. While this relieves the user of deciding on a particular microstructural model, it does make these approaches highly *unspecific*. Models of fibrous tissue, on the other hand, break down the DW-MRI signal in each imaging voxel into a sum of contributions from several compartments, where each compartment is assumed to correspond to a certain biological substrate. While these methods have to make potentially inaccurate assumptions about the nature and number of these compartments, they can provide much more *specific* information about the tissue under investigation.

Closely related to the difference between models of diffusion and models of fibrous tissue is the difference between the *diffusion ODF* and the *fiber ODF*. Models of diffusion, ranging from full-blown DSI to the much simpler DTI and DKI models, all capture the diffusion PDF, from which its simpler cousin, the diffusion ODF, can be obtained. As its name implies, the diffusion ODF captures the orientational distribution of diffusion and not of the fibers, making it ill-equipped

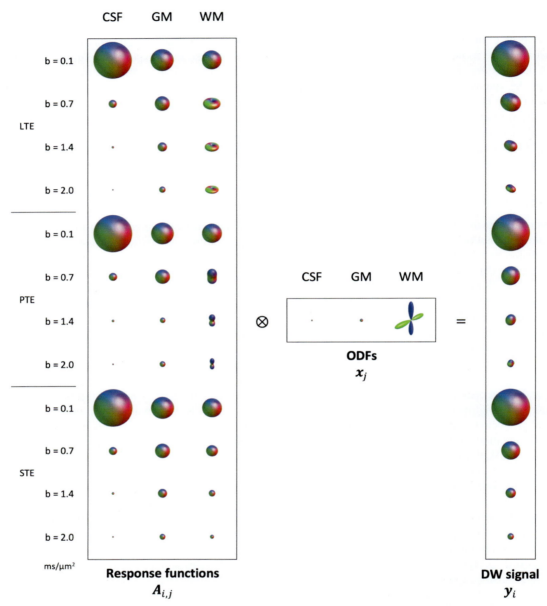

**FIG. 15** Graphical representation of multitissue spherical convolution with multiple b-tensor shapes. In this particular example, there are 12 unique b-value/b-shape combinations available in the data ($m=12$) and 3 compartments that are being modeled ($n=3$). **A** is an $m \times n$ matrix of tissue- and b-value/b-shape specific response functions; **x** is an $n \times 1$ vector of tissue-specific ODFs, and **y** is an $m \times 1$ vector of the measured DW signals for each unique b-value/b-shape combination. Note that the linear tensor encoding (LTE) response functions resemble oblate donuts, the planar tensor encoding (PTE) response functions resemble prolate peanuts, and the responses for spherical tensor encoding (STE) are spherical. *(Figure from Jeurissen, B., Szczepankiewicz, F., 2021. Multi-tissue spherical deconvolution of tensor-valued diffusion MRI. NeuroImage 245, 118717.)*

to drive fiber tracking. Models of fibrous tissue, on the other hand, typically have a *fiber ODF* associated with one or more compartments, which provides a direct estimate of the proportion of fibers running along a certain direction, making them the methods of choice for robust fiber tracking, in particular in the presence of multiple compartments and fiber orientations.

We would like to argue that multiple *b*-values are now a minimum requirement for any DW-MRI analysis, including tractography. In the past, multishell sequences were often thought of as the exclusive domain of DKI and microstructural modeling. However, it is now increasingly clear that meaningful (i.e., quantitative) fiber tracking goes hand in hand with *multicompartment* modeling of *multishell* data. Indeed, tissue- or microstructure-dependent decay curves present an

## Single-shell imaging and modeling

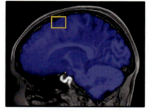

inaccurate tissue density in the case of PVE

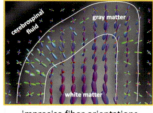

imprecise fiber orientations in the case of PVE

| measurement settings ||
|---|---|
| DW orientation | variable |
| DW strength | constant |

2D measurement space

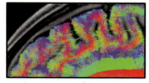

inaccurate and imprecise tractograms in the case of PVE

| modeling opportunities ||
|---|---|
| diffusion modeling | DTI |
| multiple fiber orientations | yes |
| multiple tissue types | no |
| microstructural compartments | no |

## Multi-shell imaging and modeling

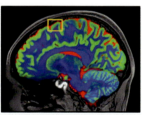

more accurate tissue density in the case of PVE

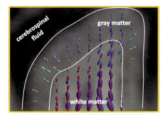

more precise fiber orientations in the case of PVE

| measurement settings ||
|---|---|
| DW orientation | variable |
| DW strength | variable |

3D measurement space

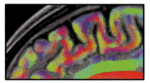

more accurate and precise tractograms in the case of PVE

| modeling opportunities ||
|---|---|
| diffusion modelling | DTI/DKI/DSI |
| multiple fiber orientations | yes+ |
| multiple tissue types | yes |
| microstructural compartments | yes- |

**FIG. 16** Graphical summary of the key differences between single- and multishell imaging and modeling.

enormous opportunity for multicompartment modeling, where the unique "weighting signature" of each compartment can be used to tease apart their contributions (see Fig. 3). This feature is not only beneficial for multicompartment modeling itself but also directly benefits downstream fiber tracking. As these approaches can provide a direct estimate of the WM/axonal fODF, they can provide specific information on the fiber orientations uncontaminated by other sources such as GM, CSF, or extracellular diffusion. Just as ignoring the possibility of multiple fiber orientations can lead to inaccurate fiber orientation estimates in voxels containing multiple underlying fiber populations (think DTI vs HARDI), ignoring the possibility of multiple compartments can lead to inaccurate estimates of the fiber densities (single- vs multishell).

## 218 PART | II Diffusion MRI

Whereas microstructural modeling and fiber tracking emerged as almost entirely different subfields within the DW-MRI community, there is now a growing realization that accurate fiber tracking goes hand in hand with accurate microstructural modeling (Girard et al., 2017; Reisert et al., 2014; Sherbondy et al., 2010).

# References

Aganj, I., Lenglet, C., Sapiro, G., 2009. ODF reconstruction in q-ball imaging with solid angle consideration. In: 2009 IEEE International Symposium on Biomedical Imaging: From Nano to Macro, pp. 1398–1401.

Alexander, D.C., 2005. Multiple-fiber reconstruction algorithms for diffusion MRI. Ann. N. Y. Acad. Sci. 1064, 113–133.

Alexander, D.C., Dyrby, T.B., Nilsson, M., Zhang, H., 2019. Imaging brain microstructure with diffusion MRI: practicality and applications. NMR Biomed. 32 (4), e3841.

Anderson, A.W., 2005. Measurement of fiber orientation distributions using high angular resolution diffusion imaging. Magn. Reson. Med. 54 (5), 1194–1206.

Assaf, Y., Basser, P.J., 2005. Composite hindered and restricted model of diffusion (CHARMED) MR imaging of the human brain. NeuroImage 27 (1), 48–58.

Baete, S.H., Yutzy, S., Boada, F.E., 2016. Radial q-space sampling for DSI. Magn. Reson. Med. 76 (3), 769–780.

Barnett, A., 2009. Theory of Q-ball imaging redux: implications for fiber tracking. Magn. Reson. Med. 62 (4), 910–923.

Basser, P.J., Mattiello, J., LeBihan, D., 1994. Estimation of the effective self-diffusion tensor from the NMR spin echo. J. Magn. Reson. Ser. B 103 (3), 247–254.

Basser, P.J., Pajevic, S., Pierpaoli, C., Duda, J., Aldroubi, A., 2000. In vivo fiber tractography using DT-MRI data. Magn. Reson. Med. 44 (4), 625–632. https://doi.org/10.1002/1522-2594(200010)44:4<625::aid-mrm17>3.0.co;2-o.

Behrens, T.E.J., Berg, H.J., Jbabdi, S., Rushworth, M.F.S., Woolrich, M.W., 2007. Probabilistic diffusion tractography with multiple fibre orientations: what can we gain? NeuroImage 34 (1), 144–155.

Behrens, T.E.J., Sotiropoulos, S.N., Jbabdi, S., 2014. MR diffusion tractography. In: Johansen-Berg, H., Behrens, T.E.J. (Eds.), Diffusion MRI, second ed. Academic Press, pp. 429–451 (Chapter 19).

Callaghan, P.T., Eccles, C.D., Xia, Y., 1988. NMR microscopy of dynamic displacements: k-space and q-space imaging. J. Phys. E Sci. Instrum. 21 (8), 820.

Catani, M., Thiebaut de Schotten, M., 2008. A diffusion tensor imaging tractography atlas for virtual in vivo dissections. Cortex 44 (8), 1105–1132.

Catani, M., Howard, R.J., Pajevic, S., Jones, D.K., 2002. Virtual in vivo interactive dissection of white matter fasciculi in the human brain. NeuroImage 17 (1), 77–94.

Christiaens, D., Sunaert, S., Suetens, P., Maes, F., 2017. Convexity-constrained and nonnegativity-constrained spherical factorization in diffusion-weighted imaging. NeuroImage 146, 507–517.

Christiaens, D., Veraart, J., Cordero-Grande, L., Price, A.N., Hutter, J., Hajnal, J.V., Tournier, J.-D., 2020. On the need for bundle-specific microstructure kernels in diffusion MRI. NeuroImage 208, 116460.

Coelho, S., Pozo, J.M., Jespersen, S.N., Jones, D.K., Frangi, A.F., 2019. Resolving degeneracy in diffusion MRI biophysical model parameter estimation using double diffusion encoding. Magn. Reson. Med. 82 (1), 395–410.

Conturo, T.E., Lori, N.F., Cull, T.S., Akbudak, E., Snyder, A.Z., Shimony, J.S., McKinstry, R.C., Burton, H., Raichle, M.E., 1999. Tracking neuronal fiber pathways in the living human brain. Proc. Natl. Acad. Sci. 96 (18), 10422–10427. https://doi.org/10.1073/pnas.96.18.10422.

Daducci, A., Dal Palú, A., Descoteaux, M., Thiran, J.-P., 2016. Microstructure informed tractography: pitfalls and open challenges. Front. Neurosci. 10, 247.

De Luca, A., Guo, F., Froeling, M., Leemans, A., 2020. Spherical deconvolution with tissue-specific response functions and multi-shell diffusion MRI to estimate multiple fiber orientation distributions (mFODs). NeuroImage 222, 117206.

Dell'Acqua, F., Tournier, J.-D., 2019. Modelling white matter with spherical deconvolution: how and why? NMR Biomed. 32 (4), e3945.

Dell'Acqua, F., Rizzo, G., Scifo, P., Clarke, R.A., Scotti, G., Fazio, F., 2007. A model-based deconvolution approach to solve fiber crossing in diffusion-weighted MR imaging. IEEE Trans. Biomed. Eng. 54 (3), 462–472.

Descoteaux, M., Deriche, R., Knösche, T.R., Anwander, A., 2009. Deterministic and probabilistic tractography based on complex fibre orientation distributions. IEEE Trans. Med. Imaging 28 (2), 269–286.

Dhital, B., Reisert, M., Kellner, E., Kiselev, V.G., 2019. Intra-axonal diffusivity in brain white matter. NeuroImage 189, 543–550.

Dhollander, T., Raffelt, D., Connelly, A., 2016. Unsupervised 3-tissue response function estimation from single-shell or multi-shell diffusion MR data without a co-registered T1 image. In: ISMRM Workshop on Breaking the Barriers of Diffusion MRI. vol. 5.

Eriksson, S., Lasic, S., Topgaard, D., 2013. Isotropic diffusion weighting in PGSE NMR by magic-angle spinning of the q-vector. J. Magn. Reson. 226, 13–18.

Fernandez-Miranda, J.C., Pathak, S., Engh, J., Jarbo, K., Verstynen, T., Yeh, F.-C., Wang, Y., Mintz, A., Boada, F., Schneider, W., Friedlander, R., 2012. High-definition fiber tractography of the human brain: neuroanatomical validation and neurosurgical applications. Neurosurgery 71 (2), 430–453.

Fieremans, E., Benitez, A., Jensen, J.H., Falangola, M.F., Tabesh, A., Deardorff, R.L., Spampinato, M.V.S., Babb, J.S., Novikov, D.S., Ferris, S.H., Helpern, J.-A., 2013. Novel white matter tract integrity metrics sensitive to Alzheimer disease progression. AJNR Am. J. Neuroradiol. 34 (11), 2105–2112.

Giraldo, D.L., Smith, R.E., Struyfs, H., Niemantsverdriet, E., De Roeck, E., Bjerke, M., Engelborghs, S., Romero, E., Sijbers, J., Jeurissen, B., 2022. Investigating tissue-specific abnormalities in Alzheimer's disease with multi-shell diffusion MRI. J. Alzheimers Dis. 90 (4), 1771–1791.

Girard, G., Daducci, A., Petit, L., Thiran, J.-P., Whittingstall, K., Deriche, R., Wassermann, D., Descoteaux, M., 2017. AxTract: toward microstructure informed tractography. Hum. Brain Mapp. 38 (11), 5485–5500.

Hosey, T., Williams, G., Ansorge, R., 2005. Inference of multiple fiber orientations in high angular resolution diffusion imaging. Magn. Reson. Med. 54 (6), 1480–1489.

Jelescu, I.O., Budde, M.D., 2017. Design and validation of diffusion MRI models of white matter. Front. Phys. 28. https://doi.org/10.3389/fphy.2017.00061.

Jensen, J.H., Helpern, J.A., 2016. Resolving power for the diffusion orientation distribution function. Magn. Reson. Med. 76 (2), 679–688. https://doi.org/10.1002/mrm.25900.

Jensen, J.H., Helpern, J.A., Ramani, A., Lu, H., Kaczynski, K., 2005. Diffusional kurtosis imaging: the quantification of non-Gaussian water diffusion by means of magnetic resonance imaging. Magn. Reson. Med. 53 (6), 1432–1440. https://doi.org/10.1002/mrm.20508.

Jespersen, S.N., Kroenke, C.D., Østergaard, L., Ackerman, J.J.H., Yablonskiy, D.A., 2007. Modeling dendrite density from magnetic resonance diffusion measurements. NeuroImage 34 (4), 1473–1486.

Jeurissen, B., Szczepankiewicz, F., 2021. Multi-tissue spherical deconvolution of tensor-valued diffusion MRI. NeuroImage 245, 118717.

Jeurissen, B., Leemans, A., Jones, D.K., Tournier, J.-D., Sijbers, J., 2011. Probabilistic fiber tracking using the residual bootstrap with constrained spherical deconvolution. Hum. Brain Mapp. 32 (3), 461–479.

Jeurissen, B., Leemans, A., Tournier, J.-D., Jones, D.K., Sijbers, J., 2013. Investigating the prevalence of complex fiber configurations in white matter tissue with diffusion magnetic resonance imaging. Hum. Brain Mapp. 34 (11), 2747–2766.

Jeurissen, B., Tournier, J.-D., Dhollander, T., Connelly, A., Sijbers, J., 2014. Multi-tissue constrained spherical deconvolution for improved analysis of multi-shell diffusion MRI data. NeuroImage 103, 411–426.

Jeurissen, B., Descoteaux, M., Mori, S., Leemans, A., 2019. Diffusion MRI fiber tractography of the brain. NMR Biomed. 32 (4), e3785.

Jian, B., Vemuri, B.C., 2007. A unified computational framework for deconvolution to reconstruct multiple fibers from diffusion weighted MRI. IEEE Trans. Med. Imaging 26 (11), 1464–1471.

Jillings, S., Van Ombergen, A., Tomilovskaya, E., Rumshiskaya, A., Litvinova, L., Nosikova, I., Pechenkova, E., Rukavishnikov, I., Kozlovskaya, I.B., Manko, O., Danilichev, S., Sunaert, S., Parizel, P.M., Sinitsyn, V., Petrovichev, V., Laureys, S., Zu Eulenburg, P., Sijbers, J., Wuyts, F.L., Jeurissen, B., 2020. Macro- and microstructural changes in cosmonauts' brains after long-duration spaceflight. Sci. Adv. 6 (36). https://doi.org/10.1126/sciadv.aaz9488.

Karan, P., Reymbaut, A., Gilbert, G., Descoteaux, M., 2022. Bridging the gap between constrained spherical deconvolution and diffusional variance decomposition via tensor-valued diffusion MRI. Med. Image Anal. 79, 102476.

Kunz, N., da Silva, A.R., Jelescu, I.O., 2018. Intra- and extra-axonal axial diffusivities in the white matter: which one is faster? NeuroImage 181, 314–322.

Kuo, L.-W., Chen, J.-H., Wedeen, V.J., Tseng, W.-Y.I., 2008. Optimization of diffusion spectrum imaging and q-ball imaging on clinical MRI system. NeuroImage 41 (1), 7–18.

Lampinen, B., Szczepankiewicz, F., Mårtensson, J., van Westen, D., Sundgren, P.C., Nilsson, M., 2017. Neurite density imaging versus imaging of microscopic anisotropy in diffusion MRI: a model comparison using spherical tensor encoding. NeuroImage 147, 517–531.

Lampinen, B., Szczepankiewicz, F., Mårtensson, J., van Westen, D., Hansson, O., Westin, C.-F., Nilsson, M., 2020. Towards unconstrained compartment modeling in white matter using diffusion-relaxation MRI with tensor-valued diffusion encoding. Magn. Reson. Med. 84 (3), 1605–1623.

Lasič, S., Szczepankiewicz, F., Eriksson, S., Nilsson, M., Topgaard, D., 2014. Microanisotropy imaging: quantification of microscopic diffusion anisotropy and orientational order parameter by diffusion MRI with magic-angle spinning of the q-vector. Front. Phys. 2, 11.

Lazar, M., Jensen, J.H., Xuan, L., Helpern, J.A., 2008. Estimation of the orientation distribution function from diffusional kurtosis imaging. Magn. Reson. Med. 60 (4), 774–781.

Mito, R., Dhollander, T., Xia, Y., Raffelt, D., Salvado, O., Churilov, L., Rowe, C.C., Brodtmann, A., Villemagne, V.L., Connelly, A., 2020. In vivo microstructural heterogeneity of white matter lesions in healthy elderly and Alzheimer's disease participants using tissue compositional analysis of diffusion MRI data. NeuroImage Clin. 28, 102479.

Morez, J., Sijbers, J., Vanhevel, F., Jeurissen, B., 2021. Constrained spherical deconvolution of nonspherically sampled diffusion MRI data. Hum. Brain Mapp. 42 (2), 521–538.

Morez, J., Szczepankiewicz, F., den Dekker, A.J., Vanhevel, F., Sijbers, J., Jeurissen, B., 2023. Optimal experimental design and estimation for q-space trajectory imaging. Hum. Brain Mapp. 44 (4), 1793–1809.

Mori, S., van Zijl, P.C., 1995. Diffusion weighting by the trace of the diffusion tensor within a single scan. Magn. Reson. Med. 33 (1), 41–52.

Mori, S., Crain, B.J., Chacko, V.P., van Zijl, P.C., 1999. Three-dimensional tracking of axonal projections in the brain by magnetic resonance imaging. Ann. Neurol. 45 (2), 265–269.

Ning, L., Bonet-Carne, E., Grussu, F., Sepehrband, F., Kaden, E., Veraart, J., Blumberg, S.B., Khoo, C.S., Palombo, M., Kokkinos, I., Alexander, D.C., Coll-Font, J., Scherrer, B., Warfield, S.K., Karayumak, S.C., Rathi, Y., Koppers, S., Weninger, L., Ebert, J., et al., 2020. Cross-scanner and cross-protocol multi-shell diffusion MRI data harmonization: algorithms and results. NeuroImage 221, 117128.

Novikov, D.S., Kiselev, V.G., Jespersen, S.N., 2018. On modeling. Magn. Reson. Med. 79 (6), 3172–3193.

Panagiotaki, E., Schneider, T., Siow, B., Hall, M.G., Lythgoe, M.F., Alexander, D.C., 2012. Compartment models of the diffusion MR signal in brain white matter: a taxonomy and comparison. NeuroImage 59 (3), 2241–2254.

Pietsch, M., Christiaens, D., Hutter, J., Cordero-Grande, L., Price, A.N., Hughes, E., Edwards, A.D., Hajnal, J.V., Counsell, S.J., Tournier, J.-D., 2019. A framework for multi-component analysis of diffusion MRI data over the neonatal period. NeuroImage 186, 321–337.

Pinto, M.S., Paolella, R., Billiet, T., Van Dyck, P., Guns, P.-J., Jeurissen, B., Ribbens, A., den Dekker, A.J., Sijbers, J., 2020. Harmonization of brain diffusion MRI: concepts and methods. Front. Neurosci. 14, 396.

Reisert, M., Kiselev, V.G., Dihtal, B., Kellner, E., Novikov, D.S., 2014. MesoFT: unifying diffusion modelling and fiber tracking. In: Medical Image Computing and Computer-Assisted Intervention: MICCAI ... International Conference on Medical Image Computing and Computer-Assisted Intervention, 17 (Pt 3), pp. 201–208.

Reisert, M., Kellner, E., Dhital, B., Hennig, J., Kiselev, V.G., 2017. Disentangling micro from mesostructure by diffusion MRI: a Bayesian approach. NeuroImage 147, 964–975.

Reisert, M., Kiselev, V.G., Dhital, B., 2019. A unique analytical solution of the white matter standard model using linear and planar encodings. Magn. Reson. Med. 81 (6), 3819–3825.

Reisert, M., Skibbe, H., Kiselev, V.G., 2014. The diffusion dictionary in the human brain is short: rotation invariant learning of basis functions. In: Schultz, T., Nedjati-Gilani, G., Venkataraman, A., O'Donnell, L., Panagiotaki, E. (Eds.), Computational Diffusion MRI and Brain Connectivity. Mathematics and Visualization. Springer, Cham. https://doi.org/10.1007/978-3-319-02475-2_5.

Sherbondy, A.J., Rowe, M.C., Alexander, D.C., 2010. MicroTrack: an algorithm for concurrent projectome and microstructure estimation. In: Medical Image Computing and Computer-Assisted Intervention: MICCAI ... International Conference on Medical Image Computing and Computer-Assisted Intervention, 13 (Pt 1), pp. 183–190.

Smith, R.E., Tournier, J.-D., Calamante, F., Connelly, A., 2013. SIFT: spherical-deconvolution informed filtering of tractograms. NeuroImage 67, 298–312.

Smith, R.E., Tournier, J.-D., Calamante, F., Connelly, A., 2015. SIFT2: enabling dense quantitative assessment of brain white matter connectivity using streamlines tractography. NeuroImage 119, 338–351.

Stejskal, E.O., Tanner, J.E., 1965. Spin diffusion measurements: spin echoes in the presence of a time-dependent field gradient. J. Chem. Phys. 42 (1), 288–292.

Szczepankiewicz, F., Lasič, S., van Westen, D., Sundgren, P.C., Englund, E., Westin, C.-F., Ståhlberg, F., Lätt, J., Topgaard, D., Nilsson, M., 2015. Quantification of microscopic diffusion anisotropy disentangles effects of orientation dispersion from microstructure: applications in healthy volunteers and in brain tumors. NeuroImage 104, 241–252.

Szczepankiewicz, F., Westin, C.-F., Nilsson, M., 2021. Gradient waveform design for tensor-valued encoding in diffusion MRI. J. Neurosci. Methods 348, 109007.

Tax, C.M.W., Jeurissen, B., Vos, S.B., Viergever, M.A., Leemans, A., 2014. Recursive calibration of the fiber response function for spherical deconvolution of diffusion MRI data. NeuroImage 86, 67–80.

Topgaard, D., 2017. Multidimensional diffusion MRI. J. Magn. Reson. 275, 98–113.

Tournier, J.-D., Calamante, F., Gadian, D.G., Connelly, A., 2004. Direct estimation of the fiber orientation density function from diffusion-weighted MRI data using spherical deconvolution. NeuroImage 23 (3), 1176–1185.

Tournier, J.-D., Calamante, F., Connelly, A., 2007. Robust determination of the fibre orientation distribution in diffusion MRI: non-negativity constrained super-resolved spherical deconvolution. NeuroImage 35 (4), 1459–1472.

Tournier, J.-D., Yeh, C.-H., Calamante, F., Cho, K.-H., Connelly, A., Lin, C.-P., 2008. Resolving crossing fibres using constrained spherical deconvolution: validation using diffusion-weighted imaging phantom data. NeuroImage 42 (2), 617–625.

Tournier, J.-D., Calamante, F., Connelly, A., 2013. Determination of the appropriate b value and number of gradient directions for high-angular-resolution diffusion-weighted imaging. NMR Biomed. 26 (12), 1775–1786.

Tuch, D.S., 2004. Q-ball imaging. Magn. Reson. Med. 52 (6), 1358–1372.

Tuch, D.S., Reese, T.G., Wiegell, M.R., Makris, N., Belliveau, J.W., Wedeen, V.J., 2002. High angular resolution diffusion imaging reveals intravoxel white matter fiber heterogeneity. Magn. Reson. Med. 48 (4), 577–582.

Tuch, D.S., Reese, T.G., Wiegell, M.R., Wedeen, V.J., 2003. Diffusion MRI of complex neural architecture. Neuron 40 (5), 885–895.

Veraart, J., Nunes, D., Rudrapatna, U., Fieremans, E., Jones, D.K., Novikov, D.S., Shemesh, N., 2020. Noninvasive quantification of axon radii using diffusion MRI. eLife 9, e49855.

Wakana, S., Jiang, H., Nagae-Poetscher, L.M., van Zijl, P.C.M., Mori, S., 2004. Fiber tract–based atlas of human white matter anatomy. Radiology 230 (1), 77–87.

Wedeen, V.J., Hagmann, P., Tseng, W.-Y.I., Reese, T.G., Weisskoff, R.M., 2005. Mapping complex tissue architecture with diffusion spectrum magnetic resonance imaging. Magn. Reson. Med. 54 (6), 1377–1386.

Wedeen, V.J., Wang, R.P., Schmahmann, J.D., Benner, T., Tseng, W.Y.I., Dai, G., Pandya, D.N., Hagmann, P., D'Arceuil, H., de Crespigny, A.J., 2008. Diffusion spectrum magnetic resonance imaging (DSI) tractography of crossing fibers. NeuroImage 41 (4), 1267–1277.

Wedeen, V.J., Rosene, D.L., Wang, R., Dai, G., Mortazavi, F., Hagmann, P., Kaas, J.H., Tseng, W.-Y.I., 2012. The geometric structure of the brain fiber pathways. Science 335 (6076), 1628–1634.

Westin, C.-F., Knutsson, H., Pasternak, O., Szczepankiewicz, F., Özarslan, E., van Westen, D., Mattisson, C., Bogren, M., O'Donnell, L.J., Kubicki, M., Topgaard, D., Nilsson, M., 2016. Q-space trajectory imaging for multidimensional diffusion MRI of the human brain. NeuroImage 135, 345–362.

Wong, E.C., Cox, R.W., Song, A.W., 1995. Optimized isotropic diffusion weighting. Magn. Reson. Med. 34 (2), 139–143.

Yeh, F.-C., Wedeen, V.J., Tseng, W.-Y.I., 2010. Generalized q-sampling imaging. IEEE Trans. Med. Imaging 29 (9), 1626–1635.

Zhan, W., Yang, Y., 2006. How accurately can the diffusion profiles indicate multiple fiber orientations? A study on general fiber crossings in diffusion MRI. J. Magn. Reson. 183 (2), 193–202.

Zhang, H., Schneider, T., Wheeler-Kingshott, C.A., Alexander, D.C., 2012. NODDI: practical in vivo neurite orientation dispersion and density imaging of the human brain. NeuroImage 61 (4), 1000–1016.

# Chapter 12

# From diffusion models to fiber orientations

Richard Stones[a,c], Maxime Descoteaux[b], and Flavio Dell'Acqua[a,c]

[a]NATBRAINLAB, Department of Neuroimaging, Institute of Psychiatry, Psychology and Neuroscience, King's College London, London, United Kingdom, [b]Sherbrooke Connectivity Imaging Laboratory (SCIL), Department of Computer Science, University of Sherbrooke, Sherbrooke, QC, Canada, [c]Department of Forensics and Neurodevelopmental Sciences, Institute of Psychiatry, Psychology and Neuroscience, King's College London, London, United Kingdom

## 1  Introduction

In the previous chapters, we have seen how diffusion MRI (dMRI) leverages the random motion of water molecules to probe the microstructural organization of biological tissues. Specifically, Chapters 10 and 11 have outlined the main diffusion models currently used in tractography applications. These models are commonly referred to as *local* because they rely exclusively on the diffusion signal at each voxel, and as a result, they provide information about the local microstructural characteristics. Crucially for tractography, these models enable the deduction of directional information about tissue structure from the anisotropic nature of the acquired diffusion signal. More specifically, they allow the extraction of the orientation of one or more fiber populations at each voxel, which can then be used as input to tractography algorithms. Currently, dMRI is the only practical in vivo and noninvasive imaging method for extracting orientational information about biological tissues, making it the primary modality for tractography.

The type of fiber orientation information inferred from these diffusion models typically falls into two main groups: a *discrete* set of fiber orientations or a *continuous* orientation distribution function (ODF). Discrete orientations are usually obtained from multicompartmental or *parametric* models that attempt to describe the diffusion process using a finite set of parameters for one or more fiber populations. Conversely, ODFs are generated from models that compute a *continuous* spherical function. This function encodes the directional distribution of fiber density or the angular information associated with the diffusion propagator.

While fiber orientations are the primary input to tractography algorithms, additional microstructural information derived from the diffusion signal can also enhance tractography reconstructions. Such information includes: anisotropy measures, fiber density, fiber dispersion, uncertainty estimates, and tissue compartment volume fractions. These can all be used in conjunction with more advanced tractography algorithms. This chapter will focus primarily on fiber orientations, exploring the main representations and operations applicable on them. Later chapters in this book will broaden the scope by exploring the use of complementary metrics and additional anatomical constraints to further improve tractography (see Chapters 14, 16, and 18).

## 2  Discrete fiber populations

As described in previous chapters, diffusion tensor imaging was one of the earliest attempts to infer the orientation of white matter fibers from the anisotropic character of the diffusion signal (Basser et al., 1994a,b). Briefly, the diffusion tensor $D$ approximates the displacement of water molecules with a trivariate Gaussian distribution. The tensor itself can be diagonalized and decomposed into three eigenvectors $v_1, v_2, v_3$ defining the principal axes of the tensor at each voxel, and three eigenvalues $\lambda_1, \lambda_2, \lambda_3$ that provide the effective diffusion coefficients along each axis. A visualization of the diffusion tensor is shown in Fig. 1A. The orientation of the eigenvector corresponding to the largest eigenvalue is often assumed to align with the dominant fiber orientation. This assumption informed many early tractography algorithms (Conturo et al., 1999; Mori et al., 1999; Basser et al., 2000; Catani et al., 2002), which often incorporated other metrics derived from the diffusion tensor. For instance, fractional anisotropy (FA) is often used to define seeding or stopping criteria during tracking.

The diffusion tensor model is conceptually easy to understand, quick to compute, and is sensitive to microstructural changes in white matter. However, its simplicity means that it averages out multiple fiber populations within each voxel leading to the well-known challenge of tracking through regions with crossing fibers. Considering that up to 90% of white matter voxels in a human brain may contain crossing fibers (Behrens et al., 2007; Descoteaux, 2008; Jeurissen et al., 2013;

---

Handbook of Diffusion MR Tractography. https://doi.org/10.1016/B978-0-12-818894-1.00035-5

Copyright © 2025 Elsevier Ltd. All rights are reserved, including those for text and data mining, AI training, and similar technologies.

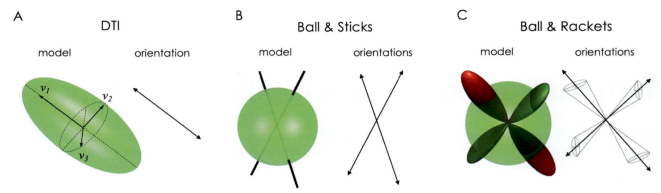

**FIG. 1** Example of discrete fiber orientations. (A) In the diffusion tensor model, the principal eigenvector, which represents the direction of the largest diffusivity, is assumed to be parallel to the main fiber orientation; (B) In the ball-and-sticks model, the orientations of two distinct fiber populations are captured by two stick-compartments; (C) In the ball-and-rackets model, each fiber population is modeled using a Bingham distribution capturing both orientation and dispersion.

Dell'Acqua et al., 2013; Volz et al., 2018) this is quite a significant limitation. Furthermore, changes in tensor metrics are not specific to particular microstructural properties of the dominant fiber population, such as fiber coherence, myelination, or axonal density. Secondary fiber populations or partial volume effects with gray matter and cerebrospinal fluid (CSF) also influence the diffusion tensor. For instance, FA, which provides only an average description of the voxel anisotropy, is known to decrease in the presence of complex fiber crossings, fanning configurations, or partial volume effects. In some brain regions it is not uncommon for tractography algorithms to terminate prematurely because FA values drop below typical stopping thresholds of FA (e.g., 0.15) even when still in deep white matter. This makes it difficult to reliably control the propagation of tractography algorithms using only information of the local diffusion tensor. One potential solution is to utilize white matter masks derived from structural MRI as a constraint for tractography, which can help to circumvent this issue. For more information, see Chapters 18 and 19.

To account for the presence of multiple fiber populations in white matter, *multitensor* approaches have been proposed as alternative local models to extract more than one fiber orientation per voxel. The general idea is to associate a separate diffusion tensor with each fiber population within a voxel, and, in some cases, additional isotropic components to model gray matter and CSF (Tuch et al., 2002; Basser and Jones, 2002; Pasternak et al., 2009). These approaches offer an intuitive way to resolve crossing fibers while maintaining the conceptual simplicity of the diffusion tensor model. However, an important limitation is the rapid increase in the number of model parameters with the addition of extra tensor components. This can result in poor model fitting, especially when modeling more than two fiber populations per voxel. The need to preselect the number of fiber populations fitted at each voxel, which may not match with the actual underlying white matter fiber organization, further complicates the fitting process and can introduce potential errors. These two factors have strongly limited the application of pure multitensor methods in tractography, but have promoted the development of simplified multitensor models with more stringent constraints on each compartment, which are more practical to use (Behrens et al., 2003; Assaf and Basser, 2005; Alexander, 2008).

One such approach that has been successfully used in tractography applications is the *ball-and-sticks model*, which consists of one isotropic compartment (the "ball") and one or more fiber populations modeled as fully anisotropic diffusion tensors (the "sticks") where $\lambda_1 = d, \lambda_2 = 0, \lambda_3 = 0$ (Behrens et al., 2003, 2007). See Fig. 1B for a visualization of the ball-and-sticks model. This approach significantly reduces the number of parameters in the model and has been shown to provide a stable fit for up to three fiber populations per voxel. Furthermore, the number of fiber populations in each voxel can also be robustly determined during the fitting process using an automatic relevance determination framework (Behrens et al., 2007). It is important to note that the ball-and-sticks model at its core is similar to a spherical convolution model where all fiber populations are described by the same signal profile or "stick" model. The main distinction is that this model provides only a discrete and limited number of fiber orientations, unlike the continuous ODF described in the following section. However, the Bayesian nature of the ball-and-sticks model does enable the quantification of uncertainty in the parameter estimation. An angular probability density function can be quantified for each fiber population based on its uncertainty estimate (Behrens and Jbabdi, 2009). These functions can then be combined to form an equivalent uncertainty ODF (or uODF) (Behrens and Jbabdi, 2009) or directly sampled at the time of tracking in bootstrapping algorithms (Berman et al., 2008; Haroon et al., 2009; Jeurissen et al., 2011). Given that both parametric models and those generating continuous ODFs can all, in principle, produce uODFs, the similarities and differences between uODFs and conventional ODFs are discussed later in this chapter. As described in more detail in Chapter 14, incorporating uncertainty is essential for several probabilistic tractography frameworks because it allows the implementation of tracking strategies in regions with high directional uncertainty and low anisotropy.

A further evolution of the *ball-and-sticks* model is the *ball-and-rackets* model, where fully anisotropic diffusion tensors are replaced with Bingham distributions (the "rackets"). These "rackets" are capable of capturing not just the orientation but also the anatomical dispersion of the estimated fiber populations (Sotiropoulos et al., 2012). As shown in Fig. 1C, the ball-and-rackets model is effectively a parametric representation of a fiber ODF, which is described in greater detail in the next section.

Even more sophisticated discrete multicompartmental models have been developed in the field of microstructure imaging in which complex analytical models for different diffusion compartments are used to explain the acquired signal (Assaf and Basser, 2005; Assaf et al., 2008; Zhang et al., 2012; Scherrer et al., 2016; Coronado-Leija et al., 2017; Palombo et al., 2020). However, these methods were developed with the objective of separating different microstructural environments and properties within each voxel (i.e., intra- and extra-axonal volume fractions, axonal radii, compartment diffusivities, etc.) rather than to accurately resolve fiber orientations. A more detailed review of these microstructural models can be found in (Panagiotaki et al., 2012; Novikov et al., 2018; Alexander et al., 2019).

# 3  Orientation distribution functions

Moving away from discrete fiber populations, the second method of representing fiber orientations is through the use of a continuous spherical function, often called an orientation distribution function (ODF). These functions, obtained from a different group of local models, are used to either describe the angular dependence of the diffusion process or to approximate the angular distribution of fiber density. Mathematically, they are represented as a function $f(\theta, \phi)$ defined on a unit sphere at each voxel, which maps orientational coordinates to a real number, i.e., $f: S^2 \to R$. The local maxima of the ODF can be associated with specific fiber populations, but since each ODF is a continuous function, it captures not only information about fiber orientations but also about their dispersion.

As discussed in the previous chapters, depending on whether the function describes a diffusion profile or the distribution of fiber density, the ODF is referred to as a *diffusion ODF* (dODF) or *fiber ODF* (fODF), respectively. Though similar in their definitions, the two types of ODFs represent conceptually different ways of characterizing the angular information contained in the dMRI signal.

## 3.1  Diffusion orientation distribution function

As shown in Chapter 11, it is theoretically possible to fully characterize the diffusion process at each voxel using the diffusion propagator $P(r, \theta, \phi, \Delta)$, which defines probability density of the water molecule displacement at distance $r$ along direction $(\theta, \phi)$ for diffusion time $\Delta$. The diffusion propagator can be approximated from the acquired diffusion MR signal using various methods such as q-space imaging (Callaghan et al., 1988), diffusion spectrum imaging (DSI) (Wedeen et al., 2005, 2008), q-ball imaging (QBI) (Tuch, 2004; Hess et al., 2006; Descoteaux et al., 2007), diffusion orientation transform (DOT) (Özarslan et al., 2006) and mean apparent propagator MRI (MAP-MRI) (Özarslan et al., 2013; Fick et al., 2016). These methods ultimately produce a dODF (see Fig. 2) containing the angular information of the water molecule displacement, obtained by projecting or approximating the radial part of the diffusion propagator as

$$dODF(\theta, \phi, \Delta) = \int_0^\infty P(r, \theta, \phi, \Delta) r^2 dr \tag{1}$$

A major advantage of using a dODF is that it generally doesn't require assumptions about the underlying white matter microstructure during reconstruction. Instead, assumptions are made later when deducing the spatial organization of the underlying white matter from the dODF. For instance, the maxima of the dODF are often interpreted as aligning with the directions of fiber populations. By identifying the local maxima of each dODF, it is possible to automatically extract the number of fiber populations, their directions and the relative amplitude as a proxy for the relative fiber density. In addition, quantitative diffusion metrics can be directly obtained from the dODF, such as the generalized fractional anisotropy (GFA)

**224 PART | II** Diffusion MRI

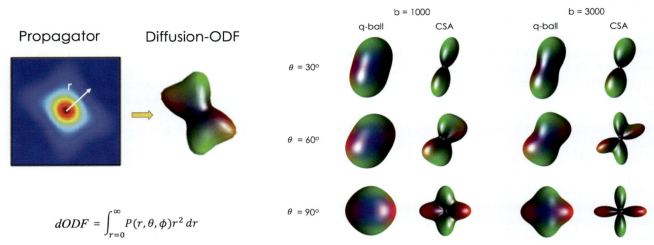

**FIG. 2** (Left) Schematic representation of the computation of the dODF profile from a propagator. (Right) Examples of dODFs calculated from a synthetic diffusion weighted imaging (DWI) signal model using the regularized q-ball method (Descoteaux et al., 2007) and the improved constant solid angle (CSA) approach (Aganj et al., 2010a) at various fiber crossing angles and $b$-values.

(Tuch, 2004), a scalar measure of anisotropy based on the dODF profile. Much like FA, GFA is also defined as ranging from zero (indicating total isotropy) to 1 (indicating maximum anisotropy). However, instead of being computed from three eigenvalues of a tensor, it is computed on the whole sphere. This approach allows GFA to capture more complex details of the intravoxel microstructural organization. GFA is computed as:

$$GFA = \frac{std(dODF)}{rms(dODF)} \quad (2)$$

where *std* is the standard deviation and *rms* is the root mean square of the dODF profile (Tuch, 2004). A common operation that is often performed on dODFs is so-called min-*max normalization*, which rescales the range of each dODF between its minimum and maximum values. This enhances the visualization of dODF peaks and normalizes all dODF amplitudes between 0 and 1. However, this process can amplify noisy dODF peaks in voxels with low or no anisotropy (e.g., gray matter and CSF regions). A more recently developed metric extracted from the dODF is the quantitative anisotropy (QA), which provides information about water spin density along individual dODF peaks (see Section 5.2 for more details about fiber specific metrics) and has been utilized to improve tractography propagation (Yeh et al., 2013).

Despite its advantages, the dODF often suffers from significant angular blurring. Even in the case of diffusion occurring only along the direction of a single fiber (as in the stick model), all diffusion gradients applied nonperpendicularly to this direction will still cause partial signal attenuation, leading to a broadened, peanut-like dODF profile instead of a sharp peak. This complicates the extraction of fiber orientation information, particularly for multiple fiber populations crossing at small angles, necessitating very high $b$-values (~4000s/mm$^2$) for acceptable angular resolutions. Recent advancements in both DSI (Canales-Rodríguez et al., 2010) and QBI (Aganj et al., 2010a; Tristán-Vega et al., 2010; Descoteaux et al., 2009) have improved angular blurring and the robustness to noise of dODFs, introducing new mathematical formulations based on spherical harmonics (SHs), as well as sharpening and deconvolution techniques. It has also been shown that modifying the expression of the dODF as described in Eq. (1) by increasing the exponent on r to higher values (e.g., $r^6$ or $r^8$) (Özarslan et al., 2013; Fick et al., 2016) enhances the dODF's sensitivity to the larger displacements or higher "radial moments" of the propagator, producing a sharper profile (Fig. 2).

Finally, dODFs have also been obtained using diffusion kurtosis imaging (DKI) (Jensen et al., 2005; Jensen and Helpern, 2010), an extension of the diffusion tensor model to better characterize the diffusion signal. This approach accounts for non-Gaussian diffusion contributions and therefore can also capture information about multiple fiber orientations. Fiber orientations may be obtained directly from the kurtosis tensor (Henriques et al., 2015) or from an approximation to the diffusion profile (Lazar et al., 2008; Jensen et al., 2014). However, because of the relatively limited number of parameters associated with the DKI model, dODFs extracted from DKI provide an even smoother representation of the underlying diffusion profile, which limits the ability to resolve small crossing angles even at large $b$-values.

## 3.2 Fiber orientation distribution function

The fODF, alternatively called the fiber orientation distribution (FOD), describes the angular distribution of fiber density on the unit sphere within each voxel. Local models producing an fODF as output can be associated with a spherical deconvolution framework, wherein the measured diffusion signal S in its simplest form can be expressed as:

$$S = fODF(\theta, \phi) \otimes R \tag{3}$$

where $R$ is a fiber response function that characterizes the signal from a voxel with a single idealized fiber population and $\otimes$ is the convolution operation. By deconvolving the measured signal S with the chosen response function, it is possible to compute an fODF. Since the fiber response accounts for the angular blurring of the diffusion signal, fODFs obtained through spherical deconvolution are generally sharper than the dODFs derived from direct modeling of the diffusion process. The higher angular resolution enables fODFs to more effectively resolve multiple crossing fibers within a voxel, especially at small crossing angles or at lower $b$-values (Fig. 3).

The fiber response function is assumed to be uniform throughout the entire brain. Although this assumption may appear simplistic, multiple studies have validated its biological plausibility. Data obtained from clinical scanners show that the profile of the fiber response doesn't change significantly for different fiber populations (Christiaens et al., 2020; Veraart et al., 2019), and in any case, errors in the fiber response have little effect on the estimated fiber orientations (Dell'Acqua and Tournier, 2019). More details can also be found in Section 5.2 and in Chapter 10.

Two common approaches for calculating the fODF are constrained spherical deconvolution (CSD) (Tournier et al., 2007) and damped Richardson-Lucy spherical deconvolution (Dell'Acqua et al., 2010). Recent advances have extended the spherical deconvolution framework to work with multishell diffusion data and accommodate multicompartmental models, including contributions from nonwhite matter compartments like gray matter and CSF (Jeurissen et al., 2014; Canales-Rodríguez et al., 2015; De Luca et al., 2020). Further developments to integrate multidimensional diffusion acquisitions have also been proposed (Jeurissen and Szczepankiewicz, 2021; Karan et al., 2022). Additionally, fODFs can be obtained by using a deconvolution operation directly on the dODF (Descoteaux et al., 2009). The term fODF will be consistently used throughout this chapter, regardless of the chosen spherical deconvolution framework, as all these techniques aim to reconstruct the ideal fODF profile representing the underlying white matter fiber density.

Similar to the dODF, the fODF profile can be used to extract the number of distinct fiber populations, their orientations, and the amplitude of each fODF lobe in each voxel. However, unlike the dODF, the absolute amplitude of fODF lobes also directly relates to the underlying fiber density and, as discussed in Section 5.2, it is useful to extract tract-specific metrics and to further assist tractography propagation.

An important aspect often overlooked is that the integral of each fODF profile over the unit sphere does *not* necessarily have to sum to one. This is due to the linearity of the convolution operation, where each fODF lobe is proportional to the signal fraction of the corresponding fiber population (i.e., fiber density). In scenarios where the total white matter signal does not account for the total signal of the voxel, such as in cases of partial volume with gray matter or CSF, the recovered fODF will not sum to one. Therefore it is advisable to interpret fODFs in terms of their absolute amplitudes, as these can

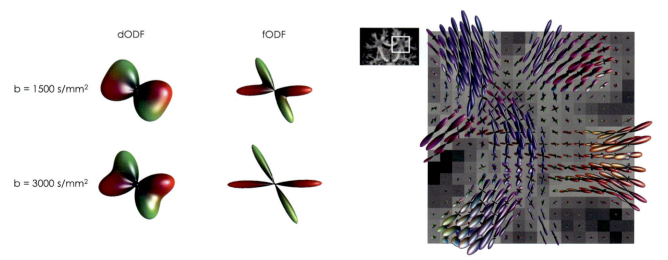

**FIG. 3** (Left) Comparison between dODF (q-ball CSA) and fODF (spherical deconvolution) computed from the same synthetic diffusion weighted imaging (DWI) data generated with different $b$-values. (Right) Large brain region with multiple crossing fibers resolved by fODFs.

more closely reflect an estimate of the underlying fiber density (Dell'Acqua and Tournier, 2019). Further details about quantitative metrics derived from the fODF are discussed later in this chapter.

## 4 ODF representation

There are several options available for the mathematical and computational representation of ODF profiles. The specific one chosen will often depend on the reconstruction method, but in general the different representations are interchangeable. The two most common representations for ODFs are either mesh and tessellation-based approaches (Tuch, 2004) or a SHs basis expansion (Frank, 2002; Alexander et al., 2002; Tournier et al., 2004).

Mesh and tessellation-based approaches, as shown in Fig. 4, are simple to implement as they directly encode the value of the ODF along a fixed number of directions or vertices at each voxel. Vertices are defined based on an approximate uniform spherical sampling, either from high-order icosahedral tessellation schemes or electrostatic repulsion strategies (Tuch, 2004; Jones, 2004; Koay, 2011), with the number of vertices typically ranging from 150 to around 800. While this representation offers some advantages in terms of efficient data vectorization and visualization, it also tends to require large amounts of memory and storage. A compromise must therefore be made between the angular resolution (or angular sampling density) of the estimated ODF and available computational resources, which may then impact the accuracy of local maxima estimates and corresponding fiber orientations. Fortunately, interpolation methods using splines, radial basis functions, or SHs can be used to upsample ODF profiles after reconstruction to a much higher number of vertices (e.g., 10,000 vertices and more) allowing precise estimation of local maxima and high-quality visualization (Tuch, 2004; Renard et al., 2010; Alaya et al., 2022; Appendix D).

SHs are another common method of encoding and representing ODF profiles, often used as part of local model reconstruction, as in analytical q-ball estimation (Hess et al., 2006; Descoteaux et al., 2007) or CSD (Tournier et al., 2004, 2007; Descoteaux et al., 2009). SH provides a compact, orthogonal set of basis functions with useful analytic properties that allow efficient computation. For instance, symmetry in the ODF profile can be enforced by selecting only even-order SHs. Additionally, by controlling the maximum order of SHs used in reconstruction, one can regularize and control noise instabilities, thereby producing smoother ODF profiles (Frank, 2002; Alexander et al., 2002; Descoteaux et al., 2006). However, one important drawback of using SHs is their susceptibility to ringing artifacts and the generation of spurious peaks introduced through truncation of the basis expansion. To mitigate this, additional regularization or filtering may be needed (Descoteaux et al., 2009; Deslauriers-Gauthier et al., 2016; Appendix D). In general, the choice of the SH order used to characterize ODF profiles depends on the reconstructed spherical function. Due to their smoothness, dODFs are often well characterized with SH up to order $l=6$ corresponding to just 28 SH coefficients (using a real symmetric SH basis).

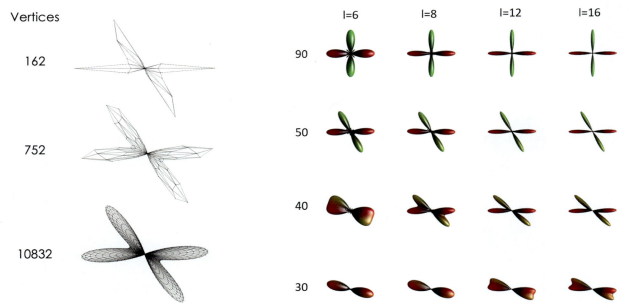

**FIG. 4** (Left) ODF representation using meshes with an increasing number of vertices. (Right) ODF representation using spherical harmonics at different orders. Higher orders can encode sharper ODF profiles, helping to resolve smaller crossing angles.

However, for fODF, while orders up to $l = 8$ are typically adopted in many applications (Tournier et al., 2007), higher orders, up to $l = 12$ (91 coefficients) or even $l = 16$ (153 coefficients), may be needed to fully capture the angular information that can be recovered by spherical deconvolution algorithms, as shown in Fig. 4.

In practice, SH and mesh/tessellation representations can be easily interchanged through simple matrix multiplications (see Appendix D). For example, an fODF can be initially estimated over a finite set of vertices with high enough sampling density to guarantee a complete SH encoding. The fODF data can then be saved compactly as SH coefficients. Finally, the SH representation can be converted back to a mesh representation at different resolutions for analysis or visualization of the fODFs.

# 5 Extracting information from the ODF

To generate streamlines from local diffusion models, we need to extract parameters that will be used as input to tracking algorithms. These parameters will determine the direction of the tracking at each step and also provide useful stopping criteria. Moreover, some of these parameters are valuable for understanding neurodevelopmental or pathological effects on white matter microstructure and may be mapped along streamlines. This section reviews the different types of information that can be extracted from the ODF.

## 5.1 Local fiber orientation

The most important feature of local diffusion models for tractography is the ability to capture orientational information about the underlying white matter fiber populations. The simplest representation for fiber orientation is that of unit vectors pointing along the directions of local maxima of each ODF. These vectors are analogous to the principal eigenvector in diffusion tensor imaging (DTI) or the orientations of the different fiber populations from multi-tensor models. This allows similar, if not identical, tractography algorithms to be applied to fiber orientations derived from different reconstruction methods.

While the diffusion tensor, ball-and-sticks model, and other multitensor approaches estimate the local fiber orientation information as part of the fitted parameters, ODF profiles do not directly provide individual fiber population orientations. The local maxima of the ODF must be identified and the orientations associated with distinct fiber populations. There are several options available for extracting local maxima depending on the ODF representation. For SHs, ODFs can be analyzed using a Newton search method (Jeurissen et al., 2013). Multiple restarts are needed with this approach to ensure full coverage of the solution space because the number of fiber populations is not known a priori for each voxel. Analytic solutions to directly extract local maxima from ODFs have also been proposed for SH representations (Aganj et al., 2010b). Alternatively, when using a mesh or tessellation-based representation, local maxima can be quickly identified by exploring all ODF vertices. However, due to the discrete angular sampling offered by this representation, estimated local maxima orientations are only as accurate as the ODF sampling density. Nevertheless, the local maxima estimates can be refined by upsampling or converting to a SH representation of the ODF and using the Newton search or equivalent method only in the vicinity of the maxima (Riffert et al., 2014). All these approaches generally also require setting a threshold on the amplitudes of each maxima in order to filter out spurious peaks introduced by signal noise or truncation of the SH basis expansion. It is worth mentioning that in some probabilistic tractography approaches, the local maxima are not needed by the tractography algorithm but, instead, the entire fODF profile is used as input for tracking. In this case, the overall shape of each fODF lobe provides a distribution of possible directions that can be sampled by the tractography algorithm. For more details, see Chapter 14 on probabilistic tractography.

## 5.2 Extracting fiber-specific information and fixels

As previously mentioned and also as discussed in Chapter 10, the fODF profile is not only useful for capturing the orientation of each fiber population and for resolving crossing fibers, but also, due to its definition over a continuous spherical function, it allows the extraction of more detailed information about each fiber population.

Due to the linearity of the convolution operation, the absolute amplitude of each fODF lobe is directly related to the diffusion signal associated with each fiber population. When assuming a convolution model with a uniform fiber response, this amplitude is associated with an estimate of the underlying fiber density. However, it is important to note that, in practice, the amplitude can be influenced by both the diffusion properties and the density of each fiber population. This has led to the idea of "Apparent Fiber Density" (AFD), originally suggested by Dell'Acqua et al. (2010) and later developed by Raffelt et al. (2012). The main realization is that within a spherical deconvolution framework using fODFs, it is possible to develop fiber-specific rather than voxel-wise metrics, providing a description of the microstructural organization of individual fiber populations and better associating changes or pathologies along individual fibers.

Building on this concept, multiple indices have been developed to practically extract this information from the fODF. The AFD, as proposed by Raffelt et al. (2012), measures the absolute amplitude of the fODF profile along any given direction; as such, it is sensitive to changes in fiber population volume fraction or to deviations of the actual response function. An analogous index to AFD is the hindrance modulated orientational anisotropy (HMOA), similarly defined as the absolute amplitude of the local maxima of each fODF lobe. This index intentionally avoids associating changes in fODFs to only fiber density (Dell'Acqua et al., 2013). Similarly to how FA is used with DTI, HMOA offers a more general and flexible terminology for describing changes and complex pathologies in white matter when multiple factors may be involved beyond just fiber density. In practice, however, these two metrics can be considered equivalent even though they may be scaled differently. Consequently, results from these two metrics in the literature should be regarded as interchangeable (Dell'Acqua and Tournier, 2019).

In a similar fashion, Riffert et al. (2014) introduced metrics like angular fiber density (also referred to as AFD) and fiber density (FD), the latter being defined as the integral of each fODF lobe parameterized as a Bingham distribution over the sphere. Interestingly, this integral can be directly computed by extracting the first coefficient of the SH basis expansion of the fODF lobe itself, making its computation very fast once parameterized as a Bingham distribution (Riffert et al., 2014; Raffelt et al., 2017). Overall, the use of fiber-specific measures instead of traditional voxel-based metrics like FA has shown to be effective in improving the specificity of mapping white matter alterations or differences along selected tracts rather than broad brain regions (Mirchandani et al., 2021; Lautarescu et al., 2022; Guberman et al., 2022). Fiber-specific metrics have also been shown to improve the general consistency and reliability of tracking and stopping criteria in tractography applications, and to serve as an effective threshold to mask spurious peaks on fODF (Dell'Acqua et al., 2013; Riffert et al., 2014).

It is important to highlight that, regardless of the metric adopted, unless a multishell spherical deconvolution approach is used (Jeurissen et al., 2013; Chapter 11), it is only at reasonably high $b$-values that the signal contribution from extra-axonal diffusion or CSF are significantly ($b > 3000\,\text{s/mm}^2$) or completely suppressed ($b > 6000\,\text{s/mm}^2$) leaving the fODF dependent primarily on the intra-axonal signal (Raffelt et al., 2012, 2015; Veraart et al., 2019). Additionally, because conventional pulse sequences on clinical scanners are not sensitive to the majority of axonal diameters, this makes the signal from each axon practically equivalent to a stick model in most cases (Dell'Acqua et al., 2013; Veraart et al., 2019). These results not only justify the widespread use of a uniform fiber response in spherical deconvolution frameworks, but they also facilitate, under these specific conditions, a straightforward association of fODF amplitudes with the underlying volume fraction of intra-axonal water. However, this assumption is valid only for perfectly impermeable axons (within the diffusion time limit of the MR pulse sequence). In the presence of pathology (such as axonal degeneration or demyelination) or white matter maturation (e.g., partial myelination), variations in permeability or other changes can still affect the diffusion properties within each fiber population, consequently influencing the obtained fODF amplitude.

A more recent concept, directly derived from the previous fiber-specific metrics, is the *fixel*. Unlike conventional neuroimaging data, which consider voxel-wise information, the fixel framework associates the information of each fiber population to its own fixel. A fixel is defined as the local maxima of the fODF (Raffelt et al., 2015, 2017). Together with the local fiber orientation, each fixel can also include local measures specific to the associated fiber population, such as estimates of FD, dispersion, and fiber cross-section (Fig. 5) (Raffelt et al., 2017).

The fixel concept has been successfully coupled with a statistical framework allowing for new fiber-specific or fixel-based analysis, in contrast to traditional voxel-wise analysis. When combined with an fODF-driven registration approach, this effectively allows for the comparison of volume fractions of distinct fiber populations within the same voxel across multiple subjects. This framework has opened the door to new analyses, enabling the investigation of white matter differences along specific tracts (Dhollander et al., 2021; Raffelt et al., 2015, 2017).

## 5.3 Fiber dispersion and structural complexity

Models of fiber bundles running in parallel straight lines through a voxel are only idealized pictures of real white matter microstructure. Compared to traditional DTI and other parametric methods returning discrete sets of fiber orientations, ODFs, and in particular fODFs, are better equipped to capture more complex white matter organization. Specifically, fODFs can be used to characterize dispersion within white matter in fanning and converging configurations. Dispersion has been quantified using microstructure imaging methods like neurite orientation density and distribution imaging (NODDI) (Zhang et al., 2012; Tariq et al., 2016), where a Bingham distribution is fit to

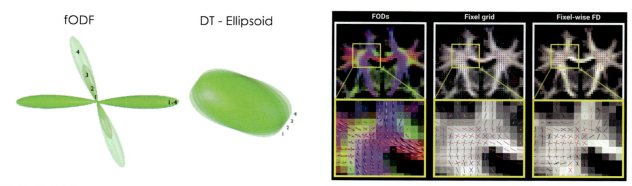

**FIG. 5** (Left) Fiber orientation distribution and diffusion tensor reconstructed using different fiber properties. (Right) Visualization of fixels showing both fiber orientation and fiber-specific metrics. *(Modified from Dhollander T., Clemente A., Singh M., Boonstra F., Civier O., Duque J.D., Egorova N., Enticott P., Fuelscher I., Gajamange S., Genc S., Gottlieb E., Hyde C., Imms P., Kelly C., Kirkovski M., Kolbe S., Liang X., Malhotra A., Fixel-based analysis of diffusion MRI: methods, applications, challenges and opportunities NeuroImage 2021 118417. https://doi.org/10.1016/j.neuroimage.2021.118417.)*

individual fiber populations. Similarly, the ball-and-rackets model (Sotiropoulos et al., 2012) uses Bingham distributions to characterize the dispersion of multiple fiber populations within the same voxel. While these models introduce Bingham distributions as part of the general model fitting, the task of fitting Bingham parameters becomes largely simplified in the presence of already computed fODFs. A large number of data points are already available for each fODF lobe, making the fitting of each Bingham distribution stable and well-posed. Each lobe can be quantified independently with a fast linear fit (Riffert et al., 2014) though nonlinear multi-Bingham fitting has also been proposed to better account for the mixing effect of overlapping fODF lobes (Henriques et al., 2016). Among the five parameters associated with the Bingham distribution (see Bingham Distribution text-box), extracting the two concentration parameters for each fODF lobe can provide a measure of its dispersion along two perpendicular axes. This approach provides a way of quantifying and comparing fiber dispersion along tracts in different brain regions or among different brains. Furthermore, each fODF lobe may show either rotationally symmetric dispersion around the fiber orientation or exhibit a flattened profile, i.e., anisotropic dispersion, along a direction perpendicular to the fiber. In this case, the two concentration parameters of the Bingham distribution can capture and provide information about these types of dispersions and can be used to directly quantify the degree of anisotropy of the dispersion (Riffert et al., 2014).

---

**Bingham distribution**

The Bingham distribution is a bivariate Gaussian distribution defined on the unit sphere (Bingham, 1975). Like standard SH and mesh-based representations of fODFs it is antipodally symmetric, making it useful as a potential representation of individual fODF lobes or fiber populations. It can be expressed in the form

$$f(\hat{\mu}) = \frac{1}{F_{k_1 k_2}} \exp\left(-k_1(\mu_1 \cdot \hat{\mu})^2 - k_2(\mu_2 \cdot \hat{\mu})^2\right)$$

where $\hat{\mu}$ is a point on the unit sphere and $F_{k_1 k_2}$ is a confluent hypergeometric function of matrix argument that normalizes the distribution. The concentration parameters $k_1$ and $k_2$ control the width of the distribution along the dispersion directions $\mu_1$ and $\mu_2$, respectively, which are orthogonal to each other and to the main orientation axis of the distribution $\mu'$, pointing, in our case, along the direction of the maximum of the fODF lobe (Fig. 6). The Bingham distribution therefore has only five free parameters: two defining the shape of the distribution and three defining the orientation (when expressed using Euler angles). The Watson distribution is a special case where $k_1 = k_2$ and the distribution becomes rotationally symmetric about the main orientation axis. The Bingham distribution has already found application in the analysis of dMRI data for modeling fiber dispersion (Kaden et al., 2007; Sotiropoulos et al., 2012; Zhang et al., 2012; Riffert et al., 2014; Tariq et al., 2016).

*Continued*

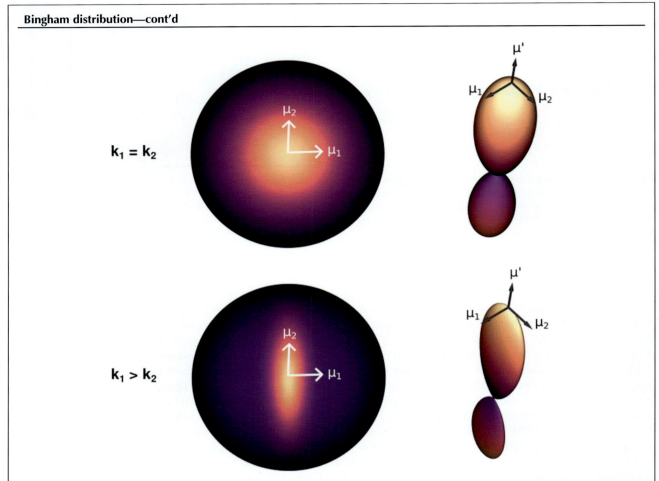

**FIG. 6** Example of Bingham distributions mapped on the unit sphere and represented as orientation distribution functions. (Top) Bingham distributions with equal concentration parameters are equivalent to Watson distributions; (Bottom) Bingham distributions with different concentration parameters capture a wider variety of fiber dispersion patterns, offering better flexibility in modeling different types of fiber configurations.

It is important to note that, while in theory the dispersion of fODF lobes should only be related to the actual dispersion of the underlying white matter, in reality the final sharpness of the fODF profile also depends on other factors. The dMRI acquisition parameters (e.g., higher $b$-values give sharper fODFs), the choice of response function and the regularization used by the spherical deconvolution algorithm all contribute to the final sharpness or smoothness of the final fODF profile. Even in the presence of perfectly straight fibers, typically estimated fODFs always present a minimum apparent dispersion of approximately 10–15 degrees (Henriques et al., 2016). These factors should be considered when interpreting dispersion results or using dispersion in a tractography algorithm.

Nevertheless, given that each fODF lobe can be well approximated by a scaled Bingham distribution, this parametrization offers, in principle, a more compact representation than the SHs representation, as it requires only six parameters per lobe. These include a scaling parameter, as well as the five parameters of the normalized Bingham distribution needed to describe each fODF lobe. Importantly, these parameters can describe white matter microstructural properties, being directly related to orientation, dispersion, and fiber density measures (Riffert et al., 2014), and can provide a set of useful fiber population or tract-specific indices for tractography analyses.

It is important to mention that fODFs, while improving over previous methods describing white matter organization, still provide a simplified and incomplete description of white matter. Small crossing angles are not resolved and are indistinguishable from dispersion. Similarly, configurations like kissing or bending are also indistinguishable from crossing or fanning configurations (Fig. 7). This is due to the symmetric nature of the diffusion signal and the symmetry of fODFs. This is an open problem that can lead to issues when tracking through regions containing these configurations (see Chapter 28 for more details). To overcome these limitations, new modeling approaches are currently being investigated and are discussed in the following sections.

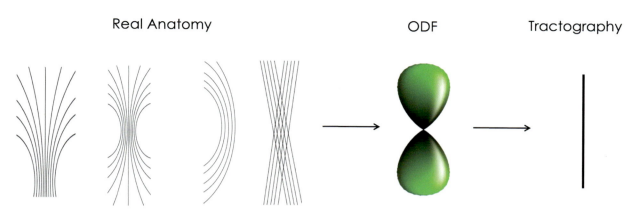

**FIG. 7** Limitation of ODFs. Fanning, kissing, bending, and crossing fibers that are not resolved can be all described by a similar ODF profile. In these situations ODFs provide only an incomplete and oversimplified description of the local white matter organization, potentially leading to incorrect tractography reconstructions.

Finally, another type of information that can be extracted from ODFs is a measure of the *local structural complexity* within a voxel. Measures like the GFA, as described in Section 3.2, provide a description of the overall ODF shape (Tuch, 2004). The total fiber density within a voxel can be directly obtained as the integral over the entire fODF profile or by simply extracting the first coefficient of its SH basis expansion (Raffelt et al., 2017; Calamante et al., 2015; Riffert et al., 2014). Riffert et al. (2014) introduced a complexity index (CX) based on how fiber density is distributed across distinct fODF lobes, ranging from $CX = 0$, when only a single fiber is present, to $CX = 1$, when all fibers within a voxel have the same fiber density. Another measure of structural complexity of white matter is related to the number of distinct fiber populations present within a voxel. This can be quantified as the number of fiber orientations (NuFO) at each voxel, after applying one or more thresholds to remove spurious fODF peaks (Dell'Acqua et al., 2013). Different authors have proposed and used similar measures, and depending on the local model adopted, it has been shown that the number of white matter voxels containing multiple fiber populations can range between 70% and 90% (Behrens et al., 2007; Descoteaux, 2008; Jeurissen et al., 2013; Dell'Acqua et al., 2013; Volz et al., 2018) with a characteristic spatial distribution across the brain. Notably, the corticospinal tract and the middle portion of the corpus callosum are the only regions where a single fiber population is clearly visible (see Fig. 8).

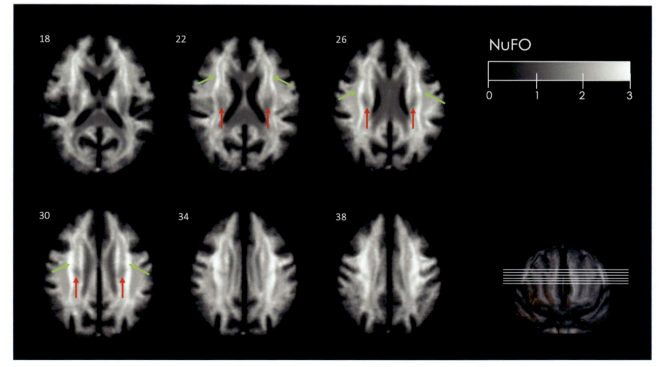

**FIG. 8** Average NuFO map showing brain regions with one, two, three, or more distinct fiber populations. *Red arrows* indicate the core of the corticospinal tract corresponding to a region with a distinct single orientation like in the body of the corpus callosum. *Green arrows* indicate regions with complex white matter organization with three or more fiber populations.

# 6 Additional orientation distribution functions

## 6.1 Uncertainty ODF

Until now we have focused our attention on the fODF and dODF. These spherical functions are the output of different local models and represent the starting point for extracting fiber orientation information to be used in tractography. However, as previously mentioned in earlier sections, another type of ODF also used in probabilistic tractography is the uncertainty ODF, or uODF.

While uODFs can be represented similarly to fODFs or dODFs, they encode conceptually different information. The uODF represents the degree of confidence in the orientation estimates and can be derived from different types of diffusion models. The computation of uODFs typically involves sampling the uncertainty of the local maxima of either fODF/dODF lobes or the uncertainty of discrete fiber orientations estimated by multicompartmental approaches, such as the ball-and-sticks or ball-and-rackets models. By applying bootstrapping or Monte Carlo methods to sample their uncertainty, each fiber orientation can then be represented by its own angular probability distribution, which captures the variability and confidence of each orientation. Combining the probability distribution for each fiber orientation into the same spherical function gives the uODF (Behrens and Jbabdi, 2009).

The resulting uODF profile usually represents highly reproducible orientations as sharp and tall peaks, concentrating high probability within a small solid angle. On the contrary, more uncertain or variable orientations will be encoded as broad uODF lobes spreading the probability of the fiber orientation over a larger area. Probabilistic tractography algorithms use this information to sample potential tracking direction based on this probability profile, along with other constraints like angle thresholds, limiting the set of possible directions, and other parameters. For more details, see Chapter 14 on probabilistic tractography.

## 6.2 Track orientation distribution function

A less common type of ODF profile is the track orientation distribution (TOD), which is generated as the angular distribution of streamlines passing through each voxel *after* fiber tracking (Dhollander et al., 2014). Because of this, it is not associated with a particular local model of diffusion or fiber response function. The TOD amplitude provides a measure of support for the existence of white matter fibers passing through a voxel in a given direction, incorporating information from neighboring voxels during the tracking process. Tractography performed on a TOD field can be used to iteratively generate further sets of TODs, resulting in increasingly smoother tractogram reconstructions. TODs have found application in bundle-specific tractography (BST) (Rheault et al., 2019), which is described in the following section on advanced processing of ODFs.

# 7 Advanced processing of ODFs

Beyond the standard methods mentioned so far, more advanced techniques exist for the enhancement and refinement of ODFs. These often involve using filtering techniques, voxel neighborhood information, or anatomical priors to spatially smooth and enhance the ODFs before performing tractography.

ODFs can be enhanced and sharpened postreconstruction using a variety of approaches. Sharpening of the ODF lobes can be applied using erosion techniques (Duits et al., 2013; Dela Haije et al., 2014). Alternatively, fODFs can be extracted from dODFs using further deconvolution steps (Descoteaux et al., 2009; Canales-Rodríguez et al., 2010) or by decomposing dODFs into sharp fODFs (Yeh and Tseng, 2013). Anatomical priors may also be introduced to create enhanced ODFs before fiber tracking. For example, in Rheault et al. (2019) a voxel-wise track orientation distribution (TOD) from a bundle-specific template is used to weight ODFs and reinforce lobes aligned with the bundle of interest (Fig. 9). This helps reconstruct tracts in sparse fanning regions compared to tracking along the original ODFs.

Voxel neighborhood information can also be used to modify ODFs, taking into account the underlying continuity of white matter fibers in the brain. One method is to use information from surrounding voxels to better estimate fiber orientations, which effectively acts as a spatial smoothing function on the ODF field, thus reducing noise. These methods can be applied after ODF reconstruction (Tax et al., 2014; Portegies et al., 2015) or through regularization applied during the reconstruction process (Canales-Rodríguez et al., 2015; Raj et al., 2011; Reisert and Kiselev, 2011).

Moreover, as we have seen in the previous section, while ODFs can capture complex white matter configurations, real white matter fiber populations may consist of even more complex arrangements, including asymmetric bending, fanning, or converging fiber configurations, which cannot be captured by the antipodal symmetry of the diffusion signal and standard

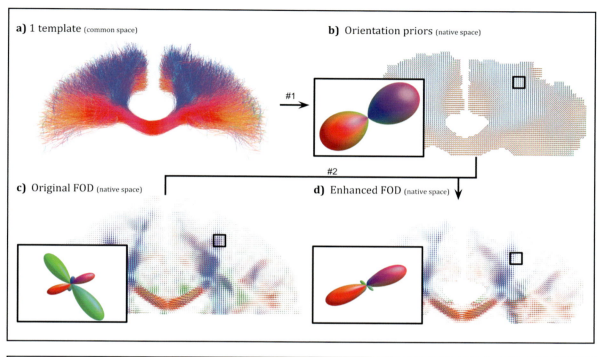

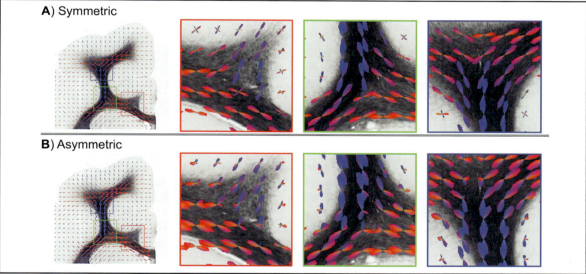

**FIG. 9** (Top) Enhanced FODs created from a tract template for performing bundle specific tractography. (Bottom) Symmetric and asymmetric FODs compared in three different brain regions. *((Top) Modified from Rheault, F., St-Onge, E., Sidhu, J., Maier-Hein, K., Tzourio-Mazoyer, N., Petit, L., Descoteaux, M., 2019. Bundle-specific tractography with incorporated anatomical and orientational priors. NeuroImage 186, 382–398. https://doi.org/10.1016/j.neuroimage.2018.11.018. (Bottom) Modified from Bastiani, M., Cottaar, M., Dikranian, K., Ghosh, A., Zhang, H., Alexander, D.C., Behrens, T.E., Jbabdi, S., Sotiropoulos, S.N., 2017. Improved tractography using asymmetric fibre orientation distributions. NeuroImage 158, 205–218. https://doi.org/10.1016/j.neuroimage.2017.06.050.)*

ODF models. Neighborhood information can once again be used to introduce asymmetry into fODF models (Fig. 9) (Bastiani et al., 2017; Karayumak et al., 2018; Poirier and Descoteaux, 2023; Reisert et al., 2012; Wu et al., 2020). This can be particularly important for modeling regions with high white matter curvature or for accurately describing the fanning of fibers within gyral blades (Wu et al., 2020). Asymmetry can be inferred by applying filters to a symmetric fODF field after reconstruction (Karayumak et al., 2018; Poirier and Descoteaux, 2023) or during the modeling process (Reisert et al., 2012; Bastiani et al., 2017; Wu et al., 2020). Although not many applications have been shown yet, tractography algorithms can, in principle, be modified to exploit the additional information provided by asymmetric fODFs and improve tracking in anatomically

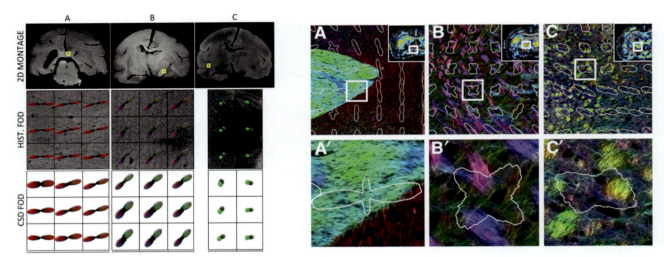

**FIG. 10** (Left) A comparison of 3D fODFs obtained from histological data, with fODFs calculated from diffusion MRI data, for three major white matter tracts in a monkey brain. (Right) 2D fiber orientation maps extracted using structural tensor analysis from histological sections. *((Left) Modified from Schilling, K., Janve, V., Gao, Y., Stepniewska, I., Landman, B.A., Anderson, A.W., 2016. Comparison of 3D orientation distribution functions measured with confocal microscopy and diffusion MRI. NeuroImage 129, 185–197. https://doi.org/10.1016/J.NEUROIMAGE.2016.01.022. (Right) Modified from Budde, M.D., Frank, J.A., 2012. Examining brain microstructure using structure tensor analysis of histological sections. NeuroImage 63 (1), 1–10. https://doi.org/10.1016/J.NEUROIMAGE.2012.06.042.)*

difficult regions (Bastiani et al., 2017). Similarly novel indices of white matter complexity can be extracted. Instead of NuFO maps based on symmetric fODF, the number of fiber directions (NuFiD) can be extracted from an asymmetric fOFD, where each lobe is counted independently and is not coupled to an equivalent antipodal direction (Poirier and Descoteaux, 2023).

## 8 fODFs from nondiffusion methods

Besides dMRI, white matter fODFs are also computed using other modalities. Histological analysis has long been used to study the brain, but recent methods have enabled quantitative comparison of high-resolution microscopy images of tissue slices with microstructural diffusion models and tractography. Fiber orientation information can be extracted from microscopy images using techniques such as manual tracing (Leergaard et al., 2010), Fourier analysis (Choe et al., 2012), or structure tensor analysis (STA) (Budde and Annese, 2013; Budde and Frank, 2012). In particular, STA has been found to work well across several microscopy modalities. It is a form of texture analysis that provides image orientation and anisotropy information at each pixel. The fODFs can be created from angular histograms of pixel orientations in piecewise regions, reflecting typical dMRI voxel sizes, and are then fit to circular (von Mises) or spherical (Bingham) probability distributions (Fig. 10). Initially applied to 2D images (Budde and Annese, 2013; Budde and Frank, 2012; Seehaus et al., 2015), STA has since been expanded to construct 3D fODFs from classical histology (Schilling et al., 2016) and advanced microscopy techniques like optical coherence tomography (Wang et al., 2015), polarized light imaging (Alimi et al., 2020), and CLARITY (Leuze et al., 2021). While these methods are only applicable for ex vivo brain tissue, they provide an important point of contact for the validation of information obtained with dMRI, at much higher resolutions than can be achieved with existing acquisition protocols and hardware (Fig. 10). For a more in-depth discussion of dMRI and tractography validation methods, refer to Chapter 26.

## 9 Conclusion

This chapter has provided a detailed overview of the various methodologies available today for extracting fiber orientation information from local diffusion models of white matter. We have covered how to obtain discrete fiber orientation estimates from the diffusion tensor or multitensor models, and how to extract continuous ODF from the diffusion propagator or spherical deconvolution approaches. This chapter has demonstrated how these methods not only facilitate the extraction of fiber orientations but also provide additional details that can be used to enhance tractography reconstructions, along-tract metrics, and information like fiber dispersion and model uncertainty. This chapter has also explored in detail the practical steps for manipulating, filtering, and visualizing fiber orientations or ODFs. In conclusion, the fiber orientation information is the main and indispensable input required by all tractography algorithms. By summarizing concepts universal to multiple diffusion models, this chapter has set the scene for the discussion on tractography algorithms that will be presented in the following chapters.

# References

Aganj, I., Lenglet, C., Sapiro, G., 2010a. ODF maxima extraction in spherical harmonic representation via analytical search space reduction. In: Lecture Notes in Computer Science (Including Subseries Lecture Notes in Artificial Intelligence and Lecture Notes in Bioinformatics).vol. 6362. pp. 84–91. https://doi.org/10.1007/978-3-642-15745-5_11.

Aganj, I., Lenglet, C., Sapiro, G., Yacoub, E., Ugurbil, K., Harel, N., 2010b. Reconstruction of the orientation distribution function in single- and multiple-shell q-ball imaging within constant solid angle. Magn. Reson. Med. 64 (2), 554–566. https://doi.org/10.1002/mrm.22365.

Alaya, I.B., Jribi, M., Ghorbel, F., Kraiem, T., 2022. Comparison analysis of local angular interpolation methods in diffusion MRI. Comput. Methods Biomech. Biomed. Eng. Imaging Vis. 10 (6), 687–696. https://doi.org/10.1080/21681163.2021.2024089.

Alexander, D.C., 2008. A general framework for experiment design in diffusion MRI and its application in measuring direct tissue-microstructure features. Magn. Reson. Med. 60 (2), 439–448. https://doi.org/10.1002/mrm.21646.

Alexander, D.C., Barker, G.J., Arridge, S.R., 2002. Detection and modeling of non-Gaussian apparent diffusion coefficient profiles in human brain data. Magn. Reson. Med. 48 (2), 331–340. https://doi.org/10.1002/mrm.10209.

Alexander, D.C., Dyrby, T.B., Nilsson, M., Zhang, H., 2019. Imaging brain microstructure with diffusion MRI: practicality and applications. NMR Biomed.. 32(4), e3841https://doi.org/10.1002/nbm.3841.

Alimi, A., Deslauriers-Gauthier, S., Matuschke, F., Müller, A., Muenzing, S.E.A., Axer, M., Deriche, R., 2020. Analytical and fast Fiber Orientation Distribution reconstruction in 3D-Polarized Light Imaging. Med. Image Anal. 65, 101760. https://doi.org/10.1016/J.MEDIA.2020.101760.

Assaf, Y., Basser, P.J., 2005. Composite hindered and restricted model of diffusion (CHARMED) MR imaging of the human brain. NeuroImage 27 (1), 48–58. https://doi.org/10.1016/j.neuroimage.2005.03.042.

Assaf, Y., Blumenfeld-Katzir, T., Yovel, Y., Basser, P.J., 2008. Axcaliber: a method for measuring axon diameter distribution from diffusion MRI. Magn. Reson. Med. 59 (6), 1347–1354. https://doi.org/10.1002/MRM.21577.

Basser, P.J., Jones, D.K., 2002. Diffusion-tensor MRI: theory, experimental design and data analysis—a technical review. In: NMR in Biomedicine. vol. 15. John Wiley & Sons, Ltd., pp. 456–467. https://doi.org/10.1002/nbm.783.

Basser, P.J., Mattiello, J., Bihan, D.L., 1994a. MR diffusion tensor spectroscopy and imaging. Biophys. J. 66 (1), 259–267. https://doi.org/10.1016/S0006-3495(94)80775-1.

Basser, P.J., Mattiello, J., Bihan, D.L., 1994b. Estimation of the effective self-diffusion tensor from the NMR spin echo. J. Magn. Reson. Ser. B 103 (3), 247–254. https://doi.org/10.1006/JMRB.1994.1037.

Basser, P.J., Pajevic, S., Pierpaoli, C., Duda, J., Aldroubi, A., 2000. In vivo fiber tractography using DT-MRI data. Magn. Reson. Med. 44 (4), 625–632. https://doi.org/10.1002/1522-2594(200010)44:4<625::AID-MRM17>3.0.CO;2-O.

Bastiani, M., Cottaar, M., Dikranian, K., Ghosh, A., Zhang, H., Alexander, D.C., Behrens, T.E., Jbabdi, S., Sotiropoulos, S.N., 2017. Improved tractography using asymmetric fibre orientation distributions. NeuroImage 158, 205–218. https://doi.org/10.1016/j.neuroimage.2017.06.050.

Behrens, T.E.J., Jbabdi, S., 2009. MR diffusion tractography. In: Johansen-Berg, H., Behrens, T.E.J. (Eds.), Diffusion MRI. Academic Press, pp. 333–351 (Chapter 15).

Behrens, T.E.J., Woolrich, M.W., Jenkinson, M., Johansen-Berg, H., Nunes, R.G., Clare, S., Matthews, P.M., Brady, J.M., Smith, S.M., 2003. Characterization and propagation of uncertainty in diffusion-weighted MR imaging. Magn. Reson. Med. 50, 1077–1088.

Behrens, T.E.J., Berg, H.J., Jbabdi, S., Rushworth, M.F.S., Woolrich, M.W., 2007. Probabilistic diffusion tractography with multiple fibre orientations: what can we gain? NeuroImage 34 (1), 144–155. https://doi.org/10.1016/j.neuroimage.2006.09.018.

Berman, J.I., Chung, S.W., Mukherjee, P., Hess, C.P., Han, E.T., Henry, R.G., 2008. Probabilistic streamline q-ball tractography using the residual bootstrap. NeuroImage 39 (1), 215–222. https://doi.org/10.1016/J.NEUROIMAGE.2007.08.021.

Bingham, C., 1975. An antipodally symmetric distribution on the sphere. Ann. Stat. 2 (6), 1201–1225.

Budde, M., Annese, J., 2013. Quantification of anisotropy and fiber orientation in human brain histological sections. Front. Integr. Neurosci.. 7https://doi.org/10.3389/fnint.2013.00003.

Budde, M.D., Frank, J.A., 2012. Examining brain microstructure using structure tensor analysis of histological sections. NeuroImage 63 (1), 1–10. https://doi.org/10.1016/J.NEUROIMAGE.2012.06.042.

Calamante, F., Smith, R.E., Tournier, J.D., Raffelt, D., Connelly, A., 2015. Quantification of voxel-wise total fibre density: investigating the problems associated with track-count mapping. NeuroImage 117, 284–293. https://doi.org/10.1016/J.NEUROIMAGE.2015.05.070.

Callaghan, P.T., Eccles, C.D., Xia, Y., 1988. NMR microscopy of dynamic displacements: k-space and q-space imaging. J. Phys. E Sci. Instrum. 21 (8), 820. https://doi.org/10.1088/0022-3735/21/8/017.

Canales-Rodríguez, E.J., Iturria-Medina, Y., Alemán-Gómez, Y., Melie-García, L., 2010. Deconvolution in diffusion spectrum imaging. NeuroImage 50 (1), 136–149. https://doi.org/10.1016/J.NEUROIMAGE.2009.11.066.

Canales-Rodríguez, E.J., Daducci, A., Sotiropoulos, S.N., Caruyer, E., Aja-Fernández, S., Radua, J., Mendizabal, J.M.Y., Iturria-Medina, Y., Melie-García, L., Alemán-Gómez, Y., Thiran, J.P., Sarró, S., Pomarol-Clotet, E., Salvador, R., 2015. Spherical deconvolution of multichannel diffusion

MRI data with non-Gaussian noise models and spatial regularization. PLoS One. 10(10), e0138910https://doi.org/10.1371/JOURNAL.PONE.0138910.

Catani, M., Howard, R.J., Pajevic, S., Jones, D.K., 2002. Virtual in vivo interactive dissection of white matter fasciculi in the human brain. NeuroImage 17 (1), 77–94. https://doi.org/10.1006/nimg.2002.1136.

Choe, A.S., Stepniewska, I., Colvin, D.C., Ding, Z., Anderson, A.W., 2012. Validation of diffusion tensor MRI in the central nervous system using light microscopy: quantitative comparison of fiber properties. NMR Biomed. 25 (7), 900–908. https://doi.org/10.1002/nbm.1810.

Christiaens, D., Veraart, J., Cordero-Grande, L., Price, A.N., Hutter, J., Hajnal, J.V., Tournier, J.D., 2020. On the need for bundle-specific microstructure kernels in diffusion MRI. NeuroImage 208, 116460. https://doi.org/10.1016/j.neuroimage.2019.116460.

Conturo, T.E., Lori, N.F., Cull, T.S., Akbudak, E., Snyder, A.Z., Shimony, J.S., McKinstry, R.C., Burton, H., Raichle, M.E., 1999. Tracking neuronal fiber pathways in the living human brain. Proc. Natl. Acad. Sci. 96 (18), 10422–10427. https://doi.org/10.1073/pnas.96.18.10422.

Coronado-Leija, R., Ramirez-Manzanares, A., Marroquin, J.L., 2017. Estimation of individual axon bundle properties by a Multi-Resolution Discrete-Search method. Med. Image Anal. 42, 26–43. https://doi.org/10.1016/j.media.2017.06.008.

De Luca, A., Guo, F., Froeling, M., Leemans, A., 2020. Spherical deconvolution with tissue-specific response functions and multi-shell diffusion MRI to estimate multiple fiber orientation distributions (mFODs). NeuroImage 222, 117206. https://doi.org/10.1016/j.neuroimage.2020.117206.

Dela Haije, T.C.J., Duits, R., Tax, C.M.W., 2014. Sharpening fibers in diffusion weighted MRI via erosion. In: Mathematics and Visualization.pp. 97–126. https://doi.org/10.1007/978-3-642-54301-2_5/FIGURES/13.

Dell'Acqua, F., Tournier, J.-D., 2019. Modelling white matter with spherical deconvolution: how and why? NMR Biomed.. 32(4), e3945https://doi.org/10.1002/nbm.3945.

Dell'Acqua, F., Scifo, P., Rizzo, G., Catani, M., Simmons, A., Scotti, G., Fazio, F., 2010. A modified damped Richardson-Lucy algorithm to reduce isotropic background effects in spherical deconvolution. NeuroImage 49 (2), 1446–1458. https://doi.org/10.1016/j.neuroimage.2009.09.033.

Dell'Acqua, F., Simmons, A., Williams, S.C.R., Catani, M., 2013. Can spherical deconvolution provide more information than fiber orientations? Hindrance modulated orientational anisotropy, a true-tract specific index to characterize white matter diffusion. Hum. Brain Mapp. 34 (10), 2464–2483. https://doi.org/10.1002/hbm.22080.

Descoteaux, M., 2008. High Angular Resolution Diffusion MRI: From Local Estimation to Segmentation and Tractography. Université Nice Sophia Antipolis.https://theses.hal.science/tel-00457458.

Descoteaux, M., Angelino, E., Fitzgibbons, S., Deriche, R., 2006. Apparent diffusion coefficients from high angular resolution diffusion imaging: estimation and applications. Magn. Reson. Med. 56 (2), 395–410. https://doi.org/10.1002/mrm.20948.

Descoteaux, M., Angelino, E., Fitzgibbons, S., Deriche, R., 2007. Regularized, fast, and robust analytical Q-ball imaging. Magn. Reson. Med. 58 (3), 497–510. https://doi.org/10.1002/MRM.21277.

Descoteaux, M., Deriche, R., Knösche, T.R., Anwander, A., 2009. Deterministic and probabilistic tractography based on complex fibre orientation distributions. IEEE Trans. Med. Imaging 28 (2), 269–286. https://doi.org/10.1109/TMI.2008.2004424.

Deslauriers-Gauthier, S., Marziliano, P., Paquette, M., Descoteaux, M., 2016. The application of a new sampling theorem for non-bandlimited signals on the sphere: improving the recovery of crossing fibers for low b-value acquisitions. Med. Image Anal. 30, 46–59. https://doi.org/10.1016/j.media.2016.01.002.

Dhollander, T., Emsell, L., Hecke, W.V., Maes, F., Sunaert, S., Suetens, P., 2014. Track orientation density imaging (TODI) and track orientation distribution (TOD) based tractography. NeuroImage 94, 312–336. https://doi.org/10.1016/j.neuroimage.2013.12.047.

Dhollander, T., Clemente, A., Singh, M., Boonstra, F., Civier, O., Duque, J.D., Egorova, N., Enticott, P., Fuelscher, I., Gajamange, S., Genc, S., Gottlieb, E., Hyde, C., Imms, P., Kelly, C., Kirkovski, M., Kolbe, S., Liang, X., Malhotra, A., et al., 2021. Fixel-based analysis of diffusion MRI: methods, applications, challenges and opportunities. NeuroImage 241, 118417. https://doi.org/10.1016/j.neuroimage.2021.118417.

Duits, R., Haije, T.D., Creusen, E., Ghosh, A., 2013. Morphological and linear scale spaces for fiber enhancement in DW-MRI. J. Math. Imaging Vision 46 (3), 326–368. https://doi.org/10.1007/s10851-012-0387-2.

Fick, R.H.J., Wassermann, D., Caruyer, E., Deriche, R., 2016. MAPL: tissue microstructure estimation using Laplacian-regularized MAP-MRI and its application to HCP data. NeuroImage 134, 365–385. https://doi.org/10.1016/j.neuroimage.2016.03.046.

Frank, L.R., 2002. Characterization of anisotropy in high angular resolution diffusion-weighted MRI. Magn. Reson. Med. 47 (6), 1083–1099. https://doi.org/10.1002/mrm.10156.

Guberman, G.I., Stojanovski, S., Nishat, E., Ptito, A., Bzdok, D., Wheeler, A.L., Descoteaux, M., 2022. Multi-tract multi-symptom relationships in pediatric concussion. eLife. 11https://doi.org/10.7554/ELIFE.70450.

Haroon, H.A., Morris, D.M., Embleton, K.V., Alexander, D.C., Parker, G.J.M., 2009. Using the model-based residual bootstrap to quantify uncertainty in fiber orientations from q-ball analysis. IEEE Trans. Med. Imaging 28 (4), 535–550. https://doi.org/10.1109/TMI.2008.2006528.

Henriques, R.N., Correia, M.M., Nunes, R.G., Ferreira, H.A., 2015. Exploring the 3D geometry of the diffusion kurtosis tensor—impact on the development of robust tractography procedures and novel biomarkers. NeuroImage 111, 85–99. https://doi.org/10.1016/J.NEUROIMAGE.2015.02.004.

Henriques, R.N., Correia, M.M., Dell'Acqua, F., 2016. Mapping fibre dispersion and tract specific metrics in multiple fibre orientation using multi Bingham distributions. Proc. Int. Soc. Magn. Reson. Med.. https://cds.ismrm.org/protected/16MProceedings/PDFfiles/3055.html.

Hess, C.P., Mukherjee, P., Han, E.T., Xu, D., Vigneron, D.B., 2006. Q-ball reconstruction of multimodal fiber orientations using the spherical harmonic basis. Magn. Reson. Med. 56 (1), 104–117. https://doi.org/10.1002/mrm.20931.

Jensen, J.H., Helpern, J.A., 2010. MRI quantification of non-Gaussian water diffusion by kurtosis analysis. NMR Biomed. 23 (7), 698–710. https://doi.org/10.1002/NBM.1518.

Jensen, J.H., Helpern, J.A., Ramani, A., Lu, H., Kaczynski, K., 2005. Diffusional kurtosis imaging: the quantification of non-Gaussian water diffusion by means of magnetic resonance imaging. Magn. Reson. Med. 53 (6), 1432–1440. https://doi.org/10.1002/MRM.20508.

Jensen, J.H., Helpern, J.A., Tabesh, A., 2014. Leading non-Gaussian corrections for diffusion orientation distribution function. NMR Biomed. 27 (2), 202–211. https://doi.org/10.1002/NBM.3053.

Jeurissen, B., Szczepankiewicz, F., 2021. Multi-tissue spherical deconvolution of tensor-valued diffusion MRI. NeuroImage 245, 118717. https://doi.org/10.1016/j.neuroimage.2021.118717.

Jeurissen, B., Leemans, A., Jones, D.K., Tournier, J.D., Sijbers, J., 2011. Probabilistic fiber tracking using the residual bootstrap with constrained spherical deconvolution. Hum. Brain Mapp. 32 (3), 461–479. https://doi.org/10.1002/HBM.21032.

Jeurissen, B., Leemans, A., Tournier, J.-D., Jones, D.K., Sijbers, J., 2013. Investigating the prevalence of complex fiber configurations in white matter tissue with diffusion magnetic resonance imaging. Hum. Brain Mapp. 34 (11), 2747–2766. https://doi.org/10.1002/hbm.22099.

Jeurissen, B., Tournier, J.D., Dhollander, T., Connelly, A., Sijbers, J., 2014. Multi-tissue constrained spherical deconvolution for improved analysis of multi-shell diffusion MRI data. NeuroImage 103, 411–426. https://doi.org/10.1016/j.neuroimage.2014.07.061.

Jones, D.K., 2004. The effect of gradient sampling schemes on measures derived from diffusion tensor MRI: a Monte Carlo study†. Magn. Reson. Med. 51 (4), 807–815. https://doi.org/10.1002/MRM.20033.

Kaden, E., Knösche, T.R., Anwander, A., 2007. Parametric spherical deconvolution: inferring anatomical connectivity using diffusion MR imaging. NeuroImage 37 (2), 474–488. https://doi.org/10.1016/j.neuroimage.2007.05.012.

Karan, P., Reymbaut, A., Gilbert, G., Descoteaux, M., 2022. Bridging the gap between constrained spherical deconvolution and diffusional variance decomposition via tensor-valued diffusion MRI. Med. Image Anal. 79, 102476. https://doi.org/10.1016/j.media.2022.102476.

Karayumak, S.C., Ozarslan, E., Unal, G., 2018. Asymmetric Orientation Distribution Functions (AODFs) revealing intravoxel geometry in diffusion MRI. Magn. Reson. Imaging 49, 145–158. https://doi.org/10.1016/J.MRI.2018.03.006.

Koay, C.G., 2011. A simple scheme for generating nearly uniform distribution of antipodally symmetric points on the unit sphere. J. Comput. Sci. 2 (4), 377–381. https://doi.org/10.1016/J.JOCS.2011.06.007.

Lautarescu, A., Bonthrone, A.F., Pietsch, M., Batalle, D., Cordero-Grande, L., Tournier, J.D., Christiaens, D., Hajnal, J.V., Chew, A., Falconer, S., Nosarti, C., Victor, S., Craig, M.C., Edwards, A.D., Counsell, S.J., 2022. Maternal depressive symptoms, neonatal white matter, and toddler social-emotional development. Transl. Psychiatry 12 (1), 1–12. https://doi.org/10.1038/s41398-022-02073-y.

Lazar, M., Jensen, J.H., Xuan, L., Helpern, J.A., 2008. Estimation of the orientation distribution function from diffusional kurtosis imaging. Magn. Reson. Med. 60 (4), 774–781. https://doi.org/10.1002/MRM.21725.

Leergaard, T.B., White, N.S., de Crespigny, A., Bolstad, I., D'Arceuil, H., Bjaalie, J.G., Dale, A.M., 2010. Quantitative histological validation of diffusion MRI fiber orientation distributions in the rat brain. PLoS One 5 (1), 1–8. https://doi.org/10.1371/journal.pone.0008595.

Leuze, C., Goubran, M., Barakovic, M., Aswendt, M., Tian, Q., Hsueh, B., Crow, A., Weber, E.M.M., Steinberg, G.K., Zeineh, M., Plowey, E.D., Daducci, A., Innocenti, G., Thiran, J.P., Deisseroth, K., McNab, J.A., 2021. Comparison of diffusion MRI and CLARITY fiber orientation estimates in both gray and white matter regions of human and primate brain. NeuroImage 228, 117692. https://doi.org/10.1016/J.NEUROIMAGE.2020.117692.

Mirchandani, A.S., Beyh, A., Lavrador, J.P., Howells, H., Dell'Acqua, F., Vergani, F., 2021. Altered corticospinal microstructure and motor cortex excitability in gliomas: an advanced tractography and transcranial magnetic stimulation study. J. Neurosurg. 134 (5), 1368–1376. https://doi.org/10.3171/2020.2.JNS192994.

Mori, S., Crain, B.J., Chacko, V.P., Zijl, P.C.M.V., 1999. Three-dimensional tracking of axonal projections in the brain by magnetic resonance imaging. Ann. Neurol. 45 (2), 265–269. https://doi.org/10.1002/1531-8249(199902)45:2<265::AID-ANA21>3.0.CO;2-3.

Novikov, D.S., Kiselev, V.G., Jespersen, S.N., 2018. On modeling. Magn. Reson. Med. 79 (6), 3172–3193. https://doi.org/10.1002/mrm.27101.

Özarslan, E., Shepherd, T.M., Vemuri, B.C., Blackband, S.J., Mareci, T.H., 2006. Resolution of complex tissue microarchitecture using the diffusion orientation transform (DOT). NeuroImage 31 (3), 1086–1103. https://doi.org/10.1016/j.neuroimage.2006.01.024.

Özarslan, E., Koay, C.G., Shepherd, T.M., Komlosh, M.E., Irfanoğlu, M.O., Pierpaoli, C., Basser, P.J., 2013. Mean apparent propagator (MAP) MRI: a novel diffusion imaging method for mapping tissue microstructure. NeuroImage 78, 16–32. https://doi.org/10.1016/j.neuroimage.2013.04.016.

Palombo, M., Ianus, A., Guerreri, M., Nunes, D., Alexander, D.C., Shemesh, N., Zhang, H., 2020. SANDI: a compartment-based model for non-invasive apparent soma and neurite imaging by diffusion MRI. NeuroImage 215, 116835. https://doi.org/10.1016/J.NEUROIMAGE.2020.116835.

Panagiotaki, E., Schneider, T., Siow, B., Hall, M.G., Lythgoe, M.F., Alexander, D.C., 2012. Compartment models of the diffusion MR signal in brain white matter: a taxonomy and comparison. NeuroImage 59 (3), 2241–2254. https://doi.org/10.1016/j.neuroimage.2011.09.081.

Pasternak, O., Sochen, N., Gur, Y., Intrator, N., Assaf, Y., 2009. Free water elimination and mapping from diffusion MRI. Magn. Reson. Med. 62 (3), 717–730. https://doi.org/10.1002/mrm.22055.

Poirier, C., Descoteaux, M., 2023. A Unified Filtering Method for Estimating Asymmetric Orientation Distribution Functions: Where and How Asymmetry Occurs in the Brain. BioRxiv, 2022.12.18.520881https://doi.org/10.1101/2022.12.18.520881.

Portegies, J.M., Fick, R.H.J., Sanguinetti, G.R., Meesters, S.P.L., Girard, G., Duits, R., 2015. Improving fiber alignment in HARDI by combining contextual PDE flow with constrained spherical deconvolution. PLoS One. 10(10), e0138122https://doi.org/10.1371/JOURNAL.PONE.0138122.

Raffelt, D., Tournier, J.D., Rose, S., Ridgway, G.R., Henderson, R., Crozier, S., Salvado, O., Connelly, A., 2012. Apparent Fibre Density: a novel measure for the analysis of diffusion-weighted magnetic resonance images. NeuroImage 59 (4), 3976–3994. https://doi.org/10.1016/j.neuroimage.2011.10.045.

Raffelt, D.A., Smith, R.E., Ridgway, G.R., Tournier, J.D., Vaughan, D.N., Rose, S., Henderson, R., Connelly, A., 2015. Connectivity-based fixel enhancement: whole-brain statistical analysis of diffusion MRI measures in the presence of crossing fibres. NeuroImage 117, 40–55. https://doi.org/10.1016/j.neuroimage.2015.05.039.

Raffelt, D.A., Tournier, J.-D., Smith, R.E., Vaughan, D.N., Jackson, G., Ridgway, G.R., Connelly, A., 2017. Investigating white matter fibre density and morphology using fixel-based analysis. NeuroImage 144, 58–73. https://doi.org/10.1016/J.NEUROIMAGE.2016.09.029.

Raj, A., Hess, C., Mukherjee, P., 2011. Spatial HARDI: improved visualization of complex white matter architecture with Bayesian spatial regularization. NeuroImage 54 (1), 396–409. https://doi.org/10.1016/j.neuroimage.2010.07.040.

Reisert, M., Kiselev, V.G., 2011. Fiber continuity: an anisotropic prior for ODF estimation. IEEE Trans. Med. Imaging 30 (6), 1274–1283. https://doi.org/10.1109/TMI.2011.2112769.

Reisert, M., Kellner, E., Kiselev, V.G., 2012. About the geometry of asymmetric fiber orientation distributions. IEEE Trans. Med. Imaging 31 (6), 1240–1249. https://doi.org/10.1109/TMI.2012.2187916.

Renard, F., Noblet, V., Grigis, A., Heinrich, C., Kremer, S., 2010. Comparison of interpolation methods for angular resampling of diffusion weighted images. In: 2010 2nd International Conference on Image Processing Theory, Tools and Applicationspp. 207–211. https://doi.org/10.1109/IPTA.2010.5586799.

Rheault, F., St-Onge, E., Sidhu, J., Maier-Hein, K., Tzourio-Mazoyer, N., Petit, L., Descoteaux, M., 2019. Bundle-specific tractography with incorporated anatomical and orientational priors. NeuroImage 186, 382–398. https://doi.org/10.1016/j.neuroimage.2018.11.018.

Riffert, T.W., Schreiber, J., Anwander, A., Knösche, T.R., 2014. Beyond fractional anisotropy: extraction of bundle-specific structural metrics from crossing fiber models. NeuroImage 100, 176–191. https://doi.org/10.1016/j.neuroimage.2014.06.015.

Scherrer, B., Schwartzman, A., Taquet, M., Sahin, M., Prabhu, S.P., Warfield, S.K., 2016. Characterizing brain tissue by assessment of the distribution of anisotropic microstructural environments in diffusion-compartment imaging (DIAMOND). Magn. Reson. Med. 76 (3), 963–977. https://doi.org/10.1002/mrm.25912.

Schilling, K., Janve, V., Gao, Y., Stepniewska, I., Landman, B.A., Anderson, A.W., 2016. Comparison of 3D orientation distribution functions measured with confocal microscopy and diffusion MRI. NeuroImage 129, 185–197. https://doi.org/10.1016/J.NEUROIMAGE.2016.01.022.

Seehaus, A., Roebroeck, A., Bastiani, M., Fonseca, L., Bratzke, H., Lori, N., Vilanova, A., Goebel, R., Galuske, R., 2015. Histological validation of high-resolution DTI in human post mortem tissue. Front. Neuroanat. 9 (July), 98. https://doi.org/10.3389/FNANA.2015.00098/BIBTEX.

Sotiropoulos, S.N., Behrens, T.E.J., Jbabdi, S., 2012. Ball and rackets: inferring fiber fanning from diffusion-weighted MRI. NeuroImage 60 (2), 1412–1425. https://doi.org/10.1016/j.neuroimage.2012.01.056.

Tariq, M., Schneider, T., Alexander, D.C., Wheeler-Kingshott, C.A.G., Zhang, H., 2016. Bingham-NODDI: mapping anisotropic orientation dispersion of neurites using diffusion MRI. NeuroImage 133, 207–223. https://doi.org/10.1016/j.neuroimage.2016.01.046.

Tax, C.M.W., Duits, R., Vilanova, A., Romeny, B.M.T.H., Hofman, P., Wagner, L., Leemans, A., Ossenblok, P., 2014. Evaluating contextual processing in diffusion MRI: application to optic radiation reconstruction for epilepsy surgery. PLoS One. 9(7), e101524https://doi.org/10.1371/JOURNAL.PONE.0101524.

Tournier, J.D., Calamante, F., Gadian, D.G., Connelly, A., 2004. Direct estimation of the fiber orientation density function from diffusion-weighted MRI data using spherical deconvolution. NeuroImage 23 (3), 1176–1185. https://doi.org/10.1016/j.neuroimage.2004.07.037.

Tournier, J.-D., Calamante, F., Connelly, A., 2007. Robust determination of the fibre orientation distribution in diffusion MRI: non-negativity constrained super-resolved spherical deconvolution. NeuroImage 35 (4), 1459–1472. https://doi.org/10.1016/j.neuroimage.2007.02.016.

Tristán-Vega, A., Westin, C.F., Aja-Fernández, S., 2010. A new methodology for the estimation of fiber populations in the white matter of the brain with the Funk-Radon transform. NeuroImage 49 (2), 1301–1315. https://doi.org/10.1016/j.neuroimage.2009.09.070.

Tuch, D.S., 2004. Q-ball imaging. Magn. Reson. Med. 52 (6), 1358–1372. https://doi.org/10.1002/mrm.20279.

Tuch, D.S., Reese, T.G., Wiegell, M.R., Makris, N., Belliveau, J.W., Wedeen, V.J., 2002. High angular resolution diffusion imaging reveals intravoxel white matter fiber heterogeneity. Magn. Reson. Med. 48 (4), 577–582. https://doi.org/10.1002/mrm.10268.

Veraart, J., Fieremans, E., Novikov, D.S., 2019. On the scaling behavior of water diffusion in human brain white matter. NeuroImage 185, 379. https://doi.org/10.1016/J.NEUROIMAGE.2018.09.075.

Volz, L.J., Cieslak, M., Grafton, S.T., 2018. A probabilistic atlas of fiber crossings for variability reduction of anisotropy measures. Brain Struct. Funct. 223 (2), 635–651. https://doi.org/10.1007/s00429-017-1508-x.

Wang, H., Lenglet, C., Akkin, T., 2015. Structure tensor analysis of serial optical coherence scanner images for mapping fiber orientations and tractography in the brain. J. Biomed. Opt. 20 (3), 036003. https://doi.org/10.1117/1.JBO.20.3.036003.

Wedeen, V.J., Hagmann, P., Tseng, W.-Y.I., Reese, T.G., Weisskoff, R.M., 2005. Mapping complex tissue architecture with diffusion spectrum magnetic resonance imaging. Magn. Reson. Med. 54 (6), 1377–1386. https://doi.org/10.1002/mrm.20642.

Wedeen, V.J., Wang, R.P., Schmahmann, J.D., Benner, T., Tseng, W.Y.I., Dai, G., Pandya, D.N., Hagmann, P., D'Arceuil, H., de Crespigny, A.J., 2008. Diffusion spectrum magnetic resonance imaging (DSI) tractography of crossing fibers. NeuroImage 41 (4), 1267–1277. https://doi.org/10.1016/j.neuroimage.2008.03.036.

Wu, Y., Hong, Y., Feng, Y., Shen, D., Yap, P.T., 2020. Mitigating gyral bias in cortical tractography via asymmetric fiber orientation distributions. Med. Image Anal. 59, 101543. https://doi.org/10.1016/j.media.2019.101543.

Yeh, F.C., Tseng, W.Y.I., 2013. Sparse solution of fiber orientation distribution function by diffusion decomposition. PLoS One. 8(10), e75747https://doi.org/10.1371/JOURNAL.PONE.0075747.

Yeh, F.C., Verstynen, T.D., Wang, Y., Fernández-Miranda, J.C., Tseng, W.Y.I., 2013. Deterministic diffusion fiber tracking improved by quantitative anisotropy. PLoS One. 8(11)https://doi.org/10.1371/journal.pone.0080713.

Zhang, H., Schneider, T., Wheeler-Kingshott, C.A., Alexander, D.C., 2012. NODDI: practical in vivo neurite orientation dispersion and density imaging of the human brain. NeuroImage 61 (4), 1000–1016. https://doi.org/10.1016/j.neuroimage.2012.03.072.

Part III

# Tractography algorithms

# Chapter 13

# Deterministic fiber tractography

**Alexander Leemans[a], Flavio Dell'Acqua[b], and Maxime Descoteaux[c]**

[a]*PROVIDI lab, Image Sciences Institute, University Medical Center Utrecht, Utrecht, The Netherlands,* [b]*NATBRAINLAB, Department of Neuroimaging,
Institute of Psychiatry, Psychology and Neuroscience, King's College London, London, United Kingdom,* [c]*Sherbrooke Connectivity Imaging Laboratory
(SCIL), Department of Computer Science, University of Sherbrooke, Sherbrooke, QC, Canada*

## 1 Historical background

Only a few years after directional dependence of the image contrast in diffusion magnetic resonance imaging (dMRI) data
was observed in vivo (Le Bihan et al., 2001; Moseley et al., 1990, 1991), the rotationally invariant diffusion tensor imaging
(DTI) framework was proposed by Basser et al. (1994a,b). This approach provided a unique way to study the microstructural organization of fibrous tissue noninvasively (Basser, 1995; Basser and Pierpaoli, 1996) and has led to the first virtual
three-dimensional (3D) reconstructions of the brain white matter (WM) architecture using a technique that is now commonly known as fiber tractography (FT) or fiber tracking (Wedeen et al., 1995; Basser, 1998; Basser et al., 2000a,b, 2002;
Mori and Barker, 1999; Mori et al., 1998, 1999a,b, 2000; Jones et al., 1998, 1999; Conturo et al., 1999; Poupon et al., 1999b,
2000; Lazar et al., 2000). Note that while the first mention of "tractography" in the context of dMRI appeared in 1993 from
Kinosada et al. (1993), these authors did not compute actual 3D trajectories of WM pathways, but merely provided qualitative maps based on maximum-intensity projections of diffusion-weighted MRI data.

Tuch et al. (2000) stated that FT methods are "... not amenable to probabilistic interpretations of the tract solutions." and
posed the question: "What is the probability associated with a particular tract solution?" To address this challenge, several
FT methods were developed that could incorporate orientation uncertainty associated with the diffusion models and/or the
data noise distributions (Lazar et al., 2001; Lazar and Alexander, 2002, 2005; Parker and Alexander, 2003; Parker et al.,
2002a, 2003; Koch et al., 2002; Behrens et al., 2002a,b, 2003a,b). These methods were then termed *probabilistic tractography* methods to differentiate them from the FT methods that did not use such a strategy. Note that only after the advent of
probabilistic FT methods was the wording *deterministic tractography* used, mainly to clearly distinguish between the two
categories (Jones and Pierpaoli, 2004). The following sections describe the key ingredients of most FT methods, as probabilistic methods are often based on their deterministic counterparts.

## 2 Key ingredients of fiber tractography

Conceptually, FT is similar to computing fluid streamlines from discrete estimates of velocity data, such as the speed of water
flowing in a river (Yeung and Pope, 1988). But instead of following fluid pathways from vector data, FT turns local estimates
of *fiber* orientations into long-range connections. While these fibers can originate from any type of tissue (heart, muscle, etc.),
we will focus in the following on WM brain tissue. In addition, although the discrete estimates of orientation can be derived
from several different modalities, including susceptibility tensor imaging (Liu, 2010), polarized light imaging (Larsen et al.,
2007), and small-angle X-ray scattering tensor tomography (Georgiadis et al., 2021), the focus here is on fiber orientation
estimates as obtained with dMRI techniques, such as DTI and spherical deconvolution (SD) methods (see Chapters 10–12).

### 2.1 Overview

In its simplest form, a dMRI-based FT algorithm works as follows. From a predefined position in the data, called a seed
point, a single tract pathway is generated by consecutively following the local fiber orientation estimates with a given step
size until some parameter thresholds are reached to terminate this propagation. As this procedure is similar to reconstructing
streamlines from a velocity field in fluid dynamics, this type of FT is also referred to as *streamline* FT. In practice, many
seed points can be defined throughout the entire dataset, aiming to reconstruct the complete wiring diagram of the brain,
also called a whole-brain tractogram.

**Handbook of Diffusion MR Tractography. https://doi.org/10.1016/B978-0-12-818894-1.00029-X**
Copyright © 2025 Elsevier Ltd. All rights are reserved, including those for text and data mining, AI training, and similar technologies.

Specifically for DTI, the datasets are often represented as discretely sampled diffusion tensors $D(r)$, that is, each voxel at position $r$ is characterized by a second-rank diffusion tensor $D$. Assuming that the first eigenvector $e_1(r)$ associated with $D(r)$ is aligned with the orientation of the underlying WM tissue at position $r$, a fiber pathway is then computed by piecing together small steps along the orientation of $e_1(r)$. Note that in this approach, the diffusion tensor field is approximated by a vector field, omitting information that is contained in the remaining degrees of freedom of the diffusion tensor. Moreover, only the orientation and not the direction of $e_1(r)$ is defined, for diffusion is inherently a center-symmetric phenomenon, that is, both $e_1(r)$ and $-e_1(r)$ are mathematically valid directions. Therefore each individual fiber pathway can be computed by propagating a line along both $e_1(r)$ and $-e_1(r)$ from the initial seed point $r_0$ (Fig. 1).

The process of connecting consecutive steps (line segments) is iterated many times until certain stopping criteria are met. For example, the tracking process can be stopped if pathways pass into areas that may not be of interest, such as cerebrospinal fluid or gray matter (GM), or if some local diffusion properties are exceeded (e.g., a diffusion anisotropy measure falls below a fixed threshold) (Jones et al., 1999; Mori et al., 1999a). In addition, it has been proposed that only a limited amount of curvature and/or torsion between consecutive segments should be tolerated, further reducing the number of potentially spurious tract pathways[a] (Jones et al., 1999; Basser et al., 2000a). In general, the FT procedure is performed for a large number of seed points $\{r_0\}$ that define a specific region of interest (ROI) or the entire dataset (Fig. 2).

Due to the nature of MRI acquisition, the diffusion tensors are only available at discrete locations, that is, the voxel positions. Although initially only taking into account these discretely sampled orientation estimates (e.g., Jones et al., 1999), it became clear that the orientation from the tracking procedure often deviated from the "true" fiber orientation, for here, the choice of orientation is limited to the 26 neighboring voxels. This problem can be avoided when tracking a continuous rather than a discrete diffusion tensor field and can be accomplished, for instance, by applying (linear) interpolation techniques and the fitting of Lagrange polynomials or B-splines (Basser, 1998; Basser et al., 2000a; Tench et al., 2002; Pajevic et al., 2002; Aldroubi and Basser, 1999). A large number of methods have been developed to further regularize and smooth the diffusion tensor field, which could improve the FT procedure (Poupon et al., 1998, 1999a, 2000, 2001; Frandsen et al., 2004; Li et al., 2004a; Coulon et al., 2001, 2004; Vemuri et al., 2001; Parker et al., 2000; McGraw et al., 2004; Wang et al., 2004).

By introducing prior anatomical knowledge, we can refine the whole-brain FT results to become more specific (Conturo et al., 1999; Catani et al., 2002). Essentially, one can define ROIs through which the tract-of-interest is known to pass, also

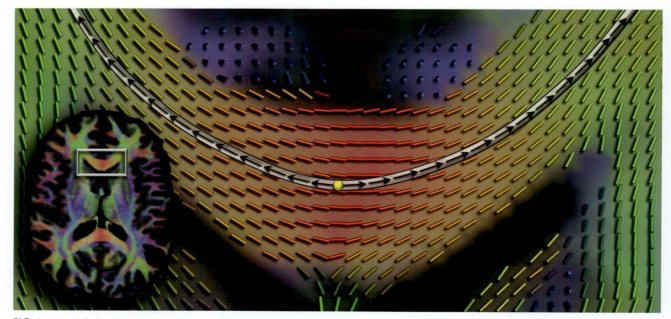

**FIG. 1** An axial view in the region of the genu of the corpus callosum, as indicated by the *white rectangle*. The colored line segments represent the first eigenvectors $e_1(r)$ associated with the diffusion tensors $D(r)$ at the voxel locations $r$. From the seed point $r_0$ (*yellow dot*), a pathway (*white line*) is computed by iteratively and consecutively piecing together these orientation estimates (*black arrows*).

---

a. Note, however, that this consideration is typically based on a priori information of the WM fiber tract geometry.

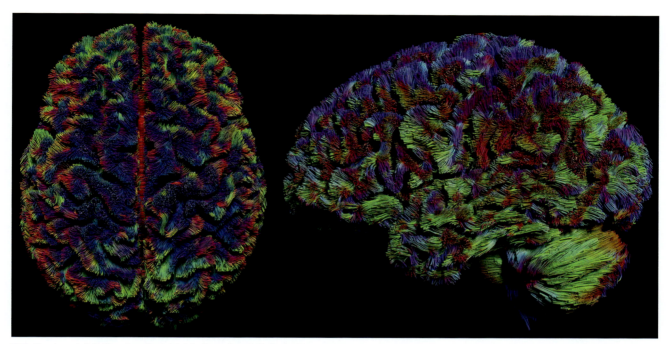

**FIG. 2** An axial (*left*) and sagittal (*right*) view of a whole-brain fiber tractography result. Here, many seed points were defined throughout the entire dataset on a 3D rectilinear grid with 1 mm equidistant spaces. The pathways are colored according to their local orientation to convey the architectural configuration more clearly (*red* = left-right, *green* = front-back, and *blue* = top-down).

known as "AND gates." Paths that enter these regions are then considered anatomically plausible, and all others are omitted. Conversely, it is also possible to define ROIs through which the tract-of-interest is known not to pass. Any trajectories that enter these ROIs, also referred to as "NOT gates," are then discarded. These methods are very powerful for identifying pathways, but they require anatomical knowledge about the tracts-of-interest and the cortical regions these are connecting to (Fig. 3).

## 2.2 Mathematical framework of streamline FT

One of the first dMRI FT frameworks was presented by Basser (1998) and Basser et al. (2000a, b, 2002). They proposed that a WM fiber pathway can be represented as a 3D space curve parameterized by its arc length $s$. In this context, the tract pathway $r(s) = [x(s)\ y(s)\ z(s)]^T$ can be characterized by the Frénet equation, that is,

$$\frac{d r(s)}{d s} = t(s), \qquad (1)$$

where $t(s)$ represents the unit vector tangential to $r(s)$ at arc length $s$. Assuming that the first eigenvector $e_1(r)$ corresponding to $D(r)$ is aligned with the tangential direction of the WM fiber trajectory, Eq. (1) can be rewritten as

$$\frac{d r(s)}{d s} = e_1[r(s)] \qquad (2)$$

or

$$r(s) = \int_{s_0} e_1[r(s)] d s, \qquad (3)$$

where $r(s = s_0) = r_0$ represents the seed point.

### 2.2.1 Euler's method

The most trivial way to perform the numerical integration of Eq. (3) is by starting at the seed point, $r_0$, which is assumed to be located on the putative fiber tract, calculating the corresponding first eigenvector, that is, $e_1(r_0)$, and following that

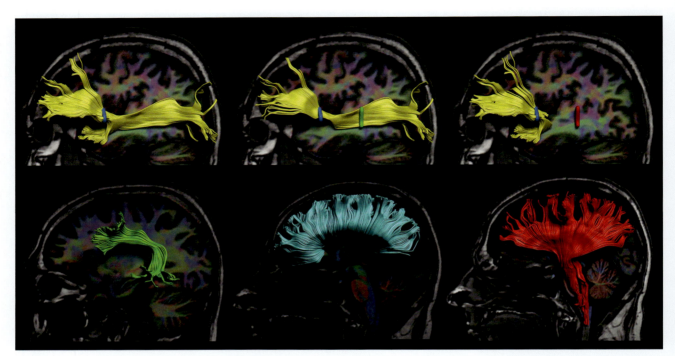

**FIG. 3** *Top row*: The *left image* shows a fiber bundle that was reconstructed with seeding ROI shown in *blue*. The *middle and right images* are obtained after applying "AND" (*green* ROI) and "NOT" (*red* ROI) gates, respectively. *Bottom row*: Trajectories related to the arcuate fasciculus (*left*), body of the corpus callosum (*middle*), and the corticospinal tract (*right*).

orientation for a short predefined distance $\Delta$ (i.e., the "step size") to obtain the next point $r_1 = r_0 + e_1(r_0)\Delta$ on the fiber pathway. With this procedure, known as Euler's method, the fiber pathway can be reconstructed by iteratively performing the aforementioned steps, that is,

$$r_{i+1} = r_i + e_1(r_i)\Delta. \tag{4}$$

This scheme has an associated propagation error of the order $\mathcal{O}\{\Delta^2\}$. With this procedure, a single fiber tract pathway can then be described as an $N \times 3$ matrix where each row represents a coordinate along its trajectory.

As previously mentioned, there is the problem of eigenvector ambiguity, that is, both $e_1$ and $-e_1$ represent the first eigenvector of $D$. This can be resolved by assuming that the integral curve does not turn more than 90 degrees between consecutive integration steps. This can be checked by determining the sign of the dot product between each intermediate tangent vector $r_i - r_{i-1}$ and eigenvector $e_1(r_i)$, that is,

$$\text{if} \quad (r_i - r_{i-1}) \cdot e_1(r_i) < 0 \quad \text{then} \quad e_1(r_i) \to -e_1(r_i). \tag{5}$$

Note, however, that from each seed point, a tract pathway must propagate in both forward and backward directions.

### 2.2.2 Runge-Kutta integration

Euler's method can suffer from a large accumulating error propagation, especially for larger step sizes $\Delta$. Therefore theoretically, higher-order numerical integration schemes are preferred, such as the second-order Runge-Kutta (RK2) integration scheme (Press et al., 1992), that is,

$$r_{i+1} = r_i + e_1\left(r_i + \frac{\Delta}{2}e_1(r_i)\right)\Delta, \tag{6}$$

which has error propagation on the order of $\mathcal{O}\{\Delta^3\}$, or the fourth-order Runge-Kutta (RK4) scheme (Press et al., 1992), that is,

$$r_{i+1} = r_i + \frac{k_1}{6} + \frac{k_2}{3} + \frac{k_3}{3} + \frac{k_4}{6}, \tag{7}$$

with

$$k_1 = e_1(r_i)\Delta$$
$$k_2 = e_1\left(r_i + \frac{k_1}{2}\right)\Delta$$
$$k_3 = e_1\left(r_i + \frac{k_2}{2}\right)\Delta \tag{8}$$
$$k_4 = e_1(r_i + k_3)\Delta.$$

The RK4 scheme has an associated error of order $\mathcal{O}\{\Delta^5\}$ and is known to be a good candidate for the numerical solution of Eq. (3). In addition, it is possible to employ adaptive step sizing to further control the amount of error introduced in each integration step. The robust RK4 integration method is implemented in several streamline FT algorithms (with or without additional heuristic modifications) and is capable of accurately integrating tract streamlines with larger step sizes $\Delta$ (Tench et al., 2002; Basser et al., 2000a; Wünsche et al., 2005; Pagani et al., 2005; Thottakara et al., 2006; Lazar and Alexander, 2003; Catani et al., 2002; Kim et al., 2003; Kuo et al., 2004; Leemans et al., 2005; Delputte et al., 2005). Note, however, that for equal step size, the RK4 approach is approximately four times slower than Euler's method. The aforementioned integration schemes can be further regularized using a variational framework for global energy minimization (Vemuri et al., 2001, 2002; Cointepas et al., 2002; Mangin et al., 2002; Poupon et al., 2001; Weng et al., 2002; McGraw et al., 2004).

### 2.2.3 Tensor deflection

An alternative approach for determining the local tract direction is to use the entire diffusion tensor $D$ to estimate the propagation direction of the trajectory (Weinstein et al., 1999), that is,

$$r_{i+1} = r_i + \Delta \frac{D(r_i) \cdot (r_i - r_{i-1})}{\|D(r_i) \cdot (r_i - r_{i-1})\|}. \tag{9}$$

This specific streamline algorithm is referred to as tensorline propagation (Weinstein et al., 1999) or tensor deflection (TEND) (Lazar et al., 2000, 2001, 2003; Lazar and Alexander, 2003). In Eq. (9), the dot product with $D$ deflects the propagation toward the first eigenvector orientation, but limits the curvature of the deflection, resulting in a smoother fiber pathway reconstruction. For instance, when the diffusion distribution profile within a voxel of interest is more planar, suggesting a higher intravoxel orientational heterogeneity, the propagation orientation is determined by the superposition of the first and second eigenvector. In the highly linear situation, the TEND approach and the "regular" streamline method will have approximately the same tract propagation orientation.

A similar approach uses the diffusion measures based on the diffusion ellipsoid's shape, differentiating between prolate, oblate, and spherical fiber distribution profiles (Westin et al., 1999, 2002; Zhang et al., 2004). Depending on these geometric measures, the propagation direction is calculated as a user-defined weighted sum of all three diffusion eigenvectors, that is, the diffusion tensor $D = D(r_i)$ in Eq. (9) is then generally defined as

$$D = \sum_{i=1}^{3} f_i(\lambda_i) e_i \cdot e_i^T, \tag{10}$$

where the $f_j$ can, for example, serve as threshold operators as a function of the eigenvalues $\lambda_i$, that is,

$$D = \begin{cases} C_L D_L & \text{if } C_L > C_P, C_S \\ C_P D_P & \text{if } C_P > C_L, C_S \\ C_S D_S & \text{if } C_S > C_L, C_P \end{cases} \tag{11}$$

In Eq. (11), the coefficients $C_L, C_P, C_S$ and tensor basis elements $D_L, D_P, D_S$ are calculated according to Westin et al. (2002):

$$D = \sum_{i=1}^{3} \lambda_i \, e_i \cdot e_i^T$$
$$= \underbrace{[\lambda_1 - \lambda_2]}_{C_L} \underbrace{e_1 \cdot e_1^T}_{D_L} + \underbrace{[\lambda_2 - \lambda_3]}_{C_P} \underbrace{\sum_{i=1}^{2} e_i \cdot e_i^T}_{D_P} + \underbrace{\lambda_3}_{C_S} \underbrace{\sum_{i=1}^{3} e_i \cdot e_i^T}_{D_S} \tag{12}$$
$$= C_L D_L + C_P D_P + C_S D_S.$$

**246 PART | III** Tractography algorithms

This propagation approach has been further regularized by minimizing a tract pathway cost model, as described in Li et al. (2004b).

## 2.3 User-defined FT settings

In the previous paragraphs, different strategies have been presented to determine the next position along a tract pathway during propagation, such as the Euler or Runge-Kutta integration. How trajectories propagate from one position to the next is also determined by the way the underlying data is interpolated. Not only the algorithmic aspect (e.g., nearest neighbors vs. cubic spline interpolation), but also the choice of which data to interpolate (e.g., diffusion-weighted images vs. diffusion tensors vs. first eigenvectors) will introduce variability in the tractography results and will provide different levels of accuracy by trading off computation complexity. In the following paragraphs, we will discuss how the typical FT settings can affect fiber pathway reconstructions. These include "seeding" choices and termination thresholds based on diffusion properties and angle deviation.

### 2.3.1 Seeding

In general, fiber pathways are reconstructed by defining the ROI that makes up the seed points from which to start the propagation procedure, that is, the seeding (or seed) ROI. So instead of having only one seed point, as shown in Fig. 1, one can use multiple starting points that are located in this ROI. If the focus is to reconstruct a specific fiber bundle, prior anatomical knowledge about its location and configuration is then required to define the seeding ROI appropriately.

There are several options to define the seeding ROI: manually by annotating the data, from an atlas-based label, or even from another imaging modality (e.g., an activation region from fMRI). Alternatively, one could also consider using the entire dataset as the seeding ROI (see Fig. 2). Specifically for the brain, one can also consider seeding in all of the WM or perhaps only at the interface between the WM and the GM, where pathways are expected to start or end.

Once the seeding ROI has been defined, one can now also choose a seeding density, that is, the number of seeding points per unit volume. For instance, consider a seeding ROI consisting of 20 voxels, where the voxel resolution is 2 mm isotropic. If only the voxel centers are defined as the seeding points, then the seeding density would be $20/2^3 = 2.5$ seed points per mm$^3$. In theory, one could define any number of seeding points per unit volume (e.g., 100 seed points per voxel or per seed ROI). Fig. 4 shows the effect of increasing the seeding density on the FT result of part of the corpus callosum. In practice, a balanced trade-off is typically made between "completeness" of the FT result and its "computation cost."

### 2.3.2 Termination thresholds

A termination threshold refers to a specific value of an FT parameter setting at which the propagation of the fiber pathway is ended. An example of such a parameter could be related to the position in the data. Specifically for the brain, one could define a 3D map (also referred to as "mask") such that the propagation stops if the next point along the pathway is located outside this mask. Different algorithms exist with a wide variety of such termination settings, incorporating complementary information from AI, neighboring voxels, or anatomical and microstructural priors (Reisert and Kiselev, 2011; Reisert et al., 2011, 2014; Mangin et al., 2013; Smith et al., 2012; Girard et al., 2014, 2017; Christiaens et al., 2015; Sherbondy et al., 2010; Chamberland et al., 2018; Neher et al., 2017). Among the most common are the angle threshold and thresholds related to diffusion properties, such as anisotropy and fiber density.

**Angle threshold**

Assuming that WM pathways do not bend sharply at the typical length scale of the voxel size, the propagation procedure is terminated if the angle $\theta$ between consecutive steps exceeds a specific threshold $\theta_t$, that is, if the pathway enters its next position at $\boldsymbol{r}_N$, tracking is stopped if

$$\theta = \big| \arccos[(\boldsymbol{r}_N - \boldsymbol{r}_{N-1}) \cdot (\boldsymbol{r}_{N-1} - \boldsymbol{r}_{N-2})] \big| > \theta_t. \tag{13}$$

Typical values for human brain data are in the range of $\theta_t \simeq$ [15 degrees–45 degrees] for a step size $\varDelta \simeq$ [0.1–1 mm]. Note that changing the step size affects the constraint of the local physical tract curvature $\kappa$. Therefore one could also apply an explicit curvature threshold $\kappa_t$ instead of $\theta_t$ (Basser et al., 2000a). Consequently, the tracking procedure is terminated if

$$\kappa = \frac{\sin \theta}{\varDelta \sin\left(\dfrac{\pi - \theta}{2}\right)} > \kappa_t. \tag{14}$$

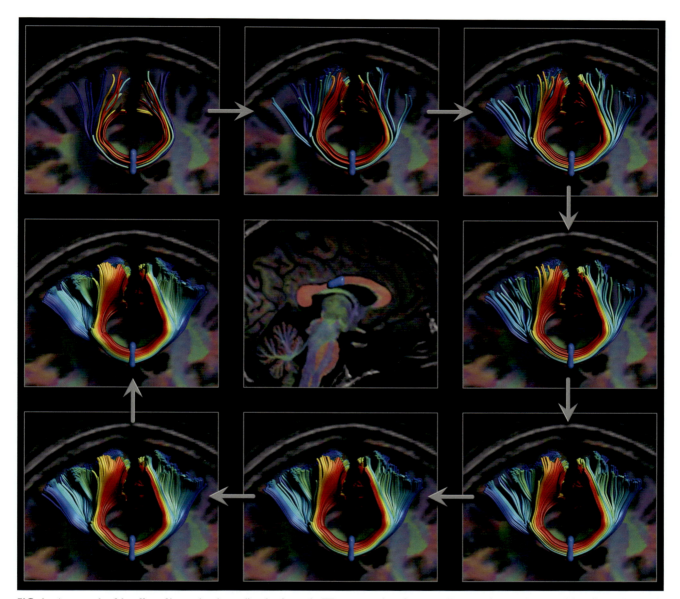

**FIG. 4** An example of the effect of increasing the seeding density on the FT reconstruction of corpus callosum pathways (the method described in Guo et al., 2020 was used). The *middle image* shows the seeding ROI (in *blue*) defined on a midsagittal slice consisting of 20 voxels. The *left top* shows a coronal view of the FT result where the center of each voxel is defined as a seed point (hence, 20 seed points). Following in a clockwise direction as indicated by the *arrows*, the number of seed points has been doubled each time (20 → 40 → ⋯ → 2560), showing a more complete representation of the pathways at each image. The color-encoding of the pathways highlights their topological configuration, with *blue-green* and *yellow-red* colors showing the medial and lateral parts of the callosal pathways, respectively.

Other pathway properties, such as torsion or length, can also be incorporated as termination thresholds to further refine the constraints imposed on the geometrical characteristics of fiber trajectories.

### Anisotropy threshold

High values of diffusion anisotropy, such as fractional anisotropy (FA) (Basser, 1995; Basser and Pierpaoli, 1996), are indicative of highly organized fiber tissue. In these regions with high FA values, the orientation used by the algorithm, that is, the first eigenvector with DTI-based FT, is then considered to be a good approximation of the orientation of the underlying fiber pathways. By contrast, in regions with low FA values, this assumption is likely not valid. A termination threshold based on the FA (i.e., $FA_t$) is therefore used to terminate pathway propagation if it drops below

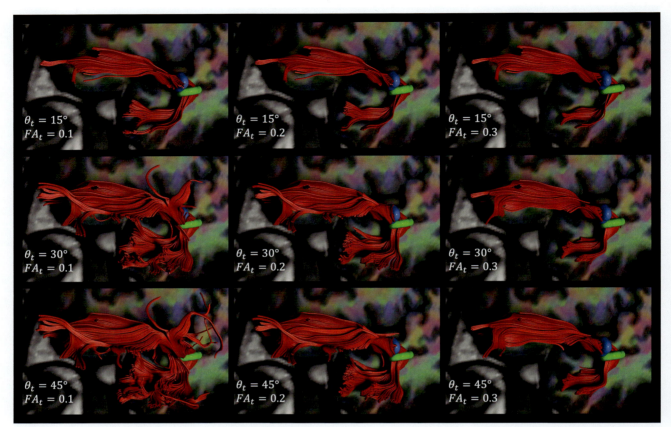

**FIG. 5** Pathways of the uncinate fasciculus reconstructed with different values for the user-defined termination threshold parameter settings $FA_t$ and $\theta_t$. The *blue and green* ROIs represent the seeding and the AND gates, respectively. A NOT gate (not shown here), similar to the one shown in the top-right image in Fig. 3, was used to leave out any pathways going to the back of the brain.

it. A value in the range of $FA_t \simeq [0.1-0.3]$ is typically used. With FT approaches based on SD techniques, a threshold based on the FOD is used, such as the "apparent fiber density" (Raffelt et al., 2012) or the "hindrance modulated orientational anisotropy" (Dell'Acqua et al., 2013).

Defining the correct termination thresholds is not trivial and may depend on the application and the fiber bundle of interest. Fig. 5 shows an example of this challenge where a fiber bundle was reconstructed with different settings for $FA_t$ (0.1, 0.2, and 0.3) and $\theta_t$ (15 degrees, 30 degrees, and 45 degrees) using the FT method described in Basser et al. (2000a) (with a step size of 1 mm). Overall, both a higher value for $FA_t$ and a lower value for $\theta_t$ will generally lead to a decreased rate of false-positive FT reconstructions, that is, one has more confidence in the validity of the computed pathways, but at the cost of an increased rate of false-negative FT reconstructions, that is, one may miss genuine (parts of) fiber pathways. This trade-off is most visible in Fig. 5 when comparing the left-bottom with the right-top image: with $FA_t = 0.1$ and $\theta_t = 45$ degrees, some parts of the streamlines may not be anatomically correct. By contrast, with $FA_t = 0.3$ and $\theta_t = 15$ degrees, the fiber bundle seems incomplete.

## 2.4 Beyond streamlines

Another way to perform FT is the use of so-called front evolution approaches, such as fast marching (FM) methods derived from level set theory (Osher and Sethian, 1988; Sethian, 1996a,b). In contrast to streamline propagation algorithms, these techniques model the evolution of an advancing 3D front through the WM by following local estimates of fiber orientations. Based on the FM algorithm of Tsitsiklis (1995), Parker et al. developed one of the first FM FT approaches (Parker, 2000; Parker et al., 2001, 2002b; Tsitsiklis, 1995). Here, the 3D front evolution was designed with a speed function proportional to the projection of the front normal to the first eigenvector. An intrinsic limitation, however, is the discrete nature of the front evolution, that is, the possible directions of propagation are limited by the number of neighboring voxels, which causes discretization errors. In the work of Tournier et al., an adaptive evolution grid was proposed to overcome this obstacle (Tournier et al., 2003).

Further developments in front-evolution-based FT approaches have been related to the description of the 3D front propagation speed function (O'Donnell et al., 2002; Campbell et al., 2002a,b; Jonasson et al., 2005; Kang et al., 2005; Duncan et al., 2004; Jackowski et al., 2004, 2005; Staempfli et al., 2006), the integration of prior knowledge and a global optimization criteria (Campbell et al., 2004a,b, 2005; Jbabdi et al., 2004), and the application of segmentation-based and directional correlation-based region growing techniques (Lin et al., 2003, 2005; Sun et al., 2001, 2003; Wang and Vemuri, 2004; Rousson et al., 2004; Feddern et al., 2003). Despite their initial promise, front-evolution-based FT approaches have not progressed into mainstream usage, partly because of their inability to provide unique point-to-point connection pathways.

## 3 Methodological considerations

Despite the numerous developments that have improved FT methods over the last decades, there are still many issues that complicate the interpretation of the results. These not only include the choice of parameter settings and differences in implementations and algorithms, but are also related to noise biases, error propagation, diffusion and fiber model selection, and the partial volume effect, among others. The reader is referred to in-depth reviews on such methodological considerations by Jones et al. (2013), Jeurissen et al. (2019), and Zhang et al. (2022). In the following, an example is presented where FT methods fail and can create so-called "phantom" pathways, that is, fiber trajectories that, based on our current knowledge of human brain anatomy, do not exist. The example also shows that even with a 1000-fold increase in spatial resolution compared to typical in vivo human brain data acquisitions to date, this issue is still present.

Let us consider the region of the pons where there are left-right oriented pontocerebellar tracts and up-down oriented corticopontine fiber pathways. The pontocerebellar pathways arise in the pontine nuclei, group together laterally, and curve dorsally to form the middle cerebellar peduncles (Naidich et al., 2009). About two-thirds of these fibers arise from the contralateral pontine nuclei and cross the midline (i.e., the basis pontis), while the remaining one-third originates from the ipsilateral pontine nuclei. The pontine nuclei also receive the corticopontine tracts from all lobes of the cerebrum. Note that the pontocerebellar pathways do not form any direct connections between the cerebellar hemispheres (Naidich et al., 2009).

Fig. 6 (top row) shows that by going to higher spatial resolutions, the thin left-right oriented pontocerebellar pathways become more visible. In addition, with higher spatial resolutions, the front-back-oriented pathways in the basis points become more visible in the area *between* the arrowheads. Notice, however, that the increasing spatial resolution does not change the left-right oriented diffusion orientations in the regions indicated by the arrowheads. In this case, the pathways connecting left cerebellum and right cerebrum are running parallel to those connecting right cerebellum and left cerebrum (Naidich et al., 2009). The observed dominant diffusion orientations are therefore originating from both sets of pathways and cannot be disambiguated at that resolution. Furthermore, in the pontine nuclei, where the corticopontine and pontocerebellar pathways end up, the up-down fiber orientations are not observed in the regions highlighted by the arrowheads.

The bottom row in Fig. 6 shows an FT reconstruction of pontocerebellar pathways *incorrectly* connecting the cerebellar hemispheres. This type of false-positive connection is extremely problematic given that it is not possible to disentangle the diffusion profiles, and consequently the fiber orientations, of the different fiber bundles.

Prior knowledge of the architectural configuration and microstructural information of fiber pathways is needed to further optimize FT reconstructions (Smith et al., 2012; Girard et al., 2014, 2017; Christiaens et al., 2015; Sherbondy et al., 2010; Chamberland et al., 2018; Neher et al., 2017; Zhylka et al., 2023). For instance, anatomical constraints have been proposed to minimize false-negative pathways (Zhylka et al., 2023). More specifically, by generating bifurcations in pathways with angles beyond the typical FT termination thresholds (e.g., angle deviations of $\theta_t \simeq 90$ degrees), a more reliable representation of known pathways can be obtained. Fig. 7 gives an example of the FT method presented in Zhylka et al. (2023), where branches are included to provide a more complete reconstruction of pathways connecting the motor cortex.

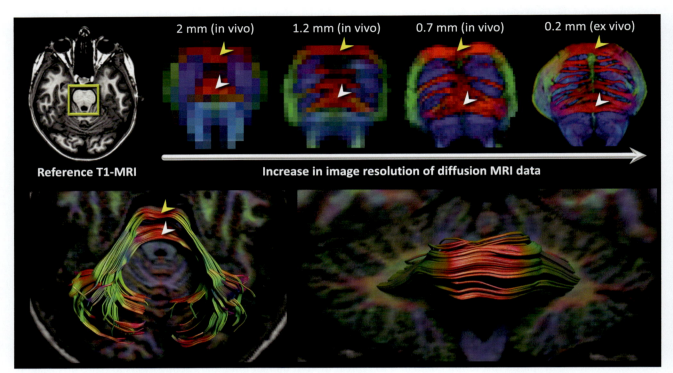

**FIG. 6** *Top row*: The *yellow square* on the leftmost image (T1-MRI) highlights the area of the pons. This region is also shown on the color-coded fractional anisotropy maps with increasingly higher spatial resolutions from left to right (2.0 mm, Leemans et al., 2006; 1.2 mm, Van Essen et al., 2013; 0.7 mm, Setsompop et al., 2018; 0.2 mm, Calabrese et al., 2015). The *red*, *green*, and *blue colors* represent the left-right, front-back, and up-down diffusion orientations, respectively, as determined by the underlying diffusion tensors. Although finer details in the pons can be seen with higher spatial resolutions, there are areas where the main diffusion orientations are not significantly changed by the increase in spatial resolution (see *arrowheads*). *Bottom row*: Superior (*left*) and frontal (*right*) views of pontocerebellar pathways connecting the left and right cerebellar regions (*arrowheads* reflecting the locations as shown in the *top row*). These are often called "phantom" pathways, as such connections do not exist as far as we know (Naidich et al., 2009).

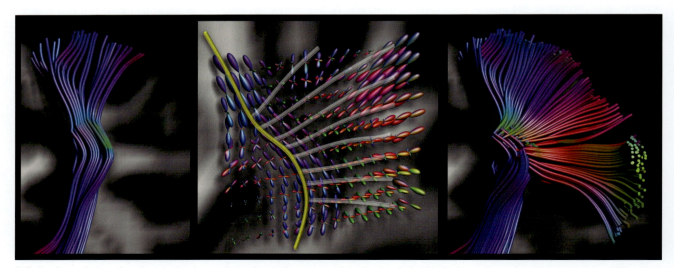

**FIG. 7** Coronal view of corticospinal pathways that end up in the cortex with seed points defined in the internal capsule using a *conventional* FT method (*left image*) and an FT method that also includes branches (*right image*). The *middle image* shows the fiber orientation distributions in that region with a schematic representation of this type of branching configuration: the *yellow line* represents one of the pathways obtained with the *conventional* FT method. The *white lines* represent the branches, which are only generated if a second peak was observed along the original (*yellow*) pathway.

# 4 Conclusion

FT is a technique for reconstructing estimates of fiber pathways in the brain. Since its inception in the late 1990s, many FT approaches have been developed and are now often classified into deterministic and probabilistic methods. While probabilistic FT methods address the uncertainty in computing fiber trajectories by incorporating data noise and model orientation uncertainty, they often rely on their deterministic counterparts (see Chapter 14). Focusing on the latter category, this chapter introduced the key concepts of FT and discussed several user-defined settings and implementation strategies.

# References

Aldroubi, A., Basser, P., 1999. Reconstruction of vector and tensor field from sampled discrete data. Contemp. Math. 247, 1–15.

Basser, P.J., 1995. Inferring microstructural features and the physiological state of tissues from diffusion-weighted images. NMR Biomed. 8 (7–8), 333–344.

Basser, P.J., 1998. Fiber-tractography via diffusion tensor MRI (DT-MRI). In: International Society for Magnetic Resonance in Medicine, p. 1226.

Basser, P.J., Pierpaoli, C., 1996. Microstructural and physiological features of tissues elucidated by quantitative-diffusion-tensor MRI. J. Magn. Reson. Ser. B 111 (3), 209–219.

Basser, P.J., Mattiello, J., Le Bihan, D., 1994a. Estimation of the effective self-diffusion tensor from the NMR spin echo. J. Magn. Reson. Ser. B 103 (3), 247–254.

Basser, P.J., Mattiello, J., Le Bihan, D., 1994b. MR diffusion tensor spectroscopy and imaging. Biophys. J. 66 (1), 259–267.

Basser, P.J., Pajevic, S., Pierpaoli, C., Duda, J., Aldroubi, A., 2000a. In vivo fiber tractography using DT-MRI data. Magn. Reson. Med. 44 (4), 625–632.

Basser, P.J., Pajevic, S., Pierpaoli, C., Aldroubi, A., Duda, J., 2000b. Fiber-tractography in human brain using diffusion tensor MRI (DT-MRI). In: International Society for Magnetic Resonance in Medicine, p. 784.

Basser, P.J., Pajevic, S., Pierpaoli, C., Aldroubi, A., 2002. Fiber tract following in the human brain using DT-MRI data. IEICE Trans. Inf. Syst. E85-D, 15–21.

Behrens, T.E.J., Jenkinson, M., Brady, J.M., Smith, S.M., 2002a. A probabilistic framework for estimating neural connectivity from diffusion weighted MRI. In: International Society for Magnetic Resonance in Medicine, p. 1142.

Behrens, T.E.J., Woolrich, M., Jenkinson, M., Brady, J.M., Smith, S.M., 2002b. Bayesian parameter estimation in diffusion weighted MRI. In: International Society for Magnetic Resonance in Medicine, p. 1160.

Behrens, T.E., Johansen-Berg, H., Woolrich, M.W., Smith, S.M., Wheeler-Kingshott, C., Barker, G.J., Boulby, P.A., Brady, J.M., Matthews, P.M., 2003a. Connectivity-based grey matter segmentation using diffusion imaging. In: International Society for Magnetic Resonance in Medicine, p. 245.

Behrens, T.E.J., Woolrich, M.W., Jenkinson, M., Johansen-Berg, H., Nunes, R.G., Clare, S., Matthews, P.M., Brady, J.M., Smith, S.M., 2003b. Characterization and propagation of uncertainty in diffusion-weighted MR imaging. Magn. Reson. Med. 50 (5), 1077–1088.

Calabrese, E., Hickey, P., Hulette, C., Zhang, J., Parente, B., Lad, S.P., Johnson, G.A., 2015. Postmortem diffusion MRI of the human brainstem and thalamus for deep brain stimulator electrode localization. Hum. Brain Mapp. 36 (8), 3167–3178. https://doi.org/10.1002/hbm.22836.

Campbell, J.S.W., Siddiqi, K., Pike, G.B., 2002a. White matter fibre tract likelihood evaluated using normalized RMS diffusion distance. In: International Society for Magnetic Resonance in Medicine, p. 1130.

Campbell, J.S.W., Siddiqi, K., Vemuri, B.C., Pike, G.B., 2002b. A geometric flow for white matter fibre tract reconstruction. In: International Symposium on Biomedical Imaging, pp. 505–508.

Campbell, J.S.W., Rymar, V., Sadikot, A.S., Siddiqi, K., Pike, G.B., 2004a. Comparison of flow- and streamline-based fibre tracking algorithms using an anisotropic diffusion phantom. In: International Society for Magnetic Resonance in Medicine, p. 1277.

Campbell, J.S.W., Siddiqi, K., Pike, G.B., 2004b. White matter fibre tractography and scalar connectivity assessment using fibre orientation likelihood distribution. In: Computer Vision and Pattern Recognition.

Campbell, J.S.W., Siddiqi, K., Rymar, V.V., Sadikot, A.F., Pike, G.B., 2005. Flow-based fiber tracking with diffusion tensor and Q-ball data: validation and comparison to principal diffusion direction techniques. NeuroImage 27 (4), 725–736.

Catani, M., Howard, R.J., Pajevic, S., Jones, D.K., 2002. Virtual in vivo interactive dissection of white matter fasciculi in the human brain. NeuroImage 17 (1), 77–94.

Chamberland, M., Tax, C.M.W., Jones, D.K., 2018. Meyer's loop tractography for image-guided surgery depends on imaging protocol and hardware. NeuroImage Clin. 2213-1582. 20, 458–465. https://doi.org/10.1016/j.nicl.2018.08.021.

Christiaens, D., Reisert, M., Dhollander, T., Sunaert, S., Suetens, P., Maes, F., 2015. Global tractography of multi-shell diffusion-weighted imaging data using a multi-tissue model. NeuroImage 1053-8119. 123, 89–101. https://doi.org/10.1016/j.neuroimage.2015.08.008.

Cointepas, Y., Mangin, J.F., Poupon, C., 2002. A spin glass based framework to reconstruct brain fiber bundles from images of the water diffusion process. Inf. Process. 2 (1), 30–36.

Conturo, T.E., Lori, N.F., Cull, T.S., Akbudak, E., Snyder, A.Z., Shimony, J.S., McKinstry, R.C., Burton, H., Raichle, M.E., 1999. Tracking neuronal fiber pathways in the living human brain. Proc. Natl. Acad. Sci. USA 96, 10422–10427.

Coulon, O., Alexander, D.C., Arridge, S.R., 2001. A regularization scheme for diffusion tensor magnetic resonance images. Lect. Notes Comput. Sci. 2082, 92–105.

Coulon, O., Alexander, D.C., Arridge, S., 2004. Diffusion tensor magnetic resonance image regularization. Med. Image Anal. 8 (1), 47–67.

# 252 PART | III Tractography algorithms

Dell'Acqua, F., Simmons, A., Williams, S.C.R., Catani, M., 2013. Can spherical deconvolution provide more information than fiber orientations? Hindrance modulated orientational anisotropy, a true-tract specific index to characterize white matter diffusion. Hum. Brain Mapp. 34 (10), 2464–2483. https://doi.org/10.1002/hbm.22080.

Delputte, S., Leemans, A., Fieremans, E., D'Asseler, Y., Lemahieu, I., Achten, R., Sijbers, J., Van de Walle, R., 2005. Density regularized fiber tractography of the brain white matter using diffusion tensor MRI. In: International Society for Magnetic Resonance in Medicine, p. 1309.

Duncan, J.S., Papademetris, X., Yang, J., Jackowski, M., Zeng, X., Staiba, L.H., 2004. Geometric strategies for neuroanatomic analysis from MRI. NeuroImage 23 (1), S34–S45.

Feddern, C., Weickert, J., Burgeth, B., 2003. Level-set methods for tensor valued images. In: IEEE Workshop on Variational, Geometric and Level Set Methods in Computer Vision, pp. 65–72.

Frandsen, J., Hobolth, A., Jensen, E.B., Vestergaard-Poulsen, P., Østergaard, L., 2004. Regularization of diffusion tensor fields in axonal fibre tracking. In: International Society for Magnetic Resonance in Medicine, p. 1221.

Georgiadis, M., Schroeter, A., Gao, Z., Guizar-Sicairos, M., Liebi, M., Leuze, C., McNab, J.A., Balolia, A., Veraart, J., Ades-Aron, B., Kim, S., Shepherd, T., Lee, C.H., Walczak, P., Chodankar, S., DiGiacomo, P., David, G., Augath, M., Zerbi, V., Sommer, S., Rajkovic, I., Weiss, T., Bunk, O., Yang, L., Zhang, J., Novikov, D.S., Zeineh, M., Fieremans, E., Rudin, M., 2021. Nanostructure-specific X-ray tomography reveals myelin levels, integrity and axon orientations in mouse and human nervous tissue. Nat. Commun. 12 (1), 2941. https://doi.org/10.1038/s41467-021-22719-7.

Girard, G., Whittingstall, K., Deriche, R., Descoteaux, M., 2014. Towards quantitative connectivity analysis: reducing tractography biases. NeuroImage 1053-8119. 98, 266–278. https://doi.org/10.1016/j.neuroimage.2014.04.074.

Girard, G., Daducci, A., Petit, L., Thiran, J.P., Whittingstall, K., Deriche, R., Wassermann, D., Descoteaux, M., 2017. AxTract: toward microstructure informed tractography. Hum. Brain Mapp. 38 (11), 5485–5500. https://doi.org/10.1002/hbm.23741.

Guo, F., Leemans, A., Viergever, M.A., Dell'Acqua, F., De Luca, A., 2020. Generalized Richardson-Lucy (GRL) for analyzing multi-shell diffusion MRI data. NeuroImage 1053-8119. 218, 116948. https://doi.org/10.1016/j.neuroimage.2020.116948.

Jackowski, M., Kao, C.Y., Qiu, M., Constable, R.T., Staib, L.H., 2004. Estimation of anatomical connectivity by anisotropic front propagation and diffusion tensor imaging. Lect. Notes Comput. Sci. 3217, 663–671.

Jackowski, M., Kao, C.Y., Qiu, M., Constable, R.T., Staib, L.H., 2005. White matter tractography by anisotropic wavefront evolution and diffusion tensor imaging. Med. Image Anal. 9 (5), 427–440.

Jbabdi, S., Bellec, P., Marrelec, G., Perlbarg, V., Benali, H., 2004. A level set method for building anatomical connectivity paths between brain areas using DTI. In: International Symposium on Biomedical Imaging, pp. 1024–1027.

Jeurissen, B., Descoteaux, M., Mori, S., Leemans, A., 2019. Diffusion MRI fiber tractography of the brain. NMR Biomed. 32 (4), e3785. https://doi.org/10.1002/nbm.3785.

Jonasson, L., Bresson, X., Hagmann, P., Cuisenaire, O., Meuli, R., Thiran, J.P., 2005. White matter fiber tract segmentation in DT-MRI using geometric flows. Med. Image Anal. 9 (3), 223–236.

Jones, D.K., Pierpaoli, C., 2004. Towards a marriage of deterministic and probabilistic tractography methods: bootstrap analysis of fiber trajectories in the human brain. In: International Society for Magnetic Resonance in Medicine, p. 1276.

Jones, D.K., Simmons, A., Williams, S.C.R., Horsfield, M.A., 1998. Noninvasive assessment of structural connectivity in white matter by diffusion tensor MRI. In: International Society for Magnetic Resonance in Medicine, p. 531.

Jones, D.K., Simmons, A., Williams, S.C.R., Horsfield, M.A., 1999. Non-invasive assessment of axonal fiber connectivity in the human brain via diffusion tensor MRI. Magn. Reson. Med. 42 (1), 37–41.

Jones, D.K., Knösche, T.R., Turner, R., 2013. White matter integrity, fiber count, and other fallacies: the do's and don'ts of diffusion MRI. NeuroImage 1053-8119. 73, 239–254. https://doi.org/10.1016/j.neuroimage.2012.06.081.

Kang, N., Zhang, J., Carlson, E.S., Gembris, D., 2005. White matter fiber tractography via anisotropic diffusion simulation in the human brain. IEEE Trans. Med. Imaging 24 (9), 1127–1137.

Kim, D.S., Kim, M., Ronen, I., Formisano, E., Kim, K.H., Ugurbil, K., Mori, S., Goebel, R., 2003. In vivo mapping of functional domains and axonal connectivity in cat visual cortex using magnetic resonance imaging. Magn. Reson. Imaging 21 (10), 1131–1140.

Kinosada, Y., Ono, M., Okuda, Y., Seta, H., Hada, Y., Hattori, T., Nomura, Y., Sakuma, H., Takeda, K., Ishii, Y., 1993. MR tractography: visualization of structure of nerve fiber system from diffusion weighted images with maximum intensity projection method. Nippon Igaku Hoshasen Gakkai Zasshi 53 (2), 171–179.

Koch, M.A., Norris, D.G., Hund-Georgiadis, M., 2002. An investigation of functional and anatomical connectivity using magnetic resonance imaging. NeuroImage 16 (1), 241–250.

Kuo, L.W., Wedeen, V.J., Weng, J.C., Reese, T.G., Chen, J.H., Tseng, W.Y.I., 2004. Mapping white matter connectivity with BOLD activated regions using diffusion spectrum imaging and fMRI. In: International Society for Magnetic Resonance in Medicine, p. 1286.

Larsen, L., Griffin, L.D., Grässel, D., Witte, O.W., Axer, H., 2007. Polarized light imaging of white matter architecture. Microsc. Res. Tech. 70 (10), 851–863. https://doi.org/10.1002/jemt.20488.

Lazar, M., Alexander, A.L., 2002. White matter tractography using random vector (RAVE) perturbation. In: International Society for Magnetic Resonance in Medicine, p. 539.

Lazar, M., Alexander, A.L., 2003. White matter tractography algorithms error analysis. NeuroImage 20 (2), 1140–1153.

Lazar, M., Alexander, A.L., 2005. Bootstrap white matter tractography (BOOT-TRAC). NeuroImage 24 (2), 524–532.

Lazar, M., Weinstein, D., Hasan, K., Alexander, A.L., 2000. Axon tractography with tensorlines. In: International Society for Magnetic Resonance in Medicine, p. 482.

Lazar, M., Hasan, K.M., Alexander, A.L., 2001. Bootstrap analysis of DT-MRI tractography techniques: streamlines and tensorlines. In: International Society for Magnetic Resonance in Medicine, p. 1527.

Lazar, M., Weinstein, D.M., Tsuruda, J.S., Hasan, K.M., Arafakinis, K., Meyerand, M.E., Badie, B., Rowley, H.A., Haughton, V., Field, A., Alexander, A.-L., 2003. White matter tractography using diffusion tensor deflection. Hum. Brain Mapp. 18 (4), 306–321.

Le Bihan, D., Mangin, J.F., Poupon, C., Clark, C.A., Pappata, S., Molko, N., Chabriat, H., 2001. Diffusion tensor imaging: concepts and applications. J. Magn. Reson. Imaging 13 (4), 534–546.

Leemans, A., Sijbers, J., De Backer, S., Vandervliet, E., Parizel, P.M., 2005. TRACT: tissue relative anisotropy based curvature thresholding for deterministic MR diffusion tensor tractography. In: European Society for Magnetic Resonance in Medicine and Biology, Basel, Switzerland, p. 289.

Leemans, A., Sijbers, J., De Backer, S., Vandervliet, E., Parizel, P., 2006. Multiscale white matter fiber tract coregistration: a new feature-based approach to align diffusion tensor data. Magn. Reson. Med. 55 (6), 1414–1423. https://doi.org/10.1002/mrm.20898.

Li, W., Tian, J., Dai, J., 2004a. Tensor field regularization for diffusion tensor MR images using nonlinear smoothing. In: International Society for Magnetic Resonance in Medicine, p. 1222.

Li, W., Tian, J., Dai, J., 2004b. White matter tractography based on minimizing the tracking cost model from diffusion tensor MRI. In: SPIE Medical Imaging, 5370, pp. 1795–1803. vol.

Lin, C.Y., Hong, C.Y., Song, S.K., Chang, C., 2003. Identify 3-dimensional white matter tracts by directional correlation based regional growing (DCRG) method. In: International Society for Magnetic Resonance in Medicine, p. 2163.

Lin, C.Y., Sun, S.W., Hong, C.Y., Chang, C., 2005. Unsupervised identification of white matter tracts in a mouse brain using a directional correlation-based region growing (DCRG) algorithm. NeuroImage 28 (2), 380–388.

Liu, C., 2010. Susceptibility tensor imaging. Magn. Reson. Med. 63 (6), 1471–1477. https://doi.org/10.1002/mrm.22482.

Mangin, J.F., Poupon, C., Cointepas, Y., Rivière, D., Papadopoulos-Orfanos, D., Clark, C.A., Régis, J., Le Bihan, D., 2002. A framework based on spin glass models for the inference of anatomical connectivity from diffusion-weighted MR data - a technical review. NMR Biomed. 15 (7–8), 481–492.

Mangin, J.F., Fillard, P., Cointepas, Y., Le Bihan, D., Frouin, V., Poupon, C., 2013. Toward global tractography. NeuroImage 1053-8119. 80, 290–296. https://doi.org/10.1016/j.neuroimage.2013.04.009.

McGraw, T., Vemuri, B.C., Chen, Y., Rao, M., Mareci, T., 2004. DT-MRI denoising and neuronal fiber tracking. Med. Image Anal. 8 (2), 95–111.

Mori, S., Barker, P.B., 1999. Diffusion magnetic resonance imaging: its principles and applications. Anat. Rec. B New Anat. 257 (3), 102–109.

Mori, S., Crain, B.J., van Zijl, P.C.M., 1998. 3D brain fiber reconstruction from diffusion MRI. In: International Conference on Functional Mapping (OHBM), p. S710.

Mori, S., Crain, B.J., Chacko, V.P., van Zijl, P.C.M., 1999a. Three-dimensional tracking of axonal projections in the brain by magnetic resonance imaging. Ann. Neurol. 45 (2), 265–269.

Mori, S., Xue, R., Crain, B.J., Solaiyappan, M., Chacko, V.P., van Zijl, P.C.M., 1999b. 3D reconstruction of axonal fibers from diffusion tensor imaging using fiber assignment by continuous tracking (FACT). In: International Society for Magnetic Resonance in Medicine, p. 320.

Mori, S., Kaufmann, W.E., Pearlson, G.D., Crain, B.J., Stieltjes, B., Solaiyappan, M., van Zijl, P.C., 2000. In vivo visualization of human neural pathways by magnetic resonance imaging. Ann. Neurol. 47 (3), 412–414.

Moseley, M.E., Cohen, Y.C., Kucharczyk, J., Asgari, H.S., Wendland, M.F., Tsuruda, J., Norman, D., 1990. Diffusion-weighted MR imaging of anisotropic water diffusion in cat central nervous system. Radiology 176 (2), 439–445.

Moseley, M.E., Kucharczyk, J., Asgari, H.S., Norman, D., 1991. Anisotropy in diffusion-weighted MRI. Magn. Reson. Med. 19 (2), 321–326.

Naidich, T.P., Duvernoy, H.M., Delman, B.N., Sorensen, A.G., Kollias, S.S., Haacke, E.M., 2009. Duvernoy's Atlas of the Human Brain Stem and Cerebellum: High-Field MRI, Surface Anatomy, Internal Structure, Vascularization and 3D Sectional Anatomy. Springer Science & Business Media.

Neher, P.F., Côté, M.A., Houde, J.C., Descoteaux, M., Maier-Hein, K.H., 2017. Fiber tractography using machine learning. NeuroImage 1053-8119. 158, 417–429. https://doi.org/10.1016/j.neuroimage.2017.07.028.

O'Donnell, L., Haker, S., Westin, C.F., 2002. New approaches to estimation of white matter connectivity in diffusion tensor MRI: elliptic PDEs and geodesics in a tensor-warped space. In: Medical Image Computing and Computer-assisted Intervention, 2489, pp. 459–466. vol.

Osher, S., Sethian, J.A., 1988. Fronts propagating with curvature-dependent speed: algorithms based on Hamilton-Jacobi formulations. J. Comput. Phys. 79 (1), 12–49.

Pagani, E., Filippi, M., Rocca, M.A., Horsfield, M.A., 2005. A method for obtaining tract-specific diffusion tensor MRI measurements in the presence of disease: application to patients with clinically isolated syndromes suggestive of multiple sclerosis. NeuroImage 26 (1), 258–265.

Pajevic, S., Aldroubi, A., Basser, P.J., 2002. A continuous tensor field approximation of discrete DT-MRI data for extracting microstructural and architectural features of tissue. J. Magn. Reson. 154 (1), 85–100.

Parker, G.J.M., 2000. Tracing fiber tracts using fast marching. In: International Society for Magnetic Resonance in Medicine, p. 85.

Parker, G.J.M., Alexander, D.C., 2003. Probabilistic Monte Carlo based mapping of cerebral connections utilising whole-brain crossing fibre information. In: Information Processing in Medical Imaging, vol. 2737, pp. 684–695.

Parker, G.J.M., Schnabel, J.A., Symms, M.R., Werring, D.J., Barker, G.J., 2000. Nonlinear smoothing for reduction of systematic and random errors in diffusion tensor imaging. J. Magn. Reson. Imaging 11 (6), 702–710.

Parker, G.J.M., Wheeler-Kingshott, C.A.M., Barker, G.J., 2001. Distributed anatomical brain connectivity derived from diffusion tensor imaging. In: Information Processing in Medical Imaging, pp. 106–120.

Parker, G.J.M., Barker, G.J., Thacker, N., Jackson, A., 2002a. A framework for a streamline-based Probabilistic Index of Connectivity (PICo) using a structural interpretation of anisotropic diffusion. In: International Society for Magnetic Resonance in Medicine, p. 1165.

Parker, G.J.M., Wheeler-Kingshott, C.A.M., Barker, G.J., 2002b. Estimating distributed anatomical connectivity using fast marching methods and diffusion tensor imaging. IEEE Trans. Med. Imaging 21 (5), 505–512.

Parker, G.J.M., Haroon, H.A., Wheeler-Kingshott, C.A.M., 2003. A framework for a streamline-based Probabilistic Index of Connectivity (PICo) using a structural interpretation of MRI diffusion measurements. J. Magn. Reson. Imaging 18 (2), 242–254.

Poupon, C., Mangin, J.F., Frouin, V., Régis, J., Poupon, F., Pachot-Clouard, M., Le Bihan, D., Bloch, I., 1998. Regularization of MR diffusion tensor maps for tracking brain white matter bundles. Lect. Notes Comput. Sci. 1496, 489–498.

Poupon, C., Clark, C.A., Frouin, V., Bloch, I., Le Bihan, D., Mangin, J.F., 1999a. Tracking white matter fascicles with diffusion tensor imaging. In: International Society for Magnetic Resonance in Medicine, p. 325.

Poupon, C., Clark, C.A., Frouin, V., Le Bihan, D., Bloch, I., Mangin, J.F., 1999b. Inferring the brain connectivity from MR diffusion tensor data. In: Taylor, C., Colchester, A. (Eds.), Medical Image Computing and Computer-Assisted Intervention - MICCAI'99, Springer, Berlin, Heidelberg, pp. 453–462.

Poupon, C., Clark, C.A., Frouin, V., Régis, J., Bloch, I., Le Bihan, D., Mangin, J.F., 2000. Regularization of diffusion-based direction maps for the tracking of brain white matter fascicles. NeuroImage 12 (2), 184–195.

Poupon, C., Mangin, J.F., Clark, C.A., Frouin, V., Régis, J., Le Bihan, D., Bloch, I., 2001. Towards inference of human brain connectivity from MR diffusion tensor data. Med. Image Anal. 5 (1), 1–15.

Press, W.H., Flannery, B.P., Teukolsky, S.A., Vetterling, W.T., 1992. Numerical Recipes in C: The Art of Scientific Computing. Cambridge University Press, New York, NY.

Raffelt, D., Tournier, J.D., Rose, S., Ridgway, G.R., Henderson, R., Crozier, S., Salvado, O., Connelly, A., 2012. Apparent fibre density: a novel measure for the analysis of diffusion-weighted magnetic resonance images. NeuroImage 1053-8119. 59 (4), 3976–3994. https://doi.org/10.1016/j.neuroimage.2011.10.045.

Reisert, M., Kiselev, V.G., 2011. Fiber continuity: an anisotropic prior for ODF estimation. IEEE Trans. Med. Imaging 30 (6), 1274–1283. https://doi.org/10.1109/TMI.2011.2112769.

Reisert, M., Mader, I., Anastasopoulos, C., Weigel, M., Schnell, S., Kiselev, V., 2011. Global fiber reconstruction becomes practical. NeuroImage 1053-8119. 54 (2), 955–962. https://doi.org/10.1016/j.neuroimage.2010.09.016.

Reisert, M., Kiselev, V.G., Dihtal, B., Kellner, E., Novikov, D.S., 2014. MesoFT: unifying diffusion modelling and fiber tracking. In: Golland, P., Hata, N., Barillot, C., Hornegger, J., Howe, R. (Eds.), Medical Image Computing and Computer-Assisted Intervention - MICCAI 2014, Springer International Publishing, Cham, pp. 201–208.

Rousson, M., Lenglet, C., Deriche, R., 2004. Level set and region based surface propagation for diffusion tensor MRI segmentation. Lect. Notes Comput. Sci. 3117, 123–134.

Sethian, J.A., 1996a. A fast marching level set method for monotonically advancing fronts. In: Proceedings of the National Academy of Sciences of the United States of America, pp. 1591–1595.

Sethian, J.A., 1996b. Level Set Methods: Evolving Interfaces in Geometry, Fluid Mechanics, Computer Vision, and Materials Science. Cambridge Monographs on Applied and Computational Mathematics, Cambridge University Press, Cambridge, UK.

Setsompop, K., Fan, Q., Stockmann, J., Bilgic, B., Huang, S., Cauley, S.F., Nummenmaa, A., Wang, F., Rathi, Y., Witzel, T., Wald, L.L., 2018. High-resolution in vivo diffusion imaging of the human brain with generalized slice dithered enhanced resolution: simultaneous multislice (gSlider-SMS). Magn. Reson. Med. 79 (1), 141–151. https://doi.org/10.1002/mrm.26653.

Sherbondy, A.J., Rowe, M.C., Alexander, D.C., 2010. MicroTrack: an algorithm for concurrent projectome and microstructure estimation. In: Jiang, T., Navab, N., Pluim, J.P.W., Viergever, M.A. (Eds.), Medical Image Computing and Computer-Assisted Intervention - MICCAI 2010, Springer, Berlin, Heidelberg, pp. 183–190.

Smith, R.E., Tournier, J.D., Calamante, F., Connelly, A., 2012. Anatomically-constrained tractography: improved diffusion MRI streamlines tractography through effective use of anatomical information. NeuroImage 1053-8119. 62 (3), 1924–1938. https://doi.org/10.1016/j.neuroimage.2012.06.005.

Staempfli, P., Jaermann, T., Crelier, G.R., Kollias, S., Valavanis, A., Boesiger, P., 2006. Resolving fiber crossing using advanced fast marching tractography based on diffusion tensor imaging. NeuroImage 30 (1), 110–120.

Sun, S.W., Song, S.K., Hong, C.Y., Chu, W.C., Chang, C., 2001. Improving relative anisotropy measurement using directional correlation of diffusion tensors. Magn. Reson. Med. 46 (6), 1088–1092.

Sun, S.W., Song, S.K., Hong, C.Y., Chu, W.C., Chang, C., 2003. Directional correlation characterization and classification of white matter tracts. Magn. Reson. Med. 49 (2), 271–275.

Tench, C.R., Morgan, P.S., Wilson, M., Blumhardt, L.D., 2002. White matter mapping using diffusion tensor MRI. Magn. Reson. Med. 47 (5), 967–972.

Thottakara, P., Lazar, M., Johnson, S.C., Alexander, A.L., 2006. Application of Brodmann's area templates for ROI selection in white matter tractography studies. NeuroImage 29 (3), 868–878.

Tournier, J.D., Calamente, F., Gadian, D.G., Connelly, A., 2003. Diffusion-weighted magnetic resonance imaging fibre tracking using a front evolution algorithm. NeuroImage 20 (1), 276–288.

Tsitsiklis, J.N., 1995. Efficient algorithms for globally optimal trajectories. IEEE Trans. Autom. Control 40, 1528–1538.

Tuch, D.S., Belliveau, J.W., Wedeen, V.J., 2000. A path integral approach to white matter tractography. In: International Society for Magnetic Resonance in Medicine, p. 791.

Van Essen, D.C., Smith, S.M., Barch, D.M., Behrens, T.E.J., Yacoub, E., Ugurbil, K., 2013. The WU-Minn human connectome project: an overview. NeuroImage 1053-8119. 80, 62–79. https://doi.org/10.1016/j.neuroimage.2013.05.041.

Vemuri, B.C., Chen, Y., Rao, M., McGraw, T., Wang, Z., Mareci, T., 2001. Fiber tract mapping from diffusion tensor MRI. In: IEEE Workshop on Variational, Geometric and Level Set Methods in Computer Vision.

Vemuri, B.C., Chen, Y., Rao, M., Wang, Z., McGraw, T., Mareci, T., Blackband, S.J., Reier, P., 2002. Automatic fiber tractography from DTI and its validation. In: International Symposium on Biomedical Imaging, pp. 501–504.

Wang, Z., Vemuri, B.C., 2004. Tensor field segmentation using region based active contour model. Lect. Notes Comput. Sci. 3024, 304–315.

Wang, Z., Vemuri, B.C., Chen, Y., Mareci, T.H., 2004. A constrained variational principle for direct estimation and smoothing of the diffusion tensor field from complex DWI. IEEE Trans. Med. Imaging 23 (8), 930–939.

Wedeen, V.J., Davis, T.L., Weisskoff, R.M., Tootell, R., Rosen, B.R., Belliveau, J.W., 1995. White matter connectivity explored by MRI. In: 1st International Conference on Functional Mapping of the Human Brain (OHBM - Paris), p. 69.

Weinstein, D., Kindlmann, G., Lundberg, E., 1999. Tensorlines: advection-diffusion based propagation through diffusion tensor fields. In: IEEE Visualization Conference, pp. 249–254.

Weng, J.C., Chen, C.M., Tseng, W.Y.I., Lin, C.P., Chen, J.H., 2002. A global approach for non-invasive axonal fiber tracking on diffusion tensor magnetic resonance image. In: International Society for Magnetic Resonance in Medicine, p. 1133.

Westin, C.F., Maier, S.E., Khidhir, B., Everett, P., Jolesz, F.A., Kikinis, R., 1999. Image processing for diffusion tensor magnetic resonance imaging. In: Medical Image Computing and Computer-assisted Intervention, pp. 441–452.

Westin, C.F., Maier, S.E., Mamata, H., Nabavi, A., Jolesz, F.A., Kikinis, R., 2002. Processing and visualization for diffusion tensor MRI. Med. Image Anal. 6 (2), 93–108.

Wünsche, B., van der Linden, J., Holmberg, N., 2005. DTI volume rendering techniques for visualising the brain anatomy. Int. Congr. Ser. 1281, 80–85.

Yeung, P., Pope, S., 1988. An algorithm for tracking fluid particles in numerical simulations of homogeneous turbulence. J. Comput. Phys. 79, 373–416.

Zhang, S., Bastin, M.E., Laidlaw, D.H., Sinha, S., Armitage, P.A., Deisboeck, T.S., 2004. Visualization and analysis of white matter structural asymmetry in diffusion tensor MRI data. Magn. Reson. Med. 51 (1), 140–147.

Zhang, F., Daducci, A., He, Y., Schiavi, S., Seguin, C., Smith, R.E., Yeh, C.H., Zhao, T., O'Donnell, L.J., 2022. Quantitative mapping of the brain's structural connectivity using diffusion MRI tractography: a review. NeuroImage 1053-8119. 249, 118870. https://doi.org/10.1016/j.neuroimage.2021.118870.

Zhylka, A., Leemans, A., Pluim, J.P.W., De Luca, A., 2023. Anatomically informed multi-level fiber tractography for targeted virtual dissection. Magn. Reson. Mater. Phys. Biol. Med. 36 (1), 79–93. https://doi.org/10.1007/s10334-022-01033-3.

# Chapter 14

# Probabilistic tractography

Gabriel Girard[a,b], Dogu Baran Aydogan[c,d,e], Flavio Dell'Acqua[f], Alexander Leemans[g], Maxime Descoteaux[h], and Stamatios N. Sotiropoulos[i,j,k]

[a]Department of Computer Science, University of Sherbrooke, Sherbrooke, QC, Canada, [b]Signal Processing Laboratory (LTS5), École Polytechnique Fédérale de Lausanne (EPFL), Lausanne, Switzerland, [c]A.I. Virtanen Institute for Molecular Sciences, University of Eastern Finland, Kuopio, Finland, [d]Department of Neuroscience and Biomedical Engineering, Aalto University School of Science, Espoo, Finland, [e]Department of Psychiatry, Helsinki University Hospital, Helsinki, Finland, [f]NATBRAINLAB, Department of Neuroimaging, Institute of Psychiatry, Psychology and Neuroscience, King's College London, London, United Kingdom, [g]PROVIDI lab, Image Sciences Institute, University Medical Center Utrecht, Utrecht, The Netherlands, [h]Sherbrooke Connectivity Imaging Laboratory (SCIL), Department of Computer Science, University of Sherbrooke, Sherbrooke, QC, Canada, [i]Sir Peter Mansfield Imaging Centre, School of Medicine, University of Nottingham, Nottingham, United Kingdom, [j]Wellcome Centre for Integrative Neuroimaging (WIN), Oxford Centre for Functional Magnetic Resonance Imaging of the Brain (FMRIB), University of Oxford, Oxford, United Kingdom, [k]NIHR Biomedical Research Centre, University of Nottingham, Nottingham University Hospitals NHS Trust, Nottingham, United Kingdom

## 1 Introduction

As outlined in the previous chapters, tractography approaches mapping of white matter bundles in the brain using diffusion MRI data. This mapping uses a model (streamlines) to calculate the paths in the brain along which diffusion of water molecules is least hindered. A fundamental assumption is then made that these least-hindrance pathways are good approximations of the underlying white matter tracts, as water diffusion is expected to be faster along rather than across tracts. While this assumption underpins all tractography methods, there are different ways of approaching the path calculation. Deterministic approaches, overviewed in the previous chapter, and probabilistic approaches that we overview here are two main groups of methods for estimating streamlines.

Probabilistic tractography, first introduced by Parker et al. (2003) and Behrens et al. (2003), performs a stochastic estimation of streamlines. Given a seed point, instead of obtaining a single streamline arising from this seed as in deterministic tracking, it estimates a spatial distribution of streamlines to reflect the fact that there is uncertainty with the estimated streamlines. To do so, it uses perturbations in the main fiber orientations in each voxel to generate streamline samples from that target spatial distribution. These perturbations are obtained as random samples from voxel-wise orientation distribution functions (ODFs) or from a distribution of a discrete number of fibers orientations, described in detail in Chapters 10–12. Why is it useful to estimate this spatial distribution in the first place? There are a number of reasons supported by the different types of probabilistic tractography approaches. The two main reasons that we will focus on here are: (i) capturing estimation uncertainty and (ii) capturing fiber orientation dispersion.

(i) The first group of methods follow in spirit the first approaches of Parker et al. (2003) and Behrens et al. (2003), and provide a spatial distribution that reflects uncertainty in the estimated streamlines (Behrens et al., 2007; Berman et al., 2008; Campbell et al., 2014; Hernandez-Fernandez et al., 2019; Kaden et al., 2007; Lazar and Alexander, 2005; Jbabdi et al., 2007, 2012; Jeurissen et al., 2011, 2014; Jones, 2003; Malcolm et al., 2010; Parker and Alexander, 2005; Pajevic and Basser, 2003; Pontabry and Rousseau, 2011; Whitcher et al., 2008; Ye and Prince, 2017; Zhang et al., 2009). Uncertainty is inherent in all features estimated from imaging, and diffusion MRI is no exception. It can be caused by a number of sources including noise in the imaging process, nonoptimal sampling, preprocessing imperfections, and modeling/algorithmic limitations. While it is very difficult to disentangle the different sources of uncertainty, probabilistic estimation provides a principled way for characterizing it. Therefore deterministic and probabilistic tractography can be seen as two sides of the same coin (model). The former aims to resolve the most likely streamline for a seed point, while the latter aims to capture the distribution of likely streamlines for a seed point given the data (i.e., the difference is effectively estimating the mean/mode of a histogram vs. the histogram itself). This distribution will capture the most likely streamline (close to what deterministic tracking will reconstruct), but it will also provide a measure of confidence around it given all the uncertainty sources; a narrow distribution provides higher confidence in the estimation, while a wider distribution denotes a lower confidence. These methods are described in Section 4.

Handbook of Diffusion MR Tractography. https://doi.org/10.1016/B978-0-12-818894-1.00030-6
Copyright © 2025 Elsevier Ltd. All rights are reserved, including those for text and data mining, AI training, and similar technologies.

(ii) A second group of methods uses the same machinery of probabilistic reconstruction (i.e., estimating a spatial histogram of streamlines per seed point), but instead of uncertainty-based perturbations, they incorporate voxel-wise measures of fiber orientation dispersion, a biophysical property of the system that is being measured (Aranda et al., 2014; Aydogan and Shi, 2021; Descoteaux et al., 2009; Girard et al., 2014; Jeurissen et al., 2014; Perrin et al., 2005; Tournier et al., 2007, 2010, 2012, 2019). These approaches aim to overcome challenges introduced by reducing complex within-voxel fiber patterns to a set of a few (1–3) discrete orientations. They do so by sampling randomly from the whole voxel-wise fiber orientation distribution functions (fODFs) directly (see Chapters 10–12). These probabilistic methods relax the constraint of deterministic tractography methods that follow the few main fiber orientations and instead follow sampled orientations from that distribution. Rather than assuming that white matter fiber populations are perfectly aligned in the main voxel-wise orientations, they assume that the fODF width provides information about fiber orientation dispersion. High dispersion (e.g., because of fiber fanning, branching) will lead to a wide distribution, while low dispersion (because of a highly coherent bundle) will lead to a narrow distribution. This group of methods is presented in Section 5.

Fig. 1 shows a toy example of probabilistic streamline tractography in voxels with different orientation distributions. The orientation distributions are assumed here to be obtained either from model uncertainty estimation or from an estimation of the voxel-wise axonal dispersion. Probabilistic tractography estimates likely trajectories of white matter fibers by sampling these voxel-wise orientation distributions. Concentrated orientation distributions (see Fig. 1A) produce a narrower bundle of streamlines, compared to wider orientation distributions (see Fig. 1B). Moreover, initiating streamlines at the same location will result in multiple bundles of streamlines when the trajectories cross voxels with multiple fiber populations (see Fig. 1C). Instead, deterministic tractography methods will generate a single streamline from the seed location. This can correspond to the trajectory with the highest confidence or fiber density, but this is not always the case. For instance, in this toy example, the reconstructed deterministic streamline will be almost identical in all three cases shown in Fig. 1. Probabilistic tractography uses the data to estimate a set of possible trajectories, which will also include this deterministic trajectory.

The two different flavors of probabilistic tractography methods (uncertainty and dispersion-based) can be linked back to the first principles of conditioning an estimated distribution, that is, given which known quantities and uncertainty sources the probabilistic histogram of streamlines is reconstructed. Uncertainty-based tracking estimates streamlines given the data and the fiber orientation model (Behrens et al., 2007; Berman et al., 2008; Campbell et al., 2014; Hernandez-Fernandez et al., 2019; Kaden et al., 2007; Lazar and Alexander, 2005; Jbabdi et al., 2007, 2012; Jeurissen et al., 2011, 2014; Jones, 2003; Malcolm et al., 2010; Parker and Alexander, 2005; Pajevic and Basser, 2003; Pontabry and Rousseau, 2011; Whitcher et al., 2008; Ye and Prince, 2017; Zhang et al., 2009). These approaches intrinsically reflect noise in the data and can be further subdivided into groups based on which different orientation features they consider the uncertainty of (for instance, uncertainty in estimating the peaks only vs. the whole fODF). On the other hand, dispersion-based tracking methods (Aranda et al., 2014; Aydogan and Shi, 2021; Descoteaux et al., 2009; Girard et al., 2014; Jeurissen et al., 2014; Perrin et al., 2005; Tournier et al., 2007, 2010, 2012, 2019) obtain streamline distributions given the fODFs and the modeled dispersion (with fODFs being either deterministically or stochastically estimated from the data).

Such different conditioning approaches will crucially produce different path estimates and answer different questions. The former group represents a stochastic estimation framework that uses estimation uncertainty in the fODF to effectively

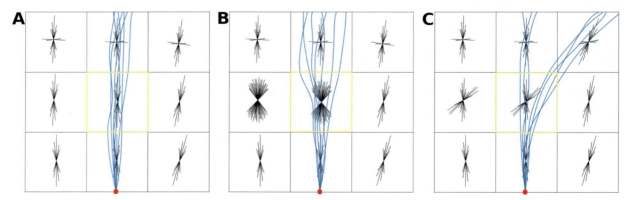

**FIG. 1** Probabilistic tractography through a voxel with one fiber population (A, B) or two fiber populations (C). The length of the *black lines* indicates the relative probability of tractography selecting the orientation for the propagation. A wide set of orientations indicate either uncertainty in the estimation of the fiber population main orientation (low confidence) or orientation dispersion estimated in the tissue being measured. Probabilistic tractography estimates likely trajectories of white matter fibers using perturbation of main fiber orientations, representing the model uncertainty or the fiber orientation dispersion. All streamlines were started at the same initial position (seed) indicated by the *red circle*.

produce a confidence interval on streamlines, as described earlier. On the other hand, the latter group aims to overcome challenges introduced by reducing complex within-voxel fiber patterns to a few discrete orientations, by sampling from the fODF directly. Both groups of methods are useful in their own right, but have been traditionally lumped under the same term of probabilistic tractography. We will make this distinction clearer throughout this chapter, but it is worth pointing out it is often very challenging to dissociate the different sources of orientation perturbations due to uncertainty and fiber dispersion, as both affect results in nontrivial ways.

In the following sections, we overview in greater detail the preceding principles. We start by discussing the sources of orientation uncertainty and orientation dispersion. Next, we describe the paradigm for probabilistic tractography and we overview the main approaches. We categorize them based on whether they are weighted toward orientation uncertainty or dispersion. We conclude the chapter by discussing interpretation and applications.

## 2 Orientation dispersion and uncertainty

As already briefly discussed, probabilistic tractography approaches utilize voxel-wise ODFs, functions defined on the sphere, to sample perturbations from the main fiber orientations in each voxel. There are two functions that are important to disentangle: the fODF that we have seen before in Chapters 10–12 and the uncertainty orientation distribution function (uODF) that will be presented in this chapter. To help describe the fODF and uODF, we define the axonal orientation distribution function (aODF) as the true distribution of the axon segment in a voxel. The aODF is a spherical function representing the proportion of the biological axon segments aligned along each orientation from the center of the voxel. The aODFs are asymmetrical, and they are defined with an infinite angular resolution and a fixed spatial resolution (voxel size). Since they represent the real biological orientation of axons, they capture both the microdispersion of axons, such as undulating fibers, and the macrodispersion of fanning, branching, and crossing fiber configurations. Dispersion will create wide aODF profiles. Although the aODF cannot be fully measured in all voxels with diffusion MRI, related information can be estimated. The estimate depends on the selected reconstruction method and on the diffusion MRI data.

The fODF is an estimate for the aODF and represents a biophysical property of the tissue—the proportion of fibers that lie along each orientation described by the aODF (see Chapter 12). It can be used to probe white matter fiber orientation dispersion caused by the lack of coherence in the fiber populations that coexist in a voxel. For instance, fibers branching out/merging in toward a direction, taking a steep bend, and/or crossing with similar incoherent patterns will lead to fiber orientation dispersion. As we have seen before in Chapters 10–12, this will lead to wide fODF profiles.

On the other hand, the uODF is not a physical property of the system that is being measured. It is a function that describes our confidence in the estimate of the aODF. It allows us to assess confidence in reconstructed tractography pathways, by directly representing our uncertainty in the measurements of fiber orientation upon which tract propagation is based. But why does such uncertainty exist in the first place? As with any estimated quantity from any model of real-world noisy data, we can never be confident about the exact orientations that we measure. Therefore estimated orientations have uncertainty associated with them, which is propagated to streamline uncertainty during tractography. This uncertainty exists regardless of whether it is formally characterized by the estimation framework (i.e., as in probabilistic tracking) or not (i.e., as in deterministic tracking).

Several sources contribute to the fiber orientation uncertainty (see Fig. 2). The most typical and less predictable source is thermal noise, which inherently corrupts diffusion MRI data. Noise gives rise to what is known as aleatoric uncertainty; repeating exactly the same imaging experiment will lead to a slightly different measurement, which in turn will lead to differences in the estimates. Although methods can partially alleviate noise effects and artifacts (see Chapter 9), other

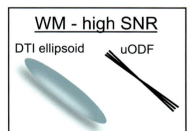 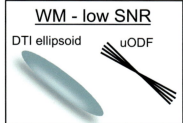 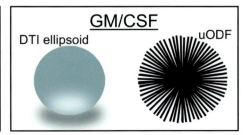

**FIG. 2** Examples of what uncertainty orientation distribution functions (uODFs) look like in two cases of highly coherent white matter (WM) and isotropic gray matter (GM) or cerebrospinal fluid (CSF)-filled areas. For the WM example, two cases are shown, one with high and one with low signal-to-noise ratio (SNR); uODFs will be wider for lower SNR (aleatoric uncertainty) and importantly for cases where a fiber orientation is not supported by the data as in GM or CSF (epistemic uncertainty).

sources of structured or so-called epistemic uncertainty exist. These include (i) acquisition parameters (e.g., nonoptimal or incomplete sampling and limited number of directions/contrast), (ii) modeling errors (e.g., simplifications in fiber orientation modeling, as partial volume and fiber dispersion due to within-voxel averaging increase the complexity of the diffusion signal), (iii) algorithmic errors (e.g., data processing choices that alter the assumed statistical properties of the data), and (iv) degeneracy in modeling parameters (i.e., multiple parameter values leading to the same signal prediction).

A good demonstration of a mixture of both effects (uncertainty and dispersion) was presented early on in Jones (2003). The study explored the variability in the fiber orientation across multiple repeats of the same diffusion MRI acquisition and visualized it with cones representing the distribution of orientations estimated across repeats. The width of these cones collectively captured one or more of the previous sources. Specifically, a bootstrapping approach was used to establish how robust the main fiber orientations, estimated from a diffusion tensor model, were across the repeated acquisitions. This robustness was represented as a 95% confidence interval (cone) in the main diffusion tensor fiber orientation. Fig. 3 shows these cones of variability estimated on in vivo human data. While the cone angle is not zero in regions of a relatively coherent single fiber population, such as the corpus callosum, it is very small, meaning that the variability of the main tensor orientation is low there. The source of this variability is likely noise and minor local fiber orientation dispersion. However, even if MRI noise is likely similar in neighboring voxels, Fig. 3 clearly shows a sharp increase in the width of the cones in voxels of the centrum semiovale, where multiple fiber populations cross. Because the diffusion tensor cannot adequately model multiple fiber populations, epistemic uncertainty in the main fiber orientation (modeling errors) increases. At the same time, fiber orientations outside the corpus callosum will likely exhibit higher geometric fiber dispersion, which also contributes to the large angles of the cones, in addition to uncertainty. Probabilistic tractography techniques can use these sources of variability in the estimation process of streamline trajectories.

It is important to point out that even if in theory we can describe sources of orientation uncertainty and orientation dispersion independently, perfectly dissociating the two is not feasible in practice. In the presence of dispersion in the aODFs, estimation uncertainty is also higher for most orientation models (we are more confident for the dominant fiber orientations than the secondary ones), so uODFs will be typically wider. On the other hand, the shape of the fODFs captures dispersion, but it also reflects a number of other factors—for instance, noise, reconstruction method, partial volume, harmonic orders in nonparametric methods, or deviation from modeling assumptions in parametric methods; see Sotiropoulos et al. (2012). In the presence of aleatoric or epistemic uncertainty sources, the width of the fODF can change in nontrivial

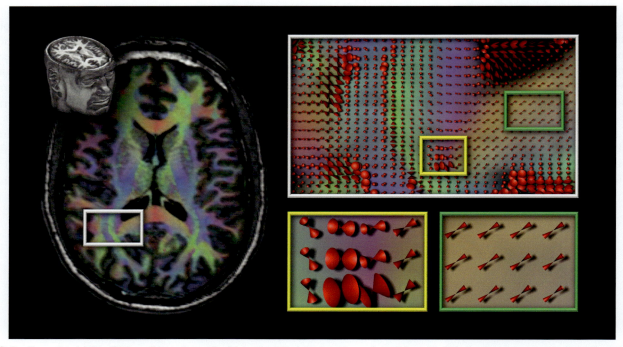

**FIG. 3** Cone of variability (95% confidence interval) of the main diffusion orientation estimated from the diffusion tensor model on in vivo human data. Voxels of the corpus callosum (*green box*) show high confidence level, while voxels of the centrum semiovale (*yellow*) show high variability (large cone angle) in the main diffusion orientation. This is likely due to the presence of multiple fiber population crossing, not adequately modeled by the diffusion tensor (epistemic uncertainty); and in addition due to the larger fiber dispersion in the centrum semiovale compared to the corpus callosum.

ways for the same aODF. As such, the orientation distributions shown in Fig. 1 can either be the results of uncertainty in the measurements, the presence of dispersion in the aODFs, or both. Similarly, the cones shown in Fig. 3 indicate more variability, both for increased orientation dispersion in the aODF and uncertainty in the data. Hence, the different tractography approaches can in practice only upweight the contribution of uncertainty or dispersion in their spatial distribution estimation, but cannot easily dissociate the two. Nevertheless, even if it is challenging to disentangle the individual effects of the multiple sources of orientation variability, probabilistic tractography approaches as a whole can formally characterize it.

## 3 The main idea of probabilistic propagation and tracking

As we saw before, probabilistic estimation starts by considering voxel-wise orientation distributions, which can be used to sample perturbations in the fiber orientations. These orientation distributions capture locally either uncertainty in the estimation (uODF) or the fiber dispersion characterized (fODF). Starting from a seed, the local information is propagated spatially with tractography. To achieve this, the probabilistic reconstruction entails an iterative process for building a spatial histogram of streamlines. In each iteration, random samples are obtained from the underlying orientation distributions (uODF or fODF) in each voxel. These orientation samples are used to propagate a new streamline. Fig. 4 depicted the estimation of a streamline from a seed location. The streamline propagates using discrete step size $\Delta s$. Moreover, the sampled orientations are constrained to be within a maximum deviation angle $\alpha$ from the previously sampled orientation, to promote smoothness of the trajectories. Similar to deterministic tracking, as orientation distributions are available on a discrete voxel grid, interpolation methods are typically used to provide estimates for intermediate locations. When propagation ends by reaching a stopping criterion, such as exiting the tracking volume, the process restarts at the seed and a new streamline gets generated. Reconstructing multiple streamlines from a seed given the considered perturbations builds up a spatial distribution of streamlines arising from that seed.

As the iterations are governed by random selection, use identical assumptions, and are independent of each other (often called in statistics the Monte Carlo method), each new streamline will be different from the ones obtained from the previous iterations. The difference depends on the level of underlying uncertainty/dispersion. Wide uODFs/fODFs along the trajectories of the targeted tract will lead to a wide streamline histogram and vice versa. This process is effectively equivalent to running many repeats of deterministic tracking but randomly perturbing the underlying orientations (i.e., peaks) before each repetition, and with the magnitude of perturbation depending on the width of the voxel-wise orientation distributions.

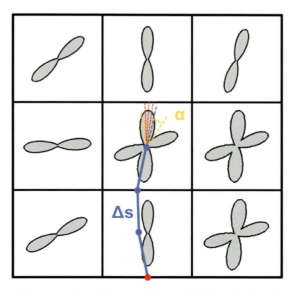

**FIG. 4** The main idea of probabilistic streamline propagation. The *gray glyphs* are depicting spherical orientation distributions representing the voxel-wise fiber dispersion (fODFs) or the uncertainty of the orientation of the fiber bundle (uODFs). Tractography iteratively propagates from the seed location (*red circle*) using discrete step size of length $\Delta s$. A propagation orientation is selected using the uODF/fODF information. Orientations with higher ODF values are more likely to be selected. Valid orientation must be within the maximum opening angle $\alpha$, defined from the previous propagation orientation (*blue dotted line*). Once selected (*red dotted lines* show likely candidate orientations), the tractography algorithm steps forward in this orientation. The process is repeated until a stopping criterion is reached.

After enough iterations, the estimated streamline distribution converges (Behrens et al., 2007; Campbell et al., 2014) and addition of new streamline samples does not offer any new information (in a similar way that representing the histogram of a normal distribution with 1000 samples instead of 100 samples does not offer much). The streamline distribution can be depicted as a spatial histogram with an arbitrary binning size (for instance, to the extreme with the track-density imaging (Calamante et al., 2010) approaches utilizing a very small binning size), but typically the original image grid is used. Individual streamlines are samples from these spatial histograms. Therefore it is important to remember that each comes with its own distribution weight that reflects whether a particular sample is close to the mode of the distribution (high weight) or the tails (low weight). Moreover, the convergence of the estimated distribution depends not only on the uncertainty and dispersion but also on the parameters of the tractography algorithm. Thus the number of streamlines will necessarily vary across applications, but the total number of experiments can be used to normalize these distributions (Behrens et al., 2007).

As in deterministic tractography, probabilistic tractography can be terminated based on stopping constraints. One difference is that strict termination criteria (e.g., anisotropy thresholds) can be relaxed with probabilistic methods, if they correspond to areas that are unlikely to considerably shift the mode of the distribution (e.g., regions where uncertainty is very high, hence propagation through there will only happen through chance and will not be replicable; see Fig. 2). Nevertheless, it is always beneficial to use anatomical constraints for any tractography approach (see Chapter 18), as the streamline model is a crude approximation regardless of the estimation method (Smith et al., 2012; Girard et al., 2014; St-Onge et al., 2018; Schilling et al., 2018; Warrington et al., 2020).

Even if the iterative propagation machinery is common to all probabilistic tractography approaches, methods differ in the ODFs and features they consider for the random sampling step, as we have seen already. Following, we overview these two main groups of methods, uncertainty-based and dispersion-based approaches.

# 4 Uncertainty-based probabilistic tractography

A key quantity to estimate for this type of approach is the uODFs for every voxel. This is calculated during the estimation of the voxel-wise orientation distribution. This uncertainty is then propagated spatially using the Monte Carlo method described earlier.

The first attempts to capture orientation uncertainty used a calibration approach. For instance, Parker et al. (2003) learned a mapping between the diffusion tensor shape/anisotropy and uncertainty in principal fiber orientations to derive uODFs; the more prolate (elongated along one axis) the tensor is, the lower the uncertainty on its principal orientation. This was extended by Parker and Alexander (2005) to consider crossing fibers in the uODFs. The authors used the persistent angular structure method, where the SNR in the image was linked to uncertainty in the main fiber orientations and used for probabilistic tractography.

Nevertheless, capturing estimation uncertainty for any signal model or reconstruction is a problem well studied in statistics. Therefore many relevant statistical approaches have been proposed over the years. Tractography methods can be grouped into further subcategories based on: (i) the way uncertainty mapping is performed and (ii) the fODF features that uncertainty is evaluated upon.

## 4.1 Bootstrapping

The most intuitive method for mapping uncertainty is assessing robustness/variability of an estimate over multiple repeats of the same experiment. Given enough repetitions (hundreds of complete diffusion MRI datasets may be needed) a uODF can be obtained in each voxel. While conceptually easy, this approach is infeasible in practice. The closest alternative to apply this concept, without needing a large number of repeats, is bootstrapping.

In its simplest form, bootstrapping needs a few (two to three) repeats, as it was done in Jones (2003). New datasets can be built from the original repeats by randomly combining them. For instance, if we have $K$ repeats of a dataset with $M$ diffusion-sensitizing directions, then we can build a new dataset (mimicking a new repeat) with $L \leq M$ volumes by randomly selecting for each volume one of the $K$ available with replacement (Pajevic and Basser, 2003; Lazar and Alexander, 2005). By repeating this random selection multiple times, we can mimic the effect of multiple repeats of the data and estimate uncertainty in the estimated parameters across the generated bootstrap samples. This process for mapping uncertainty is very generic and assumption-free, but requires extra data. Furthermore, if not enough repeats are acquired, the bootstrapped datasets can lead to an underestimation of the uncertainty compared to the uncertainty estimated from the same number of individual datasets. Simulation studies have shown that at least five data acquisition repeats are needed (O'Gorman and Jones, 2006).

An alternative bootstrapping approach that does not need repeated data acquisitions is residual bootstrap (Berman et al., 2008; Whitcher et al., 2008; Jeurissen et al., 2011; Campbell et al., 2014; Ye and Prince, 2017). This creates bootstrap samples using the residuals of a model fitted to the data. When a biophysical model for estimating orientations is fitted, the predicted signal will not fit the data precisely (due to image noise and modeling errors). The difference between the model-predicted and measured signals gives a residual for each diffusion-sensitizing measurement. Under the assumption of constant noise variance for all measurements, bootstrap samples can be created by randomly reshuffling the residuals across the measurements (i.e., adding perturbations). In the case of measured signals with different variances (as in the case of the log-transformed DTI measurements along different gradient directions, for instance), the alternative wild bootstrap can be employed (Jones, 2008; Whitcher et al., 2008), which interchanges residuals along each direction; residuals are randomly sign-flipped and added to the respective measurement points to give new bootstrap samples of the data. Uncertainty can then be obtained by assessing variability across the bootstrap samples.

## 4.2 Bayesian inference

Bayesian approaches allow formal characterization of uncertainty for any model fitted to data—see MacKay (2004), Lee (2004), and Bernardo and Smith (2000) for introductions. Uncertainty is inherently represented in the form of posterior probability density functions. Following Bayes' theorem, given data $Y$ and a model with parameters $\Theta$, the posterior probability $P(\Theta|Y)$ on particular values of model parameters that best describe the data can be calculated as a function of any prior knowledge $P(\Theta)$ and of the likelihood $P(Y|\Theta)$; with the likelihood encoding how likely it is for the observed data to be explained by particular parameter values. Once the posterior distribution is calculated, this encodes for each model parameter our belief or uncertainty on the estimated values given the observed data. Wide distributions mean high uncertainty and vice versa. Using this machinery, uODFs can be directly estimated as posterior distributions for a biophysical model that encodes fiber orientations or fODFs.

Given any biophysical model and an assumed noise distribution (e.g., Gaussian, Berman et al., 2008; Jbabdi et al., 2012; or Rician, Kaden et al., 2007; Sotiropoulos et al., 2016), the likelihood function can be defined. This must include a parametric relationship (known as a forward or generative model) that links parameters of interest (such as fiber orientations and diffusion coefficients) to the observed data (e.g., Hosey et al., 2005). The likelihood should also include assumptions about the noise structure. This is one of the strengths but at the same time weaknesses of Bayesian approaches. All modeling assumptions have to be explicitly laid down, but this allows flexibility in defining even complex forward models (e.g., Sotiropoulos et al., 2013, 2016) and formal mechanisms for mapping the uncertainty given these assumptions.

Prior distributions can provide a priori knowledge on the parameter values before observing any data. They can be completely uninformative in the inference process (i.e., estimation is driven only by the data), or they can provide some soft constraints (e.g., positivity on diffusion coefficients). A particularly interesting type of priors comes in the form of shrinkage priors (MacKay, 2004), also known as automatic relevance determination (ARD) priors. These are useful for performing on-the-fly model selection for nested models, as is the case for deciding how many crossing fiber compartments or fODF peaks exist in each voxel (Behrens et al., 2007). ARD priors impose a large prior penalty to the existence of multiple compartments, by a priori driving the respective volume fractions to zero. Model complexity can then be increased if there is strong support after observing the data through the likelihood.

Given the likelihood and priors, calculating the whole posterior distribution is effectively a stochastic model inversion process, that is, finding the ranges of parameter values that best explain the observed data given all the assumptions and prior knowledge. In the general case, this calculation can be an intractable problem. As we saw before, we only need samples from the uODFs to perform probabilistic tractography. Therefore sampling-based Bayesian methods are typically employed in practice, which overcome mathematical complexities and solve a simpler problem than estimating the whole posterior distribution of model parameters; they instead provide samples from the uODFs. Markov Chain Monte Carlo (MCMC) sampling algorithms (see Roberts and Rosenthaly, 2004, for a review) have been used extensively in diffusion MRI (Behrens et al., 2003, 2007; Hosey et al., 2005; Kaden et al., 2007; Fonteijn et al., 2007; Jbabdi et al., 2012; Sotiropoulos et al., 2016; Hernandez-Fernandez et al., 2019). For instance, FSL's Bayesian implementation (Behrens et al., 2007; Jbabdi et al., 2012; Hernandez-Fernandez et al., 2019) uses MCMC and ARD priors for model selection to estimate uODFs. MCMC is guaranteed to sample from the true posterior distribution if given enough time, but it can be computationally expensive due to its iterative nature. It can, however, be massively parallelized using modern technologies, such as GPUs, and speed-ups of three orders of magnitude are possible (Hernández et al., 2013; Hernandez-Fernandez et al., 2019). An alternative to MCMC is offered by faster approximate inference methods, such as variational Bayes (Kaden et al., 2008), or deep-learning based approaches (Cranmer et al., 2020), recently applied to diffusion MRI data (Patron et al., 2022).

## 4.3 Uncertainty of what?

All the preceding approaches evaluate uncertainty after a model or signal reconstruction is applied to the data. Uncertainty ODFs will therefore depend on that model/reconstruction and can reflect different features of the aODF (e.g., number of fiber population, curvature, shape) and potentially different sources of uncertainty.

The majority of uncertainty-based probabilistic tractography approaches consider a discrete set of fiber orientations in every voxel and evaluate uncertainty around them. They effectively follow the paradigm of deterministic tracking and focus on the uODF peaks. One type of method performs a parametric-type of spherical deconvolution against stick (i.e., no dispersion, perfect coherence) single-fiber response kernels (for instance, the ball and sticks model used in FSL bedpostX model options 1 and 2) (Behrens et al., 2007; Jbabdi et al., 2012). As fiber dispersion is not explicitly represented in the model, it will be captured as modeling error in the orientation uncertainty when present, in addition to all other uncertainty sources.

Another approach is to represent the dispersion explicitly in the deconvolution model and have a full estimate of the aODF. Many models allow doing that parametrically (e.g., Anderson, 2005; Kaden et al., 2007; Dell'Acqua et al., 2007; Sotiropoulos et al., 2012, 2016; Tariq et al., 2016). In principle, uncertainty mapped through some of these models (Kaden et al., 2007; Fonteijn et al., 2007; Berman et al., 2008; Jeurissen et al., 2011) will be lower around the peaks in a voxel with a highly dispersing fiber pattern compared to models that do not represent dispersion explicitly. In practice, the increase in model complexity and assumptions to capture dispersion compared to capturing only the peaks can have the opposite effects and lead to higher estimation uncertainty with these more complex models, as opposed to the ones that consider discrete peaks only (Jeurissen et al., 2011). Further research is needed to characterize uncertainty in both the fiber orientation and the fiber dispersion.

## 5 Dispersion-based probabilistic tractography

This family of techniques considers mostly geometric fiber dispersion, as represented by the width of the fODFs. They estimate a spatial (i.e., streamlines) distribution in a stochastic way, conditioned on voxel-wise fODFs estimated (in a deterministic way) from the data. For every voxel of the white matter, a spherical profile describing the orientation of the fibers inside that voxel is needed. The propagation direction is then sampled from that estimated fODF. Thus the probability of selecting a propagation direction is proportional to the amplitude of the fODFs for that direction. This is based on the assumption that the amplitudes of the fODFs are good estimates of the underlying aODF fiber density. Two commonly used approaches, (i) probability mass function sampling and (ii) rejection sampling methods, are detailed in the following paragraphs.

Probability mass function sampling uses a discretized set of orientations to represent the magnitude of the fODF along many orientations covering the full sphere. These can then be used as a distribution from which orientations can be sampled with probability proportional to the corresponding amplitude. Orientations not respecting the angular constraints of tractography can be set to zero before sampling the tractography propagation orientation. The orientation distribution can be computed from the amplitude of the fODF at the current tracking position only (Garyfallidis, 2012; Garyfallidis et al., 2014; Tournier et al., 2012, 2019; Girard et al., 2014), or using the product of the amplitude of the fODFs at the current and target tracking positions (Descoteaux et al., 2009).

Rejection sampling methods typically use the spherical harmonics representation of the fODF that is a continuous function on the sphere, which can be used to evaluate the fODF amplitude for any orientation. The rejection sampling method allows for sampling from a distribution that would be otherwise hard to sample directly (MacKay, 2004), as in the case of sampling the next propagation direction from fODFs. This is done by uniformly and randomly sampling an orientation respecting the angular constraints and comparing the value of the fODF for that orientation with a number randomly sampled between 0 and the estimated maximum of the fODF within the angular constraints. The randomly sampled orientation is rejected if its value is inferior to that random number. This process is repeated up to $N$ times until an orientation is accepted, and the trajectory is propagated in this orientation. The maximum can be estimated from a small set of samples of the fODF (Tournier et al., 2012, 2019).

### 5.1 Streamline propagation

Starting at the initial seed position, the probabilistic tractography iteratively propagates the streamline with an orientation sampled from the fODFs. If the fODF sampling fails to find an orientation respecting the tractography constraints (e.g., angular constraints, minimum fODF value), the propagation is stopped, assuming that the data do not support the

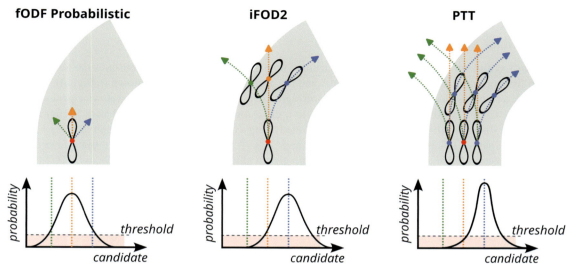

**FIG. 5** Demonstration of local fODF probabilistic, iFOD2, and PTT algorithms starting from the *red* seed point. For each algorithm, three example candidates (*green*, *orange*, and *blue*) are considered to determine the direction for the next propagation step along a bend. Local fODF probabilistic approach only takes into account the current fODF amplitudes. iFOD2 considers fODFs along arcs. PTT models cylindrical probes and considers fODFs within the probes (for simplicity, only the considered fODFs for the blue probe are marked for PTT). Representative probabilities for following the colored trajectories are shown for each algorithm. All candidates below a threshold are rejected, and any candidate above a threshold can be accepted depending on their probability. Probability distributions change depending on the fiber configuration and tractography parameters, such as the length of the arc used in iFOD2 or the probe length in PTT. While the probabilities inform about the spread of possibilities to sample from, it is important to note that tractography parameters such as step size and curvature threshold also play a role in the final spread of the streamline distribution.

propagation of the trajectory any further (Descoteaux et al., 2009; Tournier et al., 2012, 2019; Girard et al., 2014). Tournier et al. (2010) proposed modifying the probability distribution (iFOD2 algorithm) (Tournier et al., 2019), where the sampled probability is derived not only using the fODF at the current position of the trajectory but also using the fODF along the arc of a circle of fixed length and tangent to the previous propagation orientation (Fig. 5). The circle can have any radius and be rotated around the previous propagation orientation. The end of the circular arc yields the next position. The new position must respect the angular constraints of the algorithm. The iFOD2 algorithm integrates the fODF values along candidate propagation orientations, intending to capture rapid changes in the fODFs, such as in cases of high curvature and crossing configurations. The method was inspired by the Runge-Kutta method (Press et al., 1988), previously used in deterministic diffusion tensor tractography algorithms (Basser et al., 2000) to reduce trajectory error in these fiber configurations (see Appendix B). In practice, iFOD2 randomly selects a propagation orientation and then samples $N$ fODFs along the arc of the circle that is defined by the current position, the previous propagation orientation, and the candidate position. The values of the $N$ fODFs aligned with the arc of the circle are multiplied to obtain an estimate of the probability of the selected propagation orientation. This new distribution is then sampled to obtain the propagation orientation. The iFOD2 algorithm shows improvement in reconstruction trajectory with high curvature (Tournier et al., 2010) and in following the trajectory in crossing configurations.

The parallel transport tractography (PTT) algorithm (Aydogan and Shi, 2021) extends the iFOD2 strategy in two ways by using: (i) parametrized curves and (ii) probes. Similar to iFOD2, PTT samples probability using information beyond the current position of the trajectory. In PTT, this is done by probes that generalize the circular arcs used in iFOD2. A probe is a mathematical model of a cylindrical fiber bundle segment that is composed of a number of parallel lines. The approach is inspired by the topographic regularity of connections in the brain, where, in many parts, nearby regions of the brain connect to other nearby regions while preserving the spatial relationships between connections. During the propagation, several candidate probes are tried (Fig. 5), each probe is assigned a probability, and a likely trajectory is sampled. Probe probabilities are computed by adding up the fODF amplitudes along the individual parallel lines that constitute the probe. In PTT, probes are computed using parallel transport frames (Bishop, 1975), which is a curve parameterization technique akin to the Frenet-Serret approach, commonly used to study the differential geometry of curves. Parallel transport parameterization provides an analytical framework to efficiently compute candidate probes and represent the overall trajectory. With that, streamlines computed with PTT are differentiable, and they are geometrically smooth. Well-organized streamlines generated with PTT were shown to match the topographic regularity of the connections in the brain, and the use of probes was shown to mitigate noise in fODFs and improve tracking accuracy, particularly on crossing fiber regions (Aydogan and Shi, 2021).

## 6 Comparing probabilistic trajectories

### 6.1 Synthetic data

Fig. 6 shows a toy example of a single fiber bundle in a curve configuration. The blue line outlines the bundle voxel boundaries. The main fiber orientations are shown in Fig. 6A and the ground-truth aODFs are shown in Fig. 6B. The bundle is of tubular shape with a fixed radius of 10 mm. The diffusion signal was generated using the voxel-wise peaks convolved to a signal response (see Chapters 10 and 11) corresponding to a diffusion tensor, then corrupted with Rician noise (Caruyer et al., 2014). The low dispersion diffusion signal was generated using a diffusion tensor main eigenvalue $\lambda_1 = 0.0017$ and $\lambda_2 = \lambda_3 = 0.0002$ (see Chapter 10), at an average SNR = 30 and SNR = 5 (noisy scenario). To simulate fiber dispersion, the diffusion signal was generated using a diffusion tensor main eigenvalue $\lambda_1 = 0.0007$ and $\lambda_2 = \lambda_3 = 0.0002$ (Caruyer et al., 2014) (SNR = 30), increasing hindrance in the main orientation of the bundle. The diffusion MRI signal was generated at $b = 2000 mm^2/s$ on 60 uniformly distributed gradient directions, with six $b = 0 mm^2/s$ images. The streamline shown in Fig. 6C is the trajectory reconstructed using deterministic tractography on the low dispersion diffusion signal (SNR = 30), starting at the center of the red voxel.

Fig. 7 shows probabilistic estimation of 1000 trajectory samples of the bundle seeded at the center bottom of the bundle (red dot). The streamlines color corresponds to the normalized voxel-wise path density map (i.e., the spatial histogram weights). The figure shows trajectories estimated using uncertainty-based tractography, with the FSL Probtrackx2 (Behrens et al., 2007) and residual bootstrap tractography (Berman et al., 2008; Garyfallidis et al., 2014). The figure also shows dispersion-based tractography of the fODF probabilistic algorithm using on probability mass function sampling (Garyfallidis et al., 2014; Girard et al., 2014), and the iFOD2 (Tournier et al., 2012, 2019) and PTT (Aydogan and Shi, 2021) algorithms, using on rejection sampling. The first row of Fig. 7 shows the results using diffusion MRI signal at SNR = 30 and low fiber dispersion. The second row shows the same fiber dispersion, but the signal is corrupted with higher levels of noise (SNR = 5). The third row shows the diffusion MRI signal at SNR = 30, but with increased fiber dispersion. Noise and fiber dispersion increase the coverage of the estimated trajectories. The first row shows that uncertainty-based methods have consistent trajectory estimations, with little deviation. On the other hand, dispersion-based methods estimate a wider set of trajectories given the selected fiber dispersion parameter of the data, yet the obtained distributions are centered around the distributions obtained with the uODF approaches. The addition of noise to the data (second row) increases the uncertainty of the estimation, shown by wider spatial distributions. The increase in voxel-wise fiber dispersion (third row) also increases uncertainty and bundle coverage. This is especially visible on dispersion-based methods, which inherently consider dispersion to reconstruct the bundle trajectory. Although different for each method, the set of computed trajectories reflects the belief of the location of the actual fiber trajectories from the seed position, given the data, and they

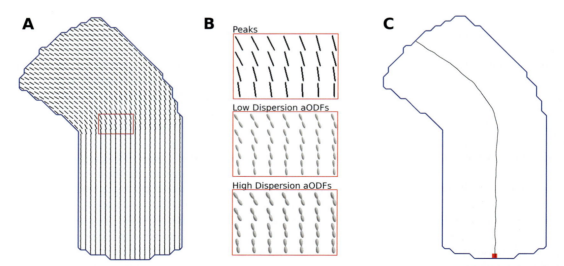

**FIG. 6** Toy example of curve bundle configuration. The main voxel-wise fiber orientations are shown in (A) and the ground-truth aODFs in (B). The bundle is of tubular shape with a fixed radius of 10 mm. The streamline shown in (C) is the trajectory reconstructed using deterministic tractography on the low dispersion diffusion signal at SNR = 30, starting at the center of the red voxel.

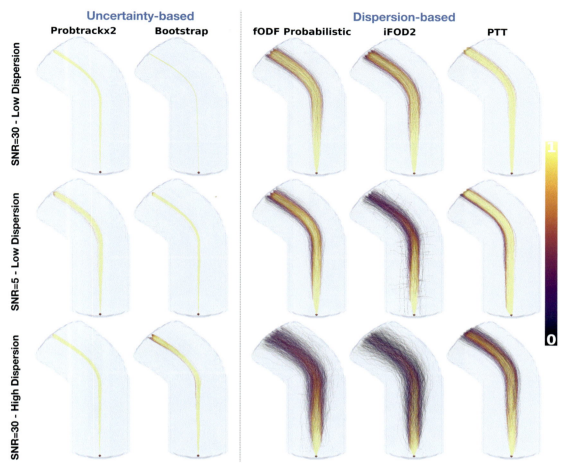

**FIG. 7** Probabilistic tractography reconstruction on a synthetic curve configuration using the Probtrackx2, bootstrap, fODF probabilistic, iFOD2, and PTT algorithms. The first row shows the results using an average SNR and low fiber dispersion scenario. The second row shows the same fiber dispersion, but when the diffusion MRI signal is corrupted significantly by noise (low SNR). The third row shows an average SNR, but high fiber dispersion scenario. The first two columns show uncertainty-based probabilistic methods, and the three last columns show dispersion-based probabilistic methods. All subfigures show 1000 reconstruction trajectories, with their color corresponding to the normalized voxel-wise path density map. The *blue outline* illustrates the bundle boundaries. The *red dot* at the bottom shows the tractography seed location.

capture either uncertainty in the estimation (uODF) or characterize the voxel-wise fiber dispersion (fODF). Two interesting patterns can be observed from these toy simulations: (a) Dissociating uncertainty and dispersion is challenging. All approaches spread more in the presence of either more noise or higher dispersion. (b) Nevertheless, uncertainty-based approaches that only consider discrete fODF peaks (e.g., Probtrackx2) seem to be more sensitive to changes in SNR than dispersion, as anticipated. And dispersion-based approaches (e.g., PTT) can be more sensitive to changes in dispersion than to changes in SNR.

## 6.2 In vivo data

The in vivo subject was obtained from the Penthera 3T dataset (Chamberland et al., 2019) (subject 001). The diffusion MRI were acquired at a 2 mm isotropic spatial resolution on a 3 Tesla MRI (Philips, Ingenia) using a single-shot echo-planar imaging sequence using three *b*-value shells, $b = [300, 1000, 2000]$mm$^2$/s with 8, 32, and 60 uniformly distributed gradient directions, respectively. The dataset contains seven $b = 0$mm$^2$/s for a total of 107 measurements. A reversed phase-encoded b0 image was also acquired. Additionally, a 1 mm isotropic resolution T1-weighted MPRAGE image was also acquired. The diffusion MRIs were corrected from Gibbs ringing artifacts (Kellner et al., 2016; Tournier et al., 2019),

**268  PART | III**  Tractography algorithms

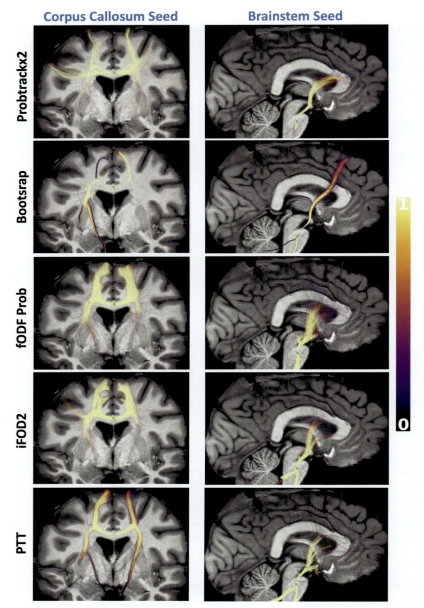

**FIG. 8** Streamline distributions obtained from a single seed position (center of the *green square*) located in the central part of the corpus callosum (*left column*) and in the brainstem (*right column*). The first two rows show uncertainty-based methods (Probtrackx2, bootstrap), and the three last rows show dispersion-based probabilistic tractography algorithms (fODF Probabilistic, iFOD2, and PTT). All subfigures show 1000 reconstructed trajectories, with their color corresponding to the normalized voxel-wise path density map.

motion artifact reduction, correction for field inhomogeneity, susceptibility-induced off-resonance field and eddy currents (Andersson et al., 2003; Andersson and Sotiropoulos, 2016; Smith et al., 2004), and spatial intensity variations (Zhang et al., 2001).

Fig. 8 displays the distributions of streamlines obtained in vivo from a single seed position using the same probabilistic tractography algorithms described and demonstrated earlier. Algorithms estimating uncertainty tend to produce more conservative estimates of the spatial histograms of streamlines, compared to dispersion-based methods. Note the figure shows the spatial histogram of streamlines from a single seed, and tractography would need to be seeded in a larger region to accurately estimate white matter bundles (e.g., corpus callosum bundle, corticospinal tract). Probabilistic approaches estimate likely trajectories of white matter fibers from the data, while deterministic approaches produce a single streamline corresponding to the most probable location of the trajectory. Therefore probabilistic methods can provide a measure of confidence given all the sources of variability in the data and model. This is shown by the color of the trajectories where bright colors (yellow and orange) show reproducible streamline portions, while dark streamlines portion (purple, black) show low reproducibility.

## 7 Interpretation: What do these probabilistic estimates mean?

As we saw before, the spatial histogram of streamlines obtained from probabilistic tractography represents the spatial distribution of diffusion paths (i.e., paths within tissue that minimally hinder water diffusion), conditioned on different features; and these features typically represent either estimation uncertainty (with uODFs) or fiber dispersion (with fODFs). The histograms can be discretized (binned) and give rise to spatial images of these distribution weights. The size of the bins is arbitrary, but voxels (either in native or standard space) provide a natural choice. As with any density function, the sum of all obtained weights can be used to normalize them in the range [0, 1]. These steps allow us to turn a continuous spatial distribution into a discrete one. An example is shown in Fig. 9, where the streamline distribution (A) is discretized into histograms calculated at the native diffusion MRI resolution (B) and at the native T1-weighted resolution (C). The discretized histograms highlight spatial density variations not visible with the spatial distribution of streamlines.

Under the assumption that governs all tractography approaches, this distribution of diffusion paths is a proxy for white matter bundles, estimated in a stochastic manner. Do the spatial histogram weights have an interpretation? It would be tempting to think of these weights as representative of white matter connectivity, that is, the number of axons passing through a region or connecting two regions. Even if probabilistic tractography weights do reflect anatomical connectivity and connectivity features to a degree (Thomas et al., 2014; van den Heuvel et al., 2015; Donahue et al., 2016; Ambrosen et al., 2020; Girard et al., 2020), they do so in a nonstraightforward way and to a moderate degree. Importantly, they also reflect a number of uninteresting-to-neuroanatomy (nuisance and nonbiophysical) factors (Jones and Cercignani, 2010; Jbabdi and Johansen-Berg, 2011), which depend on the type of probabilistic tractography employed. To understand these better, we need to look specifically into the different approaches.

### 7.1 Confidence intervals

Uncertainty-based tractography provides confidence bounds on the location of the most probable diffusion paths. Fig. 7 (first and second columns) shows the probabilistic spatial distribution, using the same seed and propagation constraints as the deterministic tractography shown in Fig. 6C (no uncertainty considered). The uODF considered here reflects only the major fiber orientations (i.e., discrete peaks), so that the two approaches utilize the same underlying orientation information (i.e., no dispersion). Probabilistic tractography shows which part of the deterministic estimation we are more confident about, given the data and modeling assumptions used. These weights will reflect anatomical features of the underlying connections. For instance, there will be more confidence in tracking stronger connections than weaker connections (as higher density of coherent axons typically results in higher diffusion anisotropy). Or the spatial distribution will spread more in the presence of fiber dispersion. However, care should be taken not to overinterpret these probabilistic weights, as they also reflect all other sources of orientation and tracking uncertainty and hence are influenced by factors such as imaging noise levels, modeling and tracking errors, path length, and voxel size. Notice that these effects are present in

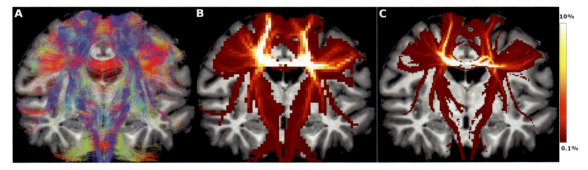

**FIG. 9** Spatial histogram of streamlines seeded in a voxel of the corpus callosum. (A) Illustration of streamline overlaid on coronal section (Probtrackx2), color-coded by orientation (*blue*: inferior-superior, *red*: left-right, *green*: posterior-anterior). The streamlines were seeded from a single voxel located in the corpus callosum (all generated streamlines were kept). (B) Discretized histogram calculated at the native diffusion MRI spatial resolution (2 mm isotropic). (C) Discretized histogram calculated at the native T1-weighted resolution (1 mm isotropic). The discretized histograms in (B) and (C) reveal density variations of streamlines, not visible with the individual streamline samples. Maximum intensity projections along the coronal plane are shown.

both types of estimation, deterministic or probabilistic tractography, as both use the same orientation information and propagation process. The probabilistic approach just provides an extra piece of information on estimation confidence around the location of the most likely path, which is not provided in the deterministic approach. It is more difficult to track a bend rather than a straight curve, a short rather than a long path, a path that goes through complex fiber patterns versus a path that does not. This difficulty is represented as reduced confidence in probabilistic tractography weights. It is also important to note that factors such as voxel size and SNR can limit what connections can be studied with tractography. Reconstructed trajectories with low confidence in this regard can be highly useful when combined with anatomical priors, and they can provide information about intricate connections in the brain, which might not be possible to study otherwise.

## 7.2 Proportion of estimated connecting fiber

If one had access to the true aODF, with a high degree of confidence, sampling from it would help us answer the question: "what is the proportion of fibers starting from A and connecting to B?" Dispersion-based tractography tries to emulate this process and provide such a measure. By assuming that the fODF along a certain orientation in a voxel provides an estimate of the proportion of axons aligned locally along this orientation, it samples from the fODF to generate a spatial histogram. In practice, however, the fODF is not always a faithful representation of the aODF. For instance, it is inherently symmetric, while the aODF may be asymmetric (Jbabdi and Johansen-Berg, 2011). This has been mitigated by estimating asymmetrical fODF profiles using information from neighboring voxels to capture complex fiber organization (Delputte et al., 2007; Reisert et al., 2012; Campbell et al., 2014; Bastiani et al., 2017; Karayumak et al., 2018; Wu et al., 2020); see Poirier and Descoteaux (2022) for a recent literature review. Moreover, the fODF can be wide in the absence of any dispersion, due to noise and modeling uncertainty (Sotiropoulos et al., 2012). A workaround to mitigate some of these effects is fODF thresholding. Typically, fODF thresholding consists of setting to zero the probability of propagation in the direction below the fODF threshold parameter. Two approaches can be used: (i) absolute fODF threshold $T_a$ and (ii) relative fODF threshold $T_r$.

The absolute fODF threshold discards all candidate directions with an fODF value lower than the threshold parameter $T_a$. This aims at removing noise in the apparent fiber density (Dell'Acqua et al., 2012; Raffelt et al., 2012; Tournier et al., 2019), focusing the propagation in orientation with higher apparent fiber density. Tractography algorithms set directions with an fODF value lower than $T_a$ to zero before sampling a direction, and stop if no direction with an fODF value above $T_a$ is found.

The relative fODF threshold approach adapts the threshold $T_r$ to the voxel-wise maximum fODF value, such as $T = T_r * \max(fODF)$. This approach similarly aims at removing noisy propagation orientation in voxels of the white matter where artifactual fODF values are relative to the maximum, such as Gibbs ringing in fODFs represented using spherical harmonics (Garyfallidis et al., 2014). In the case where the fODFs show less dominant orientations, such as in partial volume voxels (e.g., near the cortical gray matter or the ventricles) or in voxels with high dispersion (e.g., multiple pathway crossings), the fODF threshold value is low. Per contra, it increases in voxels with densely packed aligned axons, such as in the corpus callosum or the internal capsule. The fODF thresholding parameter should be selected following the noise level in the MRI data (Jeurissen et al., 2011, 2014; Parker et al., 2013; Girard et al., 2014; Tournier et al., 2019), and the modeling assumption made in the fODF estimation (see Chapters 10–12). Moreover, this parameter can be adapted to better estimate specific white matter structures (Rheault et al., 2019).

## 8 Conclusion

To summarize, probabilistic tractography provides complementary information to deterministic estimates in two crucial ways: (a) reflecting estimation uncertainty, that is, by doing a stochastic model fit versus doing a deterministic fit and/ or (b) reflecting biophysical dispersion and a continuum of within-voxel orientations versus considering only a discrete set of orientations (typically fODF peaks). Even if approaches have been devised to do both (a) and (b) (e.g., Kaden et al., 2007), the majority of commonly used methods implement either (a) or (b), due to the difficulty of completely disentangling orientation uncertainty from dispersion.

Fig. 6C shows an illustrative example of deterministic estimating, compared to different flavors of probabilistic tractography in Fig. 7. What is clear that all probabilistic approaches spread, unlike deterministic tractography. This is expected and this is what these methods are designed to do! (That is, capture a whole distribution rather than the mode of this distribution.) The degree of "spreading" depends on how the orientations are perturbed. Given similar noise levels, uncertainty-based approaches that consider only the peaks of the fODF will spread less than uncertainty-based approaches that consider the whole fODF. Also, they will tend to spread less than dispersion-based approaches for standard data, as uncertainty around the peaks tends to be small, while biophysical dispersion can be large within a macroscopic imaging voxel.

Similarly to deterministic tractography streamlines, probabilistic tractography streamlines can be used in microstructure-informed tractography methods (Sherbondy et al., 2010; Daducci et al., 2013, 2014; Pestilli et al., 2014; Smith et al., 2013, 2015; Frigo et al., 2021). Those methods require all (as much as possible) valid trajectories to adequately model the microstructure features of the tissue and filter spurious trajectories (see Chapters 16 and 18). In doing so, each streamline is weighted following the contribution of its trajectory and the microstructure model, in representing the data. These streamline weights are then combined to the spatial histogram of streamlines to provide estimates of connectivity. Probabilistic streamline tractography has shown good results, as it typically captures the full extent of white matter pathways.

Finally, it is important to note that binning the spatial histogram into weights introduces a further confound to all probabilistic approaches (uncertainty or dispersion-based), that of the voxel size. If one changes the image resolution, this will immediately change the absolute values of the weights (even if it will likely not change the spatial contrast of the discretized histogram). This is why using these values quantitatively becomes a very challenging task. An alternative to using arbitrary voxels is to use anatomical regions that have a functional meaning and are not a random choice linked to the acquisition protocol. This is the main idea underlying connectivity-based applications (see Chapters 21 and 24).

# References

Ambrosen, K.S., Eskildsen, S.F., Hinne, M., Krug, K., Lundell, H., Schmidt, M.N., van Gerven, M.A.J., Mørup, M., Dyrby, T.B., 2020. Validation of structural brain connectivity networks: the impact of scanning parameters. NeuroImage 204, 116–207. https://doi.org/10.1016/J.NEUROIMAGE.2019.116207. https://www.sciencedirect.com/science/article/pii/S1053811919307980.

Anderson, A.W., 2005. Measurement of fiber orientation distributions using high angular resolution diffusion imaging. Magn. Reson. Med. 54, 1194–1206. https://doi.org/10.1002/MRM.20667.

Andersson, J.L.R., Sotiropoulos, S.N., 2016. An integrated approach to correction for off-resonance effects and subject movement in diffusion MR imaging. NeuroImage 125, 1063. https://doi.org/10.1016/J.NEUROIMAGE.2015.10.019.

Andersson, J.L.R., Skare, S., Ashburner, J., 2003. How to correct susceptibility distortions in spin-echo echo-planar images: application to diffusion tensor imaging. NeuroImage 20, 870–888. https://doi.org/10.1016/S1053-8119(03)00336-7.

Aranda, R., Rivera, M., Ramirez-Manzanares, A., 2014. A flocking based method for brain tractography. Med. Image Anal. 18, 515–530. https://doi.org/10.1016/j.media.2014.01.009.

Aydogan, D.B., Shi, Y., 2021. Parallel transport tractography. IEEE Trans. Med. Imaging 40, 635–647. https://doi.org/10.1109/TMI.2020.3034038.

Basser, P.J., Pajevic, S., Pierpaoli, C., Duda, J., Aldroubi, A., 2000. In vivo fiber tractography using DT-MRI data. Magn. Reson. Med. 44, 625–632.

Bastiani, M., Cottaar, M., Dikranian, K., Ghosh, A., Zhang, H., Alexander, D.C., Behrens, T.E., Jbabdi, S., Sotiropoulos, S.N., 2017. Improved tractography using asymmetric fibre orientation distributions. NeuroImage 158, 205–218. https://doi.org/10.1016/J.NEUROIMAGE.2017.06.050.

Behrens, T.E.J., Woolrich, M.W., Jenkinson, M., Johansen-Berg, H., Nunes, R.G., Clare, S., Matthews, P.M., Brady, J.M., Smith, S.M., 2003. Characterization and propagation of uncertainty in diffusion-weighted MR imaging. Magn. Reson. Med. 50, 1077–1088. https://doi.org/10.1002/mrm.10609. http://www.ncbi.nlm.nih.gov/pubmed/14587019.

Behrens, T.E.J., Berg, H.J., Jbabdi, S., Rushworth, M.F.S., Woolrich, M.W., 2007. Probabilistic diffusion tractography with multiple fibre orientations: what can we gain? NeuroImage 34, 144–155. https://doi.org/10.1016/j.neuroimage.2006.09.018.

Berman, J.I., Chung, S.W., Mukherjee, P., Hess, C.P., Han, E.T., Henry, R.G., 2008. Probabilistic streamline q-ball tractography using the residual bootstrap. NeuroImage 39, 215–222. https://doi.org/10.1016/j.neuroimage.2007.08.021.

Bernardo, J.M., Smith, A.F.M., 2000. Bayesian Theory. Wiley & Sons Ltd, p. 586.

Bishop, R.L., 1975. There is more than one way to frame a curve. Am. Math. Mon. 82, 246–251.

Calamante, F., Tournier, J.D., Jackson, G.D., Connelly, A., 2010. Track-density imaging (TDI): super-resolution white matter imaging using whole-brain track-density mapping. NeuroImage 53, 1233–1243. https://doi.org/10.1016/j.neuroimage.2010.07.024.

Campbell, J.S.W., MomayyezSiahkal, P., Savadjiev, P., Leppert, I.R., Siddiqi, K., Pike, G.B., 2014. Beyond crossing fibers: bootstrap probabilistic tractography using complex subvoxel fiber geometries. Front. Neurol. 5. https://doi.org/10.3389/fneur.2014.00216.

Caruyer, E., Daducci, A., Descoteaux, M., Christophe Houde, J., Philippe Thiran, J., Verma, R., 2014. Phantomas: a flexible software library to simulate diffusion MR phantoms. In: International Symposium on Magnetic Resonance in Medicine (ISMRM'14).

Chamberland, M., Bernier, M., Girard, G., Fortin, D., Descoteaux, M., Whittingstall, K., 2019. Penthera 1.5T., https://doi.org/10.5281/ZENODO.2602022.

Cranmer, K., Brehmer, J., Louppe, G., 2020. The Frontier of simulation-based inference. Proc. Natl. Acad. Sci. USA 117, 30055–30062. https://doi.org/10.1073/PNAS.1912789117.

Daducci, A., Palu, A.D., Lemkaddem, A., Thiran, J.P., 2013. A convex optimization framework for global tractography. In: IEEE International Symposium on Biomedical Imaging, IEEE, pp. 524–527.

Daducci, A., Palu, A.D., Lemkaddem, A., Thiran, J.P., 2014. COMMIT: convex optimization modeling for micro-structure informed tractography. IEEE Trans. Med. Imaging 34, 246–257.

**272 PART | III** Tractography algorithms

Dell'Acqua, F., Rizzo, G., Scifo, P., Clarke, R.A., Scotti, G., Fazio, F., 2007. A model-based deconvolution approach to solve fiber crossing in diffusion-weighted MR imaging. IEEE Trans. Biomed. Eng. 54, 462–472. https://doi.org/10.1109/TBME.2006.888830.

Dell'Acqua, F., Simmons, A., Williams, S.C.R., Catani, M., 2012. Can spherical deconvolution provide more information than fiber orientations? Hindrance modulated orientational anisotropy, a true-tract specific index to characterize white matter diffusion. Hum. Brain Mapp. https://doi.org/10.1002/hbm.22080.

Delputte, S., Dierckx, H., Fieremans, E., D'Asseler, Y., Achten, R., Lemahieu, I., 2007. Postprocessing of brain white matter fiber orientation distribution functions. In: 2007 4th IEEE International Symposium on Biomedical Imaging: From Nano to Macro, IEEE, pp. 784–787.

Descoteaux, M., Deriche, R., Knösche, T.R., Anwander, A., 2009. Deterministic and probabilistic tractography based on complex fibre orientation distributions. IEEE Trans. Med. Imaging 28, 269–286. https://doi.org/10.1109/TMI.2008.2004424. http://www.ncbi.nlm.nih.gov/pubmed/19188114.

Donahue, C.J., Sotiropoulos, S.N., Jbabdi, S., Hernandez-Fernandez, M., Behrens, T.E., Dyrby, T.B., Coalson, T., Kennedy, H., Knoblauch, K., Essen, D.-C.V., Glasser, M.F., 2016. Using diffusion tractography to predict cortical connection strength and distance: a quantitative comparison with tracers in the monkey. J. Neurosci. 36, 6758–6770. https://doi.org/10.1523/JNEUROSCI.0493-16.2016.

Fonteijn, H.M.J., Verstraten, F.A.J., Norris, D.G., 2007. Probabilistic inference on Q-ball imaging data. IEEE Trans. Med. Imaging 26, 1515–1524. https://doi.org/10.1109/TMI.2007.907297.

Frigo, M., Zucchelli, M., Deriche, R., Deslauriers-Gauthier, S., 2021. TALON: tractograms as linear operators in neuroimaging. In: Computational Brain Connectivity Mapping (CoBCoM).

Garyfallidis, E., 2012. Toward an Accurate Brain Tractography (Ph.D. thesis). University of Cambridge.

Garyfallidis, E., Brett, M., Amirbekian, B., Rokem, A., Walt, S.V.D., Descoteaux, M., Nimmo-Smith, I., Contributors, D., 2014. Dipy, a library for the analysis of diffusion MRI data. Front. Neuroinform. 8, 1–5.

Girard, G., Whittingstall, K., Deriche, R., Descoteaux, M., 2014. Towards quantitative connectivity analysis: reducing tractography biases. NeuroImage 98, 266–278. https://doi.org/10.1016/j.neuroimage.2014.04.074.

Girard, G., Caminiti, R., Battaglia-Mayer, A., St-Onge, E., Ambrosen, K.S., Eskildsen, S.F., Krug, K., Dyrby, T.B., Descoteaux, M., Thiran, J.P., Innocenti, G.M., 2020. On the cortical connectivity in the macaque brain: a comparison of diffusion tractography and histological tracing data. NeuroImage 221, 117201. https://doi.org/10.1016/j.neuroimage.2020.117201.

Hernández, M., Guerrero, G.D., Cecilia, J.M., García, J.M., Inuggi, A., Jbabdi, S., Behrens, T.E., Sotiropoulos, S.N., 2013. Accelerating fibre orientation estimation from diffusion weighted magnetic resonance imaging using GPUs. PLoS ONE 8, e61892. https://doi.org/10.1371/JOURNAL.PONE.0061892.

Hernandez-Fernandez, M., Reguly, I., Jbabdi, S., Giles, M., Smith, S., Sotiropoulos, S.N., 2019. Using GPUs to accelerate computational diffusion MRI: from microstructure estimation to tractography and connectomes. NeuroImage 188, 598–615. https://doi.org/10.1016/J.NEUROIMAGE.2018.12.015.

Hosey, T., Williams, G., Ansorge, R., 2005. Inference of multiple fiber orientations in high angular resolution diffusion imaging. Magn. Reson. Med. 54, 1480–1489. https://doi.org/10.1002/MRM.20723.

Jbabdi, S., Johansen-Berg, H., 2011. Tractography: where do we go from here? Brain Connect. 1, 169–183. https://doi.org/10.1089/brain.2011.0033.

Jbabdi, S., Woolrich, M.W., Andersson, J.L.R., Behrens, T.E.J., 2007. A Bayesian framework for global tractography. NeuroImage 37, 116–129. https://doi.org/10.1016/j.neuroimage.2007.04.039. http://www.ncbi.nlm.nih.gov/pubmed/17543543.

Jbabdi, S., Sotiropoulos, S.N., Savio, A.M., Graña, M.G., Behrens, T.E.J., 2012. Model-based analysis of multishell diffusion MR data for tractography: how to get over fitting problems. Magn. Reson. Med. 68, 1846–1855. https://doi.org/10.1002/mrm.24204. https://pubmed.ncbi.nlm.nih.gov/22334356/.

Jeurissen, B., Leemans, A., Jones, D.K., Tournier, J.D., Sijbers, J., 2011. Probabilistic fiber tracking using the residual bootstrap with constrained spherical deconvolution. Hum. Brain Mapp. 32, 461–479. https://doi.org/10.1002/hbm.21032.

Jeurissen, B., Tournier, J.D., Dhollander, T., Connelly, A., Sijbers, J., 2014. Multi-tissue constrained spherical deconvolution for improved analysis of multi-shell diffusion MRI data. NeuroImage 103, 411–426. https://doi.org/10.1016/j.neuroimage.2014.07.061. http://www.ncbi.nlm.nih.gov/pubmed/25109526.

Jones, D.K., 2003. Determining and visualizing uncertainty in estimates of fiber orientation from diffusion tensor MRI. Magn. Reson. Med. 49, 7–12. https://doi.org/10.1002/mrm.10331. http://www.ncbi.nlm.nih.gov/pubmed/12509814.

Jones, D.K., 2008. Tractography gone wild: probabilistic fibre tracking using the wild bootstrap with diffusion tensor MRI. IEEE Trans. Med. Imaging 27, 1268–1274. https://doi.org/10.1109/TMI.2008.922191. http://www.ncbi.nlm.nih.gov/pubmed/18779066.

Jones, D.K., Cercignani, M., 2010. Twenty-five pitfalls in the analysis of diffusion MRI data. NMR Biomed. 23, 803–820. http://www.ncbi.nlm.nih.gov/pubmed/20886566.

Kaden, E., Knösche, T.R., Anwander, A., 2007. Parametric spherical deconvolution: inferring anatomical connectivity using diffusion MR imaging. NeuroImage 37, 474–488. https://doi.org/10.1016/J.NEUROIMAGE.2007.05.012.

Kaden, E., Anwander, A., Knösche, T.R., 2008. Variational inference of the fiber orientation density using diffusion MR imaging. NeuroImage 42, 1366–1380. https://doi.org/10.1016/J.NEUROIMAGE.2008.06.004. https://pubmed.ncbi.nlm.nih.gov/18603006/.

Karayumak, S.C., Özarslan, E., Unal, G., 2018. Asymmetric orientation distribution functions (AODFs) revealing intravoxel geometry in diffusion MRI. Magn. Reson. Imaging 49, 145–158. https://doi.org/10.1016/j.mri.2018.03.006.

Kellner, E., Dhital, B., Kiselev, V.G., Reisert, M., 2016. Gibbs-ringing artifact removal based on local subvoxel-shifts. Magn. Reson. Med. 76, 1574–1581. https://doi.org/10.1002/MRM.26054. https://onlinelibrary.wiley.com/doi/full/10.1002/mrm.26054.

Lazar, M., Alexander, A.L., 2005. Bootstrap white matter tractography (BOOT-TRAC). NeuroImage 24, 525–532. https://doi.org/10.1016/j.neuroimage.2004.08.050. http://www.ncbi.nlm.nih.gov/pubmed/15627594.

Lee, P.M., 2004. Bayesian Statistics: An Introduction, third ed. Arnold, p. 351.

MacKay, D.J.C., 2004. Information Theory, Inference, and Learning Algorithms. Cambridge University Press, https://doi.org/10.1017/s026357470426043x.

Malcolm, J.G., Shenton, M.E., Rathi, Y., 2010. Filtered multitensor tractography. IEEE Trans. Med. Imaging 29, 1664–1675. https://doi.org/10.1109/TMI.2010.2048121. http://www.ncbi.nlm.nih.gov/pubmed/20805043.

O'Gorman, R.L., Jones, D.K., 2006. Just how much data need to be collected for reliable bootstrap DT-MRI? Magn. Reson. Med. 56, 884–890. https://doi.org/10.1002/mrm.21014.

Pajevic, S., Basser, P.J., 2003. Parametric and non-parametric statistical analysis of DT-MRI data. J. Magn. Reson. 161, 1–14. https://doi.org/10.1016/S1090-7807(02)00178-7.

Parker, G.J.M., Alexander, D.C., 2005. Probabilistic anatomical connectivity derived from the microscopic persistent angular structure of cerebral tissue. Philos. Trans. R. Soc. Lond. Ser. B Biol. Sci. 360, 893–902. https://doi.org/10.1098/rstb.2005.1639. https://pubmed.ncbi.nlm.nih.gov/16087434/.

Parker, G.J.M., Haroon, H.A., Wheeler-Kingshott, C.A.M., 2003. A framework for a streamline-based probabilistic index of connectivity (PICo) using a structural interpretation of MRI diffusion measurements. J. Magn. Reson. Imaging 18, 242–254. https://doi.org/10.1002/jmri.10350. http://www.ncbi.nlm.nih.gov/pubmed/12884338.

Parker, G.D., Marshall, D., Rosin, P.L., Drage, N., Richmond, S., Jones, D.K., 2013. A pitfall in the reconstruction of fibre ODFs using spherical deconvolution of diffusion MRI data. NeuroImage 65, 433. https://doi.org/10.1016/J.NEUROIMAGE.2012.10.022. https://www.ncbi.nlm.nih.gov/pmc/articles/PMC3580290/.

Patron, J.P.M., Kypraios, T., Sotiropoulos, S.N., 2022. Amortised inference in diffusion MRI biophysical models using artificial neural networks and simulation-based frameworks. In: International Symposium on Magnetic Resonance in Medicilne (ISMRM'22)), p. 1644.

Perrin, M., Poupon, C., Cointepas, Y., Rieul, B., Golestani, N., Pallier, C., Rivière, D., Constantinesco, A., Bihan, D.L., Mangin, J.F., 2005. Fiber Tracking in Q-Ball Fields Using Regularized Particle Trajectories. vol. 3565. Springer Verlag, pp. 52–63. https://link.springer.com/chapter/10.1007/11505730_5.

Pestilli, F., Yeatman, J.D., Rokem, A., Kay, K.N., Wandell, B.A., 2014. Evaluation and statistical inference for human connectomes. Nat. Methods 11, 1058–1063. https://doi.org/10.1038/nmeth.3098.

Poirier, C., Descoteaux, M., 2022. A unified filtering method for estimating asymmetric orientation distribution functions: where and how asymmetry occurs in the brain. bioRxiv.

Pontabry, J., Rousseau, F., 2011. Probabilistic tractography using Q-ball modeling and particle filtering. In: International Conference on Medical Image Computing and Computer Assisted Intervention (MICCAI'11), pp. 209–216. http://www.ncbi.nlm.nih.gov/pubmed/21995031.

Press, W.H., Flannery, B.P., Teukolsky, S.A., Vatterling, W.T., 1988. Numerical Recipes Example Book C. Cambridge University Press.

Raffelt, D., Tournier, J.D., Rose, S., Ridgway, G.R., Henderson, R., Crozier, S., Salvado, O., Connelly, A., 2012. Apparent fibre density: a novel measure for the analysis of diffusion-weighted magnetic resonance images. NeuroImage 59, 3976–3994. https://doi.org/10.1016/j.neuroimage.2011.10.045. http://www.ncbi.nlm.nih.gov/pubmed/22036682.

Reisert, M., Kellner, E., Kiselev, V.G., 2012. About the geometry of asymmetric fiber orientation distributions. IEEE Trans. Med. Imaging 31, 1240–1249. https://doi.org/10.1109/TMI.2012.2187916.

Rheault, F., St-Onge, E., Sidhu, J., Maier-Hein, K., Tzourio-Mazoyer, N., Petit, L., Descoteaux, M., 2019. Bundle-specific tractography with incorporated anatomical and orientational priors. NeuroImage 186, 382–398. https://doi.org/10.1016/j.neuroimage.2018.11.018. https://linkinghub.elsevier.com/retrieve/pii/S1053811918320883.

Roberts, G.O., Rosenthaly, J.S., 2004. General state space Markov chains and MCMC algorithms. Probab. Surveys 1, 20–71. https://doi.org/10.1214/154957804100000024. https://projecteuclid.org/journals/probability-surveys/volume-1/issue-none/General-state-space-Markov-chains-and-MCMC-algorithms/10.1214/154957804100000024.full.

Schilling, K.G., Janve, V., Gao, Y., Stepniewska, I., Landman, B.A., Anderson, A.W., 2018. Histological validation of diffusion MRI fiber orientation distributions and dispersion. NeuroImage 165, 200–221. https://doi.org/10.1016/J.NEUROIMAGE.2017.10.046.

Sherbondy, A.J., Rowe, M.C., Alexander, D.C., 2010. MicroTrack: an algorithm for concurrent projectome and microstructure estimation. In: Medical Image Computing and Computer-Assisted Intervention, pp. 183–190. http://www.ncbi.nlm.nih.gov/pubmed/20879230.

Smith, S.M., Jenkinson, M., Woolrich, M.W., Beckmann, C.F., Behrens, T.E.J., Johansen-Berg, H., Bannister, P.R., Luca, M.D., Drobnjak, I., Flitney, D.E., Niazy, R.K., Saunders, J., Vickers, J., Zhang, Y., Stefano, N.D., Brady, J.M., Matthews, P.M., 2004. Advances in functional and structural MR image analysis and implementation as FSL. NeuroImage 23, S208–S219. https://doi.org/10.1016/J.NEUROIMAGE.2004.07.051.

Smith, R.E., Tournier, J.D., Calamante, F., Connelly, A., 2012. Anatomically-constrained tractography: improved diffusion MRI streamlines tractography through effective use of anatomical information. NeuroImage 63, 1924–1938. https://doi.org/10.1016/j.neuroimage.2012.06.005. http://www.ncbi.nlm.nih.gov/pubmed/22705374.

Smith, R.E., Tournier, J.D.D., Calamante, F., Connelly, A., 2013. SIFT: spherical-deconvolution informed filtering of tractograms. NeuroImage 67, 298–312. https://doi.org/10.1016/j.neuroimage.2012.11.049. http://www.ncbi.nlm.nih.gov/pubmed/23238430.

Smith, R.E., Tournier, J.D., Calamante, F., Connelly, A., 2015. SIFT2: enabling dense quantitative assessment of brain white matter connectivity using streamlines tractography. NeuroImage 119, 338–351. https://doi.org/10.1016/j.neuroimage.2015.06.092. http://www.sciencedirect.com/science/article/pii/S1053811915005972.

Sotiropoulos, S.N., Behrens, T.E.J., Jbabdi, S., 2012. Ball and rackets: inferring fiber fanning from diffusion-weighted MRI. NeuroImage 60, 1412–1425. https://doi.org/10.1016/J.NEUROIMAGE.2012.01.056.

Sotiropoulos, S.N., Jbabdi, S., Andersson, J.L., Woolrich, M.W., Ugurbil, K., Behrens, T.E.J., 2013. RubiX: combining spatial resolutions for Bayesian inference of crossing fibers in diffusion MRI. IEEE Trans. Med. Imaging 32, 969–982. https://doi.org/10.1109/TMI.2012.2231873. https://pubmed.ncbi.nlm.nih.gov/23362247/.

Sotiropoulos, S.N., Hernández-Fernández, M., Vu, A.T., Andersson, J.L., Moeller, S., Yacoub, E., Lenglet, C., Ugurbil, K., Behrens, T.E.J., Jbabdi, S., 2016. Fusion in diffusion MRI for improved fibre orientation estimation: an application to the 3T and 7T data of the Human Connectome Project. NeuroImage 134, 396–409. https://doi.org/10.1016/J.NEUROIMAGE.2016.04.014. https://pubmed.ncbi.nlm.nih.gov/27071694/.

St-Onge, E., Daducci, A., Girard, G., Descoteaux, M., 2018. Surface-enhanced tractography (SET). NeuroImage 169, 524–539. https://doi.org/10.1016/J.NEUROIMAGE.2017.12.036. https://www.sciencedirect.com/science/article/pii/S1053811917310583?via%3Dihub.

Tariq, M., Schneider, T., Alexander, D.C., Wheeler-Kingshott, C.A.G., Zhang, H., 2016. Bingham-NODDI: mapping anisotropic orientation dispersion of neurites using diffusion MRI. NeuroImage 133, 207–223. https://doi.org/10.1016/J.NEUROIMAGE.2016.01.046. https://pubmed.ncbi.nlm.nih.gov/26826512/.

Thomas, C., Ye, F.Q., Irfanoglu, M.O., Modi, P., Saleem, K.S., Leopold, D.A., Pierpaoli, C., 2014. Anatomical accuracy of brain connections derived from diffusion MRI tractography is inherently limited. Proc. Natl. Acad. Sci. USA 111, 16574–16579. https://doi.org/10.1073/PNAS.1405672111.

Tournier, J.D., Calamante, F., Connelly, A., 2007. Robust determination of the fibre orientation distribution in diffusion MRI: non-negativity constrained super-resolved spherical deconvolution. NeuroImage 35, 1459–1472. http://www.ncbi.nlm.nih.gov/pubmed/17379540.

Tournier, J.D., Calamante, F., Connelly, A., 2010. Improved probabilistic streamlines tractography by 2nd order integration over fibre orientation distributions. In: International Symposium on Magnetic Resonance in Medicine.

Tournier, J.-D., Calamante, F., Connelly, A., 2012. MRtrix: diffusion tractography in crossing fiber regions. Int. J. Imaging Syst. Technol. 22, 53–66. https://doi.org/10.1002/ima.22005.

Tournier, J.D., Smith, R., Raffelt, D., Tabbara, R., Dhollander, T., Pietsch, M., Christiaens, D., Jeurissen, B., Yeh, C.H., Connelly, A., 2019. MRtrix3: a fast, flexible and open software framework for medical image processing and visualisation. NeuroImage 202, 116137. https://doi.org/10.1016/j.neuroimage.2019.116137.

van den Heuvel, M.P., de Reus, M.A., Barrett, L.F., Scholtens, L.H., Coopmans, F.M.T., Schmidt, R., Preuss, T.M., Rilling, J.K., Li, L., 2015. Comparison of diffusion tractography and tract-tracing measures of connectivity strength in rhesus macaque connectome. Hum. Brain Mapp. 36, 3064–3075. https://doi.org/10.1002/hbm.22828.

Warrington, S., Bryant, K.L., Khrapitchev, A.A., Sallet, J., Charquero-Ballester, M., Douaud, G., Jbabdi, S., Mars, R.B., Sotiropoulos, S.N., 2020. XTRACT - standardised protocols for automated tractography in the human and macaque brain. NeuroImage 217, 116923. https://doi.org/10.1016/J.NEUROIMAGE.2020.116923.

Whitcher, B., Tuch, D.S., Wisco, J.J., Sorensen, A.G., Wang, L., 2008. Using the wild bootstrap to quantify uncertainty in diffusion tensor imaging. Hum. Brain Mapp. 29, 346–362. https://doi.org/10.1002/hbm.20395.

Wu, Y., Hong, Y., Feng, Y., Shen, D., Yap, P.T., 2020. Mitigating gyral bias in cortical tractography via asymmetric fiber orientation distributions. Med. Image Anal. 59, 101543. https://doi.org/10.1016/j.media.2019.101543.

Ye, C., Prince, J.L., 2017. Probabilistic tractography using Lasso bootstrap. Med. Image Anal. 35, 544–553. https://doi.org/10.1016/J.MEDIA.2016.08.013.

Zhang, Y., Brady, M., Smith, S., 2001. Segmentation of brain MR images through a hidden Markov random field model and the expectation-maximization algorithm. IEEE Trans. Med. Imaging 20, 45–57. https://doi.org/10.1109/42.906424. http://www.ncbi.nlm.nih.gov/pubmed/11293691.

Zhang, F., Hancock, E.R., Goodlett, C., Gerig, G., 2009. Probabilistic white matter fiber tracking using particle filtering and von Mises-Fisher sampling. Med. Image Anal. 13, 5–18. https://doi.org/10.1016/j.media.2008.05.001.

# Chapter 15

# Geodesic tractography

**Luc Florack, Rick Sengers, and Andrea Fuster**
*Department of Mathematics and Computer Science, Eindhoven University of Technology, The Netherlands*

## 1 Introduction

Differential geometry is an established framework for handling curves and surfaces. It is therefore natural to regard the problem of tractography from this perspective.[a] In *geodesic tractography*, tentative tracts are operationally defined as (locally) shortest paths relative to some data-adaptive notion of *length* or *cost*. In a broader sense, it includes families of geodesics (a.k.a. geodesic congruences), whether explicit (think of a bundle of tracts obtained tractwise) or implicit (think of a family of distance fronts transversal to such a bundle). As such, it also encompasses structural connectivity insofar as this can be inferred from such congruences.

In order to exploit the idea effectively, one needs to couple geometry to magnetic resonance imaging (MRI) data evidence to construct an efficacious notion of length or cost. The governing principle is the observation that fibrous tissue (brain white matter, muscle tissue, etc.) imparts nonrandom barriers to the diffusion of water molecules. Clinical MRI scanners are designed to probe water-bound hydrogen spins, the most abundant moiety in biological tissue.[b] By sensitizing the measured signal for the amount of diffusivity of water molecules along preselected orientations (with the help of a globally defined time-integrated magnetic gradient field $q$; see Callaghan et al., 1990 and Basser, 2002 for details), the observed *diffusion-weighted imaging* (DWI) signal reflects the amount and anisotropic nature of this process at each voxel $x$ (for a fixed diffusion time) in the form of a positive attenuation factor $0 < E(x, q) \leq 1$. Stronger attenuation is tantamount to more diffusion. The $q$-domain is usually sparsely sampled, which calls for physical $q$-space models and hypotheses,[c] of which *diffusion tensor imaging* (DTI) is probably the simplest and most familiar one (Basser, 2002; Le Bihan et al., 2001).

Mathematical rigor of geodesic tractography is offset by many unknown microstructural factors that determine the relation between the observed diffusion signal and the structure of biological tissue. Thus the problem of tractography will remain, strictly speaking, an ill-posed inverse problem. A priori knowledge and hypotheses are necessary in order to formulate a *technically* well-posed problem, but even so the solutions found cannot generally be guaranteed to correspond to biologically plausible fibers. This is the main bottleneck in geodesic tractography—and any other tractography paradigm for that matter—today. More research into the fundamental physics that governs the interplay between diffusion and tissue structure will be needed, even to the extent as to allow us to pose "the right questions," see Novikov and Kiselev (2010), Novikov et al. (2011), Liu and Özarslan (2019), Jespersen et al. (2007), Assaf et al. (2004), Stanisz et al. (1997), and Özarslan et al. (2018) for some established insights.

Meanwhile, the pragmatic approach adopted by the community is to rely on some heuristically inspired, rigorously formulated ansatz to infer geometry from the empirical evidence $E(x, q)$. In this chapter, we address some ideas, the mathematics behind them, and some important open problems. Familiarity with basic linear algebra is a requisite, notably with its pivotal concepts: vectors, linear mappings, norms, and inner products; see Appendix for a self-contained introduction. Explanations are complemented with pointers to relevant literature for details. Previous exposure to tensor calculus and differential geometry is, although helpful, not strictly necessary. For the pragmatist, the topic of geodesic tractography lends itself excellently as a concrete point of departure for endeavoring to study these topics in more generality.

---

a. Our focus is on static brain structure. An intriguing spatiotemporal differential geometric framework has been proposed in the context of brain function as well; see Le Bihan (2020).

b. The relative excess of hydrogen spins "aligned" with the static magnetic field in the scanner that can be probed to generate a signal is on the order of 1–10ppm. This low sensitivity is alleviated by the huge number of water molecules in a biological specimen, resulting in a measurable signal.

c. Physical models are explanatory, and should not be confused with physically void representations in the form of (generic or truncated) basis expansions, out of habit likewise referred to as "models," see Novikov et al. (2018).

---

**Handbook of Diffusion MR Tractography. https://doi.org/10.1016/B978-0-12-818894-1.00028-8**
Copyright © 2025 Elsevier Ltd. All rights are reserved, including those for text and data mining, AI training, and similar technologies.

**276 PART | III** Tractography algorithms

## 2 The Riemann-DTI paradigm

### 2.1 Heuristics

Geodesic tractography can be seen as a generalization of classical streamline tractography, which, at least qualitatively, remains the guiding principle, as it expresses the basic premise underlying the relation between neural tracts and observed local diffusivity patterns in DWI. In the classical approach the aim is to integrate a vector field so as to obtain a biologically meaningful tract, given an initial condition in the form of a starting point. The vector field expresses the local orientation preference. Mathematically this boils down to solving a first-order initial value problem of the form

$$\dot{x}(t) = f(x(t)) \quad \text{subject to} \quad x(0) = x_0. \tag{1}$$

In streamline tractography the vector $f(x)$ is chosen to coincide with a suitably scaled main eigenvector of the diffusion tensor at point $x$. Apart from technical problems at singularities—points on a curve at which transitions occur in eigenvalue ordering, leading to discontinuities (90 degrees turns) and ambiguities (degenerate eigenspaces)—there are (related) conceptual problems. A diffusion tensor is a highly simplified aggregate descriptor of an inherently complex diffusion pattern, which in turn is determined by the underlying tissue structure. Given that *all* (six) degrees of freedom of the diffusion tensor are determined by anatomy, one may expect that the many-to-one inverse problem of tract reconstruction from DTI should involve, besides disambiguating priors, at least all empirical evidence contained in these parameters. The theoretical conditions under which the streamline method, ignoring most of this evidence, would work as intended are not very realistic in major parts of the brain.

Assuming, therefore, that a streamline is deemed spurious by an expert, say, one is not merely confronted with a false positive (which could be rejected if recognized as such), but with a true negative as well if the initial seed point is known to lie on an actual fiber tract, since there are no alternative paths emanating from that point. In mathematical terms, the solution to Eq. (1) is unique (at least locally and under mild conditions, as discussed), which, in this scenario, is a complicating factor. Geodesic tractography radically removes this shortcoming.

In its original form, geodesic tractography was introduced at MICCAI by O'Donnell et al. (2002) and subsequently proposed at ECCV using an alternative formalism by Lenglet et al. (2004) in the context of a Riemannian paradigm for DTI. The premise is that anisotropic diffusion, regarded as an extrinsic process in Euclidean space filled with a porous medium (white matter), can be "geometrized away" by stipulating a globally diffeomorphic, vacuous Riemannian manifold, in which neural fibers follow geodesic paths. Except for the diffeomorphism,[d] this idea is reminiscent of Einstein's general relativity theory, in which Newton's extrinsic gravitational forces mysteriously acting at a distance are abolished in return for pseudo-forces arising from the local geometry of spacetime. The Riemannian 3-manifold, M say, coincides pointwise with Euclidean 3-space E (so that we may identify $M \equiv E \equiv \mathbb{R}^3$ as point sets), but is furnished with a nonstandard, spatially varying[e] inner product (or, equivalently, quadratic form), known as a *Riemannian metric*, which encases local anisotropy to lowest order. If we indicate the inner product at $x \in M$ by $(.|.)_x$, then by definition the length of a spatial vector $y \in TM_x$ (in which $TM_x$ denotes the tangent space to M at $x$; see Fig. 1) is the square root of the $y$-quadratic form[f] $(y|y)_x = g_{\mu\nu}(x)y^\mu y^\nu$. The coefficients $g_{\mu\nu}(x)$ determine the so-called Gram matrix $G$, which completely defines the inner product; see Appendix for further details.

To capture the relative alignment of (macroscopic bundles of) nerve fibers with high diffusivity pathways geometrically, the authors (*loc. cit.*) stipulate that the *inverse* Gram matrix $G^{inv}$ of the Riemannian metric, the entries of which are denoted by $g^{\mu\nu}$, coincides with the DTI matrix $D$. Put differently, the inverse DTI tensor on E is identified with the *metric tensor* of M: $g_{\mu\nu} = D_{\mu\nu}^{inv}$. Curves that are short in a Riemannian sense thus reflect pathways with relatively few diffusion obstacles and thus relatively long mean free paths. More specifically, the Riemann-geodesic paradigm (at least in its original form) stipulates that *biological fibers are geodesics relative to the DTI-induced Riemannian metric.*

Fig. 2 illustrates the DTI-induced metric in terms of an ellipsoidal gauge figure, revealing the principal diffusion orientations together with their relative apparent diffusivities (i.e., from our extrinsic point of view, recall Footnote e). The

---

d. In relativity theory the (pseudo-)Riemannian manifold renders classical spacetime obsolete; in the Riemann-DTI paradigm Riemannian and Euclidean geometries necessarily coexist, and are linked via a metric transform.

e. To signal anisotropy and spatial variation requires an extrinsic, Euclidean point of view. From an intrinsic, Riemannian point of view the metric provides a constant background skeleton (by virtue of the Levi-Civita connection, see Appendix).

f. In expressions such as here, each pair of identical upper and lower indices implies a summation over the entire index range, so that $g_{\mu\nu}(x)y^\mu y^\nu \doteq \sum_{\mu, \nu = 1, 2, 3} g_{\mu\nu}(x)y^\mu y^\nu$. This summation convention will be in effect throughout.

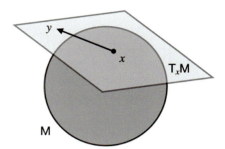

**FIG. 1** By definition, all vectors $y$ anchored at a fixed point $x$ of a manifold M span a linear space called the tangent space and denoted by $TM_x$. The manifold can generally not be thought of as a linear space, except in a Euclidean, flat world geometry (in which case all tangent spaces coincide and may be loosely said to span the manifold itself). The collection of all tangent spaces at all points of the manifold is known as the tangent bundle and denoted by TM $= \cup_{x \in M} TM_x$. One may think of the vector $y \in TM_x$ as a tangent vector of a path along M or, using the metaphor of a particle trajectory, as the instantaneous velocity at point $x \in M$. Tangent spaces are introduced for mathematical convenience, enabling local linearization of complex nonlinear spatial processes.

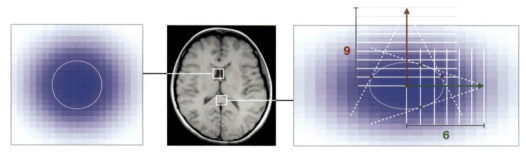

**FIG. 2** Typical diffusivity patterns captured by the DTI tensor. Bulk water in the ventricles induces isotropic diffusion (shown on the *left*). Aligned axon bundles in the corpus callosum induce anisotropic diffusion along their dominant orientation (shown on the *right*). The ellipsoids represent one level set of the quadratic form in these respective cases, $g_{\mu\nu}(x)y^\mu y^\nu = 1$, say, with fixed point $x$ somewhere inside the ventricle, respectively, along the corpus callosum. The Riemannian unit sphere, or indicatrix, $\{y \in TM_x \mid g_{\mu\nu}(x)y^\mu y^\nu = 1\}$, tells us how to construct the Riemannian length of a vector $v \in TM_x$ at base point $x \in M$. In this example, the two vector arrows are *rigidly* rotated copies. To determine their Riemannian lengths, one draws the tangent cone (in this simplified case the *dotted lines*) through the tip of each arrow to the ellipsoidal gauge figure, so as to obtain the plane (in casu *line*) containing all tangency points. Together with the parallel plane through the center of the indicatrix this determines an oriented pattern of equidistant "wave fronts" transversal to the arrow. (Generically, if vectors and DTI eigendirections are not coincidentally aligned as in this example, wave fronts and vectors will not be perpendicular.) The squared Riemannian lengths of the vectors are now given by the (fractional) number of corresponding wave fronts pierced by the arrows, so that the vertical and horizontal vectors in this example have Riemannian lengths of $\sqrt{9}$ and $\sqrt{6}$, respectively, despite equal Euclidean lengths.

figure also illustrates geometrically how such a gauge figure determines the Riemannian length of a vector depending on its orientation, while keeping its Euclidean length fixed.

Some confusion has arisen due to inadvertent reversal of the formulation of the Riemann-DTI paradigm, claiming that "geodesics are biological tracts" instead of "biological tracts are geodesics." The nature of confusion is most easily appreciated by imagining a hypothetical region consisting of similar, densely packed and perfectly straight biological fibers at microscale. Averaged over representative volume elements, such a configuration will induce a spatially homogeneous (i.e., $x$-independent) anisotropic DTI signal, and thus a ditto metric tensor field. In such a "metric affine geometry," however, *all* straight lines are geodesics, and vice versa. The preferred orientation induced by anisotropy apparently plays no role; see Fig. 3. As stated earlier, the formulation of the Riemann-geodesic paradigm ("biological tracts are geodesics") is clearly incomplete. Among all possible orientations, the preferred orientation no longer has a privileged status. Adhering to a purely intrinsic, Riemannian point of view, one apparently loses precisely the information the tractography rationale relies on: anisotropy.

It is important not to misinterpret this problem (that crucial diffusion anisotropy evidence is "geometrized away") as being related to specific features of the metric, as one might be tempted to conclude from the artificial example. Quite generally, a Riemannian manifold is said to be *geodesically complete*, which means that *any* two points can be joined by a (not necessarily unique) geodesic (the *Hopf-Rinow theorem* ensures that length-minimizing curves between any two endpoints always exist; Jost, 2011). In this sense geodesic tractography is fundamentally different from streamline tractography, as geodesics *always* emanate from any given point in *all* directions. In view of geodesic completeness, it

**278 PART | III** Tractography algorithms

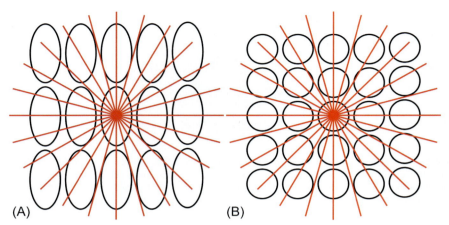

**FIG. 3** Imagine a homogeneous bundle of vertically aligned tracts, inducing a metric field illustrated by the ellipsoidal indicatrices on the *left*, revealing a vertical diffusion preference. This, however, is our extrinsic, *Euclidean* perspective. From an intrinsic, *Riemannian* perspective indicatrices are (by definition) isotropic *unit* spheres, as depicted on the *right*. If one abandons the Euclidean viewpoint altogether, one therefore loses the notion of a preferred orientation. In this simplified example geodesics are straight lines, as indicated here for a geodesic family anchored at the midpoint. Without reference to Euclidean 3-space (*left picture*) and thus to the anisotropic nature of the underlying diffusion process, one cannot tag certain geodesics as being more biologically plausible than others. For this reason the Riemann-DTI paradigm must be furnished with an explicit connectivity criterion as explained in the text. However defined, such a criterion must reflect the metric transform between Euclidean and Riemannian spaces E and M, that is, *both* viewpoints come into play in geodesic tractography. This is also reflected in Fig. 2, in which the arrows have been drawn as *rigid*, that is, *Euclidean*-rotated copies, as a result of which they can be distinguished by their Riemannian lengths, revealing a preferred direction (one may, e.g., use the ratio of Euclidean to Riemannian lengths for a hierarchical ordering along a "biological plausibility" scale).

would thus be preposterous to deem an arbitrary geodesic a biologically plausible fiber, as almost all geodesics would end up as false positives. Rather, the actual potential of geodesic tractography lies in the conjectural absence of false negatives together with the existence of an additional *connectivity criterion* to remove false or designate true positives, which must be adopted *as a constitutional part of the geodesic paradigm*.[g]

Put differently, the Riemannian manifold M does not replace the role of the original Euclidean 3-space E, and, a fortiori, the (dual) Riemannian metric tensor does not replace the diffusion tensor. We need to retain a pointwise correspondence between both base manifolds M and E and consider a so-called metric transform; recall Footnote d. A connectivity criterion must express some coupling of M and E. Simple examples are the comparisons of lengths (or other costs) of a fiducial geodesic in the geometries of M and E (average diffusivity; see Astola et al., 2007), or of the alignment of the geodesic tangent with a principal diffusion eigenvector; see Parker et al. (2002) and Lenglet et al. (2009). Without such coupling one loses the notion of anisotropy altogether.

The way to construct optimal connectivity criteria is a crucial yet unsettled case. Case specific and/or data extrinsic priors are most likely necessary. The ones mentioned previously are data induced and rather generic, expressing the basic premise of tractography, viz. the stipulated alignment of tracts and preferred diffusion channels. Such crude, evidence-based criteria lend themselves naturally to a pruning of the most conspicuous false positives. An example of how a case-specific prior might contribute to a semantic connectivity criterion (possibly via an externally driven local metric adaptation), compensating for the insufficient resolution of diffusion MRI in and near gray matter, has been proposed by St-Onge et al. (2018, 2021). Their so-called surface-enhanced tractography (SET) method relies on enriching empirical evidence by taking into account T1-weighted MRI data to better resolve the gray/white matter interface as well as a theoretical model of cortical expansion and gyrification that yields a more plausible tract continuation across that interface. A canonical orthonormal coordinate frame for gyral geometry has been presented by Cottaar et al. (2018), likewise relying on additional T1 data support for better resolution of the cortical surface. The idea of anatomically constrained tractography proposed by Smith et al. (2012) is similar; see also the application by Horbruegger et al. (2019). Also the option of adding anatomical priors for pruning (such as end-point constraints) makes geodesic tractography more versatile than first-order streamline-based methods; see Schilling et al. (2020).

More generally, recall that geodesic tractography, posed as the inverse problem to infer structure from evidence, is ill-posed. Consequently, odds are that insisting on high sensitivity (i.e., on avoiding false negatives) will incur low specificity

---

g. This applies to any metric geometry beyond the illustrated Riemannian case for which the Hopf-Rinow theorem holds.

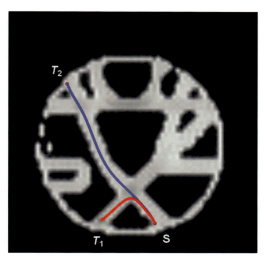

**FIG. 4** Ranking geodesics connecting fixed endpoints in the FiberCup phantom (Poupon et al., 2008, 2010; Fillard et al., 2011) based on a generic connectivity criterion—defined in this DTI-based experiment as the ratio of Euclidean to Riemannian length—is seen to prefer the seemingly "less feasible" one, connecting source point S to target point $T_1$, over the stipulated "ground truth" one, connecting S to $T_2$. Since data are nearly isotropic in the bifurcation region, both curves are likely candidates when based on relative diffusivity only, since no geometric penalty for curvature is involved. A curvature penalty may clearly alter the ranking in this case. Altering the ranking of $ST_1$ and $ST_2$ with such data-driven tractometric invariants still begs the question of which one is the physically correct one, the answer of which must lie outside (DTI) data evidence (e.g., by providing support for an effective curvature threshold in this example, or by knowledge of factual endpoint connections).

(i.e., will leave ample false positives). Fig. 4 illustrates this phenomenon on highly simplified simulation data. The a priori preference for $ST_1$ in this figure is *not* caused by the fact that it is shorter than $ST_2$, in either metric, as the connectivity criterion is scale invariant by construction. An optimal pruning to isolate true positives will ultimately require a more sophisticated classification in terms of data-intrinsic "tractometric" features, coupled to data-extrinsic knowledge to compensate for ambiguities or missing information.

## 2.2 Mathematical formulation

Geodesic tractography can be formalized in different ways. In Section 2.2.1 we consider the *Lagrangian formalism*, and in Section 2.2.2 the *Hamiltonian formalism*. There is an overarching framework based on *Hamilton-Jacobi theory*, notably the *Hamilton-Jacobi equation*, from which both formalisms may be derived, and which is of interest in itself. This is covered in Section 2.2.3. Our emphasis is on heuristic explanations, not proofs. For the latter the reader is referred to the excellent classic on this topic by Rund (1973). Appendix is a self-contained introduction to relevant mathematical concepts for readers not so familiar with linear algebra and tensor calculus, and also establishes notational conventions. In subsequent sections heuristic clarifications are provided next to formulas so as to allow the reader to safely skip mathematical details without losing the gist of the ideas presented.

### 2.2.1 Lagrangian formalism

Although named after Joseph Lagrange, the Lagrangian formalism goes back to Huygens (1690) and Pierre de Fermat; see Born and Wolf (1970). In its original form and context (optics) it states that light travels along rays minimizing travel time between any two points, a useful analogy for our purpose. The principle of Lagrange is more widely applicable and somewhat more general, seeking stationary rather than minimal solutions of a so-called action functional. In the case at hand, in which the action is a functional of a spatial curve connecting two fixed endpoints (a mapping that assigns an amplitude to each admissible curve), it stipulates that curves of interest are those that can be *arbitrarily* perturbed without significantly affecting the value attained by the functional, as long as the perturbations are sufficiently small. This principle is tremendously popular in physics, as many physical systems admit a parsimonious formulation in terms of a fundamental Lagrangian and stationary action principle.[h]

---

h. The descriptive power of an action functional goes beyond that of its stationary points; see Noether's Theorem (Neuenschwander, 2011; Noether, 1918).

**280 PART | III** Tractography algorithms

In the present context we consider a parameterized curve segment, $\gamma : [0, t] \to M \sim \mathbb{R}^3 : \tau \mapsto \gamma(\tau)$, connecting the fixed endpoints $x_0 = \gamma(0)$ and $x = \gamma(t)$ (see the two instances in Fig. 4), and define the action functional $\mathscr{L}[\gamma]$ as its length[i]:

$$\mathscr{L}[\gamma] = \int_0^t F(\gamma(\tau), \dot{\gamma}(\tau)) \, d\tau. \tag{2}$$

In this expression $F(x, y) \geq 0$ denotes the length of the tangent vector $y \in TM_x$ that "lives" at point $x \in M$, and $\tau \in [0, t]$ an *arbitrary* parameter. The reason for introducing the so-called *Finsler function*, or *Finsler norm*, $F(x, y)$, is to allow for a generalized, location-dependent notion of length, replacing the standard Euclidean length of flat space (which is independent of location $x$ and given by $F(x, y) = \|y\|$ in terms of the standard Euclidean vector norm). The rationale behind geodesic tractography is to adapt $F(x, y)$ to the diffusion MRI data evidence at $x$ in such a way that neural tracts correspond to locally shortest paths, that is, local minimizers of Eq. (2) with $x = \gamma(\tau)$ and $y = \dot{\gamma}(\tau)$. Geometrically this implies that space should no longer be conceived of as being flat. The diffusion MRI data could be said to induce a "gravitational field" causing a local bending of space, mathematically captured by a variety of curvature tensors (Bao et al., 2000; Shen and Shen, 2016), the details of which are not important for the present discussion.

An essential and natural scaling requirement for any norm is that $F(x, y)$ is absolutely homogeneous of degree one with respect to $y$,

$$F(x, \lambda y) = |\lambda| F(x, y), \tag{3}$$

which ensures that the integral in Eq. (2) is parameterization-invariant, as it should be, and also justifies the analogous terminology based on our common notion of "length" (scaling a vector by a factor $\lambda$ will make it $|\lambda|$ times as long). One may relax the requirement of parameter invariance on the condition that one clarifies the preferred parameterization used, a typical case being provided by an alternative cost in the form

$$\mathscr{C}[\gamma] = \int_0^t L(\gamma(\tau), \dot{\gamma}(\tau)) \, d\tau, \tag{4}$$

with *Lagrangian*[j]

$$L(x, y) \doteq \frac{1}{2} F^2(x, y), \tag{5}$$

and in which $\tau$ is now a so-called *affine parameter* permitting only affine reparameterizations, $\tau = A\tau' + B, A, B \in \mathbb{R}, A \neq 0$ (for only in this case stationary curves of Eqs. 2, 4 coincide). One may then interpret $\tau$ as "time," $F(\gamma(\tau), \dot{\gamma}(\tau))$ as the speed of a virtual particle moving freely within M, and $A, B$ as clock synchronization constants (unit and offset).

In Riemannian geometry, the Finsler function or homogeneous Lagrangian $F(x, y)$ is the square root of a quadratic form in $y \in TM_x$ (Pythagorean rule) at any base point $x \in M$, varying smoothly along M:

$$F(x, y) = \sqrt{g_{\mu\nu}(x) y^\mu y^\nu}. \tag{6}$$

Thus in Riemannian geometry the quadratic case (5) becomes

$$L(x, y) = \frac{1}{2} g_{\mu\nu}(x) y^\mu y^\nu. \tag{7}$$

Geodesics are, by definition, those curves that locally minimize $\mathscr{L}[\gamma]$ or $\mathscr{C}[\gamma]$ among all neighboring curves with fixed endpoints, so that $\delta\mathscr{L}[\gamma] \doteq \mathscr{L}[\gamma + \delta\gamma] - \mathscr{L}[\gamma] \approx 0$, or $\delta\mathscr{C}(\gamma) \doteq \mathscr{C}[\gamma + \delta\gamma] - \mathscr{C}[\gamma] \approx 0$, up to lowest order in variations $\delta\gamma$ of the curve $\gamma$. For a rigorous treatment, see the abundant literature on variational calculus (Abraham et al., 1988; Arnold, 1989; Lovelock and Rund, 1988; Sagan, 1992; Weinstock, 1974). In an *arbitrary* curve parameterization, this stationarity condition implies the following *geodesic equations*,[k] a system of $n = 3$ nonlinear ordinary differential equations (Bao et al., 2000):

$$\ddot{\gamma}^\mu(\tau) + \Gamma^\mu_{\nu\rho}(\gamma(\tau)) \, \dot{\gamma}^\nu(\tau) \, \dot{\gamma}^\rho(\tau) = \frac{d \ln F(\gamma(\tau), \dot{\gamma}(\tau))}{d\tau} \, \dot{\gamma}^\mu(\tau). \tag{8}$$

---

i. A (single/double) dot stands for a (first/second-order) derivative w.r.t. the curve parameter $\tau$.

j. More generally, any integrand of the type (Eq. 4), independent of the constraints (3), (5) is referred to as a Lagrangian.

k. Terminology in the literature is ambiguous to the extent that geodesics are either defined as affinely parameterized curves, excluding Eq. (8) with a nonvanishing right-hand side, or as their parameter-invariant spatial images, in which case (8) does define a geodesic (arbitrarily parameterized).

The right-hand side contributes to an acceleration along the trajectory, which has no effect on the geometry of the curve and can be effectively removed via (nonlinear) reparameterization. Any new parameter thus obtained is then an affine one in the previously introduced sense. The coefficients $\Gamma^\mu_{v\rho}(\gamma(\tau))$ are seen to generate "pseudo-forces" inducing a local acceleration at any point $\gamma(\tau)$ of the trajectory, and thus (from an extrinsic Euclidean perspective) a bending of that trajectory, which is completely determined by the metric tensor, viz.

$$\Gamma^\mu_{v\rho}(x) = \frac{1}{2} g^{\mu\lambda}(x) \left( \frac{\partial g_{v\lambda}(x)}{\partial x^\rho} + \frac{\partial g_{\lambda\rho}(x)}{\partial x^v} - \frac{\partial g_{v\rho}(x)}{\partial x^\lambda} \right). \tag{9}$$

These so-called *Christoffel symbols of the second kind* emerge by imposing two natural conditions, viz. (i) covariant constancy or metric compatibility and (ii) torsion freeness. The first basically boils down to employing units of length mediated by the local inner product (in the parlance of physics, $\Gamma^\mu_{v\rho}(x)$ is called a gauge field); recall Fig. 2. The second is a technical condition necessary to disambiguate a unique set of connection symbols and is equivalent to the symmetry condition $\Gamma^\mu_{v\rho}(x) = \Gamma^\mu_{\rho v}(x)$. This condition is, strictly speaking, redundant if one is only interested in geodesics, since Eq. (8) only depends on the symmetric part of the connection symbols with respect to their lower indices.

Recall that an affine parameter $\tau$ is one for which "speed" $F(\gamma(\tau), \dot\gamma(\tau))$ is constant along the trajectory. In that case the affinely parameterized or constant speed geodesic equation takes the form of Eq. (8) with a vanishing right-hand side; recall Footnote k. This is also the *Euler-Lagrange equation* one obtains directly by minimizing the noninvariant cost functional (Eq. 4). The irrelevance of parameterization in tractography (as opposed to spacetime physics, say, in which it relates to an observable, "proper time") is manifest in the homogeneous case (2), but not in the restricted case (4). For this reason one may want to prefer the parameter-invariant definition. Technical complications arising from this choice may then be countervailed by practical advantages. On the other hand, if one wants to avoid such complications, the nonparameter-invariant definition turns out more convenient, as long as one remains on the alert concerning the admissibility constraint on the parameterization.

### 2.2.2 Hamiltonian formalism

To fully appreciate the Riemann-DTI paradigm one needs to first understand the dual counterpart of the Lagrangian formalism as summarized in Eqs. (2)–(9), known as the Hamiltonian formalism. To this end we may in principle depart from either Eq. (2) or (4). However, subtle complications arise if one tries to adopt a unified approach encompassing both cases. This is due to the fact that, unlike the latter, the former is parameter-invariant. For this reason we consider the nonparameter-invariant case (4) only (Rund, 1973; Lovelock and Rund, 1988). The parameter-invariant case is covered elsewhere; see Rund (1959).

Let us depart from a general, not necessarily $y$-quadratic Lagrangian $L(x, y)$, excluding the case (2), (3) by assuming $L(x, \lambda y) \neq |\lambda| L(x, y)$ henceforth. Subsequently we specify results for the quadratic case of interest in the context of the Riemann-DTI paradigm. To begin with, we consider arbitrary *horizontal curves*, defined in terms of the following parameterization:

$$(x, y) = (\gamma(\tau), \dot\gamma(\tau)), \tag{10}$$

with $\dot\gamma \neq 0$. In other words, the curve $x = \gamma(\tau)$, considered as a three-dimensional spatial object in M, is "lifted" to the six-dimensional *slit tangent bundle* $TM \backslash \{0\}$, that is, the $(x, y)$-manifold obtained by including all three-dimensional tangent vectors $y \in TM_x$ at all $x \in M$, removing the zero section $y = 0$. Note that all points $(x, y) \in TM \backslash \{0\}$ can indeed be viewed implicitly as points on such lifted horizontal curves.[l] Our point of departure in the following will be to interpret the left-hand side of Eq. (10) in this way, keeping the parameterization implicit.

By analogy with classical mechanics, one may introduce the *canonical momentum* covector[m] $q \in T^*M_x$ and the *Euler-Lagrange covector* $E \in T^*M_x$, with components (recall the tacit agreement on $\tau$)

$$q_\mu \doteq q_\mu(x, y) \doteq \frac{\partial L(x, y)}{\partial y^\mu}, \tag{11}$$

respectively,

$$E_\mu \doteq E_\mu(x, y) \doteq \frac{dq_\mu(x, y)}{d\tau} - \frac{\partial L(x, y)}{\partial x^\mu}. \tag{12}$$

---

l. The terms "horizontal" and "vertical" are geometric attributes pertaining to a meaningful decomposition of the $2n$-dimensional tangent bundle, the $(x, y)$-domain, into a direct sum of two $n$-dimensional bundles, in which $y$ acts as the "vertical" coordinate. A precise definition can be found in Bao et al. (2000).

m. The notation for the momentum covector betrays a modest amount of foresight (Basser, 2002; Descoteaux et al., 2010; Tuch, 2004; Westin et al., 2016).

**282 PART | III** Tractography algorithms

One may verify that they are indeed covectors in the sense of tensor calculus, and that the affinely parameterized geodesic equations (8), with vanishing right-hand side, are equivalent to the Euler-Lagrange equations

$$E_\mu(\gamma(\tau), \dot{\gamma}(\tau)) = 0. \tag{13}$$

By virtue of the inverse function theorem we may solve Eq. (11) for $y$ as a function of $x$ and $q$, say

$$y = \phi(x, q) \quad \text{with inverse} \quad q = \psi(x, y), \tag{14}$$

provided the Jacobian of the right-hand side of Eq. (11) is nonsingular, that is,[n]

$$\det \frac{\partial^2 L(x, y)}{\partial y^\mu \partial y^\nu} \neq 0. \tag{15}$$

This condition holds for a quite general class of Lagrangians beyond the quadratic case (4), (5), except, indeed, for the 1-homogeneous case (2), (3). Violation in the latter case is a result of Euler's theorem for homogeneous functions, inducing the "subtle complications" alluded to earlier. The second-order cotensor

$$g_{\mu\nu}(x, y) \doteq \frac{\partial^2 L(x, y)}{\partial y^\mu \partial y^\nu} \tag{16}$$

is referred to in the literature as the *Riemann-Finsler metric* or *fundamental tensor* (Bao et al., 2000).

With the help of Eq. (14) we may eliminate the vector $y$ from the Lagrangian equation (5) in return for its dual momentum covector $q$, and define a scalar field commonly referred to as the *Hamiltonian*:

$$H(x, q) \doteq q_\mu \, \phi^\mu(x, q) - L(x, \phi(x, q)). \tag{17}$$

In the quadratic, Riemannian case we then obtain the Hamiltonian analog of Eq. (7) in closed form:

$$H(x, q) = \frac{1}{2} g^{\mu\nu}(x) q_\mu q_\nu. \tag{18}$$

By the same token, we may construct the one-homogeneous dual of Eq. (6), the *dual Finsler function*

$$F^*(x, q) \doteq \sqrt{g^{\mu\nu}(x) q_\mu q_\nu}, \tag{19}$$

in the context of Riemannian geometry. More generally, the 0-homogeneous dual of Eq. (16), the *dual Riemann-Finsler metric tensor*, is given by

$$g^{\mu\nu}(x, q) = \frac{\partial^2 H(x, q)}{\partial q_\mu \partial q_\nu}, \tag{20}$$

with the nondegeneracy property

$$\det \frac{\partial^2 H(x, q)}{\partial q_\mu \partial q_\nu} \neq 0, \tag{21}$$

so that

$$H(x, q) \doteq \frac{1}{2} F^{*2}(x, q); \tag{22}$$

see the Lagrangian analogs, Eqs. (5), (15). By virtue of homogeneity, Eq. (20) can be inverted, generalizing the quadratic case (18):

$$H(x, q) = \frac{1}{2} g^{\mu\nu}(x, q) \, q_\mu \, q_\nu. \tag{23}$$

One may verify that, beyond the quadratic restriction, Eqs. (16), (20) define pointwise inverse matrix fields under the one-to-one correspondence (Eq. 14) between $y$ and $q$, in the sense that

---

n. A common abuse of notation is to write $\det A_{\mu\nu}$ for the *index-free* determinant of a matrix $A$ with components $A_{\mu\nu}$.

$$g_{\mu\rho}(x, \phi(x,q)) \, g^{\rho\nu}(x,q) = g_{\mu\rho}(x,y) \, g^{\rho\nu}(x,\psi(x,y)) = \delta_\mu^\nu, \tag{24}$$

in which $\delta_\mu^\nu = 1$ if $\mu = \nu$ and 0 otherwise. The Riemann-DTI paradigm pertains to the quadratic case (7), respectively (Eq. 18). The left-hand side expressions in Eqs. (15), (16) are then independent of $y$, defining a nondegenerate metric $g_{\mu\nu}(x)$ consistent with the axioms of Riemannian geometry. Likewise, those in Eqs. (20), (21) are independent of $q$, defining the dual Riemannian metric $g^{\mu\nu}(x)$, so that Eq. (24) simplifies to

$$g_{\mu\rho}(x) \, g^{\rho\nu}(x) = \delta_\mu^\nu. \tag{25}$$

Moreover, Eq. (14) now takes the explicit form[o]

$$y^\mu = g^{\mu\nu}(x)q_\nu \quad \text{with inverse} \quad q_\mu = g_{\mu\nu}(x)y^\nu. \tag{26}$$

In the parlance of Riemannian geometry, $q \in \mathrm{T^*M}_x$ is called the dual (covector) of $y \in \mathrm{TM}_x$. It is not difficult to show that the Euler-Lagrange equations for geodesics, recall Eqs. (11)–(13), are equivalent to the first-order Hamiltonian system

$$\begin{cases} \dfrac{\partial H(x,q)}{\partial q_\mu} &= \phi^\mu(x,q) \\[2ex] \dfrac{dq_\mu}{d\tau} + \dfrac{\partial H(x,q)}{\partial x^\mu} &= 0, \end{cases} \tag{27}$$

again assuming that we are on a $\tau$-parameterized geodesic path.

### 2.2.3 Hamilton-Jacobi theory and the Hamilton-Jacobi equation

Hamilton-Jacobi theory unifies the dual formalisms of the previous sections. It stipulates the existence of a family of hypersurfaces of the form $S(x, \tau) = \text{constant}$, which is geodesically equidistant with respect to the Lagrangian[p] (Eq. 4). By this we mean that, *along a $\tau$-parameterized geodesic*, $(x, y) = (\gamma(\tau), \dot{\gamma}(\tau))$, we have $dS(x, \tau) = L(x, y) \, d\tau$, so that

$$L(x,y) = \frac{\partial S(x,\tau)}{\partial x^\mu} y^\mu + \frac{\partial S(x,\tau)}{\partial \tau}. \tag{28}$$

Note that, along a geodesic $(x, y) = (\gamma(\tau), \dot{\gamma}(\tau))$, the right-hand side is a total $\tau$-derivative, so that the value of the integral equation (4) coincides with the value $S(x, t)$ (a.k.a. the *action*) at the target point $(x, t)$, assuming $S(x_0, 0) = 0$ at the seed point[q] $(x_0, 0)$.

If we now identify a field of normals, see Eqs. (11), (28), as

$$q_\mu(x,\tau) \doteq \frac{\partial S(x,\tau)}{\partial x^\mu}, \tag{29}$$

and consider a geodesic tangent field, recall the duality relation (14); then we may rewrite Eq. (28) as follows:

$$\frac{\partial S(x,\tau)}{\partial \tau} + \frac{\partial S(x,\tau)}{\partial x^\mu} \phi^\mu(x, \nabla S(x,\tau)) - L(x, \phi(x, \nabla S(x,\tau))) = 0. \tag{30}$$

We recognize the Hamiltonian equation (17), leading to the Hamilton-Jacobi equation:

$$\frac{\partial S(x,\tau)}{\partial \tau} + H(x, \nabla S(x,\tau)) = 0. \tag{31}$$

This is a nonlinear first-order partial differential equation from which $S(x, \tau)$ may, in principle, be solved under suitable boundary conditions.

If the Hamiltonian (as in our case) does not depend explicitly on $\tau$, $S(x, \tau) - S(x, 0)$ will be linear in $\tau$, admitting the additive separation of variables

---

o. In Riemannian geometry it is common to use the *same* symbols for the components of the vector $y$ and its dual covector $q$, so that we may write $y_\mu$ instead of $q_\mu$ (respectively, $q^\mu$ instead of $y^\mu$). Loosely speaking, one may think of the lower indexed quantity as a spatial frequency, and the upper indexed one as a spatial displacement.

p. We discuss here again only the nonhomogeneous case, excluding Eqs. (2), (3); see Rund (1973) for the general case.

q. Instead of a single seed point, one may define a seed region $\Omega \subset \mathrm{M}$, such that $S(\Omega, 0) = 0$.

$$S(x, \tau) \doteq S(x) - E\tau, \tag{32}$$

by abuse of notation, for some conserved "energy" constant $E$. (To fully appreciate why this is the case, see the reference in Footnote h.) Without loss of generality we may, for our purpose, set $E = 1$, thus obtaining the following stationary form of the Hamilton-Jacobi equation for $S(x)$:

$$H(x, \nabla S(x)) = 1. \tag{33}$$

Eq. (33), subject to $S(x_0) = 0$, has a unique (generalized) *viscosity solution* (see Evans, 2010 and Arnold, 1989, 1992 for caustic phenomena related to Eq. 31). Minimal geodesics can be obtained through back-propagation in the antigradient direction after solving for $S(x)$, starting at $x$ until reaching $x_0$.

While admitting efficient numerical implementations, it is important to stress that, even if the basic conjecture holds that a neural tract corresponds to a geodesic, (back-propagation from) the viscosity solution of Eq. (33) may be incapable of delivering all existing tracts. In other words, imposing the viscosity condition for disambiguation may (and for realistic data typically will) produce false negatives. This loss of information is caused by the fact that, by the nature of this condition, only geodesics connecting $x_0$ to $x$ can be retrieved for which the overall length is *globally minimal*, that is, equals their non-Euclidean distance. In practice, false negatives tend to be (close to) *nonminimal* geodesics containing a U-shaped segment, which is short-circuited by the viscosity solution of Eq. (33). The Hamilton-Jacobi method should therefore be used with caution in geodesic tractography; see Sepasian et al. (2012) for an approach that includes nonviscosity solutions.

The "unification" alluded to at the beginning of this section entails that Eq. (31) does not merely follow from, but also implies, the geodesic equations, both in the Lagrangian as well as the Hamiltonian formalism. This underlines its fundamental importance. The reader is referred to Rund (1973) for proof and further explanation.

### 2.2.4 Operationalization

One way to implement geodesic tractography is to integrate Eq. (8) starting from two initial conditions, $\gamma(0) = x_0$ (seed point) and $\dot{\gamma}(0) = y_0$ (seed direction), say. The need for *two* conditions underlines, once again, the profound difference with first-order streamline tractography. Another marked difference is that, given a pair of initial conditions, the solution of Eq. (8) is well-defined and unique. Classical streamline tractography requires a well-defined and smoothly varying principal eigenvector of the diffusion tensor, a condition that can only be guaranteed on open neighborhoods of generic points at which there are no eigensystem degeneracies. Even if such is the case, a typical pair of endpoints has zero probability of being connected by a streamline, which in practice implies that streamline tractography tends to produce false negatives, for example, if a priori knowledge or an authoritative expert dictates that these points ought to be connected, yet fail to lie on the same streamline. By virtue of geodesic completeness, geodesic tractography will *always* produce (a) candidate tract(s). Well-posedness and geodesic completeness are the two main reasons for preferring geodesic tractography over streamline tractography. The latter may in fact be seen as a singular limit, in a precise sense, of geodesic tractography, corresponding to a degenerate metric of rank one.

A second way to implement geodesic tractography is to consider Eq. (8) furnished with fixed endpoint conditions, $\gamma(0) = x_1$, $\gamma(t) = x_2$ for some $t > 0$. In that case, one has no direct control over seed or target directions $\dot{\gamma}(0)$, respectively, $\dot{\gamma}(t)$. A brute-force method would be to temporarily ignore the target point $x_2$ and add many seed directions at $x_1$, and vice versa, integrating each resulting system as before. Despite zero probability of finding a seed direction that establishes the connection between the endpoints, slick methods could be devised that turn this idea into an interpolation scheme with control over the number of seed directions (in a trade-off with interpolation errors). The feasibility of this idea has been explored by Sepasian et al. (2008) in $n = 2$ dimensions. Extension to $n = 3$ is possible though cumbersome.

A third way to implement geodesic tractography is to exploit the Hamilton-Jacobi equation. Recall that this is a partial differential equation for a distance function $S(x; x_0)$ between an arbitrary target point $x \in M$ and a fiducial seed point $x_0 \in M$ (or region $\Omega \subset M$; recall Footnote q). This implies that geodesics intersect the level sets $S(x; x_0) = $ constant along steepest slope lines. Thus a geodesic tangent vector coincides, up to a scalar factor, with the local gradient, that is, the normal to the level sets, at every point. Beware that the concepts of "gradient" and "normal" pertain to a non-Euclidean geometry.[r] Thus in classical sense, geodesics are transversal, but typically not at right angles with the level surfaces.

A fourth way to implement geodesic tractography, again compatible with endpoint conditions, is to minimize the functional equation (2) or (4) directly. This can be done by discretizing the argument curve in terms of discrete control points, so that the functional can be approximated by a multivariate function of those control points. Local minima of this function

---

r. The Euclidean normal is proportional to the canonical momentum covector, $q = \nabla S$; recall Eqs. (11), (29). Recall that the back-propagation direction along a geodesic is governed by tangent vectors $y$ related to $q$ according to Eq. (14) and satisfying the "horizontality" condition (10).

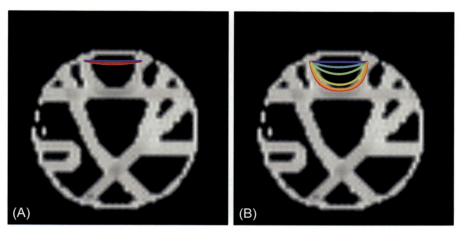

**FIG. 5** DTI tracts obtained as Riemannian geodesics in the FiberCup phantom (Poupon et al., 2008, 2010; Fillard et al., 2011) by direct numerical minimization of Eq. (2). The minimization is initialized at coarse data resolution by a straight line segment (shown in *blue*) connecting two fixed endpoints on either side of the U-shaped fiber. The *red* segments are the convergence results of a coarse-to-fine minimization algorithm, whereby data resolution is increased simultaneously with the number of control points for the multivariate approximation of the length functional (Florack and Astola, 2008; Florack and van Assen, 2012). *Left*: With the rigid identification $G = D^{\text{inv}}$ for the Gram matrix of the Riemannian metric the geodesic connection fails to follow the most plausible U-shaped path by taking a shortcut. This is the result of the fact that such a shortcut is shortest in absolute sense, even for this DTI-adjusted metric. Such shortcuts will always occur if (i) the endpoints are sufficiently close and (in a trade-off with) (ii) the DTI tensor is sufficiently "fat." *Right*: Same as left, but with an enhanced Gram matrix, elongated along its main axis relative to its minor axes with the help of a continuous stretch parameter. The value of this parameter is automatically determined so as to maximize a generic connectivity criterion (the ratio of Euclidean and Riemannian lengths for the unscaled Gram matrix). No shortcut occurs, since, at some point, the enhancement creates a shorter path via the U-turn as expected from the observations (i) and (ii) previously, which turns out to have a stronger connectivity. The same iterative procedure now indeed exhibits, shown in various rainbow colors, convergence toward a more plausible U-shaped geodesic (*red*).

then correspond to discrete approximations of geodesics; see Florack et al. (2021). Fig. 5 shows convergence results for different DTI-induced metrics, with endpoints chosen on either side of a U-shaped tract.

## 2.3 Connecting geometry to DTI and DWI

### 2.3.1 The original paradigm

It is in the Hamiltonian framework, notably Eq. (18), that one recognizes the quadratic form as it appears in the Stejskal-Tanner inspired DTI attenuation formula (for some diffusion time constant $\epsilon > 0$) (Stejskal and Tanner, 1965; Stejskal, 1965),

$$E(x, q) = \exp(-\epsilon H(x, q)), \tag{34}$$

if one identifies the dual metric with the DTI tensor, up to an irrelevant global factor, recall Eq. (18),

$$g^{\mu\nu}(x) = D^{\mu\nu}(x), \tag{35}$$

and the canonical momentum with the time-integrated diffusion sensitizing magnetic gradient field (Basser, 2002)

$$q \doteq \frac{\gamma}{2\pi} \int_0^{\Delta T} G(t)\, dt, \tag{36}$$

used in the scanner ($\gamma \approx 42.58\text{MHz/T}$, the gyromagnetic ratio of hydrogen); recall Footnote m.

Despite its elegance and simplicity one should recall that the identification expressed by Eq. (35) is heuristic. The left-hand side (arguably) relates to fiber geometry, whereas the right-hand side pertains to water diffusion. It is not clear how exactly biological fibers relate to high diffusion pathways. It has been argued that a quantitative relation—if it exists—requires a deeper physical insight into the interplay of fibrous architecture and water diffusion as well as into the coupling with Euclidean geometry in order to furnish space with meaningful tractometric connectivity measures (Astola et al., 2007; Parker et al., 2002; Lenglet et al., 2009). Also, a nonquadratic Hamiltonian might require a generalization of the DWI signal/geometry model in the form of an extension of Eq. (34). The MAP MRI representation by Özarslan et al. (2013)

**286 PART | III** Tractography algorithms

and the diffusion time extension proposed by Fick et al. (2015) may be viable starting points, although the connection with a generalized (Finslerian) Hamiltonian principle is not obvious and requires further investigation.

The DTI geodesic tractography framework is amenable to improvement, both theoretically, by taking into account relevant factors that affect our naive ansatz, notably Eqs. (18), (34), (35), and experimentally, for example, via challenges based on synthetic tractography data for parameter selection and verification (Fillard et al., 2011; Pujol et al., 2015). Although improvements have been mainly sought in sophisticated HARDI models beyond the realm of DTI, the quadratic restriction (Eq. 18) has neither been fully exploited, nor does it represent a necessary condition for geodesic tractography. Following, we discuss a few avenues for improvement.

### 2.3.2 Modifications of the original paradigm

Attempts have been made to overcome issues with the original Riemann-DTI paradigm (O'Donnell et al., 2002; Lenglet et al., 2004), such as shortcuts and false negatives (recall Fig. 5), through modification of the Riemannian metric (Eq. 35). We sketch a few directions of thoughts in ongoing research.

One option relies on tensor sharpening, either replacing the DTI matrix $D(x)$ on the right-hand side of Eq. (35) by some matrix power $D^k(x)$ for some $k > 1$ (typically a small integer[s] ), or through some form of deblurring applied to the signal function (34) (Florack et al., 2010; Dela Haije et al., 2015b).

A second option is to employ a conformal transformation or pointwise metric rescaling, replacing the right-hand side of Eq. (35) by a scalar multiple $\exp(2\alpha(x))D(x)$ for some scalar field $\alpha(x)$. Hao et al. (2011) propose a scalar field obtained by solving a Poisson equation that encourages a closer correspondence between principal eigenvector streamlines and geodesics, and verify this tendency experimentally on synthetic data. Fuster et al. (2016) impose the constraint that the Laplace-Beltrami operator for the conformally scaled metric should coincide with the diffusion operator, implying the metric to be the adjugate instead of inverse diffusion tensor, that is, $\alpha(x) = -\ln \sqrt{\det D(x)}$. Preliminary experiments seem to indicate that the induced geodesics are somewhat less likely to favor shortcuts through isotropic high diffusivity regions, such as the ventricles.

Unfortunately, any one-to-one relation between metric and diffusion tensor seems to have pros and cons. A third option is therefore to introduce a parameterized *family* of Riemannian metrics, disambiguating members on a tract-by-tract basis. The premise is that even if no effective universal relation exists between metric and diffusion tensor (35), then perhaps one could still find one *tractwise*. This approach relies on enhancing anisotropy through a relative rescaling of DTI eigenvalues governed by a priori free *control parameters*. It has analogies with both previous ones in the sense that it employs a metric transform, as well as with classical streamline tractography, which emerges in a singular limit after incorporating suitable control parameters.

The simplest scenario employs a single, global control parameter enhancing the main eigenvalue relative to the minor ones (Florack et al., 2021). This one-parameter anisotropic metric transform can be applied to either inverse or adjugate DTI-based metric, yielding, somewhat surprisingly, very similar results. The idea is to search for an "optimal" tract (given an optimality criterion in the form of a nonlinear connectivity measure[t] ) among the family of geodesics obtained for any fixed pair of endpoints induced by variation of the metric control parameter. Even in this simplest form, this control problem results in remarkable improvements over rigid methods, as can be seen on the right in Fig. 5, where it is seen to prevent a shortcut of the U-shaped bundle. In addition, the possibility of obtaining multiple optimal control parameter values (corresponding to hierarchically ordered local connectivity maxima) may provide alternative pathways, potentially preventing false negatives, in case the one found to be globally optimal turns out spurious.

### 2.3.3 Beyond DTI: Toward a comprehensive Riemann-Finsler-DWI paradigm

The options in Section 2.3.2 to improve the Riemann-DTI paradigm are motivated by the popularity of DTI and its clinical appeal (Rutten et al., 2014). In order to fully exploit state-of-the-art DWI, with its a priori unrestricted measurement degrees of freedom, we are, however, ultimately forced to abandon the quadratic constraint that defines the Riemann-DTI paradigm. Riemann-Finsler geometry is the canonical remedy, as it retains the essential notion of a norm, recall Eq. (3), while dropping the quadratic restriction (6). For this reason it has been paraphrased as "Riemannian geometry without the quadratic restriction" (Bao et al., 2000; Shen and Shen, 2016). Although this somewhat understates technical and mathematical complications—subject of ongoing and future work[u] —the geodesic tractography rationale remains essentially unaltered.

---

s. Without support from physical arguments this is no less ad hoc than the default choice $k = 1$.

t. The connectivity measure is invariably based on the untransformed metric, that is, independent of the control parameter.

u. Riemann-Finsler geometry has become an active field of research with a growing community, straddling various scientific domains; see the recently instated international conference series "New Methods in Finsler Geometry."

This section is meant to provide further motivation for this extension, with pointers to established results indicating possible avenues for future research.

The standard approach is to extend the three-dimensional spatial base manifold M to the six-dimensional slit tangent bundle TM\{0} already introduced in Section 2.2. By virtue of the homogeneity condition (3) this yields effectively a five-dimensional extended base manifold of positions and orientations, a.k.a. the *projectivized tangent bundle* or (if direction rather than orientation matters) the *sphere bundle*.[v] The Riemannian metric tensor then finds its generalization in the orientation-parameterized family (16), the Riemann-Finsler metric tensor. Early Finsler geodesic approaches along this line have been proposed by Melonakos et al. (2008), Melonakos (2009), de Boer et al. (2011), Astola (2010), Astola et al. (2011, 2014), Florack and Fuster (2014), Florack et al. (2015, 2017), Dela Haije et al. (2014, 2015a, 2019), and Dela Haije (2017).

A different approach toward Finsler geodesic tractography is based on ideas proposed by Péchaud et al. (2009), Mirebeau (2018), and Duits et al. (2014, 2016, 2018). In this case the base manifold is not the three-manifold $M \simeq \mathbb{R}^3$, but the five-dimensional manifold $\mathbb{M} = M \times U$, in which $U \doteq \{n \in \mathbb{R}^3 \mid \|n\| = 1\}$ denotes the Euclidean unit sphere in $\mathbb{R}^3$. Consequently, the norm defining Finsler function is not defined on the conventional six-dimensional slit tangent bundle TM\{0}, but on its ten-dimensional extension $T\mathbb{M}\backslash\{0\} = \{(x, n, \dot{x}, \dot{n}) \in T\mathbb{M} \mid (\dot{x}, \dot{n}) \neq (0, 0)\}$. The incommensurable natures of the vector components $\dot{x}$ and $\dot{n}$ imply that such a Finsler norm necessarily requires a coupling constant relating space and orientation, for which no a priori preferred choice presents itself. An advantage, on the other hand, is that (single-shell) DWI data can be straightforwardly interpreted as scalar fields on $\mathbb{M}$, or rather as scalar sections of $T^*\mathbb{M}$. Another potential advantage is that the geodesic tractography problem allows for simultaneous spatial localization and orientation specification at both source and target points. Recall that in the standard case one can fix either the spatial locations of both endpoints or the spatial location and initial orientation at only one of them.

Because of the mind-boggling mathematics of Finsler geometry and the lacunae in our understanding of the underlying physics of microstructural diffusion versus macrostructural fibrous tissue organization, its application to DWI calls for much future work. See also Antonelli and Zastawniak (1993, 1994, 1999) and Ma et al. (2021) for the connection between Finsler geometry, Brownian motion, and diffusion.

## 2.4 The effect of data variability on geometry

A crucial question about the effect induced by noise inevitably arises when analyzing empirical data. Among many other possible sources of uncertainty along a clinical tractography pipeline, two of them are manifestly coupled to the geodesic tractography paradigm, viz. domain and codomain uncertainties in the form of *metric uncertainty* induced by data noise, and track ambiguity due to *spatial uncertainty* in seed/target point delineation, respectively. As long as these uncertainties are "sufficiently small," their effect can be analytically described in terms of so-called *geodesic deviation*, a relative acceleration (or "tidal force" effect caused by inhomogeneities of the metric field) that quantifies the inclination of neighboring geodesics to bend toward or away from each other. To appreciate the idea, see Fig. 6, which shows (probabilistic) tubular confinement regions for perturbed geodesics as predicted by geodesic deviation along a fiducial geodesic running along the tube's central axis.

An operational definition of what it means for a perturbation to be "sufficiently small" at any given point along a geodesic would allow us to use tubes like those in Fig. 6 to represent entire volumetric bundles of neighboring tracts based on a small number of actually computed geodesics. Besides this dimensionality reduction benefit, such tubes may help to visually convey effects of noise, inter- and intrarater variability, and small misalignments by automatic segmentation algorithms.

In general relativity theory, geodesic deviation equations are used to model the effect of tidal gravitational forces on the trajectories of freely falling particles (or lightrays), and in this context they are typically formulated for an analytically stipulated metric without uncertainty; see Misner et al. (1973) for details in this context. In our case we clearly do need to account for metric perturbations, as a result of which they take a slightly more general form than found in typical textbook formulas. We present the appropriately generalized geodesic deviation equations below without proof, and omit their closed-form solution. These are provided elsewhere (Sengers et al., 2021a).

Consider a fiducial geodesic given by $(x, \dot{x}) = (\gamma(t), \dot{\gamma}(t))$, together with a two-parameter family of "infinitesimally neighboring" geodesics $(\bar{x}, \dot{\bar{x}}) = (\gamma(t; \varepsilon_1, \varepsilon_2), \dot{\gamma}(t; \varepsilon_1, \varepsilon_2))$ given by

---

v. For computational purposes a pragmatic way to handle the projectivized tangent bundle is to consider antipodally symmetric functions on the sphere bundle.

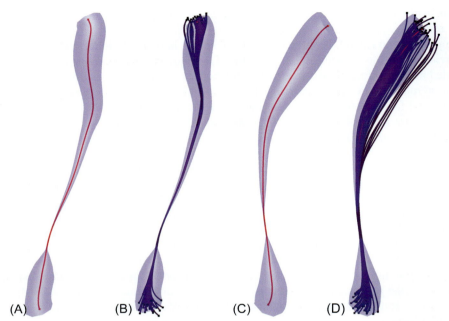

**FIG. 6** Proof of concept for geodesic tubes illustrating geodesic deviation along a fiducial geodesic, in this example generated from DTI data within the corticospinal tract (CST) using the Riemann-DTI paradigm; recall Eqs. (41)–(43). (A) Arbitrarily chosen geodesic for a given pair of seed and target point, together with its geodesic tube. (B) Same geodesic, same geodesic tube, now augmented with neighboring geodesics computed for randomly perturbed seed and target points close to the original pair for verification, (roughly) confirming the validity of the geodesic tube's capture zone. Mathematically, geodesic deviation is predicted to "work" provided perturbations are "sufficiently small," a physically void notion that will need to be operationalized for the finite perturbations occurring in practice. To illustrate this issue, consider (C) and (D), similarly obtained for another seed/target pair. Under similar perturbations as in (A) and (B), some neighboring geodesics are now seen to run partly outside the geodesic tube. Interestingly, these outliers do correspond to the most strongly perturbed seed and target points, indicating that, in this second case, some perturbations, albeit of the same order as before, are apparently no longer "sufficiently small." Thus outliers could have been prevented if only one knew how small "sufficiently small" is for each seed/target pair. See Sengers et al. (2021a) for geodesic tubes induced by DTI data noise; recall Eqs. (41), (42). Acquisition details: Original DWI scans were acquired using a Philips Achieva 3T MRI scanner, with $b = 1500$, 50 diffusion weighting directions, six $b = 0$ images, 2mm isotropic voxel size.

$$\gamma(t;\varepsilon_1,\varepsilon_2) = \gamma(t) + \varepsilon_1 y(t) + \varepsilon_2 z(t), \tag{37}$$

with $0 \leq \varepsilon_1, \varepsilon_2 \ll 1$. The vector functions $y = y(t)$ and $z = z(t)$ cannot be chosen arbitrarily, as each member of this family is required to be a geodesic under $\varepsilon_1$-perturbed metric field

$$\bar{g}_{\mu\nu}(x,\varepsilon_1) = g_{\mu\nu}(x) + \varepsilon_1 h_{\mu\nu}(x), \tag{38}$$

showing the linearized effect induced by (mild) data noise (terms of order $\mathcal{O}(\varepsilon_1^2)$ are neglected), respectively, under $\varepsilon_2$-perturbed side conditions for tract disambiguation, be it boundary conditions,

$$(\bar{x}_0, \bar{x}_T) = (x_0 + \varepsilon_2 z_0, x_T + \varepsilon_2 z_T), \tag{39}$$

with unperturbed values $x_0 = \gamma(0)$, $x_T = \gamma(T)$, or initial conditions

$$(\bar{x}_0, \bar{y}_0) = (x_0 + \varepsilon_2 z_0, y_0 + \varepsilon_2 w_0), \tag{40}$$

with unperturbed values $x_0 = \gamma(0)$, $y_0 = \dot{\gamma}(0)$, again ignoring terms of order $\mathcal{O}(\varepsilon_2^2)$. Note that $h_{\mu\nu}(x)$, $z_0$, $z_T$, and $w_0$ (and a fortiori $y(t)$, $z(t)$, and $\gamma(t;\varepsilon_1,\varepsilon_2)$) may be of a stochastic nature.

Recall the basic geodesic equation (8) governing the zeroth-order geodesic tract of interest (the unperturbed case). We assume its solution to be known in affine parameterization (zero right-hand side). The geodesic deviation equations, induced by either Eq. (38) or one of the two side conditions (39), (40), take the following generic form:

$$\frac{D^2 \xi^\mu(t)}{dt^2} + R^\mu_{\nu\rho\sigma}(\gamma(t))\,\dot{\gamma}^\nu(t)\xi^\rho(t)\dot{\gamma}^\sigma(t) = f^\mu(\gamma(t)). \tag{41}$$

In this equation, $\xi(t)$ should be identified with $y(t)$ if $\varepsilon_2 = 0$, that is, if we do not perturb the side conditions and only account for data noise, and with $z(t)$ if $\varepsilon_1 = 0$, that is, if only the side conditions are perturbed and data are considered noise-free. In the former case, there is an inhomogeneous term, viz.

$$f^\mu(\gamma(t)) = -H^\mu_{\nu\rho}(\gamma(t))\,\dot{\gamma}^\nu(t)\dot{\gamma}^\rho(t), \tag{42}$$

in which the symbol $H^\mu_{\nu\rho}$ represents the coefficient of the first-order correction term to the Christoffel symbols for the unperturbed metric, recall Eq. (9), that is, assuming $\bar{\Gamma}^\mu_{\nu\rho}(x,\varepsilon_1) = \Gamma^\mu_{\nu\rho}(x) + \varepsilon_1 H^\mu_{\nu\rho}(x)$. In the latter case, with the metric unperturbed, the geodesic deviation equation takes the more familiar homogeneous form (41) with

$$f^\mu(x) = 0. \tag{43}$$

For constructions, more explicit formulas, closed-form solutions given side condition (39) or (40), and the operational definition of the illustrated geodesic tubes, see Sengers et al. (2021a, b).

## 3 Conclusion

Geodesic tractography provides a geometric, second-order scheme as a versatile alternative to first-order streamline-based methods. Qualitatively it relies on the same heuristic principle, viz. the observation that fibrous tissue imparts nonrandom barriers to the diffusion of water molecules, inducing anisotropic local diffusion profiles that can be revealed by diffusion-weighted MRI. Its quantitative nature is rather different, though, since geodesics in general do not represent high diffusivity pathways. They are best viewed as geometric primitives for an "optimal" path between any two endpoints. Optimality entails that, given fixed endpoints, no nearby curves exist that are shorter in local geometric units tuned to the underlying anisotropy. In the case of DTI such length units are drawn from a norm induced by an inner product defined by a Gram matrix that is inversely proportional to the DTI matrix. The Hopf-Rinow theorem expresses *geodesic completeness*, ensuring existence (but not uniqueness) of such paths regardless of the choice of endpoints, a marked difference with the streamline idea.

The *Riemann-DTI paradigm* stipulates that neural tracts are optimal paths, in the preceding geodesic sense, with an *additional* optimality condition that, in its simplest form, expresses the average mean free diffusion path, or the average alignment with the principal axis of diffusion along a geodesic. The implicit need for lengths or angles in the conventional sense in the construction of such "connectivity measures" requires a coupling to Euclidean geometry.

The Riemann-DTI paradigm furthermore provides a natural mechanism for *geodesic deviation*, the tendency of neighboring geodesics to accelerate toward or away from each other due to the inhomogeneous nature of underlying DTI data. For any fiducial geodesic, geodesic deviation can be visualized in the form of a *geodesic tube*, showing the propagation of (data and/or endpoint) perturbations along the way. As such, geodesic tubes may serve to convey *uncertainty* for any geodesic, which, in turn, provides a *dimensionality reduction* of the tractography problem by virtue of their volumetric nature.

Whereas the Riemann-DTI paradigm hinges on the *quadratic restriction*, geodesic tractography as such does not, as it only requires a norm, which is not necessarily based on an inner product. The canonical extension of Riemannian geometry for norms beyond the quadratic restriction is known as Finsler geometry, which therefore provides the natural geometric framework for geodesic tractography based on more general DWI models. Such a *Finsler-DWI paradigm* calls for much future research. However, even the restrictive Riemann-DTI paradigm is, by no means, fully developed, and is of special interest in view of the robustness and clinical appeal of DTI.

## Appendix: Basic concepts from linear algebra and tensor calculus

This appendix introduces basic concepts from linear algebra and tensor calculus, and their extensions to manifolds, as far as these are relevant for geodesic tractography as outlined in this chapter.

A (real) *vector space* or *linear space* is a structured set $V$, furnished with a superposition principle enabled by *vector addition* $+: V \times V \to V$, and *scalar multiplication* $\cdot : \mathbb{R} \times V \to V$. The set $V$ is closed under these operations and satisfies the following axioms. Let $u, v, w \in V$, $\lambda, \mu \in \mathbb{R}$ henceforth; then

- $(u + v) + w = u + (v + w)$,
- there exists a neutral element $o \in V$ such that $o + u = u$,
- there exists an element $-u \in V$ such that $u + (-u) = o$,
- $u + v = v + u$,
- $\lambda \cdot (u + v) = \lambda \cdot u + \lambda \cdot v$,

**290  PART | III** Tractography algorithms

- $(\lambda + \mu) \cdot u = \lambda \cdot u + \mu \cdot u,$
- $(\lambda\mu) \cdot u = \lambda \cdot (\mu \cdot u),$
- $1 \cdot u = u.$

Another pivotal concept is that of a *linear operator*, acting between two vector spaces, $V$ and $W$, say:

- A *linear operator* $\mathcal{A} : V \to W$ is a mapping that satisfies $\mathcal{A}(\lambda \cdot v + \mu \cdot w) = \lambda \cdot \mathcal{A}(v) + \mu \cdot \mathcal{A}(w).$

Such linear operators are collectively denoted by $\mathscr{L}(V, W)$, which has a natural vector space structure:

- $\mathscr{L}(V, W)$ is a vector space, with vector addition and scalar multiplication inherited from $W$, viz. $(\lambda \cdot \mathcal{A} + \mu \cdot \mathcal{B})(v) = \lambda \cdot \mathcal{A}(v) + \mu \cdot \mathcal{B}(v)$ for all $\mathcal{A}, \mathcal{B} \in \mathscr{L}(V, W).$

The trivial case $W = \mathbb{R}$ is particularly important:

- The *dual vector space* of $V$ is $V^* = \mathscr{L}(V, \mathbb{R})$, the set of all linear functionals or *covectors*.

The ostensible asymmetry in the definition and interpretation of vectors and covectors may be removed via the introduction of the *Kronecker tensor* in the form of a bilinear operator $\langle \ , \ \rangle : V^* \times V \to \mathbb{R}$. By definition, we may identify a given covector $\omega \in V^*$ with the linear operator that arises by freezing the first argument in this bracket notation, as follows:

$$\omega \doteq \langle \omega, \ \rangle : V \to \mathbb{R}.$$

The equivalent status of vectors and covectors becomes manifest in this bracket formalism by identifying a vector $v \in V$ with the linear operator that arises by freezing the second argument:

$$v \doteq \langle \ , v \rangle : V^* \to \mathbb{R}.$$

As with a covector, a vector may thus likewise be interpreted as a linear operator (a "cocovector"), in the sense that[w] $V = \mathscr{L}(V^*, \mathbb{R}) = V^{**}$. The Kronecker bracket formalism expresses a *duality* of vectors and covectors. From the perspective of the Kronecker tensor vectors and covectors are "interacting states," each encapsulating $n$ degrees of freedom (recall Footnote w).

A *basis* of an $n$-dimensional vector space $V$ is an $n$-tuple of independent vectors, $\{\mathbf{e}_\mu\}_{\mu=1}^n$, such that each $v \in V$ can be uniquely decomposed as $v = \sum_{\mu=1}^n v^\mu \mathbf{e}_\mu$ for some $n$-tuple of scalar coefficients $\{v^\mu\}_{\mu=1}^n$. Similarly, a basis of the dual space $V^*$ is an $n$-tuple of covectors, $\{\mathbf{e}^\mu\}_{\mu=1}^n$, such that each $\omega \in V^*$ can be uniquely decomposed as $\omega = \sum_{\mu=1}^n \omega_\mu \mathbf{e}^\mu$ for some $n$-tuple of scalars $\{\omega_\mu\}_{\mu=1}^n$. A basis of $V^*$ is naturally induced by one of $V$ (vice versa), via the following *duality* principle (which, in fact, defines the bilinear Kronecker tensor):

$$\langle \mathbf{e}^\mu, \mathbf{e}_v \rangle = \delta_v^\mu.$$

Bilinearity then shows that

$$\langle \omega, v \rangle = \omega_\mu \langle \mathbf{e}^\mu, \mathbf{e}_v \rangle v^v = \omega_\mu \delta_v^\mu v^v = \omega_\mu v^\mu = \boldsymbol{\omega}^{\mathrm{T}} \mathbf{v},$$

in which boldface quantities on the right denote row and column arrays ($\mathbb{R}^n$-vectors) of (co)vector components relative to their dual bases in standard matrix notation. The index convention for the various entities is seen to be consistent with the *Einstein summation convention*, in which $\sum$-symbols are suppressed for identical upper and lower index pairs. The notation likewise reveals the $n \times n$ identity matrix as a representation of the Kronecker tensor relative to *any* choice of basis.[x] Do not confuse a contraction of a vector/covector pair with an inner product; no inner product is involved at this stage. Without an inner product neither vectors nor covectors can be regarded as "self-interacting states."

Matrix notation is convenient if one keeps track of bases and as long as objects of interest can be captured in terms of holors with no more than two labels attached. In tensor calculus, being a multivariate extension of linear algebra, one cannot get away with this, forcing us to abandon matrix notation in return for either abstract basis free notation (along with an abstract "vector bundle" terminology to distinguish the various object types) or, from a computational point of view more conveniently, one in terms of holors decorated by upper and/or lower tensor index symbols. Holors are endowed with an implicit recipe for bias removal via linear *tensor transformation* properties, expressing the relationship between holors

---

w. The identification $V = V^{**}$ requires $\dim V = \dim V^* = n < \infty$, which is tacitly assumed throughout.

x. The remarkable fact that the holor $\delta_v^\mu$ is basis independent reflects the universal nature of duality as a construct that comes for free with the axioms of a vector space. No additional structure is required.

relative to different bases. In the abstract framework tensorial equations are unbiased because no bases are involved, whereas the holor framework removes the intrinsic biases of left- and right-hand sides by virtue of identical (and invertible) linear transformations under a change of basis, rendering their validity basis independent.

An essential add-on in (metric) geodesic tractography is the notion of a *norm*. A vector space furnished with a norm is called a *normed space*. In the simplest case the norm is induced by a real *inner product*, that is, an operator of the form $(\ |\ ) : V \times V \to \mathbb{R}$, satisfying

- $(u|v) = (v|u)$,
- $(u|\lambda \cdot v + \mu \cdot w) = \lambda(u|v) + \mu(u|w)$, and
- $(u|u) \geq 0$ and $(u|u) = 0$ iff $u = o$.

A vector space $V$ furnished with an inner product (and thus with an inner product induced norm) is called an *inner product space*. The squared norm of $u$ is then defined as

$$\|u\|^2 = (u|u),$$

a special instance of a general norm (v.i.) subject to the *quadratic restriction* expressed by the right-hand side. The holor of the inner product tensor relative to a basis completely defines the inner product by virtue of bilinearity, viz.:

$$\left(\mathbf{e}_\mu \, |\mathbf{e}_v\right) = g_{\mu v}.$$

The inner product defining matrix $\mathbf{g}$, with entries $g_{\mu v}$, is called the *Gram matrix*. In terms of holors and standard matrix notation, we have, respectively,

$$(u|v) = u^\mu \left(\mathbf{e}_\mu \, |\mathbf{e}_v\right) v^v = u^\mu g_{\mu v} v^v = \mathbf{u}^{\mathrm{T}} \mathbf{g} \, \mathbf{v}.$$

This reveals another interpretation of the inner product, viz. as a toggling device for converting vectors into covectors, and vice versa. To see this, note that for any given $u \in V$ we may interpret

$$(u|\ ) : V \to \mathbb{R},$$

by virtue of linearity with respect to its single vector argument, as a covector, $\hat{u} \in V^*$, say, which is completely determined by $u$. The hat-notation for covectors in the context of an inner product space is intended to avoid confusion with vectors as well as to refer to the vectors they relate to by the equivalence $\hat{u} = (u|\ )$. In turn, $V^*$ inherits an inner product structure from $V$, with an inner product of the form $(\ |\ )_* : V^* \times V^* \to \mathbb{R}$, with associated holor

$$(\mathbf{e}^\mu \, |\mathbf{e}^v)_* = g^{\mu v},$$

the components of the *inverse Gram matrix*. This "dual metric" plays a similar role in switching between $V$ and $V^*$, since

$$(\hat{u}|\ )_* : V^* \to \mathbb{R}$$

may be identified with $u \in V^{**} = V$, with formal identification $u = (\hat{u}|\ )_*$. The inner products are related to the Kronecker tensor by the so-called Riesz representation formulas $(\hat{\ }|\hat{\ })_* = (\ |\ ) = \langle \hat{\ }, \ \rangle$.

In the holor framework the acts of converting a vector $u = u^\mu \mathbf{e}_\mu$ into its dual covector $\hat{u} = u_\mu \mathbf{e}^\mu$ are tantamount to *index lowering*:

$$u_\mu = g_{\mu v} u^v.$$

Vice versa, starting out from the (holor of a) covector, we obtain (the holor of) its corresponding vector via *index raising*:

$$u^\mu = g^{\mu v} u_v.$$

(No hats are needed on top of covector components, since index position betrays their nature.)

More generally, a norm is an operator of the form $\|\ \| : V \to \mathbb{R}$, satisfying

- $\|\lambda \cdot u\| = |\lambda| \, \|u\|$,
- $\|u + v\| \leq \|u\| + \|v\|$,
- $\|u\| \geq 0$ and $\|u\| = 0$ iff $u = o$.

Norms do not necessarily require an inner product and are much more versatile for DWI-based geodesic tractography beyond the quadratic restriction imposed by DTI. By way of analogy, one may conceive of a general norm as being induced by a *family* of inner products, $(\ |\ )_{[z]} : V \times V \to \mathbb{R}$, say, parameterized by orientation $[z] \in P(V) = \{\lambda z \mid z \in V, \ z \neq 0, \ \lambda \in \mathbb{R}\}$. The norm of a vector $u \in V$ is then measured along the *preferred orientation* $[u]$:

$$\|u\|^2 = (u|u)_{[u]} = g_{\mu\nu}([u]) \ u^\mu u^\nu,$$

in which $g_{\mu\nu}([z])$ denotes the Gram matrix associated with orientation $[z]$. In the case of an inner product space all instances of the family coincide, that is, $g_{\mu\nu}([z]) = g_{\mu\nu}$ independent of orientation. In the general case one often prefers to work with an orientation-dependent Gram matrix of the form $g_{\mu\nu}(z)$, in which a representative vector $z \in V, z \neq 0$, replaces the equivalence class $[z] \in P(V)$, with the concomitant condition of 0-homogeneity, $g_{\mu\nu}(\lambda z) = g_{\mu\nu}(z)$.

In differential geometry all concepts introduced previously are transplanted to a *differentiable manifold* M, simply by attaching a copy of the various linear spaces involved to each point $x \in$ M ("fibers"). Of particular interest is the local *tangent space*, notation $V_x \doteq TM_x$, and its dual counterpart, $V_x^* \doteq T^*M_x$. The collection $\cup_{x \in M} V_x$ of all fibers over all $x \in$ M is called a *fiber bundle*, with the (co)tangent bundles $TM = \cup_{x \in M} TM_x$ and $T^*M = \cup_{x \in M} T^*M_x$ as particularly important instances. Spatial coherence is reflected in the notion of a *section* or *field*, that is, the selection of one vector element from each fiber in such a way that vectors vary smoothly across the manifold. This requires a rigorous notion of "rate of change" so as to allow for neighborhood comparisons, which is formalized in the concept of a *connection*.

Restricting ourselves to the tangent bundle in the context of Riemannian geometry we have a *linear connection*, which is completely disambiguated by a set of (basis dependent) *connection symbols*, stating how the basis vector field $\mathbf{e}_\mu(x)$ varies to first order at $x \in$ M in the direction of $\mathbf{e}_\nu(x)$, expressed in the local basis $\{\mathbf{e}_\rho(x)\}$:

$$\nabla_{\mathbf{e}_\nu(x)} \mathbf{e}_\mu(x) = \Gamma^\rho_{\mu\nu}(x) \ \mathbf{e}_\rho(x).$$

The canonical choice is to impose natural geometric constraints of "metric compatibility" and "torsion-freeness," as a result of which the symbols are uniquely determined by the metric tensor and its first-order derivatives and known as the *Levi-Civita connection symbols*. Naturally, the connection induces *geodesics* as curves $x = \gamma(t)$ with a "constant," that is, *parallel transported* tangent vector $\dot{\gamma}(t)$: $\nabla_{\dot{\gamma}} \dot{\gamma} = 0$. This provides the basic mechanism for geodesic tractography in the Riemann-DTI paradigm. Its generalization to a Finsler-DWI paradigm requires a *nonlinear connection*. For details, see Bao et al. (2000) and Shen and Shen (2016).

## Acknowledgments

This work is part of the research program "Diffusion MRI Tractography with Uncertainty Propagation for the Neurosurgical Workflow" with project number 16338, and of the project "Bringing Tractography into Daily Neurosurgical Practice" with project number KICH1.ST03.21.004, which are (partly) financed by the Netherlands Organization for Scientific Research (NWO). We are grateful for the support received from various health technology companies and academic and regional hospitals.

The Foundation for Fundamental Research on Matter (FOM) of NWO is furthermore acknowledged for its financial support through a personal grant of A. Fuster.

The following people have provided useful contributions or suggestions for the final manuscript: Tom Dela Haije, Remco Duits, Aasa Feragen, Stephan Meesters, Pauly Ossenblok, Evren Özarslan, Faizan Siddiqui, Lars Smolders, and Anna Vilanova. Geert-Jan Rutten from the Department of Neurosurgery at ETZ Tilburg, The Netherlands, is gratefully acknowledged for clinical feedback.

## References

Abraham, R., Marsden, J.E., Ratiu, T., 1988. Manifolds, tensor analysis, and applications. In: Applied Mathematical Sciences, vol. 75. Springer-Verlag, New York.

Antonelli, P.L., Zastawniak, T.J., 1993. Diffusions on Finsler manifolds. Rep. Math. Phys. 33 (1–2), 303–315.

Antonelli, P.L., Zastawniak, T.J., 1994. Introduction to diffusion on Finsler manifolds. Math. Comput. Model. 20 (4–5), 109–116.

Antonelli, P.L., Zastawniak, T.J., 1999. Fundamentals of Finslerian diffusion with applications. In: Fundamental Theories of Physics, vol. 101. Kluwer Academic Publishers, Dordrecht, The Netherlands.

Arnold, V.I., 1989. Mathematical methods of classical mechanics. In: Graduate Texts in Mathematics, vol. 60. Springer-Verlag, New York.

Arnold, V.I., 1992. Catastrophe Theory, third, revised and expanded ed. Springer-Verlag, Berlin.

Assaf, Y., Freidlin, R.Z., Rohde, G.K., Basser, P.J., 2004. Non-mono-exponential attenuation of water and N-acetyl aspartate signals due to diffusion in brain tissue. Magn. Reson. Med. 52 (5), 965–978.

Astola, L.J., 2010. Multi-Scale Riemann-Finsler Geometry: Applications to Diffusion Tensor Imaging and High Angular Resolution Diffusion Imaging (Ph.D. thesis). Eindhoven University of Technology, Department of Mathematics and Computer Science, Eindhoven, The Netherlands.

Astola, L., Florack, L., ter Haar Romeny, B., 2007. Measures for pathway analysis in brain white matter using diffusion tensor images. In: Karssemeijer, N., Lelieveldt, B. (Eds.), Proceedings of the Twentieth International Conference on Information Processing in Medical Imaging-IPMI 2007 (Kerkrade, The Netherlands). Lecture Notes in Computer Science, vol. 4584. Springer-Verlag, Berlin, pp. 642–649.

Astola, L.J., Jalba, A.C., Balmashnova, E.G., Florack, L.M.J., 2011. Finsler streamline tracking with single tensor orientation distribution function for high angular resolution diffusion imaging. J. Math. Imaging Vis. 41 (3), 170–181.

Astola, L., Sepasian, N., Dela Haije, T., Fuster, A., Florack, L., 2014. A simplified algorithm for inverting higher order diffusion tensors. Axioms 3 (4), 369–379. https://doi.org/10.3390/axioms3040369.

Bao, D., Chern, S.S., Shen, Z., 2000. An introduction to Riemann-Finsler geometry. In: Graduate Texts in Mathematics, vol. 2000. Springer-Verlag, New York.

Basser, P.J., 2002. Relationships between diffusion tensor and q-space MRI. Magn. Reson. Med. 47 (2), 392–397.

Born, M., Wolf, E., 1970. Principles of Optics, fourth ed. Pergamon Press, Oxford.

Callaghan, P.T., MacGowan, D., Packer, K.J., Zalaya, F.O., 1990. High-resolution q-space imaging in porous structures. J. Magn. Reson. 90, 177–182.

Cottaar, M., Bastiani, M., Chen, C., Dikranian, K., Van Essen, D., Behrens, T.E., Sotiropoulos, S.N., Jbabdi, S., 2018. A gyral coordinate system predictive of fibre orientations. NeuroImage 176, 417–430. https://doi.org/10.1016/j.neuroimage.2018.04.040.

de Boer, R., Schaap, M., van der Lijn, F., Vrooman, H.A., de Groot, M., van der Lugt, A., Arfan Ikram, M., Vernooij, M.W., Breteler, M.M.B., Niessen, W.J., 2011. Statistical analysis of minimum cost path based structural brain connectivity. NeuroImage 55 (2), 557–565. https://doi.org/10.1016/j.neuroimage.2010.12.012.

Dela Haije, T.C.J., 2017. Finsler Geometry and Diffusion MRI (Ph.D. thesis). Eindhoven University of Technology, Eindhoven, The Netherlands.

Dela Haije, T.C.J., Fuster, A., Florack, L.M.J., 2014. Reconstruction of convex polynomial diffusion MRI models using semi-definite programming. In: Proceedings of the 23rd Joint Annual Meeting ISMRM-ESMRMB (Toronto, May 30–June 5, 2015). International Society for Magnetic Resonance in Medicine, p. 2821.

Dela Haije, T.C.J., Fuster, A., Florack, L.M.J., 2015a. Finslerian diffusion and the Bloch-Torrey equation. In: Hotz, I., Schultz, T. (Eds.), Visualization and Processing of Higher Order Descriptors for Multi-Valued Data. Mathematics and Visualization. Springer-Verlag, pp. 21–35.

Dela Haije, T.C.J., Sepasian, N., Fuster, A., Florack, L.M.J., 2015b. Adaptive enhancement in diffusion MRI through propagator sharpening. In: Fuster, A., Ghosh, A., Kaden, E., Rathi, Y., Reisert, M. (Eds.), MICCAI Workshop on Computational Diffusion MRI (October 9, 2015, Munich, Germany). Mathematics and Visualization, Springer-Verlag, pp. 131–143.

Dela Haije, T., Savadjiev, P., Fuster, A., Schultz, R.T., Verma, R., Florack, L., Westin, C.F., 2019. Structural connectivity analysis using Finsler geometry. SIAM J. Imaging Sci. 12 (1), 551–575. https://doi.org/10.1137/18M1209428.

Descoteaux, M., Deriche, R., Le Bihan, D., Mangin, J.F., Poupon, C., 2010. Multiple q-shell diffusion propagator imaging. Med. Image Anal. 15, 603–621.

Duits, R., Boscain, U., Rossi, F., Sachkov, Y., 2014. Association fields via cuspless sub-Riemannian geodesics in SE(2). J. Math. Imaging Vis. 49 (2), 384–417.

Duits, R., Ghosh, A., Dela Haije, T.C.J., Mashtakov, A., 2016. On sub-Riemannian geodesics in SE(3) whose spatial projections do not have cusps. J. Dyn. Control Syst. 22 (4), 771–805. https://doi.org/10.1007/s10883-016-9329-4.

Duits, R., Meesters, S., Mirebeau, J.M., Portegies, J.M., 2018. Optimal paths for variants of the 2D and 3D Reeds-Shepp car with applications in image analysis. J. Math. Imaging Vis., 1–35. Accepted for Publication in JMIV.

Evans, L.C., 2010. Partial differential equations. In: Graduate Studies in Mathematics, vol. 19. American Mathematical Society, Providence, RI.

Fick, R., Wassermann, D., Pizzolato, M., Deriche, R., 2015. A unified framework for spatial and temporal diffusion in diffusion MRI. Inf. Process. Med. Imaging 24, 167–178. https://doi.org/10.1007/978-3-319-19992-4_13.

Fillard, P., Descoteaux, M., Goh, A., Gouttard, S., Jeurissen, B., Malcolm, J., Ramirez-Manzanares, A., Reisert, M., Sakaie, K., Tensaouti, F., Yo, T., Mangin, J.F., Poupon, C., 2011. Quantitative evaluation of 10 tractography algorithms on a realistic diffusion MR phantom. NeuroImage 56 (1), 220–234. https://doi.org/10.1016/j.neuroimage.2011.01.032.

Florack, L.M.J., Astola, L.J., 2008. A multi-resolution framework for diffusion tensor images. In: Aja-Fernández, S., de Luis Garcia, R. (Eds.), CVPR Workshop on Tensors in Image Processing and Computer Vision, Anchorage, Alaska, USA, June 24–26, 2008. IEEE.

Florack, L.M.J., Fuster, A., 2014. Riemann-Finsler geometry for diffusion weighted magnetic resonance imaging. In: Westin, C.-F., Vilanova, A., Burgeth, B. (Eds.), Visualization and Processing of Tensors and Higher Order Descriptors for Multi-Valued Data. Mathematics and Visualization, Springer-Verlag, pp. 189–208.

Florack, L., van Assen, H., 2012. Multiplicative calculus in biomedical image analysis. J. Math. Imaging Vis. 42 (1), 64–75.

Florack, L., Balmashnova, E., Astola, L., Brunenberg, E., 2010. A new tensorial framework for single-shell high angular resolution diffusion imaging. J. Math. Imaging Vis. 3 (38), 171–181. https://doi.org/10.1007/s10851-010-0217-3.

Florack, L.M.J., Dela Haije, T.C.J., Fuster, A., 2015. Direction-controlled DTI interpolation. In: Hotz, I., Schultz, T. (Eds.), Mathematics and Visualization. Visualization and Processing of Higher Order Descriptors for Multi-Valued Data. Springer-Verlag, pp. 149–162.

Florack, L.M.J., Dela Haije, T.C.J., Fuster, A., 2017. Cartan scalars in Finsler-DTI for higher order local brain tissue characterization. In: Schultz, T., Özarslan, E., Hotz, I. (Eds.), Modeling, Analysis, and Visualization of Anisotropy. Mathematics and Visualization. Springer-Verlag.

Florack, L., Sengers, R., Meesters, S., Smolders, L., Fuster, A., 2021. Riemann-DTI geodesic tractography revisited. In: Özarslan, E., Schultz, T., Zhang, E., Fuster, A. (Eds.), Anisotropy Across Fields and Scales. Mathematics and Visualization. Springer-Verlag, pp. 225–243.

Fuster, A., Dela Haije, T., Tristán-Vega, A., Plantinga, B., Westin, C.F., Florack, L., 2016. Adjugate diffusion tensors for geodesic tractography in white matter. J. Math. Imaging Vis. 54 (1), 1–14. https://doi.org/10.1007/s10851-015-0586-8.

Hao, X., Whitaker, R.T., Fletcher, P.T., 2011. Adaptive Riemannian metrics for improved geodesic tracking of white matter. In: Székely, G., Hahn, H.K. (Eds.), Proceedings of the Twenty-Second International Conference on Information Processing in Medical Imaging—IPMI 2011 (Kloster Irsee, Germany). Lecture Notes in Computer Science, vol. 6801. Springer-Verlag, Berlin, pp. 13–24.

Horbruegger, M., Loewe, K., Kaufmann, J., Wagner, M., Schippling, S., Pawlitzki, M., Schoenfeld, M.A., 2019. Anatomically constrained tractography facilitates biologically plausible fiber reconstruction of the optic radiation in multiple sclerosis. NeuroImage 22 (101740), 1–12. https://doi.org/10.1016/j.nicl.2019.101740.

Huygens, C., 1690. Traité de la Lumière. Pierre van der Aa, Leiden.

Jespersen, S.N., Kroenke, C.D., Østergaard, L., Ackerman, J.J.H., Yablonskiy, D.A., 2007. Modeling dendrite density from magnetic resonance diffusion measurements. NeuroImage 34 (4), 1473–1486. https://doi.org/10.1016/j.neuroimage.2006.10.037.

Jost, J., 2011. Riemannian Geometry and Geometric Analysis. Universitext, sixth ed. Springer-Verlag, Berlin.

Le Bihan, D., 2020. On time and space in the brain: a relativistic pseudo-diffusion framework. Brain Multiphys. 1, 100016. https://doi.org/10.1016/j.brain.2020.100016.

Le Bihan, D., Mangin, J.F., Poupon, C., Clark, C.A., Pappata, S., Molko, N., Chabriat, H., 2001. Diffusion tensor imaging: concepts and applications. J. Magn. Reson. Imaging 13, 534–546.

Lenglet, C., Deriche, R., Faugeras, O., 2004. Inferring white matter geometry from diffusion tensor MRI: application to connectivity mapping. In: Pajdla, T., Matas, J. (Eds.), Proceedings of the Eighth European Conference on Computer Vision (Prague, Czech Republic, May 2004). Lecture Notes in Computer Science, vol. 3021–3024. Springer-Verlag, Berlin, pp. 127–140.

Lenglet, C., Prados, E., Pons, J.P., 2009. Brain connectivity mapping using Riemannian geometry, control theory and PDEs. SIAM J. Imaging Sci. 2 (2), 285–322.

Liu, C., Özarslan, E., 2019. Multimodal integration of diffusion MRI for better characterization of tissue biology. NMR Biomed. 32 (4), e3939. https://doi.org/10.1002/nbm.3939.

Lovelock, D., Rund, H., 1988. Tensors, Differential Forms, and Variational Principles. Dover Publications, Inc., Mineola, NY.

Ma, T., Matveev, V.S., Pavlyukevich, I., 2021. Geodesic random walks, diffusion processes and Brownian motion on Finsler manifolds. arXiv preprint arXiv:arXiv:1402.4963.

Melonakos, J., 2009. Geodesic Tractography Segmentation for Directional Medical Image Analysis (Ph.D. thesis). Georgia Institute of Technology, School of Electrical and Computer Engineering.

Melonakos, J., Pichon, E., Angenent, S., Tannenbaum, A., 2008. Finsler active contours. IEEE Trans. Pattern Anal. Mach. Intell. 30 (3), 412–423.

Mirebeau, J.M., 2018. Fast marching methods for curvature penalized shortest paths. J. Math. Imaging Vis. 60, 784–815. https://doi.org/10.1007/s10851-017-0778-5.

Misner, C.W., Thorne, K.S., Wheeler, J.A., 1973. Gravitation. Freeman, San Francisco.

Neuenschwander, D.E., 2011. Emmy Noether's Wonderful Theorem. The Johns Hopkins University Press, Baltimore.

Noether, E., 1918. Invariante Variationsprobleme. In: Nachrichten von der Königlichen Gesellschaft der Wissenschaften zu Göttingen. Mathematisch-Physikalische Klasse, Weidmannsche Buchhandlung, Berlin, pp. 235–257.

Novikov, D.S., Kiselev, V., 2010. Effective medium theory of a diffusion-weighted signal. NMR Biomed. 23, 682–697. https://doi.org/10.1002/nbm.1584.

Novikov, D.S., Fieremans, E., Jensen, J.H., Helpern, J.A., 2011. Random walks with barriers. Nature 7, 508–514.

Novikov, D.S., Kiselev, V.G., Jespersen, S.N., 2018. On modeling. Magn. Reson. Med. 79 (6), 3172–3193.

O'Donnell, L., Haker, S., Westin, C.F., 2002. New approaches to estimation of white matter connectivity in diffusion tensor MRI: elliptic PDEs and geodesics in a tensor-warped space. In: Dohi, T., Kikinis, R. (Eds.), Proceedings of the 5th International Conference on Medical Image Computing and Computer-Assisted Intervention—MICCAI 2002 (Tokyo, Japan, September 25–28, 2002). Lecture Notes in Computer Science, vol. 2488–2489. Springer-Verlag, Berlin, pp. 459–466.

Özarslan, E., Koay, C.G., Shepherd, T.M., Komlosh, M.E., İrfanoğlu, M.O., Pierpaoli, C., Basser, P.J., 2013. Mean apparent propagator (MAP) MRI: a novel diffusion imaging method for mapping tissue microstructure. NeuroImage 78, 16–32.

Özarslan, E., Yolcu, C., Herberthson, M., Knutsson, H., Westin, C.F., 2018. Influence of the size and curvedness of neural projections on the orientationally averaged diffusion MR signal. Front. Phys. 6, 17. https://doi.org/10.3389/fphy.2018.00017.

Parker, G.J.M., Wheeler-Kingshott, C.A.M., Barker, G.J., 2002. Estimating distributed anatomical connectivity using fast marching methods and diffusion tensor imaging. IEEE Trans. Med. Imaging 21 (5), 505–512.

Péchaud, M., Descoteaux, M., Keriven, R., 2009. Brain connectivity using geodesics in HARDI. In: Yang, G.-Z., Hawkes, D.J., Rueckert, D., Noble, J.A., Taylor, C.J. (Eds.), Proceedings of the 12th International Conference on Medical Image Computing and Computer Assisted Intervention—MICCAI 2009 (London, UK, September 20–24, 2009). Lecture Notes in Computer Science, vol. 5761. Springer-Verlag, Berlin, pp. 482–489.

Poupon, C., Rieul, B., Kezele, I., Perrin, M., Poupon, F., Mangin, J.F., 2008. New diffusion phantoms dedicated to the study and validation of HARDI models. Magn. Reson. Med. 60, 1276–1283.

Poupon, C., Laribiere, L., Tournier, G., Bernard, J., Fournier, D., Fillard, P., Descoteaux, M., Mangin, J.F., 2010. A diffusion hardware phantom looking like a coronal brain slice. In: Proceedings of the 18th International Society for Magnetic Resonance in Medicine, ISMRM (Stockholm, Sweden, May 1–7, 2010).

Pujol, S., et al., 2015. The DTI challenge: towards standardized evaluation of diffusion tensor imaging tractography for neurosurgery. J. Neuroimaging 25 (6), 875–882. https://doi.org/10.1111/jon.12283.

Rund, H., 1959. The Differential Geometry of Finsler Spaces. Springer-Verlag, Berlin.

Rund, H., 1973. The Hamilton-Jacobi Theory in the Calculus of Variations. Robert E. Krieger Publishing Company, Huntington, NY.

Rutten, G.J.M., Kristo, G., Pigmans, W., Peluso, J., Verheul, H.B., 2014. Het gebruik van MR-tractografie in de dagelijkse neurochirurgische praktijk (the use of MR-tractography in daily neurosurgical practice). Tijdschrift voor Neurologie & Neurochirurgie 115 (4), 204–211.

Sagan, H., 1992. Calculus of Variations. Dover Publications, Inc., New York.

Schilling, K.G., Petit, L., Rheault, F., Remedios, S., Pierpaoli, C., Anderson, A.W., Landman, B.A., Descoteaux, M., 2020. Brain connections derived from diffusion MRI tractography can be highly anatomically accurate–if we know where white matter pathways start, where they end, and where they do not go. Brain Struct. Funct. 225, 2387–2402. https://doi.org/10.1007/s00429-020-02129-z.

Sengers, R., Florack, L., Fuster, A., 2021a. Geodesic uncertainty in diffusion MRI. Front. Comput. Sci. 3, 718131. https://doi.org/10.3389/fcomp.2021.718131.

Sengers, R., Fuster, A., Florack, L., 2021b. Geodesic tubes for uncertainty quantification in diffusion MRI. In: Sommer, S., Feragen, A., Nielsen, M., Schnabel, J. (Eds.), Proceedings of the Twenty-Seventh International Conference on Information Processing in Medical Imaging—IPMI 2021 (Bornholm, Denmark). Lecture Notes in Computer Science, Springer-Verlag, Berlin, pp. 279–290.

Sepasian, N., Vilanova, A., Florack, L., ter Haar Romeny, B., 2008. A ray tracing method for geodesic based tractography in diffusion tensor images. In: Proceedings of the 9th IEEE Computer Society Workshop on Mathematical Methods in Biomedical Image Analysis, held in conjuction with the IEEE Computer Society Conference on Computer Vision and Pattern Recognition (Anchorage, Alaska, USA, June 23–28, 2008). IEEE Computer Society Press.

Sepasian, N., ten Thije Boonkkamp, J.H.M., ter Haar Romeny, B.M., Vilanova, A., 2012. Multivalued geodesic ray-tracing for computing brain connections using diffusion tensor imaging. SIAM J. Imaging Sci. 5 (2), 483–504.

Shen, Y.B., Shen, Z., 2016. Introduction to Modern Finsler Geometry. World Scientific, Singapore.

Smith, R.E., Tournier, J.D., Calamante, F., Connelly, A., 2012. Anatomically-constrained tractography: improved diffusion MRI streamlines tractography through effective use of anatomical information. NeuroImage 62, 1924–1938.

St-Onge, E., Daducci, A., Girard, G., Descoteaux, M., 2018. Surface-enhanced tractography (SET). NeuroImage 169, 524–539. https://doi.org/10.1016/j.neuroimage.2017.12.036.

St-Onge, E., Al-Sharif, N., Girard, G., Theaud, G., Descoteaux, M., 2021. Cortical surfaces integration with tractography for structural connectivity analysis. Brain Connect. https://doi.org/10.1089/brain.2020.0930.

Stanisz, G.J., Szafer, A., Wright, G.A., Henkelman, R.M., 1997. An analytical model of restricted diffusion in bovine optic nerve. Magn. Reson. Med. 37 (1), 103–111.

Stejskal, E.O., 1965. Use of spin echoes in a pulsed magnetic-field gradient to study anisotropic, restricted diffusion and flow. J. Chem. Phys. 43 (10), 3597–3603.

Stejskal, E.O., Tanner, J.E., 1965. Spin diffusion measurements: spin echoes in the presence of a time-dependent field gradient. J. Chem. Phys. 42 (1), 288–292.

Tuch, D.S., 2004. Q-ball imaging. Magn. Reson. Med. 52, 1358–1372.

Weinstock, R., 1974. Calculus of Variations With Applications to Physics & Engineering. Dover Publications, Inc., New York.

Westin, C.F., Knutsson, H., Pasternak, O., Szczepankiewicz, F., Özarslan, E., van Westen, D., Mattisson, C., Bogren, M., O'Donnell, L., Kubicki, M., Topgaard, D., Nilsson, M., 2016. Q-space trajectory imaging for multidimensional diffusion MRI of the human brain. NeuroImage 135, 345–362. https://doi.org/10.1016/j.neuroimage.2016.02.039.

Chapter 16

# Global tractography

Alessandro Daducci[a], Simona Schiavi[a,b], Daan Christiaens[c], Robert Smith[d,e], and Daniel C. Alexander[f]

[a]Department of Computer Science, University of Verona, Verona, Italy, [b]Department of Neuroscience, Rehabilitation, Ophthalmology, Genetics, Maternal and Child Health (DINOGMI), University of Genoa, Genoa, Italy, [c]Department of Electrical Engineering (ESAT/PSI), KU Leuven, Leuven, Belgium, [d]The Florey Institute of Neuroscience and Mental Health, Heidelberg, VIC, Australia, [e]The University of Melbourne, Melbourne, VIC, Australia, [f]Centre for Medical Image Computing, Department of Computer Science, University College London, London, United Kingdom

## 1 Introduction

The *ultimate goal* of any tractography algorithm is to estimate, from the acquired diffusion magnetic resonance imaging (dMRI) data, a set of macroscopic trajectories (or streamlines) that accurately mimic the intricate network of fiber tracts in the brain white matter. Thus tractography is intrinsically an *inverse problem*[a]: the term "inverse" is used to indicate that what we know, that is, the dMRI data acquired in all imaging voxels, must be somehow "inverted" to learn what we do not know, that is, the trajectories of the underlying fibers (Fig. 1A). While the transformation in the opposite direction is reasonably well defined, that is, computing the dMRI signal corresponding to a given streamlines configuration, the tractography estimation process is *ill-posed*.[b] In fact, the resolution of dMRI is much coarser than the typical size of the fibers, and different fiber configurations can generate identical dMRI profiles. Such ambiguous white-matter configurations are often called "bottlenecks" (Guevara et al., 2012; Maier-Hein et al., 2017; Girard et al., 2020; Schilling et al., 2022) and, as a consequence, multiple tractography reconstructions might be compatible with this blurred view of the underlying tissue complexity (Fig. 1B).

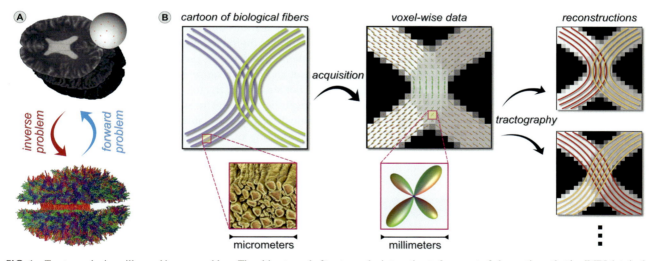

FIG. 1 Tractography is an ill-posed inverse problem. The ultimate goal of tractography is to estimate from a set of observations, that is, dMRI data in the brain (A, *top*), the macroscopic trajectories of the underlying neuronal fiber tracts that generated them (A, *bottom*). However, dMRI data contain ambiguity, as different configurations of fiber tracts (cartoonized in B, *left*) can generate very similar signal profiles (B, *middle*), and multiple tractography reconstructions are possible (B, *right*); hence, the tractography estimation process is ill-posed.

---

a. An *inverse problem* is defined as the process of estimating the causes of a phenomenon based on some observations of its effects.
b. A problem is *ill-posed* when either there are multiple equally valid yet distinct solutions or small variations to the input data, e.g., due to noise perturbations, lead to very different solutions.

The first generation of tractography algorithms presented in Chapters 13 and 14—and indeed the vast majority of those utilized in neuroscience over 20 years since its inception—attempt to solve this ill-posed inverse problem using a simple line-propagation ("streamlines") paradigm, accessing image information at the current spatial location in order to dictate the evolution of each individual trajectory independently. While the introduction of streamline tractography to the neuroscientific and clinical community generated considerable excitement given its prospective applications, described in detail in Chapters 29–33, it was also immediately recognized by the technical development community that there were a number of limitations to the technology which, if properly addressed, might further improve the robustness of conclusions that could be drawn from its use, stimulating lots of theoretical and experimental research on tractography since then.

There are two principal forms of error that may manifest in a tractography reconstruction:

- *Trajectories*: A concern that was immediately apparent was that this greedy approach *heavily suffers from local estimation inaccuracies*. At *any* spatial location along a streamline trajectory, there is scope for an error (of any magnitude and from any source) to occur in the determination of direction of propagation; further, these errors may *accumulate* as integration occurs from the seed point of a streamline along its trajectory. Hence, even a single erroneous fiber orientation estimate due to local image noise may preclude entirely the reconstruction of a macroscopic pathway despite its existence being supported by the image data on either side of that erroneous estimate, as any streamline encountering that erroneous orientation from either direction will be led astray.

- *Densities*: The fact that the number of axonal connections within a particular pathway of interest is a biological parameter of great interest, and a streamline tractography experiment yields a number of streamlines corresponding to such a pathway, makes it highly enticing to interpret the latter as being a proxy marker for the former. Such an interpretation seems, however, too simplistic, given both the wide gamut of streamline tractography reconstruction parameters that have an *undesired influence* on this emergent measure, but also the relative *absence* of *desired influence* of the local density of fibers on that measure. This means that a correlation between the numbers of connections within different pathways of the brain (or indeed within the corresponding pathway across different subjects), and the numbers of streamlines reconstructed corresponding to those pathways, is in no way guaranteed. This limitation is at the core of the common and long-standing mantra "streamline count is not quantitative" ( Jones, 2010; Jbabdi and Johansen-Berg, 2011; Jones et al., 2013; Yeh et al., 2021; Zhang et al., 2022).

The class of tractography methods colloquially referred to as "global" attempt to address these limitations by approaching tractography using a fundamentally different paradigm and introducing additional information to augment the streamlines-based reconstruction. They consider that the empirical dMRI data reflects the cumulative sum of the influence of all tissues throughout the brain on the diffusion-weighted signal (including the orientations of white matter fibers); and that therefore the goal of (implicitly whole-brain) tractography is to generate a plausible digital reconstruction of the spatial distribution of white matter fibers (and other biological constituents) that, when considered *in its entirety*, is consistent with those empirical image data. This is a drastic shift in both *implementation* and *interpretation* from the conventional streamline tractography paradigm adopted by previously described methods. For streamline tractography, the only guarantee that can be applied to the resulting tractogram is that, at each streamline vertex, there is some local image support for the presence of white matter fibers in the direction of the streamline tangent. Global tractography, instead, has the potential to provide greater confidence regarding the plausibility and biological interpretability of the resulting tractograms, not only in the *local orientations* of the reconstructed connections, but ideally their *complete trajectories*, as well as the relative *densities* and other biological attributes associated with them, throughout the image (Fig. 2).

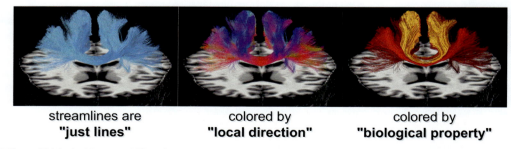

**FIG. 2** Plausibility and biological interpretability of tractograms. The output produced by any tractography algorithm is, in the end, a set of lines that describe the macroscopic trajectories of biological fibers (*left*). Local methods can only provide assurance on the *local orientations* of such lines: at any position along them, dMRI supports the presence of fibers in the direction of their local tangent (*middle*). Per contra, global methods have the potential to provide higher levels of confidence about the *plausibility of complete trajectories* as well as other *biological properties* associated with them, e.g., myelination of the corresponding tracts (*right*).

The layout of this chapter is as follows. Section 2 elucidates the meaning of the word "global" in this context, in addition to related terms that may be encountered here and/or elsewhere in the literature. In Section 3, the fundamental paradigm of global tractography is presented in a more formal fashion; in addition, this section defines and explains a key methodological attribute by which the methods proposed in this domain can be separated into two classes: *generative* and *discriminative*. Sections 4 and 5 then present the fundamental methodological frameworks of these two classes, and contextualize where those methods presented thus far in the literature fit within these frameworks. Finally, in Section 6, we discuss recent advances for integrating more information into the tractography process and the ways in which such methods can be reasonably hypothesized to evolve in the future.

## 2 What does "global" really mean?

Before the presentation of details regarding those tractography algorithms that are classified as global, and how they differ from those presented in previous chapters, it is necessary to first define carefully what is meant by "global" within this context and provide the readers with explicit distinctions between different uses of this term. In the literature there is a *sort of gray area around the expression "global tractography,"* as an explicit definition cannot be found even in the pioneering papers on this topic (Mangin et al., 2002; Kreher et al., 2008; Reisert et al., 2011). Nevertheless, there is general agreement in the community on a number of characteristics that an algorithm must have to qualify as *global tractography*, which can be summarized as follows.

1. *All dMRI data are used in the estimation process.* In the line-propagation paradigm adopted by early tracking methods, the generation of the tractography reconstruction was only ever informed by *local* image data. That is to say, for any specific vertex along the path, the decision regarding how that particular trajectory should evolve was entirely agnostic to any and all image information outside of a local sphere of influence, through, e.g., 3D image interpolation; hence the name "local tractography." With the advent of methods where the reconstruction process may be informed by *image data beyond the local interpolation neighborhood*, the community started colloquially referring to them as "global tractography." Indeed, this term was likely initially introduced as the antonym of local with the aim to identify, to a greater or lesser extent, any method involving some degree of utilization of information beyond just the local estimated fiber orientation in the decision-making process.

2. *Streamlines are estimated concurrently and are not independent.* In local tractography, each individual streamline is generated entirely independently of all others. In the global tractography paradigm, instead, these trajectories (which may confusingly still be referred to as "streamlines," despite not having been generated by a streamline algorithm) are intrinsically considered *as a whole* and the optimal configuration of streamlines is estimated *concurrently*. The estimation of all streamlines at the same time, instead of one by one, unlocks the opportunity to design the optimization such that the generation of a given trajectory is possibly influenced by all others; thus streamlines are no longer independent by design. It is postulated that this interaction among competing trajectories, as well as considering the entire tractogram as a whole, has the potential to resolve ambiguous configurations in the tractogram, and therefore to improve the anatomical accuracy of tractography reconstructions.

3. *The estimated tractogram fits the measured dMRI data.* Streamlines reconstructed with local tractography algorithms are merely infinitesimally thin trajectories through 3D space; hence, unlike biological axons, they do not occupy any volume of space nor have any biological properties associated with them. Global methods attempt instead to *attribute physical/biological properties to these trajectories* in order to explicitly recognize the fact that they are intended to represent real populations of axons. Therefore, if each reconstructed trajectory corresponds to a real bundle of axons and has ascribed to it appropriate properties reflective of the underlying bundle, then their sum total presence across the tractogram should predict the measured image data in all voxels. The assessment of the plausibility of a tractogram is driven by optimizing a *global objective function* that seeks correspondence between reconstruction and image data.

Fig. 3 highlights the main differences between "local" and "global" tractography approaches.

### 2.1 Can we ever be truly global?

In the English language, the more commonly used definition of the word global is "involving the entire globe." In more technical contexts, this term is used to denote, e.g., an algorithm or a mathematical function that considers in its formulation *all parts* of the data or situation under examination. In the context of tractography, the ambitious goal of the earliest global algorithms was for *all attributes of the resulting tractogram* to have been influenced in some manner by global image information.

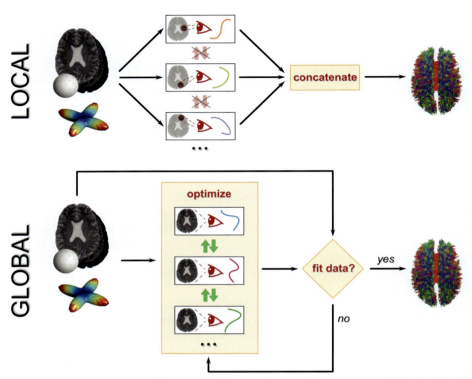

**FIG. 3** Difference between "local" and "global" tractography approaches. In a "local" tractography algorithm, each streamline trajectory is estimated independently of all others and the reconstruction is performed considering only a portion of the data (depicted by the "eye" icon); the complete tractogram is simply the concatenation of these individual trajectories, and there is *no direct comparison* between the canonical reconstruction and the dMRI data. Conversely, in a "global" tractography algorithm, the canonical reconstruction *is* compared directly to the *entire* image data, and optimization of the whole tractogram is driven by reduction of the observable differences between them.

The contents of a reconstructed tractogram can be characterized based on many attributes or features; these typically include the positions and orientations of the trajectories taken through 3D space, the locations of the terminations of these trajectories, and the number/local density of such trajectories in any particular area. However, even the earliest methods proposed in this domain were observed to come at substantial computational expense (Mangin et al., 2002; Kreher et al., 2008; Sherbondy et al., 2009, 2010), limiting their applicability in generating such reconstructions with adequate density across a large number of participants in group studies. Moreover, since the reconstructed streamlines are not "just lines" but they intrinsically represent *neuronal pathways*, it would be desirable to additionally characterize other biological features of the microenvironment associated with each pathway—for instance, myelin content or axon composition, to which advanced diffusion/quantitative MRI acquisitions are sensitive (Pike et al., 2018)—but such augmentation would exacerbate the computational tractability of such methods even further. Given these issues, alternative methods have been proposed since then that seek to provide some form of *compromise* between the fast-but-error-prone "local" approaches with the slow-but-more-accurate "global" approaches, by having only *some* of these tractogram attributes influenced by global information, with others (e.g., individual trajectories through space) being agnostic to such global information.

For this reason, and in an attempt to clear the confusion around the word "global," wherever mention to "global tractography" is made in this book, this should be inferred as a reference only to those algorithms that *meet all the three characteristics* as just described; all other approaches that satisfy only *some* of those criteria should be referred to as *semiglobal*. One notable example is represented by *geodesic methods* (Chapter 15). As they constrain both spatially distant endpoints of a trajectory by design, it is almost inevitable that any such approach to tractography will involve some degree of integration of global image information in the decision-making process. Nevertheless, the construction of each individual trajectory is not aware of the possible presence of other trajectories through the same space and is not influenced by them. Indeed, even though some published geodesic methods self-identify as "global," e.g., Jbabdi et al. (2007) and Wu et al. (2009), this is quite a distinct approach to the class of methods that are otherwise referred to as "global" in the context of tractography, highlighting the ill-posedness of this term.

**"Global" and "whole-brain" are *not* synonyms.**

When utilizing streamline tractography, one may either (i) permit the algorithm to reconstruct trajectories anywhere where the image data provide evidence for such trajectories (resulting in reconstruction of "whole-brain" connectivity in the case of brain imaging) or (ii) constrain the reconstruction to only specific regions or pathways of interest (so-called "targeted tracking"). Global tractography algorithms *must by design* deal with reconstruction of all plausible pathways within the image. Artificially constraining the reconstruction to omit specific regions or pathways would introduce an artificial discrepancy between the reconstruction and the image data, due to the biological fibers contributing to those image data being artificially excluded from the reconstruction (Smith et al., 2022), and this would therefore bias the derived quantities; see Section 3. However, referring to a tractogram reconstructed by a local streamlines algorithm as "global" simply because the set of regions/pathways included in the reconstruction was not constrained by the experimenter would be *erroneous*. It is possible to perform whole-brain fiber-tracking using a purely local streamline tractography algorithm, failing to satisfy any of the characteristics of "global" tractography enumerated earlier. The classifications "global" and "whole-brain" in the context of diffusion tractography in the brain refer to *two different concepts*, and therefore care should always be taken so as not to conflate them.

## 3  The global tractography paradigm

Global tractography algorithms *explicitly acknowledge the ill-posed nature of tractography* and formulate the reconstruction process following a global inverse problem perspective, in which *optimization* is used to find the optimal set of streamlines that best describes the measured data in all voxels as well as to explicitly capture the possible interactions between the streamlines being estimated. This reconstruction paradigm requires the definition of a *cost function*, which quantifies the compatibility between a given tractogram configuration and the acquired dMRI data (or alternatively some derivative maps obtained from it), and a *forward model*, which assigns a signal contribution to every streamline and describes how they are predicted to contribute to the image data.

While various approaches have been proposed in the literature, without loss of generality they can all be described in terms of the following *general formulation*:

$$\mathbf{T}^{*} = \underset{\mathbf{T}}{\mathrm{argmin}} \ \mathcal{E}(\mathbf{D}, \mathcal{M}(\mathbf{T})), \tag{1}$$

where

- $\mathbf{T}$ represents a *generic* tractogram configuration;
- $\mathbf{T}^{*}$ is the *optimal* tractogram configuration to be estimated;
- $\mathbf{D}$ is the empirical *dMRI data*;
- $\mathcal{M}(\cdot)$ is a *forward model* that predicts dMRI data from a tractogram configuration; and
- $\mathcal{E}(\cdot, \cdot)$ is a *cost function* that quantifies the similarity between the empirical dMRI data and that predicted from a tractogram.

Most commonly, the forward model $\mathcal{M}(\cdot)$ is some form of spherical convolution (see Chapter 10), which predicts the diffusion-weighted signal contributed from each individual streamline trajectory, both in terms of position (i.e., the set of voxels to which it contributes) and orientation (i.e., contributing more or less signal to predicted volumes with different diffusion sensitization based on the streamline tangent), based on some a priori specified model of water diffusion in axons; some methods may additionally include water pool components other than intracellular water in the forward model.

In some cases, an *alternative formulation* is adopted, where, rather than applying a forward model to the tractogram in order to compare to the empirical dMRI data, the inverse model is instead first applied to the dMRI data:

$$\mathbf{T}^{*} = \underset{\mathbf{T}}{\mathrm{argmin}} \ \mathcal{E}(\mathcal{M}^{-1}(\mathbf{D}), \mathcal{P}(\mathbf{T})), \tag{2}$$

where $\mathcal{M}^{-1}(\cdot)$ represents the inverse model (e.g., performing a spherical *de*convolution to estimate fiber orientations and densities) and $\mathcal{P}(\cdot)$ is a projection function that maps the properties of the streamlines within the tractogram onto the voxels they intersect to facilitate comparison with the diffusion model. This alternative approach has the potential to reduce the computational burden of such algorithms, but can limit the scope for integrating more complex modeling into the global optimization if the outcomes of the local diffusion model are considered immutable.

Since dMRI is a notoriously noisy acquisition modality, it may be possible for two drastically different solutions (e.g., tractograms $\mathbf{T}_1^*$ and $\mathbf{T}_2^*$) to be equally concordant with the empirical image data, or for a small perturbation of the input data to result in some method changing its solution from one to the other. For this reason, *regularization techniques* are commonly adopted to promote some kind of regularity on the solutions. The aim of regularization is to exploit all the information available in the data to resolve ambiguities and return robust and accurate estimates, as well as to encourage solutions that are considered more biologically plausible or have more desirable properties, such as smooth trajectories, sparse tractograms, penalization of statistical outliers, etc. Hence, Eqs. (1), (2) can be extended as follows:

$$\mathbf{T}^* = \underset{\mathbf{T}}{\mathrm{argmin}} \ \mathcal{E}(\mathbf{D}, \mathcal{M}(\mathbf{T})) + \Psi(\mathbf{T}) \tag{3}$$

and

$$\mathbf{T}^* = \underset{\mathbf{T}}{\mathrm{argmin}} \ \mathcal{E}(\mathcal{M}^{-1}(\mathbf{D}), \mathcal{P}(\mathbf{T})) + \Psi(\mathbf{T}), \tag{4}$$

where $\Psi(\cdot)$ represents a generic regularization function.

### 3.1 "Generative" vs. "discriminative" approaches

Global tractography algorithms can be organized in two classes depending on the approach followed to optimize Eq. (3) or (4), as illustrated in Fig. 4.

- *Generative* methods reconstruct the streamlines starting from their fundamental constituent parts, that is, small segments, each of which represents a short portion of a fiber tract, and for which it is possible to determine the corresponding signal contributions to the input dMRI data using a forward model. Continuous streamlines of greater length are formed by altering the position/shape of an *initial random collection* of these fragments and encouraging them to link together to form smooth trajectories, while simultaneously adapting to the underlying signal. They are alternatively sometimes referred to as *bottom-up* methods.
- *Discriminative* methods start from a whole-brain collection of candidate streamlines typically constructed using simpler local tractography algorithms, and attempt to globally optimize the fit of the tractogram to the empirical dMRI data by *modulating the contributions of individual streamlines*. For each trajectory, its prospective contribution to the dMRI signal in nearby voxels is computed using a forward model, based on the assumption that such contributions are constant along the length of each streamline; the optimal tractogram is then solved globally by modulating the contribution of each streamline, considering all such data concurrently when doing so. They are alternatively sometimes referred to as *top-down* methods.

In the following, we present the most relevant solutions proposed to date in the literature.

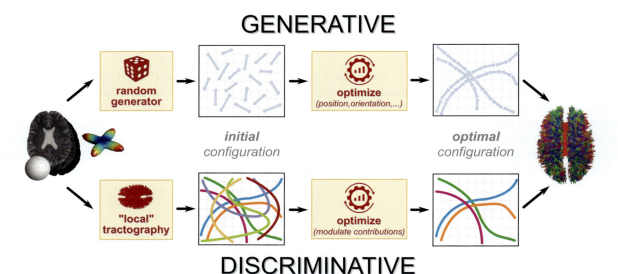

**FIG. 4** Difference between "generative" and "discriminative" approaches for global tractography. *Generative* (or alternatively "bottom-up") methods reconstruct the optimal configuration of streamlines starting from a random collection of short segments, whose signal contribution is determined using a generative model, which are encouraged to form continuous and smooth trajectories in order to explain the measured dMRI data at best. *Discriminative* (or alternatively "top-down") methods start instead from an initial collection of streamlines estimated with simpler local tractography algorithms, and attempt to modulate the contributions of individual streamlines such that the tractogram best explains the empirical dMRI data.

**Accessing a quantitative metric of connectivity.**

As discussed, trajectories reconstructed with local methods have no volume associated with them, and the correspondence between a tractography reconstruction and the diffusion image data from which they were generated only exists in the orientation domain. The relative density of *streamlines* in different voxels of the image is therefore not necessarily reflective of a difference in the density of *white matter fibers* between those locations, as the former measure is strongly influenced by other features of the reconstruction process (and indeed in most instances that reconstruction process is *agnostic* to relative white matter fiber densities in different locations) (Calamante et al., 2015). In global tractography, the reconstructed streamlines *do* contribute some nonzero white matter fiber volume along their length, that is, $\mathcal{M}(\mathbf{T})$ in Eq. (1). As such, if there is discordance in some voxels between the density of reconstructed trajectories and the density of white-matter fibers supported by the empirical image data, this manifests as a *reconstruction error*, that is, $\mathcal{E}(\mathbf{D}, \mathcal{M}(\mathbf{T}))$ in Eq. (1), for which the global tractography algorithm is responsible for *minimizing*.

The notion of "density" in this context requires careful disambiguation, as there are two domains in which that term may be invoked[1]:

- *Voxel*-wise: Within the volume in 3D space corresponding to a particular voxel on the image grid, one can define some "total amount" of content present that contributes to white matter connectivity. This could be an intraaxonal volume fraction as estimated by a diffusion model, or some measure proportional to the presence of streamlines in that region of the image.

- *Bundle*-wise: Any specific anatomical white matter bundle is defined based on some set of criteria that are satisfied for axons considered to be a part of that bundle and *not* for axons considered to *not* be a part of that bundle. Following such a selection, one could derive some measure of "density" of that bundle, such as the number of axons. Similar criteria can equally be applied to a tractogram to select streamlines from a whole-brain reconstruction that should or should not be attributed to such a bundle; and similarly, some measure of "density" could be derived, such as the number of attributed streamlines. Crucially, however, these concepts are reflective of a notion of "density" between the two *endpoints* of the bundle, which is independent of the particular trajectories taken or macroscopic shape of the bundle between those endpoints.

It is worth noting that the *dimensions* of the definition of "density" differ between these two cases. In the former case, measurement is proportional to *volume*, with dimensions $L^3$ (e.g., mm³); in the latter, this measure does *not* scale with the *length* of the bundle, and could be dimensionless in the case of a count, or may instead have dimensions $L^2$ (e.g., mm²), that is, a cross-sectional area (Smith et al., 2022).

While achieving correspondence between reconstruction and image data is generally considered beneficial by design, the practical *consequence* of doing so in this particular context is not always appreciated. *If* it is possible to generate a reconstruction where there exists correspondence in *voxel*-wise "density" between tractogram and the underlying biology (via estimates drawn from the image data), *then* it becomes possible to infer a correspondence in *bundle*-wise "density" between tractogram and the underlying biology (Smith et al., 2022; Zhang et al., 2022). As an example, if each streamline represents an intraaxonal cross-sectional area of axons, and for every image element the sum total of the contributions of all intersecting streamlines accurately replicates the local intraaxonal volume, then for any given white matter pathway reconstructed by a subset of streamlines, the sum of the contributions across those streamlines provides an estimate of the total intraaxonal cross-sectional area of that pathway: the so-called fiber bundle capacity (FBC) measure (Smith et al., 2022). This is a consequential outcome, as such bundle-wise "density" measures are almost certainly of great interest to the neuroscientific community, particularly in the context of network neuroscience (see Chapter 24), but also in hypothesis-driven interrogation of specific white matter pathways (see Chapters 21–23). Indeed, we hypothesize that the historically prolific use of alternative metrics of "connectivity," such as sampling quantitative image metrics along streamline trajectories, were likely principally driven by the inadequacy of streamline count in quantifying such.

---

1. Actually, the term "density" is also used to denote the fraction of present edges to all possible edges in the brain networks estimated with tractography; see Chapter 24.

---

# 4 Generative approaches

Generative methods for global tractography treat the inverse problem defined in Eq. (3) as one large-scale optimization problem in which, based on a forward model of the predicted dMRI signal for a given fiber configuration, a whole-brain set of streamlines is constructed and fine-tuned to best match the fiber configuration and density in the brain. The various methods that have been proposed in the literature make different choices about the forward model, the representation of the tractogram, the regularization, and the optimization techniques used. Due to the sheer size and nonlinearity of the problem, generative global tractography methods typically resort to stochastic optimization techniques.

**304 PART | III** Tractography algorithms

## 4.1 Spin glass models

A first class of methods initializes the whole-brain tractogram as chains of short, straight track *segments*, sometimes also called *particles* or *spins*. These segments are the elementary building blocks to construct the trajectories of the fiber tracts, and their position and orientation are optimized in the global tractography process. Each spin is a line segment of a short fixed length, fully defined by its position $\vec{x}$ and orientation $\vec{n}$. A spin glass is a large collection of segments $\{(\vec{x}_i, \vec{n}_i)\}$ in a disordered state. The terminology originates from condensed matter physics but is applied loosely in the context of global tractography. The core analogy is to think of the tractogram as a field of "compass needles" that align both with an external magnetic field and with the directions of other spins in their local neighborhood (Mangin et al., 2002). Starting from a random initialization, the spin glass system evolves toward a low energy, equilibrium state. In global tractography, there are typically two sources of "energy" for which the sum is minimized in pursuit of the optimal solution: the *external* energy encourages spins to align with the directions of largest diffusion, whereas the *internal* energy of the system penalizes discontinuity in orientations between adjacent segments (i.e., high curvature).

A proof of concept of spin glass tractography was demonstrated in Mangin et al. (2002) not long after the introduction of streamline tractography. In this work, the spin position was fixed to the center of each voxel (one spin per voxel), reducing the scope and complexity of the global optimization to estimate only the orientation of each spin. The external energy seeks to align spins with the principal axis of the local diffusion tensor, while the internal energy minimizes the angle between a spin and its forward and backward neighbors. In crossing and fan-shaped fiber configurations, the spin interaction model can reduce the sensitivity of tractography to the leading eigenvector of the diffusion tensor.

In more recent work, spins are no longer locked to the voxel grid (i.e., their position is manipulated in the optimization) and a single voxel can contain multiple spins (Kreher et al., 2008; Fillard et al., 2009; Reisert et al., 2011). As such, the spin glass can represent crossings and other complex fiber configurations, and can model fiber density variations throughout the image. For instance, Fillard et al. (2009) combined the DTI-based external energy used in Mangin et al. (2002) with an adaptive spin glass model. The global optimization iteratively updates each spin position and orientation to minimize the sum of internal and external energy. They demonstrated that the spin interaction potential could correctly align segments in crossing fiber regions such as the centrum semiovale, where the DTI model offers little and/or misleading guidance. This flexible spin glass model has become the standard segment-wise representation for generative global tractography. However, the large dimensionality of the parameter space (5 × the number of segments) and many local minima in the optimization landscape lead to high computational demands.

## 4.2 Gibbs tracker

A substantial breakthrough toward a practical implementation of generative global tractography was the introduction of *simulated annealing* to the global optimization process. The concept is to *propose random changes* to the segment configuration, each of which is accepted or rejected with a probability related to the external and internal energy and to the "temperature" of the system. By progressively reducing this temperature according to controlled schedule, the segment configuration gradually cools down toward a state close to the global optimum (Kreher et al., 2008).

From a Bayesian perspective, this seeks the maximum a posteriori fiber configuration $\mathbf{T}$ given the dMRI data $\mathbf{D}$, which is proportional to the product of the likelihood and the prior. Adopting a Gibbs distribution for both of them relates the likelihood to the external energy and the prior to the internal energy:

$$P(\mathbf{T}|\mathbf{D}) \propto P(\mathbf{D}|\mathbf{T})\ P(\mathbf{T}), \tag{5}$$

$$\propto e^{-U_{\text{ext}}/T}\ e^{-U_{\text{int}}/T}. \tag{6}$$

It is clear that with $U_{\text{ext}} = \mathcal{E}(\mathbf{D}, \mathcal{M}(\mathbf{T}))$ and $U_{\text{int}} = \Psi(\mathbf{T})$ this is equivalent to the cost function in Eq. (3). Simulated annealing draws random samples from the posterior, while gradually reducing the temperature according to a set cooling schedule. At low temperature, the optimization landscape becomes sharper, so the fiber configuration gradually evolves toward the mode of the distribution. Samples are generated in a Reversible Jump Markov Chain Monte Carlo (RJMCMC) process, that is, by making random changes in the current segment configuration whose effect on the internal and external energy can be evaluated efficiently. Typical examples include the addition or removal of a segment, a change in the position and orientation of a segment, and the addition or removal of a connection between two segments (Kreher et al., 2008; Reisert et al., 2011). Candidate changes are accepted with some probability (the so-called Green's ratio), depending on the energy difference between the initial and proposed state and on the temperature of the system. At lower temperature, the probability of accepting proposals decreases, effectively "freezing" the system into an equilibrium.

With smart implementation choices that minimize the computational burden of evaluating new proposals, a track configuration close to the global optimum can be reconstructed robustly, with computation times that are practical on modern desktop computers (Reisert et al., 2011). The implementation in Reisert et al. (2011) achieved top scores in an open tractography challenge on a diffusion MRI phantom (Fillard et al., 2011).

## 4.3 Multitissue and multicompartment models

The forward model in the Gibbs tracker methods described in the previous section defines the signal contribution of each segment as a hindered diffusion compartment with fixed axial and radial diffusivities and high anisotropy ($\lambda_a \gg \lambda_r$). The external energy is defined as the L2 norm between single-shell dMRI data and the sum of all segment contributions in a voxel, after subtracting the isotropic mean (Kreher et al., 2008; Reisert et al., 2011). More recent work has explored alternatives for the forward model.

For instance, MesoFT (Reisert et al., 2014) explored the potential of integrating compartment models of tissue mesostructure into the global tractography framework. They used a three-compartment model for white matter that consists of stick, zeppelin, and dot compartments (Panagiotaki et al., 2012), in which the axial diffusivities of the stick and zeppelin compartments are assumed to be equal. In addition to its position and orientation, each segment has attributed to it axial and radial diffusivities, reflecting local microstructure. The volume fractions of the different compartments, on the other hand, are properties of individual voxels (Reisert et al., 2014). The external energy is defined as the L2 distance between the empirical image data and the predicted signal, approximating the MRI noise distribution as Gaussian. The internal energy, in addition to measuring fiber curvature, is extended to also penalize sharp differences in model parameters across adjacent segments; as such, the global cost function seeks to reconstruct a tractogram with smooth mesoscopic parameter changes along fibers. The RJMCMC optimization is extended to induce random changes in the compartment model parameters. MesoFT unifies tractography and microstructure modeling, jointly reconstructing white matter fiber tracts and bundle-specific estimates of the axonal diffusivities (Reisert et al., 2014).

In Christiaens et al. (2015), a data-driven approach was chosen instead. There, the multishell multitissue model also used in constrained spherical deconvolution ( Jeurissen et al., 2014) was used, based on tissue response functions for white matter, gray matter, and cerebrospinal fluid (CSF). Rather than assuming a particular white matter model, the fiber response function is estimated empirically from the data, typically as the mean signal in voxels with high anisotropy (Tournier et al., 2004). Each segment then contributes to the predicted signal $\mathcal{M}(\mathbf{T})$ with the weighted and rotated white matter fiber response function. The isotropic gray matter and CSF response functions, also estimated from the data, contribute to the signal prediction as well. Their voxel-level signal fraction is estimated and updated using least-squares minimization on each update of the segment configuration. The results in a simulated phantom and in vivo data showed that calibrating the response functions in this data-driven way improves the accuracy of tractography (Christiaens et al., 2015).

## 4.4 Streamline-wise approaches

All approaches presented thus far in this section reconstruct the optimal configuration of streamlines starting from an initial random arrangement of small segments that need to be adjusted and assembled together to form complete and smooth trajectories. This reconstruction paradigm is very powerful and flexible but comes at a *considerable computational cost* and, for the sake of reducing complexity, no additional anatomical constraints are usually enforced on the reconstructed streamlines. For example, even though axons are known to connect neurons, streamlines are not guaranteed to start/end inside gray matter and many of them can stop prematurely in white matter; this is a well-known problem in tractography (Côté et al., 2013; Girard et al., 2014), but the ramifications that this relaxation might have in a global estimation process are often *overlooked*. In fact, even if they do not connect gray-matter regions—and thus they will not be contemplated for the quantification of connectivity—they are still considered in the global cost function and, for this reason, they can actually influence the estimation of all other streamlines (as the optimization is done concurrently) and potentially bias the final connectivity estimates.

Lemkaddem et al. (2014) proposed a solution to specifically address this limitation and to reduce the computational complexity of the reconstruction process. First, instead of modeling the trajectories as collections of small segments, the authors represent them by means of Catmull-Rom splines (Catmull and Rom, 1974); this representation has the big advantage of *drastically reducing the number of parameters to optimize*, while *conveniently enforcing smoothness* on the trajectories, as well as easily *constraining their endpoints to lie inside gray matter*. Moreover, rather than starting from a random initialization as previous methods, the algorithm creates an initial collection of streamlines using local tractography methods. This initial tractogram is then prefiltered to discard those trajectories that do not connect gray-matter regions in order to consider in the optimization only streamlines that actually represent anatomically valid connections. The global inverse problem is then solved using simulated annealing and RJMCMC as described in Section 4.2, but here the optimization is carried out considering *entire (and intrinsically smooth) streamlines*, and not constituent segments that need to be joined together, with considerable benefits in terms of computational complexity. Besides, it is worth noting that this formulation also allows the estimation of the *effective contribution of each streamline* to the measured dMRI data, which has important implications for the quantification of the connectivity (see discussion in Section 5.2). Results obtained on both synthetic and in vivo data highlighted the importance of including both anatomical priors and per-streamline contributions in the optimization process.

In the same spirit of reducing complexity, Close et al. (2015) proposed a novel mathematical representation, called Fourier tracts (FouTS), that can describe white matter bundles over a range of spatial scales, from short tract segments to whole white matter tracts, with a relatively small number of parameters. In particular, FouTS explicitly define the spatial extent and the intraaxonal volume of the white matter tracts they represent, thus reconciling correspondence to tract characteristics of interest. The posterior probability distribution over these parameters given the observed dMRI dataset is inferred within a Bayesian framework, which relates the modeled tracts directly to the observed signal. The resulting dMRI tractography method is able to unambiguously infer the posterior distribution over the spatial extent and the "Apparent Connection Strength" of white matter tracts, and therefore the resultant distributions can be directly applied to a wide range of clinical and scientific problems. A key attribute of this method is its ability to detect and penalize spatial intersections between reconstructed fascicles, in a manner akin to the BlueMatter method of Sherbondy et al. (2009) (see next section). For the majority of methods described in this chapter, multiple streamlines may contribute toward a given image voxel, and it is the sum total of these that is considered in the global optimization; but the subvoxel locations of those streamlines relative to one another are not of consequence. In FouTS, each fascicle intersecting an image voxel occupies some fraction of that voxel's volume based on the fascicle shape and position, and the optimization attempts to remove any physical overlap between different fascicles. This complicates the model and optimization considerably, but has the prospect of deriving realistic fascicle extents in the presence of partial volume effects.

## 5 Discriminative approaches

While generative methods commence the tractography reconstruction with a "clean slate" and perform the entire tractogram production as a single encapsulated algorithm, discriminative approaches involve separating the tasks of tractogram *generation* and tractogram *optimization*, as highlighted in Fig. 4. Application of these algorithms typically involves utilization of a more simple local tractography approach to obtain an initial set of *candidate streamlines*; then, they attempt to *modulate* the contributions of individual streamlines based on a given forward-model in such a way that their total predicted contributions to the dMRI signal matches that of the empirical image data. There are two principal strategies employed to perform this manipulation of the tractogram, as described hereafter.

### 5.1 Methods that filter streamlines

Algorithms in this category *remove streamlines* from the input collection of candidates that are deemed detrimental to the accuracy of the tractogram as a whole; or, expressed differently, they search an appropriate subset of the candidate streamlines that together reproduce accurately the empirical dMRI data. Due to this removal process, they are sometimes referred to as tractography "filtering methods."

#### 5.1.1 BlueMatter

BlueMatter (Sherbondy et al., 2009) is the first algorithm proposed in the literature to address the tractography reconstruction problem following a top-down strategy, and consists of two steps. In the first step, *a large collection of candidate streamlines* is generated using local tractography; the implicit assumption is that this initial tractogram must represent a superset of suitable fiber trajectories, that is, tractography is able to reconstruct all true-positive streamlines, possibly in addition to many false positives. To fulfill this requirement as well as to reduce dependency on the specific tractography algorithm chosen to create this initial tractogram, BlueMatter combines streamlines generated with various tracking techniques, e.g., deterministic, probabilistic, etc., to form a massive collection of 180 billion candidates. In the second step, then, BlueMatter searches for the *optimal subset of streamlines* from this huge set that best fit the entire image data using a parallel stochastic hill climbing algorithm. This search is based on two error terms: the first compares the predicted and measured diffusion-weighted images as other global approaches, that is, $\mathcal{E}(\,\cdot\,,\,\cdot\,)$ in Eq. (3), whereas the second term, that is, $\Psi(\cdot)$, attempts to avoid solutions with more fibers, represented by different streamlines, than a given voxel volume allows. This latter is a very important constraint as it helps resolve the ill-posed inverse problem of finding the optimal subset of streamlines that minimizes data-fitting error and to identify a *physically plausible solution*. Numerical simulations and in vivo results clearly showed the great potential of the top-down strategy to tractography introduced by BlueMatter and, even though the actual implementation was too onerous for any practical application of the technique, it stimulated a lot of research since then and many efficient alternatives have been proposed, as described in the following.

### 5.1.2 Spherical-deconvolution informed filtering of tractograms

The spherical-deconvolution informed filtering of tractograms (SIFT) method (Smith et al., 2013) seeks a more computationally tractable solution to the challenge of identifying an appropriate subset of candidate streamlines as compared to the prior BlueMatter method described in the previous section. This is achieved through the following methodological differences:

- *Optimization*: Rather than performing a stochastic search across a very large number of candidate subsets of streamlines, the method instead performs a simple gradient descent search. Commencing with the comprehensive set of candidate streamlines, it iteratively identifies and removes the streamline for which such removal results in the greatest improvement in fit to the empirical image data.
- *Projection*: As the majority of methods described in this chapter, BlueMatter employs a *forward model* to predict dMRI data from the tractogram, such that these predictions may be compared to the measured data (see Eq. 1). SIFT instead operates by first using an *inverse model* to derive estimates of local fiber densities, and then directly comparing local reconstruction densities from the tractogram to these estimates (see Eq. 2).

The SIFT method operates on the basis of *fixels*, each of which is a fiber bundle element that resides in a specific voxel (Raffelt et al., 2015). Any given image voxel may contain zero, one, or more than one fixel; each fixel has associated with it an orientation, an estimate of fiber density, and potentially other parameters depending on the model. In the typical application of SIFT, the spherical deconvolution model is used,[c] which provides a *continuous* representation of fiber orientations over the half-sphere; therefore, a segmentation algorithm is applied to these data in order to produce a finite number of *discrete* fixels in each image voxel and estimate the total fiber density associated with each (Smith et al., 2013). For each fixel, the total streamlines density from the tractogram is then calculated based on a sum of lengths of streamline intersections with that voxel that are additionally aligned with the orientation of that fixel (Smith et al., 2013).

In order to facilitate a computationally simple comparison between the fiber density and tractogram density in each fixel, the so-called *proportionality coefficient* $\mu$ is defined:

$$\mu = \frac{\sum_V \left( W_V \sum_{f \in V} FD_f \right)}{\sum_V \left( W_V \sum_{f \in V} TD_f \right)}, \tag{7}$$

where $FD_f$ and $TD_f$ are the fiber density and tractogram density in fixel $f$, respectively, $f_V$ is the set of fixels residing in voxel $V$, and $W_V$ is a weighting function that downregulates the influence of voxels that do not consist entirely of white matter. Essentially, calculation of this constant is equivalent to stating: "take the sum total of all fiber density, and the sum total of all tractogram density, and make them equal." What it facilitates is construction of a cost function that quantifies the sum total differences between the fiber densities as estimated by the diffusion model and the tractogram density, and can therefore be used to drive the initial tractogram estimate $\mathbf{T}$ toward some optimum $\mathbf{T}^*$:

$$\mathbf{T}^* = \underset{\mathbf{T}}{\mathrm{argmin}} \ \sum_V \left( W_V \sum_{f \in V} \left( \mu \cdot TD_f - FD_f \right)^2 \right). \tag{8}$$

The SIFT optimization algorithm involves calculation of the magnitude of the prospective change in the cost function that would occur for the hypothetical removal of each individual streamline, and subsequent removal of streamlines in a locally

---

c. It should be noted that, although the spherical deconvolution model forms a part of the method acronym, and is the model most associated with this method, theoretically both the SIFT model and corresponding optimization method are applicable to any diffusion model from which it is possible to derive fixel-wise (or indeed voxel-wise) estimates of fiber density.

**308 PART | III** Tractography algorithms

optimal order. As the removal of a streamline results in a decrease in the global tractogram density, parameter $\mu$ is persistently updated to reflect these removals. This filtering can terminate due to any one of a number of criteria, either predefined by the user or intrinsic to the algorithm's capability to proceed.

Application of the SIFT method has been shown to converge connection density estimates toward those estimated from postmortem dissection (Smith et al., 2015a), and to reduce the influence of the streamline seeding mechanism on connection density estimates (Yeh et al., 2016).

## 5.2 Methods that estimate streamline contributions

A key limitation of "filtering methods" is that the more strongly quantized nature of the resulting tractogram may be deleterious for downstream applications. Robust application of tractography necessitates an adequately dense reconstruction for any derivative results to possess adequate precision, and the explicit removal of streamlines directly compromises this. For this reason, instead of *removing* streamlines from the input candidates, methods in this category *modulate* their relative density contributions toward the reconstruction, such that the sum total of the candidate streamlines all modulated by their corresponding weighting factors replicates the empirical dMRI data. It is worth noting that, besides filtering out implausible trajectories, these methods *explicitly* estimate a scalar value associated with every streamline, which represents the *effective intraaxonal cross-sectional area* of the biological fibers along the corresponding trajectories. Such "weights" represent an appealing proxy measure by which to quantify the information-carrying capacity of bundles, dubbed "FBC[d]" in Smith et al. (2022), and they allow enriching the characterization of the estimated brain networks (see Chapter 24) with more quantitative and biologically informative measures (Smith et al., 2022; Zhang et al., 2022).

### 5.2.1 Convex optimization modeling for microstructure informed tractography

The convex optimization modeling for microstructure informed tractography (COMMIT) method (Daducci et al., 2013, 2015) is the first attempt to assess per-streamline cross-sectional weights, and the estimation is made possible by means of a system of linear equations that allows *complementing tractography* with local microstructural features of the neuronal tissue:

$$\mathbf{y} = \mathbf{Ax} + \eta. \tag{9}$$

In this formulation, the vector $\mathbf{y} \in \mathbb{R}_+^{n_v}$ contains the empirical dMRI data acquired in all $n_v$ imaging voxels, the columns of matrix $\mathbf{A} \in \mathbb{R}^{n_v \times n_c}$ encode the *potential* contributions to the signal in each voxel of all candidate streamlines (and possibly other tissues) according to a given multicompartment forward model, the vector $\mathbf{x} \in \mathbb{R}_+^{n_c}$ consists of the weighted contributions of these compartments that best explain the image data, and $\eta$ is the acquisition noise. The forward model implemented in Daducci et al. (2013, 2015) consists of stick, zeppelin, and ball compartments (Panagiotaki et al., 2012); however, the COMMIT formulation is very flexible and can be extended to consider more advanced biophysical models for the tissue microstructure (see Section 6.1) as well as to include multimodal data in the fitting (see Section 6.2). Experiments in both synthetic and real brain data showed the ability of COMMIT to detect and remove implausible streamlines, as well as a good agreement of the estimated compartment fractions with the expected patterns from known anatomy in the white matter (Daducci et al., 2015, 2016).

The nonnegative weights $\mathbf{x}$ that control the contributions of all the compartments, that is, columns of $\mathbf{A}$, can be efficiently estimated using regularized linear least-squares optimization:

$$\underset{\mathbf{x} \geq 0}{\operatorname{argmin}} \ \| \mathbf{Ax} - \mathbf{y} \|_2^2 + \lambda \| \mathbf{x} \|_1, \tag{10}$$

in which $\|\cdot\|_1$ and $\|\cdot\|_2$ are the standard L1- and L2-norm penalties, and parameter $\lambda$ controls the strength of the (optional) regularization term. This formulation has been recently improved by Schomburg and Hohage (2019), who suggested that although the L1 norm is appropriate to promote sparsity in the output set of streamlines, it may be suboptimal for the other voxel-wise compartments, e.g., CSF; hence, they proposed to extend this regularization by adding the Sobolev norm with the aim of promoting spatial smoothness among neighboring voxels. More advanced forms of regularization have recently been introduced (Schiavi et al., 2020; Ocampo-Pineda et al., 2021), which have enabled the possibility to include

---

d. This measure may be estimated also using the filtering methods described in Section 5.1, as they can be alternatively conceptualized as assigning a fixed weight to the retained streamlines and zero to all omitted ones; in the case of SIFT, the parameter $\mu$ can be interpreted as being proportional to the intraaxonal cross-sectional area.

anatomical information about the organization of white-matter bundles in the optimization process with the aim to further improve the anatomical accuracy of the reconstructions (see Section 6.3).

### 5.2.2 Linear fascicle evaluation

The linear fascicle evaluation (LiFE) algorithm (Pestilli et al., 2014; Caiafa and Pestilli, 2017) uses a *linear formulation* similar to Eq. (9) in order to express the measured data as a function of the input set of streamlines. The model used to generate predicted dMRI data in order to compare to the empirical data is, however, not a conventional additive multitissue signal decomposition. The empirical dMRI signal is first decomposed into its isotropic mean and the resulting anisotropic residuals. The forward model for the tractogram is further defined as an exemplar diffusion tensor, but with the mean predicted dMRI intensity subtracted; that is, the presence of a streamline modulates the predicted values of specific dMRI intensities, but its sum total influence on the predicted signal across the image is zero. The comparison between predicted and empirical dMRI data that drives the optimization process is performed based on these demeaned signal representations. The weights of the candidate streamlines are obtained by solving the system of linear equations using a nonmonotonic method for large-scale nonnegative least squares without regularization. The LiFE algorithm was recently improved by Sreenivasan et al. (2022) to include L1-norm regularization into its objective function as in Eq. (10) to obtain sparser tractograms; in addition, this implementation was the first one to introduce the use of graphics processing units to accelerate the evaluation of tractograms with LiFE.

### 5.2.3 SIFT2

The SIFT2 method (Smith et al., 2015b) was proposed as a direct successor to SIFT (see Section 5.1.2), based on the paradigm of having different streamlines within the tractogram contribute a greater or lesser density toward the reconstruction, similarly to that used in the COMMIT and LiFE methods. However, as in the SIFT method, it is based on the attribution of streamlines to fixels as estimated by a diffusion model. The tractogram density in each fixel, $TD_f$, is no longer simply a sum of streamline-voxel intersection lengths; it is augmented with per-streamline weighting factors:

$$ TD_f = \sum_s \left( |s_f| \cdot e^{F_s} \right), \tag{11} $$

where $F_s$ is a coefficient attributed to streamline $s$ that modulates the magnitude of its contribution to the reconstruction, and $|s_f|$ is the length of the intersection between streamline $s$ and the voxel occupied by fixel $f$.

Regularization is additionally utilized as per Eq. (4). Simple Tikhonov regularization can be used to encourage solutions where streamlines are not given excessively small or large contributions (i.e., $F_s \approx 0 \ni e^{F_s} \approx 1 \ \forall \ s$). More commonly, a more complex custom regularization function is utilized as defined in Smith et al. (2015b), which encourages solutions where streamlines that traverse common fixels do not obtain drastically different contribution factors; this allows streamlines within different macroscopic bundles to obtain different contribution factors in order to correct the different reconstruction biases *between* those bundles, while restricting unwanted variance in contributions for streamlines *within* bundles.

## 6 Extending the capabilities of global methods

Across the methods presented thus far in this chapter, a strong common focus is the integration of *estimation of connection densities* that are consistent with the empirical dMRI data. The paradigm of global tractography is, however, entirely flexible, and it is theoretically possible to augment such with a wide range of other sources of information: from more complex diffusion models (whether forward or inverse), to image information from modalities outside of dMRI, to a priori expectations or constraints. Here we mention a range of such methods that have been proposed thus far in the literature, though we note that at time of publication this is still a rapidly evolving domain, and there is likely scope for advancements in global tractography beyond the list provided here.

### 6.1 Adding advanced models of tissue microstructure

One promising avenue for future global tractography research is the injection of additional information about the microstructural composition of the fibers; this class of methods is commonly called *microstructure informed tractography* (Daducci et al., 2016) (see also Chapter 18). Allowing the joint optimization of both the trajectories of fiber bundles as well as their microstructural composition, e.g., caliber of the axons, can help the decision-making process to resolve

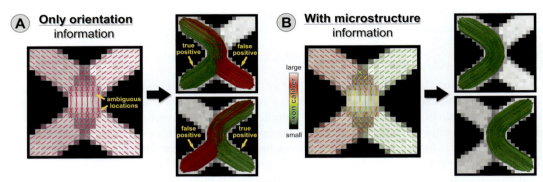

**FIG. 5** Complementing tractography with information about tissue microstructure. (A) When only the orientations of white matter fibers are available, tractography may not be able to disentangle ambiguous configurations present in the data, and erroneous trajectories may therefore be reconstructed. (B) If, instead, extra knowledge about the tissue microstructure is available, e.g., caliber of the axons in this toy example, this additional information can be exploited by tractography to resolve such ambiguities and improve the accuracy of the reconstructions.

ambiguous configurations of streamlines that standard tractography finds plausible but that are incompatible with such extra microstructural parameters (see Fig. 5).

MicroTrack (Sherbondy et al., 2010) combines the BlueMatter global tractography approach (see Section 5.1.1) with the orientationally invariant estimates of axon diameter proposed in Alexander et al. (2010) to demonstrate the potential of joint estimation to improve both tractography and microstructural parameter estimates. The algorithm assumes a constant axon diameter along each streamline and, following the BlueMatter algorithm, searches for the minimal set of streamlines from a large collection that best explains the image data. MicroTrack replaces the simple forward model in Sherbondy et al. (2009) with the model in Alexander et al. (2010), which includes an axon diameter parameter, and attaches an axon diameter parameter to each streamline. The algorithm optimizes the axon diameter parameter attached to each streamline concurrently with the selection of a subset of streamlines. The authors show in simulation that this joint-inference process (a) reduces noise in estimates of the axon diameter, by pooling information from multiple voxels along each streamline, and (b) can resolve key ambiguities that confound standard tractography, such as kissing versus crossing, by allowing only configurations that are consistent with the assumption that the axon diameter distribution remains constant along streamlines.

COMMIT$_{AxSize}$ (Barakovic et al., 2021a) also seeks to derive axon diameter index estimates at the streamline level. This method considers each streamline as consisting of a population of axons with an unknown distribution of diameters, which must be estimated. While it utilizes the same forward model component as in the original COMMIT method (see Section 5.2.1) for the extra-axonal compartment, *multiple* intraaxonal components are included in the modified model. To account for a heterogeneous composition arising from axons with distinct diameters, the authors considered *several columns for each streamline*: each corresponds to a model of parallel cylinders with fixed diameters and longitudinal diffusivity, but each column corresponds to cylinders of a different diameter. The optimization algorithm is then free to attribute whatever relative magnitudes of those columns best explain the empirical dMRI data. The results obtained through simulations and in vivo comparison with histology demonstrated that with these types of formulations it is feasible to infer *bundle-specific* biological attributes, e.g., axon diameter index, rather than local estimates as provided by voxel-wise methods (Alexander et al., 2010).

### 6.2 Adding multimodal data

Details on the microstructural composition of the fibers can be inferred from the very same dMRI acquisition used for the tracking, but a more comprehensive insight on the tissue microstructure can be obtained from other quantitative imaging modalities (see Uludağ and Roebroeck, 2014 and references therein). Incorporating multimodal data in the tractography decision-making process can enable *sensitivity to components/effects otherwise invisible to dMRI*, e.g., myelin, and this extra information has the potential to further improve the anatomical accuracy of the reconstructions as well as their biological interpretability.

COMMIT$_{T2}$ is an extension of the COMMIT formulation proposed by Barakovic et al. (2021b) that allows considering *multiecho diffusion-relaxometry data* in the optimization process and enables the estimation of multiple intraaxonal T2 relaxation times within the same voxel using tractography-based spatial regularization. As with all other discriminative methods, COMMIT$_{T2}$ assumes that T2 remains invariant along any given trajectory; but here it is permitted to vary between streamlines within the same voxel. Similarly to the way in which COMMIT$_{AxSize}$ defines multiple columns per streamline corresponding to different prospective axonal diameters, COMMIT$_{T2}$ defines multiple columns per streamline corresponding to different *prospective T2 values*. After solving the resulting global optimization problem, a vector of coefficients for every streamline is obtained, where each element quantifies the signal fraction ascribed to that streamline with the corresponding simulated T2 value, thus obtaining a distribution of T2 values for each streamline. The proof-of-concept was provided in silico demonstration as well as in vivo results, with the latter demonstrating that distinct tract-specific T2 profiles could be recovered even in the three-way crossing of the corpus callosum, arcuate fasciculus, and corticospinal tract within the centrum semiovale.

Schiavi et al. (2022) proposed another extension of COMMIT, called myelin streamline decomposition (MySD), which instead allows estimating a specific myelin content ascribed to each streamline from the total *myelin volume/water fraction* value measured in each voxel with quantitative MRI acquisitions. The weights estimated by MySD and associated with each streamline represent the myelin volume/water contribution per unit length; similarly to the case for streamline density contributions, this quantification of myelin volume/water fraction per unit length effectively yields a myelin volume/water cross-sectional area per streamline. After identifying all streamlines associated with the same macroscopic bundle, a *bundle-specific myelin fraction* for that bundle is then obtained by summing the total volume of myelin attributed to the corresponding streamlines (the product of the myelin cross-sectional area and streamline length for each), and dividing by the total macroscopic volume of the bundle. Results from in vivo data demonstrate the advantage of the bundle-specific myelin estimates provided by MySD in comparison to tractometry—that is, averaging along the path (more details in Chapter 23)—in both macroscopic white matter bundles and cortical projections of streamline values.

### 6.3 Adding anatomical information about white matter organization

Tractograms generated with global tractography can provide more biologically relevant estimates of connection density, but these estimates are *predicated on the spatial trajectories themselves being accurate* with respect to the underlying anatomy. However, recent studies have questioned the *anatomical accuracy of the reconstructions* that can be obtained with all tractography algorithms, highlighting in particular that tractograms are dominated by false positives (Thomas et al., 2014; Maier-Hein et al., 2017; Schilling et al., 2022; Maffei et al., 2022), as illustrated in Fig. 6A. Even though these spurious connections represent a major source of bias to obtain accurate estimates of connectivity (Drakesmith et al., 2015;

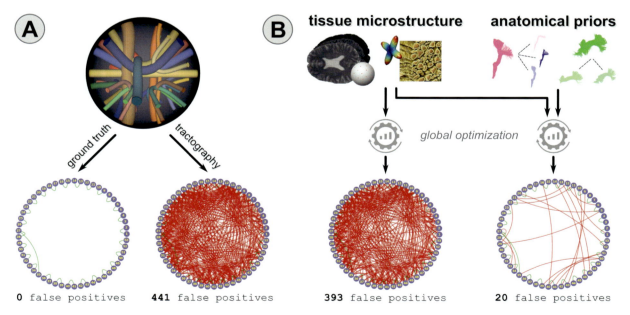

**FIG. 6** Complementing tractography with information about the organization of white matter. Tractography reconstructions are known to suffer from a high incidence of false positives, as shown in (A) using the "ISBI 2013" phantom (details in Schiavi et al., 2020). This issue is only marginally mitigated if microstructure information alone is used in the decision-making process (B, *left*), whereas the number of false positives can be drastically reduced if additional anatomical information about white matter organization is injected into the optimization (B, *right*).

Zalesky et al., 2016; Yeh et al., 2021; Zhang et al., 2022), none of the methods presented thus far has proved effective to solve this problem, as shown in the left panel of Fig. 6B.

Schiavi et al. (2020) suggested that this limitation is likely due to the fact that all existing global methods for tractography are *purely data-driven*—that is, they rely only on the empirical dMRI data in their decision-making process—and speculated that this information is not enough to disentangle all ambiguous configurations that are present in the data itself. Therefore the authors advocated the need for additional information to help tractography produce more accurate reconstructions and proposed to *incorporate in the optimization two fundamental observations* about the organization of the fiber tracts that may help resolve such ambiguities:

1. The neuronal fibers in the white matter are naturally organized as macroscopic bundles between gray-matter regions (Mandonnet et al., 2018).
2. The brain is a very expensive organ and the high metabolic costs associated with its operation do not support the existence of all fiber bundles deemed plausible by dMRI (Bullmore and Sporns, 2012).

To achieve this, the COMMIT formulation was enhanced by allowing the possibility of *grouping streamlines* according to the anatomical organization of white matter in bundles, instead of considering them as independent, and by replacing the L1 norm in Eq. (10) with the following regularization term to *promote sparsity among bundles* rather than individual streamlines:

$$\underset{\mathbf{x} \geq 0}{\operatorname{argmin}} \quad \| \mathbf{Ax} - \mathbf{y} \|_2^2 + \lambda \sum_{g \in \mathcal{G}} \| \mathbf{x}^{(g)} \|_2, \tag{12}$$

where $\mathcal{G}$ defines a partition of the streamlines into groups and $\mathbf{x}^{(g)}$ are the coefficients corresponding to the streamlines in a given group $g \in \mathcal{G}$. This novel formulation, named COMMIT2, attempts to recover the tractogram that best explains the measured dMRI data, as before, while containing a minimal number of bundles. This idea was further improved in Ocampo-Pineda et al. (2021) by allowing finer control on the streamlines grouping and introducing the possibility of defining a *multilevel partitioning* of the streamlines to be able to capture the natural organization of white-matter fibers in bundles and subbundles.

Simulations and in vivo results demonstrated that integrating both a local notion of fiber density and a global notion of bundle count can dramatically increase the anatomical accuracy of the estimated tractograms and outperforms the typical compromise between sensitivity and specificity intrinsic in standard tractography methods, as shown in the right panel of Fig. 6B.

## 7 Conclusion

The purpose of this chapter was to provide the readers with a *conceptual overview* of those tractography algorithms that are colloquially referred to as "global" and to *highlight the benefits* that such advanced formulations can provide compared to their "local" counterparts. In local algorithms, streamlines are generated *independently* one by one and the entire set of reconstructed trajectories is never directly compared to the measured dMRI data; global methods, instead, intrinsically consider all streamlines to be estimated as a whole and the estimation process is done *concurrently*, using optimization to continuously compare the streamlines configuration to the entire image data and to refine it until the optimal tractogram is found. This reconstruction strategy is able to resolve ambiguities and to identify complete trajectories, as well as potentially other associated biological attributes, for which the data offers overall support even if some voxels along the trajectory align poorly to its local tangent; thus its promise is to improve the *anatomical accuracy* and the *biological interpretability* of the tractograms compared to local approaches. Methodological research on global tractography is very active and vibrant and, in an attempt to attack the fully global problem, the formulations proposed in the literature continue to grow in sophistication and computational complexity. Fortunately, *large computational resources* such as HPC clusters and GPU farms are becoming more accessible to researchers by the day, and offer unprecedented opportunities to augment the global tractography paradigm with additional data (possibly multimodal) and prior information in the pursuit of more detailed and comprehensive characterization of the intricate architecture of white matter.

## Acknowledgments

This research was supported by the Rita Levi Montalcini Program for young researchers of the Italian Ministry of Education, University and Research; the National Health and Medical Research Council of Australia, and the Victorian Government's Operational Infrastructure Support Program; RS is a fellow of the National Imaging Facility, a National Collaborative Research Infrastructure Strategy (NCRIS) capability, at the Florey Institute of Neuroscience and Mental Health; EPSRC grants EP/M020533/1 and EP/N018702/1 and the NIHR UCLH Biomedical Research Centre support DCA's work on this topic.

# References

Alexander, D.C., Hubbard, P.L., Hall, M.G., Moore, E.A., Ptito, M., Parker, G.J.M., Dyrby, T.B., 2010. Orientationally invariant indices of axon diameter and density from diffusion MRI. NeuroImage 52, 1374–1389.

Barakovic, M., Girard, G., Schiavi, S., Romascano, D., Descoteaux, M., Granziera, C., Jones, D.K., Innocenti, G.M., Thiran, J.P., Daducci, A., 2021a. Bundle-specific axon diameter index as a new contrast to differentiate white matter tracts. Front. Neurosci. 15, 646034.

Barakovic, M., Tax, C.M.W., Rudrapatna, U., Chamberland, M., Patino, J.R., Granziera, C., Thiran, J.P., Daducci, A., Canales-Rodríguez, E.J., Jones, D.K., 2021b. Resolving bundle-specific intra-axonal T2 values within a voxel using diffusion-relaxation tract-based estimation. NeuroImage 227, 117617.

Bullmore, E., Sporns, O., 2012. The economy of brain network organization. Nat. Rev. Neurosci. 13 (5), 336–349.

Caiafa, C.F., Pestilli, F., 2017. Multidimensional encoding of brain connectomes. Sci. Rep. 7, 11491.

Calamante, F., Smith, R.E., Tournier, J.D., Raffelt, D., Connelly, A., 2015. Quantification of voxel-wise total fibre density: investigating the problems associated with track-count mapping. NeuroImage 117 (C), 284–293.

Catmull, E., Rom, R., 1974. A class of local interpolating splines. In: Computer Aided Geometric Design, Academic Press, pp. 317–326.

Christiaens, D., Reisert, M., Dhollander, T., Sunaert, S., Suetens, P., Maes, F., 2015. Global tractography of multi-shell diffusion-weighted imaging data using a multi-tissue model. NeuroImage 123, 89–101.

Close, T.G., Tournier, J.D., Johnston, L.A., Calamante, F., Mareels, I., Connelly, A., 2015. Fourier tract sampling (FouTS): a framework for improved inference of white matter tracts from diffusion MRI by explicitly modelling tract volume. NeuroImage 120, 412–427.

Côté, M.A., Girard, G., Boré, A., Garyfallidis, E., Houde, J.C., Descoteaux, M., 2013. Tractometer: towards validation of tractography pipelines. Med. Image Anal. 17 (7), 844–857.

Daducci, A., Dal Palu, A., Lemkaddem, A., Thiran, J.P., 2013. A convex optimization framework for global tractography. In: International Symposium on Biomedical Imaging, pp. 524–527.

Daducci, A., Dal Palú, A., Lemkaddem, A., Thiran, J.P., 2015. COMMIT: convex optimization modeling for microstructure informed tractography. IEEE Trans. Med. Imaging 34 (1), 246–257.

Daducci, A., Dal Palú, A., Descoteaux, M., Thiran, J.P., 2016. Microstructure informed tractography: pitfalls and open challenges. Front. Neurosci. 10, 247.

Drakesmith, M., Caeyenberghs, K., Dutt, A., Lewis, G., David, A.S., Jones, D.K., 2015. Overcoming the effects of false positives and threshold bias in graph theoretical analyses of neuroimaging data. NeuroImage 118, 313–333.

Fillard, P., Poupon, C., Mangin, J.F., 2009. A novel global tractography algorithm based on an adaptive spin glass model. In: International Conference on Medical Image Computing and Computer-Assisted Intervention, pp. 927–934.

Fillard, P., Descoteaux, M., Goh, A., Gouttard, S., Jeurissen, B., Malcolm, J., Ramirez-Manzanares, A., Reisert, M., Sakaie, K., Tensaouti, F., Yo, T., Mangin, J.F., Poupon, C., 2011. Quantitative evaluation of 10 tractography algorithms on a realistic diffusion MR phantom. NeuroImage 56 (1), 220–234.

Girard, G., Whittingstall, K., Deriche, R., Descoteaux, M., 2014. Towards quantitative connectivity analysis: reducing tractography biases. NeuroImage 98, 266–278.

Girard, G., Caminiti, R., Battaglia-Mayer, A., St-Onge, E., Ambrosen, K.S., Eskildsen, S.F., Krug, K., Dyrby, T.B., Descoteaux, M., Thiran, J.P., Innocenti, G.M., 2020. On the cortical connectivity in the macaque brain: a comparison of diffusion tractography and histological tracing data. NeuroImage 221, 117201.

Guevara, P., Duclap, D., Poupon, C., Marrakchi-Kacem, L., Fillard, P., Le Bihan, D., Leboyer, M., Houenou, J., Mangin, J.F., 2012. Automatic fiber bundle segmentation in massive tractography datasets using a multi-subject bundle atlas. NeuroImage 61 (4), 1083–1099.

Jbabdi, S., Johansen-Berg, H., 2011. Tractography: where do we go from here? Brain Connect. 1 (3), 169–183.

Jbabdi, S., Woolrich, M.W., Andersson, J.L., Behrens, T.E.J., 2007. A Bayesian framework for global tractography. NeuroImage 37 (1), 116–129.

Jeurissen, B., Tournier, J.D., Dhollander, T., Connelly, A., Sijbers, J., 2014. Multi-tissue constrained spherical deconvolution for improved analysis of multi-shell diffusion MRI data. NeuroImage 103, 411–426.

Jones, D.K., 2010. Challenges and limitations of quantifying brain connectivity in vivo with diffusion MRI. Imaging Med. 2 (3), 341–355.

Jones, D.K., Knösche, T.R., Turner, R., 2013. White matter integrity, fiber count, and other fallacies: the do's and don'ts of diffusion MRI. NeuroImage 73, 239–254.

Kreher, B.W., Mader, I., Kiselev, V.G., 2008. Gibbs tracking: a novel approach for the reconstruction of neuronal pathways. Magn. Reson. Med. 60 (4), 953–963.

Lemkaddem, A., Skiöldebrand, D., Dal Palú, A., Thiran, J.P., Daducci, A., 2014. Global tractography with embedded anatomical priors for quantitative connectivity analysis. Front. Neurol. 5, 232.

Maffei, C., Girard, G., Schilling, K.G., Aydogan, D.B., Adluru, N., Zhylka, A., Wu, Y., Mancini, M., Hamamci, A., Sarica, A., et al., 2022. Insights from the IronTract challenge: optimal methods for mapping brain pathways from multi-shell diffusion MRI. NeuroImage 257, 119327.

Maier-Hein, K.H., Neher, P.F., Houde, J.C., Côté, M.A., Garyfallidis, E., Zhong, J., Chamberland, M., Yeh, F.C., Lin, Y.C., Ji, Q., Reddick, W.E., Glass, J.O., Chen, D.Q., Feng, Y., Gao, C., Wu, Y., Ma, J., Renjie, H., Li, Q., Westin, C.F., Deslauriers-Gauthier, S., González, J.O.O., Paquette, M., St-Jean, S., Girard, G., Rheault, F., Sidhu, J., Tax, C.M.W., Guo, F., Mesri, H.Y., Dávid, S., Froeling, M., Heemskerk, A.M., Leemans, A., Boré, A., Pinsard, B., Bedetti, C., Desrosiers, M., Brambati, S., Doyon, J., Sarica, A., Vasta, R., Cerasa, A., Quattrone, A., Yeatman, J., Khan, A.R., Hodges, W., Alexander, S., Romascano, D., Barakovic, M., Auría, A., Esteban, O., Lemkaddem, A., Thiran, J.P., Cetingul, H.E., Odry, B.L., Mailhe, B., Nadar, M.S., Pizzagalli, F., Prasad, G., Villalon-Reina, J.E., Galvis, J., Thompson, P.M., Requejo, F.D.S.,

Laguna, P.L., Lacerda, L.M., Barrett, R., Dell'Acqua, F., Catani, M., Petit, L., Caruyer, E., Daducci, A., Dyrby, T.B., Holland-Letz, T., Hilgetag, C.C., Stieltjes, B., Descoteaux, M., 2017. The challenge of mapping the human connectome based on diffusion tractography. Nat. Commun. 8 (1), 1349.

Mandonnet, E., Sarubbo, S., Petit, L., 2018. The nomenclature of human white matter association pathways: proposal for a systematic taxonomic anatomical classification. Front. Neuroanat. 12, 94.

Mangin, J.F., Poupon, C., Cointepas, Y., Rivière, D., Papadopoulos-Orfanos, D., Clark, C.A., Régis, J., Le Bihan, D., 2002. A framework based on spin glass models for the inference of anatomical connectivity from diffusion-weighted MR data - a technical review. NMR Biomed. 15 (7–8), 481–492.

Ocampo-Pineda, M., Schiavi, S., Rheault, F., Girard, G., Petit, L., Descoteaux, M., Daducci, A., 2021. Hierarchical microstructure informed tractography. Brain Connect. 11, 75–88.

Panagiotaki, E., Schneider, T., Siow, B., Hall, M.G., Lythgoe, M.F., Alexander, D.C., 2012. Compartment models of the diffusion MR signal in brain white matter: a taxonomy and comparison. NeuroImage 59 (3), 2241–2254.

Pestilli, F., Yeatman, J.D., Rokem, A., Kay, K.N., Wandell, B.A., 2014. Evaluation and statistical inference for human connectomes. Nat. Methods 11 (10), 1058–1063.

Pike, B., Alexander, D.C., Stikov, N., 2018. Special Issue on "Microstructural Imaging". vol. 182 ScienceDirect, pp. 1–522.

Raffelt, D.A., Smith, R.E., Ridgway, G.R., Tournier, J.D., Vaughan, D.N., Rose, S., Henderson, R., Connelly, A., 2015. Connectivity-based fixel enhancement: whole-brain statistical analysis of diffusion MRI measures in the presence of crossing fibres. NeuroImage 117, 40–55.

Reisert, M., Mader, I., Anastasopoulos, C., Weigel, M., Schnell, S., Kiselev, V., 2011. Global fiber reconstruction becomes practical. NeuroImage 54 (2), 955–962.

Reisert, M., Kiselev, V.G., Dihtal, B., Kellner, E., Novikov, D.S., 2014. MesoFT: unifying diffusion modelling and fiber tracking. In: Lecture Notes in Computer Science. Medical Image Computing and Computer-Assisted Intervention – MICCAI 2014, vol. 8675, pp. 201–208.

Schiavi, S., Ocampo-Pineda, M., Barakovic, M., Petit, L., Descoteaux, M., Thiran, J.P., Daducci, A., 2020. A new method for accurate in vivo mapping of human brain connections using microstructural and anatomical information. Sci. Adv. 6 (31), eaba8245.

Schiavi, S., Lu, P.J., Weigel, M., Lutti, A., Jones, D.K., Kappos, L., Granziera, C., Daducci, A., 2022. Bundle myelin fraction (BMF) mapping of different white matter connections using microstructure informed tractography. NeuroImage 249, 118922.

Schilling, K.G., Tax, C.M.W., Rheault, F., Landman, B.A., Anderson, A.W., Descoteaux, M., Petit, L., 2022. Prevalence of white matter pathways coming into a single white matter voxel orientation: the bottleneck issue in tractography. Hum. Brain Mapp. 43 (4), 1196–1213.

Schomburg, H., Hohage, T., 2019. Formulation and efficient computation of $\ell_1$ - and smoothness penalized estimates for microstructure-informed tractography. IEEE Trans. Med. Imaging 38 (8), 1899–1909.

Sherbondy, A.J., Dougherty, R.F., Ananthanarayanan, R., Modha, D.S., Wandell, B.A., 2009. Think global, act local; projectome estimation with BlueMatter. In: International Conference on Medical Image Computing and Computer-Assisted Intervention, vol. 12, pp. 861–868.

Sherbondy, A.J., Rowe, M.C., Alexander, D.C., 2010. MicroTrack: an algorithm for concurrent projectome and microstructure estimation. In: International Conference on Medical Image Computing and Computer-Assisted Intervention, vol. 13, pp. 183–190.

Smith, R.E., Tournier, J.D., Calamante, F., Connelly, A., 2013. SIFT: spherical-deconvolution informed filtering of tractograms. NeuroImage 67, 298–312.

Smith, R.E., Tournier, J.D., Calamante, F., Connelly, A., 2015a. The effects of SIFT on the reproducibility and biological accuracy of the structural connectome. NeuroImage 104, 253–265.

Smith, R.E., Tournier, J.D., Calamante, F., Connelly, A., 2015b. SIFT2: enabling dense quantitative assessment of brain white matter connectivity using streamlines tractography. NeuroImage 119, 338–351.

Smith, R., Raffelt, D., Tournier, J.D., Connelly, A., 2022. Quantitative streamlines tractography: methods and inter-subject normalisation. Aperture Neuro 2, 1–25. https://doi.org/10.52294/ApertureNeuro.2022.2.NEOD9565.

Sreenivasan, V., Kumar, S., Pestilli, F., Talukdar, P., Sridharan, D., 2022. GPU-accelerated connectome discovery at scale. Nat. Comput. Sci. 2 (5), 298–306.

Thomas, C., Ye, F., Irfanoglu, O., Modi, P., Saleem, K., Leopold, D., Pierpaoli, C., 2014. Anatomical accuracy of brain connections derived from diffusion MRI tractography is inherently limited. Proc. Natl. Acad. Sci. USA 111 (46), 16574–16579.

Tournier, J.D., Calamante, F., Gadian, D.G., Connelly, A., 2004. Direct estimation of the fiber orientation density function from diffusion-weighted MRI data using spherical deconvolution. NeuroImage 23 (3), 1176–1185.

Uludağ, K., Roebroeck, A., 2014. General overview on the merits of multimodal neuroimaging data fusion. NeuroImage 102, 3–10.

Wu, X., Xu, Q., Xu, L., Zhou, J., Anderson, A.W., Ding, Z., 2009. Genetic white matter fiber tractography with global optimization. J. Neurosci. Methods 184 (2), 375–379.

Yeh, C.H., Smith, R.E., Liang, X., Calamante, F., Connelly, A., 2016. Correction for diffusion MRI fibre tracking biases: the consequences for structural connectomic metrics. NeuroImage 142, 150–162.

Yeh, C.H., Jones, D.K., Liang, X., Descoteaux, M., Connelly, A., 2021. Mapping structural connectivity using diffusion MRI: challenges and opportunities. J. Magn. Reson. Imaging 53 (6), 1666–1682.

Zalesky, A., Fornito, A., Cocchi, L., Gollo, L.L., van den Heuvel, M.P., Breakspear, M., 2016. Connectome sensitivity or specificity: which is more important? NeuroImage 142, 407–420.

Zhang, F., Daducci, A., He, Y., Schiavi, S., Seguin, C., Smith, R., Yeh, C.H., Zhao, T., O'Donnell, L.J., 2022. Quantitative mapping of the brain's structural connectivity using diffusion MRI tractography: a review. NeuroImage 249, 118870.

# Chapter 17

# Machine learning in tractography

Peter Neher[a], Philippe Poulin[b,*], Daniel Jörgens[c], Marco Reisert[d], Itay Benou[e,†], and Klaus Maier-Hein[a]

[a]*German Cancer Research Center (DKFZ) Heidelberg, Division of Medical Image Computing, Heidelberg, Germany*, [b]*Sherbrooke Connectivity Imaging Laboratory (SCIL), University of Sherbrooke, Sherbrooke, QC, Canada*, [c]*Krembil Research Institute, University Health Network, Toronto, Canada*, [d]*Department of Radiology, University Medical Center Freiburg, Freiburg, Germany*, [e]*Department of Electrical and Computer Engineering, Ben-Gurion University of the Negev, Beer-Sheva, Israel*

## 1 Introduction

The aim of diffusion MRI tractography is to noninvasively map the neuronal pathways in the human brain based on diffusion-weighted magnetic resonance imaging (dMRI). To achieve this goal, tractography algorithms aim to connect the voxel-wise discrete microstructural information with continuous structures called streamlines or tracks. This mapping involves the optimization of multiple target dimensions: (1) global connectivity between brain regions should be modeled with high sensitivity and specificity, (2) the individual connections, i.e., bundles, should be reconstructed without over- or undersegmentation, and (3) the individual reconstructed tracks in one bundle should be well aligned with the underlying real microstructure of the tissue. Depending on the use-case, a subset of these target dimensions might be more important than the rest. Most tractography algorithms only exploit the directional part of the microstructural information reflected in the dMRI signal. This step from discrete local directional information to global connectivity is a highly ambiguous and ill-posed problem.

A vast spectrum of tractography algorithms has been presented over the last two decades, ranging from local deterministic approaches (Basser, 1998; Mori et al., 1999; Lazar et al., 2003; Tournier et al., 2012) through probabilistic methods (Friman met al., 2006; Behrens et al., 2007; Zhang et al., 2007, 2013; Berman et al., 2008; Descoteaux et al., 2009; Vorburger et al., 2013) to global tractography (Jbabdi et al., 2007; Fillard et al., 2009; Reisert et al., 2011; Aganj et al., 2011; Mangin et al., 2013; Lemkaddem et al., 2014; Daducci et al., 2015), as well as a large number of variations and extensions of these broad classes (Jeurissen et al., 2019). To infer information about the complex microstructure of brain tissue and to optimally exploit the acquisition-dependent signal characteristics of diffusion-weighted images, tractography algorithms typically employ mathematical models calculated from the diffusion-weighted signal. Prominent examples include (multi-)tensor models (Basser et al., 1994; Kreher et al., 2005; Malcolm et al., 2010), spherical deconvolution (Alexander, 2005; Tournier et al., 2007; Schultz et al., 2010; Jeurissen et al., 2014), Q-ball modeling (Descoteaux et al., 2007; Aganj et al., 2009), and persistent angular structures (Jansons and Alexander, 2003; Parker and Alexander, 2005), as well as a large variety of multicompartment models (Assaf and Basser, 2005; Assaf et al., 2008; Zhang et al., 2012; Panagiotaki et al., 2012).

Classic tractography as described here suffers from a variety of serious limitations, as shown for example by Maier-Hein et al., (2017) in a tractography challenge on simulated data in the course of the 2015 ISMRM annual meeting. There is a difficult trade-off between sensitivity and specificity on a whole brain connectivity as well as on the individual bundle level. As shown in Maier-Hein et al. (2017), the average number of false positive connections included in the 96 submitted tractograms was more than four times higher than the average number of contained valid bundles. Many invalid bundles seemed coherent and visually plausible even though they were not part of the ground truth. Submissions with less invalid bundles were also less sensitive, not only on a connection level but also in terms of bundle completeness.

There are multiple potential causes for these issues, two of which are that the tractography is a difficult ill-posed problem (see Fig. 1) and that the mathematical models used to infer information about the local fiber architecture are extremely simplified abstractions of the complex microstructural reality. These models make various assumptions

---

[*] Current address: Working at Agendrix, Sherbrooke, Canada

[†] Current address: Working at Mobileye Global Inc., Jerusalem, Israel

*Handbook of Diffusion MR Tractography.* https://doi.org/10.1016/B978-0-12-818894-1.00032-X
Copyright © 2025 Elsevier Ltd. All rights are reserved, including those for text and data mining, AI training, and similar technologies.

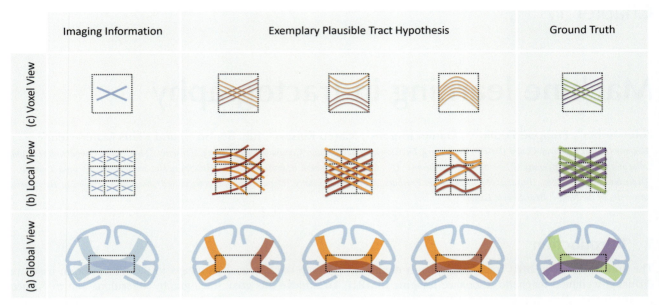

**FIG. 1** Illustration of the ambiguities that arise from the discrete imaging information that tractography is facing. Many different configurations of tracks potentially fit the present imaging information, which poses a major problem leading to a large number of false positive tracts reconstructed by current tractography methods. *(Modified from Maier-Hein, K.H., et al., 2017. The challenge of mapping the human connectome based on diffusion tractography. Nat. Commun. 8(1), Art. no. 1. https://doi.org/10.1038/s41467-017-01285-x.)*

about signal and tissue properties that show a high intra- as well as intersubject variance and are heavily dependent on the acquisition schemes.

Over the last years, machine learning (ML) and particularly deep learning (DL) has transformed many disciplines and has led to technological leaps in many research areas, for example face recognition, object detection, autonomous driving, automatic translations, and speech recognition (Goodfellow et al., 2016). ML has also shown tremendous potential in the area of medical image analysis and processing, such as the dermatologist-level classification of skin cancer with deep neural networks (Esteva et al., 2017) and many other areas.

Recently, ML-based approaches were introduced in the domain of tractography, which have several advantages and can potentially mitigate some of the limitations described earlier. ML algorithms are in theory completely data-driven and no hand-crafted simplifying modeling assumptions are necessary, which might introduce errors into the tractography process. They enable the natural and straightforward integration of additional sources of information, such as other image contrasts or prior knowledge, which have to be artificially introduced into classic model-based tractography pipelines. Further, they offer the opportunity to increasingly replace manual steps like tuning of typical tractography parameters such as thresholds on model-derived quantities like the fractional anisotropy (FA): ML approaches can directly learn the respective decisions that are based on these parameters, such as the termination of the streamline progression if the local FA value drops below a threshold, from the input data, and a manual definition becomes obsolete. ML algorithms can further naturally include an arbitrarily large neighborhood in their decision-making process, which is a promising approach for resolving fiber configurations that are locally ambiguous but become more benign to handle the larger the context.

While the use of ML techniques for tractography is therefore appealing, it comes with serious challenges, none of which are fully assessed or solved at this time. These include the availability and validity of annotated data for training and evaluation, the generalizability of approaches to new datasets with previously unseen pathologies or that were acquired with different scanners and acquisition settings, as well as the explainability of decisions made by the ML system. Nevertheless, ML methods have been successfully employed for tractography and there still is a huge potential for improvements.

In Sections 2 and 3, we give an overview of currently existing datasets and methods for ML-based tractography as well as the challenges that arise and what has to be considered in their design and creation. Section 4 reflects on the challenges that ML-based tractography is still facing, what we have learned over the last years, what could be gained, and how the methods described in Section 3 can be used. Section 5 concludes the chapter and gives a brief outlook as to which direction ML-based tractography might be heading.

## 2 Training and validation data

Data is the most crucial aspect of ML. It is the essential component that allows a model to be trained, evaluated, and compared to other models. Meanwhile, in the field of human brain tractography, a ground truth for in vivo data is impossible to obtain.

Traditional tractography methods face this problem to a lesser extent and address it by using interpretable mathematical models and assessing an algorithm's performance on phantoms and simulated datasets. Nonetheless, to benefit from contextual information and make more informed decisions, machine-learning models of various architectures have been suggested with increased frequency in the last few years, and the need for training and validation data has been growing.

Different types of data come with different advantages and disadvantages. The only way to provide actual ground truth is by using simulated datasets. Besides providing ground truth, this approach comes with the additional advantages of enabling complex phantom configurations in large numbers with relatively low effort, but at the cost of simplified diffusion properties. Indeed, simulations are not yet capable of perfectly reproducing reality, in terms of tissue properties and complexity as well as realism of the simulated MR acquisition. Only manually annotated in vivo datasets provide this realism, at the cost of trading a real ground truth for an approximate and certainly incomplete reference.

The goal of this section is to give an overview of the publicly available datasets that have been used for ML applied to tractography or that were developed with this goal in mind. Other, nonpublic, datasets have been described in the literature over the years as mentioned in Poulin et al. (2019), but are not included here since open availability is crucial for ensuring comparability and reproducibility and for being useful for this area of research in general.

### 2.1 Hardware phantom, simulated, and ex vivo data

#### 2.1.1 FiberCup

The original *FiberCup*[a] phantom dataset (Fillard, 2011) was not used directly for ML, but served as the basis for the simulated dataset, and is thus of interest to us. It was released as part of a tractography contest for the Medical Image Computing and Computer Assisted Intervention (MICCAI) conference in London in 2009. The dataset consists of a realistic diffusion MRI phantom with seven bundles of varying configurations. Diffusion images were acquired in 64 directions, at isotropic resolutions of 3 and 6 mm, using three sets of *b*-values.

Ground truth was defined by 16 manually drawn tracks, each passing through a specific seed voxel in which a single bundle is expected. Contestants were provided all six datasets (without the ground truth), and were free to apply any preprocessing of their choice. Quantitative evaluation was done by comparing a set of 16 candidate tracks to the 16 ground truth tracks using a point-based symmetric root mean square error along with three different metrics (spatial, tangent, and curve).

#### 2.1.2 Simulated FiberCup

To experiment with various acquisition parameters, Neher et al. (2014) presented a simulated version of the *FiberCup* dataset,[b] also using three sets of *b*-values ($650/1500/2000\,s/mm^2$) at a resolution of 3 mm isotropic. This way, a "real" ground truth was defined and evaluation could be performed with the same tools as the original *FiberCup*, such as the *Tractometer* (Côté et al., 2013) tool. It was used to train and test models by Neher et al. (2015, 2017). While too small to be used as a benchmark for ML models, the Simulated *FiberCup* is useful as a model selection tool or to perform quick "sanity checks" while training models (Fig. 2).

#### 2.1.3 HARDI 2013

The 2013 HARDI Challenge[c] (Caruyer et al., 2014) was organized for the 2013 ISBI conference. It focused on the local reconstruction of multiple simulated WM pathways. Both a training and a testing dataset were available, and the Tractometer tool was used to provide global metrics after applying a standard tractography algorithm on the local reconstructions of peaks or ODFs. Diffusion data was available in three sampling schemes (DTI-like, HARDI-like, and DSI-like).

The training dataset consists of 20 bundles, each containing a single fiber with a given radius, which is extremely limited in the context of data-driven tractography methods. Furthermore, the testing set provides no bundles or streamlines, only the

---

a. http://www.tractometer.org/original_fibercup/data/.

b. https://www.nitrc.org/frs/shownotes.php?release_id=2341.

c. http://hardi.epfl.ch/static/events/2013_ISBI/index.html.

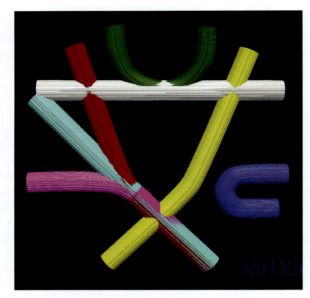

**FIG. 2** Illustration of the planar fiber configuration with seven individually colored bundles of the synthetic FiberCup replication used in Neher et al. (2017).

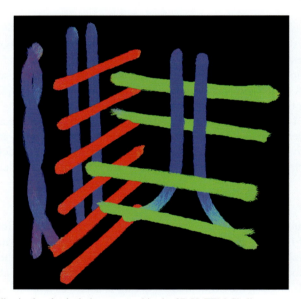

**FIG. 3** Illustration of the 16 fiber bundles in the physical phantom used in the 3D VoTEM Challenge.

ground truth orientations. Since the bundle endpoints are known, it would be possible to evaluate connectivity metrics for a candidate tractography algorithm, albeit in a limited way.

### 2.1.4 3D VoTEM phantom

The 2018 3-D VoTEM Challenge[d] offered a physical phantom with 16 bundles (see Fig. 3), along with ex vivo datasets of a macaque (with an atlas of known connections) and a squirrel monkey (with histological tracer) (Schilling et al., 2019). The ground truth of the phantom was defined as a manual delineation of the bundles, and so only voxel-wise metrics were computed (bundle overlap and overreach). On the other hand, for the macaque only spatial metrics (sensitivity and specificity) were used, while for the squirrel monkey both voxel-wise and spatial metrics were used.

---

d. https://my.vanderbilt.edu/votem/.

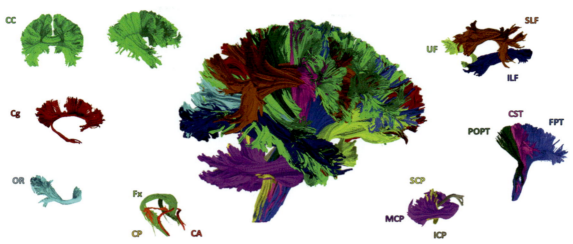

**FIG. 4** Illustration of the 25 tracts included in the phantom used for the ISMRM tractography challenge 2015 (Maier-Hein et al., 2017). *(Modified from Neher, P.F., Côté, M.-A., Houde, J.-C., Descoteaux, M., Maier-Hein, K.H., 2017. Fiber tractography using machine learning. NeuroImage 158, 417–429. https://doi.org/10.1016/j.neuroimage.2017.07.028, with permission from Elsevier.)*

### 2.1.5 ISMRM phantom

The 2015 ISMRM Tractography Challenge phantom is currently the most used benchmark for machine-learning tractography methods (Maier-Hein et al., 2017; Maier-Hein, 2015). To build the synthetic phantom, a realistic, clinical-like acquisition (including noise and artifacts) was simulated from a set of 25 bundles originally tracked and manually segmented from an HCP subject (see Fig. 4). The original data consists of a diffusion dataset with 32 $b = 1000\,\text{s/mm}^2$ images and one $b = 0\,\text{s/mm}^2$ image, with 2 mm isotropic resolution, along with a T1-like image with 1 mm isotropic resolution (Maier-Hein, 2015). Later, a high $b$-value and a high-resolution version of the phantom were published with the goal of reproducing an HCP-like protocol (Maier-Hein, 2017; Neher et al., 2017). Specifically, the newer dataset consists of 90 images for each $b$-value of 1000, 2000, and 3000 s/mm$^2$, with 1.25 mm isotropic resolution. Since the diffusion data was simulated, the ground truth is known and allows for a more straightforward evaluation of tracking methods than using a real dataset. All three datasets are available on zenodo.[e,f,g] However, the dataset consists of a single subject, which makes it impossible to use for both training and testing ML models without biasing the evaluation. Consequently, the training set was inconsistent across evaluation of different ML-based tractography methods, and the results are difficult to compare.

### 2.1.6 Random-configuration Fiberfox MRI simulations

The *Fiberfox* software (Neher et al., 2014) was also used to provide a public dataset of various simulated phantoms for tractography validation (Neher and Maier-Hein, 2019). In total, 16 phantoms are available, with varying parameters such as number of bundles (25 or 50), fiber density (50–500 streamlines per cm$^3$), bundle curvature (0–60 degrees), and bundle radius (5–30 mm). The dataset is available on zenodo.[h] Fig. 5 shows one of the randomly generated tract configurations.

While not representing human anatomy, the simulated phantoms can provide a way to quickly train and validate a data-driven architecture before moving to bigger and more realistic datasets.

### 2.1.7 The 99 simulated brains dataset

While the ISMRM 2015 Tractography Challenge phantom was used multiple times in the context of new data-driven tractography methods, it is limited in terms of only consisting of a single "subject." The new "99 simulated brains dataset"[i] can be seen as an extension of the ISMRM phantom to multiple subjects (Neher and Maier-Hein, 2020a). For 99 subjects from the HCP Young Adults database, 71 reference tracts were automatically generated using *TractSeg* (Wasserthal et al., 2018) and MITK

---

e. https://zenodo.org/record/572345.
f. https://zenodo.org/record/579933.
g. https://zenodo.org/record/1007149.
h. https://zenodo.org/record/2533250.
i. https://inrepo01.inet.dkfz-heidelberg.de/record/156611?ln=en.

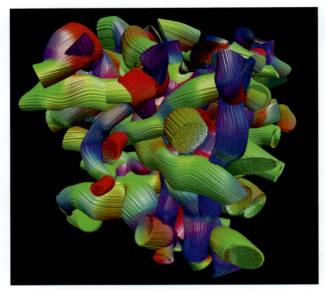

**FIG. 5** Exemplary phantom with randomized bundle configuration simulated using the Fiberfox software (Neher and Maier-Hein, 2019).

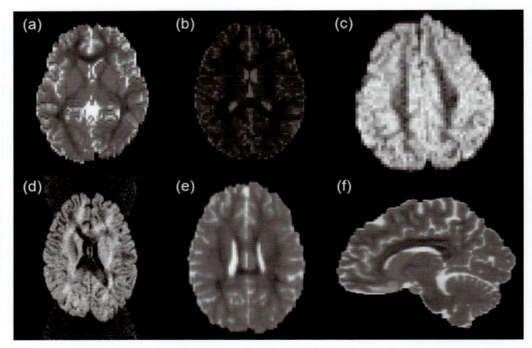

**FIG. 6** Illustration of simulated MR images with various animated artifacts (a bit excessive for illustration purposes): eddy current distortions (A), intensity drift (B), head motion and spike (C), head motion, eddy currents and noise (D), Gibbs ringing (E), inhomogeneity distortions (F).

Diffusion (Neher, 2020) (see Fig. 8 in the next section). Using these tracts with the *Fiberfox* MRI simulation software (Neher et al., 2014), realistic MR images were created with various scanner settings and artifacts (head motion, spikes, eddy currents, etc.—see Fig. 6). The dataset contains simulated MRI images (T1, T2, multiple configurations of dMRI) of 99 healthy subjects as well as corresponding reference tracts used for the simulation. Due to the large number of subjects, the dataset might be well suited for ML applications but since it has just been released recently, no publications have made use of it yet. To complement the full dataset, sample data of a single subject is also openly available and might be interesting for everyone who just wants to take a look without having to download the $\sim$400 GB of the full dataset (Neher and Maier-Hein, 2020b).

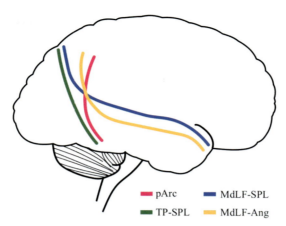

**FIG. 7** Schematic representation of four bundles connecting the posterior human cortex: the pArc *(pink)*, the TP-SPL *(green)*, the MdLF-SPL *(blue)*, and the MdLF-Ang *(yellow)* included in the HCP-minor bundle dataset. *(Reproduced with permission from Berto, G., 2020. HCP-Minor Bundle Dataset 2020. https://doi.org/10.25663/brainlife.pub.11.)*

## 2.2 In vivo human datasets

### 2.2.1 HCP-minor bundle dataset

The HCP-minor bundle dataset[j] is a collection of manually segmented white matter bundles connecting the human dorsal and posterior cortices (Berto, 2020) in 40 subjects of the HCP. The tractograms used for the manual bundle definition were created using probabilistic ensemble tractography (Takemura et al., 2016). Manual segmentations were obtained from 192 randomly selected HCP subjects using the procedure proposed by Bullock (2019). Successively, segmented bundles that were not considered plausible from the neuroanatomical point of view were removed with a semiautomatic technique remaining with 105 subjects. The actual number of subjects in the published dataset on brainlife.io is 40, which are the subjects for which all bundles exceeded a certain expert-defined quality score. The complete segmentation process is described in (Bertò, 2021). Fig. 7 illustrates the fiber bundles included in the dataset.

### 2.2.2 TractSeg

In order to train *TractSeg*, a deep-learning based WM bundle segmentation method (Wasserthal et al., 2018) (see also Section 3.3), Wasserthal and colleagues created a large-scale public tractography dataset (Wasserthal et al., 2017) available on zenodo.[k] It includes semiautomatic segmentations of 72 bundles, built on 105 subjects from the HCP Young Adults DWI database (Van Essen et al., 2013; Glasser et al., 2013). Their tractography and bundle segmentation pipeline consists of the following steps:

1. Tractography using Multi-Shell Multi-Tissue CSD (MRTrix) (Tournier et al., 2012).
2. Initial bundle extraction (TractQuerier (Wassermann et al., 2016)).
3. Bundle refinement (Manual ROIs (Stieltjes et al., 2013) + QuickBundles (Garyfallidis et al., 2012)).
4. Bundle-level quality control and cleanup.
5. Binary bundle mask generation.

Using the provided binary bundle masks, volume-oriented metrics can be computed, such as Dice score, but no streamline-oriented metrics are suggested by the authors, as *TractSeg* is a volume segmentation method. The *TractSeg* bundle database was also used in a subsequent paper to train a bundle-specific orientation mapping model (Wasserthal et al., 2019) (see Section 3.3; Fig. 8).

### 2.2.3 TractoInferno

TractoInferno is a multisite, multiprotocol tractography database of around 300 subjects, including both research and clinical-level human acquisitions, and aimed specifically at data-driven approaches. The provided preprocessed data

---

j. https://brainlife.io/pub/5e1de1371875e1ab6794cce5.
k. https://zenodo.org/record/1088278.

**322 PART | III** Tractography algorithms

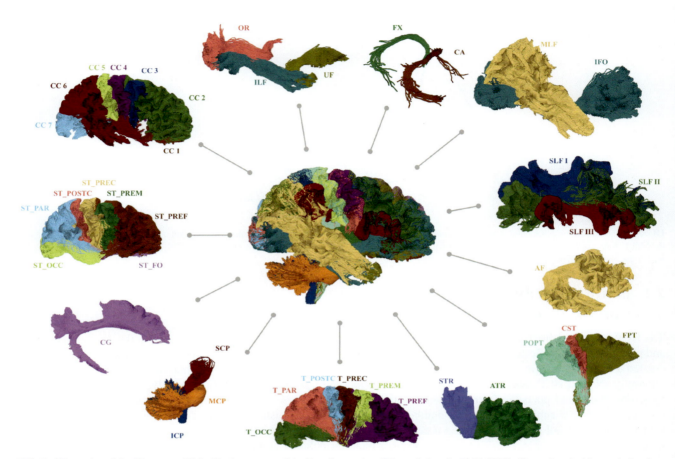

**FIG. 8** Illustration of the 72 tracts published in the context of the TractSeg project (Wasserthal et al., 2017, 2018). *(Reproduced with permission from Wasserthal, J., Neher, P., Maier-Hein, K.H., 2018. TractSeg—fast and accurate white matter tract segmentation. NeuroImage 183, 239–253. https://doi.org/10.1016/j.neuroimage.2018.07.070.)*

includes T1 and single-shell HARDI images of 3T scanners across seven sites (Philips, Siemens, and GE), with varying $b$-value (700 and 1000 s/mm$^2$), resolution (1.75, 2, and 2.3 mm), TR (7000–9200) and TE (70–93). It also contains precomputed ODFs and reference streamlines for 40 bundles. What sets this dataset apart is the fact that multiple tractography algorithms were used to generate the reference streamlines, with the idea being that different methods have different strengths and should all be included. Furthermore, this may avoid an induced bias linked to a specific tracking method.

To generate the reference streamlines, deterministic (Tournier et al., 2012) and probabilistic (Tournier et al., 2010) methods were used, along with particle-filtered tractography (Girard et al., 2014), bundle-specific tractography (Rheault et al., 2019), and surface-enhanced tractography (SET) (St-Onge et al., 2018). Furthermore, all methods used SET's surface flow to connect streamlines from the WM to the GM in an anatomically plausible way, and a streamline density subsampling process was used in order to reduce the density bias (Rheault et al., 2020a) that may affect data-driven algorithms. Specifically, the pipeline consists of the following steps:

1. Tractography with multiple independent algorithms
2. Completion of all streamline endings with *Surface Flow* (St-Onge et al., 2018)
3. Initial bundle extraction with RecoBundles (Garyfallidis et al., 2018)
4. Manual quality control and cleanup
5. Streamline density subsampling (Rheault et al., 2020a)

Using *TractoInferno*, data-driven algorithms can learn a model unbiased by a single tracking method, using a dataset as anatomically accurate as possible, but still large enough to allow large-scale learning. Furthermore, the multisite approach can also avoid overfitting related to a single-site dataset, which is a common problem with deep learning.

## 3 Current applications of machine learning in tractography

While the ML-based approach to tractography is still relatively young, various prototypes of ML-driven tractography have been presented over the last years. The differences between the individual approaches are very diverse but they can be categorized in the following groups: (1) ML-based local modeling approaches basically substitute the classic signal modeling step (tensor fit or similar) by an ML algorithm to predict the streamline progression from the data, but are otherwise relatively similar to the classic model-based streamline tractography; (2) sequence-based approaches (supervised or unsupervised) treat the streamline progression as a sequence processing task, e.g., using recurrent neural networks; (3) global approaches use ML to predict complete tracts without the iterative component of classic tractography; (4) streamline classification can be applied as a postprocessing step to filter tractograms independently of the employed tractography method.

The following sections give an overview over representative approaches published in these four categories, including their basic principles, the ML model they use, the data they work on, and how they were evaluated.

Since many of the presented models have been evaluated using Tractometer (Côté et al., 2013), particularly local and sequence-based approaches, we will quickly describe the tool and the used metrics here. *Tractometer* is used to evaluate candidate tractograms, using a bundle recognition algorithm to cluster the candidate streamlines into bundles (Garyfallidis et al., 2018), which are then compared against the ground truth bundles. The following metrics are used in the successive chapters, and are shown in Fig. 9:

- Valid bundles (VB): The number of correctly reconstructed ground truth bundles.
- Invalid bundles (IB): The number of reconstructed bundles that do not match any ground truth bundles.
- Valid connections (VC): The ratio of streamlines in valid bundles over the total number of produced streamlines.
- Bundle overlap (OL): The ratio of ground truth voxels traversed by at least one streamline over the total number of ground truth voxels.
- Bundle overreach (OR): The ratio of voxels traversed by at least one streamline that do not belong to a ground truth voxel over the total number of ground truth voxels.

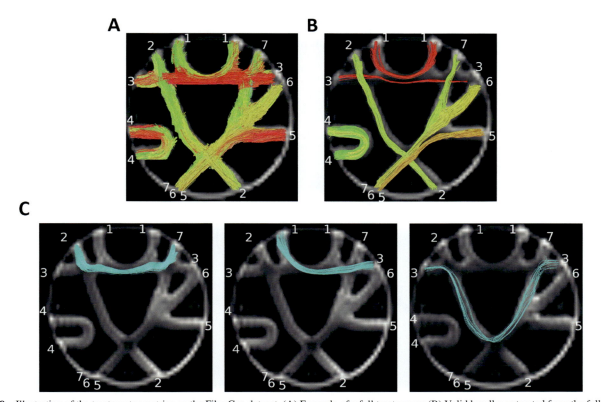

**FIG. 9** Illustration of the tractometer metrics on the FiberCup dataset. (A) Example of a full tractogram; (B) Valid bundles extracted from the full tractogram; (C) Invalid bundles extracted from the full tractogram. *(Reproduced with permission from Côté, M.-A., Girard, G., Boré, A., Garyfallidis, E., Houde, J.-C., Descoteaux, M., 2013. Tractometer: towards validation of tractography pipelines. Med. Image Anal. 17(7), 844–857. https://doi.org/10.1016/j.media.2013.03.009, with permission from Elsevier.)*

## 3.1 Local modeling

As described in Section 1, in the past virtually all tractography algorithms relied on mathematical modeling of the dMRI signal to obtain information about the voxel-wise microstructure. Recently, ML-based approaches have been introduced to mitigate limitations of the model-driven approaches, as described in the introduction. Such data-driven approaches to analyze the local microstructure and to estimate the progression direction for tractography have shown promising reconstructions of larger spatial extent of existing white matter bundles, promising reconstructions of fewer false positives, and promising robustness to known position and shape biases of current tractography techniques. On the following pages, we will describe the state of the art in these ML-based local modeling approaches for tractography.

### 3.1.1 Random forest-based local modeling

**Concept and methodology**: The first approach to employ ML for tractography was published by Neher et al., (2015, 2017). The authors used a random forest classification of the local raw diffusion-weighted signal values, in the form of the raw data or the coefficients of the spherical harmonics fit to the data, to predict the next progression direction used by the tractography algorithm to reconstruct a streamline. In this setup, 100 discrete directions are evenly distributed on the hemisphere over the plane defined by the previous progression direction. Each of these potential next progression directions is assigned a probability by the random forest classifier, facilitating deterministic as well as probabilistic tractography in a straightforward manner. The random forest-based prediction is illustrated in Fig. 10. Additionally to the local information, the method employs a sophisticated sampling scheme to exploit the information in the near neighborhood. This information is used to actively guide the streamline away from the white matter margins, thereby avoiding premature track terminations.

The previous streamline progression direction is used as a directional prior in the form of an additional input for the classifier. Furthermore, other image contrasts such as T1 or T2 anatomical image contrasts can be included as additional features in the classification process. As training data, the method employed tractograms obtained using a classic deterministic streamline approach.

**Evaluation**: The method was evaluated in silico and in vivo using a replication of the *FiberCup* phantom (Fillard, 2011) (see Fig. 2), the brain-like phantom used for the ISMRM tractography challenge 2015 (Maier-Hein et al., 2017) (see Fig. 4), as well as data from the human connectome young adult project (Van Essen et al., 2013) using multiple variations of

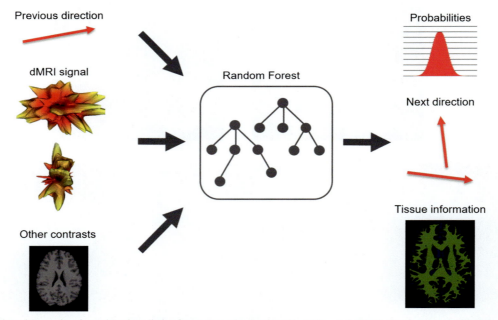

**FIG. 10** Illustration of the random forest-based prediction for tractography developed by Neher et al. (2017). Next to the raw diffusion-weighted signal in the form of the actual signal values or the coefficients of the corresponding spherical harmonics fit to the data, the approach uses the previous streamline direction and potentially additional image contrasts such as T1 or T2 weighted images as input for the random forest classifier. The output of the classifier assigns a probability to each of the 100 spherically distributed candidate directions, which enables the determination of the next progression direction in a successive step. Additionally, the approach implicitly provides tissue information by indicating if the streamline progression should proceed (i.e., white matter) or stop (i.e., outside of the white matter).

**TABLE 1** Results of the described (Neher et al. and Wegmayr et al.) local approaches on the ISMRM tractography challenge phantom as well as of two sequence-based approaches (Poulin et al. and Benou et al.) including information about the respective training data used.

| Model | Training data | VB | IB | VC (%) | OL (%) | OR (%) |
|---|---|---|---|---|---|---|
| Neher (Neher et al., 2017) | 5× HCP | 23/25 | 94 | 52 | 59 | 37 |
| NN$_{iFOD1}$ (Wegmayr et al., 2018) | 1× HCP | 22/25 | 75 | 34 | 14 | 29 |
| NN$_{iFOD3}$ (Wegmayr et al., 2018) | 3× HCP | 23/25 | 57 | 72 | 16 | 28 |
| Entrack (Wegmayr et al., 2019) | 1× HCP | 23/25 | 85 | 51 | 23 | 39 |
| Poulin (Poulin et al., 2017) | ISMRM phantom | 23/25 | 130 | 42 | 64 | 35 |
| DeepTract (Benou and Riklin Raviv, 2019) | ISMRM phantom | 23/25 | 51 | 41 | 34 | 17 |
| Track-to-Learn (Théberge et al., 2021) | – | 23/25 | 161 | 68 | 56 | n/a |

The scores are introduced in Section 2.1. It is apparent that none of the approaches yields the best result in all categories. It is further difficult to estimate whether the differences are attributable to the method itself or to a certain parameterization, since they were not analyzed jointly in a larger study.

training and test data. The quantitative results of the method trained on in vivo HCP data and tested on the ISMRM tractography challenge phantom (see Table 1) indicate that the method is competitive, but no game changer in terms of false positive reduction and bundle completeness, which are major issues in current tractography pipelines, as described in detail in Maier-Hein et al. (2017). Remarkably, the method outperformed the tractogram that was used for training.

**Availability**: The complete approach, training, and execution is openly available as source code[1] and as an executable application (Windows and Linux) included in MITK Diffusion (Neher, 2020). In MITK Diffusion, the method is embedded in the general streamline tractography framework with additional features such as anatomical constraints, directional priors, and interactive tractography. Trained models are not available but the tools required for training are included. The datasets used for training in the original publication are openly available.

### 3.1.2 Neural network based local modeling

**Concept and methodology**: In 2018, Wegmayr et al. presented their work on neural network regression for tractography (Wegmayr et al., 2018). Similar to Lazar et al. (2003), the basic idea is to predict the streamline progression direction of the tractography method on the basis of the local diffusion-weighted signal using ML. Instead of a random forest classifier, this work regresses the progression direction with a multilayer perceptron (MLP). It further directly includes the local neighborhood ($3 \times 3 \times 3$ by default) in the regression process instead of processing multiple samples of the local neighborhood independently and successively integrating the individual sampling result in a rather complex voting process. As in Neher et al. (2017), the previous streamline directions are used as additional input for the regression and serve as directional prior. Training is performed using a teacher tractography based on conventional model-based tractography. In this work the authors chose probabilistic CSD-based tractography implemented in MRtrix to create the teacher tractogram instead of the deterministic CSD tractography used in Neher et al. (2017).

Wegmayr and colleagues later extended their neural network regression for tractography from a deterministic to a probabilistic model called *Entrack* (Wegmayr et al., 2019). Instead of only predicting a single direction that is then used for the next progression step, the neural network now locally predicts the two parameters, mean direction and scalar concentration, of a Fisher-von-Mises (FvM) distribution, which is a unimodal directional distribution on the sphere (see Fig. 11). The FvM enables modeling of the aleatoric uncertainty of the measured data, i.e., the uncertainty that is intrinsic to each physical observation and cannot be eliminated completely. The predicted distribution can then be used to locally sample track directions for the probabilistic streamline tractography process.

Since neural networks are models with a high complexity, they tend to overfit without a proper regularization. This overfitting is visible in a loss that goes towards zero while at the same time driving the uncertainty to zero, i.e., the concentration parameter of the FvM towards infinity. In this case the probabilistic system collapses to a deterministic one. In

---

1. https://github.com/MIC-DKFZ/MITK-Diffusion/.

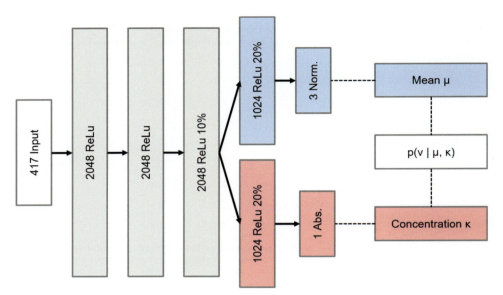

**FIG. 11** Neural network layers of Entrack. Inputs *(white box)* flow from left to right, first through shared layers *(gray boxes)*, before the two output heads for the mean *(blue boxes)* and concentration *(red boxes)* branch off. All layers are fully connected *(arrows)*. Numbers indicate the number of hidden units, followed by the layer's activation function (*ReLU*, rectified linear unit; *Norm*, normalization to unit length; *Abs*, absolute value), and dropout percentage, if applicable.

order to mitigate this issue and to enable a more robust model, the authors propose a maximum entropy regularization, the strength of which is controlled by a parameter $T$ that should reflect the noise level of the data and is determined in practice using cross-validation. They show in their experiments that the method indeed yields optimal results with $T > 0$.

An important aspect of *Entrack* is its ability to not only perform probabilistic tractography but to enable the detection and filtering of outliers since their model provides a proper probability density over tracks.

**Evaluation**: Experiments similar to the ones by Neher et al. (2017) were conducted, making the results partly comparable to the random forest-based approach. The deterministic neural network-based tractography method ($NN_{iFOD3}$) was trained on three in vivo HCP datasets and successively applied to the ISMRM tractography challenge 2015 phantom (Maier-Hein et al., 2017). The experiments showed that the deterministic neural network regression resulted in an increased specificity (in terms of less invalid bundles and more valid connections) at the cost of a decreased sensitivity (in terms of bundle completeness (overlap/overreach)) in comparison to the state of the art (see Table 1).

The probabilistic extension (*Entrack*) was evaluated with a similar setup and the results are comparable to the other state-of-the-art methods. In general, no clear tendency about any method advantages or disadvantages could be shown (see Table 1).

**Availability**: *Entrack* is open-source and available on github.[m] The python-based approach is usable on Ubuntu systems. Trained models are not available. The datasets used for training in the original publication are openly available.

### 3.1.3 Assessing the quality of neural network-based local track direction predictions

On a more general level and to abstract from the actual process of iterative track progression, Jörgens et al. conducted a study on how well a single step of streamline progression can be learned using neural networks independent of the actual tractography (Jörgens et al., 2018). The authors compare different aspects that have an effect on the result apart from the actual ML model, such as using classification or regression, using different data sampling strategies such as including neighborhood information directly in the training, only as a postprocessing step or not at all, as well as different postprocessing methods for the classification-based prediction.

The classification postprocessing consists of determining the output direction as (1) the single most likely direction, (2) probability-weighted average of all possible directions, (3) the average of all possible directions weighted with their respective probability and multiplied with the scalar product between the respective direction and the previous direction, or (4) same as (3) but at multiple neighborhood locations and a successive averaging.

---

m. https://github.com/vwegmayr/entrack.

The assessed sampling strategies comprised interpolating the image data at the current location without including neighborhood information (strategy A), interpolating the image data at eight corner points of a neighborhood cube and concatenating the data as network input (strategy B), and using a separate input layer for each corner point (B*). Each strategy was analyzed using nearest neighbor as well as linear data interpolation, resulting in a total of six network configurations.

Further, the authors analyzed the effect of varying the number of possible directions in the classification setup (100, 200, and 500) as well as using one or two previous directions as additional input for the network.

All experiments were performed using the ISMRM tractography challenge phantom and the best overall result achieved based on the average angular prediction error was approximately 3 degrees. In general, an increased number of classes lead to a decreased prediction error and classification in conjunction with postprocessing strategy (2) yielded similarly good results as a regression-based approach. The analysis of different neighborhood sampling strategies yielded no conclusive results. The inclusion of two instead of only one previous streamline direction in the prediction on the other hand clearly improved the prediction performance, which hints at the importance of incorporating curvature information in the prediction. Nevertheless, the distribution of errors with respect to the track curvature as well as the signal FA has yet to be thoroughly analyzed to exclude a sampling bias.

## 3.2 Sequence-based modeling

A likely shortcoming of local models is that they ignore possible long-term dependencies between an existing streamline path and its next direction. Indeed, the path of a streamline is very context-dependent, as shown by common tractography issues such as bottlenecks, wall effects, or narrow intersections (Rheault et al., 2020a) (see also Fig. 1). Thus some kind of sequential context might be needed in order to reduce the number of false positives resulting from those problematic regions.

Most ML-based local models propose to use a fixed sequential context of the four last streamline directions. This seems to be a good heuristic, but there is no evidence yet that indicates it is the appropriate length, or how varying the tracking step size might affect how much information is needed. For this purpose, recurrent neural networks (RNNs) provide a valuable approach to "store" information in a memory along a sequence of inputs. Thus they offer a data-driven way of incorporating long-term streamline dependencies without imposing a fixed-length context.

Specifically, sequence-based modeling suggests that diffusion data is sampled along the input streamline and fed into the network at every step. The RNN is trained to use the past information of a streamline to predict the next direction, and at test time the predicted directions are iteratively fed back into the network to build new streamlines.

### 3.2.1 Deterministic-output RNN

**Concept and methodology**: In 2017, Poulin et al. (2017) proposed using an RNN to learn the tracking process. Specifically, a gated recurrent unit (GRU) was used to learn a mapping from the diffusion signal along a streamline path to the next orientation of the streamline. As such, the model uses the spatial context of every previous step of a streamline to predict the next step. The output of the model is a 3D unit vector representing the next orientation of the streamline (as shown in Fig. 12). Consequently, it does not allow for probabilistic tracking, only deterministic.

**Evaluation**: The *2015 ISMRM Tractography Challenge* dataset was used to train, validate, and test the proposed model. Multiple architectures were tried, varying between one and four stacked GRU layers of either 500 or 1000 units, and the one with the best validation error was kept for testing.

The results show that a recurrent model is competitive (see Table 1 in Section 3.1), but it definitely did not "solve" the tractography problem. Furthermore, results were somewhat biased as the model was trained using the diffusion signal of the same subject that was used for testing (the challenge was intended for traditional algorithms and provided only one subject). While the recurrent model's performance was not state of the art, it showed that it has the capacity to learn and model this complex mapping from diffusion measurements to streamlines.

**Availability**: The code is available online,[n] but dates back to 2017, uses both Python 2 (for scoring) and 3 (for training and tracking), and depends on *Theano*, a discontinued ML library. Pretrained models are not provided. The original dataset is not provided.

---

n. https://github.com/ppoulin91/learn2track.

**328 PART | III** Tractography algorithms

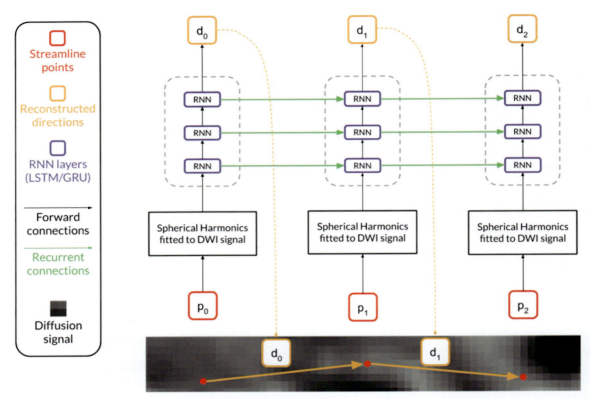

**FIG. 12** The recurrent model gets fed diffusion information for every step of the streamline, and predicts the 3D orientation of the next step given all the previous steps. Past information flows forward through the recurrent connections in order to be used for future predictions.

### 3.2.2 Probabilistic-output RNN

**Concept and methodology**: Building on the deterministic-output RNN, Benou et al. (Benou and Riklin Raviv, 2019) suggested that instead of predicting a single 3D vector, the network could predict a probability distribution over the sphere, thus facilitating both deterministic and probabilistic tractography. To this end, the tractography task is treated as a classification problem in which the unit sphere is discretized into 724 classes, corresponding to 724 evenly distributed directions. Note that unlike local probabilistic models, the output probability distribution is conditioned by the "history" of inputs along the input streamline, in the attempt to guide the model through complex fiber configurations. To account for this discretization, the target directions then become target classes. In addition, the authors suggest smoothing the output distribution using a Gaussian kernel to account for the geometric structure between nearby classes.

The training and tracking processes are illustrated in Fig. 13. The authors fixed the RNN architecture to five stacked GRU layers, with ReLU activations for all layers but the last one, which is a fully connected layer followed by a softmax. Furthermore, they incorporate an entropy-based tracking termination criteria (i.e., if the predictions are too isotropic, the tracking process should stop), which eliminates the need for a tracking mask.

**Evaluation**: Like (Poulin et al., 2017), the model is trained and tested on the *2015 ISMRM Tractography Challenge* subject, using either the challenge's ground truth streamlines or reference streamlines generated by a probabilistic algorithm as training data. Results show that *DeepTract* is competitive with other ML-based models evaluated on the ISMRM challenge (see Table 1 in Section 3.1). Particularly, it demonstrated good performance in terms of specificity (i.e., lowest overreach and number of invalid bundles), but to some extent at the expense of its sensitivity (i.e., not the highest overlap and valid connection scores).

This paper provides us with more insight into what kind of capacity might be needed to learn the tractography process using a trained model, and shows that we might be able to address the issues related to traditional probabilistic tracking. However, we still need more comprehensive experiments on both larger databases and real acquisitions to truly understand the potential of data-driven tractography.

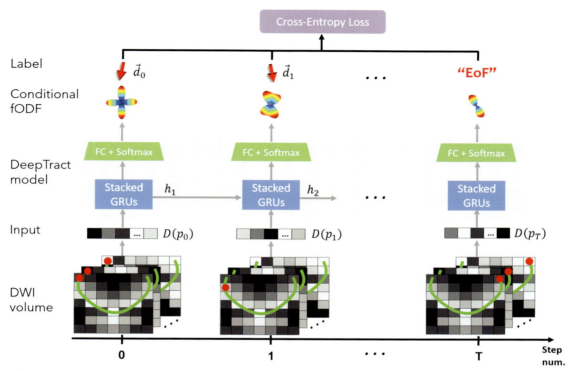

FIG. 13  The DeepTract algorithm, unrolled in time. At each step, diffusion information is fed to the network to produce conditional fODF estimations that are compared to the actual streamline directions through a cross-entropy loss. After training, tracking is performed by sequentially sampling directions from the predicted conditional fODFs. *(Modified from Benou,I. Riklin Raviv, T., 2019. DeepTract: a probabilistic deep learning framework for white matter fiber tractography. In: Medical Image Computing and Computer Assisted Intervention—MICCAI 2019, Cham, pp. 626–635. https://doi.org/10.1007/978-3-030-32248-9_70, with permission from Springer Nature.)*

**Availability**: The code is openly available on github.[o] Pretrained models are not provided. The original dataset is based on the ISMRM 2015 Tractography Challenge ground truth, which is publicly available.

### 3.2.3 Deterministic-output bundle-specific RNN

**Concept and methodology**: Using the same GRU building blocks as in Poulin et al. (2017) and Benou and Riklin Raviv (2019), Poulin et al. (2018) approached the tractography problem using a bundle-specific method. By training a different model for each individual bundle, it allows each one to "overfit" its corresponding bundle (which is a desirable trait in this case). The downside is that each model can only be used in the appropriate bundle region, as the predictions would be unreliable anywhere else. However, by focusing on a single bundle, each model has an easier task to solve.

There are multiple preliminary tasks in order to train a bundle-specific model. First, all reference streamlines must be segmented into bundles so that multiple models can be trained separately. Then, after training, new subjects need reference bundle masks to be already defined (or at least as a general area), using either manual volume segmentation or using a white matter atlas, so that each model is restricted to its own predefined bundle area. If these prerequisites are fulfilled, then bundle-specific models can be trained and used to reconstruct learned bundles on new subjects.

**Evaluation**: Using this bundle-specific approach, the authors present significant improvements over traditional tracking algorithms as seen in Table 2. Additionally, the dataset consisted of human acquisitions from 37 subjects, split into separate training and testing sets, showing that the recurrent architecture is appropriate for real data. The model architecture consisted of a single GRU layer with 512 units, and the reference streamlines were generated by a state-of-the-art, bundle-specific, probabilistic algorithm and semiautomatically segmented by a neuroanatomist expert.

Learning to track a single bundle is a much simpler task than whole-brain tractography, but given the impressive results, it might indicate that we don't yet have the necessary capacity or datasets for accurate whole-brain tractography.

**Availability**: The code or dataset are not publicly available.

---

o. https://github.com/itaybenou/DeepTract.

**TABLE 2** Evaluation of the Bundle-Wise Deep Tracker against deterministic and bundle-specific methods, using a test dataset of 12 human subjects to compute average and standard deviations for multiple metrics.

| Bundle | Method | Number of streamlines | | | Volume (mm$^3$) | | | Valid ratio (%) | | | Efficiency ratio (%) | | | Weighted Dice | | |
|---|---|---|---|---|---|---|---|---|---|---|---|---|---|---|---|---|
| AF left | Det | 1570 | ± | 381 | 17044 | ± | 2791 | 1.34 | ± | 0.33 | 11.00 | ± | 2.46 | 0.41 | ± | 0.09 |
| | BST-Det | 9623 | ± | 1586 | 37847 | ± | 2855 | 8.34 | ± | 1.24 | 45.42 | ± | 5.17 | 0.62 | ± | 0.07 |
| | BST-PFT | 17743 | ± | 2501 | 106634 | ± | 8460 | 9.59 | ± | 1.17 | 31.33 | ± | 3.72 | **0.85** | ± | **0.03** |
| | BWDT | **62223** | ± | **7080** | **122145** | ± | **5034** | **27.98** | ± | **2.03** | **64.52** | ± | **3.06** | 0.85 | ± | 0.02 |
| AF right | Det | 916 | ± | 326 | 10313 | ± | 3061 | 0.95 | ± | 0.34 | 8.07 | ± | 3.87 | 0.29 | ± | 0.13 |
| | BST-Det | 5379 | ± | 1905 | 24156 | ± | 4852 | 5.75 | ± | 2.01 | 34.87 | ± | 10.90 | 0.51 | ± | 0.12 |
| | BST-PFT | 9557 | ± | 2658 | 81231 | ± | 11008 | 6.08 | ± | 1.66 | 18.68 | ± | 5.07 | 0.80 | ± | 0.05 |
| | BWDT | **53906** | ± | **5817** | **115053** | ± | **6048** | **27.52** | ± | **3.42** | **57.94** | ± | **6.74** | **0.84** | ± | **0.02** |
| CC | Det | 2797 | ± | 916 | 19433 | ± | 4962 | 4.31 | ± | 1.26 | 27.63 | ± | 5.37 | 0.59 | ± | 0.09 |
| | BST-Det | 9379 | ± | 2281 | 33419 | ± | 6324 | 13.48 | ± | 2.61 | 44.00 | ± | 5.13 | 0.68 | ± | 0.08 |
| | BST-PFT | 12230 | ± | 3049 | **85830** | ± | **7199** | 11.31 | ± | 2.40 | 28.37 | ± | 5.55 | **0.86** | ± | **0.05** |
| | BWDT | **26746** | ± | **5170** | 66264 | ± | 6721 | **20.94** | ± | **3.62** | **46.20** | ± | **6.38** | 0.79 | ± | 0.04 |
| CST left | Det | 419 | ± | 428 | 3499 | ± | 1844 | 1.10 | ± | 1.07 | 9.52 | ± | 7.83 | 0.41 | ± | 0.20 |
| | BST-Det | 1580 | ± | 1111 | 6511 | ± | 2258 | **4.25** | ± | **2.88** | **20.90** | ± | **11.55** | 0.58 | ± | 0.16 |
| | BST-PFT | 2384 | ± | 1103 | **28468** | ± | **7185** | 3.76 | ± | 1.68 | 11.58 | ± | 4.97 | **0.85** | ± | **0.06** |
| | BWDT | **3153** | ± | **1675** | 14290 | ± | 5996 | 3.78 | ± | 2.07 | 14.91 | ± | 8.30 | 0.54 | ± | 0.14 |
| CST right | Det | 396 | ± | 315 | 3603 | ± | 1815 | 1.01 | ± | 0.80 | 9.54 | ± | 7.49 | 0.35 | ± | 0.71 |
| | BST-Det | 1498 | ± | 1039 | 6661 | ± | 2755 | **3.89** | ± | **2.73** | **19.81** | ± | **13.55** | 0.48 | ± | 0.18 |
| | BST-PFT | 1683 | ± | 749 | **25243** | ± | **6271** | 2.49 | ± | 1.12 | 7.54 | ± | 3.20 | **0.83** | ± | **0.06** |
| | BWDT | **2490** | ± | **1921** | 13422 | ± | 6575 | 2.78 | ± | 2.20 | 15.85 | ± | 12.31 | 0.47 | ± | 0.14 |

*BST-Det*, bundle-specific tractography with deterministic tracking; *BST-PFT*, bundle-specific tractography with probabilistic particle-filtered tracking; *BWDT*, bundle-wise deep tracker; *Det*: deterministic. Bold face values indicate highest score per bundle.
From Poulin, P., Rheault, F., St-Onge, E., Jodoin, P.-M., Descoteaux, M., 2018. Bundle-Wise Deep Tracker: Learning to Track Bundle-Specific Streamline Paths.

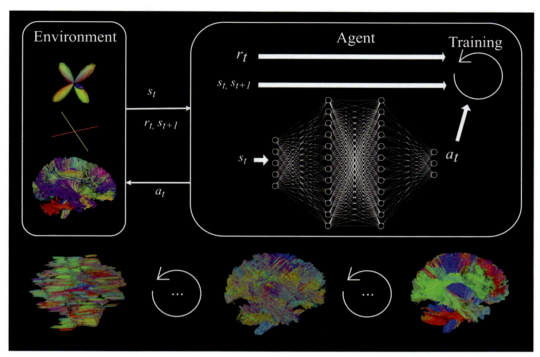

**FIG. 14** The reinforcement-learning loop as well as the evolution of produced tractograms. Top: The environment produces a state and a reward from the diffusion signal input. The reward is computed from the last tracking step and the peaks extracted from the fODFs at the streamline's head. On the basis of the state, the network proposes a new tracking step, and the reward is used for training. Bottom: Tractograms produced by agents before training begins, during training and at the end of training. *(Reproduced with permission from Théberge, A., Desrosiers, C., Descoteaux, M., Jodoin, P.-M., 2021. Track-to-Learn: a general framework for tractography with deep reinforcement learning. Med. Image Anal. 102093. https://doi.org/10.1016/j.media.2021.102093, with permission from Elsevier.)*

### 3.2.4 Deterministic-output RNN using reinforcement learning

**Concept and methodology**: All previous sequence-based methods use *supervised learning*, which means that annotated data has to be provided to the model in order to learn to reproduce it as well as possible. However, *reinforcement learning* is a call of learning that does not require annotated data, instead relying on *rewards* defined by a *reward function*. Reinforcement learning has proven to be a great way to explore the space of possibilities while learning how to accomplish a task. While defining the reward function is no easy task in the context of tractography, early work by Théberge et al. (2021) shows that it can be used successfully. Instead of reference streamlines, they define a reward function based on the alignment with the underlying fODF peaks and the length of the produced streamlines. The concept is illustrated in Fig. 14.

**Evaluation**: The method is trained and evaluated both in silico and in vivo, using three datasets: the Simulated Fibercup dataset, the ISMRM 2015 Tractography Challenge dataset, and a few subjects from the HCP Young Adults database. The Tractometer tool was used to compute performance metrics, and shows competitive results, but also shows that trained models can suffer from *reward hacking* (the model finds a way to maximize its rewards that isn't aligned with the initial goal), and need to be carefully monitored during training.

**Availability**: The datasets used in the paper are publicly available, but the code is yet to be published.

### 3.3 Global modeling

In local and memory-based modeling the progression procedure of classical streamline tractography is modeled by a general trainable mapping. In contrast to that, global methods try to estimate a streamline, or better, a bundle (as a set of streamlines), at once. Such an approach makes the track selections protocols needed in classical non-ML approaches superfluous. It can help to objectify the bundle selection and set a common ground for definitions of "gold standards." Recent research (Li, 2020) also shows that the repeatability of streamline count and volume measures can be improved by such an approach. The system learns the robustness against imaging artifacts and partial volume effects, which are common for diffusion-weighted imaging and usually disturbing classical tractographic approaches. Practically, the problem is formulated as a predictor for bundle-specific evidence maps, rather than a direct estimation of a streamline as a set of coordinates. When such evidence maps are scalar-valued, the approach may just be seen as a traditional segmentation task, where the evidence value tells how likely it is that a certain voxel is part of the bundle of interest. But evidence can also contain orientational information as in *HAMLET* (Reisert et al., 2018) or being extended by orientational information like in *TOM* (Wasserthal et al., 2019). In the following we give an overview of the existing methods.

### 3.3.1 CNN-based bundle modeling

**Concept and methodology**: There have already been a variety of approaches proposed that formulate white matter tractography as a simple direct segmentation problem. These methods are based on rather classical image-processing methods like template matching (Eckstein, 2009), Markov random field optimization (Bazin, 2011), geometric flow-based segmentation (Jonasson et al., 2005; Guo et al., 2008), surface evolution (Aganj et al., 2009; Descoteaux and Deriche, 2009), and k-NN based classification (Ratnarajah and Qiu, 2014). In 2018, Wasserthal et al. proposed to use deep concurrent neural networks to segment white matter tracts, named *TractSeg* (Wasserthal et al., 2018). A series of deep 2D U-net type networks are trained, which are applied in all three image slicings. Predictions of the slicing-associated networks are again combined by a set of CNNs. The input data is presented to the network as a set of the three most prominent directions per voxel estimated by a multishell CSD approach (Tournier et al., 2007; Jeurissen et al., 2014). By doing so, the model is robust against the actual diffusion-weighting protocol, which is an advantage. But also memory consumption is less compared to the direct use of the raw data. During training, heavy data augmentation is applied (rotation, elastic deformation, zoom, noise, contrast and brightness changes).

Wasserthal et al. later extended *TractSeg* to also predict orientations, i.e., additional to the binary decision a 3D vector is trained for every specific bundle as additional output channels (Wasserthal et al., 2019). Such tract orientation maps (TOMs) can then be directly used for simple bundle specific tractography or as prior for tractography, e.g., using MITK diffusion (Neher, 2020). Instead of using binary cross entropy as a loss function, a cosine loss is used to train the directions. Fig. 15 shows an overview over the process.

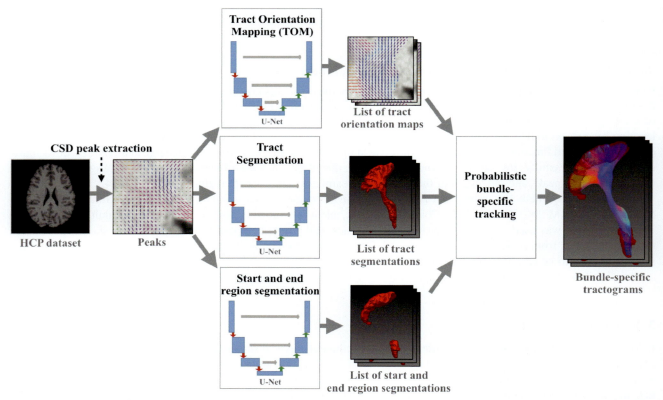

**FIG. 15** Overview of TractSeg/TOM approach. The local maximas of the FOD are stacked as triples and fed into U-net like architectures. Independently three networks are trained to learn tract presence, tract orientation, and terminal regions of the bundle of interest. Based on the prediction bundle, specific tractography can be performed. *(Reproduced with permission from Wasserthal, J., Neher, P.F., Hirjak, D., Maier-Hein, K.H., 2019. Combined tract segmentation and orientation mapping for bundle-specific tractography, Med. Image Anal. 58, 101559. https://doi.org/10.1016/j.media.2019.101559.)*

*TractSeg* predicts probability maps, binary segmentations of bundles as well as bundle endpoints, and TOMs of 72 white matter bundles.

**Evaluation**: *TractSeg* was evaluated with respect to various reference approaches (TractQuerier, WMA, RecoBundles, TRACULA, and various atlas-based approaches implemented in DIPY and MRtrix (Tournier et al., 2012; Wassermann et al., 2016; Garyfallidis et al., 2018; Yendiki, 2011; Garyfallidis, 2014)), and showed superior performance on the dataset proposed in Section 2.2. For a subset of tracts (CST, CA. IFO) it was also shown that the trained models can be transferred to other publicly available datasets. Problems appeared for small, thin structures like the anterior commissure (CA). The problems appear on high-quality (HCP) as well as on low-quality data and seem to be inherent, probably also caused by insufficiently accurate training data for these cases. The approach was evaluated on high-quality HCP data, on sub-sampled versions of these high-quality datasets to mimic a clinical-style acquisition as well as on simulated datasets created using *FiberFox* (Neher et al., 2014).

**Availability**: The dataset of 72 bundle dissections for 105 HCP subjects is openly available[p] (see Section 2.2). *TractSeg* is openly available on github, as a python package or docker container with pretrained model weights.[q]

### 3.3.2 Rotation covariant tract estimation

**Concept and methodology**: In *TractSeg* the connection between 3D space and the tensorial nature of the diffusion weighted imaging data is basically neglected. The feature dimension of the CNN has no geometrical structure and the rotational covariance properties have to be learned from data by heavy augmentation, which comes together with an enormous number of model parameters. Reisert et al. proposed an alternative approach, which incorporates the geometric character and guarantees Euclidean covariance of the finally trained model (Reisert et al., 2018). The approach (named *HAMLET*, Hierarchical hArMonic filters for LEarning Tracts) is similar to a CNN. It also consists of repeated applications of convolutions and nonlinearities; however, they all respect rotations in the way that if the input rotated the output rotates accordingly. The mathematics behind it is based on the so-called spherical tensor algebra. This intrinsic treatment of image rotation allows training without any additional data augmentation and with model complexities much less compared to what is necessary with ordinary, but much more generic, CNNs (see Fig. 16 for a sketch of the approach).

**Evaluation**: For evaluation the authors of *HAMLET* investigated the reproducibility and found *HAMLET* based bundle volumes more reproducible than a classical streamline counting approach based on global tractography (Reisert et al., 2011).

**Availability**: The code is based on the STA toolbox (Skibbe and Reisert, 2017), which is openly available on bitbucket.[r] The HAMLET code is available from the authors upon reasonable request.

## 3.4 Streamline classification

Depending on the context, not all streamlines contained in a given tractogram are of interest for the user. This might be due to the fact that current tractography methods can produce a fraction of false-positive streamlines (Maier-Hein et al., 2017), or that the application at hand focuses on particular white matter bundles (Chandio, 2020). Methods allowing for the selection of a subset of streamlines from a given tractogram are useful in such cases. Note that, as opposed to the methods introduced in Sections 3.1 and 3.2, these approaches work with *complete* streamlines as input.

In the context of ML, it is natural to formulate this streamline selection as a classification problem. Although classical approaches for tractogram filtering (Daducci et al., 2015; Smith et al., 2013, 2015; Schiavi, 2020; Pestilli et al., 2014), streamline clustering (defining clusters of coherent streamlines w.r.t. a chosen measure like distance) (Garyfallidis et al., 2012; Siless et al., 2018) or bundle segmentation (extraction of streamlines resembling the volume of a particular WM bundle) (Garyfallidis et al., 2018; Zhang, 2018) can follow various different strategies, their ML-based counterparts usually address a *streamline classification* problem in which each streamline is regarded as an individual sample.

In this section we give an overview of approaches for streamline classification that build on ML. While various flavors of model types have been employed, most build either directly on the three coordinates of sampling points along the streamlines or geometrical features derived thereof as input. Interestingly, the majority of published work has focused on the assignment of streamlines to white matter bundle labels. This reflects the inherent uncertainty about the ground truth in tractography that renders the definition of reliable binary streamline labels for tractogram filtering challenging.

---

p. https://doi.org/10.5281/zenodo.1088277.

q. https://github.com/MIC-DKFZ/TractSeg/.

r. https://bitbucket.org/skibbe/sta-toolbox/wiki/Home.

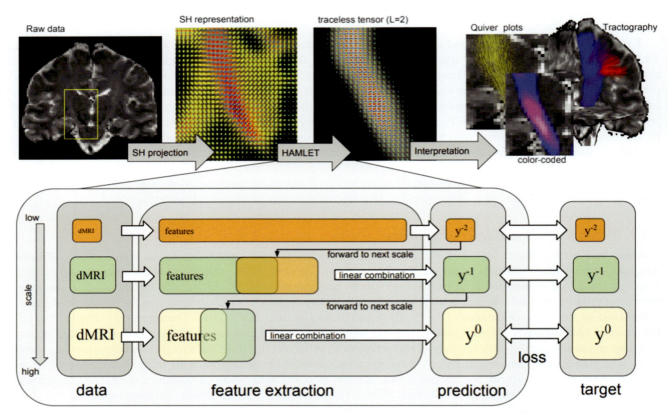

**FIG. 16** Overview of HAMLET approach. The network takes a $L=2$ spherical harmonic representation as input, which is equivalent to a traceless rank-2 Cartesian tensor. All operations of the hierarchical filter are rotations covariant, so its output, which is again an $L=2$ spherical tensor and contains all necessary information for a preceding bundle specific tractography. The SH-power of the output serves as evidence, the principal direction as orientation of the bundle.

In the following paragraphs, we group the reviewed methods based on their prediction targets (see Fig. 17), i.e., either individual labels for different white matter bundles ("Bundle prediction"), a binary classification label that could be used for acceptance or rejection of streamlines ("Dichotomous plausibility label"), or generic labels that are not explicitly defined in the model output ("Label transcription approaches").

### 3.4.1 Bundle prediction

Predicting the matching white matter bundle for each individual streamline sample is a natural formulation of the streamline classification problem. While sampling streamlines from well-defined white matter bundles is a feasible task, the definition of a rejection class, i.e., the set of streamlines that do not belong to any of the predicted bundles, is a rather challenging problem. Note that this class includes samples from bundles that are not included in the prediction labels, as well as from the set of streamlines that are artifacts of the tractography process.

#### Multiple bundle labels

**Concept and methodology**: In the simplest formulation of the classification problem, only a set of supported bundle labels is defined. With such an approach it is possible to divide a set of streamlines into clusters belonging to particular bundles. However, it is assumed that the input streamlines belong to one of the supported bundles, i.e., other bundles and false-positive streamlines are not supported in the prediction.

The tool *FiberNet* (Gupta et al., 2017a) implements this strategy building on a CNN architecture. First, a conformal mapping is employed to define a subject-specific coordinate system of the brain white matter, which accounts for individual subject anatomy. From that, spherical coordinates of the sampling points along a streamline are used as input to the CNN classifier. The final prediction is obtained from an ensemble of 20 models, each trained on a random subset of 600k streamlines from a training dataset of 4 subjects with 17 manually segmented bundles each.

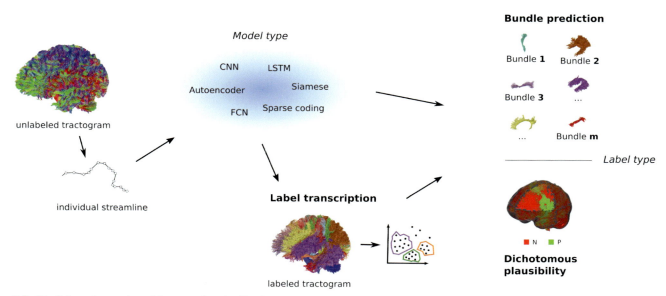

FIG. 17 Schematic overview of the streamline classification process. Individual streamline samples are assigned either to a bundle or a binary label for tractogram filtering. Different types of models have been employed to predict these labels directly as output or indirectly through a labeled template when applied.

Another idea, explored in Kumar et al. (2019) and Wu, (2020), is to assign streamlines to bundle labels based on a set of learned dictionaries that allow for a sparse representation of the streamlines in a particular bundle. In Wu (2020), it is proposed to assign an unseen streamline to the bundle that minimizes a certain cost function based on the sparse representation of that streamline in the dictionary of each considered bundle. Their method termed *TractDL* is trained on 70 subjects of the *TractSeg* bundle database to perform prediction of bilateral bundle labels, i.e., it does not distinguish between hemispheres.

**Evaluation**: Through analysis of a confusion matrix from the predictions over streamlines from 92 subjects, *FiberNet* shows a close resemblance to the streamline clustering method *autoMATE* (Jin, 2014) in terms of predicted streamline labels.

*TractDL* outperformed *RecoBundles* in terms of streamline-based accuracy in 35 subjects of the *TractSeg* bundle database (Wu, 2020).

**Availability**: The source code of *FiberNet* is available on github.[s]

## Multiple bundle labels and one rejection class

**Concept and methodology**: Augmenting the set of multiple bundle classes with an additional *rejection class* that contains streamlines not belonging to either of the supported bundles, allows for the handling of any streamlines contained in a tractogram. A basal strategy for the definition of such a rejection class from the set of streamlines that are not assigned a bundle label is direct sampling from that set (Xu et al., 2019; Gupta et al., 2017b; Ngattai et al., 2018). In order to avoid false labels in training, Ugurlu et al. try to ensure a sufficient "distance" between bundles and the rejection class through thresholding the overlap of a streamline and all considered bundles (Ugurlu et al., 2019). Zhang et al. define the rejection class as streamlines of unsupported bundles taken from an atlas as well as false-positive streamlines (Zhang et al., 2020).

This problem formulation has been investigated with different combinations of model types and input data. Lam et al. propose the tool *TRAFIC*, which consists of a neural network (two fully connected layers) applied to streamline curvature, torsion, and Euclidean distance to a set of landmarks (Ngattai et al., 2018). Ugurlu et al. use an ensemble of six networks (three fully connected layers) in their approach and for input rely on a streamline descriptor, called NRFOD (Ugurlu et al., 2018), which is based on a set of orientation histograms of local track orientations (Ugurlu et al., 2019).

The use of a CNN-based architecture is investigated in Xu et al. (2019) and Zhang et al. (2020). While Xu et al. directly use the streamline coordinates as input to their ResNet model (Xu et al., 2019), the tool *DeepWMA* builds on a hand-crafted, coordinate-based streamline descriptor, called *FiberMap*. The authors expect *FiberMap* to be robust to small anatomical variations and to allow for deep model architectures. Two interesting aspects investigated by Xu et al. are the use of

---

s. https://github.com/vikashg/FiberNet.

additional loss terms (to account for data imbalance and to improve class distances in the penultimate layer) as well as an attention mechanism with the goal to make the model predictions interpretable.

Finally, in Gupta et al. (2017b) a model with a stack of four unidirectional LSTM layers and one bidirectional LSTM layer is used in combination with the streamline coordinates after pruning points of low curvature.

**Evaluation**: The experiments in Ngattai et al. (2018) (TRAFIC) are limited to parts of the arcuate fasciculi in a small number of subjects and evaluated only qualitatively. Also the reported results in Gupta et al. (2017b) (90% accuracy, 50% recall) are obtained on a rather small dataset of three subjects with eight bundles and do not allow for a comprehensive conclusion. Ugurlu et al. compare their ensemble classifier in five HCP subjects with nine manually defined target bundles to a previous classification approach (Ugurlu et al., 2018) in terms of bundle-based minimum distance (Garyfallidis et al., 2015) and Cohen's Kappa, and find indications for slight improvements through their approach.

In Xu et al. (2019), the authors aim to showcase the utility of their model in identifying cortical regions relevant for planning pediatric epilepsy surgery. In a first experiment, they use fMRI to obtain inclusion ROIs for the definition of 64 relevant bundles in 70 healthy children. After training in 56 subjects, the best average F1 score (0.95) over 14 test subjects was obtained with the proposed loss terms and the attention mechanism. In a second experiment, the cortical endpoint regions for each prediction class are compared to the output of electrical stimulation mapping (ESM) as a gold standard. An AUC of more than 0.9 in a population-based analysis (70 children with epilepsy) of spatial overlap between the CNN model and ESM shows the potential of the approach for this application. The authors point out that the computed attention maps look different for streamlines from different classes and could be used as an additional measure to detect outliers.

Zhang et al. built a training dataset of one million streamlines in total (Zhang et al., 2020). These streamlines were sampled from 100 tractograms of HCP subjects underlying the creation of the tractography atlas in Zhang (2018) and belong either to one of 54 white matter bundles or the rejection class. The authors report a high classification accuracy of 90.99% obtained over 20% of the training data. Further, they analyzed the performance of their method in 597 subjects of six databases with different acquisition parameters, demographics as well as healthy and diseased conditions (i.e., presence of tumors). Based on the fraction of identified bundles (on average 99%) and visual inspection of different maps assessing the spatial coverage of the segmented bundles, the authors concluded that their method is robust to the mentioned sources of data variability.

**Availability**: *TRAFIC* is open source and the code is available on github.[t] The trained model and the source code of *DeepWMA* are available as part of the *SlicerDMRI* project.[u] Xu et al. published the trained model and the corresponding source code on github.[v]

### Single bundle label and one rejection class

**Concept and methodology**: Instead of training a single model for the whole classification problem, the latter can be decomposed into several binary classifications as done in Bertò (2021), Ngattai et al. (2018), Gupta et al. (2018), and Liu, (2019. In this strategy, for each target bundle an individual, bundle-specific model is trained to distinguish streamlines of that bundle from all others. With this approach it is possible to add the support for additional bundles without impacting already trained models.

The method *FiberNet2.0* (Gupta et al., 2018) implements this strategy in a two-step procedure. First, an ROI-based filtering step is employed to extract streamline candidates for all considered bundles, and after that an ensemble of CNN models aims to separate true-positive from false-positive streamlines through binary classification. In a similar fashion, the method *Classifyber* is trained to extract a streamline bundle from a crude preselection of streamline candidates (Bertò, 2021). The approach builds on a logistic regression classifier and a set of streamline-based distance measures w.r.t. to several landmarks as input.

As opposed to that, Liu et al. do not explicitly regard the definition of false-positive streamlines (Liu, 2019). Instead, they obtain the rejection class for each bundle from sampling streamlines of other bundle classes while maintaining a balance between spatially neighboring and distant bundles in their dataset. For each target bundle, a GraphCNN model, called *DeepBundle*, is trained using the streamline coordinates as input.

Finally, the tool *TRAFIC* (Ngattai et al., 2018) (see previous) is also applied in a binary classification setting with only one target bundle.

---

t. https://github.com/PrinceNgattaiLam/Trafic.

u. http://dmri.slicer.org/.

v. https://github.com/HaotianMXu/Brain-fiber-classification-using-CNNs.

**Evaluation**: The capability of *TRAFIC* to extract streamlines of the left or right arcuate fasciculus is investigated in 50 infant subjects. However, the presented evaluation allows only for limited conclusions on the actual prediction performance of the proposed model.

*FiberNet2.0* is evaluated on a set of four subjects by comparison to a human expert. The reported true-positive and true-negative rate for 10 bundles on a "validation set" seem promising. However, a clear description of the evaluation data and procedure as well as inclusion of a larger set of target bundles would be required to assess the true potential of this approach.

Liu et al. evaluate their model over 11 HCP subjects in 12 chosen bundles taken from the *TractSeg* bundle database (Wasserthal et al., 2017). The reported precision and recall for the 12 investigated bundles lies above 86% in all cases. The comparison of bundle-wise dice scores shows a better performance in their experiments compared to *RecoBundles* (Garyfallidis et al., 2018). Despite the promising results that are reported, it would be desirable to extend the presented evaluation to all 72 bundles available in the employed database. Regarding the registration-free nature of *DeepBundle* claimed by the authors, it remains to be shown in a suitable experiment if the method is fully independent of prior image space alignment of training and unseen testing subjects.

In experiments on several datasets with varying bundle size, tractography pipeline and data quality, Bertò et al. find *Classifyber* to achieve better dice scores in prediction of several bundles as compared to variants of *RecoBundles* and *TractSeg*. The authors point out the small amount of training data (15–30 subjects) required by their approach in order to reach a competitive performance.

**Availability**: While *DeepBundle* is not published open source, an implementation of the descriptions in the paper is available on github.[w] The source code of *Classifyber* also is available on github[x] and a BrainLife app exists for training[y] and testing.[z] Employed datasets are accessible online or on request to the authors (refer to Bertò, 2021 for details).

### 3.4.2 Dichotomous plausibility label

**Concept and methodology**: Aside from extracting well-defined bundle structures, the separation of a whole brain tractogram into sets of anatomically plausible and implausible streamlines is another application that can be naturally formulated as a binary classification. The central difficulty for such an approach when following a supervised learning strategy is to obtain meaningful streamline labels.

In Jörgens et al. (2019), the authors circumvent this problem by focusing on experiments based on the ISMRM 2015 tractography challenge phantom for which a ground truth is known. A CNN model relying solely on the spherical harmonics coefficients of the diffusion signal along each streamline is used to extract valid streamlines from original challenge submissions.

As opposed to that, Astolfi et al. use *dynamic edge convolutions* (DEC) to reproduce the grouping of streamlines obtained from the tool *ExTractor* (Petit et al., 2021). The approach in Astolfi (2020) treats the input streamline as a point cloud and builds in each layer on the construction of *k*NN graphs to define local neighborhood features that are mapped onto a new output representation through an MLP. In order to be sensitive to the sequential nature of the streamline sampling points, the input graph is restricted to be a bidirectional sequence graph.

**Evaluation**: In Jörgens et al. (2019), the evaluation of prediction accuracy in nine different tractograms for the ISMRM 2015 phantom indicated that nondeterministic tractograms as supervisors seem to allow for a better generalization of the trained model to other tractography approaches.

The investigated models in Astolfi (2020) are trained and evaluated on a set of 1M tractograms of 20 HCP subjects, regarding only streamlines longer than 20 mm. Based on reported accuracy, precision, and recall over three evaluation subjects, the authors find the DEC model with sequence graph input to perform best (95%–97% in all metrics). In-depth analysis of streamline characteristics showed that the sequence graph extension led to a better performance than the standard DEC approach in terms of accuracy and false-positive rate, especially for long and high-curvature streamlines.

**Availability**: The DEC method by Astolfi et al. is available open source on github[aa] and as BrainLife app.[ab] Corresponding experiment data is also provided online.[ac]

---

w. https://github.com/tueimage/DeepBundle.

x. https://github.com/FBK-NILab/app-classifyber.

y. https://doi.org/10.25663/brainlife.app.228.

z. https://doi.org/10.25663/brainlife.app.265.

aa. https://github.com/FBK-NILab/tractogram_filtering/tree/miccai2020.

ab. https://doi.org/10.25663/brainlife.app.390.

ac. https://doi.org/10.25663/brainlife.pub.13.

### 3.4.3 Label transcription approaches

**Concept and methodology**: While most of the presented methods directly predict the class label for each streamline, some approaches incorporate the streamline labels in a more indirect manner.

The authors of Jha et al. (2019) train the LSTM-based model from Gupta et al. (2017b) in a Siamese fashion in order to predict if a presented pair of streamlines belongs to the same target class or not. In combination with a bundle atlas, the method called *FS2NET* can be used to transcribe matching labels from the atlas to unlabeled streamlines. While this idea still depends on the labels in the training phase, Haitz et al. propose to use autoencoders (AEs) to learn the mapping of streamlines to a latent space from an unlabeled tractogram in order to encode geometrical similarity of streamlines by the distance between their latent representations (Legarreta et al., 2021) (see Fig. 18). This allows transcribing labels from a given tractogram to unlabeled streamlines based on a nearest-neighbor approach in the latent space.

**Evaluation**: The evaluation of *FS2NET* on the same dataset as in Gupta et al. (2017b) showed a similar performance and the same limitations apply. However, in a second experiment, the reported accuracy seems to suggest that the Siamese-type model in combination with the atlas-based labeling approach could be robust to rotations of the streamlines. However, a more thorough evaluation of this property would be needed.

The approach by Haitz et al., called *FINTA* (Legarreta et al., 2021), is shown to achieve high accuracy, sensitivity, precision, and F1 values in experiments involving the FiberCup (97%–99% in all metrics) and the ISMRM 2015 Tractography Challenge (91% in all metrics) datasets separating "plausible" from "implausible" streamlines. Further, it outperforms basic anatomy-based filtering approaches as well as the method *RecoBundles* in experiments on streamlines in the corpus callosum of a subset of the BIL&GIN database considering the same evaluation metrics.

**Availability**: These methods are at this time not publicly available.

## 4 What could be gained, what did we learn, and what challenges remain?

As described in the previous section, various prototypes of ML-driven approaches to tractography have been presented and, while they are by far not as mature and extensively tested as the classic tractography methods, a lot can already be inferred from these preliminary results.

### 4.1 General considerations on ML-based tractography

Over the last years, various groups from all over the world have shown that machine learning is indeed a suitable tool for the task of neuronal tractography. The individual classes of approaches, as described in Section 3, all proved to have certain specific advantages but also shortcomings that will be reviewed here.

Local ML-based approaches to tractography only replace a single part of the whole tractography process with an ML element, namely the inference of the next track progression direction from the underlying local image data. These approaches have shown to be competitive in comparison to the classic model based local tractography approaches, but they are no game changers. While potentially exploiting the signal more exhaustively and being more flexible in terms

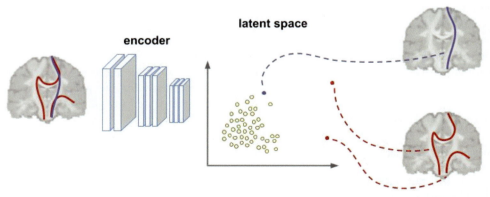

**FIG. 18** Schematic description of the autoencoder method in (Legarreta et al., 2021). Streamlines that are similar in their shape should have latent space representations that are close to each other. Streamline labeling can be performed in latent space and thus be independent of the training phase. *(Reproduced with permission from Legarreta, J.H., Petit, L., Rheault, F., Theaud, G., Lemaire, C., Descoteaux, M., Jodoin, P.M., 2021. Filtering in tractography using autoencoders (FINTA). Med. Image Anal. 72, 102126. https://doi.org/10.1016/j.media.2021.102126.)*

of input and output, they yield no drastically improved results and they suffer from the same inherent flaws as the classic approaches. Since they only take a small local window of data into consideration, they cannot resolve local ambiguities and their performance will therefore always be a trade-off between sensitivity and specificity. One aspect that has not yet been addressed and that might be an interesting focus of further research in the area of ML-based local tractography is to learn the other typical parameters of tractography instead of manually defining them, e.g., step size and curvature thresholds.

In sequence-based approaches, the previous streamline integration steps are implicitly taken into account while progressing the track locally. This memory-based strategy might mitigate some of the shortcomings of the purely local approach and in preliminary experiments some improvements could be shown. Nevertheless, these methods were never trained in vivo and their true capabilities have yet to be fully determined.

The newest and probably the best analyzed approaches are global methods. By taking the information of the complete image into account while modelling tracts, these approaches can disentangle local ambiguities and resolve even complicated configurations of multiple intersecting tracts correctly. This strength comes at the cost of a limited flexibility. Global approaches are restricted to a fixed set of bundles that were included in the training process. One could view them as a sophisticated extension of atlas-based methods, which limits their possible applications. Nevertheless, they were evaluated rather extensively and proved to yield very good reconstructions of their respective target bundles. In contrast to the local and sequence based approaches, this class of methods showed to be a real improvement over the state of the art apart from academic insights.

Streamline classification has similar advantages and drawbacks as global approaches. The subclass of methods predicting bundle labels is fixed to a predefined set of tracts and needs to be retrained from scratch to include new bundles. In this context, the idea of decoupling the definition of streamline labels from the training process is especially appealing and a promising future avenue to circumvent that limitation. A major difference compared to the global approaches is that streamline-based strategies depend heavily on the used tractography method. Since they are not generating streamlines by themselves but preprocess a given tractogram they are not capable of alleviating many of the potential limitations introduced by the choice of tractography algorithm. Nevertheless, this dependence on classic tractography does not necessarily have to be a disadvantage. Tractography can be performed with maximal sensitivity in this scenario, possibly involving multiple different algorithms, given that the ML-based streamline classification eliminates false positive streamlines.

A general issue with all these approaches is a potentially limited capability to generalize to unseen subjects that might be acquired with different scanners and acquisition-settings and to account for the corresponding intersubject variabilities, particularly also when pathologies are involved. This issue will require more research and an in-depth analysis using a consistent evaluation scheme on standardized, comprehensive, and widely recognized public datasets, which is currently not available.

A promising strategy might be to employ ML-based approaches particularly for specialized tasks and custom tailored to a specific problem. In this way, issues with generalization can be avoided and the particular strength of machine learning to optimize a certain objective can be optimally exploited. While creating such task-specific methods is promising from a performance point of view, it is challenging in terms of method design and training. Particularly, deep learning-based approaches require rather large amounts of well-curated training data, which might be hard to come by for a specialized niche problem.

Ultimately, the issue of generalization is an issue of training and validation procedures as well as datasets. While evaluation in general is a topic of its own, it has to be stated that robust comparison of biomedical image analysis methods is nontrivial, as impressively illustrated, for example, by Maier-Hein et al. (Maier-Hein, 2018) and Reinke et al. (Reinke, 2018). And it might be favorable for the tractography community to align their efforts with the guidelines provided by the Biomedical Image Analysis ChallengeS (BIAS) Initiative (Maier-Hein, 2020) to improve standardization and quality. For further details about general evaluation strategies for tractography, the interested reader is referred to Chapters 25–28. In the following, evaluation related issues that are of specific concern for the current state of the art in ML-based tractography are discussed.

## 4.2 Considerations on the state of the art in ML-based tractography validation

While multiple datasets as well as evaluation measures in the general ML and computer vision community have reached the status of de-facto standards for producing somewhat comparable results, such as the MNIST handwritten digits dataset (Lecun et al., 1998) or the CIFAR image classification dataset (Krizhevsky et al., 2009), the ML tractography community is still in its infancy and has not yet reached the point where the same metrics, strategies, tools, and datasets are used for evaluation across publications and work-groups. Consequently, it is still quite difficult to compare methods and assess their usability for specific tasks. Specifically, the ISMRM 2015 Tractography Challenge is the only dataset that has been consistently used across multiple papers. But even then the results cannot be conclusively compared because of various issues, which also generally arise when different datasets are used. In the following, we will describe and discuss the most prominent issues that should be considered in order to maximize impact and reproducibility when doing research in the area of ML-based tractography.

**Dissimilar inputs**: A characteristic feature of data-driven methods is their ability to process almost arbitrary input data, which enables a natural integration of different data sources that would be difficult to achieve with classic tractography models. If information is available to accomplish the task at hand, it will be used by the ML model to make the correct predictions. However, this has become a problem for the ML tractography community, where multiple types of input data have been used across different models, making comparisons problematic. For example, starting from the raw diffusion signal, down to the usual ODF peaks, multiple different things can be used as input:

— Raw diffusion signal in the original acquisition directions.
— Raw diffusion signal, resampled to standard directions.
— Diffusion signal fitted with spherical harmonics (X coefficients).
— Fiber ODF represented as spherical harmonics (Y coefficients).
— Fiber ODF peaks (Z vectors in three dimensions).

Additionally to the modeling, results can be influenced by the preprocessing pipeline applied to the raw signal, as is the case with classic tractography methods (e.g., eddy/topup, MRtrix's *dwidenoise* and *dwipreproc*, registration to a common space, etc.). Among all those possibilities, even if two publications use the exact same datasets, it would be hard to tell if differences in performance metrics are attributable to a proposed architecture or the choice of preprocessing or data representation.

**Varying test environments**: Tractography is a complex task and there are many variable parameters that can affect the tracking process, like the step size, the number of seeds, their spatial position (WM-GM interface vs white matter) and distribution (uniformly distributed or random), and the usage of a white matter or brain mask (or none). Changing any of those parameters can improve or worsen the results of tractography. Without a complete benchmark where all parameters are fixed, it is hard to compare results across publications. Furthermore, comparing tractography methods when variable preprocessing pipelines are involved is virtually impossible due to the large number of variables leading to a combinatorial explosion. To eliminate this issue, a standardized preprocessing pipeline must be employed that enables the comparison of the actual tractography method without additional variables. Particularly in the context of ML-based tractography, an additional challenge arises from varying training and test splits of the data. To be able to compare data-driven methods, they need to be trained and tested on the same data. A final factor of variability is the choice of evaluation metrics, which is a nontrivial task. Tractography can be assessed on a bundle, streamline, and voxel level. Each level can again be evaluated using metrics of connectivity, orientation, or spatial coverage. All of these levels of evaluation are important, particularly since they are often complementary, i.e., an improvement in one metric often comes at the expense of another, and a standardized set of evaluation metrics accepted by the whole community has yet to be defined. First steps in this direction have been taken with the Tractometer framework introduced in Section 3 (Côté et al., 2013). Finally, note that choosing the right metric might be related to a specific application, e.g., bundle reconstruction (volume-oriented) vs connectivity analysis (streamline-oriented).

**Data contamination**: Training and testing data should be similar, but must come from different sources. Models trained and tested using the same acquisition data (same diffusion signal), like the ISMRM 2015 Tractography Challenge, do not provide reliable results, as the training data was *contaminated* with testing data. The fact that few to no public datasets were available explains why some publications suffered from this. However, we should strive to do better now that data is becoming more and more publicly available.

**Disparate training data**: Reference streamlines are required to train a data-driven tractography method. However, there is no single standard way to generate streamlines, and that is reflected in the existing literature, where no proposed model was trained using the same reference streamlines. Some used deterministic methods, others used probabilistic methods. The step size and maximum angle are also not always reported and may affect the tracking process. The result is that even when the same acquisitions are used across publications, the reference streamlines may be very different, and we don't currently know how much of an impact this has on the training algorithm.

**Limited test database**: The ISMRM 2015 Tractography Challenge dataset was used for the quantitative evaluation of ML tractography in four publications, but it consisted of a single evaluation subject. Performance metrics computed from a single acquisition may be indicative of something promising, but should not necessarily be relied upon. Now that new, larger datasets are emerging, like *TractSeg*, the community should be able to get more reliable results. The next step for the community will be to use multisite databases to evaluate generalization performance of new machines and acquisition schemes.

## 4.3 Considerations on suitable datasets for ML-based tractography

As stated and discussed before, ML-based methods depend heavily on the data used for training and evaluation. While more and more datasets suitable for training and validation of ML-based tractography models are becoming available, it is worth discussing the features of an ideal dataset that can be kept in mind when choosing a dataset to train and validate one's own method or when deciding to create one's own dataset.

**Availability**: A tractography dataset should be publicly available, easy to download and provided with a clear documentation. It is too common in the tractography community that a new dataset is internally developed and not shared with the community. Due to anonymization and consent issues in the medical imaging field, it is often easier to keep a dataset private, but we should work to overcome those issues if we want to grow as a community and if we want to advance with the research field.

**Size**: New tractography datasets should always be composed of multiple subjects. This aspect is important in order to measure reliable performance metrics, along with their variability across multiple unseen test subjects (averages and standard deviations). Furthermore, having acquisitions from multiple sites is also desirable to start evaluating the generalization ability of new data-driven tractography methods. Also, the inclusion or exclusion of pathologies should be carefully considered. Taking pathologies into account is particularly relevant when thinking about the generalizability of ML models, but of course poses additional challenges, such as data availability and the creation of reference gold standards.

**In silico vs in vivo data**: As described in Section 2, in silico as well as in vivo data can be used for ML-based tractography and multiple datasets suitable for this task have already been published. Each type comes with its own limitations and advantages that have to be considered thoroughly before choosing one type for a specific application. To fully assess the capabilities of a specific tractography method, certainly a combination of both types will be necessary.

**Manual reference for in vivo data**: While a true tractography ground truth is only possible for simulated datasets, this ground truth can certainly be approximated, at least for a subset of white matter bundles, by approaching the task thoroughly. This includes working with expert neuroanatomists, radiologists, and computer scientists in an interdisciplinary and multirater setting for manually or semiautomatically annotating and segmenting bundles on the basis of large and variable in vivo imaging collectives. While certainly challenging, this approach is currently the only way to account for interrater variability and to ensure generalizability of methods trained on such a dataset (Rheault et al., 2020b).

## 5 Conclusion and take-home message

While ML-based tractography is still a young field of research, recent publications have shown that ML can add real value to brain connectivity research. Particularly global deep-learning based approaches, such as fully automatic bundle segmentation techniques, clearly outperform the previously existing methods. Other results are still inconclusive and much more methodological development and evaluation work is to be done to assess the full potential of local and sequence-based approaches to ML tractography. The many open questions in this field of research include how to optimally leverage the complete available imaging information including the micro- as well as the macrostructure of the brain, how to deal with the limited and imperfect training data, or how to improve upon it, and how to establish a common standard and processes for evaluation and quantification of tractograms.

Nevertheless, many of the developed tools are open source and usable for the community, and the number and quality of datasets suitable for training and validation of ML-based tractography is increasing. This provides a solid foundation and baseline for future research by following the concepts of open-source and open-data.

## References

Aganj, I., Lenglet, C., Sapiro, G., 2009. ODF reconstruction in q-ball imaging with solid angle consideration. In: 2009 IEEE International Symposium on Biomedical Imaging: From Nano to Macro, Jun, pp. 1398–1401, https://doi.org/10.1109/ISBI.2009.5193327.

Aganj, I., et al., 2011. A Hough transform global probabilistic approach to multiple-subject diffusion MRI tractography. Med. Image Anal. 15 (4), 414–425. https://doi.org/10.1016/j.media.2011.01.003.

Alexander, D.C., 2005. Maximum entropy spherical deconvolution for diffusion MRI. In: Information Processing in Medical Imaging, pp. 76–87, https://doi.org/10.1007/11505730_7. Berlin, Heidelberg.

Assaf, Y., Basser, P.J., 2005. Composite hindered and restricted model of diffusion (CHARMED) MR imaging of the human brain. NeuroImage 27 (1), 48–58. https://doi.org/10.1016/j.neuroimage.2005.03.042.

Assaf, Y., Blumenfeld-Katzir, T., Yovel, Y., Basser, P.J., 2008. Axcaliber: a method for measuring axon diameter distribution from diffusion MRI. Magn. Reson. Med. 59 (6), 1347–1354. https://doi.org/10.1002/mrm.21577.

**342 PART | III Tractography algorithms**

Astolfi, P., 2020. Tractogram filtering of anatomically non-plausible fibers with geometric deep learning. In: Martel, A.L., Abolmaesumi, P., Stoyanov, D., Mateus, D., Zuluaga, M.A., Zhou, S.K., Racoceanu, D., Joskowicz, L. (Eds.), Medical Image Computing and Computer Assisted Intervention—MICCAI 2020. vol. 12267. Springer International Publishing, Cham, pp. 291–301.

Basser, P.J., 1998. Fiber-tractography via diffusion tensor MRI (DT-MRI). In: Proceedings of the 6th Annual Meeting ISMRM, Sydney, Australia. vol. 1226.

Basser, P.J., Mattiello, J., Lebihan, D., 1994. Estimation of the effective self-diffusion tensor from the NMR spin echo. J. Magn. Reson. B 103 (3), 247–254. https://doi.org/10.1006/jmrb.1994.1037.

Bazin, P.-L., 2011. Direct segmentation of the major white matter tracts in diffusion tensor images. NeuroImage 58 (2), 458–468. https://doi.org/10.1016/j.neuroimage.2011.06.020.

Behrens, T.E.J., Berg, H.J., Jbabdi, S., Rushworth, M.F.S., Woolrich, M.W., 2007. Probabilistic diffusion tractography with multiple fibre orientations: what can we gain? NeuroImage 34 (1), 144–155. https://doi.org/10.1016/j.neuroimage.2006.09.018.

Benou, I., Riklin Raviv, T., 2019. DeepTract: a probabilistic deep learning framework for white matter fiber tractography. In: Medical Image Computing and Computer Assisted Intervention—MICCAI 2019, Cham, pp. 626–635, https://doi.org/10.1007/978-3-030-32248-9_70.

Berman, J.I., Chung, S., Mukherjee, P., Hess, C.P., Han, E.T., Henry, R.G., 2008. Probabilistic streamline q-ball tractography using the residual bootstrap. NeuroImage 39 (1), 215–222. https://doi.org/10.1016/j.neuroimage.2007.08.021.

Bertò, G., 2020. HCP-Minor Bundle Dataset., https://doi.org/10.25663/brainlife.pub.11.

Bertò, G., 2021. Classifyber, a robust streamline-based linear classifier for white matter bundle segmentation. NeuroImage 224, 117402. https://doi.org/10.1016/j.neuroimage.2020.117402.

Bullock, D., 2019. Associative white matter connecting the dorsal and ventral posterior human cortex. Brain Struct. Funct. 224 (8), 2631–2660. https://doi.org/10.1007/s00429-019-01907-8.

Caruyer, E., Daducci, A., Descoteaux, M., Houde, J.-C., Thiran, J.-P., Verma, R., 2014. Phantomas: a flexible software library to simulate diffusion MR phantoms.

Chandio, B.Q., 2020. Bundle analytics, a computational framework for investigating the shapes and profiles of brain pathways across populations. Sci. Rep. 10 (1), 17149. https://doi.org/10.1038/s41598-020-74054-4.

Côté, M.-A., Girard, G., Boré, A., Garyfallidis, E., Houde, J.-C., Descoteaux, M., 2013. Tractometer: towards validation of tractography pipelines. Med. Image Anal. 17 (7), 844–857. https://doi.org/10.1016/j.media.2013.03.009.

Daducci, A., Dal Palù, A., Lemkaddem, A., Thiran, J.-P., 2015. COMMIT: convex optimization modeling for microstructure informed tractography. IEEE Trans. Med. Imaging 34 (1), 246–257. https://doi.org/10.1109/TMI.2014.2352414.

Descoteaux, M., Deriche, R., 2009. High angular resolution diffusion MRI segmentation using region-based statistical surface evolution. J. Math. Imaging Vis. 33 (2), 239–252. https://doi.org/10.1007/s10851-008-0071-8.

Descoteaux, M., Angelino, E., Fitzgibbons, S., Deriche, R., 2007. Regularized, fast, and robust analytical Q-ball imaging. Magn. Reson. Med. 58 (3), 497–510. https://doi.org/10.1002/mrm.21277.

Descoteaux, M., Deriche, R., Knosche, T.R., Anwander, A., 2009. Deterministic and probabilistic tractography based on complex fibre orientation distributions. IEEE Trans. Med. Imaging 28 (2), 269–286. https://doi.org/10.1109/TMI.2008.2004424.

Eckstein, I., 2009. Active fibers: matching deformable tract templates to diffusion tensor images. NeuroImage 47, 82–89.

Esteva, A., et al., 2017. Dermatologist-level classification of skin cancer with deep neural networks. Nature 542 (7639), 7639. https://doi.org/10.1038/nature21056.

Fillard, P., 2011. Quantitative evaluation of 10 tractography algorithms on a realistic diffusion MR phantom. NeuroImage 56 (1), 220–234. https://doi.org/10.1016/j.neuroimage.2011.01.032.

Fillard, P., Poupon, C., Mangin, J.-F., 2009. A novel global tractography algorithm based on an adaptive spin glass model. In: *Medical Image Computing and Computer-Assisted Intervention—MICCAI 2009*, Berlin, Heidelberg, pp. 927–934, https://doi.org/10.1007/978-3-642-04268-3_114.

Friman, O., Farneback, G., Westin, C.-F., 2006. A Bayesian approach for stochastic white matter tractography. IEEE Trans. Med. Imaging 25 (8), 965–978. https://doi.org/10.1109/TMI.2006.877093.

Garyfallidis, E., 2014. Dipy, a library for the analysis of diffusion MRI data. Front. Neuroinform. 8, 8. https://doi.org/10.3389/fninf.2014.00008.

Garyfallidis, E., Brett, M., Correia, M.M., Williams, G.B., Nimmo-Smith, I., 2012. QuickBundles, a method for tractography simplification. Front. Neurosci. 6. https://doi.org/10.3389/fnins.2012.00175.

Garyfallidis, E., Ocegueda, O., Wassermann, D., Descoteaux, M., 2015. Robust and efficient linear registration of white-matter fascicles in the space of streamlines. NeuroImage 117, 124–140. https://doi.org/10.1016/j.neuroimage.2015.05.016.

Garyfallidis, E., et al., 2018. Recognition of white matter bundles using local and global streamline-based registration and clustering. NeuroImage 170, 283–295. https://doi.org/10.1016/j.neuroimage.2017.07.015.

Girard, G., Whittingstall, K., Deriche, R., Descoteaux, M., 2014. Towards quantitative connectivity analysis: reducing tractography biases. NeuroImage 98, 266–278. https://doi.org/10.1016/j.neuroimage.2014.04.074.

Glasser, M.F., et al., 2013. The minimal preprocessing pipelines for the Human Connectome Project. NeuroImage 80, 105–124. https://doi.org/10.1016/j.neuroimage.2013.04.127.

Goodfellow, I., Bengio, Y., Courville, A., 2016. Deep Learning. MIT Press.

Guo, W., Chen, Y., Zeng, Q., 2008. A geometric flow-based approach for diffusion tensor image segmentation. Philos. Trans. R. Soc. Lond. Math. Phys. Eng. Sci. 366 (1874), 2279–2292.

Gupta, V., Thomopoulos, S.I., Rashid, F.M., Thompson, P.M., 2017a. FiberNET: an ensemble deep learning framework for clustering white matter fibers. In: Descoteaux, M., Maier-Hein, L., Franz, A., Jannin, P., Collins, D.L., Duchesne, S. (Eds.), Medical Image Computing and Computer Assisted Intervention—MICCAI 2017. vol. 10433. Springer International Publishing, Cham, pp. 548–555.

Gupta, T., Patil, S.M., Tailor, M., Thapar, D., Nigam, A., 2017b. BrainSegNet : a segmentation network for human brain fiber tractography data into anatomically meaningful clusters. ArXiv171005158 Cs. Accessed 2017. [Online]. Available: http://arxiv.org/abs/1710.05158.

Gupta, V., Thomopoulos, S.I., Corbin, C.K., Rashid, F., Thompson, P.M., 2018. FIBERNET 2.0: an automatic neural network based tool for clustering white matter fibers in the brain. In: 2018 IEEE 15th International Symposium on Biomedical Imaging (ISBI 2018), Washington, DC, Apr, pp. 708–711, https://doi.org/10.1109/ISBI.2018.8363672.

Jansons, K.M., Alexander, D.C., 2003. Persistent angular structure: new insights from diffusion MRI data. Dummy version. In: Information Processing in Medical Imaging, Berlin, Heidelberg, pp. 672–683, https://doi.org/10.1007/978-3-540-45087-0_56.

Jbabdi, S., Woolrich, M.W., Andersson, J.L.R., Behrens, T.E.J., 2007. A Bayesian framework for global tractography. NeuroImage 37 (1), 116–129. https://doi.org/10.1016/j.neuroimage.2007.04.039.

Jeurissen, B., Tournier, J.-D., Dhollander, T., Connelly, A., Sijbers, J., 2014. Multi-tissue constrained spherical deconvolution for improved analysis of multi-shell diffusion MRI data. NeuroImage 103, 411–426. https://doi.org/10.1016/j.neuroimage.2014.07.061.

Jeurissen, B., Descoteaux, M., Mori, S., Leemans, A., 2019. Diffusion MRI fiber tractography of the brain. NMR Biomed. 32 (4), e3785. https://doi.org/10.1002/nbm.3785.

Jha, R.R., Patil, S., Nigam, A., Bhavsar, A., 2019. FS2Net: fiber structural similarity network (FS2Net) for rotation invariant brain tractography segmentation using stacked LSTM based siamese network. In: Vento, M., Percannella, G. (Eds.), Computer Analysis of Images and Patterns. vol. 11679. Springer International Publishing, Cham, pp. 459–469.

Jin, Y., 2014. Automatic clustering of white matter fibers in brain diffusion MRI with an application to genetics. NeuroImage 100, 75–90. https://doi.org/10.1016/j.neuroimage.2014.04.048.

Jonasson, L., Bresson, X., Hagmann, P., Cuisenaire, O., Meuli, R., Thiran, J.-P., 2005. White matter fiber tract segmentation in DT-MRI using geometric flows. Med. Image Anal. 9 (3), 223–236. https://doi.org/10.1016/j.media.2004.07.004.

Jörgens, D., Smedby, Ö., Moreno, R., *2018*. Learning a Single Step of Streamline Tractography Based on Neural Networks. Accessed 2018. [Online]. Available: https://www.springerprofessional.de/learning-a-single-step-of-streamline-tractography-based-on-neura/15583204.

Jörgens, D., Poulin, P., Moreno, R., Jodoin, P.-M., Descoteaux, M., 2019. Towards a Deep Learning Model for Diffusion-Aware Tractogram Filtering.

Kreher, B.W., Schneider, J.F., Mader, I., Martin, E., Hennig, J., Il'yasov, K.A., 2005. Multitensor approach for analysis and tracking of complex fiber configurations. Magn. Reson. Med. 54 (5), 1216–1225. https://doi.org/10.1002/mrm.20670.

Krizhevsky, A., Hinton, G., et al., 2009. Learning Multiple Layers of Features from Tiny Images.

Kumar, K., Siddiqi, K., Desrosiers, C., 2019. White matter fiber analysis using kernel dictionary learning and sparsity priors. Pattern Recogn. 95, 83–95. https://doi.org/10.1016/j.patcog.2019.06.002.

Lazar, M., et al., 2003. White matter tractography using diffusion tensor deflection. Hum. Brain Mapp. 18 (4), 306–321. https://doi.org/10.1002/hbm.10102.

Lecun, Y., Bottou, L., Bengio, Y., Haffner, P., 1998. Gradient-based learning applied to document recognition. Proc. IEEE 86 (11), 2278–2324. https://doi.org/10.1109/5.726791.

Legarreta, J.H., Petit, L., Rheault, F., Theaud, G., Lemaire, C., Descoteaux, M., Jodoin, P.M., 2021. Filtering in tractography using autoencoders (FINTA). Med. Image Anal. 72, 102126. https://doi.org/10.1016/j.media.2021.102126.

Lemkaddem, A., Skiöldebrand, D., Dal Palú, A., Thiran, J.-P., Daducci, A., 2014. Global tractography with embedded anatomical priors for quantitative connectivity analysis. Front. Neurol. 5. https://doi.org/10.3389/fneur.2014.00232.

Li, B., 2020. Neuro4Neuro: a neural network approach for neural tract segmentation using large-scale population-based diffusion imaging. NeuroImage 218, 116993. https://doi.org/10.1016/j.neuroimage.2020.116993.

Liu, F., 2019. DeepBundle: fiber bundle parcellation with graph convolution neural networks. In: Zhang, D., Zhou, L., Jie, B., Liu, M. (Eds.), Graph Learning in Medical Imaging. vol. 11849. Springer International Publishing, Cham, pp. 88–95.

Maier-Hein, K., May 2015. Tractography challenge ISMRM 2015 data. Zenodo. https://doi.org/10.5281/zenodo.572345.

Maier-Hein, K., May 2017. Tractography challenge ISMRM 2015 high-resolution data. Zenodo. https://doi.org/10.5281/zenodo.579933.

Maier-Hein, L., 2018. Why rankings of biomedical image analysis competitions should be interpreted with care. Nat. Commun. 9 (1), 1–13. https://doi.org/10.1038/s41467-018-07619-7.

Maier-Hein, L., 2020. BIAS: Transparent reporting of biomedical image analysis challenges. [Online]. Available: http://arxiv.org/abs/1910.04071.

Maier-Hein, K.H., et al., 2017. The challenge of mapping the human connectome based on diffusion tractography. Nat. Commun. 8 (1), 1. https://doi.org/10.1038/s41467-017-01285-x.

Malcolm, J.G., Shenton, M.E., Rathi, Y., 2010. Filtered multitensor tractography. IEEE Trans. Med. Imaging 29 (9), 1664–1675. https://doi.org/10.1109/TMI.2010.2048121.

Mangin, J.-F., Fillard, P., Cointepas, Y., Le Bihan, D., Frouin, V., Poupon, C., 2013. Toward global tractography. NeuroImage 80, 290–296. https://doi.org/10.1016/j.neuroimage.2013.04.009.

Mori, S., Crain, B.J., Chacko, V.P., Van Zijl, P.C., 1999. Three-dimensional tracking of axonal projections in the brain by magnetic resonance imaging. Ann. Neurol. 45 (2), 265–269.

Neher, P., 2020. MIC-DKFZ/MITK-Diffusion. MIC-DKFZ,.

Neher, P.F., Maier-Hein, K.H., Jan. 2019. Simulated dMRI images and ground truth of random fiber phantoms in various configurations. Zenodo. https://doi.org/10.5281/zenodo.2533250. [Online]. Available:.

Neher, P., Maier-Hein, K., 2020a. Simulated MRI images and reference fiber tracts of 99 subjects. Ger. Cancer Res. Cent. https://doi.org/10.6097/e230-20200511_1.

Neher, P., Maier-Hein, K., 2020b. Sample data of the 99 simulated brains dataset. Zenodo. https://doi.org/10.5281/zenodo.4139626.

Neher, P.F., Houde, J.-C., Descoteaux, M., Maier-Hein, K., 2017. Tractography challenge ISMRM 2015 b=3000s/mm$^2$ data. Zenodo 10. https://doi.org/10.5281/zenodo.1007149.

Neher, P.F., Laun, F.B., Stieltjes, B., Maier-Hein, K.H., 2014. Fiberfox: facilitating the creation of realistic white matter software phantoms. Magn. Reson. Med. 72 (5), 1460–1470. https://doi.org/10.1002/mrm.25045.

Neher, P.F., Götz, M., Norajitra, T., Weber, C., Maier-Hein, K.H., 2015. A machine learning based approach to fiber tractography using classifier voting. In: *Medical Image Computing and Computer-Assisted Intervention—MICCAI* 2015, Cham, pp. 45–52, https://doi.org/10.1007/978-3-319-24553-9_6.

Neher, P.F., Côté, M.-A., Houde, J.-C., Descoteaux, M., Maier-Hein, K.H., 2017. Fiber tractography using machine learning. NeuroImage 158, 417–429. https://doi.org/10.1016/j.neuroimage.2017.07.028.

Ngattai, P.D., Belhomme, G., Ferrall, J., Patterson, B., Styner, M.A., Prieto, J.C., 2018. TRAFIC: fiber tract classification using deep learning. In: *Medical Imaging* 2018: Image Processing, Houston, United States, p. 37, https://doi.org/10.1117/12.2293931.

Panagiotaki, E., Schneider, T., Siow, B., Hall, M.G., Lythgoe, M.F., Alexander, D.C., 2012. Compartment models of the diffusion MR signal in brain white matter: a taxonomy and comparison. NeuroImage 59 (3), 2241–2254. https://doi.org/10.1016/j.neuroimage.2011.09.081.

Parker, G.J.M., Alexander, D.C., 2005. Probabilistic anatomical connectivity derived from the microscopic persistent angular structure of cerebral tissue. Philos. Trans. R. Soc. B Biol. Sci. 360 (1457), 893–902. https://doi.org/10.1098/rstb.2005.1639.

Pestilli, F., Yeatman, J.D., Rokem, A., Kay, K.N., Wandell, B.A., 2014. Evaluation and statistical inference for human connectomes. Nat. Methods 11 (10), 1058–1063. https://doi.org/10.1038/nmeth.3098.

Petit, L., Rheault, F., Descoteaux, M., Tzourio-Mazoyer, N., 2021. Half of the Streamlines Built in a Whole Human Brain Tractogram is Anatomically Uninterpretable. F1000Research 10, 111. https://doi.org/10.7490/f1000research.1118488.1.

Poulin, P., et al., 2017. Learn to track: deep learning for tractography. In: *Medical Image Computing and Computer Assisted Intervention—MICCAI* 2017, Cham, pp. 540–547, https://doi.org/10.1007/978-3-319-66182-7_62.

Poulin, P., Rheault, F., St-Onge, E., Jodoin, P.-M., Descoteaux, M., 2018. Bundle-Wise Deep Tracker: Learning to Track Bundle-Specific Streamline Paths.

Poulin, P., Jörgens, D., Jodoin, P.-M., Descoteaux, M., 2019. Tractography and machine learning: current state and open challenges. Magn. Reson. Imaging 64, 37–48. https://doi.org/10.1016/j.mri.2019.04.013.

Ratnarajah, N., Qiu, A., 2014. Multi-label segmentation of white matter structures: application to neonatal brains. NeuroImage 102, 913–922.

Reinke, A., 2018. How to exploit weaknesses in biomedical challenge design and organization. In: *Medical Image Computing and Computer Assisted Intervention—MICCAI* 2018, Cham, pp. 388–395, https://doi.org/10.1007/978-3-030-00937-3_45.

Reisert, M., Mader, I., Anastasopoulos, C., Weigel, M., Schnell, S., Kiselev, V., 2011. Global fiber reconstruction becomes practical. NeuroImage 54 (2), 955–962. https://doi.org/10.1016/j.neuroimage.2010.09.016.

Reisert, M., Coenen, V.A., Kaller, C., Egger, K., Skibbe, H., 2018. HAMLET: hierarchical harmonic filters for learning tracts from diffusion MRI. ArXiv180701068 Cs. Accessed 2018. [Online]. Available: http://arxiv.org/abs/1807.01068.

Rheault, F., et al., 2019. Bundle-specific tractography with incorporated anatomical and orientational priors. NeuroImage 186, 382–398. 01 https://doi.org/10.1016/j.neuroimage.2018.11.018.

Rheault, F., Poulin, P., Caron, A.V., St-Onge, E., Descoteaux, M., 2020a. Common misconceptions, hidden biases and modern challenges of dMRI tractography. J. Neural Eng. 17 (1), 011001. https://doi.org/10.1088/1741-2552/ab6aad.

Rheault, F., et al., 2020b. Tractostorm: the what, why, and how of tractography dissection reproducibility. Hum. Brain Mapp. 41 (7), 1859–1874. https://doi.org/10.1002/hbm.24917.

Schiavi, S., 2020. A new method for accurate in vivo mapping of human brain connections using microstructural and anatomical information. Sci. Adv. 6 (31), 8245. https://doi.org/10.1126/sciadv.aba8245.

Schilling, K.G., et al., 2019. Limits to anatomical accuracy of diffusion tractography using modern approaches. NeuroImage 185, 1–11. https://doi.org/10.1016/j.neuroimage.2018.10.029.

Schultz, T., Westin, C.-F., Kindlmann, G., 2010. Multi-diffusion-tensor fitting via spherical deconvolution: a unifying framework. In: *Medical Image Computing and Computer-Assisted Intervention—MICCAI* 2010, Berlin, Heidelberg, pp. 674–681, https://doi.org/10.1007/978-3-642-15705-9_82.

Siless, V., Chang, K., Fischl, B., Yendiki, A., 2018. AnatomiCuts: hierarchical clustering of tractography streamlines based on anatomical similarity. NeuroImage 166, 32–45. https://doi.org/10.1016/j.neuroimage.2017.10.058.

Skibbe, H., Reisert, M., 2017. Spherical tensor algebra: a toolkit for 3D image processing. J. Math. Imaging Vis. 58 (3), 349–381. https://doi.org/10.1007/s10851-017-0715-7.

Smith, R.E., Tournier, J.-D., Calamante, F., Connelly, A., 2013. SIFT: spherical-deconvolution informed filtering of tractograms. NeuroImage 67, 298–312. https://doi.org/10.1016/j.neuroimage.2012.11.049.

Smith, R.E., Tournier, J.-D., Calamante, F., Connelly, A., 2015. SIFT2: enabling dense quantitative assessment of brain white matter connectivity using streamlines tractography. NeuroImage 119, 338–351. https://doi.org/10.1016/j.neuroimage.2015.06.092.

Stieltjes, B., Brunner, R.M., Fritzsche, K., Laun, F., 2013. Diffusion Tensor Imaging: Introduction and Atlas. Springer Science & Business Media.

St-Onge, E., Daducci, A., Girard, G., Descoteaux, M., 2018. Surface-enhanced tractography (SET). NeuroImage 169, 524–539. https://doi.org/10.1016/j.neuroimage.2017.12.036.

Takemura, H., Caiafa, C.F., Wandell, B.A., Pestilli, F., 2016. Ensemble tractography. PLoS Comput. Biol. 12 (2), 1004692. https://doi.org/10.1371/journal.pcbi.1004692.

Théberge, A., Desrosiers, C., Descoteaux, M., Jodoin, P.-M., 2021. Track-to-Learn: a general framework for tractography with deep reinforcement learning. Med. Image Anal., 102093. https://doi.org/10.1016/j.media.2021.102093.

Tournier, J.-D., Calamante, F., Connelly, A., 2007. Robust determination of the fibre orientation distribution in diffusion MRI: non-negativity constrained super-resolved spherical deconvolution. NeuroImage 35 (4), 1459–1472. https://doi.org/10.1016/j.neuroimage.2007.02.016.

Tournier, J.D., Calamante, F., Connelly, A., 2010. Improved probabilistic streamlines tractography by 2nd order integration over fibre orientation distributions. In: *Proceedings of the international society for magnetic resonance in medicine.* vol. 1670.

Tournier, J.-D., Calamante, F., Connelly, A., 2012. MRtrix: diffusion tractography in crossing fiber regions. Int. J. Imaging Syst. Technol. 22 (1), 53–66. https://doi.org/10.1002/ima.22005.

Ugurlu, D., Firat, Z., Türe, U., Unal, G., 2018. Neighborhood resolved fiber orientation distributions (NRFOD) in automatic labeling of white matter fiber pathways. Med. Image Anal. 46, 130–145. https://doi.org/10.1016/j.media.2018.02.008.

Ugurlu, D., Firat, Z., Ture, U., Unal, G., 2019. Supervised classification of white matter fibers based on neighborhood fiber orientation distributions using an ensemble of neural networks. In: Bonet-Carne, E., Grussu, F., Ning, L., Sepehrband, F., Tax, C.M.W. (Eds.), Computational Diffusion MRI. Springer International Publishing, Cham, pp. 143–154.

Van Essen, D.C., Smith, S.M., Barch, D.M., Behrens, T.E.J., Yacoub, E., Ugurbil, K., 2013. The WU-Minn human connectome project: an overview. NeuroImage 80, 62–79. https://doi.org/10.1016/j.neuroimage.2013.05.041.

Vorburger, R.S., Reischauer, C., Boesiger, P., 2013. BootGraph: probabilistic fiber tractography using bootstrap algorithms and graph theory. NeuroImage 66, 426–435. https://doi.org/10.1016/j.neuroimage.2012.10.058.

Wassermann, D., et al., 2016. The white matter query language: a novel approach for describing human white matter anatomy. Brain Struct. Funct. 221 (9), 4705–4721. https://doi.org/10.1007/s00429-015-1179-4.

Wasserthal, J., Neher, P.F., Maier-Hein, K.H., Dec. 2017. High quality white matter reference tracts. Zenodo. https://doi.org/10.5281/zenodo.1088278.

Wasserthal, J., Neher, P., Maier-Hein, K.H., 2018. TractSeg—fast and accurate white matter tract segmentation. NeuroImage 183, 239–253. https://doi.org/10.1016/j.neuroimage.2018.07.070.

Wasserthal, J., Neher, P.F., Hirjak, D., Maier-Hein, K.H., 2019. Combined tract segmentation and orientation mapping for bundle-specific tractography. Med. Image Anal. 58, 101559. https://doi.org/10.1016/j.media.2019.101559.

Wegmayr, V., Giuliari, G., Holdener, S., Buhmann, J., 2018. Data-driven fiber tractography with neural networks. In: *2018 IEEE 15th International Symposium on Biomedical Imaging (ISBI 2018), Apr,* pp. 1030–1033, https://doi.org/10.1109/ISBI.2018.8363747.

Wegmayr, V., Giuliari, G., Buhmann, J.M., 2019. Entrack: a data-driven maximum-entropy approach to fiber tractography. In: *Pattern Recognition,* Cham, pp. 232–244, https://doi.org/10.1007/978-3-030-33676-9_16.

Wu, Y., 2020. Tract dictionary learning for fast and robust recognition of fiber bundles. In: Martel, A.L., Abolmaesumi, P., Stoyanov, D., Mateus, D., Zuluaga, M.A., Zhou, S.K., Racoceanu, D., Joskowicz, L. (Eds.), Medical Image Computing and Computer Assisted Intervention—MICCAI 2020. vol. 12267. Springer International Publishing, Cham, pp. 251–259.

Xu, H., Dong, M., Lee, M.-H., O'Hara, N., Asano, E., Jeong, J.-W., 2019. Objective detection of eloquent axonal pathways to minimize postoperative deficits in pediatric epilepsy surgery using diffusion tractography and convolutional neural networks. IEEE Trans. Med. Imaging 38 (8), 1910–1922. https://doi.org/10.1109/TMI.2019.2902073.

Yendiki, A., 2011. Automated probabilistic reconstruction of white-matter pathways in health and disease using an atlas of the underlying anatomy. Front. Neuroinform. 5 (23), 12–23.

Zhang, F., 2018. An anatomically curated fiber clustering white matter atlas for consistent white matter tract parcellation across the lifespan. NeuroImage 179, 429–447. https://doi.org/10.1016/j.neuroimage.2018.06.027.

Zhang, F., Goodlett, C., Hancock, E., Gerig, G., 2007. Probabilistic fiber tracking using particle filtering. In: *Medical Image Computing and Computer-Assisted Intervention—MICCAI* 2007, Berlin, Heidelberg, pp. 144–152, https://doi.org/10.1007/978-3-540-75759-7_18.

Zhang, H., Schneider, T., Wheeler-Kingshott, C.A., Alexander, D.C., 2012. NODDI: practical in vivo neurite orientation dispersion and density imaging of the human brain. NeuroImage 61 (4), 1000–1016. https://doi.org/10.1016/j.neuroimage.2012.03.072.

Zhang, M., Sakaie, K.E., Jones, S.E., 2013. Logical foundations and fast implementation of probabilistic tractography. IEEE Trans. Med. Imaging 32 (8), 1397–1410. https://doi.org/10.1109/TMI.2013.2257179.

Zhang, F., Karayumak, S.C., Hoffmann, N., Rathi, Y., Golby, A.J., O'Donnell, L.J., 2020. Deep white matter analysis (DeepWMA): fast and consistent tractography segmentation. Med. Image Anal. 65, 101761. https://doi.org/10.1016/j.media.2020.101761.

# Chapter 18

# Improving tractography using anatomical priors and multimodal integration

Etienne St-Onge[a], Gabriel Girard[b], Kurt G. Schilling[c], Alessandro Daducci[d], Samuel Deslauriers-Gauthier[e], Laurent Petit[f], and Maxime Descoteaux[g]

[a]*Department of Computer Science and Engineering, Université du Québec en Outaouais, Saint-Jérôme, QC, Canada,* [b]*Department of Computer Science, University of Sherbrooke, Sherbrooke, QC, Canada,* [c]*Department of Radiology and Radiological Sciences, Vanderbilt University Medical Center, Nashville, TN, United States,* [d]*Department of Computer Science, University of Verona, Verona, Italy,* [e]*CRONOS, Inria Centre at Université Côte d'Azur, France,* [f]*Groupe d'Imagerie Neurofonctionnelle, Institut des Maladies Neurodégénératives (GIN-IMN), UMR5293, CNRS, CEA, Université de Bordeaux, Bordeaux, France,* [g]*Sherbrooke Connectivity Imaging Laboratory (SCIL), Department of Computer Science, University of Sherbrooke, Sherbrooke, QC, Canada*

## 1 Introduction

Enhancing diffusion magnetic resonance imaging (dMRI) tractography with anatomical knowledge as priors can significantly improve the reconstruction of white matter (WM) architecture. This anatomical knowledge can encompass various levels of organization, ranging from macroscopic features such as gyri, sulci, gray nuclei, and WM bundles, to the microarchitecture at the level of neurons. By incorporating this valuable information, tractography algorithms can achieve more accurate and precise reconstructions of the complex WM pathways in the brain, overcoming the inherent limitations of dMRI, such as lower spatial resolution and complexities associated with crossing, kissing, and fanning fibers (Jbabdi and Johansen-Berg, 2011). Without such prior information, the reconstruction is prone to generating a large number of "false positive" streamlines and incomplete reconstruction of "true positive" WM bundles (Maier-Hein et al., 2017; Schilling et al., 2019b; Girard et al., 2020). This integration of anatomical knowledge as priors in dMRI tractography holds great promise in advancing our understanding of the intricate connectivity patterns in the brain and their role in brain function and neurological disorders.

As evident in Chapters 13, 14, 16, and 17, various approaches have been developed to enhance the precision of tractography algorithms. In this chapter, we elucidate how incorporating information such as tissue segmentation, geometrical constraints, microstructure, and functional MRI (fMRI) as priors into the tracking process can significantly improve tractography reconstruction.

## 2 Guiding tractography with tissue maps

As seen in previous chapters, the effectiveness of tractography algorithms heavily depends on the precise characterization of brain tissues to ensure accurate tracking through the spatial exploration process. Therefore brain tissue maps of WM, but also gray matter (GM) and cerebrospinal fluid (CSF), are needed to guide the tractography algorithm, in particular to determine where streamlines should start and stop (Smith et al., 2012; Girard et al., 2014). These maps are often named tissue "masks" when depicting the location of a single structure with a Boolean value (e.g., 0 for the background and 1 for the WM), or "probabilistic maps" when representing it with a probability or distribution. A frequent way to obtain a WM mask is by thresholding the DTI fractional anisotropy (FA) map around 0.15 (Lazar et al., 2003). However, FA thresholding has its limitations and is often imprecise in crossing fiber regions, in partial volume (PV) regions at tissue boundaries and at the boundaries of the brain mask. To address FA thresholding limitations, tissue maps estimated from a T1-weighted or T2-weighted image are generally a good option (see Fig. 1), since they are commonly acquired along dMRI and have typically higher spatial resolution than dMRI. Alternatively, WM maps can be estimated from the dMRI data using the fODF value (Tournier et al., 2019; Raffelt et al., 2012; Jeurissen et al., 2014) or obtained through dMRI compartment modeling (Canales-Rodríguez et al., 2015). This anatomical information is necessary to reliably assess structural connectivity and estimate WM architecture. Multiple approaches to integrate tissue maps into tractography algorithms are described in this

---

Handbook of Diffusion MR Tractography. https://doi.org/10.1016/B978-0-12-818894-1.00021-5
Copyright © 2025 Elsevier Ltd. All rights reserved, including those for text and data mining, AI training, and similar technologies.

**348 PART | III** Tractography algorithms

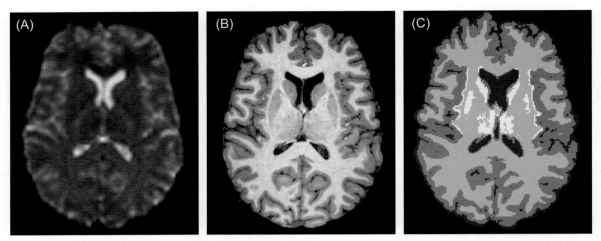

**FIG. 1** (A) $b = 0s/mm^2$ image, 2 mm isotropic resolution, (B) T1-weighted image, 1 mm isotropic, and (C) tissue segmentation from the T1w: subcortical GM (*white*), WM (*light gray*), GM (*gray*), and CSF (*black*).

section, such as anatomically constrained tractography (ACT) and particle filtering tractography (PFT) (Smith et al., 2012; Girard et al., 2014).

## 2.1 Binary tissue masks

To constrain the reconstruction of tractography inside WM voxels, tractography needs an input on valid tracking regions. One approach is to provide binary masks to describe where it can go and where it cannot. Anatomically, axons initiate and terminate in GM regions and propagate inside WM structures (Chapters 1 and 2). Thus both WM and GM binary masks can be provided to specify where streamlines, estimated with tractography, can go through (WM) and where they should stop (GM). These masks constitute a direct condition for every iteration of a streamline's reconstruction. Streamlines trajectory will progress inside WM and stop once they leave it. Different conditions can be applied based on endpoint location: for example, if both endpoints reach GM the pathway is considered "valid," and otherwise it is rejected.

### 2.1.1 Seeding mask

In addition to tissue segmentation, a seeding mask is often inputted to tractography algorithms to specify the starting location of streamlines. The starting point of each streamline, also called the "seed," is randomly chosen inside this seeding mask. This seeding mask can be directly derived from the previous WM or GM masks. Tractography's seeding approaches are separated into two families: WM seeding and WM/GM interface seeding. WM seeding methods initialize streamlines from a point inside WM and propagate them in two opposite directions. In opposition, GM seeding techniques start streamlines inside GM and only propagate them toward the WM. Analogously, WM-GM interface seeding can be seen as a GM seeding, since it starts along the GM and behaves in a similar way. These seeding approaches are illustrated in Fig. 2.

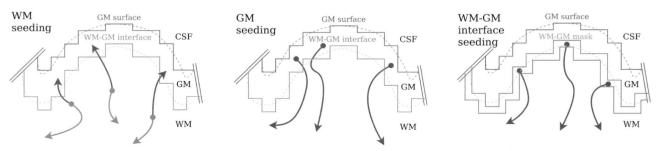

**FIG. 2** Comparison of streamline initialization from seed positions, represented with a *gray dot*, inside a chosen mask: WM, GM, and WM-GM interface seeding approaches. *(Reproduced with permission from St-Onge, E., 2021. Analyse de L'architecture de la Matière Blanche et Projection de Mesures sur la Surface Corticale (Ph.D. thesis).)*

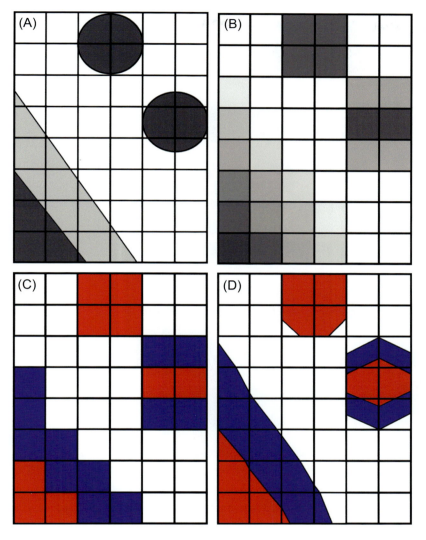

FIG. 3 (A) Toy example with three different tissue types. (B) Estimated image, where the partial volume (PV) effect is due to the averaging of different structures at each voxel. (C) Voxel-wise segmentation with an observable staircase artifact. (D) Surface-based segmentation using level-set.

### 2.1.2 Limitations of binary mask

Binary masks are prone to PV effects and a so-called staircase artifact caused by the image discretization. Illustrated in Fig. 3, these discretization artifacts impact the spatial precision of tractography near the interface of two different tissue structures. This PV effect impacts the accuracy of the segmentation, since multiple tissues values are averaged together. The staircase artifact affects the precision of the mask due to the domain discretization of an image, creating a cube-shaped profile, similar to a rasterization (Shirley et al., 2009). This type of artifact can be reduced with continuous approaches such as level-set (Osher and Sethian, 1988; see Fig. 3D).

## 2.2 Probabilistic tissue maps

State-of-the-art tractography methods, such as ACT (Smith et al., 2012) and PFT (Girard et al., 2014), take advantage of probabilistic brain tissue segmentation (see Fig. 4). These probabilistic segmentation maps, also called PV estimate maps, further increase the precision and facilitate image interpolation. The integration of probabilistic segmentation in tractography reduces PV effect, especially in-between tissues, such as the WM-GM interface and near the ventricles. It also enables the filtering of streamlines based on each pathway probability. The incorporation of these probabilistic maps further improves the tractography algorithm spatial precision when obtained from a higher resolution T1w image.

The ACT method interpolates probabilistic tissue maps to define tractography reconstruction behavior. At each position, PV estimate maps are describing the proportion of WM, GM, and CSF, such that $PV^{WM} + PV^{GM} + PV^{CSF} = 1$. If the current

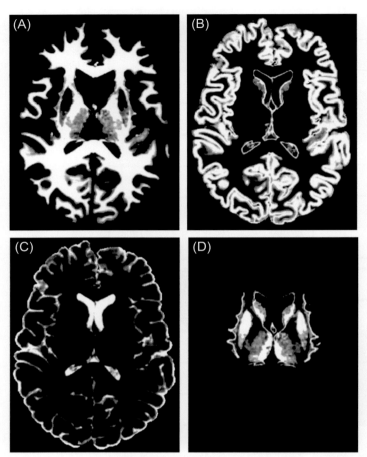

**FIG. 4** Probabilistic tissue maps of the (A) white matter (WM), (B) gray matter (GM), (C) cerebrospinal fluid (CSF), and (D) subcortical structures (SCs).

tracking location is constituted of a majority of WM ($PV^{WM} > 0.5$), the tractography continues normally. However, if a position with $PV^{GM} > 0.5$ is reached, the streamline propagation is stopped and this streamline is included in the tractogram. Otherwise, all streamlines containing *CSF* voxels are excluded from the final tractogram (regions where the $PV^{CSF}$ has the highest proportion). In the end, the resulting tractogram contains only streamlines going through WM with both endpoints in regions with $PV^{GM} > 0.5$ or with one endpoint exiting the brainstem, suitable for connectivity analysis.

For PFT, a continuous map criterion approach provides similar behavior by replacing threshold values (of 0.5) with probabilities. Thus going through a higher $PV^{GM}$ estimate results in a higher likelihood that a streamline propagation will stop and be included in the final tractogram. Similarly, in regions of high $PV^{CSF}$ the trajectories are likely to be excluded. Moreover, multiple tractography steps in voxels with low CSF or GM PVs are likely to result in stopping the streamline propagation. The probability of continuing the propagation at a position $p$ is defined as

$$P_p^{continue} = (1 - PV_p^{GM} - PV_p^{CSF})^{s/l},$$

where $s$ is the step size and $l$ is the average length of a voxel edge of PV maps. Streamlines stopping without reaching GM, with $PV^{GM} > 0.5$, are also rejected from the final tractogram. When propagation stops, the probability of including a streamline in the final tractography is defined by

$$P_p^{include} = \frac{PV_p^{GM}}{PV_p^{GM} + PV_p^{CSF}}.$$

When a streamline reaches a location leading to a trajectory rejection, ACT tries to find an alternative trajectory using an iterative backtracking approach. The tractography restarts at a previous location, with each rejection increasing the backtracking distance by one step. If this process leads to backtracking to the initial seed location, the streamline is removed from the tractogram. By avoiding invalid regions, this backtracking strategy allows the tractography algorithm to follow WM in narrow corridors or high curvature that would otherwise be challenging to reconstruct with conventional tractography algorithms.

Both ACT and CMC can be generalized to use other tissue volume fraction or probability maps to guide streamline propagation, such as including lesion maps to help determine whether the tracking should stop or continue, and if streamlines should be included or excluded. Also, subcortical structures (SCs) are often segmented as GM along with the cortex; however, some segmentation methods separate them into different labels (Fig. 4). Having a distinct label for SC can be useful when distinct starting or stopping criteria are needed (Smith et al., 2012).

### 2.2.1 Limitations of probabilistic mask

Nonetheless, incorporating tissue segmentation maps is still highly dependent on the precision of the segmentation algorithm. For example, voxels at the interface between WM and CSF are often mislabeled as GM due to PV effect, that is, averaging a low (CSF) and high (WM) intensity results in a value similar to GM. This phenomenon is depicted in Fig. 3 and can be observed in Fig. 4 GM around the ventricles, specifically in the GM segmentation. Recent advanced segmentation methods, employing atlases or machine learning, could reduce those limitations, resulting in a better reconstruction (Cabezas et al., 2011; Avants et al., 2011; Despotović et al., 2015; Dadar and Collins, 2021).

### 2.2.2 Tissue maps from dMRI

Instead of using T1- or T2-weighted images, tissue maps could be obtained directly from dMRI data. This reduces scanning time by removing the requirement on an anatomical image acquisition. Moreover, extracting different tissue maps directly from the dMRI also removes the necessity of a registration step between the diffusion volume and the anatomical space (Liu et al., 2007; Yap et al., 2015; Nie et al., 2018; Ciritsis et al., 2018; Cheng et al., 2020; Little and Beaulieu, 2021; Zhang et al., 2021; Theaud et al., 2022). However, doing a tissue segmentation only using the dMRI is challenging because it has a low spatial resolution, a lower contrast, and more distortions. To alleviate these problems, alternative approaches of registration and segmentation could combine both diffusion and anatomical data.

### 2.2.3 Segmentation recommendations

In general, segmenting tissues from a high resolution anatomical (T1) image results in better delineation of brain tissues; however, it requires the alignment of this anatomical image to the dMRI space. This registration can introduce blurring and a misalignment of brain structures, reducing the precision of those PV estimates maps. For this, the quality of the segmentation relies not only on a good registration but also on a good distortion correction (eddy and top-up) so that the geometry of brain in dMRI matches structural images. In opposition to binary masks that are limited to nearest-neighbor interpolation, probabilistic maps with scalar values can be continuously interpolated and transformed with minimal loss of precision.

## 2.3 Cortical surfaces

Recently, multiple surface-based tractography algorithms were proposed to improve the precision of the tractography near WM-GM interfaces (Cottaar et al., 2015; St-Onge et al., 2015). Since meshes are not restricted to the voxel grid, they can be created with finer details (from a higher resolution image) and avoid the staircase artifact (see Fig. 3). Cortical surfaces of the pial and the WM-GM interface are generally estimated from a high resolution T1w image using FreeSurfer (Dale et al., 1999), or other mesh reconstruction methods such as CLASP (MacDonald et al., 2000), CIVET (Kim et al., 2005), or BrainVisa (Auzias et al., 2013). Depicted in Fig. 5, the segmentation precision near the GM is essential for tractography methods to reconstruct pathways that reach the cortex accurately.

Cortical meshes were initially incorporated in the previously mentioned tractography algorithms (ACT and PFT), resulting in *Mesh-ACT* (MACT) (Yeh et al., 2017) and *surface-enhanced tractography* (SET) (St-Onge et al., 2018). Surfaces were also utilized in global tractography to better constrain streamline projection near the cortex (Teillac et al., 2017). Moreover, SCs and ventricles can be included for more advanced filtering or connectivity analysis (St-Onge et al., 2021).

### 2.3.1 Geometrical modeling

Geometrical approaches can be employed to mitigate tractography endpoint biases, such as gyral bias, since previous tractography algorithms overestimate the number of terminations along gyri (Van Essen et al., 2013; Reveley et al., 2015; Schilling et al., 2018). Along with the integration of cortical surfaces, St-Onge et al. (2018) and Cottaar et al. (2021) suggested estimating the WM fanning structure in the superficial WM. These geometrical models recreate the fanning structure of axons near the cortex, based on the shape of the brain (curvature, gyrification, and folding). Similar geometrical approaches are also used to estimate the cortical thickness and boundary between layers (Tallinen et al., 2016). This modeling further improves tractography reconstructions near the cortex and helps streamlines to reach GM regions. It

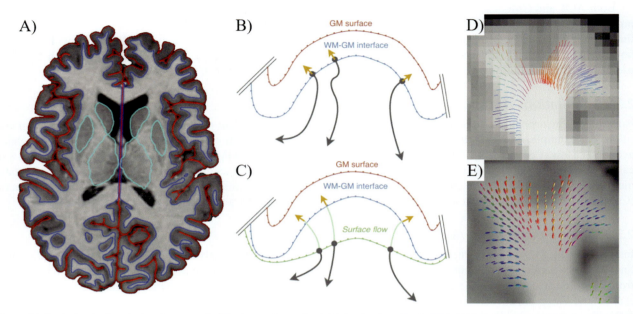

**FIG. 5** (A) Cortical and subcortical surfaces over the T1w image. Streamline initialization from (B) the WM-GM interface mesh, (C) with a geometrical flow initialization. This geometrical modeling of the fanning structure near the gray matter can be represented with (D) line trajectories or (E) local orientation distribution. *(Reproduced with permission from St-Onge, E., 2021. Analyse de L'architecture de la Matière Blanche et Projection de Mesures sur la Surface Corticale (Ph.D. thesis).)*

can also distribute connections more evenly along the cortex (St-Onge et al., 2021), mitigating gyral bias effect and improving the overall cortical coverage. Illustrated in Fig. 5, this can be used to both initialize pathways from the cortex toward the deep WM, and to project streamlines to the cortex.

### 2.3.2 Limitations of surfaces

The main limitation of a surface-based tractography algorithm is the requirement of an accurate reconstruction of cortical and subcortical surfaces. These estimated meshes frequently have geometrical imperfections; thus the precision of the resulting interface may vary from one reconstruction to another. Other types of meshing error (flipped facet, degenerate triangles) will also directly impact any geometrical modeling, which are sensitive to local estimation (area, curvature). However, advanced meshes processing can reduce or remove these issues. New algorithms, or further improvements to existing cortical surface reconstruction methods, could significantly improve mesh quality and precision, resulting in more accurate modeling (Ren et al., 2022; Huo et al., 2016; Rickmann et al., 2022; Ma et al., 2023; Bongratz et al., 2022; Gopinath et al., 2021). Finally, neurosurgical cases and abnormal anatomy can also be quite challenging for surface-based mesh reconstructions that should therefore be used with care on pathological anatomy (Chapter 19).

## 3 Anatomical constraints

Anatomical priors based on pathway shapes (length, curvature, and loop) or positions (endpoint location, traversing invalid regions) are a useful way to constrain or filter streamlines. Some of these priors can be integrated directly into the tractography algorithm or applied afterward as a filtering step. This is done especially for assessing brain pathology along a specific fascicle or for neurosurgical planning around a particular WM area (Urbanski et al., 2008; Assaf and Pasternak, 2008; Dayan et al., 2015). Streamline filtering approaches are well detailed in Chapters 21 and 22. This section will focus on methods that combine anatomical priors directly into the tractography algorithms.

### 3.1 ROI-based tractography

When looking for a specific bundle or region of interest (ROI), reconstructing a whole-brain tractogram is not always the best practice. In fact, the seeding map (described in Section 2.1) can be easily modified to focus on a specified ROI, such as the insula, for example, Ghaziri et al. (2017). Additionally, other tractography maps (tissue maps or atlases) can be adapted based on the target reconstruction, and then used to constrain streamlines (Wassermann et al., 2016). ROI-based tractography reconstruction reduces the computation time and increases the specificity of the WM reconstruction by reducing unwanted streamlines.

ROIs are often used to filter tractography with automatic algorithms or manual segmentation. This can be done to constrain or group WM streamlines based on their general location, that is, selecting specific way points or ending regions. When this knowledge is used prior to tractography, it limits the extent of the WM reconstruction. Most of these approaches often heavily depend on anatomical knowledge of WM fascicles and cortical regions, obtained from postmortem dissection or tractography virtual dissection from neuroanatomists.

This type of approach is employed by ExTractor to extract anatomically plausible streamlines of the human brain (Petit et al., 2019, 2023). From the most basic point of view, the human WM is organized in fiber tracts, that are aggregations of axons running in close apposition to each other and sharing cortical and/or subcortical origins and destinations. With this goal, ExTractor aims to reduce the complexity of connections and pathways from a whole-brain tractography employing a series of rules to organize the streamlines (Schmahmann and Pandya, 2006; Nieuwenhuys et al., 2007):

**(A)** Every cortical area is linked with other cortical and subcortical areas by pathways grouped into three distinct categories originally defined by Meynert (1885), namely the association, commissural, and projection fibers (Chapter 1).

**(B)** According to Schmahmann and Pandya (2006): commissural, projection, and long-range association fibers travel within the central WM part of the core of a gyrus, forming a convergence of fibers also known as "the cord." In other words, these fibers enter or leave a gyrus via a gyral stem and therefore never from the side of a gyrus by traversing a sulcus.

**(C)** Shorter U-shaped association fibers connect adjacent gyri by running in a thin band immediately beneath the GM constituting the most external part of the gyral stem.

Illustrated in Fig. 6, ExTractor employs these rules with ROIs from the Johns Hopkins University template (Oishi et al., 2009) to define anatomical filtering rules respecting the principles of brain organization listed earlier. The resulting filtered and partitioned tractogram, obtained from ROIs and anatomical rules, can be used as a set of plausible streamlines for further analysis.

Other specialized approaches adapt local diffusion models or tractography parameters in specific ROIs to improve the reconstruction. This can be used in edema or near glial tumor, where tractography pipelines need to be adapted to work properly (Deslauriers-Gauthier et al., 2018; Parker et al., 2018; Vanderweyen et al., 2020).

## 3.2 Bundle-specific tractography

Some dMRI and streamlines analysis often focus on a specific bundle of interest (BOI) instead of a whole-brain tractogram for connectivity analysis. The implementation is similar to a ROI-based tractography, where the tractography is adapted to reconstruct a specific bundle. With this method, each BOI can be reconstructed from a series of anatomical rules. These bundle priors can be manually given or automatically extracted from a template.

This idea of reconstructing each bundle separately from anatomical priors was recently implemented in XTRACT (Warrington et al., 2020). By incorporating tractography masks and seeding regions adapted to specific WM pathways, XTRACT is able to reliably reconstruct 42 WM bundles. BOI-specific rules are based on anatomical knowledge from previous WM study and analysis from both human and macaque, detailed in the supplementary material of Warrington et al. (2020). A BOI-based tractography can be described in a few series of operations:

**(1)** Generate seeding, termination, and tracking maps from prior anatomical studies, WM analysis, or brain atlases.

**(2)** For each subject, compute the bundle-specific tractography (BST) with the previously generated maps, and by adapting tractography parameters (maximum curvature, streamline lengths, etc.) for each WM pathway.

**(3)** Filter generated streamlines based on the prescribed template, using anatomical descriptors (location) and filtering methods.

When adapting local model orientations, BOI-based tractography can overcome some typical reconstruction limitations (Jbabdi and Johansen-Berg, 2011), such as traversing crossing regions (Rheault et al., 2019). In addition, this single bundle reconstruction has the flexibility to change and optimize the tractography algorithm or parameters for the desired bundle (Girard et al., 2014; Takemura et al., 2016). Recently, a few tractography methods have used directional enhancement to better reconstruct a specific WM bundle: MAGNEtic Tractography (MAGNET) (Chamberland et al., 2017), BST (Rheault et al., 2019), and tract segmentation and orientation mapping (TrackSeg) (Wasserthal et al., 2019). First, MAGNET was proposed to overcome a difficult fiber crossing configuration in tracking the optic radiations from thalamus, by manually

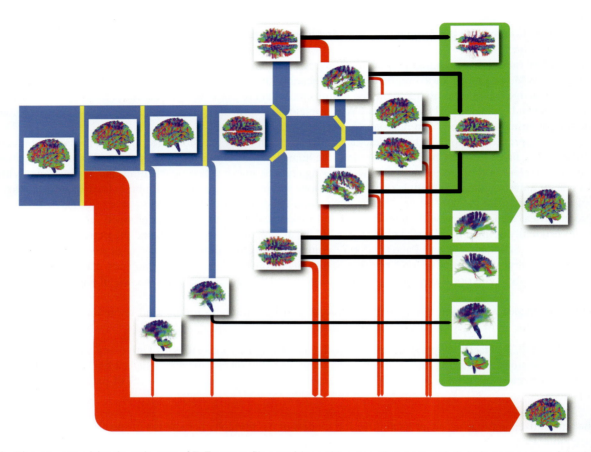

**FIG. 6** Diagram summarizing the main steps of ExTractor to filter out false-positive streamlines from a whole-brain tractogram by following brain neuroanatomy organizational principles. In this example, an original tractogram of a single subject initially composed of more than 3.5 million streamlines provides a final whole-brain tractogram of around 590,000 anatomically plausible streamlines once a series of filters (anatomical rules schematically represented with *yellow lines*) are applied. *(Reproduced with permission from Petit, L., Ali, K.M., Rheault, F., Boré, A., Cremona, S., Corsini, F., De Benedictis, A., Descoteaux, M., Sarubbo, S., 2023. The structural connectivity of the human angular gyrus as revealed by microdissection and diffusion tractography. Brain Struct. Funct. 228 (1), 103–120.)*

encouraging streamlines to curve more and be attracted by local directions pointing toward the occipital lobe. In opposition, BST and TrackSeg automatically generate directional priors and adapt tissue maps based on a template bundle without any manual intervention. For BST, directional priors are estimated using Track Orientation Density Imaging (Dhollander et al., 2014), whereas TrackSeg employs deep-learning techniques to generate those directions.

These methods are able to enhance the local model directions for a chosen BOI, by reconstructing a single bundle at a time with orientational priors (see Fig. 7). This type of approach significantly reduces the complexity of the tractography reconstruction by focusing on a single pathway, thus avoiding the crossing/kissing/branching problem. Thus BOI-based tractography can greatly improve tractography reconstruction in selected key areas. This can be more significant in areas where tissue segmentations have difficulty (e.g., narrow regions with high PV effect), since it also incorporates tracking maps (described in Section 3.2). This approach is comparable to a data-driven streamline registration, where the WM structure is reconstructed if the subject directions are coherent with the atlas.

### 3.2.1 Region-based and bundle-specific tractography

Overall, adapting tracking maps and masks in tractography serves to minimize, or even eliminate, local uncertainty in the tracking process. However, both ROI and BOI-based approaches heavily depend on strong anatomical priors and rely on atlases and templates. Additionally, they are optimal for only a limited number of WM pathways with established anatomical knowledge from postmortem dissection and tracing validation. In future tractography algorithms, the integration of additional anatomical rules could further enhance the reconstruction of WM pathways. These methods significantly reduce reconstruction variability and are particularly useful when reconstructing specific bundles or connections.

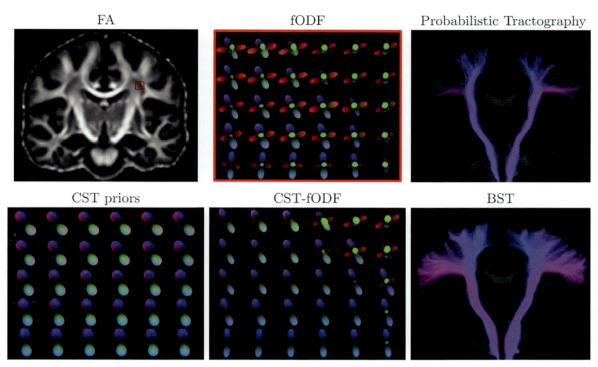

**FIG. 7** Bundle-specific tractography (BST) reconstruction of the corticospinal tract (CST) employing enhanced CST-fODF, combining fODF with CST prior orientations. *(Reproduced with permission from Rheault, F., St-Onge, E., Sidhu, J., Chenot, Q., Petit, L., Descoteaux, M., 2018. Bundle-specific tractography. In: Computational Diffusion MRI: MICCAI Workshop, Québec, Canada, September 2017, pp. 129–139.)*

## 4 Tractography with microstructure information

Recent work in microstructural mapping, in particular axon radii mapping (Veraart et al., 2020), suggests that while microstructural properties are not perfectly homogeneous along a pathway, they are at least continuous and slowly varying. These WM properties, often obtained from diffusion models, can be exploited to regularize tractography connectivity or orientations.

### 4.1 Filtering tractograms with microstructure information

Proposed by Daducci et al. (2015), the COMMIT framework exploits microstructural properties in a top-down approach by associating a microstructure model of the neuronal tissue to streamlines to improve the anatomical accuracy of the estimated tractogram and the biological interpretability of the reconstructions. COMMIT is detailed in Chapter 16 as well as other tractogram filtering techniques such as SIFT and LIFE (Smith et al., 2015; Qi et al., 2016). Combining tractography with microstructure information allows filtering from an input tractogram those streamlines with trajectories that are inconsistent with the imaging data and the chosen model, as well as estimating the actual contribution, or weight, of the plausible ones. In the original formulation (Daducci et al., 2015), a stick model represents the intraaxonal dMRI signal, but COMMIT was later extended to consider multimodal data and more advanced microstructure models. For instance, in Schiavi et al. (2022) the candidate collection of streamlines is fit to a myelin water fraction map, thus allowing one to infer the myelin content of fibers along the plausible trajectories. In addition, Barakovic et al. (2021a) proposed an extension to allow COMMIT to use a more advanced biophysical model for microstructure and map the axon diameter index along the trajectories, whereas in Barakovic et al. (2021b) the tractogram is fit to multiecho diffusion data to estimate the distribution of T2 relaxation times of each streamline.

It is important to note that being able to weigh streamlines based on microstructural measures provides a way to quantitatively map microstructure to the streamlines, and provides an alternative to using streamline count for quantitative connectivity analysis (Chapters 16 and 24).

## 4.2 Guiding trajectories using microstructure information

Girard et al. (2017) proposed a bottom-up approach to microstructure-informed tractography mapping with the AxTract method, which extends deterministic streamline propagation methods (Chapter 13) by adding the active use of the local axon diameter information to guide trajectories. In this specific work, the AMICOX (Auría et al., 2015) framework was employed to estimate the diameter information for every peak of the fODF at the tracking location. Classical FOD-based deterministic streamline tractography methods iteratively follow the local peaks forming the smallest angle with the previous tracking orientation among all peaks of the voxels. AxTract relaxes this constraint by allowing the tracking algorithm to select any peaks within a predefined maximum angle. Instead of solely minimizing the angular deviation, AxTract method selectively chooses the local WM orientation based on estimated diameter information that closely aligns with the diameter information of the previously tracked orientation. Assuming a microstructure estimation method provides a reliable estimate of the axon diameter differences from the dMRI data, AxTract can change the propagation direction to resolve complex WM configurations (Girard et al., 2017).

Although AxTract employs axon diameter measures, this concept could be extended to other microstructure properties that are constant or change smoothly along the trajectory of WM bundles, such as myelin or T2 relaxation time. Even though it relies on the estimation of such diffusion properties for each peak independently, increasing the complexity of the model fit, it could improve the anatomical accuracy of the reconstructions by increasing its coherence with the underlying microstructure environment (Daducci et al., 2016).

## 5 Tractography with functional maps

In a parallel to dMRI applications that map the structural connectivity of the brain, functional connectivity is used and interpreted as mapping the functional activations of the brain. With the intuition that brain structure guides brain function, functional signal contrast seems a logical modality to guide or inform tractography algorithms. fMRI relies on detecting small signal changes often associated with neural activity, an observation known as the blood oxygenation level-dependent (BOLD) effect (Friston et al., 1994; Attwell and Iadecola, 2002; Logothetis, 2003; Poldrack et al., 2011). For example, an experiment may include presenting a stimulus or performing a task (such as a visual stimulus or motor task), which leads to neural activity, hemodynamic changes, and a detectable MRI signal change. With appropriate analysis tools, locations of signal changes may be related to specific functions. Additionally, rather than using a stimulus, recording the brain at rest (i.e., resting state fMRI), allows calculating correlations in time between MRI signals from different regions, which are widely used to reveal neural circuits (e.g., motor, language, or default mode networks).

In comparison to fMRI, electroencephalography (EEG) and magnetoencephalography (MEG) provide a direct measure of the electrical brain activity at a millisecond resolution (Baillet, 2017; Baillet et al., 2001; Schomer and da Silva, 2017). However, because the sensors are located at the surface of the scalp or near the head, an ill-posed linear problem must be solved to extract brain activity from the recordings. Solving this problem typically involves the computation of a volume conductor model from anatomical MRI and in some cases dMRI. This volume conductor is then used to solve Maxwell's equations, referred to as solving the forward problem. Given the forward solution, brain activity can be recovered by inverting the model under suitable constraints, for example, minimum norm or smoothness.

### 5.1 Functional regions

Functional areas revealed through functional activations have been used to inform, guide, and filter tractography. A simple, and intuitive, example is utilizing functionally defined ROIs to initialize or filter tractography streamlines. This is employed to further explore the link between brain function and structure. It also allows a direct visualization of the structural substrate (i.e., the WM pathways) associated with areas of the activated cortex. This has been applied in both tasks and resting state fMRI, for example, generating the underlying structural connections of resting state networks (Preti et al., 2012; Chamberland et al., 2015). Additionally, a streamline filtering step can be performed by extracting the BOLD time series from the endpoints of streamlines and performing correlations with the rest of the brain, keeping only those for which there is evidence of functional connections.

A similar idea can also be applied to more invasive imaging modalities like electrocorticography. Here, the location of the electrodes placed directly on the cortical surface is recorded, in addition to the location of electrical stimulation sites. By using the locations of every stimulation and electrode pair as regions of interest, the underlying WM fibers supporting the functional activations can be identified (Conner et al., 2011). This allows the in vivo quantification of physiological parameters of WM bundles, for example, their conduction velocity (Filipiak et al., 2021) and to relate them to microstructural

Improving tractography using anatomical priors and multimodal integration **Chapter | 18 357**

information provided by dMRI. The limitation of these approaches is their extreme invasiveness, allowing the procedure to be performed only on patients already scheduled for neurosurgery, for example, during tumor resection.

In general, this approach is similar to ROI-based tractography (described in Section 3), where regions are not obtained from anatomical knowledge but from subject-wise activation maps. This can further improve tractography reconstruction by integrating those functional maps. It can also segment and compare resulting streamlines based on activation correlation. These analyses may provide insight into structure-function relationships and a subject-specific basis and may facilitate neurosurgical applications where the underlying structural network associated with a stimulated cortex is of paramount importance.

## 5.2 Functional connectivity

Structural connectivity, obtained from tractography along with diffusion measures, is often complemented with additional functional measures. For example, the functional correlations matrix can be used in addition to the dMRI connectivity to simultaneously map structure and function (Honey et al., 2009; Calamante et al., 2013; Abdelnour et al., 2018; Becker et al., 2018; Deslauriers-Gauthier et al., 2020b).

Recently, a way to combine functional signals from fMRI with tractography to gain insight into functional WM has been introduced, called the functionnectome (Nozais et al., 2021). Here, tractography-derived anatomical priors that describe the connectivity from every voxel to every other brain voxel are used to project GM fMRI signals into the WM. Traditional fMRI analysis can then be performed on this whole-brain BOLD time series. This framework has been used to analyze the joint GM and WM contribution to functional networks for motor systems, working memory, and language functions or to analyze the joint contribution to resting state networks of the brain and shows promise in using tractography to increase statistical power of fMRI in the WM (Nozais et al., 2021). Currently the functionnectome is based on a population-average tractogram in a standard space, but projecting functional data into WM can also be done for subject-specific tractograms.

Strategies to inject WM connectivity into the MEG and EEG inverse problem, which estimates cortical activity from recordings, have also been proposed. In addition to improving the ill-posed nature of the problem by limiting the space of solutions, these approaches also allow the projection of functional information onto the WM substrate. For example, because of the millisecond temporal resolution provided by EEG and MEG, the WM introduces observable delays into MEG and EEG data (Drakesmith and Jones, 2019). This provides a unique opportunity to estimate the directionality of WM connections, which is not possible from dMRI alone. A first strategy is to represent brain network dynamics using a multivariate autoregressive (MAR) model with constraints on the coefficients being driven by both dMRI and fMRI. The parameters of the model are then estimated using a variational Bayesian algorithm to explain the MEG data, yielding an estimate of the effective connectivity (Fukushima et al., 2015; Filatova et al., 2018). A second strategy is to model the possible interactions between brain regions at different times using a Bayesian network derived from dMRI. In addition to being subject specific, this Bayesian network also encodes tract-specific information, such as delays. By introducing EEG or MEG measurements as evidence into this model, it is possible to infer information flow in WM connections at each time sample for both the afferent and efferent directions (Deslauriers-Gauthier et al., 2019, 2020a). Finally, more straightforward linear approaches have also been proposed to constrain the EEG and MEG inverse problem using structural connectivity (Hammond et al., 2013; Kojčić et al., 2021).

The methods mentioned make use of functional information to complement, enhance, or prune the structural connectivity captured by the tractogram, but never reconsider its relation to the dMRI data. In contrast, it is possible to fit the dMRI data in a manner similar to COMMIT, but simultaneously promoting fiber bundles that are coherent with resting state fMRI measurements. The resulting tractogram is not only coherent with the structural data, but also in agreement with the functional connectivity matrix (Frigo et al., 2019).

Furthermore, mapping streamline endpoints along cortical surfaces (presented in Section 2.3) is a compelling way to apply multimodal research. This integration of cortical surfaces with meshes facilitates the analysis of tractography and other diffusion measures along the cortex. Meshes (vertices) can be used as a domain to analyze cortical measures (thickness, area, and volume), structural connectivity (tractography and dMRI metrics), and functional activations (fMRI, EEG, and MEG). A similar approach was recently employed by Cole et al. (2021) to analyze structural and functional connectivity with a continuous representation over cortical surfaces.

## 5.3 Functional local orientation

In the WM, fMRI BOLD contrast has traditionally been ignored, or considered as false positive results. This is because its origins may be unclear, with lower blood flow in WM, and it is not clear if there is a hemodynamic response to required

**358 PART | III** Tractography algorithms

**FIG. 8** Resting state signals have local anisotropic correlations within neighborhoods, with directionality that largely follows the underlying white matter structure. (A) T1-weighted structural image, (B) local functional correlations, (C) diffusion tensors in the same region, and (D) an RGB colormap obtained without any diffusion weighting, and using only correlations in resting state fluctuations. *(Reproduced with permission from Ding, Z., Xu, R., Bailey, S.K., Wu, T.L., Morgan, V.L., Cutting, L.E., Anderson, A.W., Gore, J.C., 2016. Visualizing functional pathways in the human brain using correlation tensors and magnetic resonance imaging. Magn. Reson. Imaging 34 (1), 8–17.)*

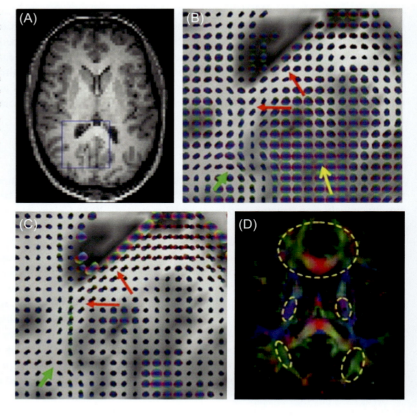

energy changes. However, recent studies have suggested that although BOLD effects are weaker in the WM, using appropriate detection and analysis methods they are robustly detectable in both resting and task fMRI. First, resting state signals have been shown to have local anisotropic correlation within neighborhoods (i.e., correlation with neighbors may be stronger in anterior-posterior directions than superior-inferior, for example) prompting fitting of local correlation coefficients to tensor fits (in a technique dubbed function tensor imaging) (Ding et al., 2016; Schilling et al., 2019a). Displayed in Fig. 8, these maps look very similar to what is expected from diffusion tensor imaging and diffusion HARDI techniques, and have been used for "functional" tractography, which suggests that these measures of BOLD signal directionality may be used to inform connectivity measures.

Over and above orientation information, task-induced BOLD activations have been shown to occur in WM, suggesting a WM hemodynamic response function analogous to that in GM (Li et al., 2019). This could potentially identify or segment regions of WM based on response similarity, or response to specific stimuli, and be used to inform tractography algorithms. One could imagine using a motor task, in combination with activation regions and mapping of hemodynamic response functions, to cluster, filter, or propagate streamlines. Alternatively, correlations between GM cortical regions and specific WM regions may be a way to identify or segment specific patterns of activated regions that could be used for subsequent filtering. However, these directional correlations with functional signals could also be due to vascular drainage effects or correlations caused by susceptibility fields (Andersson et al., 2020). Overall, studies of both WM and GM bold signals may not only provide new insights into brain function, but present opportunities to understand how function and structure are linked.

## 6 Conclusion

Several ways exist to improve tractography by integrating anatomical priors. The use of segmentation maps or surfaces extracted from high-resolution anatomical images is crucial for improving tractography accuracy, particularly near the WM-GM interface. Surface-based modeling can further enhance the quality of streamlines near the cortex and superficial WM, decreasing end-point biases. However, aligning these segmentation maps to diffusion volumes perfectly is challenging.

Anatomical constraints should be used, when applicable, to decrease the number of invalid streamlines. The difficulty of using anatomical constraints is often increased in brain disease and pathological brains. Care should be taken in such applications. As such, when studying specific WM pathways, using ROIs can significantly reduce erroneous streamlines in the analysis, focusing results on the desired pathway and trajectory. In addition, exploiting presegmented tractography bundles from a WM atlas is a powerful way to enhance local orientations and filter streamlines with automatic procedures. Future neuroanatomical studies focusing on WM organization will provide more priors and better validations for current tractography methods.

Microstructure measures, from diffusion MRI or any other acquisitions, along WM trajectory provides regularization that can be exploited for both quantitative connectivity matrix weighting. Some information could also be extracted from functional imaging, providing tractography with temporal data. These additional features provide unique connectivity characteristics that could alleviate the complexity of the brain structural network. These applications remain preliminary and should be further researched to fully take advantage of their potential.

# References

Abdelnour, F., Dayan, M., Devinsky, O., Thesen, T., Raj, A., 2018. Functional brain connectivity is predictable from anatomic network's Laplacian Eigenstructure. NeuroImage 172, 728–739.

Andersson, M., Kjer, H.M., Rafael-Patino, J., Pacureanu, A., Pakkenberg, B., Thiran, J.P., Ptito, M., Bech, M., Dahl, A.B., Dahl, V.A., et al., 2020. Axon morphology is modulated by the local environment and impacts the noninvasive investigation of its structure-function relationship. Proc. Natl. Acad. Sci. USA 117 (52), 33649–33659.

Assaf, Y., Pasternak, O., 2008. Diffusion tensor imaging (DTI)-based white matter mapping in brain research: a review. J. Mol. Neurosci. 34, 51–61.

Attwell, D., Iadecola, C., 2002. The neural basis of functional brain imaging signals. Trends Neurosci. 25 (12), 621–625.

Auría, A., Romascano, D., Canales-Rodriguen, E., Wiaux, Y., Dirby, T.B., Alexander, D., Thiran, J.P., Daducci, A., 2015. Accelerated microstructure imaging via convex optimisation for regions with multiple fibres (AMICOx). In: 2015 IEEE International Conference on Image Processing (ICIP), pp. 1673–1676.

Auzias, G., Lefevre, J., Le Troter, A., Fischer, C., Perrot, M., Régis, J., Coulon, O., 2013. Model-driven harmonic parameterization of the cortical surface: HIP-HOP. IEEE Trans. Med. Imaging 32 (5), 873–887.

Avants, B.B., Tustison, N.J., Wu, J., Cook, P.A., Gee, J.C., 2011. An open source multivariate framework for n-tissue segmentation with evaluation on public data. Neuroinformatics 9 (4), 381–400.

Baillet, S., 2017. Magnetoencephalography for brain electrophysiology and imaging. Nat. Neurosci. 20 (3), 327–339.

Baillet, S., Mosher, J.C., Leahy, R.M., 2001. Electromagnetic brain mapping. IEEE Signal Process. Mag. 18 (6), 14–30.

Barakovic, M., Girard, G., Schiavi, S., Romascano, D., Descoteaux, M., Granziera, C., Jones, D.K., Innocenti, G.M., Thiran, J.P., Daducci, A., 2021a. Bundle-specific axon diameter index as a new contrast to differentiate white matter tracts. Front. Neurosci. 15, 687.

Barakovic, M., Tax, C.M.W., Rudrapatna, U., Chamberland, M., Rafael-Patino, J., Granziera, C., Thiran, J.P., Daducci, A., Canales-Rodríguez, E.J., Jones, D.K., 2021b. Resolving bundle-specific intra-axonal T2 values within a voxel using diffusion-relaxation tract-based estimation. NeuroImage 227, 117617.

Becker, C.O., Pequito, S., Pappas, G.J., Miller, M.B., Grafton, S.T., Bassett, D.S., Preciado, V.M., 2018. Spectral mapping of brain functional connectivity from diffusion imaging. Sci. Rep. 8 (1), 1–15.

Bongratz, F., Rickmann, A.M., Pölsterl, S., Wachinger, C., 2022. Vox2cortex: fast explicit reconstruction of cortical surfaces from 3D MRI scans with geometric deep neural networks. In: Proceedings of the IEEE/CVF Conference on Computer Vision and Pattern Recognition, pp. 20773–20783.

Cabezas, M., Oliver, A., Lladó, X., Freixenet, J., Cuadra, M.B., 2011. A review of atlas-based segmentation for magnetic resonance brain images. Comput. Methods Programs Biomed. 104 (3), e158–e177.

Calamante, F., Masterton, R.A.J., Tournier, J.D., Smith, R.E., Willats, L., Raffelt, D., Connelly, A., 2013. Track-weighted functional connectivity (TW-FC): a tool for characterizing the structural-functional connections in the brain. NeuroImage 70, 199–210.

Canales-Rodríguez, E.J., Daducci, A., Sotiropoulos, S.N., Caruyer, E., Aja-Fernández, S., Radua, J., Mendizabal, J.M.Y., Iturria-Medina, Y., Melie-García, L., Alemán-Gómez, Y., et al., 2015. Spherical deconvolution of multichannel diffusion MRI data with non-Gaussian noise models and spatial regularization. PLoS One 10 (10), e0138910.

Chamberland, M., Bernier, M., Fortin, D., Whittingstall, K., Descoteaux, M., 2015. 3D interactive tractography-informed resting-state fMRI connectivity. Front. Neurosci. 9, 275.

Chamberland, M., Scherrer, B., Prabhu, S.P., Madsen, J., Fortin, D., Whittingstall, K., Descoteaux, M., Warfield, S.K., 2017. Active delineation of Meyer's loop using oriented priors through MAGNEtic tractography (MAGNET). Hum. Brain Mapp. 38 (1), 509–527.

Cheng, H., Newman, S., Afzali, M., Fadnavis, S.S., Garyfallidis, E., 2020. Segmentation of the brain using direction-averaged signal of DWI images. Magn. Reson. Imaging 69, 1–7.

Ciritsis, A., Boss, A., Rossi, C., 2018. Automated pixel-wise brain tissue segmentation of diffusion-weighted images via machine learning. NMR Biomed. 31 (7), e3931.

Cole, M., Murray, K., St-Onge, E., Risk, B., Zhong, J., Schifitto, G., Descoteaux, M., Zhang, Z., 2021. Surface-based connectivity integration: an atlas-free approach to jointly study functional and structural connectivity. Hum. Brain Mapp. 42 (11), 3481–3499.

Conner, C.R., Ellmore, T.M., DiSano, M.A., Pieters, T.A., Potter, A.W., Tandon, N., 2011. Anatomic and electro-physiologic connectivity of the language system: a combined DTI-CCEP study. Comput. Biol. Med. 41 (12), 1100–1109.

Cottaar, M., Jbabdi, S., Glasser, M.F., Dikranian, K., Van Essen, D.C., Behrens, T.E., Sotiropoulos, S.N., 2015. A generative model of white matter axonal orientations near the cortex. In: International Society of Magnetic Resonance in Medicine (ISMRM).

Cottaar, M., Bastiani, M., Boddu, N., Glasser, M.F., Haber, S., Van Essen, D.C., Sotiropoulos, S.N., Jbabdi, S., 2021. Modelling white matter in gyral blades as a continuous vector field. NeuroImage 227, 117693.

Dadar, M., Collins, D.L., 2021. BISON: brain tissue segmentation pipeline using T1-weighted magnetic resonance images and a random forest classifier. Magn. Reson. Med. 85 (4), 1881–1894.

Daducci, A., Canales-Rodríguez, E.J., Zhang, H., Dyrby, T.B., Alexander, D.C., Thiran, J.P., 2015. Accelerated microstructure imaging via convex optimization (AMICO) from diffusion MRI data. NeuroImage 105, 32–44.

Daducci, A., Dal Palú, A., Descoteaux, M., Thiran, J.P., 2016. Microstructure informed tractography: pitfalls and open challenges. Front. Neurosci. 10, 247.

Dale, A.M., Fischl, B., Sereno, M.I., 1999. Cortical surface-based analysis: I. Segmentation and surface reconstruction. NeuroImage 9 (2), 179–194.

Dayan, M., Munoz, M., Jentschke, S., Chadwick, M.J., Cooper, J.M., Riney, K., Vargha-Khadem, F., Clark, C.A., 2015. Optic radiation structure and anatomy in the normally developing brain determined using diffusion MRI and tractography. Brain Struct. Funct. 220 (1), 291–306.

Deslauriers-Gauthier, S., Parker, D., Rheault, F., Deriche, R., Brem, S., Descoteaux, M., Verma, R., 2018. Edema-informed anatomically constrained particle filter tractography. In: International Conference on Medical Image Computing and Computer-Assisted Intervention, pp. 375–382.

Deslauriers-Gauthier, S., Lina, J.M., Butler, R., Whittingstall, K., Gilbert, G., Bernier, P.M., Deriche, R., Descoteaux, M., 2019. White matter information flow mapping from diffusion MRI and EEG. NeuroImage 201, 116017.

Deslauriers-Gauthier, S., Costantini, I., Deriche, R., 2020a. Non-invasive inference of information flow using diffusion MRI, functional MRI, and MEG. J. Neural Eng. 17 (4), 045003.

Deslauriers-Gauthier, S., Zucchelli, M., Frigo, M., Deriche, R., 2020b. A unified framework for multimodal structure-function mapping based on eigenmodes. Med. Image Anal. 66, 101799.

Despotović, I., Goossens, B., Philips, W., 2015. MRI segmentation of the human brain: challenges, methods, and applications. Comput. Math. Methods Med. 2015, 450341.

Dhollander, T., Emsell, L., Van Hecke, W., Maes, F., Sunaert, S., Suetens, P., 2014. Track orientation density imaging (TODI) and track orientation distribution (TOD) based tractography. NeuroImage 94, 312–336.

Ding, Z., Xu, R., Bailey, S.K., Wu, T.L., Morgan, V.L., Cutting, L.E., Anderson, A.W., Gore, J.C., 2016. Visualizing functional pathways in the human brain using correlation tensors and magnetic resonance imaging. Magn. Reson. Imaging 34 (1), 8–17.

Drakesmith, M., Jones, D.K., 2019. Mapping axon conduction delays in vivo from microstructural MRI. bioRxiv, 503763.

Filatova, O.G., Yang, Y., Dewald, J.P.A., Tian, R., Maceira-Elvira, P., Takeda, Y., Kwakkel, G., Yamashita, O., Van der Helm, F.C.T., 2018. Dynamic information flow based on EEG and diffusion MRI in stroke: a proof-of-principle study. Front. Neural Circuits 12, 79.

Filipiak, P., Almairac, F., Papadopoulo, T., Fontaine, D., Mondot, L., Chanalet, S., Deriche, R., Clerc, M., Wassermann, D., 2021. Towards linking diffusion MRI based macro- and microstructure measures with cortico-cortical transmission in brain tumor patients. NeuroImage 226, 117567.

Frigo, M., Costantini, I., Deriche, R., Deslauriers-Gauthier, S., 2019. Resolving the crossing/kissing fiber ambiguity using functionally informed commit. In: International Conference on Medical Image Computing and Computer-Assisted Intervention, pp. 335–343.

Friston, K.J., Jezzard, P., Turner, R., 1994. Analysis of functional MRI time-series. Hum. Brain Mapp. 1 (2), 153–171.

Fukushima, M., Yamashita, O., Knösche, T.R., Sato, M.A., 2015. MEG source reconstruction based on identification of directed source interactions on whole-brain anatomical networks. NeuroImage 105, 408–427.

Ghaziri, J., Tucholka, A., Girard, G., Houde, J.C., Boucher, O., Gilbert, G., Descoteaux, M., Lippé, S., Rainville, P., Nguyen, D.K., 2017. The cortico-cortical structural connectivity of the human insula. Cereb. Cortex 27 (2), 1216–1228.

Girard, G., Whittingstall, K., Deriche, R., Descoteaux, M., 2014. Towards quantitative connectivity analysis: reducing tractography biases. NeuroImage 98, 266–278.

Girard, G., Daducci, A., Petit, L., Thiran, J.P., Whittingstall, K., Deriche, R., Wassermann, D., Descoteaux, M., 2017. Ax tract: toward microstructure informed tractography. Hum. Brain Mapp. 38 (11), 5485–5500.

Girard, G., Caminiti, R., Battaglia-Mayer, A., St-Onge, E., Ambrosen, K.S., Eskildsen, S.F., Krug, K., Dyrby, T.B., Descoteaux, M., Thiran, J.P., et al., 2020. On the cortical connectivity in the macaque brain: a comparison of diffusion tractography and histological tracing data. NeuroImage 221, 117201.

Gopinath, K., Desrosiers, C., Lombaert, H., 2021. Segrecon: learning joint brain surface reconstruction and segmentation from images. In: Medical Image Computing and Computer Assisted Intervention—MICCAI 2021: 24th International Conference, Strasbourg, France, September 27–October 1, 2021, Proceedings, Part VII 24, pp. 650–659.

Hammond, D.K., Scherrer, B., Warfield, S.K., 2013. Cortical graph smoothing: a novel method for exploiting DWI-derived anatomical brain connectivity to improve EEG source estimation. IEEE Trans. Med. Imaging 32 (10), 1952–1963.

Honey, C.J., Sporns, O., Cammoun, L., Gigandet, X., Thiran, J.P., Meuli, R., Hagmann, P., 2009. Predicting human resting-state functional connectivity from structural connectivity. Proc. Natl. Acad. Sci. USA 106 (6), 2035–2040.

Huo, Y., Plassard, A.J., Carass, A., Resnick, S.M., Pham, D.L., Prince, J.L., Landman, B.A., 2016. Consistent cortical reconstruction and multi-atlas brain segmentation. NeuroImage 138, 197–210.

Jbabdi, S., Johansen-Berg, H., 2011. Tractography: where do we go from here? Brain Connect. 1 (3), 169–183.

Jeurissen, B., Tournier, J.D., Dhollander, T., Connelly, A., Sijbers, J., 2014. Multi-tissue constrained spherical deconvolution for improved analysis of multi-shell diffusion MRI data. NeuroImage 103, 411–426.

Kim, J.S., Singh, V., Lee, J.K., Lerch, J., Ad-Dab'bagh, Y., MacDonald, D., Lee, J.M., Kim, S.I., Evans, A.C., 2005. Automated 3-D extraction and evaluation of the inner and outer cortical surfaces using a Laplacian map and partial volume effect classification. NeuroImage 27 (1), 210–221.

Kojčić, I., Papadopoulo, T., Deriche, R., Deslauriers-Gauthier, S., 2021. Incorporating transmission delays supported by diffusion MRI in MEG source reconstruction. In: 2021 IEEE 18th International Symposium on Biomedical Imaging (ISBI), pp. 64–68.

Lazar, M., Weinstein, D.M., Tsuruda, J.S., Hasan, K.M., Arfanakis, K., Meyerand, M.E., Badie, B., Rowley, H.A., Haughton, V., Field, A., et al., 2003. White matter tractography using diffusion tensor deflection. Hum. Brain Mapp. 18 (4), 306–321.

Li, M., Newton, A.T., Anderson, A.W., Ding, Z., Gore, J.C., 2019. Characterization of the hemodynamic response function in white matter tracts for event-related fMRI. Nat. Commun. 10 (1), 1–11.

Little, G., Beaulieu, C., 2021. Automated cerebral cortex segmentation based solely on diffusion tensor imaging for investigating cortical anisotropy. NeuroImage 237, 118105.

Liu, T., Li, H., Wong, K., Tarokh, A., Guo, L., Wong, S.T., 2007. Brain tissue segmentation based on DTI data. NeuroImage 38 (1), 114–123.

Logothetis, N.K., 2003. The underpinnings of the BOLD functional magnetic resonance imaging signal. J. Neurosci. 23 (10), 3963–3971.

Ma, Q., Li, L., Robinson, E.C., Kainz, B., Rueckert, D., Alansary, A., 2023. CortexODE: learning cortical surface reconstruction by neural ODEs. IEEE Trans. Med. Imaging 42, 430–443.

MacDonald, D., Kabani, N., Avis, D., Evans, A.C., 2000. Automated 3-D extraction of inner and outer surfaces of cerebral cortex from MRI. NeuroImage 12 (3), 340–356.

Maier-Hein, K.H., Neher, P.F., Houde, J.C., Côté, M.A., Garyfallidis, E., Zhong, J., Chamberland, M., Yeh, F.C., Lin, Y.C., Ji, Q., et al., 2017. The challenge of mapping the human connectome based on diffusion tractography. Nat. Commun. 8 (1), 1–13.

Meynert, T., 1885. Psychiatry; A Clinical Treatise on Diseases of the Fore-Brain Based Upon a Study of Its Structure, Functions, and Nutrition. GP Putnam's Sons.

Nie, D., Wang, L., Adeli, E., Lao, C., Lin, W., Shen, D., 2018. 3-D fully convolutional networks for multimodal isointense infant brain image segmentation. IEEE Trans. Cybern. 49 (3), 1123–1136.

Nieuwenhuys, R., Voogd, J., Van Huijzen, C., 2007. The Human Central Nervous System: A Synopsis and Atlas. Springer Science & Business Media.

Nozais, V., Forkel, S.J., Foulon, C., Petit, L., de Schotten, M.T., 2021. Functionnectome as a framework to analyse the contribution of brain circuits to fMRI. Commun. Biol. 4 (1), 1–12.

Oishi, K., Faria, A., Jiang, H., Li, X., Akhter, K., Zhang, J., Hsu, J.T., Miller, M.I., van Zijl, P.C.M., Albert, M., et al., 2009. Atlas-based whole brain white matter analysis using large deformation diffeomorphic metric mapping: application to normal elderly and Alzheimer's disease participants. NeuroImage 46 (2), 486–499.

Osher, S., Sethian, J.A., 1988. Fronts propagating with curvature-dependent speed: algorithms based on Hamilton-Jacobi formulations. J. Comput. Phys. 79 (1), 12–49.

Parker, D., Desiderios, L., Brem, S., Verma, R., 2018. Tracking through edema: enhanced neurosurgical planning using advanced diffusion modeling of the peritumoral tissue microstructure. In: Neuro-Oncology, vol. 20. Oxford University Press, Cary, NC, pp. 63–64.

Petit, L., Rheault, F., Descoteaux, M., Tzourio-Mazoyer, N., 2019. Half of the streamlines built in a whole human brain tractogram is anatomically uninterpretable. In: Proceedings OHBM, p. Th785.

Petit, L., Ali, K.M., Rheault, F., Boré, A., Cremona, S., Corsini, F., De Benedictis, A., Descoteaux, M., Sarubbo, S., 2023. The structural connectivity of the human angular gyrus as revealed by microdissection and diffusion tractography. Brain Struct. Funct. 228 (1), 103–120.

Poldrack, R.A., Mumford, J.A., Nichols, T.E., 2011. Handbook of Functional MRI Data Analysis. Cambridge University Press.

Preti, M.G., Makris, N., Lagana, M.M., Papadimitriou, G., Baglio, F., Griffanti, L., Nemni, R., Cecconi, P., Westin, C.F., Baselli, G., 2012. A novel approach of fMRI-guided tractography analysis within a group: construction of an fMRI-guided tractographic atlas. In: 2012 Annual International Conference of the IEEE Engineering in Medicine and Biology Society, pp. 2283–2286.

Qi, S., Meesters, S., Nicolay, K., ter Haar Romeny, B.M., Ossenblok, P., 2016. Structural brain network: what is the effect of LiFE optimization of whole brain tractography? Front. Comput. Neurosci. 10, 12.

Raffelt, D., Tournier, J.D., Rose, S., Ridgway, G.R., Henderson, R., Crozier, S., Salvado, O., Connelly, A., 2012. Apparent fibre density: a novel measure for the analysis of diffusion-weighted magnetic resonance images. NeuroImage 59 (4), 3976–3994.

Ren, J., Hu, Q., Wang, W., Zhang, W., Hubbard, C.S., Zhang, P., An, N., Zhou, Y., Dahmani, L., Wang, D., et al., 2022. Fast cortical surface reconstruction from MRI using deep learning. Brain Inform. 9 (1), 6.

Reveley, C., Seth, A.K., Pierpaoli, C., Silva, A.C., Yu, D., Saunders, R.C., Leopold, D.A., Frank, Q.Y., 2015. Superficial white matter fiber systems impede detection of long-range cortical connections in diffusion MR tractography. Proc. Natl. Acad. Sci. USA 112 (21), E2820–E2828.

Rheault, F., St-Onge, E., Sidhu, J., Maier-Hein, K., Tzourio-Mazoyer, N., Petit, L., Descoteaux, M., 2019. Bundle-specific tractography with incorporated anatomical and orientational priors. NeuroImage 186, 382–398.

Rickmann, A.M., Bongratz, F., Pölsterl, S., Sarasua, I., Wachinger, C., 2022. Joint reconstruction and parcellation of cortical surfaces. In: Machine Learning in Clinical Neuroimaging: 5th International Workshop, MLCN 2022, Held in Conjunction with MICCAI 2022, Singapore, September 18, 2022, Proceedings, pp. 3–12.

Schiavi, S., Lu, P.J., Weigel, M., Lutti, A., Jones, D.K., Kappos, L., Granziera, C., Daducci, A., 2022. Bundle myelin fraction (BMF) mapping of different white matter connections using microstructure informed tractography. NeuroImage 249, 118922.

**362 PART | III** Tractography algorithms

Schilling, K., Gao, Y., Janve, V., Stepniewska, I., Landman, B.A., Anderson, A.W., 2018. Confirmation of a gyral bias in diffusion MRI fiber tractography. Hum. Brain Mapp. 39 (3), 1449–1466.

Schilling, K.G., Gao, Y., Li, M., Wu, T.L., Blaber, J., Landman, B.A., Anderson, A.W., Ding, Z., Gore, J.C., 2019a. Functional tractography of white matter by high angular resolution functional-correlation imaging (HARFI). Magn. Reson. Med. 81 (3), 2011–2024.

Schilling, K.G., Nath, V., Hansen, C., Parvathaneni, P., Blaber, J., Gao, Y., Neher, P., Aydogan, D.B., Shi, Y., Ocampo-Pineda, M., et al., 2019b. Limits to anatomical accuracy of diffusion tractography using modern approaches. NeuroImage 185, 1–11.

Schmahmann, J.D., Pandya, D.N., 2006. Fiber Pathways of the Brain. Oxford University Press, New York.

Schomer, D.L., da Silva, F.H.L., 2017. Niedermeyer's Electroencephalography: Basic Principles, Clinical Applications, and Related Fields. Oxford University Press.

Shirley, P., Ashikhmin, M., Marschner, S., 2009. Fundamentals of Computer Graphics. AK Peters/CRC Press.

Smith, R.E., Tournier, J.D., Calamante, F., Connelly, A., 2012. Anatomically-constrained tractography: improved diffusion MRI streamlines tractography through effective use of anatomical information. NeuroImage 62 (3), 1924–1938.

Smith, R.E., Tournier, J.D., Calamante, F., Connelly, A., 2015. SIFT2: enabling dense quantitative assessment of brain white matter connectivity using streamlines tractography. NeuroImage 119, 338–351.

St-Onge, E., Girard, G., Whittingstall, K., Descoteaux, M., 2015. Surface tracking from the cortical mesh complements diffusion MRI fiber tracking near the cortex. In: International Society of Magnetic Resonance in Medicine (ISMRM).

St-Onge, E., Daducci, A., Girard, G., Descoteaux, M., 2018. Surface-enhanced tractography (set). NeuroImage 169, 524–539.

St-Onge, E., Al-Sharif, N., Girard, G., Theaud, G., Descoteaux, M., 2021. Cortical surfaces integration with tractography for structural connectivity analysis. Brain Connect. 11 (7), 505–517.

Takemura, H., Caiafa, C.F., Wandell, B.A., Pestilli, F., 2016. Ensemble tractography. PLoS Comput. Biol. 12 (2), e1004692.

Tallinen, T., Chung, J.Y., Rousseau, F., Girard, N., Lefèvre, J., Mahadevan, L., 2016. On the growth and form of cortical convolutions. Nat. Phys. 12 (6), 588–593.

Teillac, A., Beaujoin, J., Poupon, F., Mangin, J.F., Poupon, C., 2017. A novel anatomically-constrained global tractography approach to monitor sharp turns in gyri. In: Medical Image Computing and Computer Assisted Intervention—MICCAI 2017: 20th International Conference, Quebec City, QC, Canada, September 11–13, 2017, Proceedings, Part I, pp. 532–539.

Theaud, G., Edde, M., Dumont, M., Zotti, C., Zucchelli, M., Deslauriers-Gauthier, S., Deriche, R., Jodoin, P.M., Descoteaux, M., 2022. DORIS: a diffusion MRI-based 10 tissue class deep learning segmentation algorithm tailored to improve anatomically-constrained tractography. Front. Neuroimaging 1, 917806.

Tournier, J.D., Smith, R., Raffelt, D., Tabbara, R., Dhollander, T., Pietsch, M., Christiaens, D., Jeurissen, B., Yeh, C.H., Connelly, A., 2019. MRtrix3: a fast, flexible and open software framework for medical image processing and visualisation. NeuroImage 202, 116137.

Urbanski, M., De Schotten, M.T., Rodrigo, S., Catani, M., Oppenheim, C., Touzé, E., Chokron, S., Méder, J.F., Lévy, R., Dubois, B., et al., 2008. Brain networks of spatial awareness: evidence from diffusion tensor imaging tractography. J. Neurol. Neurosurg. Psychiatry 79 (5), 598–601.

Van Essen, D.C., Jbabdi, S., Sotiropoulos, S.N., Chen, C., Dikranian, K., Coalson, T., Harwell, J., Behrens, T.E., Glasser, M.F., 2013. Mapping connections in humans and non-human primates: aspirations and challenges for diffusion imaging. In: Johansen-Berg, H., Behrens, T.E.J. (Eds.), Diffusion MRI, pp. 337–358.

Vanderweyen, D.C., Theaud, G., Sidhu, J., Rheault, F., Sarubbo, S., Descoteaux, M., Fortin, D., 2020. The role of diffusion tractography in refining glial tumor resection. Brain Struct. Funct. 225 (4), 1413–1436.

Veraart, J., Nunes, D., Rudrapatna, U., Fieremans, E., Jones, D.K., Novikov, D.S., Shemesh, N., 2020. Noninvasive quantification of axon radii using diffusion MRI. eLife 9, e49855.

Warrington, S., Bryant, K.L., Khrapitchev, A.A., Sallet, J., Charquero-Ballester, M., Douaud, G., Jbabdi, S., Mars, R.B., Sotiropoulos, S.N., 2020. Xtract-standardised protocols for automated tractography in the human and macaque brain. NeuroImage 217, 116923.

Wassermann, D., Makris, N., Rathi, Y., Shenton, M., Kikinis, R., Kubicki, M., Westin, C.F., 2016. The white matter query language: a novel approach for describing human white matter anatomy. Brain Struct. Funct. 221, 4705–4721.

Wasserthal, J., Neher, P.F., Hirjak, D., Maier-Hein, K.H., 2019. Combined tract segmentation and orientation mapping for bundle-specific tractography. Med. Image Anal. 58, 101559.

Yap, P.T., Zhang, Y., Shen, D., 2015. Brain tissue segmentation based on diffusion MRI using $\ell_0$ sparse-group representation classification. In: International Conference on Medical Image Computing and Computer-Assisted Intervention, pp. 132–139.

Yeh, C., Smith, R.E., Dhollander, T., Connelly, A., 2017. Mesh-based anatomically-constrained tractography for effective tracking termination and structural connectome construction. In: International Society of Magnetic Resonance in Medicine (ISMRM).

Zhang, F., Breger, A., Cho, K.I.K., Ning, L., Westin, C.F., O'Donnell, L.J., Pasternak, O., 2021. Deep learning based segmentation of brain tissue from diffusion MRI. NeuroImage 233, 117934.

# Chapter 19

# Tractography in pathological anatomy: Some general considerations

Guillaume Theaud[a], Manon Edde[a], Alexander Leemans[b], Flavio Dell'Acqua[c,e], Joseph Yuan-Mou Yang[d,*], and Maxime Descoteaux[f,*]

[a]Sherbrooke Connectivity Imaging Laboratory (SCIL), University of Sherbrooke, Sherbrooke, QC, Canada, [b]PROVIDI lab, Image Sciences Institute, University Medical Center Utrecht, Utrecht, The Netherlands, [c]NATBRAINLAB, Department of Neuroimaging, Institute of Psychiatry, Psychology and Neuroscience, King's College London, London, United Kingdom, [d]Neuroscience Advanced Clinical Imaging Service (NACIS), Department of Neurosurgery, The Royal Children's Hospital, Melbourne, VIC, Australia, [e]Department of Forensics and Neurodevelopmental Sciences, Institute of Psychiatry, Psychology and Neuroscience, King's College London, London, United Kingdom, [f]Sherbrooke Connectivity Imaging Laboratory (SCIL), Department of Computer Science, University of Sherbrooke, Sherbrooke, QC, Canada

## 1 Introduction

In most clinical applications and in selected research settings, there is a need to perform tractography in pathological brains. In research, it is acceptable to spend hours to perform state-of-art diffusion magnetic resonance imaging (dMRI) preprocessing (e.g., performing data denoising, and motion and echo-planar imaging (EPI) distortion corrections), and to optimize white matter (WM) modeling and tracking processing to ensure the tractography results are of highest quality possible. In contrast, computation time allowed in clinical applications, such as tractography informed presurgical planning in acute neurosurgical settings (Farquharson et al., 2013; Jeurissen et al., 2019; Vanderweyen et al., 2020; Yang et al., 2021), is limited. The dMRI acquisition, tractography reconstruction, and integration into presurgical planning must be completed within a very short time (typically in less than an hour) to avoid interrupting clinical workflow and to reduce potential risks to patient safety. Thus clinical tractography processing faces the challenges of *lack of time* and *lack of data* and *data quality*. Typical ways to address such challenges include reducing the data acquisition time, and incorporating fast or real-time, interactive tractography reconstruction and visualization techniques.

The readers should note that most recommendations and state-of-the-art techniques described in this book have been developed based on using high-quality MRI datasets from healthy adult volunteers in well-controlled research settings. These recommendations and techniques will need to be adjusted to suit different clinical applications. Clinical MRI data are often of lower quality than those used for research method development, partly due to limited acquisition time. The process was made more challenging due to the need to consider WM modeling and tracking accuracy in and adjacent to pathology.

The purpose of this chapter is to highlight both the technical and practical considerations to make anatomically plausible tractography reconstructions possible, especially in challenging clinical situations with *lack of time* and *lack of data* and *data quality*. We will revisit some of the key concepts of tractography algorithms and make practical suggestions.

## 2 Technical considerations for tractography within pathological brains

A first important prerequisite to successful tractography reconstruction is to ensure the tracking mask covers all brain regions. Lesional areas from the presence of aforementioned pathological processes are often excluded from the tracking mask, which can then indirectly affect the tracking accuracy.

Depending on the tractography algorithm, the tracking mask can range from a *"simple"* diffusion tensor imaging (DTI)-based fractional anisotropy (FA) mask (Chapter 10), to more *"complex"* masks specific to different brain tissue types—that is, the gray matter (GM), WM, cerebrospinal fluid (CSF), and the different GM nuclei or even pial surface meshes (Smith et al., 2012; Girard et al., 2014; Chapter 18). These tissue-specific masks are typically derived from structural MRI data, such as the T1- or T2-weighted image, which introduces additional anatomical priors to the tractography processing. The latter gave rise

---

*Cosenior authors. These senior authors have contributed equally.

**364 PART | III** Tractography algorithms

to a family of so-called *"anatomically constrained tractography (ACT)"* algorithms based on using precise GM, WM, and CSF masks (Smith et al., 2012; Girard et al., 2014; Chapter 18).

A trade-off to adding anatomical priors to tractography processing in pathological anatomy is that an error in estimating this anatomical prior, or misregistration of this anatomical image to the dMRI data, will lead to erroneous tractography outputs and introduce bias to quantitative DWI microstructural metrics derived from the tractography. In principle, all ACT-based algorithms require accurate voxelwise tissue segmentation in the DWI space, which is very difficult to achieve in pathology affected brain regions. This is illustrated using an example multiple sclerosis (MS) patient in Fig. 1, where the tracking accuracy within the demyelinating lesions is affected by erroneous anatomical priors introduced to the tracking masks.

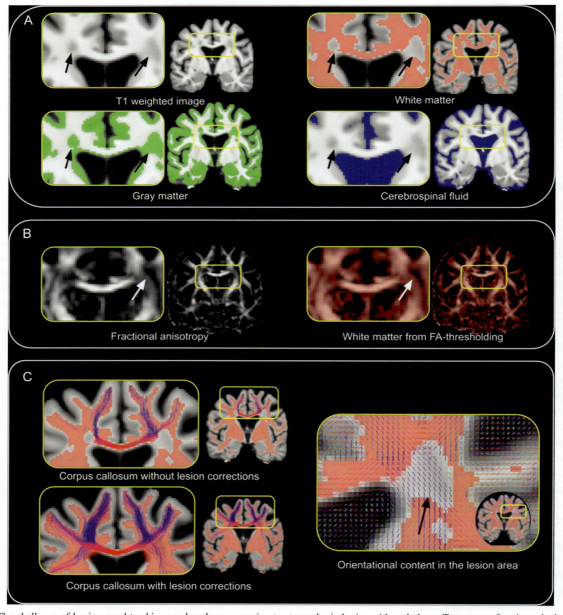

**FIG. 1** The challenge of having good tracking masks when processing tractography in brains with pathology. Two areas of periventricular multiple sclerosis (MS) lesions are marked by the *arrows*. In (A), the white matter (WM) segmented mask (*red*) from a T1-weighted image excludes the MS lesions. These excluded lesional areas are considered as gray matter (*green*) by the tissue segmentation algorithm. In (B), the brain mask derived from using a fractional anisotropy-thresholding at 0.15 also misses these areas due to the presence of three-way crossing fibers. These segmentation errors will lead to tracking errors over the lesional areas. In (C), the *right panel* shows the expected fiber orientation distribution function (fODF) peaks (maximum amplitude of the fODF) are preserved within the lesional areas. In the *top left panel*, the corpus callosum is extracted from a tractogram using a WM mask (in A) without lesion correction. In the *bottom left panel*, the corpus callosum is extracted from a tractogram using the lesion mask added to the WM mask in (A). Both tractograms used the same fODF-based probabilistic tracking algorithm and the same local tracking parameters: five seeds per voxel, seeding from the WM mask, angle 45 degrees, step size of 0.5 mm.

Although the dMRI signal is affected by the presence of pathology, in certain pathology, it remains possible to explore the diffusion signal with tractography. For example, in Fig. 1C, within an MS lesional area (red pixels) containing low FA or hypointense T1-weighted signal, coherent fiber orientation distribution function (fODF) peaks are preserved, which leads to streamlines traversing the lesion, with plausible anatomical paths. A well-defined tracking mask that's inclusive of all lesions is an important prerequisite to avoid premature tracking termination over the lesion and tracking mask interface (see Fig. 1C).

In other instances, incoherent or noisy fODF is estimated within the pathology. For presurgical planning, the dMRI signal within such lesions cannot be reliably explored using tractography. It is important to caution the neurosurgeons to not depend solely on the tractography results to guide surgical resections. In such instances, direct cortical and subcortical electrostimulation and functional brain mapping should be considered to validate the functional relevance of the intralesional tracking results. A case example is shown in Fig. 2; also see Section 6 for a more in-depth discussion of this point.

# 3 Practical considerations for tractography within pathological brains

Here, we describe three categories of tractography pipelines that can be used to address the clinical challenges of *lack of time* and *lack of data* and *data quality*. For each tractography pipeline, we review the *"why"* and the *"how"* of utilizing each pipeline in clinics and highlight its practical considerations, including the need to consider the trade-offs between dMRI acquisition time, computation time, the complexity of local WM modeling techniques, and the quality of tractography reconstructions. We use brain tumor patients and an MS patient as case illustrations. Data-processing details for all cases are included in the Appendix.

## 3.1 Real-time tractography as a visualization and data exploration tool

The first category of tractography pipeline is real-time tractography (RTT; Chamberland et al., 2014). In this pipeline, **no** streamline is computed offline or before visualization. This accelerates computation time significantly as only local orientation information, from the DTI or higher-order reconstruction methods (fODF, multitensors, etc.), need to be computed before the visualization. Both the tracking masks and tracking parameters can be updated and tweaked on-the-fly, in the visualization software, as illustrated in Fig. 3. This enables tracking of a WM bundle that is displaced by the brain tumor.

### 3.1.1 Why?

The ability to rapidly visualize and explore the tractography results using different tracking region-of-interest (ROI) strategies and different tracking parameters is particularly useful when the processing time is limited (e.g., under an hour), making it well suited for presurgical planning in the acute neurosurgical settings, and is potentially useful for intraoperative RTT updates, using data acquired from intraoperative MRI scanners. RTT permits users to explore the tracking results in specific brain regions impacted by the pathology, such as in the peritumoral WM edema, without the need to reconstruct a whole brain tractogram. The default tracking parameters, which are usually set for healthy and typically developing brains, can be adjusted on-the-fly, in real-time, with updated streamline displays and instant visual feedback. For example, the user may lower the FA threshold, to 0 or 0.01, to "force" the tractography to explore tracking within the tumor and the peritumoral WM edema.

### 3.1.2 How?

Some tractography algorithms are directly implemented in the visualization software (Chamberland et al., 2014; Rheault et al., 2016; Aydogan et al., 2021; Leemans et al., 2009). These algorithms are based on fODF peaks (i.e., maximum amplitude of the fODF) or the principal DTI eigenvector. In the software (see Fig. 3), an ROI can be placed to initiate tracking streamlines that propagate through the precomputed diffusion peak fODF field. The tracking mask used, for example, based on an FA map with a predefined tracking threshold, would determine whether the streamlines can be generated and where they terminate (Vanderweyen et al., 2020). A limitation of this real-time approach is that the streamlines generated may contain a large amount of false positives, as streamlines are generated purely by following the local orientations without additional a priori anatomical information, such as the tract shape, to constrain the tracking results.

## 3.2 The diffusion tensor imaging tractography pipeline

For this pipeline, DTI is used to model the WM microstructure and principal fiber orientation (i.e., eigenvector) at every voxel. The tracking streamlines follow the step-by-step, voxelwise principal eigenvector using a deterministic tracking algorithm, such as the Fiber Assignment by Continuous Tracking (FACT) algorithm, based on a preselected FA threshold (usually around 0.15 for healthy adult brains; Christidi et al., 2016; Sinke et al., 2018; Hagler et al., 2009).

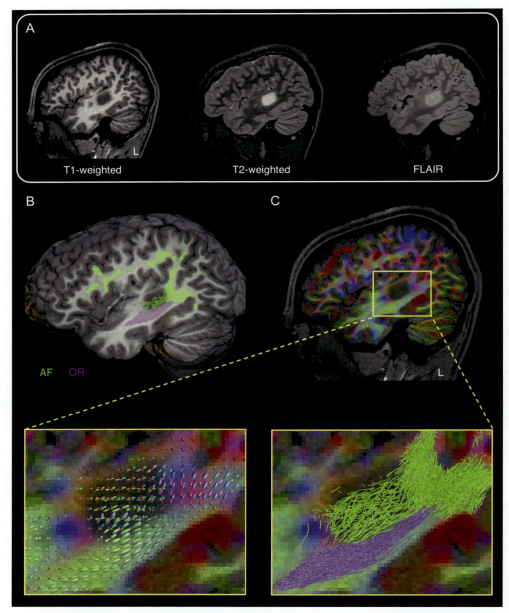

**FIG. 2** Incoherent fiber orientation distribution function (fODF) and tracking streamlines within the lesion. This is a 14-year-old, right-handed boy with a left posterior temporal subcortical white matter (WM) lesion in his language dominant hemisphere, identified on initial MRI scan (A). The radiological differential diagnoses include both demyelination and brain tumors, thus a diagnostic open brain biopsy for this lesion is planned for. In (B), a 3D rendered tangential cut-away view shows the left arcuate fasciculus (AF) traverses through the lesion, and the optic radiation (OR) is displaced and abuts the inferior border of the lesion. In (C), the *top right panel* shows an fODF-based directionally encoded color map in a selected sagittal plane. The *bottom panels* are close-up views highlighted in *yellow outlined rectangle blocks*. The *bottom left* shows the fODF glyphs, and the *bottom right* shows the AF (in *green*) and OR (in *purple*) tracking streamlines. Incoherent and noisy fODF glyphs and tracking streamlines are demonstrated within the lesion. Since there are no anatomical or biological certainties of the estimated lesional fODF, the AF tracking result could not be accurately relied upon for presurgical planning. Added functional information is required to guide the open biopsy. An awake craniotomy and functional brain mapping with direct brain electrostimulation is planned for this case. On the other hand, the temporal lobe WM adjacent to the lesion shows coherent fODF glyphs, consistent with the expected OR anatomy course. Both AF and OR tractographies were reconstructed using the multitissue constrained spherical deconvolution model, *lmax* = 8, an iFOD-2 probabilistic algorithm, WM anatomy-guided tracking ROIs, fiber orientation distribution threshold = 0.1, max angle 45 degrees, step size of 1.15 mm.

### *3.2.1 Why?*

The deterministic DTI tractography pipeline is the most widely available "press-button" solution on all MRI systems. It is also the only tractography method that's cleared by the US Food and Drug Administration and incorporated by the surgical image-guidance (neuronavigation) vendors for presurgical planning and for intrasurgical visualization purposes

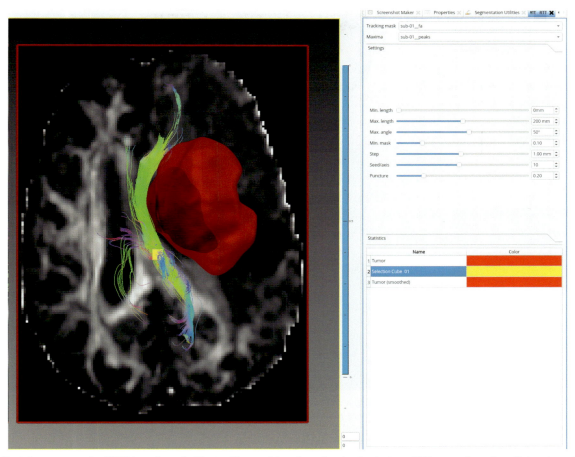

**FIG. 3** Real-time tractography (RTT) is performed with a tracking mask based on a fractional anisotropy (FA) map and a region of interest seen in *yellow* ("Selection Cube 01" in the *right panel*). This visualization tool enables virtual dissection of the right cingulum bundle in real-time, showing displacement around the tumor. The *right panel* allows the user to modify the tracking parameters, including the FA threshold used for WM mask, the step size, and angular constraint of the tracking process.

(Vanderweyen et al., 2020; Chamberland et al., 2014; Theaud et al., 2019). In practice, the deterministic DTI tracking is "easy-to-run" because it is fast, widely available, and compatible with most dMRI acquisition with more than six diffusion directions. Acquiring dMRI data for DTI modeling ($b = 700-1000 mm^2/s$, 12–32 diffusion directions) takes less than 3–5 min on a typical clinical MRI scanner. Generating a whole brain tractogram using such a tractography algorithm is typically not time consuming (see Fig. 4), making it clinically feasible. On an average computer with 8 CPU cores and 16GB RAM or more, the DTI tractography pipeline takes approximately 30 min or less to run.

### 3.2.2 How?

The DTI tractography pipeline requires a tracking mask and the field of diffusion tensors or principal eigenvector (also called peak). The tracking mask is usually the thresholded FA map computed from DTI at $FA \geq 0.15$. If pathology or WM abnormalities are present, the mask should be corrected or adjusted before tractography to account for the presence of pathology, as shown in Fig. 4A (Theaud et al., 2017, 2020).

Using the deterministic DTI tractography pipeline to explore tracking in and around the pathology needs to be done with care and awareness of the related technical limitations, especially when the tracking is performed over the crossing fiber WM regions. The DTI model cannot accurately estimate WM fODF containing crossing fiber or other complex fiber configurations. Together with a step-wise deterministic tracking approach, this leads to underestimation of the displayed WM anatomy (i.e., a false negative tracking problem). Examples of the observed tracking failure include poor or partial streamline spatial coverage, and premature streamline terminations, with the reconstruction missing parts of the expected WM tract morphology. The centrum semiovale, a subcortical WM region, superior and lateral to the lateral ventricle, is a

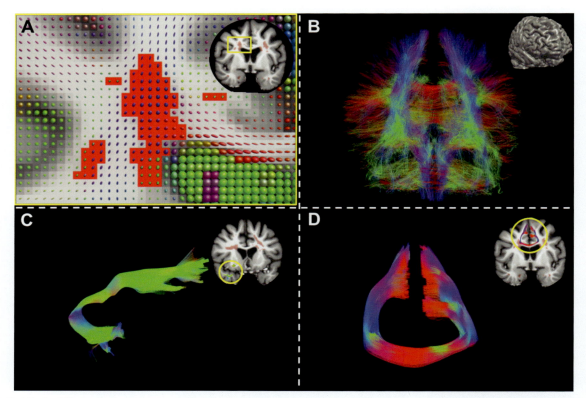

**FIG. 4** In (A), the diffusion tensors are represented as glyphs around a lesion area. This lesion area is manually segmented by an expert. In (B), a DTI-based whole brain tractogram is shown. This tractogram was generated within minutes with deterministic tracking, using five seeds per voxels and a tracking mask based on a fractional anisotropy threshold of 0.1. The step size used was 0.5 mm with a theta of 60 degrees. The uncinate fasciculus (in C) and the corpus callosum (in D) were extracted using this whole brain tractogram.

good example of a crossing fiber region, containing several WM fiber populations with different orientations. Here, the DTI tractography cannot reconstruct the lateral and low convexity face/tongue fibers of the corticospinal tract and the radiation fibers of the corpus callosum (see Fig. 4D), resulting in partial tract reconstructions. On this point, the deterministic DTI tracking can be seen as "hard-to-use." Substituting a deterministic with a probabilistic tracking algorithm, combining with DTI, also cannot fully address the crossing fiber tracking problem given the limitation is inherent to the DTI model itself.

Nonetheless, in selected pathology, especially when the preserved principal fODF is in line with the expected tract anatomy, it remains possible to reliably reconstruct major WM bundles, such as the arcuate fasciculus and the uncinate fasciculus (see Fig. 4C), using the deterministic DTI tractography pipeline.

### 3.3 Tractography pipelines robust to crossing fibers

In Chapters 13–18, we have already described a plethora of tractography algorithms beyond the DTI tractography, from deterministic to probabilistic, to global, to geodesic, and to machine learning tractography. In clinical applications, tractography pipelines robust to crossing fibers have mostly been implemented around the *"local"* deterministic and probabilistic tractography techniques: *"local"* because the algorithm implements a step-by-step approach based on a local field of orientations available at every MRI voxel. As opposed to global tractography approaches (Chapter 16), the tracking algorithm only takes local decisions as it steps through the WM field of orientations. In its simplest form, this pipeline is an extension of the DTI tractography pipeline just described, but uses a field of local orientations robust to crossing fibers (see Chapters 10–12) instead of the diffusion tensors.

#### 3.3.1 Why?

Local tractography robust to crossing fibers is becoming frequently used in research applications, thanks to increasingly available open-source software solutions. Contrasting the DTI tractography pipeline, the advantage of this tractography pipeline is the use of fODFs or an equivalent local reconstruction of multiorientation fields to generate streamlines. The fODFs make it possible to take into account the fiber crossings (see Fig. 5) and to reconstruct WM bundles with more complete cortical coverage and projections, compared to those generated by the DTI tractography pipeline (Chapters 13 and 14). This pipeline is thus computationally more demanding than the previous two pipelines described, although the running time still remain acceptable in certain clinical settings (i.e., around 1–2 h or less on an average computer having 8 CPU cores and 16GB of RAM or more). Depending on the implementation, one can expect more or less 1 million streamlines generated within an hour or less.

While there remains a need to improve the technical efficiency of advanced methods to be comparable with acute neurosurgical time, it is worth noting that application of advanced tractography methods is highly feasible for many lesion-based neurosurgeries performed on either a semiurgent or an elective basis. This includes most of the benign intrinsic brain lesions, such as low-grade gliomas, developmental brain tumors, and the majority of lobectomy and lesionectomy-based epilepsy surgeries. Additionally, continual advances in MRI hardware and dMRI acquisition schemes, such as simultaneous multislice imaging, have reduced the acquisition time for high angular resolution diffusion imaging data (HARDI data; $b \geq 3000$ mm$^2$/s, $\geq 45$ directions) and multishell DWI data to clinically acceptable times, facilitating the clinical adoption of higher-order dMRI models. There is also growing research evidence, based on selected brain tumor surgical series, demonstrating improved anatomical plausibility of the reconstructed tracts using probabilistic fODF-based tractography and other higher-order multifiber approaches, as compared to the DTI-based tractography (Farquharson et al., 2013; Bucci et al., 2013; Kuhnt et al., 2013; Mormina et al., 2015). The benefits of probabilistic fODF-based tractography are most pronounced when reconstructing WM bundles near the pathology or in the presence of perilesional WM edema (Kuhnt et al., 2013). In one study, this was validated using direct subcortical electrostimulation confirming correspondence between the reconstructed tractography results and the true WM bundles' location (Bucci et al., 2013).

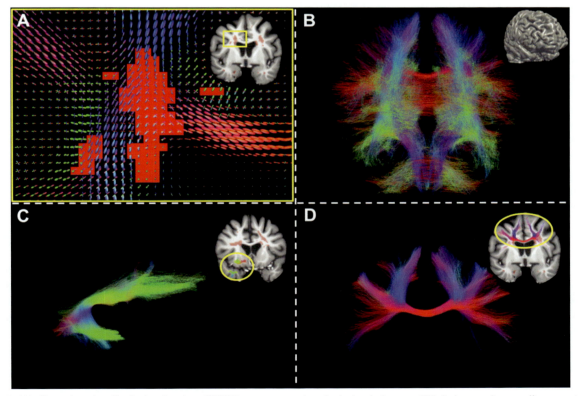

**FIG. 5** In (A), fiber orientation distribution functions (fODFs) are represented as glyphs in a lesion area. This lesion area is manually segmented by an expert. In (B), an fODF-based whole brain tractogram is shown. This tractogram was generated using five seeds per voxels and a tracking mask based on editing a white matter mask by filling the lesion holes. The step size used was 0.5 mm with a theta of 20 degrees. The uncinate fasciculus (in C) and the corpus callosum (in D) were dissected from this whole brain tractogram, using the same regions-of-interest as in Fig. 4.

### 3.3.2 How?

This tractography pipeline requires seeding and tracking masks as well as the field of fODF or multiorientation local information. Similar to the DTI tractography pipeline, the tracking mask can be a thresholded FA map, a mask defined from the fODF field, or a WM mask obtained from an anatomical image or even a template to deal with lesions caused by abnormal anatomy (Theaud et al., 2020). Most of the available open-source software can perform this tractography pipeline using either a deterministic or a probabilistic tracking algorithm. Unfortunately, at this time, this technique is rarely available on clinical MRI scanners and is not widely available on standard commercial surgical neuronavigation systems. Many practical challenges and barriers need to be overcome first. For example, this pipeline is typically set up using personal or research computers and is implemented in parallel to the MRI or the neuronavigation systems. Data flow between all systems can be challenging. If the data flow is to be between different hospitals and research institutions, there are also added ethical considerations, including patient data anonymity, cybersecurity concern on data transfer, and logistic issues, such as the interinstitutional differences in computation capacity and infrastructure, affecting the efficiency and completeness of the data transfer.

Despite such practical challenges, there are emerging open-source tools being developed and now made available in the public domain to transfer advanced tractography and import them into selected surgical neuronavigation platforms, enabling intraoperative image-guidance by integrating tractography visualizations with other image data (Beare et al., 2022). Alternatively, there are novel tools (e.g., OpenIGLink; Tokuda et al., 2009) to integrate open-source viewing platforms and real-time surgical neuronavigation data. This is an area of ongoing research development with the potential of significant clinical impact. There are now emerging eloquent-region brain tumor and epilepsy surgical series reported from selected neurosurgical centers with colocated research institution, using advanced tractography techniques robust to crossing fibers to inform presurgical planning and intraoperative neuronavigation (Yang et al., 2019, 2020; Macdonald-Laurs et al., 2021). These surgical series demonstrated promising surgical outcomes including postoperative functional preservation without compromising the extent of lesion resection in eloquent brain and reported good seizure outcomes.

## 4 Challenges of implementing tractography pipelines in pathological brains

Across the different tractography pipelines, the common challenging questions when tracking in or adjacent to pathology are: *What tracking mask do I use? and How can I compute it given the abnormal anatomy of my patient?* In the previous sections, we presented the FA thresholded tracking mask, which although not fully robust to all cases, remains useful for interactive inquiry and rapid visualization of the tracking results. To perform WM tract-based analysis such as tractometry or connectomics, more efforts have to be deployed to obtain a high-quality WM mask adapted to explore a subject with pathology or WM anomalies (Lipp et al., 2019; Theaud et al., 2017). Adequate inclusion of the lesional area within the tracking mask is critical for accurate tract-based quantitative analysis by considering the lesion effect on diffusion signals. In this section, two types of WM mask correction techniques are presented here: (i) semiautomatic and manual lesion filling and (ii) atlas-based WM masking.

### 4.1 Semiautomatic and manual lesion filling

#### 4.1.1 Why?

A lesion mask can be obtained automatically or with manual lesion segmentation. Automatic segmentation algorithms are rarely perfect and may require manual editing, hence the term *semiautomatic*. This lesion mask is useful when we want to quantify diffusion measures of the lesion areas, or compute the lesion numbers or a lesion volume load as a proportion of the whole brain WM or specific WM bundle volume. It is also important to quantify diffusion measures outside the lesion areas, the so-called *normal appearing white matter*, to evaluate the extent of disease impact, such as in MS cases. The emergence and improvements in machine-learning-based image segmentation techniques have the potential to improve tracking masks used for clinical tractography applications.

#### 4.1.2 How?

Manual segmentation is usually performed by an expert in neuroanatomy and the processes can be tedious and time consuming, especially when segmenting large lesions with complex contrasts and poorly defined lesion margin, and the need for batch data processes from a large number of subjects. There are increasing numbers of both supervised and unsupervised automatic machine-learning based lesion segmentation algorithms being made available in the public domain. They were usually developed to segment one particular lesion type (e.g., MS lesions, WM hyperintensity lesions, specific brain tumor type, etc.). Currently, they remain primarily research tools and have not been fully evaluated and validated for clinical tractography applications. The segmentation outputs should be quality checked by, preferably, a neuroanatomy expert. If deemed suboptimal, manual editing and correction of the segmentation outputs should still be carried out.

Once a lesion mask or a "pathology" mask is obtained, it can be added to the WM mask to make sure the tractography algorithm is allowed to explore all brain WM. This manual, semiautomatic lesion filling strategy often permits the tractography process to traverse lesion areas, as seen in Figs. 4A and 5A.

## 4.2 Atlas-based WM masking

### 4.2.1 Why?

If manual segmentation is impossible and automatic segmentation does not work correctly, an alternative option is to use an atlas-based WM mask that bypasses the need to consider lesion effects on the WM mask segmentation, thus allowing the tractography pipeline to explore the lesion areas. This approach is faster than performing manual lesion segmentation and does not require an expert neuroanatomist input.

### 4.2.2 How?

The WM atlases, templates, or outputs from open-source tools, such as Freesurfer (among others) developed based on using healthy and typically developing subjects, can be registered to the dMRI data (Fischl, 2012; Fan et al., 2016; Avants et al., 2009). The WM mask quality should be carefully evaluated, on a case-by-case basis, to check for imaging registration accuracy, and the completeness of including all lesional area within the WM mask.

## 5 What can I do with my tractogram in practice?

### 5.1 Interactive visualization

Interactive visualization of the tractogram output is an important quality control step. It allows validity check of the whole brain tractogram and using the tracking ROIs, one can validate the presence or absence of major WM bundles (see Fig. 3; Theaud and Descoteaux, 2022). It is important to note that, once the whole brain tractogram is computed, one cannot change the tracking parameters on-the-fly. If this is the intended use, one should consider using the RTT method, as mentioned previously in Section 3.1.

### 5.2 Virtual white matter bundle dissection

From a whole brain tractogram, it is possible to apply automatic segmentation algorithms to dissect specific WM bundles (Chapters 21 and 22). These bundles can then be qualitatively visualized and quantitatively assessed using tractometry (Chapter 23).

### 5.2.1 Qualitative assessment

This includes assessing both the position and morphological appearance of a WM bundle as impacted by the pathology. For example, a portion of the bundle adjacent to the lesion can be compressed or displaced from its usual position. Alternatively, the tracking streamlines may be observed to traverse through the perilesional WM edema (see Figs. 6, 8, and 10). This is critically useful for presurgical planning, working out a safe surgical corridor by displaying the spatial relationship between the intended resection lesion and the adjacent WM bundles, thus avoiding postoperative functional deficits due to inadvertent WM injuries (Vanderweyen et al., 2020; Tax et al., 2015; Yang et al., 2021).

### 5.2.2 Quantitative assessment

Quantitative assessments can be performed on the selected WM bundles using the tractometry technique (Chapter 23). Tractometry enables diffusion-based measures as WM microstructural surrogates, to be extracted along the profile of one or multiple WM bundles. Tractometry has been used to locate WM anomalies and to assess the microstructural properties of the studied WM bundles impacted by high-grade gliomas and in drug-resistant epilepsy patients with focal cortical dysplasia (FCD; Fekonja et al., 2021; Chamberland et al., 2021).

In the MS context (see Fig. 7), a WM bundle can traverse through several demyelinating lesions. Diffusion measures as well as other imaging-based measures can be quantified from within the normal-appearing WM or from these lesional areas within the WM bundle.

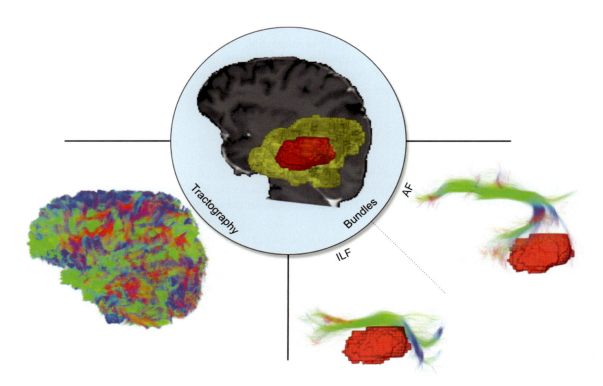

**FIG. 6** The *yellow volume* represents edema and the *red volume* represents tumor core. Selected white matter bundles were virtually dissected from the whole brain tractogram. The inferior longitudinal fasciculus (ILF) and the posterior component of arcuate fasciculus (AF) are both displaced and affected by the tumor core. *(Adapted from Vanderweyen, D.C., Theaud, G., Sidhu, J., Rheault, F., Sarubbo, S., Descoteaux, M., Fortin, D., 2020. The role of diffusion tractography in refining glial tumor resection. Brain Struct. Funct. 225 (4), 1413–1436.)*

### 5.3 Connectomic analysis

A structural connectome (Chapter 24) can be obtained from a whole brain tractogram reconstructed using both healthy subjects and subjects with brain pathology. Based on the chosen brain atlas with a cortical and subcortical parcellation scheme suited for clinical applications, a matrix composed of streamline counts or any other diffusion measures could be extracted between every pair of parcellated brain regions from the atlas. These pairwise connectomic measures can then be compared between healthy and disease individuals, using graph-theory analysis. Measures derived from structural connectomes reconstructed using drug-resistant epilepsy patients have been shown as promising biomarkers estimating seizure outcomes following surgery (Bonilha et al., 2015; Gleichgerrcht et al., 2020). Alternatively, virtual lesioned structural connectome can be reconstructed to mimic the effects of glioma surgical resection. Here, the connectome is reconstructed by removing the streamlines affected by the lesion. This lesioned connectome then serves as the basis to build a simulated brain dynamic model, representing the effects of surgery (Aerts et al., 2020).

## 6 Which tractography pipeline can I use in my application?

Answers to this question depend on the application context (i.e., for research or for clinical purposes), the capability of local MRI hardware, the available tractography technique from the MRI and surgical neuronavigation vendors, and the accessibility of local research expertise to implement advanced tractography methods in the clinical realm. Together, these contextual factors would largely determine the data quality, type of local WM modeling technique, and tractography pipelines used, and the processing time allowed to optimize the tractography outputs. Differences in clinical urgency are also an important practical consideration. For example, it is currently feasible to implement higher-order WM models and advanced tractography techniques to presurgical planning in semiacute and elective surgical settings, but not feasible in emergency surgical settings due to the limited clinical time.

Additionally, the lesion types (i.e., cortical or subcortical lesions), lesion characteristics (i.e., solitary or cystic lesions; presence or absence of intralesional hematoma), and the severity of perilesional WM changes (i.e., WM edema or gliosis)

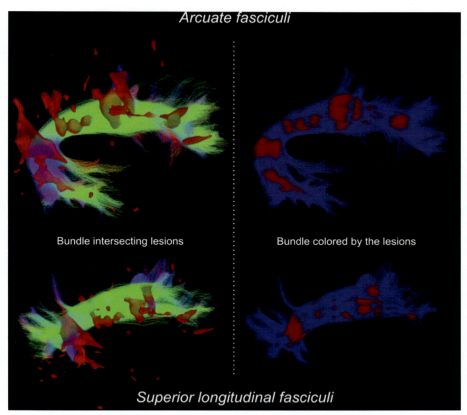

FIG. 7 The *top and bottom rows* show the arcuate fasciculi and the superior longitudinal fasciculi, respectively, reconstructed from using the same whole brain tractogram presented in Fig. 5B. The *left column* images show the distribution of demyelinating lesions in this patient with multiple sclerosis (in *red*). In the *right column*, the tracking streamlines for both WM bundles are colored in *red* if they traverse through the lesions.

can present different challenges to tractography reconstructions. This section makes general suggestions depending on the selected lesion types and the perilesional effects.

## 6.1 Disorders consisting of white matter lesions

Disorders that resulted in multiple isolated WM lesions spread across the brain can create holes in the tracking mask, which may prematurely stop the tracking streamlines. Example conditions include WM hyperintensity lesions in the aging brains, in individuals with Alzheimer's disease, and both active and degenerated WM lesions in MS. The general recommendation in these conditions is to use the local tractography pipeline with a lesion mask derived from either the manual or automatic segmentation, or from an atlas-based approach as described previously. Using the local tractography pipeline permits the tracking process to explore as much of the lesional areas as possible that will help optimize the tractography results.

Low-grade gliomas typically appear as a solitary WM lesion on MRI. Uncertainties of intratumoral fODF estimation would affect tracking accuracy. In this case, it is important not to rely solely on tractography, and functional imaging data should be used to aid WM bundle dissections (also see the following sections).

## 6.2 Disorders consisting of cortical-based lesions

Many surgical remediable epileptic developmental brain lesions in children fall under this category. Common histopathology encountered includes FCD, cortical tubers in tuberous sclerosis complex, and developmental brain tumors (e.g., dysembryoplastic neuroepithelial tumor, and ganglioglioma). These lesions are slow growing and larger ones tend to cause mechanical tract displacement or infiltrate the WM bundles (see Fig. 8A). Developmental vascular lesions, such as cavernous hemangioma, can be broadly considered within this category, as micro- or macrohemorrhage from the vascular nidus tends to displace the adjacent WM bundles. Any aforementioned tractography pipeline can be applied in this instance,

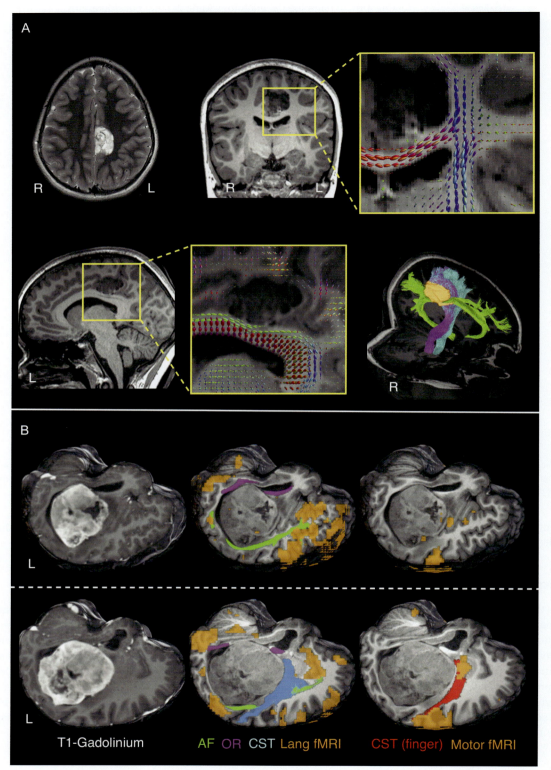

**FIG. 8** Tractography in low-grade brain tumor and extraaxial brain tumor with marked mass effect. In (A) is a 17-year-old boy with drug-resistant epilepsy referable to a left cingulate developmental brain tumor. Close-up views (*yellow outlined squares*) of perilesional fODF glyphs show preserved orientations. The *bottom right* is a 3D render showing the cingulum (in *green*), corticospinal tract (in *cyan*), and primary somatosensory tract (in *pink*) and their spatial relations with the tumor (in *orange*). In (B) is a 13.5-year-old, right-handed boy, presented with headache referable to a large left intraventricular meningioma with marked brain compression. The 3D rendered images are displayed in surgical orientation. Both the left corticospinal tract (CST, in *blue*) and optic radiation (OR, in *purple*) can be reconstructed using anatomy-based tracking ROIs. The CST abuts the anterior tumor margin, and the OR is displaced inferiorly by the meningioma. Conversely, anatomy-based ROIs could not be adequately delineated for arcuate fasciculus (AF). In this case, eloquent cortices (*orange*) localized by language functional MRI (Lang fMRI) are used as tracking regions-of-interest. The AF wraps around the superior and posterior tumor margins. The CST finger fibers (in *red*) are also dissected using a finger motor fMRI. Tractography in both cases used the multitissue constrained spherical deconvolution model, *lmax* = 8, an iFOD-2 probabilistic algorithm, fiber orientation distribution threshold = 0.1 (in A) and 0.07 (in B), max angle 45 degrees, step size of 1.15 mm.

although one must exercise caution and consider the technical limitation of each local WM modeling techniques on the tractography results. In some cases, there may be associated subcortical WM signal changes deep to the cortical lesion. The accuracy of local fODF estimates in this WM region should be checked for.

## 6.3 Disorders with marked brain anatomy distortion

Examples in this category include encephalomalacia and cerebral gliosis following acquired brain injuries, such as stroke, traumatic brain injury, and anatomy altered by previous brain surgeries; and large intraaxial lesions (e.g., thalamic or basal ganglia tumors) and extraaxial lesions (e.g., meningioma), causing marked mechanical compression of the brain parenchyma (Hou et al., 2012). To dissect the WM bundle of interest in these instances, the general principle is to place the tracking ROIs only in areas with recognizable anatomy, and avoid placing precise ROIs in areas affected by pathology, which would introduce tracking bias due to ambiguous a priori anatomical knowledge. When the brain anatomy cannot be accurately delineated, functional imaging data should be used to aid WM bundle dissections (see Fig. 8B; also see the next section).

## 6.4 Disorders with microscopic disease infiltrates and perilesional WM changes

A classic example in this category is high-grade gliomas (e.g., glioblastoma multiforme and anaplastic astrocytoma) with surrounding vasogenic brain edema. Another example is treatment-related peritumoral WM injury/necrosis following cranial irradiation therapy. Acute hemorrhage from vascular lesions, such as cavernous hemangioma, or arteriovenous malformation can also be associated with perilesional edema. In these cases, important information needed from tractography includes: *Is the adjacent WM bundle displaced or disrupted by the lesion? Does the WM bundle traverse through the perilesional WM edema, gliosis, or areas infiltrated by microscopic tumor deposits or metastasis?* Generally speaking, these lesion-tissue effects introduce uncertainties in WM modeling based on the diffusion signal (Provenzale et al., 2004). They typically introduce noisier and less anisotropic diffusion signal, leading to inaccurate local modeling of orientation estimates, and premature tracking terminations (see cases illustrated in Fig. 9 and also Fig. 2C; Roberts et al., 2005).

The development of edema correction tractography methods is an active area of dMRI research, detailed descriptions of which are beyond the scope of this chapter. The readers are referred to relevant sections of recent reviews for more details (Vanderweyen et al., 2020; Yang et al., 2021). Based on one author's own surgical tractography experience (JYMY), combining a probabilistic tracking algorithm with a local tractography pipeline robust to crossing fibers can partly overcome tracking difficulties in the perilesional WM edema, resulting in anatomically plausible tracking results (see Fig. 10). One should acknowledge there are no anatomical or biological certainties of the local fODF orientation estimates derived from within the lesion and the perilesional WM anomalies, including microscopic tumor deposits. It remains unclear what anatomical rules to follow for the tracking streamlines. The readers are to be reminded that tractography is a mathematical computing process that has no direct quantitative physical and biological attributes. To this end, it is critical to introduce functional brain imaging modalities, such as functional MRI and navigated transcranial magnetic stimulation, to localize the eloquent cortex, and to guide the tracking ROI definition. This will lead to functionally relevant tractography reconstructions that are useful for surgical image-guidance (Sanvito et al., 2020; Yang et al., 2020; Kleiser et al., 2010). The RTT pipeline can be used here for real-time interactive tracking visualization as presented in Section 3.1. The accuracy of tracking results can also be confirmed intraoperatively using direct cortical and subcortical electrostimulation, which is

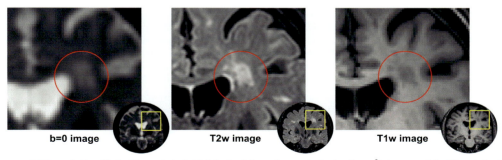

**FIG. 9** An edematous WM area is identified with a *red circle*. This lesion is hyperintense in the $b = 0$ mm$^2$/s image and the T2-weighted (T2w) image, but hypointense in the T1-weighted (T1w) image. The less anisotropic fiber orientation distribution function here makes it hard for tracking streamlines to traverse through the edema areas.

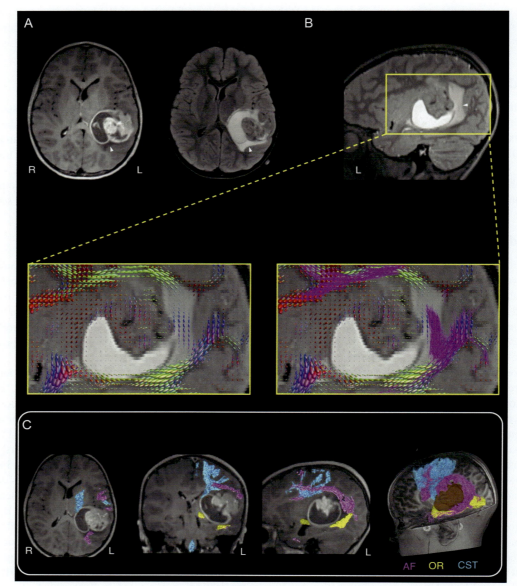

**FIG. 10** Peritumoral edema tractography in high-grade brain tumor. This is an 8-year-old boy with a large left parietal anaplastic ganglioglioma. The initial MRI (A) shows the tumor has mixed solid and cystic components and marked peritumoral WM edema (*arrowheads*) posteriorly in the deep temporoparietal WM. In (B) are close-up views (*yellow outlined panels*) of the fODF glyphs and arcuate fasciculus (AF, in *pink*) tracking streamlines shown as overlays on T2-weighted sagittal images. The multifiber WM modeling technique used is able to recover the principal fODF orientation profiles in the edematous WM. This enables AF to track through the edema, resulting in anatomically plausible tractography reconstruction. Image (C) shows both 2D orthogonal and 3D rendered views of the reconstructed left AF (in *pink*), corticospinal tract (CST, in *cyan*), and optic radiation (OR, in *yellow*) and their spatial relationship with the tumor (in *brown*). Tractography in this case used the multitissue constrained spherical deconvolution model, $lmax = 8$, an iFOD-2 probabilistic algorithm, WM anatomy-guided ROIs, fiber orientation distribution threshold = 0.07, max angle 45 degrees, step size of 1.15 mm. *((C) Adapted and modified from Yang, J.Y.M., Yeh, C.H., Poupon, C., Calamante, F., 2021. Diffusion MRI tractography for neurosurgery: the basics, current state, technical reliability and challenges. Phys. Med. Biol. 66 (15), 15TR01, with CC BY 4.0 copyright permission.)*

the surgical gold standard for localizing the eloquent cortex and confirming the WM bundle position and its functional relevance (Berger and Hadjipanayis, 2007; Duffau, 2015).

## 7 Conclusion

This chapter highlights some technical and practical considerations when performing tractography in pathological brains. Even in the context of *lack of time* and *lack of data* and *data quality*, there remain feasible options to make tractography possible in these challenging situations. We revisited some of the key considerations regarding selections of the tractography pipelines and made practical suggestions for different neurological and neurosurgical conditions. With pathology, it is also essential to challenge the default tracking parameters and default tracking masks that were defined based on the healthy and typically developing brains.

# Appendix

### Acquisition information about the MS subject used for illustration purpose

This acquisition is provided by the Sherbrooke Connectivity Imaging Laboratory. Whole brain MRI data were acquired on a Philips Healthcare Ingenia 3T MRI scanner using a 32-channel head coil. The acquisition consisted of a 3D T1-weighted image and multishell diffusion-weighted images (DWI). The 3D T1-weighted MPRAGE image was acquired at 1.0 mm isotropic following a repetition time of 7.9 ms, an echo time of 3.5 ms, and an inversion time of 950 ms. The field-of-view used was $224 \times 224$ mm$^2$ yielding 150 slices, and the flip angle used was 8 degrees. The multishell DWI images were acquired with single-shot spin echo. The resolution used was 2 mm isotropic with a repetition time of 4800 ms and an echo time of 92 ms. The field-of-view and flip angle used were $112 \times 112$ mm$^2$ yielding 66 slices and 90 degrees. The DWIs were acquired using 100 uniform directions spread over three shells at $b = 300$ mm$^2$/s (8 directions), $b = 1000$ mm$^2$/s (32 directions), $b = 2000$ mm$^2$/s (60 directions), and seven $b = 0$ mm$^2$/s images. A reversed phase encoded $b = 0$ image was also acquired to correct EPI distortion in the same resolution as the DWI. The whole acquisition time is approximately 30 min.

### Acquisition information about the patients used for Figs. 2, 8, and 10

The MRI data acquisition, DWI preprocessing, tractography reconstruction, and combined multimodal imaging display for surgical visualization were provided by the Neuroscience Advanced Clinical Imaging Service (NACIS), the Royal Children's Hospital, Melbourne, Australia. The MRI data processing and multimodal imaging displays were performed and generated by Dr. Sila Genc and Dr. Joseph Yuan-Mou Yang. The 3D renders displayed in Figs. 2 and 8B were generated based on using an in-house technique developed by Dr. Bonnie Alexander. The 3D renders for cases used in Figs. 8A and 10 were generated using the MR view volume render function in MRtrix3 (Tournier et al., 2019). Whole brain MRI data were acquired on a Siemens Magnetom Prisma fit 3T MRI scanner using a 32-channel head coil. Different combinations of structural sequences were used for different case illustrations. They included: 3D T1-weighted images, 3D T1-weighted postgadolinium contrast, 2D or 3D T2-weighted images, and 2D fluid attenuated inversion recovery (FLAIR) images.

The multishell DWI data were acquired using single-shot spin echo and with multiband acceleration in all cases. For cases illustrated in Figs. 2, 8B, and 10, the sequence resolution used was 2.3 mm isotropic with a repetition time of 3500 ms, and an echo time of 77 ms. The field-of-view used was $258 \times 258$ mm$^2$ yielding 64 slices, and the flip angle used was 90 degrees. The DWI was acquired with an anterior-to-posterior phase encode direction, using 90 uniform directions spread over 2 diffusion-weighted shells at $b = 1000$ mm$^2$/s (30 directions), $b = 3000$ mm$^2$/s (60 directions), and 11 interleaved $b = 0$ mm$^2$/s images. The acquisition time was 6.09 min. For the case illustrated in Fig. 8A, a legacy multishell DWI sequence from our institution was used, with each of the three diffusion-weighted shells acquired separately: $b = 1000$ mm$^2$/s (25 directions and 6 interleaved $b = 0$ mm$^2$/s images), $b = 2000$ mm$^2$/s (45 directions and 6 interleaved $b = 0$ mm$^2$/s images), and $b = 2800$ mm$^2$/s (60 directions and 10 interleaved $b = 0$ mm$^2$/s images). For all shells, the resolution used was 2.3 mm isotropic with a repetition time of 3200 ms and an echo time of 110 ms. The field-of-view used was $258 \times 258$ mm$^2$ yielding 64 slices, and the flip angle used was 90 degrees. The combined total acquisition time for all three shells was around 10.50 min. In both multishell DWI sequences, a pair of anterior-to-posterior and posterior-to-anterior phase encode $b = 0$ mm$^2$/s images were acquired to correct EPI distortion in the same resolution as the DWI.

# References

Aerts, H., Schirner, M., Dhollander, T., Jeurissen, B., Achten, E., Van Roost, D., Ritter, P., Marinazzo, D., 2020. Modeling brain dynamics after tumor resection using The Virtual Brain. NeuroImage 213, 116738.

Avants, B.B., Tustison, N., Song, G., 2009. Advanced normalization tools (ANTS). Insight J. 2 (365), 1–35.

Aydogan, D.B., Souza, V.H., Lioumis, P., Ilmoniemi, R.J., 2021. Towards real-time tractography-based TMS neuronavigation. Brain Stimul. 14 (6), 1609.

Beare, R., Alexander, B., Warren, A., Kean, M., Seal, M., Wray, A., Maixner, W., Yang, J.Y.M., 2022. Karawun: a software package for assisting evaluation of advances in multimodal imaging for neurosurgical planning and intraoperative neuronavigation. Int. J. Comput. Assist. Radiol. Surg. 18 (1), 171–179.

Berger, M.S., Hadjipanayis, C.G., 2007. Surgery of intrinsic cerebral tumors. Neurosurgery 61 (Suppl 1), SHC-279.

Bonilha, L., Jensen, J.H., Baker, N., Breedlove, J., Nesland, T., Lin, J.J., Drane, D.L., Saindane, A.M., Binder, J.R., Kuzniecky, R.I., 2015. The brain connectome as a personalized biomarker of seizure outcomes after temporal lobectomy. Neurology 84 (18), 1846–1853.

Bucci, M., Mandelli, M.L., Berman, J.I., Amirbekian, B., Nguyen, C., Berger, M.S., Henry, R.G., 2013. Quantifying diffusion MRI tractography of the corticospinal tract in brain tumors with deterministic and probabilistic methods. NeuroImage 3, 361–368.

Chamberland, M., Whittingstall, K., Fortin, D., Mathieu, D., Descoteaux, M., 2014. Real-time multi-peak tractography for instantaneous connectivity display. Front. Neuroinform. 8, 59.

Chamberland, M., Genc, S., Tax, C.M.W., Shastin, D., Koller, K., Raven, E.P., Cunningham, A., Doherty, J., van den Bree, M.B.M., Parker, G.D., et al., 2021. Detecting microstructural deviations in individuals with deep diffusion MRI tractometry. Nat. Comput. Sci. 1 (9), 598–606.

Christidi, F., Karavasilis, E., Samiotis, K., Bisdas, S., Papanikolaou, N., 2016. Fiber tracking: a qualitative and quantitative comparison between four different software tools on the reconstruction of major white matter tracts. Eur. J. Radiol. Open 3, 153–161.

Duffau, H., 2015. Stimulation mapping of white matter tracts to study brain functional connectivity. Nat. Rev. Neurol. 11 (5), 255–265.

Fan, L., Li, H., Zhuo, J., Zhang, Y., Wang, J., Chen, L., Yang, Z., Chu, C., Xie, S., Laird, A.R., et al., 2016. The human brainnetome atlas: a new brain atlas based on connectional architecture. Cereb. Cortex 26 (8), 3508–3526.

Farquharson, S., Tournier, J.D., Calamante, F., Fabinyi, G., Schneider-Kolsky, M., Jackson, G.D., Connelly, A., 2013. White matter fiber tractography: why we need to move beyond DTI. J. Neurosurg. 118 (6), 1367–1377.

Fekonja, L.S., Wang, Z., Aydogan, D.B., Roine, T., Engelhardt, M., Dreyer, F.R., Vajkoczy, P., Picht, T., 2021. Detecting corticospinal tract impairment in tumor patients with fiber density and tensor-based metrics. Front. Oncol. 10, 622358.

Fischl, B., 2012. FreeSurfer. NeuroImage 62 (2), 774–781.

Girard, G., Whittingstall, K., Deriche, R., Descoteaux, M., 2014. Towards quantitative connectivity analysis: reducing tractography biases. NeuroImage 98, 266–278.

Gleichgerrcht, E., Keller, S.S., Drane, D.L., Munsell, B.C., Davis, K.A., Kaestner, E., Weber, B., Krantz, S., Vandergrift, W.A., Edwards, J.C., et al., 2020. Temporal lobe epilepsy surgical outcomes can be inferred based on structural connectome hubs: a machine learning study. Ann. Neurol. 88 (5), 970–983.

Hagler Jr., D.J., Ahmadi, M.E., Kuperman, J., Holland, D., McDonald, C.R., Halgren, E., Dale, A.M., 2009. Automated white-matter tractography using a probabilistic diffusion tensor atlas: application to temporal lobe epilepsy. Hum. Brain Mapp. 30 (5), 1535–1547.

Hou, Y., Chen, X., Xu, B., 2012. Prediction of the location of the pyramidal tract in patients with thalamic or basal ganglia tumors. PLoS One 7 (11), e48585.

Jeurissen, B., Descoteaux, M., Mori, S., Leemans, A., 2019. Diffusion MRI fiber tractography of the brain. NMR Biomed. 32 (4), e3785.

Kleiser, R., Staempfli, P., Valavanis, A., Boesiger, P., Kollias, S., 2010. Impact of fMRI-guided advanced DTI fiber tracking techniques on their clinical applications in patients with brain tumors. Neuroradiology 52, 37–46.

Kuhnt, D., Bauer, M.H.A., Sommer, J., Merhof, D., Nimsky, C., 2013. Optic radiation fiber tractography in glioma patients based on high angular resolution diffusion imaging with compressed sensing compared with diffusion tensor imaging-initial experience. PLoS One 8 (7), e70973.

Leemans, A., Jeurissen, B., Sijbers, J., Jones, D.K., 2009. ExploreDTI: a graphical toolbox for processing, analyzing, and visualizing diffusion MR data. Proc. Int. Soc. Mag. Reson. Med. 17 (1), 3537.

Lipp, I., Parker, G.D., Tallantyre, E., Goodall, A., Grama, S., Patitucci, E., Heveron, P., Tomassini, V., Jones, D.K., 2019. Tractography in the presence of white matter lesions in multiple sclerosis. bioRxiv, 559708.

Macdonald-Laurs, E., Maixner, W.J., Bailey, C.A., Barton, S.M., Mandelstam, S.A., Yang, J.Y.M., Warren, A.E.L., Kean, M.J., Francis, P., MacGregor, D., et al., 2021. One-stage, limited-resection epilepsy surgery for bottom-of-sulcus dysplasia. Neurology 97 (2), e178–e190.

Mormina, E., Longo, M., Arrigo, A., Alafaci, C., Tomasello, F., Calamuneri, A., Marino, S., Gaeta, M., Vinci, S.L., Granata, F., 2015. MRI tractography of corticospinal tract and arcuate fasciculus in high-grade gliomas performed by constrained spherical deconvolution: qualitative and quantitative analysis. Am. J. Neuroradiol. 36 (10), 1853–1858.

Provenzale, J.M., McGraw, P., Mhatre, P., Guo, A.C., Delong, D., 2004. Peritumoral brain regions in gliomas and meningiomas: investigation with iso-tropic diffusion-weighted MR imaging and diffusion-tensor MR imaging. Radiology 232 (2), 451–460.

Rheault, F., Houde, J.C., Goyette, N., Morency, F., Descoteaux, M., 2016. MI-brain, a software to handle tractograms and perform interactive virtual dissection. In: Proceedings of the ISMRM Diffusion Study Group Workshop, Lisbon.

Roberts, T.P.L., Liu, F., Kassner, A., Mori, S., Guha, A., 2005. Fiber density index correlates with reduced fractional anisotropy in white matter of patients with glioblastoma. Am. J. Neuroradiol. 26 (9), 2183–2186.

Sanvito, F., Caverzasi, E., Riva, M., Jordan, K.M., Blasi, V., Scifo, P., Iadanza, A., Crespi, S.A., Cirillo, S., Casarotti, A., et al., 2020. fMRI-targeted high-angular resolution diffusion MR tractography to identify functional language tracts in healthy controls and glioma patients. Front. Neurosci. 14, 225.

Sinke, M.R.T., Otte, W.M., Christiaens, D., Schmitt, O., Leemans, A., van der Toorn, A., Sarabdjitsingh, R.A., Joëls, M., Dijkhuizen, R.M., 2018. Diffusion MRI-based cortical connectome reconstruction: dependency on tractography procedures and neuroanatomical characteristics. Brain Struct. Funct. 223 (5), 2269–2285.

Smith, R.E., Tournier, J.D., Calamante, F., Connelly, A., 2012. Anatomically-constrained tractography: improved diffusion MRI streamlines tractography through effective use of anatomical information. NeuroImage 62 (3), 1924–1938.

Tax, C.M.W., Chamberland, M., van Stralen, M., Viergever, M.A., Whittingstall, K., Fortin, D., Descoteaux, M., Leemans, A., 2015. Seeing more by showing less: orientation-dependent transparency rendering for fiber tractography visualization. PLoS One 10 (10), e0139434.

Theaud, G., Descoteaux, M., 2022. dMRIQCpy: a python-based toolbox for diffusion MRI quality control and beyond. In: 31th Scientific Meeting of the International Society for Magnetic Resonance in Medicine.

Theaud, G., Dilharreguy, B., Catheline, G., Descoteaux, M., 2017. Impact of white-matter hyperintensities on tractography. In: 25th Annual Meeting of the International Society for Magnetic Resonance in Medicine (ISMRM), International Society for Magnetic Resonance in Medicine, Honolulu.

Theaud, G., Fortin, D., Morency, F., Descoteaux, M., 2019. Brain tumors: a challenge for tracking algorithms. In: 27th Annual Meeting of the International Society for Magnetic Resonance in Medicine (ISMRM), International Society for Magnetic Resonance in Medicine, Honolulu.

Theaud, G., Houde, J.C., Boré, A., Rheault, F., Morency, F., Descoteaux, M., 2020. TractoFlow-ABS (Atlas-Based Segmentation). bioRxiv.

Tokuda, J., Fischer, G.S., Papademetris, X., Yaniv, Z., Ibanez, L., Cheng, P., Liu, H., Blevins, J., Arata, J., Golby, A.J., et al., 2009. OpenIGTLink: an open network protocol for image-guided therapy environment. Int. J. Med. Robot. Comput. Assist. Surg. 5 (4), 423–434.

Tournier, J.D., Smith, R., Raffelt, D., Tabbara, R., Dhollander, T., Pietsch, M., Christiaens, D., Jeurissen, B., Yeh, C.H., Connelly, A., 2019. MRtrix3: a fast, flexible and open software framework for medical image processing and visualisation. NeuroImage 202, 116137.

Vanderweyen, D.C., Theaud, G., Sidhu, J., Rheault, F., Sarubbo, S., Descoteaux, M., Fortin, D., 2020. The role of diffusion tractography in refining glial tumor resection. Brain Struct. Funct. 225 (4), 1413–1436.

Yang, J.Y.M., Beare, R., Wu, M.H., Barton, S.M., Malpas, C.B., Yeh, C.H., Harvey, A.S., Anderson, V., Maixner, W.J., Seal, M., 2019. Optic radiation tractography in pediatric brain surgery applications: a reliability and agreement assessment of the tractography method. Front. Neurosci. 13, 1254.

Yang, J.Y., Menon, R., Barton, S., Mandelstam, S.A., Kerr, R., Wrennall, J., Bailey, C., Freeman, J., Maixner, W.J., Harvey, A.S., 2020. One-stage, language-dominant, opercular-insular epilepsy surgery with multimodal structural and functional neuroimaging evaluation. In: International Society for Magnetic Resonance in Medicine Conference.

Yang, J.Y.M., Yeh, C.H., Poupon, C., Calamante, F., 2021. Diffusion MRI tractography for neurosurgery: the basics, current state, technical reliability and challenges. Phys. Med. Biol. 66 (15), 15TR01.

# Chapter 20

# Tractography visualization

**Maxime Chamberland[a], Charles Poirier[b], Tom Hendriks[a], Dmitri Shastin[c], Anna Vilanova[a,d], and Alexander Leemans[e]**

[a]*Eindhoven University of Technology, Eindhoven, The Netherlands,* [b]*Sherbrooke Connectivity Imaging Lab, University of Sherbrooke, Sherbrooke, Canada,* [c]*Cardiff University Brain Research Imaging Centre (CUBRIC), School of Psychology, Cardiff University, Cardiff, United Kingdom,* [d]*Delft University of Technology, Delft, The Netherlands,* [e]*PROVIDI lab, Image Sciences Institute, University Medical Center Utrecht, Utrecht, The Netherlands*

## 1 Introduction

The main objective of medical scientific visualization is to clearly and effectively represent the data obtained in scientific research (Comaniciu et al., 2016; Zhou et al., 2022). This data is often represented as a continuous field. The main difference between scientific visualization and other visualization domains is that *scivis* deals with continuous data (that has been sampled) often representing spatial data (i.e., in the form of scalar, tensor, vector fields, or multivariate fields). With the ever-growing size and dimensionality of data acquired nowadays, it quickly becomes a real challenge to display such large datasets. Scientific visualization, therefore, has also a mandate to not only allow the intelligent display of this data but also ensure its perception and understanding. In other words, visualization must not sacrifice any critical information present in the data for the sake of simplicity or esthetics.

As seen through this book, one can reconstruct the three-dimensional (3D) architecture of the human brain via diffusion magnetic resonance imaging (MRI) streamline tractography ( Jeurissen et al., 2019). From a scientific visualization point-of-view, the notion of streamline is a concept that exists from fluid dynamics (Helman and Hesselink, 1991). From a computer graphics point-of-view, streamlines (or polylines) can aid in the display of tensor fields (Delmarcelle and Hesselink, 1992; Westin et al., 2002). Fiber tractography produces dense datasets that are formed of multiple interdigitating 3D polylines, which makes direct visualization challenging (Rheault et al., 2017). Indeed, when visualizing whole-brain tractograms, visual clutter occurs. To address this problem, various approaches were proposed in an attempt to better visualize tractography data in the context of medical imaging.

This chapter aims to provide an overview of the current state of research on tractography visualization. While there have been numerous works on this topic (Vilanova et al., 2006; Leemans, 2010; Margulies et al., 2013; Isenberg, 2015; Schultz and Vilanova, 2019; Chen et al., 2018), this chapter aims to synthesize and build upon the existing knowledge in a way that is accessible to a wide audience. By focusing only on tractogram and streamline visualization, we hope to provide a clear and up-to-date overview of the field of tractography. Methods associated with local reconstruction (e.g., glyphs visualization) and postprocessing of tractograms (e.g., graph theory and connectomics) are therefore not covered in this chapter. All in all, this chapter will be a valuable resource for researchers and practitioners alike who are interested in gaining a deeper understanding of tractography visualization.

## 2 Tractography rendering from a scientific visualization point-of-view

### 2.1 The many flavors of streamline rendering

Streamlines derived from diffusion MRI tractography are commonly visualized using polylines or streamtubes. In the context of tractography, streamlines are zero-radius polylines consisting of a list of points in 3D space, connected by line segments. In this chapter, the notion of polyline and streamline are used interchangeably for simplicity. Starting from a seed point, the next point of the line is calculated using the available diffusion MRI data (e.g., simply following the principal direction of diffusion for a set distance (Conturo et al., 1999), or using more advanced regularized approaches (Zhukov and Barr, 2002)). A streamline ends when a terminating condition is met (e.g., the anisotropy of diffusion falls below a threshold, or an anatomical end region is reached (Part III). The trajectory of a streamline only captures information

---

*Handbook of Diffusion MR Tractography.* https://doi.org/10.1016/B978-0-12-818894-1.00002-1
Copyright © 2025 Elsevier Ltd. All rights are reserved, including those for text and data mining, AI training, and similar technologies.

about the principal direction of diffusion, which is often emphasized by coloring line segments based on their orientation in 3D space. Spiral curves can capture additional information about secondary and tertiary directions of diffusion (Peeters et al., 2006).

Hyperstreamlines, or streamtubes, are 3D tubes whose trajectories are determined by following the major eigenvectors in a tensorfield, while stretching their cross-sections in accordance with the two orthogonal eigenvectors (Delmarcelle and Hesselink, 1992). In diffusion MRI, this results in the trajectory of the streamtube following the principal direction of diffusion, and the cross-section visualizing the ratio between the secondary and tertiary directions of diffusion (Zhang et al., 2003). Alternatively, other properties (e.g., orientation distribution functions [ODFs], or uncertainty maps) can be captured in the shape of the hyperstreamline (Vos et al., 2013; Fig. 1). While hyperstreamlines contain more information, they are also more complex to render. Tube imposters can be used as a computationally more efficient method of rendering 3D tubes (Nesi et al., 2017; Petrovic et al., 2007). Overall, the use of streamtubes (combined with lights and shadows) provide improved 3D perception of the scene (Zhang et al., 2003).

From a computational point-of-view, whole-brain tractograms are usually visualized by rendering thousands, or even millions, of polylines, which often leads to dense, cluttered views. To clarify the visualization of diffusion tractography, a number of methods have been proposed. Shading techniques can be used to simulate lighting effects (Yuan et al., 2011), such as diffuse and specular reflections, and can be combined with other visualization techniques, such as transparency, to create more complex and realistic renderings. Illuminating streamlines can improve the spatial perception (Zockler et al., 1996; Peeters et al., 2006). Similarly, illustrative visualization techniques (e.g., focus and context visualization, depth-dependent halos, etc.) can improve the comprehension of the data (Everts et al., 2009; Otten et al., 2010; Isenberg, 2015). Ambient occlusion effects also improve the spatial and structural perception of (hyper)streamlines (Eichelbaum et al., 2012; Schott et al., 2012) by adding shadows. Examples of some of these techniques can be seen in Fig. 2 and a description can be found in Table 1.

To reduce the cluttering of streamlines, polylines of the same bundle can be combined into a single abstraction using a variety of methods. Moving the streamlines together creates dense tracts surrounded by empty space (Everts et al., 2015). Replacing streamlines with a single hyperstreamline allows for bundle-based measurements and intersubject comparison (O'Donnell et al., 2009). Simplifying tracts in both geometric and image space also results in clearer images with less occlusion (Bouma et al., 2020). Wrapping streamlines in hulls[a] can visualize bundles with different fiber densities,

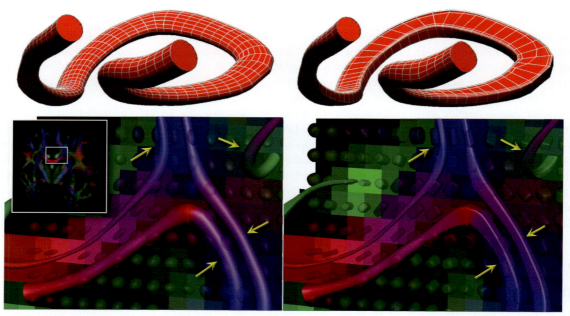

**FIG. 1** Hyperstreamline visualization in the centrum semiovale region (*inset white border*). *Left*: Conventional tubular-shaped streamlines rendered as a mesh (*top*) and overlayed on top of a diffusion tensor field (*bottom*). *Right*: Superquadric hyperstreamlines (Wiens et al., 2014) rendered as a mesh (*top*) and overlayed on top of a superquadrics tensor field (Kindlmann, 2004) (*bottom*), which can be used for uncertainty mapping. Streamlines and tensors are color-coded by direction (*red*: left-right, *green*: anteroposterior, *blue*: top-down). The *yellow arrows* highlight the differences between the two techniques. Generated using ExploreDTI (Leemans et al., 2009).

---

a. A simplified 3D geometric shape that tightly encompasses a more complex object or set of objects (e.g., streamlines), similar to an envelope.

Tractography visualization **Chapter | 20** 383

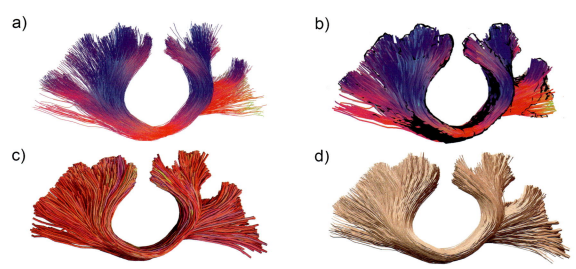

**FIG. 2** Various types of tractography visualization approaches applied to the midbody of the corpus callosum. (A) Streamline rendering combined with local orientation coloring and ambient shading. (B) Illustrative silhouette rendering combined with local orientation coloring and ambient shading. (C) Streamtube rendering combined with global orientation coloring and Phong shading. (D) Photorealistic rendering simulating the brain's white matter. Generated using FiberNavigator (Chamberland et al., 2014) and Surf Ice (Rorden, 2021).

**TABLE 1** Visualization terms employed in computer graphics.

| Term | Description |
| --- | --- |
| Focus-and-context | A technique used to emphasize a certain region of interest in an image while also showing the surrounding context. It is often used to deal with cluttered or complex scenes |
| Ambient occlusion | A shading technique used to create the effect of shadows in areas where objects are close together, resulting in a more realistic image |
| Halos | A postprocessing effect that creates a glowing aura around objects in an image, making them stand out more |
| Lensing | A technique used to distort the image to simulate the effect of a lens. It is often used to create a fish-eye effect or to zoom in on a specific area |
| Shading | The process of adding colors to an object in a way that simulates how it would appear in different lighting conditions |
| Depth-of-field | A technique used to simulate the effect of a camera lens in which objects in the foreground or background appear blurred, while the main focus remains sharp |
| Texturing | The process of adding detail to an object's surface to create the appearance of a specific material (e.g., brain tissue) |
| Field-of-view | The angle of the scene that is visible to the viewer. It is often used to show or hide elements of the scene |
| Illustrative rendering | This nonphotorealistic technique uses abstraction to convey the relevant information while minimizing less important details |

depending on the distance of the hull from the core of the bundle (Enders et al., 2005; e.g., similar to Fig. 6D). These hulls can be interactively dissected to show the streamlines in areas of interest (Röttger et al., 2012). Illustrative techniques for the visualization of streamline bundles have also been proposed (Otten et al., 2010).

Aside from polylines, other shapes have been suggested to represent fiber bundles. A hybrid form using textured triangle strips and point sprites efficiently renders a visualization similar to tubes (Merhof et al., 2006). Hyperstream ribbons capture only the peaks of the ODF but are much more efficient to render than hyperstreamlines (Vos et al., 2013). Ridge surfaces, based on fractional anisotropy creases, can be used to delineate white matter structure (Kindlmann et al., 2007). Finally, streamsurfaces (or sheets) can be used to visualize the complex interwoven structure of fibers that occurs at some parts of the brain (Wedeen et al., 2012; Tax et al., 2016; Ankele and Schultz, 2018). An overview of existing methods is given in Schultz and Vilanova (2019).

**384 PART | III** Tractography algorithms

What has been covered so far pertained to the 3D aspect of streamlines. Another set of approaches consists of mapping 3D streamlines onto a two-dimensional (2D) texture. One approach, called track density imaging (TDI, Calamante et al., 2010), produces high-quality images of the white matter, with high anatomical contrast using tractography as input. The approach works by summing the total number of streamlines present in each voxel of a user-defined grid. By using grid elements that are smaller than the original diffusion MRI voxels, one can achieve superresolution. TDI is, however, limited by the ongoing challenges associated with fiber count (Jones et al., 2013) and should be used qualitatively or combined with quantitative measures of density (Calamante et al., 2015). Another type of texture-based tractography visualization relies on line integral convolution (Cabral and Leedom, 1993), a widely used approach for visualizing local vector fields. This method produces an image representation of a vector field by *blurring* a noisy texture with a given vector field (e.g., main diffusion directions extracted from a tensor field (Deoni et al., 2003; Schultz et al., 2019) or multiple fiber orientations (Höller et al., 2014)).

## 2.2 Leveraging the graphical processing unit

As in many computer graphics applications, graphical processing units (GPUs) play an important part in tractography visualization. Not only are they essential for performing the basic rendering of streamlines, but their capabilities can also be harnessed to increase the visual perception of the data and to further accelerate the rendering. This section provides an overview of techniques using the GPU to enhance tractogram visualization, either in terms of computation times or memory requirements.

A naive polygon-based approach would consist of copying each streamline as an array of vertices to GPU memory and then drawing each pair of consecutive points as a line segment, using depth testing (z-buffer) to deal with overlapping segments. However, because tractograms can contain millions of streamlines, this method is expensive both computationally and in terms of memory usage. Compressing streamlines before or during loading (Rheault et al., 2017; Schertzer et al., 2022) is a simple workaround to cut down memory usage. Also, when the compressed tractogram is obtained by subsampling the streamlines (e.g., Presseau et al., 2015), the reduction in the number of vertices can render results in an increased frame rate. Even greater performances can be achieved by taking advantage of the highly parallel architecture of the GPU. For example, the GPU can be used to efficiently generate the geometry to render on-the-fly (Reina et al., 2006; Petrovic et al., 2007; Köhn et al., 2009). This enables applications to efficiently render more complex geometries such as streamtubes without increasing the amount of memory and preprocessing required to store a triangulated set of vertices by copying only the parameters describing the geometry to GPU memory. The use of a more complex geometry enables the visualization application to convey more useful information on the data. For instance, hyperstreamlines (Reina et al., 2006) are a multimodal, tube-like representation describing the properties of the diffusion tensor at each step along a fiber trajectory. To increase visual perception of the tracks, GPU-accelerated ambient occlusion has been proposed (Eichelbaum et al., 2012; Schertzer et al., 2022). As we see in Fig. 3, ambient occlusion simplifies the interpretation of tractograms by better capturing the spatial arrangement of the rendered fibers while maintaining interactive frame rates.

To further accelerate the rendering of tractograms, the use of level of detail has been suggested (Petrovic et al., 2007). Using occlusion queries, the authors determine the precision required to correctly describe the pathways to be rendered given the quantity of information that can be represented for a given screen resolution. The overlapping property of tractograms—that is, many of the rendered pathways are occluded by other pathways—has also been used to design more efficient rendering strategies. In Schertzer et al. (2022), the authors describe a method for culling the occluded geometry depending on the position of the observer and the arrangement of the fibers. The occluded fiber segments can therefore be discarded before going through the entire rendering pipeline.

The GPU has also been used to enable real-time tracking by distributing the reconstruction of fiber pathways across GPU threads (Jeong et al., 2007; McGraw and Nadar, 2007; Köhn et al., 2009; Van Aart et al., 2011). In such applications, because the data is generated on the GPU, it is readily available for rendering. Depending on the nature of the tracking algorithm implemented, different information can be displayed. For instance, Jeong et al. (2007) use the GPU to compute and visualize volumetric pathways connecting two user-defined regions of interest (ROIs) in a DTI volume, and McGraw and Nadar (2007) use it to render probabilistic heat maps describing the probability of reaching any region from some initial seeding position.

## 3 Interacting with tractograms

Interacting with streamlines is an essential aspect of tractography data visualization. In the context of virtual dissection, there are two main ways to interact with streamlines. The first is to perform the fiber tracking in *real time* (Mittmann et al., 2011; Golby et al., 2011; Chamberland et al., 2014; e.g., by providing the user with a seed region that can be moved in 3D),

FIG. 3 Tractography visualization with ambient occlusion (as done in Eichelbaum et al., 2012). View: Coronal cut.

and the second is to precompute a large number of streamlines, and then filter the relevant ones interactively via selection objects (Catani et al., 2002; e.g., cubes, spheres, anatomical regions, etc.). These interaction techniques can be combined in various ways to create a rich and engaging user experience and can be implemented using a variety of different software tools and technologies. Research into the design and implementation of interactive tractography visualizations using visual analytics is still an active field of research (Chen et al., 2009; Jianu et al., 2009; Xu et al., 2020). Following are some use cases related to common interaction techniques.

- *Navigation*: This allows the user to move around the tractogram and change their point of view, in order to explore the data from different angles and perspectives. This can be achieved using techniques such as panning, zooming, and rotating.
- *Selection*: This allows the user to filter the data in the tractogram, in order to remove or hide certain tracts or to highlight specific features. This can be achieved using techniques such as thresholding, clustering, and masking.

## 3.1 Selecting streamlines with regions of interest

When streamlines are rendered as a whole, it is impossible to correctly distinguish the paths that each individual streamline takes due to cluttering. The use of ROIs brings a solution to this problem (Fig. 4). ROIs can be manually defined or extracted from an atlas (Wakana et al., 2004). Selection ROIs can be seen as filters (Chen et al., 2009; Vaillancourt et al., 2010), thereby reducing the amount of information displayed on the screen. Some binary operations can then be applied between ROIs such as inclusion, exclusion, union, and intersection. To add interactivity, spherical and cuboid ROIs can also be moved by the user. The selection of streamlines is generally done by testing where at least one vertex is found within the ROI. To accelerate the process, streamline points can be assigned to predefined portions of the imaging volume upon loading via the use of a space-partitioning data structure (e.g., octree, kd-tree; Blaas et al., 2005). One can then simply look up in which quadrant of the volume the selection ROI is currently, and then only query streamline points from that quadrant.

## 3.2 Interactively slicing tractograms via selection planes

Whole-brain tractograms can also be visualized in 2D by means of slicing planes (Fig. 5A). This can be achieved by (1) testing whether streamline points fall inside a given plane boundary (the thickness of the plane being defined by the user as seen in Fig. 5B, typically between 1–3 slices) and (2) only rendering those streamline points. This approach can also be combined with, for example, opacity rendering to show translucency, where an α value is assigned to each streamline. Combining the alpha value between overlapping streamlines results in a spectrum of translucency (0: black, 1: white) which can then be used to weigh the color of each tract segment (Fig. 5C). The opacity of tractography can also be set by the general orientation of a given bundle in relation to the scene camera (e.g., viewing angle; Osman et al., 2023;

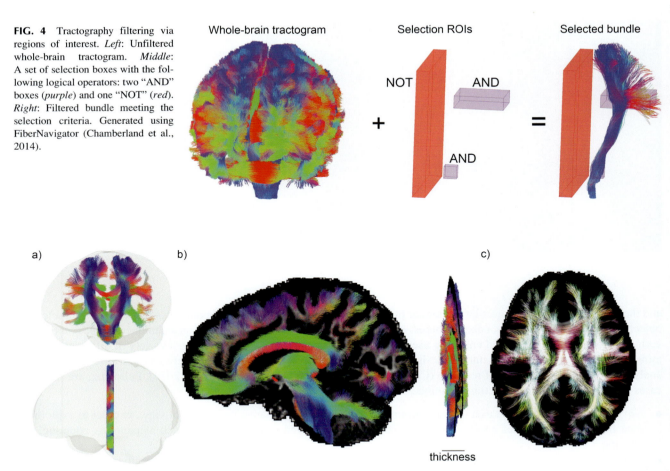

**FIG. 4** Tractography filtering via regions of interest. *Left*: Unfiltered whole-brain tractogram. *Middle*: A set of selection boxes with the following logical operators: two "AND" boxes (*purple*) and one "NOT" (*red*). *Right*: Filtered bundle meeting the selection criteria. Generated using FiberNavigator (Chamberland et al., 2014).

**FIG. 5** Tractography slicing. (A) Coronal slice and brain surface. (B) Sagittal slice overlaying an FA map. (C) Axial slice combined with an opacity blending function. Generated using FiberNavigator (Chamberland et al., 2014).

Tax et al., 2015). Slicing techniques have been used to compare tractography data with postmortem dissections (Catani et al., 2012).

### 3.3 Real-time tractography

Typically, fiber tractography is performed with a fixed set of parameters, either by launching seeds from the gray matter/white matter interface, or from everywhere in the white matter based on a tracking threshold (e.g., fractional anisotropy, fiber orientation distribution function amplitude), or even starting from a specific ROI. How to efficiently cover the whole-brain space with seed points has been a field of research in flow visualization and diffusion MRI (Vilanova et al., 2004). Previous studies have investigated the feasibility of visualization based on a set of precomputed offline streamlines using accelerated techniques (Peeters et al., 2006; Petrovic et al., 2007; Reina et al., 2006; Hlawitschka et al., 2008; Eichelbaum et al., 2012; Köhn et al., 2009; Jeong et al., 2007; McGraw and Nadar, 2007; Van Aart et al., 2011). These methods focus on how to efficiently render a set of streamlines generated by offline tractography algorithms and are for visualization only. From a neurosurgical point-of-view, the produced fiber tracts are typically then shown to a neurosurgeon inside a visualization tool. However, it is well known that fiber tractography is sensitive to the tractography parameters used. Moreover, in neurosurgery applications, the input diffusion MRI datasets are very different from one subject to the other and, thus, parameters may be subject-dependent (e.g., due to pathology; Chamberland et al., 2015; Takemura et al., 2016). Hence, an expert usually has to prepare the fiber tract datasets by manually editing and segmenting the tracts of interests, which can be time-consuming. As the name suggests, real-time fiber tracking (Mittmann et al., 2011; Golby et al., 2011; Chamberland et al., 2014) provides information *on-the-fly* about the uncertainty associated with tracking parameters and limitations that are well-known (Maier-Hein et al., 2017), thereby avoiding the use of tracking algorithms as "black boxes." This solution, which links the computation and visualization stages, at once permits the user to instantaneously compute and visualize streamlines interactively.

With this method, every time a user interacts with the application, fiber tracking is automatically carried out. For example, fiber tracking is carried out when a user is dragging an ROI on the screen, from which fiber trajectories (i.e., seeds) are launched, or when any of the typical tractography parameters are changed (e.g., masking, step size, maximum angle, weights on input and current directions, etc.). Early approaches were, however, limited to relatively small datasets (i.e., native diffusion dimensions) and were also solely based on tensor fields with a single main direction per voxel, which does not take into account the known limitations of DTI in areas of curved and crossings fibers (Tournier et al., 2011; Chapters 10 and 11). However, it can be computationally heavy to extract multiple tracking directions from the set of local fODF glyphs in a real-time fashion. Therefore, to achieve such real-time performance, the authors of Chamberland et al. (2014) proposed a novel evolution equation based on the preextracted and upsampled principal directions, also called peaks. Real-time tractography can also be used to guide neuronavigation in transcranial magnetic stimulation (Aydogan et al., 2021).

## 3.4 Placing tractography in context

In-context visualization of tractography refers to the use of additional information, such as brain surfaces (Schultz et al., 2008; Svetachov et al., 2010; Schurade et al., 2010) or other MRI modalities (e.g., functional MRI), to provide additional context and to improve the interpretability of the tractography data (Fig. 6). Some examples of in-context visualization techniques include:

- *Surface rendering*: This involves rendering the tractogram jointly with a surface representation of the brain, such as a cortical or subcortical mesh, in order to show how the tracts relate to the underlying anatomy. This can provide a more intuitive and immersive way to visualize the data.

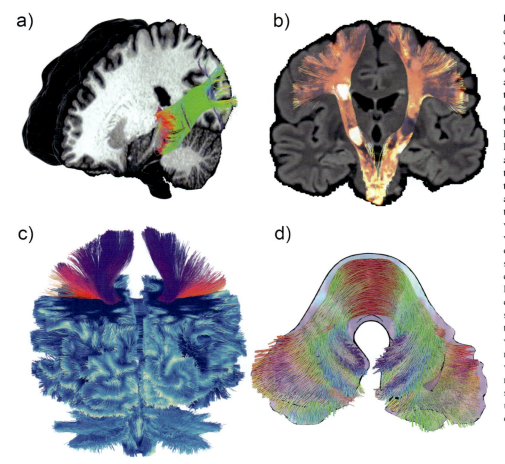

**FIG. 6** In-context visualization of diffusion MRI tractography. (A) 3D view of the optic radiation, overlaid on an anatomical T1w image. The optic radiation is shown in *green*, and the T1w image provides additional context and orientation. (B) Coronal view of the corticospinal tract (*orange*, bilateral), overlaid on a FLAIR image. Multiple sclerosis lesions are also mapped onto the tract, and are shown in *white* for hyperintense lesions and *purple* for hypointense lesions. Transparency has been applied to the tractogram to allow the underlying FLAIR image to be visible. (C) Whole-brain tractogram visualization, where the top portion of the brain has been cut out using a selection plane. This reveals the corpus callosum, which connects the left and right motor cortices. The cut-out plane provides a different perspective on the tractogram, allowing the corpus callosum to be more clearly visible. (D) A set of tracts forming the middle cerebellar peduncle placed within its envelope, delineating the region of interest, which is rendered semiopaque (*purple*). Generated using FiberNavigator (Chamberland et al., 2014).

**388 PART | III** Tractography algorithms

- *Overlay*: This involves combining the tractogram with other MRI data, such as functional or structural images, in order to provide additional information about the brain and to highlight specific features or regions. This can be useful for comparing the tractography data with other data modalities and for identifying relationships between different aspects of brain function and structure.

## 4 Advanced tractography visualization

In this section, we briefly touch upon other types of visualizations related to tractography. Our goal is to introduce those concepts in a concise manner and recommend further reading material to interested readers.

### 4.1 Along-tract mapping and profiling

One way to visualize information over a given bundle is to sample diffusion MRI measures at every 3D point (i.e., vertex) of a streamline by means of an interpolation technique. This process is typically referred to as *tractometry* (De Santis et al., 2014) and can provide a more comprehensive assessment of the underlying white matter microstructure—by mapping information along specific white matter tracts (Chapter 23). Streamlines can be color-coded using various methods. The most basic approach is the use of a uniform color scheme, which can be useful to differentiate various bundles. Alternatively, streamlines can be color-coded according to their main orientation (e.g., by mapping the 3D vector formed by their start and ending points to a set of RGB values: $x \rightarrow R, y \rightarrow G, z \rightarrow B$). This principle can also be applied locally, where each line segment forming a streamline is color coded. Quantitative indices derived from multimodal acquisitions may also be mapped onto streamlines, by sampling the underlying values at each vertex. This of course requires that both the streamlines and the underlying map are coregistered in the same space. A look-up table is then commonly used to map the measures to gray-scale of color-scale values. Those colormaps may be interactively tailored to focus on specific aspects of the data (e.g., highlight regions of high anisotropy, detect the presence of anomalies in the white matter, etc.). This has been the focus of a recent work where the authors studied the ability of the human visual system to extract statistical information from features in the context of streamline color mapping (Chen et al., 2019). From a computational point-of-view, it quickly becomes intractable for existing analysis pipelines to process multiple measurements at each voxel and at each vertex forming a streamline, highlighting the need for new ways to visualize and analyze such high-dimensional data (e.g., by using dimensionality reduction approaches and the direct mapping of components onto the streamlines; Chamberland et al., 2019).

### 4.2 Web-based visualization

Web-based visualization applications can be an effective way to share and interact with tractography data. These applications allow users to access and explore the data from any device with a web browser, without the need to install specialized software. Additionally, web-based visualization can also be useful for educational and outreach purposes, as it can enable the dissemination of fiber tractography results to a wider audience. Several research papers have explored the use of web-based applications for tractography visualization, and together these approaches have contributed to online brain connectivity visualization overall (Prados et al., 2007; Haehn et al., 2014, 2020; Ledoux et al., 2017; Yeatman et al., 2018; Franke and Haehn, 2020; Franke et al., 2021). These papers describe the technical challenges and solutions involved in creating web-based tractography visualization applications and can serve as useful references for those interested in developing similar tools.

### 4.3 Uncertainty mapping

Uncertainty visualization has gained popularity in recent years, especially in medicine, where enormous volumes of data are collected. Communicating uncertainty is critical to ensuring effective treatment decisions (Gillmann et al., 2021a, b). In the context of tractography and surgical planning, uncertainty visualization refers to the display of the degree of confidence or accuracy of reconstructed fiber tracts. This can be achieved by showing the variability or the spread of estimated fiber pathways, which reflect the uncertainty in the reconstruction process. This can be useful in identifying regions where fiber tracking is less reliable and in guiding the interpretation of the results. The methods used for uncertainty visualization in fiber tractography can vary, but they typically involve the use of statistical or probabilistic models to estimate the distribution of possible fiber pathways and to represent the uncertainty through a visual representation, such as color maps or transparency. For example, by using volumetric hulls that encompass a certain amount of streamlines (Enders et al., 2005).

For more information on uncertainty visualization in the context of diffusion MRI and tractography, see, for example, Jones (2003, 2008), von Kapri et al. (2010), Wiens et al. (2014), Schultz et al. (2014), Gruen et al. (2022), Brecheisen et al. (2013), Siddiqui et al. (2021) and Chapter 14.

### 4.4 Immersive virtuality

Interactive 3D renderings of key anatomical and diseased features may improve surgical intervention, radiation therapy, and minimally invasive procedures. With 3D representations, the spatial relationships between diseased regions and eloquent structures at risk may be properly assessed. For example, diffusion MRI tractography can also be combined with augmented reality (AR) technology to superimpose the tractography data onto a real-time video feed of the subject's brain. This can provide a more immersive and interactive way to visualize the tractography data, allowing the user to see how the white matter tracts relate to the underlying anatomy of the brain (Yang et al., 2021; Luzzi et al., 2021). This approach has the potential to be especially useful for surgical planning, as it can provide real-time guidance to the surgeon during the procedure. Virtual reality (VR), on the other hand, can provide an immersive and interactive 3D environment where diffusion MRI tractography can be visualized as virtual objects, enabling users to explore and manipulate the data from different perspectives (Rick et al., 2011; Qiu et al., 2010; Lo et al., 2007; Pinter et al., 2020; Ard et al., 2017). Research into the use of AR/VR for diffusion MRI tractography is still in its early stages, but early work in the field can provide valuable insights into the potential applications and challenges of combining AR and diffusion MRI tractography. AR/VR can also be useful for educational and research purposes, allowing students and researchers to interact with the tractography data in a more intuitive and engaging way (Rick et al., 2011; Ille et al., 2021).

### 4.5 Photorealistic rendering

To generate stunning high-quality tractography images, similar to photorealistic or cinematic rendering (Comaniciu et al., 2016; Lakhani and Deib, 2022), it is necessary to use advanced rendering techniques and algorithms that can create highly detailed and realistic visualizations of the data. This can involve using a mix of techniques previously described, such as global illumination (Banks and Westin, 2008; Kanzler et al., 2018; Kern et al., 2020), ambient occlusion (Eichelbaum et al., 2012; Schott et al., 2012), and depth of field to create more realistic lighting and shading effects (Comaniciu et al., 2016), and using advanced shading models, such as physically based rendering (Banks and Westin, 2008; Garyfallidis et al., 2021) or raytracing (McGraw, 2020) to simulate the interaction of light with the surface of the brain tissue (Fig. 7). For example, the work of Lakhani and Deib (2022) depicts photorealistic models of intracranial tumors derived from conventional 3D-T1 anatomical MRI. Additionally, it is important to use high-resolution data and appropriate tractography techniques to ensure that the resulting images are as detailed and accurate as possible. This can involve using high-quality diffusion MRI data and carefully selecting the streamlines to be visualized, in order to create the most visually striking and informative images.

**FIG. 7** Photorealistic rendering of the short association fibers network (Shastin et al., 2022) associated with the arcuate fasciculus pathway (lateral view). Streamlines are rendered as tubes and color-coded with a brain-looking texture, combined with ambient occlusion (e.g., shadows inside the sulci). Generated using Surf Ice (Rorden, 2021).

## 5 Conclusion

In conclusion, tractography visualization is a powerful tool for gaining a deeper understanding of diffusion MRI data. By allowing us to visualize and localize significant white matter pathways, this technique can be used to improve diagnosis and surgical planning in a variety of neurosurgical interventions. As the field of tractography visualization continues to evolve, we can expect to see the development of new and more advanced techniques, particularly those that incorporate machine-learning approaches. As these techniques continue to be refined, we believe that they will become an increasingly indispensable tool in the field of neurosurgery and beyond.

## References

Ankele, M., Schultz, T., 2018. DT-MRI streamsurfaces revisited. IEEE Trans. Vis. Comput. Graph. 25 (1), 1112–1121.

Ard, T., Krum, D.M., Phan, T., Duncan, D., Essex, R., Bolas, M., Toga, A., 2017. NIVR: neuro imaging in virtual reality. In: 2017 IEEE Virtual Reality (VR), pp. 465–466.

Aydogan, D.B., Souza, V.H., Lioumis, P., Ilmoniemi, R.J., 2021. Towards real-time tractography-based TMS neuronavigation. Brain Stimul. 14 (6), 1609.

Banks, D.C., Westin, C.F., 2008. Global illumination of white matter fibers from DT-MRI data. In: Visualization in Medicine and Life Sciences, pp. 173–184.

Blaas, J., Botha, C.P., Peters, B., Vos, F.M., Post, F.H., 2005. Fast and reproducible fiber bundle selection in DTI visualization. In: VIS 05. IEEE Visualization, 2005, pp. 59–64.

Bouma, S., Hurter, C., Telea, A., 2020. Structure-aware trail bundling for large DTI datasets. Algorithms 13 (12), 316.

Brecheisen, R., Platel, B., ter Haar Romeny, B.M., Vilanova, A., 2013. Illustrative uncertainty visualization of DTI fiber pathways. Vis. Comput. 29, 297–309.

Cabral, B., Leedom, L.C., 1993. Imaging vector fields using line integral convolution. In: Proceedings of the 20th Annual Conference on Computer Graphics and Interactive Techniques, pp. 263–270.

Calamante, F., Tournier, J.D., Jackson, G.D., Connelly, A., 2010. Track-density imaging (TDI): super-resolution white matter imaging using whole-brain track-density mapping. NeuroImage 53 (4), 1233–1243.

Calamante, F., Smith, R.E., Tournier, J.D., Raffelt, D., Connelly, A., 2015. Quantification of voxel-wise total fibre density: investigating the problems associated with track-count mapping. NeuroImage 117, 284–293.

Catani, M., Howard, R.J., Pajevic, S., Jones, D.K., 2002. Virtual in vivo interactive dissection of white matter fasciculi in the human brain. NeuroImage 17 (1), 77–94.

Catani, M., Dell'Acqua, F., Vergani, F., Malik, F., Hodge, H., Roy, P., Valabregue, R., De Schotten, M.T., 2012. Short frontal lobe connections of the human brain. Cortex 48 (2), 273–291.

Chamberland, M., Whittingstall, K., Fortin, D., Mathieu, D., Descoteaux, M., 2014. Real-time multi-peak tractography for instantaneous connectivity display. Front. Neuroinform. 8, 59.

Chamberland, M., Bernier, M., Fortin, D., Whittingstall, K., Descoteaux, M., 2015. 3D interactive tractography-informed resting-state fMRI connectivity. Front. Neurosci. 9, 275.

Chamberland, M., Raven, E.P., Genc, S., Duffy, K., Descoteaux, M., Parker, G.D., Tax, C.M.W., Jones, D.K., 2019. Dimensionality reduction of diffusion MRI measures for improved tractometry of the human brain. NeuroImage 200, 89–100.

Chen, W., Zhang, S., MacKay-Brandt, A., Correia, S., Qu, H., Crow, J.A., Tate, D.F., Yan, Z., Peng, Q., et al., 2009. A novel interface for interactive exploration of DTI fibers. IEEE Trans. Vis. Comput. Graph. 15 (6), 1433–1440.

Chen, W., Shi, L., Chen, W., 2018. A survey of macroscopic brain network visualization technology. Chin. J. Electron. 27 (5), 889–899.

Chen, J., Zhang, G., Chiou, W., Laidlaw, D.H., Auchus, A.P., 2019. Measuring the effects of scalar and spherical colormaps on ensembles of dMRI tubes. IEEE Trans. Vis. Comput. Graph. 26 (9), 2818–2833.

Comaniciu, D., Engel, K., Georgescu, B., Mansi, T., 2016. Shaping the future through innovations: from medical imaging to precision medicine. Med. Image Anal. 33, 19–26.

Conturo, T.E., Lori, N.F., Cull, T.S., Akbudak, E., Snyder, A.Z., Shimony, J.S., McKinstry, R.C., Burton, H., Raichle, M.E., 1999. Tracking neuronal fiber pathways in the living human brain. Proc. Natl. Acad. Sci. USA 96 (18), 10422–10427.

De Santis, S., Drakesmith, M., Bells, S., Assaf, Y., Jones, D.K., 2014. Why diffusion tensor MRI does well only some of the time: variance and covariance of white matter tissue microstructure attributes in the living human brain. NeuroImage 89, 35–44.

Delmarcelle, T., Hesselink, L., 1992. Visualization of second order tensor fields and matrix data. In: Proceedings Visualization '92, pp. 316–317.

Deoni, S.C.L., Rutt, B.K., Peters, T.M., 2003. Visualization of neural DTI vector fields using line integral convolution. In: Medical Image Computing and Computer-Assisted Intervention—MICCAI 2003: 6th International Conference, Montréal, Canada, November 15–18, 2003. Proceedings 6, pp. 207–214.

Eichelbaum, S., Hlawitschka, M., Scheuermann, G., 2012. LineAO-improved three-dimensional line rendering. IEEE Trans. Vis. Comput. Graph. 19 (3), 433–445.

Enders, F., Sauber, N., Merhof, D., Hastreiter, P., Nimsky, C., Stamminger, M., 2005. Visualization of white matter tracts with wrapped streamlines. VIS 05. IEEE Visualization, Minneapolis, MN, USA. pp. 51–58.

Everts, M.H., Bekker, H., Roerdink, J.B.T.M., Isenberg, T., 2009. Depth-dependent halos: illustrative rendering of dense line data. IEEE Trans. Vis. Comput. Graph. 15 (6), 1299–1306.

Everts, M.H., Begue, E., Bekker, H., Roerdink, J.B.T.M., Isenberg, T., 2015. Exploration of the brain's white matter structure through visual abstraction and multi-scale local fiber tract contraction. IEEE Trans. Vis. Comput. Graph. 21 (7), 808–821.

Franke, L., Haehn, D., 2020. Modern scientific visualizations on the web. Informatics 7 (4), 37.

Franke, L., Weidele, D.K.I., Zhang, F., Cetin-Karayumak, S., Pieper, S., O'Donnell, L.J., Rathi, Y., Haehn, D., 2021. Fiberstars: visual comparison of diffusion tractography data between multiple subjects. In: 2021 IEEE 14th Pacific Visualization Symposium (PacificVis), pp. 116–125.

Garyfallidis, E., Koudoro, S., Guaje, J., Côté, M.A., Biswas, S., Reagan, D., Anousheh, N., Silva, F., Fox, G., Contributors, F., 2021. FURY: advanced scientific visualization. J. Open Source Softw. 6 (64), 3384. https://doi.org/10.21105/joss.03384.

Gillmann, C., Saur, D., Wischgoll, T., Scheuermann, G., 2021a. Uncertainty-aware visualization in medical imaging—a survey. Comput. Graph. Forum 40 (3), 665–689.

Gillmann, C., Smit, N.N., Gröller, E., Preim, B., Vilanova, A., Wischgoll, T., 2021b. Ten open challenges in medical visualization. IEEE Comput. Graph. Appl. 41 (5), 7–15.

Golby, A.J., Kindlmann, G., Norton, I., Yarmarkovich, A., Pieper, S., Kikinis, R., 2011. Interactive diffusion tensor tractography visualization for neurosurgical planning. Neurosurgery 68 (2), 496–505.

Gruen, J., van der Voort, G., Schultz, T., 2022. Model averaging and bootstrap consensus-based uncertainty reduction in diffusion MRI tractography. Comput. Graph. Forum 42 (1), 217–230.

Haehn, D., Rannou, N., Ahtam, B., Grant, E., Pienaar, R., 2014. Neuroimaging in the browser using the x toolkit. Front. Neuroinform. 101.

Haehn, D., Franke, L., Zhang, F., Cetin-Karayumak, S., Pieper, S., O'Donnell, L.J., Rathi, Y., 2020. Trako: efficient transmission of tractography data for visualization. In: International Conference on Medical Image Computing and Computer-Assisted Intervention, pp. 322–332.

Helman, J.L., Hesselink, L., 1991. Visualizing vector field topology in fluid flows. IEEE Comput. Graph. Appl. 11 (3), 36–46.

Hlawitschka, M., Eichelbaum, S., Scheuermann, G., 2008. Fast and memory efficient GPU-based rendering of tensor data. IADIS Comput. Graph. Vis. 2008, 28–30.

Höller, M., Otto, K.M., Klose, U., Groeschel, S., Ehricke, H.H., 2014. Fiber visualization with LIC maps using multidirectional anisotropic glyph samples. Int. J. Biomed. Imaging 2014, 401819.

Ille, S., Ohlerth, A.K., Colle, D., Colle, H., Dragoy, O., Goodden, J., Robe, P., Rofes, A., Mandonnet, E., Robert, E., et al., 2021. Augmented reality for the virtual dissection of white matter pathways. Acta Neurochir. 163 (4), 895–903.

Isenberg, T., 2015. A survey of illustrative visualization techniques for diffusion-weighted MRI tractography. In: Visualization and Processing of Higher Order Descriptors for Multi-Valued Data, Springer, pp. 235–256.

Jeong, W.K., Fletcher, P.T., Tao, R., Whitaker, R., 2007. Interactive visualization of volumetric white matter connectivity in DT-MRI using a parallel-hardware Hamilton-Jacobi solver. IEEE Trans. Vis. Comput. Graph. 13 (6), 1480–1487.

Jeurissen, B., Descoteaux, M., Mori, S., Leemans, A., 2019. Diffusion MRI fiber tractography of the brain. NMR Biomed. 32 (4), e3785.

Jianu, R., Demiralp, C., Laidlaw, D., 2009. Exploring 3D DTI fiber tracts with linked 2D representations. IEEE Trans. Vis. Comput. Graph. 15 (6), 1449–1456.

Jones, D.K., 2003. Determining and visualizing uncertainty in estimates of fiber orientation from diffusion tensor MRI. Magn. Reson. Med. 49 (1), 7–12.

Jones, D.K., 2008. Tractography gone wild: probabilistic fibre tracking using the wild bootstrap with diffusion tensor MRI. IEEE Trans. Med. Imaging 27 (9), 1268–1274.

Jones, D.K., Knösche, T.R., Turner, R., 2013. White matter integrity, fiber count, and other fallacies: the do's and don'ts of diffusion MRI. NeuroImage 73, 239–254.

Kanzler, M., Rautenhaus, M., Westermann, R., 2018. A voxel-based rendering pipeline for large 3D line sets. IEEE Trans. Vis. Comput. Graph. 25 (7), 2378–2391.

Kern, M., Neuhauser, C., Maack, T., Han, M., Usher, W., Westermann, R., 2020. A comparison of rendering techniques for 3D line sets with transparency. IEEE Trans. Vis. Comput. Graph. 27 (8), 3361–3376.

Kindlmann, G., 2004. Superquadric tensor glyphs. In: Proceedings of the Sixth Joint Eurographics—IEEE TCVG Conference on Visualization, pp. 147–154.

Kindlmann, G., Tricoche, X., Westin, C.F., 2007. Delineating white matter structure in diffusion tensor MRI with anisotropy creases. Med. Image Anal. 11 (5), 492–502.

Köhn, A., Klein, J., Weiler, F., Peitgen, H.O., 2009. A GPU-based fiber tracking framework using geometry shaders. In: Medical Imaging 2009: Visualization, Image-Guided Procedures, and Modeling, vol. 7261, pp. 508–517.

Lakhani, D.A., Deib, G., 2022. Photorealistic depiction of intracranial tumors using cinematic rendering of volumetric 3T MRI data. Acad. Radiol. 29 (10), e211–e218.

Ledoux, L.P., Morency, F.C., Cousineau, M., Houde, J.C., Whittingstall, K., Descoteaux, M., 2017. Fiberweb: diffusion visualization and processing in the browser. Front. Neuroinform. 11, 54.

Leemans, A., 2010. Visualization of diffusion MRI data. In: Diffusion MRI, Oxford University Press, Oxford, pp. 354–379.

Leemans, A., Jeurissen, B., Sijbers, J., Jones, D.K., 2009. ExploreDTI: a graphical toolbox for processing, analyzing, and visualizing diffusion MR data. Proc. Int. Soc. Mag. Reson. Med. 17 (1), 3537.

Lo, C.Y., Chao, Y.P., Chou, K.H., Guo, W.Y., Su, J.L., Lin, C.P., 2007. DTI-based virtual reality system for neurosurgery. In: 2007 29th Annual International Conference of the IEEE Engineering in Medicine and Biology Society, pp. 1326–1329.

Luzzi, S., Lucifero, A.G., Martinelli, A., Del Maestro, M., Savioli, G., Simoncelli, A., Lafe, E., Preda, L., Galzio, R., 2021. Supratentorial high-grade gliomas: maximal safe anatomical resection guided by augmented reality high-definition fiber tractography and fluorescein. Neurosurg. Focus 51 (2), E5.

Maier-Hein, K.H., Neher, P.F., Houde, J.C., Côté, M.A., Garyfallidis, E., Zhong, J., Chamberland, M., Yeh, F.C., Lin, Y.C., Ji, Q., et al., 2017. The challenge of mapping the human connectome based on diffusion tractography. Nat. Commun. 8 (1), 1349.

Margulies, D.S., Böttger, J., Watanabe, A., Gorgolewski, K.J., 2013. Visualizing the human connectome. NeuroImage 80, 445–461.

McGraw, T., 2020. High-quality real-time raycasting and raytracing of streamtubes with sparse voxel octrees. In: 2020 IEEE Visualization Conference (VIS), pp. 21–25.

McGraw, T., Nadar, M., 2007. Stochastic DT-MRI connectivity mapping on the GPU. IEEE Trans. Vis. Comput. Graph. 13 (6), 1504–1511.

Merhof, D., Sonntag, M., Enders, F., Nimsky, C., Hastreiter, P., Greiner, G., 2006. Hybrid visualization for white matter tracts using triangle strips and point sprites. IEEE Trans. Vis. Comput. Graph. 12 (5), 1181–1188.

Mittmann, A., Nobrega, T.H.C., Comunello, E., Pinto, J.P.O., Dellani, P.R., Stoeter, P., Von Wangenheim, A., 2011. Performing real-time interactive fiber tracking. J. Digit. Imaging 24 (2), 339–351.

Nesi, L.L., Rorden, C., Munsell, B.C., 2017. A simple and efficient cylinder imposter approach to visualize DTI fiber tracts. In: International Workshop on Connectomics in Neuroimaging, pp. 89–97.

O'Donnell, L.J., Westin, C.F., Golby, A.J., 2009. Tract-based morphometry for white matter group analysis. NeuroImage 45 (3), 832–844.

Osman, B., Pereira, M., van de Wetering, H., Chamberland, M., 2023. Voxlines: streamline transparency through voxelization and view-dependent line orders. In: International Workshop on Computational Diffusion MRI. Cham, Springer Nature Switzerland, pp. 92–103.

Otten, R., Vilanova, A., Van De Wetering, H., 2010. Illustrative white matter fiber bundles. Comput. Graph. Forum 29 (3), 1013–1022.

Peeters, T.H.J.M., Vilanova, A., ter Haar Romeny, R., 2006. Visualization of DTI fibers using hair-rendering techniques. In: Proceedings of ASCI, pp. 66–73.

Petrovic, V., Fallon, J., Kuester, F., 2007. Visualizing whole-brain DTI tractography with GPU-based tuboids and LoD management. IEEE Trans. Vis. Comput. Graph. 13 (6), 1488–1495.

Pinter, C., Lasso, A., Choueib, S., Asselin, M., Fillion-Robin, J.C., Vimort, J.B., Martin, K., Jolley, M.A., Fichtinger, G., 2020. SlicerVR for medical intervention training and planning in immersive virtual reality. IEEE Trans. Med. Robot. Bionics 2 (2), 108–117. https://doi.org/10.1109/TMRB.2020.2983199.

Prados, F., Boada, I., Feixas, M., Prats, A., Blasco, G., Pedraza, S., Puig, J., 2007. DTIWeb: a web-based framework for DTI data visualization and processing. In: International Conference on Computational Science and Its Applications, pp. 727–740.

Presseau, C., Jodoin, P.M., Houde, J.C., Descoteaux, M., 2015. A new compression format for fiber tracking datasets. NeuroImage 109, 73–83. https://doi.org/10.1016/j.neuroimage.2014.12.058.

Qiu, T.M., Zhang, Y., Wu, J.S., Tang, W.J., Zhao, Y., Pan, Z.G., Mao, Y., Zhou, L.F., 2010. Virtual reality presurgical planning for cerebral gliomas adjacent to motor pathways in an integrated 3-D stereoscopic visualization of structural MRI and DTI tractography. Acta Neurochir. 152, 1847–1857.

Reina, G., Bidmon, K., Enders, F., Hastreiter, P., Ertl, T., 2006. GPU-based hyperstreamlines for diffusion tensor imaging. In: EuroVis, vol. 6, pp. 35–42.

Rheault, F., Houde, J.C., Descoteaux, M., 2017. Visualization, interaction and tractometry: dealing with millions of streamlines from diffusion MRI tractography. Front. Neuroinform. 11, 42.

Rick, T., von Kapri, A., Caspers, S., Amunts, K., Zilles, K., Kuhlen, T., 2011. Visualization of probabilistic fiber tracts in virtual reality. In: Medicine Meets Virtual Reality 18, IOS Press, pp. 486–492.

Rorden, C., 2021. Surf Ice: Main Page. NITRC, pp. 1–57.

Röttger, D., Merhof, D., Müller, S., 2012. The BundleExplorer: A Focus and Context Rendering Framework for Complex Fiber Distributions. The Eurographics Association.

Schertzer, J., Mercier, C., Rousseau, S., Boubekeur, T., 2022. Fiblets for real-time rendering of massive brain tractograms. Comput. Graph. Forum 41 (2), 447–460.

Schott, M., Martin, T., Grosset, A.V.P., Smith, S.T., Hansen, C.D., 2012. Ambient occlusion effects for combined volumes and tubular geometry. IEEE Trans. Vis. Comput. Graph. 19 (6), 913–926.

Schultz, T., Vilanova, A., 2019. Diffusion MRI visualization. NMR Biomed. 32 (4), e3902.

Schultz, T., Sauber, N., Anwander, A., Theisel, H., Seidel, H.P., 2008. Virtual klingler dissection: putting fibers into context. Comput. Graph. Forum 27 (3), 1063–1070.

Schultz, T., Vilanova, A., Brecheisen, R., Kindlmann, G., 2014. Fuzzy fibers: uncertainty in dMRI tractography. Sci. Vis., 79–92.

Schultz, T., Klose, U., Hauser, T.K., Ehricke, H., 2019. Anatomy-Focused Volume Line Integral Convolution for Brain White Matter Visualization. Václav Skala-UNION Agency.

Schurade, R., Hlawitschka, M., Hamann, B., Scheuermann, G., Knösche, T.R., Anwander, A., 2010. Visualizing white matter fiber tracts with optimally fitted curved dissection surfaces. In: Bartz, D., Botha, C., Hornegger, J., Machiraju, R., Wiebel, A., Preim, B. (Eds.), Eurographics Workshop on Visual Computing for Biology and Medicine, The Eurographics Association.

Shastin, D., Genc, S., Parker, G.D., Koller, K., Tax, C.M.W., Evans, J., Hamandi, K., Gray, W.P., Jones, D.K., Chamberland, M., 2022. Surface-based tracking for short association fibre tractography. NeuroImage 260, 119423.

Siddiqui, F., Höllt, T., Vilanova, A., 2021. A progressive approach for uncertainty visualization in diffusion tensor imaging. Comput. Graph. Forum 40 (3), 411–422.

Svetachov, P., Everts, M.H., Isenberg, T., 2010. DTI in context: illustrating brain fiber tracts in situ. Comput. Graph. Forum 29 (3), 1023–1032.

Takemura, H., Caiafa, C.F., Wandell, B.A., Pestilli, F., 2016. Ensemble tractography. PLoS Comput. Biol. 12 (2), e1004692.

Tax, C.M.W., Chamberland, M., van Stralen, M., Viergever, M.A., Whittingstall, K., Fortin, D., Descoteaux, M., Leemans, A., 2015. Seeing more by showing less: orientation-dependent transparency rendering for fiber tractography visualization. PLoS One 10 (10), e0139434.

Tax, C.M.W., Haije, T.D., Fuster, A., Westin, C.F., Viergever, M.A., Florack, L., Leemans, A., 2016. Sheet probability index (SPI): characterizing the geometrical organization of the white matter with diffusion MRI. NeuroImage 142, 260–279.

Tournier, J.D., Mori, S., Leemans, A., 2011. Diffusion tensor imaging and beyond. Magn. Reson. Med. 65 (6), 1532.

Vaillancourt, O., Boré, A., Girard, G., Descoteaux, M., 2010. A fiber navigator for neurosurgical planning (neuroplanningnavigator). In: IEEE Visualization, vol. 231.

Van Aart, E., Sepasian, N., Jalba, A., Vilanova, A., 2011. CUDA-accelerated geodesic ray-tracing for fiber tracking. Int. J. Biomed. Imaging 2011, 698908.

Vilanova, A., Berenschot, G., Van Pul, C., 2004. DTI visualization with streamsurfaces and evenly-spaced volume seeding. In: Proceedings of the Sixth Joint Eurographics—IEEE TCVG conference on Visualization, pp. 173–182.

Vilanova, A., Zhang, S., Kindlmann, G., Laidlaw, D., 2006. An introduction to visualization of diffusion tensor imaging and its applications. In: Visualization and Processing of Tensor Fields. Mathematics and Visualization, Springer, Berlin, Heidelberg, pp. 121–153.

von Kapri, A., Rick, T., Caspers, S., Eickhoff, S.B., Zilles, K., Kuhlen, T., 2010. Evaluating a visualization of uncertainty in probabilistic tractography. In: Medical Imaging 2010: Visualization, Image-Guided Procedures, and Modeling, vol. 7625, pp. 986–997.

Vos, S.B., Viergever, M.A., Leemans, A., 2013. Multi-fiber tractography visualizations for diffusion MRI data. PLoS One 8 (11), e81453.

Wakana, S., Jiang, H., Nagae-Poetscher, L.M., Van Zijl, P.C.M., Mori, S., 2004. Fiber tract-based atlas of human white matter anatomy. Radiology 230 (1), 77–87.

Wedeen, V.J., Rosene, D.L., Wang, R., Dai, G., Mortazavi, F., Hagmann, P., Kaas, J.H., Tseng, W.Y.I., 2012. The geometric structure of the brain fiber pathways. Science 335 (6076), 1628–1634.

Westin, C.F., Maier, S.E., Mamata, H., Nabavi, A., Jolesz, F.A., Kikinis, R., 2002. Processing and visualization for diffusion tensor MRI. Med. Image Anal. 6 (2), 93–108.

Wiens, V., Schlaffke, L., Schmidt-Wilcke, T., Schultz, T., 2014. Visualizing uncertainty in HARDI tractography using superquadric streamtubes. EuroVis (Short Papers) 5.

Xu, C., Neuroth, T., Fujiwara, T., Liang, R., Ma, K.L., 2020. A predictive visual analytics system for studying neurodegenerative disease based on DTI fiber tracts. arXiv preprint arXiv:2010.07047.

Yang, J.Y.M., Yeh, C.H., Poupon, C., Calamante, F., 2021. Diffusion MRI tractography for neurosurgery: the basics, current state, technical reliability and challenges. Phys. Med. Biol. 66 (15), 15TR01.

Yeatman, J.D., Richie-Halford, A., Smith, J.K., Keshavan, A., Rokem, A., 2018. A browser-based tool for visualization and analysis of diffusion MRI data. Nat. Commun. 9 (1), 1–10.

Yuan, Z., Dai, F., Xu, D., 2011. Improved visualization of fiber tracts in whole brain using illuminated line rendering. In: 2011 4th International Conference on Biomedical Engineering and Informatics (BMEI), vol. 1, pp. 328–332.

Zhang, S., Demiralp, C., Laidlaw, D.H., 2003. Visualizing diffusion tensor MR images using streamtubes and streamsurfaces. IEEE Trans. Vis. Comput. Graph. 9 (4), 454–462.

Zhou, L., Fan, M., Hansen, C., Johnson, C.R., Weiskopf, D., 2022. A review of three-dimensional medical image visualization. Health Data Sci. 2022, 9840519.

Zhukov, L., Barr, A.H., 2002. Oriented tensor reconstruction: tracing neural pathways from diffusion tensor MRI. In: IEEE Visualization, 2002. VIS 2002, pp. 387–394.

Zockler, M., Stalling, D., Hege, H.C., 1996. Interactive visualization of 3D-vector fields using illuminated stream lines. In: Proceedings of Seventh Annual IEEE Visualization'96, pp. 107–113.

Part IV

# From streamlines to tracts

# Chapter 21

# Dissecting white matter pathways: A neuroanatomical approach

Stephanie J. Forkel[a,b,c], Cesare Bortolami[b], Lilit Dulyan[a], Rachel L.C. Barrett[d], and Ahmad Beyh[d,e]

[a]Donders Institute for Brain Cognition Behaviour, Radboud University, Nijmegen, The Netherlands, [b]Brain Connectivity and Behaviour Laboratory, Sorbonne Universities, Paris, France, [c]Centre for Neuroimaging Sciences, Department of Neuroimaging, Institute of Psychiatry, Psychology and Neuroscience, King's College London, London, United Kingdom, [d]Department of Forensics and Neurodevelopmental Sciences, Institute of Psychiatry, Psychology and Neuroscience, King's College London, London, United Kingdom, [e]Laboratory of Neurobiology, Department of Cell and Developmental Biology, University College London, London, United Kingdom

The brain is the most magnificent structure, and we are only at the cusp of unraveling some of its complexity. Neuroanatomy is the best tool to map the brain's structural complexity. As such, neuroanatomy is not just an academic exercise; it serves our fundamental understanding of the neurobiology of cognition and improves clinical practice (Dulyan et al., 2024). A deepened anatomical understanding has advanced our conceptual grasp of the evolution of the brain, interindividual variability of cognition in health and disease, and the conceptual shift toward the emergence of cognition. For the past 20 years, diffusion imaging tractography has dramatically facilitated these advances by enabling the study of the delicate networks that orchestrate brain processes (for review, see Park and Friston, 2013; Thiebaut de Schotten and Forkel, 2022).

Historically, the bedrock of neuroanatomy is formed by clinical cases presenting significant changes in behavior and cognition after suffering a lesion to the brain (Broca, 1865; Corkin, 2002; Macmillan, 2000; Thiebaut de Schotten et al., 2015; Wernicke, 1874). These famous cases were assessed with the methods available at the time, and in some circumstances, the brain (Dronkers et al., 2007; Signoret et al., 1984), skull (Damasio et al., 1994; Ratiu et al., 2004; Van Horn et al., 2012), or digital data (Annese et al., 2014; Augustinack et al., 2014; Corkin et al., 1997) were preserved and reanalyzed with modern technologies. With the emergence of novel methods, neuroscience's famous cases are regularly reevaluated (Dronkers et al., 2007; Signoret et al., 1984; Thiebaut de Schotten et al., 2015), most recently using diffusion-weighted tractography to visualize the networks damaged by lesions (Dronkers et al., 2007; Signoret et al., 1984; Thiebaut de Schotten et al., 2015). While neurosciences have learned a lot from individual patients, the availability of magnetic resonance imaging (MRI) introduced a shift toward studying the brain in health and disease in large populations and across time (i.e., longitudinal study designs). There are many methods to study the brain's surface (gyri and sulci, e.g., Lerch et al., 2017; Naidich et al., 1992; Ono et al., 1990; Ribas, 2010; Yousry et al., 1997), its cross-sectional anatomy (subcortical structures, e.g., Lerch et al., 2017), cytoarchitectonics (e.g., Amunts et al., 1999; Uylings et al., 2005), receptorarchitecture (e.g., Froudist-Walsh et al., 2023; Haghir et al., 2023) and postmortem connectional neuroanatomy (Akeret et al., 2022; Dziedzic et al., 2021; Kadri et al., 2017; Klingler, 1935; Lawes et al., 2008; Thiebaut de Schotten et al., 2011; Türe et al., 2000; Vergani et al., 2014b; Yendiki et al., 2022). However, there is only one method that allows the investigation of the structural networks of the living human brain—diffusion-weighted imaging (DWI) tractography (e.g., Assaf et al., 2019; Basser et al., 1994; Catani and Thiebaut de Schotten, 2008; Dell'Acqua and Tournier, 2019; Jbabdi and Johansen-Berg, 2011; Pierpaoli et al., 1996).

With tractography, it is possible to map the connectional anatomy of the brain across large numbers of healthy participants (Catani and Thiebaut de Schotten, 2008; Mori et al., 2005; Rojkova et al., 2016; Thiebaut de Schotten et al., 2010) and various clinical cohorts (Ciccarelli et al., 2008; Forkel et al., 2014a; Thiebaut de Schotten et al., 2005), associate white matter properties with cognitive correlates (Catani et al., 2012; Forkel et al., 2022a), and study inter- and intraindividual variability (Croxson et al., 2018; Forkel et al., 2014a, 2020; Scholz et al., 2009; Thiebaut de Schotten and Forkel, 2022). A recent metaanalysis showed that studying white matter is a reliable measure of variability in anatomy and that this variability is related to cognitive differences in health and disease (Forkel et al., 2022a). We therefore dedicate this chapter to studying the network neuroanatomy of the human brain.

Handbook of Diffusion MR Tractography. https://doi.org/10.1016/B978-0-12-818894-1.00012-4
Copyright © 2025 Elsevier Ltd. All rights are reserved, including those for text and data mining, AI training, and similar technologies.

**398 PART | IV From streamlines to tracts**

**FIG. 1** Comparison of Klingler postmortem dissection (A), diffusion tensor imaging (DTI) tractography (B), and spherical deconvolution (C) in vivo dissections. *(Personal data.)*

Despite tractography being a relatively novel method, it has rapidly developed over the past four decades resulting in 6880 studies today (PubMed search term *tractography*, 08/11/2022). Anatomically guided manual dissections often rely on deterministic tracking algorithms that have been optimized to best match the connectional anatomy of the human brain as defined by Klingler postmortem dissections (see Part III: Tractography algorithms, Fig. 1).

Deterministic tractography is commonly used as a whole-brain analysis tool (Chapter 13). The resulting tractogram, or connectome (i.e., whole-brain white matter), is reconstructed from the diffusion signal before individual pathways or networks are segmented using anatomically defined regions of interest (ROIs). The placement of ROIs follows specific anatomical criteria but may vary depending on each individual's brain anatomy (Croxson et al., 2018). While a given set of ROIs might be consistently applied in healthy participants, these regions may need adjustments in clinical cohorts due to brain tissue displacement or loss of tissue. The advantage of using tractography to segment white matter pathways based on anatomically guided manual dissections is that it can be applied to any species (Barrett et al., 2020; Croxson et al., 2018; Friedrich et al., 2021; Mars et al., 2018), population, or pathology, which automatic algorithms may not easily achieve. As the dissections are performed in the participant space and brain-by-brain, tractography accounts for interindividual variability; see for review Forkel et al. (2022a).

When analyzing tractography data, several steps are consistent across all studied populations and brain states (health/disease). We will discuss various considerations for dissections across populations and give practical tips on common pitfalls and features to improve the visualization of the dissections. We briefly discuss specific considerations for manual dissections in nonhuman primates (NHPs). Lastly, we provide an atlas of ROIs for the most commonly delineated white matter connections.

# 1 Principles of anatomically guided manual dissections

## 1.1 Optimal diffusion maps for ROI delineation

Before placing a ROI, it is crucial to locate prominent anatomical features (e.g., midsection of the corpus callosum (CC), hand knob area in the precentral gyrus, or the anterior commissure (AC)) to orient oneself. Identifying anatomical features not only depends on the expertise of the tractographer but also varies significantly with the resolution of the data and the choice of brain map (e.g., fractional anisotropy (FA) map, anisotropic power (AP) map, T1-weighted scan; Fig. 2). Diffusion data is acquired rapidly using a spin-echo sequence (Chapters 6 and 7), which prioritizes a fast acquisition over spatial resolution (classically around 2–2.5 mm); however, recent developments have pushed spatial resolution in the living human brain (1.25 mm in 3T Human Connectome Project (HCP), Fig. 2A) and in postmortem imaging (~500 μm). The higher the spatial resolution, the easier it is (especially for novices) to identify anatomical structures. As such, high-resolution in vivo data is optimal for anatomical studies (e.g., 7T HCP dataset, ~1 mm).

Traditionally, a grayscale FA map is used to delineate ROIs (Fig. 2B). Often, the color-coded FA map is even more helpful, as it indicates the direction of connections in a standard Red-Green-Blue (RGB) system representing commissural-association-projection connections. However, FA maps lose signal in areas closer to the cortical ribbon and when two or more connections cross. This can lead to erroneous interpretations by the algorithms and interpreters.

More recently, software and preprocessing pipelines have facilitated the coregistration of a high-resolution structural scan to the tractogram (Fig. 2C). Other approaches rely on the use of diffusion-derived maps, such as the AP map, which has a similar contrast as a structural T1-weighted scan but additionally contains information about the diffusion signal and can visualize anisotropy closer to the cortex compared to a classical FA map (Fig. 2B) (Dell'Acqua et al., 2014).

All these maps have their advantages and limitations when delineating ROIs. These differences are amplified when studying clinical populations and should be carefully considered (Forkel and Catani, 2018). The clinical usability of these

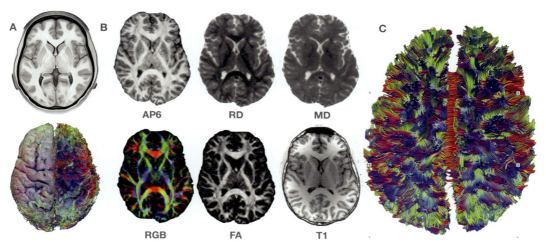

**FIG. 2** Anatomical resolution in the HCP dataset (1.25 mm, A) and a healthy control dataset (1.8 mm, B). Panel A details the HCP anatomy of subcortical and white matter structures. Panel B shows the range of possible maps used for manual delineations. Panel C shows a dorsal view of the tractogram (i.e., whole brain white matter) obtained from a healthy participant. *AP6*, anisotropy power map; *FA*, grayscale fractional anisotropy map; *MD*, mean diffusivity; *RD*, radial diffusivity; *RGB*, color-coded (red-green-blue) fractional anisotropy map; *T1*, T1-weighted structural scan.

contrasts varies in their sensitivity to lesion onset (e.g., in stroke), image resolution, and how long it takes to acquire the image (e.g., motion artifacts with longer sequences). Depending on the clinical pathology, an added layer of complexity might be introduced by the lesion's appearance on a scan (e.g., pseudonormalization). For example, diffusion indices are highly sensitive to stroke lesions that suddenly alter the brain's anatomy and microstructure. However, brain tumors can gradually shift and displace neuronal tissue, including white matter fibers (Zhang et al., 2022; Zhylka et al., 2021). In clinical settings, it is vital to choose a map that represents the lesion best and be aware of the interactions between a given pathology and diffusion-derived maps for ROI delineation.

## 1.2 Tractography in clinical populations—Stroke and neurosurgical patients

Diffusion-derived maps are sensitive to early microstructural changes in the immediate aftermath of a stroke (Moseley et al., 1990; Pierpaoli et al., 1996). This makes them better candidates than structural maps in the acute stages of stroke if one is interested in delineating ROIs surrounding a stroke or assessing the extent to which certain white matter tracts are affected by the injury. However, it is crucial to remember that this effect varies over time. In the immediate aftermath (minutes) of a stroke, the diffusion signal changes in the affected tissue and remains in this altered state. Despite the injury, the affected tissue slowly and temporarily regains its original signal in a process known as pseudonormalization (Schlaug et al., 1997). It may take several days for longer-term changes to start pushing the diffusion signal away from its normal range again (Warach et al., 1995). This phenomenon is highly relevant if one studies the brains of acute stroke patients but is less relevant in the chronic stages (Forkel and Catani, 2018).

It is essential to keep this information in mind if one is interested in performing dissections in stroke patients. The altered diffusion metrics might affect both the diffusion-derived maps on which ROIs are delineated and the tracking algorithms. For example, edema can cause diffusion-derived metrics such as FA to drop, which may be interpreted as a change in the underlying white matter fibers. However, despite this reduced FA, the white matter fibers passing through edematous tissue may not (yet) be directly affected. Consequently, tractography algorithms may stop tracking when they reach an area of low FA and produce a false negative result where the connection is not traced through edema (Chapter 19). Advanced tractography algorithms and multishell acquisitions can, however, partially overcome these limitations. In stroke patients, advanced tractography methods have emerged as reliable tools to predict long-term recovery based on acute neuroimaging by looking at the anatomy of the contralesional hemisphere (Forkel et al., 2014a) or the pattern of white matter disconnections caused by a lesion (Dulyan et al., 2022; Hope et al., 2024; Pacella et al., 2019; Salvalaggio et al., 2020; Souter et al., 2022; Talozzi et al., 2021).

In cases where injury to the brain is caused by a mass (e.g., brain tumors), the mass may have compressed or displaced neighboring white matter fibers causing their shape and location to deviate from the expected anatomy. For example, a lateral frontal lobe tumor can compress and push a part of the corticospinal tract (CST) toward the medial wall. However, this will only affect the portion of the CST immediately neighboring the tumor, altering its overall trajectory (Fig. 3).

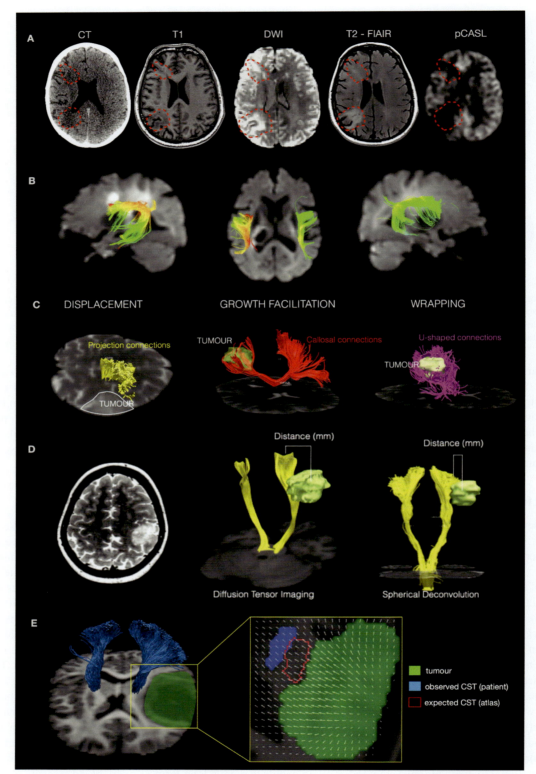

**FIG. 3** Tractography in stroke and brain tumors. Panel A shows the computer tomography (CT), structural (T1, T2 FLAIR), diffusion, and perfusion (pCASL) images acquired in a single patient with acute stroke and aphasia. The frontal lesion is barely visible on conventional image contrasts (e.g., CT, T1). By contrast, diffusion and perfusion imaging immediately highlight the vast lesion impact. While all these scans indicate that the white matter is affected by the stroke, none of those maps allow us to visualize white matter pathways along their entire course. Panel B shows the individual reconstruction of the arcuate fasciculus in an acute stroke patient with the scalar FA value plotted along the pathway. In the left hemisphere, the lesion reduces the FA index along the arcuate fasciculus, while the right hemisphere still harbors a healthy pathway. Panel C shows the various possible configurations that the presence of a tumor might have on the anatomical architecture of the white matter. Panel D shows the impact tractography algorithms (e.g., tensor-based or spherical deconvolution) can have on estimating the distance between a lesion and the white matter of the corticospinal tract (CST). Panel E shows the displacement *(white vectors)* of the CST caused by a tumor *(green)* in reference to the atlas-based location of the CST *(red)* in a healthy brain. This shows how the location of the CST, as observed *(blue)*, has been pushed medially due to the compressing and displacing forces exerted by the growing mass on the surrounding tissue. *((D) Courtesy of Francesco Vergani and Henrietta Howells (personal data).)*

Importantly, this will be specific to each patient, so there are no other means of anticipating the exact shape of the CST without visualizing it in each patient using tractography. Such clinical uses of tractography have made it a pivotal tool in the neurosurgical theater (Cochereau et al., 2020; Dragoy et al., 2020; Duffau, 2008; Kemerdere et al., 2016; Leclercq et al., 2010; Mirchandani et al., 2020; Sollmann et al., 2020; Teichmann et al., 2015; Thiebaut de Schotten et al., 2005).

## 1.3 Anatomical placement of ROIs

ROIs are typically placed in "bottleneck" regions, constituting obligatory passages (Catani et al., 2002; Hau et al., 2016; Schilling et al., 2022). For example, the inferior fronto-occipital fasciculus (IFOF) fans out in the frontal and occipital lobes (Forkel et al., 2014b; Schmahmann and Pandya, 2007). Still, it is compressed into a compact bundle while traveling through the external capsule. Placing a single ROI in this bottleneck region will capture most of the streamlines associated with the IFOF. However, it is best to use three ROIs to separate connections; in this case, the IFOF is dissected as part of the ventral triade with the uncinate fasciculus (UF) and the inferior longitudinal fasciculus (ILF) (Fig. 4). Another example is the frontotemporal arcuate fasciculus (AF), which can be segmented by placing a single ROI in the inferior parietal white matter where the connection consolidates, rather than in the frontal and temporal cortices where the pathway fans out as it reaches its cortical targets (Catani et al., 2005).

Not every connection, however, passes through a bottleneck region. In these cases, a multiple ROI approach defines a connection as passing between two brain regions (e.g., all streamlines between the frontal and parietal lobes) (Thiebaut de Schotten et al., 2011). When multiple ROIs are used, each can serve as an inclusion or an exclusion region using Boolean operators ("AND," "OR," "NOT"). In addition, the operator chosen for an ROI can be applied to any part of a tractography streamline (the "waypoint" rule) or strictly to its terminations (the "endpoint" rule). In this context, a "waypoint" ROI is used to visualize (or exclude) streamlines that pass through it, while an "endpoint" ROI is used to visualize (or exclude) streamlines that terminate in it ("either end," "both ends," "no end"). A particular use of the endpoint option is "both ends," which is best suited when dissecting short U-shaped fibers that connect neighboring gyri. Only streamlines that begin and end within the same ROI are visualized (Fig. 4). It is possible to examine nearly every white matter connection with these few rules.

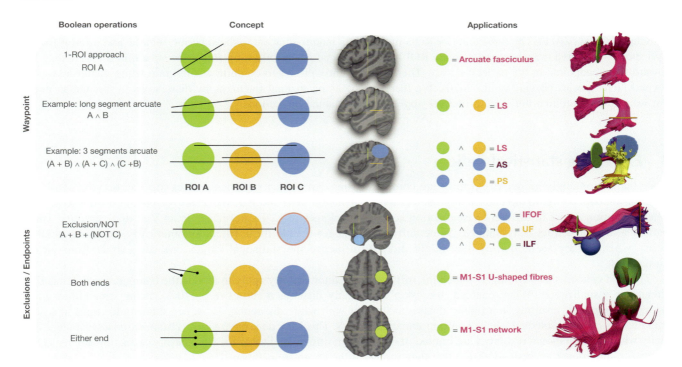

**FIG. 4** Region of interest (ROI) operations to virtually reconstruct white matter connections using waypoint ROIs and exclusion regions in the combinations of "AND" and "NOT," as well as endpoint rules such as "both ends" or "either end." *AS*, anterior segment; *IFOF*, inferior fronto-occipital fasciculus; *ILF*, inferior longitudinal fasciculus; *LS*, long segment; *M1*, motor cortex; *PS*, posterior segment; *ROI*, region of interest; *S1*, somatosensory cortex; *UF*, uncinate fasciculus.

## 2 Anatomical delineation in nonhuman primates

Most methods available for studying the connectional anatomy of the brain are invasive by nature and, as such, are only applicable to nonhuman brains (e.g., axonal tracing) or postmortem specimens (e.g., histological staining) (for review, see Yendiki et al., 2022).

Tract dissection in NHPs offers a unique opportunity to compare white matter anatomy across species (Barrett et al., 2020; Croxson et al., 2005; Dyrby et al., 2007; Jbabdi et al., 2013; Knösche et al., 2015; Li et al., 2013; Rilling et al., 2008; Roumazeilles et al., 2020; Warrington et al., 2020) and validate findings from tractography by comparing directly with axonal tracing (Azadbakht et al., 2015; Calabrese et al., 2015; Dauguet et al., 2007; Donahue et al., 2016; Gao et al., 2013; Jbabdi et al., 2013; Schilling et al., 2019; Schmahmann et al., 2007; Thomas et al., 2014; van den Heuvel et al., 2015).

DWI is the only method available for comparative studies to apply to living human and NHP brains. The last decade has seen a steep increase in comparative neuroimaging studies aiming to map evolution through connectivity (Friedrich et al., 2021). This comparison allowed for the computation of deformation fields between species' brains (Mars et al., 2018). These deformation fields capture similarities and differences between species. Comparative studies assume that similarities between species can be traced back to a common ancestor and account for preserving specific functions across evolution (Croxson et al., 2018). Recent comparative work revealed one of the first comprehensive maps of the phylogenetic organization of brain regions (Mars et al., 2018). Such technical advances in comparative neuroimaging will allow targeted studies that better match human brain mechanisms to their phylogenetic counterparts. Further, they may also help discover and mimic neuroprotective mechanisms in animals that could potentially translate to improving human disease models and therapeutics (Royo et al., 2021).

The approach to the manual delineation of connections in NHPs is very similar to what has been described earlier for humans, except that anatomical references will differ. The anatomical guidance for ROI placement in monkeys is most commonly based on axonal tracing (Schmahmann and Pandya, 2006; Thiebaut de Schotten et al., 2012), although other methods, including Klingler dissections, may also be referred to. For nonhuman species, axonal tracing is often considered the gold standard of the methods available for describing white matter anatomy, as it is robust to false positives and able to distinguish between mono- and polysynaptic pathways (Petrides and Pandya, 2002; Schmahmann et al., 2007). As with human tractography dissections, ROIs can be drawn manually (Thiebaut de Schotten et al., 2012) or imported from atlases based on cortical parcellations (Paxinos et al., 2000; Rohlfing et al., 2012). For example, the three branches of the fronto-parietal network of the superior longitudinal fasciculus (SLFI–III) have been dissected using four waypoint ROIs (Thiebaut de Schotten et al., 2011). The three anterior ROIs in the white matter adjacent to the superior, middle, and inferior frontal gyri specify the three branches, and a single ROI has been placed posteriorly to intercept the streamlines before they fan toward their endpoints in the parietal cortex. There are currently efforts to compile whole-brain white matter atlases of several NHP species (Bryant et al., 2020). We therefore focus on manual tractography dissections in the human brain, but want to reiterate that the same principles apply to NHP manual dissections.

## 3 Extracting statistical indices

Once the anatomically guided dissections are completed, one can extract statistical values that can be voxel-based or tract-based, depending on the chosen approach (Part II: Diffusion MRI and Part III: Tractography algorithms). Several filtering options exist in different visualization software packages (e.g., ExploreDTI, TrackVis) that can alter which part of a dissected tract is visualized. Some software packages enable these filters automatically, so it is crucial to verify whether they are active as they might directly influence the statistical values one can extract from virtual dissections.

The first concerns a "*skip filter*," which the software may automatically apply to reduce the computational demand of streamline visualization. When enabled, the "skip" filter only displays a subset (e.g., 70%) of the streamlines while "skipping" the rest, resulting in a lighter dataset to interact with while dissecting. However, this would result in a 30% data loss if it remains selected when exporting statistical values. This filter does have a clear visualization advantage, especially when computational resources are limited. It is, however, critical to disable the filter before extracting any measurements for subsequent analysis to ensure that they represent the complete pathway.

Another filter that can be applied is a "*length threshold*." While dissecting, we may apply a filter to the displayed streamlines using a length threshold. This can be especially useful for removing spurious streamlines that follow unrealistic trajectories (e.g., crossing sulcal boundaries). However, caution should be applied while using this filter as it does not discriminate between anatomically plausible and implausible (i.e., artifacts) streamlines. Therefore extra care should be taken when the chosen length threshold is close to the length of a target pathway. It would also be necessary to consider interindividual differences in brain size and pathway length that are likely to affect the chosen length threshold for each individual.

After the filters have been checked, various voxel-based and tract-based statistical measures can be extracted. The available indices will depend on the chosen processing pipeline (Chapter 23). The most commonly reported index is FA (50% of studies, see Forkel et al., 2022a), which reflects the anisotropy of water diffusion in a given voxel based on the diffusion tensor model (Basser and Jones, 2002; Catani and Forkel, 2019; Jones et al., 2013b; Jones, 2008). Such values are then exported into statistical programs and used for further analysis (for a detailed example, see Forkel and Catani, 2018).

Statistical analysis can be performed with values extracted from individual dissections in each participant's native space or at the group level based on data in a common standard space. For the latter, the final virtual dissection can be exported as a tract density map to create an overlay map across a group of interest. This allows the creation, for example, of percentage overlay maps that encode the interindividual variability across the group for each connection (Fig. 5).

Many features exist to visualize virtual reconstructions and render them akin to axonal connections. The most impactful rendering option is visualizing tubes instead of streamlines, giving more volume to each reconstruction. This allows the visualization software to treat streamlines as 3D objects that reflect light and produce shadows, adding clarity to their 3D trajectory. This is especially helpful for understanding the anatomy of white matter bundles as it creates a depth effect in the image. However, it is essential to remember that this is simply a visualization tool that does not bear any meaning on the actual thickness or size of the dissected connection and does not convey direct information about the underlying axonal populations.

Before producing the final image of the dissected pathway, it is worth verifying whether the visualization software has an *antialiasing* feature, which is effectively a low-pass filter that is applied to the rendered image. Aliasing is an artifact that occurs when the visualization software simplifies the graphical representation of the image for efficiency, making smooth lines appear jagged. The antialiasing filter ensures that the final rendered image is smooth and free of such artifacts.

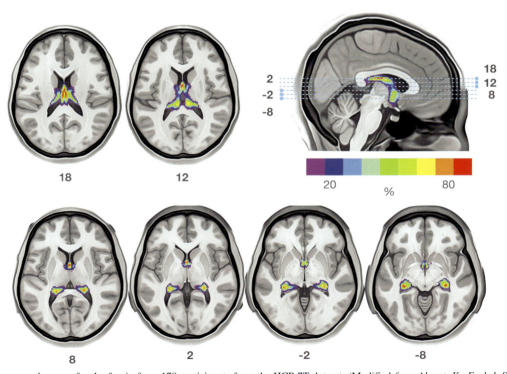

**FIG. 5** Percentage overlay map for the fornix from 178 participants from the HCP 7T dataset. (*Modified from Akeret, K., Forkel, S.J., Buzzi, R.M., Vasella, F., Amrein, I., Colacicco, G., Serra, C., Krayenbühl, N., 2022. Multimodal anatomy of the human forniceal commissure. Commun. Biol. 5.*)

## 4 Atlas of neuroanatomical dissections

For ease of viewing and anatomical references, the ROIs will be shown in one hemisphere while the other is used for anatomical labeling. We offer the inclusion and exclusion ROIs and detail how to combine them during the virtual dissections. The ROIs described later are based on previous atlases that used a tensor-based algorithm (Catani and de Schotten, 2012; Catani and Thiebaut de Schotten, 2008) and an advanced spherical deconvolution algorithm (Rojkova et al., 2016). Details about the advantages and limitations of these algorithms can be found in Chapters 10–12.

### 4.1 Superior longitudinal fasciculus

The three branches of the SLF (SLFI–III, Fig. 6D) were first described in the living human brain in 2011 and are intrahemispheric frontoparietal association connections (Thiebaut de Schotten et al., 2011). The SLFs are in both hemispheres with a slight rightward prominence. Each branch is dissected using a two-ROI approach (Fig. 6A and B) plus an exclusion ROI in the left and right temporal lobes (Fig. 6C).

### 4.2 Cingulum

The cingulum (Fig. 7C) is an interlobar association connection on the medial aspect of the brain. Its white matter runs within the cingulate gyrus that surrounds the corpus callosum (CC). The most extended cingulate connections link the orbitofrontal cortex with the anterior temporal lobe. Several shorter connections in the cingulum branch out to the medial frontal, parietal, occipital, and temporal lobes. Anatomically, it is considered part of the limbic system (Catani et al., 2013).

Using postmortem dissection techniques and tractography, the results often indicate a dorsal and ventral part of the cingulum (Jones et al., 2013a). The separation between the two occurs at the level of the parieto-occipital fissure. Due to its proximity to the commissural connections of the CC, an exclusion region needs to be used (Fig. 7B).

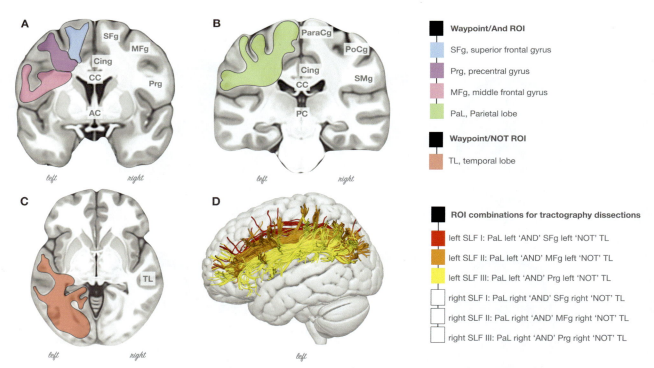

**FIG. 6** ROIs and their combinations of the three branches of the SLF. (A) Three waypoint ROIs are placed in the frontal lobe covering the superior frontal gyrus (SFg), middle frontal gyrus (MFg), and precentral gyrus (Prg). (B) A single posterior waypoint region is placed in the left parietal lobe (PaL). (C) A single exclusion ROI is placed around the white matter of the temporal lobe to exclude the connections belonging to the fronto-temporal arcuate fasciculus in either hemisphere. (D) Virtual reconstruction of the three branches of the SLF in the left hemisphere. *AC*, anterior commissure; *CC*, corpus callosum; *Cing*, cingulum; *MFg*, middle frontal gyrus; *PaL*, parietal lobe; *ParaCg*, paracentral gyrus; *PC*, posterior commissure; *PoCg*, postcentral gyrus; *Prg*, precentral gyrus; *SFg*, superior frontal gyrus; *SLF*, superior longitudinal fasciculus; *TL*, temporal lobe.

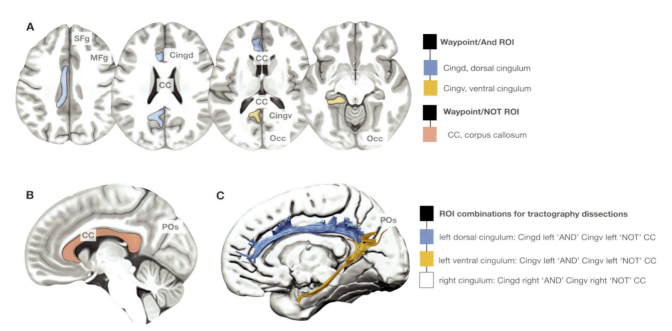

FIG. 7  ROIs and their combinations for the dissection of the dorsal and ventral cingulum. (A) Two waypoint ROIs are placed to define the cingulum dorsal to the parieto-occipital sulcus (POs) and ventral to it. (B) A single waypoint exclusion region is placed in the corpus callosum (CC). (C) Virtual reconstruction of the full cingulum in the left hemisphere. *CC*, corpus callosum; *Cingd*, dorsal cingulum; *Cingv*, ventral cingulum; *MFg*, middle frontal gyrus; *Occ*, occipital lobe; *POs*, parieto-occipital sulcus; *SFg*, superior frontal gyrus.

### 4.3  Uncinate fasciculus

The UF (Fig. 8C) is a frontotemporal association connection and is considered part of the limbic network and the language system (Catani et al., 2013). It is also part of a ventral fronto-temporal-occipital system often dissected in a three-ROIs approach (Fig. 8A). The uncinate consolidates and runs in the ventral floor of the external capsule and radiates into the orbitofrontal and anterior temporal lobes.

### 4.4  Inferior longitudinal fasciculus

The ILF (Fig. 8D) is an occipital-temporal association connection composed of long-range medial connections and shorter lateral connections linking early and late visual cortices and branching into the occipital and temporal lobes (Catani et al., 2003). The ILF connects occipital visual areas to the amygdala and hippocampus and higher-order visual areas in the temporal lobe.

### 4.5  Inferior fronto-occipital fasciculus

As its name suggests, the IFOF (Fig. 8B) is a white matter pathway that connects occipital and frontal cortices. . As the IFOF leaves the occipital lobe and enters the temporal stem, its fibers gather at the level of the external/extreme capsule level above the UF. These fibers form a thin sheet in the frontal lobe that curves toward the inferior frontal gyrus, lateral orbitofrontal region, and frontal pole. This pathway has a convoluted history; for review, see Forkel et al. (2014b), Schmahmann and Pandya (2007), and Türe et al. (1997). A remnant of its history is indicated in its name, as labeling the pathway "inferior" implies that there is a "superior" fronto-occipital fasciculus. Even though this label is still available in some atlases (e.g., Johns Hopkins University (JHU) atlas (Mori et al., 2005)), several studies using different methods reject its existence (Forkel et al., 2014b; Liu et al., 2020; Meola et al., 2015).

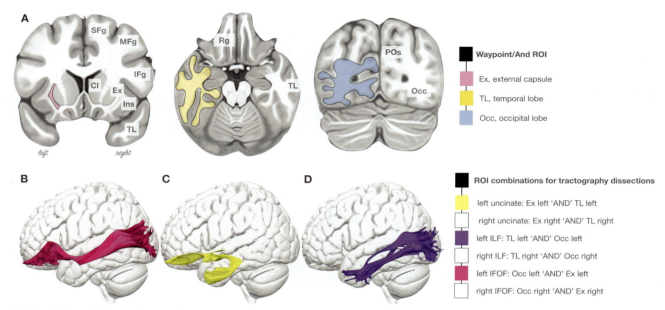

**FIG. 8** ROIs and their combinations for the dissection of the "ventral network" composed of the inferior fronto-occipital fasciculus (IFOF, B), the uncinate fasciculus (C), and the inferior longitudinal fasciculus (ILF, D). The three connections share some ROIs (see Fig. 4) and are therefore shown together. (A) Three-waypoint ROIs are placed to define the subcortical external capsule, the temporal lobe white matter (TL), and the occipital lobe white matter (Occ). (B–D) Virtual reconstruction of the ventral network in the left hemisphere. *IC*, internal capsule; *Ex*, external capsule; *IFg*, inferior frontal gyrus; *IFOF*, inferior fronto-occipital fasciculus; *ILF*, inferior longitudinal fasciculus; *Ins*, insula; *MFg*, middle frontal gyrus; *Occ*, occipital lobe; *POs*, parieto-occipital sulcus; *Rg*, gyrus rectus; *SFg*, superior frontal gyrus; *TL*, temporal lobe.

### 4.6 Arcuate fasciculus

The AF (Fig. 9B) is an interlobar (fronto-parietal-temporal) association connection that can be separated into three segments that link the inferior frontal gyrus, inferior parietal lobule (supramarginal gyrus (SMg), angular gyrus (Ag)), and posterior temporal lobe (superior temporal gyrus (STg), middle temporal gyrus (MTg; Fig. 9A). The long segment of the AF (or arcuate *sensu strictu*) is a frontotemporal connection (Catani et al., 2005). The nomenclature of the additional components varies in the literature. There is a consensus that the additional segments are more lateral than the frontotemporal connection, one being a frontoparietal (also referred to as horizontal or anterior) segment and the other being a parietal-temporal (also referred to as vertical or posterior) segment (see Catani et al., 2005; Forkel et al., 2022a; Frey et al., 2008; Giampiccolo and Duffau, 2022; Kaplan et al., 2010). The exact terminations of each component are still a matter of debate and vary depending on the methods applied (Giampiccolo and Duffau, 2022).

Originating from early anatomical papers, we are still faced with a body of literature that uses the terms SLF and AF interchangeably. While there is some overlap between both networks, for example, between the SLF-III and the anterior segment of the AF, other branches (SLF-I and SLF-II) and segments (long and posterior arcuate) are distinct. From an anatomical and etymological perspective, the SLF should be considered to be solely those fibers connecting frontal and parietal regions (i.e., superior and longitudinal; Thiebaut de Schotten et al., 2011), whereas the long AF *sensu strictu* should be considered to be the frontotemporal connection (i.e., "arcuatus" translates to "arching" around the Sylvian fissure) (Catani et al., 2005).

### 4.7 Frontal aslant tract

The frontal aslant tract (FAT) (Fig. 10B) is an intralobar connection in the frontal lobe connecting the inferior frontal gyrus and the superior frontal gyrus (Catani et al., 2011; Vergani et al., 2014a). A two-ROI approach can delineate the FAT (Fig. 10A).

### 4.8 Fronto-insular tracts

The fronto-insular tracts (FITs) are a group of short connections between the inferior frontal gyrus (incl. orbitofrontal cortex, pars triangularis, pars opercularis, precentral gyrus, and subcentral gyrus) and the three anterior short gyri and the two posterior long gyri of the insula cortex (Fig. 11A). The FITs are numbered (1–5) from anterior to posterior (Fig. 11B).

Dissecting white matter pathways: A neuroanatomical approach **Chapter | 21** **407**

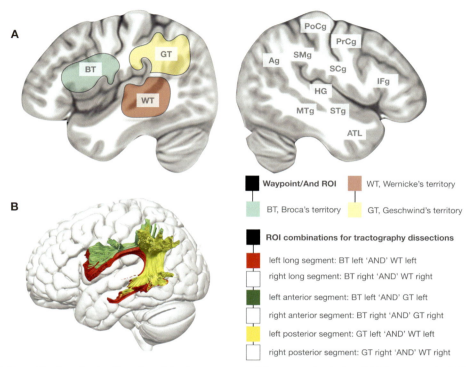

**FIG. 9** ROIs and their combinations for the dissection of the three segments of the arcuate fasciculus. (A) Three waypoint ROIs are placed to define Broca's territory (BT) in the inferior frontal gyrus and the ventral premotor cortex, Geschwind's territory (GT) covering the inferior parietal lobe (SMg and Ag), and Wernicke's territory (WT) in the posterior temporal lobe (STg and MTg). (B) Virtual reconstruction of the three segments of the arcuate fasciculus in the left and right hemispheres. *Ag*, angular gyrus; *ATL*, anterior temporal lobe; *BT*, Broca's territory; *GT*, Geschwind's territory; *HG*, Heschl's gyrus; *IFg*, inferior frontal gyrus; *MTg*, middle temporal gyrus; *PoCg*, postcentral gyrus; *PrCg*, precentral gyrus; *SCg*, subcentral gyrus; *SMg*, supramarginal gyrus; *STg*, superior temporal gyrus; *WT*, Wernicke's territory.

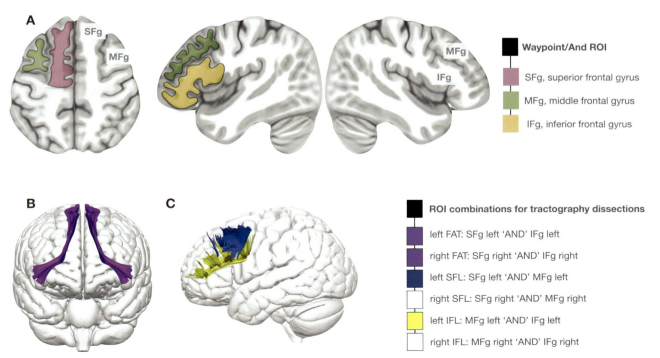

**FIG. 10** ROIs and their combinations for the dissection of the frontal aslant tract (FAT), inferior and superior frontal longitudinal fasciculi (IFL, SFL). (A) A total of three waypoint ROIs are placed on axial slices to define the superior frontal gyrus (SFg), middle frontal gyrus (MFg), and inferior frontal gyrus (IFg, including the pars opercularis, triangularis, and orbitalis). Virtual reconstruction of the bilateral FAT (B), SFL, and IFL (C). *FAT*, frontal aslant tract; *IFg*, inferior frontal gyrus; *IFL*, inferior frontal longitudinal fasciculus; *MFg*, middle frontal gyrus; *SFg*, superior frontal gyrus; *SFL*, superior frontal longitudinal fasciculus.

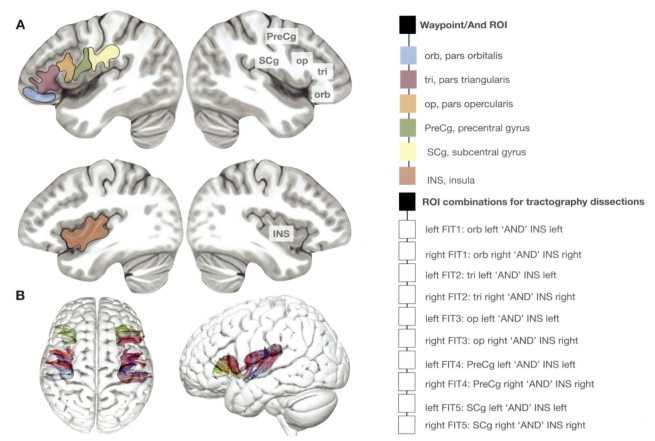

**FIG. 11** ROIs and their combinations for the dissection of the fronto-insular tracts (FITs). (A) A total of six waypoint ROIs are placed on sagittal slices to define the insula cortex (INS), orbitofrontal cortex (orb), pars triangularis (tri), pars opercularis (op), ventral precentral gyrus (PreCg), and the subcentral gyrus (SC). (B) Virtual reconstruction of the FITs 1–5. *FIT*, fronto-insular tract; *INS*, insula; *op*, pars opercularis; *orb*, pars orbitalis; *PreCg*, precentral gyrus; *SCg*, subcentral gyrus; *tri*, pars triangularis.

### 4.9 Corticospinal tract

The CST (Fig. 12B) is part of an extensive projection system between the cortex, the internal capsule, and the brainstem. The CST is the strand connecting to the motor cortex in the precentral gyrus (Fig. 12A). The CST is the most commonly studied pathway in neurological patients (Forkel et al., 2022a).

### 4.10 Anterior thalamic radiations

Thalamocortical projections (Fig. 13B) are the primary drivers of cortical activity in sensory areas and associative brain regions. The thalamus is densely connected to the entire brain, often called the brain's gatekeeper. The only sensory input that is not filtered by the thalamus is olfaction. As a result, the thalamic projections radiate into the cortex. Here we visualize only anterior thalamic projections into the frontal cortex (Fig. 13B). Still, the same logic can be applied to the rest of the brain by combining the thalamus ROI with different lobar ROIs (Fig. 13A).

### 4.11 Frontostriatal projections

Frontostriatal projections (Fig. 13C) connect frontal lobe regions with subcortical nuclei, including the nucleus accumbens, caudate nucleus, and the putamen. The circuit mediates motor, cognitive, and behavioral functions within the brain. The striatal projections form a delicate brain circuit that radiates into the frontal, parietal, and occipital cortices.

Dissecting white matter pathways: A neuroanatomical approach **Chapter | 21 409**

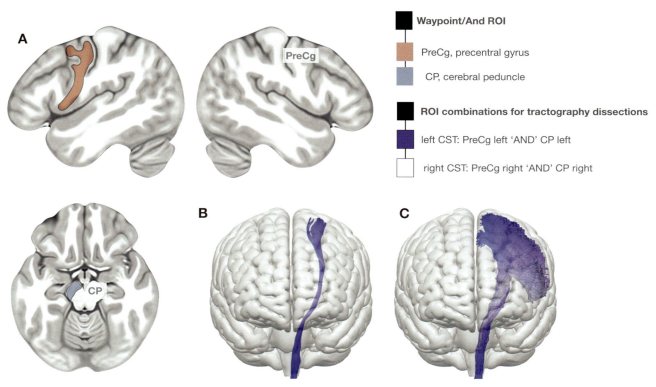

**FIG. 12** ROIs and their combinations for the dissection of the corticospinal tract (CST). (A) Two waypoint ROIs are placed on sagittal and axial slices to define the motor cortex (PreCg) and the cerebral peduncle in the brainstem (CP). Virtual reconstruction of the CST using a deterministic tensor-based algorithm (B) and a probabilistic spherical deconvolution algorithm (C) on the same data with the identical preprocessing. *CP*, cerebral peduncle; *CST*, corticospinal tract; *PreCg*, precentral gyrus.

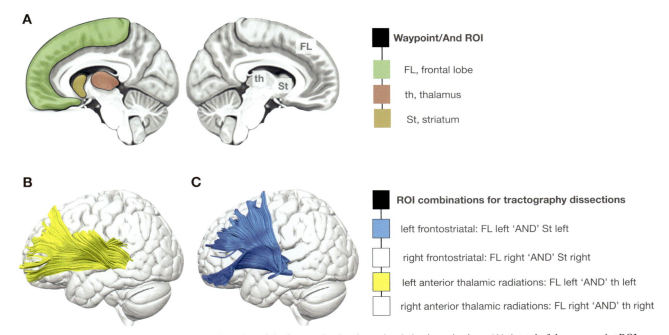

**FIG. 13** ROIs and their combinations for the dissection of the frontostriatal and anterior thalamic projections. (A) A total of three waypoint ROIs are placed in the frontal lobe (FL), the thalamus (th), and the striatum (St). Virtual reconstruction of the anterior thalamic projections (B) and the frontostriatal projections (C). *FL*, frontal lobe; *St*, striatum; *th*, thalamus.

## 4.12 Precentral and postcentral U-shaped fibers

The precentral and postcentral gyri (Fig. 14B) are functionally considered the motor and sensory cortices. A close interaction between both structures is essential, especially for the hand (e.g., fine finger movements) and face (e.g., mouth movements) (Catani et al., 2011). Fig. 14 details the ROIs to dissect the short U-shaped fibers between both structures to isolate the connections of the paracentral gyrus, the dorsal and ventral hand knob region, and the motor-sensory face areas. These connections can be dissected using a two-ROI approach, as shown in Fig. 14A. It is possible to use a sphere as ROI and select a filter specifying that the connections should start and terminate within the sphere. This is often called the "both ends" filter and is a valuable option for short U-shaped connections.

## 4.13 Optic radiation (geniculocalcarine fasciculus)

The optic radiation (OR) (Fig. 15B) connects the thalamus's lateral geniculate nucleus (LGN) with calcarine and pericalcarine cortex in the occipital lobe—hence, it is also known as the geniculocalcarine fasciculus. The OR loops from the LGN anteriorly, forming Meyer's loop near the temporal horn of the lateral ventricle, before it takes a posterior turn toward the occipital lobe; its occipital terminations span the striate cortex around the calcarine sulcus as well as several other pericalcarine visual regions (Fig. 15A). Some features of the OR, such as Meyer's loop or the full extent of its occipital terminations, might be challenging to dissect with stringent angular thresholds and with the diffusion tensor model, and can benefit from advanced modeling and tracking algorithms (Chamberland et al., 2017) (Chapter 18).

## 4.14 Medial occipital longitudinal tract

The medial occipital longitudinal tract (MOLT) (Fig. 16B) is a recently defined occipital-temporal connection that links the cuneus and lingual gyrus on one end with the medial temporal cortex, namely the parahippocampal gyrus, on the other end

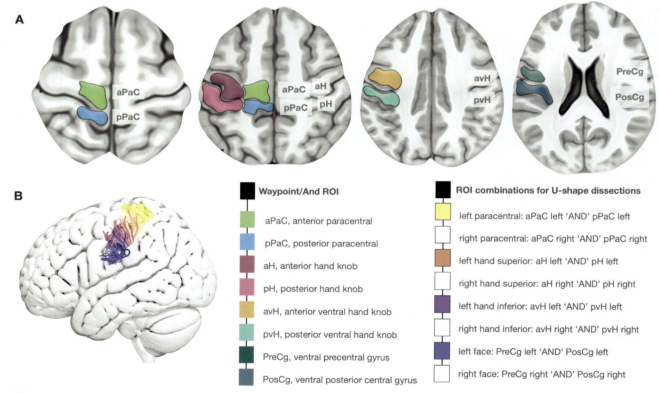

FIG. 14 ROIs and their combinations for the dissection of the paracentral, hand superior, hand inferior, and face U-shaped connections. (A) A total of eight waypoint ROIs are placed in pre- and postcentral gyri from dorsal to ventral. (B) Virtual reconstruction of the short interlobar U-shaped connections for the paracentral gyrus, dorsal, and ventral hand region, and the face area. *aH*, anterior hand knob; *aPaC*, anterior paracentral gyrus; *avH*, anterior ventral hand knob; *pH*, posterior hand knob; *PosCg*, ventral posterior central gyrus; *pPaC*, posterior paracentral gyrus; *PreCg*, ventral precentral gyrus; *pvH*, posterior ventral hand knob.

Dissecting white matter pathways: A neuroanatomical approach **Chapter | 21** **411**

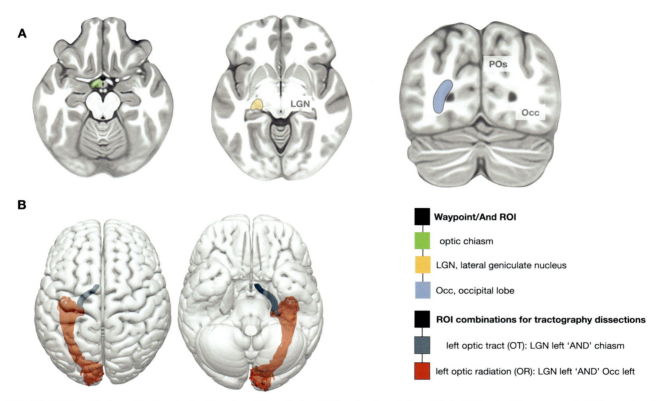

**FIG. 15** ROIs and their combinations for the dissection of the optic tract (OT) and optic radiations (ORs). (A) For the OT, two waypoint ROIs are placed around the lateral geniculate nucleus (LGN) and the optic chiasm; for the OR, the same LGN ROI is combined with another waypoint ROI placed in the white matter of the occipital lobe. (B) Virtual reconstruction of the OT and OR between the LGN and visual cortex. Note that tractography reconstructions of the OT are difficult, especially in the region of the optic chiasm, due to image distortions caused by the acquisition sequence and by eye movements; the latter cause severe shifts in the location of the optic nerve (anterior to the chiasm), and by association cause movements in the chiasm as well. *LGN*, lateral geniculate nucleus; *Occ*, occipital lobe; *POs*, parieto-occipital sulcus.

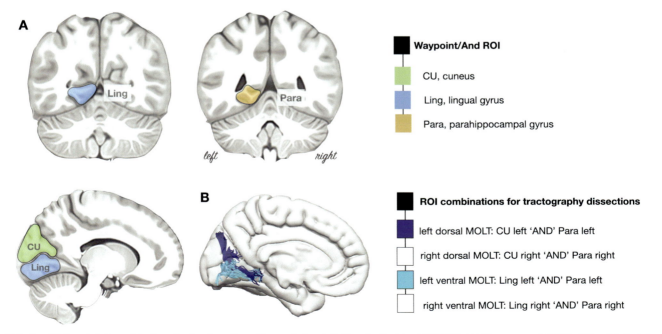

**FIG. 16** ROIs and their combinations for the dissection of the medial occipital longitudinal tract (MOLT). (A) Three-waypoint ROIs define the cuneus and the white matter of the lingual and parahippocampal gyri. (B) Virtual reconstruction of the MOLT between medial temporal and occipital cortices. *CU*, cuneus; *Ling*, lingual gyrus; *MOLT*, medial occipital longitudinal tract; *Para*, parahippocampal gyrus.

(Beyh et al., 2022). The MOLT's ventral (lingual) component is larger than the dorsal (cuneus) component, and the tract shows slight right lateralization. Functionally, the MOLT has been implicated in the visuospatial processing domain, particularly in the encoding of spatial configuration.

### 4.15 Corpus callosum

The CC is the most prominent white matter tract in the human brain that connects the frontoparietal, occipital, and temporal cortices of the two hemispheres (Fig. 17). The only part of the brain not connected by the CC is the anterior temporal lobes associated with the anterior commissure.

### 4.16 Anterior commissure

The AC is a well-known landmark to orient neuroimaging scans with stereotactic atlases (Fig. 18). On cross-sectional sagittal slices, it is easily identified as a small white structure. Its 3D anatomy is, however, slightly less established. The AC is an interhemispheric white matter pathway that primarily connects the anterior temporal lobes and parts of the visual cortex (Catani and Thiebaut de Schotten, 2008; Peltier et al., 2011). The fibers of the AC cross the interhemispheric midline between the anterior and posterior columns of the fornix (Fx) (Akeret et al., 2022) just above the intersection of the optic nerves. Significant interspecies differences have been reported, whereby the AC is smaller in the human brain (Barrett et al., 2020; Foxman et al., 1986).

### 4.17 Frontomarginal and fronto-orbito-polar tracts

Two intralobar frontal pole connections have been identified (Fig. 19B), namely the frontomarginal tract (FMT) and the fronto-orbito-polar tract (FOP) (Catani et al., 2011; Rojkova et al., 2016). The FMT runs beneath the frontomarginal sulcus from the medial to the lateral part of the frontal pole (BA10) (Rojkova et al., 2016). The FOP connects the posterior orbital gyrus (BA12) to the anterior orbital gyrus (BA11) and the ventromedial region of the frontal pole (Fig. 19A).

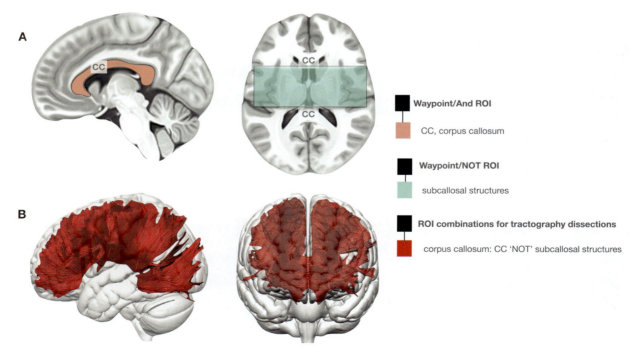

**FIG. 17** ROIs and their combinations for the dissection of the corpus callosum (CC). (A) Three-waypoint ROIs define the cuneus and the white matter of the lingual and parahippocampal gyri. (B) Virtual reconstruction of the MOLT between medial temporal and occipital cortices. *CC*, corpus callosum.

Dissecting white matter pathways: A neuroanatomical approach **Chapter | 21** **413**

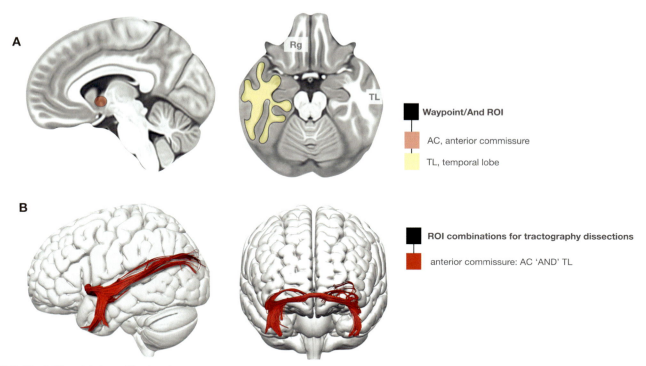

**FIG. 18** ROIs and their combinations for the dissection of the anterior commissure (AC). (A) Two-waypoint ROIs define the AC on a sagittal plane just off the midline (due to the crossing with the fornix). This ROI can be combined with a left and right temporal ROI for a cleaner dissection. The posterior projections are consistently dissected with tractography but are often removed in publications. (B) Virtual reconstruction of the AC connecting the left and right anterior temporal lobes. *AC*, anterior commissure; *Rg*, gyrus rectus; *TL*, temporal lobe.

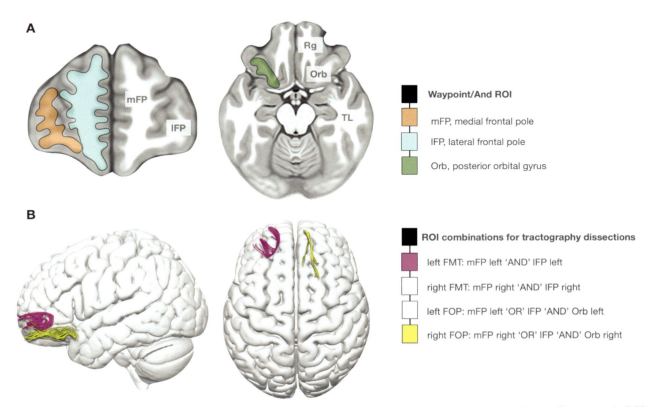

**FIG. 19** ROIs and their combinations for dissection of the frontomarginal tract (FMT) and the fronto-orbito-polar tract (FOP). (A) Three-waypoint ROIs define the lateral and medial frontal pole and the posterior orbitofrontal cortex. (B) Virtual reconstruction of the FMT and FOP. *FMT*, frontomarginal tract; *FOP*, fronto-orbito-polar tract; *lFP*, lateral frontal pole; *mFP*, medial frontal pole; *Orb*, orbitofrontal cortex; *Rg*, gyrus rectus; *TL*, temporal lobe.

## 4.18 Fornix

The Fx (Fig. 20B) is a complex structure underneath the CC and connects the mammillary bodies to the medial temporal lobe (Dogan et al., 2022). Anteriorly, it engulfs the AC between its anterior and posterior columns. The body of the Fx is abutting the CC and can sometimes be split. Posteriorly, the Fx separates into two branches known as the fimbriae. Owing to its anatomy, automatic methods that rely on priors of commissural-projection-association connections might miss the connection in its entirety due to its commissural (body), projection (fimbria), and longitudinal orientations along its course. The Fx in either hemisphere can be dissected using a single ROI (Fig. 20A). Functionally, the Fx is considered part of the limbic system (Catani et al., 2013). A multimodal approach to the forniceal anatomy recently revisited the existence of the forniceal commissure and concluded that it is a thin membrane rather than a commissure (Akeret et al., 2022).

## 4.19 Vertical occipital fasciculus

The vertical occipital fasciculus (VOF) (Fig. 21B) connects the dorsolateral and ventrolateral visual cortex (Takemura et al., 2017; Vergani et al., 2014b; Yeatman et al., 2014), including the visual word form area (VWFA), a part of the ventral occipitotemporal cortex specialized in processing visual formation of words and reading (Dehaene et al., 2002; Forkel et al., 2022b; Wandell et al., 2012). The first comprehensive description of the occipital lobe white matter was provided in the atlas *The white matter of the human cerebrum. Part 1. The Occipital Lobe* by Heinrich Sachs (Forkel, 2015; Forkel et al., 2015). Sachs's mentor Carl Wernicke was an enthusiastic advocate of his anatomical insights and encouraged his trainee to further pursue this research. The atlas was, in fact, intended to be a multivolume project, in which subsequent books would have been dedicated to the function and clinical correlates of each tract. In recent years, and with the help of tractography and Klingler dissections, the occipital white matter was revisited (Bugain et al., 2021; Vergani et al., 2014b). To identify the VOF, two ROIs need to be placed, one in the ventral and one in the dorsal occipital lobe. The ROIs are delineated on axial

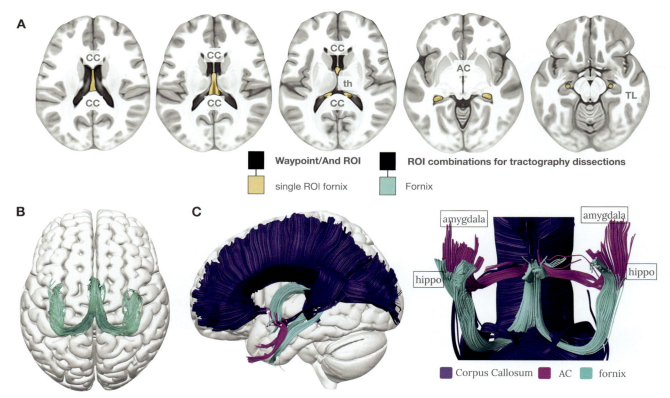

**FIG. 20** ROIs and their combinations for the dissection of the fornix. (A) A single waypoint ROI is defined on axial slices around the columns, body, and fimbriae. (B) Virtual reconstruction of the fornix (dorsal view). (C) The fornix in relation to the corpus callosum and the anterior commissure (lateral and ventral view). *AC*, anterior commissure; *CC*, corpus callosum; *hippo*, hippocampus; *th*, thalamus; *TL*, temporal lobe.

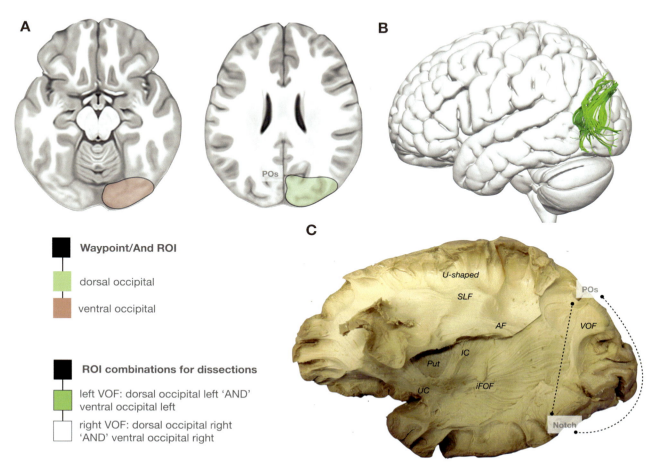

**FIG. 21** ROIs (A) and their combinations for the dissection of the vertical occipital fasciculus (VOF, B). (C) The reconstruction compared to a Klingler dissection of the VOF (personal data). *AF*, arcuate fasciculus; *IC*, internal capsule; *IFOF*, inferior fronto-occipital fasciculus; *Put*, putamen; *SLF*, superior longitudinal fasciculus; *UC*, uncinate fasciculus; *VOF*, vertical occipital fasciculus.

slides to be perpendicular to the vertical fiber system. The dorsal ROI is placed just behind the parieto-occipital sulcus (POs), and the ventral ROI is delineated on axial slides behind the occipital notch. The most lateral aspect of the occipital region contains the preoccipital (or temporo-occipital) notch, which demarcates the border between the inferior temporal gyrus and the ventral surface of the inferior occipital gyrus. Sometimes this landmark is in continuation with the inferior temporal sulcus. The borders of the occipital lobe are drawn between the landmarks of the POs and the notch to separate the occipital from the temporal and parietal lobes.

## 4.20 Accumbofrontal pathway

The accumbofrontal pathway is a white matter pathway that connects the nucleus accumbens, a brain region involved in reward and pleasure, with the orbital/prefrontal cortex, a brain region involved in decision-making and executive function (Karlsgodt et al., 2015). The accumbofrontal pathway is thought to play a role in regulating reward-related behavior and may be involved in the development of addiction (Feltenstein and See, 2008). The accumbofrontal pathway is in close vicinity to other pathways, such as UF and anterior thalamic radiation, but is located more medially (Karlsgodt et al., 2015; Pascalau et al., 2018) (Fig. 22).

In conclusion, manual virtual dissections are essential to capture the magnitude of interindividual variability associated with the brain's white matter. This is particularly important in the presence of brain lesions that amplify the variability in anatomy by disrupting, disconnecting, and displacing brain structures.

We hope that this dissection manual and atlas of 34 pathways will assist the next generation of tractographers in mapping the brain's white matter comprehensively and accurately.

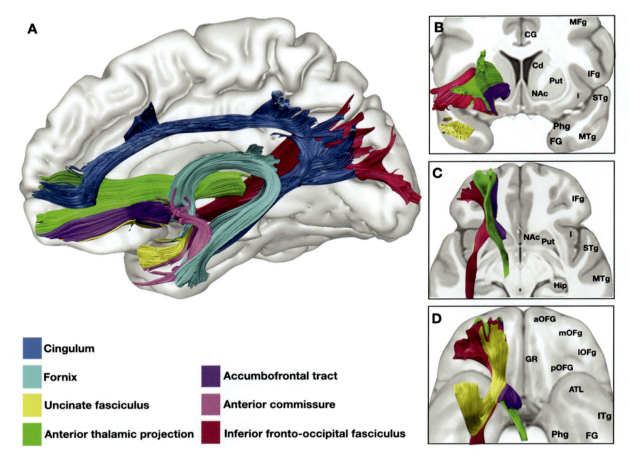

**FIG. 22** The accumbofrontal pathway in relation to other white matter pathways (A–D). The limbic network (cingulum in *blue*, fornix in *teal*), accumbofrontal pathway *(purple)*, anterior frontostriatal pathway *(green)*, anterior commissure *(pink)*, and the inferior longitudinal fasciculus (ILF, *magenta*). *am*, amygdala; *aOFG*, anterior orbitofrontal gyrus; *ATL*, anterior temporal lobe; *Cd*, caudate nucleus; *CG*, cingulate gyrus; *FG*, fusiform gyrus; *GP*, globus pallidus; *GR*, gyrus rectus; *Hip*, hippocampus; *I*, insular cortex; *IFg*, inferior frontal gyrus; *ITg*, inferior temporal gyrus; *lOFg*, lateral orbitofrontal gyrus; *MFg*, middle frontal gyrus; *mOFg*, medial orbitofrontal gyrus; *MTg*, middle temporal gyrus; *NAc*, nucleus accumbens; *OFC*, orbitofrontal cortex; *OLF*, olfactory cortex; *Phg*, parahippocampal gyrus; *pOFG*, posterior orbitofrontal gyrus; *Put*, putamen; *SFG*, superior frontal gyrus; *STG*, superior temporal gyrus.

The general principles of anatomically guided manual dissections that apply regardless of the studied population (health/pathology), species (e.g., NHPs), or the type of data (e.g., FA/RGB) are the following: (i) know your anatomy, (ii) be consistent in your approach (e.g., applying thresholds), and (iii) fully document your approach to improve reproducibility across studies.

## Funding

This project has received funding from the European Union's Horizon 2020 research and innovation program under the Marie Skłodowska-Curie grant agreement No. 101028551 (SJF, PERSONALISED) and the Donders Mohrmann Fellowship No. 2401515 (SJF, NEUROVARIABILITY). We thank all the students who used this guide in their work and offered constructive feedback, in particular Jiska Koemans, Janniek Wester, and Milou Huijsmans.

Linked document: https://docs.google.com/document/d/1jAZ587noDAkOXTYIdHkHjPCT2XvpyNqW0YJXTiFQDMo/edit?usp=sharing.

## References

Akeret, K., Forkel, S.J., Buzzi, R.M., Vasella, F., Amrein, I., Colacicco, G., Serra, C., Krayenbühl, N., 2022. Multimodal anatomy of the human forniceal commissure. Commun. Biol. 5.

Amunts, K., Schleicher, A., Bürgel, U., Mohlberg, H., Uylings, H.B., Zilles, K., 1999. Broca's region revisited: cytoarchitecture and intersubject variability. J. Comp. Neurol. 412, 319–341.

Annese, J., Schenker-Ahmed, N.M., Bartsch, H., Maechler, P., Sheh, C., Thomas, N., Kayano, J., Ghatan, A., Bresler, N., Frosch, M.P., Klaming, R., Corkin, S., 2014. Postmortem examination of patient H.M.'s brain based on histological sectioning and digital 3D reconstruction. Nat. Commun. 5, 3122.

Assaf, Y., Johansen-Berg, H., Thiebaut de Schotten, M., 2019. The role of diffusion MRI in neuroscience. NMR Biomed. 32, e3762.

Augustinack, J.C., van der Kouwe, A.J.W., Salat, D.H., Benner, T., Stevens, A.A., Annese, J., Fischl, B., Frosch, M.P., Corkin, S., 2014. H.M.'s contributions to neuroscience: a review and autopsy studies. Hippocampus 24, 1267–1286.

Azadbakht, H., Parkes, L.M., Haroon, H.A., Augath, M., Logothetis, N.K., de Crespigny, A., D'Arceuil, H.E., Parker, G.J.M., 2015. Validation of high-resolution tractography against in vivo tracing in the macaque visual cortex. Cereb. Cortex 25, 4299–4309.

Barrett, R.L.C., Dawson, M., Dyrby, T.B., Krug, K., Ptito, M., D'Arceuil, H., Croxson, P.L., Johnson, P.J., Howells, H., Forkel, S.J., Dell'Acqua, F., Catani, M., 2020. Differences in frontal network anatomy across primate species. J. Neurosci. 40, 2094–2107.

Basser, P.J., Jones, D.K., 2002. Diffusion-tensor MRI: theory, experimental design and data analysis—a technical review. NMR Biomed. 15, 456–467.

Basser, P.J., Mattiello, J., LeBihan, D., 1994. MR diffusion tensor spectroscopy and imaging. Biophys. J. 66, 259–267.

Beyh, A., Dell'Acqua, F., Cancemi, D., Requejo, F.D.S., Ffytche, D., Catani, M., 2022. The medial occipital longitudinal tract supports early stage encoding of visuospatial information. Commun. Biol. 5, 318.

Broca, P., 1865. Sur le siège de la faculté du langage articulé. Bull. Mém. Soc. Anthropol. Paris 6, 377–393.

Bryant, K.L., Li, L., Eichert, N., Mars, R.B., 2020. A comprehensive atlas of white matter tracts in the chimpanzee. PLoS Biol. 18, e3000971.

Bugain, M., Dimech, Y., Torzhenskaya, N., Thiebaut de Schotten, M., Caspers, S., Muscat, R., Bajada, C.J., 2021. Occipital Intralobar fasciculi: a description, through tractography, of three forgotten tracts. Commun. Biol. 4, 433.

Calabrese, E., Badea, A., Cofer, G., Qi, Y., Johnson, G.A., 2015. A diffusion MRI tractography connectome of the mouse brain and comparison with neuronal tracer data. Cereb. Cortex 25, 4628–4637.

Catani, M., de Schotten, M.T., 2012. Atlas of Human Brain Connections. Oxford University Press.

Catani, M., Forkel, S.J., 2019. Diffusion imaging methods in language sciences. In: The Oxford Handbook of Neurolinguistics. Oxford University Press.

Catani, M., Thiebaut de Schotten, M., 2008. A diffusion tensor imaging tractography atlas for virtual in vivo dissections. Cortex 44, 1105–1132.

Catani, M., Howard, R.J., Pajevic, S., Jones, D.K., 2002. Virtual in vivo interactive dissection of white matter fasciculi in the human brain. NeuroImage 17, 77–94.

Catani, M., Jones, D.K., Donato, R., Ffytche, D.H., 2003. Occipito-temporal connections in the human brain. Brain 126, 2093–2107.

Catani, M., Jones, D.K., Ffytche, D.H., 2005. Perisylvian language networks of the human brain. Ann. Neurol. 57, 8–16.

Catani, M., Dell'Acqua, F., Vergani, F., Malik, F., Hodge, H., Roy, P., Valabregue, R., Thiebaut de Schotten, M., 2011. Short frontal lobe connections of the human brain. Cortex 48, 273–291.

Catani, M., Dell'Acqua, F., Bizzi, A., Forkel, S.J., Williams, S.C., Simmons, A., Murphy, D.G., Thiebaut de Schotten, M., 2012. Beyond cortical localization in clinico-anatomical correlation. Cortex 48, 1262–1287.

Catani, M., Dell'Acqua, F., Thiebaut de Schotten, M., 2013. A revised limbic system model for memory, emotion and behaviour. Neurosci. Biobehav. Rev. 37 (8), 1724–1737. https://doi.org/10.1016/j.neubiorev.2013.07.001 (Epub 2013 Jul 9. PMID: 23850593).

Chamberland, M., Scherrer, B., Prabhu, S.P., Madsen, J., Fortin, D., Whittingstall, K., Descoteaux, M., Warfield, S.K., 2017. Active delineation of Meyer's loop using oriented priors through MAGNEtic tractography (MAGNET). Hum. Brain Mapp. 38, 509–527.

Ciccarelli, O., Catani, M., Johansen-Berg, H., Clark, C., Thompson, A., 2008. Diffusion-based tractography in neurological disorders: concepts, applications, and future developments. Lancet Neurol. 7, 715–727.

Cochereau, J., Lemaitre, A.-L., Wager, M., Moritz-Gasser, S., Duffau, H., Herbet, G., 2020. Network-behavior mapping of lasting executive impairments after low-grade glioma surgery. Brain Struct. Funct. 225, 2415–2429.

Corkin, S., 2002. What's new with the amnesic patient H.M.? Nat. Rev. Neurosci. 3, 153–160.

Corkin, S., Amaral, D.G., Gilberto González, R., Johnson, K.A., Hyman, B.T., 1997. H. M.'s medial temporal lobe lesion: findings from magnetic resonance imaging. J. Neurosci. 7 (10), 3964–3979.

Croxson, P.L., Johansen-Berg, H., Behrens, T.E.J., Robson, M.D., Pinsk, M.A., Gross, C.G., Richter, W., Richter, M.C., Kastner, S., Rushworth, M.F.S., 2005. Quantitative investigation of connections of the prefrontal cortex in the human and macaque using probabilistic diffusion tractography. J. Neurosci. 25, 8854–8866.

Croxson, P.L., Forkel, S.J., Cerliani, L., Thiebaut de Schotten, M., 2018. Structural variability across the primate brain: a cross-species comparison. Cereb. Cortex 28, 3829–3841.

Damasio, H., Grabowski, T., Frank, R., Galaburda, A.M., Damasio, A.R., 1994. The return of Phineas Gage: clues about the brain from the skull of a famous patient. Science 264, 1102–1105.

Dauguet, J., Peled, S., Berezovskii, V., Delzescaux, T., Warfield, S.K., Born, R., Westin, C.-F., 2007. Comparison of fiber tracts derived from in-vivo DTI tractography with 3D histological neural tract tracer reconstruction on a macaque brain. NeuroImage 37, 530–538.

Dehaene, S., Le Clec'H, G., Poline, J.-B., Le Bihan, D., Cohen, L., 2002. The visual word form area: a prelexical representation of visual words in the fusiform gyrus. Neuroreport 13, 321–325.

Dell'Acqua, F., Tournier, J.-D., 2019. Modelling white matter with spherical deconvolution: how and why? NMR Biomed. 32, e3945.

Dell'Acqua, F., Lacerda, L., Catani, M., Simmons, A., 2014. Anisotropic power maps: a diffusion contrast to reveal low anisotropy tissues from HARDI data. Proc. Int. Soc. Magn. Reson. Med., 29960–29967.

Dogan, E., Gungor, A., Dogulu, F., Türe, U., 2022. The historical evolution of the fornix and its terminology: a review. Neurosurg. Rev. 45, 979–988.

Donahue, C.J., Sotiropoulos, S.N., Jbabdi, S., Hernandez-Fernandez, M., Behrens, T.E., Dyrby, T.B., Coalson, T., Kennedy, H., Knoblauch, K., Van Essen, D.C., Glasser, M.F., 2016. Using diffusion tractography to predict cortical connection strength and distance: a quantitative comparison with tracers in the monkey. J. Neurosci. 36, 6758–6770.

**418** **PART | IV** From streamlines to tracts

Dragoy, O., Zyryanov, A., Bronov, O., Gordeyeva, E., Gronskaya, N., Kryuchkova, O., Klyuev, E., Kopachev, D., Medyanik, I., Mishnyakova, L., Pedyash, N., Pronin, I., Reutov, A., Sitnikov, A., Stupina, E., Yashin, K., Zhirnova, V., Zuev, A., 2020. Functional linguistic specificity of the left frontal aslant tract for spontaneous speech fluency: evidence from intraoperative language mapping. Brain Lang. 208, 104836.

Dronkers, N.F., Plaisant, O., Iba-Zizen, M.T., Cabanis, E.A., 2007. Paul Broca's historic cases: high resolution MR imaging of the brains of Leborgne and Lelong. Brain 130, 1432–1441.

Duffau, H., 2008. The anatomo-functional connectivity of language revisited. New insights provided by electrostimulation and tractography. Neuropsychologia 46, 927–934.

Dulyan, L., Talozzi, L., Pacella, V., Corbetta, M., Forkel, S.J., Thiebaut de Schotten, M., 2022. Longitudinal prediction of motor dysfunction after stroke: a disconnectome study. Brain Struct. Funct. 227, 3085–3098.

Dulyan, L., Guzmán Chacón, E., Forkel, S.J., 2024. Navigating neuroanatomy. In: Encyclopedia of the Human Brain, second ed. Elsevier. In press.

Dyrby, T.B., Søgaard, L.V., Parker, G.J., Alexander, D.C., Lind, N.M., Baaré, W.F.C., Hay-Schmidt, A., Eriksen, N., Pakkenberg, B., Paulson, O.B., Jelsing, J., 2007. Validation of in vitro probabilistic tractography. NeuroImage 37, 1267–1277.

Dziedzic, T.A., Balasa, A., Jeżewski, M.P., Michałowski, Ł., Marchel, A., 2021. White matter dissection with the Klingler technique: a literature review. Brain Struct. Funct. 226, 13–47.

Feltenstein, M.W., See, R.E., 2008. The neurocircuitry of addiction: an overview. Br. J. Pharmacol. 154, 261–274.

Forkel, S.J., 2015. Heinrich Sachs (1863–1928). J. Neurol. 262, 498–500. https://doi.org/10.1007/s00415-014-7517-2.

Forkel, S.J., Catani, M., 2018. Structural neuroimaging. In: de Groot, A.M.B., Hagoort, P. (Eds.), Research Methods in Psycholinguistics and the Neurobiology of Language: A Practical Guide. Wiley Blackwell, pp. 288–309.

Forkel, S.J., Catani, M., 2018. Lesion mapping in acute stroke aphasia and its implications for recovery. Neuropsychologia 115, 88–100.

Forkel, S.J., Thiebaut de Schotten, M., Dell'Acqua, F., Kalra, L., Murphy, D.G.M., Williams, S.C.R., Catani, M., 2014a. Anatomical predictors of aphasia recovery: a tractography study of bilateral perisylvian language networks. Brain 137, 2027–2039.

Forkel, S.J., Thiebaut de Schotten, M., Kawadler, J.M., Dell'Acqua, F., Danek, A., Catani, M., 2014b. The anatomy of fronto-occipital connections from early blunt dissections to contemporary tractography. Cortex 56, 73–84.

Forkel, S.J., Mahmood, S., Vergani, F., Catani, M., 2015. The white matter of the human cerebrum: part I the occipital lobe by Heinrich Sachs. Cortex 62, 182–202.

Forkel, S.J., Rogalski, E., Drossinos Sancho, N., D'Anna, L., Luque Laguna, P., Sridhar, J., Dell'Acqua, F., Weintraub, S., Thompson, C., Mesulam, M.-M., Catani, M., 2020. Anatomical evidence of an indirect pathway for word repetition. Neurology 94, e594–e606.

Forkel, S.J., Friedrich, P., Thiebaut de Schotten, M., Howells, H., 2022a. White matter variability, cognition, and disorders: a systematic review. Brain Struct. Funct. 227, 529–544.

Forkel, S.J., Labache, L., Nachev, P., de Schotten, M.T., Hesling, I., 2022b. Stroke disconnectome decodes reading networks. bioRxiv.

Foxman, B.T., Oppenheim, J., Petito, C.K., Gazzaniga, M.S., 1986. Proportional anterior commissure area in humans and monkeys. Neurology 36, 1513–1517.

Frey, S., Campbell, J.S.W., Pike, G.B., Petrides, M., 2008. Dissociating the human language pathways with high angular resolution diffusion fiber tractography. J. Neurosci. 28, 11435–11444.

Friedrich, P., Forkel, S.J., Amiez, C., Balsters, J.H., Coulon, O., Fan, L., Goulas, A., Hadj-Bouziane, F., Hecht, E.E., Heuer, K., Jiang, T., Latzman, R.D., Liu, X., Loh, K.K., Patil, K.R., Lopez-Persem, A., Procyk, E., Sallet, J., Toro, R., Vickery, S., Weis, S., Wilson, C.R.E., Xu, T., Zerbi, V., Eickoff, S.B., Margulies, D.S., Mars, R.B., Thiebaut de Schotten, M., 2021. Imaging evolution of the primate brain: the next frontier? NeuroImage 228, 117685.

Froudist-Walsh, S., Xu, T., Niu, M., Rapan, L., Zhao, L., Margulies, D.S., Zilles, K., Wang, X.J., Palomero-Gallagher, N., 2023. Gradients of neurotransmitter receptor expression in the macaque cortex. Nat. Neurosci. 26 (7), 1281–1294. https://doi.org/10.1038/s41593-023-01351-2 (Epub 2023 Jun 19. PMID: 37336976; PMCID: PMC10322721).

Gao, Y., Choe, A.S., Stepniewska, I., Li, X., Avison, M.J., Anderson, A.W., 2013. Validation of DTI tractography-based measures of primary motor area connectivity in the squirrel monkey brain. PLoS One 8, e75065.

Giampiccolo, D., Duffau, H., 2022. Controversy over the temporal cortical terminations of the left arcuate fasciculus: a reappraisal. Brain 145 (4), 1242–1256. https://doi.org/10.1093/brain/awac057 (PMID: 35142842).

Haghir, H., Kuckertz, A., Zhao, L., Hami, J., Palomero-Gallagher, N., 2023. A new map of the rat isocortex and proisocortex: cytoarchitecture and M2 receptor distribution patterns. Brain Struct. Funct. https://doi.org/10.1007/s00429-023-02654-7.

Hau, J., Sarubbo, S., Perchey, G., Crivello, F., Zago, L., Mellet, E., Jobard, G., Joliot, M., Mazoyer, B.M., Tzourio-Mazoyer, N., Petit, L., 2016. Cortical terminations of the inferior fronto-occipital and uncinate fasciculi: anatomical stem-based virtual dissection. Front. Neuroanat. 10, 58.

Hope, T.M.H., Neville, D., Talozzi, L., Foulon, C., Forkel, S.J., de Schotten, M.T., Price, C.J., 2024. Testing the disconnectome symptom discoverer model on out-of-sample post-stroke language outcomes. Brain 147 (2), e11–e13. https://doi.org/10.1093/brain/awad352. PMID: 37820032; PMCID: PMC10834246.

Jbabdi, S., Johansen-Berg, H., 2011. Tractography: where do we go from here? Brain Connect. 1, 169–183.

Jbabdi, S., Lehman, J.F., Haber, S.N., Behrens, T.E., 2013. Human and monkey ventral prefrontal fibers use the same organizational principles to reach their targets: tracing versus tractography. J. Neurosci. 33, 3190–3201.

Jones, D.K., 2008. Studying connections in the living human brain with diffusion MRI. Cortex 44, 936–952.

Jones, D.K., Christiansen, K.F., Chapman, R.J., Aggleton, J.P., 2013a. Distinct subdivisions of the cingulum bundle revealed by diffusion MRI fibre tracking: implications for neuropsychological investigations. Neuropsychologia 51, 67–78.

Jones, D.K., Knösche, T.R., Turner, R., 2013b. White matter integrity, fiber count, and other fallacies: the do's and don'ts of diffusion MRI. NeuroImage 73, 239–254.

Kadri, P.A.S., de Oliveira, J.G., Krayenbühl, N., Türe, U., de Oliveira, E.P.L., Al-Mefty, O., Ribas, G.C., 2017. Surgical approaches to the temporal horn: an anatomic analysis of white matter tract interruption. Oper. Neurosurg. (Hagerstown) 13, 258–270.

Kaplan, E., Naeser, M.A., Martin, P.I., Ho, M., Wang, Y., Baker, E., Pascual-Leone, A., 2010. Horizontal portion of arcuate fasciculus fibers track to pars opercularis, not pars triangularis, in right and left hemispheres: a DTI study. NeuroImage 52, 436–444.

Karlsgodt, K.H., John, M., Ikuta, T., Rigoard, P., Peters, B.D., Derosse, P., Malhotra, A.K., Szeszko, P.R., 2015. The accumbofrontal tract: diffusion tensor imaging characterization and developmental change from childhood to adulthood. Hum. Brain Mapp. 36, 4954–4963.

Kemerdere, R., de Champfleur, N.M., Deverdun, J., Cochereau, J., Moritz-Gasser, S., Herbet, G., Duffau, H., 2016. Role of the left frontal aslant tract in stuttering: a brain stimulation and tractographic study. J. Neurol. 263, 157–167.

Klingler, J., 1935. Erleichterung der makroskopischen Präparation des Gehirn durch den Gefrierprozess. Schweiz. Arch. Neurol. Psychiatr. 36, 247–256.

Knösche, T.R., Anwander, A., Liptrot, M., Dyrby, T.B., 2015. Validation of tractography: comparison with manganese tracing. Hum. Brain Mapp. 36, 4116–4134.

Lawes, I.N.C., Barrick, T.R., Murugam, V., Spierings, N., Evans, D.R., Song, M., Clark, C.A., 2008. Atlas-based segmentation of white matter tracts of the human brain using diffusion tensor tractography and comparison with classical dissection. NeuroImage 39, 62–79.

Leclercq, D., Duffau, H., Delmaire, C., Capelle, L., Gatignol, P., Ducros, M., Chiras, J., Lehéricy, S., 2010. Comparison of diffusion tensor imaging tractography of language tracts and intraoperative subcortical stimulations. J. Neurosurg. 112, 503–511.

Lerch, J.P., van der Kouwe, A.J.W., Raznahan, A., Paus, T., Johansen-Berg, H., Miller, K.L., Smith, S.M., Fischl, B., Sotiropoulos, S.N., 2017. Studying neuroanatomy using MRI. Nat. Neurosci. 20, 314–326.

Li, L., Hu, X., Preuss, T.M., Glasser, M.F., Damen, F.W., Qiu, Y., Rilling, J., 2013. Mapping putative hubs in human, chimpanzee and rhesus macaque connectomes via diffusion tractography. NeuroImage 80, 462–474.

Liu, X., Kinoshita, M., Shinohara, H., Hori, O., Ozaki, N., Nakada, M., 2020. Does the superior fronto-occipital fascicle exist in the human brain? Fiber dissection and brain functional mapping in 90 patients with gliomas. NeuroImage Clin. 25, 102192.

Macmillan, M., 2000. Restoring Phineas Gage: a 150th retrospective. J. Hist. Neurosci. 9, 46–66.

Mars, R.B., Sotiropoulos, S.N., Passingham, R.E., Sallet, J., Verhagen, L., Khrapitchev, A.A., Sibson, N., Jbabdi, S., 2018. Whole brain comparative anatomy using connectivity blueprints. eLife 7, e35237.

Meola, A., Comert, A., Yeh, F.-C., Stefaneanu, L., Fernandez-Miranda, J.C., 2015. The controversial existence of the human superior fronto-occipital fasciculus: connectome-based tractographic study with microdissection validation. Hum. Brain Mapp. 36, 4964–4971.

Mirchandani, A.S., Beyh, A., Lavrador, J.P., Howells, H., Dell'Acqua, F., Vergani, F., 2020. Altered corticospinal microstructure and motor cortex excitability in gliomas: an advanced tractography and transcranial magnetic stimulation study. J. Neurosurg., 1–9.

Mori, S., Wakana, S., van Zijl, P.C.M., Nagae-Poetscher, L.M., 2005. MRI Atlas of Human White Matter. Elsevier.

Moseley, M.E., Cohen, Y., Mintorovitch, J., Chileuitt, L., Shimizu, H., Kucharczyk, J., Wendland, M.F., Weinstein, P.R., 1990. Early detection of regional cerebral ischemia in cats: comparison of diffusion- and T2-weighted MRI and spectroscopy. Magn. Reson. Med. 14, 330–346.

Naidich, P., Valavanis, G., Kubik, S., 1992. Corretta localizzazione anatomica delle principali circonvoluzioni e dei principali solchi lungo la convessità inferiore e mediana con RM in sezione sagittale. Riv. Neuroradiol. 5, 299–307.

Ono, M., Kubik, S., Abernathey, C.D., 1990. Atlas of the Cerebral Sulci. Thieme.

Pacella, V., Foulon, C., Jenkinson, P.M., Scandola, M., Bertagnoli, S., Avesani, R., Fotopoulou, A., Moro, V., Thiebaut de Schotten, M., 2019. Anosognosia for hemiplegia as a tripartite disconnection syndrome. eLife 8.

Park, H.J., Friston, K., 2013. Structural and functional brain networks: from connections to cognition. Science 342. https://doi.org/10.1126/science.1238411.

Pascalau, R., Stănilă, R.P., Sfrângeu, S., Szabo, B., 2018. Anatomy of the limbic white matter tracts as revealed by fiber dissection and tractography. World Neurosurg. 113, e672–e689. https://doi.org/10.1016/j.wneu.2018.02.121 (Epub 2018 Mar 1. PMID: 29501514).

Paxinos, G., Huang, X.-F., Toga, A.W., 2000. The Rhesus Monkey Brain in Stereotaxic Coordinates. Faculty of Health and Behavioural Sciences—Papers (Archive).

Peltier, J., Verclytte, S., Delmaire, C., Pruvo, J.-P., Havet, E., Le Gars, D., 2011. Microsurgical anatomy of the anterior commissure: correlations with diffusion tensor imaging fiber tracking and clinical relevance. Neurosurgery 69, ons241–ons246. discussion ons246–ons247.

Petrides, M., Pandya, D.N., 2002. Comparative cytoarchitectonic analysis of the human and the macaque ventrolateral prefrontal cortex and corticocortical connection patterns in the monkey. Eur. J. Neurosci. 16, 291–310.

Pierpaoli, C., Jezzard, P., Basser, P.J., Barnett, A., Di Chiro, G., 1996. Diffusion tensor MR imaging of the human brain. Radiology 201, 637–648.

Ratiu, P., Talos, I.-F., Haker, S., Lieberman, D., Everett, P., 2004. The tale of Phineas Gage, digitally remastered. J. Neurotrauma 21, 637–643.

Ribas, G.C., 2010. The cerebral sulci and gyri. Neurosurg. Focus. 28, E2.

Rilling, J.K., Glasser, M.F., Preuss, T.M., Ma, X., Zhao, T., Hu, X., Behrens, T.E.J., 2008. The evolution of the arcuate fasciculus revealed with comparative DTI. Nat. Neurosci. 11, 426–428.

Rohlfing, T., Kroenke, C.D., Sullivan, E.V., Dubach, M.F., Bowden, D.M., Grant, K.A., Pfefferbaum, A., 2012. The INIA19 template and NeuroMaps atlas for primate brain image parcellation and spatial normalization. Front. Neuroinform. 6, 27.

Rojkova, K., Volle, E., Urbanski, M., Humbert, F., Dell'Acqua, F., Thiebaut de Schotten, M., 2016. Atlasing the frontal lobe connections and their variability due to age and education: a spherical deconvolution tractography study. Brain Struct. Funct. 221, 1751–1766.

**420** **PART** | **IV** From streamlines to tracts

Roumazeilles, L., Eichert, N., Bryant, K.L., Folloni, D., Sallet, J., Vijayakumar, S., Foxley, S., Tendler, B.C., Jbabdi, S., Reveley, C., Verhagen, L., Dershowitz, L.B., Guthrie, M., Flach, E., Miller, K.L., Mars, R.B., 2020. Longitudinal connections and the organization of the temporal cortex in macaques, great apes, and humans. PLoS Biol. 18, e3000810.

Royo, J., Forkel, S.J., Pouget, P., Thiebaut de Schotten, M., 2021. The squirrel monkey model in clinical neuroscience. Neurosci. Biobehav. Rev. 128, 152–164.

Salvalaggio, A., De Filippo De Grazia, M., Zorzi, M., Thiebaut de Schotten, M., Corbetta, M., 2020. Post-stroke deficit prediction from lesion and indirect structural and functional disconnection. Brain 143, 2173–2188.

Schilling, K.G., Gao, Y., Stepniewska, I., Janve, V., Landman, B.A., Anderson, A.W., 2019. Anatomical accuracy of standard-practice tractography algorithms in the motor system—a histological validation in the squirrel monkey brain. Magn. Reson. Imaging 55, 7–25.

Schilling, K.G., Tax, C.M.W., Rheault, F., Landman, B.A., Anderson, A.W., Descoteaux, M., Petit, L., 2022. Prevalence of white matter pathways coming into a single white matter voxel orientation: the bottleneck issue in tractography. Hum. Brain Mapp. 43, 1196–1213.

Schlaug, G., Siewert, B., Benfield, A., Edelman, R.R., Warach, S., 1997. Time course of the apparent diffusion coefficient (ADC) abnormality in human stroke. Neurology 49, 113–119.

Schmahmann, J.D., Pandya, D.N., 2006. Fiber Pathways of the Brain. Oxford University Press.

Schmahmann, J.D., Pandya, D.N., 2007. The complex history of the fronto-occipital fasciculus. J. Hist. Neurosci. 16, 362–377.

Schmahmann, J.D., Pandya, D.N., Wang, R., Dai, G., D'Arceuil, H.E., de Crespigny, A.J., Wedeen, V.J., 2007. Association fibre pathways of the brain: parallel observations from diffusion spectrum imaging and autoradiography. Brain 130, 630–653.

Scholz, J., Klein, M.C., Behrens, T.E.J., Johansen-Berg, H., 2009. Training induces changes in white-matter architecture. Nat. Neurosci. 12, 1370–1371.

Signoret, J.L., Castaigne, P., Lhermitte, F., Abelanet, R., Lavorel, P., 1984. Rediscovery of Leborgne's brain: anatomical description with CT scan. Brain Lang. 22, 303–319.

Sollmann, N., Zhang, H., Fratini, A., Wildschuetz, N., Ille, S., Schröder, A., Zimmer, C., Meyer, B., Krieg, S.M., 2020. Risk assessment by presurgical tractography using navigated TMS maps in patients with highly motor- or language-eloquent brain tumors. Cancers 12.

Souter, N.E., Wang, X., Thompson, H., Krieger-Redwood, K., Halai, A.D., Lambon Ralph, M.A., Thiebaut de Schotten, M., Jefferies, E., 2022. Mapping lesion, structural disconnection, and functional disconnection to symptoms in semantic aphasia. Brain Struct. Funct. 227, 3043–3061.

Takemura, H., Pestilli, F., Weiner, K.S., Keliris, G.A., Landi, S.M., Sliwa, J., Ye, F.Q., Barnett, M.A., Leopold, D.A., Freiwald, W.A., Logothetis, N.K., Wandell, B.A., 2017. Occipital white matter tracts in human and macaque. Cereb. Cortex 27, 3346–3359.

Talozzi, L., Forkel, S.J., Pacella, V., Nozais, V., Corbetta, M., Nachev, P., De Schotten, M.T., 2021. Latent disconnectome prediction of long-term cognitive symptoms in stroke.

Teichmann, M., Rosso, C., Martini, J.-B., Bloch, I., Brugières, P., Duffau, H., Lehéricy, S., Bachoud-Lévi, A.-C., 2015. A cortical-subcortical syntax pathway linking Broca's area and the striatum. Hum. Brain Mapp. 36, 2270–2283.

Thiebaut de Schotten, M., Forkel, S.J., 2022. The emergent properties of the connected brain. Science 378, 505–510. https://doi.org/10.1126/science.abq2591.

Thiebaut de Schotten, M., Urbanski, M., Duffau, H., Volle, E., Lévy, R., Dubois, B., Bartolomeo, P., 2005. Direct evidence for a parietal-frontal pathway subserving spatial awareness in humans. Science 309, 2226–2228.

Thiebaut de Schotten, M., Ffytche, D.H., Bizzi, A., Dell'Acqua, F., Allin, M., Walshe, M., Murray, R., Williams, S.C., Murphy, D.G.M., Catani, M., 2010. Atlasing location, asymmetry and inter-subject variability of white matter tracts in the human brain with MR diffusion tractography. NeuroImage 54, 49–59.

Thiebaut de Schotten, M., Dell'Acqua, F., Forkel, S.J., Simmons, A., Vergani, F., Murphy, D.G.M., Catani, M., 2011. A lateralized brain network for visuospatial attention. Nat. Neurosci. 14, 1245–1246.

Thiebaut de Schotten, M., Dell'Acqua, F., Valabregue, R., Catani, M., 2012. Monkey to human comparative anatomy of the frontal lobe association tracts. Cortex 48, 82–96.

Thiebaut de Schotten, M., Dell'Acqua, F., Ratiu, P., Leslie, A., Howells, H., Cabanis, E., Iba-Zizen, M.T., Plaisant, O., Simmons, A., Dronkers, N.F., Corkin, S., Catani, M., 2015. From Phineas Gage and Monsieur Leborgne to H.M.: revisiting disconnection syndromes. Cereb. Cortex 25, 4812–4827.

Thomas, C., Ye, F.Q., Irfanoglu, M.O., Modi, P., Saleem, K.S., Leopold, D.A., Pierpaoli, C., 2014. Anatomical accuracy of brain connections derived from diffusion MRI tractography is inherently limited. Proc. Natl. Acad. Sci. U. S. A. 111, 16574–16579.

Türe, U., Yaşargil, M.G., Pait, T.G., 1997. Is there a superior occipitofrontal fasciculus? A microsurgical anatomic study. Neurosurgery 40, 1226–1232.

Türe, U., Yaşargil, M.G., Friedman, A.H., Al-Mefty, O., 2000. Fiber dissection technique: lateral aspect of the brain. Neurosurgery 47, 417–426. discussion 426–427.

Uylings, H.B.M., Rajkowska, G., Sanz-Arigita, E., Amunts, K., Zilles, K., 2005. Consequences of large interindividual variability for human brain atlases: converging macroscopical imaging and microscopical neuroanatomy. Anat. Embryol. 210, 423–431.

van den Heuvel, M.P., de Reus, M.A., Barrett, L.F., Scholtens, L.H., Coopmans, F.M.T., Schmidt, R., Preuss, T.M., Rilling, J.K., Li, L., 2015. Comparison of diffusion tractography and tract-tracing measures of connectivity strength in rhesus macaque connectome. Hum. Brain Mapp. 36 (8), 3064–3075. https://doi.org/10.1002/hbm.22828 (Epub 2015 Jun 9. PMID: 26058702; PMCID: PMC6869766).

Van Horn, J.D., Irimia, A., Torgerson, C.M., Chambers, M.C., Kikinis, R., Toga, A.W., 2012. Mapping connectivity damage in the case of Phineas Gage. PLoS One 7, e37454.

Vergani, F., Lacerda, L., Martino, J., Attems, J., Morris, C., Mitchell, P., Thiebaut de Schotten, M., Dell'Acqua, F., 2014a. White matter connections of the supplementary motor area in humans. J. Neurol. Neurosurg. Psychiatry 85, 1377–1385.

Vergani, F., Mahmood, S., Morris, C.M., Mitchell, P., Forkel, S.J., 2014b. Intralobar fibres of the occipital lobe: a post mortem dissection study. Cortex 56, 145–156.

Wandell, B.A., Rauschecker, A.M., Yeatman, J.D., 2012. Learning to see words. Annu. Rev. Psychol. 63, 31–53.

Warach, S., Gaa, J., Siewert, B., 1995. Acute human stroke studied by whole brain echo planar diffusion-weighted magnetic resonance imaging. Ann. Neurol. 37 (2), 231–241. https://doi.org/10.1002/ana.410370214 (PMID: 7847864.weN).

Warrington, S., Bryant, K.L., Khrapitchev, A.A., Sallet, J., Charquero-Ballester, M., Douaud, G., Jbabdi, S., Mars, R.B., Sotiropoulos, S.N., 2020. XTRACT—standardised protocols for automated tractography in the human and macaque brain. NeuroImage 217, 116923.

Wernicke, C., 1874. Der aphasische Symptomencomplex: eine psychologische Studie auf anatomischer Basis. Cohn & Weigert.

Yeatman, J.D., Weiner, K.S., Pestilli, F., Rokem, A., Mezer, A., Wandell, B.A., 2014. The vertical occipital fasciculus: a century of controversy resolved by in vivo measurements. Proc. Natl. Acad. Sci. U. S. A. 111, E5214–E5223.

Yendiki, A., Aggarwal, M., Axer, M., Howard, A.F.D., van Walsum, A.-M.v.C., Haber, S.N., 2022. Post mortem mapping of connectional anatomy for the validation of diffusion MRI. NeuroImage 256, 119146.

Yousry, T.A., Schmid, U.D., Alkadhi, H., Schmidt, D., Peraud, A., Buettner, A., Winkler, P., 1997. Localization of the motor hand area to a knob on the precentral gyrus. A new landmark. Brain 120 (Pt 1), 141–157.

Zhang, W., Ille, S., Schwendner, M., Wiestler, B., Meyer, B., Krieg, S.M., 2022. Tracking motor and language eloquent white matter pathways with intraoperative fiber tracking versus preoperative tractography adjusted by intraoperative MRI-based elastic fusion. J. Neurosurg., 1–10.

Zhylka, A., Sollmann, N., Kofler, F., Radwan, A., De Luca, A., Gempt, J., Wiestler, B., Menze, B., Krieg, S.M., Zimmer, C., Kirschke, J.S., Sunaert, S., Leemans, A., Pluim, J.P.W., 2021. Tracking the corticospinal tract in patients with high-grade glioma: clinical evaluation of multi-level fiber tracking and comparison to conventional deterministic approaches. Front. Oncol. 11, 761169.

# Chapter 22

# Dissecting white matter pathways: Automatic and semiautomatic approaches

Eleftherios Garyfallidis[a], Marc-Alexandre Côté[b], Francois Rheault[c], and Emanuele Olivetti[d]

[a]*Department of Intelligent Systems Engineering, Luddy School of Informatics, Computing and Engineering, Indiana University Bloomington, Bloomington, IN, United States,* [b]*Microsoft Research, Montreal, QC, Canada,* [c]*Department of Electrical and Computer Engineering, Vanderbilt University, Nashville, TN, United States,* [d]*Fondazione Bruno Kessler, Trento, Italy*

## 1 Introduction

The human brain contains billions of axons that bundle together in tracts and fasciculi. These can be reconstructed in vivo by collecting diffusion magnetic resonance imaging data (Basser et al., 1994; Le Bihan et al., 2001; Alexander et al., 2007) and deploying tractography algorithms (Farquharson et al., 2013; Catani and De Schotten, 2008; Gong et al., 2008; Mori and Van Zijl, 2002). The products of tractography algorithms are called tractograms. These tractograms are represented digitally using streamlines, which are three-dimensional (3D) curves traversing the brain. Whole brain tractograms are densely populated with millions of streamlines, which makes it difficult to visually and computationally inspect and characterize brain pathways. When streamlines of similar shape and characteristics travel together through the white matter, they are called bundles. These bundles approximate the white matter fiber bundles that connect distant parts of the brain to each other and are also known as tracts or fasciculi. These bundles of axons carry crucial information between cortical and/ or subcortical areas and potential damage to these bundles, for example, from surgery, trauma, or disease, can have tremendous consequences for the patient's cognitive function and quality of life. Every bundle has a different functional association or set of functional associations. For example, the arcuate fasciculus is involved in the understanding of language (Geschwind, 1970) while the optic radiation is involved primarily with visual processing (Leuret, 1857; Schmahmann et al., 2009).

Manual virtual dissection (Catani et al., 2002; Wang et al., 2007; Chamberland et al., 2014) and automatic bundle extraction (Garyfallidis et al., 2017; Wasserthal et al., 2018; Guevara et al., 2012; Lawes et al., 2008; Jonasson et al., 2005; Bertò et al., 2021) of white matter bundles have enabled the scientific community to gather thousands of extracted fiber tract exemplars from source tractograms and visualize them in vivo. With the rise of machine learning in the field of neuroimaging, we are able to automate complex and unwieldy tasks such as white matter fiber bundle segmentation from whole brain tractograms. Segmenting bundles has become convenient, efficient, and fast (Garyfallidis et al., 2012, 2017; Bertò et al., 2021), and this practice can accommodate the generation and processing of exceedingly large datasets (Garyfallidis et al., 2019b; Vu et al., 2015; Smith et al., 2015; Maier-Hein et al., 2017).

**Why is automatic segmentation a challenging problem**? Tractograms are different data structures than what the medical community is used to working with. Tractograms contain streamlines that belong in the category of unstructured data in contrast to images that are traditionally used. For example, in a 3D image if we know the position $(i, j, k)$ of a voxel we have quick access to its neighboring voxels with a simple change of the coordinate position, that is, $(i - 1, j, k)$. In a tractogram, we do not have access to such information, and therefore we do not know which streamlines are neighbors to each other. This information is not provided from the data structure itself and needs to be calculated. In addition, we do not know how the points composing the streamline are ordered. Two streamlines that look identical could be represented as two different sequences of points, that is, listing the points starting from different ends. An additional difficulty arises from the fact that tractograms grow faster in size than the underlying images because their coordinates are saved in submillimeter (subvoxel) resolution. These aforementioned issues increase computational complexity of algorithms dealing with tractograms.

These are not the only issues that one will face when building or using methods for automatically segmenting anatomical bundles from tractograms. One of the major issues is that neuroanatomists often disagree on the definition of many

---

*Handbook of Diffusion MR Tractography. https://doi.org/10.1016/B978-0-12-818894-1.00013-6*
Copyright © 2025 Elsevier Ltd. All rights are reserved, including those for text and data mining, AI training, and similar technologies.

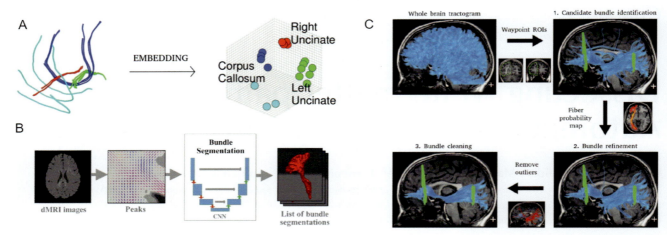

**FIG. 1** Examples of different strategies that can be used to dissect a tractogram in an automatic or semiautomatic way. (A) Spectral embedding. (B) Deep learning-based segmentation with masked ROIs. (C) Apparent fiber quantification (AFQ) ROI-based segmentation.

tracts (Vavassori et al., 2021). Also, there are bundles in the tractogram that may be unknown to anatomists or simply just false positives (Rheault et al., 2020b; Maier-Hein et al., 2017). Additionally, there are not many labeled datasets available to drive commonly available machine-learning techniques. However, these issues are improving, driven by large datasets and new methods that help improve the labeling process (Garyfallidis and Chandio, 2021; Yeh et al., 2018).

We should note here that, although this chapter focuses primarily on streamline-based representations, tractograms and bundles are not always represented using streamlines. They can be represented as surfaces (external boundary), regions of interest (ROIs) (collections of internal voxels), or graphs (connecting endpoints) (Fig. 1).

In order to segment a tractogram and obtain bundles of interest, it is crucial to define a way to compare streamlines. This seemingly simple question is in fact a very challenging one. Here, we face two schools of thoughts: the feature-based versus distance-based approach. In machine learning, a feature describes an object using its properties. For a streamline representation, features can include torsion, dispersion, curvature, and length. Streamlines can then be compared using their features. Similar features assume similar streamlines. A distance-based approach will directly use the points of the streamlines, and those with a small distance will be considered similar, potentially belonging to the same bundle. The concept of distance, a.k.a. distance function or metric, is well-defined in mathematics and has to respect a list of properties. Only a few distance functions between streamlines have such properties.

## 2 Why is it important?

There are three key reasons why automated segmentation is a crucial step in current tractography analysis pipelines.

(a) Manual segmentation takes time, and humans can be consciously or accidentally biased (Rheault et al., 2020a). Additionally, the number of datasets and size of datasets are increasing every year. Therefore the time needed for manual segmentation also increases. (b) Bundles have functional meaning and therefore are important for building theories about how the brain works. The more precise the methods, the more accurate the explanations will be. (c) By extracting anatomically valid bundles, we can save ourselves from wasting time dealing with false positives (Rheault et al., 2020b).

## 3 What is a streamline?

A tractogram $T$ is represented as a set $T = \{s_1, ..., s_n, ..., s_N\}$ where $s_n$ denotes an individual streamline, $n \in [1, N]$ denotes the index of the streamline within the tractogram, and $N$ is the number of streamlines in the tractogram. Each streamline $s_n$ can be thought of as a $K_n \times 3$ matrix of 3D points where $K_n$ is the number of points of the $n$th streamline (i.e., $s_n$). For example, assume a streamline with only four points; then this would be represented as

$$s_n = \begin{pmatrix} x_{11} & x_{12} & x_{13} \\ x_{21} & x_{22} & x_{23} \\ x_{31} & x_{32} & x_{33} \\ x_{41} & x_{42} & x_{43} \end{pmatrix}$$

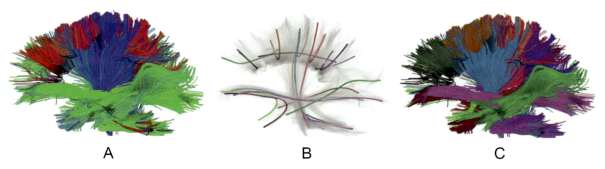

**FIG. 2** A QuickBundles example. In (A) an input tractogram is shown colored with average streamline orientation. Note that this is a reduced-in-size tractogram used for visual clarity. In (B) the centroids of the method are shown. In (C) the corresponding clusters of (B) are shown. Note the improvement in conspicuity. A single threshold at 15 mm suffices to provide this result. However, the larger and denser the tractogram is, the harder and the more challenging the problem becomes. (A) Original tractogram, (B) QuickBundles centroids, (C) QuickBundles clusters.

where each row is a different point $x_i \in \mathbb{R}^3$. Note that each streamline can have a different number of points. Also, note that visually $s_n$ and

$$s'_n = \begin{pmatrix} x_{41} & x_{42} & x_{43} \\ x_{31} & x_{32} & x_{33} \\ x_{21} & x_{22} & x_{23} \\ x_{11} & x_{12} & x_{13} \end{pmatrix}$$

are indistinguishable. Note also that visually any permutation (without replacement) of the streamlines in the tractogram will have no noticeable difference. For example, assume $T_1 = \{s_1, s_2, s_3, s_4\}$, $T_2 = \{s_3, s_2, s_1, s_4\}$, and $T_3 = \{s_1, s_4, s_3, s_2\}$. All three tractograms are identical for the user, and there is no computationally economical way to find that they are stored differently. Nonetheless, every streamline is a sequence of ordered 3D points, and this allows for many properties to be computed. For example, by selecting consecutive pairs of points an average orientation vector can be calculated. This is commonly used for coloring purposes (see Fig. 2A). Other statistical or geometrical properties can be calculated such as curvature, torsion, and mean deviation. Besides calculating these properties, many labs often store any associated anatomical data with each bundle. For example, they interpolate FA values at each point of the streamline. Another piece of information that is often stored with the tractogram is the coordinate system, allowing the data to appear in native coordinates or world coordinates with a neurological or radiological convention.

To make processing of tractograms easier, we can set all streamlines to have the same number of points. One way to do that is by specifying the number of points and resampling the original streamline along its length. The result provides an approximation of the original streamline that is often as good as the original. For example, by keeping only 20 equidistant points on the streamlines, most streamlines are well represented. This was shown in Garyfallidis et al. (2012). An alternative approach is to build a vector representation of a streamline using other streamlines as reference (Olivetti et al., 2012). In that approach, we can select some streamlines randomly or from bundle atlases as prototypes. A vector can be built where each element refers to a distance from the current streamline to a prototype streamline. This is called dissimilarity projection and is a common approach in pattern recognition (Pekalska and Duin, 2005). After this encoding procedure is created, we can apply the most commonly used vector-based methods to process our data since most machine-learning methods expect feature vectors as input.

## 4 Streamline distance functions

To be truly useful when comparing streamlines, a metric needs to have excellent characteristics. For example: (a) The metric should be robust to small deformations and displacements, which are omnipresent in tractograms. (b) The metric should satisfy the following three mathematical axioms: (1) the identity of indiscernibles, (2) the symmetry and hopefully, and (3) the triangle inequality. If these properties are not satisfied, the domain can have singular locations. (c) The metric should be simple to interpret. (d) The metric should be fast to compute since tractograms are massive unstructured datasets. For example, a tractogram with 1 million streamlines can easily have half a billion points.

Given two streamlines $s_a$ and $s_b$, here we provide the mathematical formulation of some of the streamline distance functions frequently used in the literature. The idea is that streamlines belonging to the same anatomical structure are close to each other, that is, lie in small distances.

**1.** Mean of closest distances (Corouge et al., 2004):

$$d_{\mathrm{MC}}(s_a, s_b) = \frac{d_m(s_a, s_b) + d_m(s_b, s_a)}{2} \tag{1}$$

where $d_m(s_a, s_b) = \frac{1}{|s_a|} \sum_{\mathbf{x}_i \in s_a} \min_{\mathbf{x}_j \in s_b} \| \mathbf{x}_i - \mathbf{x}_j \|_2$

**2.** Shorter mean of closest distances (Zhang et al., 2008):

$$d_{\mathrm{SC}}(s_a, s_b) = \min(d_m(s_a, s_b), d_m(s_b, s_a)) \tag{2}$$

**3.** Longer mean of closest distances (Zhang et al., 2008):

$$d_{\mathrm{LC}}(s_a, s_b) = \max(d_m(s_a, s_b), d_m(s_b, s_a)) \tag{3}$$

**4.** Minimum average direct-flip (MDF) distance (Garyfallidis et al., 2012), after resampling each streamline to a given number of points $p$, such as $s_a = \{\mathbf{x}_1^a, \ldots, \mathbf{x}_p^a\}$ and $s_b = \{\mathbf{x}_1^b, \ldots, \mathbf{x}_p^b\}$:

$$d_{\mathrm{MDF}}(s_a, s_b) = \min(d_{\mathrm{direct}}(s_a, s_b), d_{\mathrm{flipped}}(s_a, s_b)) \tag{4}$$

where $d_{\mathrm{direct}}(s_a, s_b) = \frac{1}{p} \sum_{i=1}^{p} \| \mathbf{x}_i^a - \mathbf{x}_i^b \|_2$ and $d_{\mathrm{flipped}}(s_a, s_b) = \frac{1}{p} \sum_{i=1}^{p} \| \mathbf{x}_i^a - \mathbf{x}_{p-i+1}^b \|_2$

**5.** Point density model (PDM) (Siless et al., 2013):

$$d_{\mathrm{PDM}}^2(s_a, s_b) = \langle s_a, s_a \rangle_{pdm} + \langle s_b, s_b \rangle_{pdm} - 2\langle s_a, s_b \rangle_{pdm} \tag{5}$$

where

$$\langle s_a, s_b \rangle_{pdm} = \frac{1}{|s_a||s_b|} \sum_{i=1}^{|s_a|} \sum_{j=1}^{|s_b|} K_\sigma(\mathbf{x}_i^a, \mathbf{x}_j^b) \tag{6}$$

and $K_\sigma(\mathbf{x}_i^a, \mathbf{x}_j^b) = \exp\left(-\frac{\|\mathbf{x}_i^a - \mathbf{x}_j^b\|_2^2}{\sigma^2}\right)$ is a Gaussian kernel between two 3D points.

**6.** Varifolds distance (Charon and Trouvé, 2013), which is the nonoriented version of the currents distance (Gori et al., 2016), does not need streamlines $s_a$ and $s_b$ to have a consistent orientation:

$$d_{\mathrm{varifolds}}^2(s_a, s_b) = \langle s_a, s_a \rangle_{var} + \langle s_b, s_b \rangle_{var} - 2\langle s_a, s_b \rangle_{var} \tag{7}$$

where

$$\langle s_a, s_b \rangle_{var} = \sum_{i=1}^{|s_a|-1} \sum_{j=1}^{|s_b|-1} K_\sigma(\mathbf{p}_i^a, \mathbf{p}_j^b) K_n(\mathbf{n}_i^a, \mathbf{n}_j^b) \| \mathbf{n}_i^a \|_2 \| \mathbf{n}_j^b \|_2 \tag{8}$$

with $K_n(\mathbf{n}_i^a, \mathbf{n}_j^b) = \left(\frac{(\mathbf{n}_i^a)^T \mathbf{n}_j^b}{\|\mathbf{n}_i^a\|_2 \|\mathbf{n}_j^b\|_2}\right)^2$ where $\mathbf{p}_i^a$ (resp. $\mathbf{p}_j^b$) and $\mathbf{n}_i^a$ (resp. $\mathbf{n}_j^b$) are the center and tangent vector of segment $i$ (resp. $j$) of streamline $s_a$ (resp. $s_b$). The endpoints of segment $i$ are $\mathbf{x}_i$ and $\mathbf{x}_{i+1}$ for $i \in [1, \ldots, n-1]$.

Notice that $d_{\mathrm{DPM}}$ and $d_{\mathrm{varifolds}}$ have higher computational complexity than the other distance functions. Moreover, $d_{\mathrm{MDF}}$ is the fastest to compute because it requires only a number of Euclidean distances between 3D points, which scale linearly with the number of points of the streamlines. Differently, all other distances scale quadratically or worse.

## 4.1 Comparing distances

In Olivetti et al. (2017), the authors designed a way to compare streamline distances for the task of bundle segmentation. The proposed procedure starts with a curated segmentation of a given bundle, like the corticospinal tract, in the tractogram of one subject, for example, to segment the same bundle in the tractogram of another target subject. After registering the target tractogram to the example tractogram, for each streamline of the example bundle, its nearest streamline is computed within the target tractogram according to the desired streamline distance function. The resulting set of nearest streamlines is an approximation of the desired target bundle by means of nearest neighbor (NN) segmentation through the selected streamline distance. The quality of the NN segmentation is measured by comparing the degree of voxel overlap, called the Dice similarity coefficient (DSC), between the resulting target bundle and a curated segmentation of that target bundle.

Dissecting white matter pathways: Automatic and semiautomatic approaches **Chapter | 22** **427**

**TABLE 1** Mean DSC voxel table across the 90 pairs of subjects and 9 different bundles, for each of the different distance functions (columns) considered.

|  | $d_{MC}$ | $d_{SC}$ | $d_{LC}$ | $d_{MDF,12}$ | $d_{MDF,20}$ | $d_{MDF,32}$ | $d_{PDM}$ | $d_{varifolds}$ |
|---|---|---|---|---|---|---|---|---|
| Means | 0.53 | 0.53 | 0.53 | 0.53 | 0.53 | 0.53 | 0.54 | 0.55 |

**TABLE 2** Time to compute 90,000 pairwise streamline distances.

|  | $d_{MC}$ | $d_{SC}$ | $d_{LC}$ | $d_{MDF,12}$ | $d_{MDF,20}$ | $d_{MDF,32}$ | $d_{PDM}$ | $d_{varifolds}$ |
|---|---|---|---|---|---|---|---|---|
| Time (s) | 0.5 | 0.5 | 0.5 | 0.03 | 0.04 | 0.05 | 16 | 28 |

By adopting different streamline distance functions, it is possible to observe if the quality of segmentation improves in terms of DSC, that is, whether or not distance functions play an important role in segmentation.

In Table 1, we report the result presented in Olivetti et al. (2017), that is, the mean DSC obtained with the described bundle segmentation procedure across 90 pairs of subjects, where one subject in the pair is used as example and the other as target. For each pair of subjects, nine different bundles[a] are considered and the six different distance functions described previously,[b] in Section 4. For each bundle and distance function, the standard deviation of DSC is approximately 0.10, which corresponds to a standard deviation of the mean of 0.01. This value includes the variances due to the anatomical variability across subjects and the limitations of the segmentation technique used as ground truth.

In Table 2, we report the time (in seconds) required by a standard workstation[c] to compute 90,000 pairwise streamline comparisons for every distance function considered. The differences in time are due to both the different computational cost of the formulas in Section 4 and their implementation. Here, $d_{MC}$, $d_{SC}$, $d_{LC}$, and $d_{MDF}$, available from DIPY, were implemented in Cython; $d_{PDM}$ and $d_{varifolds}$ were implemented using NumPy. The main reason behind the difference in computation time between $d_{PDM}$ and $d_{varifolds}$ with respect to the other distance is the increased computational complexity of their algorithms.

These results show that there are no major differences in segmentation accuracy, measured as DSC, when using different distance functions. However, the computational time of the distance functions can be very different; for example, the MDF distance is faster by a $10\times$ factor. This is the reason why the MDF distance should be preferred unless there are no other specific reasons.

# 5 Overview of state-of-the-art methods

This section will present a limited overview of methods for segmentation. Algorithms were not chosen for their popularity, but mostly for their differences from each other.

A categorization of the different approaches using machine-learning terminology can be divided between unsupervised, supervised, and semisupervised methods. An additional categorization can be created depending on how much the methods depend on selecting ROIs in images. For example, there are methods that are completely tractography based (streamline-based) and others that require integration with an image template, or an existing cortical surface or parcellation.

## 5.1 Unsupervised and semisupervised methods

This section discusses methods that do not require any labels (unsupervised) or only minimal labeling (semisupervised).

---

a. Cingulum bundle (left and right), inferior fronto-occipital fasciculus (left and right), uncinate fasciculus (left and right), segments 2 and 7 of corpus callosum, and left arcuate fasciculus.

b. The MDF distance is computed using different resamplings of the streamlines: at 12, 20, and 32 points.

c. 3.2 GHz quad-core workstation with 16GB RAM.

### 5.1.1 QuickBundles

**Description**

QuickBundles (QB) (Garyfallidis et al., 2012) available with DIPY is the first algorithm that demonstrated that it is possible to cluster streamlines at very high speeds and low memory usage. This method goes through the data once, building the clusters in a single pass. The distance used is the minimum direct-flip distance (MDF). MDF is also the first fast streamline distance used in the field that provided good accuracy (see comparison in Table 2). See example of use in Fig. 2A–C.

QuickBundlesX (QBX) (Garyfallidis et al., 2016) is a dynamic version of QB where clustering can take place at multiple thresholds. The method was devised as a way to speed up execution at lower thresholds using additional higher thresholds (a divide and conquer strategy) often surpassing 25× speedups (one CPU core) over QB with a single threshold.

**Advantages/disadvantages**

The main advantages of QB are: (a) It has only one parameter, a distance threshold with a physical interpretation, (b) it is fast, and (c) it has a low memory footprint. QBX has more thresholds (often three are recommended) that reduce execution time. Additionally, it provides as output a tree with access to the clusters at each of the three levels. Moreover, it is again single pass. A disadvantage of these methods is that they do not always guarantee the same solution. They are stochastic and the result can depend on the order the streamlines were selected. This problem can be easily handled by running QB multiple times using different random selections, as shown in Fig. 3. The theoretical benefits of such methods have been studied extensively using generic data in a recent publication by Garyfallidis et al. (2021) showing remarkable results (both in speed and accuracy) over well-established methods such as *k*-means.

**Quality assurance**

For human adult brains 10–15 mm thresholds (assuming 1 mm$^3$ space) are expected to provide consistent bundles. These recommended thresholds are expected to be different for different animals or pediatric brains because their size differs. Note that such unsupervised algorithms are not aiming to provide bulletproof anatomical results but provide outcomes with similar geometry: for example, by grouping streamlines that are adjacent and have similar shape. In addition, deciding the level of accuracy depends on the application at hand. Because tractograms are very large datasets often containing millions of streamlines, QB can be used as a simplification (Garyfallidis et al., 2012) of the tractogram to provide a fast search. QB can reduce searching speed up to 1000× with a negligible memory footprint.

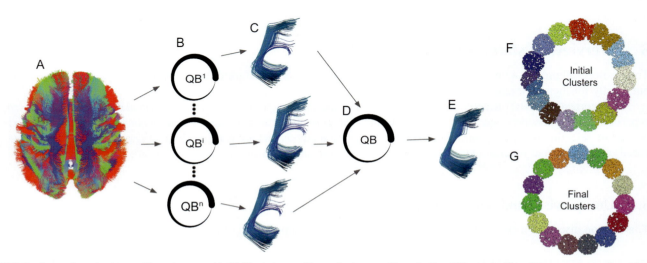

**FIG. 3** Improving robustness with randomness. (A–E) Clustering quality can be improved by selecting different shuffles of the same data and running QuickBundles (QB) in parallel for each shuffle. Then, all the centroids are provided as input to a new QB. Finally, all streamlines of the input tractogram are reassigned to the closest centroids obtained by that last QB. This rather simple technique can have dramatic improvement on the final labels. See, for example, a comparison between a single QB run in (F) and the randomized approach in (G). This example uses uniform disks on a ring, which is hard for most algorithms to cluster correctly.

### 5.1.2 FFClust

**Description**

FFClust (Vázquez et al., 2020) is another algorithm that focuses on speed and uses a similar distance as QB MDF, but rather than the minimum, it uses the maximum between direct and flipped distances. In addition, the algorithm proceeds in the following four steps: (1) building point clusters, (2) generating preliminary streamline clusters, (3) reassigning small preliminary streamline clusters, and (4) merging candidate streamline clusters using a graph system. The idea of separating the tractography clustering problem into smaller problems is a concept that has been investigated considerably in Guevara et al. (2010, 2012, 2017) and Labra et al. (2017).

**Advantages/disadvantages**

The main advantage of this approach is an increase in speed and parallelism. In addition, breaking the problem into smaller problems can increase clustering quality. However, the disadvantage of FFClust is that it introduces more parameters to tune, and the algorithm becomes harder to understand and predict.

**Quality assurance**

The Davies-Bouldin index, a well-known approach for comparing clusters, is reported by the authors. In addition, the authors provide an atlas of 100 reproducible bundles in Guevara et al. (2017) that facilitates quality control for people who want to segment bundles that have some relevance to the anatomy and the original authors' results.

### 5.1.3 Spectral clustering

**Description**

Spectral clustering uses the $k$-most salient eigenvalues of the similarity matrix of the data to project the data points onto a low-dimension embedding space before applying $k$-means to cluster them. Once done, the derived projection matrix can be used to project new data points onto the same spectral space before assigning each point to its closest cluster (i.e., closest centroid). O'Donnell and Westin (2007) make use of spectral clustering to devise a multisubject atlas of streamlines that can be used to segment streamlines of new subjects. The similarity matrix is obtained by computing the pairwise distance between all streamlines, then converting them to affinity measure using a Gaussian kernel. Different streamline distances can be used, for example, symmetrized Hausdorff distance, mean closest point distance, and others.

**Advantages/disadvantages**

The main advantage of this method lies within the spectral embedding where similarity relationships are represented by simple spatial coordinates, which make them easier to cluster. However, spectral clustering is computationally expensive since it needs the similarity matrix $W$ between all pairs of streamlines. As done in O'Donnell and Westin (2007), one can use the Nystrom method (Fowlkes et al., 2004) to approximate $W$, hence reducing the amount of computation needed. However, recent research (Fowlkes et al., 2004) pointed out that Nystrom approximation could lead to problematic $W$ (e.g., not element-wisely positive). Spectral clustering also inherits common disadvantages of $k$-means, such as the need to know in advance how many clusters (i.e., $k$) to look for, and it is sensitive to the distance function used.

**Quality assurance**

Segmented bundles of streamlines can be visualized with standard tools. The quality of segmentation can be measured as the degree of overlap with expert-made segmentation. Multiple graph theoretic techniques can be used to analyze the similarity matrix.

## 5.2 Supervised methods

In this section, we present methods that require more labels but more importantly try also to move beyond identifying geometrical similarities to actual tools that come close to manual segmentations by expert neuroanatomists. In addition, the methods described in this section are created by using two or more subjects, in contrast to the previous section that uses single subject data. For this reason, notice that some sort of registration is a mandatory step.

## 5.2.1 XTRACT

### Description

XTRACT, available with FSL, provides a segmentation framework based on a standard space dissection protocol. It uses anatomical knowledge from the standard space seeding strategy and inclusion, exclusion ROIs and termination regions are utilized to isolate pathways of interest. As for now, it comprises 42 bundles in 2 species (human and macaque). This automatic method was shown to be robust to a range of data quality and to preserve intersubject variability.

### Advantages/disadvantages

This method provides well-defined pathways in their voxel representation (density maps). While this is an advantage for reproducibility, this method does not create streamline representation, which can limit shape analysis further down the road. The analogous bundles between species could provide an interesting basis for future studies. It is important to mention that a possible limitation is that XTRACT expects (as input) preprocessed data using FSL bedpostx and will subsequently run FSL probtrackx2. This does not allow the user to provide data processed with other tools.

### Quality assurance

The output of the method can be in either native diffusion space or standard space. An easy way to verify quality is to look at the average pathways (across a population) in standard space. FSL does provide FSL XTRACT viewer in combination with FSLEYES to display bundles with a specific colormap (with matching left/right coloring) to facilitate Q/A.

## 5.2.2 Tracula

### Description

Tracula, available in FreeSurfer, allows for automatic reconstruction of pathways using anatomical priors and global tractography. By using information from neighboring anatomical structure from FreeSurfer cortical parcellation and subcortical segmentation of the subject, the global tractography algorithm can be constrained and pathways isolated. The anatomical priors' distributions were obtained from manual segmentation of training data and Bayesian probability estimation. The method was shown to have good test-retest reproducibility as well as working well with a variety of datasets (acquisition quality, aging or epilepsy, etc.).

### Advantages/disadvantages

The tool provides a framework for longitudinal analysis as well as bundle statistics. Configuration can be somewhat difficult at first, but it is worth it to join reconstruction and analysis of longitudinal datasets (or with multiple acquisition). Similarly to XTRACT, Tracula does not reconstruct streamlines and outputs only voxel representations of pathways (probability maps), which can limit any forthcoming shape analysis. Because it uses FreeSurfer, some preprocessing is necessary (intra-/intersubject registration, templates creation, etc.) and the underlying ball-and-stick model must be estimated (using bedpostX) and must all be computed before Tracula can be launched. Therefore the total processing time is rather long.

### Quality assurance

Freeview allows the visualization of the voxel-representation, and the volumetric distributions can be loaded and interacted with as isosurfaces using various parameters.

## 5.2.3 TractQuerier

### Description

TractQuerier is an open-source tool to describe white matter tracts in a human-readable language (i.e., the White Matter Query Language [WMQL]) (Wassermann et al., 2016). Tract descriptions are written as logical statements that involve predefined WMQL sets of streamlines related to different anatomical regions extracted by FreeSurfer. TractQuerier is considered as a voxel-based segmentation since those WMQL's sets are defined by streamlines traversing or ending in certain regions of the brain.

### Advantages/disadvantages

TractQuerier makes streamline segmentation more user-friendly and interpretable. For instance, anatomists can extend bundle definitions for finer-grained segmentations by describing which areas the bundles pass through or start/end. TractQuerier implicitly requires T1 image registration (via FreeSurfer) to obtain the anatomical regions. A disadvantage of the method is with clinical brains (e.g., TBI data). If the T1 is not well parcellated in affected regions, then TractQuerier cannot extract correct bundles connecting these regions.

### Quality assurance

TractQuerier (through WQML) unifies the bundle definitions which can be used on across different subjects. The resulting output can be visualized with most available viewers. Note that some bundles may not be available in some datasets and therefore it is expected to not find all the described languages in a bundle.

### 5.2.4 RecoBundles

#### Description

This method greatly simplifies the segmentation process by extracting specific bundles from the otherwise intractable volume of streamlines. Here using RecoBundles (RB) which stands for bundle recognition (Garyfallidis et al., 2017), the approach goes a step further from automated bundle extraction as it requires no training. This method learns from single examples of bundles also known as model bundles, provided by an atlas of bundles or a single subject dataset.

In this approach, in order to extract white matter fiber tracts from whole brain tractogram, the authors first register an input target tractogram (A.a) to a model tractogram (A.b) using streamline-based linear (affine) registration (SLR) (Garyfallidis et al., 2015) as shown in Fig. 4. After the target tractogram is transformed from native space to common space, we begin extracting bundles.

The model bundle is used as a reference bundle to find corresponding streamlines in the target tractogram, ignoring unrealistic streamlines. The model bundle is part of the atlas used for registering the target tractogram to common space (see Fig. 4). Both target tractogram and model bundle are in common space. The first step in RB is to reduce the search space and find neighboring areas for the model bundle in the target tractogram. This is achieved by a process called far pruning. That is to say, the streamlines that are too different (by a streamline distance) from the streamlines of the model bundle are excluded from our search space. After local registration of streamlines using (SLR), local pruning of neighbor streamlines is performed in a similar manner as far as pruning. Neighbor streamlines whose MDF distance with model bundle streamlines is greater than the pruning threshold are discarded. An autocalibration step was used in the latest version of the algorithm, as described in Chandio et al. (2020), that improves recognition capabilities especially for shorter bundles. Autocalibration was found to be especially useful when dealing with noisy data or when extracting small bundles like the

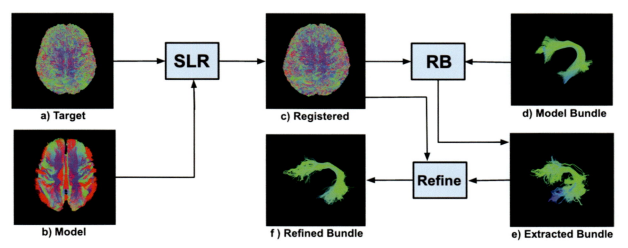

**FIG. 4** Overview of RecoBundles (RB) with autocalibration and registration. The input target tractogram (a) is registered to a model tractogram (b) using streamline-based linear registration (SLR). The target tractogram is now in model's space, extraction can begin. The input to RB is the registered target tractogram (c) and a model bundle (d) from the model tractogram (b). RB generates an extracted bundle (e) that looks as closely as possible to the model bundle (d). The autocalibration step refines further the extracted bundle by rerunning RB with the extracted bundle as input. The result is shown in (f).

**432 PART | IV** From streamlines to tracts

uncinate fasciculus (UF) from whole brain tractograms. During the autocalibration step, the final extracted bundle output of standard RB becomes our new model bundle and the process restarts.

### Advantages/disadvantages

Whole brain tractograms contain a large number of false positives, for example, and unrealistically small or extremely long streamlines. Nonetheless, RB in principle only requires one or a few model bundles to find the bundles of interest. The authors also provide access to a curated bundle atlas (Garyfallidis et al., 2019b; Yeh and Tseng, 2011; Yeh et al., 2018). In addition, it does not require diffeomorphic registration, which is hard for clinical datasets, for example, with tumors. RB is a fast method and can directly connect to BUAN for statistical analysis. Also if the datasets are already in the same space (warped), it can directly take that into account and disable local registration. The method has been applied successfully in thousands of subjects and hundreds of bundles with more than 98% accuracy. In addition, the method is not tied to a species or age group.

### Quality assurance

Multiple methods can be used to estimate quality of outcome including bundle adjacency (BA) (see Section 6). Furthermore, the method does try to extract a bundle similar to what the user has provided. RB is robust to small to medium deformations but the user should make sure that the provided model bundle corresponds to appropriate neuroanatomical definitions. For example, an undersampled model bundle may extract undersampled bundles. A thicker model bundle will most likely generate thicker recognized bundles. Finally, RB was used to evaluate the tractography challenge reported in Maier-Hein et al. (2017) and therefore is highly recommended for evaluation/validation purposes.

### 5.2.5 TractSeg

#### Description

This recently proposed method (Wasserthal et al., 2018) is based on the analysis of voxel-based DW-MRIs to jointly segment 72 different bundles with deep learning techniques. First, as a preprocessing step, a new brain volume is created using the peak values of the orientation distribution function estimated via constrained spherical deconvolution from raw data. Coronal, axial, and sagittal slices of the resulting volume are resampled at a prespecified resolution of $144 \times 144 \times 144$. These are then given as input each to a dedicated two-dimensional (2D) fully convolutional neural network (FCNN) to obtain as output a five-dimensional volume. Similarly, this output is used as input for a second FCNN, which operates on coronal, axial, and sagittal. The output of this second network is then a $144 \times 144 \times 144 \times 72$ volume where the 72 values in each voxel represent the estimated probabilities for each bundle. As an optional postprocessing step, TractSeg provides means to track streamlines in the voxel masks computed from the previous step.

### Advantages/disadvantages

The experiments reported in Wasserthal et al. (2018) show that the quality of bundles segmented by TractSeg outperforms a large number of other automatic segmentation methods when used on research-quality dMRI data. This is the first algorithm using deep learning for the segmentation task. In Bertò et al. (2021), TractSeg is compared to other methods in more challenging contexts, such as probabilistic tracking or clinical-quality data, showing that the quality of segmentation substantially decreases when the shape of bundles is nonsmooth/irregular. TractSeg is fast to run on new data and can be retrained on new expert-made examples. In such a case, the training requires very long computations. Nonetheless, an important property is that TractSeg shows a level of invariance to rotation differences between subjects.

### Quality assurance

The results of segmentation are voxel masks of the bundles and, optionally, streamlines of those bundles can be visualized with standard tools. The quality of segmentations was quantitatively analyzed in Wasserthal et al. (2018) and Bertò et al. (2021) by measuring the degree of overlap with expert-made segmentations. Care needs to be taken with such approaches, as multiple streamlines (often 1000) can belong to a single voxel. Therefore users should employ streamline-based approaches to have a more accurate understanding of the result.

### 5.2.6 DeepWMA

**Description**

DeepWMA (Zhang et al., 2020) is a deep-learning tractography segmentation method that utilized a 2D multichannel feature descriptor (FiberMap). FiberMap encodes spatial coordinates along streamlines in a certain way in order to be robust to streamlines orientation and slight displacement. Training was performed using labeled streamlines (categorized into bundles of interest) as well as an extra category for false positives. The streamlines are converted into their FiberMap representation and used to train a convolution neural network.

**Advantages/disadvantages**

As for now, the code does not seem to be easily available. The method was shown to be robust to a variety of datasets acquisition quality, as well as pathological data, and only requires a volume-based affine registration. Sensitivity and specificity computation performance are impressive.

**Quality assurance**

The results are in the streamlines representation and can be visualized with a variety of tools. However, there are no Q/A tools specifically provided by DeepWMA. A comparison to the atlas used for training could be done to identify outliers.

### 5.2.7 LAP

**Description**

In Sharmin et al. (2018), the authors describe a streamline-based bundle segmentation method built on the concept of the linear assignment problem (LAP). The goal of this segmentation method is to find which streamlines of the tractogram of interest correspond to those of expert-made examples of that bundle from other subjects. The correspondence between example streamlines and target tractogram is computed as a combinatorial optimization problem by minimizing the distances of the example and target streamlines with a one-to-one constraint: in other words, by solving the LAP.

**Advantages/disadvantages**

The experiments in Sharmin et al. (2018) and Bertò et al. (2021) show that LAP produces very high-quality bundles; however, the cost of the computation is very high and prohibitive for very large tractograms of 1 million streamlines or more.

**Quality assurance**

The results of segmentation are the streamlines of the segmented bundle, which can be visualized with standard tools. The quality of segmentations was quantitatively analyzed in Sharmin et al. (2018) and Bertò et al. (2021) by measuring the degree of overlap with expert-made segmentations.

### 5.2.8 Classifyber

**Description**

Classifyber (Bertò et al., 2021) is a streamline-based bundle segmentation method. It consists of a set of simple and fast linear classifiers, one for each bundle of interest, for example, one classifier for the Arcuate Fasciculus, one for the IFOF, etc. Each linear classifier takes as input a single streamline and predicts as output whether such a streamline belongs or not to the bundle of interest. Given a tractogram of a new subject/patient as input, Classifyber loops over the streamlines and predicts each of them. Classifyber is based on creating a vector representation for the input streamline, of approximately 300 values. Such values describe the geometrical properties and connectivity pattern of the streamline with respect to a set of global landmark streamlines, local streamlines, and atlas-based ROIs that characterize the bundle of interest. In other words, the novelty of Classifyber is that it jointly exploits information from geometry, connectivity, and anatomy of the streamlines, previously addressed only individually by the automatic segmentation methods in the literature (Fig. 5).

**Advantages/disadvantages**

The extensive experiments reported in Bertò et al. (2021) show that Classifyber consistently provides very high-quality segmented bundles across different scenarios, such as deterministic tracking, probabilistic tracking, small versus large

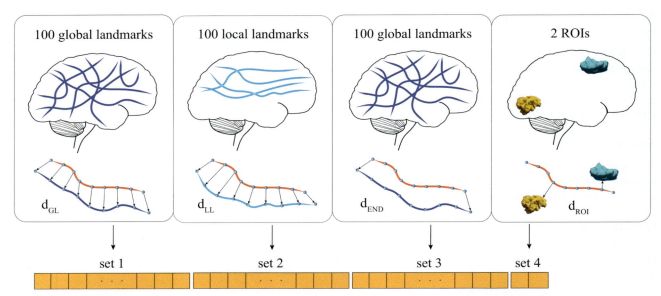

**FIG. 5** Classifyber feature selection. The feature vector used in Classifyber is built with distances that take into account the overall length of the streamlines, the endpoints' distances, and distances from specific connecting ROIs. Notice that both local and global or connectivity information is being deployed.

bundles, and smooth and irregular shaped bundles. Classifyber significantly outperforms all main bundle segmentation methods in all those scenarios on a very large set of bundles. The algorithm is fast in segmenting bundles in new subjects and the linear models can be easily retrained for new kinds of bundles or data-quality using a small amount of expert-made examples. A disadvantage of the method is that for training it requires a selection of local and global prototype streamlines and connecting regions for specific bundles. However, by including this higher level of prior information it can also boost accuracy.

### Quality assurance

The results of segmentation are the streamlines of the segmented bundle, which can be visualized with standard tools such as Horizon, Trackvis, etc. The quality of segmentations was quantitatively analyzed in Bertò et al. (2021) by measuring the degree of overlap with expert-made segmentations and qualitatively by expert neuroanatomists.

## 6 Shape similarities using bundle adjacency and fractal dimensions

In the last section, we discussed how to compare streamlines with each other and reviewed different segmentation approaches. To compare two bundles, we rely on the BA metric (Garyfallidis et al., 2012) that calculates the shape similarity between the same type of bundles across subjects and groups. The higher the BA value, the higher the similarity between the shapes of the two bundles.

BA is calculated between two bundles. BA uses an MDF distance (Garyfallidis et al., 2012; Chandio et al., 2020) to get the distance between two streamlines; 0 BA means no match (or bundles too far) and 1 means a perfect match.

When it comes to bundle comparison, it is likely that more than one metric is required to get the full picture. For example, BA by itself does not quantify overlap. Therefore it is impossible to differentiate a case where bundles perfectly overlap at the core with a large overreach at extremities and a case where bundles are shifted from each other. Dice score can provide the overlap and BA can provide the average distance (overlapping or not), thus helping to disentangle the previous mentioned cases. Another example is when the number of streamlines is different between bundles, this "density" information is lost to the Dice coefficient or BA. This is why using a mixture of Dice coefficient, weighted-Dice coefficient (Cousineau et al., 2017), and BA is useful to interpret exactly how different the bundles are (Rheault et al., 2020a).

## 6.1 The fractal dimension of a bundle mask

The degree of irregularity of the shape of a bundle seems to deeply affect some bundle segmentation methods while others are insensitive, as explained in Bertò et al. (2021). The fractal dimension (FD) is a numerical quantity that characterizes a shape: for simple shapes like a line, a 2D flat square, or a 3D cube, FD is 1, 2, and 3, respectively. For more complex shapes, FD assumes fractional values. For example, a convoluted 3D shape with holes has an FD between 2 and 3. The FD of the voxel mask of the entire white matter has been estimated in Zhang et al. (2006) to be between 2.1 and 2.5 and is used, for example, to characterize patients with multiple sclerosis (Esteban et al., 2007).

Convolutional neural networks (CNNs) are known to be biased toward rounded segmentations, because of the max-pooling operation (Sabour et al., 2017). Based on this fact, Bertò et al. (2021) explained why TractSeg—which is based on CNNs—shows a poor quality of segmentation when segmenting bundles with low FD, which have an irregular shape. Differently, streamline-based segmentation methods seem unaffected by this issue. The FD of a voxel mask can be approximated by means of the *box-counting dimension*.

## 6.2 Streamline-based bundle atlases

In this chapter, we used a publicly available streamline-based bundle atlas to produce some of the figures. Our atlas was a reduced form of the HCP-842 template (Yeh and Tseng, 2011; Yeh et al., 2018). The atlas was curated by removing the bundles that had unrealistic streamlines and then transforming them to MNI 152 space; 30 bundles of interest were selected for this project. The final atlas can be downloaded from DIPY (Garyfallidis et al., 2014). The atlas contains a whole brain tractogram and the corresponding subset of the 30 up to 80 bundles. Similar atlases are currently being made available from multiple labs (Yeh et al., 2018; Zhang et al., 2018; Chenot et al., 2019; Poulin et al., 2022) (https://dmri.mgh.harvard.edu/tract-atlas/; https://zenodo.org/records/4630660#.YJvmwXVKhdU; https://doi.org/10.6084/m9.figshare.12089652.v5).

## 7 The impact of different pipeline choices

Here, we present some important notes for caution. Often users think the problem is with the segmentation but most likely the issue is at a prior step. For example, most segmentation methods will be affected by lack of quality-controlled preprocessing. In addition, researchers should check if any registration is done correctly, or check the images for any obvious pathologies. Now, streamline-based segmentation can be more robust to the latter, but still these steps should be checked in case of concerns. Also, often neuroscientists may wrongly assume that a segmentation method should provide the same results for different types of tractography. However, the different types of tractography (e.g., deterministic, probabilistic, or global) will have a measurable difference in the final output of the segmentation methods. For instance, extracted bundles from deterministic tractography are expected to be smoother than probabilistic. In relation to postprocessing, the audience should also keep in mind that classification/regression techniques, such as TractSeg, DeepWMA, or Classifyber, do learn specific types of data. Therefore the recommendation is to train and test them with the same type of tractography. A good and carefully executed segmentation strategy can easily reduce the number of false streamlines and bundles in tractograms, hence helping any further analysis to focus on important parts of the organ under study.

## 8 Visualizing the virtual dissections

Horizon (Garyfallidis et al., 2019a) is a 3D visualization tool available in DIPY that allows users to switch between data, clusters, and centroids interactively. In addition, it allows segmented bundles to be seen in native, world coordinates and a gallery. In similar fashion, there are multiple visualization tools to quality-assure segmentation results, including Trackvis (Wang et al., 2007), Fibernavigator (Vaillancourt et al., 2010), ABrainVis (Osorio et al., 2021), AFQ Browser (Yeatman et al., 2018), and FiberStars (Franke et al., 2021). Most software allow simultaneous visualization of streamlines and anatomical information, for example, T1 or FA images. Often the issue is not the lack of visualization but more the difficulty of converting the file formats from one software type to another so that they are supported in the visualization tool of interest. Nibabel (Brett et al., 2018) provides a way to alleviate that problem, as it supports the conversion of multiple file formats, including TRK, TCK, and others.

# 9 Summary

The domain of virtual dissection has improved exponentially in the last few years, providing multiple approaches that allow for reliable automatic segmentation of well-known bundles. More work is needed in providing high-quality training data and community-approved definitions of bundles. But even if such definitions take time to resolve, the algorithms today can adapt and generate results for multiple definitions. However, the neuroscientists and medical practitioners need to communicate parameter choices and atlases used in their publications. People should describe what definitions were used, and which atlases or datasets were used during training, validation, testing, or deployment. Researchers should also report any unexpected changes due to hyperparameter tuning. The readers should be aware that today there are multiple open-source implementations available that adhere to excellent engineering principles, and that the communities that built these tools welcome any feedback from the scientists to improve upon future versions of their software. Finally, a well-executed segmentation strategy using the tools described here can easily reduce the number of false streamlines and bundles in tractograms and help any forthcoming analyses.

# References

Alexander, A.L., Lee, J.E., Lazar, M., Field, A.S., 2007. Diffusion tensor imaging of the brain. Neurotherapeutics 4 (3), 316–329.

Basser, P.J., Mattiello, J., LeBihan, D., 1994. MR diffusion tensor spectroscopy and imaging. Biophys. J. 66 (1), 259–267.

Bertò, G., Bullock, D., Astolfi, P., Hayashi, S., Zigiotto, L., Annicchiarico, L., Corsini, F., De Benedictis, A., Sarubbo, S., Pestilli, F., Avesani, P., Olivetti, E., 2021. Classifyber, a robust streamline-based linear classifier for white matter bundle segmentation. NeuroImage 224, 117402. https://doi.org/10.1016/j.neuroimage.2020.117402.

Brett, M., Hanke, M., Markiewicz, C., Côté, M.A., McCarthy, P., Ghosh, S., Wassermann, D., et al., 2018. nipy/nibabel: 2.3. 0. 1287921. https://doi.org/10.5281/zenodo.

Catani, M., De Schotten, M.T., 2008. A diffusion tensor imaging tractography atlas for virtual in vivo dissections. Cortex 44 (8), 1105–1132.

Catani, M., Howard, R.J., Pajevic, S., Jones, D.K., 2002. Virtual in vivo interactive dissection of white matter fasciculi in the human brain. NeuroImage 17 (1), 77–94.

Chamberland, M., Whittingstall, K., Fortin, D., Mathieu, D., Descoteaux, M., 2014. Real-time multi-peak tractography for instantaneous connectivity display. Front. Neuroinform. 8, 59.

Chandio, B.Q., Risacher, S.L., Pestilli, F., Bullock, D., Yeh, F.C., Koudoro, S., Rokem, A., Harezlak, J., Garyfallidis, E., 2020. Bundle analytics, a computational framework for investigating the shapes and profiles of brain pathways across populations. Sci. Rep. 10 (1), 1–18.

Charon, N., Trouvé, A., 2013. The varifold representation of nonoriented shapes for diffeomorphic registration. SIAM J. Imaging Sci. 6 (4), 2547–2580. https://doi.org/10.1137/130918885.

Chenot, Q., Tzourio-Mazoyer, N., Rheault, F., et al., 2019. A population-based atlas of the human pyramidal tract in 410 healthy participants. Brain Struct. Funct. 224 (2), 599–612. https://link.springer.com/article/10.1007/s00429-018-1798-7.

Corouge, I., Gouttard, S., Gerig, G., 2004. Towards a shape model of white matter fiber bundles using diffusion tensor MRI. In: IEEE International Symposium on Biomedical Imaging: Nano to Macro, 2004, vol. 1. IEEE, pp. 344–347, https://doi.org/10.1109/isbi.2004.1398545.

Cousineau, M., Jodoin, P.M., Garyfallidis, E., Côté, M.A., Morency, F.C., Rozanski, V., Grand'Maison, M., Bedell, B.J., Descoteaux, M., 2017. A test-retest study on Parkinson's PPMI dataset yields statistically significant white matter fascicles. NeuroImage 16, 222–233.

Esteban, F.J., Sepulcre, J., de Mendizábal, N.V., Goñi, J., Navas, J., de Miras, J.R., Bejarano, B., Masdeu, J.C., Villoslada, P., 2007. Fractal dimension and white matter changes in multiple sclerosis. NeuroImage 36 (3), 543–549. https://doi.org/10.1016/j.neuroimage.2007.03.057.

Farquharson, S., Tournier, J.D., Calamante, F., Fabinyi, G., Schneider-Kolsky, M., Jackson, G.D., Connelly, A., 2013. White matter fiber tractography: why we need to move beyond DTI. J. Neurosurg. 118 (6), 1367–1377.

Fowlkes, C., Belongie, S., Chung, F., Malik, J., 2004. Spectral grouping using the Nystrom method. IEEE Trans. Pattern Anal. Mach. Intell. 26 (2), 214–225.

Franke, L., Weidele, D.K.I., Zhang, F., Cetin-Karayumak, S., Pieper, S., O'Donnell, L.J., Rathi, Y., Haehn, D., 2021. FiberStars: visual comparison of diffusion tractography data between multiple subjects. In: 2021 IEEE 14th Pacific Visualization Symposium (PacificVis), pp. 116–125.

Garyfallidis, E., Chandio, B., 2021. Atlas of 30 human brain bundles in MNI space., https://doi.org/10.6084/m9.figshare.12089652.v5.

Garyfallidis, E., Brett, M., Correia, M.M., Williams, G.B., Nimmo-Smith, I., 2012. QuickBundles, a method for tractography simplification. Front. Neurosci. 6, 175. https://doi.org/10.3389/fnins.2012.00175.

Garyfallidis, E., Brett, M., Amirbekian, B., Rokem, A., van der Walt, S., Descoteaux, M., Nimmo-Smith, I., Contributors, Dipy, 2014. Dipy, a library for the analysis of diffusion MRI data. Front. Neuroinform. 8, 8.

Garyfallidis, E., Ocegueda, O., Wassermann, D., Descoteaux, M., 2015. Robust and efficient linear registration of white-matter fascicles in the space of streamlines. NeuroImage 117, 124–140.

Garyfallidis, E., Côté, M.A., Rheault, F., Descoteaux, M., 2016. QuickBundlesX: sequential clustering of millions of streamlines in multiple levels of detail at record execution time. In: International Society for Magnetic Resonance Imaging (ISMRM), p. 1.

Garyfallidis, E., Côté, M.A., Rheault, F., Sidhu, J., Hau, J., Petit, L., Fortin, D., Cunanne, S., Descoteaux, M., 2017. Recognition of white matter bundles using local and global streamline-based registration and clustering. NeuroImage 170, 283–295.

Garyfallidis, E., Côté, M.A., Chandio, B.Q., Fadnavis, S., Guaje, J., Aggarwal, R., St-Onge, E., Juneja, K.S., Koudoro, S., Reagan, D., 2019a. DIPY horizon: fast, modular, unified and adaptive visualization. In: International Society of Magnetic Resonance in Medicine (ISMRM), p. 1.

Garyfallidis, E., Nahla, E., Wu, Y.C., 2019b. Superresolved HYDI dataset., https://doi.org/10.6084/m9.figshare.10266194.v1.

Garyfallidis, E., Fadnavis, S., Park, J.S., Chandio, B.Q., Guaje, J., Koudoro, S., Anousheh, N., 2021. Theta-fast and robust clustering via a distance parameter. arXiv preprint arXiv:2102.07028.

Geschwind, N., 1970. The organization of language and the brain. Science 170 (3961), 940–944.

Gong, G., He, Y., Concha, L., Lebel, C., Gross, D.W., Evans, A.C., Beaulieu, C., 2008. Mapping anatomical connectivity patterns of human cerebral cortex using in vivo diffusion tensor imaging tractography. Cereb. Cortex 19 (3), 524–536.

Gori, P., Colliot, O., Marrakchi-Kacem, L., Worbe, Y., De Vico Fallani, F., Chavez, M., Poupon, C., Hartmann, A., Ayache, N., Durrleman, S., 2016. Parsimonious approximation of streamline trajectories in white matter fiber bundles. IEEE Trans. Med. Imaging. http://view.ncbi.nlm.nih.gov/pubmed/27416589.

Guevara, P., Poupon, C., Rivière, D., Cointepas, Y., Descoteaux, M., Thirion, B., Mangin, J.F., 2010. Robust clustering of massive tractography datasets. NeuroImage 54 (3), 1975–1993. https://doi.org/10.1016/j.neuroimage.2010.10.028.

Guevara, P., Duclap, D., Poupon, C., Marrakchi-Kacem, L., Fillard, P., Le Bihan, D., Leboyer, M., Houenou, J., Mangin, J.F., 2012. Automatic fiber bundle segmentation in massive tractography datasets using a multi-subject bundle atlas. NeuroImage 61 (4), 1083–1099.

Guevara, M., Román, C., Houenou, J., Duclap, D., Poupon, C., Mangin, J.F., Guevara, P., 2017. Reproducibility of superficial white matter tracts using diffusion-weighted imaging tractography. NeuroImage 147, 703–725.

Jonasson, L., Bresson, X., Hagmann, P., Cuisenaire, O., Meuli, R., Thiran, J.P., 2005. White matter fiber tract segmentation in DT-MRI using geometric flows. Med. Image Anal. 9 (3), 223–236.

Labra, N., Guevara, P., Duclap, D., Houenou, J., Poupon, C., Mangin, J.F., Figueroa, M., 2017. Fast automatic segmentation of white matter streamlines based on a multi-subject bundle atlas. Neuroinformatics 15, 71–78.

Lawes, I.N.C., Barrick, T.R., Murugam, V., Spierings, N., Evans, D.R., Song, M., Clark, C.A., 2008. Atlas-based segmentation of white matter tracts of the human brain using diffusion tensor tractography and comparison with classical dissection. NeuroImage 39 (1), 62–79.

Le Bihan, D., Mangin, J.F., Poupon, C., Clark, C.A., Pappata, S., Molko, N., Chabriat, H., 2001. Diffusion tensor imaging: concepts and applications. J. Magn. Reson. Imaging 13 (4), 534–546.

Leuret, F., 1857. Anatomie Comparée du Système Nerveux: Considéré Dans ses Rapports Avec L'intelligence. vol. 2 J.-B. Baillière et fils.

Maffei, C., Lee, C., Planich, M., Ramprasad, M., Ravi, N., Trainor, D., Urban, Z., Kim, M., Jones, R.J., Henin, A., Hofmann, S.G., Pizzagalli, D.A., Auerbach, R.P., Gabrieli, J.D.E., Whitfield-Gabrieli, S., Greve, D.N., Haber, S.N., Yendiki, A., 2021. Using diffusion MRI data acquired with ultra-high gradient strength to improve tractography in routine-quality data. NeuroImage 245, 118706. https://dmri.mgh.harvard.edu/tract-atlas/.

Maier-Hein, K.H., Neher, P.F., Houde, J.C., Côté, M.A., Garyfallidis, E., Zhong, J., Chamberland, M., Yeh, F.C., Lin, Y.C., Ji, Q., et al., 2017. The challenge of mapping the human connectome based on diffusion tractography. Nat. Commun. 8 (1), 1–13.

Maier-Hein, K.H., Neher, P.F., Houde, J.C., Cote, M.A., Garyfallidis, E., Zhong, J., Chamberland, M., Yeh, F.C., Lin, Y.C., Ji, Q., Reddick, W.E., Glass, J.-O., Chen, D.Q., Feng, Y., Gao, C., Wu, Y., Ma, J., Renjie, H., Li, Q., Westin, C.F., Deslauriers-Gauthier, S., Gonzalez, J.O.O., Paquette, M., St-Jean, S., Girard, G., Rheault, F., Sidhu, J., Tax, C.M.W., Guo, F., Mesri, H.Y., David, S., Froeling, M., Heemskerk, A.M., Leemans, A., Bore, A., Pinsard, B., Bedetti, C., Desrosiers, M., Brambati, S., Doyon, J., Sarica, A., Vasta, R., Cerasa, A., Quattrone, A., Yeatman, J., Khan, A.R., Hodges, W., Alexander, S., Romascano, D., Barakovic, M., Auria, A., Esteban, O., Lemkaddem, A., Thiran, J.P., Cetingul, H.E., Odry, B.L., Mailhe, B., Nadar, M.-S., Pizzagalli, F., Prasad, G., Villalon-Reina, J.E., Galvis, J., Thompson, P.M., Requejo, F.D.S., Laguna, P.L., Lacerda, L.M., Barrett, R., Dell'Acqua, F., Catani, M., Petit, L., Caruyer, E., Daducci, A., Dyrby, T.B., Holland-Letz, T., Hilgetag, C.C., Stieltjes, B., Descoteaux, M., 2017. The challenge of mapping the human connectome based on diffusion tractography. Nat. Commun. 8, 1349.

Mori, S., Van Zijl, P.C.M., 2002. Fiber tracking: principles and strategies—a technical review. NMR Biomed. 15 (7–8), 468–480.

O'Donnell, L.J., Westin, C.F., 2007. Automatic tractography segmentation using a high-dimensional white matter tracts. IEEE Trans. Med. Imaging 26 (11), 1562–1575.

Olivetti, E., Nguyen, T.B., Garyfallidis, E., 2012. The approximation of the dissimilarity projection. In: 2012 International Workshop on Pattern Recognition in NeuroImaging (PRNI), pp. 85–88.

Olivetti, E., Bertò, G., Gori, P., Sharmin, N., Avesani, P., 2017. Comparison of distances for supervised segmentation of white matter tractography. In: 2017 International Workshop on Pattern Recognition in Neuroimaging (PRNI), IEEE, pp. 1–4, https://doi.org/10.1109/prni.2017.7981502.

Osorio, I., Guevara, M., Bonometti, D., Carrasco, D., Descoteaux, M., Poupon, C., Mangin, J.F., Hernández, C., Guevara, P., 2021. ABrainVis: an android brain image visualization tool. Biomed. Eng. Online 20 (1), 72. https://doi.org/10.1186/s12938-021-00909-0.

Pekalska, E., Duin, R., 2005. The Dissimilarity Representation for Pattern Recognition - Foundations and Applications. World Scientific Publishing, Singapore.

Poulin, P., Theaud, G., Rheault, F., St-Onge, E., Bore, A., Renauld, E., de Beaumont, L., Guay, S., Jodoin, P.M., Descoteaux, M., 2022. TractoInferno—a large-scale, open-source, multi-site database for machine learning dMRI tractography. Sci. Data 9 (1), 725. https://www.nature.com/articles/s41597-022-01833-1.

Rheault, F., De Benedictis, A., Daducci, A., Maffei, C., Tax, C.M.W., Romascano, D., Caverzasi, E., Morency, F.C., Corrivetti, F., Pestilli, F., et al., 2020a. Tractostorm: the what, why, and how of tractography dissection reproducibility. Hum. Brain Mapp. 41 (7), 1859–1874.

Rheault, F., Poulin, P., Caron, A.V., St-Onge, E., Descoteaux, M., 2020b. Common misconceptions, hidden biases and modern challenges of dMRI tractography. J. Neural Eng. 17 (1), 011001.

Sabour, S., Frosst, N., Hinton, G.E., 2017. Dynamic routing between capsules. In: NIPS'17. Proceedings of the 31st International Conference on Neural Information Processing Systems, Curran Associates Inc., pp. 3859–3869. http://dl.acm.org/citation.cfm?id=3294996.3295142.

Schmahmann, J.D., Schmahmann, J., Pandya, D., 2009. Fiber Pathways of the Brain. Oxford University Press.

Sharmin, N., Olivetti, E., Avesani, P., 2018. White matter tract segmentation as multiple linear assignment problems. Front. Neurosci. 11. https://doi.org/10.3389/fnins.2017.00754.

Siless, V., Medina, S., Varoquaux, G., Thirion, B., 2013. A comparison of metrics and algorithms for fiber clustering. In: 2013 International Workshop on Pattern Recognition in Neuroimaging, IEEE, pp. 190–193, https://doi.org/10.1109/prni.2013.56.

Smith, R.E., Tournier, J.D., Calamante, F., Connelly, A., 2015. SIFT2: enabling dense quantitative assessment of brain white matter connectivity using streamlines tractography. NeuroImage 119, 338–351.

Vaillancourt, O., Boré, A., Girard, G., Descoteaux, M., 2010. A fiber navigator for neurosurgical planning (neuroplanningnavigator). In: IEEE Visualization, vol. 231.

Vavassori, L., Sarubbo, S., Petit, L., 2021. Hodology of the superior longitudinal system of the human brain: a historical perspective, the current controversies, and a proposal. Brain Struct. Funct. 226, 1363–1384.

Vázquez, A., López-López, N., Sánchez, A., Houenou, J., Poupon, C., Mangin, J.F., Hernández, C., Guevara, P., 2020. FFClust: fast fiber clustering for large tractography datasets for a detailed study of brain connectivity. NeuroImage 220, 117070.

Vu, A.T., Auerbach, E., Lenglet, C., Moeller, S., Sotiropoulos, S.N., Jbabdi, S., Andersson, J., Yacoub, E., Ugurbil, K., 2015. High resolution whole brain diffusion imaging at 7 T for the human connectome project. NeuroImage 122, 318–331.

Wang, R., Benner, T., Sorensen, A.G., Wedeen, V.J., 2007. Diffusion toolkit: a software package for diffusion imaging data processing and tractography. In: Proceedings of the International Society for Magnetic Resonance in Medicine, vol. 15, p. 3720.

Wassermann, D., Makris, N., Rathi, Y., Shenton, M., Kikinis, R., Kubicki, M., Westin, C.F.F., 2016. The white matter query language: a novel approach for describing human white matter anatomy. Brain Struct. Funct. 221, 4705–4721. http://view.ncbi.nlm.nih.gov/pubmed/26754839.

Wasserthal, J., Neher, P., Maier-Hein, K.H., 2018. Tractseg-fast and accurate white matter tract segmentation. NeuroImage 183, 239–253.

Yeatman, J.D., Richie-Halford, A., Smith, J.K., Keshavan, A., Rokem, A., 2018. A browser-based tool for visualization and analysis of diffusion MRI data. Nat. Commun. 9 (1), 940. https://doi.org/10.1038/s41467-018-03297-7.

Yeh, F.C., Tseng, W.Y.I., 2011. NTU-90: a high angular resolution brain atlas constructed by q-space diffeomorphic reconstruction. NeuroImage 58 (1), 91–99.

Yeh, F.C., Panesar, S., Fernandes, D., Meola, A., Yoshino, M., Fernandez-Miranda, J.C., Vettel, J.M., Verstynen, T., 2018. Population-averaged atlas of the macroscale human structural connectome and its network topology. NeuroImage 178, 57–68. https://pubmed.ncbi.nlm.nih.gov/29758339/.

Zhang, L., Liu, J.Z., Dean, D., Sahgal, V., Yue, G.H., 2006. A three-dimensional fractal analysis method for quantifying white matter structure in human brain. J. Neurosci. Methods 150 (2), 242–253. https://doi.org/10.1016/j.jneumeth.2005.06.021.

Zhang, S., Correia, S., Laidlaw, D.H., 2008. Identifying white-matter fiber bundles in DTI data using an automated proximity-based fiber-clustering method. IEEE Trans. Vis. Comput. Graph. 14 (5), 1044–1053. https://doi.org/10.1109/tvcg.2008.52.

Zhang, F., Wu, Y., Norton, I., Rigolo, L., Rathi, Y., Makris, N., O'Donnell, L.J., 2018. An anatomically curated fiber clustering white matter atlas for consistent white matter tract parcellation across the lifespan. NeuroImage 179, 429–447. https://www.sciencedirect.com/science/article/abs/pii/S1053811918305342?via%3Dihub.

Zhang, F., Karayumak, S.C., Hoffmann, N., Rathi, Y., Golby, A., O'Donnell, L., 2020. Deep white matter analysis (DeepWMA): fast and consistent tractography segmentation. Med. Image Anal. 65, 101761.

# Chapter 23

# Methods and statistics for diffusion MRI tractometry

Maxime Chamberland[a], Samuel St-Jean[b,c], Derek K. Jones[d], Maxime Descoteaux[e], and Alexander Leemans[f]

[a]*Eindhoven University of Technology, Eindhoven, The Netherlands,* [b]*Department of Biomedical Engineering, Faculty of Medicine and Dentistry, University of Alberta, Edmonton, AB, Canada,* [c]*Clinical Sciences Lund, Lund University, Lund, Sweden,* [d]*Cardiff University Brain Research Imaging Centre (CUBRIC), Cardiff University, Cardiff, United Kingdom,* [e]*Sherbrooke Connectivity Imaging Laboratory (SCIL), Department of Computer Science, University of Sherbrooke, Sherbrooke, QC, Canada,* [f]*PROVIDI lab, Image Sciences Institute, University Medical Center Utrecht, Utrecht, The Netherlands*

## 1  Introduction

As seen in this book, one can derive microstructural properties of the white matter and visualize its structural architecture using diffusion magnetic resonance imaging (dMRI). For years, however, the analysis of diffusion data was largely divided into these two categories: (i) voxel-wise mapping of microstructural properties and (ii) tractography reconstructions of pathways. As in "photography," in which one makes a qualitative *picture* composed of *relative* intensities, so tractography makes a qualitative picture of the white matter pathways. In contrast, *photometry* measures and therefore *quantifies* the intensity of light. This linkage between qualitative pictures and quantification led us naturally to propose the term *tractometry*, which is not just making a picture of the course of a bundle, but quantifying important aspects of the bundle.

The term was not officially coined until 2011, in an abstract presented at the International Society for Magnetic Resonance in Imaging (ISMRM) by Bells et al. (2011), where the emphasis was on quantifying multiple attributes of microstructure (e.g., myelination, axon density) along the same bundle. The basic technology underlying the approach, however, was proposed at least 6 years earlier, for example, profiling parameters along a specific tractography-reconstructed streamline ( Jones et al., 2005b), and computing a representative summary statistic (e.g., mean or median) along the whole bundle ( Jones et al., 2006; de Gervai et al., 2014; Corouge et al., 2006; Yeatman et al., 2012, 2018; Colby et al., 2012; Cousineau et al., 2017).

Along-streamline profiling based on dMRI tractometry is a framework that maps summary measures of the diffusion data (e.g., fractional anisotropy) at multiple points along a given pathway, resulting in a so-called *bundle-profile*. Along-streamline profiling has been applied previously to investigate neurodevelopment (Yeatman et al., 2014; Travis et al., 2015; Groeschel et al., 2014) and aging (Yeatman et al., 2014), as well as various brain conditions such as multiple sclerosis (Mezer et al., 2013; Dayan et al., 2015; Blecher et al., 2019), Parkinson's disease (Cousineau et al., 2017; Chandio et al., 2020), amyotrophic lateral sclerosis (ALS) (Sarica et al., 2017), Alzheimer's disease (Dou et al., 2020), schizophrenia (Sun et al., 2015; Xiao et al., 2018), and traumatic brain injury (Main et al., 2017; Yeh et al., 2017).

This chapter introduces the main concepts relevant to tractometry and provides an overview of the computational steps involved for along-streamline profiling of microstructural MRI measures. This includes the segmentation of white matter bundles, the extraction of a core representative streamline, the mapping of microstructural tissue properties along the core, and the statistical analysis of bundle-profiles. Finally, the advantages of tractometry are discussed with respect to more traditional frameworks like voxel-based morphometry (VBM; Ashburner and Friston, 2000) and tract-based spatial statistics (TBSS; Smith et al., 2006).

## 2  Tractometry: Along-streamline analysis

Tractometry involves many steps, which are summarized in Fig. 1. This typically includes, but is not limited to, extracting bundles from a whole-brain tractogram (Chapters 21 and 22) (A), finding a descriptive center pathway for each bundle (B), dividing the bundles into segments (C), and mapping quantitative features along bundles of interest (D) to create a so-called *bundle-profile* (Yeatman et al., 2012). As is usually the case with most processing pipelines, this multistage process

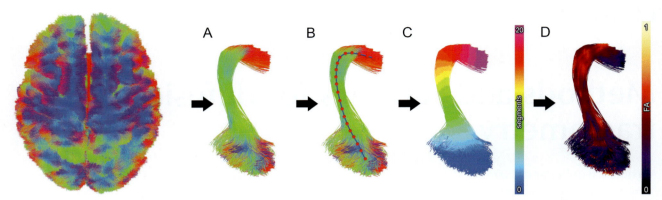

**FIG. 1** Overview of the tractometry pipeline. (A) A bundle is first extracted from a whole-brain tractogram. (B) The core representative centerline is then derived. (C) Each vertex is then assigned to the closest core centerline point. (D) Measures are sampled at each vertex location and averaged within each segment.

increases the chance of errors to occur at various stages of the tractometry pipeline. Fortunately, most of these steps are currently thoroughly implemented and fully automated in various open-source software tools made freely available to the community (Yendiki et al., 2011; Leemans et al., 2009; Yeatman et al., 2012; Garyfallidis et al., 2014; Cousineau et al., 2017; Wasserthal et al., 2018, 2020; Chandio et al., 2020; Maffei et al., 2021). Yet, it is important that the future users of these tools are aware of the many underlying assumptions behind each of these steps, as this may have a direct impact on interpretation of their results. In this section, we will go through the most basic steps required to perform tractometry. From this point on, we also assume that the reader is familiar with data preprocessing procedures to remove artifacts (Tax et al., 2021) (Chapter 9), as well as with the various tractography algorithms (Zhang et al., 2022) (Part III).

## 2.1 Bundle segmentation

As seen in Chapter 22, the main objective of bundle segmentation is to classify or assign tractography-derived streamlines to anatomically defined white matter tracts. A streamline is traditionally represented by a set of points $s = \{p_1, p_2, ..., p_n\}$, $p_i \in \mathbb{R}^3$. Similarly, a bundle represents a combination of multiple streamlines given by $b = \{s_1, s_2, ..., s_m\}$. Bundles are the bread and butter of tractometry and various approaches have been proposed to segment the white matter. For tractometry, the use of automated approaches gives the added benefit of reducing the number of false-positive streamlines, which can introduce variability between profiles. Indeed, automated bundle segmentation algorithms implicitly perform a preliminary removal of outliers. This is an important step, since working with *unpruned* bundles may bias the results down the road. Although tractography reconstructions have become more reproducible over the years, it is important to keep in mind that the technique still faces significant challenges (Maier-Hein et al., 2017; Jones et al., 2013; Schilling et al., 2019, 2021). Therefore the task of bundle segmentation remains an open challenge to the community, both from an anatomical and computational point of view (Chapter 22).

## 2.2 Streamline ordering

At this stage, we assume that we have now obtained a set of delineated tracts for each subject within our cohort. By default, streamlines generated from tracking algorithms do not have any directional information encoded in them. For example, for a given bundle $b_i$ connecting regions A and B of the brain, we have no way of determining if all streamlines $s$ forming the bundle were initiated from region A, region B, or even somewhere else in the brain. All we know is that they connect A and B. This is due to the fact that information about the seeding technique that was employed to generate $b_i$ is discarded at this stage. Indeed, seeds could have been placed randomly inside a white matter mask, a set of ROIs, or at the interface between white and gray matters. Therefore one has to reorder streamlines within a bundle to ensure that their starting and ending points are coherent within that same bundle. Metrics like the mean distance flip (O'Donnell and Westin, 2007; O'Donnell et al., 2009; Garyfallidis et al., 2012; Leemans et al., 2006) typically offer an easy solution to that problem. Assuming that endpoints are now coherent *within* a bundle, it is also crucial that they coherently match *across* subjects. This is to ensure that one is comparing the same anatomical location between subjects. Simply put, imagine a scenario where half of the subjects in a group study have their streamlines running from left to right in the corpus callosum, while the rest of them have their streamlines encoded from right to left. This simple incoherence could entirely mask out the effect of interest one is looking for. Therefore one needs to reorder global bundle orientations based on a fixed reference axis across subjects. For this, the recommendation is to simply employ the same convention across all subjects (e.g., association pathways could always be coded anteroposteriorly, superoinferiorly for projection pathways, and from left to right for commissural pathways). However, do keep in mind that the main orientation of some white matter bundles may still be challenging to define; for instance, association bundles like the arcuate fasciculus and uncinate fasciculus (UF) have their starting and ending points running along the Z-axis due to their curved aspect. Another example is the cingulum bundle, which may not entirely loop down the parahypoccampal gyrus in all subjects, hence the need for quality assurance beforehand.

## 2.3 Core streamline definition

At the core of along-streamline profiling lies the concept of a *representative streamline* (O'Donnell et al., 2009), which is used to project measures along the course of a given bundle. This is typically done by resampling all streamlines forming the bundle to $n$ points and by averaging their spatial coordinates in a point-wise fashion (O'Donnell et al., 2009; Colby et al., 2012; Yeatman et al., 2012). Typically, $n$ should be fixed for all subjects to ensure correspondence. A reasonable value for $n$ is to divide the average streamline length of the fiber bundle of interest by the length scale of the voxel size. For example, if the voxel size is $2 \times 2 \times 2$ mm$^3$ and the average length of a given bundle equals 70 mm, then $n = 70/2 = 35$ points would generally suffice. Longer bundles will, of course, necessitate more points. However, it is important to keep in mind that sampling more points than the length scale of the voxel size will not add any new information. Conversely, sampling fewer points along the streamline will result in averaging microstructural values over larger segments.

This resampling technique will produce a reasonable estimate of the average trajectory when there is very little branching and dispersion between the streamlines forming the pathway. However, streamlines within a given bundle can vary both in length and orientation, making it inappropriate to directly average their coordinates (O'Donnell et al., 2009; Chamberland et al., 2018). Indeed, if the underlying streamlines are even slightly dispersed from each other, the resulting representative streamline obtained by traditional averaging of the coordinates can misrepresent the overall shape of the bundle (Fig. 2). Incorrect assignment of vertices to their closest core point can also occur. In extreme scenarios, the core streamline may even extend beyond the shape of the bundle. Not only does this representation become anatomically implausible, but also it can directly hamper further steps down the tractometry pipeline, that is, when averaging measures along different sections of the pathway. A common solution to overcome this problem is to perform tractometry by excluding data from the extremities, which tend to include fanning (Yeatman et al., 2012; Glozman et al., 2018; St-Jean et al., 2019). This approach greatly helps to (1) quickly obtain a representative streamline and (2) mitigate variability between subjects since bundles are essentially reduced to a simpler representation. However, discarding data from both extremities of a bundle inherently limits the benefits conferred by state-of-the-art tractography techniques that can recover fanning and branching portions of white matter bundles, as well as the study of cortical endings. In some cases, pathways

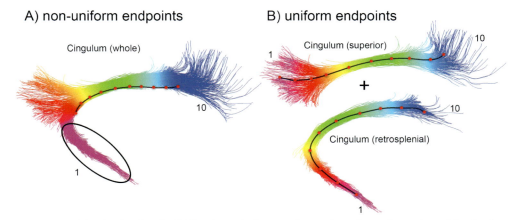

**FIG. 2** Centerline extraction of the cingulum bundle. (A) Fanning endpoints near the anterior and posterior portions of the cingulum create ambiguous core definition and point assignment (*circled area*). This is caused by the incorrect averaging of space- and length-varying streamlines. (B) Uniform endpoints and anatomically correct core streamlines are achieved by separating the branching portions of the bundle into two separate bundles (e.g., superior and retrosplenial cingulum).

**442 PART | IV** From streamlines to tracts

may be better represented using streamsurfaces rather than core streamlines (Vilanova et al., 2004; Zhang et al., 2003; Kindlmann et al., 2007; Yushkevich et al., 2009; Schultz et al., 2009; Qiu et al., 2010; Chen et al., 2016; Tax et al., 2016).

## 2.4 Measure assignment

The next step is to assign vertex values from each individual streamline to the previously defined core streamline. A simple approach is to evaluate the measure of interest (e.g., fractional anisotropy) at $n$ equidistant points along each streamline. Those values can then be averaged per point $n$. However, this can lead to inconsistencies due to varying streamline lengths within a bundle. A more robust approach is to rely on a distance metric (e.g., by using the Euclidean or geodesic distance; Cousineau et al., 2017; Chamberland et al., 2018; Chandio et al., 2020; Wasserthal et al., 2020) to assign each vertex to a point along the core, creating sharper transitions between segments (Fig. 1D). This approach also allows users to weight the measure of interest, based on how far a point is from the core. For example, it could be advantageous to add more weight to points that are closer to the core than to those that are further away from it, to mitigate the effect of partial volume (Vos et al., 2011). Furthermore, a postprocessing step can be applied to the resulting bundle-profiles, to ensure better alignment between subjects. This can be done by aligning bundle-profiles (St-Jean et al., 2019; Glozman et al., 2018; Benou et al., 2019) directly. Note that this alignment can also be performed a priori, by directly aligning tracts of all subjects (Garyfallidis et al., 2015; Parker et al., 2016; Leemans et al., 2006).

# 3 Statistical analysis

## 3.1 Choice of space for along-streamline analysis

The first step is to choose a space to analyze the along-streamline data of the subjects. Choosing the correct subspace allows one to extract only information that is necessary for the analysis, reducing confounding effects at the same time by excluding redundant information. In this section, we show some examples of subspaces that are commonly used to perform along-streamline analyses.

### 3.1.1 Bundle average

The simplest way to analyze the data consists of averaging all the values collected along bundles for each subject separately (Bells et al., 2011; De Santis et al., 2014). Doing so provides higher statistical power since multiple points are collapsed to a single value, but at the expense of spatial specificity since the location along the bundle is discarded during the averaging process. This approach is sensible only when averaging and comparing the same measure in all subjects, and as such this averaging step needs to be repeated for every measure of interest.

### 3.1.2 Bundle-profiles

There is a middle ground between keeping every point created from the tractography and averaging tracts to a single value: that is keeping only a few points capturing the key features of a bundle segment (O'Donnell et al., 2009). This type of analysis study refers to what was previously described in the aforementioned section by deriving streamline-profiles for each subject (Jones et al., 2005b; Colby et al., 2012; Yeatman et al., 2012). At this step, we assume that the extracted measures are representative of the original bundle segment as previously discussed in Section 2.4. Additional correction to ensure optimal correspondence between subjects is available by realigning the measures after extraction of the streamline profiles (St-Jean et al., 2019). Another available option is to perform functional streamline-profiling analysis by mapping the discretized points into continuous functions and to study these functions instead. In the next section, we will go over methods for the statistical analysis of streamline-profiles.

## 3.2 Statistical tests for along-streamline analysis

As we saw previously, there are multiple ways to conduct an along-streamline analysis and, unsurprisingly, multiple statistical tests are available to draw conclusions about the studied dataset. Hypothesis testing assesses the plausibility of two opposing statements given representative samples, usually denoted as $H_0$ (the null hypothesis) and $H_1$ (the alternative hypothesis), inferring properties at the population level. Each test also has various assumptions (e.g., equality of variance

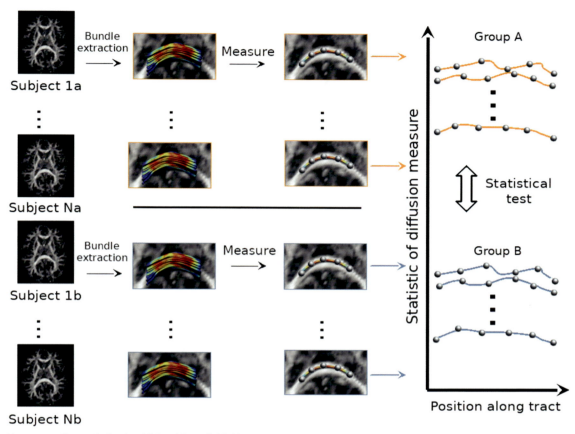

**FIG. 3** Along-streamline analysis of multiple subjects divided into two groups. A bundle of interest is first virtually dissected. A measure to study (e.g., FA) is then chosen and extracted at keypoints along the bundle for each subject. A statistical test is then conducted on each point along the bundle with a statistic representing each group, such as the mean or the median of every subject.

between groups, assumption of normally distributed data) that needs to be verified *before* applying the chosen statistical test. These topics are briefly discussed in the following paragraphs and illustrated in Fig. 3.

### 3.2.1 Hypothesis testing in the context of along-streamline analysis

Generally speaking, two hypotheses are stated when looking for statistical difference along bundles between groups. The $H_0$ hypothesis states that the test statistic (e.g., the mean, the median) is the same in each group, while $H_1$ states that the statistic in at least one of those groups is different from the rest. The probability of choosing to reject $H_0$ or failing to reject $H_0$ (i.e., there is not enough evidence statistically to dismiss $H_0$) is controlled by balancing two types of error. We distinguish type I errors (often referred to as false positives, i.e., claiming that there is an effect of interest while there was none by wrongly rejecting $H_0$), from type II errors, or false negatives, dismissing an existing effect mistakenly by failing to reject $H_0$. Type I errors are controlled by increasing the confidence level (usually denoted $\alpha$) while type II errors (usually denoted $\beta$) depend on multiple factors. One way to reduce $\beta$ is by increasing the power, which is given by $(1 - \beta)$, such as by increasing the sample size. The confidence level $\alpha$ is chosen by the researcher based on the research question and the quality of the available data, with conventional ad hoc values being $\alpha = 0.05$ and $\beta = 0.20$ (Cohen, 1992). Note that these values are generic guidelines and can be adjusted depending on the research question and the quantity and quality of available datasets. Controlling the acceptable trade-off between the two types of errors is therefore done by carefully balancing $\alpha$, $\beta$ and the sample size according to the difficulty of collecting additional data, but also depends on the (unknown) effect of size of the study. For example, lower values of $\beta$ come at the cost of higher values of $\alpha$, but a higher false negative rate might be desirable for detecting a genuine group difference in an along-streamline analysis, as it also provides a lower rate of false negatives. The sample size can then be chosen to strike the desirable balance between $\alpha$ and $\beta$.

**444** **PART | IV** From streamlines to tracts

### 3.2.2 Statistical tests for identifying along-streamline differences between groups

Several tests are available to identify samples belonging to different populations and fall into two broad categories: parametric and nonparametric tests. Parametric tests assume a known probability distribution for the population of the data samples and compare representative values of their tendency, such as the mean in a normal distribution. Nonparametric tests instead rely on properties that are independent of the distribution, without explicitly specifying the distribution, such as the median. The trade-off for using nonparametric tests is the resulting higher $P$-value, but nonparametric tests usually require fewer assumptions than parametric tests.

Among the *parametric tests* we find Student's $t$-test (assuming equal variance), the Welch's $t$-test (assuming unequal variance), the one-way analysis of variance (ANOVA, assuming errors are i.i.d.), with the goal of determining if the means of the populations are significantly different given representative samples by assuming that the mean is a good representation of the central tendency of the population itself.

For the *nonparametric tests*, a few common examples are the Mann-Whitney $U$ test, the Kolmogorov-Smirnov test, or the Kruskal-Wallis test. There is no direct equivalence between parametric and nonparametric tests since different tests use different statistics to draw conclusions. The hypotheses and the statistic used by the chosen test should therefore be chosen according to their practical meaningfulness in the context of the study.

Other tests also exist in a paired alternative (e.g., paired $t$-test, repeated ANOVA, Wilcoxon rank test), which offers more power for longitudinal studies than the nonpaired counterpart by assuming that the underlying data is the same, but subject to a different effect across groups. A practical example would be to use hypothesis testing to study a change or improvement in the same cohort of subjects from measurements through time. These tests can be applied on the whole bundle, identifying which tracts are different from their counterpart in the studied groups, or on a point-by-point basis along the streamlines, identifying the spatial locations that are different in each bundle. A point-by-point analysis requires more power than considering the whole bundle at once and therefore requires a larger sample of subjects to locally detect a potential effect.

It is important to mention that the presented tests do *not* provide a guarantee that the underlying assumptions of the model nor the conclusion are 100% valid; they provide a framework where the researcher must balance the probability of false positive and false negative results, allowing a risk-reward analysis with a probable outcome at a level predetermined by the researcher.

### 3.2.3 Correcting for multiple comparisons

The examples we have presented so far do not necessarily take into account that the tests may be repeatable, such as when testing every point along several bundles. Indeed, it may be possible to obtain a false positive or false negative result simply because of the sheer number of tests realized concurrently. The multiple comparison framework provides correction to the $P$-values of the statistics to remedy this issue. To name a few popular possibilities for controlling the type I error rate, known as the family-wise error rate (FWER; Nichols and Holmes, 2002), we have the Bonferroni correction, which provides a $P$-value so that no type I error occurs.

As an FWER correction might be too conservative by providing artificially low values of $\alpha$ and therefore lower power, another approach is to control the *probability* of making a type I error. Another approach is to control the false discovery rate, available through the Benjamini-Hochberg procedure for example, where the researcher instead allows some of the hypotheses to be incorrectly not rejected, but this will have more power than by using an FWER correction. It is, however, impossible to know which of the hypotheses were not wrongly rejected by the procedure. Note that some omnibus tests, such as the ANOVA, already include mitigation of the multiple comparison issue directly into their statistical framework.

Other frameworks are available, such as Bayesian analysis, but the multiple comparison theory is not ready yet for large samples (Keysers et al., 2020) as encountered in along-streamline analysis. Other approaches include permutation testing (Good, 2005), where distribution of *any* statistic can be constructed in addition to providing a confidence interval. This is done by exchanging samples between groups and recomputing the statistic of interest, over and over, with every possible swap of the datasets. While permutation testing does not necessarily assume any underlying distribution of the data, the computational cost of this approach can be cumbersome for large datasets, as the number of tests that needs to be realized is approximately factorial. An idea to mitigate the multiple comparisons issue could be to first test the whole bundle using an omnibus test, such as the ANOVA, as it accounts in its design for the multiple comparisons issue. If any difference is detected, a subsequent local test (e.g., a $t$-test) can be used on every point of the segment, accounting for multiple testing afterwards, to identify if any part of the segment is statistically different between groups. This only works if the hypothesis

$H_0$ of the omnibus test is a generalization of the hypothesis used for the local test, however. We refer the reader to Evans and Rosenthal (2010, Chapters 5, 9, and 10) and Motulsky (2017, Parts D–F and H) for more information about statistical inference in general.

### 3.2.4 Continuous representations and functional profile analysis

Another approach is to map the collected discrete samples into continuous functions by choosing and fitting the data to a set of well-chosen basis functions. The field of functional data analysis allows the combination and modeling of multiple diffusion measures and covariates of interest (e.g., age, gender, height, response to a treatment) in the same unified framework. This flexibility in modeling, however, comes at the cost of additional complexity and ad hoc choices in the study design, such as choosing an adequate set of continuous functions to be used (e.g., a linear combination of low degree polynomials, spline functions, and their degrees) or fitting the basis functions and the type of regularization to use. There is also the design of statistical tests (either locally on a segment or globally for the whole bundle) and hypothesis testing since the data is now represented in a continuous fashion (see, e.g., Zhu et al., 2010, 2011 for applications in diffusion MRI and Ramsay and Silverman, 2005 for theoretical details in general).

## 3.3 Normative modeling and machine learning

The previous approaches rely on comparing statistics at the population level using estimates from representative samples. However, this methodology may not translate well when only a few subjects of a specific group are available (Yeh et al., 2017), or if the group of interest is highly heterogeneous (Marquand et al., 2016). One way to circumvent this issue is to assemble data from a normative population (e.g., healthy controls), and then compare individuals to that normative representation. Normative modeling is an emerging statistical framework that aims to capture variability by comparing individuals to a normative population (Marquand et al., 2016). Current efforts to apply normative modeling in neuroimaging, however, have so far relied on voxel-based methods (Marquand et al., 2019), which can be suboptimal for white matter. Tractography, on the other hand, offers a more intuitive manifold (Chamberland et al., 2021). By defining a threshold based on a specific $z$-score, one can assess whether each point along the bundle-profile falls within a normal range or not (Yeatman et al., 2012; Yeh et al., 2017). Indeed, with tractometry, one can assemble a normative range of bundle-profiles based on a healthy population, as shown in Fig. 4.

### 3.3.1 Dimensionality reduction

Dimensionality reduction refers to the process of reducing the number of features in your data while preserving as much information as possible from the original data. The space in which the new features are represented is often referred to as the *latent* space. The analysis spaces previously discussed typically only extract information from one diffusion measure at a time, requiring the user to process each measure independently before statistical testing. However, this has the potential to obscure key relationships *between* different tracts or different measures. In addition, when analyzing multiple measures (even when derived within the same bundle), statistical analysis is hampered by: (i) the multiple comparisons problem

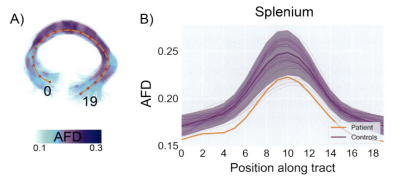

**FIG. 4** Along-streamline profiling overview of the posterior aspect of the corpus callosum (splenium). (A) Superior axial view of the splenium color-coded with the apparent fiber density (AFD; Raffelt et al., 2012). The core streamline (*orange*) is overlaid and consist of $n = 20$ points. (B) Example application where the individual profile of the subject of interest (*orange*) is compared to a normative population of healthy controls (*purple lines*). The population average is depicted in *bold* with the *shaded area* representing 1.65 standard deviations ($Z \pm 1$). In this case, the subject's bundle-profile falls outside of the defined normative range.

**446 PART | IV** From streamlines to tracts

and (ii) any covariance between measurements (De Santis et al., 2014; Chamberland et al., 2019; Geeraert et al., 2020). Alternatively, multivariate approaches can increase statistical power by combining the sensitivity profiles of independent measures (Dean et al., 2017; Chamberland et al., 2019; Taylor et al., 2020). To address the multivariate nature of dMRI data and circumvent the *curse of dimensionality*, dimensionality reduction approaches based on principal component analysis (PCA), or deep autoencoders for larger datasets, were recently proposed (Dean et al., 2017; Chamberland et al., 2019, 2021; Taylor et al., 2020). When combined with clustering, unsupervised dimensionality reduction approaches offer great ways to visualize the complex space of diffusion bundle-profiles (Legarreta et al., 2021; Attye et al., 2021). This can allow data scientists to identify several subgroups or patterns within the data, potentially revealing important insights otherwise hidden by the multidimensional aspect of the data. For example, by projecting bundle-profiles into a two-dimensional (2D) space, one can train a classifier to distinguish between controls and patients with ALS (Yeatman et al., 2018). From a machine learning point-of-view, bundle-profiles can therefore be seen as *features* and can be used to solve classification and regression problems (Sarica et al., 2017; Dou et al., 2020; Richie-Halford et al., 2021; Chamberland et al., 2021; Attye et al., 2021).

## 4 Comparison with other frameworks

Considerable effort has gone into designing methods for dMRI analysis: voxel-based analysis (VBA), TBSS, and, as seen in this chapter, the tractometry framework.

VBA for dMRI compares each brain voxel on a group level (Hecke et al., 2016). This method, which was inspired by structural- (e.g., T1-weighted; Ashburner and Friston, 2000) and functional-MRI analyses, requires datasets to be spatially aligned; each voxel is statistically evaluated to detect differences between populations. However, VBA faces major challenges, for example, misalignments and partial volume effects (Jones et al., 2005a). Moreover, dMRI has orientational information (Leemans and Jones, 2009), and it is unclear how to propagate this through nonlinear deformations. Finally, since one dataset contains hundreds of thousands of voxels, statistical power is hampered by multiple comparisons (Jones et al., 2005a) when correcting for false positives (Bennett et al., 2009).

TBSS attempts to mitigate VBA problems and relies on the projection of maximum FA values onto a template FA skeleton (Smith et al., 2006), followed by standard voxel-based statistics on each skeleton voxel. Although referring to fiber tracts, this framework does not operate on state-of-the-art tractography outputs, but on voxel-wise FA maps. While more robust to registration errors than typical VBA, it is not devoid of limits (e.g., potential bias around white matter (WM) abnormalities; Bach et al., 2014).

Tractometry has several advantages over the more ubiquitous voxel-based or region-of-interest (ROI)-based analyses. First and foremost, the white matter is not a homogeneous mass, but comprises distinct anatomical pathways that may pass very close to each other (e.g., in adjacent voxels), but connect very different regions of the brain. For example, the UF, which interconnects frontal and temporal cortex, runs immediately adjacent to the inferior fronto-occipital fasciculus (IFOF), which interconnects frontal and occipital cortex. Without the inference of continuous trajectories afforded by tractography, it can be impossible to ascertain which particular pathway a voxel should be assigned to, and therefore how to ensure that, across participants, one is comparing "like for like." Note that, despite the name, the current most popular framework for voxel-based analyses, that is, TBSS (Smith et al., 2006), does not incorporate any such information and so cannot ascertain whether two adjacent voxels correspond to the same bundle (which is actually rather important when it comes to voxel cluster-based inference). In summary, the first benefit of tractometry is that it provides enhanced anatomical specificity.

Second, if one is to compute a summary statistic along a given bundle (e.g., the mean anisotropy of the left UF), then it is not necessary to spatially normalize the data to a common reference space before analysis (as is essential for voxel-based analyses). Rather, one needs to just reconstruct the left UF of each participant and compute the mean. This therefore obviates the need for any spatial interpolation step, which introduces a level of smoothing and may influence the statistical inference.

Third, one *could* adopt an ROI-based approach, where—with good prior anatomical knowledge—an expert observer can manually define a 2D or three-dimensional (3D) region on the quantitative maps, and compute a summary statistic (e.g., the mean) and compare between individuals/groups. While this removes the need for both spatial normalization and tractography, it is only really applicable to a small number of anatomical regions where the anatomical boundaries are clearly defined, for example, the midsagittal section of the corpus callosum, or the cerebral peduncles. When the boundaries are less well defined (as in the aforementioned UF and IFOF), or when streamlines take a tortuous route in and out of the plane of the screen, manual delineation becomes problematic. In these situations, the tractography provides a semiautomatic

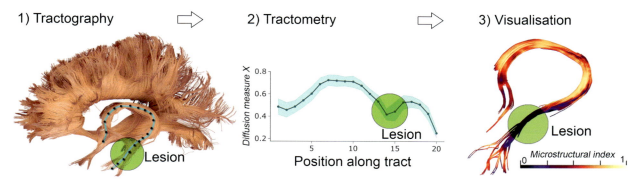

FIG. 5 Along-streamline profiling in the presence of a white matter lesion reveals a local change in tissue microstructure.

creation of a 3D region of interest. Note that the manual segmentation can be replaced by the registration of ROIs from an existing atlas (e.g., in MNI space).

Finally, in applications where highly localized pathology is not suspected (e.g., in studying neurodevelopmental disorders, neurodevelopment itself, or perhaps a group of healthy individuals), in comparison to a pure voxel-wise analysis (which effectively discretizes a bundle into small, independent cubes of tissue), the tractography approach confers an SNR advantage in that it averages across voxels belonging to the same anatomical structure, and can therefore not only increase the anatomical specificity but improve the sensitivity. Self-evidently, if highly localized pathologies are expected, then taking a bundle-wise global average risks obscuring the pathology, and so *along-streamline profiling* is recommended (Fig. 5). Furthermore, there are multiple differently oriented fiber pathways present within a single voxel (Jeurissen et al., 2013). In these "crossing fiber" regions, voxel-based measures like the FA cannot be attributed to any of these pathways in an unambiguous way. Interpretation of bundle-profiles in this case should be taken with great care. Whenever possible, we recommend the use of fiber-specific measures, for example, apparent fiber density (Raffelt et al., 2012; Dell'Acqua et al., 2013), or any microstructural measure that can be related to a specific fiber population (Assaf and Basser, 2005; Alexander et al., 2017).

## 5 Conclusion

In this chapter, we presented the main ingredients of the tractometry pipeline. First, after choosing a specific dMRI method and concomitant tractography approach, a manual or automated bundle segmentation is required. Then, an along-streamline spatial sampling strategy is needed to provide the framework for the statistical inference of the diffusion measures of interest. Despite the numerous methodological considerations and parameter choices in tractometry, it is becoming increasingly popular as it combines the best of two worlds: it provides the ability to locally investigate white matter properties, such as is also the case in voxel-based analyses, *and* it takes advantage of the increase in spatial specificity in the white matter by incorporating long-distance information from fiber tractography.

## Acknowledgments

MC was supported by the Radboud University Excellence Initiative Fellowship at the time of writing. SStJ was supported by the Natural Sciences and Engineering Research Council of Canada (NSERC) (funding reference number BP-546283-2020) and the Fonds de recherche du Québec—Nature et technologies (FRQNT) (Dossier 290978).

## References

Alexander, D.C., Dyrby, T.B., Nilsson, M., Zhang, H., 2017. Imaging brain microstructure with diffusion MRI: practicality and applications. NMR Biomed. 32 (4), e3841.

Ashburner, J., Friston, K.J., 2000. Voxel-based morphometry—the methods. NeuroImage 11 (6), 805–821. https://doi.org/10.1006/nimg.2000.0582.

Assaf, Y., Basser, P.J., 2005. Composite hindered and restricted model of diffusion (CHARMED) MR imaging of the human brain. NeuroImage 27 (1), 48–58.

Attye, A., Renard, F., Baciu, M., Roger, E., Lamalle, L., Dehail, P., Cassoudesalle, H., Calamante, F., 2021. TractLearn: a geodesic learning framework for quantitative analysis of brain bundles. NeuroImage 233, 117927.

**448 PART | IV From streamlines to tracts**

Bach, M., Laun, F.B., Leemans, A., Tax, C.M.W., Biessels, G.J., Stieltjes, B., Maier-Hein, K.H., 2014. Methodological considerations on tract-based spatial statistics (TBSS). NeuroImage 100, 358–369. https://doi.org/10.1016/j.neuroimage.2014.06.021.

Bells, S., Cercignani, M., Deoni, S., Assaf, Y., Pasternak, O., Evans, C.J., Leemans, A., Jones, D.K., 2011. Tractometry-comprehensive multi-modal quantitative assessment of white matter along specific tracts. In: Proceedings of ISMRM, 678. vol.

Bennett, C.M., Wolford, G.L., Miller, M.B., 2009. The principled control of false positives in neuroimaging. Soc. Cogn. Affect Neurosci. 4 (4), 417–422.

Benou, I., Veksler, R., Friedman, A., Raviv, T.R., 2019. Combining white matter diffusion and geometry for tract-specific alignment and variability analysis. NeuroImage 200, 674–689. https://doi.org/10.1016/j.neuroimage.2019.05.003.

Blecher, T., Miron, S., Schneider, G.G., Achiron, A., Ben-Shachar, M., 2019. Association between white matter microstructure and verbal fluency in patients with multiple sclerosis. Front. Psychol. 10, 1607.

Chamberland, M., St-Jean, S., Tax, C.M.W., Jones, D.K., 2018. Obtaining representative core streamlines for white matter tractometry of the human brain. In: International Conference on Medical Image Computing and Computer-Assisted Intervention, pp. 359–366.

Chamberland, M., Raven, E.P., Genc, S., Duffy, K., Descoteaux, M., Parker, G.D., Tax, C.M.W., Jones, D.K., 2019. Dimensionality reduction of diffusion MRI measures for improved tractometry of the human brain. NeuroImage 200, 89–100.

Chamberland, M., Genc, S., Tax, C.M.W., Shastin, D., Koller, K., Raven, E.P., Cunningham, A., Doherty, J., van den Bree, M., Parker, G.D., et al., 2021. Detecting microstructural deviations in individuals with deep diffusion MRI tractometry. Nat. Comput. Sci. 1 (9), 598–606.

Chandio, B.Q., Risacher, S.L., Pestilli, F., Bullock, D., Yeh, F.C., Koudoro, S., Rokem, A., Harezlak, J., Garyfallidis, E., 2020. Bundle analytics, a computational framework for investigating the shapes and profiles of brain pathways across populations. Sci. Rep. 10 (1), 1–18.

Chen, Z., Zhang, H., Yushkevich, P.A., Liu, M., Beaulieu, C., 2016. Maturation along white matter tracts in human brain using a diffusion tensor surface model tract-specific analysis. Front. Neuroanat. 10, 9.

Cohen, J., 1992. A power primer. Psychol. Bull. 112 (1), 155–159. https://doi.org/10.1037/0033-2909.112.1.155.

Colby, J.B., Soderberg, L., Lebel, C., Dinov, I.D., Thompson, P.M., Sowell, E.R., 2012. Along-tract statistics allow for enhanced tractography analysis. NeuroImage 59 (4), 3227–3242.

Corouge, I., Fletcher, P.T., Joshi, S., Gouttard, S., Gerig, G., 2006. Fiber tract-oriented statistics for quantitative diffusion tensor MRI analysis. Med. Image Anal. 10 (5), 786–798.

Cousineau, M., Jodoin, P.-M., Garyfallidis, E., Cote, M.A., Morency, F.C., Rozanski, V., Grand Maison, M., Bedell, B.J., Descoteaux, M., 2017. A test-retest study on Parkinson's PPMI dataset yields statistically significant white matter fascicles. NeuroImage 16, 222–233.

Dayan, M., Munoz, M., Jentschke, S., Chadwick, M.J., Cooper, J.M., Riney, K., Vargha-Khadem, F., Clark, C.A., 2015. Optic radiation structure and anatomy in the normally developing brain determined using diffusion MRI and tractography. Brain Struct. Funct. 220 (1), 291–306.

de Gervai, P.D., Sboto-Frankenstein, U.N., Bolster, R.B., Thind, S., Gruwel, M.L.H., Smith, S.D., Tomanek, B., 2014. Tractography of Meyer's Loop asymmetries. Epilepsy Res. 108 (5), 872–882.

De Santis, S., Drakesmith, M., Bells, S., Assaf, Y., Jones, D.K., 2014. Why diffusion tensor MRI does well only some of the time: variance and covariance of white matter tissue microstructure attributes in the living human brain. NeuroImage 89, 35–44.

Dean, D.C., Lange, N., Travers, B.G., Prigge, M.B., Matsunami, N., Kellett, K.A., Freeman, A., Kane, K.L., Adluru, N., Tromp, D.P.M., et al., 2017. Multivariate characterization of white matter heterogeneity in autism spectrum disorder. NeuroImage 14, 54–66.

Dell'Acqua, F., Simmons, A., Williams, S.C.R., Catani, M., 2013. Can spherical deconvolution provide more information than fiber orientations? Hindrance modulated orientational anisotropy, a true-tract specific index to characterize white matter diffusion. Hum. Brain Mapp. 34 (10), 2464–2483.

Dou, X., Yao, H., Feng, F., Wang, P., Zhou, B., Jin, D., Yang, Z., Li, J., Zhao, C., Wang, L., An, N., Liu, B., Zhang, X., Liu, Y., 2020. Characterizing white matter connectivity in Alzheimer's disease and mild cognitive impairment: an automated fiber quantification analysis with two independent datasets. Cortex 129, 390–405. https://doi.org/10.1016/j.cortex.2020.03.032.

Evans, M., Rosenthal, J.S., 2010. Probability and Statistics: The Science of Uncertainty, second ed. W.H. Freeman and Co., p. 773.

Garyfallidis, E., Brett, M., Correia, M.M., Williams, G.B., Nimmo-Smith, I., 2012. Quickbundles, a method for tractography simplification. Front. Neurosci. 6, 175.

Garyfallidis, E., Brett, M., Amirbekian, B., Rokem, A., Van Der Walt, S., Descoteaux, M., Nimmo-Smith, I., Contributors, Dipy, 2014. Dipy, a library for the analysis of diffusion MRI data. Front. Neuroinform. 8, 8.

Garyfallidis, E., Ocegueda, O., Wassermann, D., Descoteaux, M., 2015. Robust and efficient linear registration of white-matter fascicles in the space of streamlines. NeuroImage 117, 124–140.

Geeraert, B.L., Chamberland, M., Lebel, R.M., Lebel, C., 2020. Multimodal principal component analysis to identify major features of white matter structure and links to reading. PLoS One 15 (8), e0233244.

Glozman, T., Bruckert, L., Pestilli, F., Yecies, D.W., Guibas, L.J., Yeom, K.W., 2018. Framework for shape analysis of white matter fiber bundles. NeuroImage 167, 466–477.

Good, P.I., 2005. Permutation, Parametric and Bootstrap Tests of Hypotheses. Springer Series in Statistics, Springer-Verlag, New York, p. 315, https://doi.org/10.1007/b138696.

Groeschel, S., Tournier, J.D., Northam, G.B., Baldeweg, T., Wyatt, J., Vollmer, B., Connelly, A., 2014. Identification and interpretation of microstructural abnormalities in motor pathways in adolescents born preterm. NeuroImage 87, 209–219.

Hecke, W.V., Leemans, A., Emsell, L., 2016. DTI analysis methods: voxel-based analysis. In: Diffusion Tensor Imaging, Springer, pp. 183–203.

Jeurissen, B., Leemans, A., Tournier, J.D., Jones, D.K., Sijbers, J., 2013. Investigating the prevalence of complex fiber configurations in white matter tissue with diffusion magnetic resonance imaging. Hum. Brain Mapp. 34 (11), 2747–2766.

Jones, D.K., Symms, M.R., Cercignani, M., Howard, R.J., 2005a. The effect of filter size on VBM analyses of DT-MRI data. NeuroImage 26 (2), 546–554.

Jones, D.K., Travis, A.R., Eden, G., Pierpaoli, C., Basser, P.J., 2005b. PASTA: pointwise assessment of streamline tractography attributes. Magn. Reson. Med. 53 (6), 1462–1467.

Jones, D.K., Catani, M., Pierpaoli, C., Reeves, S.J.C., Shergill, S.S., O'Sullivan, M., Golesworthy, P., McGuire, P., Horsfield, M.A., Simmons, A., et al., 2006. Age effects on diffusion tensor magnetic resonance imaging tractography measures of frontal cortex connections in schizophrenia. Hum. Brain Mapp. 27 (3), 230–238.

Jones, D.K., Knösche, T.R., Turner, R., 2013. White matter integrity, fiber count, and other fallacies: the do's and don'ts of diffusion MRI. NeuroImage 73, 239–254. https://doi.org/10.1016/j.neuroimage.2012.06.081.

Keysers, C., Gazzola, V., Wagenmakers, E.J., 2020. Using Bayes factor hypothesis testing in neuroscience to establish evidence of absence. Nat. Neurosci. 23 (7), 788–799. https://doi.org/10.1038/s41593-020-0660-4.

Kindlmann, G., Tricoche, X., Westin, C.F., 2007. Delineating white matter structure in diffusion tensor MRI with anisotropy creases. Med. Image Anal. 11 (5), 492–502.

Leemans, A., Jones, D.K., 2009. The B-matrix must be rotated when correcting for subject motion in DTI data. Magn. Reson. Med. 61 (6), 1336–1349.

Leemans, A., Sijbers, J., De Backer, S., Vandervliet, E., Parizel, P., 2006. Multiscale white matter fiber tract coregistration: a new feature-based approach to align diffusion tensor data. Magn. Reson. Med. 55 (6), 1414–1423. https://doi.org/10.1002/mrm.20898.

Leemans, A., Jeurissen, B., Sijbers, J., Jones, D.K., 2009. ExploreDTI: a graphical toolbox for processing, analyzing, and visualizing diffusion MR data. Proc. Int. Soc. Mag. Reson. Med. 17 (1), 3537.

Legarreta, J.H., Petit, L., Rheault, F., Theaud, G., Lemaire, C., Descoteaux, M., Jodoin, P.M., 2021. Filtering in tractography using autoencoders (FINTA). Med. Image Anal. 72, 102126.

Maffei, C., Lee, C., Planich, M., Ramprasad, M., Ravi, N., Trainor, D., Urban, Z., Kim, M., Jones, R.J., Henin, A., et al., 2021. Using diffusion MRI data acquired with ultra-high gradient strength to improve tractography in routine-quality data. NeuroImage 245, 118706.

Maier-Hein, K.H., Neher, P.F., Houde, J.C., Côté, M.A., Garyfallidis, E., Zhong, J., Chamberland, M., Yeh, F.C., Lin, Y.C., Ji, Q., et al., 2017. The challenge of mapping the human connectome based on diffusion tractography. Nat. Commun. 8, 1349.

Main, K.L., Soman, S., Pestilli, F., Furst, A., Noda, A., Hernandez, B., Kong, J., Cheng, J., Fairchild, J.K., Taylor, J., et al., 2017. DTI measures identify mild and moderate TBI cases among patients with complex health problems: a receiver operating characteristic analysis of US veterans. NeuroImage 16, 1–16.

Marquand, A.F., Rezek, I., Buitelaar, J., Beckmann, C.F., 2016. Understanding heterogeneity in clinical cohorts using normative models: beyond case-control studies. Biol. Psychiatry 80 (7), 552–561. https://doi.org/10.1016/j.biopsych.2015.12.023.

Marquand, A.F., Kia, S.M., Zabihi, M., Wolfers, T., Buitelaar, J.K., Beckmann, C.F., 2019. Conceptualizing mental disorders as deviations from normative functioning. Mol. Psychiatry 24 (10), 1415–1424. https://doi.org/10.1038/s41380-019-0441-1.

Mezer, A., Yeatman, J.D., Stikov, N., Kay, K.N., Cho, N.J., Dougherty, R.F., Perry, M.L., Parvizi, J., Hua, L.H., Butts-Pauly, K., et al., 2013. Quantifying the local tissue volume and composition in individual brains with magnetic resonance imaging. Nat. Med. 19 (12), 1667–1672.

Motulsky, H., 2017. Intuitive Biostatistics: A Nonmathematical Guide to Statistical Thinking, fourth ed. Oxford University, Toronto, p. 608.

Nichols, T.E., Holmes, A.P., 2002. Nonparametric permutation tests for functional neuroimaging: a primer with examples. Hum. Brain Mapp. 15 (1), 1–25.

O'Donnell, L.J., Westin, C.F., 2007. Automatic tractography segmentation using a high-dimensional white matter atlas. IEEE Trans. Med. Imaging 26 (11), 1562–1575.

O'Donnell, L.J., Westin, C.F., Golby, A.J., 2009. Tract-based morphometry for white matter group analysis. NeuroImage 45 (3), 832–844.

Parker, G.D., LLoyd, D., Jones, D.K., 2016. The best of both worlds: combining the strengths of TBSS and tract-specific measurements for group-wise comparison of white matter microstructure. In: International Society on Magnetic Resonance in Medicine (ISMRM'16).

Qiu, A., Oishi, K., Miller, M.I., Lyketsos, C.G., Mori, S., Albert, M., 2010. Surface-based analysis on shape and fractional anisotropy of white matter tracts in Alzheimer's disease. PLoS One 5 (3), e9811.

Raffelt, D., Tournier, J.D., Rose, S., Ridgway, G.R., Henderson, R., Crozier, S., Salvado, O., Connelly, A., 2012. Apparent fibre density: a novel measure for the analysis of diffusion-weighted magnetic resonance images. NeuroImage 59 (4), 3976–3994.

Ramsay, J.O., Silverman, B.W., 2005. Functional Data Analysis. Springer, New York, https://doi.org/10.1007/b98888.

Richie-Halford, A., Yeatman, J., Simon, N., Rokem, A., 2021. Multidimensional analysis and detection of informative features in human brain white matter. PLoS Comput. Biol. 17 (6), e1009136.

Sarica, A., Cerasa, A., Valentino, P., Yeatman, J., Trotta, M., Barone, S., Granata, A., Nisticò, R., Perrotta, P., Pucci, F., et al., 2017. The corticospinal tract profile in amyotrophic lateral sclerosis. Hum. Brain Mapp. 38 (2), 727–739.

Schilling, K.G., Nath, V., Hansen, C., Parvathaneni, P., Blaber, J., Gao, Y., Neher, P., Aydogan, D.B., Shi, Y., Ocampo-Pineda, M., Schiavi, S., Daducci, A., Girard, G., Barakovic, M., Rafael-Patino, J., Romascano, D., Rensonnet, G., Pizzolato, M., Bates, A., Fischi, E., Thiran, J.P., Canales-Rodríguez, E.J., Huang, C., Zhu, H., Zhong, L., Cabeen, R., Toga, A.W., Rheault, F., Theaud, G., Houde, J.C., Sidhu, J., Chamberland, M., Westin, C.F., Dyrby, T.B., Verma, R., Rathi, Y., Irfanoglu, M.O., Thomas, C., Pierpaoli, C., Descoteaux, M., Anderson, A.W., Landman, B.A., 2019. Limits to anatomical accuracy of diffusion tractography using modern approaches. NeuroImage 185, 1–11. https://doi.org/10.1016/j.neuroimage.2018.10.029.

Schilling, K.G., Rheault, F., Petit, L., Hansen, C.B., Nath, V., Yeh, F.C., Girard, G., Barakovic, M., Rafael-Patino, J., Yu, T., et al., 2021. Tractography dissection variability: what happens when 42 groups dissect 14 white matter bundles on the same dataset? NeuroImage 243, 118502.

Schultz, T., Theisel, H., Seidel, H.P., 2009. Crease surfaces: from theory to extraction and application to diffusion tensor MRI. IEEE Trans. Vis. Comput. Graph. 16 (1), 109–119.

Smith, S.M., Jenkinson, M., Johansen-Berg, H., Rueckert, D., Nichols, T.E., Mackay, C.E., Watkins, K.E., Ciccarelli, O., Cader, M.Z., Matthews, P.M., Behrens, T.E.J., 2006. Tract-based spatial statistics: voxelwise analysis of multi-subject diffusion data. NeuroImage 31 (4), 1487–1505. https://doi.org/10.1016/j.neuroimage.2006.02.024.

St-Jean, S., Chamberland, M., Viergever, M.A., Leemans, A., 2019. Reducing variability in along-tract analysis with diffusion profile realignment. NeuroImage 199, 663–679.

Sun, H., Lui, S., Yao, L., Deng, W., Xiao, Y., Zhang, W., Huang, X., Hu, J., Bi, F., Li, T., et al., 2015. Two patterns of white matter abnormalities in medication-naive patients with first-episode schizophrenia revealed by diffusion tensor imaging and cluster analysis. JAMA Psychiatry 72 (7), 678–686.

Tax, C.M.W., Haije, T.D., Fuster, A., Westin, C.F., Viergever, M.A., Florack, L., Leemans, A., 2016. Sheet Probability Index (SPI): characterizing the geometrical organization of the white matter with diffusion MRI. NeuroImage 142, 260–279.

Tax, C.M.W., Bastiani, M., Veraart, J., Garyfallidis, E., Irfanoglu, M.O., 2021. What's new and what's next in diffusion MRI preprocessing. NeuroImage 249, 118830.

Taylor, P.N., da Silva, N.M., Blamire, A., Wang, Y., Forsyth, R., 2020. Early deviation from normal structural connectivity: a novel intrinsic severity score for mild TBI. Neurology 94 (10), e1021–e1026.

Travis, K.E., Leitner, Y., Feldman, H.M., Ben-Shachar, M., 2015. Cerebellar white matter pathways are associated with reading skills in children and adolescents. Hum. Brain Mapp. 36 (4), 1536–1553.

Vilanova, A., Berenschot, G., Van Pul, C., 2004. DTI visualization with streamsurfaces and evenly-spaced volume seeding. In: Proceedings of the Sixth Joint Eurographics-IEEE TCVG Conference on Visualization, pp. 173–182.

Vos, S.B., Jones, D.K., Viergever, M.A., Leemans, A., 2011. Partial volume effect as a hidden covariate in DTI analyses. NeuroImage 55 (4), 1566–1576.

Wasserthal, J., Neher, P., Maier-Hein, K.H., 2018. TractSeg—fast and accurate white matter tract segmentation. NeuroImage 183, 239–253.

Wasserthal, J., Maier-Hein, K.H., Neher, P.F., Northoff, G., Kubera, K.M., Fritze, S., Harneit, A., Geiger, L.S., Tost, H., Wolf, R.C., et al., 2020. Multiparametric mapping of white matter microstructure in catatonia. Neuropsychopharmacology 45 (10), 1750–1757.

Xiao, Y., Sun, H., Shi, S., Jiang, D., Tao, B., Zhao, Y., Zhang, W., Gong, Q., Sweeney, J.A., Lui, S., 2018. White matter abnormalities in never-treated patients with long-term schizophrenia. Am. J. Psychiatry 175 (11), 1129–1136.

Yeatman, J.D., Dougherty, R.F., Myall, N.J., Wandell, B.A., Feldman, H.M., 2012. Tract profiles of white matter properties: automating fiber-tract quantification. PLoS One 7 (11), e49790.

Yeatman, J.D., Wandell, B.A., Mezer, A.A., 2014. Lifespan maturation and degeneration of human brain white matter. Nat. Commun. 5 (1), 1–12.

Yeatman, J.D., Richie-Halford, A., Smith, J.K., Keshavan, A., Rokem, A., 2018. A browser-based tool for visualization and analysis of diffusion MRI data. Nat. Commun. 9 (1), 1–10.

Yeh, P.H., Guan Koay, C., Wang, B., Morissette, J., Sham, E., Senseney, J., Joy, D., Kubli, A., Yeh, C.H., Eskay, V., et al., 2017. Compromised neurocircuitry in chronic blast-related mild traumatic brain injury. Hum. Brain Mapp. 38 (1), 352–369.

Yendiki, A., Panneck, P., Srinivasan, P., Stevens, A., Zöllei, L., Augustinack, J., Wang, R., Salat, D., Ehrlich, S., Behrens, T., et al., 2011. Automated probabilistic reconstruction of white-matter pathways in health and disease using an atlas of the underlying anatomy. Front. Neuroinform. 5, 23.

Yushkevich, P.A., Zhang, H., Simon, T.J., Gee, J.C., 2009. Structure-specific statistical mapping of white matter tracts. In: Visualization and Processing of Tensor Fields, Springer, pp. 83–112.

Zhang, S., Demiralp, C., Laidlaw, D.H., 2003. Visualizing diffusion tensor MR images using streamtubes and streamsurfaces. IEEE Trans. Vis. Comput. Graph. 9 (4), 454–462.

Zhang, F., Daducci, A., He, Y., Schiavi, S., Seguin, C., Smith, R., Yeh, C.H., Zhao, T., O'Donnell, L.J., 2022. Quantitative mapping of the brain's structural connectivity using diffusion MRI tractography: a review. NeuroImage 249, 118870.

Zhu, H., Styner, M., Tang, N., Liu, Z., Lin, W., Gilmore, J.H., 2010. FRATS: functional regression analysis of DTI tract statistics. IEEE Trans. Med. Imaging 29 (4), 1039–1049. https://doi.org/10.1109/TMI.2010.2040625.

Zhu, H., Kong, L., Li, R., Styner, M., Gerig, G., Lin, W., Gilmore, J.H., 2011. FADTTS: functional analysis of diffusion tensor tract statistics. NeuroImage 56 (3), 1412–1425. https://doi.org/10.1016/j.neuroimage.2011.01.075.

# Chapter 24

# Connectivity and connectomics

Andrew Zalesky[a,b], Stamatios N. Sotiropoulos[c,d], Saad Jbabdi[c], and Alex Fornito[e]

[a]*Department of Biomedical Engineering, The University of Melbourne, Melbourne, VIC, Australia,* [b]*Melbourne Neuropsychiatry Centre, Department of Psychiatry, The University of Melbourne, Melbourne, VIC, Australia,* [c]*Wellcome Centre for Integrative Neuroimaging (WIN), Oxford Centre for Functional Magnetic Resonance Imaging of the Brain (FMRIB), University of Oxford, Oxford, United Kingdom,* [d]*Sir Peter Mansfield Imaging Centre, School of Medicine, University of Nottingham, Nottingham, United Kingdom,* [e]*Turner Institute for Brain and Mental Health, School of Psychological Sciences, and Monash Biomedical Imaging, Monash University, Melbourne, VIC, Australia*

## 1 Introduction

The connectome refers to a comprehensive structural description of the network elements and connections forming a nervous system (Sporns et al., 2005). Diffusion MRI tractography is one of the most important techniques for mapping connectomes and is the only technique available for mapping connectomes in vivo. The basic approach to mapping connectomes with tractography has remained largely unchanged since the first human connectomes were mapped in 2005 (Hagmann, 2005; Sporns et al., 2005). After performing tractography in the whole brain, the number of tractography streamlines interconnecting each pair of regions comprising a predefined parcellation atlas is enumerated and used to populate a connectivity matrix (Fig. 1). However, tractography was never originally developed to map connectomes and the prolific use of tractography for this purpose nowadays has given rise to numerous challenges and misconceptions. For example, despite emerging as the de facto measure of structural connectivity in diffusion MRI, the interpretation of interregional streamline counts remains elusive and often misconstrued. Another challenge is the tendency for some tractography algorithms to reconstruct large numbers of spurious connections, referred to as false positives.

This chapter introduces the basics of tractography-based connectome mapping and describes recent advances developed to address some of these challenges.

The chapter begins with an introduction to the fundamentals of connectome mapping with diffusion MRI tractography. The first step in mapping a connectome is node delineation. A *node* is simply a discrete brain region. Section 2 reviews the key properties of brain nodes and considers parcellation methods commonly used to delineate a set of nodes covering the whole cortex. Once a set of nodes is defined, the next step is to determine which pairs of nodes are structurally connected. A connection between a pair of nodes is known as an *edge* in network parlance. The use of tractography as a technique to infer connectome edges is introduced in Section 3. Evidence of connectivity inferred from tractography is indirect and thus validation studies employing invasive techniques and connectome phantoms are vital. Connectome validation is considered in Section 4. The final section of this chapter introduces the burgeoning field of network neuroscience and details some of the recent discoveries in this field about network principles governing brain organization. The chapter thus covers node delineation, mapping connectivity and connectomes, connectome validation, and network analysis.

Tractography-derived connectomes are macroscale connectomes. Connections reconstructed with tractography represent large-scale bundles of coherently organized axons. Although these connections constitute less than 10% of the human brain's total number of neuronal connections (Schuz, 2002), they form complex networks that are efficiently wired to integrate information, as discussed in Section 5. While not considered in this chapter, it is important to remark that connectomes can be mapped at finer spatial scales using invasive techniques, albeit not in humans. At the finest conceivable scale, the definition of connectome nodes (neurons) and edges (synapses) becomes unequivocal, although mapping connectomes at this scale is practically and technically infeasible for large nervous systems. The complete microscale connectome has been mapped for the nematode *C. elegans* (Cook et al., 2019; White et al., 1986). Using invasive tract tracing techniques to map mesoscale connectomes in rodents (Oh et al., 2014) and nonhuman primates (Markov et al., 2014) is an active area of research. This chapter specifically focuses on mapping macroscale connectomes with diffusion MRI tractography. We emphasize that while this is the only technique available to map human connectomes in vivo, the field of connectomics is much broader than the topics covered in this chapter and some of the most important advances in the field have been facilitated by invasive connectome mapping.

**Handbook of Diffusion MR Tractography. https://doi.org/10.1016/B978-0-12-818894-1.00007-0**
Copyright © 2025 Elsevier Ltd. All rights are reserved, including those for text and data mining, AI training, and similar technologies.

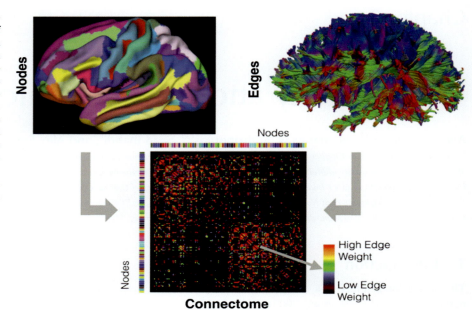

FIG. 1 Connectome mapping with diffusion MRI tractography. The nodes of the connectome are first delineated. Each cortical region comprising a parcellation atlas is a distinct node (*top, left*). Whole-brain tractography is used to infer the presence or absence of structural connections between pairs of nodes (*top, right*). Connections are referred to as edges. The tractography streamline count is often used as an index of connectivity strength (edge weight). Connectivity strengths are represented in a connectivity matrix (*bottom*). Tractography cannot resolve fiber directionality and thus the connectivity matrix is symmetric.

## 2 Delineating nodes

In science as in engineering, understanding a system often begins with breaking it down into its component subunits and analyzing how these units interact. Over the past century, neuroscience has taken this same "divide and conquer" approach to understanding the brain. At the microscopic level, the natural component units are cells (neurons and glia), and their interactions are electrochemical. In systems neuroscience, the study of the brain as a whole, the component units are loosely referred to as "brain areas" (sometimes with the addition of an adjective, such as "functional," to hint at a more precise meaning). This chapter is about modeling the brain as a macroscopic network, a collection of nodes and edges, where the nodes of the network are brain areas. But what are brain areas?

### 2.1 What is a brain area?

Anatomists have always labored to subdivide the brain into different parts. This ranges from gross delineations into large areas based on their phylogenetics (e.g., neocortex vs allocortex), on their nearest divisions of the skull (cortical lobes), or on the location and appearance of cortical folds (e.g., superior frontal gyrus, etc.). Arguably the most important definition of a brain area is in terms of function and computation. In his monumental "Histological Studies on the Localisation of Cerebral Function," Alfred W. Campbell clearly aims to establish the functional organization of cortex. But even then, c. 1905, he knew that he could not rely exclusively on cortical folding. As he noted, while some primary functions are correlated with the location of primary folds:

> "the visual area is associated in the closest manner possible with the calcarine fissure, the olfactory with the hippocampal, the motor and the common sensory with the Rolandic and the auditory with the Sylvian;"

This does not generally hold true:

> "Save these 'primary' fissures there is really no practical naked-eye guide to the localisation of the different areas, for all the 'so-called' secondary fissures bear a very inconstant relation to given fields and cannot be utilised as fixed boundary lines."

A compelling confirmation of this view comes from more recent studies combining modern MRI techniques and histology (Fischl et al., 2008). So what constitutes a brain area? Although there is no agreed-upon definition, an identifiable brain area is typically associated with one or more of the following properties:

(i) To be in the gray matter (obviously)—as it is the doctrine that neurons perform the computations in the brain, although axons also play an important and nontrivial role (Alcami and El Hady, 2019). However, as we will see later, information guiding the delineation of gray matter areas does not need to rely on local information alone, but can also be informed by data from other parts of the brain, including white matter.

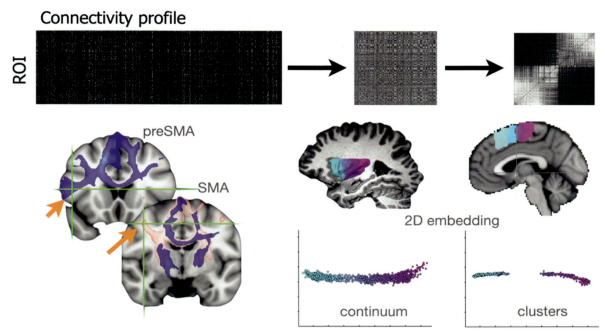

FIG. 2 Connectivity-based parcellation. *Top*: One approach to parcellation takes the connectivity profiles of every voxel in an ROI and builds a similarity matrix, which is then reordered to reveal cluster structure (in this case the ROI is the medial prefrontal cortex containing SMA and pre-SMA). *Bottom-left*: Tractography from SMA and pre-SMA shows distinguishing features (*orange arrows*), such as connectivity to the inferior lateral prefrontal lobule (pre-SMA) and the lateral premotor cortex (SMA). *Bottom-right*: Two-dimensional nonlinear embedding of the features can reveal whether the connectivity data is highly clustered (as is the case for the SMA and pre-SMA) or forms a continuum (insula). *SMA*, supplementary motor area. *(Bottom-right figure adapted from Cerliani, L., Thomas, R.M., Jbabdi, S., Siero, J.C., Nanetti, L., Crippa, A., Gazzola, V., D'Arceuil, H., Keysers, C., 2012. Probabilistic tractography recovers a rostrocaudal trajectory of connectivity variability in the human insular cortex. Hum. Brain Mapp. 33, 2005–2034.)*

(ii) To have a distinctive cellular organization—for example, different parts of cerebral cortex can be distinguished by the appearance of their constituent cortical layers, which in turn are defined according to the types, sizes, and morphologies of their constituent cells and neurites. The cerebellar cortex, on the other hand, has a striking layer organization but with very little differentiation between its constituent parts (Voogd and Glickstein, 1998).

(iii) To have a distinctive distribution of receptors. Different cortical layers are known to have strikingly different densities of receptors, such as NMDA and GABA receptors (Palomero-Gallagher and Zilles, 2019). These density patterns can vary considerably between different cytoarchitectonic areas (Fig. 2). However, due to the large densities of receptors on dendrites, which can reach across different areas tangentially, receptor patterns may sometimes cross cytoarchitectonic areal boundaries (Palomero-Gallagher and Zilles, 2019).

(iv) To have a distinctive pattern of connectivity to the rest of the brain. Indeed, the operations that can be performed by an area are in part determined by its extrinsic input and output (Passingham et al., 2002).

(v) To have a distinctive function and computational role. While this criterion is arguably the most central, it is also the most loosely defined and difficult to assess in practice. The most direct method is task functional localizers, where a stimulus-evoked activation is contrasted against a suitable baseline to isolate brain areas preferentially activated. However, this presupposes knowledge of what the relevant computations are (Kanwisher, 2010).

An additional criterion often used in practice is that brain areas are spatially contiguous patches. In a recent paper summarizing advances in brain parcellations in humans and nonhuman animals, Glasser and colleagues neatly formalize the definition of what constitutes a cortical area, adding areal topography to account for known sensory maps in cortex (Glasser et al., 2016):

> *A cortical area is a distinctive region of cortex that differs reliably from neighboring areas in one or more neurobiological properties from four basic categories:* **function, architecture, connectivity, and/or topographic organization**. *As shorthand, we will sometimes refer to these as the "FACT" categories.*

In sum, a brain area must have some distinctive properties. These properties are not always in agreement with one another. A cortical fold does not necessarily indicate a boundary between functional areas. As we already said, changes in receptor profiles do not always align with cytoarchitecture changes. Therefore the relevant features for defining a brain

**454 PART | IV** From streamlines to tracts

area are necessarily dependent on the question at hand. This chapter is about studying the brain as an interconnected network. Thus an important property for distinguishing brain areas is its *connectivity* features, or "fingerprints."

## 2.2 Connectivity fingerprints

The term *connectivity fingerprint* refers to the pattern of connections of a given brain location (voxel, vertex, or area). In human in vivo techniques, there is a distinction between structural/anatomical connectivity (as inferred by diffusion tractography) and functional/effective connectivity (as inferred through modeling of functional measurements with FMRI or electrophysiology). Given a measure of connectivity, a fingerprint may be defined as the connectivity to a predefined set of brain areas, to the rest of the gray matter, or as the pattern of associated white matter connections (Mars et al., 2018a).

In one of the earliest studies of connectional fingerprints in humans (Johansen-Berg et al., 2004), the pattern of tractography streamlines from medial prefrontal cortex was used as a fingerprint to separate medial premotor area 6 (SMA) from medial prefrontal area 8 (pre-SMA). The former was characterized by a predominance of projections to the corticospinal tract and lateral premotor cortex, while the latter had strong rostral prefrontal connectivity and strong projections to the inferior frontal gyrus (Fig. 2). This white matter fingerprint not only strongly dissociated the two areas but their boundary was aligned with the expected functional subdivision based on task FMRI. This type of white matter fingerprint has subsequently been used extensively to identify many areas of cortex, including subdivisions of the parietal lobes (Mars et al., 2011), the frontal lobes (Sallet et al., 2013), the temporal lobes (Roumazeilles et al., 2020), the insular cortex (Tian and Zalesky, 2018), and the occipital lobe (Thiebaut de Schotten et al., 2014). Thus, equipped with a suitable connectivity fingerprint, we can parcellate the entire brain. But why would we want to?

## 2.3 Brain parcellation—Why parcellate the brain?

Admittedly, the first question we should have asked in this section is why parcellate at all? Why not preserve the graininess or spatial resolution of the measurements and model the brain as a dense connectome, i.e., a point-to-point graph? There are two types of answers to this question, a practical answer and a conceptual one. With current imaging technology, working with a dense connectome in practice means $\sim 10^5$ nodes (voxels), and thus $\sim 10^{10}$ connections. Some methods are amenable to work on graphs of this scale (e.g., spectral methods), whereas other methods would struggle (e.g., finding shortest paths in the graph). But the measurement atoms (the voxels) bear no relation to the brain organization. Some brain areas may span a large number of voxels with much redundancy between the signals of different voxels, whereas other functional units may have subvoxel scale (e.g., cortical-columns). Parcellating the brain forces us to make sense of the data by finding the right level of abstraction, averaging out irrelevant fine-grained details and extracting essential features.

## 2.4 Brain parcellation—The methods

Many approaches to parcellation (segmentation, clustering, community detection, etc.) of the brain have been proposed, but they can all be segregated into two broad categories: agglomerative approaches and border-mapping approaches (Eickhoff et al., 2018).

Agglomerative methods attempt to group together voxels (or surface vertices) that have similar features. Border-mapping methods search for sharp transitions in the features between adjacent gray matter locations. Unlike agglomerative methods, border-mapping assumes that areas are topologically contiguous, although this assumption can also be explicitly incorporated into an agglomerative method (Blumensath et al., 2013; Tian and Zalesky, 2018). Both types of methods require the definition of a similarity function, i.e., a scalar function that quantitatively compares the features of two brain locations. Agglomerative methods further require a *cost function*, i.e., another scalar function that quantifies the quality of the clustering. For example, in the case of the $k$-means clustering, this function quantifies within cluster and between cluster similarities, where the objective is to maximize the former and minimize the latter.

Once all voxels have been assigned a cluster label, it is often useful to visualize the quality of the clustering by representing the voxel-wise features in a lower dimensional space, where we can simultaneously visualize the clustering assignment and the feature dissimilarities (Fig. 2). This type of visualization can also help address the old question of how many clusters are supported by the data.

Similarly, border-mapping methods provide a natural way to visualize both the clustering and the quality of the clustering: the sharpness of the borders are a measure of how much the data supports a division of two parcels. This naturally leads to the question of where to draw the line (pardon the pun) between a sharp border and a smooth transition between areas. This question can be addressed with both border-mapping and low-dimensional embedding.

Another important distinction between parcellation approaches is in the amount of anatomical knowledge that is injected into the algorithm. At one extreme, which can be called "hypothesis-driven," one can look for parcels that are highly connected with specific predefined brain areas. For example, in their pioneering work, Behrens and colleagues parcellated the thalamus into different subregions that preferentially connect to specific cortical lobes or parcels, on the basis that thalamic nuclei have been shown from tracer studies to have preferential cortical connections. At the other extreme, which can be called "data-driven," many proposed cortical parcellation methods utilize the connectivity profile to the entire brain to drive the parcellation (Johansen-Berg et al., 2004). This is useful when there is no a priori knowledge about the connections, such as is often the case in humans. Less common, but perhaps even more powerful, are approaches that are a combination of data-driven and hypothesis-driven. For example, Saygin parcellated the amygdala based on carefully crafted anatomical hypotheses on the underlying structural connections (Saygin et al., 2011). This helped reduce false-positive/negative connections but still left enough freedom for the data to drive the parcellation. This approach is a promising avenue for connectivity-driven parcellations, where information from nonhuman primate tracer data can be used to refine the connectivity profiles that guide the parcellations.

## 2.5 Connectivity-based parcellation

Parcellation requires features. We have already discussed the many different types of regional properties that can be used to define a brain area (and thus drive parcellation), such as cortical morphology, cytoarchitecture, receptor-architecture, function, and connectivity. We will focus here on methods that are available in vivo, since variability between individual brains limits the utility of atlases derived from postmortem data.

Among in vivo measures, there are arguments for favoring connectivity as the feature that is driving the parcellation. One argument refers to the general principle that we are interested in breaking down the brain into component units and studying how they interact. It is natural to *define* these units based on how they interact. There is a nice symmetry in this approach: the interaction (connectivity) defines the units, and the units are the things that interact. Another practical, yet important, argument is that since our goal is to do network modeling, it is important to not have heterogeneity in the connectivity within a node. Such heterogeneity can greatly bias the estimation of the network edges. For example, Smith et al. (2011) used extensive simulations of FMRI-like functional connectivity in small networks. They showed that when estimating network edges, even a small amount of between-area mixing (inducing heterogeneity within regions) can have a dramatic impact on network edge estimation. However, network analyses do not always require accurate estimation of edge strength. In some cases, where summary measures of network structure are derived, the exact definition of the nodes can have less impact (Zalesky et al., 2010b).

Nevertheless, it is natural, and in fact common, to use connectivity measures to drive brain parcellations. But there is also an argument to be made in favor of multimodal parcellations.

## 2.6 Multimodal parcellation

Although we have argued for connectivity as the main driver for informing parcellation, multimodal approaches are also becoming popular. By multimodal we mean combining multiple facets of regional differentiation, such as microstructure measurements (myelin-dependent contrast being the most utilized in MRI), cortical folding, and functional connectivity. Together, these multiple types of measures can serve as proxies that cover all the FACT categories mentioned earlier.

One advantage of multimodal parcellation is that all measurements have some degree of inaccuracy and bias. For example, in the next section we discuss in detail the biases in using diffusion MR tractography to infer precise point-to-point connectivity in convoluted cortex, which limits the border-mapping methods that use structural connectivity. These biases are absent when measuring functional connectivity. Thus combining multiple contrast mechanisms and features can help reduce biases. A compelling example of a multimodal parcellation of the human brain was proposed by Glasser et al. (2016). The primary modalities driving their parcellations were resting-state functional connectivity and myelin-sensitive mapping using T1- and T2-weighted MRI. The myelin-sensitive contrast gives a great delineation of primary areas, whereas functional connectivity was useful for finding parcels in association cortex.

However, a clear problem with a multimodal approach is: what to do when the modalities disagree? If there is a border in functional connectivity features but no border in microstructural features, should we parcellate or not? In some cases the answer is simple: myelin has very good contrast to distinguish between primary and association cortex, but very little contrast between association areas. Thus we should favor other types of features for this part of the brain in order to decide where to draw the borders. But the general question of how to combine information from multiple modalities to drive parcellation is still open.

## 2.7 Gradients and soft parcellations

Thus far, we have talked about brain areas as if they were homogeneous entities with well-defined boundaries, in the same way that we can talk about the liver or heart or any other body organ. This is of course an oversimplification. Often, rather than sharp borders, there is a smooth transition (gradient) between brain areas. There may be heterogeneity and internal organization *within* a well-defined brain area. For example, many visual areas contain a 2D retinotopic map, and there are stronger connections among these visual areas for locations that have similar receptive fields. Thus there is internal heterogeneity in both function and connectivity. Generalizing this to nonsensory cortex is difficult, as we do not know what maps to look for.

Soft parcellation approaches, such as independent component analysis (Hyvärinen, 1999), have been developed to allow a certain degree of overlap, or partial volume, between different areas. These approaches have predominantly been used to analyze functional MRI connectivity (Beckmann and Smith, 2004) but have also been used on structural connectivity (O'Muircheartaigh and Jbabdi, 2018; Thompson et al., 2020). Another, more recent set of approaches has emerged that embraces the idea of internal areal organization and explicitly searches for a representation of these organizations, referred to as connectivity gradients (Huntenburg et al., 2018). They generally do so by mapping the dense connectivity information into a lower dimensional space, where the axes of this embedding space correspond to the principal gradients of connectivity. A similar approach was used to represent the SMA and Insula connectivity in Fig. 2. Model selection can be used to quantitatively evaluate whether continuous gradients or discrete boundaries provide the most parsimonious representation across the cortex, yielding hybrid gradient-boundary representations (Tian and Zalesky, 2018). This approach has recently been applied to parcellate the human subcortex (Tian et al., 2020).

## 2.8 Looking forward

While this chapter has a general focus on building connectomes and analyzing them as a graph, brain parcellation is a research area in its own right with applications that go beyond finding nodes for graph theory. Here we finish this section by looking at recent developments and open questions for brain parcellation.

One powerful insight is that assigning features to brain locations equips these locations with a *representation*. For example, by describing the connectivity fingerprints of a brain area, we can analyze the data in this feature space, independent of other potentially irrelevant properties such as area size, extent, shape, or physical location. This allows comparisons between areas across the brain of an individual, across the brains of different individuals, and across the brains of different species. For example, Mars et al. (2018b) assign connectional fingerprints to all cortical locations in humans and macaques. Importantly, these fingerprints are matched between the species, because they are defined in terms of connections to major white matter tracts which are known to be homologs (Warrington et al., 2020). This connectional representation allows mapping the brains of these different species into the same feature space, thus enabling us to address comparative/evolutionary neuroscience questions.

Another path of ongoing development is the idea of internal areal organization, or gradients. An interesting question is whether we can identify multiple patterns of connectional organization within a given area (Haak and Beckmann, 2020; Jbabdi et al., 2013). The coexistence of spatial patterns of connections within a single cytoarchitectonic area suggests that neuronal activity may also follow spatial patterns dictated by superimposed connectivities. Overlapping connection patterns may interact in complex ways to enable more elaborate functions. For example, a region that contains two or more overlapping sensory maps, such as the superior parietal cortex (Sereno and Huang, 2006), can efficiently calculate transformations between these maps that a region with a single map cannot. It has also been suggested that overlapping maps may be useful for representing complex multidimensional stimuli or actions onto the two-dimensional cortical surface (Graziano and Aflalo, 2007). Identifying spatial patterns of connectivity may allow us to have a new interpretation of functional organization. The concept of a functional unit being constrained to a cytoarchitectonic area may be extended by considering spatial anatomical patterns that can give rise to predictable brain activity (Saygin et al., 2012).

Finally, one may ask: what is the right parcellation? The answer is probably that there is no single "right" parcellation. Different parcellations will be obtained with different modalities and different features. As a case in point, consider the somatosensory motor cortex (Fig. 3). By using diffusion tractography to produce whole-brain connectivity fingerprints, we can subdivide this region according to the somatotopy axis. But if we restrict the connectional fingerprint to only include connections to the thalamus, the main axis of organization is now anterior-posterior (i.e., sensory vs motor). This is because such an axis of organization is mirrored in the thalamus. Of course, both are valid *functional* axes; the second additionally overlaps with an axis of cellular organization. Thus deciding which features to utilize for defining the parcels is ultimately guided by the aims of the investigator.

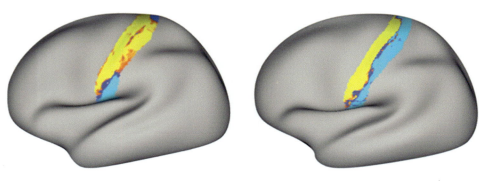

**FIG. 3** Feature selection. Parcellation of the sensorimotor strip leads to separation along different axes depending on the features used. *(Figure courtesy of Michiel Cottaar.)*

## 3 Mapping connectivity and connectomes

Connectome edges between pairs of nodes reflect the underlying network wiring that allows information transfer between brain regions and determines what the brain can and cannot compute (Mill et al., 2017; Van Essen, 2013). Using MRI, edges cannot be measured directly; they are inferred from the data (Behrens and Sporns, 2012). This *estimation* process can be performed using different neuroimaging modalities, deriving in each case regional pairwise connectivity, in slightly different (yet complementary) ways (Jbabdi et al., 2015).

For instance, synchronization in recorded spontaneous activity between regions, as measured by resting-state functional MRI or MEG, can be thought of as belonging to the same subnetwork and is therefore "connected," directly or indirectly (Smith et al., 2013b). The degree of synchrony or association of resting-state activity can be used as a proxy for functional connectivity. The degree of similarity in regional responses to the same task conditions and stimuli can be used as another estimate of functional connectivity (Toro et al., 2008). Similarity of cytoarchitectonics (for instance, cortical thickness as measured using anatomical MRI) has also been used as a different proxy for the degree of connectivity between regions (Evans, 2013).

A more intuitive approach to estimate connections from MRI is to probe the fiber pathways that connect gray matter regions through white matter. A network edge between two regions then corresponds to the fiber pathway that connects them. Diffusion MRI inherently provides this capability and is the focus of this section. Axonal fiber bundles are organized coherently such that water diffusion occurs preferentially along the orientations of least hindrance, which are typically parallel to the fibers. In contrast, diffusion is maximally hindered in the perpendicular direction. The preferred diffusion orientations can be indirectly mapped to fiber orientations at a voxel-wise level (Dell'Acqua and Tournier, 2019; Seunarine and Alexander, 2009). Tractography approaches then integrate the voxel-wise information at a global scale and propagate curves that are maximally tangential to the local preferred diffusion orientations (Basser et al., 2000). These curves provide estimates of white matter bundles (Catani and Thiebaut de Schotten, 2008).

Despite being seemingly well suited for this task, diffusion MRI is an indirect approach for probing brain connections; it measures the diffusion scatter pattern of water molecules within brain tissue. Connectivity inferences are made upon the assumption that paths of least hindrance to diffusion of water molecules are proxies for coherent tissue structure, such as the one expected along axonal bundles. This assumption works reasonably well for localizing the core of major white matter tracts (Catani and Thiebaut de Schotten, 2008; Garyfallidis et al., 2018; Warrington et al., 2020; Wassermann et al., 2016; Yendiki et al., 2011). However, when using tractography to map brain networks and connectomes, an additional set of challenges exists—tracking through the "superficial" white matter (Van Essen, 2013) (Fig. 4A), i.e., some tractography paradigms are not well suited to navigating the interface with cortex and subcortex. Here, we overview some of these challenges and potential solutions.

### 3.1 Deep white matter tracking

Conventional tractography approaches are optimized for reconstructing the core of fiber bundles within deep white matter. In the absence of more robust and direct alternatives, they are the method of choice for estimating connectomes.

Tractography reconstructs pathways that provide the least hindrance to diffusion. The local (voxel-wise) preferred diffusion orientations are typically mapped into fiber orientations through a deconvolution process; either parametric (Anderson, 2005; Behrens et al., 2007; Dell'Acqua et al., 2007) or nonparametric (Jeurissen et al., 2014; Tournier et al., 2007). This information can be utilized in three main ways: (a) *Local* propagation (Basser et al., 2000;

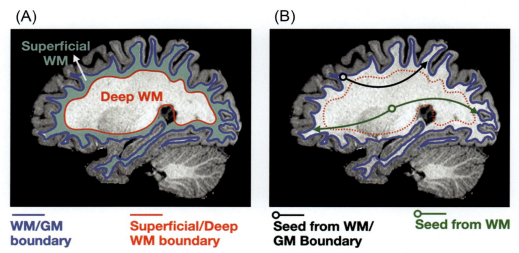

**FIG. 4** (A) Deep and superficial white matter with respect to the white matter/gray matter boundary. (B) Different tractography seeding strategies for obtaining a connectome either from many locations at the WM/GM boundary (*black arrow*) or from many locations in white matter (*green arrow*). The former approach sets the starting point of an edge and seeks the termination point. The latter strategy looks for both endpoints of an edge.

Behrens et al., 2003b; Mori et al., 1999; Parker et al., 2003), which in a greedy, step-by-step fashion propagates curves (or *streamlines*) that are tangent to vector fields representing fiber orientations. (b) *Global* tracking, for instance (Campbell et al., 2005; Daducci et al., 2015; Iturria-Medina et al., 2007; Jbabdi et al., 2007; Kreher et al., 2008; Reisert et al., 2011; Sotiropoulos et al., 2010b), which estimates paths that are optimally fitted to the entirety of available fiber orientations, rather than on a voxel-by-voxel basis. Such paths are not necessarily tangent at every point of their route to the local fiber orientations and, in principle, they are more immune to local errors. (c) *Pseudo-global* approaches, such as (Pestilli et al., 2014; Smith et al., 2013a), which evaluate the goodness of fit of previously obtained local streamlines to some global function of the dMRI data post hoc. These approaches aim to improve the biological validity of connection weights under the assumption that the global cost minimization ensures neuroanatomical fidelity.

Local streamline methods have been more widely used for connectome reconstruction than the other approaches. They can be further subdivided into deterministic and probabilistic techniques, depending on whether they perform a deterministic or stochastic estimation. Deterministic methods (Basser et al., 2000; Mori et al., 1999) provide a point estimate of the path of least hindrance to diffusion between two points. Probabilistic methods (Behrens et al., 2003b; Parker and Alexander, 2003) estimate a spatial distribution for this path. Several studies have evaluated the test-retest reliability of these methods for connectome reconstruction (Bonilha et al., 2015; Buchanan et al., 2014; Prckovska et al., 2016). Estimates derived from probabilistic tractography generally show greater connectome reproducibility than deterministic methods, reduce the effect of residual spatial misalignment errors, and potentially improve some of the statistical properties of the sampled paths (i.e., normality). However, probabilistic tractography can reconstruct numerous spurious trajectories, leading to poor specificity (Zalesky et al., 2016). In particular, probabilistic tractography yields greater spatial dispersion in streamline trajectories, which may lead to more spurious connections. Connectomes derived from deterministic tractography generally comprise fewer edges, but show greater variation within and across individuals, particularly in data with low angular resolution or low signal-to-noise ratio (see Section 4 in this chapter).

The *seeding strategy* for tractography can impact connectome reconstruction (Li et al., 2012). Typically, streamlines can be initiated from all white matter voxels and the streamlines intersecting pairs of nodes are mapped to the respective edge (Fig. 4B). This "brute force" approach is suggested to be more sensitive at detecting long white matter bundles (Hagmann et al., 2008). An alternative is to seed from the boundary between white and gray matter (Girard et al., 2014; Smith et al., 2012). Boundary seeding may be associated with fewer biases, including less gyral bias (Schilling et al., 2018), better predictions of path length distributions, and better sensitivity and specificity (Donahue et al., 2016; Smith et al., 2015a). However, reconstructing long fibers can be challenging with boundary seeding.

Regardless of the significant implementation differences across different methods and strategies, tractography's neuroanatomical fidelity relies on a number of conditions: (i) There is adequate anisotropy in the signal to extract reliable *orientational* information from the diffusion scatter pattern; (ii) it is possible to *map* preferred *diffusion* orientations to fiber orientations; (iii) the relevant *architecture* and structure of the underlying fiber bundles *can be recovered at the imaged spatial resolution*. There are clear indications and neuroanatomical scenarios where these conditions are reasonably met; for instance, separating functionally different, but spatially neighboring, parts of the superior longitudinal fasciculus

(Thiebaut de Schotten et al., 2011) or distinguishing connectivity profiles of neighboring gray matter regions (Johansen-Berg et al., 2004; Mars et al., 2011). However, in the general case of connectome estimation, fulfillment of these conditions is challenging, as we discuss in the next section.

## 3.2 Path termination and superficial white matter tracking

While robust identification of tractography endpoints in gray matter can assist delineation of large bundles in white matter (Wassermann et al., 2016), it is not necessary for such a task, as large bundles can be defined based on gross anatomical guiding principles. However, accurate tractography of the gray matter is an essential component of connectome reconstruction and is inherently linked to the spatial fidelity and resolution of connectome reconstructions. It is this aspect that challenges tractography algorithms compared to more conventional WM bundle localization applications.

Identifying fiber terminations is inherently difficult with tractography (Jbabdi and Johansen-Berg, 2011) because propagation cannot be terminated in an unsupervised manner and ad-hoc heuristics are needed to determine endpoints. This makes termination criteria an important choice and is the reason why anatomically driven rules can improve reliability of connectomes (Girard et al., 2014; Hernandez-Fernandez et al., 2019; Smith et al., 2012). For instance, crossing the white/gray matter boundary multiple times is likely to yield spurious trajectories and propagation within the cortex may be prone to greater errors due to low anisotropy in gray matter.

In the vicinity of cortex, another obstacle that biases the estimation of termination points is the presence of passing white matter fibers, such as the U-fibers that run parallel to the WM/GM boundary (Reveley et al., 2015). The density of these fibers is high at the sulcal fundi, resulting in diffusion MRI-estimated principal fiber orientations that run parallel to the sulcal surface. It is therefore difficult for tractography to traverse the boundary and escape white matter in these regions. This underrepresentation of tractography streamlines at the sulci compared to gyri was first described in Van Essen et al. (2013) and later confirmed in other studies (Schilling et al., 2018) as *gyral bias* (Fig. 5). This bias has two effects on tractography (Cottaar et al., 2021): (a) Streamlines propagating toward cortex tend to terminate at the gyral crowns, (b) Streamlines propagating away from the cortex tend to remain near the gyral walls. These effects can create artificial termination density patterns and will be more evident for finer parcellation schemes and a large number of nodes. Yet, they can also introduce a confound for coarser parcellations, particularly when average curvature and sulcal depth profiles vary considerably across nodes.

For these reasons, several recent tractography approaches attempt to address challenges linked with superficial white matter tracking. Estimation of fiber orientations can include more explicit models of partial volume compartments and multishell acquisition schemes to provide a better separation of GM and WM components at the vicinity of their boundary (De Luca et al., 2020; Jeurissen et al., 2014; Sotiropoulos et al., 2016). Higher spatial and angular resolution data can also assist (Sotiropoulos and Zalesky, 2019). However, as accurate orientation information on its own cannot provide a full solution (Yeh et al., 2019), new tractography paradigms have also been proposed. In one approach, fiber orientations within the cortical ribbon are constrained to be radial to the WM/GM boundary surface extracted from an anatomical image, and flow preservation rules are used to successfully track into/out of various locations along the boundary (St-Onge et al., 2018). Another approach uses a semiglobal generative model for fiber orientation within the gyral blades as a continuous vector field (Cottaar, 2020). This field can be fitted globally to the diffusion data, but can also incorporate constraints on fiber

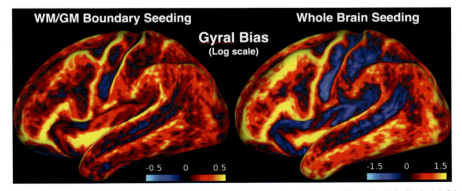

**FIG. 5** Gyral bias in tractography termination points on the WM/GM boundary surface for boundary (*left*) and whole-brain (*right*) seeding strategies. The number of streamlines on a dense grid of surface vertices was enumerated. The average streamline density for zero-curvature points on the surface was then computed and used as a reference. The relative streamline density at every surface vertex was calculated with respect to the zero-curvature density. This relative density is presented in log-scale. High numbers at gyral crowns correspond to a higher number of streamline termination points than zero-curvature locations. Low numbers at sulcal fundi correspond to a lower number of streamline termination points than zero-curvature locations. The gyral bias is smaller for WM/GM boundary seeding.

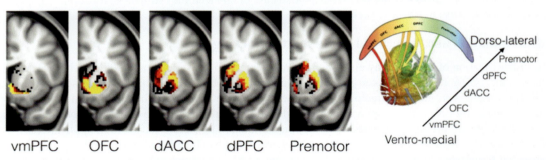

**FIG. 6** Pattern of corticostriatal connectivity revealed by tractography. Seed locations were in different cortical regions, including vmPFC (ventromedial prefrontal cortex), OFC (orbitofrontal cortex), dACC (dorsal anterior cingulate cortex), dPFC (dorsal prefrontal cortex), and premotor cortex. Path probabilities (*yellow*: high, *red*: intermediate, *black*: low) were obtained using probabilistic tractography on Human Connectome Project data and are shown within the striatum. These reveal a pattern from ventral to dorsal striatum preferentially connecting to different parts of the cortex. Note the strong similarity to patterns of tracer-based connectivity reported in the macaque monkey, shown in the sketch on the *right*. (*Reproduced with permission from Haber, S.N., Behrens, T.E.J., 2014. The neural network underlying incentive-based learning: implications for interpreting circuit disruptions in psychiatric disorders. Neuron 83, 1019–1039.*)

orientation and density at the cortical boundary. The field obtained provides a one-to-one mapping from the highly convoluted WM/GM boundary to a much flatter superficial/deep white matter boundary (blue vs red in Fig. 4A), from which conventional deep white matter tracking can be less challenging (Cottaar et al., 2021). This also allows better cortical termination constraints and rules to be globally fitted to data. In Wu et al. (2020), a reduction in gyral bias is achieved by encouraging a smooth transition between the radial fiber orientation in the gray matter and the tangential orientation underneath using asymmetric fiber orientation distribution functions (Bastiani et al., 2017).

Density constraints on streamline endpoints can also be added as part of streamline filtering approaches (Daducci et al., 2016), such as Contrack (Sherbondy et al., 2010), SIFT/SIFT2 (Smith et al., 2013a), LiFE (Pestilli et al., 2014), or COMMIT (Daducci et al., 2015). These algorithms have in common that they filter or assign weights to streamlines produced by local tractography algorithms to represent their relative contribution. While these weights are only fitted to the diffusion MRI data, density of termination endpoints could be added as an additional constraint. However, this does require having a sufficient population of streamlines connecting to the sulcal walls and fundi to begin with, i.e., these methods cannot assist in the cases of local tractography false negatives. For instance, streamlines connecting sulcal fundi at both ends are so rare in tractography (but not in brain anatomy) that even after postprocessing they may be underrepresented in the final fiber population. Postprocessing approaches might achieve a reduction of gyral bias simply by upweighting the fundi-to-crown connections and not including the many fundi-fundi connections. Proper superficial white matter tracking models can reconstruct such connections (Cottaar et al., 2021).

In the subcortex and cerebellar nuclei, termination problems also exist, in the sense that using tractography to find synaptic termination locations within these nuclei is challenging. A difference, however, with cortical termination is that the inherently higher anisotropy within major subcortical structures allows topographical organization and connection patterns to be probed within their volumes; for instance, see Behrens et al. (2003a) and Tziortzi et al. (2014) and the example in Fig. 6. In the case of cortex, due to low anisotropy, contrast, and the inherent resolution limits, sensitivity is greatest to connectional patterns *along* the cortical sheet (WM/GM boundary).

In summary, there are many challenges to address in achieving accurate termination with relatively high spatial resolution and this is a field of active research. Regardless of the unmet challenges, however, current approaches for estimating brain connectivity provide unique "unobserved imaging contrasts" that can probe brain architecture in ways impossible without diffusion MRI (Jbabdi et al., 2015; Mars et al., 2018a). The depth and breadth of information that can be extracted and encoded is reflected in the quantifiability of the edges.

## 3.3 Quantifying edges

Ideally, the weights of the connectome edges should represent something quantitative about the connections, such as the number of synapses or cells receiving/sending long-range axons. While such information can be gleaned for mesoscale connectomes obtained using chemical tracing (Markov et al., 2014; Oh et al., 2014), diffusion MRI tractography cannot provide such direct measures (Jbabdi and Johansen-Berg, 2011; Jones et al., 2013). Instead, tractography allows estimation of edge weights that indirectly reflect some relevant quantitative properties. These range from simple binary values, denoting the presence or absence of an edge, to approximations of biophysical properties of connections, reflecting micro- or macrostructure.

The likelihood of a connection or "connection strength" is typically quantified using some function of *streamline counts*, or the number of tractography streamlines intersecting a pair of regions. Diffusion path probabilities, obtained from probabilistic tractography, reflect normalized conditionals of streamline counts given the orientation model, seeding strategy, and termination/counting criteria (Behrens et al., 2007). Streamline counts and functions thereof can be enumerated for all pairs of regions to populate the cells of a connectivity matrix (Hagmann et al., 2008; Sporns, 2011). Streamlines that are permitted to propagate within gray matter can intersect more than two distinct regions, in which case they contribute to the streamline count for multiple pairs. Anatomical constraints can be imposed to avoid such scenarios, as discussed before (Hernandez-Fernandez et al., 2019; Smith et al., 2012).

Streamline counts can be symmetrized, normalized or transformed in various ways (Iturria-Medina et al., 2007), which aims to reduce the effect of confounds reflecting algorithmic choices and potentially ensures better consistency across subjects. Logarithmic transformations are sometimes applied to the streamline counts before analysis to achieve approximate normality. Normalization by node sizes (Hagmann et al., 2008) can be used to account for volume/area variability in the chosen gray-matter parcellation. While it may be that larger brain regions are indeed more strongly connected by virtue of anatomy, a greater number of streamlines is likely to terminate in regions with a larger interface between gray and white matter due to the tractography process (Girard et al., 2014; Zalesky et al., 2010b). Considered further in the next section of the chapter, streamline filtering (Daducci et al., 2015; Pestilli et al., 2014; Smith et al., 2013a) provides an indirect way of modifying raw streamline counts, which increases reproducibility and in certain scenarios the biological relevance of the obtained edge weights (Smith et al., 2015a).

Alternative metrics that reflect *microstructural* properties along edges can be considered as edge weights. For instance, voxel-specific measures of anisotropy can be averaged over all voxels traversed by a path that is assigned to a particular pair of nodes (Lemkaddem et al., 2014). The resulting tract-averaged measure thus characterizes the anisotropy of a connectome edge. Other microstructural measures can also be used (Hagmann et al., 2010), such as measures of axonal myelin derived from images of magnetization transfer ratio (Mandl et al., 2010; Van den Heuvel et al., 2010). Different weighting functions can be employed for the averaging process to give greater weight to different parts of the tract (e.g., using streamline counts/probabilities in each voxel can give greater weight in the main tract core relative to its periphery). Features that reflect tract *macrostructure* are another possibility for edge quantification. In fact, volume and cross-sectional area of paths intuitively relate more directly to connection strength than their microstructure counterparts; they are, however, more challenging to accurately measure and new tractography paradigms may be needed (Close et al., 2015).

## 3.4 Quantification biases

A major shortcoming for quantifying connectome edges is that none of the preceding approaches provide an interregional measure of the number of connecting axons, which is a desirable measure of connectivity strength in many applications. Tract-averaged microstructural measures may provide an interpretable biophysical property per edge. However, it is questionable how informative such properties are when treating connectomes as networks, which would require some proxy of connectivity. Some investigators report no correlation between such microstructural measures and axonal strengths measured by tracers (van den Heuvel et al., 2015). On the other hand, functions of streamline counts may be more relevant in such a network context (Donahue et al., 2016). However, factors that reflect data quality, algorithmic choices, and inherent limitations bias these measures (Jbabdi and Johansen-Berg, 2011; Sotiropoulos and Zalesky, 2019).

The *gyral bias,* discussed earlier, induces an unrealistic spatial distribution of connections along the cortex concentrated at the gyral crowns. It remains to be seen whether superficial white matter tracking approaches can provide a comprehensive solution to this challenge. In addition, a *distance bias* is reflected in connectome weights. Streamline counts between distant regions, interconnected by longer tracts, are often smaller than counts between neighboring regions. Algorithmic limitations contribute to this pattern; for instance longer tracts are more difficult to reconstruct with tractography because streamlines must be propagated for a longer distance and each propagation step provides an opportunity for "wrong turns" (Zalesky, 2008; Zalesky and Fornito, 2009). However, connection strengths as measured by tracers follow an exponential decay with connection length, with the majority of connections being short and strong and the long connections being weak, comprising fewer axons (Beul et al., 2015; Ercsey-Ravasz et al., 2013). Tracers of course have their own sources of error and the extent to which the algorithmic distance bias of tractography reflects the biology remains to be explored.

The *seeding* strategy can also have a considerable impact on the edge weights. When streamlines are seeded from all white matter, longer tracts are inevitably sampled more abundantly because they occupy a greater volume than shorter tracts. To compensate for this bias toward long tracts, the streamline count can be normalized by the average length of

**462 PART | IV** From streamlines to tracts

the streamlines contributing to the count (Roberts et al., 2016). However, given that many tracts are sheet-like and vary considerably in cross-sectional area and morphology, simple normalization factors such as the streamline length might not adequately correct for the oversampling of tracts occupying greater volumes. Initiating streamlines from the WM/GM boundary interface is an alternative that overcomes this limitation and can potentially provide more realistic path length distribution (Donahue et al., 2016; Girard et al., 2014; Smith et al., 2015a), but this seeding approach can have difficulties in tracing out long bundles.

### 3.5 Summary

Tractography was originally developed to virtually dissect particular fiber bundles of interest with diffusion MRI. The now widespread use of tractography as a high-throughput, quantitative technique to infer the strength of connectivity between potentially thousands of pairs of regions presents numerous challenges for tractography in the era of connectomics. We outlined several of these challenges in this section and briefly discussed new approaches that have been developed to overcome them. Key challenges include the difficulty of identifying fiber terminations, particularly due to the bias for streamlines to terminate in gyral crowns, as well as the elusive and often misconstrued interpretation of interregional streamline counts. Streamline counts were the first proposed measure of structural connectivity and they have remained the default measure ever since. While enumerating the number of streamlines between a pair of regions lacks sound theoretical justification as a measure of interregional connectivity, streamline counts have proven to be incredibly useful in practice and yield connectomes that recapitulate key network properties found in nervous systems mapped using alternative neuroimaging modalities, as discussed further in the next section.

Advances in diffusion MRI acquisition protocols and hardware, next-generation tractography algorithms, and novel streamline filters are some of the ways forward to overcome these challenges. We considered the pros and cons of global and local tractography algorithms in the context of connectome mapping and highlighted the importance of minimizing the reconstruction of spurious connections. Selecting a tractography algorithm to map connectomes should be guided by numerous factors and it is difficult to provide a universal recommendation. Global tractography algorithms are an attractive choice, yet they are complex, computationally burdensome, and not widely used. The choice between deterministic and probabilistic methods depends to a large extent on whether the reconstruction of false positive or false negative connections is deemed most detrimental for the application at hand. Probabilistic tractography is more prone to reconstructing false positive connections, although we briefly introduced normalization procedures, thresholding, and postreconstruction filters that aim to reduce the impact of spurious connections and improve connectome validity. Streamline filters and thresholding will be considered in greater detail in the next section of this chapter. Deterministic tractography is a simple and reliable approach that yields relatively sparse connectomes, but it is often criticized for overlooking connections.

It is important to remark that algorithms for assigning streamline endpoints to specific nodes of the connectome are an important consideration, which is distinct from the choice of tractography algorithm. Streamline-to-node assignment procedures are based on heuristics and they can substantially influence the final connectome (Yeh et al., 2019), yet the choice of heuristic is usually given much less consideration than the choice of tractography algorithm and brain parcellation atlas. This underscores the complexity of connectome mapping and the wealth of choices that confront an investigator when establishing an algorithmic workflow.

## 4 Connectome accuracy and validation

The interregional streamline counts that most commonly populate connectivity matrices derived from tractography do not necessarily provide a direct measure of connectivity strength (Jones et al., 2013). They rather characterize the probability with which routes of least resistance to the diffusion of water molecules can be reconstructed in white matter (Jbabdi and Johansen-Berg, 2011; Johansen-Berg and Behrens, 2006). While these routes recapitulate the spatial trajectories of axonal fiber bundles quite well (Dauguet et al., 2007; Schilling et al., 2019), mapping connectomes and connectivity usually only utilizes the number of streamlines interconnecting pairs of regions, discarding the rich spatial information that tractography was originally designed to recover.

Streamline counts might have been considered a byproduct of tractography experiments in the early days. But in the connectomics era, they have become one of the primary outputs. The extent to which streamline counts can provide a meaningful measure of connectivity strength for network neuroscientists remains an open question, and it is unlikely that tractography pioneers would have envisaged the prolific use of streamline counts today. Given that there is no principled reason why modeling routes of least resistance to water diffusion would furnish a valid measure of structural connectivity strength, the need to validate the accuracy of connectomes mapped with tractography is imperative.

Connectome validation is usually undertaken by comparing tractography-derived connectomes to connectomes for which the ground truth is known with high confidence, either by construction, in the case of physical and numerical connectome phantoms, or through the use of invasive tract tracing techniques in animal models and postmortem brains. Fortunately, the correspondence between ground truth and tractography-derived connectivity measures is fair to quite good, depending on the tractography algorithm, diffusion model, and type of ground truth (Sotiropoulos and Zalesky, 2019). Validation experiments can also assist with evaluating model accuracy and quantifying the extent to which well-known tractography biases influence streamline counts. While streamline counts are undoubtedly affected by numerous methodological biases and lack sound theoretical justification as a measure of connectivity, in practice, they provide accurate predictors of clinical response (Horn et al., 2017), facilitate circuit-based localization of treatment targets (Sammartino et al., 2016), and significantly associate with age (Betzel et al., 2014), behavioral measures (Zalesky et al., 2011), and functional connectivity (Suárez et al., 2020).

In this section, we first examine efforts to validate the accuracy of tractography-derived connectomes using tract-tracing experiments and connectome phantoms. We then consider postconstruction methods to improve connectome accuracy, including streamline thresholding and filtering techniques. Finally, we conclude this section with a discussion of potential ways forward to improve connectome accuracy and validity.

## 4.1 In vivo validation

The anatomical gold standard in tractography validation is classical fiber dissection (Klingler, 1935) in postmortem brains (Martino et al., 2011). However, fiber dissection is painstaking work and most suited to validating the trajectories and spatial extent of fiber tracts, rather than quantifying their connectivity strength. Tract tracing is a more scalable and quantitative validation technique than dissection. A chemical tracer is injected at a specific brain site, where it permeates into neuronal cell bodies at the site and is transferred to peripheral axonal terminals via active axonal transport. Retrograde tracers enable tracing in the reverse direction, from synapse to soma. Interregional connectivity strength between the injection site and distant sites of tracer uptake can be quantified based on the fraction of neurons labeled by the chemical tracer (Bohland et al., 2009). This process can be repeated for several injection sites to build up a whole-brain connectivity matrix (Markov et al., 2013; Oh et al., 2014; Zingg et al., 2014). Tractography-derived connectivity matrices can be quantitatively compared to their tracer-derived counterparts (Fig. 7A). The agreement is generally modest to quite good and certainly above chance levels (Sotiropoulos and Zalesky, 2019).

Connectivity strengths derived from tracers span up to five orders of magnitude across fiber lengths (Markov et al., 2013) and streamline counts recapitulate this variation to a large extent (Donahue et al., 2016). Most studies report correlations between streamline counts and tracer-derived connectivity strengths that range between 0.4 and 0.8 (Calabrese et al., 2015; Donahue et al., 2016; van den Heuvel et al., 2015). However, the devil is in the details and the extent of agreement depends on the tractography pipeline and other key experimental factors (Ambrosen et al., 2020; Chen et al., 2015). Deterministic tractography algorithms guided by the diffusion tensor often overlook connections uncovered by tract-tracing experiments, whereas probabilistic algorithms coupled with unconstrained fiber orientation models tend to reconstruct spurious connections. Many tracer studies thus report a trade-off between connectome sensitivity and specificity (Azadbakht et al., 2015; Calabrese et al., 2015; Knösche et al., 2015; Schilling et al., 2019; Thomas et al., 2014). Probabilistic methods often yield reconstructed connectomes with high connection densities (Girard et al., 2020).

Tracer and histological validations are not without their limitations. There is an inherent mismatch in spatial resolution between connectivity inferred from tractography and invasive tract tracing. Whereas tractography is a macroscale technique capable of resolving fiber bundles up to a millimeter resolution, tracers can in principle map individual axonal projections that are invisible to tractography (Kennedy et al., 2013). Tracers also reveal the direction of axonal projections and provide directed connectivity matrices, whereas tractography cannot resolve directionality. Furthermore, the injected tracer can be taken up by axons traversing, but not synapsing, with the injected site, leading to reconstruction of spurious connections. The replicability of tracer experiments also remains to be established, with one study reporting up to two orders of magnitude variation in connectivity strength between replication experiments (Markov et al., 2013). Hence, while tracers have validated tractography-derived connectomes to a certain extent, these validations are confounded by several challenges. Some of these challenges can be overcome with connectome phantoms.

## 4.2 Connectome phantoms

Fiber phantoms can be constructed by arranging lengths of synthetic fibers such as nylon into tightly packed cylindrical bundles, usually encased in water (Fieremans et al., 2008; Perrin et al., 2005; Poupon et al., 2008; Tournier et al., 2008;

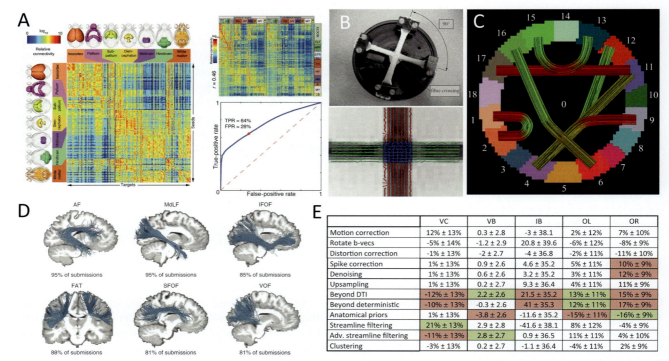

**FIG. 7** Validation of tractography-derived connectomes with tract-tracing studies and connectome phantoms. (A) The tractography-derived mouse connectome (*left*, 296 regions) recapitulates the tracer-derived connectivity matrix (*top-right*) with reasonable accuracy ($r = 0.46$). Anterograde viral tracers were injected at approximately 500 unique sites. The receiver operating characteristic curve (*bottom-right*) exemplifies the inherent trade-off between false positives and false negatives. Thresholding enables arbitration of this trade-off. (B) Physical phantom comprising two crossing bundles of rayon fibers (*top*). Fibers are 17 μm in diameter and enclosed in a water-filled glass container with a square cross-section of 10 mm². Fiber orientations were estimated using q-ball imaging (*bottom*). Notice that two orientations are estimated in the fiber-crossing area. (C) Seven fiber bundles comprising the Fiber Cup phantom and node demarcations (1–18) to facilitate connectivity matrix mapping. (D) Six examples of spurious fibers reconstructed for a brain-like in silico connectome phantom generated with Fiberfox. The percentage of tractography pipelines that reconstructed each spurious fiber is indicated. (E) Impact of pipeline variations on reconstruction accuracy. Valid connections (VC), valid bundles (VB), and invalid bundles (IB) are most relevant to connectivity matrix accuracy. The overlap (OL) and overreach (OR) metrics are less relevant to connectomics. *Green and red squares* indicate variations imparting a significant positive and negative impact on accuracy, respectively. (*Images adapted from Calabrese, E., Badea, A., Cofer, G., Qi, Y., Johnson, G.A., 2015. A diffusion MRI tractography connectome of the mouse brain and comparison with neuronal tracer data. Cereb. Cortex 25, 4628–4637 (A), Perrin, M., Poupon, C., Rieul, B., Leroux, P., Constantinesco, A., Mangin, J.F., Lebihan, D., 2005. Validation of q-ball imaging with a diffusion fibre-crossing phantom on a clinical scanner. Philos. Trans. R. Soc. Lond. Ser. B Biol. Sci. 360, 881–891 (B), Neher, P.F., Laun, F.B., Stieltjes, B., Maier-Hein, K.H., 2014. Fiberfox: facilitating the creation of realistic white matter software phantoms. Magn. Reson. Med. 72, 1460–1470 (C) and Maier-Hein, K.H., Neher, P.F., Houde, J.C., Côté, M.A., Garyfallidis, E., Zhong, J., Chamberland, M., Yeh, F.C., Lin, Y.C., Ji, Q., Reddick, W.E., Glass, J.O., Chen, D.Q., Feng, Y., Gao, C., Wu, Y., Ma, J., He, R., Li, Q., Westin, C.F., Deslauriers-Gauthier, S., González, J.O.O., Paquette, M., St-Jean, S., Girard, G., Rheault, F., Sidhu, J., Tax, C.M.W., Guo, F., Mesri, H.Y., Dávid, S., Froeling, M., Heemskerk, A.M., Leemans, A., Boré, A., Pinsard, B., Bedetti, C., Desrosiers, M., Brambati, S., Doyon, J., Sarica, A., Vasta, R., Cerasa, A., Quattrone, A., Yeatman, J., Khan, A.R., Hodges, W., Alexander, S., Romascano, D., Barakovic, M., Auría, A., Esteban, O., Lemkaddem, A., Thiran, J.P., Cetingul, H.E., Odry, B.L., Mailhe, B., Nadar, M.S., Pizzagalli, F., Prasad, G., Villalon-Reina, J.E., Galvis, J., Thompson, P.M., Requejo, F.S., Laguna, P.L., Lacerda, L.M., Barrett, R., Dell'Acqua, F., Catani, M., Petit, L., Caruyer, E., Daducci, A., Dyrby, T.B., Holland-Letz, T., Hilgetag, C.C., Stieltjes, B., Descoteaux, M., 2017. The challenge of mapping the human connectome based on diffusion tractography. Nat. Commun. 8, 1349 (D,E).*)

Yanasak and Allison, 2006). These synthetic configurations can be imaged to produce diffusion MRI data with a known ground truth to validate tractography reconstructions (Fig. 7B). Connectome phantoms can be constructed by integrating multiple such fibers in crossing and kissing configurations. As shown in Fig. 7C, the Fiber Cup phantom is perhaps the most well-known physical connectome phantom and comprises seven acrylic fibers simulating a coronal section of the human brain (Poupon et al., 2010). The diffusion MRI data produced by this phantom was released to the public and 10 research teams submitted tractography algorithms to compete in reconstructing the ground truth. Competitors were ranked based on how well they reconstructed the spatial extent and curvature of the acrylic fibers (Fillard et al., 2011). With the rise of connectomics, these spatial metrics were largely supplanted by new metrics aimed at assessing the accuracy of connectivity matrix reconstruction, such as the number of valid and invalid connections (Côté et al., 2012). To enable systematic validation with these new metrics, online tractography and connectome validation systems such as Tractometer were developed (Côté et al., 2012). Validation studies using these tools report that global tractography algorithms and deterministic tractography steered by a crossing-fiber diffusion model yield the most accurate connectome reconstructions of the Fiber Cup phantom (Neher et al., 2015). In contrast, probabilistic methods better characterize the spatial extent of fibers.

Physical connectome phantoms are limited by the extent to which they can recapitulate complex white matter architecture. While providing an important advance toward validation, the Fiber Cup phantom and its predecessors might be considered not particularly brain-like. Synthetic fibers are also usually larger in size than the white matter architecture that they intend to model, introducing further confounds (Jbabdi and Johansen-Berg, 2011). Some of these limitations can be overcome with numerical phantoms, where established diffusion MRI signal models are used to simulate diffusion MRI data corresponding to connectome architectures that are defined in silico (Daducci et al., 2014; Leemans et al., 2005; Lori et al., 2002). This provides flexibility to rapidly prototype and construct phantom connectomes that are more realistic and brain-like than phantoms instantiated in hardware. Software tools such as Fiberfox provide a convenient framework for phantom construction and prototyping (Close et al., 2009; Neher et al., 2014).

Maier-Hein and colleagues constructed a numerical phantom comprising 25 realistic fiber bundles and used the phantom to evaluate almost 100 tractography pipelines contributed by multiple research teams (Maier-Hein et al., 2017). While most pipelines successfully reconstructed the 25 ground truth fibers, several pipelines also reconstructed numerous invalid connections between regions that were not connected by a fiber in the ground truth. These invalid connections are termed false positives (Catani, 2007) and represent a particularly detrimental confound for graph-theoretic analyses (Zalesky et al., 2016). Six examples of false positive fiber reconstructions are shown in Fig. 7D. Consistent with earlier hardware-based validation experiments, pipelines based on deterministic tractography were generally found to be most suited to connectome mapping and crossing-fiber models substantially improved connectome sensitivity relative to the diffusion tensor. Correcting data for spikes, motion, and other distortions as well as seeding strategies did not significantly impact connectome mapping accuracy (Maier-Hein et al., 2017). The impact of common pipeline variations on mapping accuracy is tabulated in Fig. 7E. While criticisms based on the extent to which numerical phantoms resemble the brain are warranted, these same conclusions are borne out for larger phantoms comprising crossing and kissing fiber configurations chosen to match the complexity of the human brain (Sarwar et al., 2019).

In summary, validation experiments suggest that relatively basic tractography pipelines can adequately reconstruct connectomes with reasonable accuracy, particularly if the goal is to infer the presence or absence of connections. Determining connectivity strength is, however, more challenging, as is the reconstruction of long-range fibers (Sotiropoulos and Zalesky, 2019). While advanced tractography algorithms can improve the reconstruction accuracy of fiber curvature and spatial extent, this does not necessarily translate to improvements in connectivity matrix accuracy (Maier-Hein et al., 2017; Schilling et al., 2019). An important consideration is the trade-off between connectome sensitivity and specificity and the impact of spurious connections on connectome integrity. Global tractography (Jbabdi et al., 2007; Sotiropoulos et al., 2010a; Zalesky, 2008) and deep-learning methods (Poulin et al., 2019; Sarwar et al., 2020) provide a promising way forward, although they are often computationally demanding and the latter approaches mandate large sample sizes.

## 4.3 Other evidence of connectome validity

Streamline counts often provide the structural backbone in dynamic network models of large-scale brain activity (Cabral et al., 2014; Deco et al., 2011). The functional connectivity patterns computed using the activity simulated from these models matches empirical patterns from functional MRI quite well (Honey et al., 2009; Messé et al., 2014; Zalesky et al., 2014), which would not be the case for inaccurately mapped or randomized connectomes (Preti and Van De Ville, 2019). Reasonable agreement is also evident with modalities other than functional MRI, including positron emission topography (Tziortzi et al., 2014) and electrophysiological recordings (Elias et al., 2012). Structural connectivity derived from tractography is also increasingly used to assist in determining optimal targets for deep brain stimulation therapies for numerous neuropsychiatric and neurological disorders (Azriel et al., 2020; Pouratian et al., 2011). Some studies report that streamline counts between subcortical stimulation sites and key cortical regions can predict an individual's clinical response to treatment (Horn et al., 2017). Finally, as discussed later in this chapter, tractography-derived connectomes comprise hubs, rich clubs, and other topological properties that are synonymous with biological networks. While this catalog of evidence is indirect, it provides further validation for tractography-derived connectomes from a diverse number of modalities and experimental designs.

### 4.4 Connectome thresholding

As discussed already, validation studies indicate that the accuracy of tractography-derived connectomes is modest. In this section, we investigate postconstruction methods that can be applied to reconstructed connectomes with the goal of improving accuracy and biological validity. The simplest of these is connectome thresholding, sometimes referred to as filtering. Thresholding involves removing the connections comprising the fewest streamlines from a connectivity matrix (Fig. 8A). This is justified on the basis that connections comprising the fewest streamlines are more likely to be spurious, and thus removing them can potentially improve connectome specificity. However, real connections that are weak in strength can also comprise relatively few streamlines and thresholding indiscriminately eliminates these and spurious connections. The severity of thresholding arbitrates a trade-off between connectome sensitivity and specificity, with severer thresholds yielding higher specificity at the expense of lower sensitivity. As shown in Fig. 8B, experiments performed on connectome phantoms indicate that thresholding is particularly beneficial for improving the accuracy of connectomes reconstructed with probabilistic tractography (Sarwar et al., 2019).

Nevertheless, there is much debate about whether thresholding is warranted and the extent to which connectomes should be thresholded, if at all. The reason for thresholding is not only to improve accuracy. Thresholding can improve network visualizations of dense connectomes derived from probabilistic tractography, which typically resemble a bird's nest if not thresholded. Thresholding can also aid the interpretation and computation of some graph-theoretic measures (Fornito et al., 2013). For high-resolution connectomes mapped at the resolution of vertices and voxels, thresholding becomes mandatory to reduce the enormous storage and computational burden. Without thresholding, structural connectivity matrices mapped at the resolution of vertices comprising the standard cortical surface used by the Human Connectome Project can occupy upward of 10 GB of memory per individual. Finally, thresholding can strengthen connectome associations with age and behavioral measures (Buchanan et al., 2020).

Opponents of thresholding argue that removing connections comprising the fewest streamlines is largely inconsequential and thus thresholding is an unnecessary step that needlessly adds to workflow complexity (Civier et al., 2019). Indeed, even if not explicitly removed during thresholding, the contribution of weak connections is inherently downplayed in most weighted network analyses. Another key limitation of thresholding is the choice of threshold value, which is usually selected arbitrarily and conclusions may be contingent on the precise threshold choice (Garrison et al., 2015; van Wijk et al., 2010). Given the arbitrariness in threshold choice, it is important to test the sensitivity of analyses across a range of thresholds. This can be achieved with summary measures such as the area-under-curve (Rubinov et al., 2009) and multi-threshold permutation correction (Drakesmith et al., 2015).

Many thresholding methods are available. The two most used methods in network neuroscience are *density* and *weight* thresholding, also known as *proportional* and *absolute* thresholding, respectively (Rubinov and Sporns, 2010). With density-based thresholding, connections are removed until a desired connection density is achieved, where density is defined as the number of remaining connections divided by the total possible number of connections. Weight-based

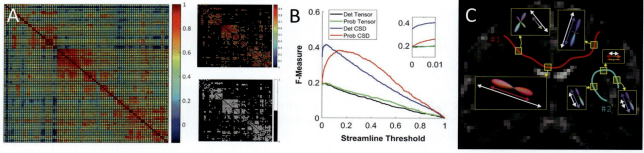

**FIG. 8** Postconstruction methods to improve connectome accuracy. (A) The connectivity matrix (*left*) is first thresholded (*top-right*) and then binarized (*bottom-right*). Thresholding involves removing the connections comprising the fewest streamlines. Binary connectivity matrices only encode the absence or presence of connections. (B) Connectome accuracy (F-measure) as a function thresholding severity. Thresholding improves the accuracy of deterministic (*blue line*) and probabilistic (*red line*) tractography algorithms steered by a cross fiber model (constrained spherical deconvolution, CSD), although excessive thresholding is detrimental. (C) The SIFT method is a postconstruction filter that selectively remove streamlines with the goal of minimizing the discrepancy between streamline counts and white matter microstructure. For voxels traversed by the *red streamline*, the streamline counts (commensurate to length of *white arrows*) exceed the FOD lobe integrals, and thus the *red streamline* is a good candidate for removal. This is not the case for the *green streamline*. FOD, fiber orientation distribution. *(Images reproduced from Sarwar, T., Ramamohanarao, K., Zalesky, A., 2019. Mapping connectomes with diffusion MRI: deterministic or probabilistic tractography? Magn. Reson. Med. 81, 1368–1384 and Smith, R.E., Tournier, J.D., Calamante, F., Connelly, A., 2013a. SIFT: spherical-deconvolution informed filtering of tractograms. Neuroimage 67, 298–312. https://doi.org/10.1016/j.neuroimage.2012.11.049 [Epub 2012 Dec 11. PMID: 23238430].)*

thresholding involves removing connections that comprise fewer than a predefined number of streamlines. These two methods are equivalent in the sense that a one-to-one correspondence exists between density and weight thresholds. However, important practical differences emerge between them in group studies.

Consider comparing connectomes between a group of patients with globally weaker structural connectivity than a healthy comparison group. Removing connections comprising fewer than a fixed number of streamlines in both groups would create between-group differences in connection density, confounding between-group comparisons of the thresholded connectomes. While density-based thresholding would certainly circumvent this confound, to do so, a greater number of weak and potentially spurious connections would be admitted into the thresholded patient connectomes to make up the numbers required to achieve the desired connection density. This introduces a different confound and potentially explains the subtle randomization of connectome topology that has been reported in schizophrenia (Fornito et al., 2012). Spurious connections are more likely to be randomly positioned and diverge from established topological wiring principles, which can conflate disease-related phenomena with thresholding confounds. Heuristics have been developed to disambiguate some of these confounds from real effects (van den Heuvel et al., 2017).

Advanced thresholding methods are available to preserve certain connectome properties and overcome limitations associated with the two global methods described earlier. Streamline counts and other measures of connectivity strength generally decrease with increasing fiber distance. While short-range axonal connections are indisputably more abundant in nervous systems than long fibers (Buzsáki et al., 2004), the extent of this abundance is overestimated in tractography-derived connectomes due to the difficulty in reconstructing long fibers and the prevalent use of thresholding methods that do not factor in the effect of fiber distance. Distance-dependent thresholding methods aim to overcome this bias by keeping connections that are strong for their length, rather than just strong overall (Roberts et al., 2017). This can reduce the over-representation of short-range connections in thresholded connectomes.

Consensus thresholding methods aim to keep connections that are consistently found across a group of individuals (de Reus and van den Heuvel, 2013a). Rather than keeping strong connections, connections that are found in more than a predefined proportion of individuals are kept, irrespective of their strength. However, consensus methods are vulnerable to systematic tractography biases that consistently localize to the same connections across individuals. Consensus thresholding methods can also be applied in a distance-dependent manner, ensuring that consensus-thresholded connectomes preserve the distribution of connection distances (Betzel et al., 2019). Some thresholding methods focus on preserving connections that are deemed strong within their topological locale, even though they might not be strong at the level of the whole network. While local methods such as the disparity filter (Serrano et al., 2009) are uncommon in connectomics, they can potentially unveil more detailed topologically structure than commonly used global methods (Kale et al., 2018) and have been used to suppress outliers (Tian and Zalesky, 2018).

Finally, connectome thresholding can be approached from a statistical perspective. By repeatedly randomizing fiber orientations estimated for each voxel and reperforming tractography, an empirical null distribution can be built up for streamline counts (Morris et al., 2008). This enables computation of $P$ values for each connection and thresholding based on a desired statistical significance level. The alternative hypothesis is that streamline counts are greater than expected when streamlines are steered in random directions. Functional brain networks are more amenable to statistical thresholding, given that the null hypothesis can be defined and tested more straightforwardly (Afyouni et al., 2019; Váša et al., 2018).

Binarization is an additional postconstruction step that can optionally be performed after thresholding. It can also be directly performed on the sparse connectivity matrices derived from deterministic tractography. Binary connectomes only encode information about the presence or absence of connections, but not variation in connectivity strength. Streamline counts span several orders of magnitude and lumping connections comprising 10 and 10,000 streamlines into the same category potentially squanders valuable and biologically meaningful information. However, some of this variation is inevitably confounded by noise, methodological limitations, and tractography biases. Binarization is a blunt tool for eliminating these confounds, albeit at the cost of also eliminating true variation. Binary connectomes are also more amenable to some network analyses, compared to their weighted counterparts. For example, Maslov-Sneppen rewiring of binary connectomes, which is commonly used to generate surrogate networks for statistical inference (discussed in the following), is computationally efficient and exact (Maslov and Sneppen, 2002), whereas rewiring weighted networks mandates the use of expensive and approximate heuristics (Rubinov and Sporns, 2011).

## 4.5 Connectome filters

The simplest kind of connectome filter is thresholding. We have already seen that thresholding is a simple postconstruction step that can improve connectome accuracy, aid network visualization, and strengthen connectome-behavior associations. However, thresholding is fundamentally a heuristic approach that is typically implemented using arbitrarily chosen thresholds. In recent years, more principled connectome filters have been developed to improve the biological validity, quantifiability, and accuracy of connectome reconstructions. Most of these filters operate by iteratively removing or downweighting streamlines to optimize a global cost function. The cost function specifies a theoretical relationship between streamline counts and white matter microstructural properties derived from the diffusion-weighted MRI acquisition. By optimizing the cost function using numerical optimization techniques, the filtered streamline counts become more consistent with the underlying white matter microstructure and diffusion signal at each voxel.

Numerous connectome filters following these principles have been developed, including SIFT (Smith et al., 2013a), LiFE (Pestilli et al., 2014), COMMIT (Daducci et al., 2015; Schiavi et al., 2020), and ReAl-LiFE (Kumar et al., 2019). The most salient differences between these filters are the cost function and optimization methodology. SIFT seeks to remove streamlines to minimize the discrepancy at each white matter voxel between the scaled streamline counts and the integral of the relevant lobe of the fiber orientation distribution function, where streamline counts are scaled by a global proportionality constant (Fig. 8C). In contrast, COMMIT fits a weight to each streamline using nonnegative least-squares so that the sum of streamline weights within each white matter voxel is consistent with the restricted and hindered components of the local diffusion signal. LiFE operates on a similar principle.

Some of the developers of these filters have undertaken validation experiments demonstrating that advanced filtering can improve the reproducibility and quantifiability of tractography-derived structural connectivity (Smith et al., 2015a). However, independent validations of SIFT performed with tract tracing experiments in rats (Sinke et al., 2018), connectome phantoms (Maier-Hein et al., 2017), and ex vivo brain specimens (Schilling et al., 2019) indicate that advanced filtering provides no appreciable improvement in accuracy. Indeed, pre- and postfiltered connectomes are very highly correlated within individuals. Some advanced filtering techniques can also lead to erroneous inference in pathological connectomes and caution is warranted when using them for between-group comparisons (Zalesky et al., 2020).

Advanced filtering methods are relatively new in the field of connectomics. The jury is still out on the utility and potential benefits of currently available filters. Incorporating functional, anatomical, and microstructural priors into the filtering process may be a promising way forward and it is likely that more sophisticated multimodal connectome filters will be developed in the future.

## 4.6 Summary and future directions

This section focused on the validation of tractography-derived connectomes and introduced several postconstruction methods that can be used to improve connectome accuracy. Early validation studies were based on in vivo tract-tracing experiments, whereas more recent validation work benefits from the flexibility and scalability of physical and in silico connectome phantoms. While both validation approaches suffer shortcomings, they both suggest that tractography does a reasonable job at reconstructing connectomes, although the margin for improvement is significant. Relatively basic connectome mapping pipelines that incorporate a robust crossing-fiber model with deterministic tractography perform quite well and the potential benefits of building more sophisticated methods into the mapping pipeline remain unclear (Maier-Hein et al., 2017). While probabilistic tractography can reconstruct fiber curvature and spatial extent more accurately than deterministic methods, faithfully reconstructing these attributes does not necessarily translate to improved connectivity matrix estimation. Probabilistic methods are also prone to mapping spurious connections, whereas deterministic methods are more susceptible to overlooking fibers. We discussed a range of postconstruction filters and thresholding methods. Thresholding enables arbitration of the inherent trade-off between the sensitivity and specificity of tractography-derived connectomes. More advanced filters aim to achieve consistency between streamline counts and microstructural tissue properties.

Using interregional streamline counts to quantify structural connectivity lacks sound theoretical justification. There is no principled reason why the probability with which routes of least resistance to the diffusion of water molecules would necessarily provide a valid measure of structural connectivity. Despite this lack of theoretical support, numerous validation experiments have demonstrated that streamline counts are practically useful and yield accurate connectivity measures.

While new connectome validation tools such as Fiberfox enable systematic evaluation of connectome mapping pipelines, the in silico connectome phantoms on which they are based remain relatively simplistic and do not recapitulate many of the fiber complexities of the human connectome. It will be important to develop more realistic phantoms in the future. Assessing the reproducibility and reliability of connectome mapping pipelines together with more traditional accuracy-based validation metrics is also emerging as an important area of future research (Rheault et al., 2020).

# 5 Graph analysis of the connectome

Despite the challenges outlined previously, diffusion MRI-based tractography remains the most efficient tool available for reconstructing whole-brain connectomes at macroscopic resolution. Once connectivity between each pair of brain regions has been mapped, the results can be succinctly summarized in matrix form, where each row and column of the matrix corresponds to a distinct network node and each matrix element, $C_{ij}$, of the matrix encodes the type and strength of connectivity between node pairs. The matrix, $C$, can also be represented as a graph of nodes connected by edges, where the nodes correspond to the rows/columns of $C$ and the edges correspond to the $C_{ij}$ elements. This equivalence allows us to use the mathematics of graph theory to characterize a diverse range of network properties, offering a rich repertoire of tools for understanding connectome structure and function (Albert and Barabasi, 2002; Boccaletti et al., 2006; Fornito et al., 2016; Newman, 2003, 2010).

In connectomics, network nodes could represent individual neurons, populations of neurons, or macroscopic brain regions, and edges represent some type of structural or functional interaction between nodes. For connectomes generated with diffusion MRI, edges represent some tractography-derived estimate of interregional connectivity and nodes represent macroscopic brain regions defined according to a particular parcellation. The resulting graph model is a remarkably simple abstraction, rendering the complex organization of the brain into a network of elements and their connections. Doing so allows one to focus on the essential elements of the system and offers tremendous flexibility, providing a common framework for modeling diverse neuroscientific data, including connectomes constructed using different methods, at different resolutions, and in different species. Graphs also allow us to embed these neuroscientific models within a much broader theoretical framework for understanding complex systems more generally, aiding the identification of universal principles of network organization across superficially diverse systems, ranging from the physical and biological to technological and social (Bullmore et al., 2009; Bullmore and Sporns, 2009; Newman, 2003).

This section will cover some practical considerations when conducting graph analysis of connectomes derived from diffusion MRI data. It will overview the types of analyses and inferences that are possible and summarize current knowledge about the core organizational properties of brain networks.

## 5.1 Building a graph model

The simplest possible graph model is a binary, undirected graph. This type of graph only tells us which pairs of nodes are connected and provides no information about the strength or direction of connectivity. The corresponding matrix representation is symmetric about the diagonal, such that $C_{ij} = C_{ji}$ and $C_{ij} = 1$ if nodes $i$ and $j$ are connected and $C_{ij} = 0$ otherwise (Fig. 9, left).

At first glance, this model seems like a gross oversimplification for connectomes, since we know that brain connectivity is intrinsically directed (i.e., each anatomical projection has an origin and target) and that connection strengths can vary over several orders of magnitude (Markov et al., 2014; Oh et al., 2014). We should thus expect that a weighted, directed graph will provide a more accurate model. In such graphs, connectivity strength is represented by variations in edge thickness and directionality is represented by appending arrowheads to the edges (Fig. 9, right). The corresponding connectivity matrix is not symmetric about the diagonal, meaning that elements $C_{ij}$ and $C_{ji}$ are not necessarily equivalent. Outgoing connections can be listed along either the matrix rows or columns. Being a weighted graph, the matrix elements correspond to the connectivity weights derived from the measurement procedure (for example, streamline counts or a tract-averaged estimate of fractional anisotropy).

Weighted, directed graphs may still provide simplistic models of the connectome. These models assume that all network nodes are the same, but brain regions vary in terms of their intrinsic microcircuitry, molecular architecture, and dynamics (Huntenburg et al., 2018; Wang, 2020). Moreover, while the majority of long-range interregional projections are excitatory, their net effect on the target neuronal population will depend on local synaptic properties, such that some may have a net excitatory effect whereas others may have a net inhibitory effect. Annotated and signed graphs can be used to model node and edge heterogeneity, respectively. However, they have rarely been used in connectomics. For exceptions, see Gal et al. (2017), Murphy et al. (2016), and Rubinov and Sporns (2011). Moreover, the inability of diffusion MRI to resolve directionality means that most human connectomic studies use undirected (binary or weighted) graph models.

The choice between a binary or weighted graph depends on the research question. A binary graph allows one to focus on the basic properties of pairwise connectivity in the network but ignores variations in connectivity strength. As discussed earlier, precisely how one should interpret connectivity weights estimated with diffusion MRI is a matter of debate, given that the technique can only provide an indirect estimate of the connection strength. In some cases, a

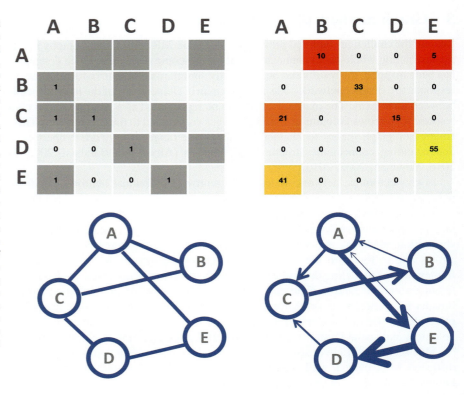

FIG. 9 Matrices and graphs are equivalent representations of a network. *Top left* shows a matrix representation of a binary, undirected network comprising five nodes. In this matrix, $C_{ij}=1$ if a connection is found between regions $i$ and $j$ and $C_{ij}=0$ otherwise. Since the network is undirected, the matrix is symmetric about the diagonal. The corresponding graph representation is shown below. *Top right* shows a matrix representation of a weighted, directed network. The values in each matrix element, $C_{ij}$, correspond to connectivity weights estimated for a given pair of nodes. *Colors* in the matrix represent variations in the weights (higher weights in *yellow*). In this matrix, outgoing connections for each node are listed down the columns. Note that the matrix is no longer symmetric about the diagonal. Asymmetries in the *upper and lower triangles* of the matrix represent the directionality of interregional connectivity. The corresponding graph is shown below. In this graph, *arrowheads* indicate the source and target of each connection, and edge thickness corresponds to connectivity strength. It is also possible to construct weighted, undirected networks and binary, directed networks.

transformation of the weights can be applied to ensure that any quantities computed on the network are not dominated by outlying values. Other remappings can also be applied depending on the specific network property being investigated (Fornito et al., 2016; Goñi et al., 2014).

A final decision before analyzing network properties concerns whether a threshold should be applied to the graph. We have already considered the strengths and weaknesses of different thresholding methods. Here, we note that thresholding can pose specific problems for the estimation of some network properties. For example, connectomes generated using probabilistic tractography are often very dense, with a nonzero connectivity value obtained for nearly every pair of nodes; i.e., $C_{ij}>0$ for nearly all connections, meaning that a binary analysis will be inappropriate. Specific methods for thresholding have been devised for such dense connectomes (Roberts et al., 2017). In contrast, deterministic tractography typically yields sparser connectomes, but it is still common practice to apply a threshold to remove weak or inconsistently found connections (de Reus and van den Heuvel, 2013a). If the threshold is too stringent, the network may fragment into subsets of nodes that are disconnected from each other. An excessively stringent threshold may also produce isolated nodes that have no connections to other regions. These two phenomena of network fragmentation and node isolation are biologically implausible and can complicate the estimation and interpretation of some network properties (Newman, 2003). Thus thresholds, if used, should be applied carefully to ensure that the resulting network forms a single connected component, such that all nodes have at least one connection and a path can be found between any pair of nodes in the network (there are no disjoint sets of nodes).

Having settled on an appropriate graph model for the data, it is possible to analyze the network in terms of its connectivity or topology. Analyses of connectivity focus on variations in the type and strength of connectivity at each edge. Analyses of topology focus on how connections are arranged with respect to each other in the network and can provide insight into overarching organizational principles. We consider each of these analyses in the following sections.

### 5.2 Analyzing brain network connectivity

Given a set of reconstructed connectomes, it is often desirable to map which specific connections covary with some perturbation (whether through development, disease, or experiment) or other phenotype of interest (e.g., cognitive performance). Such analyses pose a considerable multiple comparisons problem since, for an undirected network comprising

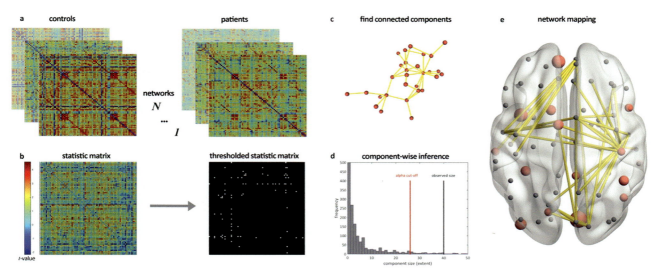

FIG. 10 Example workflow for the network-based statistic (NBS). (A) Example connectivity matrices for a hypothetical group of patients and healthy controls. (B) A *t*-statistic is estimated at each edge to quantify the difference between groups in mean connectivity values, resulting in a matrix of *t*-statistics. This matrix is then thresholded at some nominal level, $\tau$, to generate a binary matrix of supra-threshold *t*-statistics. (C) The sizes of the connected components in this thresholded *t*-statistic matrix are then estimated. (D) The patient and control labels are shuffled many times and steps B–C are repeated under the null hypothesis of random group membership. The maximal component size obtained at each iteration is stored to generate an empirical null distribution to estimate the statistical significance of the observed component sizes. (E) Any significant components can then be mapped on the brain and their anatomical and functional properties interrogated. *(Figure reproduced from Fornito, A., Zalesky, A., Bullmore, E.T., 2016. Fundamentals of Brain Network Analysis. Academic Press, Inc, San Diego, USA with permission.)*

$N$ nodes, a comprehensive mass-univariate mapping involves $(N-1)^2/2$ statistical tests. As connectomes constructed with diffusion MRI typically comprise $10^2-10^3$ nodes, the resulting number of tests is often very large and traditional correction methods such as the Bonferroni or Benjamini-Hochberg (Benjamini and Hochberg, 1995) procedures will be very conservative.

Several solutions to this multiple comparison problem have been proposed, using various elements of multivariate statistics (Mišić et al., 2016; Shehzad et al., 2014; Smith et al., 2015b) and machine learning (He et al., 2020; Maglanoc et al., 2020; Shen et al., 2017; Sui et al., 2020). One approach that specifically relies on graph theoretic methods is the network-based statistic (NBS) (Zalesky et al., 2010a, 2012). The NBS can be used in conjunction with any univariate statistical model estimated for an ensemble of edges, but it is most easily understood by considering an example experiment comparing connectivity values between two groups, say patients and healthy controls (Fig. 10). In this application, the first step of the NBS is to estimate a *t*-statistic at each edge to quantify the mean difference in connectivity strength between groups (Fig. 10A and B). Second, the resulting matrix of *t*-statistics is thresholded at some nominal value, $\tau$, and binarized (Fig. 10B). Third, the connected components of the resulting graph are identified (Fig. 10C). Recall that a connected component is a subset of nodes that can be linked by a set of supra-threshold edges, the assumption being that edges comprising any such component will share a common effect of disease. Fourth, the number of edges in each of these components is estimated and stored. Fifth, the group labels of the original data are shuffled many times and the same analysis steps are applied to each permutation of the data. The largest component size obtained for each permutation is then retained to generate a distribution of maximal component sizes obtained under the null hypothesis. Finally, a *P* value is obtained for each observed component by computing the fraction of null components that are larger (Fig. 10D and E). Critically, because the null distribution consists of only the maximal component size obtained for each permutation of the data, the resulting *P* values control the family-wise error rate (Nichols and Hayasaka, 2003).

An arbitrary choice in conducting the NBS is the selection of the initial component-forming threshold, $\tau$, which is similar to cluster-wise correction methods for voxel-wise activation mapping of functional MRI data (Woo et al., 2014). More liberal thresholds will be sensitive to weak effects distributed across multiple edges, whereas more stringent thresholds will be sensitive to strong effects distributed over smaller subsets of edges. Some heuristics have been proposed to circumvent this choice of threshold (Baggio et al., 2018).

### 5.3 Analyzing brain network topology

Topological properties of a network are those properties that are invariant to any continuous, spatial transformation of the system. For example, the basic pattern of pairwise connectivity between nodes in the graphs shown in Fig. 9 would remain constant if we stretched, rotated, or resized the graphs. The topology of the network thus defines the specific arrangement and organization of connections between node pairs.

Graph theory offers numerous metrics for quantifying diverse aspects of network topology. An exhaustive treatment of these measures is beyond the scope of this chapter and comprehensive summaries have been provided elsewhere (Fornito et al., 2016; Newman, 2003, 2010; Rubinov and Sporns, 2010). Here, we consider the general overarching organizational principles of connectomes that have been revealed through the application of these metrics. Indeed, graph theory has been used to identify a remarkable consistency of these principles across connectomes mapped in different species, from *C. elegans* to human, at resolutions ranging from the level of individual neurons and synapses to large-scale brain regions, and using techniques as diverse as electron microscopy, tract tracing, and diffusion MRI (Bullmore et al., 2009; Bullmore and Sporns, 2012; Fornito et al., 2016; van den Heuvel et al., 2016). A recurring theme in this work is that connectivity is not distributed uniformly across the individual nodes of different nervous systems; that is, connectomes are distinctly nonrandom. Some key properties relevant to this discussion are shown in Fig. 11.

One way in which connectomes differ from random networks is in their degree distribution. The degree of a node is simply the number of connections to which it is attached (Fig. 11A). The degree distribution of a network thus defines how degree (i.e., connectivity) is distributed across nodes. In a random graph, the degree distribution is binomial, such that most nodes show a characteristic average degree with a low probability of identifying nodes with degrees much higher or lower than this mean. Seminal work by Barabasi and Albert (1999) showed that many real-world networks are characterized by a scale-free power-law degree distribution of the form $P(\text{degree}=k) \sim k^{-\gamma}$. In these networks, most nodes have low degree,

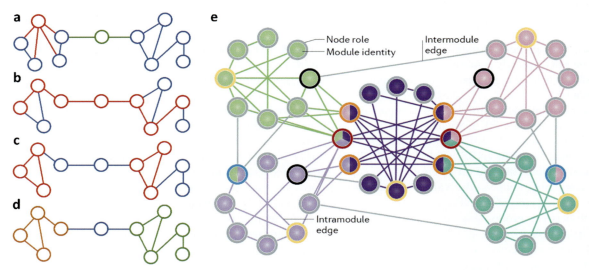

**FIG. 11** Key topological properties of brain networks. (A–D) A toy graph showing some commonly studied topological properties. (A) Network hubs can be simply defined as nodes with many connections; i.e., nodes with high degree (e.g., *red node*). However, alternative definitions are possible. For example, all shortest paths between the cluster of nodes on the left and the cluster on the right pass through the *green node*, so one could argue that the *green node* is more influential than the *red node*. Diverse metrics are available for capturing these distinct notions of network centrality. (B) The shortest path between two nodes at opposite ends of the network is shown in *red*. (C) A network with a high clustering coefficient comprises many *closed triangles*, as shown in *red*. (D) Example of two modules, shown in *orange and green*. The nodes within a module have stronger connectivity with each other than with other nodes. In this case, the assignment of the node in the middle to a single module is ambiguous, as it has equal connectivity to the *orange and green modules*. This example demonstrates the limitations of a hard module partition. Under a fuzzy or overlapping partition, this node could be assigned to both modules. (E) Simplified representation of key topological properties identified for the human connectome. *Colors* represent different network modules. Here, we allow for overlapping module structure, such that nodes with multiple colors belong to more than one module. Each specialized module has one or more hubs, which form a rich-club (*dark purple*) arranged much like a module that overlaps with the others. Modular architecture supports specialization and segregation of function. Rich-club organization supports functional integration across modules. *Yellow borders* identify nodes that acts as within-module (provincial) hubs; *orange and red borders* identify nodes that connect many diverse modules (also known as connector hubs). This schematic is a simplification, as the brain's modular architecture is hierarchical, containing submodules nested within modules across several resolution scales. *(Panel (E) reproduced from Fornito, A., Zalesky, A., Breakspear, M., 2015. The connectomics of brain disorders. Nat. Rev. Neurosci. 16, 159–172 with permission.)*

but the probability of finding nodes with very high degree is much greater than in a random graph. These high-degree nodes represent putative network hubs. Early studies, principally in high-resolution functional connectivity networks, found evidence for a similar power-law form in degree distributions of brain networks (Hayasaka and Laurienti, 2010; van den Heuvel et al., 2008), but the bulk of evidence suggests that any power-law scaling is only present over a limited range and becomes truncated at very high values of degree (Achard et al., 2006; Amaral et al., 2000; Gong et al., 2009). This result implies that it is unlikely to find nodes with a very large degree in brain networks, which is consistent with metabolic and physical limits on the number of connections that can be attached to any single node. Thus brain networks do appear to have hubs, although these hubs may not be as influential as those found in pure scale-free networks. We do note, however, that rigorous fitting and model comparison of different functional forms to the degree distributions of connectomes (Clauset et al., 2009) has not yet been conducted.

Another way in which connectomes were initially found to differ from random graphs is in their small-worldness. The small-worldness of a network is defined with respect to two properties: its characteristic path length and its clustering coefficient. The characteristic path length is the average shortest path between any two nodes in the network (Fig. 11B) and is related to the efficiency and integration of the network (Latora and Marchiori, 2001). The clustering coefficient of a network is the fraction of closed triangles (fully connected triplets) attached to a node (Fig. 11C), averaged across all nodes, and is taken as an index of locally cliquish connectivity. In their pioneering work, Watts and Strogatz (1998) showed that many real-world networks, including the neuron-and-synapse connectome of *C. elegans*, show a comparable path length to a random graph, implying a high degree of integration, coupled with much higher clustering, suggesting locally structured and cliquish connectivity. The same observations have since been made in many other connectomes, reviewed in Bassett and Bullmore (2006). Although the small-worldness of nervous systems has been questioned (Markov et al., 2013), careful analysis reveals that connectome topology is indeed compatible with small-world topology (Bassett and Bullmore, 2017).

Clustered connectivity in brain networks is largely attributable to a modular organization. Network modules are subsets of nodes that are more strongly connected with each other than with the rest of the network. In the brain, modules are organized hierarchically, such that modules contain nested submodules across multiple levels of resolution (Bassett et al., 2010; Meunier et al., 2009). This organization is thought to support specialization and segregation of function across different spatial scales (Sporns and Betzel, 2016). In anatomical brain networks, network modules tend to be spatially localized, comprising regions that are located near each other in physical space. This organization likely arises from the well-known distance-dependence of brain connectivity, in which the probability of connection between two nodes decays approximately as exponential function of the distance between them (Arnatkeviciute et al., 2018; Ercsey-Ravasz et al., 2013; Fulcher and Fornito, 2016; Horvát et al., 2016). This predominance of short-range connectivity helps to limit wiring costs and minimize metabolic expenditure, but connectomes possess more long-range connections than expected under a pure cost-minimization model (Bullmore and Sporns, 2012; Chen et al., 2013; Kaiser and Hilgetag, 2006). A large fraction of these long-range projections connect hubs that reside in different topological modules, resulting in a dense interconnectivity between hubs, called a rich-club, that supports coordinated dynamics across the network (de Reus and van den Heuvel, 2013b; van den Heuvel and Sporns, 2011). The strategic placement of these long-range projections allows brains to economically satisfy competitive selection pressures to minimize network wiring costs and support integrated, complex, and adaptive function (Bullmore and Sporns, 2012). Indeed, the wiring cost of connectomes appears to be near-minimal given their level of topological complexity (Bassett et al., 2010). Critically, many of these topological properties are not just shared across the brains of different individuals and species, but with many other natural and man-made systems in which connections are formed at some cost (Bullmore and Sporns, 2012; Fornito et al., 2016; van den Heuvel et al., 2016). Fig. 11E provides a schematic demonstration of these properties.

Graph theoretic studies have shed important light on some of the fundamental principles that govern connectome organization. However, there are many practical considerations when conducting such analyses. For instance, an initial choice concerns whether one should focus on binary or weighted aspects of network topology. Weighted graphs can be more informative. For example, consider two nodes, each with three connections that have weights $A = \{5, 7, 8\}$ and $B = \{23, 56, 72\}$. The average weight of connections attached to node B is much higher, so we should naturally conclude that node B plays a more influential role in network function. If we relied on a binary representation, we would only know that each node has three connections and we would conclude that nodes A and B are equally influential.

Conversely, some weighted measures may be overly sensitive to outlying values. For instance, many topological metrics rely on finding shortest paths between node pairs. In binary networks, the shortest path is the path that traverses the fewest connections (Fig. 11B). In weighted networks, the shortest path is the path with the lowest aggregate weight. For connectomes, the raw connectivity weights are usually remapped prior to such an analysis so that highly weighted edges have lower values (for a discussion, see Fornito et al., 2016). Assuming that such a remapping has been applied, algorithms for finding weighted shortest paths can be biased to identify paths that traverse edges with very low remapped weights, since these paths will tend to have the lowest aggregate weight after remapping, even if the paths traverse many different edges. In other words, the algorithms may prefer routes that cross many edges if the total aggregate weight of the path is lower than a path traversing fewer edges. Although heuristic solutions to this problem have been proposed (Opsahl et al., 2010), this example demonstrates that the choice between binary or weighted analysis depends on the research question at hand.

Another complicating factor in graph theoretic studies is that there are often multiple different methods for quantifying a specific topological property, and the specific method used can lead to different results and interpretations. For instance, Fig. 11A shows a toy modular network in which a node with a small number of connections can exert a greater influence on network function than a node with many more connections, because the low-degree node acts as a bridge between two otherwise disconnected subsets of nodes. This example highlights the multifaceted nature of network centrality, which can be defined as the degree to which a node influences, or is influenced by, other nodes by virtue of its connection topology (Fornito et al., 2016). Numerous measures of node centrality have been proposed, each being sensitive to different aspects of what it means for a node to be influential and each making different assumptions about how communication unfolds on the network (Borgatti, 2005). It is therefore essential to choose a given centrality measure that aligns with a realistic model for network communication. This has proven to be a particular challenge in connectomics, where many studies have relied on centrality measures that assume information propagates along the shortest topological paths between nodes. This is an unrealistic assumption, as it requires each node to have global knowledge of the network to find such a path.

Alternative communication processes have been proposed. These can be categorized along a spectrum (Avena-Koenigsberger et al., 2018; Fornito et al., 2016): at one end lie routing strategies like those relying on shortest paths, which require some knowledge of network structure to propagate a message from source to target; at the other end lie diffusion-like strategies that propagate messages according to local rules that are blind to global topology and in which the source or target of a message may be undefined. Centrality measures based on the latter may be more appropriate for modeling neural communication. Indeed, when computed on a structural network, they are more predictive of empirical patterns of functional connectivity (Avena-Koenigsberger et al., 2019; Goñi et al., 2014; Mišić et al., 2015; Seguin et al., 2018, 2019). For some applications, such as understanding patterns of communication and information flow in the network, these distinctions are vitally important. For others, such as identifying topologically central network nodes, they may be less critical, given evidence that many commonly used centrality measures are highly correlated when applied to real-world networks (Oldham et al., 2019).

Similar challenges face attempts to characterize the modular organization of the brain, with a large number of different algorithms available for partitioning a network into modules (Fortunato, 2010). By far, the most commonly used approach in neuroscience involves finding a partition of the network that maximizes a quality function, $Q$, which is defined as the difference between the observed degree of intramodule connectivity and the degree of intramodule connectivity observed by chance (Newman and Girvan, 2004). Finding a partition that maximizes $Q$ is an NP-complete problem (Brandes et al., 2006), so numerous heuristic algorithms have been developed to identify high-$Q$ partitions. Many of these algorithms are stochastic, resulting in an optimal partition that may vary from run to run. Moreover, the optimization landscape is highly degenerate, meaning that it is possible to identify many different partitions with similar values of $Q$ (Good et al., 2010). As a result, investigators often iterate the algorithm multiple times and use consensus clustering approaches to arrive at a single summary partition (Lancichinetti and Fortunato, 2012). There are two further complications. First, many algorithms include a tunable resolution parameter that can be used to adjust the size of the detected modules. On the one hand, this complicates analysis as there is no single optimum resolution for a hierarchical network such as the brain; on the other hand, it offers the flexibility to derive a more comprehensive multiscale characterization of brain architecture. Second, the modules that one obtains very much depend on how chance-expected connectivity is defined in the $Q$-function. The most commonly used definition relies on the so-called configuration model, which defines the expectations for an ensemble of random graphs with the same degree sequence as the empirical network. Alternative definitions are possible that can be used to identify distinct types of modules, such as those that show stronger internal connectivity than predicted by a simple distance-dependent connectivity rule (Bazzi et al., 2014; Betzel et al., 2017; Expert et al., 2011).

Critically, one may also obtain a very different result when using an alternative partitioning algorithm. For instance, $Q$-optimization procedures typically yield a hard partition of the network such that nodes can only belong to one module, but other algorithms allow recovery of overlapping partitions in which individual nodes can belong to multiple modules (reviewed in Xie et al., 2013). Moreover, some modeling approaches allow for a more flexible definition of modules, such as distinctions between assortative (connections are mainly formed within the same subsets) and disassortative (connections are mainly formed between different subsets) node communities (Betzel et al., 2018). Benchmarking studies have been performed to evaluate the relative performance of some of these module-detection algorithms (Lancichinetti and Fortunato, 2012; Xie et al., 2013) and can provide some guidance for deciding which methods are more accurate for a given purpose.

## 5.4 The importance of null models

Assuming that a researcher has settled on a specific method for quantifying a given topological property, the next question that arises is whether that property is expressed in the brain to any notable degree. For instance, assume that we have measured a given property, $T$, and we obtain a value of 5. Is 5 large or small? This question can be answered with respect to a null model that establishes expectations for observing such a value by chance. By far the most widely used null model in connectomics is based on the Maslov-Sneppen (MS) rewiring procedure (Maslov and Sneppen, 2002), which iteratively and randomly rewires the connections of the empirical network while preserving the same degree sequence, such that each node in the rewired network has the exact same degree as the original network, with all other properties randomized. This constraint is preferred over a completely random graph because the degree distribution of a network can drive much of its topological structure. Variants of this approach have been developed to deal with weighted and signed networks (Rubinov and Sporns, 2011). In each case, many rewired networks are generated and the property $T$ is estimated at each iteration to construct a null distribution. This distribution is then used to evaluate the likelihood of obtaining the observed value of $T$ under the null hypothesis that $T$ is consistent with expectations in a randomized network with the same degree sequence as the empirical data.

The degree sequence of the network is just one constraint that can be used when defining null graphs. Alternative constraints can be specified to allow tests of specific hypotheses. For instance, several studies suggest that simple models in which network connections are formed at random, but subject to a similar distance-dependence as observed in the brain (Henderson and Robinson, 2014), possess many topological properties that mimic those observed in empirical data. We may thus seek to determine whether our observed property $T$ assumes a higher value than expected by a simple distance-dependent random connection rule. This hypothesis can be tested using algorithms that preserve the empirically observed relationship between distance and connectivity, while also optionally preserving other features such as degree sequences and distributions (Roberts et al., 2016; Samu et al., 2014). These, and other methods (Rubinov, 2016), can be used to test increasingly specific hypotheses about the network characteristics that may or may not explain a given topological property observed in the brain.

## 5.5 Summary

Graphs offer extraordinary flexibility for quantifying different connectome properties. As discussed here, they should be used judiciously, with due consideration of the limitations of each method and algorithm. More generally, the results of any graph theoretic analysis should be interpreted with respect to the limitations of the measurement process itself. This chapter has highlighted several limitations of diffusion MRI tractography, and these limitations constrain the inferences that can be drawn from any subsequent analysis. It is nonetheless reassuring that many properties identified in such analyses are also observed in connectomes generated with other methods, and at different resolution scales, suggesting that they are indeed robust. The precision with which we can estimate these properties will only improve in line with the accuracy of our measurement tools.

Over the next few years, we expect to see further developments in graph theoretic methods specifically tailored to neuroscientific questions. Early applications simply adapted existing measures developed for other applications, such as centrality measures based on shortest paths, that have limited applicability for brain networks. As methods develop for characterizing diverse kinds of communication processes on networks, we expect a more accurate understanding of the roles that different network elements play in shaping brain dynamics.

A critical challenge for the field will involve shifting from merely describing network properties to building useful models. Graphs offer great flexibility in this regard. Generative models of network wiring, in which networks are grown in silico according to specific wiring rules, are elucidating some of the core principles and candidate selection pressures on connectome organization, while also providing an efficient means for summarizing diverse aspects of network topology (Betzel et al., 2016; Vértes et al., 2012; Zhang et al., 2021). The recent development of network control theory offers an elegant framework for understanding how brain networks respond to different perturbations and is finding diverse neuroscientific applications (Gu et al., 2015; Lynn and Bassett, 2019; Yan et al., 2017). Multilayer network analysis offers a powerful framework for understanding how networks evolve through time (Bassett et al., 2011; Kivelä et al., 2014) and for linking diverse kinds of connectivity information acquired in the same network (Bentley et al., 2016). Moreover, the combination of connectomics with information on regional measures of microstructure and molecular function is affording new opportunities to understand how different scales of brain organization relate to each other (Fornito et al., 2019; Scholtens and van den Heuvel, 2018).

**476 PART | IV From streamlines to tracts**

# References

Achard, S., Salvador, R., Whitcher, B., Suckling, J., Bullmore, E., 2006. A resilient, low-frequency, small-world human brain functional network with highly connected association cortical hubs. J. Neurosci. 26, 63–72.

Afyouni, S., Smith, S.M., Nichols, T.E., 2019. Effective degrees of freedom of the Pearson's correlation coefficient under autocorrelation. Neuroimage 199, 609–625.

Albert, R., Barabasi, A.L., 2002. Statistical mechanics of complex networks. Rev. Mod. Phy. 74, 47–97.

Alcami, P., El Hady, A., 2019. Axonal computations. Front. Cell. Neurosci. 13, 413.

Amaral, L.A., Scala, A., Barthelemy, M., Stanley, H.E., 2000. Classes of small-world networks. Proc. Natl. Acad. Sci. U. S. A. 97, 11149–11152.

Ambrosen, K.S., Eskildsen, S.F., Hinne, M., Krug, K., Lundell, H., Schmidt, M.N., van Gerven, M.A.J., Mørup, M., Dyrby, T.B., 2020. Validation of structural brain connectivity networks: the impact of scanning parameters. Neuroimage 204, 116207.

Anderson, A.W., 2005. Measurement of fiber orientation distributions using high angular resolution diffusion imaging. Magn. Reson. Med. 54, 1194–1206.

Arnatkeviciute, A., Fulcher, B.D., Pocock, R., Fornito, A., 2018. Hub connectivity, neuronal diversity, and gene expression in the Caenorhabditis elegans connectome. PLoS Comput. Biol. 14, e1005989.

Avena-Koenigsberger, A., Mišić, B., Sporns, O., 2018. Communication dynamics in complex brain networks. Nat. Neurosci. 19, 17–33.

Avena-Koenigsberger, A., Yan, X., Kolchinsky, A., van den Heuvel, M.P., Hagmann, P., Sporns, O., 2019. A spectrum of routing strategies for brain networks. PLoS Comput. Biol. 15, e1006833.

Azadbakht, H., Parkes, L.M., Haroon, H.A., Augath, M., Logothetis, N.K., de Crespigny, A., D'Arceuil, H.E., Parker, G.J., 2015. Validation of high-resolution tractography against in vivo tracing in the macaque visual cortex. Cereb. Cortex 25, 4299–4309.

Azriel, A., Farrand, S., Di Biase, M., Zalesky, A., Lui, E., Desmond, P., Evans, A., Awad, M., Moscovici, S., Velakoulis, D., Bittar, R.G., 2020. Tractography-guided deep brain stimulation of the anteromedial globus pallidus internus for refractory obsessive-compulsive disorder: case report. Neurosurgery 86, E558–E563.

Baggio, H.C., Abos, A., Segura, B., Campabadal, A., Garcia Diaz, A., Uribe, C., Compta, Y., Marti, M.J., Valldeoriola, F., Junqué, C., 2018. Statistical inference in brain graphs using threshold-free network-based statistics. Hum. Brain Mapp. 39, 2289–2302.

Barabasi, A., Albert, R., 1999. Emergence of scaling in random networks. Science 286, 509–512.

Basser, P.J., Pajevic, S., Pierpaoli, C., Duda, J., Aldroubi, A., 2000. In vivo fiber tractography using DT-MRI data. Magn. Reson. Med. 44, 625–632.

Bassett, D.S., Bullmore, E., 2006. Small-world brain networks. Neuroscientist 12, 512–523.

Bassett, D.S., Bullmore, E.T., 2017. Small-world brain networks revisited. Neuroscientist 23, 499–516.

Bassett, D.S., Greenfield, D.L., Meyer-Lindenberg, A., Weinberger, D.R., Moore, S.W., Bullmore, E.T., 2010. Efficient physical embedding of topologically complex information processing networks in brains and computer circuits. PLoS Comput. Biol. 6, e1000748.

Bassett, D.S., Wymbs, N.F., Porter, M.A., Mucha, P.J., Carlson, J.M., Grafton, S.T., 2011. Dynamic reconfiguration of human brain networks during learning. Proc. Natl. Acad. Sci. U. S. A. 108, 7641–7646.

Bastiani, M., Cottaar, M., Dikranian, K., Ghosh, A., Zhang, H., Alexander, D.C., Behrens, T.E., Jbabdi, S., Sotiropoulos, S.N., 2017. Improved tractography using asymmetric fibre orientation distributions. Neuroimage 158, 205–218.

Bazzi, M., Porter, M.A., Williams, S., McDonald, M., 2014. Community detection in temporal multilayer networks, and its application to correlation networks. arXiv. 1501.00040v2.

Beckmann, C.F., Smith, S.M., 2004. Probabilistic independent component analysis for functional magnetic resonance imaging. IEEE Trans. Med. Imaging 23, 137–152.

Behrens, T.E., Sporns, O., 2012. Human connectomics. Curr. Opin. Neurobiol. 22, 144–153.

Behrens, T.E., Johansen-Berg, H., Woolrich, M.W., Smith, S.M., Wheeler-Kingshott, C.A., Boulby, P.A., Barker, G.J., Sillery, E.L., Sheehan, K., Ciccarelli, O., Thompson, A.J., Brady, J.M., Matthews, P.M., 2003a. Non-invasive mapping of connections between human thalamus and cortex using diffusion imaging. Nat. Neurosci. 6, 750–757.

Behrens, T.E., Woolrich, M.W., Jenkinson, M., Johansen-Berg, H., Nunes, R.G., Clare, S., Matthews, P.M., Brady, J.M., Smith, S.M., 2003b. Characterization and propagation of uncertainty in diffusion-weighted MR imaging. Magn. Reson. Med. 50, 1077–1088.

Behrens, T.E., Berg, H.J., Jbabdi, S., Rushworth, M.F., Woolrich, M.W., 2007. Probabilistic diffusion tractography with multiple fibre orientations: what can we gain? Neuroimage 34, 144–155.

Benjamini, Y., Hochberg, Y., 1995. Controlling the false discovery rate: a practical and powerful approach to multiple testing. J. R. Stat. Soc. Series B 57, 289–300.

Bentley, B., Branicky, R., Barnes, C.L., Chew, Y.L., Yemini, E., Bullmore, E.T., Vértes, P.E., Schafer, W.R., 2016. The multilayer connectome of Caenorhabditis elegans. PLoS Comput. Biol. 12, e1005283.

Betzel, R.F., Byrge, L., He, Y., Goñi, J., Zuo, X.N., Sporns, O., 2014. Changes in structural and functional connectivity among resting-state networks across the human lifespan. Neuroimage 102 (Pt. 2), 345–357.

Betzel, R.F., Avena-Koenigsberger, A., Goñi, J., He, Y., de Reus, M.A., Griffa, A., Vértes, P.E., Mišić, B., Thiran, J.-P., Hagmann, P., van den Heuvel, M., Zuo, X.-N., Bullmore, E.T., Sporns, O., 2016. Generative models of the human connectome. Neuroimage 124, 1054–1064.

Betzel, R.F., Medaglia, J.D., Papadopoulos, L., Baum, G.L., Gur, R., Gur, R., Roalf, D., Satterthwaite, T.D., Bassett, D.S., 2017. The modular organization of human anatomical brain networks: accounting for the cost of wiring. Netw. Neurosci. 1, 42–68.

Betzel, R.F., Medaglia, J.D., Bassett, D.S., 2018. Diversity of meso-scale architecture in human and non-human connectomes. Nat. Commun., 1–14.

Betzel, R.F., Griffa, A., Hagmann, P., Mišić, B., 2019. Distance-dependent consensus thresholds for generating group-representative structural brain networks. Netw. Neurosci. 3, 475–496.

Beul, S.F., Grant, S., Hilgetag, C.C., 2015. A predictive model of the cat cortical connectome based on cytoarchitecture and distance. Brain Struct. Funct. 220, 3167–3184.

Blumensath, T., Jbabdi, S., Glasser, M.F., Van Essen, D.C., Ugurbil, K., Behrens, T.E., Smith, S.M., 2013. Spatially constrained hierarchical parcellation of the brain with resting-state fMRI. Neuroimage 76, 313–324.

Boccaletti, S., Latora, V., Moreno, Y., Chavez, M., Hwang, D., 2006. Complex networks: structure and dynamics. Phys. Rep. 424, 175–308.

Bohland, J.W., Wu, C., Barbas, H., Bokil, H., Bota, M., Breiter, H.C., Cline, H.T., Doyle, J.C., Freed, P.J., Greenspan, R.J., Haber, S.N., Hawrylycz, M., Herrera, D.G., Hilgetag, C.C., Huang, Z.J., Jones, A., Jones, E.G., Karten, H.J., Kleinfeld, D., Kötter, R., Lester, H.A., Lin, J.M., Mensh, B.D., Mikula, S., Panksepp, J., Price, J.L., Safdieh, J., Saper, C.B., Schiff, N.D., Schmahmann, J.D., Stillman, B.W., Svoboda, K., Swanson, L.W., Toga, A.-W., Van Essen, D.C., Watson, J.D., Mitra, P.P., 2009. A proposal for a coordinated effort for the determination of brainwide neuroanatomical connectivity in model organisms at a mesoscopic scale. PLoS Comput. Biol. 5, e1000334.

Bonilha, L., Gleichgerrcht, E., Fridriksson, J., Rorden, C., Breedlove, J.L., Nesland, T., Paulus, W., Helms, G., Focke, N.K., 2015. Reproducibility of the structural brain connectome derived from diffusion tensor imaging. PLoS One 10, e0135247.

Borgatti, S.P., 2005. Centrality and network flow. Soc. Networks 27, 55–71.

Brandes, U., Delling, D., Gaertler, M., Görke, R., 2006. Maximizing modularity is hard. arXiv. 0608255v2.

Buchanan, C.R., Pernet, C.R., Gorgolewski, K.J., Storkey, A.J., Bastin, M.E., 2014. Test-retest reliability of structural brain networks from diffusion MRI. Neuroimage 86, 231–243.

Buchanan, C.R., Bastin, M.E., Ritchie, S.J., Liewald, D.C., Madole, J.W., Tucker-Drob, E.M., Deary, I.J., Cox, S.R., 2020. The effect of network thresholding and weighting on structural brain networks in the UK Biobank. Neuroimage 211, 116443.

Bullmore, E., Sporns, O., 2009. Complex brain networks: graph theoretical analysis of structural and functional systems. Nat. Rev. Neurosci. 10, 186–198.

Bullmore, E., Sporns, O., 2012. The economy of brain network organization. Nat. Rev. Neurosci. 13, 336–349.

Bullmore, E., Barnes, A., Bassett, D.S., Fornito, A., Kitzbichler, M., Meunier, D., Suckling, J., 2009. Generic aspects of complexity in brain imaging data and other biological systems. Neuroimage 47, 1125–1134.

Buzsáki, G., Geisler, C., Henze, D.A., Wang, X.J., 2004. Interneuron diversity series: circuit complexity and axon wiring economy of cortical interneurons. Trends Neurosci. 27, 186–193.

Cabral, J., Kringelbach, M.L., Deco, G., 2014. Exploring the network dynamics underlying brain activity during rest. Prog. Neurobiol. 114, 102–131.

Calabrese, E., Badea, A., Cofer, G., Qi, Y., Johnson, G.A., 2015. A diffusion MRI tractography connectome of the mouse brain and comparison with neuronal tracer data. Cereb. Cortex 25, 4628–4637.

Campbell, J.S., Siddiqi, K., Rymar, V.V., Sadikot, A.F., Pike, G.B., 2005. Flow-based fiber tracking with diffusion tensor and q-ball data: validation and comparison to principal diffusion direction techniques. Neuroimage 27, 725–736.

Catani, M., 2007. From hodology to function. Brain 130, 602–605.

Catani, M., Thiebaut de Schotten, M., 2008. A diffusion tensor imaging tractography atlas for virtual in vivo dissections. Cortex 44, 1105–1132.

Chen, Y., Wang, S., Hilgetag, C.C., Zhou, C., 2013. Trade-off between multiple constraints enables simultaneous formation of modules and hubs in neural systems. PLoS Comput. Biol. 9, e1002937.

Chen, H., Liu, T., Zhao, Y., Zhang, T., Li, Y., Li, M., Zhang, H., Kuang, H., Guo, L., Tsien, J.Z., Liu, T., 2015. Optimization of large-scale mouse brain connectome via joint evaluation of DTI and neuron tracing data. Neuroimage 115, 202–213.

Civier, O., Smith, R.E., Yeh, C.H., Connelly, A., Calamante, F., 2019. Is removal of weak connections necessary for graph-theoretical analysis of dense weighted structural connectomes from diffusion MRI? Neuroimage 194, 68–81.

Clauset, A., Shalizi, C.R., Newman, M.E.J., 2009. Power-law distributions in empirical data. SIAM Rev. 51, 661–703.

Close, T.G., Tournier, J.D., Calamante, F., Johnston, L.A., Mareels, I., Connelly, A., 2009. A software tool to generate simulated white matter structures for the assessment of fibre-tracking algorithms. Neuroimage 47, 1288–1300.

Close, T.G., Tournier, J.D., Johnston, L.A., Calamante, F., Mareels, I., Connelly, A., 2015. Fourier Tract Sampling (FouTS): a framework for improved inference of white matter tracts from diffusion MRI by explicitly modelling tract volume. Neuroimage 120, 412–427.

Cook, S.J., Jarrell, T.A., Brittin, C.A., Wang, Y., Bloniarz, A.E., Yakovlev, M.A., Nguyen, K.C.Q., Tang, L.T., Bayer, E.A., Duerr, J.S., Bülow, H.E., Hobert, O., Hall, D.H., Emmons, S.W., 2019. Whole-animal connectomes of both Caenorhabditis elegans sexes. Nature 571, 63–71.

Côté, M.A., Boré, A., Girard, G., Houde, J.C., Descoteaux, M., 2012. Tractometer: online evaluation system for tractography. Med. Image Comput. Comput. Assist. Interv. 15, 699–706.

Cottaar, M., Bastiani, M., Boddu, N., Glasser, M.F., Haber, S., Van Essen, D.C., Sotiropoulos, S.N., Jbabdi, S., 2021. Modelling white matter in gyral blades as a continuous vector field. Neuroimage 227, 117693. https://doi.org/10.1016/j.neuroimage.2020.117693 (Epub 2020 Dec 30. PMID: 33385545; PMCID: PMC7610793).

Daducci, A., Canales-Rodríguez, E.J., Descoteaux, M., Garyfallidis, E., Gur, Y., Lin, Y.C., Mani, M., Merlet, S., Paquette, M., Ramirez-Manzanares, A., Reisert, M., Reis Rodrigues, P., Sepehrband, F., Caruyer, E., Choupan, J., Deriche, R., Jacob, M., Menegaz, G., Prčkovska, V., Rivera, M., Wiaux, Y., Thiran, J.P., 2014. Quantitative comparison of reconstruction methods for intra-voxel fiber recovery from diffusion MRI. IEEE Trans. Med. Imaging 33, 384–399.

Daducci, A., Dal Palù, A., Lemkaddem, A., Thiran, J.P., 2015. COMMIT: convex optimization modeling for microstructure informed tractography. IEEE Trans. Med. Imaging 34, 246–257.

Daducci, A., Dal Palu, A., Descoteaux, M., Thiran, J.P., 2016. Microstructure informed tractography: pitfalls and open challenges. Front. Neurosci. 10, 247.

Dauguet, J., Peled, S., Berezovskii, V., Delzescaux, T., Warfield, S.K., Born, R., Westin, C.F., 2007. Comparison of fiber tracts derived from in-vivo DTI tractography with 3D histological neural tract tracer reconstruction on a macaque brain. Neuroimage 37, 530–538.

Deco, G., Jirsa, V.K., McIntosh, A.R., 2011. Emerging concepts for the dynamical organization of resting-state activity in the brain. Nat. Rev. Neurosci. 12, 43–56.

Dell'Acqua, F., Tournier, J.D., 2019. Modelling white matter with spherical deconvolution: how and why? NMR Biomed. 32, e3945.

Dell'Acqua, F., Rizzo, G., Scifo, P., Clarke, R.A., Scotti, G., Fazio, F., 2007. A model-based deconvolution approach to solve fiber crossing in diffusion-weighted MR imaging. IEEE Trans. Biomed. Eng. 54, 462–472.

De Luca, A., Guo, F., Froeling, M., Leemans, A., 2020. Spherical deconvolution with tissue-specific response functions and multi-shell diffusion MRI to estimate multiple fiber orientation distributions (mFODs). Neuroimage 222, 117206. https://doi.org/10.1016/j.neuroimage.2020.117206 (Epub 2020 Aug 1. PMID: 32745681).

de Reus, M.A., van den Heuvel, M.P., 2013a. Estimating false positives and negatives in brain networks. Neuroimage 70, 402–409.

de Reus, M.A., van den Heuvel, M.P., 2013b. Rich club organization and intermodule communication in the cat connectome. J. Neurosci. 33, 12929–12939.

Donahue, C.J., Sotiropoulos, S.N., Jbabdi, S., Hernandez-Fernandez, M., Behrens, T.E., Dyrby, T.B., Coalson, T., Kennedy, D., Knoblauch, K., Van Essen, D.C., Glasser, M.F., 2016. Using diffusion tractography to predict cortical connection strength and distance: a quantitative comparison with tracers in the monkey. J. Neurosci. 36, 6758–6770.

Drakesmith, M., Caeyenberghs, K., Dutt, A., Lewis, G., David, A.S., Jones, D.K., 2015. Overcoming the effects of false positives and threshold bias in graph theoretical analyses of neuroimaging data. Neuroimage 118, 313–333.

Eickhoff, S.B., Yeo, B.T.T., Genon, S., 2018. Imaging-based parcellations of the human brain. Nat. Rev. Neurosci. 19, 672–686.

Elias, W.J., Zheng, Z.A., Domer, P., Quigg, M., Pouratian, N., 2012. Validation of connectivity-based thalamic segmentation with direct electrophysiologic recordings from human sensory thalamus. Neuroimage 59, 2025–2034.

Ercsey-Ravasz, M., Markov, N.T., Lamy, C., Van Essen, D.C., Knoblauch, K., Toroczkai, Z., Kennedy, H., 2013. A predictive network model of cerebral cortical connectivity based on a distance rule. Neuron 80, 184–197.

Evans, A.C., 2013. Networks of anatomical covariance. Neuroimage 80, 489–504.

Expert, P., Evans, T.S., Blondel, V.D., Lambiotte, R., 2011. Uncovering space-independent communities in spatial networks. Proc. Natl. Acad. Sci. U. S. A. 108, 7663–7668.

Fieremans, E., De Deene, Y., Delputte, S., Ozdemir, M.S., Achten, E., Lemahieu, I., 2008. The design of anisotropic diffusion phantoms for the validation of diffusion weighted magnetic resonance imaging. Phys. Med. Biol. 53, 5405–5419.

Fillard, P., Descoteaux, M., Goh, A., Gouttard, S., Jeurissen, B., Malcolm, J., Ramirez-Manzanares, A., Reisert, M., Sakaie, K., Tensaouti, F., Yo, T., Mangin, J.F., Poupon, C., 2011. Quantitative evaluation of 10 tractography algorithms on a realistic diffusion MR phantom. Neuroimage 56, 220–234.

Fischl, B., Rajendran, N., Busa, E., Augustinack, J., Hinds, O., Yeo, B.T., Mohlberg, H., Amunts, K., Zilles, K., 2008. Cortical folding patterns and predicting cytoarchitecture. Cereb. Cortex 18, 1973–1980.

Fornito, A., Zalesky, A., Pantelis, C., Bullmore, E.T., 2012. Schizophrenia, neuroimaging and connectomics. Neuroimage 62, 2296–2314.

Fornito, A., Zalesky, A., Breakspear, M., 2013. Graph analysis of the human connectome: promise, progress, and pitfalls. Neuroimage 80, 426–444.

Fornito, A., Zalesky, A., Bullmore, E.T., 2016. Fundamentals of Brain Network Analysis. Academic Press, Inc, San Diego, USA.

Fornito, A., Arnatkeviciute, A., Fulcher, B.D., 2019. Bridging the gap between connectome and transcriptome. Trends Cogn. Sci. 23, 34–50.

Fortunato, S., 2010. Community detection in graphs. Phys. Rep. 486, 75–174.

Fulcher, B.D., Fornito, A., 2016. A transcriptional signature of hub connectivity in the mouse connectome. Proc. Natl. Acad. Sci. U. S. A. 113, 1435–1440.

Gal, E., London, M., Globerson, A., Ramaswamy, S., Reimann, M.W., Muller, E., Markram, H., Segev, I., 2017. Rich cell-type-specific network topology in neocortical microcircuitry. Nat. Neurosci. 20, 1004–1013.

Garrison, K.A., Scheinost, D., Finn, E.S., Shen, X., Constable, R.T., 2015. The (in)stability of functional brain network measures across thresholds. Neuroimage 118, 651–661.

Garyfallidis, E., Cote, M.A., Rheault, F., Sidhu, J., Hau, J., Petit, L., Fortin, D., Cunanne, S., Descoteaux, M., 2018. Recognition of white matter bundles using local and global streamline-based registration and clustering. Neuroimage 170, 283–295.

Girard, G., Whittingstall, K., Deriche, R., Descoteaux, M., 2014. Towards quantitative connectivity analysis: reducing tractography biases. Neuroimage 98, 266–278.

Girard, G., Caminiti, R., Battaglia-Mayer, A., St-Onge, E., Ambrosen, K.S., Eskildsen, S.F., Krug, K., Dyrby, T.B., Descoteaux, M., Thiran, J.P., Innocenti, G.M., 2020. On the cortical connectivity in the macaque brain: a comparison of diffusion tractography and histological tracing data. Neuroimage 221, 117201.

Glasser, M.F., Coalson, T.S., Robinson, E.C., Hacker, C.D., Harwell, J., Yacoub, E., Ugurbil, K., Andersson, J., Beckmann, C.F., Jenkinson, M., Smith, S.-M., Van Essen, D.C., 2016. A multi-modal parcellation of human cerebral cortex. Nature 536 (7615), 171–178. https://doi.org/10.1038/nature18933 (Epub 2016 Jul 20. PMID: 27437579; PMCID: PMC4990127).

Gong, G., He, Y., Concha, L., Lebel, C., Gross, D.W., Evans, A.C., Beaulieu, C., 2009. Mapping anatomical connectivity patterns of human cerebral cortex using in vivo diffusion tensor imaging tractography. Cereb. Cortex 19, 524–536.

Goñi, J., van den Heuvel, M.P., Avena-Koenigsberger, A., Velez de Mendizabal, N., Betzel, R.F., Griffa, A., Hagmann, P., Corominas-Murtra, B., Thiran, J.-P., Sporns, O., 2014. Resting-brain functional connectivity predicted by analytic measures of network communication. Proc. Natl. Acad. Sci. U. S. A. 111, 833–838.

Good, B.H., de Montjoye, Y.A., Clauset, A., 2010. Performance of modularity maximization in practical contexts. Phys. Rev. E Stat. Nonlin. Soft Matter Phys. 81, 046106.

Graziano, M.S., Aflalo, T.N., 2007. Rethinking cortical organization: moving away from discrete areas arranged in hierarchies. Neuroscientist 13, 138–147.

Gu, S., Pasqualetti, F., Cieslak, M., Telesford, Q.K., Yu, A.B., Kahn, A.E., Medaglia, J.D., Vettel, J.M., Miller, M.B., Grafton, S.T., Bassett, D.S., 2015. Controllability of structural brain networks. Nat. Commun. 6, 8414.

Haak, K.V., Beckmann, C.F., 2020. Understanding brain organisation in the face of functional heterogeneity and functional multiplicity. Neuroimage 220, 117061.

Hagmann, P., 2005. From Diffusion MRI to Brain Connectomics. EPFL.

Hagmann, P., Cammoun, L., Gigandet, X., Meuli, R., Honey, C.J., Wedeen, V.J., Sporns, O., 2008. Mapping the structural core of human cerebral cortex. PLoS Biol. 6, e159.

Hagmann, P., Sporns, O., Madan, N., Cammoun, L., Pienaar, R., Wedeen, V.J., Meuli, R., Thiran, J.P., Grant, P.E., 2010. White matter maturation reshapes structural connectivity in the late developing human brain. Proc. Natl. Acad. Sci. U. S. A. 107, 19067–19072.

Hayasaka, S., Laurienti, P.J., 2010. Comparison of characteristics between region-and voxel-based network analyses in resting-state fMRI data. Neuroimage 50, 499–508.

He, T., Kong, R., Holmes, A.J., Nguyen, M., Sabuncu, M.R., Eickhoff, S.B., Bzdok, D., Feng, J., Yeo, B.T.T., 2020. Deep neural networks and kernel regression achieve comparable accuracies for functional connectivity prediction of behavior and demographics. Neuroimage 206, 116276.

Henderson, J.A., Robinson, P.A., 2014. Relations between the geometry of cortical gyrification and white-matter network architecture. Brain Connect. 4, 112–130.

Hernandez-Fernandez, M., Reguly, I., Jbabdi, S., Giles, M., Smith, S., Sotiropoulos, S.N., 2019. Using GPUs to accelerate computational diffusion MRI: from microstructure estimation to tractography and connectomes. Neuroimage 188, 598–615.

Honey, C.J., Sporns, O., Cammoun, L., Gigandet, X., Thiran, J.P., Meuli, R., Hagmann, P., 2009. Predicting human resting-state functional connectivity from structural connectivity. Proc. Natl. Acad. Sci. U. S. A. 106, 2035–2040.

Horn, A., Reich, M., Vorwerk, J., Li, N., Wenzel, G., Fang, Q., Schmitz-Hübsch, T., Nickl, R., Kupsch, A., Volkmann, J., Kühn, A.A., Fox, M.D., 2017. Connectivity predicts deep brain stimulation outcome in Parkinson disease. Ann. Neurol. 82, 67–78.

Horvát, S., Gămănuţ, R., Ercsey-Ravasz, M., Magrou, L., Gămănuţ, B., Van Essen, D.C., Burkhalter, A., Knoblauch, K., Toroczkai, Z., Kennedy, H., 2016. Spatial embedding and wiring cost constrain the functional layout of the cortical network of rodents and primates. PLoS Biol. 14, e1002512.

Huntenburg, J.M., Bazin, P.-L., Margulies, D.S., 2018. Large-scale gradients in human cortical organization. Trends Cogn. Sci. 22, 21–31.

Hyvärinen, A., 1999. Fast and robust fixed-point algorithms for independent component analysis. IEEE Trans. Neural Netw. 10, 626–634.

Iturria-Medina, Y., Canales-Rodriguez, E.J., Melie-Garcia, L., Valdes-Hernandez, P.A., Martinez-Montes, E., Aleman-Gomez, Y., Sanchez-Bornot, J.M., 2007. Characterizing brain anatomical connections using diffusion weighted MRI and graph theory. Neuroimage 36, 645–660.

Jbabdi, S., Johansen-Berg, H., 2011. Tractography: where do we go from here? Brain Connect. 1, 169–183.

Jbabdi, S., Woolrich, M.W., Andersson, J.L., Behrens, T.E., 2007. A Bayesian framework for global tractography. Neuroimage 37, 116–129.

Jbabdi, S., Sotiropoulos, S.N., Behrens, T.E., 2013. The topographic connectome. Curr. Opin. Neurobiol. 23, 207–215.

Jbabdi, S., Sotiropoulos, S.N., Haber, S.N., Van Essen, D.C., Behrens, T.E., 2015. Measuring macroscopic brain connections in vivo. Nat. Neurosci. 18, 1546–1555.

Jeurissen, B., Tournier, J.D., Dhollander, T., Connelly, A., Sijbers, J., 2014. Multi-tissue constrained spherical deconvolution for improved analysis of multi-shell diffusion MRI data. Neuroimage 103, 411–426.

Johansen-Berg, H., Behrens, T.E., 2006. Just pretty pictures? What diffusion tractography can add in clinical neuroscience. Curr. Opin. Neurol. 19, 379–385.

Johansen-Berg, H., Behrens, T.E., Robson, M.D., Drobnjak, I., Rushworth, M.F., Brady, J.M., Smith, S.M., Higham, D.J., Matthews, P.M., 2004. Changes in connectivity profiles define functionally distinct regions in human medial frontal cortex. Proc. Natl. Acad. Sci. U. S. A. 101, 13335–13340.

Jones, D.K., Knosche, T.R., Turner, R., 2013. White matter integrity, fiber count, and other fallacies: the do's and don'ts of diffusion MRI. Neuroimage 73, 239–254.

Kaiser, M., Hilgetag, C.C., 2006. Nonoptimal component placement, but short processing paths, due to long-distance projections in neural systems. PLoS Comput. Biol. 2, e95.

Kale, P., Zalesky, A., Gollo, L.L., 2018. Estimating the impact of structural directionality: how reliable are undirected connectomes? Netw. Neurosci. 2, 259–284.

Kanwisher, N., 2010. Functional specificity in the human brain: a window into the functional architecture of the mind. Proc. Natl. Acad. Sci. U. S. A. 107, 11163–11170.

Kennedy, H., Knoblauch, K., Toroczkai, Z., 2013. Why data coherence and quality is critical for understanding interareal cortical networks. Neuroimage 80, 37–45.

Kivelä, M., Arenas, A., Barthelemy, M., Gleeson, J.P., Moreno, Y., Porter, M.A., 2014. Multilayer networks. J. Complex Netw. 2, 203–271.

Klingler, J., 1935. Erleichterung der makroskopischen Praeparation des Gehirns durch den Gefrierprozess. Schweiz. Arch. Neurol. Psychiatr. 36, 247–256.

Knösche, T.R., Anwander, A., Liptrot, M., Dyrby, T.B., 2015. Validation of tractography: comparison with manganese tracing. Hum. Brain Mapp. 36, 4116–4134.

Kreher, B.W., Mader, I., Kiselev, V.G., 2008. Gibbs tracking: a novel approach for the reconstruction of neuronal pathways. Magn. Reson. Med. 60, 953–963.

Kumar, S., Sreenivasan, V., Talukdar, P., Pestilli, F., Sridharan, D., 2019. ReAl-LiFE: accelerating the discovery of individualized brain connectomes on GPUs. Proc. AAAI Conf. Artif. Intell. 33, 630–638.

Lancichinetti, A., Fortunato, S., 2012. Consensus clustering in complex networks. Sci. Rep. 2, 336.

Latora, V., Marchiori, M., 2001. Efficient behavior of small-world networks. Phys. Rev. Lett. 87, 198701.

Leemans, A., Sijbers, J., Verhoye, M., Van der Linden, A., Van Dyck, D., 2005. Mathematical framework for simulating diffusion tensor MR neural fiber bundles. Magn. Reson. Med. 53, 944–953.

Lemkaddem, A., Daducci, A., Kunz, N., Lazeyras, F., Seeck, M., Thiran, J.P., Vulliemoz, S., 2014. Connectivity and tissue microstructural alterations in right and left temporal lobe epilepsy revealed by diffusion spectrum imaging. Neuroimage Clin. 5, 349–358.

Li, L., Rilling, J.K., Preuss, T.M., Glasser, M.F., Hu, X., 2012. The effects of connection reconstruction method on the interregional connectivity of brain networks via diffusion tractography. Hum. Brain Mapp. 33, 1894–1913.

Lori, N.F., Akbudak, E., Shimony, J.S., Cull, T.S., Snyder, A.Z., Guillory, R.K., Conturo, T.E., 2002. Diffusion tensor fiber tracking of human brain connectivity: aquisition methods, reliability analysis and biological results. NMR Biomed. 15, 494–515.

Lynn, C.W., Bassett, D.S., 2019. The physics of brain network structure, function and control. Nat. Rev. Phys., 1–15.

Maglanoc, L.A., Kaufmann, T., van der Meer, D., Marquand, A.F., Wolfers, T., Jonassen, R., Hilland, E., Andreassen, O.A., Landrø, N.I., Westlye, L.T., 2020. Brain connectome mapping of complex human traits and their polygenic architecture using machine learning. Biol. Psychiatry 87, 717–726.

Maier-Hein, K.H., Neher, P.F., Houde, J.C., Côté, M.A., Garyfallidis, E., Zhong, J., Chamberland, M., Yeh, F.C., Lin, Y.C., Ji, Q., Reddick, W.E., Glass, J.-O., Chen, D.Q., Feng, Y., Gao, C., Wu, Y., Ma, J., He, R., Li, Q., Westin, C.F., Deslauriers-Gauthier, S., González, J.O.O., Paquette, M., St-Jean, S., Girard, G., Rheault, F., Sidhu, J., Tax, C.M.W., Guo, F., Mesri, H.Y., Dávid, S., Froeling, M., Heemskerk, A.M., Leemans, A., Boré, A., Pinsard, B., Bedetti, C., Desrosiers, M., Brambati, S., Doyon, J., Sarica, A., Vasta, R., Cerasa, A., Quattrone, A., Yeatman, J., Khan, A.R., Hodges, W., Alexander, S., Romascano, D., Barakovic, M., Auría, A., Esteban, O., Lemkaddem, A., Thiran, J.P., Cetingul, H.E., Odry, B.L., Mailhe, B., Nadar, M.-S., Pizzagalli, F., Prasad, G., Villalon-Reina, J.E., Galvis, J., Thompson, P.M., Requejo, F.S., Laguna, P.L., Lacerda, L.M., Barrett, R., Dell'Acqua, F., Catani, M., Petit, L., Caruyer, E., Daducci, A., Dyrby, T.B., Holland-Letz, T., Hilgetag, C.C., Stieltjes, B., Descoteaux, M., 2017. The challenge of mapping the human connectome based on diffusion tractography. Nat. Commun. 8, 1349.

Mandl, R.C., Schnack, H.G., Luigjes, J., van den Heuvel, M.P., Cahn, W., Kahn, R.S., Hulshoff Pol, H.E., 2010. Tract-based analysis of magnetization transfer ratio and diffusion tensor imaging of the frontal and frontotemporal connections in schizophrenia. Schizophr. Bull. 36, 778–787.

Markov, N.T., Ercsey-Ravasz, M., Lamy, C., Ribeiro Gomes, A.R., Magrou, L., Misery, P., Giroud, P., Barone, P., Dehay, C., Toroczkai, Z., Knoblauch, K., Van Essen, D.C., Kennedy, H., 2013. The role of long-range connections on the specificity of the macaque interareal cortical network. Proc. Natl. Acad. Sci. U. S. A. 110, 5187–5192.

Markov, N.T., Ercsey-Ravasz, M.M., Ribeiro Gomes, A.R., Lamy, C., Magrou, L., Vezoli, J., Misery, P., Falchier, A., Quilodran, R., Gariel, M.A., Sallet, J., Gamanut, R., Huissoud, C., Clavagnier, S., Giroud, P., Sappey-Marinier, D., Barone, P., Dehay, C., Toroczkai, Z., Knoblauch, K., Van Essen, D.C., Kennedy, H., 2014. A weighted and directed interareal connectivity matrix for macaque cerebral cortex. Cereb. Cortex 24, 17–36.

Mars, R.B., Jbabdi, S., Sallet, J., O'Reilly, J.X., Croxson, P.L., Olivier, E., Noonan, M.P., Bergmann, C., Mitchell, A.S., Baxter, M.G., Behrens, T.E., Johansen-Berg, H., Tomassini, V., Miller, K.L., Rushworth, M.F., 2011. Diffusion-weighted imaging tractography-based parcellation of the human parietal cortex and comparison with human and macaque resting-state functional connectivity. J. Neurosci. 31, 4087–4100.

Mars, R.B., Passingham, R.E., Jbabdi, S., 2018a. Connectivity fingerprints: from areal descriptions to abstract spaces. Trends Cogn. Sci. 22, 1026–1037.

Mars, R.B., Sotiropoulos, S.N., Passingham, R.E., Sallet, J., Verhagen, L., Khrapitchev, A.A., Sibson, N., Jbabdi, S., 2018b. Whole brain comparative anatomy using connectivity blueprints. eLife 7.

Martino, J., De Witt Hamer, P.C., Vergani, F., Brogna, C., de Lucas, E.M., Vázquez-Barquero, A., García-Porrero, J.A., Duffau, H., 2011. Cortex-sparing fiber dissection: an improved method for the study of white matter anatomy in the human brain. J. Anat. 219, 531–541.

Maslov, S., Sneppen, K., 2002. Specificity and stability in topology of protein networks. Science 296, 910–913.

Messé, A., Rudrauf, D., Benali, H., Marrelec, G., 2014. Relating structure and function in the human brain: relative contributions of anatomy, stationary dynamics, and non-stationarities. PLoS Comput. Biol. 10, e1003530.

Meunier, D., Lambiotte, R., Fornito, A., Ersche, K.D., Bullmore, E.T., 2009. Hierarchical modularity in human brain functional networks. Front. Neuroinform. 3, 37.

Mill, R.D., Ito, T., Cole, M.W., 2017. From connectome to cognition: the search for mechanism in human functional brain networks. Neuroimage 160, 124–139.

Mišić, B., Betzel, R.F., Nematzadeh, A., Goñi, J., Griffa, A., Hagmann, P., Flammini, A., Ahn, Y.-Y., Sporns, O., 2015. Cooperative and competitive spreading dynamics on the human connectome. Neuron 86, 1518–1529.

Mišić, B., Betzel, R.F., de Reus, M.A., van den Heuvel, M.P., Berman, M.G., McIntosh, A.R., Sporns, O., 2016. Network-level structure-function relationships in human neocortex. Cereb. Cortex 26, 3285–3296.

Mori, S., Crain, B.J., Chacko, V.P., van Zijl, P.C., 1999. Three-dimensional tracking of axonal projections in the brain by magnetic resonance imaging. Ann. Neurol. 45, 265–269.

Morris, D.M., Embleton, K.V., Parker, G.J., 2008. Probabilistic fibre tracking: differentiation of connections from chance events. Neuroimage 42, 1329–1339.

Murphy, A.C., Gu, S., Khambhati, A.N., Wymbs, N., Grafton, S.T., Satterthwaite, T.D., Bassett, D.S., 2016. Explicitly linking regional activation and function connectivity: community structure of weighted networks with continuous annotation. arXiv. arXiv:1611.07962v1.

Neher, P.F., Laun, F.B., Stieltjes, B., Maier-Hein, K.H., 2014. Fiberfox: facilitating the creation of realistic white matter software phantoms. Magn. Reson. Med. 72, 1460–1470.

Neher, P.F., Descoteaux, M., Houde, J.C., Stieltjes, B., Maier-Hein, K.H., 2015. Strengths and weaknesses of state of the art fiber tractography pipelines—a comprehensive in-vivo and phantom evaluation study using Tractometer. Med. Image Anal. 26, 287–305.

Newman, M.E.J., 2003. The structure and function of complex networks. SIAM Rev. 45, 167–256.

Newman, M.E.J., 2010. Networks: An Introduction. Oxford University Press, Oxford.

Newman, M.E.J., Girvan, M., 2004. Finding and evaluating community structure in networks. Phys. Rev. E Stat. Nonlin. Soft Matter Phys. 69, 026113.

Nichols, T., Hayasaka, S., 2003. Controlling the familywise error rate in functional neuroimaging: a comparative review. Stat. Methods Med. Res. 12, 419–446.

Oh, S.W., Harris, J.A., Ng, L., Winslow, B., Cain, N., Mihalas, S., Wang, Q., Lau, C., Kuan, L., Henry, A.M., Mortrud, M.T., Ouellette, B., Nguyen, T.N., Sorensen, S.A., Slaughterbeck, C.R., Wakeman, W., Li, Y., Feng, D., Ho, A., Nicholas, E., Hirokawa, K.E., Bohn, P., Joines, K.M., Peng, H., Hawrylycz, M.J., Phillips, J.W., Hohmann, J.G., Wohnoutka, P., Gerfen, C.R., Koch, C., Bernard, A., Dang, C., Jones, A.R., Zeng, H., 2014. A mesoscale connectome of the mouse brain. Nature 508, 207–214.

Oldham, S., Fulcher, B., Parkes, L., Arnatkevicˇiūtė, A., Suo, C., 2019. Consistency and differences between centrality measures across distinct classes of networks. PLoS One 14 (7), e0220061. https://doi.org/10.1371/journal.pone.0220061 (PMID: 31348798; PMCID: PMC6660088).

O'Muircheartaigh, J., Jbabdi, S., 2018. Concurrent white matter bundles and grey matter networks using independent component analysis. Neuroimage 170, 296–306.

Opsahl, T., Agneessens, F., Skvoretz, J., 2010. Node centrality in weighted networks: generalizing degree and shortest paths. Soc. Networks 32, 245–251.

Palomero-Gallagher, N., Zilles, K., 2019. Cortical layers: cyto-, myelo-, receptor- and synaptic architecture in human cortical areas. Neuroimage 197, 716–741.

Parker, G.J., Alexander, D.C., 2003. Probabilistic Monte Carlo based mapping of cerebral connections utilising whole-brain crossing fibre information. Inf. Process. Med. Imaging 18, 684–695.

Parker, G.J., Haroon, H.A., Wheeler-Kingshott, C.A., 2003. A framework for a streamline-based probabilistic index of connectivity (PICo) using a structural interpretation of MRI diffusion measurements. J. Magn. Reson. Imaging 18, 242–254.

Passingham, R.E., Stephan, K.E., Kotter, R., 2002. The anatomical basis of functional localization in the cortex. Nat. Rev. Neurosci. 3, 606–616.

Perrin, M., Poupon, C., Rieul, B., Leroux, P., Constantinesco, A., Mangin, J.F., Lebihan, D., 2005. Validation of q-ball imaging with a diffusion fibre-crossing phantom on a clinical scanner. Philos. Trans. R. Soc. Lond. Ser. B Biol. Sci. 360, 881–891.

Pestilli, F., Yeatman, J.D., Rokem, A., Kay, K.N., Wandell, B.A., 2014. Evaluation and statistical inference for human connectomes. Nat. Methods 11, 1058–1063.

Poulin, P., Jörgens, D., Jodoin, P.M., Descoteaux, M., 2019. Tractography and machine learning: current state and open challenges. Magn. Reson. Imaging 64, 37–48.

Poupon, C., Rieul, B., Kezele, I., Perrin, M., Poupon, F., Mangin, J.F., 2008. New diffusion phantoms dedicated to the study and validation of high-angular-resolution diffusion imaging (HARDI) models. Magn. Reson. Med. 60, 1276–1283.

Poupon, C., Guevara, P., Laribière, L., Tournier, J.D., Bernard, A., Fillard, P., Descoteaux, M., Mangin, J.F., 2010. A human-like coronal tractography phantom. In: Proceedings of the Joint ISMRM/ESMRMB Conference.

Pouratian, N., Zheng, Z., Bari, A.A., Behnke, E., Elias, W.J., Desalles, A.A., 2011. Multi-institutional evaluation of deep brain stimulation targeting using probabilistic connectivity-based thalamic segmentation. J. Neurosurg. 115, 995–1004.

Prckovska, V., Rodrigues, P., Puigdellivol Sanchez, A., Ramos, M., Andorra, M., Martinez-Heras, E., Falcon, C., Prats-Galino, A., Villoslada, P., 2016. Reproducibility of the structural connectome reconstruction across diffusion methods. J. Neuroimaging 26, 46–57.

Preti, M.G., Van De Ville, D., 2019. Decoupling of brain function from structure reveals regional behavioral specialization in humans. Nat. Commun. 10, 4747.

Reisert, M., Mader, I., Anastasopoulos, C., Weigel, M., Schnell, S., Kiselev, V., 2011. Global fiber reconstruction becomes practical. Neuroimage 54, 955–962.

Reveley, C., Seth, A.K., Pierpaoli, C., Silva, A.C., Yu, D., Saunders, R.C., Leopold, D.A., Ye, F.Q., 2015. Superficial white matter fiber systems impede detection of long-range cortical connections in diffusion MR tractography. Proc. Natl. Acad. Sci. U. S. A. 112, E2820–E2828.

Rheault, F., De Benedictis, A., Daducci, A., Maffei, C., Tax, C.M.W., Romascano, D., Caverzasi, E., Morency, F.C., Corrivetti, F., Pestilli, F., Girard, G., Theaud, G., Zemmoura, I., Hau, J., Glavin, K., Jordan, K.M., Pomiecko, K., Chamberland, M., Barakovic, M., Goyette, N., Poulin, P., Chenot, Q., Panesar, S.S., Sarubbo, S., Petit, L., Descoteaux, M., 2020. Tractostorm: the what, why, and how of tractography dissection reproducibility. Hum. Brain Mapp. 41, 1859–1874.

Roberts, J.A., Perry, A., Lord, A.R., Roberts, G., Mitchell, P.B., Smith, R.E., Calamante, F., Breakspear, M., 2016. The contribution of geometry to the human connectome. Neuroimage 124, 379–393.

Roberts, J.A., Perry, A., Roberts, G., Mitchell, P.B., Breakspear, M., 2017. Consistency-based thresholding of the human connectome. Neuroimage 145, 118–129.

Roumazeilles, L., Eichert, N., Bryant, K.L., Folloni, D., Sallet, J., Vijayakumar, S., Foxley, S., Tendler, B.C., Jbabdi, S., Reveley, C., Verhagen, L., Dershowitz, L.B., Guthrie, M., Flach, E., Miller, K.L., Mars, R.B., 2020. Longitudinal connections and the organization of the temporal cortex in macaques, great apes, and humans. PLoS Biol. 18, e3000810.

Rubinov, M., 2016. Constraints and spandrels of interareal connectomes. Nat. Commun. 7, 13812.

Rubinov, M., Sporns, O., 2010. Complex network measures of brain connectivity: uses and interpretations. Neuroimage 52, 1059–1069.

Rubinov, M., Sporns, O., 2011. Weight-conserving characterization of complex functional brain networks. Neuroimage 56, 2068–2079.

**482 PART | IV** From streamlines to tracts

Rubinov, M., Knock, S.A., Stam, C.J., Micheloyannis, S., Harris, A.W., Williams, L.M., Breakspear, M., 2009. Small-world properties of nonlinear brain activity in schizophrenia. Hum. Brain Mapp. 30, 403–416.

Sallet, J., Mars, R.B., Noonan, M.P., Neubert, F.X., Jbabdi, S., O'Reilly, J.X., Filippini, N., Thomas, A.G., Rushworth, M.F., 2013. The organization of dorsal frontal cortex in humans and macaques. J. Neurosci. 33, 12255–12274.

Sammartino, F., Krishna, V., King, N.K., Lozano, A.M., Schwartz, M.L., Huang, Y., Hodaie, M., 2016. Tractography-based ventral intermediate nucleus targeting: novel methodology and intraoperative validation. Mov. Disord. 31, 1217–1225.

Samu, D., Seth, A.K., Nowotny, T., 2014. Influence of wiring cost on the large-scale architecture of human cortical connectivity. PLoS Comput. Biol. 10, e1003557.

Sarwar, T., Ramamohanarao, K., Zalesky, A., 2019. Mapping connectomes with diffusion MRI: deterministic or probabilistic tractography? Magn. Reson. Med. 81, 1368–1384.

Sarwar, T., Seguin, C., Ramamohanarao, K., Zalesky, A., 2020. Towards deep learning for connectome mapping: a block decomposition framework. Neuroimage 212, 116654.

Saygin, Z.M., Osher, D.E., Augustinack, J., Fischl, B., Gabrieli, J.D., 2011. Connectivity-based segmentation of human amygdala nuclei using probabilistic tractography. Neuroimage 56, 1353–1361.

Saygin, Z.M., Osher, D.E., Koldewyn, K., Reynolds, G., Gabrieli, J.D., Saxe, R.R., 2012. Anatomical connectivity patterns predict face selectivity in the fusiform gyrus. Nat. Neurosci. 15, 321–327.

Schiavi, S., Ocampo-Pineda, M., Barakovic, M., Petit, L., Descoteaux, M., Thiran, J.P., Daducci, A., 2020. A new method for accurate in vivo mapping of human brain connections using microstructural and anatomical information. Sci. Adv. 6, eaba8245.

Schilling, K., Gao, Y., Janve, V., Stepniewska, I., Landman, B.A., Anderson, A.W., 2018. Confirmation of a gyral bias in diffusion MRI fiber tractography. Hum. Brain Mapp. 39, 1449–1466.

Schilling, K.G., Nath, V., Hansen, C., Parvathaneni, P., Blaber, J., Gao, Y., Neher, P., Aydogan, D.B., Shi, Y., Ocampo-Pineda, M., Schiavi, S., Daducci, A., Girard, G., Barakovic, M., Rafael-Patino, J., Romascano, D., Rensonnet, G., Pizzolato, M., Bates, A., Fischi, E., Thiran, J.P., Canales-Rodríguez, E.J., Huang, C., Zhu, H., Zhong, L., Cabeen, R., Toga, A.W., Rheault, F., Theaud, G., Houde, J.C., Sidhu, J., Chamberland, M., Westin, C.F., Dyrby, T.B., Verma, R., Rathi, Y., Irfanoglu, M.O., Thomas, C., Pierpaoli, C., Descoteaux, M., Anderson, A.W., Landman, B.A., 2019. Limits to anatomical accuracy of diffusion tractography using modern approaches. Neuroimage 185, 1–11.

Scholtens, L.H., van den Heuvel, M.P., 2018. Multimodal connectomics in psychiatry: bridging scales from micro to macro. Biol. Psychiatry Cogn. Neurosci. Neuroimaging 3, 767–776.

Schuz, A.B.V., 2002. The human cortical white matter: quantitative aspects of cortico-cortical long-range connectivity. In: Shuez, A., Miller, R. (Eds.), Cortical Areas: Unity and Diversity. Taylor & Francis, London, pp. 377–384.

Seguin, C., van den Heuvel, M.P., Zalesky, A., 2018. Navigation of brain networks. Proc. Natl. Acad. Sci. U. S. A. 115, 6297–6302.

Seguin, C., Razi, A., Zalesky, A., 2019. Inferring neural signalling directionality from undirected structural connectomes. Nat. Commun. 10, 4289.

Sereno, M.I., Huang, R.S., 2006. A human parietal face area contains aligned head-centered visual and tactile maps. Nat. Neurosci. 9, 1337–1343.

Serrano, M.A., Boguñá, M., Vespignani, A., 2009. Extracting the multiscale backbone of complex weighted networks. Proc. Natl. Acad. Sci. U. S. A. 106, 6483–6488.

Seunarine, K.K., Alexander, D.C., 2009. Multiple fibers: beyond the diffusion tensor. In: Johansen-Berg, H., Behrens, T.E. (Eds.), Diffusion MRI: From Quantitative Measurement to In-Vivo Neuroanatomy. Elsevier, pp. 55–72.

Shehzad, Z., Kelly, C., Reiss, P.T., Cameron Craddock, R., Emerson, J.W., McMahon, K., Copland, D.A., Castellanos, F.X., Milham, M.P., 2014. A multivariate distance-based analytic framework for connectome-wide association studies. Neuroimage 93 (Pt. 1), 74–94.

Shen, X., Finn, E.S., Scheinost, D., Rosenberg, M.D., Chun, M.M., Papademetris, X., Constable, R.T., 2017. Using connectome-based predictive modeling to predict individual behavior from brain connectivity. Nat. Protoc. 12, 506–518.

Sherbondy, A.J., Rowe, M.C., Alexander, D.C., 2010. MicroTrack: an algorithm for concurrent projectome and microstructure estimation. Med. Image Comput. Comput. Assist. Interv. 13, 183–190.

Sinke, M.R.T., Otte, W.M., Christiaens, D., Schmitt, O., Leemans, A., van der Toorn, A., Sarabdjitsingh, R.A., Joëls, M., Dijkhuizen, R.M., 2018. Diffusion MRI-based cortical connectome reconstruction: dependency on tractography procedures and neuroanatomical characteristics. Brain Struct. Funct. 223, 2269–2285.

Smith, S.M., Miller, K.L., Salimi-Khorshidi, G., Webster, M., Beckmann, C.F., Nichols, T.E., Ramsey, J.D., Woolrich, M.W., 2011. Network modelling methods for FMRI. Neuroimage 54, 875–891.

Smith, R.E., Tournier, J.D., Calamante, F., Connelly, A., 2012. Anatomically-constrained tractography: improved diffusion MRI streamlines tractography through effective use of anatomical information. Neuroimage 62, 1924–1938.

Smith, R.E., Tournier, J.D., Calamante, F., Connelly, A., 2013a. SIFT: spherical-deconvolution informed filtering of tractograms. Neuroimage 67, 298–312.

Smith, S.M., Vidaurre, D., Beckmann, C.F., Glasser, M.F., Jenkinson, M., Miller, K.L., Nichols, T.E., Robinson, E.C., Salimi-Khorshidi, G., Woolrich, M.-W., Barch, D.M., Ugurbil, K., Van Essen, D.C., 2013b. Functional connectomics from resting-state fMRI. Trends Cogn. Sci. 17, 666–682.

Smith, R.E., Tournier, J.D., Calamante, F., Connelly, A., 2015a. The effects of SIFT on the reproducibility and biological accuracy of the structural connectome. Neuroimage 104, 253–265 (Epub 2012 Dec 11. PMID: 23238430).

Smith, S.M., Nichols, T.E., Vidaurre, D., Winkler, A.M., Behrens, T.E.J., Glasser, M.F., Uğurbil, K., Barch, D.M., Van Essen, D.C., Miller, K.L., 2015b. A positive-negative mode of population covariation links brain connectivity, demographics and behavior. Nat. Neurosci. 18, 1565–1567.

Sotiropoulos, S.N., Zalesky, A., 2019. Building connectomes using diffusion MRI: why, how and but. NMR Biomed. 32, e3752.

Sotiropoulos, S.N., Bai, L., Morgan, P.S., Constantinescu, C.S., Tench, C.R., 2010a. Brain tractography using Q-ball imaging and graph theory: improved connectivities through fibre crossings via a model-based approach. Neuroimage 49, 2444–2456.

Sotiropoulos, S.N., Jones, D.E., Bai, L., Kypraios, T., 2010b. Exact and analytic Bayesian inference for orientation distribution functions. In: 7th IEEE International Symposium on Biomedical Imaging: From Nano to Macro, pp. 1189–1192.

Sotiropoulos, S.N., Hernández-Fernández, M., Vu, A.T., Andersson, J.L., Moeller, S., Yacoub, E., Lenglet, C., Ugurbil, K., Behrens, T.E., Jbabdi, S., 2016. Fusion in diffusion MRI for improved fibre orientation estimation: an application to the 3T and 7T data of the human connectome project. Neuroimage 134, 396–409. https://doi.org/10.1016/j.neuroimage.2016.04.014 (Epub 2016 Apr 9. PMID: 27071694; PMCID: PMC6318224).

Sporns, O., 2011. The human connectome: a complex network. Ann. N. Y. Acad. Sci. 1224, 109–125.

Sporns, O., Betzel, R.F., 2016. Modular brain networks. Annu. Rev. Psychol. 67, 613–640.

Sporns, O., Tononi, G., Kotter, R., 2005. The human connectome: a structural description of the human brain. PLoS Comput. Biol. 1, e42.

St-Onge, E., Daducci, A., Girard, G., Descoteaux, M., 2018. Surface-enhanced tractography (SET). Neuroimage 169, 524–539.

Suárez, L.E., Markello, R.D., Betzel, R.F., Misic, B., 2020. Linking structure and function in macroscale brain networks. Trends Cogn. Sci. 24, 302–315.

Sui, J., Jiang, R., Bustillo, J., Calhoun, V., 2020. Neuroimaging-based individualized prediction of cognition and behavior for mental disorders and health: methods and promises. Biol. Psychiatry 88 (11), 818–828. https://doi.org/10.1016/j.biopsych.2020.02.016 (Epub 2020 Feb 27. PMID: 32336400; PMCID: PMC7483317).

Thiebaut de Schotten, M., Dell'Acqua, F., Forkel, S.J., Simmons, A., Vergani, F., Murphy, D.G., Catani, M., 2011. A lateralized brain network for visuo-spatial attention. Nat. Neurosci. 14, 1245–1246.

Thiebaut de Schotten, M., Urbanski, M., Valabregue, R., Bayle, D.J., Volle, E., 2014. Subdivision of the occipital lobes: an anatomical and functional MRI connectivity study. Cortex 56, 121–137.

Thomas, C., Ye, F.Q., Irfanoglu, M.O., Modi, P., Saleem, K.S., Leopold, D.A., Pierpaoli, C., 2014. Anatomical accuracy of brain connections derived from diffusion MRI tractography is inherently limited. Proc. Natl. Acad. Sci. U. S. A. 111, 16574–16579.

Thompson, E., Mohammadi-Nejad, A.R., Robinson, E.C., Andersson, J.L.R., Jbabdi, S., Glasser, M.F., Bastiani, M., Sotiropoulos, S.N., 2020. Non-negative data-driven mapping of structural connections with application to the neonatal brain. Neuroimage 222, 117273. https://doi.org/10.1016/j.neuroimage.2020.117273 (Epub 2020 Aug 18. PMID: 32818619; PMCID: PMC7116021).

Tian, Y., Zalesky, A., 2018. Characterizing the functional connectivity diversity of the insula cortex: subregions, diversity curves and behavior. Neuroimage 183, 716–733.

Tian, Y., Margulies, D.S., Breakspear, M., Zalesky, A., 2020. Topographic organization of the human subcortex unveiled with functional connectivity gradients. Nat. Neurosci. 23 (11), 1421–1432. https://doi.org/10.1038/s41593-020-00711-6 (Epub 2020 Sep 28. PMID: 32989295).

Toro, R., Fox, P.T., Paus, T., 2008. Functional coactivation map of the human brain. Cereb. Cortex 18, 2553–2559.

Tournier, J.D., Calamante, F., Connelly, A., 2007. Robust determination of the fibre orientation distribution in diffusion MRI: non-negativity constrained super-resolved spherical deconvolution. Neuroimage 35, 1459–1472.

Tournier, J.D., Yeh, C.H., Calamante, F., Cho, K.H., Connelly, A., Lin, C.P., 2008. Resolving crossing fibres using constrained spherical deconvolution: validation using diffusion-weighted imaging phantom data. Neuroimage 42, 617–625.

Tziortzi, A.C., Haber, S.N., Searle, G.E., Tsoumpas, C., Long, C.J., Shotbolt, P., Douaud, G., Jbabdi, S., Behrens, T.E., Rabiner, E.A., Jenkinson, M., Gunn, R.N., 2014. Connectivity-based functional analysis of dopamine release in the striatum using diffusion-weighted MRI and positron emission tomography. Cereb. Cortex 24, 1165–1177.

van den Heuvel, M.P., Sporns, O., 2011. Rich-club organization of the human connectome. J. Neurosci. 31, 15775–15786.

van den Heuvel, M.P., Stam, C.J., Boersma, M., Pol, H.E.H., 2008. Small-world and scale-free organization of voxel-based resting-state functional connectivity in the human brain. Neuroimage 43, 528–539.

Van den Heuvel, M.P., Mandl, R.C., Stam, C.J., Kahn, R.S., Hulshoff Pol, H.E., 2010. Aberrant frontal and temporal complex network structure in schizophrenia: a graph theoretical analysis. J. Neurosci. 30, 15915–15926.

van den Heuvel, M.P., de Reus, M.A., Feldman Barrett, L., Scholtens, L.H., Coopmans, F.M., Schmidt, R., Preuss, T.M., Rilling, J.K., Li, L., 2015. Comparison of diffusion tractography and tract-tracing measures of connectivity strength in rhesus macaque connectome. Hum. Brain Mapp. 36, 3064–3075.

van den Heuvel, M.P., Bullmore, E.T., Sporns, O., 2016. Comparative connectomics. Trends Cogn. Sci. 20, 345–361.

van den Heuvel, M.P., de Lange, S.C., Zalesky, A., Seguin, C., Yeo, B.T.T., Schmidt, R., 2017. Proportional thresholding in resting-state fMRI functional connectivity networks and consequences for patient-control connectome studies: issues and recommendations. Neuroimage 152, 437–449.

Van Essen, D.C., 2013. Cartography and connectomes. Neuron 80, 775–790.

Van Essen, D.C., Jbabdi, S., Sotiropoulos, S.N., Chen, C., Dikranian, K., Coalson, T., Harwell, J., Behrens, T.E.J., Glasser, M.F., 2013. Mapping connections in humans and nonhuman primates: aspirations and challenges for diffusion imaging. In: Johansen-Berg, H., Behrens, T.E. (Eds.), Diffusion MRI: From Quantitative Measurement to In-Vivo Neuroanatomy, second ed. Elsevier, pp. 337–358.

van Wijk, B.C., Stam, C.J., Daffertshofer, A., 2010. Comparing brain networks of different size and connectivity density using graph theory. PLoS One 5, e13701.

Váša, F., Bullmore, E.T., Patel, A.X., 2018. Probabilistic thresholding of functional connectomes: application to schizophrenia. Neuroimage 172, 326–340.

Vértes, P.E., Alexander-Bloch, A.F., Gogtay, N., Giedd, J.N., Rapoport, J.L., Bullmore, E.T., 2012. Simple models of human brain functional networks. Proc. Natl. Acad. Sci. U. S. A. 109, 5868–5873.

Voogd, J., Glickstein, M., 1998. The anatomy of the cerebellum. Trends Neurosci. 21, 370–375.

Wang, X.J., 2020. Macroscopic gradients of synaptic excitation and inhibition in the neocortex. Nat. Rev. Neurosci. 21 (3), 169–178. https://doi.org/10.1038/s41583-020-0262-x (Epub 2020 Feb 6. PMID: 32029928; PMCID: PMC7334830).

Warrington, S., Bryant, K.L., Khrapitchev, A.A., Sallet, J., Charquero-Ballester, M., Douaud, G., Jbabdi, S., Mars, R.B., Sotiropoulos, S.N., 2020. XTRACT—standardised protocols for automated tractography in the human and macaque brain. Neuroimage 217, 116923.

Wassermann, D., Makris, N., Rathi, Y., Shenton, M., Kikinis, R., Kubicki, M., Westin, C.F., 2016. The white matter query language: a novel approach for describing human white matter anatomy. Brain Struct. Funct. 221, 4705–4721.

Watts, D.J., Strogatz, S.H., 1998. Collective dynamics of 'small-world' networks. Nature 393, 440–442.

White, J.G., Southgate, E., Thomson, J.N., Brenner, S., 1986. The structure of the nervous system of the nematode Caenorhabditis elegans. Philos. Trans. R. Soc. Lond. Ser. B Biol. Sci. 314, 1–340.

Woo, C.W., Krishnan, A., Wager, T.D., 2014. Cluster-extent based thresholding in fMRI analyses: pitfalls and recommendations. Neuroimage 91, 412–419. https://doi.org/10.1016/j.neuroimage.2013.12.058 (Epub 2014 Jan 8. PMID: 24412399; PMCID: PMC4214144).

Wu, Y., Hong, Y., Feng, Y., Shen, D., Yap, P.-T., 2020. Mitigating gyral bias in cortical tractography via asymmetric fiber orientation distributions. Med. Image Anal. 59, 101543.

Xie, J., Kelley, S., Szymanski, B.K., 2013. Overlapping community detection in networks. ACM Comput. Surv. 45, 1–35.

Yan, G., Vértes, P.E., Towlson, E.K., Chew, Y.L., Walker, D.S., Schafer, W.R., Barabási, A.-L., 2017. Network control principles predict neuron function in the Caenorhabditis elegans connectome. Nature 550, 519–523.

Yanasak, N., Allison, J., 2006. Use of capillaries in the construction of an MRI phantom for the assessment of diffusion tensor imaging: demonstration of performance. Magn. Reson. Imaging 24, 1349–1361.

Yeh, C.-H., Smith, R.E., Dhollander, T., Calamante, F., Connelly, A., 2019. Connectomes from streamlines tractography: assigning streamlines to brain parcellations is not trivial but highly consequential. Neuroimage 199, 160–171.

Yendiki, A., Panneck, P., Srinivasan, P., Stevens, A., Zollei, L., Augustinack, J., Wang, R., Salat, D., Ehrlich, S., Behrens, T., Jbabdi, S., Gollub, R., Fischl, B., 2011. Automated probabilistic reconstruction of white-matter pathways in health and disease using an atlas of the underlying anatomy. Front. Neuroinform. 5, 23.

Zalesky, A., 2008. DT-MRI fiber tracking: a shortest paths approach. IEEE Trans. Med. Imaging 27, 1458–1471.

Zalesky, A., Fornito, A., 2009. A DTI-derived measure of cortico-cortical connectivity. IEEE Trans. Med. Imaging 28, 1023–1036.

Zalesky, A., Fornito, A., Bullmore, E.T., 2010a. Network-based statistic: identifying differences in brain networks. Neuroimage 53, 1197–1207.

Zalesky, A., Fornito, A., Harding, I.H., Cocchi, L., Yucel, M., Pantelis, C., Bullmore, E.T., 2010b. Whole-brain anatomical networks: does the choice of nodes matter? Neuroimage 50, 970–983.

Zalesky, A., Fornito, A., Seal, M.L., Cocchi, L., Westin, C.F., Bullmore, E.T., Egan, G.F., Pantelis, C., 2011. Disrupted axonal fiber connectivity in schizophrenia. Biol. Psychiatry 69, 80–89.

Zalesky, A., Cocchi, L., Fornito, A., Murray, M.M., Bullmore, E., 2012. Connectivity differences in brain networks. Neuroimage 60, 1055–1062.

Zalesky, A., Fornito, A., Cocchi, L., Gollo, L.L., Breakspear, M., 2014. Time-resolved resting-state brain networks. Proc. Natl. Acad. Sci. U. S. A. 111, 10341–10346.

Zalesky, A., Fornito, A., Cocchi, L., Gollo, L.L., van den Heuvel, M.P., Breakspear, M., 2016. Connectome sensitivity or specificity: which is more important? Neuroimage 142, 407–420.

Zalesky, A., Sarwar, T., Ramamohanarao, K., 2020. A cautionary note on the use of SIFT in pathological connectomes. Magn. Reson. Med. 83, 791–794.

Zhang, X., Braun, U., Harneit, A., Zang, Z., Geiger, L.S., Betzel, R.F., Chen, J., Schweiger, J.I., Schwarz, K., Reinwald, J.R., Fritze, S., Witt, S., Rietschel, M., Nöthen, M.M., Degenhardt, F., Schwarz, E., Hirjak, D., Meyer-Lindenberg, A., Bassett, D.S., Tost, H., 2021. Generative network models of altered structural brain connectivity in schizophrenia. Neuroimage 225, 117510. https://doi.org/10.1016/j.neuroimage.2020.117510 (Epub 2020 Nov 5. PMID: 33160087).

Zingg, B., Hintiryan, H., Gou, L., Song, M.Y., Bay, M., Bienkowski, M.S., Foster, N.N., Yamashita, S., Bowman, I., Toga, A.W., Dong, H.W., 2014. Neural networks of the mouse neocortex. Cell 156, 1096–1111.

# Chapter 25

# Tractography validation Part 1: Foundations, numerical simulations, and phantom models

Tim B. Dyrby[a,b], Els Fieremans[c], Francois Rheault[d], Adam W. Anderson[e], Marco Palombo[f,g], Silvio Sarubbo[h], Peter Neher[i], and Kurt G. Schilling[j]

[a]*Danish Research Centre for Magnetic Resonance, Centre for Functional and Diagnostic Imaging and Research, Copenhagen University Hospital—Amager and Hvidovre, Copenhagen, Denmark,* [b]*Department of Applied Mathematics and Computer Science, Technical University of Denmark (DTU), Kongens Lyngby, Denmark,* [c]*Bernard and Irene Schwartz Center for Biomedical Imaging, Department of Radiology, New York University Grossman School of Medicine, New York, NY, United States,* [d]*Department of Computer Science, University of Sherbrooke, Sherbrooke, QC, Canada,* [e]*Biomedical Engineering, Vanderbilt University, Nashville, TN, United States,* [f]*Cardiff University Brain Research Imaging Centre (CUBRIC), School of Psychology, Cardiff University, Cardiff, United Kingdom,* [g]*School of Computer Science and Informatics, Cardiff University, Cardiff, United Kingdom,* [h]*Department of Neurosurgery, "S. Chiara" Hospital, Trento, Italy,* [i]*German Cancer Research Center (DKFZ) Heidelberg, Division of Medical Image Computing, Heidelberg, Germany,* [j]*Department of Radiology and Radiological Sciences, Vanderbilt University Medical Center, Nashville, TN, United States*

## 1 Anatomy, tractography, and validation

Diffusion MRI fiber tractography as a tool to study the human connectome and fiber pathways of the brain has proven valuable in modeling normal white matter anatomy, segmenting the brain into structural subareas, understanding brain development and neurological and psychiatric disease, and has found application in preoperative surgical planning. However, the process from acquiring diffusion MRI data to generating and interpreting these 3D maps of fiber connectivity is a multistep procedure with numerous assumptions and uncertainties that can affect the ability of tractography to faithfully represent the true axonal connections of the brain. Because of this, validation of tractography is critical for these techniques to become useful biomedical tools. Tractography validation is the act of assessing and quantifying the relationship between anatomy and tractography. Validation enables the ability to assess the strengths and limitations of the fiber tractography process, and aims to ultimately understand, improve, and refine the process of fiber tractography.

## 1.1 Anatomical length scales

The anatomy of brain networks and descriptions of structural connectivity can span several anatomical scales (Fig. 1, left column) (Kjer et al., 2024). At the micrometer length (μm) scale, we have the **axon**, which is the long fiber-like part of a nerve cell along which electrical impulses are conducted. The cross-sectional diameters of axons in the human central nervous system are typically on the order of 1 μm, ranging from 0.1 to 12 μm, with larger axons able to conduct signals to axon terminals at a faster speed (Dyrby et al., 2018). The length of axons can be as short as a few millimeters for typical intracortical connections, to as long as tens to several hundreds of millimeters for those projecting via the white matter and connecting distal functional regions. At the millimeter length (mm) scale, a group of axons in the central nervous system is called a **tract**. Synonymous with tract, a **fiber bundle** also refers to a group of fibers, or axons, with an anatomical or functional meaning, for example, the arcuate fasciculus, fornix, and optic radiations. It is important to emphasize that the use of the term tract does not necessarily mean that all axons of the tract start and end within the same functional regions. Rather tracts, or fiber bundles, can be imagined as "highways" of the white matter, where

---

Handbook of Diffusion MR Tractography. https://doi.org/10.1016/B978-0-12-818894-1.00017-3
Copyright © 2025 Elsevier Ltd. All rights are reserved, including those for text and data mining, AI training, and similar technologies.

**486 PART | IV From streamlines to tracts**

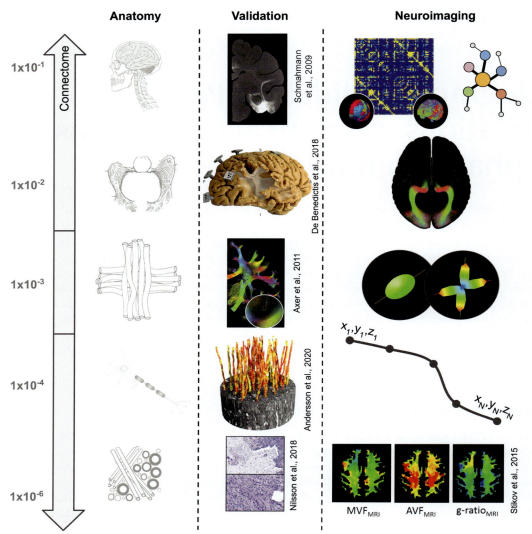

**FIG. 1** Validation is the act of assessing and quantifying the relationship between anatomy and neuroimaging results. In diffusion fiber tractography, validation must be performed at spatial scales covering several orders of magnitude. The relevant anatomy starts at the scale of tissue microstructure, with groups of axons forming tracts or fiber bundles with distinct anatomical or functional characteristics. The human connectome describes the mapping of all structural connections of the brain. Our neuroimaging, diffusion MRI, is sensitive to the tissue microstructure, and processing generates estimates of fiber orientation on the scale of imaging voxels, from which streamlines are generated. Tractography can be used to quantify features of groups of streamlines, bundles, or the connections of the entire human brain, the connectome. For validation, we must choose the experiment and modalities that enable direct visualization and quantification of the relevant spatial scale. *(Figures reproduced with the permission from Axer, H., Beck, S., Axer, M., Schuchardt, F., Heepe, J., Flücken, A., Axer, M., Prescher, A., Witte, O.W., 2011. Microstructural analysis of human white matter architecture using polarized light imaging: views from neuroanatomy. Front. Neuroinform. 5. https://doi.org/10.3389/fninf.2011.00028. Andersson, M., Kjer, H.M., Rafael-Patino, J., Pacureanu, A., Pakkenberg, B., Thiran, J.-P., Ptito, M., et al., 2020. Axon morphology is modulated by the local environment and impacts the noninvasive investigation of its structure-function relationship. Proc. Natl. Acad. Sci. U. S. A. 117 (52), 33649–33659. Stikov, N., Campbell, J.S.W., Stroh, T., Lavelée, M., Frey, S., Novek, J., Nuara, S., et al., 2015. In vivo histology of the myelin g-ratio with magnetic resonance imaging. NeuroImage 118, 397–405. Nilsson, M., Englund, E., Szczepankiewicz, F., van Westen, D., Sundgren, P.C., 2018. Imaging brain tumour microstructure. NeuroImage 182, 232–250. De Benedictis, A., Nocerino, E., Menna, F., Remondino, F., Barbareschi, M., Rozzanigo, U., Corsini, F., et al., 2018. Photogrammetry of the human brain: a novel method for three-dimensional quantitative exploration of the structural connectivity in neurosurgery and neurosciences. World Neurosurg. 115, e279–e291.)*

functionally related tracts enter and leave this highway as they travel to and from their target regions. Finally, at the larger scale of the whole brain (tens of centimeters), a **brain network** describes the collection of fiber bundle connections between different functional regions, and the **human connectome** describes the entire structural connections of the human brain. The connectome is typically visualized and analyzed as a connectivity matrix that captures how different anatomical and functional regions are connected.

Tractography validation Part 1: Foundations, numerical simulations, and phantom models **Chapter | 25 487**

## 1.2  Neuroimaging length scales

Just as anatomy has multiple scales of complexity and organization, so does tractography (Fig. 1, right column). At the finest scale, tractography is derived from contrast on diffusion MRI (dMRI), which is sensitive to the motion of water on the scale of tens of micrometers. Fortunately, this scale is very much in line with the size of restrictions, boundaries, and diameters (i.e., **tissue microstructure**) of axons that form the connections of the brain. Tractography, then, is the attempt to reconstruct and quantify these connections of the brain. At the next scale, diffusion MRI data is acquired and organized into **MRI imaging voxels**. While the average axon diameter is on the order of 1 μm, MRI imaging voxels are on the order of 1–2.5 mm; thus each voxel contains millions of axons, with potentially various geometries and orientations contained within them. The next step of the tractography process aims to represent the directionality and **orientation of axons** within a voxel. Next, voxel-wise measurements of orientation are connected through some **streamline** generation process to create virtual connections throughout regions of the brain. Thus a streamline is a digital representation of a curve in space approximating the underlying axonal fiber. The final scale of tractography is the manipulation, interpretation, and quantification of groups of streamlines, or tracts. Generally, the main applications or uses of tractography can be divided into the fields of bundle segmentation and connectomics. As a tool for **bundle segmentation**, tractography can be used to virtually dissect a set of streamlines, i.e., a tract or bundle, and used to quantify shape, location, trajectory, or biophysical properties along the bundle. Thus, as a tool for **connectomics**, tractography allows us to quantify the brain as a complex, interconnected network, acting as a tool to describe the connection topology of the brain. The field of connectomics promises insight into how brain networks develop and age and respond to pathological and disease perturbations.

## 1.3  What needs to be validated?

Thus we are using a multiscale and multistep neuroimaging process to represent anatomy composed of multiple scales of complexity, which may not correspond in a one-to-one fashion. This insight is key to validation studies and allows us to answer the question: "What needs to be validated?" We propose that the answer is twofold: we need to validate (1) every step in the tractography process, and (2) every quantitative metric that we interpret.

First, we need to validate every step in the tractography process. Fig. 2 illustrates the various steps in this process and the ambiguity that accumulates with each step. We start with the diffusion acquisition, which has uncertainties similar to any parametric imaging—including motion and distortion artifacts, signal-to-noise ratio, scanner, and resolution effects. Also, choices and differences in preprocessing contribute to ambiguities in the pipeline (Tax et al., 2022). Next is the local reconstruction step, which is performed for each individual voxel, and may be dependent on which reconstruction method is utilized, how complicated the geometry of axons is within the voxel, the number of sampled diffusion directions, and the diffusion weighting. Next, the tracking process itself depends on how you start and stop streamlines, and what logic is used to move through the 3D field of orientations. The final step is the interpretation and quantification of the data, and any decisions used to quantify "connectivity" or "connection strength." Ultimately, all the ambiguities obscure the final use cases of diffusion MRI-based tractography: making inferences on the connections of a single subject, i.e., single subject inference, or determining differences between populations or cohorts, i.e., population inference.

In addition to validating every step in the process, we also need all quantitative metrics that we aim to interpret (Fig. 1). Again, this starts at the level of tissue microstructure as measures of volume fractions or axon diameters, or the geometry and orientation of structures within a voxel. Next, the location and trajectories of streamlines must be validated or compared against ground truth. The location, existence, and connectivity profile of fiber bundles must be demonstrated by alternative contrasts or measures, and the derived properties and topology of the connectome must be validated against orthogonal ground truths.

## 1.4  Bridging anatomy and tractography

As described earlier, validation is the act of bridging anatomy and the neuroimaging under investigation (i.e., tractography) (Fig. 1, middle column). Remember that diffusion MRI is an indirect measure of anatomical features because of its limited image resolution in the millimeter range. In tractography, we therefore model fiber orientations and the microstructure within a voxel as illustrated in Fig. 3, and link these together to map a tract. Validation is performed at a length scale that allows a quantification or direct visualization of an anatomical feature using an independent imaging modality. Diffusion MRI is not an independent modality because it is used for tractography—any other image modality can be used—also MRI techniques.

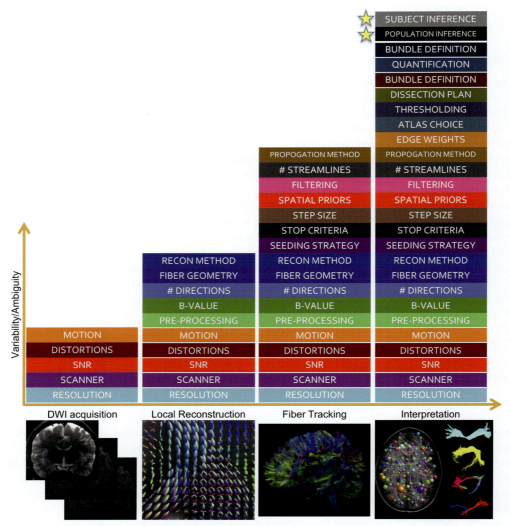

**FIG. 2** The process from data acquisition to quantitative analysis of fiber pathways is a multistep procedure, with several decisions and assumptions that can affect the ability of tractography to represent the true structural connections of the brain. Ultimately, these variabilities and ambiguities obscure and hinder the use of tractography as a biomedical tool for single-subject and population-based inferences of neuroimaging studies of the brain.

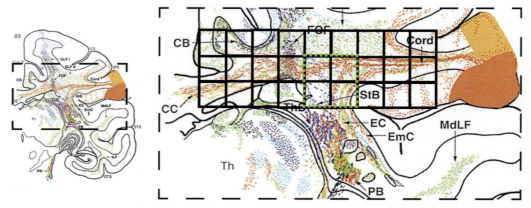

**FIG. 3** Single invasive anterograde tracer injections have been made in a set of 36 monkey brains in total. The labeled axons as well as their orientations have been manually visualized. Axonal trajectories from several injection sites (and monkeys) have been merged onto a standard brain with a set of coronal slices, one color per injection site (Schmahmann and Pandya, 2009). In this material, we get a unique insight into the complexity of a tractogram that we try to restore with noninvasive tractography. Also, we can observe the kind of topology that tracts follow when interfacing with each other. Due to the limited image resolution of dMRI, one can place a voxel grid representing the typical voxel size of ex vivo dMRI in Old World monkeys, i.e., isotropic 500 μm (*black grid*, left). The *striped green square* indicates a crossing fiber region. Clearly, and as expected, most MRI "voxels" contain complex fiber configurations including kissing, deviations, merging, crossing, and parallel. Interestingly, the tracer labeled material also indicates the existence of a complex fiber configuration emanating from the same injection site (Right, *orange labeled tract*). (Modified from figure 6-13 slice 89 in Schmahmann, J.D., Pandya, D., 2009. Fiber Pathways of the Brain. Oxford University Press.)

Validation may utilize one or more independent measures to image and assess an anatomical feature to be compared with the feature that a diffusion MRI-based model is assumed to map. Typically, the independent techniques or image modalities are in such a high image resolution that we can directly "see" the structure that we assume to model via diffusion MRI. As the structural features to be validated exist at different anatomical scales, so do the validation techniques. For example, to validate fiber tracking using tractography, we would need to use a modality that can map the whole tract, such as histological slices with invasive tracers, microdissection photographics, or similar. In these cases, we sacrifice the microscopic image resolution to be able to quantify the fiber bundle trajectory. Likewise, to validate a crossing fiber model one sacrifices the whole-brain field of view to zoom onto a smaller field of view at the subvoxel level to visualize the organization of axonal fibers or to visualize the trajectory of individual axons. Ideally one would like to have an imaging modality that provides both submicrometer image resolution and whole-brain coverage. Today's imaging techniques cannot really achieve this within a reasonable scan time and the amount of data becomes a challenge to handle.

Image modalities including MRI and those used for validation have an image contrast weighted toward one or more specific anatomical feature(s). A single image modality cannot visualize all features. This challenges the validation of diffusion MRI models, because we wish to verify if our model correlates well with the "ground truth" of the anatomical feature we assume the diffusion model to be sensitive to. But if the validation image modality uses tissue staining to enhance one type of anatomical feature over others, like myelinated axons over cells, is it then the "ground truth" we are comparing against? Not really! Validation is therefore basically a procedure to confirm our initial assumption of our diffusion MRI model, but most often we explore some kinds of discrepancies. Attention to the discrepancies is important to get insight into what they are, because they are the ones that resulted in deviations from our initial model assumptions. Hopefully, we then start reflecting: "There might be something that we did not know…." The reflection often leads to the acceptance of a discrepancy between anatomy and the model, or it leads to an adaptation of the assumptions as well as of the model (Dyrby et al., 2018). We have achieved new knowledge and adapted to it. We then start a new iteration of the validation process based on the updated assumptions and the modified model and do the reflections and so on. Hence, validation is an iterative process of "lessons learned" where we achieve new knowledge when comparing independent modalities to diffusion MRI models. Several examples in the literature demonstrate that it is the "lesson learned" validation process that can push the frontiers of a research field.

In summary, we have a highly multidimensional space of anatomies that we wish to represent and quantify, and we need to ensure the accuracy of every quantitative diffusion-derived metric and assess and verify every step that leads us to that metric. Thus each validation "study" that is performed must carefully consider what feature of tractography is to be validated—whether the presence or absence of connections, the weight of connections between regions, or the spatial extent of fiber pathways—and consider what aspects of the process will be investigated (acquisition, reconstruction, tractography, interpretation), and must also carefully choose the validation strategy (or strategies) described in the following sections that will let the investigator appropriately answer the questions under investigation.

# 2 Numerical simulations

Numerical models for the purpose of evaluating fiber tractography try to approximate the following aspects of in vivo dMRI as realistically as possible using computer simulations: (I) a rich and complex micro- and macrostructure, (II) diffusion properties similar to the human brain, as well as (III) the actual magnetic resonance (MR) acquisition step, including all its effects on the final imaging result. An ideal simulation would reproduce all these aspects of reality perfectly.

Due to the enormous complexity of the central nervous system across anatomical length scales, such a perfect reproduction of reality with a simulation at the neuronal level is currently not possible. To make this problem more tractable, it is typically approached with a relatively coarse macroscopic white matter bundle model that can be enriched with microstructural models, mimicking cellular fingerprints along the macroscopic bundle model, as well as simplified models of the MR acquisition. Such simulations have shown to be very advantageous due to their unique capability of delivering real ground truth and being fully controllable, and they have been used extensively in the area of fiber tractography.

In the following, we will first describe the steps typically involved in designing and constructing such a numerical phantom for fiber tractography validation. Then we describe the next finer step in numerical simulations by adding simulations of the microstructure features that are responsible for forming the brain network. Although still in its beginning, the microstructural features can be used to inform tractography algorithms.

## 2.1 Simulating the brain network

Simulating the brain network typically involves three primary decisions that must be made to create numerical diffusion MRI images (Fig. 4). The first decision defines the tissue that will be investigated, including the shape and trajectory of

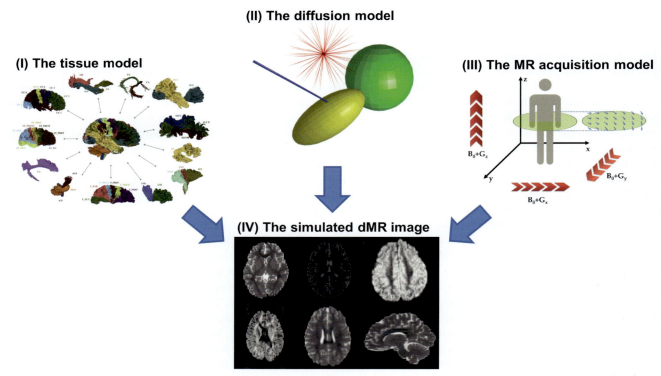

**FIG. 4** Illustration of the components involved in simulating artificial dMR images (IV): A structural tissue model (I), a model of the diffusion process to simulate tissue attenuation (II), a model of the actual MR experiment (III). *(Subfigure I is reprinted from Wasserthal, J., Neher, P., Maier-Hein, K.H., 2018. TractSeg—fast and accurate white matter tract segmentation. NeuroImage 183, 239–253.)*

white matter (WM) bundles that will exist within the simulated dataset. Second, the diffusion model must be chosen, which describes how the diffusion process relates to the underlying substrate chosen in the tissue. Finally, the MR acquisition must be modeled, which will determine not only the diffusion signal based on acquisition parameters but may also include typical acquisition artifacts.

**(I)** The tissue model: Shape and trajectory of whole-brain connections

Numerical phantoms to evaluate fiber tractography are based on white matter models of the long-range brain connections and their microstructural features, which can be imagined as a simplified "drawing" of the brain anatomy of interest. In the early stages of numerical phantom creation, simple geometric shapes abstractly mimicking the centerlines of brain connections, such as lines, circles, ellipses, helices, and crossings, were modeled using mathematical functions (Gössl et al., 2002; Tournier et al., 2002; Lori et al., 2002; Lazar and Alexander, 2003; Kang et al., 2005; Kreher et al., 2005; Staempfli et al., 2006; Batchelor et al., 2006; Aganj et al., 2011; Cetingul et al., 2012; Wu et al., 2012). These fiber models based on mathematical models evolved to manually delineated white matter tracts, which are more flexible and allow the creation of more variable shapes at the cost of a higher manual design effort (Leemans et al., 2005; Close et al., 2009; Neher et al., 2014). The most recent long-range brain connection models rely on the automatic randomized generation of variable and complex trajectories and shapes or directly use curated tractograms obtained from real dMRI datasets as structural fiber models (Neher et al., 2014).

The latter approach, while being far from perfect, enables the creation of complex brain connections of an unprecedented level of realism in terms of similarity to the trajectory and shape of real long-range brain connections. Typically, curated tractograms are created by manually or automatically extracting a clean set of well-known WM fiber bundles from a whole brain tractogram (Wasserthal et al., 2018; Maier-Hein et al., 2017). Other tissue types, such as gray matter (GM) or corticospinal fluid (CSF), can be added to the WM model in the form of scalar volume fraction maps, obtainable from real images by using, e.g., FSL or MRtrix (Jenkinson et al., 2012; Smith et al., 2012; Tournier et al., 2012), specifying their relative amount per voxel.

**(II)** The diffusion model: Generating the diffusion signal of MRI volumes

To obtain a simulated MR image of the numerical phantom, the tissue model described in the previous section (I) must be translated to a voxel-wise dMRI signal as if the model were acquired using an MRI scanner. The receipt for creating such a 3D image volume with a diffusion-weighted contrast based on the tissue model typically includes the following steps:

**(1)** Define the basic parameters of the simulation, such as the image resolution, the field of view of the MRI volume as well as the parameters specific for dMRI (the $b$-value, the number of gradient directions, and their distribution on the sphere).

**(2)** Discretize the tissue model to the image grid defined in step (1), calculate the voxel-wise tissue compartment fractions and, in the case of the WM tissue compartment, the voxel-wise fiber orientations. Typical compartments are intra- and extraaxonal WM, GM, and CSF.

**(3)** Define diffusion models and their diffusivity parameters mimicking the diffusion processes in the different tissue types. A simple and commonly used diffusion model is the diffusion tensor model. See, for example Panagiotaki et al. (2012) for a comprehensive overview of common parametric diffusion models.

**(4)** Generate a voxel-wise diffusion signal by sampling the model defined for each tissue in step (3) using the dMRI parameters specified in step (1).

**(5)** Finally, the synthetic dMRI image is created by averaging the voxel-wise diffusion signal contributions for each tissue type weighted by their respective volume fraction.

Notably, in step (2) we exemplified the simple tensor model as a diffusion model. While this is a very rough approximation of brain tissue diffusion, it is flexible and fast enough to quickly generate whole-brain dMRI datasets. This model is often combined with other simple parametric models (Panagiotaki et al., 2012). Alternatively, more accurate microstructural compartment models can be generated using Monte Carlo simulations, as described in Section 2.2. Since Monte Carlo simulations in such complex tissues are computationally extremely challenging, these types of phantoms typically comprise only a very limited number of voxels and their application in the context of fiber tractography is a matter of future research.

**(III)** The MR acquisition model: Adding MRI artifacts and limited signal-to-noise ratio (SNR)

The diffusion-weighted images created using the tissue model (I) and the diffusion model (II) are lacking the tissue specific T1 and T2 contrasts as well as artifacts typically occurring in real MR images and are therefore not realistic. Since all these aspects introduce imperfections into the reconstructed MR image and add to the accumulated error that all tractography methods suffer from (Fig. 2), they have to be accounted for in the tractography pipeline.

Simulating such effects can be approached by introducing artifacts and noise into an existing image volume a posteriori. In this way, no complex simulation of the actual MR acquisition step is required. This approach is relatively simple, but it only enables a limited and often insufficient realism of the introduced effects. Examples of this approach are the model-based introduction of Rician noise or of eddy currents, other distortions, and head motion via simple transformations of the image or the approximation of signal relaxation by decreasing the voxel values by predefined amounts based on the tissue model.

To increase the realism of the simulated effects, it is necessary to model the actual MR acquisition step, which enables the introduction of basically arbitrary and realistic effects, such as eddy currents, inhomogeneity-induced distortions, ghosts, head motion, T1 and T2 relaxation, and many more. Depending on the research target and what effect one wants to analyze, this approach can be followed in many different degrees of realism and complexity, e.g., by transforming the images to $k$-space via a customized Fourier transform and then transforming them back to image space after introducing changes that occur in real $k$-space acquisitions (Neher et al., 2014) or by Bloch equation-based simulations of MR acquisition (Neher et al., 2014; Drobnjak et al., 2006, 2010a,b).

### 2.1.1 Tips, tricks, and available tools

When approaching the task of creating a phantom for your specific needs, many aspects must be considered. The first is to clearly define your requirements. Part of this is deciding whether the focus of your planned project is a comprehensive analysis of tractography performance in a brain-like situation or whether you are rather interested in a specific aspect, such as performance in fiber crossings or the influence of a specific artifact on the tractography process. In this context, it is also important to decide which metrics you want to analyze using which tools. For general information on metrics for evaluating tractography results in comparison to ground truth, please refer to Chapter 26.

If an already published and openly available phantom is suitable to be reused for your experiments, it is always recommended to do so instead of simulating a new one, since it makes your results much more comparable to the state of the art

**492** PART | IV From streamlines to tracts

**TABLE 1** List of openly available numerical dMRI phantoms and corresponding image data.

| Dataset | Description | Generated with | References |
|---|---|---|---|
| DiSCo[a] | A numerical phantom that mimics complex anatomical fiber pathway trajectories while also accounting for microstructural features such as axonal diameter distribution, myelin presence, and variable packing densities. | Numerical Fiber Generator (used for parts of the simulation) | Rafael-Patino et al. (2021) |
| Simulated FiberCup[b] | A series of synthetic diffusion weighted images with a structure similar to the well-known FiberCup hardware phantom. | Fiberfox | Neher et al. (2014) |
| 99 simulated brains[c] | Whole brain dMRIs with varying parameters and artifacts of 99 subjects | Fiberfox | – |
| ISMRM tractography challenge 2015[d,e,f] | Whole brain dMRI with varying parameters and multiple artifacts of a single subject | Fiberfox | Maier-Hein et al. (2017) |
| ISBI HARDI Reconstruction Challenge[g] | Single abstractly brain-like ISBI phantom with varying parameters and noise levels. No simulation of artifacts or the MR acquisition. | Phantomas | Caruyer et al. (2014) |
| Simple Phantomas phantoms[h] | Phantomas config files for various crossings, kissings, and fannings | Phantomas | – |
| Randomized Fiberfox phantoms[i] | Simulated dMRI images and ground truth of random fiber phantoms in various configurations | Fiberfox | Fritzsche et al. (2012), Neher et al. (2014) |
| DW-POSSUM dataset[j] | dMRI data with highly realistic eddy current, motion, and susceptibility artifacts | POSSUM | Graham et al. (2016), Graham et al. (2017) |

[a]*https://data.mendeley.com/datasets/fgf86jdfg6/1.*
[b]*https://www.nitrc.org/frs/shownotes.php?release_id=2341.*
[c]*https://doi.org/10.5281/zenodo.4139626.*
[d]*https://doi.org/10.5281/zenodo.572345.*
[e]*https://doi.org/10.5281/zenodo.1007149.*
[f]*https://doi.org/10.5281/zenodo.579933.*
[g]*http://hardi.epfl.ch/static/events/2013_ISBI/training_data.html.*
[h]*https://github.com/ecaruyer/phantomas/tree/master/examples.*
[i]*https://doi.org/10.5281/zenodo.2533250.*
[j]*https://www.nitrc.org/projects/diffusionsim/.*

that might have used the same phantom for evaluation. A list of some openly available simulated datasets is compiled in Table 1. In general, it is also advisable to include more than just one image in your analysis to obtain credible results. If you decide that none of the existing phantoms suit your needs, then the next step is to decide which tool to use for simulating a new dataset. Using the three components described here, namely a tissue model (I), a diffusion model (II), and a model of the MR acquisition (III), it is possible to simulate rather realistic approximations of real diffusion-weighted images. To this end, a series of openly available tools has been published, such as Numerical Fiber Generator,[a] Phantomas,[b] Fiberfox,[c] and POSSUM[d]:

- The **Numerical Fiber Generator** (Close et al., 2009) randomly generates numerical fiber structures and simulates corresponding dMR images. The focus of this method is the automatic optimization of the created structure to obtain

---

a. https://www.nitrc.org/projects/nfg/.

b. http://www.emmanuelcaruyer.com/phantomas.php.

c. https://github.com/MIC-DKFZ/MITK-Diffusion/.

d. https://fsl.fmrib.ox.ac.uk/fsl/fslwiki/POSSUM.

Tractography validation Part 1: Foundations, numerical simulations, and phantom models **Chapter | 25** **493**

densely packed bundles with certain constraints on the overlaps between the bundles as well as with manually placed spherical isotropic regions.

- **Phantomas** (Caruyer et al., 2014) enables the manual definition of tubular fiber bundles as a tissue model and the simulation of corresponding dMR images using a composite hindered and restricted model of diffusion (CHARMED) model without a sophisticated MR acquisition model in a lightweight web interface. The source code of Phantomas is also available online on GitHub[e] and was used to create numerical phantoms for the testing and training data of the 2nd HARDI Reconstruction Challenge,[f] organized at ISBI 2013.

- **Fiberfox** (Neher et al., 2014) uses a tissue model based on streamlines that can be manually defined, automatically generated with certain parameters, or directly obtained from real tractograms. This structural model can then be used in conjunction with a flexible multicompartment diffusion model and a simulation of the actual MR acquisition to simulate brain-like images including realistic image properties and typical dMRI artifacts. Fiberfox is available open-source and as a binary application included in MITK Diffusion.[g] It has been used in various tractography evaluation studies, for example for the ISMRM tractography challenge 2015 (Maier-Hein et al., 2017).

- **POSSUM (Physics-Oriented Simulated Scanner for Understanding MRI)** (Drobnjak et al., 2006, 2010a) focuses on a realistic simulation of the MR acquisition process. The diffusion-weighting and the tissue model are directly obtained from a real acquisition without using a diffusion model (Graham et al., 2016) in the form of volume fraction maps and spherical harmonics interpolations of the dMRI signal. The diffusion-weighted images are combined with a full Bloch equation-based simulation of MRI images. Its focus is rather the analysis of MR-related artifacts and effects other than the validation of fiber tractography, but it is the most sophisticated simulator for whole brain MRI acquisitions to date. Nevertheless, it is possible in a straightforward way to enable POSSUM for use in the context of tractography by using the tissue and diffusion model of Fiberfox to simulate the diffusion-weighted image that serves as the input of the MR acquisition model of POSSUM.

Pipelines following Fig. 4 for creating multiple realistic phantoms in various configurations can involve many steps and they tend to get rather complex. It is therefore advisable to thoroughly document your experiments, potentially using a logging tool that allows you to exactly reproduce and understand everything you have done.

Furthermore, and particularly when using a phantom generation tool that relies on the manual extraction of individual fiber bundles from a real tractogram, it is advisable to use automatic or semiautomatic tools that assist with this process, such as TractSeg (Wasserthal et al., 2018, 2019), Quickbundles (Garyfallidis et al., 2012), or Tract Querier (Wassermann et al., 2016), since this avoids inconsistencies across subjects and the process can otherwise be extremely time consuming. In general, try to automate as much of your process as possible to make it consistent and reproducible. When extracting white matter bundles from a real tractogram for the purpose of using them as white matter tissue models for the simulation, watch out for gaps in your tissue model arising from voxels not fully occupied by your compartments, i.e., where the sum of volume fractions of the individual compartments is not 1.0. Due to the bundle extraction process, the white matter will shrink in some places, but a GM volume fraction map obtained from the original image will stay the same, resulting in a gap between the two compartments that have to be closed, e.g., using some heuristic.

The final choice you must make is the suitable parameterization of the chosen simulation tool. Many parameters are a direct result of your research question, for example, if you want to analyze complex microstructural models you will have to choose a sufficient number of diffusion-encoding gradients and a suitable $b$-value. Other parameters can potentially be taken directly from scientific literature, such as relaxation constants or diffusivities. Nevertheless, in some cases it can be useful to tune your parameters using automatic optimization methods such as Optuna,[h] Optunity,[i] Hyperopt,[j] or Hyppopy[k] (Bergstra et al., 2015; Agrawal, 2021) to improve the realism of the resulting dMRI images.

After having obtained your phantom dataset, existing or newly simulated, it can be used for running your tractography experiment and for the successive evaluation procedure. Particularly the Fiberfox-based datasets can be readily used, for example using the Tractometer[l] tractography validation suite that calculates a set of connectivity and overlap-based metrics

---

e. https://github.com/ecaruyer/phantomas.

f. http://hardi.epfl.ch/static/events/2013_ISBI/.

g. https://github.com/MIC-DKFZ/MITK-Diffusion/.

h. https://optuna.org/.

i. https://optunity.readthedocs.io/en/latest/.

j. http://hyperopt.github.io/hyperopt/.

k. https://github.com/MIC-DKFZ/Hyppopy.

l. http://tractometer.org.

for tractography results in comparison to reference tractograms (Côté et al., 2013). Like the recommendation of reusing existing datasets to enable comparability to the state of the art, the same is true for the used evaluation metrics and tools. Always try to make your results comparable to previously published results. Of course, you are not limited to the metrics used in the state of the art and extensions are often necessary. One of the most established numerical dMRI phantoms used for validation in recent tractography publications (Neher et al., 2017; Poulin et al., 2017; Théberge et al., 2021; Benou and Raviv, 2019) is the ISMRM tractography challenge 2015 phantom, which is a whole brain simulated dataset of 25 association, projection, and commissural bundles.

While of course being far from perfect, the currently available tools for creating numerical dMRI phantoms and images have proven to be very useful for the validation of tractography. One challenge for future developments will be to make dMRI simulations tractable that use a realistic Monte Carlo model of the diffusion processes in conjunction with a complex tissue model representative of the human brain and a comprehensive approximation of a real MR acquisition. One of the first steps in this direction was done by Rafael-Patino et al. with the DiSCo dataset (see Table 1).

In general, when approaching the task of fiber tractography validation with numerical phantoms, it is crucial to keep in mind that you are only validating the performance on the phantom, which is perfect for validating how an approach behaves in certain situations and in comparison to other approaches. Numerical phantoms are invaluable tools to quantify, for example, which methods produce more false positive connections in a certain white matter model or which methods are most capable of reproducing the fiber densities of the ground truth. Nevertheless, if method A can produce a lower number of false positives in a phantom than method B, it does not mean that A will produce an equally low number of false positives in a real dataset, but it is an indicator that A will produce a lower number of false positives than B. Any transfer of the obtained results must be treated with caution though, because many of the assumptions made in the phantom are based on how we *think* the real anatomy looks from various observations.

## 2.2 Simulating tissue microstructure along the brain network

A few tractography methods such as Convex Optimization for Microstructure Informed Tractography (COMMIT) (Daducci et al., 2015) take an additional step to improve the accuracy of streamline tractography by including tract-related microstructure features along the brain connection that MRI also can detect. Microstructural features can both be specific anatomical related, e.g., axon diameter and density estimation (Alexander et al., 2010; Zhang et al., 2011, 2012; Veraart et al., 2020) or unspecific, such as tissue relaxation, diffusivity, etc. A numerical verification of such methods is challenging because the tractography method links anatomical information across several length scales from brain connections to the microstructural length scale. To simplify the amount of data to process, numerical verification focuses on voxels containing anatomical configurations of fiber architecture. This could be straight, bending, or crossing fibers where each streamline or tract is given a ground truth microstructural fingerprint. In this section, we give a brief introduction to numerical microstructural simulations. The grand challenge to generate realistic synthetic dMRI signals is to generate synthetic realistic 3D microstructural environments that can be used to inform tractography.

### 2.2.1 Simulating the microscopic diffusion process

Concerning the simulation of the diffusion process at the microscopic scale, the most widely used approaches can be divided into two major classes: Monte Carlo (MC) methods and finite element/difference (FEM/FDM) methods. MC methods use a microscopic statistical description of the diffusion process (e.g., random walk) and calculate the corresponding dMRI signal by averaging the accumulated phase over many particles (Hall and Alexander, 2009; Fieremans and Lee, 2018); instead, the FEM/FDM methods solve numerically the differential equations of motion and the dMRI signal is computed by solving the corresponding Bloch-Torrey equations given some boundary conditions, dictated by the tissue geometry. There are several open-source tools freely available that implement these methods for dMRI simulations. For instance, CAMINO (http://camino.cs.ucl.ac.uk) (Cook et al., 2006), Monte Carlo Diffusion and Collision simulator (MC/DC) (github.com/jonhrafe/MCDC_Simulator_public) (Rafael-Patino et al., 2020), DW-MRI Random Walk Simulator (www.NITRC.org: project name: "DW-MRI Random Walk Simulator") (Landman et al., 2010), DIFSIM (http://csci.ucsd.edu/software/difsim.html) (Balls and Frank, 2009), Diffusion Microscopist Simulator (Yeh et al., 2013), and the Realistic Monte Carlo Simulator (RMS) https://github.com/NYU-DiffusionMRI/monte-carlo-simulation-3D-RMS (Lee et al., 2021) based on MC; SPINDOCTOR (https://github.com/jingrebeccali/SpinDoctor) (Lee et al., 2021) based on FEM/FDM.

Here, we focus on generating synthetic substrates for MC simulations that include a user-defined microstructure geometry mimicking, e.g., a WM voxel. MC simulations are widely used because of their simplicity, flexibility, and power in terms of simulating different MRI signals in any user-defined microstructure geometry. In MC simulations, particles are initiated in the synthetic substrate and the diffusion process is simulated as a *random walk* of these particles in space. Cellular membranes are modeled in the synthetic substrate as barriers to the random walk and they can be defined by mathematical relations defining the boundaries of simple geometries (e.g., spheres, cylinders, ellipsoids, etc.) or as three-dimensional surface meshes of general shapes. The latter provides more flexibility for the design of realistic substrates and it is the most commonly used approach in advanced MC simulations (Andersson et al., 2020, 2022; Lee et al., 2020a,b). However, because of the absolute size of the substrate (i.e., voxel), some parts of a microstructure geometry may cross the boundary (i.e., voxel sides), which causes a problem for the MC simulation. Therefore a substrate is typically designed using periodic boundary conditions, which ensure that if a part of a cell or axon geometry is placed outside the substrate boundary, the geometry is simply copied by mirroring it to the opposite side of the substrate so the "outside" part now is inside the substrate. Hence, the periodic boundary conditions ensure that particles crossing a boundary will be mapped back into the geometry placed on the opposite side within the substrate and the MC will not fail (Hall and Alexander, 2009).

Given the stochastic nature of the random walk approximation of the diffusion process, it is necessary to properly set up the simulations to avoid undesired biases in the simulated MRI signal. For example, the number of walkers $N$ and the step size $dr$ must be carefully chosen. For a complete list of criteria and recommendations on how to set up and proof check MC simulations, we refer the reader to (Hall and Alexander, 2009; Fieremans and Lee, 2018; Rafael-Patino et al., 2020).

### 2.2.2 Generative algorithms to construct simple cylinder-based WM substrates

The most challenging aspect of simulating dMRI signals in realistic WM microstructure is the generation of synthetic substrates that sufficiently mimic the microstructural environment of real WM tissue. For many practical applications, it is often not necessary to simulate WM microstructure in all its complexity. Indeed, a too simple synthetic environment can introduce a mismatch to the ground truth as seen with the axon diameter estimation model (Lee et al., 2019; Veraart et al., 2020; Andersson et al., 2020). For these reasons, simulations of tissue microstructure can vary over a wide range of complexity and realism based on the intended experimental design (Fig. 5).

A simple substrate of WM microstructure is assumed to be a densely packed nontouching cylinder substrate. The cylinders may contain a distribution of diameters typically obtained from histology. The cylinders are assumed to be parallel and therefore are invariant along the fiber axis, which simplifies the diffusion simulation to a problem in two dimensions (2D) (Fig. 5A). The dense packing of axons is thus equivalent to the generation of random 2D packing of disks. There are

**WM digital substrates**

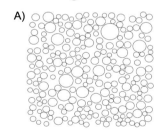

A) Dense 2D packing of circles representing cylinders

B) Dense 3D packing of complex axon-like morphologies

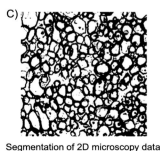

C) Segmentation of 2D microscopy data

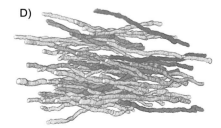

D) Accurate reconstruction of complex axonal morphologies from 3D Electron Microscopy

**Complexity & Realism** →

FIG. 5 Digital substrates mimicking the WM microstructure at increasing levels of complexity and realism (from left to right), by either generative modeling approaches (first row) or three-dimensional reconstructions from microscopy data of real tissue (second row). In addition to more intricate geometries, features of exchange and detailed axonal morphology are new developments.

several open-source software suites to generate such cylinder substrates. For example, MC simulators such as CAMINO and MC/DC can generate substrates of densely packed cylinders with gamma distributed radii. After setting the parameters of the chosen gamma distribution and the expected cylinder volume fraction, circles of diameter drawn from the chosen gamma distribution are randomly displaced in the plane of the substrate until the desired volume fraction is reached. Typically, the substrate generator algorithm is based on a random sampling of each circle position, guaranteeing that circles in the substrate do not overlap. Given the stochastic nature of the algorithm, it can become computationally expensive and very time demanding when volume fractions higher than 0.60 are targeted. To overcome this limitation, other packing algorithms have been proposed, such as the AxonPacking algorithm (https://github.com/neuropoly/axonpacking) (Mingasson et al., 2017). Here, the different steps of process packing are the following: first, the diameters of $N$ disks are randomly drawn from a chosen gamma distribution. Then, the positions of disks are initialized on a grid, and they migrate toward the center of the packing area until the maximum disk volume fraction is achieved. The migration is simulated using a molecular dynamics algorithm, providing each circle with a migration velocity, and guaranteeing that two circles never cross each other when they collide.

To increase complexity, periodic or randomly placed enlargements along cylinders, or axonal beading (Fig. 5B), may be added to the simulation to mimic axonal mechanical stress pathologies or traumatic injuries (Budde and Frank, 2010; Palombo et al., 2017). Axonal beading substrates can be constructed by starting from the dense packing of straight cylinders and varying the radius of each cylinder along its main axis following the geometrical path traced by a point on an ellipse as it rolls along an axis, e.g., according to the eqs. 11 and 12 in Brabec et al. (2020). Given the 3D mesh of each cylinder and its central line (the main axis), this is achieved by changing the position of each vertex of the mesh to match the distance from the central line computed according to the desired geometrical path. Care should be taken to guarantee that the modified surface of one cylinder does not overlap with any of the neighboring cylinders. This can be achieved by, for example, keeping the distance between the vertices of meshes of different cylinders constant.

Studies have shown that real axons are not straight (Andersson et al., 2020, 2022; Lee et al., 2019). Simple substrates of nonstraight axons can be generated as an undulating cylinder with amplitude along the main axis. It enables the quantification of axonal microscopic orientation dispersion on the estimated apparent axonal diameter (Nilsson et al., 2012; Brabec et al., 2020). The substrates can be constructed by starting from the dense packing of straight cylinders and varying the direction of the center line (the main axis) following a sinusoidal path, e.g., following eq. 1 in Nilsson et al. (2012). The center line can be partitioned into short subsegments, each one oriented according to the chosen undulation path, and defining a small subcylinder. Then, the whole undulating cylinder is obtained by connecting all the subcylinders. Once again, care should be taken to guarantee that the modified surface of one cylinder does not overlap with any of the neighboring cylinders. Of course, the procedures to create beading and undulating cylinders can be combined to obtain more complex substrates of beading and undulating cylinders.

### 2.2.3 Generative algorithms to construct realistic synthetic substrates of WM

Increasing the complexity and realism of synthetic WM substrates, going beyond models of straight or undulating cylinders, is quite challenging. Two possible solutions to this challenge have been so far investigated: one involves the definition of generative algorithms able to generate substrates of high enough complexity and realism (Fig. 5B); the other is to use as substrates the meshes of reconstruction at a micrometric resolution of the three-dimensional structure of real WM tissue samples from 2D or 3D microscopy imaging techniques, such as electron microscopy (EM) or X-ray synchrotron imaging also known as X-ray holographic nanotomography (XNH) (Fig. 5C and D).

Concerning the generative algorithms, the difficulty mostly lies in packing with realistically high enough density fiber geometries with also controllable and realistic orientation dispersion. Ideally, to be a realistic representation of the WM structure, we would need to achieve values of volume fraction occupied by the intra-axonal space between 50% and 70% (Syková and Nicholson, 2008) and orientation dispersion between 10 and 35 degrees (Lee et al., 2019; Andersson et al., 2020). Two generative algorithms recently proposed can generate substrates of packed fibers of complex morphologies that satisfy these requirements. These are MEDUSA (Ginsburger et al., 2019) and ConFig (Callaghan et al., 2020). Using MEDUSA, it is possible to generate substrates where axons are modeled as cylinders with variable caliber along their main axis. Nodes of Ranvier can be added, and it is possible to reach high degrees of orientation dispersion and fiber crossing at high volume fractions (Ginsburger et al., 2018, 2019). The ConFig algorithm offers the generation of similar anatomical features and allows control of subtle details of fiber morphologies, such as bulging/beading and irregular cross-sectional area. The peculiar anatomical features of real axonal morphologies are inspired by the growth-cone process of axons in the central nervous system (CNS) (Callaghan et al., 2019, 2020, review Dyrby et al., 2018).

### 2.2.4 3D microscopy data of real WM to construct synthetic WM substrate

The challenges involved in the generation of complex and realistic digital substrates using generative algorithms can be overcome by methods based on 3D reconstructions of real WM microstructure obtained from microscopy data. However, this advantage comes with some other challenges and considerations.

First, it requires a microscopy platform that enables 3D imaging of tissue. Recently, EM or two photon imaging has enabled the collection of 3D data typically by taking an image of the sample surface and then cutting away the image slice and then taking a new surface image. Synchrotron XNH imaging enables a 3D scan of the intact sample without cutting (review Dyrby et al., 2018). Typically, EM provides higher in-plane image resolution than synchrotron imaging, which can go down to about 50 nm isotropic. However, synchrotron imaging can easily cover a large field of view (FOV) similar to an MRI voxel within a day, whereas EM takes days. Indeed, the image resolution limits the size of geometries that can be segmented for quantification like axon diameters (Abdollahzadeh et al., 2019; Lee et al., 2019; Andersson et al., 2020). Notably, if the image resolution becomes too low to quantify microstructure geometries, one can consider using the structure-tensor analysis to analyze fiber orientation distributions—see Chapter 26.

Secondly, accurate segmentation algorithms are needed to properly segment microstructural compartments of interest such as axons, cell bodies, blood vessels and vacuoles, extracellular space, etc. (Andersson et al., 2020; Abdollahzadeh et al., 2019). Often segmentation results need additional manual segmentation to reach a quality level sufficient to produce smooth and continuous 3D surface meshes to be used as substrates for MC simulations.

Third, substrates based on real data do not guarantee the periodic boundary conditions as for the generative methods. The problem originates from a limited FOV and image resolution of the microscopic techniques. Therefore one must consider the diffusion times to be simulated in relation to the length of the 3D geometrical meshes to be simulated. This is a minor problem for spherical (restricted) cell geometries but not for axons. Axons appear as hollow tube-like structures extending the whole voxel and can produce a simulation bias because particles risk leaving the axon ends for longer diffusion times. Indeed, one can close the mesh's ends, but closed ends also can introduce biases if too many particles reach the end and contribute to a significant bouncing back signal. One can reduce such biases by initiating particles at the middle of the axon and/or simply extending the length of the axon by mirroring the axon (Fieremans and Lee, 2018). However, the original axon needs to be imaged for a certain length for the MC simulation results to represent the real anatomy (Fig. 5D); 3D meshes of real axons only, or a combination of cells and vessels, are shared, which allows the generation of real WM substrates for MC simulation.

For example, 3D EM data of WM axonal microstructure for mouse corpus callosum for an FOV of $22 \times 22 \times 22\,\mu m$ and the segmented axonal 3D surface meshes are freely available at http://cai2r.net/resources/software together with open-source code for segmenting the EM data, preparing the digital substrate, and performing the MC simulations available at https://github.com/NYU-DiffusionMRI. Also, 3D synchrotron X-ray nano-holotomography (XNH) data of monkey corpus callosum and crossing fiber region with a large FOV of $150 \times 150 \times 150\,\mu m$. The meshes of axons, cells, and vacuoles and their corresponding dMRI scan can be downloaded here: https://www.drcmr.dk/axon-morphology-dataset.

## 3 Physical phantoms

Physical or hardware phantoms can be used to understand tractography methods and verify their performance. We define them here as well-characterized objects in terms of size and composition that emulate brain white matter (or other fibrous tissue) to evaluate the accuracy and precision of diffusion MRI and tractography. In their simplest form, phantoms are liquids or gels that are visible with (diffusion) MRI and exhibit isotropic diffusion. For testing tractography, these MRI visible liquids or gels are then submerged into nuclear magnetic resonance (NMR) invisible materials to induce microstructural properties including anisotropy.

Here we review both simple isotropic phantoms and more sophisticated microstructure phantoms, considering their usefulness for calibrating tractography sequences and validating tractography algorithms. We will discuss challenges in the choice of materials and construction to achieve sufficient SNR and fractional anisotrophy (FA) values similar to brain white matter. For a detailed cookbook describing and discussing the wide range of phantoms proposed for brain microstructure MRI, we refer to (Fieremans and Lee, 2018).

### 3.1 Isotropic liquids

**Pure liquids** in a test tube or container exhibit free isotropic Gaussian diffusion, characterized by a well-established (time-independent or constant) diffusion coefficient, and are therefore useful as a standard for quality assurance. Water is the most popular main component of diffusion phantoms, due to its properties of being inert, stable, cheap, and readily available. While water has a relatively high diffusion coefficient $D$ at room temperature ($2\,mm^2/ms$) as compared to those observed in the brain, ice water at 0°C has a $D$ of $1.1\,mm^2/ms$ and has been shown useful in multicenter studies to study the coefficient of variation of $D$ among different scanners, sites, field strengths, and vendors (Chenevert et al., 2011; Grech-Sollars et al., 2015; Malyarenko et al., 2016; Mulkern et al., 2015).

More pure liquids (Tofts et al., 2000; Komlosh et al., 2017; Wagner et al., 2017) as well as several aqueous solutions (e.g., with polyvinylpyrrolidone (Pierpaoli et al., 2009; Palacios et al., 2017; Wagner et al., 2017; Wang et al., 2017), sucrose (Laubach et al., 1998; Delakis et al., 2004; Lavdas et al., 2013; Hara et al., 2014; Wang et al., 2017), and albumin (Fukuzaki et al., 1995)), and mixtures of acetone and water (Wang et al., 2017) have been proposed to create diffusion phantoms with a range of diffusion coefficients $D$, all listed in Table 1 in Fieremans and Lee (2018). All these are useful as standards for quality assurance for evaluating diffusion sequences, evaluating scanner stability, and calibration of the diffusion weighting or $b$-value.

## 3.2 Anisotropic phantoms for diffusion tractography

**Microstructure phantoms**, as used for tractography, consist of (NMR-visible) liquids that are immersed in microstructurally confined (NMR-invisible) phantom materials to create anisotropy and/or fiber crossings. Water is particularly well suited as the main component in such diffusion phantoms because of several appealing MR properties: its single NMR spectral line is a reference for NMR and MRI and is free from chemical shift artifacts (i.e., shifts in echo-planar imaging (EPI) images). In addition, the long intrinsic T1- and T2-values of (distilled) water help achieve reasonable SNR in diffusion phantoms, as the proton density is typically low due to the presence of other NMR invisible materials, and surface relaxation and local susceptibility differences causing additional T2(*) relaxation (Fieremans and Lee, 2018). Alternatively, 1% agarose gel could be used instead of water. As water-based phantoms potentially have motion issues if used without settling time, agarose gels provide a good alternative, as it has Gaussian diffusion properties with $D$ being relatively independent of the agarose concentration (up to 3%), similar to water (Fieremans and De Deene, 2020).

A wide range of anisotropic phantoms has been proposed to evaluate tractography, and they are reviewed here in Table 2 adapted from Fieremans and Lee (2018). In particular, fibrous materials with micrometer-sized geometries are used to induce the (desired) anisotropy either mimicking intra-, or extra-axonal space or a combination.

Here we review the fibrous phantom materials proposed so far. Importantly, their (often unknown) material properties determine the SNR and could potentially result in artifacts or biases when evaluating for tractography, as explained next. These materials are then enclosed in containers that minimize field distortions, e.g., with cylindrical shapes or tubes that are aligned with the static field of the magnet, and made of a material with susceptibility matching the liquid (Wapler et al., 2014).

### 3.2.1 Plants

Plants, in particular monocotyl plants or monocots such as asparagus (Fig. 6) and celery (Gruwel et al., 2013) have been used for tractography. They exhibit anisotropic diffusion in the vascular bundles (xylem), with inner diameter (I.D.) $\sim$100 mm. While they are widely available and require minimal preparation, drawbacks of these phantoms are the low degree of anisotropy (FA $\sim$0.15 in asparagus), and the presence of air causing susceptibility artifacts, in addition to limited preservability because of season availability and dependence on storage and delivery conditions.

Recently, wood has been proposed as an alternative anisotropic phantom with durability evaluated up to 9 months (Suzuki et al., 2019). Phantom preparation includes placing it in water and then boiling to remove air and fill up the wood with water. Anisotropy originates then from water inside the vessels and potentially also from inside wood fibers, resulting in FA values ranging from 0.4 to 0.7, depending on the wood type.

### 3.2.2 Plain fiber phantoms

Plain fiber phantoms made of solid, nonhollow fibers with a diameter of $1$–$20\,\mu m$ are commonly used as diffusion phantoms to simulate diffusion in the extra-axonal space. The diffusion properties of fiber phantoms depend on the fiber packing density as well as the fiber diameter (Fieremans et al., 2008a): To obtain sufficient anisotropy, fibers with a diameter $<20\,\mu m$ are needed as the spaces in between the fibers are on the order of the diffusion length of the measurement. In addition, the fibers should be packed sufficiently densely to obtain FA values $>0.2$, where fiber phantoms with FA $\sim$0.4 correspond to fiber density or volume fraction of $\sim$0.6 (see Fig. 6 in Fieremans et al., 2008a). Practically, much higher densities are difficult to obtain (cf. the highest density of randomly packed uniform nonoverlapping circles in a plane is 0.82

**TABLE 2** Anisotropic phantoms for diffusion tractography.

| | Phantom material | Sizes (μm) | Applications | References |
|---|---|---|---|---|
| Plants | Asparagus | ~100 (vascular bundle—xylem) | Modeling diffusion Sequence testing | Latt et al. (2007), Panagiotaki et al. (2010), Sigmund and Song (2006), Boujraf (2001), Baete et al. (2015) |
| | Celery | | Tractography Sequence testing Cell size mapping | Gruwel et al. (2013), Gaggl et al. (2014), Özarslan et al. (2011) |
| | Wood | 5–12 (wood fibers) 42–205 (vessels) | Diffusion tensor imaging (DTI) and diffusion kurtosis imaging (DKI) phantom | Suzuki et al. (2019) |
| Plain fiber phantoms mimicking extra-axonal space | Rayon fibers | O.D.: 17 | Crossing fibers | Perrin et al. (2005) |
| | Dyneema (polyethylene) fibers | O.D.: 17 | Characterization, biophysical modeling of $D(t)$, quality assurance (QA), effect of susceptibility and surface relaxation on diffusion, crossing fibers | Burcaw et al. (2015), Farrher et al. (2012), Fieremans et al. (2008a,b), Lemberskiy et al. (2017), Lorenz et al. (2008), Jeurissen (2012), Fieremans (2008) |
| | Tsunooga (polyethylene) fibers | O.D.: 12 | DTI and DKI phantom | Suzuki et al. (2019) |
| | Acrylic fibers | O.D.: 20 | Crossing fibers, tractography validation | Fillard et al. (2011), Lichenstein et al. (2016), Poupon et al. (2008) |
| | Polyamide fibers | O.D.: 15 | Modeling effect of susceptibility on diffusion | Laun et al. (2009) |
| | Polyester fibers | O.D.: 10 | Crossing fibers | Pullens et al. (2010) |
| | Polypropylene fibers | O.D.: 25 | Fiber tractography | Schilling et al. (2019a,b), Whitton et al. (2016) |
| Capillary phantoms mimicking intra-axonal space | Glass capillaries (Schott North America, Southbridge, MA, United States) | I.D. only: 23, 48, 82 | DTI phantom | Yanasak and Allison (2006), Yanasak et al. (2009) |
| | Glass capillary wafers or glass capillary array (GCA) (Photonis, Sturbridge, MA, United States) | I.D. only: 2.5, 10, 15, 13.7 | Modeling restricted diffusion, pore size mapping with double pulsed field gradient (dPFG) | Benjamini et al. (2014), Komlosh et al. (2011) |
| | Fused silica capillaries (Polymicro Technologies—Molex) | I.D.: 1–50 O.D.: 75, 150 | Modeling restricted diffusion, diameter mapping, crossing fibers | Avram et al. (2004), Bar-Shir et al. (2008), Li et al. (2014), Milne and Conradi (2009), Morozov et al. (2013), Shemesh et al. (2010), Siow et al. (2012), Tournier et al. (2008), Vellmer et al. (2017) |

*Continued*

**TABLE 2** Anisotropic phantoms for diffusion tractography—cont'd

| | Phantom material | Sizes (μm) | Applications | References |
|---|---|---|---|---|
| | Plastic capillary sheets (polytetrafluoroethylene) (Cole-Parmer Instrument) | I.D: 50 O.D.: 350 | Crossing fibers | Lin et al. (2003), von dem Hagen and Henkelman (2002) |
| Biomimetic fiber phantoms | Taxons, hollow polypropylene yarns | I.D: 11.8±1.2 O.D.: 33.5±2.3 | Characterization, fiber tracking, diameter mapping Crossing fiber phantom | Fan et al. (2018), Guise et al. (2016), Schneider et al. (2019), Fan et al. (2018), Pathak et al. (2020), Baete et al. (2021) |
| | Hollow polycaprolactone (PCL) microfibers | I.D.: 0.5–13.4 (tunable) | Characterization, diameter mapping | Zhou et al. (2018), Hubbard et al. (2015), Grech-Sollars et al. (2018), Huang et al. (2021) |

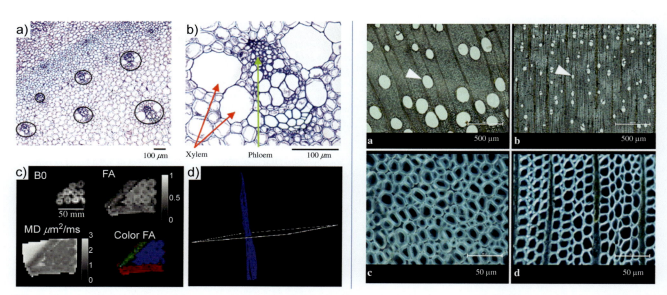

**FIG. 6** Plants have been proposed as anisotropic diffusion phantoms: Left: Microscopic images of asparagus with zoom-in on the vascular bundle, consisting of long tubular vessels with a thick wall (xylem) and smaller cells (phloem) that are either tubular and filled with air or spherical containing water; the resulting parametric maps and fiber tracking results also shown (FA=0.15). Right: Microscopic images of wood showing the vessels (top) with zoom-in on the wood fibers (bottom), illustrating how vessels differ in two types of woods: A, C: *Gleditsia triacanthos*, FA=0.38 and B, D *Acer palmatum*, with FA=0.79. *(Left: Figure A, B, D reproduced with permission from Fieremans, E., 2008. Validation Methods for Diffusion Weighted Magnetic Resonance Imaging in Brain White Matter. Ghent University. https://biblio.ugent.be/publication/891081 and Figure C made based on data from Baete, S.H., Cho, G., Sigmund, E.E., 2013. Multiple-echo diffusion tensor acquisition technique (MEDITATE) on a 3T clinical scanner. NMR Biomed. https://doi.org/10.1002/nbm.2978. Right: Figure reproduced with permission from Suzuki, M., Moriya, S., Hata, J., Tachibana, A., Senoo, A., Niitsu, M., 2019. Development of anisotropic phantoms using wood and fiber materials for diffusion tensor imaging and diffusion kurtosis imaging. MAGMA 32 (5), 539–547.)*

(Kausch et al., 1971)) and also result in low SNR due to the corresponding low proton density. Several methods have been described to pack the fibers densely, including compressing fibers inside a custom-designed mold or matrix after submersion in water (Poupon et al., 2008) or winding the fibers on a spindle (Moussavi-Biugui et al., 2011). Alternatively, dense fiber packings can be obtained by placing fiber bundles in water inside shrinking tubes that subsequently shrink by heating the water (causing shrinkage), as explained in detail in Fieremans et al. (2008a). Automated systems to count the fibers inside each bundle help to increase the reproducibility of the phantom manufacturing (Fieremans, 2008; Moussavi-Biugui et al., 2011).

Aside from fiber diameter, the fiber material properties of surface relaxation and magnetic susceptibility need to be considered for determining SNR and minimizing confounding effects on diffusion quantification in fiber phantoms: while surface relaxation is often overlooked, it causes additional T2 relaxation (proportionate to the S/V of the packing) and the resulting SNR loss can be significant, as illustrated in Fig. 5(a) in Fieremans et al. (2008a), comparing R2 of phantoms made of fibers with different surface relaxivities. Hydrophobic fiber materials have minimal surface relaxation with water, yet such fiber phantoms are prone to contain air bubbles. However, the air bubbles can be minimized by making the phantoms underwater and squeezing the fibers by hand or using tweezers to remove bubbles, as well as using vacuum chambers and ultrasonic baths (Fieremans et al., 2008a).

In addition, differences in magnetic susceptibility between fiber and fluid in the phantom may cause distortion artifacts and result in internal gradients that induce additional T2 relaxation (Fieremans et al., 2008a) and potentially may compromise the diffusion quantification. While these unwanted effects can be minimized by aligning phantom fibers to the static B0-field, they come into play in crossing fiber phantoms, or other phantoms with multiple fiber orientations (e.g., in Fig. 8B).

Plain fibers, while mimicking extra-axonal diffusion only, are commonly used to create anisotropic diffusion phantoms, because of their small radius (20 µm and less, Table 2) and their shown value as test objects. They have been used to create fiber crossings (Perrin et al., 2005; Reischauer et al., 2009; Pullens et al., 2010; Moussavi-Biugui et al., 2011) to study the value of beyond diffusion tensor imaging (DTI) fiber tracking methods, as illustrated in Fig. 7A. They also have been used in several anthropomorphic head phantoms, as shown in Fig. 8.

### 3.2.3 Capillary phantoms

Capillary phantoms mimicking the intra-axonal space, are made of microcapillaries, capillary arrays, or capillary sheets, consisting of hollow fibers with I.D.s varying from 1 to 50 µm, and outer diameters (O.D.s) of 75 µm or more. They have a low ratio of I.D./O.D. or no fluid in-between and therefore are often used for evaluating pore size and axon diameter mapping methods (Siow et al., 2012; Shemesh et al., 2010; Morozov et al., 2013, 2014; Benjamini et al., 2014). They have also been proposed as DTI phantoms (Yanasak and Allison, 2006) and for evaluating tractography (Lin et al., 2003; Tournier et al., 2008). Compared to plain fibers, phantoms made from capillaries can in principle obtain much higher FA values if all MRI-visible protons are inside the capillaries and the I.D. is small as compared to the diffusion length of the experiment. However, experimentally, these phantoms often have low proton density and corresponding SNR due to the low ratio of I.D./O.D (or no fluid in-between), and the presence of air bubbles may cause susceptibility artifacts. Air bubbles may be reduced by using under pressure.

Even more so than in plain fiber phantoms, filling capillaries with water while avoiding air bubbles can be challenging. For that, capillary action is needed, which occurs when the adhesion to the surface material is stronger than the cohesive forces between the water molecules and is limited by surface tension and gravity. For hydrophilic surfaces, surface tension is small, which will enhance capillary action and get water into the tube, while hydrophobic surfaces increase surface tension, hence will make it harder for water to get into the capillaries. From this perspective, it is beneficial to have capillaries made of a hydrophilic material, which makes it relatively straightforward to fill them with water and remove air bubbles. On the other hand, hydrophilic materials also have increased NMR surface relaxation (Chen et al., 2006), resulting in a decrease in T2 relaxation for the water inside hydrophilic capillaries, which will lower the overall SNR of the dMRI experiment, as well as potentially affect the diffusion properties, as discussed earlier for plain fibers and in Fieremans and Lee (2018).

Capillaries have been used to create fiber crossings to evaluate higher-order diffusion methods, as illustrated in Fig. 7B and C. Of note, these capillaries have I.D.s that are larger than axons (~1 µm) and therefore cannot be considered as sticks, i.e., cylinders with zero radii, for standard diffusion measurements with diffusion times on the order of 10–100 ms (Vellmer et al., 2017). They typically also have much thicker walls as compared to the myelin thickness of axons, prompting further pursuit of more realistic phantoms mimicking axons in brain white matter.

### 3.2.4 Biomimetic phantoms

Recently, a new class of so-called biomimetic phantoms has been proposed to address the limitations of the phantoms described already and resemble more closely axons in the brain. Using co-electron spinning, hollow polycaprolactone (PCL) microfibers can be created where the fiber cross-sectional features are influenced by production time, solution flow rate, and x-y stage translation speed (Zhou et al., 2011; Hubbard et al., 2015), with size distribution varying from ~1.2 µm (Zhou et al., 2011), 9.5 µm up to 13.4 µm (Hubbard et al., 2015). Phantoms consisting of aligned or randomly oriented hollow microfibers filled with cyclohexane were produced to respectively mimic brain white and GM with FA values

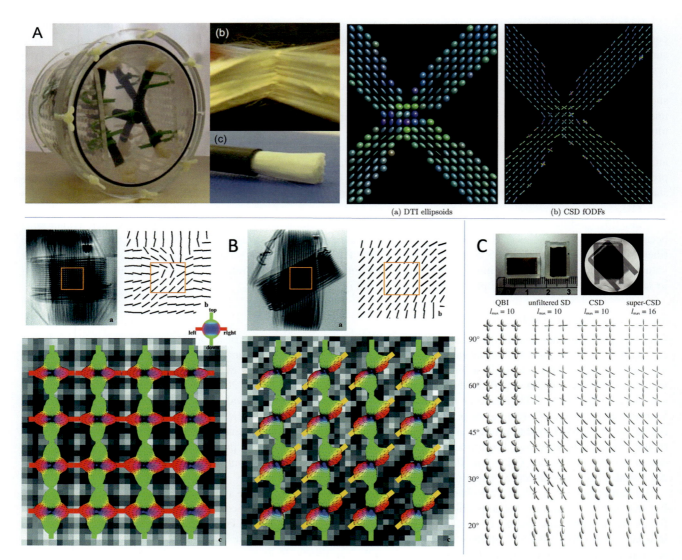

**FIG. 7** Examples of crossing fiber phantoms made from plain (A) and capillary (B and C) fibers showing the inability of DTI to capture crossings and benefit of higher-order methods: (A) Crossing fiber phantom of Dyneema fibers where shrinking tubes ensure adequate tightening of the fibers within each bundle (c). While DTI reports oblate diffusion tensor (DT) ellipsoids at the crossing (a), CSD reports very sharp fiber orientation distribution functions (fODF)'s at the fiber crossing, with the fODF maxima along the expected fiber orientations (b); (B) Sheets of plastic (PTFE) capillaries with I.D. of 50 mm and O.D. of 350 mm placed at crossing angles of 90 degrees (left) and 45 degrees (right), with corresponding high-resolution T2w-MRI. Maps plotting the direction of the first eigenvector show the inability of DTI to capture the fiber crossing, while DSI-derived ODFs determine the orientation correctly. (C) Sheets of plastic capillaries (fused silica) with I.D. of 20 mm and O.D. of 90 mm placed at varying crossing angles. The measured fiber orientation distribution (FOD)/ODF at $b=4000$ s/mm$^2$ from Q-ball imaging (QBI), unfiltered spherical deconvolution (SD), constrained SD (CSD), and super-CSD. All techniques resolve crossing at 45 degrees and above, while only unfiltered SD and super-CSD reliably resolve crossings at 30 degrees. ((A) Figure reproduced with permission from Fieremans, E., 2008. Validation Methods for Diffusion Weighted Magnetic Resonance Imaging in Brain White Matter. Ghent University. https://biblio.ugent.be/publication/891081. Jeurissen, B., 2012. Improved Analysis of Brain Connectivity Using High Angular Resolution Diffusion MRI. Thesis. https://visielab.uantwerpen.be/sites/default/files/jeurissen-phdthesis-2012.pdfhttps://search.proquest.com/openview/95fe8990d0a2c1a22c081d3a4237a74a/1?pq-origsite=gscholar&cbl=18750https://visielab.uantwerpen.be/sites/default/files/jeurissen-phdthesis-2012.pdf. (B) Figure reproduced with permission from Lin, C.-P., Wedeen, V.J., Chen, J.-H., Yao, C., Tseng, W.-Y.I., 2003. Validation of diffusion spectrum magnetic resonance imaging with manganese-enhanced rat optic tracts and ex vivo phantoms. NeuroImage 19 (3), 482–495. (C) Figure reproduced with permission from Tournier, J.-D., Yeh, C.-H., Calamante, F., Cho, K.-H., Connelly, A., Lin, C.-P., 2008. Resolving crossing fibres using constrained spherical deconvolution: validation using diffusion-weighted imaging phantom data. NeuroImage 42 (2), 617–625.)

Tractography validation Part 1: Foundations, numerical simulations, and phantom models **Chapter | 25** **503**

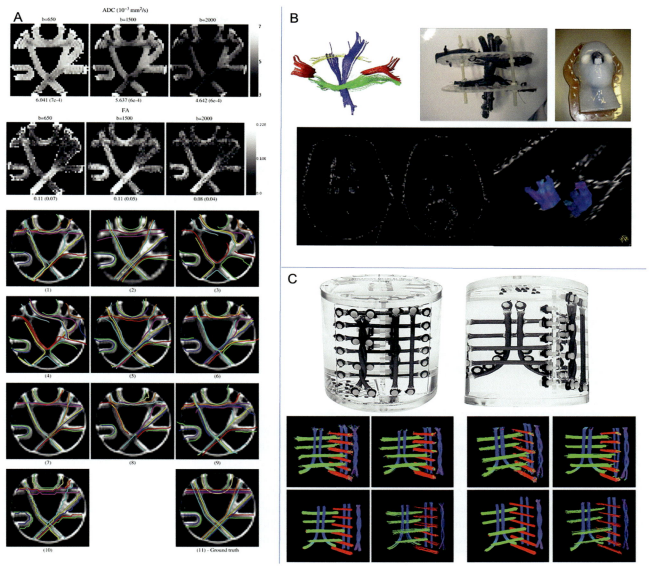

**FIG. 8** Examples of anthropomorphic head phantoms made from plain fibers used to evaluate fiber tractography: (A) Fibercup09 phantom made of acrylic fibers (Poupon et al., 2008) mimicking a coronal section of the human brain (Fillard et al., 2011): apparent diffusion coefficient (ADC) and FA maps derived from DTI based on different *b*-values are shown. Note the overall low FA values ~0.1. Different contributions of the fibercup09 contests are shown along with the ground truth. (B) Prototype head phantom mimicking several brain white matter bundles made of Dyneema fibers surrounded by shrinking tubes and immersed in agar gel. The flexible bundles are shaped according to the following tracts: corticospinal tracts—*blue*, forceps major and minor of the corpus callosum—*red*, optical tracts—*yellow*, fronto-occipital tracts—*green*, and secured in 3D by attaching them to a grid with zip ties and placed inside a head phantom filled with gel and a hollow pipe representing the trachea. Axial FA maps and fiber tracking results of the corticospinal tracts are shown. While the agar gel was doped with Gd-DTPA to match T1 and T2 of brain white matter, susceptibility mismatches between fiber and gel and the lower proton density caused additional T2 relaxation, and resulting low SNR, particularly for bundles perpendicular to the static B0-field; (C) The Anisotropic Diffusion Phantom (Synaptive Medical, Toronto, ON) contains 16 flexible fiber bundles, creating pathways aligned in orthogonal planes, as well as in curved (both 90 degrees and helical curving), and kissing geometries to mimic complex nerve fibers of the brain, with bundle dimensions of magnitudes comparable to major white matter pathways in the human brain, ranging from 2 mm up to 6 mm diameter bundles (Schilling et al., 2019b). The phantom is filled with distilled water, resulting in a very long intrinsic T2 that is shortened by surface relaxation and internal gradients. *((A) Figure reproduced with permission from Fillard, P., Descoteaux, M., Goh, A., Gouttard, S., Jeurissen, B., Malcolm, J., Ramirez-Manzanares, A., et al., 2011. Quantitative evaluation of 10 tractography algorithms on a realistic diffusion MR phantom. NeuroImage 56 (1), 220–234. (B) Figure modified from Fieremans, E., De Deene, Y., Delputte, S., Ozdemir, M.S., Achten, E., Lemahieu, I., 2008. The design of anisotropic diffusion phantoms for the validation of diffusion weighted magnetic resonance imaging. Phys. Med. Biol. 53 (19), 5405–5419.)*

of 0.4 or 0.6 in the white matter phantom (for different fiber types) or less than 0.1 in the GM phantom with phantom stability confirmed over 33 months (Grech-Sollars et al., 2018). Surfactants in the PCL solution improved the hydrophilicity of the PCL, which could be beneficial for phantom filling (Zhou et al., 2018). PCL microfibers have been recently used to create a spinal cord phantom (Zhou et al., 2020), to validate models for time-dependent diffusion (Lundell et al., 2019) and to validate pore size mapping on a Connectom scanner with gradients of 300 mT/m (Huang et al., 2021).

Similarly, using melt-spinning techniques borrowed from the textile industry, hollow polypropylene (PP) yarns, also dubbed "taxons," with an I.D. of about 12 μm and O.D. of 34 μm have been proposed (Schneider et al., 2019; Pathak et al., 2020). PP have thicker walls than PCL fibers (that may be permeable), making the phantoms suitable for mimicking both intra- and extra-axonal space, They have been used to study the effect of dispersion and crossings on fiber tracking (Guise et al., 2016), as well as to validate compartment size and fraction measurements (Fan et al., 2018). An anisotropic phantom based on taxons with I.D. of 0.8 μm and packing density matching human axon histology is under development aiming to advance cross-laboratory research and calibration (Schneider et al., 2019; Pathak et al., 2020). Most recently, this phantom has been used as ground truth to evaluate orientation distribution function (ODF) fingerprinting (Baete et al., 2021; Filipiak et al., 2021).

### 3.3 Which phantoms for which experiment?

As listed previously, and in Table 2, several phantoms with increasing complexity have been proposed as test objects for diffusion MRI and tractography methods, as follows.

Test tubes or containers shaped as cylinders or spheres, filled with isotropic liquids, are the most accurate and well-characterized test systems. They are the go-to phantoms and are useful for initial diffusion sequence, tractography testing, and b-value calibration. They are also ideal phantoms for multicenter and longitudinal studies. Several liquids and mixtures have been proposed covering a wide range of diffusivities (Table 1 in Fieremans and Lee, 2018). Ice water and PVP solutions have diffusion coefficients similar to the brain. Other liquids, in particular alkanes, have lower viscosity, but also potential drawbacks, including toxicity or the presence of resonance frequencies different from water. The latter may result either in chemical shift artifacts (for EPI-based sequences) or suppressed signals when "fat saturation" is on.

Anisotropic phantoms may be useful for evaluating gradient directions or testing particular tractography applications. Simple phantoms such as asparagus require minimal preparation and may be helpful for one-time testing. More stable and durable options include man-made anisotropic phantoms based on either plain fibers or capillaries. Such phantoms are more challenging to manufacture (e.g., to minimize air bubbles) and typically not very well characterized, but can be kept over time (albeit often by adding small concentrations of toxic compounds). Such phantoms are used to mimic fiber crossings (Fig. 7) and several prototype head phantoms have been proposed for evaluating tractography (Fig. 8). Of note, such phantoms consist of fibers that are either not hollow, impermeable, or they have an inner and/or outer diameter far outside the range of axons in the brain. Numerical phantoms offer the possibility of more realistic phantoms.

Finally, there is an ongoing search for biomimetic phantoms that display realistic brain microstructure features. Promising candidates include hollow PCL microfibers and taxons. Such phantoms require advanced manufacturing, but may become more accessible with the advent of 3D printing and gain importance in future applications.

# References

Abdollahzadeh, A., Belevich, I., Jokitalo, E., Tohka, J., Sierra, A., 2019. Automated 3D axonal morphometry of white matter. Sci. Rep. 9 (1), 6084.

Aganj, I., Lenglet, C., Jahanshad, N., Yacoub, E., Harel, N., Thompson, P.M., Sapiro, G., 2011. A Hough transform global probabilistic approach to multiple-subject diffusion MRI tractography. Med. Image Anal. 15 (4), 414–425.

Agrawal, T., 2021. Optuna and AutoML. In: Hyperparameter Optimization in Machine Learning., https://doi.org/10.1007/978-1-4842-6579-6_5.

Alexander, D.C., Hubbard, P.L., Hall, M.G., Moore, E.A., Ptito, M., Parker, G.J.M., Dyrby, T.B., 2010. Orientationally invariant indices of axon diameter and density from diffusion MRI. NeuroImage 52 (4), 1374–1389.

Andersson, M., Kjer, H.M., Rafael-Patino, J., Pacureanu, A., Pakkenberg, B., Thiran, J.-P., Ptito, M., et al., 2020. Axon morphology is modulated by the local environment and impacts the noninvasive investigation of its structure-function relationship. Proc. Natl. Acad. Sci. U. S. A. 117 (52), 33649–33659.

Andersson, M., Pizzolato, M., Kjer, H.M., Skodborg, K.F., Lundell, H., Dyrby, T.B., 2022. Does powder averaging remove dispersion bias in diffusion MRI diameter estimates within real 3D axonal architectures? NeuroImage 248, 118718.

Avram, L., Assaf, Y., Cohen, Y., 2004. The effect of rotational angle and experimental parameters on the diffraction patterns and micro-structural information obtained from q-space diffusion NMR: implication for diffusion in white matter fibers. J. Magn. Reson., 30–38.

Baete, S.H., Cho, G.Y., Sigmund, E.E., 2015. Dynamic diffusion-tensor measurements in muscle tissue using the single-line multiple-echo diffusion-tensor acquisition technique at 3T. NMR Biomed. 28 (6), 667–678.

Baete, S.H., Filipiak, P., Lee, B., Zuccolotto, A., Lin, Y.-C., Placantonakis, D.G., Shepherd, T., Schneider, W., Boada, F.E., 2021. A closer look at diffusion and fiber ODFs in a ground truth crossing fiber phantom. Proc. Int. Soc. Magn. Reson. Med., 3409.

Balls, G.T., Frank, L.R., 2009. A simulation environment for diffusion weighted MR experiments in complex media. Magn. Reson. Med. 62, 771–778. https://doi.org/10.1002/mrm.22033.

Bar-Shir, A., Avram, L., Özarslan, E., Basser, P.J., Cohen, Y., 2008. The effect of the diffusion time and pulse gradient duration ratio on the diffraction pattern and the structural information estimated from q-space diffusion MR: experiments and simulations. J. Magn. Reson. 194 (2), 230–236.

Batchelor, P.G., Calamante, F., Tournier, J.-D., Atkinson, D., Hill, D.L.G., Connelly, A., 2006. Quantification of the shape of fiber tracts. Magn. Reson. Med. 55 (4), 894–903.

Benjamini, D., Komlosh, M.E., Basser, P.J., Nevo, U., 2014. Nonparametric pore size distribution using D-PFG: comparison to S-PFG and migration to MRI. J. Magn. Reson. 246, 36–45.

Benou, I., Raviv, T.R., 2019. DeepTract: a probabilistic deep learning framework for white matter fiber tractography. Lect. Notes Comput. Sci. https://doi.org/10.1007/978-3-030-32248-9_70.

Bergstra, J., Komer, B., Eliasmith, C., Yamins, D., Cox, D.D., 2015. Hyperopt: a python library for model selection and hyperparameter optimization. Comput. Sci. Discov. https://doi.org/10.1088/1749-4699/8/1/014008.

Boujraf, S., 2001. Echo planar magnetic resonance imaging of anisotropic diffusion in asparagus stems. Magn. Reson. Mater. Phys. Biol. Med. https://doi.org/10.1016/s1352-8661(01)00132-6.

Brabec, J., Lasič, S., Nilsson, M., 2020. Time-dependent diffusion in undulating thin fibers: impact on axon diameter estimation. NMR Biomed. 33 (3), e4187.

Budde, MD, Frank, JA, 2010. Neurite beading is sufficient to decrease the apparent diffusion coefficient after ischemic stroke. Proc. Natl. Acad. Sci. U. S. A. 107 (32), 14472–14477. https://doi.org/10.1073/pnas.1004841107. Epub 2010 Jul 26 PMID: 20660718. PMCID: PMC2922529.

Burcaw, L.M., Fieremans, E., Novikov, D.S., 2015. Mesoscopic structure of neuronal tracts from time-dependent diffusion. NeuroImage 114, 18–37.

Callaghan, R, Alexander, DC, Zhang, G, Palombo, M, 2019. Contextual fibre growth to generate realistic axonal packing for diffusion MRI simulation. Proceedings of the 26th International Conference on Information Processing in Medical Imaging (IPMI). https://doi.org/10.1007/978-3-030-20351-1_33.

Callaghan, R., Alexander, D.C., Palombo, M., Zhang, H., 2020. ConFiG: contextual fibre growth to generate realistic axonal packing for diffusion MRI simulation. NeuroImage 220, 117107.

Caruyer, E., Daducci, A., Descoteaux, M., Houde, J.-C., Thiran, J.-P., Verma, R., 2014. Phantomas: a flexible software library to simulate diffusion MR phantoms. Proc. Int. Soc. Magn. Reson. Med. 2666. https://hal.inria.fr/hal-00944644/.

Cetingul, H.E., Cetingül, H.E., Nadar, M., Thompson, P., Sapiro, G., Lenglet, C., 2012. Simultaneous ODF estimation and tractography in HARDI. IEEE, https://doi.org/10.1109/EMBC.2012.6345877.

Chen, J., Hirasaki, G.J., Flaum, M., 2006. NMR wettability indices: effect of OBM on wettability and NMR responses. J. Pet. Sci. Eng. https://doi.org/10.1016/j.petrol.2006.03.007.

Chenevert, T.L., Galbán, C.J., Ivancevic, M.K., Rohrer, S.E., Londy, F.J., Kwee, T.C., Meyer, C.R., Johnson, T.D., Rehemtulla, A., Ross, B.D., 2011. Diffusion coefficient measurement using a temperature-controlled fluid for quality control in multicenter studies. J. Magn. Reson. Imaging 34 (4), 983–987.

Close, T.G., Tournier, J.-D., Calamante, F., Johnston, L.A., Mareels, I., Connelly, A., 2009. A software tool to generate simulated white matter structures for the assessment of fibre-tracking algorithms. NeuroImage 47 (4), 1288–1300.

Cook, P.A., Bai, Y., Nedjati-Gilani, S., Seunarine, K.K., Hall, M.G., Parker, G.J., 2006. Camino: open-source diffusion-MRI reconstruction and processing. Proc. Int. Soc. Magn. Reson. Med., 2759.

Côté, M.-A., Girard, G., Boré, A., Garyfallidis, E., Houde, J.-C., Descoteaux, M., 2013. Tractometer: towards validation of tractography pipelines. Med. Image Anal. 17 (7), 844–857.

Daducci, A., Dal Palù, A., Lemkaddem, A., Thiran, J.-P., 2015. COMMIT: convex optimization modeling for microstructure informed tractography. IEEE Trans. Med. Imaging 34 (1), 246–257.

Delakis, I., Moore, E.M., Leach, M.O., De Wilde, J.P., 2004. Developing a quality control protocol for diffusion imaging on a clinical MRI system. Phys. Med. Biol. 49 (8), 1409–1422.

Drobnjak, I., Gavaghan, D., Süli, E., Pitt-Francis, J., Jenkinson, M., 2006. Development of a functional magnetic resonance imaging simulator for modeling realistic rigid-body motion artifacts. Magn. Reson. Med. 56 (2), 364–380.

Drobnjak, I., Pell, G.S., Jenkinson, M., 2010a. Simulating the effects of time-varying magnetic fields with a realistic simulated scanner. Magn. Reson. Imaging 28 (7), 1014–1021.

Drobnjak, I., Siow, B., Alexander, D.C., 2010b. Optimizing gradient waveforms for microstructure sensitivity in diffusion-weighted MR. J. Magn. Reson. 206 (1), 41–51.

Dyrby, T.B., Innocenti, G.M., Bech, M., Lundell, H., 2018. Validation strategies for the interpretation of microstructure imaging using diffusion MRI. NeuroImage 182, 62–79.

Fan, Q., Nummenmaa, A., Wichtmann, B., Witzel, T., Mekkaoui, C., Schneider, W., Wald, L.L., Huang, S.Y., 2018. Validation of diffusion MRI estimates of compartment size and volume fraction in a biomimetic brain phantom using a human MRI scanner with 300 mT/m maximum gradient strength. NeuroImage 182, 469–478.

Farrher, E., Joachim Kaffanke, A., Celik, A., Stöcker, T., Grinberg, F., Jon Shah, N., 2012. Novel multisection design of anisotropic diffusion phantoms. Magn. Reson. Imaging 30 (4), 518–526.

Fieremans, E., 2008. Validation Methods for Diffusion Weighted Magnetic Resonance Imaging in Brain White Matter. Ghent University. https://biblio.ugent.be/publication/891081.

**506 PART | IV From streamlines to tracts**

Fieremans, E., De Deene, Y., 2020. Gel phantoms for diffusion MRI studies. In: NMR and MRI of Gels. Royal Society of Chemistry, pp. 379–400 (Chapter 11).

Fieremans, E., Lee, H.-H., 2018. Physical and numerical phantoms for the validation of brain microstructural MRI: a cookbook. NeuroImage 182, 39–61.

Fieremans, E., De Deene, Y., Delputte, S., Ozdemir, M.S., Achten, E., Lemahieu, I., 2008a. The design of anisotropic diffusion phantoms for the validation of diffusion weighted magnetic resonance imaging. Phys. Med. Biol. 53 (19), 5405–5419.

Fieremans, E., De Deene, Y., Delputte, S., Ozdemir, M.S., D'Asseler, Y., Vlassenbroeck, J., Deblaere, K., Achten, E., Lemahieu, I., 2008b. Simulation and experimental verification of the diffusion in an anisotropic fiber phantom. J. Magn. Reson. 190 (2), 189–199.

Filipiak, P., Lee, B., Zuccolotto, A., Lin, Y.-C., Placantonakis, D.G., Shepherd, T., Schneider, W., Boada, F.E., Baete, S.H., 2021. ODF-fingerprinting improves reconstruction of fibers crossing at shallow angles: a study on diffusion phantom. Proc. Int. Soc. Magn. Reson. Med., 4285.

Fillard, P., Descoteaux, M., Goh, A., Gouttard, S., Jeurissen, B., Malcolm, J., Ramirez-Manzanares, A., et al., 2011. Quantitative evaluation of 10 tractography algorithms on a realistic diffusion MR phantom. NeuroImage 56 (1), 220–234.

Fritzsche, K.H., Neher, P.F., Reicht, I., van Bruggen, T., Goch, C., Reisert, M., Nolden, M., et al., 2012. MITK diffusion imaging. Methods Inf. Med. 51 (5), 441.

Fukuzaki, M., Miura, N., Shinyashiki, N., Kurita, D., Shioya, S., Haida, M., Mashimo, S., 1995. Comparison of water relaxation time in serum albumin solution using nuclear magnetic resonance and time domain reflectometry. J. Phys. Chem. 99 (1), 431–435.

Gaggl, W., Jesmanowicz, A., Prost, R.W., 2014. High-resolution reduced field of view diffusion tensor imaging using spatially selective RF pulses. Magn. Reson. Med. 72 (6), 1668–1679.

Garyfallidis, E., Brett, M., Correia, M.M., Williams, G.B., Nimmo-Smith, I., 2012. QuickBundles, a method for tractography simplification. Front. Neurosci. https://doi.org/10.3389/fnins.2012.00175.

Ginsburger, K., Poupon, F., Beaujoin, J., Estournet, D., Matuschke, F., Mangin, J.-F., Axer, M., Poupon, C., 2018. Improving the realism of white matter numerical phantoms: a step toward a better understanding of the influence of structural disorders in diffusion MRI. Front. Phys. 6, 1–18. https://doi.org/10.3389/fphy.2018.00012.

Ginsburger, K., Matuschke, F., Poupon, F., Mangin, J.-F., Axer, M., Poupon, C., 2019. MEDUSA: a GPU-based tool to create realistic phantoms of the brain microstructure using tiny spheres. NeuroImage 193, 10–24.

Gössl, C., Fahrmeir, L., Pütz, B., Auer, L.M., Auer, D.P., 2002. Fiber tracking from DTI using linear state space models: detectability of the pyramidal tract. NeuroImage 16 (2), 378–388.

Graham, M.S., Drobnjak, I., Zhang, H., 2016. Realistic simulation of artefacts in diffusion MRI for validating post-processing correction techniques. NeuroImage 125, 1079–1094.

Graham, M.S., Drobnjak, I., Jenkinson, M., Zhang, H., 2017. Quantitative assessment of the susceptibility artefact and its interaction with motion in diffusion MRI. PLoS One 12 (10), e0185647.

Grech-Sollars, M., Hales, P.W., Miyazaki, K., Raschke, F., Rodriguez, D., Wilson, M., Gill, S.K., et al., 2015. Multi-centre reproducibility of diffusion MRI parameters for clinical sequences in the brain. NMR Biomed. 28 (4), 468–485.

Grech-Sollars, M., Zhou, F.-L., Waldman, A.D., Parker, G.J.M., Hubbard, P.L., Cristinacce., 2018. Stability and reproducibility of co-electrospun brain-mimicking phantoms for quality assurance of diffusion MRI sequences. NeuroImage. https://doi.org/10.1016/j.neuroimage.2018.06.059.

Gruwel, M.L.H., Latta, P., Sboto-Frankenstein, U., Gervai, P., 2013. Visualization of water transport pathways in plants using diffusion tensor imaging. Progress Electromagn. Res. C. https://doi.org/10.2528/pierc12110506.

Guise, C., Fernandes, M.M., Nóbrega, J.M., Pathak, S., Schneider, W., Fangueiro, R., 2016. Hollow polypropylene yarns as a biomimetic brain phantom for the validation of high-definition fiber tractography imaging. ACS Appl. Mater. Interfaces 8 (44), 29960–29967.

Hall, M.G., Alexander, D.C., 2009. Convergence and parameter choice for Monte-Carlo simulations of diffusion MRI. IEEE Trans. Med. Imaging. https://doi.org/10.1109/tmi.2009.2015756.

Hara, M., Kuroda, M., Ohmura, Y., Matsuzaki, H., Kobayashi, T., Murakami, J., Katashima, K., Ashida, M., Ohno, S., Asaumi, J.-I., 2014. A new phantom and empirical formula for apparent diffusion coefficient measurement by a 3 Tesla magnetic resonance imaging scanner. Oncol. Lett. 8 (2), 819–824.

Huang, C.-c., Hsu, C.-c.H., Zhou, F.-l., Kusmia, S., Drakesmith, M., Parker, G.J.M., Lin, C.-p., Jones, D.K., 2021. Validating pore size estimates in a complex microfiber environment on a human MRI system. Magn. Reson. Med. https://doi.org/10.1002/mrm.28810.

Hubbard, P.L., Zhou, F.-L., Eichhorn, S.J., Parker, G.J.M., 2015. Biomimetic phantom for the validation of diffusion magnetic resonance imaging. Magn. Reson. Med. 73 (1), 299–305.

Jenkinson, M., Beckmann, C.F., Behrens, T.E.J., Woolrich, M.W., Smith, S.M., 2012. FSL. NeuroImage 62 (2), 782–790.

Jeurissen, B., 2012. Improved Analysis of Brain Connectivity Using High Angular Resolution Diffusion MRI. Thesis. https://visielab.uantwerpen.be/sites/default/files/jeurissen-phdthesis-2012.pdf.

Kang, N., Zhang, J., Carlson, E.S., Gembris, D., 2005. White matter fiber tractography via anisotropic diffusion simulation in the human brain. IEEE Trans. Med. Imaging 24 (9), 1127–1137.

Kausch, H.H., Fesko, D.G., Tschoegl, N.W., 1971. The random packing of circles in a plane. J. Colloid Interface Sci. 37 (3), 603–611.

Kjer, H.M., Andersson, M., He, Y., Pacureanu, A., Daducci, A., Pizzolato, M., Salditt, T., Robisch, A.-L., Eckermann, M., Toepperwien, M., Dahl, A.B., Elkjær, M.L., Illes, Z., Ptito, M., Dahl, V.A., Dyrby, T.B., 2024. Bridging the 3D geometrical organisation of white matter pathways across anatomical length scales and species. eLife 13, RP94917. https://doi.org/10.7554/eLife.94917.1.

Komlosh, M.E., Özarslan, E., Lizak, M.J., Horkay, F., Schram, V., Shemesh, N., Cohen, Y., Basser, P.J., 2011. Pore diameter mapping using double pulsed-field gradient MRI and its validation using a novel glass capillary array phantom. J. Magn. Reson. 208 (1), 128–135.

Komlosh, M.E., Benjamini, D., Barnett, A.S., Schram, V., Horkay, F., Avram, A.V., Basser, P.J., 2017. Anisotropic phantom to calibrate high-Q diffusion MRI methods. J. Magn. Reson. 275, 19–28.

Kreher, B.W., Schneider, J.F., Mader, I., Martin, E., Hennig, J., Il'Yasov, K.A., 2005. Multitensor approach for analysis and tracking of complex fiber configurations. Magn. Reson. Med. 54 (5), 1216–1225.

Landman, BA, Farrell, JA, Smith, SA, Reich, DS, Calabresi, PA, van Zijl, PC, 2010. Complex geometric models of diffusion and relaxation in healthy and damaged white matter. NMR Biomed., 152–162. https://doi.org/10.1002/nbm.1437.

Latt, J., Nilsson, M., Malmborg, C., Rosquist, H., Wirestam, R., Stahlberg, F., Topgaard, D., Brockstedt, S., 2007. Accuracy of q-space related parameters in MRI: simulations and phantom measurements. IEEE Trans. Med. Imaging 26 (11), 1437–1447.

Laubach, H.J., Jakob, P.M., Loevblad, K.O., Baird, A.E., Bovo, M.P., Edelman, R.R., Warach, S., 1998. A phantom for diffusion-weighted imaging of acute stroke. J. Magn. Reson. Imaging 8 (6), 1349–1354.

Laun, F.B., Huff, S., Stieltjes, B., 2009. On the effects of dephasing due to local gradients in diffusion tensor imaging experiments: relevance for diffusion tensor imaging fiber phantoms. Magn. Reson. Imaging 27 (4), 541–548.

Lavdas, I., Behan, K.C., Papadaki, A., McRobbie, D.W., Aboagye, E.O., 2013. A phantom for diffusion-weighted MRI (DW-MRI). J. Magn. Reson. Imaging 38 (1), 173–179.

Lazar, M., Alexander, A.L., 2003. An error analysis of white matter tractography methods: synthetic diffusion tensor field simulations. NeuroImage 20 (2), 1140–1153.

Lee, H.-H., Yaros, K., Veraart, J., Pathan, J.L., Liang, F.-X., Kim, S.G., Novikov, D.S., Fieremans, E., 2019. Along-axon diameter variation and axonal orientation dispersion revealed with 3D Electron microscopy: implications for quantifying brain white matter microstructure with histology and diffusion MRI. Brain Struct. Funct. 224 (4), 1469–1488.

Lee, H.-H., Jespersen, S.N., Fieremans, E., Novikov, D.S., 2020a. The impact of realistic axonal shape on axon diameter estimation using diffusion MRI. NeuroImage 223, 117228.

Lee, H.-H., Papaioannou, A., Kim, S.-L., Novikov, D.S., Fieremans, E., 2020b. A time-dependent diffusion MRI signature of axon caliber variations and beading. Commun. Biol. 3 (1), 354.

Lee, H.-H., Fieremans, E., Novikov, D.S., 2021. Realistic microstructure simulator (RMS): Monte Carlo simulations of diffusion in three-dimensional cell segmentations of microscopy images. J. Neurosci. Methods 350, 109018. https://doi.org/10.1016/j.jneumeth.2020.109018.

Leemans, A., Sijbers, J., Verhoye, M., Van der Linden, A., Van Dyck, D., 2005. Mathematical framework for simulating diffusion tensor MR neural fiber bundles. Magn. Reson. Med. 53 (4), 944–953.

Lemberskiy, G., Baete, S.H., Cloos, M.A., Novikov, D.S., Fieremans, E., 2017. Validation of surface-to-volume ratio measurements derived from oscillating gradient spin echo on a clinical scanner using anisotropic fiber phantoms. NMR Biomed. 30 (5), e3708.

Li, H., Gore, J.C., Xu, J., 2014. Fast and robust measurement of microstructural dimensions using temporal diffusion spectroscopy. J. Magn. Reson. 242, 4–9.

Lichenstein, S.D., Bishop, J.H., Verstynen, T.D., Yeh, F.-C., 2016. Diffusion capillary phantom vs. human data: outcomes for reconstruction methods depend on evaluation medium. Front. Neurosci. 10, 407.

Lin, C.-P., Wedeen, V.J., Chen, J.-H., Yao, C., Tseng, W.-Y.I., 2003. Validation of diffusion spectrum magnetic resonance imaging with manganese-enhanced rat optic tracts and ex vivo phantoms. NeuroImage 19 (3), 482–495.

Lorenz, R., Bellemann, M.E., Hennig, J., Il'Yasov, K.A., 2008. Anisotropic phantoms for quantitative diffusion tensor imaging and Fiber-tracking validation. Appl. Magn. Reson. 33 (4), 419–429.

Lori, N.F., Akbudak, E., Shimony, J.S., Cull, T.S., Snyder, A.Z., Guillory, R.K., Conturo, T.E., 2002. Diffusion tensor fiber tracking of human brain connectivity: aquisition methods, reliability analysis and biological results. NMR Biomed. 15, 494–515.

Lundell, H., Nilsson, M., Dyrby, T.B., Parker, G.J.M., Hubbard Cristinacce, P.L., Zhou, F.-L., Topgaard, D., Lasič, S., 2019. Multidimensional diffusion MRI with spectrally modulated gradients reveals unprecedented microstructural detail. Sci. Rep. 9 (1), 9026.

Maier-Hein, K.H., Neher, P.F., Houde, J.-C., Côté, M.-A., Garyfallidis, E., Zhong, J., Chamberland, M., et al., 2017. The challenge of mapping the human connectome based on diffusion tractography. Nat. Commun. 8 (1), 1349.

Malyarenko, D.I., Newitt, D., Wilmes, L.J., Tudorica, A., Helmer, K.G., Arlinghaus, L.R., Jacobs, M.A., et al., 2016. Demonstration of nonlinearity Bias in the measurement of the apparent diffusion coefficient in multicenter trials. Magn. Reson. Med. 75 (3), 1312–1323.

Milne, M.L., Conradi, M.S., 2009. Multi-exponential signal decay from diffusion in a single compartment. J. Magn. Reson. 197 (1), 87–90.

Mingasson, T., Duval, T., Stikov, N., Cohen-Adad, J., 2017. AxonPacking: an open-source software to simulate arrangements of axons in white matter. Front. Neuroinform. 11, 5.

Morozov, D., Bar, L., Sochen, N., Cohen, Y., 2013. Modeling of the diffusion MR signal in calibrated model systems and nerves. NMR Biomed. 26 (12), 1787–1795.

Morozov, D., Bar, L., Sochen, N., Cohen, Y., 2014. Microstructural information from angular double-pulsed-field-gradient NMR: from model systems to nerves. Magn. Reson. Med. 74 (1), 25–32.

Moussavi-Biugui, A., Stieltjes, B., Fritzsche, K., Semmler, W., Laun, F.B., 2011. Novel spherical phantoms for Q-ball imaging under in vivo conditions. Magn. Reson. Med. https://doi.org/10.1002/mrm.22602.

Mulkern, R.V., Ricci, K.I., Vajapeyam, S., Chenevert, T.L., Malyarenko, D.I., Kocak, M., Poussaint, T.Y., 2015. Pediatric Brain Tumor Consortium multisite assessment of apparent diffusion coefficient z-axis variation assessed with an ice–water phantom. Acad. Radiol. 22 (3), 363–369.

Neher, P.F., Laun, F.B., Stieltjes, B., Maier-Hein, K.H., 2014. Fiberfox: facilitating the creation of realistic white matter software phantoms. Magn. Reson. Med. 72 (5), 1460–1470. https://doi.org/10.1002/mrm.25045.

Neher, P.F., Côté, M.-A., Houde, J.-C., Descoteaux, M., Maier-Hein, K.H., 2017. Fiber tractography using machine learning. NeuroImage. https://doi.org/10.1016/j.neuroimage.2017.07.028.

Nilsson, M., Lätt, J., Ståhlberg, F., van Westen, D., Hagslätt, H., 2012. The importance of axonal undulation in diffusion MR measurements: a Monte Carlo simulation study. NMR Biomed. 25 (5), 795–805.

Özarslan, E., Komlosh, M.E., Lizak, M.J., Horkay, F., Basser, P.J., 2011. Double pulsed field gradient (double-PFG) MR imaging (MRI) as a means to measure the size of plant cells. Magn. Reson. Chem. 49 (Suppl. 1), S79–S84.

Palacios, E.M., Martin, A.J., Boss, M.A., Ezekiel, F., Chang, Y.S., Yuh, E.L., Vassar, M.J., et al., 2017. Toward precision and reproducibility of diffusion tensor imaging: a multicenter diffusion phantom and traveling volunteer study. AJNR Am. J. Neuroradiol. 38 (3), 537–545.

Palombo, M, Ligneul, C, Hernandez-Garzon, E, Valette, J, 2017. Can we detect the effect of spines and leaflets on the diffusion of brain intracellular metabolites? NeuroImage 182, 283–293. https://doi.org/10.1016/j.neuroimage.2017.05.003. PMID: 28495635.

Panagiotaki, E., Hall, M.G., Zhang, H., Siow, B., Lythgoe, M.F., Alexander, D.C., 2010. High-Fidelity meshes from tissue samples for diffusion MRI simulations. In: Medical Image Computing and Computer-Assisted Intervention: MICCAI .. International Conference on Medical Image Computing and Computer-Assisted Intervention. vol. 13 (Pt 2), pp. 404–411.

Panagiotaki, E., Schneider, T., Siow, B., Hall, M.G., Lythgoe, M.F., Alexander, D.C., 2012. Compartment models of the diffusion MR signal in brain white matter: a taxonomy and comparison. NeuroImage 59 (3), 2241–2254.

Pathak, S., Schneider, W., Zuccolotto, A., Huang, S., Fan, Q., Witzel, T., Wald, L., Fieremans, E., Komlosh, M.E., Benjamini, D., Avram, A.V., Basser, P.-J., 2020. Diffusion ground truth quantification of axon scale phantom: limits of diffusion MRI on 7T, 3T and connectome 1.0. Proc. Int. Soc. Magn. Reson. Med., 0737.

Perrin, M., Poupon, C., Rieul, B., Leroux, P., Constantinesco, A., Mangin, J.-F., Lebihan, D., 2005. Validation of q-ball imaging with a diffusion fibre-crossing phantom on a clinical scanner. Philos. Trans. R. Soc. Lond. B Biol. Sci. 360 (1457), 881–891.

Pierpaoli, C., Sarlls, J., Nevo, U., Basser, P.J., Horkay, F., 2009. Polyvinylpyrrolidone (PVP) water solutions as isotropic phantoms for diffusion MRI studies. Proc. Int. Soc. Magn. Reson. Med., 1414.

Poulin, P., Côté, M.-A., Houde, J.-C., Petit, L., Neher, P.F., Maier-Hein, K.H., Larochelle, H., Descoteaux, M., 2017. Learn to track: deep learning for tractography. bioRxiv.

Poupon, C., Rieul, B., Kezele, I., Perrin, M., Poupon, F., Mangin, J.-F., 2008. New diffusion phantoms dedicated to the study and validation of high-angular-resolution diffusion imaging (HARDI) models. Magn. Reson. Med. 60 (6), 1276–1283.

Pullens, P., Roebroeck, A., Goebel, R., 2010. Ground truth hardware phantoms for validation of diffusion-weighted MRI applications. J. Magn. Reson. Imaging 32 (2), 482–488.

Rafael-Patino, J., Romascano, D., Ramirez-Manzanares, A., Canales-Rodríguez, E.J., Girard, G., Thiran, J.-P., 2020. Robust Monte-Carlo simulations in diffusion-MRI: effect of the substrate complexity and parameter choice on the reproducibility of results. Front. Neuroinform. https://doi.org/10.3389/fninf.2020.00008.

Rafael-Patino, J., Girard, G., Truffet, R., Pizzolato, M., Caruyer, E., Thiran, J.-P., 2021. The diffusion-simulated connectivity (DiSCo) dataset. Data Brief 38, 107429.

Reischauer, C., Staempfli, P., Jaermann, T., Boesiger, P., 2009. Construction of a temperature-controlled diffusion phantom for quality control of diffusion measurements. J. Magn. Reson. Imaging 29 (3), 692–698.

Schilling, K.G., Daducci, A., Maier-Hein, K., Poupon, C., Houde, J.-C., Nath, V., Anderson, A.W., Landman, B.A., Descoteaux, M., 2019a. Challenges in diffusion MRI tractography—lessons learned from international benchmark competitions. Magn. Reson. Imaging 57, 194–209.

Schilling, K.G., Nath, V., Hansen, C., Parvathaneni, P., Blaber, J., Gao, Y., Neher, P., et al., 2019b. Limits to anatomical accuracy of diffusion tractography using modern approaches. NeuroImage 185, 1–11.

Schmahmann, J.D., Pandya, D., 2009. Fiber Pathways of the Brain. Oxford University Press.

Schneider, W., Pathak, S., Wu, Y., Busch, D., Dzikiy, J., 2019. Taxon anisotropic phantom delivering human scale parametrically controlled diffusion compartments to advance cross laboratory research and calibration. Proc. Int. Soc. Magn. Reson. Med., 3634.

Shemesh, N., Ozarslan, E., Adiri, T., Basser, P.J., Cohen, Y., 2010. Noninvasive bipolar double-pulsed-field-gradient NMR reveals signatures for pore size and shape in polydisperse, randomly oriented, inhomogeneous porous media. J. Chem. Phys. 133 (4), 044705.

Sigmund, E.E., Song, Y.-Q., 2006. Multiple echo diffusion tensor acquisition technique. Magn. Reson. Imaging 24 (1), 7–18.

Siow, B., Drobnjak, I., Chatterjee, A., Lythgoe, M.F., Alexander, D.C., 2012. Estimation of pore size in a microstructure phantom using the optimised gradient waveform diffusion weighted NMR sequence. J. Magn. Reson. 214 (1), 51–60.

Smith, R.E., Tournier, J.-D., Calamante, F., Connelly, A., 2012. Anatomically-constrained tractography: improved diffusion MRI streamlines tractography through effective use of anatomical information. NeuroImage 62 (3), 1924–1938.

Staempfli, P., Jaermann, T., Crelier, G.R., Kollias, S., Valavanis, A., Boesiger, P., 2006. Resolving fiber crossing using advanced fast marching tractography based on diffusion tensor imaging. NeuroImage 30 (1), 110–120.

Suzuki, M., Moriya, S., Hata, J., Tachibana, A., Senoo, A., Niitsu, M., 2019. Development of anisotropic phantoms using wood and fiber materials for diffusion tensor imaging and diffusion kurtosis imaging. MAGMA 32 (5), 539–547.

Syková, E., Nicholson, C., 2008. Diffusion in brain extracellular space. Physiol. Rev. https://doi.org/10.1152/physrev.00027.2007.

Tax, C.M.W., Bastiani, M., Veraart, J., Garyfallidis, E., Okan Irfanoglu, M., 2022. What's new and what's next in diffusion MRI preprocessing. NeuroImage 249, 118830.

Théberge, A., Desrosiers, C., Descoteaux, M., Jodoin, P.-M., 2021. Track-to-learn: a general framework for tractography with deep reinforcement learning. Med. Image Anal. 72, 102093.

Tofts, P.S., Lloyd, D., Clark, C.A., Barker, G.J., Parker, G.J., McConville, P., Baldock, C., Pope, J.M., 2000. Test liquids for quantitative MRI measurements of self-diffusion coefficient in vivo. Magn. Reson. Med. 43 (3), 368–374.

Tournier, J.-D., Calamante, F., King, M.D., Gadian, D.G., Connelly, A., 2002. Limitations and requirements of diffusion tensor fiber tracking: an assessment using simulations. Magn. Reson. Med. 47 (4), 701–708.

Tournier, J.-D., Yeh, C.-H., Calamante, F., Cho, K.-H., Connelly, A., Lin, C.-P., 2008. Resolving crossing fibres using constrained spherical deconvolution: validation using diffusion-weighted imaging phantom data. NeuroImage 42 (2), 617–625.

Tournier, J.-D., Calamante, F., Connelly, A., 2012. MRtrix: diffusion tractography in crossing fiber regions. Int. J. Imaging Syst. Technol. https://doi.org/10.1002/ima.22005.

Vellmer, S., Edelhoff, D., Suter, D., Maximov, I.I., 2017. Anisotropic diffusion phantoms based on microcapillaries. J. Magn. Reson. 279, 1–10.

Veraart, J., Nunes, D., Rudrapatna, U., Fieremans, E., Jones, D.K., Novikov, D.S., Shemesh, N., 2020. Noninvasive quantification of axon radii using diffusion MRI. eLife 9. https://doi.org/10.7554/eLife.49855.

von dem Hagen, E.A.H., Henkelman, R.M., 2002. Orientational diffusion reflects fiber structure within a voxel. Magn. Reson. Med. https://doi.org/10.1002/mrm.10250.

Wagner, F., Laun, F.B., Kuder, T.A., Mlynarska, A., Maier, F., Faust, J., Demberg, K., et al., 2017. Temperature and concentration calibration of aqueous polyvinylpyrrolidone (PVP) solutions for isotropic diffusion MRI phantoms. PLoS One 12 (6), e0179276.

Wang, X., Reeder, S.B., Hernando, D., 2017. An acetone-based phantom for quantitative diffusion MRI. J. Magn. Reson. Imaging 46 (6), 1683–1692.

Wapler, M.C., Leupold, J., Dragonu, I., von Elverfeld, D., Zaitsev, M., Wallrabe, U., 2014. Magnetic properties of materials for MR engineering, micro-MR and beyond. J. Magn. Reson. 242, 233–242.

Wassermann, D., Makris, N., Rathi, Y., Shenton, M., Kikinis, R., Kubicki, M., Westin, C.-F., 2016. The white matter query language: a novel approach for describing human white matter anatomy. Brain Struct. Funct. 221 (9), 4705–4721.

Wasserthal, J., Neher, P., Maier-Hein, K.H., 2018. TractSeg—fast and accurate white matter tract segmentation. NeuroImage 183, 239–253.

Wasserthal, J., Neher, P.F., Hirjak, D., Maier-Hein, K.H., 2019. Combined tract segmentation and orientation mapping for bundle-specific tractography. Med. Image Anal. https://doi.org/10.1016/j.media.2019.101559.

Whitton, G., Gmeiner, T., Harris, C.T., Kerins, F., 2016. A novel diffusion tensor imaging phantom that simulates complex neuro-architecture for potential validation of DTI processing. Proc. Int. Soc. Magn. Reson. Med., 3061.

Wu, X., Xie, M., Zhou, J., Anderson, A.W., Gore, J.C., Ding, Z., 2012. Globally optimized fiber tracking and hierarchical clustering—a unified framework. Magn. Reson. Imaging. https://doi.org/10.1016/j.mri.2011.12.017.

Yanasak, N., Allison, J., 2006. SU-FF-I-58: demonstration of the use of a capillary phantom to monitor DTI image processing: dyadic sorting of tensor eigenvalues. Med. Phys. 33 (6Part3), 2010.

Yanasak, N.E., Allison, J.D., Hu, T.C.C., Zhao, Q., 2009. The use of novel gradient directions with DTI to synthesize data with complicated diffusion behavior. Med. Phys. 36 (5), 1875–1885.

Yeh, CH, Schmitt, B, Le Bihan, D, Li-Schlittgen, JR, Lin, CP, Poupon, C, 2013. Diffusion microscopist simulator: a general Monte Carlo simulation system for diffusion magnetic resonance imaging. PLoS One 8 (10), e76626. https://doi.org/10.1371/journal.pone.0076626.

Zhang, H., Hubbard, P.L., Parker, G.J.M., Alexander, D.C., 2011. Axon diameter mapping in the presence of orientation dispersion with diffusion MRI. NeuroImage. https://doi.org/10.1016/j.neuroimage.2011.01.084.

Zhang, H., Schneider, T., Wheeler-Kingshott, C.A., Alexander, D.C., 2012. NODDI: practical in vivo neurite orientation dispersion and density imaging of the human brain. NeuroImage. https://doi.org/10.1016/j.neuroimage.2012.03.072.

Zhou, F.-L., Hubbard, P.L., Eichhorn, S.J., Parker, G.J.M., 2011. Jet deposition in near-field electrospinning of patterned polycaprolactone and sugar-polycaprolactone core–shell fibres. Polymer. https://doi.org/10.1016/j.polymer.2011.06.002.

Zhou, F.-L., Li, Z., Gough, J.E., Cristinacce, P.L.H., Parker, G.J.M., 2018. Axon mimicking hydrophilic hollow polycaprolactone microfibres for diffusion magnetic resonance imaging. Mater. Des. 137, 394–403.

Zhou, F., Grussu, F., Li, Z., Parker, G., 2020. Co-electrospun spinal cord phantom for diffusion MRI. Proc. Int. Soc. Magn. Reson. Med., 4507.

Chapter 26

# Tractography validation Part 2: The use of anatomical model systems and measures for validation

Tim B. Dyrby[a,b], Silvio Sarubbo[h], Francois Rheault[d], Els Fieremans[c], Adam W. Anderson[e], Marco Palombo[f,g], Peter Neher[i], Kathleen S. Rockland[k], and Kurt G. Schilling[j]

[a]Danish Research Centre for Magnetic Resonance, Centre for Functional and Diagnostic Imaging and Research, Copenhagen University Hospital—Amager and Hvidovre, Copenhagen, Denmark, [b]Department of Applied Mathematics and Computer Science, Technical University of Denmark (DTU), Kongens Lyngby, Denmark, [c]Bernard and Irene Schwartz Center for Biomedical Imaging, Department of Radiology, New York University Grossman School of Medicine, New York, NY, United States, [d]Department of Computer Science, University of Sherbrooke, Sherbrooke, QC, Canada, [e]Biomedical Engineering, Vanderbilt University, Nashville, TN, United States, [f]Cardiff University Brain Research Imaging Centre (CUBRIC), School of Psychology, Cardiff University, Cardiff, United Kingdom, [g]School of Computer Science and Informatics, Cardiff University, Cardiff, United Kingdom, [h]Department of Neurosurgery, "S. Chiara" Hospital, Trento, Italy, [i]German Cancer Research Center (DKFZ) Heidelberg, Division of Medical Image Computing, Heidelberg, Germany, [j]Department of Radiology and Radiological Sciences, Vanderbilt University Medical Center, Nashville, TN, United States, [k]Department of Anatomy & Neurobiology, Boston University, Boston, MA, United States

## 1 Anatomical model systems

There are a great number of advantages to validation studies when tractography is applied to a real anatomical system and acquired in a "real" MRI scanner. Here the only assumptions applied are those incorporated into the tractography models, meaning that if results deviate from histology, then we need to find an explanation for why. As an anatomical model system, we can use different animal species, which allows us to use independent histological techniques such as tracers or staining techniques as the ground truth. A wide range of validation techniques exist. Here we cover a representative spectrum of the techniques to give some practical insight into how they work and what to be aware of. Indeed, these validation methods also have their limitations to be considered when used as ground truth, especially to align the expectations of what tractography methods reflect and what anatomy actually looks like.

### 1.1 Species and validation considerations

#### 1.1.1 Species differences

Animal brains are an important resource that allows for some measure of direct histological-imaging validation, as well as invasive studies. However, it is important to consider interspecies differences when choosing anatomical features to be validated (Fig. 1) and how the conclusions can be translated to the human brain. For example, species that share closer homologies to humans may be necessary for the translation and validation of complex brain networks, whereas more simple murine models may be sufficient to validate tissue microstructure. Species differences include not only differences in brain size, but also tissue constituents, the complexity of connections, and folding patterns.

Fundamental species differences occur in **tissue volume fractions**. For example, cortical cell density is greater in mice than in rats, and greater in rats than in humans (Defelipe, 2011), and there is increased morphological complexity of glia in the human brain (Oberheim et al., 2006). Obviously, from a volumetric point of view, there is relatively more white matter (WM) in humans and nonhuman primates (NHPs), a lesser degree in hoofed animals, and little in rodents. The organization of WM may also vary. In animals with less WM, a proportion of the fibers may travel in the deeper cortical layers, without forming distinct bundles in the WM per se. For WM tracts, therefore, across species, the source area, the target area, and the

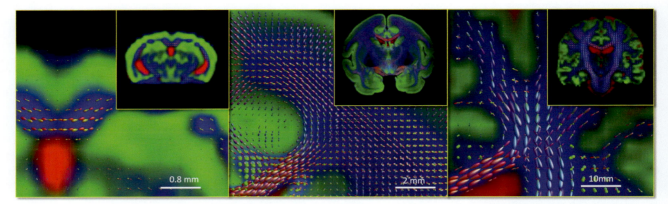

**FIG. 1** Species differences may occur in brain size, tissue volume fractions, gyrification, and existence and trajectory of fiber pathways, and they must be considered when selecting validation models of tractography. From left to right is shown a mouse, squirrel monkey, and human brain, colored by tissue components estimated using multishell multitissue spherical deconvolution (Jeurissen et al., 2014). Cerebrospinal fluid (CSF) *(red)*, GM *(green)*, and WM *(blue)*. (Mouse data was kindly provided by Ben Jeurissen (Jillings, S., Morez, J., Vidas-Guscic, N., Van Audekerke, J., Wuyts, F., Verhoye, M., Sijbers, J., Jeurissen, B., 2020. Multi-tissue constrained spherical deconvolution in a murine brain. In: International Society for Magnetic Resonance in Medicine, 0900. https://archive.ismrm.org/2020/0900.html).)

extended interconnected cortico-cortical network are likely to have significant structural and functional differences (Assaf et al., 2020). While major WM tracts are recognizable across species—corpus callosum, anterior commissure, corticospinal tract, fornix, cingulum—the structure and, even more so, the function will be highly variable. Biological properties of the constituent axons are also likely to differ across species, including axon diameter distributions, degree of myelination, organization of nodes of Ranvier, etc., all of which impact conduction velocity, among other parameters (Drakesmith et al., 2019).

**Gyrencephalic and lissencephalic** brains are obviously different, and the larger gyrencephalic animal brains are likely to provide a closer model for human brain organization. Macaque or vervet monkeys have traditionally been the gyrencephalic animal model of choice, but other, more readily available, species with gyrencephalic brains are attracting attention: for example, the minipig (Lind et al., 2007; Bech et al., 2020). An interesting proposal is that gyrencephaly is actually the ancestral condition (Llinares-Benadero and Borrell, 2019).

In summary, it is critical to consider which species to choose for validation studies. If we want a model that faces the same tractography challenges as the human connectome, it is necessary to choose a brain with similar complexity and gyrification, for example, NHPs or hoofed brains. Similarly, if studying the location and trajectory of specific pathways, NHP is the best match for humans, whereas other species (hoofed) may have similar pathways, but functional areas can be shifted in location. For understanding challenges related to fiber orientation estimation, NHP and hoof brains may be more appropriate, as their brains contain multiple overlapping pathways leading to complex fiber configurations (i.e., crossing, kissing, branching pathways) as expected in the human brain. For microstructure modeling, most brains will contain relevant tissue constituents, albeit with differences in axon diameters, and myelination appropriate for tractography validation studies.

### 1.1.2 Ethics and the 3 Rs

In animal research, ethics are constantly reviewed to enhance animal welfare when used in experiments. The ethical requirements are centralized around the 3 Rs: Replacement, Reduction, and Refinement. As described earlier, while the NHP (and especially the old-world monkeys) are often seen as the "gold standard" in tractography validation frameworks, it is important to justify their use and reduce the number of animals required. Other species may be appropriate depending on the feature to be validated. Comparing tractography within the same brain with preinjected tracers is defined as *direct validation* whereas *indirect validation* is when diffusion (d)MRI and histology are not applied to the same brain. Naturally, direct validation is the strongest type of validation since anatomical concerns about locational variation between applied techniques can be ignored.

In relation to animal ethics and especially the 3 Rs, one can see that animal models are required to validate tractography in real, complex anatomy, but we can strongly argue the need to reduce the number of animals sacrificed: we can in diffusion MRI replace and refine validation by using synthetic data simulations and phantoms to get basic insights into tractography methods before including animals. We can reduce the number of animals by using fixed brain tissue for ex vivo dMRI imaging. A distinct advantage of ex vivo animal models is that if tissue is well preserved and well stored, it can be rescanned over many years (Dyrby et al., 2011).

### 1.1.3 Preclinical ex vivo dMRI

The use of stronger preclinical MRI scanners in combination with smaller or cryo-coil radiofrequency (RF) coils generally allows us to obtain much higher image resolutions for animals as compared to clinical scanners. Since tractography is a measure of brain structure, we can further benefit from collecting dMRI data ex vivo over in vivo. With ex vivo dMRI, one can gain even higher image quality and higher image resolution, given the higher signal-to-noise ratio. The latter allows us to explore the performance of tractography methods at different levels of partial volume effects to provide better insights into the actual performance of tractography. However, since our target brain for tractography validation is the human brain, which is many times larger than the laboratory animal brains, the image resolution should be made relative to the brain size or WM volume. Hence, the use of a higher image resolution is no guarantee of providing the needed insight into the methodological challenges of a method if the brain volume ratio or relative brain size differences in anatomical structures are not discussed between human and animal (Dyrby et al., 2007, 2014; Ambrosen et al., 2020). For example, very high image resolution dMRI is needed to obtain insights into the relative performance of tractography in humans when performed in mice, although mice can offer a unique 3D reconstructed tracer material (Oh et al., 2014).

Practicalities for preclinical in vivo and ex vivo dMRI experiments will not be covered in this chapter and we refer the reader to Schilling et al. (2023) and Jelescu et al. (2024). However, a few basic principles for ensuring the collections of ex vivo MRI that can be reliably used for validation are provided. A detailed description of high-quality ex vivo imaging and its four steps (Fixation, Storage, Preparation, and MRI scanning) are described in Dyrby et al. (2011) and in a review by Dyrby et al. (2018) (Fig. 2). First of all, ensuring that the microstructure environment is close to the in vivo state is crucial to performing reliable tractography validations. The postmortem interval must be as close to zero as possible, meaning that the tissue should not have time to "discover" its death to avoid autolytic processes (degeneration of tissue) (D'Arceuil and de Crespigny, 2007; Hukkanen and Röyttä, 1987). Therefore it is preferred that tissue is perfusion fixed where the fixative is flushed through the vascular system transcardially, while the animal is under deep anesthesia (Cahill et al., 2012). The penetration of the fixative into the tissue relies on the diffusion process and takes roughly about 1 h/mm. Therefore immersion fixation where tissue is soaked into the fixative should only be done on small samples, so that tissue becomes fixed before it starts degeneration. After perfusion fixation, postfixation takes additional time for the tissue to be well fixed and ready for dMRI ($>$1 week depending on sample size). A trick to check the quality of fixed tissue is to collect a dMRI dataset for diffusion tensor imaging (DTI) (one shell with a $b$-value of about 3000 s/mm$^2$) and generate an fractional anisotropy (FA) map. The FA values should be the same as, or similar to, those of in vivo tissue, because tissue anisotropy is preserved. If FA values are clearly lower and/or appear noisy, then there is a risk that the tissue is not well preserved and should not be used for validation. When performing a comparison between in vivo and ex vivo, be aware of differences in partial volume effects at boundaries (e.g., corpus callosum or gray matter (GM) structures) because differences in image resolution can introduce FA differences. Such differences in FA should not be intermingled with tissue quality. Note that the diffusivity ex vivo is about three to four times lower than in vivo, so correspondingly higher $b$-values are needed for ex vivo dMRI (Sun et al., 2005). Since preclinical MRI scanners have much stronger gradient hardware than clinical, the use of higher $b$-values is not a problem.

Ex vivo imaging (as well as phantom scanning) allows for long scan sessions over days and even weeks, but the scanning time is not infinite because the magnetic field drifts over time (B0-drift). A magnetic drift will be seen as a translation of the object in the image. How much magnetic drift occurs depends on the magnet quality, and the manufacturer can provide the drift information. The number of hours it takes for one voxel drift can be calculated when knowing the B0-drift (Hz/h), the bandwidth (BW (Hz)), and matrix size in the read-out direction (R0): hours = BW/(R0*B0-drift). A trick to minimize the impact of B0-drift is to split up a long dMRI protocol into smaller blocks of 8–15 h and perform a frequency adjustment at the beginning of each block. Validation is always to ensure good data quality, and critical visual inspection is essential.

## 1.2 Anatomical model systems: Microdissection

The investigation of the structural connectivity of the human brain has a long history. Starting from the early observations of the subcortical WM carried out throughout the 17th century, progressive technical improvements furthered understanding

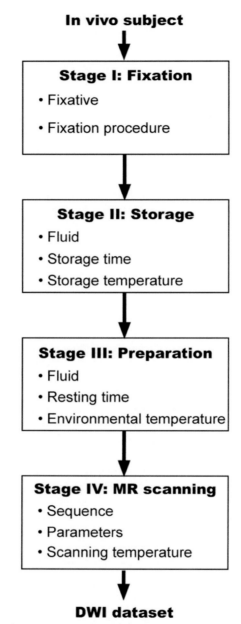

**FIG. 2** Flow chart of the four stages going from in vivo to the collection of high-quality ex vivo diffusion MRI datasets on postmortem brains. Following the four steps will ensure minimizing the risk of introducing autolytic effects degrading tissue quality (stages I and II) and preparing and collecting ex vivo dMRI (stages III and IV). *(From figure 1 in Dyrby, T.B., Baaré, W.F.C., Alexander, D.C., Jelsing, J., Garde, E. and Søgaard, L.V., 2011. An ex vivo imaging pipeline for producing high-quality and high-resolution diffusion-weighted imaging datasets. Hum. Brain Mapp. 32, 544–563.)*

of the microscopic distribution patterns of brain connections. These first efforts led to the delineation of the general organization of the human brain's WM, with the classification of the main fiber pathways into commissural, projection, and association based on their course and topography (Schmahmann et al., 2007).

### 1.2.1 Freezing/thawing enables the separation of axonal bundles

In 1935, the postmortem exploration of the WM anatomy was reshaped by the introduction of a new technique for specimen preparation based on repeated sessions of freezing/defrosting of previously formalin-fixed brains (Klingler, 1935). Joseph Klingler's (1888–1966) revolution exploited a basic property of ice crystals, which, while expanding, separate WM fibers

Tractography validation: The use of anatomical model systems and measures for validation **Chapter | 26 515**

through the disruption of glial cells within the extracellular matrix. Still, myelin sheaths are preserved by this process and the dissection of the fascicles is therefore facilitated (Zemmoura et al., 2016). The development of this method marked the beginning of a new era, characterized by more accurate and systematic investigations of WM bundles' structural features and their reciprocal relationships. Klingler's dissection technique represents an important reference tool for the validation of in vivo WM pathways reconstruction from tractography (Lawes et al., 2008). Klingler's method envisages the fixation of each cerebral hemisphere in a 10% formalin solution for 40 days, followed by a freezing process for 40 days at $-20°C$, or lower. After gradual defrosting of the specimens, the blood vessels, the arachnoid mater, and the pia mater are removed, and the hemispheres are refrozen at $-15°C$ for at least 15 days (De Benedictis et al., 2014; Sarubbo et al., 2013, 2015).

### 1.2.2 The microdissection procedure

A typical microdissection procedure is illustrated in Fig. 3. After carefully evaluating the anatomy of the cortical surface, the standard dissection workflow starts from the lateral convexity. The ideal tools for this procedure are wooden spatulas with different dimensions and consistencies, and surgical microscopes or loops glasses. The first step of dissection usually consists in the removal of the cortical layer. Still, since this procedure may lead to difficulties in characterizing the cortical distribution of terminal fibers, a "cortex-sparing" approach can be adopted, with the preservation of the gyral gray matter during the decortication phase (Martino et al., 2011). A sequential "layer-by-layer" exposition of the associative and projection pathways is then classically performed, allowing progressive isolation of the "U-shaped" intergyral fibers, the two adjacent components of the superior longitudinal fascicle (SLF II, SLF III), and the arcuate fascicle, the vertical occipital fascicle, the inferior frontooccipital fascicle (IFOF), the inferior longitudinal fascicle, the optic radiation, and eventually the corona radiata (De Benedictis et al., 2012, 2014; Sarubbo et al., 2015). To better understand the relationships between association fibers and projection tracts, the basal ganglia region can also be evaluated (De Benedictis et al., 2014). Considering the fragility of WM fibers, the dissection requires the highest level of accuracy to maximally preserve the integrity of each fascicle all along its course (Sarubbo et al., 2013, 2015).

The standard lateral-to-medial approach ensures a proper dissection of the main association pathways of the brain, but this procedure can be modified depending on the specific WM structures that need to be investigated. For instance, if the analysis is focused on interhemispheric connectivity, the best approach for the isolation of the corpus callosum consists of starting the dissection from the vertex and then proceeding following a vertical course (De Benedictis et al., 2016). Instead, when the goal of the observation is to investigate the WM organization of a given region, for example, according to a specific surgical perspective, the dissection can be exempt from performing a systematic representation of the entire course of each pathway, and rather focus on the reciprocal topographical relationships among bundles (Dziedzic et al., 2021; De Benedictis et al., 2012, 2014).

### 1.2.3 3D microdissection for tractography validation

Given the current relevance of in vivo tractography, there is an increasing necessity for an accurate definition of the effective organization of WM fiber pathways represented within an image volume, i.e., in 3D. Today, microdissection uses photographs taken at different steps during the dissection procedure. This is to document the trajectory of the dissected axonal bundles, categorize the cortical territories of terminations, and collect information about the anatomical relationships with other, cortical and subcortical, structures encountered during the progressive dissecting-exploration of each structure. Typically, the photographs are key to documenting the spatial and topological organization of anatomy and the agreement with methods like tractography. For example, Sarubbo et al. (2019) performed in the same monkey brain tractography and microdissection to demonstrate the topological agreement; as validation they overlaid the two images as shown in Fig. 4. However, this validation approach does not allow us to quantify the spatial agreement between microdissection and tractography, simply because the pictures do not currently allow a quantification.

This first approach allowed the validation of tractography results in terms of confirmation of pathway trajectory, terminations, and even the existence of different associative and projection WM bundles. A second possible use of classical microdissection results was developed, for instance, to validate, refine, or improve the acquisition methods of tractography for reconstructing WM structures with high-critical features (i.e., high-curving shape, high-crossing/kissing regions, etc.). Maffei et al., for example, allowed the setting of an ultrahigh $b$-value for obtaining realistic reconstructions (i.e., comparable to microdissection results) of the acoustic radiation (Maffei et al., 2018). This enabled the development of a reliable atlas of these critical projection fibers (Maffei et al., 2019).

Finally, microdissection results, and in particular the evidence regarding the stem (high-density region collecting the whole axonal fibers of a WM bundle) of certain WM fascicles, led to the development of new techniques for tracking (e.g.,

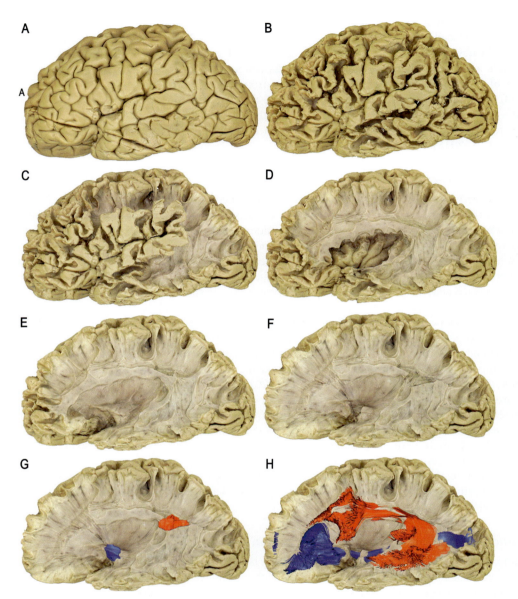

**FIG. 3** A picture collection showing a progressive lateromedial human brain convexity (A) layer-by-layer microdissection with cortex-sparing technique. Starting from the first two steps, i.e., the decortication of the gray matter (B) became spongy because of the Klingler preparation and the exposition of the shorter and more superficial U-fibers (C), up to the deeper white matter associative and projection fibers of the dorsal and ventral streams (C–F). These pictures are extracted from a photogrammetrist 3D model that was aligned to the MNI-152 space and merged with tractography atlas reconstructions, fitting with the dissection steps for both locations of the stems of arcuate fascicle *(red)* and inferior frontooccipital fascicle *(blue)* and course of the fibers of these respective bundles.

stem-based approach) (Hau et al., 2016) and validating the automatic system for segmentation of WM fibers (Bertò et al., 2021).

Nevertheless, obtaining quantitative validation results with microdissection is still an open challenge, because one has to convert a series of photographs for each dissection step into a 3D tract image volume to be compared with tractography. Currently, there exist two photogrammetry approaches that enable converting high-resolution photographs from postmortem dissection of real specimens into a 3D image volume representation. De Benedictis et al. (2018) used a set of fiducial markers placed on the surface of the postmortem brain, which allowed them to "decode" the spatial information

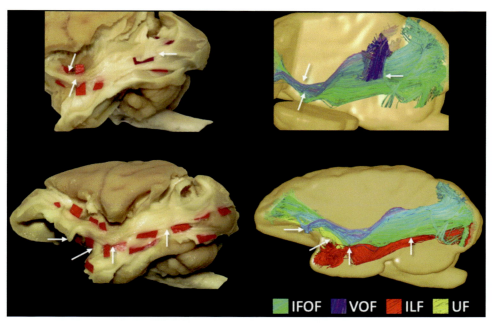

**FIG. 4** Microdissection of the different association tracts in the monkey brain and comparison with 3D rendered tractography results applied to the same brain using ex vivo diffusion MRI. Four tracts were delineated using microdissection (left) and tractography: the inferior fronto-occipital fascicle (IFOF) *(green)*, vertical occipital fascicle (VOF) *(purple)*, the inferior longitudinal fascicle (ILF) *(red)*, and the uncinate fascicle (UF) *(yellow)*. *(Modified from figure 2 in Sarubbo, S., Petit, L., De Benedictis, A., Chioffi, F., Ptito, M., Dyrby, T.B., 2019. Uncovering the inferior fronto-occipital fascicle and its topological organization in non-human primates: the missing connection for language evolution. Brain Struct. Funct. 224 (4), 1553–1567.)*

of multilayered microdissection photographs to be overlayed into a 3D volume representation. Preliminary results demonstrated that this method might have many possible applications, including interactive visualization of WM tracts, quantitative data analysis (volumes, linear measures, etc.), and integration between data coming from multilayer microdissection and tractography (De Benedictis et al., 2018). Zemmoura et al. (2014) also used fiducial markers and combined different techniques to map the surfaces of multilayer microdissection into an ex vivo magnetic resonance (MR) image volume. First, an ex vivo MRI with fiducial markers was collected. Then a laser scanner was used to scan the surfaces of each microdissection step, and these were reconstructed into a 3D volume and mapped into the MRI scan. They demonstrated how different pathways of the same brain could be dissected and mapped into the MRI voxel space (Zemmoura et al., 2014).

### 1.2.4 Considerations

The main limitations of the classical microdissection are: (1) typically, only visual comparisons with tractography are possible; (2) risk of false negatives as a consequence of the progressive layer-by-layer dissection of the WM; (3) the learning curve for WM anatomy for reliable dissection and the time-consuming aspect; and (4) the lack of digital comparison and possible quantifications with other image modalities. Therefore integrating the application of photogrammetry microdissection as shown in Fig. 4 into a common MRI-based radiological atlas space is a way to overcome some of the limitations. Further, it will allow a kind of validation of tractography on the group level (Bürgel et al., 2006). Interestingly, Bürgel et al. (2006) dissected a number of well-known tracts from a group of human brains. Mapped into a common atlas space, they could quantify tract-specific anatomical variations along the tracts. Such information provides new insights into how tracts can be expected to vary along their trajectory at the group level, as well as what can be expected by tractography at the group level. On the other hand, there is a risk of false negatives from microdissection appearing at the interface between tracts with complex shapes, particularly in high-density and crossing areas (see for instance David et al., 2019), such as in the WM underneath the central region where high-density projection fibers and associative ones are strongly intermingled.

**FIG. 5** An injection site of the anterograde tracer biotinylated dextran amine (BDA) in macaque area V4 (**). Local intrinsic connections are visible as a less dense "haze." Labeled axon fragments exiting in the white matter are seen at progressively higher magnifications of a single coronal section (50 μm), as denoted by *white and black arrows*.

### 1.3 Anatomical model systems: Neuronal tracers

Anatomical results in NHPs and other laboratory animals are commonly taken as gold-standard validation for tractography in the human brain. For connections, the preponderance of data is from injections of anterograde or retrograde tracers made in the cortical or subcortical gray matter (Fig. 5). These reveal brain-wide distributed neurons sending terminations to the injected tissue (retrograde tracers), or brain-wide distributed axons and terminations originating from neurons in the injected tissue (anterograde tracers). These primarily address the gray matter to gray matter connectivity. Only a small minority of studies, such as those with intracellular fills or serially reconstructed axons, directly pertain to WM organization; and all these experiments need careful interpretation—all the more, since the fuller complexity of what is meant by "connections" is not straightforward. Do we mean area-to-area connections? Neuron-to-neuron? What degree of reciprocity? What about connectional weights? In this section, we begin with an overview of what tracer injections can say about connections. For additional background, see Rockland (2015), Dyrby et al. (2018) and Grier et al. (2020).

A network analysis basically combines the validation of more brain connections between cortical regions individually delineated with tracer injections, as described in the previous section. An early systematization of cortical connections in the macaque has depicted these as a source-target connectivity matrix, "CoCoMac" being an early example (Bezgin et al., 2012), followed by the quantitative weights database of Markov-Kennedy (Markov et al., 2014), and a more recent database and simulation platform, "The Virtual Brain" (Shen et al., 2019). Tracer connectivity matrices focus on quantifying the number of labeled neurons at target gray matter sites in relation to the injection site. What the connectivity matrix reflects in relation to tractography (streamline count) is often an extrapolation and requires careful interpretation. As the trajectory of axonal bundles is not apparent from retrograde tracers, validation by network analysis is limited. More recent anatomical tools emphasize high-resolution in vivo MRI templates, supported by improved analysis pipelines (e.g., Seidlitz et al., 2018); and further developments, especially utilizing three-dimensional visualizations, are rapidly becoming available (Feng et al., 2017; Saleem et al., 2021).

In this section, we address what tracer studies actually reflect in terms of network analysis and bundle organization, and end by coming back to the broader issue of what are connections. We aim to shed some light on how connectivity "weights," used to compare with tractography, can be interpreted. Bear in mind that, as opposed to tractography, neuronal tracers are directionally dependent.

#### 1.3.1 What do neuronal tracers give us?

**Connectional weights and "numbers"**

The tractography connectivity matrix is generated by counting the number of streamlines that reach different target regions from a seed region. Depending on the output format of the tracer connectivity matrix, the streamline count is thresholded and often binarized. If not a binarized connectivity matrix as in Markov et al. (2014), several studies show a correlation between streamline count and labeled neurons. But how can a streamline count that has no biological meaning except to map the most probable axonal trajectory based on a tissue anisotropy measure have any relation to the strength of connectivity? What does the weighting in a network analysis mean?

An explanation can be that different tracts have different cross-sectional areas, meaning that more streamlines are likely to follow those with larger areas, and these might be the ones that contain more neurons. However, it is the axon morphology and myelination that are the main factors influencing conduction velocity along brain connections (Rockland and DeFelipe, 2016; Innocenti et al., 2019). How do we interpret connectional weights in network analysis, when not estimating conduction velocity?

For the tracers, a common practice is to equate the efficacy of connections with connectional "weights," either as a number of retrogradely labeled neurons or the number and size of synapses. While this is to some extent reasonable, numbers alone have to be regarded with some skepticism in this context. One conspicuous example is the corticothalamic projections, to the pulvinar and mediodorsal thalamus, from the associational cortex. These arise from neurons in layers 5 and 6. The neurons in layer 6 outnumber those in layer 5, but owing to a less-strategic distal postsynaptic location—among a few other factors—they are considered to be less efficacious ("modulatory vs driving"; table 7-1 in Sherman and Guillery, 2006; but see Bickford, 2015). Another issue is that populations of retrogradely labeled neurons are not homogeneous, structurally or, presumably, functionally. The inference is that there is differential "efficacy" across the 10s or 100s of neurons within an injected site.

Similarly, the number alone of geniculocortical synapses, as often remarked, is a small proportion of the total synapses in layer 4, even though they are efficacious in "driving" receptive field properties (Rockland, 2015). Thus a more complete evaluation of synaptic "weight" requires data on the numbers of presynaptic neurons, the numbers and location of synapses, and convergence with other projections.

From the perspective of anterograde labeling, the determination of "connectional weight" by discrete counts is more difficult, and to some extent arbitrary. What do we count? The number of synapses from one presynaptic neuron to a given postsynaptic neuron? The total number of synapses in a projection zone, resulting from a discrete injection site? Labeled axons, even though these branch extensively at their distal portions in the gray matter? At the distal termination in gray matter, axons typically have multiple terminal arbors that are spatially dispersed over a volume of several millimeters. These are not stereotyped, but usually vary in size and number of synapses, and sometimes in laminar targets (if cortical). In other words, one axon, originating from one neuron, can differentially influence postsynaptic neuron assemblies differing in number and spatial location of neurons, even though in the same target area (leading to the suggestion of "principal" and "secondary" components; review in Rockland, 2020).

## Axonal branching/morphology

The not-uncommon occurrence of axon branching (Rockland, 2013) suggests possibly coherent but differential influence in the several target structures. Branching axons are called collateral and have been shown in both mice (Rodriguez-Moreno et al., 2020; Han et al., 2018) and NHPs (Rockland, 2013) (Fig. 6). Evidence suggests that the collateral axons do not necessarily have the same morphology and targets as the main axon from which the branch comes. Collateral axons may differ in diameter, degree of myelination, and synapses (Zhong and Rockland, 2003; Rodriguez-Moreno et al., 2020), suggesting that the morphological features are dependent on or influenced by the target area as shown by Innocenti et al. (2014) for the main axons. However, when characterizing axon morphology or the trajectory of axonal bundles, we currently do not differentiate or assign morphology to collateral axons vs main axons. Much is unknown about axonal branching, including how commonly branching occurs, as branching applies to only a subpopulation of all axons. It is also unknown where branching typically occurs, but it has been documented in thalamocortical (TC) axons (Rodriguez-Moreno et al., 2020), and axons of the parietal and temporal association cortices (Rockland, 2013). Branching occurs in both the superficial and deep WM (Rockland, 2002). Improvements in histologically identifying and characterizing branching are needed to better understand this dynamic anatomical phenomenon and its biological relevance.

## Topography

All tracts have topological organization, and in fact this is a phylogenetically conserved principle (Suryanarayana et al., 2022). Some degree of topographic organization within bundles is a general rule. This can be demonstrated by comparing multiple animals with the results of single injections of anterograde tracer if the originating area is sufficiently large and differentiated so that injections can be spatially separated.

Despite extensive investigations, even the corpus callosum data are still incomplete. We know, from single injections of anterograde tracers in cortical areas, that there is a discernible topography of callosal projections (e.g., Caminiti et al., 2013); but, significantly, callosal axons project not only to mirror corresponding contralateral loci ("homotopic"), but also can have contralateral projections to the same regions that receive ipsilateral ("heterotopic"). The heterotopic projections

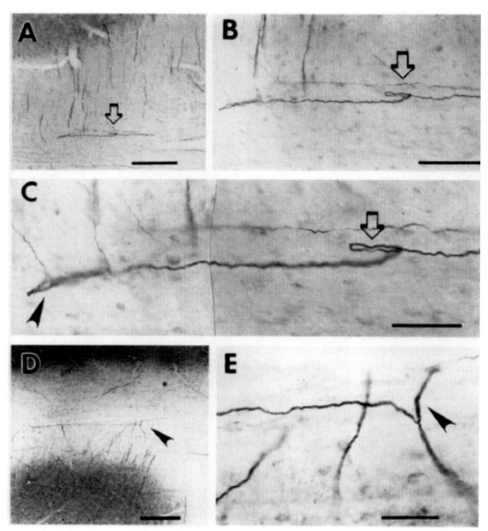

**FIG. 6** Anterogradely labeled axon segments in single histology sections (squirrel monkey, 50 μm thick) with several examples of branched axon segments. (A, B, and (higher magnification) C) One feedback axon from V2 to V1 with a "hairpin" turn and two thin perpendicular branches, also making a small "loop" (*hollow arrow* in C); (D and (higher magnification) E) feedback axon from V2 to V1, with a quasiright angle branch in the white matter below V1. Scale bars = 200 μm in A, 100 μm in B, 300 μm in D, 50 μm in C, E. *(Reproduced with permission from Rockland, K.S., Knutson, T., 2000. Feedback connections from area MT of the squirrel monkey to areas V1 and V2. J. Comp. Neurol. 425 (3), 345–368 (their figure 22).)*

are supposed to be less dense, but the distal trajectories to the heterotopic targets are poorly understood. Double tracer injections, placed in corresponding locations in the two hemispheres, allow us to more clearly decipher the organization of reciprocating ipsi- and contralateral projections and have shown a layered topology of the corpus callosum in rodents (Zhou et al., 2013).

The optic tract has been exceptionally conducive to topographic studies of tract order and target specificity. Studies in mice demonstrate that retinofugal axons lose topographic organization at the optic chiasm, but then re-sort as they enter the optic tract (Sitko and Mason, 2016). Three coexisting mechanisms have been identified: namely, ordered topography, with a dorsal-ventral switch as the axons approach the lateral geniculate nucleus; chronotopic order, where developmentally younger axons course superficially to older axons; and eye-specific order, where ipsilaterally derived axons are superficial in relation to those from the contralateral eye (Figure 13.1 in Sitko and Mason, 2016).

Other studies have used neuronal tracers in macaque monkeys to study the topography of corticobulbar projections from motor areas and in the internal capsule (Morecraft et al., 2017), and in subcortical projections from the prefrontal cortex (Lehman et al., 2011).

Tractography validation: The use of anatomical model systems and measures for validation **Chapter | 26** **521**

## Heterogeneity of axon diameters

As stated earlier, axonal bundles are composed of axons originating from multiple neuronal subtypes. While the majority of parent neurons, for corticofugal bundles, are excitatory pyramidal cells, these will differ even within the same source layer according to soma size, dendritic morphology, receptor distribution, number and distribution of local intrinsic axon collaterals, and number, distribution, and identity of inputs. Not surprisingly, tract-specific axon diameters, visualized by tracers, are also nonhomogeneous. For example, a heterogeneity of axon diameters has been reported in the corpus callosum, in the macaque corticospinal tract, and in the connections from area V1. Different diameters of axons are significant in the context of conduction velocity and are postulated to reflect functional specialization (Firmin et al., 2014). Recent results from the corticospinal tract suggest that the thinner, more slowly conducting axons are not reflected as expected in electrophysiological experiments due to recording bias (Kraskov et al., 2020).

Recent studies show that axon morphology is not like uniformly sized cylinders (Andersson et al., 2020; Lee et al., 2019). Andersson et al. (2020) show that along-axon variation comes from the local environmental structures (vessels, cell clusters, vacuoles, and crossing axons) shaping the axon morphology. This means that the diameter of a single axon cannot be captured by a single cross-sectional measurement; rather, axons have a variety of diameters along their length. Likewise, variation of the g-ratio along an axon was dominated by the axon diameter variation, while the myelin thickness was constant along the myelin segment (Andersson et al., 2020).

## Reciprocity

The "directionality" of an axonal bundle is important; many but not all connections are bidirectional in the global sense. For example, corticostriatal and corticocollicular projections are unidirectional. Cortical feedback connections are also often nonreciprocal; area V4 reliably projects back to V1, but area V1 projects to V4 only in the central visual field representation (Rockland et al., 1994).

At a finer level, it is unknown to what extent reciprocating connections terminate back on the originating neurons in the source area. Relevant to this is the overriding connectivity principle of connectional divergence and convergence in the gray matter. By divergence: one neuron has 100s or 1000s of synapses with postsynaptic neurons in one or more target structures. By convergence: one and the same neuron will receive 100s or 1000s of synapses from extrinsic sources.

## Weak (auxiliary) connections

Sparse connections are often overlooked but have been repeatedly demonstrated, both by retrograde and anterograde tracers. Examples are the direct auditory and parietal projections to area V1 in the peripheral visual field representation (Rockland and Ojima, 2003; Borra and Rockland, 2011). The WM trajectory of these "weaker" (sparse) projections has not been mapped.

## Neurochemical visualization (immunocytochemistry)

Neurochemically distinct axon populations can be visualized in postmortem tissue reacted for appropriate antibodies. Thus histological processing will reveal populations positive for serotonin, tyrosine hydroxylase, choline acetyltransferase, and (with more difficulty) gamma-aminobutyric acid (GABA), among others (see Collection 6 in MacBrain Resource Center: https://macbraingallery.yale.edu/collection6/#mbr-collection6). For example, in primates (but not rodents), the TC projections are positive for parvalbumin or calbindin (Jones, 2001), and can be easily identified (Fig. 7) in the WM (in gray matter, there is potential confound with inhibitory GABAergic neurons, which also use parvalbumin or calbindin). Since the label is global, it cannot by itself give information on the source structure; for example, parvalbumin-positive TC projections to the occipital lobe might include fibers from the pulvinar as well as the lateral geniculate nucleus.

## *1.3.2 Resources*

A number of resources in anatomical model systems have been used, or may be useful, for tractography validation. Here, we summarize three resources of interest: the connectivity matrices of (1) the CoCoMac database, (2) the Markov connectivity database, and (3) the Allen Brain Atlas, and list further resources in Table 1.

The CoCoMac database is a dedicated website that annotates and classifies connections in the macaque brain (Bakker et al., 2011). The data are based on extracted tracer information from the literature that is then mapped onto a common space atlas. Unfortunately, the tabulation of connections adopts a binary "pairwise" connectivity matrix, although many connections collateralize to multiple targets. Another limitation is that CoCoMac, like many other databases, prioritizes

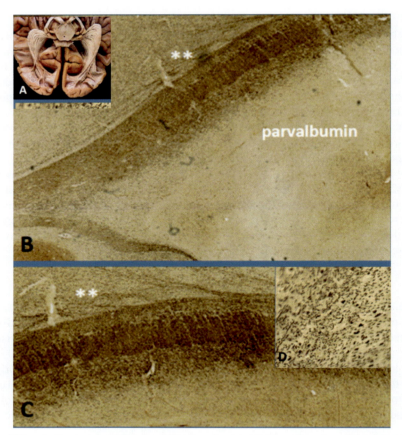

**FIG. 7** The optic radiations (TC projections) are parvalbumin-positive in primates. (A) Gross dissection of the optic radiations, ventral surface of a human brain. (B and C) (higher magnification from **) histological preparation (50 μm thick) reacted for parvalbumin. Note the nonhomogeneous, slab-like microstructure. (D) Higher magnification of the geniculocortical axons, mostly in oblique or cross section.

**TABLE 1** Resources and anatomical models for tractography validation.

| Resource | Name | Information | Location/citation |
|---|---|---|---|
| Connectivity matrix | CoCoMac Database | Binary connectivity matrix of cortico-cortical connections from tracers in macaque | http://cocomac.g-node.org/main/index.php |
| Connectivity matrix | Markov Macaque atlas | Weighted connectivity matrix from retrograde tracers in macaque | http://core-nets.org |
| Connectivity matrix | Allen Brain Atlas | Tracer connectivity matrix in mouse from >2900 injection sites | Oh et al. (2014) |
| Simulations | The Virtual Brain | Macaque connectome informed by tracers + tractography | Shen et al. (2019) |
| 3D atlas + histology | adult and infant macaque: MacBrain Resource Center | Tracers, EM, histochemistry, immunohistochemistry, and tissue blocks of macaques of various ages | https://medicine.yale.edu/neuroscience/macbrain/ |
| 3D atlas + histology | National Chimpanzee Brain Resource | Collection and distribution of chimpanzee neuroimaging data and postmortem brain tissue | https://www.chimpanzeebrain.org/ |
| 3D atlas + histology | Macaque digital atlas | Population-averaged macaque brain atlas with high-resolution ex vivo DTI integrated into in vivo space | Feng et al. (2017) |

# Tractography validation: The use of anatomical model systems and measures for validation Chapter | 26 523

**TABLE 1** Resources and anatomical models for tractography validation—cont'd

| Resource | Name | Information | Location/citation |
|---|---|---|---|
| 3D atlas +histology | Macaque digital atlas | High-resolution digital atlas of subcortical regions using mean apparent propagator (MAP)-MRI and histology | Saleem et al. (2021) |
| 3D atlas +histology | Macaque digital template | Population MRI brain template and analysis tools for the macaque | Seidlitz et al. (2018) |
| 3D atlas +histology | Human, rat, and mouse brain atlases | Human, rat, and mouse brain atlases and navigation tools | https://ebrains.eu/services/atlases/ |
| 3D atlas +histology | Squirrel monkey digital atlas+histology | Squirrel monkey template, atlas, and histology | Schilling et al. (2017b) |
| Beyond tracers | Electrical microstimulation with functional MRI | Electrical microstimulation with functional MRI in the macaque prefrontal cortex | Xu et al. (2022) |
| Beyond tracers | Consortium groups | The PRIMatE Data Exchange (PRIME-DE) Global Collaboration Workshop and Consortium | PRIMatE Data Exchange (PRIME-DE) Global Collaboration Workshop and Consortium (2020) |

Resources are divided into connectivity matrices, simulations, 3D atlases and histology, and ongoing experimental work beyond tracers, which are well suited to dense mapping connections of the brain.

cortico-cortical connections, largely to the exclusion of connections with subcortical areas. It also prioritizes ipsilateral cortical connections (http://cocomac.g-node.org/main/index.php).

Markov et al. (2014) presented a connectivity matrix for macaque (Markov et al., 2014). Unlike CoCoMac, the data are both weighted by connection strength and described by the directionality of connections (http://core-nets.org). The intra-hemispheric interareal connectivity of the macaque cerebral cortex is quantified using retrograde tracer injections in 29 regions. The weights in the graph are the extrinsic fraction of labeled neurons (FLNe) in gray matter areas projecting to the injection site. The FLNe value of an area is estimated by the number of labeled neurons in that area relative to the total number of labeled neurons, excluding the labeled neurons intrinsic to the injected area (Markov et al., 2014).

The Allen Brain Atlas provides a data portal of the mouse brain where a tracer connectivity matrix can be obtained from many injection sites (Oh et al., 2014). For each injection, the whole mouse brain has been imaged with serial two-photon microscopy, meaning that axonal trajectory information can be extracted from the dataset: https://connectivity.brain-map.org/. The atlas currently contains data for more than 2900 injection sites and projection data and is continuously growing.

## 1.3.3 Tracer injections considerations

While basic technical issues of tracer sensitivity—false positives or false negatives—are relatively easy to spot, there is a long list of other pitfalls important to take into consideration.

**(1)** Tracer injections are typically delimited ($\sim$0.5–3.0 mm in diameter for NHPs), and, except for small structures, do not cover the full extent of a given target, i.e., they are subtotal.

**(2)** Since target structures are typically anatomically and functionally nonuniform, tracer labeling may only be representative for the particular subdivision that is injected. Good examples of this are premotor cortex (e.g., Marconi et al., 2003), primary visual cortex (Rockland and Ojima, 2003; Borra and Rockland, 2011; Palmer and Rosa, 2006), and visual area V2 (Abel et al., 2000), where injected areas have significant internal topographic organization. Thus the standard terminology "area A projects to area B" should, strictly speaking, be rephrased more precisely: "An injection of x mm in y region of area A results in a projection focus of Y mm in target Z."

**(3)** Injections are not 100% standardized, and strict reproducibility can be problematic. This is due to inherently variable parameters of the injection (exact size, shape, and location of the needle or pipette tip; proximity to pulsating blood vessels; anesthetic state of the animal; metabolism and other physiological variables of the animal) and idiosyncratic tissue properties that can influence tracer uptake and transport (e.g., neuronal density, degree of myelination, extra-cellular space).

(4) Defining the injection site is not straightforward. Typically, neuronal uptake will be uneven, especially with progressive distance from the injection core or center toward the fringe or edge, what has often been called the "halo." At the fringe or halo, depending on the tracer, uptake may be strictly local, with little or no longer-distance transport, i.e., the effective injection may be smaller than what is qualitatively judged "by eye." For injections in primary sensory areas, evaluating transport to or from a well-characterized structure with sensory or motor maps of known dimensions can offer some confirmation of qualitative judgment of the injection site, but this calibration is less possible for cortical association areas.

(5) Tracer injections are still usually extracellular and relatively large (>0.5 mm in diameter in NHPs). Cellular identities and spatial arrangements are obscured within the injection. For cortical areas, the extrinsic projections originate overwhelmingly from excitatory pyramidal neurons, but these are heterogeneous by laminar position, dendritic morphology, transcriptional profile, and other criteria, and will vary in an area-specific manner. Callosally projecting neurons, for example, originate from layer 3 in early visual and other areas, but from layers 3 and 5 in most association areas (Hedreen and Yin, 1981; Kennedy et al., 1986). Visual feedback connections originate from neurons in layers 2, upper 3, lower 5, and 6 (review: Vanni et al., 2020). Transcriptomic methods, moreover, are providing new evidence for finer subdivisions within projectionally identified pyramidal cell populations (in mice: Tasic et al., 2016; Cembrowski and Spruston, 2019; humans: Berg et al., 2021).

(6) Extracellular injections of anterograde or retrograde tracers are essentially incomplete, "guillotine" experiments. Knowing which *neurons* in multiple targets project to an injection site (retrograde tracers) leaves unanswered: axon trajectory, axon diameter, number of presynaptic synapses, and identity and spatial distribution of postsynaptic neurons. Knowing the targets and density of anterogradely labeled *terminations* leaves unanswered: the identity and spatial distribution of the neurons of origin, the dendritic location on the postsynaptic neurons, and unless serial sections are processed without gaps—in histological series or in whole brain volumes—the axon trajectories will not be visualized in detail. These limitations can be circumscribed to some extent by intracellular or juxtacellular injections, which reveal the exact parent cell, the trajectory of its axon, collateralization if any, and distribution and a number of presynaptic terminations. Such whole cell visualizations are particularly effective in mapping out divergence of connections (NHP: Parent and Parent, 2006; rat: Kita and Kita, 2012), but they carry the limitation of small sample size. In the mouse brain, whole-brain reconstructions of thy-1 or enhanced green fluorescent protein (eGFP)-expressing neurons are now relatively common and can provide detailed whole-axon maps (Winnubst et al., 2019). It is now possible, in rodent, NHP, and hoofed animal models, to use cell-specific tracers to target excitatory or inhibitory-specific neurons and reconstruct and map the entire trajectory of cell-specific brain connections.

(7) From the anatomical perspective, there are several obvious pitfalls to be avoided in discussing "connectivity." One, as already said, is that connections are not "pairwise." One neuron projects divergently to hundreds or thousands of other neurons (Fig. 8). For neurons in the association cortex, there are often collateral branches to multiple spatially dispersed

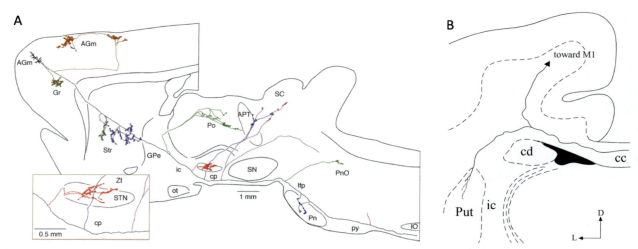

**FIG. 8** (A) Divergence. Sagittal section of the rat brain, with a filled neuron in the motor cortex. The full axon arborization includes intrinsic and extrinsic cortical projections and a large number of subcortical projections. (B) Reconstruction of an anterogradely labeled axon of a neuron in the macaque motor cortex. Note branched collaterals to the putamen and contralateral cortex (through the corpus callosum). *(From Kita, T., Kita, H., 2012. The subthalamic nucleus is one of multiple innervation sites for long-range corticofugal axons: a single-axon tracing study in the rat. J. Neurosci. Off. J. Soc. Neurosci. 32 (17), 5990–5999 (A), and Parent, M., Parent, A., 2006. Single-axon tracing study of corticostriatal projections arising from primary motor cortex in primates. J. Comp. Neurol. 496 (2), 202–213, with permission.)*

targets. Since most cortical pyramidal neurons have extensive local collaterals, even for those neurons projecting to a single target, the basic architecture is four ways: extrinsic projection, intrinsic collateral, extrinsic postsynaptic target neurons, and the local collaterals of the target neuron. Of these, the least is known anatomically about the portion of an axon ("extrinsic projections") in the WM. This elaborate machinery of anterograde and retrograde axonal transport and cytoskeletal elements (e.g., Mikhaylova et al., 2020)—whose job it is to secure the "infrastructure" of the axon for optimal conduction velocity and chemical signaling at the dendrites—is largely uncharacterized.

Second, to say that x area "has WM connections" with another area Y is arguably jargon. Synapses—a property of gray matter—have to be taken into account. "Connectivity" at a minimum involves both the WM and gray matter components. Connectivity properties specific to WM are already in evidence (as, "trans-axonal signaling" in development, but also in adults (Spead and Poulain, 2020); and the glia-axon unit (Wang and Marquardt, 2013)). Thus it may be useful to consider: (1) gray matter alone (neuron-to-neuron) synaptic connectivity, with the prominent role of synaptic convergence and divergence, (2) WM alone (for which there are still only scant data at the physiological level), and (3) GM/WM interactions as a third compartment.

## 1.4 Anatomical model systems: Validating fiber orientation

Validating the local fiber orientation is a prerequisite for accurate streamline propagation. While fiber orientation can be validated using numerical simulation or phantoms, they often fail to capture the true geometric complexity seen in living systems. To address this, validation can be performed directly in a real tissue environment. This can be done on ultra-high resolution microscopic imaging techniques to directly image the anisotropic microstructure features that diffusion is assumed to be sensitive to, i.e., axons. Since many staining methods robustly can stain for myelin, myelin is widely targeted for the validation of axonal fiber orientations. Principally, any microstructure feature such as the cell membrane of axons or the intra-axonal neurofilaments that represent the axon direction can be used.

In this section, we cover two widely used methodologies to extract fiber orientation: those based on indirect image processing-derived orientation and those that directly measure orientation (Fig. 9). Both enable a quantitative comparison of dMRI and histology by first allowing in vivo and/or ex vivo dMRI prior to the tissue processing for microscopic imaging.

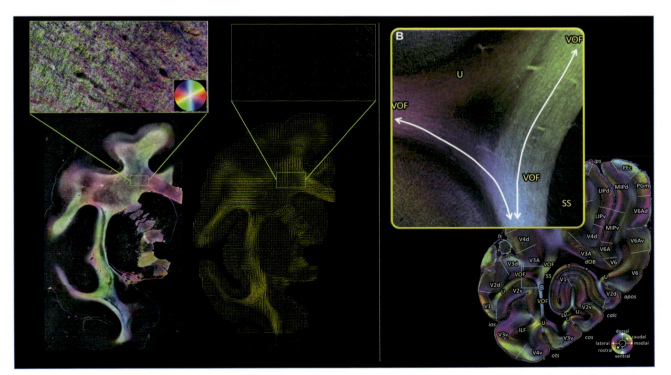

**FIG. 9** Indirect (left) and direct (right) measures of fiber orientation. Indirect measures are derived from histological samples and subsequent image processing techniques, for example, structure tensor analysis from light microscopy of myelin-stained samples (Schilling et al., 2017a). In contrast, direct measures are derived from optical imaging directly sensitive to tissue orientation, for example, polarized light imaging (Takemura et al., 2020).

### 1.4.1 Indirect measures of fiber orientation

The most prevalent method to derive fiber orientation information is from optical imaging of histological samples and subsequent image processing. Here, because optical imaging is not directly sensitive to orientation, we rely on digital image processing to extract this information; hence we refer to these as indirect measures of fiber orientation. Several digital processing methods have been used to analyze stained tissue sections including manual tracing (Leergaard et al., 2010), and Fourier-based filters (Choe et al., 2012). However, the most common method for diffusion validation experiments has been **structure tensor analysis**.

The structure tensor was introduced in the late 1980s for point and edge detection (Bigun and Granlund, 1987; Harris and Stephens, 1988), and can be applied to any imaging modality to extract the orientation of the imaged structures. The structure tensor is based on the image intensity gradients of the image, which can be calculated using any desired image filter—for example, Sobel or Prewitt gradient operators, or more commonly Gaussian derivative filters. This image gradient operation is illustrated for an example simulated cylinder in Fig. 10A, which might represent neuronal structures seen in 3D optical imaging methods. Ideally, the image gradients will be orthogonal to the fiber at all points (Fig. 10B). Next, an object known as the gradient square tensor is calculated for each point in the image by taking the dyadic product of the gradient vector with itself. At this point, we have a tensor for every pixel/voxel in the histological image. This tensor is a 3-by-3 symmetric, semipositive definite, rank-two tensor that has several parallels with the diffusion tensor. This matrix has three positive eigenvalues and can be visualized as an ellipsoid (Fig. 10C). However, unlike in DTI where one is typically interested in the largest eigenvalue and eigenvector (which points in the direction of greatest diffusivity and is parallel to the fiber orientation within the MR voxel), for structure tensor analysis one is typically interested in the smallest eigenvector and eigenvalue. This is because the image intensity gradients are strongest perpendicular to the fibers, which means that the

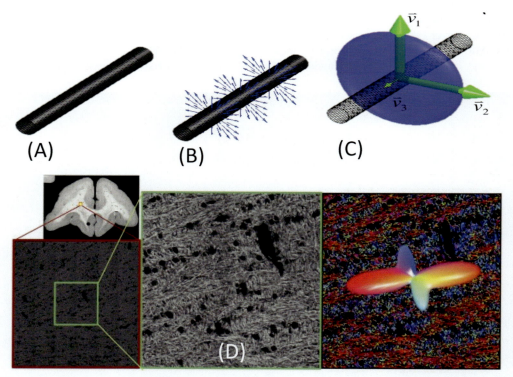

**FIG. 10** Structure tensor analysis illustrated on a simulated cylindrical fiber (A). The structure tensor is derived from the image intensity gradients (B), which point orthogonal to the fiber at all points. The Cartesian product of the gradient vector with itself is taken and averaged over a local neighborhood to derive the structure tensor, which has three eigenvalues and three eigenvectors (C). The tertiary eigenvector, v3, will point in the direction of minimum intensity variation—parallel to the fiber. Now, all fiber orientations for every voxel in the histological image can be cumulated to create the ground truth histological fiber orientation distribution (D) visualized in a 3D confocal image. *((A–C) Modified from Schilling, K., Janve, V., Gao, Y., Stepniewska, I., Landman, B.A., Anderson, A.W., 2016. Comparison of 3D orientation distribution functions measured with confocal microscopy and diffusion MRI. NeuroImage 129, 185–197; Khan, A.R., Cornea, A., Leigland, L.A., Kohama, S.G., Jespersen, S.N., Kroenke, C.D., 2015. 3D structure tensor analysis of light microscopy data for validating diffusion MRI. NeuroImage 111, 192–203. (D) Reproduced from Schilling, K.G., Janve, V., Gao, Y., Stepniewska, I., Landman, B.A., Anderson, A.W., 2018. Histological validation of diffusion MRI fiber orientation distributions and dispersion. NeuroImage 165, 200–221.)*

Tractography validation: The use of anatomical model systems and measures for validation **Chapter | 26 527**

two largest eigenvectors will also be perpendicular to these structures. Hence, the direction of minimal intensity variation is parallel to the fiber orientation in each image pixel, which is given by the tertiary eigenvector.

The next step in this process is averaging these orientations over a "patch" or neighborhood. These fiber orientation estimates for every pixel in the image can be accumulated over all pixels within a selected field of view, or neighborhood (often equivalent to an MRI voxel) to generate the "ground truth" fiber orientation distribution within that region. An example of this is illustrated in Fig. 10D.

## Decisions to be made in structure tensor analysis

Structure tensor analysis can be applied to any microscopic imaging modality to extract the orientation of the imaged structures. Validation studies have utilized a number of different types of stainings such as light microscopy (Leergaard et al., 2010; Budde and Frank, 2012), confocal microscopy (Schilling et al., 2016, 2018; Khan et al., 2015), optical coherence tomography (Wang et al., 2015), electron microscopy (EM) (Salo et al., 2018), X-ray synchrotron imaging (Kjer et al., 2024; Salo et al., 2018; Teh et al., 2017), microcomputed tomography (CT) (Walsh et al., 2021), and light-sheet fluorescence imaging of cleared brains (Leuze et al., 2021). Common for all these modalities is that images contain a structure that can be related to the fiber orientation of densely packed axons such as myelin, intra-axonal neurofilaments, or cell membranes of axons.

For validating diffusion MRI, it is important to consider the dimensionality of the histological data; 2D slices imaged with brightfield microscopy, for example, represent a projection of all fiber orientations onto a 2D plane, which makes comparison with 3D diffusion fiber orientation distribution (FOD) challenging. Additionally, care must be taken when running structure tensor analysis on 3D modalities, for example, those that acquire images with anisotropic resolution (higher in-plane than through-plane resolution) or anisotropic point spread functions, or those with limited fields of view much smaller than the MRI voxel. Several pipelines have been developed to not bias structure tensor measurements, including correcting for image intensity attenuation, image deblurring/deconvolution for anisotropic point spread functions, and interpolation to isotropic resolution (Schilling et al., 2016, 2018; Khan et al., 2015).

Finally, image processing decisions include the size of the filter to determine image intensity gradients and the size of the patch to average orientations over. When using a Gaussian derivative filter, we recommend using a filter that is approximately the size of the structures in the image you want to detect. The patch, or neighborhood size, is traditionally the size of the corresponding MRI voxels to make direct comparisons of MRI and histology. However, these can be any desired size to investigate local fiber orientation distributions and their relationship to resolution (Kjer et al., 2024; Budde and Frank, 2012; Schilling et al., 2017a).

## Direct measures of fiber orientation

The second method to measure histological fiber orientation is to utilize optical imaging that is directly sensitive to tissue orientation. To map fiber orientations optically, light imaging techniques explore changes in the transmitted light after penetrating through an unstained histology-prepared tissue slice placed on a glass plate. Similar to a diffusion experiment where molecular motion is imaged in different directions to map local fiber directions, a series of transmitted light images are collected with the light source stepwise rotated in relation to the tissue slices and so to the tissue orientation. Depending on the light imaging technique used, voxel-wise in-plan fiber direction and to a certain degree the inclination angle (out of plan) representation of the true 3D fiber orientation for single and crossing fiber can be estimated. Basically, two types of light imaging techniques exist to measure fiber orientation.

The first is **polarized light Imaging** (PLI) (Axer et al., 2011a). Here, the light source is linearly polarized light, which makes the transmitted light orientationally dependent on densely packed axons. Polarized light imaging is obtained by placing a polarizer in front of the light source, which can be stepwise rotated (0–180 degrees), and for each step, an image of the transmitted light is collected by the camera in the microscope (see for detailed experimental setup description (Axer et al., 2011a,b)). A local fiber model is fitted to the rotated series of transmitted light amplitudes and finds the 2D in-plane fiber orientation. To map the fiber orientation in 3D for validation of diffusion MRI fiber models and to perform tractography, the inclination angle, i.e., fibers pointing out-of-plan of densely packed axons, is needed too. However, 2D PLI methods were not sensitive and did not generate contrast, for out-of-plane axons. To improve the 3D estimation, Axer et al. (2011c) introduced 3D-PLI by collecting an extra series of images where the stage holder of the histological slice was slightly tilted in relation to the in-plane rotations of polarized light. Also, Wang et al. (2014) obtained a 2D polarization image contrast to fiber orientations by using an optical coherence scanner.

The image resolution of the PLI is basically determined by the in-plane image resolution of the microscope in combination with the slice thickness ranging 4–5 μm and 100–30 μm, respectively. So, despite the high in-plane image resolution, one should be aware that the transmitted light measured is weighted toward the anisotropic tissue volume and so is the local fiber orientation and orientation distribution. Although the very high image resolution crossing fiber regions still does exist at this level, the PLI shows limited contrast to separate crossing fiber regions. The crossing-fiber problem can be observed in regions at the interface between two tracts with different trajectory directions where a thin band of dark pixel values can be seen. Similarly, tract-interface effects can be observed in directionally color-coded DTI images at the interface of two tracts where low FA values are to be found due to crossing fibers that DTI cannot model.

The second light imaging method is the recently introduced **scattering light imaging** (SLI) (Menzel et al., 2020). A similar microscopic experimental setup to that for PLI is used, but without the use of polarized light. Instead, a disk mask with small holes in the circumference of the mask is placed in front of the light source and rotated in steps. The holes are placed as a circle with a constant angle and radius from the (rotation) center of the disk mask. For each rotation step, light will shine through a hole and the transmitted light measured by a microscope camera will be a scattered light. The amplitude of the scattered light will, like the polarized light, depend on the 2D in-plane orientation of the anisotropic tissue modeled. A local fiber model including the distance information is then fitted to the series of scattered signal images (for details on the experimental setup for SLI, see Menzel et al., 2021). Because the transmitted light depends on a scattering effect, its sensitivity to tissue anisotropy is not limited to densely packed myelinated axons in the same way as PLI. This means it has better sensitivity to fiber orientations in both cortex and WM, and especially to detect crossing fibers in WM, as shown in Fig. 11. So far, the SLI has low sensitivity to estimate the inclination angle of fiber that points out-of-plane. This means that SLI only provides the 2D in-plane component of a 3D fiber orientation, so one should be careful when comparing it with 3D diffusion fiber models. However, since SLI uses a similar setup as PLI, the two techniques can be combined to map crossing fibers and to have better sensitivity to the inclination angle of local fibers.

### 1.4.2 Challenges and opportunities of histology

**Registration/alignment**

One particular challenge when using histology is aligning the histology to MRI for quantitative comparisons. This is particularly important for tractography validation, where it is critical to compare voxel-wise estimates of orientation or the presence/absence of tracer and tractography. The challenge lies in the fact that very often the histology is composed of 2D slices that must be mapped into a 3D brain volume, or histology may be a small field of view (for example, EM) that is only a small fraction of a full MRI voxel.

This alignment problem can be solved in several ways. Most commonly, the use of an intermediate modality, usually referred to as a block-face image, can be used (Kim et al., 1997) (Fig. 12). Block-face images are usually digital photographs of the tissue block as it is being sectioned. These undistorted block-face images can be stacked into a 3D volume and registered to the 3D MRI directly. Because each 2D digital photograph can be mapped to a specific 2D histological slice, 2D registration can be performed to align histology to the appropriate block-face image. Now, the concatenation of these 2D-to-2D and 3D-to-3D registration fields can be used to associate any histological location to 3D MRI space. This technique has been performed in mice, monkeys, and humans (Mancini et al., 2020; Annese et al., 2006; Lebenberg et al., 2010; Choe et al., 2011) and used for several optical imaging modalities, including brightfield microscopy (Dauguet et al., 2007a, b,c; Gao et al., 2013), confocal microscopy (Schilling et al., 2016, 2018), EM (Dauguet et al., 2007a), and PLI (Palm et al., 2010), to validate both fiber orientation estimates and fiber tractography location.

Other options include 3D-printed molds as sample holders. Here, a cursory scan is performed and used to design a mold that will hold the sample for subsequent scans. Importantly, the mold can be designed with cutting guides, or slots, that will be used when the tissue is sectioned or blocked, which facilitates the identification of the histological slice within the 3D image volume (Baldi et al., 2019; Luciano et al., 2016; Bourne et al., 2017). Finally, several studies have been performed after manual alignment between MRI and histology, for example validating microstructure measures (Jespersen et al., 2010), fiber orientation (Jespersen et al., 2010; Leergaard et al., 2010; Kjer et al., 2024), or tractography (Thomas et al., 2014).

**Staining and specificity**

Notably, staining methods only allow image contrast to a specific anatomical feature. This means that nonaxonal structures such as cell bodies and their processes do not directly contribute to the fiber orientation information extracted, which can cause a bias in anisotropy compared with diffusion MRI. Similarly, the tissue preparation for staining and microscopic imaging uses a dehydration step that can introduce a shrinkage effect. A shrinkage effect is not expected to have a major impact on the fiber orientation information, but it can impact anisotropy-like metrics, because the shape of, e.g., ST eigenvalues changes.

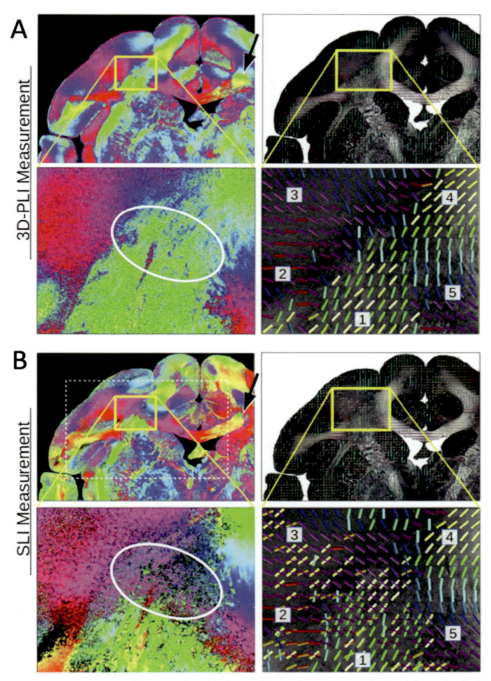

FIG. 11 Fiber reconstruction from 3D-PLI and SLI data collected in the deep WM region centrum semiovale of the monkey brain. The WM region is known to include crossing fibers. (A) 3D-PLI and (B) SLI reconstructions in the same histological slice. Clearly, 3D-PLI cannot map crossing fibers but SLI can. *(Modified from figure 8 in Menzel, M., Reuter, J.A., Gräßel, D., Huwer, M., Schlömer, P., Amunts, K., Axer, M., 2021. Scattered light imaging: resolving the substructure of nerve fiber crossings in whole brain sections with micrometer resolution. NeuroImage 233, 117952.)*

## Obtaining a large field of view in micrometer image resolution

One of the challenges with microscopic imaging techniques is their limited field of view. They provide micrometer image resolution but, similarly, the field of view is limited to the pixel matrix size of the camera, which cannot cover a whole brain in high image resolution. Moreover, most imaging modalities rely on stacking 2D images into an image volume needed to validate tractography or to validate the local fiber reconstruction models. A few techniques such as synchrotron imaging

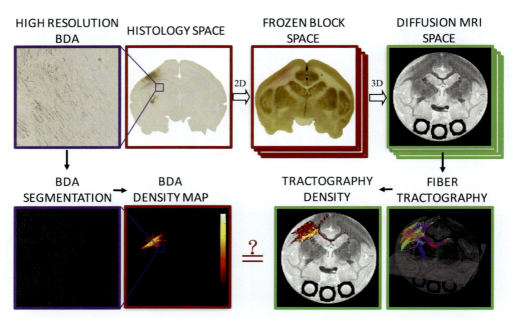

**FIG. 12** Alignment of histology and MRI using intermediate block-face images. In this example, high-resolution 2D BDA micrographs can be registered to the corresponding digital photograph of the frozen tissue block, which is registered to the 3D diffusion MRI volume. Now direct comparisons can be made between histology derived measures (for example, BDA density or orientation) and MRI-derived measures (for example, tractography). *(Figure from Schilling, K.G., Gao, Y., Stepniewska, I., Janve, V., Landman, B.A., Anderson, A.W., 2019. Anatomical accuracy of standard-practice tractography algorithms in the motor system—a histological validation in the squirrel monkey brain. Magn. Reson. Imaging 55, 7–25.)*

i.e., X-ray nanoholotomography (XNH) allow nondestructured 3D images of a sample, but EM can also produce 3D images (Dyrby et al., 2018). Nevertheless, three challenges remain to obtain a large field of view (FOV) that covers the whole brain while keeping a high image resolution. Firstly, to extend the FOV we stitch together small FOVs determined by the camera, referred to as tiles. Depending on how stable the imaging setup is, each tile often needs a certain percentage of overlap with neighboring tiles to ensure robust registration. Secondly, imaging modalities that image a 2D histological slice at the time will need to stack these into an image volume. Slicing and handling of the tissue, when placed on glass plates, risk introducing nonlinear deformations to the tissue in relation to the neighboring slices. This is a complex nonlinear image registration process, but can be solved with today's image registration tools. Lastly, creating a whole-brain image volume in micrometer image resolution generates enormous datasets to be handled. This requires big computer facilities with large storage and memory capabilities, as well as considerations on how best to handle the processing of these datasets. Axer et al. (2011c) have succeeded in realizing such a data-processing pipeline to create whole-brain image volumes for their PLI setup and presented impressive PLI tractography results on the monkey and human brains.

## 2 Empirical validations

### 2.1 What is empirical validation?

In cases where the ground truth is not known, experiments can still establish the reproducibility of tractography and identify the major contributions to variability. The empirical validation is mostly applied to in vivo humans where ground truth image modalities are rarely available. The focus of this section is to present methods for quantifying the impact of factors that limit reproducibility, leaving aside the question of bias, that is, reproducible errors in tractography. Of course, this is only a small part of the overall goal of reliable tractography: it is always possible to make very reproducible pathways that have low accuracy (Dyrby et al., 2007). Nevertheless, quantifying reproducibility has a critical role in the development and application of new methods. As more accurate tractography algorithms are developed, there will always be a need to characterize their precision (independent of accuracy) and understand the sources of test-retest variability, to optimize imaging and analysis protocols and interpret the results of experiments. *Empirical validation, then, is the act of observing the results that occur when changes are made to the tractography pipeline*, and because there is no ground truth, it is the *changes* in connectivity or variability that are of primary interest.

## 2.2 What can be assessed?

As outlined in Chapter 25, choices made at each stage of data acquisition and data processing contribute to potential sources of errors in the final tractography results: MR scanning, data processing, local fiber orientation distribution estimations, tract generation, and subsequent choices in quantification. A straightforward strategy for evaluating the relative importance of these factors is to quantify the variability of tractography that results from sampling the parameter space of possibilities for each contributing factor.

Empirical validation is typically assessed in three ways. The first method, as described earlier, involves **experimentally isolating and changing one step of the tractography process**. An example of this is scanning the same subject on the same scanner to quantify systematic variability in scanner performance, scanning the same subject on different scanners to investigate variability due to scanner hardware/software, or changing a parameter in the tractography process (starting/stopping criterion, propagation method, filtering, etc.) to quantify sensitivity to experimental choices. Here, no ground truth exists, but the changes observed offer insight into the reliability of different decisions in the tractography process. Second, it is common to acquire or create an **immensely oversampled dataset** and investigate how combinations of subsamples of the dataset, or noisy realizations of the dataset, affect outcome measures. A common example here is acquiring a high spatial resolution, high angular resolution, multiple $b$-values, and repeated acquisition scans that could potentially be used to study the effects of resolution, the number of diffusion directions, $b$-values, or signal-to-noise ratio (SNR) on desired measures. Here, again, there is no ground truth, rather, specific subsets of the data may be compared to the fully sampled dataset that might represent a best-case scenario. Third, empirical validation may involve **comparing against prior anatomical knowledge**. This may take the form of knowing the trajectory or connectivity of spatial pathways, or where in the brain features are expected to occur (for example, crossing fiber locations or low anisotropy in different tissue types). Following, we describe how empirical studies have been utilized to study local reconstruction, bundle segmentation, and connectome quantification.

## 2.3 Empirical validation: Connectomes

Empirical validation of connectomes is usually performed by observing results, change in results, and variability when varying choices are made in the connectome-generation process. Often, connectomes are quantified and summarized as a single scalar graph theoretical measure, and variation across acquisitions and methodologies may be quantified. Using a fixed analysis framework, connectomic measures have shown excellent reproducibility and intrasubject repeatability; however, nearly every step in the process of generating a connectome alters these measures, which may create large discrepancies in the literature. For a thorough review, see Qi et al. (2015). Empirical validation makes it possible to optimize methodologies to reduce variance (Qi et al., 2015), enabling studies of sensitivity to brain parcellations (Zalesky et al., 2010; Bassett et al., 2011), acquisition schemes (Vaessen et al., 2010), reconstruction (Rodrigues et al., 2013), tractography algorithm (Bastiani et al., 2012), and the number of streamlines (Yeh et al., 2016, 2018; Moyer et al., 2019). While most studies have focused on graph theory summary statistics, it is also possible to study the consistency of nodes and individual edges in the connectome, to gain insight into why variability exists and how it may be mitigated.

## 2.4 Empirical validation: Fiber bundles

Regarding bundle segmentation, empirical validation has been most often used to study where variability is introduced in the tractography and virtual dissection processes. A useful fiber segmentation protocol will result in a bundle and quantification of properties of that bundle (i.e., size, shape, or microstructural properties) that are reproducible when performed on different scanners, with different acquisitions, by different people (researcher, MRI technician, clinician), or when repeated by the same person or algorithm.

As an example, the exceptional study by Wakana and colleagues studied the precision and reproducibility of tractography measurements, which is necessary to determine whether tractography can detect differences in tracts between two groups (i.e., disease or disorder) (Wakana et al., 2007). Here they measured the reproducibility of tractography bundle segmentation when performed by the same rater (i.e., person performing segmentation) and by different raters; variability of mean diffusivity (MD), FA, T2, and tract volume were quantified, enabling power analysis for detecting potential group differences. Similar studies have focused on individual pathways (Chenot et al., 2019), quantifying variability at different levels from individual streamlines to volumetric overlap (Rheault et al., 2020), characterizing manual dissection protocol differences due to different datasets and data quality (Rheault et al., 2022), or observing differences that may result when different groups segment the same bundle using different protocols (Schilling et al., 2021a).

**532** **PART | IV** From streamlines to tracts

Several carefully designed public databases can be used for these purposes (in addition to local reconstruction). The Multiple Acquisitions for Standardization of Structural Imaging Validation and Evaluation (MASSIVE) database consists of >8000 diffusion volumes, with $b$-values up to $9000 \, \text{s/mm}^2$, on a single subject over several imaging sessions (Froeling et al., 2017). Other test-retest databases include The Penthera 1.5T and 3T datasets of 10 subjects scanned four times at 1.5T and 13 subjects scanned six times at 3T (Paquette et al., 2019); the multisite, multiscanner, and multisubject acquisitions for studying variability in diffusion weighted MRI (MASiVAR) database of 300+ sessions on 90+ subjects on different scanners (Cai et al., 2021); the Human Connectome Project (HPC) test-retest data on 46 subjects; or the computational diffusion MRI (CDMRI) harmonization database of 15 subjects on 3 scanners with varied protocols (spatial resolution, $b$-values, diffusion directions) to study effects across scanners and across protocols (Tax et al., 2019; Ning et al., 2020). These datasets have been utilized to investigate intrasession and intrascanner reproducibility (Nath et al., 2020), and have shown that bundle segmentation method are dependent on scanner effects, vendor effects, acquisition resolution, $b$-value, number of diffusion directions, and the method/protocol used to segment the pathways (Schilling et al., 2021b).

A number of surgically relevant pathways have been studied, and methodology validated, through empirical means. For many procedures, visualization, localization, and coordinates of the pathway are critical for planning, for avoiding during procedures, or as targets for stimulation. Differences in the acquisition, software packages, and decisions in the tractography process (number of streamlines, choices in seeding strategies), have been observed in small cranial nerves (Jacquesson et al., 2019), the corticospinal tract (Rheault et al., 2020), and the dentatorubrothalamic tract (Nowacki et al., 2018). Here, the ground truth is inaccessible, or hard to quantify in vivo, and decisions and validation can be based on prior anatomical knowledge. In a study on optic radiations, a pathway that is necessary to localize to avoid surgical visual field defects, Chamberland et al. (2018) utilized empirical validation by observing and measuring simple geometric features of the reconstructed pathways (distance from Meyer's loop to the temporal pole) and the variation that results from changing acquisition parameters, to make informed choices regarding spatial and angular resolution for adequate tractography of these pathways.

## 2.5 Empirical validation: Local reconstruction

Empirical validation of local reconstruction of fiber orientation techniques can take many forms. One common approach is to acquire a comprehensive and oversampled dataset, possibly with multiple diffusion weightings, diffusion times, and multiple repeats; this acquisition can be subsampled to observe the effects of choices in acquisition on experimental outcomes. A series of studies on DTI from Farrell and Landman et al. exemplifies this idea well (Farrell et al., 2007; Landman et al., 2007). By scanning a single subject over several sessions and repeating the same acquisition 45 times, they were able to investigate the effects of SNR on DTI scalar and orientation values or investigate whether an increasing directional resolution is preferable to repeating observations (while keeping scan time constant), while making comparisons to the results observed when using the high SNR 45–repeat acquisition. Their findings, including biases in FA and lower precision of eigenvectors at low SNR, and biases in nonoptimal direction schemes, well-matched other literature using alternative validation methods (i.e., Monte Carlo simulations) (Jones and Basser, 2004). Heavily oversampled and high SNR datasets have been similarly utilized to investigate and observe the effects of the number of directions, $b$-values, and combinations of acquisition schemes for reconstructing high angular resolution diffusion data.

Related to local reconstruction, multicompartment models have been similarly validated utilizing empirically acquired data and observing results of fitting to different models. A series of studies by Ferizi et al. (2014, 2015, 2017) and Panagiotaki et al. (2012) exemplifies this well. Starting in 2012 on an ex vivo rat model, they acquired data with a wide range of imaging parameters (diffusion sensitization, diffusion times, diffusion directions), and created a taxonomy of 47 possible multicompartment models (Panagiotaki et al., 2012). Using the full dataset, and combinations of subsampled data, they investigated and ranked the models based on the ability to explain the measured data, the stability of fit, and the observed compartment parameters. Extending this to human in vivo data in 2014, they similarly used the data itself to rank models, further informing the choices of modeling and acquisition in the human brain (Ferizi et al., 2014). Finally, this was done again in 2017 with the aim to compare results from $60 \, \text{mT/m}$ and $300 \, \text{mT/m}$ gradient systems (Ferizi et al., 2017), and finally results were presented as a community challenge in the same year (Ferizi et al., 2017). This strategy of ranking models based on observations has been further explored to investigate modeling in the spinal cord and optimal sampling strategies for multicompartment models.

Other notable examples of using observations to inform local reconstruction algorithms may include comparing against anatomical knowledge—for example, when introducing Q-ball and the Forecast (spherical deconvolution) models, Tuch (2004) and Anderson (2005), respectively, validated their approaches by showing expected complex, crossing fiber patterns where fiber pathways were known to cross in the brain. Similar empirical observations may be used to identify limitations of tractography, for example observing where crossing fibers occur and their prevalence at different spatial resolutions

(Schilling et al., 2017a; Jeurissen et al., 2013). Finally, observing the diffusion signal itself, and characterizing its magnitude and angular features, have been used to determine the minimal number of sampling directions for reconstruction approaches (Tournier et al., 2013).

## 2.6 Considerations in empirical validation

Although the approach just outlined is conceptually simple, it has some limitations. Importantly, it overlooks the fact that factors at different stages in the measurement can interact: for example, an aim of data preprocessing is to mitigate sources of variation in data acquisition and the same is true for fiber tracking algorithms and postprocessing. Hence, to get an accurate idea of the contributions to variability in a specific data acquisition and analysis pipeline, ideally, all measurements should be made varying only one factor at a time in that pipeline. This is impractical due to the large space of parameter choices. An alternative approach is to compare the results obtained with completely different pipelines. International challenges that provide the same data to many groups, each using their own pipeline(s), generate valuable data on variability across pipelines (Schilling et al., 2019). Finally, despite high reproducibility and consistent reconstruction, bundles or connectomes do not imply anatomical accuracy, and biases due to systematic effects and variance across methods may confound our interpretation of results (Calamante, 2019).

## 3 Measures for validation of tractography

It is necessary to quantify the agreement between tractography and our gold standard or ground truth to assign values to the accuracy and reliability of the neuroimaging methods, to compare methods, and to identify (and overcome) limitations. Because both neuroimaging (tractography) and anatomy are represented across several anatomical length scales, a number of measures have been proposed for the validation of tractography. In the next several sections, we describe commonly employed measures for the quantifying of similarity between neuroimaging and ground truth, ranging from tissue microstructure to the connectome. For each, we focus on what the measure quantifies, what features of anatomy it aims to capture, how it may be interpreted, and when it might be used.

A common theme at all spatial scales is the "confusion matrix," or "error matrix," used in statistical classification, which will describe the relationship between the dMRI metric and the independent validation "ground truth." As visualized in Fig. 13 each column represents instances of the ground truth class that are classified as actually positive and actually negative, and each column represents instances of the dMRI class that are classified as predicted to be positive and predicted to be negative. These positive and negative classifications can be the presence/absence of a fiber orientation estimation, a streamline or bundle connection, or an index or value in the connectome matrix. Thus the confusion matrix elements are the True-Positive (TP), True-Negative (TN), False-Positive (FP), and False-Negative (FN) classifications. The elements of the confusion matrix can then be used to derive useful evaluation metrics, in the dMRI literature most commonly sensitivity, specificity, precision, and accuracy.

## 3.1 Measures for validating connections/connectomes

A number of validation metrics have been proposed to capture the accuracy of the connectome, i.e., the matrix of region-to-region connectivity generated by tractography (Gouttard et al., 2012; Côté et al., 2013). Here, it is important to appreciate the number of ways that this matrix can be generated and represented, and the multiscale analysis that is possible. A ground

|  | Actually Positive (1) | Actually Negative (0) |
|---|---|---|
| Predicted Positive (1) | True Positives (TP) | False Positives (FP) |
| Predicted Negative (0) | False Negatives (FN) | True Negatives (TN) |

FIG. 13 The confusion matrix is the fundament for establishing evaluation metrics to validate the spatial agreement between the tractography results and an expected ground truth. True-Positive: Tractography results agree with ground truth. False-Positive: Tractography results find something not existing in the ground truth. False-Negative: Tractography did not find anything, but ground truth suggests it does exist. True-Negative: Both tractography and ground truth agree there should be no finding.

truth connectivity matrix, whether generated from accumulating tracer studies, physical phantoms, or numerically simulated connectomes, may have a "strength of connectivity" or a "number of connections" as a scalar value for each edge of the connectome matrix, or maybe a simple binary "connection" or "no connection." Subsequently, each edge-to-edge connection can be associated with a pathway that has its own unique location, shape, and connections. The connectome, then, can be validated at the level of graph theory summary statistics, node-based metrics, or edge-based analysis.

Some common validation metrics of a binarized connectivity matrix against the ground truth are to quantify **valid bundles** and **invalid bundles**, i.e., the percentage of bundles or edges connecting expected regions of interest (ROIs) or connecting unexpected ROIs. Similarly, classification measures of **sensitivity** and **specificity** have commonly described connectome classification accuracy. For weighted connectomes, i.e., scalar values as edges, a threshold could be applied for binarization, and the preceding metrics can be similarly calculated. Alternatively, other measures, such as **Jensen-Shannon Divergence (JSD)**, **mutual information**, or cross-entropy (measures that describe the similarity or information content of the tractography compared to ground truth) have been used to validate or compare connectomes, with the advantage that they enable a histogram-based similarity metric that may provide insights into connection strength estimates.

Finally, whole sets of streamlines can be compared to a ground truth set of bundles of streamlines. From this, much like connectome validation measures, **valid connections, invalid connections**, or the percentage of streamlines that connect expected (or unexpected) regions can be quantified, in addition to calculating **no connections**, or the percentage of streamlines that do not connect two ROIs. Together, these measures give insight into the trade-offs in the tractography process.

## 3.2 Measures for validating bundles

There are similarly a large number of measures that have been used to describe the anatomical accuracy of fiber bundles, describing agreements and disagreements in volume or spatial overlap. These measures are typically ROI-based or voxel-wise accuracy measures (Côté et al., 2013). The ROI-based measures characterize the validity of tract region-to-region connectivity, while the voxel-wise measures assess the spatial extent of reconstructed pathways on the scale of individual voxels. For both, it is common to measure **sensitivity** and **specificity**. Again, sensitivity measures the true proportion of positives (regions or voxels that are occupied by ground truth) that are correctly identified as such using tractography. Specificity measures the proportion of negatives (regions or voxels that do not contain the ground truth) that are correctly identified as not containing streamlines. These measures can be summarized using other common classification measures, including **accuracy**, **precision**, and, in several studies, **Youden's $J$ statistic** (Sensitivity + Specificity − 1), where a value of 1 indicates a perfect "test," or tractography algorithm.

Other common measures, often applied on the scale of voxels, include the dice overlap coefficient, bundle overlap, and bundle overreach (Côté et al., 2013). The **dice overlap** describes the overall volume similarity and is calculated as twice the intersection of two bundles (one ground truth, the other tractography) divided by the union. A value of 1 indicates perfect spatial overlap. **Bundle overlap** describes the proportion of voxels that contain the ground truth volume that is also traversed by at least one streamline (a sensitivity measure). The **bundle overreach** is calculated as the number of voxels containing streamlines that are outside the ground truth volume divided by the total volume of the ground truth bundle (a specificity measure), where a perfect bundle has full overlap (value of 1), and no overreach (value of 0). When using such overlap similarity metrics, one must be aware that these are sensitive to the volume-to-surface ratio of the two structures being compared. This means that two spheres are more likely to have a good agreement than two planes where a shift of one voxel can lead to a score of 0. Often, having a low volume-to-surface ratio, the results can show low spatial overlap agreement which simply is because the objects compared to have a low volume, i.e., few voxels.

The preceding measures compare one volume to another, with little-to-no consideration of individual streamlines. Alternative quantification may be used if tractography and ground truth are in the form of streamlines. This may include a **density correlation** measure, calculated as the correlation coefficient between the voxel-wise densities. **Measures of disagreement** may be calculated by quantifying average distances between streamlines/volumes using one of many distance metrics (average mean distance, Hausdorff distance, etc.). Finally, some studies have quantified, or ranked, bundles by **qualitative visualization** based on 3D viewing of tracts rated by anatomists or diffusion experts.

## 3.3 Measures for validation of local fiber orientation distribution

Accurate estimates of axonal fiber orientation in each voxel are a necessary prerequisite for the generation of streamlines that adequately follow these fibers, and hence, are a necessary prerequisite for accurate tractography. Fiber orientation may be represented as a continuous probability distribution over a sphere (similar to an orientational histogram), or in the form of a 3D vector, or multiple 3D vectors, usually representing local peaks in the continuous fiber orientation distribution. There are various ways to measure agreement between the estimated fiber orientation distribution and ground truth orientation(s). In general, most studies evaluate the agreement between dMRI and histology estimates of orientation by focusing on (a) angular error in estimation (b) correct assessment of the number of discrete "fiber populations" or peaks in each voxel, and (c) overall agreement in shape between histological and diffusion spherical functions.

**Angular error** is intuitively, the angular difference between two vectors, one from the validation data (whether it be histologically derived, simulation, or a phantom) and one from the diffusion data (Leergaard et al., 2010). A smaller angular difference designates better agreement, with a value of 90 degrees being the worst-case difference. While this is a simple measure for the comparison of two vectors, it quickly becomes complicated when multiple orientations exist within either the diffusion data or the ground truth data. Typically, diffusion orientations are compared to the nearest orientation that exists within the ground truth. However, this does not completely describe the accuracy of the diffusion data. For this reason, angular error, and the correct assessment of a number of orientations, or peaks, is often quantified (Daducci et al., 2014). The number of true and false positive peaks, or true and false negative peaks, derived from dMRI can be assessed, and measures of **sensitivity**, **specificity**, and **accuracy** have been quantified using the confusion matrix described above. Moreover, the **success rate** has often been used, where a fiber reconstruction algorithm is successful if it successfully identifies all peaks that exist within the ground truth.

So far, measures have been described to compare discrete peaks. If both the ground-truth and dMRI data are represented as continuous orientation distribution functions, the overall similarity of shape in these can be quantified (Daducci et al., 2014). For example, a similarity metric between two orientation distribution functions (ODFs) can use an **angular correlation coefficient** (ACC, Anderson, 2005) or **Jensen-Shannon divergence** (JSD, Cohen-Adad et al., 2011). Here, an ACC of 1, or JSD of 0, represents perfect relationships between dMRI and ground truth.

## 3.4 Measures for validating microstructure

Microstructure is not directly used in most current tractography pipelines. However, with advances in fiber orientation analysis, microstructure-informed tractography, and bundle-specific tractography—techniques that attempt to incorporate microstructural features into the streamline generation and filtering process—it is likely that microstructure validation will be necessary for structural connectivity in the future (Daducci et al., 2016; Girard et al., 2017). Example microstructure measures that are currently estimated from dMRI may include tissue volume fractions (for example, axon or myelin fractions), sizes of structures (soma or axon diameters), or densities of components (cellular or axonal density).

Typically, measures of tissue microstructure are validated by quantifying **accuracy**, **precision**, and **correlation** of the dMRI estimated feature as the ground truth. This may be performed on the scale of individual voxels, or over regions of interest. Examples might be comparing dMRI estimated diameters and volume fractions to simulations (Palombo et al., 2020), comparing diameters to physical phantoms (Fan et al., 2018), correlating volume fractions to histological axon-staining intensities (Jespersen et al., 2010), or comparing axon diameters to XNH or EM or known patterns expected throughout the brain (Lee et al., 2019; Assaf et al., 2008; Andersson et al., 2020). In all cases, *accuracy* may be quantified as the absolute difference between predicted and ground truth; the *correlation* between predicted and the truth captures the strength of the relationship between the two; the *precision* identifies variations in the estimation procedure. Together, this help to identify **biases** in estimates and cases where microstructural modeling strategies are (or are not) successful.

It is important to emphasize that most microstructure validation studies look at, and assess, the dMRI model and microstructure in isolation and without relation to tractography. Concepts such as network topology being balanced using microstructure priors, microstructure-weighted connectivity matrices, or pathways with microstructure signatures (global or along its length) are much harder to quantify and validate. It would be crucial to not only quantify the precision and accuracy of a feature, but also to ensure that feature is specific to the pathway, node, or streamline(s) under investigation.

## 3.5 Quantification considerations

It is crucial to remember what anatomical scale of interest is being investigated, what it is being compared to, and what interpretation can be drawn from it. An accurate local orientation does not necessarily lead to an accurate reconstruction of major WM pathways, or an accurate reconstruction of major WM pathways does not necessarily lead to an accurate connectome. While publications related to validation often focus on a particular scale and provide crucial insight, the agreement between dMRI and tractography and the true underlying anatomy has not yet reached a holistic/integrated/comprehensive/aggregate level.

## References

Abel, P.L., O'Brien, B.J., Olavarria, J.F., 2000. Organization of callosal linkages in visual area V2 of macaque monkey. J. Comp. Neurol. 428 (2), 278–293.

Ambrosen, K.S., Eskildsen, S.F., Hinne, M., Krug, K., Lundell, H., Schmidt, M.N., van Gerven, M.A.J., Mørup, M., Dyrby, T.B., 2020. Validation of structural brain connectivity networks: the impact of scanning parameters. NeuroImage 204, 116207.

Anderson, A.W., 2005. Measurement of fiber orientation distributions using high angular resolution diffusion imaging. Magn. Reson. Med. 54 (5), 1194–1206.

Andersson, M., Kjer, H.M., Rafael-Patino, J., Pacureanu, A., Pakkenberg, B., Thiran, J.-P., Ptito, M., et al., 2020. Axon morphology is modulated by the local environment and impacts the noninvasive investigation of its structure-function relationship. Proc. Natl. Acad. Sci. U. S. A. 117 (52), 33649–33659.

Annese, J., Sforza, D.M., Dubach, M., Bowden, D., Toga, A.W., 2006. Postmortem high-resolution 3-dimensional imaging of the primate brain: blockface imaging of perfusion stained tissue. NeuroImage 30 (1), 61–69.

Assaf, Y., Blumenfeld-Katzir, T., Yovel, Y., Basser, P.J., 2008. AxCaliber: a method for measuring axon diameter distribution from diffusion MRI. Magn. Reson. Med. 59 (6), 1347–1354. https://doi.org/10.1002/mrm.21577. PMID: 18506799. PMCID: PMC4667732.

Assaf, Y., Bouznach, A., Zomet, O., Marom, A., Yovel, Y., 2020. Conservation of brain connectivity and wiring across the mammalian class. Nat. Neurosci. 23 (7), 805–808.

Axer, M., Amunts, K., Grässel, D., Palm, C., Dammers, J., Axer, H., Pietrzyk, U., Zilles, K., 2011a. A novel approach to the human connectome: ultra-high resolution mapping of fiber tracts in the brain. NeuroImage. https://doi.org/10.1016/j.neuroimage.2010.08.075.

Axer, H., Beck, S., Axer, M., Schuchardt, F., Heepe, J., Flücken, A., Axer, M., Prescher, A., Witte, O.W., 2011b. Microstructural analysis of human white matter architecture using polarized light imaging: views from neuroanatomy. Front. Neuroinform. 5. https://doi.org/10.3389/fninf.2011.00028.

Axer, M., Grässel, D., Kleiner, M., Dammers, J., Dickscheid, T., Reckfort, J., Hütz, T., et al., 2011c. High-resolution fiber tract reconstruction in the human brain by means of three-dimensional polarized light imaging. Front. Neuroinform. 5, 34.

Bakker, R., Potjans, T.C., Wachtler, T., Diesmann, M., 2011. Macaque structural connectivity revisited: CoCoMac 2.0. BMC Neurosci. https://doi.org/10.1186/1471-2202-12-s1-p72.

Baldi, D., Aiello, M., Duggento, A., Salvatore, M., Cavaliere, C., 2019. MR imaging-histology correlation by tailored 3D-printed slicer in oncological assessment. Contrast Media Mol. Imaging 2019, 1071453.

Bassett, D.S., Brown, J.A., Deshpande, V., Carlson, J.M., Grafton, S.T., 2011. Conserved and variable architecture of human white matter connectivity. NeuroImage 54 (2), 1262–1279.

Bastiani, M., Shah, N.J., Goebel, R., Roebroeck, A., 2012. Human cortical connectome reconstruction from diffusion weighted MRI: the effect of tractography algorithm. NeuroImage 62 (3), 1732–1749.

Bech, J., Orlowski, D., Glud, A.N., Dyrby, T.B., Sørensen, J.C.H., Bjarkam, C.R., 2020. Ex vivo diffusion-weighted MRI tractography of the Göttingen minipig limbic system. Brain Struct. Funct. 225 (3), 1055–1071.

Berg, J., Sorensen, S.A., Ting, J.T., Miller, J.A., Chartrand, T., Buchin, A., Bakken, T.E., et al., 2021. Human neocortical expansion involves glutamatergic neuron diversification. Nature 598 (7879), 151–158.

Bertò, G., Bullock, D., Astolfi, P., Hayashi, S., Zigiotto, L., Annicchiarico, L., Corsini, F., et al., 2021. Classifyber, a robust streamline-based linear classifier for white matter bundle segmentation. NeuroImage 224, 117402.

Bezgin, G., Vakorin, V.A., John van Opstal, A., McIntosh, A.R., Bakker, R., 2012. Hundreds of brain maps in one atlas: registering coordinate-independent primate neuro-anatomical data to a standard brain. NeuroImage 62 (1), 67–76.

Bickford, M.E., 2015. Thalamic circuit diversity: modulation of the driver/modulator framework. Front. Neural Circuits 9, 86.

Bigun, J., Granlund, G.H., 1987. Optimal orientation detection of linear symmetry. In: ICCV, London, June 8–11. IEEE Computer Society, pp. 433–438.

Borra, E., Rockland, K.S., 2011. Projections to early visual areas v1 and v2 in the calcarine fissure from parietal association areas in the macaque. Front. Neuroanat. 5, 35.

Bourne, R.M., Bailey, C., Johnston, E.W., Pye, H., Heavey, S., Whitaker, H., Siow, B., et al., 2017. Apparatus for histological validation of in vivo and ex vivo magnetic resonance imaging of the human prostate. Front. Oncol. 7, 47.

Budde, M.D., Frank, J.A., 2012. Examining brain microstructure using structure tensor analysis of histological sections. NeuroImage 63 (1), 1–10.

Bürgel, U., Amunts, K., Hoemke, L., Mohlberg, H., Gilsbach, J.M., Zilles, K., 2006. White matter fiber tracts of the human brain: three-dimensional mapping at microscopic resolution, topography and intersubject variability. NeuroImage 29 (4), 1092–1105.

Cahill, L.S., Laliberté, C.L., Ellegood, J., Spring, S., Gleave, J.A., van Eede, M.C., Lerch, J.P., Mark Henkelman, R., 2012. Preparation of fixed mouse brains for MRI. NeuroImage 60 (2), 933–939.

Cai, L.Y., Yang, Q., Kanakaraj, P., Nath, V., Newton, A.T., Edmonson, H.A., Luci, J., et al., 2021. MASiVar: multisite, multiscanner, and multisubject acquisitions for studying variability in diffusion weighted MRI. Magn. Reson. Med. 86 (6), 3304–3320.

Calamante, F., 2019. The seven deadly sins of measuring brain structural connectivity using diffusion MRI streamlines fibre-tracking. Diagnostics (Basel, Switzerland) 9 (3). https://doi.org/10.3390/diagnostics9030115.

Caminiti, R., Carducci, F., Piervincenzi, C., Battaglia-Mayer, A., Confalone, G., Visco-Comandini, F., Pantano, P., Innocenti, G.M., 2013. Diameter, length, speed, and conduction delay of callosal axons in macaque monkeys and humans: comparing data from histology and magnetic resonance imaging diffusion tractography. J. Neurosci. Off. J. Soc. Neurosci. 33 (36), 14501–14511.

Cembrowski, M.S., Spruston, N., 2019. Heterogeneity within classical cell types is the rule: lessons from hippocampal pyramidal neurons. Nat. Rev. Neurosci. 20 (4), 193–204.

Chamberland, M., Tax, C.M.W., Jones, D.K., 2018. Meyer's loop tractography for image-guided surgery depends on imaging protocol and hardware. NeuroImage Clin. 20, 458–465.

Chenot, Q., Tzourio-Mazoyer, N., Rheault, F., et al., 2019. A population-based atlas of the human pyramidal tract in 410 healthy participants. Brain Struct. Funct. 224, 599–612. https://doi.org/10.1007/s00429-018-1798-7.

Choe, A.S., Gao, Y., Li, X., Compton, K.B., Stepniewska, I., Anderson, A.W., 2011. Accuracy of image registration between MRI and light microscopy in the ex vivo brain. Magn. Reson. Imaging 29 (5), 683–692.

Choe, A.S., Stepniewska, I., Colvin, D.C., Ding, Z., Anderson, A.W., 2012. Validation of diffusion tensor MRI in the central nervous system using light microscopy: quantitative comparison of fiber properties. NMR Biomed. 25 (7), 900–908.

Cohen-Adad, J., Descoteaux, M., Wald, L.L., 2011. Quality assessment of high angular resolution diffusion imaging data using bootstrap on Q-ball reconstruction. J. Magn. Reson. Imaging 33 (5), 1194–1208.

Côté, M.-A., Girard, G., Boré, A., Garyfallidis, E., Houde, J.-C., Descoteaux, M., 2013. Tractometer: towards validation of tractography pipelines. Med. Image Anal. 17 (7), 844–857.

D'Arceuil, H., de Crespigny, A., 2007. The effects of brain tissue decomposition on diffusion tensor imaging and tractography. NeuroImage 36 (1), 64–68.

Daducci, A., Canales-Rodríguez, E.J., Descoteaux, M., Garyfallidis, E., Gur, Y., Lin, Y.-C., Mani, M., et al., 2014. Quantitative comparison of reconstruction methods for intra-voxel fiber recovery from diffusion MRI. IEEE Trans. Med. Imaging 33 (2), 384–399.

Daducci, A., Dal Palú, A., Descoteaux, M., Thiran, J.-P., 2016. Microstructure informed tractography: pitfalls and open challenges. Front. Neurosci. 10, 247.

Dauguet, J., Davi, B., Clay Reid, R., Warfield, S.K., 2007a. Alignment of large image series using cubic B-splines tessellation: application to transmission electron microscopy data. In: Medical Image Computing and Computer-Assisted Intervention: MICCAI .. International Conference on Medical Image Computing and Computer-Assisted Intervention. vol. 10 (Pt 2), pp. 710–717.

Dauguet, J., Delzescaux, T., Condé, F., Mangin, J.-F., Ayache, N., Hantraye, P., Frouin, V., 2007b. Three-dimensional reconstruction of stained histological slices and 3D non-linear registration with in-vivo MRI for whole baboon brain. J. Neurosci. Methods 164 (1), 191–204.

Dauguet, J., Peled, S., Berezovskii, V., Delzescaux, T., Warfield, S.K., Born, R., Westin, C.-F., 2007c. Comparison of fiber tracts derived from in-vivo DTI tractography with 3D histological neural tract tracer reconstruction on a macaque brain. NeuroImage 37 (2), 530–538.

David, S., Heemskerk, A.M., Corrivetti, F., Thiebaut, M., de Schotten, S., Sarubbo, F.C., De Benedictis, A., et al., 2019. The superoanterior fasciculus (SAF): a novel White matter pathway in the human brain? Front. Neuroanat. https://doi.org/10.3389/fnana.2019.00024.

De Benedictis, A., Sarubbo, S., Duffau, H., 2012. Subcortical surgical anatomy of the lateral frontal region: human white matter dissection and correlations with functional insights provided by intraoperative direct brain stimulation: laboratory investigation. J. Neurosurg. 117 (6), 1053–1069.

De Benedictis, A., Duffau, H., Paradiso, B., Grandi, E., Balbi, S., Granieri, E., et al., 2014. Anatomo-functional study of the temporo-parieto-occipital region: dissection, tractographic and brain mapping evidence from a neurosurgical perspective. J. Anat. https://doi.org/10.1111/joa.12204.

De Benedictis, A., Petit, L., Descoteaux, M., Marras, C.E., Barbareschi, M., Corsini, F., Dallabona, M., Chioffi, F., Sarubbo, S., 2016. New insights in the homotopic and heterotopic connectivity of the frontal portion of the human corpus callosum revealed by microdissection and diffusion tractography. Hum. Brain Mapp. 37 (12), 4718–4735.

De Benedictis, A., Nocerino, E., Menna, F., Remondino, F., Barbareschi, M., Rozzanigo, U., Corsini, F., et al., 2018. Photogrammetry of the human brain: a novel method for three-dimensional quantitative exploration of the structural connectivity in neurosurgery and neurosciences. World Neurosurg. 115, e279–e291.

Defelipe, J., 2011. The evolution of the brain, the human nature of cortical circuits, and intellectual creativity. Front. Neuroanat. 5, 29.

Drakesmith, M., Harms, R., Rudrapatna, S.U., Parker, G.D., John Evans, C., Jones, D.K., 2019. Estimating axon conduction velocity in vivo from microstructural MRI. NeuroImage 203, 116186.

Dyrby, T.B., Søgaard, L.V., Parker, G.J., Alexander, D.C., Lind, N.M., Baaré, W.F.C., Hay-Schmidt, A., et al., 2007. Validation of in vitro probabilistic tractography. NeuroImage 37 (4), 1267–1277.

Dyrby, T.B., Baaré, W.F.C., Alexander, D.C., Jelsing, J., Garde, E., Søgaard, L.V., 2011. An ex vivo imaging pipeline for producing high-quality and high-resolution diffusion-weighted imaging datasets. Hum. Brain Mapp. 32, 544–563. https://doi.org/10.1002/hbm.21043.

Dyrby, T.B., Lundell, H., Burke, M.W., Reislev, N.L., Paulson, O.B., Ptito, M., Siebner, H.R., 2014. Interpolation of diffusion weighted imaging datasets. NeuroImage 103, 202–213.

Dyrby, T.B., Innocenti, G.M., Bech, M., Lundell, H., 2018. Validation strategies for the interpretation of microstructure imaging using diffusion MRI. NeuroImage 182, 62–79.

Dziedzic, T.A., Balasa, A., Jeżewski, M.P., Michałowski, Ł., Marchel, A., 2021. White matter dissection with the Klingler technique: a literature review. Brain Struct. Funct. https://doi.org/10.1007/s00429-020-02157-9.

Fan, Q., Nummenmaa, A., Wichtmann, B., Witzel, T., Mekkaoui, C., Schneider, W., Wald, L.L., Huang, S.Y., 2018. Validation of diffusion MRI estimates of compartment size and volume fraction in a biomimetic brain phantom using a human MRI scanner with 300 mT/m maximum gradient strength. NeuroImage 182, 469–478.

Farrell, J.A.D., Landman, B.A., Jones, C.K., Smith, S.A., Prince, J.L., van Zijl, P.C.M., Mori, S., 2007. Effects of signal-to-noise ratio on the accuracy and reproducibility of diffusion tensor imaging-derived fractional anisotropy, mean diffusivity, and principal eigenvector measurements at 1.5 T. J. Magn. Reson. Imaging 26 (3), 756–767.

Feng, L., Jeon, T., Qiaowen, Y., Ouyang, M., Peng, Q., Mishra, V., Pletikos, M., et al., 2017. Population-averaged macaque brain atlas with high-resolution ex vivo DTI integrated into in vivo space. Brain Struct. Funct. 222 (9), 4131–4147.

Ferizi, U., Schneider, T., Panagiotaki, E., Nedjati-Gilani, G., Zhang, H., Wheeler-Kingshott, C.A.M., Alexander, D.C., 2014. A ranking of diffusion MRI compartment models with in vivo human brain data. Magn. Reson. Med. 72 (6), 1785–1792.

**538 PART | IV** From streamlines to tracts

Ferizi, U., Schneider, T., Witzel, T., Wald, L.L., Zhang, H., Wheeler-Kingshott, C.A.M., Alexander, D.C., 2015. White matter compartment models for in vivo diffusion MRI at 300mT/m. NeuroImage 118, 468–483.

Ferizi, U., Scherrer, B., Schneider, T., Alipoor, M., Eufracio, O., Fick, R.H.J., Deriche, R., et al., 2017. Diffusion MRI microstructure models with in vivo human brain connectome data: results from a multi-group comparison. NMR Biomed. 30 (9). https://doi.org/10.1002/nbm.3734.

Firmin, L., Field, P., Maier, M.A., Kraskov, A., Kirkwood, P.A., Nakajima, K., Lemon, R.N., Glickstein, M., 2014. Axon diameters and conduction velocities in the macaque pyramidal tract. J. Neurophysiol. 112 (6), 1229–1240.

Froeling, M., Tax, C.M.W., Vos, S.B., Luijten, P.R., Leemans, A., 2017. 'MASSIVE' brain dataset: multiple acquisitions for standardization of structural imaging validation and evaluation. Magn. Reson. Med. 77 (5), 1797–1809.

Gao, Y., Choe, A.S., Stepniewska, I., Li, X., Avison, M.J., Anderson, A.W., 2013. Validation of DTI tractography-based measures of primary motor area connectivity in the squirrel monkey brain. PLoS One 8 (10), e75065.

Girard, G., Daducci, A., Petit, L., Thiran, J.-P., Whittingstall, K., Deriche, R., Wassermann, D., Descoteaux, M., 2017. AxTract: toward microstructure informed tractography. Hum. Brain Mapp. 38 (11), 5485–5500.

Gouttard, S., Goodlett, C.B., Kubicki, M., Gerig, G., 2012. Measures for validation of DTI tractography. Proc. SPIE Int. Soc. Opt. Eng. 8314, 83140J.

Grier, M.D., Zimmermann, J., Heilbronner, S.R., 2020. Estimating brain connectivity with diffusion-weighted magnetic resonance imaging: promise and peril. Biol. Psychiatry Cognit. Neurosci. Neuroimaging 5 (9), 846–854.

Han, Y., Kebschull, J.M., Campbell, R.A.A., Cowan, D., Imhof, F., Zador, A.M., Mrsic-Flogel, T.D., 2018. The logic of single-cell projections from visual cortex. Nature 556 (7699), 51–56.

Harris, C., Stephens, M., 1988. A combined corner and edge detector. Proceedings of the Alvey Vision Conference, p. 23.1. https://doi.org/10.5244/C.2.23.

Hau, J., Sarubbo, S., Perchey, G., Crivello, F., Zago, L., Mellet, E., Jobard, G., et al., 2016. Cortical terminations of the inferior fronto-occipital and uncinate fasciculi: anatomical stem-based virtual dissection. Front. Neuroanat. 10, 58.

Hedreen, J.C., Yin, T.C., 1981. Homotopic and heterotopic callosal afferents of caudal inferior parietal lobule in *Macaca mulatta*. J. Comp. Neurol. 197 (4), 605–621.

Hukkanen, V., Röyttä, M., 1987. Autolytic changes of human white matter: an electron microscopic and electrophoretic study. Exp. Mol. Pathol. 46 (1), 31–39.

Innocenti, G.M., Vercelli, A., Caminiti, R., 2014. The diameter of cortical axons depends both on the area of origin and target. Cereb. Cortex. https://doi.org/10.1093/cercor/bht070.

Innocenti, G.M., Dyrby, T.B., Girard, G., St-Onge, E., Thiran, J.-P., Daducci, A., Descoteaux, M., 2019. Topological principles and developmental algorithms might refine diffusion tractography. Brain Struct. Funct. 224 (1), 1–8.

Jacquesson, T., Frindel, C., Kocevar, G., Berhouma, M., Jouanneau, E., Attyé, A., Cotton, F., 2019. Overcoming challenges of cranial nerve tractography: a targeted review. Neurosurgery 84 (2), 313–325.

Jelescu, I.O., Grussu, F., Ianus, A., Hansen, B., Barrett, R.L.C., Aggarwal, M., Michielse, S., Nasrallah, F., Syeda, W., Wang, N., Veraart, J., Roebroeck, A., Bagdasarian, A.F., Eichner, C., Sepehrband, F., Zimmermann, J., Soustelle, L., Bowman, C., Tendler, B.C., Hertanu, A., Jeurissen, B., Verhoye, M., Frydman, L., van de Looij, Y., Hike, D., Dunn, J.F., Miller, K., Landman, B.A., Shemesh, N., Anderson, A., McKinnon, E., Farquharson, S., Acqua, F.D., Pierpaoli, C., Drobnjak, I., Leemans, A., Harkins, K.D., Descoteaux, M., Xu, D., Huang, H., Santin, M.D., Grant, S.C., Obenaus, A., Kim, G.S., Wu, D., Bihan, D.L., Blackband, S.J., Ciobanu, L., Fieremans, E., Bai, R., Leergaard, T.B., Zhang, J., Dyrby, T.B., Johnson, G.A., Cohen-Adad, J., Budde, M.D., Schilling, K.G., 2024. Recommendations and guidelines from the ISMRM Diffusion Study Group for preclinical diffusion MRI: part 1—in vivo small-animal imaging. arXiv. https://arxiv.org/abs/2209.12994.

Jespersen, S.N., Bjarkam, C.R., Nyengaard, J.R., Mallar Chakravarty, M., Hansen, B., Vosegaard, T., Østergaard, L., Yablonskiy, D., Nielsen, N.C., Vestergaard-Poulsen, P., 2010. Neurite density from magnetic resonance diffusion measurements at ultrahigh field: comparison with light microscopy and electron microscopy. NeuroImage 49 (1), 205–216.

Jeurissen, B., Leemans, A., Tournier, J.-D., Jones, D.K., Sijbers, J., 2013. Investigating the prevalence of complex fiber configurations in white matter tissue with diffusion magnetic resonance imaging. Hum. Brain Mapp. 34 (11), 2747–2766.

Jeurissen, B., Tournier, J.-D., Dhollander, T., Connelly, A., Sijbers, J., 2014. Multi-tissue constrained spherical deconvolution for improved analysis of multi-shell diffusion MRI data. NeuroImage 103, 411–426.

Jones, E.G., 2001. The thalamic matrix and thalamocortical synchrony. Trends Neurosci. 24 (10), 595–601.

Jones, D.K., Basser, P.J., 2004. 'Squashing peanuts and smashing pumpkins': how noise distorts diffusion-weighted MR data. Magn. Reson. Med. 52 (5), 979–993.

Kennedy, H., Dehay, C., Bullier, J., 1986. Organization of the callosal connections of visual areas V1 and V2 in the macaque monkey. J. Comp. Neurol. 247 (3), 398–415.

Khan, A.R., Cornea, A., Leigland, L.A., Kohama, S.G., Jespersen, S.N., Kroenke, C.D., 2015. 3D structure tensor analysis of light microscopy data for validating diffusion MRI. NeuroImage 111, 192–203.

Kim, B., Boes, J.L., Frey, K.A., Meyer, C.R., 1997. Mutual information for automated unwarping of rat brain autoradiographs. NeuroImage 5 (1), 31–40.

Kita, T., Kita, H., 2012. The subthalamic nucleus is one of multiple innervation sites for long-range corticofugal axons: a single-axon tracing study in the rat. J. Neurosci. Off. J. Soc. Neurosci. 32 (17), 5990–5999.

Kjer, H.M., Andersson, M., He, Y., Pacureanu, A., Daducci, A., Pizzolato, M., Salditt, T., Robisch, A.-L., Eckermann, M., Toepperwien, M., Dahl, A.B., Elkjær, M.L., Illes, Z., Ptito, M., Dahl, V.A., Dyrby, T.B., 2024. Bridging the 3D geometrical organisation of white matter pathways across anatomical length scales and species. eLife 13, RP94917. https://doi.org/10.7554/eLife.94917.1.

Klingler, J., 1935. Erleichterung Der Makrokopischen Präparation Des Gehirns Durch Den Gefrierprozess. Orell Füssli.

Kraskov, A., Soteropoulos, D.S., Glover, I.S., Lemon, R.N., Baker, S.N., 2020. Slowly-conducting pyramidal tract neurons in macaque and rat. Cereb. Cortex 30 (5), 3403–3418.

Landman, B.A., Farrell, J.A.D., Jones, C.K., Smith, S.A., Prince, J.L., Mori, S., 2007. Effects of diffusion weighting schemes on the reproducibility of DTI-derived fractional anisotropy, mean diffusivity, and principal eigenvector measurements at 1.5T. NeuroImage 36 (4), 1123–1138.

Lawes, I.N.C., Barrick, T.R., Murugam, V., Spierings, N., Evans, D.R., Song, M., Clark, C.A., 2008. Atlas-based segmentation of white matter tracts of the human brain using diffusion tensor tractography and comparison with classical dissection. NeuroImage. https://doi.org/10.1016/j.neuroimage.2007.06.041.

Lebenberg, J., Hérard, A.-S., Dubois, A., Dauguet, J., Frouin, V., Dhenain, M., Hantraye, P., Delzescaux, T., 2010. Validation of MRI-based 3D digital atlas registration with histological and autoradiographic volumes: an anatomofunctional transgenic mouse brain imaging study. NeuroImage 51 (3), 1037–1046.

Lee, H.-H., Yaros, K., Veraart, J., Pathan, J.L., Liang, F.-X., Kim, S.G., Novikov, D.S., Fieremans, E., 2019. Along-axon diameter variation and axonal orientation dispersion revealed with 3D Electron microscopy: implications for quantifying brain white matter microstructure with histology and diffusion MRI. Brain Struct. Funct. 224 (4), 1469–1488.

Leergaard, T.B., White, N.S., de Crespigny, A., Bolstad, I., D'Arceuil, H., Bjaalie, J.G., Dale, A.M., 2010. Quantitative histological validation of diffusion MRI fiber orientation distributions in the rat brain. PLoS One 5 (1), e8595.

Lehman, J.F., Greenberg, B.D., McIntyre, C.C., Rasmussen, S.A., Haber, S.N., 2011. Rules ventral prefrontal cortical axons use to reach their targets: implications for diffusion tensor imaging tractography and deep brain stimulation for psychiatric illness. J. Neurosci. Off. J. Soc. Neurosci. 31 (28), 10392–10402.

Leuze, C., Goubran, M., Barakovic, M., Aswendt, M., Tian, Q., Hsueh, B., Crow, A., et al., 2021. Comparison of diffusion MRI and clarity fiber orientation estimates in both gray and white matter regions of human and primate brain. NeuroImage 228, 117692.

Lind, N.M., Moustgaard, A., Jelsing, J., Vajta, G., Cumming, P., Hansen, A.K., 2007. The use of pigs in neuroscience: modeling brain disorders. Neurosci. Biobehav. Rev. 31 (5), 728–751.

Llinares-Benadero, C., Borrell, V., 2019. Deconstructing cortical folding: genetic, cellular and mechanical determinants. Nat. Rev. Neurosci. 20 (3), 161–176.

Luciano, N.J., Sati, P., Nair, G., Guy, J.R., Ha, S.-K., Absinta, M., Chiang, W.-Y., et al., 2016. Utilizing 3D printing technology to merge MRI with histology: a protocol for brain sectioning. J. Vis. Exp. 118. https://doi.org/10.3791/54780.

Maffei, C., Jovicich, J., De Benedictis, A., Corsini, F., Barbareschi, M., Chioffi, F., Sarubbo, S., 2018. Topography of the human acoustic radiation as revealed by ex vivo fibers micro-dissection and in vivo diffusion-based tractography. Brain Struct. Funct. 223 (1), 449–459.

Maffei, C., Sarubbo, S., Jovicich, J., 2019. Diffusion-based tractography atlas of the human acoustic radiation. Sci. Rep. 9 (1), 4046.

Mancini, M., Casamitjana, A., Peter, L., Robinson, E., Crampsie, S., Thomas, D.L., Holton, J.L., Jaunmuktane, Z., Iglesias, J.E., 2020. A multimodal computational pipeline for 3D histology of the human brain. Sci. Rep. 10 (1), 13839.

Marconi, B., Genovesio, A., Giannetti, S., Molinari, M., Caminiti, R., 2003. Callosal connections of dorso-lateral premotor cortex. Eur. J. Neurosci. 18 (4), 775–788.

Markov, N.T., Ercsey-Ravasz, M.M., Ribeiro Gomes, A.R., Lamy, C., Magrou, L., Vezoli, J., Misery, P., et al., 2014. A weighted and directed interareal connectivity matrix for macaque cerebral cortex. Cereb. Cortex 24 (1), 17–36.

Martino, J., De Witt, P.C., Hamer, F.V., Brogna, C., de Lucas, E.M., Vázquez-Barquero, A., García-Porrero, J.A., Duffau, H., 2011. Cortex-sparing fiber dissection: an improved method for the study of white matter anatomy in the human brain. J. Anat. https://doi.org/10.1111/j.1469-7580.2011.01414.x.

Menzel, M., Axer, M., De Raedt, H., Costantini, I., Silvestri, L., 2020. Toward a high-resolution reconstruction of 3D nerve fiber architectures and crossings in the brain using light scattering measurements and finite-difference time-domain simulations. Phys. Rev. X 10, 021002.

Menzel, M., Reuter, J.A., Gräßel, D., Huwer, M., Schlömer, P., Amunts, K., Axer, M., 2021. Scattered light imaging: resolving the substructure of nerve fiber crossings in whole brain sections with micrometer resolution. NeuroImage 233, 117952.

Mikhaylova, M., Rentsch, J., Ewers, H., 2020. Actomyosin contractility in the generation and plasticity of axons and dendritic spines. Cells 9 (9). https://doi.org/10.3390/cells9092006.

Morecraft, R.J., Binneboese, A., Stilwell-Morecraft, K.S., Ge, J., 2017. Localization of orofacial representation in the corona radiata, internal capsule and cerebral peduncle in *Macaca mulatta*. J. Comp. Neurol. 525 (16), 3429–3457.

Moyer, D.C., Thompson, P., Ver Steeg, G., 2019. Measures of tractography convergence. In: Computational Diffusion MRI. Springer International Publishing, pp. 295–307.

Nath, V., Schilling, K.G., Parvathaneni, P., Huo, Y., Blaber, J.A., Hainline, A.E., Barakovic, M., et al., 2020. Tractography reproducibility challenge with empirical data (TraCED): the 2017 ISMRM diffusion study group challenge. J. Magn. Reson. Imaging 51 (1), 234–249.

Ning, L., Bonet-Carne, E., Grussu, F., Sepehrband, F., Kaden, E., Veraart, J., Blumberg, S.B., et al., 2020. Cross-scanner and cross-protocol multi-shell diffusion MRI data harmonization: algorithms and results. NeuroImage 221, 117128.

Nowacki, A., Schlaier, J., Debove, I., Pollo, C., 2018. Validation of diffusion tensor imaging tractography to visualize the dentatorubrothalamic tract for surgical planning. J. Neurosurg. 130 (1), 99–108.

Oberheim, N.A., Wang, X., Goldman, S., Nedergaard, M., 2006. Astrocytic complexity distinguishes the human brain. Trends Neurosci. 29 (10), 547–553.

Oh, S.W., Harris, J.A., Ng, L., Winslow, B., Cain, N., Mihalas, S., Wang, Q., et al., 2014. A mesoscale connectome of the mouse brain. Nature 508 (7495), 207–214.

Palm, C., Axer, M., Gräßel, D., Dammers, J., Lindemeyer, J., Zilles, K., Pietrzyk, U., Amunts, K., 2010. Towards ultra-high resolution fibre tract mapping of the human brain—registration of polarised light images and reorientation of fibre vectors. Front. Hum. Neurosci. 4, 9.

**540** **PART | IV** From streamlines to tracts

Palmer, S.M., Rosa, M.G.P., 2006. A distinct anatomical network of cortical areas for analysis of motion in far peripheral vision. Eur. J. Neurosci. 24 (8), 2389–2405.

Palombo, M., Ianus, A., Guerreri, M., Nunes, D., Alexander, D.C., Shemesh, N., Zhang, H., 2020. SANDI: a compartment-based model for non-invasive apparent soma and neurite imaging by diffusion MRI. NeuroImage 215, 116835.

Panagiotaki, E., Schneider, T., Siow, B., Hall, M.G., Lythgoe, M.F., Alexander, D.C., 2012. Compartment models of the diffusion MR signal in brain white matter: a taxonomy and comparison. NeuroImage 59 (3), 2241–2254.

Paquette, M., Gilbert, G., Descoteaux, M., 2019. Penthera 3T. https://doi.org/10.5281/zenodo.2602049.

Parent, M., Parent, A., 2006. Single-axon tracing study of corticostriatal projections arising from primary motor cortex in primates. J. Comp. Neurol. 496 (2), 202–213.

PRIMatE Data Exchange (PRIME-DE) Global Collaboration Workshop and Consortium, 2020. Accelerating the evolution of nonhuman primate neuro-imaging. Neuron 105 (4), 600–603.

Qi, S., Meesters, S., Nicolay, K., Ter Haar Romeny, B.M., Ossenblok, P., 2015. The influence of construction methodology on structural brain network measures: a review. J. Neurosci. Methods 253, 170–182.

Rheault, F., De Benedictis, A., Daducci, A., Maffei, C., Tax, C.M.W., Romascano, D., Caverzasi, E., et al., 2020. Tractostorm: the what, why, and how of tractography dissection reproducibility. Hum. Brain Mapp. 41 (7), 1859–1874.

Rheault, F., Bayrak, R.G., Wang, X., Schilling, K.G., Greer, J.M., Hansen, C.B., Kerley, C., et al., 2022. TractEM: evaluation of protocols for deterministic tractography white matter atlas. Magn. Reson. Imaging 85, 44–56.

Rockland, K.S., 2002. Non-uniformity of extrinsic connections and columnar organization. J. Neurocytol. 31 (3–5), 247–253.

Rockland, K.S., 2013. Collateral branching of long-distance cortical projections in monkey. J. Comp. Neurol. 521 (18), 4112–4123.

Rockland, K.S., 2015. About connections. Front. Neuroanat. 9, 61.

Rockland, K.S., 2020. What we can learn from the complex architecture of single axons. Brain Struct. Funct. 225 (4), 1327–1347.

Rockland, K.S., DeFelipe, J., 2016. Neuroanatomy for the XXIst century. Front. Res. Top. https://doi.org/10.3389/978-2-88919-916-7.

Rockland, K.S., Ojima, H., 2003. Multisensory convergence in calcarine visual areas in macaque monkey. Int. J. Psychophysiol. 50 (1–2), 19–26.

Rockland, K.S., Saleem, K.S., Tanaka, K., 1994. Divergent feedback connections from areas V4 and TEO in the macaque. Vis. Neurosci. 11 (3), 579–600.

Rodrigues, P., Prats-Galino, A., Gallardo-Pujol, D., Villoslada, P., Falcon, C., Prckovska, V., 2013. Evaluating structural connectomics in relation to different Q-space sampling techniques. vol. 16 (Pt 1). International Conference on Medical Image Computing and Computer-Assisted Intervention, pp. 671–678.

Rodriguez-Moreno, J., Porrero, C., Rollenhagen, A., Rubio-Teves, M., Casas-Torremocha, D., Alonso-Nanclares, L., Yakoubi, R., et al., 2020. Area-specific synapse structure in branched posterior nucleus axons reveals a new level of complexity in thalamocortical networks. J. Neurosci. Off. J. Soc. Neurosci. 40 (13), 2663–2679.

Saleem, K.S., Avram, A.V., Glen, D., Yen, C.C.-C., Ye, F.Q., Komlosh, M., Basser, P.J., 2021. High-resolution mapping and digital atlas of subcortical regions in the macaque monkey based on matched MAP-MRI and histology. NeuroImage 245, 118759.

Salo, R.A., Belevich, I., Manninen, E., Jokitalo, E., Gröhn, O., Sierra, A., 2018. Quantification of anisotropy and orientation in 3D electron microscopy and diffusion tensor imaging in injured rat brain. NeuroImage 172, 404–414.

Sarubbo, S., De Benedictis, A., Maldonado, I.L., Basso, G., Duffau, H., 2013. Frontal terminations for the inferior fronto-occipital fascicle: anatomical dissection, DTI study and functional considerations on a multi-component bundle. Brain Struct. Funct. 218 (1), 21–37.

Sarubbo, S., De Benedictis, A., Milani, P., Paradiso, B., Barbareschi, M., Rozzanigo, U., Colarusso, E., et al., 2015. The course and the anatomo-functional relationships of the optic radiation: a combined study with 'post mortem' dissections and 'in vivo' direct electrical mapping. J. Anat. https://doi.org/10.1111/joa.12254.

Sarubbo, S., Petit, L., De Benedictis, A., Chioffi, F., Ptito, M., Dyrby, T.B., 2019. Uncovering the inferior fronto-occipital fascicle and its topological organization in non-human primates: the missing connection for language evolution. Brain Struct. Funct. 224 (4), 1553–1567.

Schilling, K., Janve, V., Gao, Y., Stepniewska, I., Landman, B.A., Anderson, A.W., 2016. Comparison of 3D orientation distribution functions measured with confocal microscopy and diffusion MRI. NeuroImage 129, 185–197.

Schilling, K., Gao, Y., Janve, V., Stepniewska, I., Landman, B.A., Anderson, A.W., 2017a. Can increased spatial resolution solve the crossing fiber problem for diffusion MRI? NMR Biomed. 30 (12). https://doi.org/10.1002/nbm.3787.

Schilling, K.G., Gao, Y., Stepniewska, I., Tung-Lin, W., Wang, F., Landman, B.A., Gore, J.C., Chen, L.M., Anderson, A.W., 2017b. The VALiDATe29 MRI based multi-channel atlas of the squirrel monkey brain. Neuroinformatics 15 (4), 321–331.

Schilling, K.G., Janve, V., Gao, Y., Stepniewska, I., Landman, B.A., Anderson, A.W., 2018. Histological validation of diffusion MRI fiber orientation distributions and dispersion. NeuroImage 165, 200–221.

Schilling, K.G., Daducci, A., Maier-Hein, K., Poupon, C., Houde, J.-C., Nath, V., Anderson, A.W., Landman, B.A., Descoteaux, M., 2019. Challenges in diffusion MRI tractography—lessons learned from international benchmark competitions. Magn. Reson. Imaging 57, 194–209.

Schilling, K.G., Rheault, F., Petit, L., Hansen, C.B., Nath, V., Yeh, F.-C., Girard, G., et al., 2021a. Tractography dissection variability: what happens when 42 groups dissect 14 white matter bundles on the same dataset? NeuroImage 243, 118502.

Schilling, K.G., Tax, C.M.W., Rheault, F., Hansen, C., Yang, Q., Yeh, F.-C., Cai, L., Anderson, A.W., Landman, B.A., 2021b. Fiber tractography bundle segmentation depends on scanner effects, vendor effects, acquisition resolution, diffusion sampling scheme, diffusion sensitization, and bundle segmentation workflow. NeuroImage 242, 118451.

Schilling, K.G., Grussu, F., Ianus, A., Hansen, B., Barrett, R.L.C., Aggarwal, M., Michielse, S., Nasrallah, F., Syeda, W., Wang, N., Veraart, J., Roebroeck, A., Bagdasarian, A.F., Eichner, C., Sepehrband, F., Zimmermann, J., Soustelle, L., Bowman, C., Tendler, B.C., Hertanu, A., Jeurissen, B., Frydman, L., van de Looij, Y., Hike, D., Dunn, J.F., Miller, K., Landman, B.A., Shemesh, N., Anderson, A., McKinnon, E., Farquharson, S., Acqua, F.D., Pierpaoli, C., Drobnjak, I., Leemans, A., Harkins, K.D., Descoteaux, M., Xu, D., Huang, H., Santin, M.D., Grant, S.C., Obenaus, A., Kim, G.S., Wu, D., Bihan, D.L., Blackband, S.J., Ciobanu, L., Fieremans, E., Bai, R., Leergaard, T., Zhang, J., Dyrby, T.B., Johnson, G.-A., Cohen-Adad, J., Budde, M.D., Jelescu, I.O., 2023. Recommendations and guidelines from the ISMRM Diffusion Study Group for preclinical diffusion MRI: Part 2—ex vivo imaging. arXiv. https://arxiv.org/abs/2209.13371.

Schmahmann, J.D., Pandya, D.N., Wang, R., Dai, G., D'Arceuil, H.E., de Crespigny, A.J., Wedeen, V.J., 2007. Association fibre pathways of the brain: parallel observations from diffusion spectrum imaging and autoradiography. Brain J. Neurol. 130 (Pt 3), 630–653.

Seidlitz, J., Sponheim, C., Glen, D., Ye, F.Q., Saleem, K.S., Leopold, D.A., Ungerleider, L., Messinger, A., 2018. A population MRI brain template and analysis tools for the macaque. NeuroImage 170, 121–131.

Shen, K., Bezgin, G., Schirner, M., Ritter, P., Everling, S., McIntosh, A.R., 2019. A macaque connectome for large-scale network simulations in TheVirtualBrain. Sci. Data 6 (1), 123.

Sherman, S.M., Guillery, R.W., 2006. Exploring the Thalamus and Its Role in Cortical Function, second ed. MIT Press, p. 484.

Sitko, A.A., Mason, C.A., 2016. Organization of axons in their tracts. In: Axons and Brain Architecture. https://doi.org/10.1016/b978-0-12-801393-9.00013-x.

Spead, O., Poulain, F.E., 2020. Trans-axonal signaling in neural circuit wiring. Int. J. Mol. Sci. 21 (14). https://doi.org/10.3390/ijms21145170.

Sun, S.-W., Neil, J.J., Liang, H.-F., He, Y.Y., Schmidt, R.E., Hsu, C.Y., Song, S.-K., 2005. Formalin fixation alters water diffusion coefficient magnitude but not anisotropy in infarcted brain. Magn. Reson. Med. 53 (6), 1447–1451.

Suryanarayana, S.M., Pérez-Fernández, J., Robertson, B., Grillner, S., 2022. The lamprey forebrain—evolutionary implications. Brain Behav. Evol. 96 (4–6), 318–333.

Takemura, H., Palomero-Gallagher, N., Axer, M., Gräßel, D., Jorgensen, M.J., Woods, R., Zilles, K., 2020. Anatomy of nerve fiber bundles at micrometer-resolution in the vervet monkey visual system. eLife 9. https://doi.org/10.7554/eLife.55444.

Tasic, B., Menon, V., Nguyen, T.N., Kim, T.K., Jarsky, T., Yao, Z., Levi, B., et al., 2016. Adult mouse cortical cell taxonomy revealed by single cell transcriptomics. Nat. Neurosci. 19 (2), 335–346.

Tax, C.M., Grussu, F., Kaden, E., Ning, L., Rudrapatna, U., Evans, C.J., St-Jean, S., et al., 2019. Cross-scanner and cross-protocol diffusion MRI data harmonisation: a benchmark database and evaluation of algorithms. NeuroImage 195, 285–299.

Teh, I., McClymont, D., Zdora, M.-C., Whittington, H.J., Davidoiu, V., Lee, J., Lygate, C.A., Rau, C., Zanette, I., Schneider, J.E., 2017. Validation of diffusion tensor MRI measurements of cardiac microstructure with structure tensor synchrotron radiation imaging. J Cardiovasc Magn Reson 19 (1), 31.

Thomas, C., Ye, F.Q., Okan Irfanoglu, M., Modi, P., Saleem, K.S., Leopold, D.A., Pierpaoli, C., 2014. Anatomical accuracy of brain connections derived from diffusion MRI tractography is inherently limited. Proc. Natl. Acad. Sci. U. S. A. 111 (46), 16574–16579.

Tournier, J.-D., Calamante, F., Connelly, A., 2013. Determination of the appropriate B value and number of gradient directions for high-angular-resolution diffusion-weighted imaging. NMR Biomed. 26 (12), 1775–1786.

Tuch, D.S., 2004. Q-ball imaging. Magn. Reson. Med. 52 (6), 1358–1372.

Vaessen, M.J., Hofman, P.A.M., Tijssen, H.N., Aldenkamp, A.P., Jansen, J.F.A., Backes, W.H., 2010. The effect and reproducibility of different clinical DTI gradient sets on small world brain connectivity measures. NeuroImage 51 (3), 1106–1116.

Vanni, S., Hokkanen, H., Werner, F., Angelucci, A., 2020. Anatomy and physiology of macaque visual cortical areas V1, V2, and V5/MT: bases for biologically realistic models. Cereb. Cortex 30 (6), 3483–3517.

Wakana, S., Caprihan, A., Panzenboeck, M.M., Fallon, J.H., Perry, M., Gollub, R.L., Hua, K., Zhang, J., Jiang, H., Dubey, P., Blitz, A., van Zijl, P., Mori, S., 2007. Reproducibility of quantitative tractography methods applied to cerebral white matter. NeuroImage 1053-8119. 36 (3), 630–644. https://doi.org/10.1016/j.neuroimage.2007.02.049.

Walsh, C.L., Tafforeau, P., Wagner, W.L., Jafree, D.J., Bellier, A., Werlein, C., Kühnel, M.P., et al., 2021. Imaging intact human organs with local resolution of cellular structures using hierarchical phase-contrast tomography. Nat. Methods 18 (12), 1532–1541.

Wang, L., Marquardt, T., 2013. What axons tell each other: axon-axon signaling in nerve and circuit assembly. Curr. Opin. Neurobiol. 23 (6), 974–982.

Wang, H., Zhu, J., Reuter, M., Vinke, L.N., Yendiki, A., Boas, D.A., Fischl, B., Akkin, T., 2014. Cross-validation of serial optical coherence scanning and diffusion tensor imaging: a study on neural fiber maps in human medulla oblongata. NeuroImage 100, 395–404.

Wang, H., Lenglet, C., Akkin, T., 2015. Structure tensor analysis of serial optical coherence scanner images for mapping fiber orientations and tractography in the brain. J. Biomed. Opt. 20 (3), 036003.

Winnubst, J., Bas, E., Ferreira, T.A., Zhuhao, W., Economo, M.N., Edson, P., Arthur, B.J., et al., 2019. Reconstruction of 1,000 projection neurons reveals new cell types and organization of long-range connectivity in the mouse brain. Cell 179 (1), 268–281.e13.

Xu, R., Bichot, N.P., Takahashi, A., Desimone, R., 2022. The cortical connectome of primate lateral prefrontal cortex. Neuron 110 (2), 312–327.e7.

Yeh, C.-H., Smith, R.E., Liang, X., Calamante, F., Connelly, A., 2016. Correction for diffusion MRI fibre tracking biases: the consequences for structural connectomic metrics. NeuroImage 142, 150–162.

Yeh, C.-H., Smith, R.E., Liang, X., Calamante, F., Connelly, A., 2018. Investigating the streamline count required for reproducible structural connectome construction across a range of brain parcellation resolutions. Proc. Int. Soc. Magn. Reson. Med. p. 1558.

Zalesky, A., Fornito, A., Harding, I.H., Cocchi, L., Yücel, M., Pantelis, C., Bullmore, E.T., 2010. Whole-brain anatomical networks: does the choice of nodes matter? NeuroImage 50 (3), 970–983.

Zemmoura, I., Serres, B., Andersson, F., Barantin, L., Tauber, C., Filipiak, I., Cottier, J.-P., Venturini, G., Destrieux, C., 2014. FIBRASCAN: a novel method for 3D white matter tract reconstruction in MR space from cadaveric dissection. NeuroImage 103, 106–118.

Zemmoura, I., Blanchard, E., Raynal, P.-I., Rousselot-Denis, C., Destrieux, C., Velut, S., 2016. How Klingler's dissection permits exploration of brain structural connectivity? An electron microscopy study of human white matter. In: Brain Structure and Function. https://doi.org/10.1007/s00429-015-1050-7.

Zhong, Y.-M., Rockland, K.S., 2003. Inferior parietal lobule projections to anterior inferotemporal cortex (area TE) in macaque monkey. Cereb. Cortex 13 (5), 527–540.

Zhou, J., Wen, Y., She, L., Sui, Y.-N., Liu, L., Richards, L.J., Poo, M.-M., 2013. Axon position within the corpus callosum determines contralateral cortical projection. Proc. Natl. Acad. Sci. U. S. A. 110 (29), E2714–E2723.

# Chapter 27

# Tractography validation part 3: Lessons learned through validation studies

**Kurt G. Schilling[a], Francois Rheault[b], and Tim B. Dyrby[c,d]**

[a]*Department of Radiology and Radiological Sciences, Vanderbilt University Medical Center, Nashville, TN, United States,* [b]*Department of Computer Science, University of Sherbrooke, Sherbrooke, QC, Canada,* [c]*Danish Research Centre for Magnetic Resonance, Centre for Functional and Diagnostic Imaging and Research, Copenhagen University Hospital—Amager and Hvidovre, Copenhagen, Denmark,* [d]*Department of Applied Mathematics and Computer Science, Technical University of Denmark (DTU), Kongens Lyngby, Denmark*

Diffusion MRI fiber tractography has become a mainstay in neuroimaging studies due to its ability to noninvasively map the structural connectivity of the brain, revealing fundamental insight into brain function, development, and cognition, as well as complex brain disorders. However, the process from acquiring data to generating and interpreting 3D maps of brain connections is a multistep procedure with numerous assumptions and decisions that affect the ability of tractography to represent the true structural connectivity of the brain. Clearly, validation of tractography is critical for it to become a useful biomedical tool. Through the years, using simulations, physical phantoms, anatomical model systems, and empirical validation, we have gained a tremendous understanding of the successes and limitations of tractography, leading to an improved ability to develop new techniques and interpret tractography findings. While the focus of the previous two chapters was to act as a "cookbook" for different validation strategies, describing best practices and how and when to use different validation strategies, the focus of this chapter is to highlight "what we have learned" through validation studies, with a focus on how these lessons were learned and their implications for tractography.

## 1 Anatomy can be more complex than our models

Brain anatomy is enormously complex. Complexity occurs not only in structure, geometry, and architecture across spaces with scales ranging from axons to fiber pathways, but also in the nomenclature and terminology used to describe this anatomy. **Thus the first lesson learned is that our current tractography neuroimaging methods may paint an overly simplistic picture of the structural connections of the brain.**

First, axons are not perfect sticks or cylinders. Axons exhibit heterogeneity of axon diameters along individual axons, across axons within the same functional tract, and across tracts in the white matter. Diameter variations along the axon (Fig. 1) have been observed in 3D electron microscopy of mice (Lee et al., 2019) and 3D synchrotron X-ray imaging in monkeys (Andersson et al., 2020), and indirectly validated through empirical measurements of unique time-dependent diffusion signatures (Lee et al., 2020b). Not only do individual axons have varying diameters along the trajectory, but also within the same pathway, where the distribution can change as axonal tracts can enter and depart along a pathway. This has been visualized through neuronal tracers (Call and Bergles, 2021; Basu et al., 2023) and electron microscopy in the macaque pyramidal tract (Firmin et al., 2014). Finally, axon diameter distributions vary widely across pathways in the human brain, again observed through histological studies (Aboitiz et al., 1992; Liewald et al., 2014; Innocenti et al., 2015) and observed empirically with specialized diffusion measures (Huang et al., 2020; Veraart et al., 2020, 2021). Thus axonal diameter varies along axons, within axons of the same fiber pathways, and across different fiber pathways of the brain. Despite this apparent complexity of describing both along- and across-axon variation, most diffusion weighted measurements (i.e., low to medium gradient strength systems) are insensitive to axon diameters (Dyrby et al., 2013b; McKinnon et al., 2017; Veraart et al., 2019), which supports the use of microstructure and fiber orientation reconstruction methods that model axons as an array of zero-radius sticks (Behrens et al., 2003). Further, this phenomenon means that certain measures (e.g., high b-value data) are exquisitely sensitive to the intra-axonal volume fraction and fiber dispersion, which can (and have) been utilized to improve tractography accuracy or facilitate quantitative tractography.

---

**Handbook of Diffusion MR Tractography. https://doi.org/10.1016/B978-0-12-818894-1.00004-5**
Copyright © 2025 Elsevier Ltd. All rights are reserved, including those for text and data mining, AI training, and similar technologies.

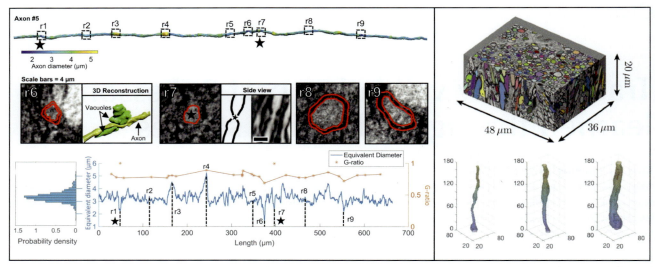

**FIG. 1** Axons are not perfect sticks/cylinders with a single diameter, but along the axon, they have a distribution of diameters. (Left) A single axon (top) identified through synchrotron X-ray imaging of a vervet monkey corpus callosum from Andersson et al. (2020). The axon diameter variation along the axon (bottom, *blue line*) can be caused by the local environment such as cell bodies, blood vessels, and vacuoles as in (r6) or at the region of the nodes of Ranvier (r7). (Right) Intra-axonal space of several axons segmented from 3D scanning electron microscopy of a mouse corpus callosum from Lee et al. (2020a) shows changes in cross-sectional area (diameter) and also changes in directionality of axons. The mean diameter is ∼1 μm. The units along *x*-, *y*-, and *z*-axis are 0.1 μm.

Second, axons may exhibit trajectories with complexities that may be hard to capture adequately with diffusion tractography. Axons may exhibit a wavelike undulating course (Fig. 2A), with undulations occurring over the scale of tens of micrometers, an effect observed in light microscopy from tracer studies or axonal staining, confocal microscopy, polarized light image (PLI), and in electron microscopy (Nilsson et al., 2012, 2013). These undulations may infer important biomechanical properties to tissue (Sunderland and Bradley, 1961; Nilsson et al., 2012; Dyrby et al., 2013a), but may complicate tractography by introducing intravoxel fiber orientation dispersion (Mollink et al., 2017). These undulations do not need to be periodic, and any deviation of axons from their main axis might be better described with the term *microdispersion* (Andersson et al., 2020). Beyond microdispersion, axonal branching may also occur, where axon bundles that have traveled together over a certain distance separate toward potentially different targets. Further, branching may occur along a single axon, with collateral axons potentially having different targets from the main axonal brain (see Chapter 26). Both observations of this branching phenomena are routinely seen in tracer studies (Fig. 2B) and can be detected with diffusion MRI

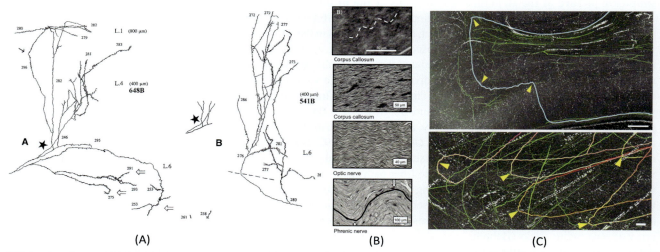

**FIG. 2** Complex axonal trajectories include branching (A), undulation (B), and complex curvatures (C) derived from single axon tracing experiments and confocal and brightfield microscopy. (A) From Rockland, (2020). (B) From Brabec et al., (2020) and Dyrby et al., (2013a). (C) From Xu et al., (2021).

(Lundell et al., 2011). Finally, other complex trajectories can be observed with histology, including 180-degree or right-angle turns (Fig. 2C) (Xu et al., 2021), at the interface of gray and white matter or within the deep white matter. Yet, while the trajectories of axons can be complex, they are not chaotic. Most of the association, projection, and commissural pathways that we intend to study with tractography are composed of a confluence of well-ordered axons sharing similar trajectories, volume, and orientations (and function) as other axons within the tract. In fact, comparisons between micro-dissection and streamlines are often so visually similar (Lawes et al., 2008) that tractography is described by many researchers as a virtual in vivo dissection.

Third, even with the seminal work of the Dejerine-Klumpke (1895) dating to more than a century ago, there is not yet a full consensus on the name or anatomical definitions of many white matter pathways. Thus the nomenclature is also complicated. Historically, multiple discoveries happened in parallel across nations, which led to differences in the descriptions and names of white matter pathways (Bullock et al., 2022). A classic example is the superior longitudinal fasciculus (SLF) (Janelle et al., 2022). The SLF has long been recognized as the largest associative fiber bundle system in the brain, connecting the frontal lobe with other areas of the ipsilateral hemisphere (Martino et al., 2013). Yet, despite its apparent simplicity, various descriptions can give rise to confusion in anatomical classification and subdivisions of this pathway. Traditionally, the SLF is divided into three segments (SLFI, SLFII, SLFIII), described in monkeys using tracer studies (Schmahmann and Pandya, 2006; Schmahmann et al., 2007), in humans using microdissection (Wang et al., 2016), and in humans using diffusion tractography (Thiebaut de Schotten et al., 2011; Catani and Thiebaut de Schotten, 2012; Schurr et al., 2020); yet the literature also contains convincing descriptions of the SLF having anywhere from two segments (typically an anterior and posterior segment) (Zhang et al., 2010; De Benedictis et al., 2014) to as many as five (Kamali et al., 2014), with differences in the descriptions of their exact connections (see Table 1 of Janelle et al., 2022 for a summary of SLF subdivisions). Moreover, even though the SLF and the arcuate fasciculus (AF) are widely recognized as distinct entities both anatomically and functionally, there are still descriptions of these bundles that suggest possible overlap in connections and locations (Catani and Thiebaut de Schotten, 2012; De Benedictis et al., 2014; Zemmoura et al., 2014; Wu et al., 2016; Mandonnet et al., 2018), originating as early as Dejerine using the SLF and AF as synonyms (Dejerine-Klumpke, 1895). Further examples exist, suggesting the SLFI is indistinguishable from another bundle, the cingulum (Wang et al., 2016). Another example involves the anatomy of fronto-occipital connectivity. While there is no doubt that communication exists between the primary/secondary visual cortex of the occipital lobe and frontal cortex, whether these are direct (monosynaptic) or indirect (polysynaptic) pathways is controversial. The inferior fronto-occipital fasciculus has been demonstrated in monkeys using gross dissection and tractography (Sarubbo et al., 2013, 2019; Hau et al., 2016), neither of which are able to distinguish mono- versus polysynaptic pathways; meanwhile, these direct connections were not demonstrated in tract-tracing studies (Schmahmann et al., 2007), where authors declared a direct IFOF "spurious" and "nonexistent" and rather composed of indirect pathways coursing through the extreme capsule (Schmahmann and Pandya, 2006, 2007). Yet again, recent tracer studies have found sparse direct connections in the monkey brain (Gerbella et al., 2010), and careful dissections have demonstrated fibers coursing between occipital and frontal lobes that are distinct from others in the extreme capsule system (Sarubbo et al., 2019). Overall, these highlight not only the difficulty in extrapolating results from nonhuman to human and vice versa (Rilling et al., 2008; Forkel et al., 2014; Bullock et al., 2022), but also the complexity in describing white matter pathways.

Fourth, these small variations in terminology and definition may lead to different tractography representations of the same pathway. The pyramidal tract (PyT) is a good example of a well-known pathway with a variety of similar, almost equivalent anatomical definitions. First described by Dejerine-Klumpke (1895), it is now defined as starting from the cortex, going through the internal capsule, then through the anterior region of the brainstem, and finally to the medulla oblongata within the pyramids (Chenot et al., 2019). It is considered to include both the corticospinal tract (CST) and the corticobulbar tract (CBT). Since the PyT is made of motor neurons, some definitions mention explicitly the termination in the precentral and postcentral gyri, while others only specify that any neurons that originated at the cortex and continue past the lower brainstem are part of the PyT (Ebeling and Reulen, 1992). One can easily see multiple ways to describe the almost exact same pathway while using a variation of the same definition, resulting in possibly different experimental results. This has been observed empirically in several studies, for example where tractography experts, anatomists, or clinical users were asked to perform bundle segmentation on various pathways of the brain on the same datasets and observed large variability due to differences in anatomical definition and segmentation protocol (Fig. 3). Such differences among experts in neuroanatomy are common (Forkel et al., 2014; Mandonnet et al., 2018; Panesar and Fernandez-Miranda, 2019; Schilling et al., 2021a); even though an agreement may have been reached on a general anatomical definition, small inconsistency can lead to major deviations in the results.

Finally, the sheer number of axons and the combinatorial number of possible cortico-cortical connections makes fully characterizing the human connectome a challenging endeavor. Tracer studies have provided an excellent resource for

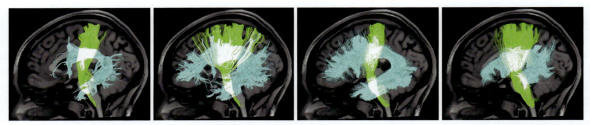

FIG. 3 Differences in anatomical definition may lead to different bundle segmentation protocols and different representations of the same intended pathways. Here, differences in size, shape, and connections are easily observed in the corticospinal tract and arcuate fasciculus when segmentation is performed by different groups or protocols (Schilling et al., 2021a).

validating tractography, yet, as described in Chapter 26, these tracers are typically an extremely sparse representation of the total number of possible connections in the brain. For example, commonly used macaque "ground truths" are composed of 29 injection sites (Markov et al., 2014), or 36 injection sites (Schmahmann and Pandya, 2006). Optimistically, recent Collations of Connectivity data of the Macaque (CoCoMac) now enable large-scale collations of thousands of experimental findings of anterograde and retrograde tracer experiments and measures of histologically derived "connection strength" metadata (see Chapter 26). Further, work in other species including the Allen Brain Atlas now includes axonal projections from thousands of injection sites covering nearly the entire mouse brain (Oh et al., 2014). However, despite these resources, our knowledge of neuronal connectivity remains incomplete for nearly all species, particularly when compared to the complete wiring diagram of the 302 neurons of *C. elegans* obtained through electron microscopy (White et al., 1986; Emmons, 2015).

In short, due to the sheer complexity of anatomical model systems, we emphasize that it is important to consider these cadaveric and histological studies not as ground truth, but as a "silver standard" for comparison (Knösche et al., 2015). Even more so, it is important to critically assess the features of these complex systems (orientation, number of connections, the density of connections, etc.) that we are trying to replicate, and/or simplify, through the tractography process.

## 2 Reconstruction techniques capture fiber orientation distribution but are limited in extracting discrete peaks and orientation

Because estimates of the fiber orientation form the basis of nearly all tractography algorithms, validating the ability to characterize the geometry and orientations is necessary to understand the limitations of tractography. One of the first major recognized limitations was the crossing fiber problem. Validation studies using tractography based on DTI highlighted areas where pathways were not accurately tracked, noting this occurred in regions where multiple pathways crossed (Dauguet et al., 2007; Gao et al., 2013a,b; Jbabdi et al., 2013). This challenge became known as the crossing fiber problem, and through empirical measurements using diffusion MRI (Jeurissen et al., 2013) and histological quantification (Schilling et al., 2017), has been shown to affect a majority of voxels in the brain. This challenge spurred the development of several fiber reconstruction algorithms to overcome this challenge. **Overall, many of these fiber reconstruction algorithms well-capture the orientation distribution of white matter within MRI voxels**.

These algorithms are most often validated through simulations, which can quickly identify successes and limitations across a range of fiber crossing angles, acquisition choices, noise characteristics, and algorithmic choices (Alexander, 2005). This has allowed comparisons across algorithms, and optimization of techniques and updates in the model assumptions (Canales-Rodriguez et al., 2015, 2018). Notably, a community challenge on a field of simulated fiber orientations allowed comparisons of 20 different algorithms (Fig. 4), concluding that none outperformed the others in every experimental condition. However, many were successful with different diffusion weightings, diffusion directions, and complicated ground truth orientation distributions (Daducci et al., 2014).

As described in the previous chapter, anatomical model systems in combination with microscopy techniques can be compared to diffusion-estimated orientations. Most commonly, two-dimensional light microscopy in combination with manual tracing/annotation (Leergaard et al., 2010), filter matching (Choe et al., 2012), or structure tensor analysis (Budde and Frank, 2012) allow quantification of tissue orientation, which has been shown to generally align well with the diffusion tensor. Extending validation to three dimensions with multiphoton (Kamagata et al., 2016), confocal microscopy (Khan et al., 2015; Schilling et al., 2016, 2018a), PLI (Axer et al., 2016), and OCT (Wang et al., 2015; Jones et al., 2020) has allowed a full characterization of the geometrical structures within volumes the size of MRI voxels. Even while the histological fiber orientation distributions are much more complex than the simulations described

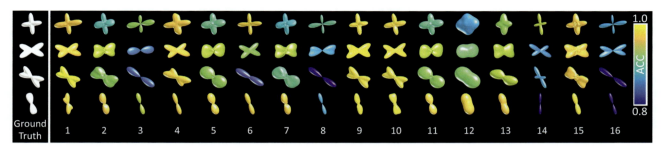

**FIG. 4** Reconstruction techniques well-capture the fiber orientation distribution, with trade-offs in false positives and false negatives in identifying discrete fiber populations. This figure is derived from data from Daducci et al. (2014) using a simulation framework for validation at varying crossing angles and with varying SNR. Fiber orientation distribution glyphs are shown for 16 reconstruction algorithms (as columns), compared to ground truth fiber orientations (left, white glyphs), and colored based on the angular correlation coefficient (ACC) with ground truth. A higher ACC indicates greater overall correlation and agreement with ground truth. Visually, the ability to successfully resolve crossing fibers (success rate) and numbers of false positive and false negative fiber populations vary as a function of cross angle, and vary across algorithms. For details on quantitative metrics (ACC, success rate, false positive and false negative peaks), see Chapter 26. *(Data from Alessandro Daducci.)*

previously, histological analysis confirms simulation results, with no model outperforming others in every quality metric evaluated (see Chapter 26 for quantitative comparison metrics of fiber orientations). In fact, all methods describe the overall shape of the fiber distributions quite well (Schilling et al., 2018a; Jones et al., 2020). However, histological evaluation highlights limitations in extracting discrete measures from the distribution (e.g., the number of peaks, or discrete orientations), as well as the inability to resolve fiber populations that cross at sharp (acute) angles due to partial volume effects between these two fiber populations. These studies have highlighted trade-offs in acquisition and sampling schemes, resolution, and reconstruction methods, but also highlight the fact that real brain tissue has complex geometries (see Section 1, Anatomy can be more complicated than our models), with crossing, curving, and undulating/dispersion of fiber orientations throughout the voxel.

In addition to the directionality and number of fiber populations, validating the dispersion, or alternatively, the coherence, of fibers is relevant for both tractography algorithms and microstructural modeling with diffusion MRI. Quantification of histology has well-demonstrated that white matter is not simply parallel sets of axons but exhibits fiber dispersions of approximately 15–25 degrees in rodents (Leergaard et al., 2010), primates (Schilling et al., 2018a; Andersson et al., 2020), and human samples (Lontis et al., 2009; Ronen et al., 2014). Validating against expected anatomy (i.e., high dispersion in and near the neocortex), as well as against anatomical models in the spinal cord and brain, has shown that many representations of the fiber orientation are able to adequately capture fiber-orientated dispersion effects within a voxel.

Rather than extracting and validating the FOD directly based on histology, several studies have empirically assessed the diffusion signal itself. This includes investigating acquisition schemes, modeling, and SNR on the observed directionality, reproducibility, or angular information in the diffusion signal (Jones, 2004; Jones and Basser, 2004; Farrell et al., 2007; Landman et al., 2007). As a classic example, Farrell et al. (2007) and Landman et al. (2007) take a highly oversampled human dataset and empirically assess the effects of SNR and diffusion weighting schemes on precision and accuracy of DTI measures by comparing subsampled data to the fully sampled dataset. Together, these have shown that the optimal acquisition for capturing orientation depends on the model used, and the optimal angular resolution is inherently linked to the diffusion weighting and reconstruction used (Tournier et al., 2013).

Finally, it is important to stress that accurate fiber orientation estimates alone do not fully characterize the fiber geometry. The underlying FOD is often treated as a compass needle centered at the center of the voxel. This means that spatial information on how pathways are organized at the subvoxel level is lost when estimating the FOD. Several geometric configurations of fibers (crossing, "kissing," branching, turning, etc.) can result in the same FOD (Alexander and Seunarine, 2010). This loss of spatial information of fiber orientations within a voxel, and across neighboring voxels, can be a major challenge in the specificity of tractography and contributes to several biases (described in Section 8).

## 3 Tractography can reconstruct known WM anatomy—Valid path, shape, and position of WM bundles

As a tool to analyze specific white matter bundles of the brain, it is critical that tractography bundle segmentation is able to extract the correct location, shape, and endpoints of the bundles. Thus we must validate not only the streamline generation process but also the techniques and decisions used to create and/or segment the fiber bundle from the rest of the brain. **Towards this end, a great number of studies have demonstrated that tractography can indeed reconstruct known WM bundles, delineating the overall path and shape of the true anatomy.**

Early on, several exploratory and validation studies shared a quite optimistic view of the ability to map major white matter bundles. Many of the pioneers who first developed the constraints used to dissect bundles relied on prior knowledge to determine the regions through which streamlines must start, end, or pass through, at the time creating tractography segmentation protocols (Wakana et al., 2004, 2007) and white matter atlases (Wakana et al., 2004; Catani and Thiebaut de Schotten, 2008; Mori et al., 2008; Oishi et al., 2008) for broad dissemination and replication. Many noted, although qualitatively, the obvious correspondence between tractography and both early and contemporary texts of dissection, leading to the moniker of a "virtual dissection." One of the first studies to demonstrate a clear comparison of the two modalities (Lawes et al., 2008) visualized DTI tractography atlases and their variability across a population against that of photographs of cadaveric dissections (Fig. 5). Remarkably, these comparisons showed similar orientations and locations, and fine-scale similarities including curvatures and gyral terminations. Early comparisons against injected tracers in the pig brain (Dyrby et al., 2007) and monkey brain (Dauguet et al., 2006, 2007) have also shown good overall agreement in overlap (specifically, Dice overlap coefficient) of volumes between tracer and tractography, with variability noted due to tractography parameter settings and reconstruction method (both DTI and more advanced methods). Overall, tracer studies have shown that tractography bundle segmentation can replicate major features of white matter bundles (Schmahmann et al., 2007) and represent the gross geometrical organization of these pathways, with most studies focusing on major association pathways.

The striking similarities between tractography and expected anatomy have led to the use of tractography to evaluate the qualitative observations of microdissections, and as a complementary tool to improve the understanding and location of pathways. Examples include using the quantitative nature of tractography to classify the inferior fronto-occipital fasciculus into subcomponents (Sarubbo et al., 2013), or using tractography to provide evidence of the existence and uniqueness of this pathway (Forkel et al., 2014; Sarubbo et al., 2019) that has been controversial in the literature. Similar studies have used tractography to quantify dissection observations, for example, to classify and distinguish components of the SLF (Wang et al., 2016), define the stem and shape of the uncinate fasciculus (Hau et al., 2016), or provide 3D descriptions and relationships of the acoustic radiations with other structures (Maffei et al., 2019) that are otherwise hard to do through the dissection process. In a similar manner, using tractography to verify tracer-derived organization principles (e.g., the spatial

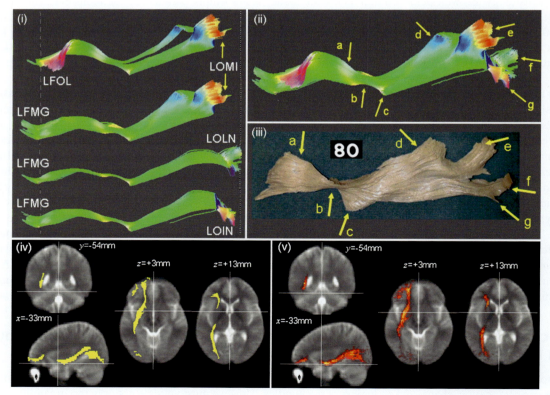

**FIG. 5** Tractography can reconstruct known white matter anatomy, with valid paths, shapes, and positions of bundles. This figure shows a qualitative comparison of dissection to tractography (Lawes et al., 2008) of the inferior frontal-occipital fasciculus.

Tractography validation part 3: Lessons learned through validation studies **Chapter | 27 549**

organization and rotation of trajectories from different connections), Jbabdi et al. (2013) investigated projections from the frontal cortex, showing that tractography provides converging evidence of the connections of this system of fibers.

More confidence in the ability to delineate known pathways has been shown in physical phantoms. Of note, the FiberCup dataset (Fillard et al., 2011) was created to represent a coronal slice of the brain with crossing, curving, and fanning structures (see Chapter 25 for more details). Several acquisition schemes, reconstruction algorithms, and tractography methods have been investigated using these techniques, with strong evidence that many algorithms can reconstruct the known bundles of this phantom, with high-volume bundle coverage. In the future, the ability to create complex phantoms and/or simulations (Neher et al., 2014; Sarwar et al., 2019) and analyze a combinatorial number of pipelines (Cote et al., 2013) has the potential to elucidate more limitations of existing algorithms and lead to improvements in future ones.

Finally, further evidence of the ability of tractography to segment known bundles has been provided in the form of visual evaluation by experts. Notably, the "DTI Challenge" (Pujol et al., 2015) was a public contest to reconstruct the pyramidal tract in neurosurgical cases presenting with gliomas near the motor cortex. Many algorithms reconstructed bundles that were graded highly by neurosurgeons, even in the presence of tumors. This means most algorithms were able to find a majority of the expected projections. However, many were not able to trace specific connections (the lateral projections of the hand, face, and tongue area) that require strong curvature and passing through a region of complex fiber configurations. Here, it was noted that a large variability exists in the same tract reconstructed using different algorithms. Despite limitations, tractography on clinical-quality data and relatively simple model fitting (diffusion tensor, two-tensor models) was able to provide clinically useful information to neurosurgeons in pathological tissues.

## 4  Tractography is a fair, but far from perfect, predictor of the presence of connections: There is an inherent sensitivity/specificity trade-off

A major emphasis of many early validation studies was on the sensitivity of algorithms to detect known anatomy (i.e., true connections); however, it is also critical that tractography have a high specificity (i.e., the ability to correctly identify when no connections exist) to provide anatomically accurate maps of the brain. Thus, while it was found that tractography can reasonably segment specific bundles with the use of prior anatomical knowledge, it is necessary to also validate its ability to estimate region-to-region connections of the (potentially unknown) bundles of the brain.

One of the first validation studies to quantify both sensitivity and specificity was performed by Thomas et al. (2014). Here, they acquired high-quality data with both high resolution and a large number of diffusion directions, in an ex vivo macaque brain, and compared tractography to maps of known axonal projections from previous tracer studies. Dividing the brain into white matter and gray matter regions and designating them as tracer-positive or tracer-negative (i.e., the ground truth), they were able to quantify tractography passing through (or not passing through) the same regions as either true positives or false positives (or true negatives or false negatives). Despite the exceptional quality of the diffusion data, none of the tested tractography pipelines demonstrated high anatomical accuracy, showing a fundamental trade-off in sensitivity and specificity (Fig. 6). Those that had the highest sensitivity showed the lowest specificity, and vice versa.

These findings of limited ability to achieve both high sensitivity and high specificity at the same time were confirmed in pig brains using manganese tracing of the same brains (Knösche et al., 2015), in mouse brains by comparing against atlases from a large number of mice brains (Aydogan et al., 2018), and in other primates where tracer and tractography were performed in the same brain (Schilling et al., 2019b). In an additional community challenge (Schilling et al., 2018b) (utilizing both phantom data and the same datasets as in Thomas et al. (2014) and Schilling et al. (2019b)) allowing researchers to implement their own algorithms with any desired settings, most algorithms were found to lie on the extreme ends of the sensitivity and specificity curve (i.e., high sensitivity and low specificity, or vice-versa). Quantifying volume-based measures of accuracy, those that resulted in high overlap with ground truth consistently had high overreach, and vice versa. Again, tractography algorithms can be susceptible to both false-positives and false-negatives.

Thus this sensitivity and specificity trade-off has been demonstrated in several brains for a number of pathways. The overall accuracy was shown to be dependent on the fiber reconstruction methods and tracking algorithm, as well as seed regions and how the seed is utilized, dependent on distance from the seed, and worse for voxel-based comparisons than for larger region-based analysis (Thomas et al., 2014; Aydogan et al., 2018; Schilling et al., 2019b). **Most studies concluded that there exists an inherent limitation(s) in mapping long-range connections that cannot be solved with better data quality or better modeling of orientation alone**.

While the preceding studies have been focused on one or a few bundles of the brain, others have quantified the sensitivity and specificity of mapping connection region-to-region connectivity of the many cortical connections—typically using the (sparse) ground truth connectomes from macaque tracer studies. These have again illustrated similar trade-offs in tractography's ability to delineate connections. Using tracer connectivity matrices accumulated through many injections of

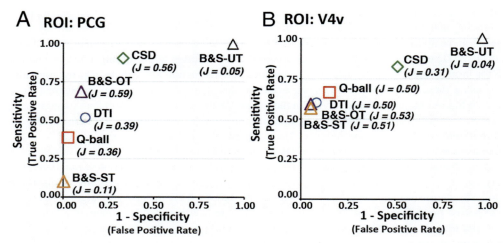

FIG. 6 Tractography is a fair, but far from perfect, predictor of the presence of connection with an inherent sensitivity/specificity trade-off. Figure from Thomas et al. (2014), where ground truth was defined from tracer injections in two regions (PCG and V4v) from Schmahmann and Pandya (2006). Here, various reconstruction and tractography methods show different ability to reconstruct true positive connections while avoiding true negative connections, but there is a trade-off in sensitivity and specificity. Modified from Thomas et al., (2014).

the macaque brain, Azadbakht et al. (2015) showed that tractography is a fair predictor of the presence or absence of a connection between two regions (with a focus on the connections of the visual cortex), and is able to identify a majority, 74%, of the expected connections in this area. However, the accuracy was again dependent on the length of the connections (Liptrot et al., 2014), curvature and stopping constraints, and thresholding (for both tracer and tractography)—all highlighting the inherent trade-off in sensitivity and specificity. A final international collaborative effort to validate the region-to-region connections using simulated human data (Maier-Hein et al., 2017) found that many algorithms produced tractograms containing over 90% of the ground truth bundles. However, while true connections existed, algorithms contained an even larger number of invalid bundles. These false-positive connections looked real, were coherently oriented, and occurred systematically across many algorithms. These challenges were attributed to certain *bottleneck* regions of the brain that cause ambiguities in pathway selections for tractography (see Section 8).

## 5 The strength of connections has useful predictive power

In the validation studies described previously, a cluster of streamlines representing a fiber pathway was treated as a binary entity, specifying the existence or lack of a connection from one place to another. However, ideally, we would be able to quantify some measure of connectivity to and from these regions representative of some feature of the neuroanatomical "connectivity." As described in Chapter 26, the idea of neuroanatomical connectivity is not well-defined. One intuitive measure would simply be the number of axonal projections between two regions: the more axons, the more connected these regions may be. It could also be some measure of the density of axonal connections. However, connectivity is more than simply a structural component, and connectivity may also be considered as some measure of the capacity of information transfer, which necessitates the inclusion of some microstructural attributes of or along the pathway (i.e., saltatory conduction velocity is related to the diameter of axons, intermodal spacing, and myelination (Drakesmith et al., 2019)). Ultimately, the purpose of studying connectivity is to better understand brain function, and whatever image-derived measures should relate to some variation in brain function.

Attempts at parallel measures using tractography often assume that measures of streamline counts, streamline counts weighted by connecting areas (or streamline lengths), or counts weighted by some microstructural measure (i.e., commonly FA) can be treated as proxies for true anatomical features. Clearly, understanding how the strength of connections from tractography varies with true anatomical features is necessary to understand and quantify connectivity in the brain using these techniques.

Validating the strength of connections is commonly done using histological tracers. Instead of comparing the percent of correctly determined binary connections (as described earlier), tractography measures can be compared against the number of labeled neurons in a region (Donahue et al., 2016) or the total fraction of labeled neurons (Ambrosen et al., 2020). Several examples of quantifying histological tracers include (1) counting the number of pixels with tracer signal within a region (i.e., density); (2) measuring the total tracer signal intensity per pixel (Calabrese et al., 2015); (3) counting the stained axons crossing a given boundary or region (Gao et al., 2013a,b); or (4) categorizing connectivity as weak, intermediate, or strong based on visual observation (van den Heuvel et al., 2015). Several histological databases or atlases in murine and primate models exist with quantitative measures of tracer-based "connection strength" (Stephan et al., 2001; Bakker et al., 2012; Calabrese et al., 2015; Dell et al., 2019). Many of these are compiled from tens or hundreds of published studies, and not necessarily from the same animal as that analyzed with tractography.

The main takeaway from several studies comparing tracer and tractography is that tractography-derived strength of connections has useful predictive power and can provide region-to-region connectivity that modestly correlates with neuroanatomical connectivity. This is interesting because the correlation is not related to saltatory conduction velocity, merely a geometrical feature of the tract like cross-sectional size and its lengths. A range of streamline-correlation coefficients has been observed, due to differences in connectivity measure quantified, possibly specimen-to-specimen variation, and differences in acquisition and tractography algorithms. In a study using BDA tracers in the motor cortex of a squirrel monkey (Gao et al., 2013a,b), DTI-based tractography streamlines and BDA fibers in the same brain resulted in correlation coefficients between 0.44 and 0.93, with results dependent on algorithms used and seed regions. Tractography was less reliable at mapping the fine-scale connections at the scale of individual voxels. Similar findings were observed in the mouse brain, comparing tractography against the Allen Brain Atlas neuronal tracer data from a set of 488 tracer injection experiments (Calabrese et al., 2015). Again, tractography varied based on the coarsity of the brain parcellation, with correlations between the number of streamlines and tracer density ranging from 0.46 for fine scale (469 injections by 592 regions), to 0.77 for midlevel (structural level with 296 regions), to 0.99 at a very coarse level (13 regions).

A number of validation studies have investigated the more complex macaque brain, utilizing the CoCoMac database (a whole brain parcellation classified as strong, moderate, weak connections) and the Markov Kennedy database (a single hemisphere quantification of label tracers) to investigate the relationship between streamline counts and anatomical connectivity. Here, various tractography streamline densities and the number of streamlines has shown low to moderate positive correlations with tracer connection strengths ranging from 0.25 to 0.31 (van den Heuvel et al., 2015), 0.59 (Donahue et al., 2016), and 0.70 (Shen et al., 2019a) (Fig. 7). Thus tractography is a modest predictor of weight of neuronal connections. Further investigations found that tractography underestimates connection weights for pathways with high tracer connection weights, and correlations with tracer are much less for weaker connected pathways, yet still perform better than chance (Donahue et al., 2016). Again, different analysis methods, different data, angular and spatial resolution, SNR, and different algorithms lead to different prediction abilities. As an example, in Shen et al. (2019a), algorithm parameters were varied to optimize the correlation between tractography and the relatively sparse tracer knowledge to create an optimized full connectome (which could be used in subsequent validation studies). Continuing with these datasets, this same group also showed that several connectome graph theory measures are well captured by tractography (i.e., degree and modularity), whereas classification of network centrality, participation coefficients, and core periphery was not as predictive of that from the relatively incomplete tracer data (Shen et al., 2019b).

As an alternative to a simple correlation coefficient, it is possible that information theory metrics may lend insight into the relationship and information content in tractography. Towards this end, Ambrosen et al. (2020) utilized tractography and the Markov tracer dataset to measure cross entropy as a metric for the ability of tractography to properly encode tracer strengths. Here, they again confirmed that tractography is moderately predictive of the ground truth. Moreover, they found that (1) high angular resolution and high signal-to-noise ratio are important to estimate connectivity; (2) tractography derived from moderate image resolution is in better agreement with the tracer network than that derived from very low resolution, and (3) surprisingly, moderate image resolution results are in better agreement with tracer than very high-resolution data, which means there is a significant relationship between the voxel size and the size of the bundles we are attempting to capture.

Finally, the prediction power of tractography has also been tested in the ferret. Again, tractography connection strength is highly correlated with tract-tracing data ($r = 0.67$–$0.91$) (Delettre et al., 2019) comparable to the macaque. In agreement with Donahue et al. (2016) they found that correlations decreased or were no longer significant after regressing out path-length distances due to path-length biases in tractography (Liptrot et al., 2014) (see following discussion on tractography biases).

## 6 Methods are reproducible, but there is significant variance across methods

Every step of the tractography process leads to variability and ambiguity in the final tractogram (see Fig. 1 in Chapter 25). As described earlier, validation studies have shown that differences in these steps can lead to differences in orientations, bundles, and connectivity. While not strictly validation studies, empirical observations of changes introduced in the final quantification due to differences in acquisition, reconstruction, and protocols offer further insight into the tractography process. **To date, most studies have shown that with a fixed acquisition, fixed voxel-wise reconstruction, and fixed tractography algorithm, results from connectivity studies, bundle segmentation, and bundle microstructure features**

**552 PART | IV From streamlines to tracts**

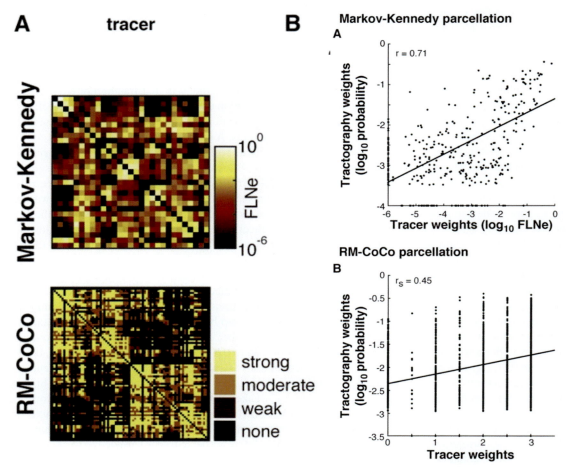

**FIG. 7** Strength of connections has useful predictive power. The strength of connections quantified from histological tracer studies from two different macaque resources is well captured using diffusion tractography (Shen et al., 2019b). On the left are two different tracer-derived connectivity matrices, and the right are tractography weights plotted against tracer-weights, resulting in modest-to-high correlations.

**are fairly reproducible across scanners and repeated scans; however, different methods and procedures can vary dramatically**. Following, we describe (mostly empirical) studies of reproducibility of (1) bundle segmentation, (2) segmentation of surgically relevant pathways, and (3) connectomics and graph theory measures. A detailed review of reproducibility of microstructure measures (i.e., DTI-based metrics) is given in Chapter 33.

For bundle segmentation, several studies have shown that users are consistently able to delineate regions of interest to segment bundles that lead to high reproducibility of microstructure measures and high-volume overlap (Wakana et al., 2007). In-depth analysis of individual pathways has shown that, while volume overlap is similar, interpreting individual streamlines may be highly variable (Rheault et al., 2020a), as well as measures of microstructure along pathways, due to differences in region placement and user decisions in the process. These volume and streamline differences are dependent on experience with anatomy (Rheault et al., 2020a), dependent on pathways (larger, well-defined pathways are more reproducible) (Bayrak et al., 2019), and vary depending on data quality. Intrasession reproducibility is slightly higher than reproducibility across sessions and scanners, yet tractography remains fairly reproducible across scanners (Vishwesh Nath et al., 2018).

Very few, often surgically relevant, pathways have been studied in detail. The larger corticospinal tract was highly reproducible when segmentation was performed by both experts and nonexperts in neuroanatomy (Rheault et al., 2020a); however, depending on how the segmentation is utilized, reproducibility of the overall volume of the pathway is higher than the reproducibility of streamline-based quantification. Visualization of the dentatorubrothalamic tract, a possible substrate for deep brain stimulation-induced tremor alleviation, was also sensitive to different tracking methodologies (Andreas et al., 2018), which could lead to different localizations, coordinates in atlas space, and brain targets. For the smaller cranial nerves, differences in acquisition and software packages led to dramatic differences in the ability to

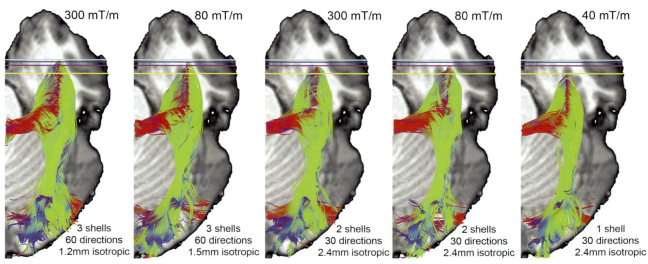

**FIG. 8** Methods are reproducible, but there is significant variance across methods. Here, differences in hardware and data acquisition may lead to differences in tractography of the optic radiations for image-guided surgery; however, for a given acquisition and bundle segmentation technique, the location and trajectory of this pathway are highly reproducible (Chamberland et al., 2018). Note that differences in hardware (gradients) and acquisition (shells, directions, resolution) are shown as text at top and bottom of image, respectively.

successfully track nerves, and visual differences in the resulting streamlines (Jacquesson et al., 2019). In a study on the optic radiations, a pathway that is necessary to localize to avoid surgical visual field defects, Chamberland et al. (2018), found that data acquisition choices (i.e., resolution, b-value, number of diffusion directions) lead to substantial variance in this pathway. By measuring simple geometrical features of this pathway (distance from Meyer's loop to the temporal pole) variations as much as 16 mm were observed. While the variance in acquisition leads to variance in tractography, higher spatial and angular resolution led to a qualitatively more complete delineation of the pathway (Fig. 8). Finally, by inviting many groups to dissect the same fiber bundles on the same datasets (Schilling et al., 2020b), it was found that bundle segmentation variability due to segmentation protocol differences is larger than variability within a protocol and variability across subjects. This variance stems from not only the process of tractography but also the definitions used to describe the anatomy of a bundle and the subsequent constraints used to dissect it (see Section 1). In a follow-up study investigating and ranking various sources of variability, Schilling et al., (2021b) found that differences due to the choice of bundle segmentation workflows are larger than any other studied confound, followed by the effects of acquisition protocol, in particular acquisition image resolution, resulting in the lowest reproducibility of tractography. This was followed by vendor effects, scanner effects, and finally diffusion scheme and b-value effects, which had similar reproducibility to scan-rescan variation. The main point is that experimental factors greatly impact the reliability of tractography, and these confounds need to be considered when interpreting conclusions or results across sites, acquisitions, or workflows.

Parallel findings have been shown for the quantification of structural connectivity networks, with highly reproducible connectomes and graph theoretical measures, but significant variations across acquisitions and methodologies. Using a fixed analysis framework, connectomics measures show greater intrasubject similarity than across subjects (Girard et al., 2015), with generally excellent reproducibility (Roine et al., 2019). However, nearly every step in the process of generating a connectome alters these measures, which may create large discrepancies in the literature. For a thorough review, see Chapter 24 and also Qi et al. (2015), where the sensitivity of graph measures to brain parcellations (Zalesky et al., 2010; Bassett et al., 2011), acquisition schemes (Vaessen et al., 2010), reconstruction (Rodrigues et al., 2013), and tractography algorithms (Bastiani et al., 2012) is summarized. Again, without a gold standard for comparison, and no way to eliminate these influences, Qi et al. (2015) suggest quantification of variance, and optimization of methodologies to reduce variance is necessary to make comparable inferences in the literature. A final source of variability is the number of streamlines, which has been studied empirically and found to be dependent on the number of nodes in the brain parcellation, reconstruction, and tractography techniques (Liptrot et al., 2014; Donahue et al., 2016; Yeh et al., 2016, 2018; Moyer et al., 2019). The number of streamlines to ensure adequate information content and maintain reproducible metrics can vary widely from tens of thousands to tens of millions depending on subsequent analysis.

Despite high reproducibility, consistent tractography does not imply reliable tractography, and biases due to systematic effects and variance across methods may confound our interpretation of results (Calamante, 2019).

## 7 There exists no "optimal" combination of acquisition, reconstruction, or tractography parameters

Currently, there is no consensus on an "optimal" pipeline for tractography (see the review by Tax et al. (2022) for thorough discussions on artifacts, preprocessing steps, and practical considerations in diffusion MRI), and no single step in the process, from acquisition to reconstruction to tractography, has been shown to consistently lead to more anatomically accurate results than any others. **Thus, while we know that different choices in acquisition, preprocessing, and tractography lead to different results (Section 6), there is no tractography pipeline that is "optimal" for all acquisitions, pathways, and experimental questions.** In general, it is intuitive that higher SNR, and more sampling directions, lead to better estimates of fiber orientation, as shown empirically (Farrell et al., 2007; Landman et al., 2007), in simulations (Alexander and Barker, 2005; Cook et al., 2007; Daducci et al., 2014; Canales-Rodriguez et al., 2018), phantoms (Cho et al., 2008; Tournier et al., 2008), and histological comparisons (Schilling et al., 2016, 2018a; Ambrosen et al., 2020), although high SNR and more data are not always feasible due to time constraints. Similarly, a higher b-value is known to result in larger angular contrast (Jones and Basser, 2004; Tournier et al., 2013), although at the expense of SNR. Interestingly, pushing the image resolution to the scale of tens of microns will still not solve the crossing fiber problem, as verified both ex vivo and through the histological analysis (Schilling et al., 2017). Overall, for orientation estimation, many reconstruction methods adequately estimate orientations, crossing fibers, and dispersion of fibers, yet all are more (or less) successful with varying acquisition conditions and different fiber geometries (Daducci et al., 2014).

For tractography, the choice of method parameters greatly influences the final results. Again, each parameter choice represents the fundamental specificity/sensitivity trade-off (Knösche et al., 2015), where parameters that let false-positive tracts disappear will inevitably lead to the removal of true positives. Similarly, measures of overlap, overreach, and overall tractography-to-ground truth disagreement are all affected by choices made in the tractography process (Schilling et al., 2019b). Nearly all parameters (Fig. 9) have been shown to significantly affect results, with differences due to parameters comparable in effect to differences between subjects (Aydogan et al., 2018). No combination of parameters is always optimal in all situations (Sarwar et al., 2019), with variations occurring across methods, deterministic and probabilistic differences, influences of stopping criteria (Dauguet et al., 2007; Bastiani et al., 2012; Azadbakht et al., 2015), curvature (Dauguet et al., 2007; Bastiani et al., 2012), integration steps, and thresholding on both bundles and overall connectivity (Bastiani et al., 2012). In general, more directions lead to an overall improvement in accuracy; improvements in SNR (more averages, more directions) are always beneficial, and gross brain structures are less susceptible to variations in parameters (Ambrosen et al., 2020). Overall, again, no single method outperforms all others, for all pathways of the brain (Knösche et al., 2015; Maier-Hein et al., 2017; Schilling et al., 2018b).

## 8 Limitations: Obstacles, biases, and challenges to overcome

A detailed description of biases and limitations of tractography is given in Chapter 28. Here, we briefly summarize these limitations, but with a focus on how this knowledge was gained through validation.

The first sets of biases deal with the shape of pathways. Short-range, straight, and larger (cross-sectionally) pathways are more reliably detected than long-range, bending, and narrow fibers, which results in the underrepresentation of certain pathways and creates an interaction between path length and sensitivity/specificity. The sources of errors have been investigated using simple numerical simulations that show the combination of noise and integration errors on generating streamlines leads to the propagation of error over distance (Anderson, 2001; Lazar and Alexander, 2003), and this was confirmed on phantoms, which highlights that the longer, more complex, pathways require more decisions through potentially complex regions to accurately determine a connection (Cote et al., 2013). Direct comparisons against tracers show that most false-negative connections are long-range projections (Azadbakht et al., 2015; Schilling et al., 2019b), while tractography (streamline) connectivity strength has been shown to be highly dependent on the distance (Liptrot et al., 2014; Donahue et al., 2016; Delettre et al., 2019). Distance is also a biological principle for the preference of connections between two regions (Hilgetag et al., 2016; Hilgetag and Goulas, 2020). Thus attempts to reduce this tractography bias may be hard to disentangle from biological truth.

Next, tractography and subsequent quantification are sensitive to seeding strategies. In contrast to the harder to reconstruct long pathways (due to path-length dependency), seeding from all white matter voxels may cause an overrepresentation of the large and long bundles (Smith et al., 2012; Girard et al., 2014), because more streamlines would be initiated in the longer and larger bundles. Seeding from the interface of white and gray matter may partially alleviate this problem; however, this introduces partial volume effects at the interface itself and may hinder the ability to dissect pathways that have few voxels at the interface (i.e., fornix), or must pass through narrow white matter regions (inferior fronto-occipital fasciculus) (Rheault et al., 2020b).

Tractography validation part 3: Lessons learned through validation studies Chapter | 27 555

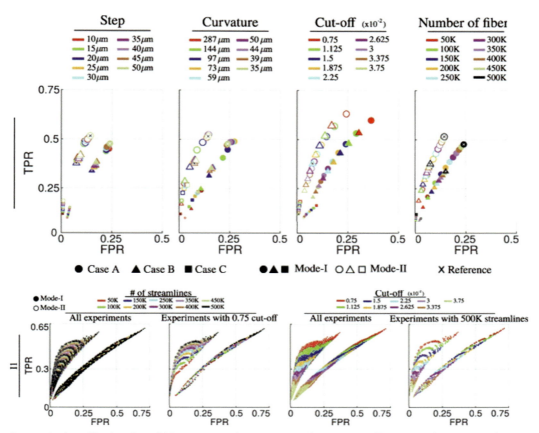

FIG. 9 There is no optimal combination of acquisition, reconstruction, or tractography parameters. Here, comparing tracers against tractography in the mouse brain, Aydogan et al. show that the accuracy of tractography varies based on every decision made in the tractography process (including step size, curvature and stopping thresholds, number of streamlines, quantification procedures, among others) (Aydogan et al., 2018). Plots show the true positive rate (TPR) plotted against false positive rate (FPR) for three different tractography algorithms (cases A, B, C) and two quantification strategies (mode-I and mode-II, which include anatomical constraints in the quantification process), and a number of parameters (colored, with legends above plots).

These limitations were largely discovered from qualitative observations of streamline endpoints, comparisons of streamline count based on different tractography strategies and methodologies, and assessing the effects of seeding on connectomes.

The next set of obstacles is due to partial volume effects. Overlap of white and gray matter voxels, or even overlap of pathways within voxels or within regions, can contribute to inaccurate results. Crossing fibers have been shown to be extremely prevalent in the brain using both diffusion MRI itself (Jeurissen et al., 2013) and through histology (Schilling et al., 2017), with the first location of the discrepancy between tractography and tracer often occurring immediately adjacent to the gray matter (Gao et al., 2013a,b; Girard et al., 2020). Additional partial volume effects occur in bottleneck regions, where multiple pathways converge towards a single area, sharing a common orientation, before remerging towards their own unique endpoints. Tractography is then faced with a combinatorial number of options for starting points and endpoints, which leads to a high risk for false-positive connections. This was first highlighted in an international challenge on a simulated brain dataset with known bundles (Maier-Hein et al., 2017), where the false-positive bundles occurred nearly four times as often as true positives, but these false positives were consistent across algorithms and were often associated with bottleneck regions of tractography overlap. By simply observing the overlap of the known association, projection, and commissural fiber pathways, it has been shown that a majority of white matter voxels present as bottlenecks for tractography, which may lead to incorrect or erroneous estimates of connectivity (Schilling et al., 2022). However, optimistically, incorporating anatomical knowledge into the tractography process can overcome these bottleneck obstacles and lead to highly accurate tractography (Schilling et al., 2020a). By utilizing the same validation datasets that have previously highlighted sensitivity/specificity trade-offs in tractography (Thomas et al., 2014), it was shown that prior knowledge (in the form of manually placed or template-driven constraints) can achieve a high specificity and high sensitivity simultaneously, overcoming many of the limitations described earlier (Schilling et al., 2020a).

**556 PART | IV** From streamlines to tracts

A number of validation studies have shown that accuracy decreases from a coarse to a fine spatial scale. For example, comparisons of tracers in both bundle segmentation and connectomics have consistently shown tractography can measure region-to-region connectivity at a coarse scale, whereas voxel-to-voxel connections are unreliable (Gao et al., 2013a,b; Calabrese et al., 2015; Ambrosen et al., 2020). Similarly, phantom studies have shown that measuring volume overlap and the presence or absence of a bundle is more anatomically accurate than streamline-based measures of accuracy (i.e., measures derived from valid/invalid bundles led to higher fidelity metrics than those derived from valid/invalid streamlines; see (Fillard et al., 2011; Cote et al., 2013) for details on bundle-based and streamline-based accuracy measures for the Tractometer dataset). Similarly, reproducibility is dependent on scale, with bundle volume and location more reproducible than streamline locations (Rheault et al., 2020a), larger pathways more reproducible than smaller ones (Vishwesh Nath et al., 2018), and connectivity measures more consistent with fewer nodes (Zalesky et al., 2010; Bassett et al., 2011).

Finally, we have learned that tractography has limitations in determining where the white matter pathways end and insert into the gray matter. This limitation is often referred to as the gyral bias, which refers to the overrepresentation of pathways connecting the gyral crowns, while missing many of the connections to the walls of the gyrus or the sulcal fundi. This bias was observed empirically using tractography itself, where it was hypothesized that it could be a true anatomical finding (Nie et al., 2012; Chen et al., 2013; Kleinnijenhuis et al., 2015). However, when compared to the true connectivity revealed through the myelin stains (Reveley et al., 2015; Schilling et al., 2018c), tracer data (Reveley et al., 2015), or polarized light imaging (Mollink et al., 2017), tractography was shown to overestimate connections at gyral crowns. This gyral bias is dependent on the seeding strategy (discussed earlier), curvature constraints, and quantification of connectivity (Schilling et al., 2018c). The cause of this bias was attributed to two biological phenomena. First, the high curvature of fibers as they enter the cortex, compared to the relatively straight fibers of the crown (Mollink et al., 2017; Schilling et al., 2018c), makes the crown the preferred route of travel for most algorithms. Second, a dense set of superficial fibers exists directly underneath many sulcal fundi (Reveley et al., 2015), traveling tangential to the gray matter and presenting orientation-based obstacles for streamlines to reach the cortex. Thus, while tractography can delineate the correct location, shape, and trajectory of white matter pathways (Section 3), one must be cautious when interpreting endpoints, or the exact cortical connections derived from streamlines.

## 9 Bridging anatomy and tractography is challenging

Two decades of validation studies have provided unprecedented insight into the challenges and limitations of diffusion tractography and spurred development in not only image acquisition, reconstruction, tractography, and bundle segmentation, but also dissemination of reproducible workflows. To summarize lessons learned: tractography has a proven ability to reproducibly reconstruct known white matter anatomy, identify the presence or absence of connections, and reflect measures that are proportional to biological measures of axonal connections and density. Despite successes, there are still challenges associated with two challenging inverse problems: reconstruction of fiber geometries from the diffusion signal, and reconstructing continuous pathways from discrete estimates of orientation (Schilling et al., 2019a).

### 9.1 Bridging anatomy and tractography

As described in Chapter 25, validation is the act of bridging anatomy and the neuroimaging technique under investigation (i.e., tractography) to identify limitations and improve the technique. The major challenge with diffusion tractography lies in the multiple length scales covered by this technique, from the physics of diffusion at the scale of micrometers, to the averaging effect of diffusion over voxels at the scale of millimeters, to the streamline-generation process over centimeters across the brain. Thus validation is required at multiple scales—and fundamental trade-offs or limitations exist when using alternative techniques (physical phantoms, simulations, dissection, microscopy, and/or tracer data) to image the anatomy that will often serve as our "ground truth" or "gold standard" for comparison with tractography. Examples include mismatches in the fields of view being compared between MRI voxels and high-resolution optical imaging modalities (i.e., electron microscopy); tissue and spatial distortions occurring during tissue preparation and optical imaging (tissue dehydration, sectioning, etc.); differences in anatomy between the human and validation model system (murine, nonhuman primate); and possibly oversimplistic assumptions and generation of phantom and/or simulation datasets. Yet, progress is continually being made on all fronts—including large-scale multimodality datasets of mice (Oh et al., 2014) and nonhuman primates (Xu et al., 2021; Howard et al., 2022), quantification of the traditionally qualitative microdissection process (Zemmoura et al., 2014; De Benedictis et al., 2018), improvements in 3D mapping of orientation with polarized light imaging (Menzel et al., 2020), optical coherence tomography (Magnain et al., 2019), polarization-sensitive OCT (Jones et al., 2021), light-sheet imaging of cleared tissue (Leuze et al., 2021), synchrotron imaging (Andersson et al., 2020;

Georgiadis et al., 2021), large field of view 3D electron microscopy (Eberle and Zeidler, 2018), improvements to histology-MRI correspondence (Huszar et al., 2019; Casamitjana et al., 2022; Perens et al., 2023), and complex simulations utilizing existing histology (Ginsburger et al., 2019; Palombo et al., 2019; Lee et al., 2021). Together, these improvements and technologies will help us better bridge the gap between tractography and the true anatomy.

## 9.2 Validation as an iterative process

With these challenges in mind, it is important to emphasize that tractography validation is an iterative process. In this process, we not only validate our tractography toolset and algorithms, but also extend our knowledge about the ground truth anatomy. This improved knowledge leads us to change and refine our methodology, assumptions, and techniques, which can then be compared to the ground truth again. As described in a recent review by Dyrby et al. (2018) on diffusion MRI microstructural modeling validation, we "loop until our method's results agree with the gold standard, and/or until the updated knowledge of the ground truth can explain the discrepancies observed" (Dyrby et al., 2018). In tractography, we start with the theory—theory of how the diffusion process affects the observed signal, the theory about how the observed signal reflects tissue geometry, and the theory about how the anatomy of the bundles and connections of the brain can be recreated from discrete voxel-wise measures of fiber orientation. We then select one or several independent modalities to serve as a gold standard(s) for validating this new method (described in Chapters 25 and 26). We then describe and understand any deviations between our tractography (signal, FOD, bundles, connectome) and the validation method. Finally, we adjust and improve our tractography pipeline or our understanding of the ground truth. This process nicely highlights the fact that bridging the gap between anatomy and tractography is not a one-way bridge. Indeed, diffusion MRI and tractography can inform our histology measurements, and this is the way to push forward advances in brain mapping. By representing the same feature with different modalities, at different length scales or different resolutions, it may become possible to perform what could be called a hybrid-tractography, correlative tractography, or joint-tractography by merging technologies and knowledge across image resolution scales and across modalities to improve estimates of connectivity.

Together, with the combination of gathering anatomical knowledge of the systems we intend to investigate and studying the influence and effects they have on the tractography algorithms, we have been able to identify the successes and limitations of tractography. And understanding their limitations, and identifying problems for future generations of algorithms to solve, is the first step towards improving the methods and advancing their use in studying brain connectivity.

# References

Aboitiz, F., Scheibel, A.B., Fisher, R.S., Zaidel, E., 1992. Fiber composition of the human corpus callosum. Brain Res. 598 (1-2), 143–153.

Alexander, D.C., 2005. Multiple-fiber reconstruction algorithms for diffusion MRI. Ann. N. Y. Acad. Sci. 1064 (1), 113–133.

Alexander, D.C., Barker, G.J., 2005. Optimal imaging parameters for fiber-orientation estimation in diffusion MRI. NeuroImage 27 (2), 357–367.

Alexander, D.C., Seunarine, K.K., 2010. Mathematics of crossing fibers. In: Jones, P.D.K. (Ed.), Diffusion MRI: Theory, Methods, and Applications. Oxford University Press, p. 451.

Ambrosen, K.S., Eskildsen, S.F., Hinne, M., Krug, K., Lundell, H., Schmidt, M.N., van Gerven, M.A.J., Morup, M., Dyrby, T.B., 2020. Validation of structural brain connectivity networks: the impact of scanning parameters. NeuroImage 204, 116207.

Anderson, A.W., 2001. Theoretical analysis of the effects of noise on diffusion tensor imaging. Magn. Reson. Med. 46 (6), 1174–1188.

Andersson, M., Kjer, H.M., Rafael-Patino, J., Pacureanu, A., Pakkenberg, B., Thiran, J.P., Ptito, M., Bech, M., Bjorholm Dahl, A., Andersen Dahl, V., Dyrby, T.B., 2020. Axon morphology is modulated by the local environment and impacts the noninvasive investigation of its structure-function relationship. Proc. Natl. Acad. Sci. USA 117 (52), 33649–33659.

Andreas, N., Jürgen, S., Ines, D., Claudio, P., 2018. Validation of diffusion tensor imaging tractography to visualize the dentatorubrothalamic tract for surgical planning. J. Neurosurg. 130 (1), 99–108.

Axer, M., Strohmer, S., Grassel, D., Bucker, O., Dohmen, M., Reckfort, J., Zilles, K., Amunts, K., 2016. Estimating fiber orientation distribution functions in 3D-polarized light imaging. Front. Neuroanat. 10, 40.

Aydogan, D.B., Jacobs, R., Dulawa, S., Thompson, S.L., Francois, M.C., Toga, A.W., Dong, H., Knowles, J.A., Shi, Y., 2018. When tractography meets tracer injections: a systematic study of trends and variation sources of diffusion-based connectivity. Brain Struct. Funct. 223 (6), 2841–2858.

Azadbakht, H., Parkes, L.M., Haroon, H.A., Augath, M., Logothetis, N.K., de Crespigny, A., D'Arceuil, H.E., Parker, G.J., 2015. Validation of high-resolution tractography against in vivo tracing in the macaque visual cortex. Cereb. Cortex 25 (11), 4299–4309.

Bakker, R., Wachtler, T., Diesmann, M., 2012. CoCoMac 2.0 and the future of tract-tracing databases. Front. Neuroinform. 6 (30).

Bassett, D.S., Brown, J.A., Deshpande, V., Carlson, J.M., Grafton, S.T., 2011. Conserved and variable architecture of human white matter connectivity. NeuroImage 54 (2), 1262–1279.

Bastiani, M., Shah, N.J., Goebel, R., Roebroeck, A., 2012. Human cortical connectome reconstruction from diffusion weighted MRI: the effect of tractography algorithm. NeuroImage 62 (3), 1732–1749.

Basu, K., Appukuttan, S., Manchanda, R., Sik, A., 2023. Difference in axon diameter and myelin thickness between excitatory and inhibitory callosally projecting axons in mice. Cereb. Cortex 33 (7), 4101–4115. https://doi.org/10.1093/cercor/bhac329.

Bayrak, R.G., Schilling, K.G., Greer, J.M., Hansen, C.B., Greer, C.M., Blaber, J.A., Williams, O., Beason-Held, L.L., Resnick, S.M., Rogers, B.P., Landman, B.A., 2019. TractEM: fast protocols for whole brain deterministic tractography-based white matter atlas. bioRxiv: 651935.

Behrens, T.E., Woolrich, M.W., Jenkinson, M., Johansen-Berg, H., Nunes, R.G., Clare, S., Matthews, P.M., Brady, J.M., Smith, S.M., 2003. Characterization and propagation of uncertainty in diffusion-weighted MR imaging. Magn. Reson. Med. 50 (5), 1077–1088.

Brabec, J., Lasic, S., Nilsson, M., 2020. Time-dependent diffusion in undulating thin fibers: impact on axon diameter estimation. NMR Biomed 33 (3), e4187.

Budde, M.D., Frank, J.A., 2012. Examining brain microstructure using structure tensor analysis of histological sections. NeuroImage 63 (1), 1–10.

Bullock, D.N., Hayday, E.A., Grier, M.D., Tang, W., Pestilli, F., Heilbronner, S.R., 2022. A taxonomy of the brain's white matter: twenty-one major tracts for the 21st century. Cereb. Cortex 32 (20), 4524–4548.

Calabrese, E., Badea, A., Cofer, G., Qi, Y., Johnson, G.A., 2015. A diffusion MRI tractography connectome of the mouse brain and comparison with neuronal tracer data. Cereb. Cortex 25 (11), 4628–4637.

Calamante, F., 2019. The seven deadly sins of measuring brain structural connectivity using diffusion MRI streamlines fibre-tracking. Diagnostics (Basel, Switzerland) 9 (3), 115.

Call, C.L., Bergles, D.E., 2021. Cortical neurons exhibit diverse myelination patterns that scale between mouse brain regions and regenerate after demyelination. Nat. Commun. 12 (1), 4767.

Canales-Rodriguez, E.J., Daducci, A., Sotiropoulos, S.N., Caruyer, E., Aja-Fernandez, S., Radua, J., Yurramendi Mendizabal, J.M., Iturria-Medina, Y., Melie-Garcia, L., Aleman-Gomez, Y., Thiran, J.P., Sarro, S., Pomarol-Clotet, E., Salvador, R., 2015. Spherical deconvolution of multichannel diffusion MRI data with non-Gaussian noise models and spatial regularization. PLoS One 10 (10), e0138910.

Canales-Rodriguez, E.J., Legarreta, J.H., Pizzolato, M., Rensonnet, G., Girard, G., Patino, J.R., Barakovic, M., Romascano, D., Aleman-Gomez, Y., Radua, J., Pomarol-Clotet, E., Salvador, R., Thiran, J.P., Daducci, A., 2018. Sparse wars: a survey and comparative study of spherical deconvolution algorithms for diffusion MRI. NeuroImage 184, 140–160.

Casamitjana, A., Lorenzi, M., Ferraris, S., Peter, L., Modat, M., Stevens, A., Fischl, B., Vercauteren, T., Iglesias, J.E., 2022. Robust joint registration of multiple stains and MRI for multimodal 3D histology reconstruction: application to the Allen human brain atlas. Med. Image Anal. 75, 102265.

Catani, M., Thiebaut de Schotten, M., 2008. A diffusion tensor imaging tractography atlas for virtual in vivo dissections. Cortex 44 (8), 1105–1132.

Catani, M., Thiebaut de Schotten, M., 2012. Atlas of Human Brain Connections. Oxford University Press, Oxford, UK.

Chamberland, M., Tax, C.M.W., Jones, D.K., 2018. Meyer's loop tractography for image-guided surgery depends on imaging protocol and hardware. Neuroimage Clin. 20, 458–465.

Chen, H., Zhang, T., Guo, L., Li, K., Yu, X., Li, L., Hu, X., Han, J., Hu, X., Liu, T., 2013. Coevolution of gyral folding and structural connection patterns in primate brains. Cereb. Cortex 23 (5), 1208–1217.

Chenot, Q., Tzourio-Mazoyer, N., Rheault, F., Descoteaux, M., Crivello, F., Zago, L., Mellet, E., Jobard, G., Joliot, M., Mazoyer, B., Petit, L., 2019. A population-based atlas of the human pyramidal tract in 410 healthy participants. Brain Struct. Funct. 224 (2), 599–612.

Cho, K.H., Yeh, C.H., Tournier, J.D., Chao, Y.P., Chen, J.H., Lin, C.P., 2008. Evaluation of the accuracy and angular resolution of q-ball imaging. NeuroImage 42 (1), 262–271.

Choe, A.S., Stepniewska, I., Colvin, D.C., Ding, Z., Anderson, A.W., 2012. Validation of diffusion tensor MRI in the central nervous system using light microscopy: quantitative comparison of fiber properties. NMR Biomed. 25 (7), 900–908.

Cook, P.A., Symms, M., Boulby, P.A., Alexander, D.C., 2007. Optimal acquisition orders of diffusion-weighted MRI measurements. J. Magn. Reson. Imaging 25 (5), 1051–1058.

Cote, M.A., Girard, G., Bore, A., Garyfallidis, E., Houde, J.C., Descoteaux, M., 2013. Tractometer: towards validation of tractography pipelines. Med. Image Anal. 17 (7), 844–857.

Daducci, A., Canales-Rodríguez, E.J., Descoteaux, M., Garyfallidis, E., Gur, Y., Lin, Y.C., Mani, M., Merlet, S., Paquette, M., Ramirez-Manzanares, A., Reisert, M., Reis Rodrigues, P., Sepehrband, F., Caruyer, E., Choupan, J., Deriche, R., Jacob, M., Menegaz, G., Prčkovska, V., Rivera, M., Wiaux, Y., Thiran, J.P., 2014. Quantitative comparison of reconstruction methods for intra-voxel fiber recovery from diffusion MRI. IEEE Trans. Med. Imaging 33 (2), 384–399.

Dauguet, J., Peled, S., Berezovskii, V., Delzescaux, T., Warfield, S.K., Born, R., Westin, C.F., 2006. 3D histological reconstruction of fiber tracts and direct comparison with diffusion tensor MRI tractography. Med. Image Comput. Comput. Assist. Interv. 9 (Pt 1), 109–116.

Dauguet, J., Peled, S., Berezovskii, V., Delzescaux, T., Warfield, S.K., Born, R., Westin, C.F., 2007. Comparison of fiber tracts derived from in-vivo DTI tractography with 3D histological neural tract tracer reconstruction on a macaque brain. NeuroImage 37 (2), 530–538.

De Benedictis, A., Duffau, H., Paradiso, B., Grandi, E., Balbi, S., Granieri, E., Colarusso, E., Chioffi, F., Marras, C.E., Sarubbo, S., 2014. Anatomo-functional study of the temporo-parieto-occipital region: dissection, tractographic and brain mapping evidence from a neurosurgical perspective. J. Anat. 225 (2), 132–151.

De Benedictis, A., Nocerino, E., Menna, F., Remondino, F., Barbareschi, M., Rozzanigo, U., Corsini, F., Olivetti, E., Marras, C.E., Chioffi, F., Avesani, P., Sarubbo, S., 2018. Photogrammetry of the human brain: a novel method for three-dimensional quantitative exploration of the structural connectivity in neurosurgery and neurosciences. World Neurosurg. 115, e279–e291.

Dejerine-Klumpke, D.J.D.M., 1895. Anatomie des Centres Nerveux. vol. 1.

Delettre, C., Messe, A., Dell, L.A., Foubet, O., Heuer, K., Larrat, B., Meriaux, S., Mangin, J.F., Reillo, I., de Juan Romero, C., Borrell, V., Toro, R., Hilgetag, C.C., 2019. Comparison between diffusion MRI tractography and histological tract-tracing of cortico-cortical structural connectivity in the ferret brain. Netw. Neurosci. 3 (4), 1038–1050.

Dell, L.-A., Innocenti, G.M., Hilgetag, C.C., Manger, P.R., 2019. Cortical and thalamic connectivity of occipital visual cortical areas 17, 18, 19, and 21 of the domestic ferret (Mustela putorius furo). J. Comp. Neurol. 527 (8), 1293–1314.

Donahue, C.J., Sotiropoulos, S.N., Jbabdi, S., Hernandez-Fernandez, M., Behrens, T.E., Dyrby, T.B., Coalson, T., Kennedy, H., Knoblauch, K., Van Essen, D.C., Glasser, M.F., 2016. Using diffusion tractography to predict cortical connection strength and distance: a quantitative comparison with tracers in the monkey. J. Neurosci. 36 (25), 6758–6770.

Drakesmith, M., Harms, R., Rudrapatna, S.U., Parker, G.D., Evans, C.J., Jones, D.K., 2019. Estimating axon conduction velocity in vivo from microstructural MRI. NeuroImage 203, 116186.

Dyrby, T.B., Søgaard, L.V., Parker, G.J., Alexander, D.C., Lind, N.M., Baaré, W.F., Hay-Schmidt, A., Eriksen, N., Pakkenberg, B., Paulson, O.B., Jelsing, J., 2007. Validation of in vitro probabilistic tractography. NeuroImage 37 (4), 1267–1277.

Dyrby, T.B., Burke, M.W., Alexander, D.C., Ptito, M., 2013a. Undulating and Crossing Axons in the Corpus Callosum May Explain the Overestimation of Axon Diameters with ActiveAx.

Dyrby, T.B., Sogaard, L.V., Hall, M.G., Ptito, M., Alexander, D.C., 2013b. Contrast and stability of the axon diameter index from microstructure imaging with diffusion MRI. Magn. Reson. Med. 70 (3), 711–721.

Dyrby, T.B., Innocenti, G.M., Bech, M., Lundell, H., 2018. Validation strategies for the interpretation of microstructure imaging using diffusion MRI. NeuroImage 182, 62–79.

Ebeling, U., Reulen, H.J., 1992. Subcortical topography and proportions of the pyramidal tract. Acta Neurochir. 118 (3-4), 164–171.

Eberle, A.L., Zeidler, D., 2018. Multi-beam scanning electron microscopy for high-throughput imaging in connectomics research. Front. Neuroanat. 12, 112.

Emmons, S.W., 2015. The beginning of connectomics: a commentary on White et al. (1986) "The structure of the nervous system of the nematode Caenorhabditis elegans". Philos. Trans. R. Soc. Lond. Ser. B Biol. Sci. 370 (1666).

Farrell, J.A., Landman, B.A., Jones, C.K., Smith, S.A., Prince, J.L., van Zijl, P.C., Mori, S., 2007. Effects of signal-to-noise ratio on the accuracy and reproducibility of diffusion tensor imaging-derived fractional anisotropy, mean diffusivity, and principal eigenvector measurements at 1.5 T. J. Magn. Reson. Imaging 26 (3), 756–767.

Fillard, P., Descoteaux, M., Goh, A., Gouttard, S., Jeurissen, B., Malcolm, J., Ramirez-Manzanares, A., Reisert, M., Sakaie, K., Tensaouti, F., Yo, T., Mangin, J.F., Poupon, C., 2011. Quantitative evaluation of 10 tractography algorithms on a realistic diffusion MR phantom. NeuroImage 56 (1), 220–234.

Firmin, L., Field, P., Maier, M.A., Kraskov, A., Kirkwood, P.A., Nakajima, K., Lemon, R.N., Glickstein, M., 2014. Axon diameters and conduction velocities in the macaque pyramidal tract. J. Neurophysiol. 112 (6), 1229–1240.

Forkel, S.J., Thiebaut de Schotten, M., Kawadler, J.M., Dell'Acqua, F., Danek, A., Catani, M., 2014. The anatomy of fronto-occipital connections from early blunt dissections to contemporary tractography. Cortex 56, 73–84.

Gao, Y., Choe, A.S., Li, X., Stepniewska, I., Anderson, A.W., 2013a. Comparison of In Vivo and Ex Vivo DTI Cortical Connectivity Measurements in the Squirrel Monkey Brain. International Society for Magnetic Resonance in Medicine, Salt Lake City, Utah, USA.

Gao, Y., Choe, A.S., Stepniewska, I., Li, X., Avison, M.J., Anderson, A.W., 2013b. Validation of DTI tractography-based measures of primary motor area connectivity in the squirrel monkey brain. PLoS One 8 (10), e75065.

Georgiadis, M., Schroeter, A., Gao, Z., Guizar-Sicairos, M., Liebi, M., Leuze, C., McNab, J.A., Balolia, A., Veraart, J., Ades-Aron, B., Kim, S., Shepherd, T., Lee, C.H., Walczak, P., Chodankar, S., DiGiacomo, P., David, G., Augath, M., Zerbi, V., Sommer, S., Rajkovic, I., Weiss, T., Bunk, O., Yang, L., Zhang, J., Novikov, D.S., Zeineh, M., Fieremans, E., Rudin, M., 2021. Nanostructure-specific X-ray tomography reveals myelin levels, integrity and axon orientations in mouse and human nervous tissue. Nat. Commun. 12 (1), 2941.

Gerbella, M., Belmalih, A., Borra, E., Rozzi, S., Luppino, G., 2010. Cortical connections of the macaque caudal ventrolateral prefrontal areas 45A and 45B. Cereb. Cortex 20 (1), 141–168.

Ginsburger, K., Matuschke, F., Poupon, F., Mangin, J.F., Axer, M., Poupon, C., 2019. MEDUSA: A GPU-based tool to create realistic phantoms of the brain microstructure using tiny spheres. NeuroImage 193, 10–24.

Girard, G., Whittingstall, K., Deriche, R., Descoteaux, M., 2014. Towards quantitative connectivity analysis: reducing tractography biases. NeuroImage 98, 266–278.

Girard, G., Whittingstall, K., Deriche, R., Descoteaux, M., 2015. Structural Connectivity Reproducibility Through Multiple Acquisitions. Organization for Human Brain Mapping, Honolulu, United States.

Girard, G., Caminiti, R., Battaglia-Mayer, A., St-Onge, E., Ambrosen, K.S., Eskildsen, S.F., Krug, K., Dyrby, T.B., Descoteaux, M., Thiran, J.-P., Innocenti, G.M., 2020. On the cortical connectivity in the macaque brain: a comparison of diffusion tractography and histological tracing data. NeuroImage 221, 117201.

Hau, J., Sarubbo, S., Perchey, G., Crivello, F., Zago, L., Mellet, E., Jobard, G., Joliot, M., Mazoyer, B.M., Tzourio-Mazoyer, N., Petit, L., 2016. Cortical terminations of the inferior fronto-occipital and uncinate fasciculi: anatomical stem-based virtual dissection. Front. Neuroanat. 10, 58.

Hilgetag, C.C., Goulas, A., 2020. 'Hierarchy' in the organization of brain networks. Philos. Trans. R. Soc. B Biol. Sci. 375 (1796), 20190319.

Hilgetag, C.C., Medalla, M., Beul, S.F., Barbas, H., 2016. The primate connectome in context: principles of connections of the cortical visual system. NeuroImage 134, 685–702.

Howard, A.F., Huszar, I.N., Smart, A., Cottaar, M., Daubney, G., Hanayik, T., Khrapitchev, A.A., Mars, R.B., Mollink, J., Scott, C., 2022. The BigMac dataset: an open resource combining multi-contrast MRI and microscopy in the macaque brain. bioRxiv: 2022.2009. 2008.506363.

Huang, S.Y., Tian, Q., Fan, Q., Witzel, T., Wichtmann, B., McNab, J.A., Daniel Bireley, J., Machado, N., Klawiter, E.C., Mekkaoui, C., Wald, L.L., Nummenmaa, A., 2020. High-gradient diffusion MRI reveals distinct estimates of axon diameter index within different white matter tracts in the in vivo human brain. Brain Struct. Funct. 225 (4), 1277–1291.

Huszar, I.N., Pallebage-Gamarallage, M., Foxley, S., Tendler, B.C., Leonte, A., Hiemstra, M., Mollink, J., Smart, A., Bangerter-Christensen, S., Brooks, H., Turner, M.R., Ansorge, O., Miller, K.L., Jenkinson, M., 2019. Tensor image registration library: automated non-linear registration of sparsely sampled histological specimens to post-mortem MRI of the whole human brain. bioRxiv: 849570.

Innocenti, G.M., Caminiti, R., Aboitiz, F., 2015. Comments on the paper by Horowitz et al. (2014). Brain Struct. Funct. 220 (3), 1789–1790.

Jacquesson, T., Frindel, C., Kocevar, G., Berhouma, M., Jouanneau, E., Attyé, A., Cotton, F., 2019. Overcoming challenges of cranial nerve tractography: a targeted review. Neurosurgery 84 (2), 313–325.

Janelle, F., Iorio-Morin, C., D'Amour, S., Fortin, D., 2022. Superior longitudinal fasciculus: a review of the anatomical descriptions with functional correlates. Front. Neurol. 13, 794618.

Jbabdi, S., Lehman, J.F., Haber, S.N., Behrens, T.E., 2013. Human and monkey ventral prefrontal fibers use the same organizational principles to reach their targets: tracing versus tractography. J. Neurosci. 33 (7), 3190–3201.

Jeurissen, B., Leemans, A., Tournier, J.D., Jones, D.K., Sijbers, J., 2013. Investigating the prevalence of complex fiber configurations in white matter tissue with diffusion magnetic resonance imaging. Hum. Brain Mapp. 34 (11), 2747–2766.

Jones, D.K., 2004. The effect of gradient sampling schemes on measures derived from diffusion tensor MRI: a Monte Carlo study. Magn. Reson. Med. 51 (4), 807–815.

Jones, D.K., Basser, P.J., 2004. "Squashing peanuts and smashing pumpkins": how noise distorts diffusion-weighted MR data. Magn. Reson. Med. 52 (5), 979–993.

Jones, R., Grisot, G., Augustinack, J., Magnain, C., Boas, D.A., Fischl, B., Wang, H., Yendiki, A., 2020. Insight into the fundamental trade-offs of diffusion MRI from polarization-sensitive optical coherence tomography in ex vivo human brain. NeuroImage, 116704.

Jones, R., Maffei, C., Augustinack, J., Fischl, B., Wang, H., Bilgic, B., Yendiki, A., 2021. High-fidelity approximation of grid- and shell-based sampling schemes from undersampled DSI using compressed sensing: post mortem validation. NeuroImage 244, 118621.

Kamagata, K., Kerever, A., Yokosawa, S., Otake, Y., Ochi, H., Hori, M., Kamiya, K., Tsuruta, K., Tagawa, K., Okazawa, H., Aoki, S., Arikawa-Hirasawa, E., 2016. Quantitative histological validation of diffusion tensor MRI with two-photon microscopy of cleared mouse brain. Magn. Reson. Med. Sci. 15 (4), 416–421.

Kamali, A., Flanders, A.E., Brody, J., Hunter, J.V., Hasan, K.M., 2014. Tracing superior longitudinal fasciculus connectivity in the human brain using high resolution diffusion tensor tractography. Brain Struct. Funct. 219 (1), 269–281.

Khan, A.R., Cornea, A., Leigland, L.A., Kohama, S.G., Jespersen, S.N., Kroenke, C.D., 2015. 3D structure tensor analysis of light microscopy data for validating diffusion MRI. NeuroImage 111, 192–203.

Kleinnijenhuis, M., van Mourik, T., Norris, D.G., Ruiter, D.J., van Cappellen van Walsum, A.M., Barth, M., 2015. Diffusion tensor characteristics of gyrencephaly using high resolution diffusion MRI in vivo at 7T. NeuroImage 109, 378–387.

Knösche, T.R., Anwander, A., Liptrot, M., Dyrby, T.B., 2015. Validation of tractography: comparison with manganese tracing. Hum. Brain Mapp. 36 (10), 4116–4134.

Landman, B.A., Farrell, J.A., Jones, C.K., Smith, S.A., Prince, J.L., Mori, S., 2007. Effects of diffusion weighting schemes on the reproducibility of DTI-derived fractional anisotropy, mean diffusivity, and principal eigenvector measurements at 1.5T. NeuroImage 36 (4), 1123–1138.

Lawes, I.N., Barrick, T.R., Murugam, V., Spierings, N., Evans, D.R., Song, M., Clark, C.A., 2008. Atlas-based segmentation of white matter tracts of the human brain using diffusion tensor tractography and comparison with classical dissection. NeuroImage 39 (1), 62–79.

Lazar, M., Alexander, A.L., 2003. An error analysis of white matter tractography methods: synthetic diffusion tensor field simulations. NeuroImage 20 (2), 1140–1153.

Lee, H.H., Yaros, K., Veraart, J., Pathan, J.L., Liang, F.X., Kim, S.G., Novikov, D.S., Fieremans, E., 2019. Along-axon diameter variation and axonal orientation dispersion revealed with 3D electron microscopy: implications for quantifying brain white matter microstructure with histology and diffusion MRI. Brain Struct. Funct. 224 (4), 1469–1488. https://doi.org/10.1007/s00429-019-01844-6.

Lee, H.H., Jespersen, S.N., Fieremans, E., Novikov, D.S., 2020a. The impact of realistic axonal shape on axon diameter estimation using diffusion MRI. NeuroImage 223, 117228.

Lee, H.H., Papaioannou, A., Kim, S.L., Novikov, D.S., Fieremans, E., 2020b. A time-dependent diffusion MRI signature of axon caliber variations and beading. Commun. Biol. 3 (1), 354.

Lee, H.H., Fieremans, E., Novikov, D.S., 2021. Realistic microstructure simulator (RMS): Monte Carlo simulations of diffusion in three-dimensional cell segmentations of microscopy images. J. Neurosci. Methods 350, 109018.

Leergaard, T.B., White, N.S., de Crespigny, A., Bolstad, I., D'Arceuil, H., Bjaalie, J.G., Dale, A.M., 2010. Quantitative histological validation of diffusion MRI fiber orientation distributions in the rat brain. PLoS One 5 (1), e8595.

Leuze, C., Goubran, M., Barakovic, M., Aswendt, M., Tian, Q., Hsueh, B., Crow, A., Weber, E.M.M., Steinberg, G.K., Zeineh, M., Plowey, E.D., Daducci, A., Innocenti, G., Thiran, J.P., Deisseroth, K., McNab, J.A., 2021. Comparison of diffusion MRI and CLARITY fiber orientation estimates in both gray and white matter regions of human and primate brain. NeuroImage 228, 117692.

Liewald, D., Miller, R., Logothetis, N., Wagner, H.J., Schuz, A., 2014. Distribution of axon diameters in cortical white matter: an electron-microscopic study on three human brains and a macaque. Biol. Cybern. 108 (5), 541–557.

Liptrot, M.G., Sidaros, K., Dyrby, T.B., 2014. Addressing the path-length-dependency confound in white matter tract segmentation. PLoS One 9 (5), e96247.

Lontis, E.R., Nielsen, K., Struijk, J.J., 2009. In vitro magnetic stimulation of pig phrenic nerve with transverse and longitudinal induced electric fields: analysis of the stimulation site. IEEE Trans. Biomed. Eng. 56 (2), 500–512.

Lundell, H., Nielsen, J.B., Ptito, M., Dyrby, T.B., 2011. Distribution of collateral fibers in the monkey cervical spinal cord detected with diffusion-weighted magnetic resonance imaging. NeuroImage 56 (3), 923–929.

Maffei, C., Sarubbo, S., Jovicich, J., 2019. Diffusion-based tractography atlas of the human acoustic radiation. Sci. Rep. 9 (1), 4046.

Magnain, C., Augustinack, J.C., Tirrell, L., Fogarty, M., Frosch, M.P., Boas, D., Fischl, B., Rockland, K.S., 2019. Colocalization of neurons in optical coherence microscopy and Nissl-stained histology in Brodmann's area 32 and area 21. Brain Struct. Funct. 224 (1), 351–362.

Maier-Hein, K.H., Neher, P.F., Houde, J.C., Cote, M.A., Garyfallidis, E., Zhong, J., Chamberland, M., Yeh, F.C., Lin, Y.C., Ji, Q., Reddick, W.E., Glass, J.-O., Chen, D.Q., Feng, Y., Gao, C., Wu, Y., Ma, J., Renjie, H., Li, Q., Westin, C.F., Deslauriers-Gauthier, S., Gonzalez, J.O.O., Paquette, M., St-Jean, S., Girard, G., Rheault, F., Sidhu, J., Tax, C.M.W., Guo, F., Mesri, H.Y., David, S., Froeling, M., Heemskerk, A.M., Leemans, A., Bore, A., Pinsard, B., Bedetti, C., Desrosiers, M., Brambati, S., Doyon, J., Sarica, A., Vasta, R., Cerasa, A., Quattrone, A., Yeatman, J., Khan, A.R., Hodges, W., Alexander, S., Romascano, D., Barakovic, M., Auria, A., Esteban, O., Lemkaddem, A., Thiran, J.P., Cetingul, H.E., Odry, B.L., Mailhe, B., Nadar, M.-S., Pizzagalli, F., Prasad, G., Villalon-Reina, J.E., Galvis, J., Thompson, P.M., Requejo, F.S., Laguna, P.L., Lacerda, L.M., Barrett, R., Dell'Acqua, F., Catani, M., Petit, L., Caruyer, E., Daducci, A., Dyrby, T.B., Holland-Letz, T., Hilgetag, C.C., Stieltjes, B., Descoteaux, M., 2017. The challenge of mapping the human connectome based on diffusion tractography. Nat. Commun. 8 (1), 1349.

Mandonnet, E., Sarubbo, S., Petit, L., 2018. The nomenclature of human white matter association pathways: proposal for a systematic taxonomic anatomical classification. Front. Neuroanat. 12, 94.

Markov, N.T., Ercsey-Ravasz, M.M., Ribeiro Gomes, A.R., Lamy, C., Magrou, L., Vezoli, J., Misery, P., Falchier, A., Quilodran, R., Gariel, M.A., Sallet, J., Gamanut, R., Huissoud, C., Clavagnier, S., Giroud, P., Sappey-Marinier, D., Barone, P., Dehay, C., Toroczkai, Z., Knoblauch, K., Van Essen, D.C., Kennedy, H., 2014. A weighted and directed interareal connectivity matrix for macaque cerebral cortex. Cereb. Cortex 24 (1), 17–36.

Martino, J., De Witt Hamer, P.C., Berger, M.S., Lawton, M.T., Arnold, C.M., de Lucas, E.M., Duffau, H., 2013. Analysis of the subcomponents and cortical terminations of the perisylvian superior longitudinal fasciculus: a fiber dissection and DTI tractography study. Brain Struct. Funct. 218 (1), 105–121.

McKinnon, E.T., Jensen, J.H., Glenn, G.R., Helpern, J.A., 2017. Dependence on b-value of the direction-averaged diffusion-weighted imaging signal in brain. Magn. Reson. Imaging 36, 121–127.

Menzel, M., Axer, M., De Raedt, H., Costantini, I., Silvestri, L., Pavone, F.S., Amunts, K., Michielsen, K., 2020. Toward a high-resolution reconstruction of 3D nerve fiber architectures and crossings in the brain using light scattering measurements and finite-difference time-domain simulations. Phys. Rev. X 10 (2), 021002.

Mollink, J., Kleinnijenhuis, M., Cappellen van Walsum, A.V., Sotiropoulos, S.N., Cottaar, M., Mirfin, C., Heinrich, M.P., Jenkinson, M., Pallebage-Gamarallage, M., Ansorge, O., Jbabdi, S., Miller, K.L., 2017. Evaluating fibre orientation dispersion in white matter: comparison of diffusion MRI, histology and polarized light imaging. NeuroImage 157, 561–574.

Mori, S., Oishi, K., Jiang, H., Jiang, L., Li, X., Akhter, K., Hua, K., Faria, A.V., Mahmood, A., Woods, R., Toga, A.W., Pike, G.B., Neto, P.R., Evans, A., Zhang, J., Huang, H., Miller, M.I., van Zijl, P., Mazziotta, J., 2008. Stereotaxic white matter atlas based on diffusion tensor imaging in an ICBM template. NeuroImage 40 (2), 570–582.

Moyer, D.C., Thompson, P., Steeg, G.V., 2019. Measures of tractography convergence. In: Computational Diffusion MRI. Springer International Publishing, Cham.

Neher, P.F., Laun, F.B., Stieltjes, B., Maier-Hein, K.H., 2014. Fiberfox: facilitating the creation of realistic white matter software phantoms. Magn. Reson. Med. 72 (5), 1460–1470.

Nie, J., Guo, L., Li, K., Wang, Y., Chen, G., Li, L., Chen, H., Deng, F., Jiang, X., Zhang, T., Huang, L., Faraco, C., Zhang, D., Guo, C., Yap, P.T., Hu, X., Li, G., Lv, J., Yuan, Y., Zhu, D., Han, J., Sabatinelli, D., Zhao, Q., Miller, L.S., Xu, B., Shen, P., Platt, S., Shen, D., Hu, X., Liu, T., 2012. Axonal fiber terminations concentrate on gyri. Cereb. Cortex 22 (12), 2831–2839.

Nilsson, M., Latt, J., Stahlberg, F., van Westen, D., Hagslatt, H., 2012. The importance of axonal undulation in diffusion MR measurements: a Monte Carlo simulation study. NMR Biomed. 25 (5), 795–805.

Nilsson, M., van Westen, D., Ståhlberg, F., Sundgren, P.C., Lätt, J., 2013. The role of tissue microstructure and water exchange in biophysical modelling of diffusion in white matter. MAGMA 26 (4), 345–370.

Oh, S.W., Harris, J.A., Ng, L., Winslow, B., Cain, N., Mihalas, S., Wang, Q., Lau, C., Kuan, L., Henry, A.M., Mortrud, M.T., Ouellette, B., Nguyen, T.N., Sorensen, S.A., Slaughterbeck, C.R., Wakeman, W., Li, Y., Feng, D., Ho, A., Nicholas, E., Hirokawa, K.E., Bohn, P., Joines, K.M., Peng, H., Hawrylycz, M.J., Phillips, J.W., Hohmann, J.G., Wohnoutka, P., Gerfen, C.R., Koch, C., Bernard, A., Dang, C., Jones, A.R., Zeng, H., 2014. A mesoscale connectome of the mouse brain. Nature 508 (7495), 207–214.

Oishi, K., Zilles, K., Amunts, K., Faria, A., Jiang, H., Li, X., Akhter, K., Hua, K., Woods, R., Toga, A.W., Pike, G.B., Rosa-Neto, P., Evans, A., Zhang, J., Huang, H., Miller, M.I., van Zijl, P.C., Mazziotta, J., Mori, S., 2008. Human brain white matter atlas: identification and assignment of common anatomical structures in superficial white matter. NeuroImage 43 (3), 447–457.

Palombo, M., Alexander, D.C., Zhang, H., 2019. A generative model of realistic brain cells with application to numerical simulation of the diffusion-weighted MR signal. NeuroImage 188, 391–402.

Panesar, S.S., Fernandez-Miranda, J., 2019. Commentary: the nomenclature of human white matter association pathways: proposal for a systematic taxonomic anatomical classification. Front. Neuroanat. 13, 61.

Perens, J., Salinas, C.G., Roostalu, U., Skytte, J.L., Gundlach, C., Hecksher-Sorensen, J., Dahl, A.B., Dyrby, T.B., 2023. Multimodal 3D mouse brain atlas framework with the skull-derived coordinate system. Neuroinformatics 21 (2), 269–286. https://doi.org/10.1007/s12021-023-09623-9 (Epub 2023 Feb 21).

Pujol, S., Wells, W., Pierpaoli, C., Brun, C., Gee, J., Cheng, G., Vemuri, B., Commowick, O., Prima, S., Stamm, A., Goubran, M., Khan, A., Peters, T., Neher, P., Maier-Hein, K.H., Shi, Y., Tristan-Vega, A., Veni, G., Whitaker, R., Styner, M., Westin, C.F., Gouttard, S., Norton, I., Chauvin, L., Mamata, H., Gerig, G., Nabavi, A., Golby, A., Kikinis, R., 2015. The DTI challenge: toward standardized evaluation of diffusion tensor imaging tractography for neurosurgery. J. Neuroimaging 25 (6), 875–882.

Qi, S., Meesters, S., Nicolay, K., Romeny, B.M.T.H., Ossenblok, P., 2015. The influence of construction methodology on structural brain network measures: a review. J. Neurosci. Methods 253, 170–182.

Reveley, C., Seth, A.K., Pierpaoli, C., Silva, A.C., Yu, D., Saunders, R.C., Leopold, D.A., Ye, F.Q., 2015. Superficial white matter fiber systems impede detection of long-range cortical connections in diffusion MR tractography. Proc. Natl. Acad. Sci. USA 112 (21), E2820–E2828.

Rheault, F., De Benedictis, A., Daducci, A., Maffei, C., Tax, C.M.W., Romascano, D., Caverzasi, E., Morency, F.C., Corrivetti, F., Pestilli, F., Girard, G., Theaud, G., Zemmoura, I., Hau, J., Glavin, K., Jordan, K.M., Pomiecko, K., Chamberland, M., Barakovic, M., Goyette, N., Poulin, P., Chenot, Q., Panesar, S.S., Sarubbo, S., Petit, L., Descoteaux, M., 2020a. Tractostorm: the what, why, and how of tractography dissection reproducibility. Hum. Brain Mapp. 41 (7), 1859–1874. https://doi.org/10.1002/hbm.24917.

Rheault, F., Poulin, P., Valcourt Caron, A., St-Onge, E., Descoteaux, M., 2020b. Common misconceptions, hidden biases and modern challenges of dMRI tractography. J. Neural Eng. 17 (1), 011001.

Rilling, J.K., Glasser, M.F., Preuss, T.M., Ma, X., Zhao, T., Hu, X., Behrens, T.E., 2008. The evolution of the arcuate fasciculus revealed with comparative DTI. Nat. Neurosci. 11 (4), 426–428.

Rockland, K.S., 2020. What we can learn from the complex architecture of single axons. Brain Struct. Funct. 225 (4), 1327–1347.

Rodrigues, P., Prats-Galino, A., Gallardo-Pujol, D., Villoslada, P., Falcon, C., Prčkovska, V., 2013. Evaluating structural connectomics in relation to different q-space sampling techniques. In: Medical Image Computing and Computer-Assisted Intervention—MICCAI 2013. Springer, Berlin, Heidelberg.

Roine, T., Jeurissen, B., Perrone, D., Aelterman, J., Philips, W., Sijbers, J., Leemans, A., 2019. Reproducibility and intercorrelation of graph theoretical measures in structural brain connectivity networks. Med. Image Anal. 52, 56–67.

Ronen, I., Budde, M., Ercan, E., Annese, J., Techawiboonwong, A., Webb, A., 2014. Microstructural organization of axons in the human corpus callosum quantified by diffusion-weighted magnetic resonance spectroscopy of N-acetylaspartate and post-mortem histology. Brain Struct. Funct. 219 (5), 1773–1785.

Sarubbo, S., De Benedictis, A., Maldonado, I.L., Basso, G., Duffau, H., 2013. Frontal terminations for the inferior fronto-occipital fascicle: anatomical dissection, DTI study and functional considerations on a multi-component bundle. Brain Struct. Funct. 218 (1), 21–37.

Sarubbo, S., Petit, L., De Benedictis, A., Chioffi, F., Ptito, M., Dyrby, T.B., 2019. Uncovering the inferior fronto-occipital fascicle and its topological organization in non-human primates: the missing connection for language evolution. Brain Struct. Funct. 224 (4), 1553–1567.

Sarwar, T., Ramamohanarao, K., Zalesky, A., 2019. Mapping connectomes with diffusion MRI: deterministic or probabilistic tractography? Magn. Reson. Med. 81 (2), 1368–1384.

Schilling, K., Janve, V., Gao, Y., Stepniewska, I., Landman, B.A., Anderson, A.W., 2016. Comparison of 3D orientation distribution functions measured with confocal microscopy and diffusion MRI. NeuroImage 129, 185–197.

Schilling, K., Gao, Y., Janve, V., Stepniewska, I., Landman, B.A., Anderson, A.W., 2017. Can increased spatial resolution solve the crossing fiber problem for diffusion MRI? NMR Biomed 30 (12). https://doi.org/10.1002/nbm.3787.

Schilling, K.G., Janve, V., Gao, Y., Stepniewska, I., Landman, B.A., Anderson, A.W., 2018a. Histological validation of diffusion MRI fiber orientation distributions and dispersion. NeuroImage 165, 200–221.

Schilling, K.G., Nath, V., Hansen, C., Parvathaneni, P., Blaber, J., Gao, Y., Neher, P., Aydogan, D.B., Shi, Y., Ocampo-Pineda, M., Schiavi, S., Daducci, A., Girard, G., Barakovic, M., Rafael-Patino, J., Romascano, D., Rensonnet, G., Pizzolato, M., Bates, A., Fischi, E., Thiran, J.P., Canales-Rodriguez, E.J., Huang, C., Zhu, H., Zhong, L., Cabeen, R., Toga, A.W., Rheault, F., Theaud, G., Houde, J.C., Sidhu, J., Chamberland, M., Westin, C.-F., Dyrby, T.B., Verma, R., Rathi, Y., Irfanoglu, M.O., Thomas, C., Pierpaoli, C., Descoteaux, M., Anderson, A.W., Landman, B.A., 2018b. Limits to anatomical accuracy of diffusion tractography using modern approaches. NeuroImage 185, 1–11.

Schilling, K., Gao, Y., Janve, V., Stepniewska, I., Landman, B.A., Anderson, A.W., 2018c. Confirmation of a gyral bias in diffusion MRI fiber tractography. Hum. Brain Mapp. 39 (3), 1449–1466.

Schilling, K.G., Daducci, A., Maier-Hein, K., Poupon, C., Houde, J.C., Nath, V., Anderson, A.W., Landman, B.A., Descoteaux, M., 2019a. Challenges in diffusion MRI tractography—Lessons learned from international benchmark competitions. Magn. Reson. Imaging 57, 194–209.

Schilling, K.G., Gao, Y., Stepniewska, I., Janve, V., Landman, B.A., Anderson, A.W., 2019b. Anatomical accuracy of standard-practice tractography algorithms in the motor system—a histological validation in the squirrel monkey brain. Magn. Reson. Imaging 55, 7–25.

Schilling, K.G., Petit, L., Rheault, F., Remedios, S., Pierpaoli, C., Anderson, A.W., Landman, B.A., Descoteaux, M., 2020a. Brain connections derived from diffusion MRI tractography can be highly anatomically accurate-if we know where white matter pathways start, where they end, and where they do not go. Brain Struct. Funct. 225 (8), 2387–2402.

Schilling, K.G., Rheault, F., Petit, L., Hansen, C.B., Nath, V., Yeh, F.-C., Girard, G., Barakovic, M., Rafael-Patino, J., Yu, T., Fischi-Gomez, E., Pizzolato, M., Ocampo-Pineda, M., Schiavi, S., Canales-Rodríguez, E.J., Daducci, A., Granziera, C., Innocenti, G., Thiran, J.-P., Mancini, L., Wastling, S., Cocozza, S., Petracca, M., Pontillo, G., Mancini, M., Vos, S.B., Vakharia, V.N., Duncan, J.S., Melero, H., Manzanedo, L., Sanz-Morales, E., Peña-Melián, Á., Calamante, F., Attyé, A., Cabeen, R.P., Korobova, L., Toga, A.W., Vijayakumari, A.A., Parker, D., Verma, R., Radwan, A., Sunaert, S., Emsell, L., De Luca, A., Leemans, A., Bajada, C.J., Haroon, H., Azadbakht, H., Chamberland, M., Genc, S., Tax, C.M.W., Yeh, P.-H., Srikanchana, R., McKnight, C., Yang, J.Y.-M., Chen, J., Kelly, C.E., Yeh, C.-H., Cochereau, J., Maller, J.J., Welton, T., Almairac, F., Seunarine, K.K., Clark, C.A., Zhang, F., Makris, N., Golby, A., Rathi, Y., O'Donnell, L.J., Xia, Y., Aydogan, D.B., Shi, Y., Fernandes, F.G., Raemaekers, M., Warrington, S., Michielse, S., Ramírez-Manzanares, A., Concha, L., Aranda, R., Meraz, M.R., Lerma-Usabiaga, G., Roitman, L., Fekonja, L.S., Calarco, N., Joseph, M., Nakua, H., Voineskos, A.N., Karan, P., Grenier, G., Legarreta, J.H., Adluru, N., Nair, V.A., Prabhakaran, V., Alexander, A.L., Kamagata, K., Saito, Y., Uchida, W., Andica, C., Masahiro, A., Bayrak, R.G., Gandini Wheeler-Kingshott, C.A.M., D'Angelo, E., Palesi, F., Savini, G., Rolandi, N., Guevara, P., Houenou, J., López-López, N., Mangin, J.-F., Poupon, C., Román, C., Vázquez,

A., Maffei, C., Arantes, M., Andrade, J.P., Silva, S.M., Raja, R., Calhoun, V.D., Caverzasi, E., Sacco, S., Lauricella, M., Pestilli, F., Bullock, D., Zhan, Y., Brignoni-Perez, E., Lebel, C., Reynolds, J.E., Nestrasil, I., Labounek, R., Lenglet, C., Paulson, A., Aulicka, S., Heilbronner, S., Heuer, K., Anderson, A.W., Landman, B.A., Descoteaux, M., 2020b. Tractography dissection variability: what happens when 42 groups dissect 14 white matter bundles on the same dataset? bioRxiv: 2020.2010.2007.321083.

Schilling, K.G., Rheault, F., Petit, L., Hansen, C.B., Nath, V., Yeh, F.C., Girard, G., Barakovic, M., Rafael-Patino, J., Yu, T., Fischi-Gomez, E., Pizzolato, M., Ocampo-Pineda, M., Schiavi, S., Canales-Rodriguez, E.J., Daducci, A., Granziera, C., Innocenti, G., Thiran, J.P., Mancini, L., Wastling, S., Cocozza, S., Petracca, M., Pontillo, G., Mancini, M., Vos, S.B., Vakharia, V.N., Duncan, J.S., Melero, H., Manzanedo, L., Sanz-Morales, E., Pena-Melian, A., Calamante, F., Attye, A., Cabeen, R.P., Korobova, L., Toga, A.W., Vijayakumari, A.A., Parker, D., Verma, R., Radwan, A., Sunaert, S., Emsell, L., De Luca, A., Leemans, A., Bajada, C.J., Haroon, H., Azadbakht, H., Chamberland, M., Genc, S., Tax, C.M.W., Yeh, P.H., Srikanchana, R., McKnight, C.D., Yang, J.Y., Chen, J., Kelly, C.E., Yeh, C.H., Cochereau, J., Maller, J.J., Welton, T., Almairac, F., Seunarine, K.K., Clark, C.A., Zhang, F., Makris, N., Golby, A., Rathi, Y., O'Donnell, L.J., Xia, Y., Aydogan, D.B., Shi, Y., Fernandes, F.G., Raemaekers, M., Warrington, S., Michielse, S., Ramirez-Manzanares, A., Concha, L., Aranda, R., Meraz, M.R., Lerma-Usabiaga, G., Roitman, L., Fekonja, L.S., Calarco, N., Joseph, M., Nakua, H., Voineskos, A.N., Karan, P., Grenier, G., Legarreta, J.H., Adluru, N., Nair, V.A., Prabhakaran, V., Alexander, A.L., Kamagata, K., Saito, Y., Uchida, W., Andica, C., Abe, M., Bayrak, R.G., Wheeler-Kingshott, C., D'Angelo, E., Palesi, F., Savini, G., Rolandi, N., Guevara, P., Houenou, J., Lopez-Lopez, N., Mangin, J.F., Poupon, C., Roman, C., Vazquez, A., Maffei, C., Arantes, M., Andrade, J.P., Silva, S.M., Calhoun, V.D., Caverzasi, E., Sacco, S., Lauricella, M., Pestilli, F., Bullock, D., Zhan, Y., Brignoni-Perez, E., Lebel, C., Reynolds, J.E., Nestrasil, I., Labounek, R., Lenglet, C., Paulson, A., Aulicka, S., Heilbronner, S.R., Heuer, K., Chandio, B.Q., Guaje, J., Tang, W., Garyfallidis, E., Raja, R., Anderson, A.W., Landman, B.A., Descoteaux, M., 2021a. Tractography dissection variability: what happens when 42 groups dissect 14 white matter bundles on the same dataset? NeuroImage 243, 118502.

Schilling, K.G., Tax, C.M.W., Rheault, F., Hansen, C., Yang, Q., Yeh, F.C., Cai, L., Anderson, A.W., Landman, B.A., 2021b. Fiber tractography bundle segmentation depends on scanner effects, vendor effects, acquisition resolution, diffusion sampling scheme, diffusion sensitization, and bundle segmentation workflow. NeuroImage 242, 118451.

Schilling, K.G., Tax, C.M.W., Rheault, F., Landman, B.A., Anderson, A.W., Descoteaux, M., Petit, L., 2022. Prevalence of white matter pathways coming into a single white matter voxel orientation: the bottleneck issue in tractography. Hum. Brain Mapp. 43 (4), 1196–1213. https://doi.org/10.1002/hbm.25697.

Schmahmann, J.D., Pandya, D.N., 2006. Fiber Pathways of the Brain. Oxford University Press, Oxford; New York.

Schmahmann, J.D., Pandya, D.N., 2007. The complex history of the fronto-occipital fasciculus. J. Hist. Neurosci. 16 (4), 362–377.

Schmahmann, J.D., Pandya, D.N., Wang, R., Dai, G., D'Arceuil, H.E., de Crespigny, A.J., Wedeen, V.J., 2007. Association fibre pathways of the brain: parallel observations from diffusion spectrum imaging and autoradiography. Brain 130 (Pt 3), 630–653.

Schurr, R., Zelman, A., Mezer, A.A., 2020. Subdividing the superior longitudinal fasciculus using local quantitative MRI. NeuroImage 208, 116439.

Shen, K., Bezgin, G., Schirner, M., Ritter, P., Everling, S., McIntosh, A.R., 2019a. A macaque connectome for large-scale network simulations in TheVirtualBrain. Sci. Data 6 (1), 123.

Shen, K., Goulas, A., Grayson, D.S., Eusebio, J., Gati, J.S., Menon, R.S., McIntosh, A.R., Everling, S., 2019b. Exploring the limits of network topology estimation using diffusion-based tractography and tracer studies in the macaque cortex. NeuroImage 191, 81–92.

Smith, R.E., Tournier, J.D., Calamante, F., Connelly, A., 2012. Anatomically-constrained tractography: improved diffusion MRI streamlines tractography through effective use of anatomical information. NeuroImage 62 (3), 1924–1938.

Stephan, K.E., Kamper, L., Bozkurt, A., Burns, G.A., Young, M.P., Kotter, R., 2001. Advanced database methodology for the collation of connectivity data on the macaque brain (CoCoMac). Philos. Trans. R. Soc. Lond. Ser. B Biol. Sci. 356 (1412), 1159–1186.

Sunderland, S., Bradley, K.C., 1961. Stress-Strain Phenomena In Human Peripheral Nerve trunks. Brain 84 (1), 102–119.

Tax, C.M.W., Bastiani, M., Veraart, J., Garyfallidis, E., Okan Irfanoglu, M., 2022. What's new and what's next in diffusion MRI preprocessing. NeuroImage 249, 118830.

Thiebaut de Schotten, M., Dell'Acqua, F., Forkel, S.J., Simmons, A., Vergani, F., Murphy, D.G., Catani, M., 2011. A lateralized brain network for visuospatial attention. Nat. Neurosci. 14 (10), 1245–1246.

Thomas, C., Ye, F.Q., Irfanoglu, M.O., Modi, P., Saleem, K.S., Leopold, D.A., Pierpaoli, C., 2014. Anatomical accuracy of brain connections derived from diffusion MRI tractography is inherently limited. Proc. Natl. Acad. Sci. USA 111 (46), 16574–16579.

Tournier, J.D., Yeh, C.H., Calamante, F., Cho, K.H., Connelly, A., Lin, C.P., 2008. Resolving crossing fibres using constrained spherical deconvolution: validation using diffusion-weighted imaging phantom data. NeuroImage 42 (2), 617–625.

Tournier, J.D., Calamante, F., Connelly, A., 2013. Determination of the appropriate b value and number of gradient directions for high-angular-resolution diffusion-weighted imaging. NMR Biomed. 26 (12), 1775–1786.

Vaessen, M.J., Hofman, P.A.M., Tijssen, H.N., Aldenkamp, A.P., Jansen, J.F.A., Backes, W.H., 2010. The effect and reproducibility of different clinical DTI gradient sets on small world brain connectivity measures. NeuroImage 51 (3), 1106–1116.

van den Heuvel, M.P., de Reus, M.A., Feldman Barrett, L., Scholtens, L.H., Coopmans, F.M., Schmidt, R., Preuss, T.M., Rilling, J.K., Li, L., 2015. Comparison of diffusion tractography and tract-tracing measures of connectivity strength in rhesus macaque connectome. Hum. Brain Mapp. 36 (8), 3064–3075.

Veraart, J., Fieremans, E., Novikov, D.S., 2019. On the scaling behavior of water diffusion in human brain white matter. NeuroImage 185, 379–387.

Veraart, J., Nunes, D., Rudrapatna, U., Fieremans, E., Jones, D.K., Novikov, D.S., Shemesh, N., 2020. Nonivasive quantification of axon radii using diffusion MRI. elife 9.

Veraart, J., Raven, E.P., Edwards, L.J., Weiskopf, N., Jones, D.K., 2021. The variability of MR axon radii estimates in the human white matter. Hum. Brain Mapp. 42 (7), 2201–2213.

Vishwesh Nath, K.G.S., Hainline, A.E., Huo, Y., Parvathaneni, P., Blaber, J.A., Rowe, M., Rodrigues, P., Prchkovska, V., Aydogan, D.B., Sun, W., Shi, Y., Parker, W.A., Ismail, A.A.O., Verma, R., Cabeen, R.P., Toga, A.W., Newton, A.T., Wasserthal, J., Neher, P., Maier-Hein, K., Savini, G., Palesi, F., Kaden, E., Wu, Y., He, J., Fen, Y., Barakovic, M., Romascano, D., Rafael-Patino, J., Frigo, M., Girard, G., Daducci, A., Thiran, J.P., Paquette, M., Rheault, F., Sidhu, J., Lebel, C., Leemans, A., Descoteaux, M., Dyrby, T.B., Landman, B.A., 2018. Tractography Reproducibility Challenge with Empirical Data (TraCED): The 2017 ISMRM Diffusion Study Group Challenge.

Wakana, S., Jiang, H., Nagae-Poetscher, L.M., van Zijl, P.C., Mori, S., 2004. Fiber tract-based atlas of human white matter anatomy. Radiology 230 (1), 77–87.

Wakana, S., Caprihan, A., Panzenboeck, M.M., Fallon, J.H., Perry, M., Gollub, R.L., Hua, K., Zhang, J., Jiang, H., Dubey, P., Blitz, A., van Zijl, P., Mori, S., 2007. Reproducibility of quantitative tractography methods applied to cerebral white matter. NeuroImage 36 (3), 630–644.

Wang, H., Lenglet, C., Akkin, T., 2015. Structure tensor analysis of serial optical coherence scanner images for mapping fiber orientations and tractography in the brain. J. Biomed. Opt. 20 (3), 036003.

Wang, X., Pathak, S., Stefaneanu, L., Yeh, F.C., Li, S., Fernandez-Miranda, J.C., 2016. Subcomponents and connectivity of the superior longitudinal fasciculus in the human brain. Brain Struct. Funct. 221 (4), 2075–2092.

White, J.G., Southgate, E., Thomson, J.N., Brenner, S., 1986. The structure of the nervous system of the nematode Caenorhabditis elegans. Philos. Trans. R. Soc. Lond. Ser. B Biol. Sci. 314 (1165), 1–340.

Wu, Y., Sun, D., Wang, Y., Wang, Y., Wang, Y., 2016. Tracing short connections of the temporo-parieto-occipital region in the human brain using diffusion spectrum imaging and fiber dissection. Brain Res. 1646, 152–159.

Xu, F., Shen, Y., Ding, L., Yang, C.Y., Tan, H., Wang, H., Zhu, Q., Xu, R., Wu, F., Xiao, Y., Xu, C., Li, Q., Su, P., Zhang, L.I., Dong, H.W., Desimone, R., Xu, F., Hu, X., Lau, P.M., Bi, G.Q., 2021. High-throughput mapping of a whole rhesus monkey brain at micrometer resolution. Nat. Biotechnol. 39 (12), 1521–1528.

Yeh, C.H., Smith, R.E., Liang, X., Calamante, F., Connelly, A., 2016. Correction for diffusion MRI fibre tracking biases: the consequences for structural connectomic metrics. NeuroImage 142, 150–162.

Yeh, C.-H., Smith, R.E., Liang, X., Calamante, F., Connelly, A., 2018. Investigating the Streamline Count Required for Reproducible Structural Connectome Construction Across a Range of Brain Parcellation Resolutions. International Society of Magnetic Resonance in Medicine, Paris, France.

Zalesky, A., Fornito, A., Harding, I.H., Cocchi, L., Yücel, M., Pantelis, C., Bullmore, E.T., 2010. Whole-brain anatomical networks: does the choice of nodes matter? NeuroImage 50 (3), 970–983.

Zemmoura, I., Serres, B., Andersson, F., Barantin, L., Tauber, C., Filipiak, I., Cottier, J.P., Venturini, G., Destrieux, C., 2014. FIBRASCAN: a novel method for 3D white matter tract reconstruction in MR space from cadaveric dissection. NeuroImage 103, 106–118.

Zhang, Y., Zhang, J., Oishi, K., Faria, A.V., Jiang, H., Li, X., Akhter, K., Rosa-Neto, P., Pike, G.B., Evans, A., Toga, A.W., Woods, R., Mazziotta, J.C., Miller, M.I., van Zijl, P.C., Mori, S., 2010. Atlas-guided tract reconstruction for automated and comprehensive examination of the white matter anatomy. NeuroImage 52 (4), 1289–1301.

# Chapter 28

# Current challenges and opportunities for tractography

Francois Rheault[a,b], Philippe Poulin[a], Alex Valcourt Caron[a], Etienne St-Onge[c], Kurt G. Schilling[b], Laurent Petit[d], Flavio Dell'Acqua[e,f], Alexander Leemans[g], and Maxime Descoteaux[h]

[a]*Sherbrooke Connectivity Imaging Laboratory (SCIL), University of Sherbrooke, Sherbrooke, QC, Canada,* [b]*Vanderbilt University Institute of Imaging Science, Vanderbilt University Medical Center, Nashville, TN, United States,* [c]*Department of Computer Science and Engineering, Université du Québec en Outaouais, Saint-Jérôme, QC, Canada,* [d]*Groupe d'Imagerie Neurofonctionnelle, Institut des Maladies Neurodégénératives (GIN-IMN), UMR5293, CNRS, CEA, Université de Bordeaux, Bordeaux, France,* [e]*NATBRAINLAB, Department of Neuroimaging, Institute of Psychiatry, Psychology and Neuroscience, King's College London, London, United Kingdom,* [f]*Department of Forensics and Neurodevelopmental Sciences, Institute of Psychiatry, Psychology and Neuroscience, King's College London, London, United Kingdom,* [g]*PROVIDI lab, Image Sciences Institute, University Medical Center Utrecht, Utrecht, The Netherlands,* [h]*Sherbrooke Connectivity Imaging Laboratory (SCIL), Department of Computer Science, University of Sherbrooke, Sherbrooke, QC, Canada*

## 1 The rise of tractography in neuroscience

The field of tractography has emerged from close collaboration between mathematicians, physicists, computer scientists, biomedical engineers, neuroanatomists, and clinicians, all with their complementary expertise when studying the brain and its *connectome*. This multidisciplinary nature means that researchers studying similar questions are often from laboratories with different backgrounds and expertise. Additionally, researchers interested in tractography have multiple options to perform their research, and the choices are difficult for new users. However, the end-user of tractography must be familiar with the limitations of underlying algorithms used by the tool they picked in order to properly interpret any results they obtained. Diffusion magnetic resonance imaging (dMRI) tractography is particularly prone to these challenges, since many misconceptions, biases, and pitfalls can arise at various steps during processing.

Various reviews of existing challenges in dMRI already exist, but most focus on challenges related to image acquisition, local models, microstructure, or statistical analysis (Jbabdi and Johansen-Berg, 2011; Jones et al., 2013; O'Donnell and Pasternak, 2015; Sotiropoulos and Zalesky, 2019; Maier-Hein et al., 2017). However, the purpose of this chapter is to focus on dMRI tractography processing, validation, and interpretation.

The popularity of dMRI tractography in clinical research has risen tremendously since its origin more than 20 years ago. It started as a computationally heavy technical tool and has become a clinical and neurosurgical tool. Over the years dMRI tractography became a widespread technological tool available to *all* and it is clearly growing in popularity (see Fig. 1). However, as more and more neuroscience and clinical journal publications are adopting tractography, the associated limitations need to be discussed.

## 2 Virtual reconstruction versus underlying anatomy: What is and what isn't?

The main challenges of dMRI tractography reside in the fact that it is expected to reconstruct the brain connectivity using only local diffusion information (Maier-Hein et al., 2017; Jeurissen et al., 2017). This problem cannot be easily solved by tractography alone. Tractography can generate streamlines that look valid but have no basis in the underlying anatomy, or streamlines that obviously look invalid, but until verified cannot be discarded. While dMRI tractography can indeed reconstruct correctly known WM anatomy (Maffei et al., 2018; Hau et al., 2017; Benedictis et al., 2016; Chenot et al., 2019), this problem makes it difficult to use tractography for exploratory investigations. This section will detail some of the known challenges when it comes to tractogram reconstruction. Note that, despite numerous examples, this list is not exhaustive and does not cover all existing algorithms.

Handbook of Diffusion MR Tractography. https://doi.org/10.1016/B978-0-12-818894-1.00037-9
Copyright © 2025 Elsevier Ltd. All rights are reserved, including those for text and data mining, AI training, and similar technologies.

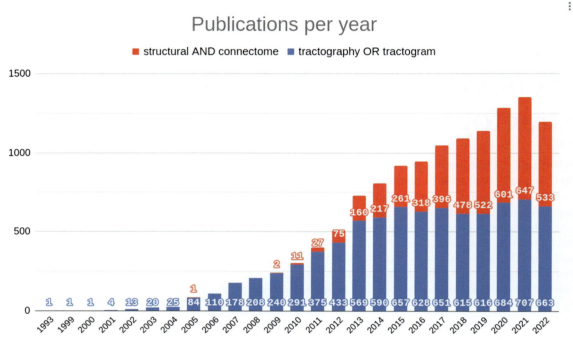

**FIG. 1** Slowly but surely: The rise in the number of publications including the reference to the keywords "structural connectome," "tractography," or "tractogram" (PubMed).

## 2.1 Local signal challenges

Despite advances in the field of dMRI tractography and local modeling over the last few years, it is important to reiterate that the problem of tractography is fundamentally ill-posed (Maier-Hein et al., 2017; Schilling et al., 2022). Estimating global connectivity from local information about diffusion will inherently generate both false positives and true positives. Even global tractography algorithms are not immune to this, as they use the local voxel-wise information to connect segments together in order to generate plausible streamlines (Maier-Hein et al., 2017; Jeurissen et al., 2017; Daducci et al., 2016).

### 2.1.1 Bottleneck effect

While the name may vary, the bottleneck effect is probably the most challenging limitation of dMRI tractography. When multiple fasciculi, or a single fasciculus with multiple origins, converge toward a small region, the observed diffusion signal will lead to a single orientation within the same voxel. During tractography, reconstruction at the bottleneck exit will have all equally probable choices of direction. Deterministic tractography will pick the straightest path, while probabilistic tractography will pick all possible paths. Both types of algorithms will lead to invalid terminations as shown in Fig. 2. This phenomenon can be observed in the temporal stem, the corpus callosum (CC), or in the internal capsule (IC). This was demonstrated in more detail in Maier-Hein et al. (2017) and Schilling et al. (2022) by using a synthetic reconstruction of a subset of the long-ranging WM bundles.

Without advanced knowledge of the underlying anatomy, explicit rules, and priors, this situation cannot be solved with *orientation information* alone. Simply stepping forward in a directions field (i.e., local models) is not enough to avoid this pitfall (see Fig. 3). Even when the streamlines are most likely all plausible (as in Fig. 3), the spatial distribution or the number of streamlines may not be anatomically relevant. Bundle-specific tractography recently made an appearance in the literature, either using manually placed priors (Chamberland et al., 2017), using bundle templates (Rheault et al., 2019) or machine learning (Wasserthal et al., 2018; Poulin et al., 2017). All are a direct attempt to solve this problem and inform tractography how to untangle ambiguous signals by injecting information about the origin and the target of a streamline.

This effect is not unique to deterministic or probabilistic tractography algorithms; in a bottleneck region, even global tractography algorithms struggle to correctly piece together into a streamline. There is still a combinatorial number of possibilities when it comes to connecting segments that go through a single-direction region.

Current challenges and opportunities for tractography **Chapter | 28 567**

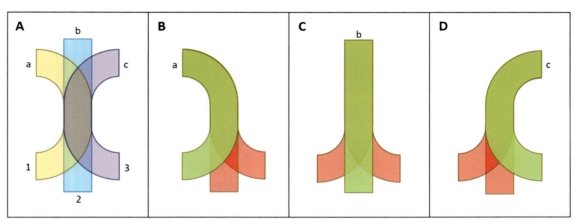

**FIG. 2** The bottleneck effect. Three (3) *bundles* with independent endpoints have an overlapping region (A), the bottleneck in *shaded gray*. In this region, all bundles run parallel and, upon exit, the direction to take is unknown. In such a case, with three origins and three destinations, there are nine potential outcomes (B–C–D), despite having only three valid configurations in the underlying architecture. This configuration will generate more invalid connections (IC) than valid connections (VC), as shown in Maier-Hein et al. (2017).

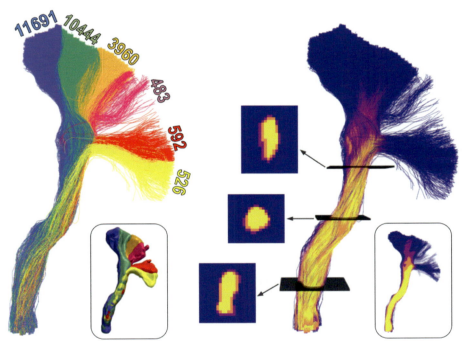

**FIG. 3** Segmentation of the pyramidal tract (PyT), cleaned to represent a typical anatomical definition; the six subgroups of streamlines were divided using their terminations. This figure does not show the thousands of streamlines going anteriorly or posteriorly. In this scenario, the PyT bottleneck makes it difficult to disentangle the origin/target of streamlines; hence the spatial configuration of the streamlines cannot easily be linked to the underlying anatomy. At each of the three cross-sections, the number of subbundles overlapping (0–6) is shown; the core of the bottleneck is more often than not an overlap of the six subbundles. Despite a constant overlap, in a region with a single-fiber population, along the majority of their length and having a common starting region, they do not have a common ending region.

### 2.1.2 Wall effect

One of the lesser-known challenges of tractography is encountered when a pathway crosses another at a low angle or in a kissing region. In this case, the reconstructed streamlines often look like they simply bounced off each other. In regions where kissing pathways are expected, at the junction where they touch or overlap at a low angle, the reconstructed local model is not as sharp as it needs to be (or is simply reconstructed as a single direction) to disentangle both paths. When incoming streamlines take the curve and reach the overlapping region, they straighten their path due to the blurred

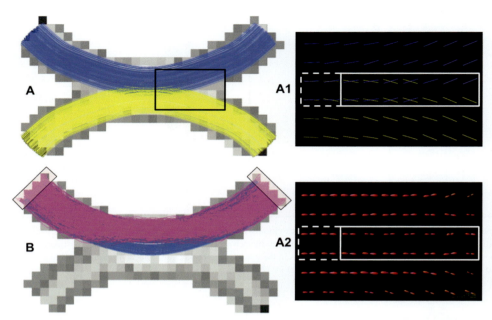

**FIG. 4** Representation of the wall effect using the open-source framework Fiberfox (Neher et al., 2014) and an in-house geometry generator. In (A), the *ground truth* is composed of two bundles kissing. ROI A1 represents the two separate vector fields from the beginning of the overlap to the point where they run parallel. ROI A2 shows the reconstructed fODF field, using CSD on the simulated data from Fiberfox. In both A1 and A2, the elongated *white boxes* show where the wall effect occurs, one could expect to see more than one fiber orientation, but the wall effect prevents it. The *dotted boxes* show where the fODF are expected to contain a single orientation. This is due to the fact that the ground-truth vectorfields are parallel at this location. In (B), the pink streamlines generated using probabilistic tractography show a clear decrease in volume when compared to the blue streamlines (ground truth). The lower portion of the arc is completely missing due to the wall effect.

single-direction local models until they exit the region. Instead of actually having a kissing or small angle crossing region, the reconstruction appears to be separated by a *wall* as shown in Fig. 4. In such a case, termination of the streamlines could be considered valid, but the overall spatial extent is reduced. However, it is possible this effect could send streamlines off-path if they stay in a straight line following the *wall*.

### 2.1.3 Narrow intersection effect

This effect is what happens when a streamline follows the wrong local direction in a crossing. When a streamline reaches a position where multiple directions are considered valid, it is possible to simply pick the wrong one. The same can happen at the next step, and so forth, with the errors accumulating until the crossing region is over. Contrary to the previously described effect, this is not due to the fact that the local model is too simplistic when pathways overlap. In this case, the crossing can be correctly resolved, in terms of local modeling, and it is simply due to the tractography algorithm picking the wrong direction when more than one choice seems valid.

In the case of (some) deterministic and probabilistic tractography algorithms that use constrained spherical deconvolution (CSD) (Tournier et al., 2007; Descoteaux et al., 2007), the amplitude of the fiber orientation distribution function (fODF) can lead to more error. The choice is sometimes influenced by the amplitude of the local model as well as its direction, this leads to another bias where dense fasciculi will increase their signal weight compared to others, and thus generate fODF with unequal lobe sizes (Raffelt et al., 2012). This will lead to streamlines *choosing* more easily these directions if they are within their valid options, since the lobe amplitude is superior to the alternatives (see top row of Fig. 5).

The narrow intersection effect can be observed in the centrum semiovale, a well-known region often pointed out as a major pitfall for diffusion tensor imaging (DTI). Local models that support more than one direction per voxel have their own limitations in this area. When entering the region perpendicular to the other underlying fiber populations, there is no problem. However, as the incoming angle decreases, over the course of a few voxels, some streamlines will follow the easiest path available (e.g., smallest angle) and sometimes go along the wrong direction, effectively *switching pathways*. In the centrum semiovale region, the CC and PyT are fanning and the contribution of the AF becomes bigger, making the reconstruction of the CC and PyT more difficult. Within this crossing, multiple streamlines from the CC and PyT take a sharp turn along the anterior-posterior axis (see bottom row of Fig. 5).

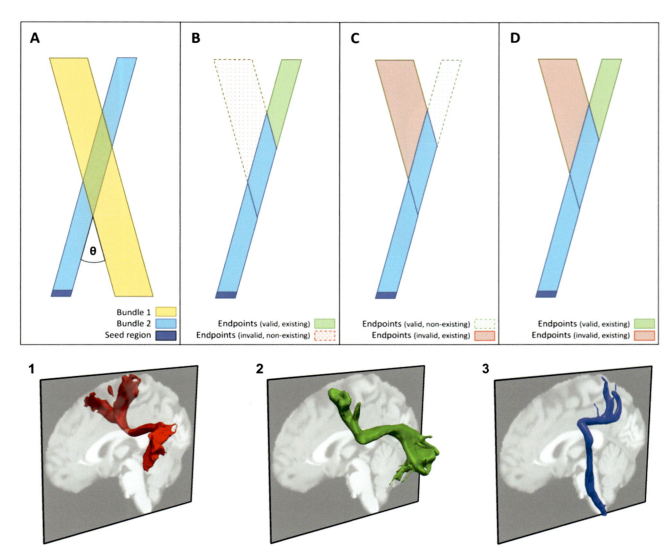

**FIG. 5** The narrow intersection effect. Two bundles crossing each other at a narrowangle $\theta$ (A). In this scenario, the reconstructed local models correctly contain two independent lobes/peaks in the crossing region. The seeding region is shown using a *dark blue box*. A tractography algorithm varying the cone aperture ($\theta$) (Chapter 14) at each step will generate scenario A if the angle is low enough, while scenario C is also possible in the case of a high angle parameter setting. Most likely, scenario D will be the result of most tractography algorithms and the proportion of streamlines at both terminations will vary with the curvature parameters, stepsize, and the dimension and angle of the crossing. Various regions of the brain are prone to this effect, such as the centrum semiovale. Pathways (1), (2), and (3) show reconstructed connections. It is harder to evaluate if they should be considered valid or invalid based on our knowledge of neuroanatomy.

Global tractography algorithms can also be vulnerable to the narrow intersection effect, but the turns are usually less pronounced because of the use of parameters penalizing broken lines and promoting smoothness. If the crossing happens at a low angle over a somewhat large region, streamlines can still *switch pathways* and in such a case invalid streamlines are even harder to distinguish from valid streamlines.

## 2.2 Path generation challenge

Challenges due to local models are the main source of error for dMRI tractography, but it is important to consider local modeling on a larger scale. Decisions made at the voxel level are hard to inspect, but their impact on the scale of the whole brain is more easily understood. Accumulation of tracking errors along the whole path can lead to implausible pathways or streamlines with unique shapes, despite being locally well-supported by the local models. Overall, the effects

570 PART | IV From streamlines to tracts

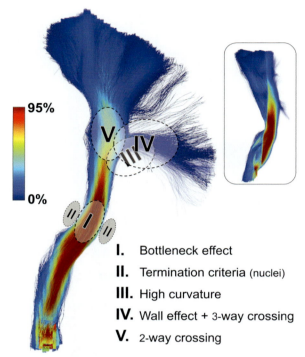

**FIG. 6** Local density of streamlines projected on the pyramidal tract (PyT) showing the major *obstacles* along the way; on the right is a lateral (cut) view from the right showing the density within the core. The internal capsule region is a single fiber population bottleneck (I) squeezed in between a partial volume effect of the surrounding nuclei (II). The region fanning laterally is first restricted by a sharp turn (III) and then a wall effect caused by the contribution of the arcuate fasciculus (AF) (IV), while also being a three-way crossing. This can explain why the number of streamlines reaching the medial/superior region at the cortex is higher, encountering only one difficult region, that is, the two-way crossing from the corpus callosum (CC) (V).

previously described can be seen as *difficult regions* or even *obstacles*. When looking at a reconstructed bundle and the number of streamlines in it, it is crucial to consider the number of such obstacles along its way. One of the reasons why not all bundles converge similarly to a stable white matter coverage is that different bundles have different obstacle variety along their path (a more complete description is available in Section 2.4, or Fig. 8). Bundles with a constant curving path passing through crossing regions will be harder to reconstruct than straight bundles. This is one of the reasons why the fanning of the PyT is harder to reconstruct than the core. The fanning region includes a crossing area in the centrum semiovale and it is easier to follow the straight path of the IC to the hand region than the constant curve from IC to the leg region. This explains why reconstructing the full spatial extent of the PyT is a challenging task even for multidirection local models, as shown in Fig. 6.

Furthermore, the longer a bundle is, the higher the probability of encountering obstacles will be. In practice, this means that more *good* local decisions are required to avoid mistakes. In the case of the PyT, the streamlines all have to navigate through areas of partial volume effects in the brainstem, traverse the IC surrounded by nuclei, and finally follow the complex fanning structure into a large area of the cortex. As a result, streamline density in a bundle is often strongly linked to distance because of these encounters with difficult regions (Sotiropoulos and Zalesky, 2019; Rheault et al., 2019). These biases are strictly due to processing, reconstruction, and algorithmic choice and not related to the true underlying anatomy.

Postprocessing algorithms such as SIFT, LIFE, and COMMIT (Smith et al., 2015; Pestilli et al., 2014; Daducci et al., 2015) attempt to fix this disparity in density that does not represent the local signal. By assigning a weight to each streamline to explain the observed diffusion signal, streamlines can be discarded if they do not provide a significant contribution. However, poor initial reconstruction will lead to suboptimal filtering results. These algorithms necessitate tens of millions of streamlines to fully cover white matter space (see Section 2.4), which leads to the emergence of somewhat dense false positives. Dense anatomically implausible groupings of streamlines can often be supported by the underlying dMRI signal better than a few plausible streamlines. These methods require a good initialization; optimizing tractogram reconstruction before postprocessing (i.e., filtering) is always a good practice to increase the quality of the previously mentioned algorithms.

Improvements over these methods or machine learning techniques attempted to quantify invalid paths in order to identify outliers (Schiavi et al., 2019; Poulin et al., 2019), but the wild variations in possible shapes make it difficult to identify automatically. Injecting more anatomical priors into the tractography algorithms themselves is probably necessary to decrease the number of invalid streamlines, which will make them more easily identifiable. For more details, see Chapter 18.

## 2.3 Origin and termination

Determination of the termination criteria is one of the most influential decisions to make during a dMRI tractography project (along with the local model and the tracking algorithm). Decisions need to be made when embarking on a tractography study. Is tracking allowed to traverse or does it need to stop in deep subcortical structures? Are the streamlines forced to finish at the cortex? Can streamlines go into the cerebellum or should they stop at the level of pontine nuclei? Should the seeds be placed in the whole WM or just the WM/GM interface, and should the nuclei be included? Even when these questions can be answered with confidence, the origin and termination of streamlines is still subject to challenges that cannot be easily solved. One of the most well-known termination biases for dMRI tractography is the gyral bias (Van Essen et al., 2014; Reveley et al., 2015), which was extensively covered in Chapter 18.

### 2.3.1 Deep nuclei

The deep nuclei provide another challenge for streamline termination. Their complex structure often renders dMRI tractography uninterpretable within a nucleus, despite well-organized underlying WM pathways passing through the surrounding regions and along the surface. Short connections between pairs of nuclei are also very hard to interpret, despite being more validated than other long-ranging WM pathways. Due to the simplicity of the local decisions during tractography, streamlines can (and often will) stop around the nuclei due to partial volume effects and poorly defined termination criteria. This blurs the line between the anatomically plausible streamlines and methodological artifacts, as shown in Fig. 7. However, some of these streamlines could simply be passing by and running parallel for a short distance. More advanced algorithms using surfaces could prevent such terminations, but the lack of confirmation from anatomical validation makes it difficult to establish clear rules on what streamlines can and cannot do around the deep nuclei.

### 2.3.2 Anatomical priors

Injecting more anatomical priors is a good way to tackle some of the challenges of tractography. This is often done by including tissue segmentation maps (WM, GM, and CSF) in the tractography process, also called anatomically constrained tractography (ACT) (Chapter 18). For example, FSL-FAST (Woolrich et al., 2009) often classifies the deep nuclei as a mix of WM and GM, which will be a challenge for tractography. ANTS-ATROPOS (Avants et al., 2011) classifies deep nuclei as GM only and any streamline coming near them will stop. Most segmentation tools inadvertently create a ring of GM around the ventricles due to partial volume effects between CSF and WM. Such a layer of GM wrongfully allows termination and leads to *broken streamlines*, which are perfectly valid in terms of shape, but abruptly stop in what should be WM (Smith et al., 2012; Girard et al., 2014; St-Onge et al., 2018). Furthermore, the lack of differentiation between the cortical GM and the deep nuclei GM does not allow for spatially variant rules for tractography. Even if this were possible, such rules would need to be designed and confirmed by neuroanatomists.

Moreover, proximity between GM regions and partial volume effects can reduce the area of WM *passage* or even completely block it. This is common in the temporal stem, where GM *leaks* into WM, which prevents the inferior fronto-occipital fasciculus (IFOF) and uncinate fasciculus (UF) from being reconstructed completely (Hau et al., 2017). In pathological cases with white matter lesions, tissue segmentation is often very challenging, where thresholding a diffusion metric such as FA could be an option. However, if the FA threshold is too high (above 0.15), numerous crossing regions and most nuclei will not be included in the mask, and this could result in incomplete reconstructions of perfectly healthy bundles simply due to the nature of DTI and poor specificity of FA. These challenges will be revisited in Chapter 19. Examples of how tissue segmentation algorithms impact tractography reconstruction are shown in Fig. 7.

## 2.4 Tractography and spatial coverage

DMRI tractography can be seen as a way to explore the white matter and each streamline adds to the overall spatial coverage of the whole tractogram. This is a stochastic process, in which each streamline is a new sample to add to the

**572 PART | IV From streamlines to tracts**

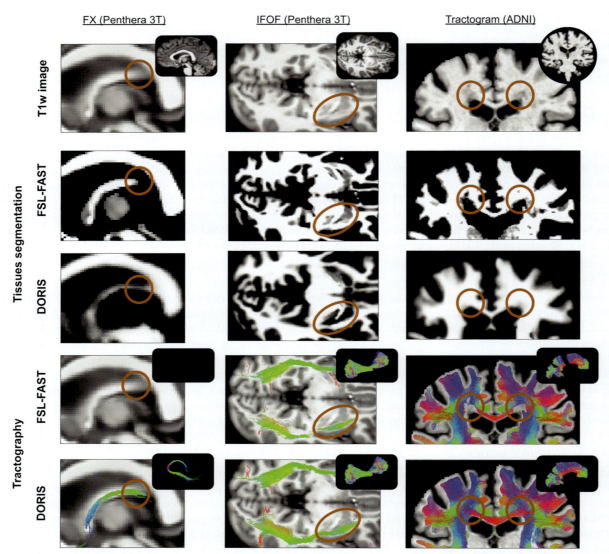

**FIG. 7** The uncertainty around termination criteria is challenging for tractography. Segmentation algorithms have their own quasiunique tissue segmentation output, with varying levels of partial volume effect (PVE), illustrated here on a dataset from a healthy control multishell dMRI public database (Penthera 3T—Columns 1–2, Paquette et al., 2019; Beaudoin et al., 2021) and single-shell dMRI acquisition from ADNI (Column 3, Petersen et al., 2010). This is shown in this comparison of FST-FAST (Woolrich et al., 2009) (Rows 2 and 4) and DORIS (Theaud et al., 2022) (Rows 3 and 5). The regions around the thalamic nuclei involving the body of the fornix (FX, Column 1) and the stem of the inferior fronto-occipital fasciculus (IFOF, Column 2) are often contaminated by overextended PVE. Even a small decrease in PVE can increase the spatial coverage of nearby bundles. Inadequate classification due to lesions near the corpus callosum (CC, Row 3) in an aging dataset leads to a similar challenge. Tissue segmentation has a high impact on the quality and spatial extent of tractography reconstruction and represents an open challenge.

tractogram. This is why it is usually recommended to generate hundreds of thousands (if not millions) of streamlines to obtain tractogram reconstructions that reconstruct most WM configurations. If whole brain coverage is desired, it is not always realistic to achieve a reconstruction with enough streamlines. An obvious interrogation can be raised: How many streamlines are enough (Moyer et al., 2018; Gauvin, 2016)? The answer is more complex than it seems at first glance. The answer to this question will change depending on the way tractography is seeded or the algorithm used. The actual consequence of not generating enough streamlines depends on the intended use, as shown in Fig. 8. For example, the node weight in a deterministic connectome matrix will stabilize faster than a probabilistic connectome matrix, but not all nodes will stabilize at the same speed (Yeh et al., 2018). Some connections in the connectome are harder to track than others. Overall, poor seeding can easily lead to unobserved pathways, while seeding too much can lead to the observation of a large number of uninterpretable pathways.

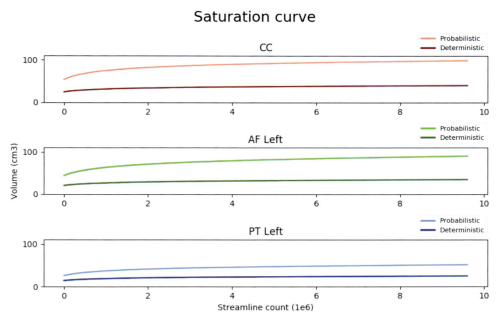

**FIG. 8** Bundle saturation effect. Saturation curves for three major WM bundles, the corpus callosum (CC, *red*), arcuate fasciculus (AF, *green*), and pyramidal tract (PyT, *blue*) using probabilistic (*lighter*) and deterministic (*darker*) tractography. On the *X*-axis, the number of generated streamlines in the tractogram is shown, up to 10M streamlines. On the *Y*-axis, the volume of the bundle of interest is given in cubic centimeters. Probabilistic tractography reconstructs a larger volume and converges/stabilizes slower than deterministic tractography. Volume in probabilistic tractography is greatly influenced by the cone aperture threshold and convergence will change as a direct consequence. In this example, the aperture of the cone was set to 20 degrees. For example, the probabilistic CC at 100k and 200k streamlines has 54.1 and 58.2cm$^3$ respectively, while at 9.9M and 10M streamlines the volume is at 98.9 and 99.0cm$^3$, respectively. The scaling of the *X*-axis is compressed for visualization, giving the impression of ever-increasing volume.

### 2.4.1 Seeding strategies

Seeding strategies are crucial to consider when interpreting the resulting tractograms. Seeding strategies that initialize streamlines from WM voxels will reconstruct major bundles (large volume) more easily than WM-GM interface seeding strategies, as they occupy more space and therefore more initiated streamlines will participate in its reconstruction (Smith et al., 2012; Girard et al., 2014). This is sometimes referred to as spatial coverage (or volume) convergence (or saturation). These expressions mean that the volume of a reconstructed bundle does not change even if more streamlines are generated, as the volume of the bundle has converged. It means that "enough" streamlines were generated for that bundle, but not necessarily for others.

This is a potential bias that needs to be considered in connectomics; as a potential solution connectomics studies tend to use seeding methods from the WM/GM interface (or more recently, surfaces) to reduce this bias (Girard et al., 2014; St-Onge et al., 2018). As opposed to being biased toward bundles with large volume, streamline distribution is linked to the area at the WM/GM interface. This strategy is not as problematic due to a squared growth (of the surface, i.e., cortex) instead of a cubic growth (of the volume, i.e., white matter) in streamline count. This is known as the *square-cube law* and can be adapted to our context as follows: as a bundle grows in size, its volume grows faster than its surface area. This means seeding from its volume (WM) will generate more streamlines than seeding from its cortical surface.

### 2.4.2 Impact on bundle segmentation and connectomics

Not all bundle volume will converge (or be saturated) at rate, and this is the reason why it is nearly impossible to say when a tractogram is saturated or not. This can lead to an underestimation of the spatial extent for some connections or the overestimation of others. This can be problematic for the interpretation of results since each connection may have reached its maximum volume coverage (or not). Deterministic tractography algorithms will *converge* much faster than probabilistic ones as they *sample* streamlines from a much smaller distribution. In this context, smaller distribution means that due to the underlying generation algorithm of deterministic tractography there are fewer possible *choices* of streamline path. This inherently leads to smaller volumes that get covered faster. Global tractography is not immune to this issue. Global tractography algorithms are simply exploring the space differently at their own speed. This means that, just like the classical approaches, factors such as segment length, voxel size, and number of seeds to initialize will influence the number of streamlines needed and will influence how slow/fast the spatial coverage of connections will be achieved.

## 3 Pathways and connectomes: How to interpret them?

Once a tractogram has been generated, it is quite challenging to interpret the findings from it. Clinically oriented research typically uses one of the two most widespread approaches: bundle segmentation to perform tractometry (Chapters 21 and 22) or connectomics (Chapter 24). This section will briefly summarize factors that need to be carefully thought out before starting a project involving one of these approaches.

### 3.1 Bundle segmentation

#### 3.1.1 Manual approach

Despite being heavily time-consuming, manual segmentation is still very much used by neuroanatomy experts (Chapter 21). Evidently, all previously mentioned challenges will influence how the tractogram is reconstructed and any bundles dissected from it. However, the act of performing the segmentation comes with its own bias. Intrarater and interrater variability can create a surprisingly large variation in WM pathways obtained from tractography (Gwet, 2012, as seen in Fig. 9). Since dMRI tractography is a virtual method, experimental variability can be assessed. The same anatomical definition, on the same tractogram, from the same rater at different time points will inevitably produce slightly different results, which may also happen if the segmentation is done by more than one rater (Rheault et al., 2020, 2022a, c). This bias could become a major pitfall if the set of rules is not well clarified in advance. As shown in Section 4.2 interpretation of a definition can be challenging. Observations and interpretations based on a single segmentation could change if performed by a different rater, especially if this variability is not accounted for in the statistical analysis. In such a case, conclusions could be different.

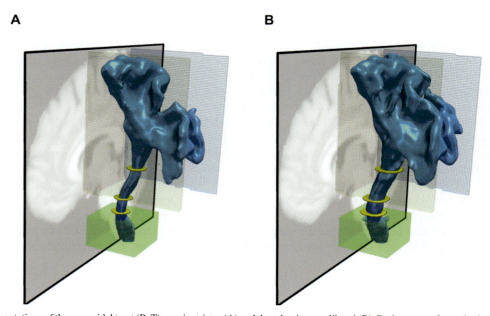

**FIG. 9** Two segmentations of the pyramidal tract (PyT): one is stricter (A) and the other is more liberal (B). Both were performed using rules from Chenot et al. (2019) to isolate the path. Three disk regions of passage (internal capsule, midbrain, and medulla oblongata) and three regions of exclusion (midline, posterior to postcentral gyrus, and anterior of precentral gyrus) were used. The conservative version used planar disks of 5 mm radius and the liberal version used planar disks of 7.5 mm radius (disks were enlarged for visualization), while all other ROIs were identical. The spatial extent of the fanning increases to completely cover both gyri and the core of the bundle becomes thicker in the liberal segmentation. The volume of the conservative version (A) is 36.2cm$^3$ and the liberal version (B) reaches 58.0cm$^3$. Such a difference can be expected when changing the anatomical definition used for the segmentation. Statistical comparisons or conclusions should be adapted if the discrepancy emerges from raters' variability (intrarater or interrater).

### 3.1.2 Automatic approach

While automatic segmentation methods do not have intrarater and interrater variability, they, however, rely on algorithms that are often not free of their own challenges. It is important to consider that most automatic methods claiming to reconstruct anatomically relevant bundles rely on some sort of registration algorithm, which can be strongly affected by datasets containing pathological or aging brains. Such datasets always present a challenge, but using automatic methods designed on healthy datasets can have unexpected results.

Another common challenge with automatic methods is that they heavily rely on predetermined bundle definitions. Such definitions are sometimes not fully agreed upon and could even be considered wrong (mentioned in Section 4.2). Sometimes developed by researchers with no background in anatomy, new methods are presented purely for algorithmic consideration, and can add "noise" to the literature when it comes to bundle segmentation/definition. If (when) the lack of consensus in anatomical definitions is addressed in the future, any automatic method that would follow the agreed-upon definition would be a very interesting asset for the field of dMRI tractography. In the meantime, when using an automatic segmentation method, the resulting bundle of interest (BOI) should be considered with caution.

Finally, each segmentation approach has its advantages and disadvantages, and depending on the intended goal they can be very useful or problematic (Chapter 22). This section will present a limited overview of known methods, where algorithms were not chosen for their popularity but rather their differences from each other, which highlight various considerations. Fig. 10 highlights differences between submissions (on an identical dataset) from a project where 42 groups performed their own segmentation of 14 different bundles (Schilling et al., 2021).

Large-scale open datasets are now being created and shared to facilitate the evaluation of methods or the reproducibility of diffusion MRI (and subsequently tractography) (Cai et al., 2021; Koller et al., 2021; Duchesne et al., 2019). The TractoInferno dataset (Poulin et al., 2021) was created specifically for the training and evaluation of machine learning tractography algorithms. Multicenter efforts to quantify variability in tools producing streamlines or bundles segmentation methods are aiming to better understand algorithmic variability (Schilling et al., 2021).

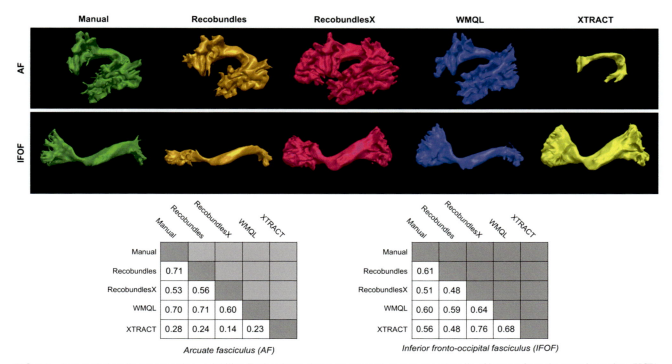

**FIG. 10** Automatic bundle-segmentation variability. Each column shows an approach to segment a bundle of interest from the same input data (HCP). The rows show two association pathways: the arcuate fasciculus (AF) and the inferior fronto-occipital fasciculus. The methods do not follow the same anatomical definition and therefore do not reconstruct the same spatial extent. In order, Manual segmentation, Recobundles (Garyfallidis et al., 2018), RecobundlesX (Rheault, 2020; Garyfallidis et al., 2018), White Matter Query Language (WMQL) (Wasserthal et al., 2018), and XTRACT (Warrington et al., 2020). Data for the figure is from submissions of Schilling et al. (2021) on identical probabilistic tractogram. The table below shows the Dice coefficient between all pairs of bundles to highlight the difference in their spatial extend (0—no overlap, 1—complete overlap). The figure shows 3D rendering of surfaces simply for visualization purposes; the initial representation was streamline.

**576 PART | IV From streamlines to tracts**

## 3.2 Effects on connectomics

All the tractography challenges previously mentioned will also influence the connectome matrix computed from the tractogram. However, when it comes to building a connectome matrix, the tractogram is only half of the equation, and the remaining factors are as important (Qi et al., 2015), as mentioned in the Chapter 24. The choice of the segmentation atlas, which will determine the size of the parcels and their boundaries, has an important effect on the way the tractogram is filtered to obtain the connectivity matrix.

There is increasing effort put toward structural connectomics analysis and open-source tools are now more common (Seguin et al., 2022; Cruces et al., 2022; Serin et al., 2021; Baggio et al., 2018; Rheault et al., 2021). International challenges such as the diffusion-simulated connectivity (DiSCo) ISMRM challenge show that structural connectome reconstruction can be accurate despite the potential biases that may occur during processing.

# 4 Challenges and opportunities for the dMRI tractography community

## 4.1 Nomenclature and terminology

A key component of modern science is the common agreement among experts in a field on the terminology to use when describing their experiments, results, and conclusions (Sageder, 2010). However, in tractography, discrepancies in definition will always exist, due to the intrinsic multidisciplinarity of the topic (physics, biology, mathematics, computer science, medicine, chemistry, etc.). Here, each field has its own terminology and the choice of words of researchers in one field can sometimes convey a different meaning for researchers in another field.

While this section is not an attempt to redefine or propose a common scientific terminology for the field of dMRI tractography, we explicitly present examples of common words or expressions that may lead to misunderstandings. By acknowledging the variety of definitions and their risks, misunderstandings can be avoided and an agreement can be more easily reached over time. This represents an opportunity for developers and neuroanatomists to join forces and set terminology standards (see Mandonnet et al., 2018).

### 4.1.1 Concept of a single fiber

Track, tract, fibre (UK) and fiber (US), streamlines, pathways, and trajectories are terms used interchangeably but may not always refer to the same thing. The easiest to define is "streamline," because it is specific to virtual reconstruction: a streamline is a polyline, an individual element generated by tractography. This is a mathematical representation and is not necessarily specific to tractography. "Track" is often considered analogous to streamline. The term "tract" is often used to describe a set of streamlines with similar spatial characteristics. However, it can sometimes refer to an anatomical group of axons (see Section 4.2). "Fibre"/"fiber" are used in tractography as well as in anatomy. However, it was recommended that "fibers" should be reserved to reference the underlying fibrous tissues (Jones et al., 2013; Côté et al., 2013) and they can also apply to cardiac MRI with the fibrous texture of muscles. On the other end, terms like "axon" should be avoided and not associated with tractography reconstruction.

### 4.1.2 Concept of a group of fibers

Another concept often leading to confusion is that of a "group of fibers." In its digital representation as well as in the real brain, we tend to classify things and refer to them as a group for simplicity. The terms "tracks"/"tracts," "bundle," "fascicle"/"fasciculus," "cluster," and "pathway" are sometimes used as synonymous, but again their definitions can vary considerably depending on the context. Overall, the definition of "fascicle" and "pathway" is somewhat clearer, typically describing a large group of axons that respect an anatomical definition. However, these terms are used for both tractography representations as well as for neuroanatomy. The terms "bundle" and "tract" are similarly used for both tractography and the underlying anatomy. However, while some studies have used "bundle" to group streamlines with a known anatomical counterpart (i.e., entire tracts), others have used the term "bundle" to refer simply to a group of streamlines without a clear anatomical definition or to refer to a fraction or subcomponent of an anatomically defined tract. The term "cluster" is more specific to neuroimaging applications, and is used to describe a group of morphological and spatially similar streamlines, but does not necessarily represent a known anatomical structure.

In this chapter, we will always refer to the anatomical structure as a "fascicle," and always use "bundle" to refer to a group of streamlines. The term "pathway" is used in both contexts to describe the general path of known WM structures.

### 4.1.3 Concept of the brain structural connectivity

The last concept is related to the overall brain structural connections. The terms "connection," "tractogram," "connectome," and "connectivity" are not always used in the same way by everyone. A "tractogram" usually refers to a whole brain output of tractography (usually composed of hundreds of thousands to millions of streamlines). A "connectome," in tractography, defines all connections between pairs of regions (Hagmann, 2005), but in neuroscience and neuroanatomy fields, "connectome" rather describes the general concept of the entire multiscale mapping of the human brain connectivity, in some cases, down to the axonal level (Sporns et al., 2005). Finally, "connection" and "connectivity" are words referring to white matter connecting specific brain regions and are similarly used by neuroimaging as well as neuroanatomy.

## 4.2 Reaching consensus on anatomical definitions

Electrical stimulation during surgery to guide resection (Szelényi et al., 2010; Suess et al., 2006), stroke location to predict deficits (lesions mapping) (Mazzocchi and Vignolo, 1979; Robinson et al., 1984), and deep brain stimulation for Parkinson's (Benabid, 2003) or epilepsy (Sprengers et al., 2017) are examples that indicate that our understanding of the human brain anatomy is quite advanced. However, these examples highlight the knowledge of cortical regions and deep nuclei, not the white matter connections linking them. Even with the seminal work of Dejerine and Dejerine-Klumpke (1895) dating to more than a century ago, the current state of consensus on anatomical definitions of WM pathways is not perfect and is an opportunity for the community to come up with adequate and modern white matter anatomy definitions.

### 4.2.1 Debate is common in neuroimaging

The connectivity of the brain is a complex subject; there is debate when it comes to terminology, anatomical definition, and even the existence (or nonexistence) of pathways. Historically, multiple discoveries happened in parallel across nations that led to various differences in descriptions and names of white matter pathways. Small variations in terminology can explain most observable differences in pathways having the same name, such as the case of even well-known pathways sometimes having a variety of similar, almost equivalent, anatomical definitions. A summary of those debates and challenges for the neuroanatomy community is available in the third portion of the validation chapters (Chapter 27).

All the debates and controversies are by no means specific to the field of structural connectivity or dMRI tractography. Well-known brain structures like the hippocampus have been at the center of neuroanatomical discussions for years. Descriptions of the structure have varied through time (position, shape, extent), but protocols on how to perform a 3D virtual segmentation are even more heterogeneous. A review of the existing hippocampus segmentation protocols showed that the agreement across protocols was far from perfect (Boccardi et al., 2011). There was no clear consensus in neuroimaging/radiology on the anatomical definition of the hippocampus. Steps were taken to create a harmonized protocol that would minimize the chance of error, maximize reproducibility, and reduce confusion (Boccardi et al., 2015). Such protocol allows researchers to aggregate data across studies to perform metaanalyses and facilitate literature reviews (Frisoni et al., 2015). A similar approach is needed for every major WM pathway of the human brain (Rheault et al., 2022b).

## 4.3 Limitations in anatomical validation

From the perspective of the neuroimaging field, validations with histological staining, tracing, postmortem dissection, or, more recently, methods such as polarized light imaging (PLI), optical coherence tomography (OCT), or clarity are often considered *ground truth* to our reconstruction methods (Axer et al., 2016; Mollink et al., 2017; Wang et al., 2011, 2015; Chang et al., 2017; Morawski et al., 2018). Due to the complexity of these validation methods, they should be considered as *gold standard* rather than ground truth, and even with this consideration a few caveats must be acknowledged (Chapters 25 and 26).

As is the case for every experimental approach, there is an inherent variability of results (Gwet, 2012). This variability cannot be truly evaluated, since a procedure cannot be performed twice on the same brain. The designation *ground truth* is not adequate, and a comparison between any of these approaches with dMRI tractography will likely overestimate the error from tractography (since the experimental variability is not taken into account). Any of these techniques have their specific biases, pitfalls, and biological or algorithm limitations. It is important to consider them as a *gold standard*, as they can only capture a portion of the real anatomy.

**578 PART | IV From streamlines to tracts**

# 5 Conclusion

The purpose of this chapter was to put into perspective the existing challenges for the field of dMRI tractography. Further details of the challenges covered in this chapter can be found in the following reviews (Jbabdi and Johansen-Berg, 2011; Jones et al., 2013; O'Donnell and Pasternak, 2015; Sotiropoulos and Zalesky, 2019; Maier-Hein et al., 2017). Understanding the limitations of dMRI tractography is as important as understanding its strengths. This chapter therefore should be seen as a constructive critique for dMRI tractography and a list of open challenges, offering an opportunity for new research and tractography development.

# References

Avants, B.B., Tustison, N.J., Wu, J., Cook, P.A., Gee, J.C., 2011. An open source multivariate framework for n-tissue segmentation with evaluation on public data. Neuroinformatics 9 (4), 381–400.

Axer, M., Strohmer, S., Gräßel, D., Bücker, O., Dohmen, M., Reckfort, J., Zilles, K., Amunts, K., 2016. Estimating fiber orientation distribution functions in 3D-polarized light imaging. Front. Neuroanat. 10, 40.

Baggio, H.C., Abos, A., Segura, B., Campabadal, A., Garcia-Diaz, A., Uribe, C., Compta, Y., Marti, M.J., Valldeoriola, F., Junque, C., 2018. Statistical Inference in Brain Graphs Using Threshold-Free Network-Based Statistics. Wiley Online Library.

Beaudoin, A.M., Rheault, F., Theaud, G., Laberge, F., Whittingstall, K., Lamontagne, A., Descoteaux, M., 2021. Modern technology in multi-shell diffusion MRI reveals diffuse white matter changes in young adults with relapsing-remitting multiple sclerosis. Front. Neurosci. 15, 665017.

Benabid, A.L., 2003. Deep brain stimulation for Parkinson's disease. Curr. Opin. Neurobiol. 13 (6), 696–706.

Benedictis, A., Petit, L., Descoteaux, M., Marras, C.E., Barbareschi, M., Corsini, F., Dallabona, M., Chioffi, F., Sarubbo, S., 2016. New insights in the homotopic and heterotopic connectivity of the frontal portion of the human corpus callosum revealed by microdissection and diffusion tractography. Hum. Brain Mapp. 37 (12), 4718–4735.

Boccardi, M., Ganzola, R., Bocchetta, M., Pievani, M., Redolfi, A., Bartzokis, G., Camicioli, R., Csernansky, J.G., De Leon, M.J., deToledo Morrell, L., et al., 2011. Survey of protocols for the manual segmentation of the hippocampus: preparatory steps towards a joint EADC-ADNI harmonized protocol. J. Alzheimers Dis. 26 (S3), 61–75.

Boccardi, M., Bocchetta, M., Apostolova, L.G., Barnes, J., Bartzokis, G., Corbetta, G., DeCarli, C., Firbank, M., Ganzola, R., Gerritsen, L., et al., 2015. Delphi definition of the EADC-ADNI Harmonized Protocol for hippocampal segmentation on magnetic resonance. Alzheimers Dement. 11 (2), 126–138.

Cai, L.Y., Yang, Q., Kanakaraj, P., Nath, V., Newton, A.T., Edmonson, H.A., Luci, J., Conrad, B.N., Price, G.R., Hansen, C.B., et al., 2021. MASiVar: multisite, multiscanner, and multisubject acquisitions for studying variability in diffusion weighted MRI. Magn. Reson. Med. 86 (6), 3304–3320.

Chamberland, M., Scherrer, B., Prabhu, S.P., Madsen, J., Fortin, D., Whittingstall, K., Descoteaux, M., Warfield, S.K., 2017. Active delineation of Meyer's loop using oriented priors through MAGNEtic tractography (MAGNET). Hum. Brain Mapp. 38 (1), 509–527.

Chang, E.H., Argyelan, M., Aggarwal, M., Chandon, T.S.S., Karlsgodt, K.H., Mori, S., Malhotra, A.K., 2017. The role of myelination in measures of white matter integrity: combination of diffusion tensor imaging and two-photon microscopy of CLARITY intact brains. NeuroImage 147, 253–261.

Chenot, Q., Tzourio-Mazoyer, N., Rheault, F., Descoteaux, M., Crivello, F., Zago, L., Mellet, E., Jobard, G., Joliot, M., Mazoyer, B., et al., 2019. A population-based atlas of the human pyramidal tract in 410 healthy participants. Brain Struct. Funct. 224 (2), 599–612.

Côté, M.A., Girard, G., Boré, A., Garyfallidis, E., Houde, J.C., Descoteaux, M., 2013. Tractometer: towards validation of tractography pipelines. Med. Image Anal. 17 (7), 844–857.

Cruces, R.R., Royer, J., Herholz, P., Larivière, S., de Wael, R.V., Paquola, C., Benkarim, O., Park, B.Y., Degré-Pelletier, J., Nelson, M., et al., 2022. Micapipe: a pipeline for multimodal neuroimaging and connectome analysis. NeuroImage 263, 119612.

Daducci, A., Dal Palù, A., Lemkaddem, A., Thiran, J.P., 2015. COMMIT: convex optimization modeling for microstructure informed tractography. IEEE Trans. Med. Imaging 34 (1), 246–257.

Daducci, A., Dal Palú, A., Descoteaux, M., Thiran, J.P., 2016. Microstructure informed tractography: pitfalls and open challenges. Front. Neurosci. 10, 247.

Dejerine, J., Dejerine-Klumpke, A., 1895. Anatomie des centres nerveux: Méthodes générales d'étude-embryologie-histogénèse et histologie. Anatomie du cerveau. vol. 1 Rueff.

Descoteaux, M., Angelino, E., Fitzgibbons, S., Deriche, R., 2007. Regularized, fast, and robust analytical Q-ball imaging. Magn. Reson. Med. 58 (3), 497–510.

Duchesne, S., Chouinard, I., Potvin, O., Fonov, V.S., Khademi, A., Bartha, R., Bellec, P., Collins, D.L., Descoteaux, M., Hoge, R., et al., 2019. The Canadian dementia imaging protocol: harmonizing national cohorts. J. Magn. Reson. Imaging 49 (2), 456–465.

Frisoni, G.B., Jack Jr., C.R., Bocchetta, M., Bauer, C., Frederiksen, K.S., Liu, Y., Preboske, G., Swihart, T., Blair, M., Cavedo, E., et al., 2015. The EADC-ADNI harmonized protocol for manual hippocampal segmentation on magnetic resonance: evidence of validity. Alzheimers Dement. 11 (2), 111–125.

Garyfallidis, E., Côté, M.A., Rheault, F., Sidhu, J., Hau, J., Petit, L., Fortin, D., Cunanne, S., Descoteaux, M., 2018. Recognition of white matter bundles using local and global streamline-based registration and clustering. NeuroImage 170, 283–295.

Gauvin, A., 2016. Assurance Qualité en Dissection Virtuelle des Faisceaux de la Matière Blanche par Tractographie. (Master's thesis). Universite de Sherbrooke.

Girard, G., Whittingstall, K., Deriche, R., Descoteaux, M., 2014. Towards quantitative connectivity analysis: reducing tractography biases. NeuroImage 98, 266–278.

Gwet, K.L., 2012. Handbook of Inter-Rater Reliability: The Definitive Guide to Measuring the Extent of Agreement Among Multiple Raters. Advanced Analytics, LLC.

Hagmann, P., 2005. From Diffusion MRI to Brain Connectomics. EPFL.

Hau, J., Sarubbo, S., Houde, J.C., Corsini, F., Girard, G., Deledalle, C., Crivello, F., Zago, L., Mellet, E., Jobard, G., et al., 2017. Revisiting the human uncinate fasciculus, its subcomponents and asymmetries with stem-based tractography and microdissection validation. Brain Struct. Funct. 222 (4), 1645–1662.

Jbabdi, S., Johansen-Berg, H., 2011. Tractography: where do we go from here? Brain Connect. 1 (3), 169–183.

Jeurissen, B., Descoteaux, M., Mori, S., Leemans, A., 2017. Diffusion MRI fiber tractography of the brain. NMR Biomed. 32 (4), e3785.

Jones, D.K., Knösche, T.R., Turner, R., 2013. White matter integrity, fiber count, and other fallacies: the do's and don'ts of diffusion MRI. NeuroImage 73, 239–254.

Koller, K., Rudrapatna, U., Chamberland, M., Raven, E.P., Parker, G.D., Tax, C.M.W., Drakesmith, M., Fasano, F., Owen, D., Hughes, G., et al., 2021. MICRA: microstructural image compilation with repeated acquisitions. NeuroImage 225, 117406.

Maffei, C., Jovicich, J., De Benedictis, A., Corsini, F., Barbareschi, M., Chioffi, F., Sarubbo, S., 2018. Topography of the human acoustic radiation as revealed by ex vivo fibers micro-dissection and in vivo diffusion-based tractography. Brain Struct. Funct. 223 (1), 449–459.

Maier-Hein, K.H., Neher, P.F., Houde, J.C., Côté, M.A., Garyfallidis, E., Zhong, J., Chamberland, M., Yeh, F.C., Lin, Y.C., Ji, Q., et al., 2017. The challenge of mapping the human connectome based on diffusion tractography. Nat. Commun. 8 (1), 1349.

Mandonnet, E., Sarubbo, S., Petit, L., 2018. The nomenclature of human white matter association pathways: proposal for a systematic taxonomic anatomical classification. Front. Neuroanat. 12, 94.

Mazzocchi, F., Vignolo, L.A., 1979. Localisation of lesions in aphasia: clinical-CT scan correlations in stroke patients. Cortex 15 (4), 627–653.

Mollink, J., Kleinnijenhuis, M., van Walsum, A.M.V.C., Sotiropoulos, S.N., Cottaar, M., Mirfin, C., Heinrich, M.P., Jenkinson, M., Pallebage-Gamarallage, M., Ansorge, O., et al., 2017. Evaluating fibre orientation dispersion in white matter: comparison of diffusion MRI, histology and polarized light imaging. NeuroImage 157, 561–574.

Morawski, M., Kirilina, E., Scherf, N., Jäger, C., Reimann, K., Trampel, R., Gavriilidis, F., Geyer, S., Biedermann, B., Arendt, T., et al., 2018. Developing 3D microscopy with CLARITY on human brain tissue: towards a tool for informing and validating MRI-based histology. NeuroImage 182, 417–428.

Moyer, D.C., Thompson, P., Ver Steeg, G., 2018. Measures of tractography convergence. In: International Conference on Medical Image Computing and Computer-Assisted Intervention, pp. 295–307.

Neher, P.F., Laun, F.B., Stieltjes, B., Maier-Hein, K.H., 2014. Fiberfox: facilitating the creation of realistic white matter software phantoms. Magn. Reson. Med. 72 (5), 1460–1470.

O'Donnell, L.J., Pasternak, O., 2015. Does diffusion MRI tell us anything about the white matter? An overview of methods and pitfalls. Schizophr. Res. 161 (1), 133–141.

Paquette, M., Gilbert, G., Descoteaux, M., 2019. Penthera 3T., https://doi.org/10.5281/ZENODO.2602049.

Pestilli, F., Yeatman, J.D., Rokem, A., Kay, K.N., Wandell, B.A., 2014. Evaluation and statistical inference for human connectomes. Nat. Methods 11 (10), 1058.

Petersen, R.C., Aisen, P.S., Beckett, L.A., Donohue, M.C., Gamst, A.C., Harvey, D.J., Jack, C.R., Jagust, W.J., Shaw, L.M., Toga, A.W., et al., 2010. Alzheimer's disease neuroimaging initiative (ADNI) clinical characterization. Neurology 74 (3), 201–209.

Poulin, P., Cote, M.A., Houde, J.C., Petit, L., Neher, P.F., Maier-Hein, K.H., Larochelle, H., Descoteaux, M., 2017. Learn to track: deep learning for tractography. In: International Conference on Medical Image Computing and Computer-Assisted Intervention, pp. 540–547.

Poulin, P., Jorgens, D., Jodoin, P.M., Descoteaux, M., 2019. Tractography and machine learning: current state and open challenges. Magn. Reson. Imaging 64, 37–48.

Poulin, P., Theaud, G., Rheault, F., St-Onge, E., Bore, A., Renauld, E., de Beaumont, L., Guay, S., Jodoin, P.M., Descoteaux, M., 2021. TractoInferno: a large-scale, open-source, multi-site database for machine learning dMRI tractography. Sci. Data 9 (1), 725.

Qi, S., Meesters, S., Nicolay, K., ter Haar Romeny, B.M., Ossenblok, P., 2015. The influence of construction methodology on structural brain network measures: a review. J. Neurosci. Methods 253, 170–182.

Raffelt, D., Tournier, J.D., Rose, S., Ridgway, G.R., Henderson, R., Crozier, S., Salvado, O., Connelly, A., 2012. Apparent fibre density: a novel measure for the analysis of diffusion-weighted magnetic resonance images. NeuroImage 59 (4), 3976–3994.

Reveley, C., Seth, A.K., Pierpaoli, C., Silva, A.C., Yu, D., Saunders, R.C., Leopold, D.A., Frank, Q.Y., 2015. Superficial white matter fiber systems impede detection of long-range cortical connections in diffusion MR tractography. Proc. Natl. Acad. Sci. USA 112 (21), E2820–E2828.

Rheault, F., 2020. Analyse et Reconstruction de Faisceaux de la Matière Blanche (Ph.D. thesis). Universite de Sherbrooke. https://savoirs.usherbrooke.ca/handle/11143/17255.

Rheault, F., St-Onge, E., Sidhu, J., Maier-Hein, K., Tzourio-Mazoyer, N., Petit, L., Descoteaux, M., 2019. Bundle-specific tractography with incorporated anatomical and orientational priors. NeuroImage 186, 382–398.

Rheault, F., De Benedictis, A., Daducci, A., Maffei, C., Tax, C.M.W., Romascano, D., Caverzasi, E., Morency, F.C., Corrivetti, F., Pestilli, F., et al., 2020. Tractostorm: the what, why, and how of tractography dissection reproducibility. Hum. Brain Mapp. 41, 1859–1874.

Rheault, F., Houde, J.C., Sidhu, J., Obaid, S., Guberman, G., Daducci, A., Descoteaux, M., 2021. Connectoflow: a cutting-edge nextflow pipeline for structural connectomics. In: Proceedings of ISMRM.

Rheault, F., Bayrak, R.G., Wang, X., Schilling, K.G., Greer, J.M., Hansen, C.B., Kerley, C., Ramadass, K., Remedios, L.W., Blaber, J.A., et al., 2022a. TractEM: evaluation of protocols for deterministic tractography white matter atlas. Magn. Reson. Imaging 85, 44–56.

Rheault, F., Schilling, K.G., Obaid, S., Begnoche, J., Cutting, L.E., Descoteaux, M., Landman, B.A., Petit, L., 2022b. The influence of regions of interest on tractography virtual dissection protocols: general principles to learn and to follow. Brain Struct. Funct. 227, 2191–2207.

Rheault, F., Schilling, K.G., Valcourt-Caron, A., Théberge, A., Poirier, C., Grenier, G., Guberman, G.I., Begnoche, J., Legarreta, J.H., Y Cai, L., et al., 2022c. Tractostorm 2: optimizing tractography dissection reproducibility with segmentation protocol dissemination. Hum. Brain Mapp. 43, 2134–2147.

Robinson, R.G., Kubos, K.L., Starr, L.B., Rao, K., Price, T.R., 1984. Mood disorders in stroke patients: importance of location of lesion. Brain 107 (1), 81–93.

Sageder, D., 2010. Terminology today: a science, an art or a practice?: some aspects on terminology and its development. Brno Stud. Eng. 36 (1), 123–134.

Schiavi, S., Barakovic, M., Ocampo-Pineda, M., Descoteaux, M., Thiran, J.P., Daducci, A., 2019. Reducing false positives in tractography with microstructural and anatomical priors. bioRxiv, 608349.

Schilling, K.G., Rheault, F., Petit, L., Hansen, C.B., Nath, V., Yeh, F.C., Girard, G., Barakovic, M., Rafael-Patino, J., Yu, T., et al., 2021. Tractography dissection variability: what happens when 42 groups dissect 14 white matter bundles on the same dataset? NeuroImage 243, 118502.

Schilling, K.G., Tax, C.M.W., Rheault, F., Landman, B.A., Anderson, A.W., Descoteaux, M., Petit, L., 2022. Prevalence of white matter pathways coming into a single white matter voxel orientation: the bottleneck issue in tractography. Hum. Brain Mapp. 43 (4), 1196–1213.

Seguin, C., Smith, R.E., Zalesky, A., et al., 2022. Connectome spatial smoothing (CSS): concepts, methods, and evaluation. NeuroImage 250, 118930.

Serin, E., Zalesky, A., Matory, A., Walter, H., Kruschwitz, J.D., 2021. NBS-predict: a prediction-based extension of the network-based statistic. NeuroImage 244, 118625.

Smith, R.E., Tournier, J.D., Calamante, F., Connelly, A., 2012. Anatomically-constrained tractography: improved diffusion MRI streamlines tractography through effective use of anatomical information. NeuroImage 62 (3), 1924–1938.

Smith, R.E., Tournier, J.D., Calamante, F., Connelly, A., 2015. SIFT2: enabling dense quantitative assessment of brain white matter connectivity using streamlines tractography. NeuroImage 119, 338–351.

Sotiropoulos, S.N., Zalesky, A., 2019. Building connectomes using diffusion MRI: why, how and but. NMR Biomed. 32 (4), e3752.

Sporns, O., Tononi, G., Kötter, R., 2005. The human connectome: a structural description of the human brain. PLoS Comput. Biol. 1 (4), e42.

Sprengers, M., Vonck, K., Carrette, E., Marson, A.G., Boon, P., 2017. Deep brain and cortical stimulation for epilepsy. Cochrane Database Syst. Rev. 7, CD008497.

St-Onge, E., Daducci, A., Girard, G., Descoteaux, M., 2018. Surface-enhanced tractography (SET). NeuroImage 169, 524–539.

Suess, O., Suess, S., Brock, M., Kombos, T., 2006. Intraoperative electrocortical stimulation of Brodman area 4: a 10-year analysis of 255 cases. Head Face Med. 2 (1), 20.

Szelényi, A., Bello, L., Duffau, H., Fava, E., Feigl, G.C., Galanda, M., Neuloh, G., Signorelli, F., Sala, F., 2010. Intraoperative electrical stimulation in awake craniotomy: methodological aspects of current practice. Neurosurg. Focus 28 (2), E7.

Theaud, G., Edde, M., Dumont, M., Zotti, C., Zucchelli, M., Deslauriers-Gauthier, S., Deriche, R., Jodoin, P.M., Descoteaux, M., 2022. DORIS: a diffusion MRI-based 10 tissue class deep learning segmentation algorithm tailored to improve anatomically-constrained tractography. Front. Neuroimaging 1, 917806.

Tournier, J.D., Calamante, F., Connelly, A., 2007. Robust determination of the fibre orientation distribution in diffusion MRI: non-negativity constrained super-resolved spherical deconvolution. NeuroImage 35 (4), 1459–1472.

Van Essen, D.C., Jbabdi, S., Sotiropoulos, S.N., Chen, C., Dikranian, K., Coalson, T., Harwell, J., Behrens, T.E.J., Glasser, M.F., 2014. Mapping connections in humans and non-human primates: aspirations and challenges for diffusion imaging. In: Diffusion MRI, Elsevier, pp. 337–358.

Wang, X., Grimson, W.E.L., Westin, C.-F., 2011. Tractography segmentation using a hierarchical Dirichlet processes mixture model. NeuroImage 54 (1), 290–302.

Wang, H., Lenglet, C., Akkin, T., 2015. Structure tensor analysis of serial optical coherence scanner images for mapping fiber orientations and tractography in the brain. J. Biomed. Opt. 20 (3), 036003.

Warrington, S., Bryant, K.L., Khrapitchev, A.A., Sallet, J., Charquero-Ballester, M., Douaud, G., Jbabdi, S., Mars, R.B., Sotiropoulos, S.N., 2020. XTRACT-standardised protocols for automated tractography in the human and macaque brain. NeuroImage 217, 116923.

Wasserthal, J., Neher, P.F., Maier-Hein, K.H., 2018. Tract orientation mapping for bundle-specific tractography. arXiv preprint arXiv:1806.05580.

Woolrich, M.W., Jbabdi, S., Patenaude, B., Chappell, M., Makni, S., Behrens, T., Beckmann, C., Jenkinson, M., Smith, S.M., 2009. Bayesian analysis of neuroimaging data in FSL. NeuroImage 45 (1), S173–S186.

Yeh, C.H., Smith, R.E., Liang, X., Calamante, F., 2018. Investigating the streamline count required for reproducible structural connectome construction across a range of brain parcellation resolutions. In: Proceedings of ISMRM, p. 1558.

Part V

# Tractography applications

# Chapter 29

# Tractography: Applications to neurodevelopment, aging, and plasticity

Catherine Lebel[a,b], David Salat[c,d], and Jason Yeatman[e,f]

[a]Department of Radiology, University of Calgary, Calgary, AB, Canada, [b]Alberta Children's Hospital Research Institute and Hotchkiss Brain Institute, Calgary, AB, Canada, [c]Department of Radiology, Harvard Medical School, Boston, MA, United States, [d]Massachusetts General Hospital, Boston, MA, United States, [e]Department of Pediatrics, Stanford University, Stanford, CA, United States, [f]Maternal & Child Health Research Institute, Palo Alto, CA, United States

## 1 Introduction

Throughout life, the brain undergoes structural changes that are closely linked with behavioral and cognitive changes, as well as genetic, physiological, and environmental factors. Diffusion imaging provides a sensitive and versatile noninvasive brain mapping method to quantify tissue changes and has contributed greatly to our understanding of patterns and mechanisms of brain development, aging, and plasticity. Diffusion tractography offers a unique way to examine age-related changes throughout the lifespan, as it allows us to interrogate specific brain pathways and understand regional variation. Indeed, tractography, which virtually reconstructs the brain pathways ("tracts") to produce volumes for visualization and/or analysis ("tracks"), has provided substantial information to help understand patterns of brain change in childhood and adulthood, and how they relate to cognition and behavior. In this chapter, we aim to provide a high-level overview of the application of diffusion tractography to study neural and cognitive development, aging, and plasticity with some historic review and selected recent literature. A thorough description of the technical aspects of diffusion imaging and tractography can be found elsewhere in this book and in prior technical reviews. However, the application of this family of imaging techniques to the study of development, aging, and plasticity requires a general understanding of technical variations in data acquisition and processing, and the strengths and limitations of each. We therefore provide a very broad set of concepts to consider when reviewing prior literature or applying these procedures in new research or interpreting findings from previously published results. The bulk of this chapter is divided into three sections: one that describes tractography's contributions to understanding neurodevelopment, one focused on brain changes in aging, and one section discussing how tractography has advanced our understanding of brain plasticity. These processes are inherently intertwined; for example, there is no clear delineation between brain development and brain aging. We separated them here using somewhat arbitrary boundaries simply for ease of reading. We primarily focus on typical development and aging, but also include studies examining common aberrant conditions in each section. Finally, this chapter will highlight current limitations, knowledge gaps, and potential future directions of the application of tractography to neurodevelopment, aging, and plasticity.

### 1.1 Tractography measures

First, it is important to provide context to the multiple measures derived from diffusion imaging used to understand development, aging, and plasticity in the reviewed studies. Diffusion imaging uses diffusion sensitizing gradients to provide information about the magnitude of water diffusion in different directions. By far the most commonly used model in tractography studies of neurodevelopment and aging is the diffusion tensor model, from diffusion tensor imaging (DTI). Fractional anisotropy (FA) and mean diffusivity (MD) are typically reported by DTI studies, sometimes accompanied by axial diffusivity (AD; also called parallel or longitudinal diffusivity) and radial diffusivity (RD; also called perpendicular or transverse diffusivity). The tensor model has known limitations, perhaps most notable of which is its inability to model more than one fiber population per voxel (Jones and Cercignani, 2010; Tournier et al., 2011). Furthermore, while FA and MD are highly sensitive measures of brain changes, they lack specificity. AD and RD are often interpreted as slightly more specific measures of axonal changes and myelination/axon packing, respectively, than FA and MD (Song et al., 2002, 2003, 2005), but remain nonspecific. Many more advanced diffusion models have emerged recently with potential for

Handbook of Diffusion MR Tractography. https://doi.org/10.1016/B978-0-12-818894-1.00009-4
Copyright © 2025 Elsevier Ltd. All rights reserved, including those for text and data mining, AI training, and similar technologies.

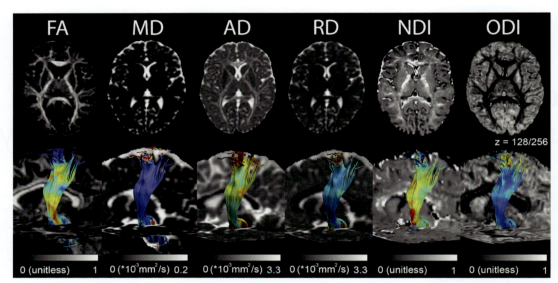

**FIG. 1** Maps of key parameters derived from the tensor model (FA, MD, AD, RD) and the NODDI model (NDI, ODI) shown in the corticospinal tract of a typically developing child. *AD*, axial diffusivity; *FA*, fractional anisotropy; *MD*, mean diffusivity; *NDI*, neurite density index; *ODI*, orientation dispersion index; *RD*, radial diffusivity. *(Modified from Geeraert, B.L., Reynolds, J.E., Lebel, C., 2018. Diffusion imaging perspectives on brain development in childhood and adolescence. In: Kadosh, K. (Ed.), The Oxford Handbook of Developmental Cognitive Neuroscience. Oxford Press.)*

revealing more specific neural correlates of age-related brain changes. Diffusion kurtosis imaging (DKI) overcomes the Gaussian limitation of the tensor model and incorporates non-Gaussian signal behavior caused by restricted diffusion (Jensen and Helpern, 2010; Jensen et al., 2005). The metrics produced by DKI include mean kurtosis (MK) as well as axial kurtosis (AK) and radial kurtosis (RK), all analogous to DTI measures MD, AD, and RD, respectively. Myelinated axons and other membranes lead to non-Gaussian diffusion and, thus, the DKI model offers more nuanced insights into the underlying biology of diffusion changes. Indeed, models such as the white matter tract integrity model posit specific biophysical interpretations of the DKI model (Fieremans et al., 2011; Jelescu et al., 2015). The composite hindered and restricted model of diffusion (CHARMED) model (Assaf and Basser, 2005) allows separation of intra- and extra-axonal diffusion compartments for quantification of the "restricted signal fraction" (FR) of diffusion within the brain.

Neurite orientation dispersion and density imaging (NODDI) is a diffusion technique that models three water compartments: intracellular, extracellular, and cerebrospinal fluid (Zhang et al., 2012). The most common parameters reported by NODDI are neurite density index (NDI), which reflects the intracellular volume fraction as modeled by NODDI, and orientation dispersion index (ODI), which represents the angular variation in neurite orientation. The classic NODDI acquisition requires two *b*-values in addition to the nondiffusion weighted scans, but new formulations eliminate the cerebrospinal fluid (CSF) compartment and estimate NDI and ODI parameters from a single *b*-value acquisition (Edwards et al., 2017). Fig. 1 shows maps of the diffusion parameters derived from DTI and from NODDI models.

Acquisition schemes with high angular resolution (e.g., Wedeen et al., 2005) can model multiple fiber orientations within each voxel. Fixel-based analysis is one way to analyze multiple fiber orientations; it separates populations of fibers within the same voxel into "fixels," which can then be characterized (Raffelt et al., 2015). Measures such as apparent fiber density (AFD), fiber cross section (FC), and fiber density (FD) are commonly assessed within the fixel-based framework (Raffelt et al., 2012), as seen in Chapters 10 and 12.

## 1.2 Types of analysis

The numerous approaches for delineating tracks are covered elsewhere in this book (see Chapters 21 and 22). Here, we briefly review analysis methods relevant to this chapter. One very common approach is to average diffusion parameters across the entire volume of a track, creating one single value of each metric (e.g., FA) per bundle. However, microstructure can vary substantially along the length of a tract (Salat et al., 2010; Yeatman et al., 2012b) (e.g., see Fig. 2) due to fiber density or organization, crossing fiber bundles, and other tissue features. Furthermore, in cases such as brain injury or infarct, damage may be localized to a particular section of a tract. It is therefore not always appropriate to average across the whole length of a bundle, and several different along-track analyses have been proposed to overcome this limitation and provide more detail about variation within bundles. One approach is to subdivide bundles into equidistant units along their

Tractography: Applications to neurodevelopment, aging, and plasticity **Chapter | 29** **585**

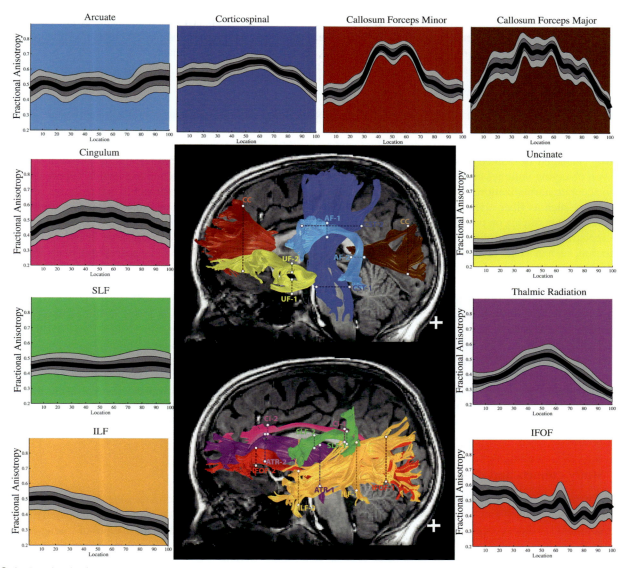

**FIG. 2** Fractional anisotropy (FA) values are shown along the length of major white matter fibers in the brain. All bundles show substantial variation along their lengths, with values tending to be lower where tracks show more fanning. It is important to note that profiles of overlapping tracks, such as the inferior longitudinal fasciculus (ILF, *orange*), inferior fronto-occipital fasciculus (IFOF, *red*; called IFO elsewhere in this chapter), and uncinate fasciculus *(yellow)*, show differences from each other. Tractography is a valuable technique that can help separate the profiles of distinct but overlapping tracks, such as these. *(From Yeatman, J.D., Dougherty, R.F., Myall, N.J., Wandell, B.A., Feldman, H.M., 2012. Tract profiles of white matter properties: automating fiber-tract quantification. PLoS One 7, e49790.)*

length, creating an average for a small cross-section, rather than the entire bundle (De Santis et al., 2014); various versions of this have been implemented in the literature, including Automated Fiber Quantification (AFQ) (Yeatman et al., 2012b), a MATLAB toolbox for along-track statistics (Colby et al., 2012), and TRActs Constrained by UnderLying Anatomy (TRACULA) (Yendiki et al., 2011). A surface-based approach has also been used to study brain maturation (Chen et al., 2016). However, there are also limitations to this type of analysis that must be considered. For example, along-track methods test multiple points along each bundle, and thus can lead to high rates of false positives. Whole bundle averaging or along-track statistics may each be appropriate, depending on the study design and questions.

Tractometry is an analysis approach that uses tracks defined by tractography as volumes of interest for analysis of other metrics (Bells et al., 2011). For example, studies have used tractography to delineate major white matter bundles and then assess age-related changes in not only FA and MD, but other diffusion metrics such as AFD and NDI, as well as

**586 PART | V** Tractography applications

nondiffusion metrics such as myelin water fraction and g-ratio (Chamberland et al., 2019; Geeraert et al., 2019). One benefit of tractometry is that microstructural properties of each fiber bundle can be summarized in tidy tables (Wickham, 2014), facilitating data sharing and reproducibility (Yeatman et al., 2018). Please see also Chapter 23.

## 1.3 Confounds

Brain volume roughly doubles in the first year of life, followed by steady increases until approximately age 6 years (Holland et al., 2014; Matsuzawa et al., 2001). Throughout life, the gray matter-white matter ratio changes, and after middle age, total brain volume declines, even in healthy individuals (Good et al., 2001; Taubert et al., 2020). These volume differences can be easily managed in manual analyses, and also do not generally cause problems for studies with narrow age ranges. However, changing volumes can create challenges for studies that span wide age ranges and use analysis approaches that rely on accurate normalization. One key advantage of tractography over other analysis approaches is that it uses additional information, such as fiber orientation, FA and/or MD thresholds, anatomical information, and tissue priors to delineate tracks, rather than relying on perfect normalization between scans. These additional pieces of information can help mitigate small registration errors or differences in local volumes. In contrast, methods that rely much more heavily on accurate registration, such as voxel-wise analysis or automated region-of-interest approaches, are more susceptible to errors caused by different sized brains and misalignment. New methods, such as automated tractography adapted for the infant brain (Zollei et al., 2019), will help further address some of these challenges.

Children and older adults are more likely to move during an MRI scan than young or middle-aged adults (Roalf et al., 2016; Savalia et al., 2017). Motion can produce blurring or artifacts in images. Relevant to the developmental context, motion during diffusion scans tends to cause underestimation of age-related changes in children and youth (Roalf et al., 2016). Motion can be mitigated at the acquisition stage, for example by allowing children to watch a movie (Thieba et al., 2018; Vanderwal et al., 2015), by preparing participants in a mock scanner (de Bie et al., 2010; Hallowell et al., 2008; Raschle et al., 2009), and/or by ensuring participant comfort and using head padding. Motion problems can also be mitigated on the processing side, using freely available software packages like DTIPrep (Oguz et al., 2014) or RESTORE (Chang et al., 2012), or by visually inspecting data and removing bad volumes or bad datasets (Walton et al., 2018). However, small amounts of motion can bias estimates of diffusion properties, making it essential to carefully control for motion and report quality metrics (Roalf et al., 2016).

## 1.4 Study cohorts and design

In addition to image data acquisition, it is important to consider the population studied when assessing results across studies. Studies of "development" and "aging" may refer to effects across the entire lifespan, effects in children compared to adults, young compared to older cohorts, or the effects of age limited to a specific age range. Each of these types of studies could be expected to produce disparate results regarding the degree and spatial nature of age-associated microstructural changes. Additionally, with regard to aging, it should be expected that, unless explicitly screened with advanced biomarkers, a portion of adults in the later decades will harbor some degree of Alzheimer's disease neuropathology as well as vascular pathology, even when enrolled as "typically aging," "healthy aging," and/or "cognitively intact." Finally, examples of differences in cross-sectional compared to longitudinal results demonstrate that care must be taken when inferring effects of age from cross-sectional data, as selection biases and other cohort effects may taint such results. For example, a selection bias is likely in the study of aging where differing developmental environment (e.g., better diet in the younger cohort) as well as a selection bias for lower-risk older adults (e.g., a survival bias linked to reduced genetic or other risk factors).

## 2 Brain development

The brain matures rapidly from infancy to childhood, with protracted development continuing into young adulthood. Broad developmental changes in white matter have been recently reviewed in detail elsewhere (Dubois et al., 2014; Lebel and Deoni, 2018; Lebel et al., 2019; Ouyang et al., 2019; Tamnes et al., 2018). Overall, three key trends have emerged in brain white matter development: development is nonlinear, it includes substantial regional variation, and it continues into early adulthood.

Early studies of postnatal white matter development tended to have relatively small sample sizes and/or narrow age ranges, and modeled development linearly (Berman et al., 2005; Clayden et al., 2012; Dubois et al., 2006; Eluvathingal et al., 2007; Jeon et al., 2015; Scherf et al., 2014; Yeo et al., 2014) or compared different age groups (Clayden et al., 2012; Loenneker et al., 2011) or different time points within subjects (Brouwer et al., 2012; Giorgio et al., 2010). With larger sample sizes and wider age ranges, it has become clear that development patterns are not linear, and larger studies now typically use nonlinear models to characterize white matter changes more accurately and/or in more detail (Chen et al., 2016; Lebel et al., 2008b; Lynch et al., 2020; Reynolds et al., 2019). Nonlinear patterns allow studies to describe various features of development, such as peaks or plateaus, and to examine differential rates of change at different ages; however, it is important to note that the choice of model, the particular track, as well as the age range included, will influence any measures of peak, plateau, or other developmental trajectory features (Fjell et al., 2010; Lebel et al., 2019).

Tractography studies show that brain maturation proceeds very rapidly in infancy, with increases on the order of ~1%–2% FA per week in the first months of life (Dubois et al., 2008), then 1%–2% per year across early childhood (Dimond et al., 2020b; Lynch et al., 2020; Reynolds et al., 2019), with continued changes through later childhood that are generally <1%/year (Brouwer et al., 2012; Chen et al., 2016; Clayden et al., 2012; Eluvathingal et al., 2007; Jeon et al., 2015; Lebel and Beaulieu, 2011; Lebel and Deoni, 2018; Scherf et al., 2014). MD and RD follow similar patterns of decreases. NDI increases by ~15%/year over infancy, and ~1%–2%/year through early childhood, while ODI shows no changes (Dimond et al., 2020a; Lynch et al., 2020). NDI also continues to increase during late childhood and adolescence showing slightly larger changes than FA, while ODI remains stable (Geeraert et al., 2018, 2019; Lynch et al., 2020; Mah et al., 2017). Tractography may provide more specific and interpretable information about tract development, but the trends of increasing FA and NDI, decreasing MD, and stable ODI are consistent with studies using voxel-based and region-of-interest approaches (see, for example, the following reviews: Dubois et al., 2014; Lebel and Deoni, 2018; Lebel et al., 2019; Ouyang et al., 2019; Tamnes et al., 2018).

White matter development displays consistent regional variation, with frontal connections demonstrating more prolonged maturation than posterior tracks, and association tracks maturing after commissural and projection fibers. Evidence for this comes from studies showing steeper changes in frontal tracts in childhood and adolescence (Brouwer et al., 2012; Mah et al., 2017), from tracts modeling nonlinear changes and showing later peaks/plateaus in frontal connections (Dimond et al., 2020b; Lebel and Beaulieu, 2011; Lebel et al., 2008b; Lynch et al., 2020; Reynolds et al., 2019), and from along-track studies demonstrating more prominent changes in anterior portions of tracts (Chen et al., 2016; Yeatman et al., 2012a). Infant studies provide further support for these patterns by demonstrating lower FA and higher MD of microstructure in frontal connections than in posterior and projection fibers (Dubois et al., 2008).

White matter development continues into young adulthood. While total brain volume reaches adult values by age ~6 years (Matsuzawa et al., 2001) and cortical volume peaks between ~10–12 years (Giedd et al., 1999; Lenroot et al., 2007), the total white matter volume peaks in the mid-30s (Lebel et al., 2012; Westlye et al., 2010). Measures of microstructure undergo considerable changes into early adulthood. Cross-sectional studies have shown significant changes in late adolescence/early adulthood (Jeon et al., 2015; Scherf et al., 2014), peaks/plateaus during the 20s or early 30s (Chen et al., 2016; Lebel et al., 2008b) in frontal connections (see regional variation earlier). The most compelling evidence for continued maturation into young adulthood comes from longitudinal studies demonstrating changes within individuals into their 20s (Giorgio et al., 2010; Lebel and Beaulieu, 2011).

## 2.1 Contributions of tractography

The preceding features of white matter development have been shown by tractography studies, as well as numerous other diffusion image analysis techniques. However, tractography has made several unique contributions to our understanding of white matter development that would not have been revealed with other methods. We focus on three examples in the following discussion: tracking fibers in utero in the fetal brain, using tractography to separate development patterns in overlapping tracts, and graph theory analysis of brain development.

## 2.2 In utero development of white matter tracts

Myelination begins in the second or third trimester, though it is largely a postnatal event (Barkovich et al., 1988; Yakovlev and Lecours, 1967). Despite the lack of myelin in utero, however, axonal membranes influence the diffusion sufficiently to quantify anisotropy and allow for fiber tracking (Beaulieu, 2002). Tractography has been used to identify white matter fiber bundles in the postmortem fetal brain as young as 13 weeks, showing that tracts increase substantially in length, fullness, and extent from 13 to 19 weeks (Huang et al., 2009). Regional variation is notable, with limbic, cerebellar, as well as some commissural and projection fibers identifiable as early as 13 weeks, the corpus callosum around 15 weeks, and association fibers emerging between 15 and 19 weeks (Huang et al., 2006, 2009). Tractography has also been used to show a shift from radial organization to a more corticocortical connectivity during fetal brain development in postmortem studies, and a general posterior-to-anterior pattern of maturation (Takahashi et al., 2012). Fig. 3 shows examples of emerging brainstem, projection, limbic, commissural, and association fibers during fetal and infant development (Ouyang et al., 2019). The sequence of maturation shows clear regional variation. Cerebellar tracts, the fornix, internal capsule, and a small portion of the central corpus

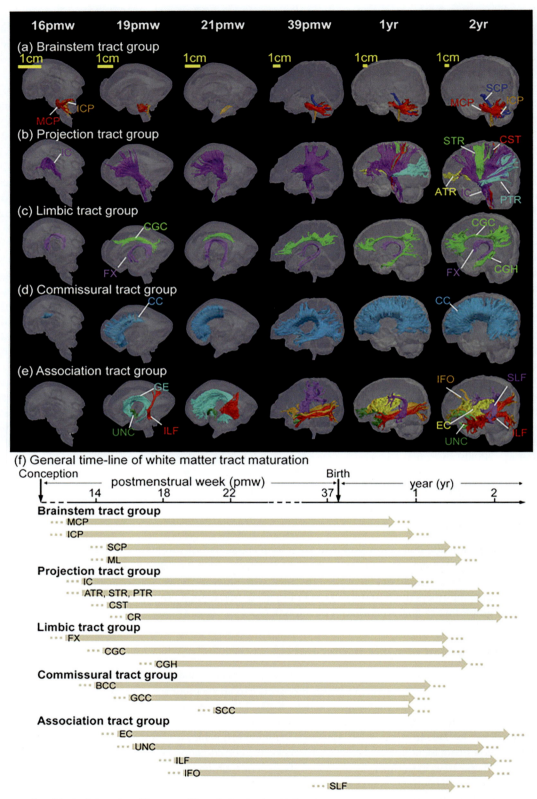

FIG. 3 Tractography of the brain's major white matter fibers during fetal and infant development. Tractography shows clear development in the extent and complexity of tracks from midpregnancy to the end of the second year of life, as well as differences in the timing of emergence of certain bundles (e.g., most association tracks emerge in late pregnancy). *(Modified from Ouyang, M., Dubois, J., Yu, Q., Mukherjee, P., Huang, H., 2019. Delineation of early brain development from fetuses to infants with diffusion MRI and beyond. NeuroImage 185, 836–850.)*

callosum are evident at 16 weeks' gestation. On the other hand, the external capsule and the superior longitudinal fasciculus demonstrate later maturation and are not really apparent until after birth. This figure also illustrates just how rapid brain development is in utero and during the first year, as changes over just a few weeks are evident to the naked eye.

Advanced methods now exist for in utero tractography (Hunt et al., 2020; Pontabry et al., 2013; Song et al., 2018), and have shown its feasibility as young as 18 weeks gestation (Kasprian et al., 2008). One study used in utero tractography to describe three phases of brain maturation corresponding to changes in diffusion parameters (Zanin et al., 2011). The first stage was characterized by large increases in AD with small increases of RD (resulting in increases of both MD and FA), suggested to correspond with the organization of axons into coherent bundles before ~25–28 weeks' gestation. The second phase consists of small decreases in both AD and RD (leading to decreases of MD and stable FA). During this phase (from 25–28 weeks to 32–35 weeks), oligodendrocyte precursors increase and begin to organize. The final phase of in utero development (from ~32–35 weeks until birth) sees slight decreases in AD alongside fast decreases of RD (leading to increasing FA) and corresponding to the beginning of myelination. Callosal regions were the first to switch from the first to second phases (~25 weeks), but also persisted in the premyelination phase the longest (~36 weeks or later). The corticospinal tracts and optic radiation had shorter second phases, lasting from ~26–27 weeks until ~33–35 weeks (Zanin et al., 2011).

## 2.3 Understanding differential maturation in overlapping tracks

Many brain areas have overlapping or crossing fiber bundles that cannot be well separated with voxel-based analysis or regions-of-interest. For example, the inferior fronto-occipital fasciculus (IFO) shares most of its posterior trajectory with the inferior longitudinal fasciculus (ILF) and most if its anterior trajectory with the uncinate fasciculus (UF). While these bundles share similar trajectories, they have different profiles of diffusion parameters along their length/surface, even in adjacent/overlapping areas (Chen et al., 2016; Yeatman et al., 2012b). Fig. 2 illustrates the along-track profiles of FA for several major white matter tracks including the IFO, ILF, and UF. Note the substantial variation along the length of the tracks and the subtle but evident differences between overlapping sections of the different tracks.

In early childhood (2–8 years), tractography shows three distinct regional patterns of white matter tract development (Reynolds et al., 2019). The corpus callosum and pyramidal tracts demonstrate relatively slow maturation with high initial (~2 years) FA values. Frontal-temporal connections including the UF also show relatively slow maturation, but have low initial FA values. A third group, including the inferior fronto-occipital and longitudinal fasciculi (IFO, ILF), had the most rapid changes of FA and MD across the age range. Similar patterns have been confirmed using fixel-based analysis in an independent group of girls (Dimond et al., 2020b).

Studies in later childhood further separate patterns in these tracks, showing earlier peaks in the ILF and the IFO than the uncinate (Chen et al., 2016; Lebel and Beaulieu, 2011; Lebel et al., 2008a). For example, Fig. 4 shows maturation along the surface of the ILF, IFO, and UF. The ages of peak FA values are similar in neighboring areas, but variation is

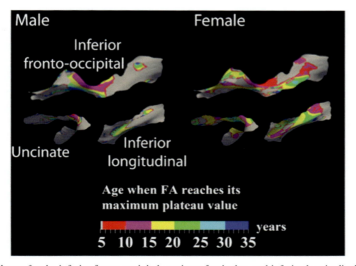

**FIG. 4** Maturation patterns are shown for the inferior fronto-occipital, uncinate fasciculus, and inferior longitudinal fasciculi. Note the different maturation patterns, even in overlapping segments of the tracks. *(Modified from Chen, Z., Zhang, H., Yushkevich, P.A., Liu, M., Beaulieu, C., 2016. Maturation along white matter tracts in human brain using a diffusion tensor surface model tract-specific analysis. Front. Neuroanat. 10, 9.)*

evident between the tracks. For example, the posterior and superior portion of the ILF shows peak ages of ~15–20 years in males and ~10–15 years in females, but the inferior posterior portion of the IFO does not show significant changes in the same region. Lifespan tractography studies provide further support to these differential development rates, with peak FA values occurring in the ILF before the IFO, and both occurring before peaks in the UF (Hasan et al., 2010; Lebel et al., 2012).

Studies have used tractometry to extend these findings to other white matter measures. In general, similar patterns emerge, with the ILF demonstrating the most rapid/early maturation in NDI values, the IFO peaking or continuing to develop slightly later, and the UF demonstrating the most protracted maturation (Geeraert et al., 2019; Mah et al., 2017), though one study found relatively early maturation of the uncinate (Lynch et al., 2020). Increases of myelin water fraction have also been noted in these tracks, again with more prolonged changes in the UF compared to the ILF and IFO (Geeraert et al., 2019), suggesting ongoing maturation of myelin into adolescence. Using a principal component analysis to reduce dimensionality across multiple diffusion metrics, one study showed increases in the principal component related to hindrance/restriction in the IFO and UF (among other tracks) with age in a group of children and adolescents (Chamberland et al., 2019) without accompanying changes in the ILF, suggesting more protracted changes in those tracks. This principal component included contributions from AFD, RD, and restricted signal fraction, and was attributed largely to axonal packing (hindrance/restriction). The second principal component, including contributions from the number of fiber orientations, as well as AD and MD, increased with age in the superior longitudinal fasciculus but not the ILF, IFO, or UF. Interestingly, most fiber bundles did not show significant relationships between individual diffusion parameters and age, suggesting that principal components derived from this type of dimensionality reduction may be more interpretable and more sensitive to age-related changes than individual diffusion metrics (Chamberland et al., 2019).

## 2.4 Graph theory analysis

Tractography has enabled the study of the brain's structural connectome, by allowing the measurement of connections (edges) between separate brain regions (nodes). The human brain, throughout life, shows a small world network arrangement, which means it has high local clustering and short average path length. Neighboring nodes tend to be well-connected, and both short- and long-range communication can be accomplished efficiently (Bassett and Bullmore, 2017). This overall network arrangement remains relatively consistent across ages (Lim et al., 2015), though more specific network properties undergo changes with development. For more on graph theory, see Chapter 24.

In general, infant development involves rapid increases in efficiency and decreases in path length (Hagmann et al., 2010; Huang et al., 2015; Yap et al., 2011), followed by slower, more steady changes into early adulthood (Baum et al., 2017; Chen et al., 2013b; Dennis et al., 2013a). However, mixed findings have been noted and one study showed decreasing global and local efficiency across childhood (Lim et al., 2015). Modularity refers to the clustering, or community structure, of a network (Newman, 2006). Higher modularity generally indicates smaller, segregated clusters that are more densely connected within the group than they are to other clusters. Some studies have shown increasing modularity with age, with a shift from more local connectivity pattern to more distributed connectivity pattern with distinct clusters (Baum et al., 2017; Hagmann et al., 2010; Wierenga et al., 2018); see Fig. 5 for an example. Other studies find relatively stable modularity with age (Chen et al., 2016; Lim et al., 2015), or both increases and decreases (Dennis et al., 2013a). Part of the discrepancy is likely related to the definitions of modules, since some studies have used automatic detection of modules (Chen et al., 2016), while others use a priori defined modules based on functional brain segmentation (Baum et al., 2017). Finally, the human brain displays a rich club organization, in which network hubs are densely connected with each other (Colizza et al., 2006). This rich club organization of the structural brain connectome increases during adolescence, and undergoes a shift from subcortical hubs to more frontal hubs (Baker et al., 2015; Dennis et al., 2013b; Wierenga et al., 2018).

Graph theory has enabled a detailed characterization of the network structure of the brain and its changes over time. This would not be possible without tractography, which facilitates the measurement and assessment of the connections (edges) between distinct nodes.

# 3 Aging

Numerous studies have described substantial changes in diffusion measures both globally and regionally with aging (e.g., Bhagat and Beaulieu, 2004; Head et al., 2004; Madden et al., 2004; Nomura et al., 1994; Nusbaum et al., 2001; O'Sullivan et al., 2001; Salat et al., 2005a; Sullivan et al., 2001). These studies highlight the vulnerability of anterior white matter (Head et al., 2004; Pfefferbaum et al., 2005; Salat et al., 2005a) and have contributed to hypothesized models including

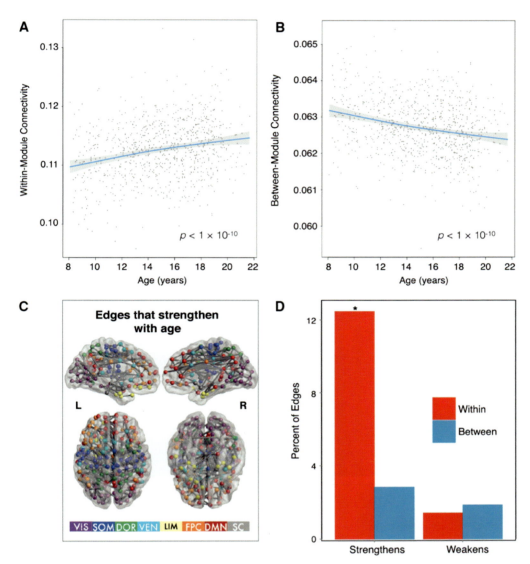

**FIG. 5** Analysis of the brain's structural connectome indicates increasing within-module connectivity (A) and decreasing between-module connectivity with age (B). Edges that strengthen with age are shown in (C). Most of the edges that strengthen with age are within-modules (D). *(Figure from Baum, G.L., Ciric, R., Roalf, D.R., Betzel, R.F., Moore, T.M., Shinohara, R.T., Kahn, A.E., Vandekar, S.N., Rupert, P.E., Quarmley, M., Cook, P.A., Elliott, M.A., Ruparel, K., Gur, R.E., Gur, R.C., Bassett, D.S., Satterthwaite, T.D., 2017. Modular segregation of structural brain networks supports the development of executive function in youth. Curr. Biol. 27, 1561–1572 e1568.)*

"anterior to posterior gradient" (Pfefferbaum et al., 2005) and "last in first out" (Bender et al., 2016). This anterior-posterior effect, as well as an "outer-to-inner" effect, is apparent in the corpus callosum (Lebel et al., 2010), a structure that has been examined in detail in diffusion studies of aging (Abe et al., 2002; Hasan et al., 2008; McLaughlin et al., 2007; Ota et al., 2006; Pietrasik et al., 2020; Sullivan et al., 2010) (Fig. 6).

However, it has been noted that these types of anatomical generalizations are limited, and in some cases somewhat conflicting. For example, certain studies suggest no relationship between developmental patterns and degeneration patters, at least at the track level (Yeatman et al., 2014) (Fig. 7). Furthermore, age-related changes in frontal brain regions show a relatively stronger effect than posterior areas; however, there are many regional exceptions (Coutu et al., 2014).

It is likely that the effects of age are based more on anatomical properties that have not yet been examined in detail, such as ventricular and vascular anatomy, as opposed to a simple gradient along a primary axis of the brain. For example, although the prefrontal white matter is thought to be generally vulnerable, effects of age are greatest in white matter on the ventricular borders (which deteriorate with age) and effects resemble the anatomy of vascular border zones (e.g., medial orbitofrontal white matter (Salat et al., 2005b)) where vascular physiology may be compromised.

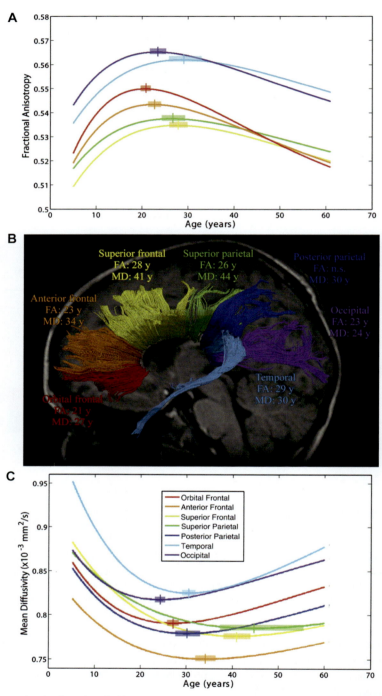

**FIG. 6** Development and age-associated trajectories of white matter across several major callosal subsections. The outer-to-inner pattern of age-related changes is apparent across subdivisions. *(From Lebel, C., Caverhill-Godkewitsch, S., Beaulieu, C., 2010. Age-related regional variations of the corpus callosum identified by diffusion tensor tractography. NeuroImage 52, 20–31.)*

Similar to studies in development, most studies of aging have used simple linear models. However, several studies have demonstrated that aging trajectories (just like developmental trajectories) are better captured by higher-order models (Hsu et al., 2010; Pietrasik et al., 2020). As discussed earlier, limited work to date has explored lifespan transitions from development into later aging. There is substantial regional heterogeneity in the age of peak white matter development (Lebel et al., 2012) (Fig. 8). Such studies are critical for understanding when the transition occurs from constructive to destructive processes within the nervous system (Lebel et al., 2012). Longitudinal studies of aging are generally in agreement with

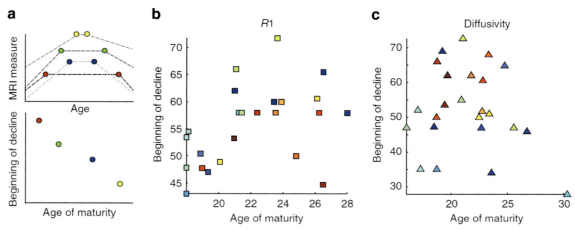

**FIG. 7** Using piecewise linear models, a "last-in-first-out" pattern of white matter aging was not detected. Panel A demonstrates theoretical models of such a relationship. Panels B and C show actual data using R1 values (B) and diffusivity values (C) demonstrating limited correspondence between the time that a track begins to decline and the age of maturity of the track. *(From Yeatman, J.D., Wandell, B.A., Mezer, A.A., 2014. Lifespan maturation and degeneration of human brain white matter. Nat. Commun. 5, 4932.)*

cross-sectional trends, though provide an enhanced ability to detect nonlinear trajectories, such as how the rate of white matter changes accelerates with age (i.e., aging processes happen faster in older adults) (Sexton et al., 2014). For example, when assessing diffusion changes ~3.5 years apart in older adults, Nicolas reported a regional increase in MD with strong effects in frontal white matter that correlated with verbal free recall scores (Nicolas et al., 2020). Longitudinal work has provided evidence that white matter changes accelerate with increasing age.

### 3.1 Contributions of tractography

When examining over the extent of a track, certain paths seem to be more vulnerable to aging than others. As expected, this work generally shows, similarly to voxel-based analyses, that aging effects are widespread and anterior tracks are vulnerable to aging. Recent tractography work described the greatest age effects in the posterior and anterior thalamic radiations (Li et al., 2020) (Fig. 9).

More advanced diffusion models possible through high gradient diffusion imaging can provide additional mechanistic insights and sensitivity. For example, such procedures allow for modeled estimates of axon diameter and axon density in tissue, demonstrating a potential increase in the proportion of large diameter axons in the anterior callosum with aging (Fan et al., 2019) (Fig. 10). Additional microstructural modeling advances provide hope that future uses of diffusion imaging will include specific quantification of histological and pathologic properties within tissue.

### 3.2 Cognitive associations

It is now well-established that diffusion brain metrics are associated with cognition across the lifespan. Seminal work in 2000 demonstrated that diffusion MRI measures of white matter tracks are correlated with reading skills and differ in people with dyslexia (Klingberg et al., 2000). This observation—that individual differences in white matter properties are related to individual differences in performance—opened the floodgates to a flurry of papers investigating the relationships between various cognitive functions, microstructural properties of different white matter tracks, and the developmental time-course of these effects. For example, between 2005 and 2006 three different labs published papers extending Klingberg's observation to children with dyslexia (Beaulieu et al., 2005; Deutsch et al., 2005; Niogi and McCandliss, 2006), establishing a relationship between structural properties of temporoparietal white matter and reading skills in children and adults. Since then, numerous studies have linked diffusion metrics during brain development to emerging cognitive skills; see for example reviews of diffusion studies of reading (Geeraert et al., 2018; Vandermosten et al., 2012b), executive function (Goddings et al., 2020), and mathematics (Matejko and Ansari, 2015). A smaller, but growing, literature of longitudinal studies shows that changes over time in diffusion parameters are linked to cognitive skills during development, showing that structural brain changes parallel functional improvements. For example, changes over time in reading ability are linked to changes of FA (Yeatman et al., 2012a) and MD (Treit et al., 2013) during childhood and adolescence.

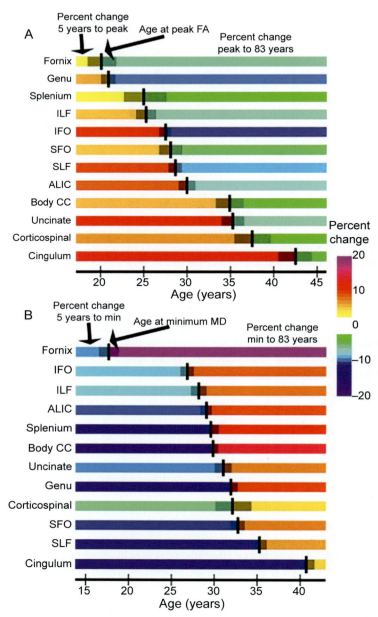

**FIG. 8** Assessment of the transition from "development" to "senescence" across multiple brain regions and tracks. There is a wide variation in the age at peak FA across regions. Trends in FA are only partially mirrored by MD but do differ substantially. *(From Lebel, C., Gee, M., Camicioli, R., Wieler, M., Martin, W., Beaulieu, C., 2012. Diffusion tensor imaging of white matter tract evolution over the lifespan. NeuroImage 60, 340–352.)*

On the other end of the age spectrum, diffusion metrics have been associated with executive function in prefrontal white matter during aging (Grieve et al., 2007). Similarly, Ziegler and colleagues found a dissociation in regional associations with anterior/frontal white matter being related to cognitive control and temporal/parietal white matter being related to episodic memory (Ziegler et al., 2010). Similarly, Kennedy and Raz reported differential associations between diffusion metrics and cognitive performance with anterior age-effects associated with processing speed and working memory and inhibition, task switching costs associated with variation in posterior regions, and episodic memory associated with aging of central white matter (Kennedy and Raz, 2009).

Certain studies suggest that white matter microstructure measured with diffusion imaging is more sensitive to cognitive variation and aging compared to multiple other neuroimaging indicators, including gray matter morphometry (Ziegler et al., 2010) as well as in measures of functional connectivity (Fjell et al., 2017). For example, an age-associated decline of FA in selective anatomical pathways was associated with time-based prospective memory reduction, whereas this association was

Tractography: Applications to neurodevelopment, aging, and plasticity **Chapter | 29** 595

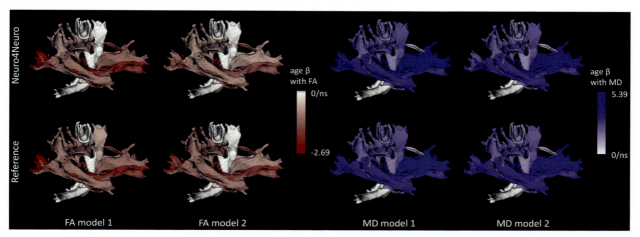

**FIG. 9** Age effects across multiple tracks using FA (left panel) and MD (right panel). *(From Li, B., de Groot, M., Steketee, R.M.E., Meijboom, R., Smits, M., Vernooij, M.W., Ikram, M.A., Liu, J., Niessen, W.J., Bron, E.E., 2020. Neuro4Neuro: a neural network approach for neural tract segmentation using large-scale population-based diffusion imaging. NeuroImage 218, 116993.)*

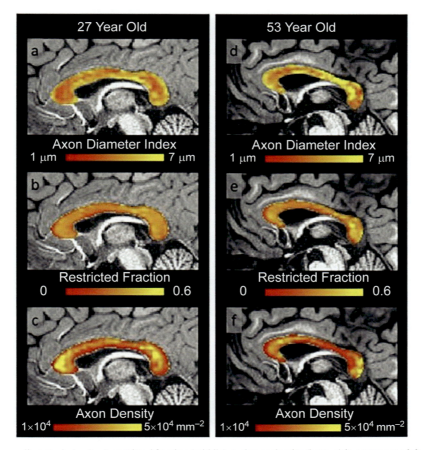

**FIG. 10** Comparison of axon diameter index (top), restricted fraction (middle), and axon density (bottom) in a younger adult (left) compared to an older adult (right). *(From Fan, Q., Tian, Q., Ohringer, N.A., Nummenmaa, A., Witzel, T., Tobyne, S.M., Klawiter, E.C., Mekkaoui, C., Rosen, B.R., Wald, L.L., Salat, D.H., Huang, S.Y., 2019. Age-related alterations in axonal microstructure in the corpus callosum measured by high-gradient diffusion MRI. NeuroImage 191, 325–336.)*

**596 PART | V** Tractography applications

not found for gray matter morphometry (Morand et al., 2021). However, these types of results are greatly influenced by the imaging and image-processing parameters as well as the cognitive tests and scoring utilized and should be interpreted in this context.

Cognitive associations may be more related to global changes in white matter as opposed to changes within a specific track, especially when accounting for other pathologies such as brain amyloid (Rabin et al., 2019). Potentially more interesting is the utilization of microstructural and morphometric information in a combined manner or in complementary analyses. Richard and colleagues recently demonstrated a "tissue specific brain age" approach that may provide novel information about heterogeneity in brain aging and sensitivity to cognitive associations (Richard et al., 2018).

## 4  Sex differences

Most tractography studies do not explicitly test sex differences in development or aging, and several studies report no differences in development trends or aging trajectories between males and females (Kodiweera et al., 2016; Krogsrud et al., 2016; Lebel et al., 2008b, 2012). Several studies show more protracted development in males, suggesting earlier development in females (Bava et al., 2010; Pohl et al., 2016; Seunarine et al., 2016; Simmonds et al., 2014). Some support for this is provided by recent tractography studies. For example, a tractography study of development from childhood to early adulthood showed earlier ages of peak FA values in females compared to males when the entire track was averaged, as well as more localized regional differences between sexes in the along-track surface analysis (see Fig. 4) (Chen et al., 2016). A recent tractography study in young children showed limited sex differences, with slightly larger decreases of MD in males in the fornix and superior and inferior longitudinal fasciculi, and slightly steeper increases in FA of the pyramidal tract in females (Reynolds et al., 2019). A study of FA, MD, and additional white matter measures found only one age-sex interaction, with males showing steeper declines of MD in the UF during adolescence (Geeraert et al., 2019).

In aging, some diffusion studies report accelerated age-related changes in men compared to women (Abe et al., 2010). It is possible that such effects are more apparent with advanced imaging protocols. For example, Toschi and colleagues reported that age-related white matter changes may initiate earlier in men compared to women, but this finding was only apparent with the multishell CHARMED procedure that separates intra- and extra-axonal diffusion compartments, and not with conventional diffusion metrics (Toschi et al., 2020).

Overall, more research is necessary to determine the timing, regional variation, and magnitude of sex differences in white matter development and aging. Longitudinal studies in particular will help answer the question of whether male and female brains develop and age differently.

## 5  Atypical populations

### 5.1  Developmental disorders

There are numerous examples of atypical white matter structure in children with developmental disorders. A growing recent body of literature also supports the idea that maturation is different in children with atypical conditions.

Autism spectrum disorder (ASD) is a neurodevelopmental disorder characterized by repetitive behaviors and social communication difficulties; symptoms typically appear in early childhood (Lord et al., 2020). Tractography studies of young children at high risk of autism show that children who eventually are diagnosed with ASD have initially higher FA, but slower development of several white matter bundles than high-risk children who are not later diagnosed with ASD (Wolff et al., 2012). These infants also show less efficient structural brain networks, and efficiency predicted symptom severity at age 2 years (Lewis et al., 2017). Similar patterns of atypical development have been observed in toddlers, with higher initial FA and slower development in toddlers with ASD compared to controls (Solso et al., 2016); see Fig. 11.

Another example of atypical development is children with fetal alcohol spectrum disorder (FASD). FASD is a neurodevelopmental disorder caused by prenatal alcohol exposure (PAE) that affects ~4% of children in North America (Flannigan et al., 2018; May et al., 2018). FASD is associated with widespread cognitive, behavioral, and neurological abnormalities (Lebel et al., 2011; Mattson et al., 2011; Riley et al., 2011; Tsang et al., 2016; Wozniak and Muetzel, 2011). PAE and FASD also lead to atypical brain development, which appears to follow similar patterns to children with ASD: higher FA at younger ages and slower development through childhood (Kar et al., 2022). However, an interesting study of older children and adolescents with PAE also noted faster decreases of MD in several tracks, perhaps suggesting a later "catch-up" phase (Treit et al., 2013). The brain connectome in FASD tends to show a similar network structure in infancy (Roos et al., 2018), but reduced efficiency, centrality, and between-network connectivity emerges in later childhood and adolescence (Long et al., 2020), suggesting altered network development.

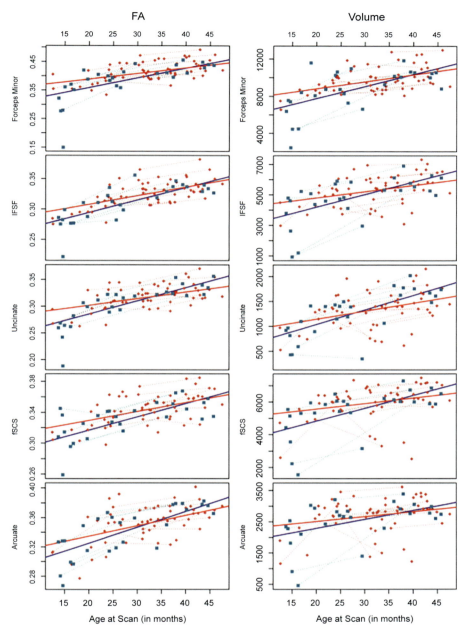

**FIG. 11** Fractional anisotropy (FA) is initially higher in several tracks (uncinate and arcuate fasciculi shown here) in children with autism spectrum disorder (ASD; *red*) vs children with typical development *(blue)*. However, development rates are slower in children with ASD, and at older ages, they demonstrate lower FA in these same tracks. *(Modified from Solso, S., Xu, R., Proudfoot, J., Hagler, D.J., Jr., Campbell, K., Venkatraman, V., Carter Barnes, C., Ahrens-Barbeau, C., Pierce, K., Dale, A., Eyler, L., Courchesne, E., 2016. Diffusion tensor imaging provides evidence of possible axonal overconnectivity in frontal lobes in autism spectrum disorder toddlers. Biol. Psychiatry 79 (8), 676–684.)*

## 5.2 Cerebrovascular and Alzheimer's disease

Cerebrovascular disease and Alzheimer's disease, two highly prevalent conditions of aging, result in characteristic disturbances in white matter that are beyond the normal effects of age; an appreciable portion of the effects reported for "typical" aging are likely accounted for by subclinical aspects of these conditions that are highly prevalent in older adults. More generally, brain aging can be considered the cumulative effects of variation in a range of health parameters that span the subclinical to disease states and should not be considered a distinct phenomenon unrelated to health and other conditions that lead to disease. For example, blood pressure is associated with white matter microstructure in middle-aged and older adults, even when limited to individuals in the prehypertensive range (Salat et al., 2012) (Fig. 12).

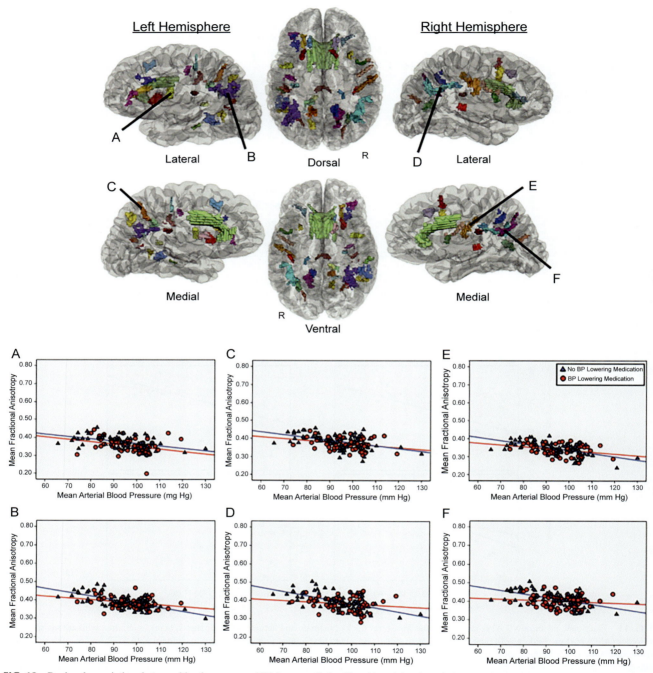

**FIG. 12** Regional associations between blood pressure and FA in generally healthy older adults. Associations were apparent even in the prehypertensive range, suggesting that white matter is influenced by "risk" and even variation in vascular health as opposed to solely impacted by overt cerebrovascular disease. *(From Salat, D.H., Williams, V.J., Leritz, E.C., Schnyer, D.M., Rudolph, J.L., Lipsitz, L.A., McGlinchey, R.E., Milberg, W.P., 2012. Interindividual variation in blood pressure is associated with regional white matter integrity in generally healthy older adults. NeuroImage 59, 181–192.)*

Similar effects are found for insulin resistance and other markers of vascular and systemic health (Ryu et al., 2014, 2017; Salat et al., 2012). Cerebrovascular disease and risk is associated with white matter lesions (Awad et al., 1986a,b,c; Fazekas et al., 1988; Gerard and Weisberg, 1986; Lechner et al., 1988), typically quantified with FLAIR and T2 structural imaging as "white matter hyperintensities" (but also quantifiable from T1 weighted imaging as "hypointensities" (Coutu et al., 2016; Dadar et al., 2018), potentially providing greater sensitivity to more damaged and clinically relevant tissue (Riphagen et al., 2018)). White matter lesion burden is not independent of the health of the normal-appearing white

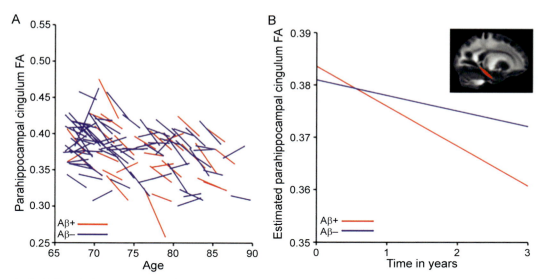

**FIG. 13** Longitudinal effects in the parahippocampal cingulum are most significant in amyloid positive individuals. *(From Rieckmann, A., Van Dijk, K. R., Sperling, R.A., Johnson, K.A., Buckner, R.L., Hedden, T., 2016. Accelerated decline in white matter integrity in clinically normal individuals at risk for Alzheimer's disease. Neurobiol. Aging 42, 177–188.)*

matter, as total lesion burden is strongly associated with microstructural properties of white matter outside of lesioned tissue (Leritz et al., 2014).

Alzheimer's disease has prominent effects on parahippocampal/hippocampal-cingulum white matter (Lee et al., 2015; Salat et al., 2010; Zavaliangos-Petropulu et al., 2019). As noted previously, effects of age and age-associated disease do not tend to be apparent along the full extent of the fiber pathway. In the case of Alzheimer's disease, effects are more prominent in anterior sections of the bundle, bordering regions susceptible to primary Alzheimer's pathology such as the entorhinal cortex (Salat et al., 2010). These diffusion effects are associated with hippocampal volume (Lee et al., 2015) and longitudinal changes are greatest in individuals with higher baseline amyloid burden (Rieckmann et al., 2016) (Fig. 13), suggesting that changes may be linked to primary pathology of this condition and may contribute to further isolation of the hippocampal formation from neocortical communication.

Although white matter lesions are present in both typical aging and Alzheimer's disease, lesion burden is more severe in patients with Alzheimer's disease (e.g., Lindemer et al., 2017a,b). Additionally, damage to tissue within lesions is more severe in patients with Alzheimer's disease as quantified by greater disruption of diffusion microstructure within lesions (Coutu et al., 2016), demonstrating that diffusion tissue measures can be used to quantify lesion burden beyond traditional volumetrics.

Diffusion measures are associated with Alzheimer's biomarkers. For example, free water elimination diffusion measures throughout white matter are associated with cerebrospinal fluid pTau181/Aβ42 levels (Hoy et al., 2017). Similarly, diffusion tractography has been used to predict regional tau accumulation based on structural connectivity patterns (Jacobs et al., 2018).

### 5.3 Modifiable risk factors

A critical question regarding the aging brain is whether interventions can slow or reverse the changes resulting from typical aging. Several modifiable risk factors are related to white matter change measures with diffusion imaging in older adults. For example, blood pressure/hypertension (Leritz et al., 2010; Sabisz et al., 2019; Salat et al., 2012), obesity, insulin resistance/diabetes, smoking, physical activity, nutrition, cholesterol levels, and other factors have all been linked to white matter integrity. These risk factors seem to be associated with changes in white matter regions that also show most prominent age effects, potentially being the primary causal factors for aging white matter. It is possible that several of these risk factors have downstream effects on cerebral blood flow, which is also strongly associated with white matter microstructure in older adults (Chen et al., 2013a).

The impact of modifiable risk factors on white matter with aging was recently comprehensively reviewed (Wassenaar et al., 2019) (Fig. 14). This assessment of the existing data concluded that the evidence from cross-sectional studies of

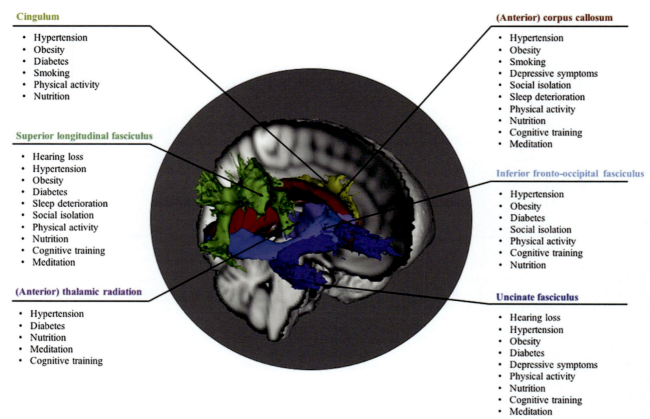

**FIG. 14** White matter bundles that are vulnerable to modifiable risk factors. It is possible that treatment of individual factors can impact changes in specific fibers over time. *(From Wassenaar, T.M., Yaffe, K., van der Werf, Y.D., Sexton, C.E., 2019. Associations between modifiable risk factors and white matter of the aging brain: insights from diffusion tensor imaging studies. Neurobiol. Aging 80, 56–70.)*

treated risk factors and interventional studies of protective factors is supportive of the idea that modification of factors may be protective of white matter integrity.

Recent work in the Systolic Blood Pressure Intervention Trial Memory and Cognition in Decreased Hypertension (SPRINT-MIND) study demonstrated that intensive blood pressure treatment slowed the rate of white matter lesion progression in older adults with vascular risk factors in a randomized clinical trial (The SPRINT MIND Investigators for the SPRINT Research Group et al., 2019). Although this study did not describe results from diffusion imaging, it provides a proof of concept that white matter degeneration with aging may be modifiable and such studies will be of great interest in the coming years. Additionally, diffusion measures are associated with cardiorespiratory fitness (Hayes et al., 2015; Johnson et al., 2012; Marks et al., 2007) and may be impacted by exercise in older adults (Fuhrmann et al., 2019). However, more work is necessary in this area to determine whether such interventions will diminish the cumulative effects of age-associated factors contributing to white matter deterioration.

## 6 Plasticity

Traditionally, white matter was viewed as static infrastructure of the brain; research on plasticity and learning generally focused on the synapse. However, a series of observations have spurred a sea change in how neuroscientists view white matter. First, oligodendrocytes, the glial cells that are responsible for myelination in white matter, actively monitor neural activity through signaling mechanisms that sense nerve discharges, allowing them to change properties of white matter tissue in response to neural activity (Barres and Raff, 1993; Ishibashi et al., 2006). Second, diffusion-weighted magnetic resonance imaging (dMRI) measurements of tract microstructural properties are correlated with most, if not all, aspects of cognitive function, predict concurrent and future behavioral measurements, and are related to individual differences in behavioral performance and learning (Fields, 2008; Wandell and Yeatman, 2013). Third, studies in humans and animal models have revealed a surprising capacity for experience-dependent white matter plasticity throughout the lifespan and demonstrated that this previously ignored form of plasticity is critical for learning (Fields, 2015; Gibson et al., 2014; Hofstetter et al., 2013; Huber et al., 2018; Sampaio-Baptista and Johansen-Berg, 2017). Indeed, it is now largely appreciated that white matter not only plays an essential role in behavior, but also that white matter plasticity is essential for learning. The goal of this section is to provide an overview of what has been learned about human white matter plasticity from diffusion MRI and tractography. We highlight three areas of the literature that exemplify different approaches to the study of plasticity: (1) correlational studies relating cognitive development to individual differences in white matter development; (2) studies of expert populations, such as musicians, that have had different and specialized experience not characteristic of the general population; (3) intervention studies that have systematically manipulated environmental factors (e.g., juggling, reading instruction, meditation) and studied the impacts on white matter development.

## 6.1 Cognitive development is correlated with white matter properties

Previous work linking white matter microstructure with cognitive development inspired two important questions. First, which specific white matter bundles are related to which aspects of cognition? Early work by Klingberg and others employed voxel-based approaches. The effect they discovered was at the intersection of three different tracks: the arcuate fasciculus, superior longitudinal fasciculus, and corticospinal tract (Ben-Shachar et al., 2007). Subsequent metaanalyses (Vandermosten et al., 2012b), reviews (Wandell and Yeatman, 2013), and empirical studies (Vandermosten et al., 2012a; Yeatman et al., 2011, 2012a) capitalized on evolving tractography approaches to better localize the microstructural difference to specific anatomical connections. These studies revealed effects localized to the arcuate fasciculus (Vandermosten et al., 2012a; Yeatman et al., 2011, 2012a) but also suggested there are differences in the macroanatomical configuration of these major fascicles in people with dyslexia. For example, increased interhemispheric connectivity through the posterior callosum could displace the arcuate fasciculus and cortical-spinal tract leading to voxel-based differences at the intersection of tracks in the temporoparietal cortex (Ben-Shachar et al., 2007; Schwartzman et al., 2005).

The second question of key interest was, do correlations between white matter structure and cognition reflect a static property of the brain that constrains cognitive abilities, or do correlations emerge because of environmental influences on brain development (i.e., plasticity)? Most studies relating white matter to cognitive development have interpreted group differences and behavioral correlations as reflecting static deficits. The predominant assumption has been that white matter is not sufficiently plastic over short time-scales to explain effects that are present in early childhood. For example, the aforementioned studies on white matter and reading in young children led many to assume that these white matter differences were characteristics of the dyslexic brain over the lifespan. However, longitudinal investigations revealed a dynamic interplay between an individual's white matter development and their learning trajectory (Wang et al., 2017; Yeatman et al., 2012b). Moreover, recent experiments in model organisms and human subjects have called this assumption into question (Fields, 2015). It is now widely accepted that white matter is dynamic and its tissue properties change over rapid timescales to adapt to environmental demands (Huber et al., 2018; Sampaio-Baptista and Johansen-Berg, 2017). These findings will be covered in more detail in the section on experimental approaches.

## 6.2 Long-term impacts of childhood experience: White matter development in expert musicians

Cross-sectional studies comparing subjects with expertise in a specific domain (e.g., expert musicians) to subjects without expertise are a common methodology to study how different life experiences shape white matter development (Zatorre et al., 2012). For example, studies of musical expertise have revealed higher FA values in the corpus callosum in expert musicians compared to nonexperts. Steele et al. (2013) used tractography to show that the region of the corpus callosum associated with musical expertise contains connections between the left and right hemisphere sensorimotor cortex. FA values in this region were correlated with performance on a sensorimotor synchronization task. Interestingly, the effect of expertise on FA values in the corpus callosum was only present for expert musicians that had learned early in life— before the age of 7—suggesting a sensitive period, early in development, when musical training can shape white matter development in the corpus callosum. Indeed, a compendium of behavioral studies supports the notion that musical expertise is better developed when training begins before 7 years of age. The sensitive period hypothesis suggests a link between the enhanced ability to learn a musical instrument when training begins before 7 years of age and the enhanced capacity of relevant white matter pathways to be sculpted by environmental influences during this developmental window. Beyond musical training, studies of expert athletes (Roberts et al., 2013), children raised in orphanages (Nelson et al., 2007), congenitally blind (Aguirre et al., 2016), and other special populations make it clear that a child's environment and early experiences, both positive and negative, have a profound and long-lasting impact on white matter development.

## 6.3 The causal influence of environmental factors on white matter development: Intervention studies

The most highly cited experiments showing that learning a new skill induces measurable structural brain changes used longitudinal MRI to study adults learning to juggle. An initial study found an increase in gray matter volume localized to posterior parietal cortex for subjects in the juggling training group, but not those in the control group (Draganski et al., 2004). Follow-up diffusion MRI work also revealed training-induced changes in the underlying white matter bundles connecting to posterior parietal cortex (Scholz et al., 2009). These seminal studies on white matter plasticity and motor learning invigorated a heightened interest into the mechanisms of white matter plasticity and defined an important new direction of research linking experience, learning, and changes in white matter.

Following the influential results from motor learning paradigms, research turned to the question of whether a purely mental act, like teaching a child to read, perform mental arithmetic, or focus their thoughts, can induce similar changes in white matter structure. The ultimate test of whether purely mental experiences are sufficient to cause changes in white matter is meditation. Tang et al. (2010) conducted a study in which college students were randomly assigned to relaxation training or meditation training and diffusion MRI was used to measure changes in white matter properties. They found that training and practice with meditation led to increased FA in the cingulum, a tract that is particularly important for self-regulation (Tang et al., 2010). Thus the purely mental act of focusing one's thoughts has a causal impact on white matter structure.

Educational interventions have been another major source of inquiry into the dynamics of white matter plasticity. Educational interventions offer a particularly compelling tool to support causal inferences as researchers can manipulate specific aspects of a child's learning environment and measure the impact of the environmental factor on white matter structure. For example, Keller and Just (2009) enrolled children with dyslexia in a supplemental reading instruction program and found that after the 6-month training period there was an increase in FA. This training-related increase in FA was localized to voxels in the centrum semiovale, encompassing a region that showed differences between the dyslexic and control groups prior to intervention (but anterior to the original region reported by Klingberg and colleagues). Jolles et al. (2016) used diffusion MRI and tractography to relate white matter plasticity to math learning over the course of a tutoring program for children with dyscalculia. Focusing on subcomponents of the superior longitudinal fasciculus, they found that, on average, there was no change in diffusion properties over the intervention. However, there was substantial variability among subjects and those with the largest gains in math scores showed an increase in FA localized to the arcuate fasciculus. Those with the smallest gains in math scores showed a decrease in FA. These data suggest a relationship between white matter plasticity and individual differences in math learning.

Following up on the extensive literature linking white matter properties to reading abilities, Huber et al. (2018) enrolled a group of struggling readers in an 8-week, intensive (4 hours a day), one-on-one reading intervention program and collected diffusion MRI data at 2.5-week intervals. Using tractography to examine white matter tracks that are conventionally associated with reading (left arcuate fasciculus and left ILF), they reported significant changes in MD and FA emerging within the first couple of weeks of instruction. Individual time-courses of MD change tracked reading improvements over the 8-week intervention period. Surprisingly, beyond the plasticity observed within the core reading circuitry, they observed large-scale changes (effect sizes ranging from $d = 0.5$ to $d = 1.0$) throughout an extensive network of tracts in the left and right hemisphere (Fig. 15). These data highlight the surprising capacity for rapid and extensive remodeling of white matter in response to environmental changes (such as an improved educational environment). More broadly, the compendium of studies reviewed here emphasizes the critical importance of environmental factors in sculpting white matter architecture over the lifespan.

## 7 Conclusions and future directions

A great number of studies have described the effects of development, aging, and plasticity on the brain's white matter pathways, though several areas of investigation remain unexplored. One area is in realizing the full promise of the exquisite microstructural information provided through diffusion imaging to better understand the mechanistically complex tissue changes that occur with the aging process. For example, a range of microstructural parameters can be derived from diffusion procedures, yet only very limited work has aimed to integrate information across domains for a more comprehensive understanding of the microstructural effects of age (Geeraert et al., 2020; Konukoglu et al., 2016). Such multimodal approaches are important for differentiating the less specific effects from any individual measure. For example, combining multiple diffusion metrics into principal components can provide more sensitive measures of age-related changes across childhood (Chamberland et al., 2019; Geeraert et al., 2020). Moreover, combining diffusion MRI and tractography with other quantitative MRI measures, such as myelin water fraction, macromolecular tissue volume, R1, and magnetization transfer, can

Tractography: Applications to neurodevelopment, aging, and plasticity Chapter | 29 603

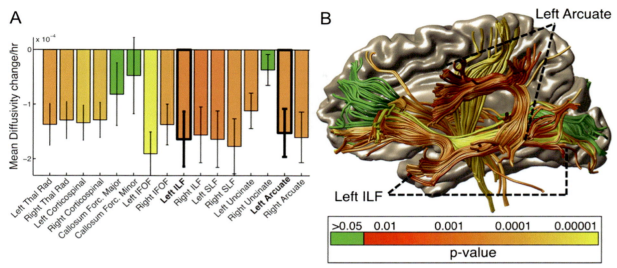

FIG. 15 Reading intervention induces changes in mean diffusivity. Changes in mean diffusivity over the course of the reading intervention in Huber et al. were estimated with a linear mixed effects model. (A) The magnitude (±1 standard error) of change in mean diffusivity as a function of intervention time is shown for each white matter track. (B) The colormap shows the amount of change for each left hemisphere white matter track as estimated with a linear mixed effects model. Plasticity was distributed over an extensive network of tracks. Two key tracks in the reading circuitry, the left hemisphere arcuate and ILF, are highlighted for reference.

lead to more detailed and biologically specific models of white matter development, aging, and plasticity (Deoni et al., 2012; Geeraert et al., 2019; Yeatman et al., 2014).

More sophisticated diffusion models are emerging that can provide more specific and sensitive measures. For example, recent studies demonstrate that age-associated effects on diffusion tensor microstructural measures are attenuated when the free water compartment is eliminated (Chad et al., 2018). We speculate that advanced microstructural modeling will provide novel mapping procedures to differentiate among different types of structural changes within the brain. For more information, please see Chapters 5–7.

As discussed previously, high spatial resolution diffusion imaging could be used in quantification of tissue changes in smaller regions of anatomy, such as subcortical u-fibers, or in cortical gray matter. Such goals may be achieved by using hybrid imaging schemes that allow both high angular and high spatial resolution diffusion imaging (Fan et al., 2017). Finally, novel tractography procedures will continue to be developed that provide optimal possible quantification of tissue properties at critical anatomical scales for understanding functional variation in older adults (e.g., Siless et al., 2020). These types of acquisition and analysis are further enabled by emerging advances in imaging techniques that offer higher spatial resolution, faster imaging, and/or higher signal-to-noise ratios.

Another promising development is the availability of public datasets, which provide the opportunity to explore the impact of age on the brain through diffusion imaging. One notable dataset is the Human Connectome Project (HCP), which includes high-resolution multishell diffusion acquisitions across multiple arms: the Lifespan Developing HCP (1500 participants 20–44 weeks postconception) (Bastiani et al., 2019), the Baby Connectome Project (500 children aged 0–5 years) (Howell et al., 2019), HCP Development (1250 participants aged 5–21 years) (Harms et al., 2018), and HCP Aging (1200 participants aged 36–100 years) (Bookheimer et al., 2019; Harms et al., 2018). Another notable publicly available dataset of relevance to brain development is the Adolescent Brain and Cognitive Development (ABCD) project, which has recruited over 10,000 9–10-year-olds across 21 centers in the United States and plans to follow them for 10 years with longitudinal neuroimaging and comprehensive neuropsychological assessments (Casey et al., 2018; Jernigan et al., 2018). These types of open datasets offer a wealth of possibilities for investigators to study brain development and aging without needing to collect new data.

In conclusion, diffusion tractography offers a sensitive and versatile tool to study brain changes across the lifespan. Since its first applications to neurodevelopment, aging, and brain plasticity over 20 years ago, it has offered much insight into structural changes of the brain's major pathways. In this chapter, we have described developmental and aging trends at a high level, as well as provided some key examples of specific instances where tractography has enhanced our understanding of brain maturation and aging. With the advent of new hardware and software to enable more detailed assessment of brain white matter, many promising discoveries are likely to emerge in the coming years.

# References

Abe, O., Aoki, S., Hayashi, N., Yamada, H., Kunimatsu, A., Mori, H., Yoshikawa, T., Okubo, T., Ohtomo, K., 2002. Normal aging in the central nervous system: quantitative MR diffusion-tensor analysis. Neurobiol. Aging 23, 433–441.

Abe, O., Yamasue, H., Yamada, H., Masutani, Y., Kabasawa, H., Sasaki, H., Takei, K., Suga, M., Kasai, K., Aoki, S., Ohtomo, K., 2010. Sex dimorphism in gray/white matter volume and diffusion tensor during normal aging. NMR Biomed. 23, 446–458.

Aguirre, G.K., Datta, R., Benson, N.C., Prasad, S., Jacobson, S.G., Cideciyan, A.V., Bridge, H., Watkins, K.E., Butt, O.H., Dain, A.S., Brandes, L., Gennatas, E.D., 2016. Patterns of individual variation in visual pathway structure and function in the sighted and blind. PLoS One 11, e0164677.

Assaf, Y., Basser, P.J., 2005. Composite hindered and restricted model of diffusion (CHARMED) MR imaging of the human brain. NeuroImage 27, 48–58.

Awad, I., Modic, M., Little, J.R., Furlan, A.J., Weinstein, M., 1986a. Focal parenchymal lesions in transient ischemic attacks: correlation of computed tomography and magnetic resonance imaging. Stroke 17, 399–403.

Awad, I.A., Johnson, P.C., Spetzler, R.F., Hodak, J.A., 1986b. Incidental subcortical lesions identified on magnetic resonance imaging in the elderly. II. Postmortem pathological correlations. Stroke 17, 1090–1097.

Awad, I.A., Spetzler, R.F., Hodak, J.A., Awad, C.A., Carey, R., 1986c. Incidental subcortical lesions identified on magnetic resonance imaging in the elderly. I. Correlation with age and cerebrovascular risk factors. Stroke 17, 1084–1089.

Baker, S.T., Lubman, D.I., Yucel, M., Allen, N.B., Whittle, S., Fulcher, B.D., Zalesky, A., Fornito, A., 2015. Developmental changes in brain network hub connectivity in late adolescence. J. Neurosci. 35, 9078–9087.

Barkovich, A.J., Kjos, B.O., Jackson Jr., D.E., Norman, D., 1988. Normal maturation of the neonatal and infant brain: MR imaging at 1.5 T. Radiology 166, 173–180.

Barres, B.A., Raff, M.C., 1993. Proliferation of oligodendrocyte precursor cells depends on electrical activity in axons. Nature 361, 258–260.

Bassett, D.S., Bullmore, E.T., 2017. Small-world brain networks revisited. Neuroscientist 23, 499–516.

Bastiani, M., Andersson, J.L.R., Cordero-Grande, L., Murgasova, M., Hutter, J., Price, A.N., Makropoulos, A., Fitzgibbon, S.P., Hughes, E., Rueckert, D., Victor, S., Rutherford, M., Edwards, A.D., Smith, S.M., Tournier, J.D., Hajnal, J.V., Jbabdi, S., Sotiropoulos, S.N., 2019. Automated processing pipeline for neonatal diffusion MRI in the developing Human Connectome Project. NeuroImage 185, 750–763.

Baum, G.L., Ciric, R., Roalf, D.R., Betzel, R.F., Moore, T.M., Shinohara, R.T., Kahn, A.E., Vandekar, S.N., Rupert, P.E., Quarmley, M., Cook, P.A., Elliott, M.A., Ruparel, K., Gur, R.E., Gur, R.C., Bassett, D.S., Satterthwaite, T.D., 2017. Modular segregation of structural brain networks supports the development of executive function in youth. Curr. Biol. 27, 1561–1572. e1568.

Bava, S., Thayer, R., Jacobus, J., Ward, M., Jernigan, T.L., Tapert, S.F., 2010. Longitudinal characterization of white matter maturation during adolescence. Brain Res. 1327, 38–46.

Beaulieu, C., 2002. The basis of anisotropic water diffusion in the nervous system—a technical review. NMR Biomed. 15, 435–455.

Beaulieu, C., Plewes, C., Paulson, L.A., Roy, D., Snook, L., Concha, L., Phillips, L., 2005. Imaging brain connectivity in children with diverse reading ability. NeuroImage 25 (4), 1266–1271.

Bells, S., Cercignani, M., Deoni, S., Assaf, Y., Pasternak, O., Evans, C.J., Leemans, A., Jones, D.K., 2011. Tractometry—comprehensive multi-modal quantitative assessment of white matter along specific tracts. In: International Society for Magnetic Resonance in Medicine Annual Conference, Montreal, PQ.

Bender, A.R., Volkle, M.C., Raz, N., 2016. Differential aging of cerebral white matter in middle-aged and older adults: a seven-year follow-up. NeuroImage 125, 74–83.

Ben-Shachar, M., Dougherty, R.F., Wandell, B.A., 2007. White matter pathways in reading. Curr. Opin. Neurobiol. 17, 258–270.

Berman, J.I., Mukherjee, P., Partridge, S.C., Miller, S.P., Ferriero, D.M., Barkovich, A.J., Vigneron, D.B., Henry, R.G., 2005. Quantitative diffusion tensor MRI fiber tractography of sensorimotor white matter development in premature infants. NeuroImage 27, 862–871.

Bhagat, Y.A., Beaulieu, C., 2004. Diffusion anisotropy in subcortical white matter and cortical gray matter: changes with aging and the role of CSF-suppression. J. Magn. Reson. Imaging 20, 216–227.

Bookheimer, S.Y., Salat, D.H., Terpstra, M., Ances, B.M., Barch, D.M., Buckner, R.L., Burgess, G.C., Curtiss, S.W., Diaz-Santos, M., Elam, J.S., Fischl, B., Greve, D.N., Hagy, H.A., Harms, M.P., Hatch, O.M., Hedden, T., Hodge, C., Japardi, K.C., Kuhn, T.P., Ly, T.K., Smith, S.M., Somerville, L.H., Ugurbil, K., van der Kouwe, A., Van Essen, D., Woods, R.P., Yacoub, E., 2019. The lifespan Human Connectome Project in aging: an overview. NeuroImage 185, 335–348.

Brouwer, R.M., Mandl, R.C., Schnack, H.G., van Soelen, I.L., van Baal, G.C., Peper, J.S., Kahn, R.S., Boomsma, D.I., Hulshoff Pol, H.E., 2012. White matter development in early puberty: a longitudinal volumetric and diffusion tensor imaging twin study. PLoS One 7, e32316.

Casey, B.J., Cannonier, T., Conley, M.I., Cohen, A.O., Barch, D.M., Heitzeg, M.M., Soules, M.E., Teslovich, T., Dellarco, D.V., Garavan, H., Orr, C.A., Wager, T.D., Banich, M.T., Speer, N.K., Sutherland, M.T., Riedel, M.C., Dick, A.S., Bjork, J.M., Thomas, K.M., Chaarani, B., Mejia, M.H., Hagler Jr., D.J., Daniela Cornejo, M., Sicat, C.S., Harms, M.P., Dosenbach, N.U.F., Rosenberg, M., Earl, E., Bartsch, H., Watts, R., Polimeni, J.R., Kuperman, J.M., Fair, D.A., Dale, A.M., the ABCD Imaging Acquisition Workgroup, 2018. The Adolescent Brain Cognitive Development (ABCD) study: imaging acquisition across 21 sites. Dev. Cogn. Neurosci. 32, 43–54.

Chad, J.A., Pasternak, O., Salat, D.H., Chen, J.J., 2018. Re-examining age-related differences in white matter microstructure with free-water corrected diffusion tensor imaging. Neurobiol. Aging 71, 161–170.

Chamberland, M., Raven, E.P., Genc, S., Duffy, K., Descoteaux, M., Parker, G.D., Tax, C.M.W., Jones, D.K., 2019. Dimensionality reduction of diffusion MRI measures for improved tractometry of the human brain. NeuroImage 200, 89–100.

Chang, L.C., Walker, L., Pierpaoli, C., 2012. Informed RESTORE: a method for robust estimation of diffusion tensor from low redundancy datasets in the presence of physiological noise artifacts. Magn. Reson. Med. 68, 1654–1663.

Chen, J.J., Rosas, H.D., Salat, D.H., 2013a. The relationship between cortical blood flow and sub-cortical white-matter health across the adult age span. PLoS One 8, e56733.

Chen, Z., Liu, M., Gross, D.W., Beaulieu, C., 2013b. Graph theoretical analysis of developmental patterns of the white matter network. Front. Hum. Neurosci. 7, 716.

Chen, Z., Zhang, H., Yushkevich, P.A., Liu, M., Beaulieu, C., 2016. Maturation along white matter tracts in human brain using a diffusion tensor surface model tract-specific analysis. Front. Neuroanat. 10, 9.

Clayden, J.D., Jentschke, S., Munoz, M., Cooper, J.M., Chadwick, M.J., Banks, T., Clark, C.A., Vargha-Khadem, F., 2012. Normative development of white matter tracts: similarities and differences in relation to age, gender, and intelligence. Cereb. Cortex 22, 1738–1747.

Colby, J.B., Soderberg, L., Lebel, C., Dinov, I.D., Thompson, P.M., Sowell, E.R., 2012. Along-tract statistics allow for enhanced tractography analysis. NeuroImage 59, 3227–3242.

Colizza, V., Flammiini, A., Serrano, M., Vespignani, A., 2006. Detecting rich-club ordering in complex networks. Nat. Phys. 2, 110–115.

Coutu, J.P., Chen, J.J., Rosas, H.D., Salat, D.H., 2014. Non-Gaussian water diffusion in aging white matter. Neurobiol. Aging 35, 1412–1421.

Coutu, J.P., Goldblatt, A., Rosas, H.D., Salat, D.H., Alzheimer's Disease Neuroimaging Initiative, 2016. White matter changes are associated with ventricular expansion in aging, mild cognitive impairment, and Alzheimer's disease. J. Alzheimers Dis. 49, 329–342.

Dadar, M., Maranzano, J., Ducharme, S., Carmichael, O.T., Decarli, C., Collins, D.L., Alzheimer's Disease Neuroimaging Initiative, 2018. Validation of T1w-based segmentations of white matter hyperintensity volumes in large-scale datasets of aging. Hum. Brain Mapp. 39, 1093–1107.

de Bie, H.M., Boersma, M., Wattjes, M.P., Adriaanse, S., Vermeulen, R.J., Oostrom, K.J., Huisman, J., Veltman, D.J., Delemarre-Van de Waal, H.A., 2010. Preparing children with a mock scanner training protocol results in high quality structural and functional MRI scans. Eur. J. Pediatr. 169, 1079–1085.

Dennis, E.L., Jahanshad, N., McMahon, K.L., de Zubicaray, G.I., Martin, N.G., Hickie, I.B., Toga, A.W., Wright, M.J., Thompson, P.M., 2013a. Development of brain structural connectivity between ages 12 and 30: a 4-Tesla diffusion imaging study in 439 adolescents and adults. NeuroImage 64, 671–684.

Dennis, E.L., Jahanshad, N., Toga, A.W., McMahon, K.L., de Zubicaray, G.I., Hickie, I., Wright, M.J., Thompson, P.M., 2013b. Development of the "rich club" in brain connectivity networks from 438 adolescents & adults aged 12 to 30. In: Proceedings/IEEE International Symposium on Biomedical Imaging, pp. 624–627.

Deoni, S.C., Dean 3rd, D.C., O'Muircheartaigh, J., Dirks, H., Jerskey, B.A., 2012. Investigating white matter development in infancy and early childhood using myelin water fraction and relaxation time mapping. NeuroImage 63, 1038–1053.

De Santis, S., Drakesmith, M., Bells, S., Assaf, Y., Jones, D.K., 2014. Why diffusion tensor MRI does well only some of the time: variance and covariance of white matter tissue microstructure attributes in the living human brain. Neuroimage 89, 35–44.

Deutsch, G.K., Dougherty, R.F., Bammer, R., Siok, W.T., Gabrieli, J.D., Wandell, B., 2005. Children's reading performance is correlated with white matter structure measured by diffusion tensor imaging. Cortex 41 (3), 354–363.

Dimond, D., Heo, S., Ip, A., Rohr, C.S., Tansey, R., Graff, K., Dhollander, T., Smith, R.E., Lebel, C., Dewey, D., Connelly, A., Bray, S., 2020a. Maturation and interhemispheric asymmetry in neurite density and orientation dispersion in early childhood. NeuroImage 221, 117168.

Dimond, D., Rohr, C.S., Smith, R.E., Dhollander, T., Cho, I., Lebel, C., Dewey, D., Connelly, A., Bray, S., 2020b. Early childhood development of white matter fiber density and morphology. NeuroImage 210, 116552.

Draganski, B., Gaser, C., Busch, V., Schuierer, G., Bogdahn, U., May, A., 2004. Neuroplasticity: changes in grey matter induced by training. Nature 427, 311–312.

Dubois, J., Hertz-Pannier, L., Dehaene-Lambertz, G., Cointepas, Y., Le Bihan, D., 2006. Assessment of the early organization and maturation of infants' cerebral white matter fiber bundles: a feasibility study using quantitative diffusion tensor imaging and tractography. NeuroImage 30, 1121–1132.

Dubois, J., Dehaene-Lambertz, G., Perrin, M., Mangin, J.F., Cointepas, Y., Duchesnay, E., Le Bihan, D., Hertz-Pannier, L., 2008. Asynchrony of the early maturation of white matter bundles in healthy infants: quantitative landmarks revealed noninvasively by diffusion tensor imaging. Hum. Brain Mapp. 29, 14–27.

Dubois, J., Dehaene-Lambertz, G., Kulikova, S., Poupon, C., Huppi, P.S., Hertz-Pannier, L., 2014. The early development of brain white matter: a review of imaging studies in fetuses, newborns and infants. Neuroscience 276, 48–71.

Edwards, L.J., Pine, K.J., Ellerbrock, I., Weiskopf, N., Mohammadi, S., 2017. NODDI-DTI: estimating neurite orientation and dispersion parameters from a diffusion tensor in healthy white matter. Front. Neurosci. 11, 720.

Eluvathingal, T.J., Hasan, K.M., Kramer, L., Fletcher, J.M., Ewing-Cobbs, L., 2007. Quantitative diffusion tensor tractography of association and projection fibers in normally developing children and adolescents. Cereb. Cortex 17, 2760–2768.

Fan, Q., Nummenmaa, A., Polimeni, J.R., Witzel, T., Huang, S.Y., Wedeen, V.J., Rosen, B.R., Wald, L.L., 2017. HIgh b-value and high Resolution Integrated Diffusion (HIBRID) imaging. NeuroImage 150, 162–176.

Fan, Q., Tian, Q., Ohringer, N.A., Nummenmaa, A., Witzel, T., Tobyne, S.M., Klawiter, E.C., Mekkaoui, C., Rosen, B.R., Wald, L.L., Salat, D.H., Huang, S.Y., 2019. Age-related alterations in axonal microstructure in the corpus callosum measured by high-gradient diffusion MRI. NeuroImage 191, 325–336.

Fazekas, F., Niederkorn, K., Schmidt, R., Offenbacher, H., Horner, S., Bertha, G., Lechner, H., 1988. White matter signal abnormalities in normal individuals: correlation with carotid ultrasonography, cerebral blood flow measurements, and cerebrovascular risk factors. Stroke 19, 1285–1288.

Fields, R.D., 2008. White matter in learning, cognition and psychiatric disorders. Trends Neurosci. 31, 361–370.

Fields, R.D., 2015. A new mechanism of nervous system plasticity: activity-dependent myelination. Nat. Rev. Neurosci. 16, 756–767.

Fieremans, E., Jensen, J.H., Helpern, J.A., 2011. White matter characterization with diffusional kurtosis imaging. NeuroImage 58, 177–188.

**606 PART | V** Tractography applications

Fjell, A.M., Walhovd, K.B., Westlye, L.T., Ostby, Y., Tamnes, C.K., Jernigan, T.L., Gamst, A., Dale, A.M., 2010. When does brain aging accelerate? Dangers of quadratic fits in cross-sectional studies. NeuroImage 50, 1376–1383.

Fjell, A.M., Sneve, M.H., Grydeland, H., Storsve, A.B., Walhovd, K.B., 2017. The disconnected brain and executive function decline in aging. Cereb. Cortex 27, 2303–2317.

Flannigan, K., Unsworth, K., Harding, K., 2018. The Prevalence of Fetal Alcohol Spectrum Disorder. CanFASD, Vancouver, BC.

Fuhrmann, D., Nesbitt, D., Shafto, M., Rowe, J.B., Price, D., Gadie, A., Cam, C.A.N., Kievit, R.A., 2019. Strong and specific associations between cardiovascular risk factors and white matter micro- and macrostructure in healthy aging. Neurobiol. Aging 74, 46–55.

Geeraert, B.L., Reynolds, J.E., Lebel, C., 2018. Diffusion imaging perspectives on brain development in childhood and adolescence. In: Kadosh, K. (Ed.), The Oxford Handbook of Developmental Cognitive Neuroscience. Oxford Press.

Geeraert, B.L., Lebel, R.M., Lebel, C., 2019. A multi-parametric analysis of white matter maturation during late childhood and adolescence. Hum. Brain Mapp. 40 (15), 4345–4356.

Geeraert, B.L., Chamberland, M., Lebel, R.M., Lebel, C., 2020. Multimodal principal component analysis to identify major features of white matter structure and links to reading. PLoS One 15, e0233244.

Gerard, G., Weisberg, L.A., 1986. MRI periventricular lesions in adults. Neurology 36, 998–1001.

Gibson, E.M., Purger, D., Mount, C.W., Goldstein, A.K., Lin, G.L., Wood, L.S., Inema, I., Miller, S.E., Bieri, G., Zuchero, J.B., Barres, B.A., Woo, P.J., Vogel, H., Monje, M., 2014. Neuronal activity promotes oligodendrogenesis and adaptive myelination in the mammalian brain. Science 344, 1252304.

Giedd, J.N., Blumenthal, J., Jeffries, N.O., Castellanos, F.X., Liu, H., Zijdenbos, A., Paus, T., Evans, A.C., Rapoport, J.L., 1999. Brain development during childhood and adolescence: a longitudinal MRI study. Nat. Neurosci. 2, 861–863.

Giorgio, A., Watkins, K.E., Chadwick, M., James, S., Winmill, L., Douaud, G., De Stefano, N., Matthews, P.M., Smith, S.M., Johansen-Berg, H., James, A.C., 2010. Longitudinal changes in grey and white matter during adolescence. NeuroImage 49, 94–103.

Goddings, A.-L., Roalf, D.R., Lebel, C., Tamnes, C.K., 2020. Development of white matter microstructure and executive functions during childhood and adolescence: A review of diffusion MRI studies. PsyArXiv, https://doi.org/10.31234/osf.io/kyxbq.

Good, C.D., Johnsrude, I.S., Ashburner, J., Henson, R.N., Friston, K.J., Frackowiak, R.S., 2001. A voxel-based morphometry study of ageing in 465 normal adult human brains. NeuroImage 14, 21–36.

Grieve, S.M., Williams, L.M., Paul, R.H., Clark, C.R., Gordon, E., 2007. Cognitive aging, executive function, and fractional anisotropy: a diffusion tensor MR imaging study. AJNR Am. J. Neuroradiol. 28, 226–235.

Hagmann, P., Sporns, O., Madan, N., Cammoun, L., Pienaar, R., Wedeen, V.J., Meuli, R., Thiran, J.P., Grant, P.E., 2010. White matter maturation reshapes structural connectivity in the late developing human brain. Proc. Natl. Acad. Sci. U. S. A. 107, 19067–19072.

Hallowell, L.M., Stewart, S.E., de Amorim, E.S.C.T., Ditchfield, M.R., 2008. Reviewing the process of preparing children for MRI. Pediatr. Radiol. 38, 271–279.

Harms, M.P., Somerville, L.H., Ances, B.M., Andersson, J., Barch, D.M., Bastiani, M., Bookheimer, S.Y., Brown, T.B., Buckner, R.L., Burgess, G.C., Coalson, T.S., Chappell, M.A., Dapretto, M., Douaud, G., Fischl, B., Glasser, M.F., Greve, D.N., Hodge, C., Jamison, K.W., Jbabdi, S., Kandala, S., Li, X., Mair, R.W., Mangia, S., Marcus, D., Mascali, D., Moeller, S., Nichols, T.E., Robinson, E.C., Salat, D.H., Smith, S.M., Sotiropoulos, S.N., Terpstra, M., Thomas, K.M., Tisdall, M.D., Ugurbil, K., van der Kouwe, A., Woods, R.P., Zollei, L., Van Essen, D.C., Yacoub, E., 2018. Extending the Human Connectome Project across ages: imaging protocols for the Lifespan Development and Aging projects. NeuroImage 183, 972–984.

Hasan, K.M., Kamali, A., Kramer, L.A., Papnicolaou, A.C., Fletcher, J.M., Ewing-Cobbs, L., 2008. Diffusion tensor quantification of the human midsagittal corpus callosum subdivisions across the lifespan. Brain Res. 1227, 52–67.

Hasan, K.M., Kamali, A., Abid, H., Kramer, L.A., Fletcher, J.M., Ewing-Cobbs, L., 2010. Quantification of the spatiotemporal microstructural organization of the human brain association, projection and commissural pathways across the lifespan using diffusion tensor tractography. Brain Struct. Funct. 214, 361–373.

Hayes, S.M., Salat, D.H., Forman, D.E., Sperling, R.A., Verfaellie, M., 2015. Cardiorespiratory fitness is associated with white matter integrity in aging. Ann. Clin. Transl. Neurol. 2, 688–698.

Head, D., Buckner, R.L., Shimony, J.S., Williams, L.E., Akbudak, E., Conturo, T.E., McAvoy, M., Morris, J.C., Snyder, A.Z., 2004. Differential vulnerability of anterior white matter in nondemented aging with minimal acceleration in dementia of the Alzheimer type: evidence from diffusion tensor imaging. Cereb. Cortex 14, 410–423.

Hofstetter, S., Tavor, I., Tzur Moryosef, S., Assaf, Y., 2013. Short-term learning induces white matter plasticity in the fornix. J. Neurosci. 33, 12844–12850.

Holland, D., Chang, L., Ernst, T.M., Curran, M., Buchthal, S.D., Alicata, D., Skranes, J., Johansen, H., Hernandez, A., Yamakawa, R., Kuperman, J.M., Dale, A.M., 2014. Structural growth trajectories and rates of change in the first 3 months of infant brain development. JAMA Neurol. 71, 1266–1274.

Howell, B.R., Styner, M.A., Gao, W., Yap, P.T., Wang, L., Baluyot, K., Yacoub, E., Chen, G., Potts, T., Salzwedel, A., Li, G., Gilmore, J.H., Piven, J., Smith, J.K., Shen, D., Ugurbil, K., Zhu, H., Lin, W., Elison, J.T., 2019. The UNC/UMN Baby Connectome Project (BCP): an overview of the study design and protocol development. NeuroImage 185, 891–905.

Hoy, A.R., Ly, M., Carlsson, C.M., Okonkwo, O.C., Zetterberg, H., Blennow, K., Sager, M.A., Asthana, S., Johnson, S.C., Alexander, A.L., Bendlin, B.B., 2017. Microstructural white matter alterations in preclinical Alzheimer's disease detected using free water elimination diffusion tensor imaging. PLoS One 12, e0173982.

Hsu, J.L., Van Hecke, W., Bai, C.H., Lee, C.H., Tsai, Y.F., Chiu, H.C., Jaw, F.S., Hsu, C.Y., Leu, J.G., Chen, W.H., Leemans, A., 2010. Microstructural white matter changes in normal aging: a diffusion tensor imaging study with higher-order polynomial regression models. NeuroImage 49, 32–43.

Huang, H., Zhang, J., Wakana, S., Zhang, W., Ren, T., Richards, L.J., Yarowsky, P., Donohue, P., Graham, E., van Zijl, P.C., Mori, S., 2006. White and gray matter development in human fetal, newborn and pediatric brains. Neuroimage 33 (1), 27–38.

Huang, H., Xue, R., Zhang, J., Ren, T., Richards, L.J., Yarowsky, P., Miller, M.I., Mori, S., 2009. Anatomical characterization of human fetal brain development with diffusion tensor magnetic resonance imaging. J. Neurosci. 29, 4263–4273.

Huang, H., Shu, N., Mishra, V., Jeon, T., Chalak, L., Wang, Z.J., Rollins, N., Gong, G., Cheng, H., Peng, Y., Dong, Q., He, Y., 2015. Development of human brain structural networks through infancy and childhood. Cereb. Cortex 25, 1389–1404.

Huber, E., Donnelly, P.M., Rokem, A., Yeatman, J.D., 2018. Rapid and widespread white matter plasticity during an intensive reading intervention. Nat. Commun. 9, 2260.

Hunt, D., Dighe, M., Gatenby, C., Studholme, C., 2020. Automatic, age consistent reconstruction of the corpus callosum guided by coherency from in utero diffusion-weighted MRI. IEEE Trans. Med. Imaging 39 (3), 601–610.

Ishibashi, T., Dakin, K.A., Stevens, B., Lee, P.R., Kozlov, S.V., Stewart, C.L., Fields, R.D., 2006. Astrocytes promote myelination in response to electrical impulses. Neuron 49, 823–832.

Jacobs, H.I.L., Hedden, T., Schultz, A.P., Sepulcre, J., Perea, R.D., Amariglio, R.E., Papp, K.V., Rentz, D.M., Sperling, R.A., Johnson, K.A., 2018. Structural tract alterations predict downstream tau accumulation in amyloid-positive older individuals. Nat. Neurosci. 21, 424–431.

Jelescu, I.O., Veraart, J., Adisetiyo, V., Milla, S.S., Novikov, D.S., Fieremans, E., 2015. One diffusion acquisition and different white matter models: how does microstructure change in human early development based on WMTI and NODDI? NeuroImage 107, 242–256.

Jensen, J.H., Helpern, J.A., 2010. MRI quantification of non-Gaussian water diffusion by kurtosis analysis. NMR Biomed. 23, 698–710.

Jensen, J.H., Helpern, J.A., Ramani, A., Lu, H., Kaczynski, K., 2005. Diffusional kurtosis imaging: the quantification of non-Gaussian water diffusion by means of magnetic resonance imaging. Magn. Reson. Med. 53, 1432–1440.

Jeon, T., Mishra, V., Ouyang, M., Chen, M., Huang, H., 2015. Synchronous changes of cortical thickness and corresponding white matter microstructure during brain development accessed by diffusion MRI tractography from parcellated cortex. Front. Neuroanat. 9, 158.

Jernigan, T.L., Brown, S.A., Dowling, G.J., 2018. The adolescent brain cognitive development study. J. Res. Adolesc. 28, 154–156.

Johnson, N.F., Kim, C., Clasey, J.L., Bailey, A., Gold, B.T., 2012. Cardiorespiratory fitness is positively correlated with cerebral white matter integrity in healthy seniors. NeuroImage 59, 1514–1523.

Jolles, D., Ashkenazi, S., Kochalka, J., Evans, T., Richardson, J., Rosenberg-Lee, M., Zhao, H., Supekar, K., Chen, T., Menon, V., 2016. Parietal hyperconnectivity, aberrant brain organization, and circuit-based biomarkers in children with mathematical disabilities. Dev. Sci. 19, 613–631.

Jones, D.K., Cercignani, M., 2010. Twenty-five pitfalls in the analysis of diffusion MRI data. NMR Biomed. 23, 803–820.

Kar, P., Reynolds, J.E., Gibbard, W.B., McMorris, C., Tortorelli, C., Lebel, C., 2022. Trajectories of brain white matter development in young children with prenatal alcohol exposure. Hum. Brain Mapp. 43 (13), 4145–4157.

Kasprian, G., Brugger, P.C., Weber, M., Krssak, M., Krampl, E., Herold, C., Prayer, D., 2008. In utero tractography of fetal white matter development. NeuroImage 43, 213–224.

Keller, T.A., Just, M.A., 2009. Altering cortical connectivity: remediation-induced changes in the white matter of poor readers. Neuron 64, 624–631.

Kennedy, K.M., Raz, N., 2009. Aging white matter and cognition: differential effects of regional variations in diffusion properties on memory, executive functions, and speed. Neuropsychologia 47, 916–927.

Klingberg, T., Hedehus, M., Temple, E., Salz, T., Gabrieli, J.D., Moseley, M.E., Poldrack, R.A., 2000. Microstructure of temporo-parietal white matter as a basis for reading ability: evidence from diffusion tensor magnetic resonance imaging. Neuron 25, 493–500.

Kodiweera, C., Alexander, A.L., Harezlak, J., McAllister, T.W., Wu, Y.C., 2016. Age effects and sex differences in human brain white matter of young to middle-aged adults: a DTI, NODDI, and q-space study. NeuroImage 128, 180–192.

Konukoglu, E., Coutu, J.P., Salat, D.H., Fischl, B., Alzheimer's Disease Neuroimaging Initiative, 2016. Multivariate statistical analysis of diffusion imaging parameters using partial least squares: application to white matter variations in Alzheimer's disease. NeuroImage 134, 573–586.

Krogsrud, S.K., Fjell, A.M., Tamnes, C.K., Grydeland, H., Mork, L., Due-Tonnessen, P., Bjornerud, A., Sampaio-Baptista, C., Andersson, J., Johansen-Berg, H., Walhovd, K.B., 2016. Changes in white matter microstructure in the developing brain—a longitudinal diffusion tensor imaging study of children from 4 to 11 years of age. NeuroImage 124, 473–486.

Lebel, C., Beaulieu, C., 2011. Longitudinal development of human brain wiring continues from childhood into adulthood. J. Neurosci. 31, 10937–10947.

Lebel, C., Deoni, S., 2018. The development of brain white matter microstructure. Neuroimage 182, 207–218.

Lebel, C., Rasmussen, C., Wyper, K., Walker, L., Andrew, G., Yager, J., Beaulieu, C., 2008a. Brain diffusion abnormalities in children with fetal alcohol spectrum disorder. Alcohol. Clin. Exp. Res. 32, 1732–1740.

Lebel, C., Treit, S., Beaulieu, C., 2019. A review of diffusion MRI of typical white matter development from early childhood to young adulthood. NMR Biomed. 32 (4), e3778.

Lebel, C., Walker, L., Leemans, A., Phillips, L., Beaulieu, C., 2008b. Microstructural maturation of the human brain from childhood to adulthood. NeuroImage 40, 1044–1055.

Lebel, C., Caverhill-Godkewitsch, S., Beaulieu, C., 2010. Age-related regional variations of the corpus callosum identified by diffusion tensor tractography. NeuroImage 52, 20–31.

Lebel, C., Roussotte, F., Sowell, E.R., 2011. Imaging the impact of prenatal alcohol exposure on the structure of the developing human brain. Neuropsychol. Rev. 21 (2), 102–118.

Lebel, C., Gee, M., Camicioli, R., Wieler, M., Martin, W., Beaulieu, C., 2012. Diffusion tensor imaging of white matter tract evolution over the lifespan. NeuroImage 60, 340–352.

Lechner, H., Schmidt, R., Bertha, G., Justich, E., Offenbacher, H., Schneider, G., 1988. Nuclear magnetic resonance image white matter lesions and risk factors for stroke in normal individuals. Stroke 19, 263–265.

Lee, S.H., Coutu, J.P., Wilkens, P., Yendiki, A., Rosas, H.D., Salat, D.H., Alzheimer's Disease Neuroimaging Initiative, 2015. Tract-based analysis of white matter degeneration in Alzheimer's disease. Neuroscience 301, 79–89.

Lenroot, R.K., Gogtay, N., Greenstein, D.K., Wells, E.M., Wallace, G.L., Clasen, L.S., Blumenthal, J.D., Lerch, J., Zijdenbos, A.P., Evans, A.C., Thompson, P.M., Giedd, J.N., 2007. Sexual dimorphism of brain developmental trajectories during childhood and adolescence. NeuroImage 36, 1065–1073.

Leritz, E.C., Salat, D.H., Milberg, W.P., Williams, V.J., Chapman, C.E., Grande, L.J., Rudolph, J.L., Schnyer, D.M., Barber, C.E., Lipsitz, L.A., McGlinchey, R.E., 2010. Variation in blood pressure is associated with white matter microstructure but not cognition in African Americans. Neuropsychology 24, 199–208.

Leritz, E.C., Shepel, J., Williams, V.J., Lipsitz, L.A., McGlinchey, R.E., Milberg, W.P., Salat, D.H., 2014. Associations between T1 white matter lesion volume and regional white matter microstructure in aging. Hum. Brain Mapp. 35, 1085–1100.

Lewis, J.D., Evans, AC., Pruett J.R., Jr., Botteron, K.N., McKinstry, R.C., Zwaigenbaum, L., Estes, A.M., Collins, D.L., Kostopoulos, P., Gerig, G., Dager, S.R., Paterson, S., Schultz, R.T., Styner, M.A., Hazlett, H.C., Piven, J., Infant Brain Imaging Study Network, 2017. The emergence of network inefficiencies in infants with autism spectrum disorder. Biol. Psychiatry 82 (3), 176–185.

Li, B., de Groot, M., Steketee, R.M.E., Meijboom, R., Smits, M., Vernooij, M.W., Ikram, M.A., Liu, J., Niessen, W.J., Bron, E.E., 2020. Neuro4Neuro: a neural network approach for neural tract segmentation using large-scale population-based diffusion imaging. NeuroImage 218, 116993.

Lim, S., Han, C.E., Uhlhaas, P.J., Kaiser, M., 2015. Preferential detachment during human brain development: age- and sex-specific structural connectivity in diffusion tensor imaging (DTI) data. Cereb. Cortex 25, 1477–1489.

Lindemer, E.R., Greve, D.N., Fischl, B., Augustinack, J.C., Salat, D.H., Alzheimer's Disease Neuroimaging Initiative, 2017a. Differential regional distribution of juxtacortical white matter signal abnormalities in aging and Alzheimer's disease. J. Alzheimers Dis. 57, 293–303.

Lindemer, E.R., Greve, D.N., Fischl, B.R., Augustinack, J.C., Salat, D.H., 2017b. Regional staging of white matter signal abnormalities in aging and Alzheimer's disease. NeuroImage Clin. 14, 156–165.

Loenneker, T., Klaver, P., Bucher, K., Lichtensteiger, J., Imfeld, A., Martin, E., 2011. Microstructural development: organizational differences of the fiber architecture between children and adults in dorsal and ventral visual streams. Hum. Brain Mapp. 32, 935–946.

Long, X., Little, G., Treit, S., Beaulieu, C., Gong, G., Lebel, C., 2020. Altered brain white matter connectome in children and adolescents with prenatal alcohol exposure. Brain Struct. Funct. 225 (3), 1123–1133.

Lord, C., Brugha, T.S., Charman, T., Cusack, J., Dumas, G., Frazier, T., Jones, E.J.H., Jones, R.M., Pickles, A., State, M.W., Taylor, J.L., Veenstra-VanderWeele, J., 2020. Autism spectrum disorder. Nat. Rev. Dis. Primers 6 (1), 5.

Lynch, K.M., Cabeen, R.P., Toga, A.W., Clark, K.A., 2020. Magnitude and timing of major white matter tract maturation from infancy through adolescence with NODDI. NeuroImage 116672.

Madden, D.J., Whiting, W.L., Huettel, S.A., White, L.E., MacFall, J.R., Provenzale, J.M., 2004. Diffusion tensor imaging of adult age differences in cerebral white matter: relation to response time. NeuroImage 21, 1174–1181.

Mah, A., Geeraert, B.L., Lebel, C., 2017. Detailing neuroanatomical development in late childhood and early adolescence using NODDI. PLoS One 12, e0182340.

Marks, B.L., Madden, D.J., Bucur, B., Provenzale, J.M., White, L.E., Cabeza, R., Huettel, S.A., 2007. Role of aerobic fitness and aging on cerebral white matter integrity. Ann. N. Y. Acad. Sci. 1097, 171–174.

Matejko, A.A., Ansari, D., 2015. Drawing connections between white matter and numerical and mathematical cognition: a literature review. Neurosci. Biobehav. Rev. 48, 35–52.

Matsuzawa, J., Matsui, M., Konishi, T., Noguchi, K., Gur, R.C., Bilker, W., Miyawaki, T., 2001. Age-related volumetric changes of brain gray and white matter in healthy infants and children. Cereb. Cortex 11, 335–342.

Mattson, S.N., Crocker, N., Nguyen, T.T., 2011. Fetal alcohol spectrum disorders: neuropsychological and behavioral features. Neuropsychol. Rev. 21 (2), 81–101.

May, P.A., Chambers, C.D., Kalberg, W.O., Zellner, J., Feldman, H., Buckley, D., Kopald, D., Hasken, J.M., Xu, R., Honerkamp-Smith, G., Taras, H., Manning, M.A., Robinson, L.K., Adam, M.P., Abdul-Rahman, O., Vaux, K., Jewett, T., Elliott, A.J., Kable, J.A., Akshoomoff, N., Falk, D., Arroyo, J.A., Hereld, D., Riley, E.P., Charness, M.E., Coles, C.D., Warren, K.R., Jones, K.L., Hoyme, H.E., 2018. Prevalence of fetal alcohol spectrum disorders in 4 US communities. JAMA 319 (5), 474–482.

McLaughlin, N.C., Paul, R.H., Grieve, S.M., Williams, L.M., Laidlaw, D., DiCarlo, M., Clark, C.R., Whelihan, W., Cohen, R.A., Whitford, T.J., Gordon, E., 2007. Diffusion tensor imaging of the corpus callosum: a cross-sectional study across the lifespan. Int. J. Dev. Neurosci. 25, 215–221.

Morand, A., Segobin, S., Lecouvey, G., Gonneaud, J., Eustache, F., Rauchs, G., Desgranges, B., 2021. Brain substrates of time-based prospective memory decline in aging: a voxel-based morphometry and diffusion tensor imaging study. Cereb. Cortex 31 (1), 396–409.

Nelson 3rd, C.A., Zeanah, C.H., Fox, N.A., Marshall, P.J., Smyke, A.T., Guthrie, D., 2007. Cognitive recovery in socially deprived young children: the Bucharest Early Intervention Project. Science 318, 1937–1940.

Newman, M.E., 2006. Modularity and community structure in networks. Proc. Natl. Acad. Sci. U. S. A. 103, 8577–8582.

Nicolas, R., Hiba, B., Dilharreguy, B., Barse, E., Baillet, M., Edde, M., Pelletier, A., Periot, O., Helmer, C., Allard, M., Dartigues, J.F., Amieva, H., Peres, K., Fernandez, P., Catheline, G., 2020. Changes over time of diffusion MRI in the white matter of aging brain, a good predictor of verbal recall. Front. Aging Neurosci. 12, 218.

Niogi, S.N., McCandliss, B.D., 2006. Left lateralized white matter microstructure accounts for individual differences in reading ability and disability. Neuropsychologia 44 (11), 2178–2188.

Nomura, Y., Sakuma, H., Takeda, K., Tagami, T., Okuda, Y., Nakagawa, T., 1994. Diffusional anisotropy of the human brain assessed with diffusion-weighted MR: relation with normal brain development and aging. AJNR Am. J. Neuroradiol. 15, 231–238.

Nusbaum, A.O., Tang, C.Y., Buchsbaum, M.S., Wei, T.C., Atlas, S.W., 2001. Regional and global changes in cerebral diffusion with normal aging. AJNR Am. J. Neuroradiol. 22, 136–142.

Oguz, I., Farzinfar, M., Matsui, J., Budin, F., Liu, Z., Gerig, G., Johnson, H.J., Styner, M., 2014. DTIPrep: quality control of diffusion-weighted images. Front. Neuroinform. 8, 4.

O'Sullivan, M., Jones, D.K., Summers, P.E., Morris, R.G., Williams, S.C., Markus, H.S., 2001. Evidence for cortical "disconnection" as a mechanism of age-related cognitive decline. Neurology 57, 632–638.

Ota, M., Obata, T., Akine, Y., Ito, H., Ikehira, H., Asada, T., Suhara, T., 2006. Age-related degeneration of corpus callosum measured with diffusion tensor imaging. NeuroImage 31, 1445–1452.

Ouyang, M., Dubois, J., Yu, Q., Mukherjee, P., Huang, H., 2019. Delineation of early brain development from fetuses to infants with diffusion MRI and beyond. NeuroImage 185, 836–850.

Pfefferbaum, A., Adalsteinsson, E., Sullivan, E.V., 2005. Frontal circuitry degradation marks healthy adult aging: evidence from diffusion tensor imaging. NeuroImage 26, 891–899.

Pietrasik, W., Cribben, I., Olsen, F., Huang, Y., Malykhin, N.V., 2020. Diffusion tensor imaging of the corpus callosum in healthy aging: investigating higher order polynomial regression modelling. NeuroImage 213, 116675.

Pohl, K.M., Sullivan, E.V., Rohlfing, T., Chu, W., Kwon, D., Nichols, B.N., Zhang, Y., Brown, S.A., Tapert, S.F., Cummins, K., Thompson, W.K., Brumback, T., Colrain, I.M., Baker, F.C., Prouty, D., De Bellis, M.D., Voyvodic, J.T., Clark, D.B., Schirda, C., Nagel, B.J., Pfefferbaum, A., 2016. Harmonizing DTI measurements across scanners to examine the development of white matter microstructure in 803 adolescents of the NCANDA study. Neuroimage 130, 194–213.

Pontabry, J., Rousseau, F., Oubel, E., Studholme, C., Koob, M., Dietemann, J.L., 2013. Probabilistic tractography using Q-ball imaging and particle filtering: application to adult and in-utero fetal brain studies. Med. Image Anal. 17, 297–310.

Rabin, J.S., Perea, R.D., Buckley, R.F., Neal, T.E., Buckner, R.L., Johnson, K.A., Sperling, R.A., Hedden, T., 2019. Global white matter diffusion characteristics predict longitudinal cognitive change independently of amyloid status in clinically normal older adults. Cereb. Cortex 29, 1251–1262.

Raffelt, D., Tournier, J.D., Rose, S., Ridgway, G.R., Henderson, R., Crozier, S., Salvado, O., Connelly, A., 2012. Apparent Fibre Density: a novel measure for the analysis of diffusion-weighted magnetic resonance images. NeuroImage 59, 3976–3994.

Raffelt, D.A., Smith, R.E., Ridgway, G.R., Tournier, J.D., Vaughan, D.N., Rose, S., Henderson, R., Connelly, A., 2015. Connectivity-based fixel enhancement: whole-brain statistical analysis of diffusion MRI measures in the presence of crossing fibres. NeuroImage 117, 40–55.

Raschle, N.M., Lee, M., Buechler, R., Christodoulou, J.A., Chang, M., Vakil, M., Stering, P.L., Gaab, N., 2009. Making MR imaging child's play— pediatric neuroimaging protocol, guidelines and procedure. J. Vis. Exp. (29), 1309.

Reynolds, J.E., Grohs, M.N., Dewey, D., Lebel, C., 2019. Global and regional white matter development in early childhood. NeuroImage 196, 49–58.

Richard, G., Kolskar, K., Sanders, A.M., Kaufmann, T., Petersen, A., Doan, N.T., Monereo Sanchez, J., Alnaes, D., Ulrichsen, K.M., Dorum, E.S., Andreassen, O.A., Nordvik, J.E., Westlye, L.T., 2018. Assessing distinct patterns of cognitive aging using tissue-specific brain age prediction based on diffusion tensor imaging and brain morphometry. PeerJ 6, e5908.

Rieckmann, A., Van Dijk, K.R., Sperling, R.A., Johnson, K.A., Buckner, R.L., Hedden, T., 2016. Accelerated decline in white matter integrity in clinically normal individuals at risk for Alzheimer's disease. Neurobiol. Aging 42, 177–188.

Riley, E.P., Infante, M.A., Warren, K.R., 2011. Fetal alcohol spectrum disorders: an overview. Neuropsychol. Rev. 21 (2), 73–80.

Riphagen, J.M., Gronenschild, E., Salat, D.H., Freeze, W.M., Ivanov, D., Clerx, L., Verhey, F.R.J., Aalten, P., Jacobs, H.I.L., 2018. Shades of white: diffusion properties of T1- and FLAIR-defined white matter signal abnormalities differ in stages from cognitively normal to dementia. Neurobiol. Aging 68, 48–58.

Roalf, D.R., Quarmley, M., Elliott, M.A., Satterthwaite, T.D., Vandekar, S.N., Ruparel, K., Gennatas, E.D., Calkins, M.E., Moore, T.M., Hopson, R., Prabhakaran, K., Jackson, C.T., Verma, R., Hakonarson, H., Gur, R.C., Gur, R.E., 2016. The impact of quality assurance assessment on diffusion tensor imaging outcomes in a large-scale population-based cohort. NeuroImage 125, 903–919.

Roberts, R.E., Bain, P.G., Day, B.L., Husain, M., 2013. Individual differences in expert motor coordination associated with white matter microstructure in the cerebellum. Cereb. Cortex 23, 2282–2292.

Ryu, S.Y., Coutu, J.P., Rosas, H.D., Salat, D.H., 2014. Effects of insulin resistance on white matter microstructure in middle-aged and older adults. Neurology 82, 1862–1870.

Roos, A., Fouche, J.P., Ipser, J.C., Narr, K., Woods, R., Zar, H.J., Stein, D., Donald, K.A., 2018. Structural and Functional Brain Network Connectivity in Prenatal Alcohol Exposed Neonates as Assessed by Multimodal Brain Imaging. Research Society on Alcoholism Annual Scientific Meeting, San Diego, CA.

Ryu, C.W., Coutu, J.P., Greka, A., Rosas, H.D., Jahng, G.H., Rosen, B.R., Salat, D.H., 2017. Differential associations between systemic markers of disease and white matter tissue health in middle-aged and older adults. J. Cereb. Blood Flow Metab. 37, 3568–3579.

Sabisz, A., Naumczyk, P., Marcinkowska, A., Graff, B., Gasecki, D., Glinska, A., Witkowska, M., Jankowska, A., Konarzewska, A., Kwela, J., Jodzio, K., Szurowska, E., Narkiewicz, K., 2019. Aging and hypertension—independent or intertwined white matter impairing factors? Insights from the quantitative diffusion tensor imaging. Front. Aging Neurosci. 11, 35.

Salat, D.H., Tuch, D.S., Greve, D.N., van der Kouwe, A.J., Hevelone, N.D., Zaleta, A.K., Rosen, B.R., Fischl, B., Corkin, S., Rosas, H.D., Dale, A.M., 2005a. Age-related alterations in white matter microstructure measured by diffusion tensor imaging. Neurobiol. Aging 26, 1215–1227.

Salat, D.H., Tuch, D.S., Hevelone, N.D., Fischl, B., Corkin, S., Rosas, H.D., Dale, A.M., 2005b. Age-related changes in prefrontal white matter measured by diffusion tensor imaging. Ann. N. Y. Acad. Sci. 1064, 37–49.

Salat, D.H., Tuch, D.S., van der Kouwe, A.J., Greve, D.N., Pappu, V., Lee, S.Y., Hevelone, N.D., Zaleta, A.K., Growdon, J.H., Corkin, S., Fischl, B., Rosas, H.D., 2010. White matter pathology isolates the hippocampal formation in Alzheimer's disease. Neurobiol. Aging 31, 244–256.

Salat, D.H., Williams, V.J., Leritz, E.C., Schnyer, D.M., Rudolph, J.L., Lipsitz, L.A., McGlinchey, R.E., Milberg, W.P., 2012. Inter-individual variation in blood pressure is associated with regional white matter integrity in generally healthy older adults. NeuroImage 59, 181–192.

Sampaio-Baptista, C., Johansen-Berg, H., 2017. White matter plasticity in the adult brain. Neuron 96, 1239–1251.

Savalia, N.K., Agres, P.F., Chan, M.Y., Feczko, E.J., Kennedy, K.M., Wig, G.S., 2017. Motion-related artifacts in structural brain images revealed with independent estimates of in-scanner head motion. Hum. Brain Mapp. 38, 472–492.

Scherf, K.S., Thomas, C., Doyle, J., Behrmann, M., 2014. Emerging structure-function relations in the developing face processing system. Cereb. Cortex 24, 2964–2980.

Scholz, J., Klein, M.C., Behrens, T.E., Johansen-Berg, H., 2009. Training induces changes in white-matter architecture. Nat. Neurosci. 12, 1370–1371.

Schwartzman, A., Dougherty, R.F., Taylor, J.E., 2005. Cross-subject comparison of principal diffusion direction maps. Magn. Reson. Med. 53, 1423–1431.

Seunarine, K.K., Clayden, J.D., Jentschke, S., Munoz, M., Cooper, J.M., Chadwick, M.J., Banks, T., Vargha-Khadem, F., Clark, C.A., 2016. Sexual dimorphism in white matter developmental trajectories using tract-based spatial statistics. Brain Connect. 6 (1), 37–47.

Sexton, C.E., Walhovd, K.B., Storsve, A.B., Tamnes, C.K., Westlye, L.T., Johansen-Berg, H., Fjell, A.M., 2014. Accelerated changes in white matter microstructure during aging: a longitudinal diffusion tensor imaging study. J. Neurosci. 34, 15425–15436.

Siless, V., Davidow, J.Y., Nielsen, J., Fan, Q., Hedden, T., Hollinshead, M., Beam, E., Vidal Bustamante, C.M., Garrad, M.C., Santillana, R., Smith, E.E., Hamadeh, A., Snyder, J., Drews, M.K., Van Dijk, K.R.A., Sheridan, M., Somerville, L.H., Yendiki, A., 2020. Registration-free analysis of diffusion MRI tractography data across subjects through the human lifespan. NeuroImage 214, 116703.

Song, S.K., Sun, S.W., Ramsbottom, M.J., Chang, C., Russell, J., Cross, A.H., 2002. Dysmyelination revealed through MRI as increased radial (but unchanged axial) diffusion of water. NeuroImage 17, 1429–1436.

Song, S.K., Sun, S.W., Ju, W.K., Lin, S.J., Cross, A.H., Neufeld, A.H., 2003. Diffusion tensor imaging detects and differentiates axon and myelin degeneration in mouse optic nerve after retinal ischemia. NeuroImage 20, 1714–1722.

Song, S.K., Yoshino, J., Le, T.Q., Lin, S.J., Sun, S.W., Cross, A.H., Armstrong, R.C., 2005. Demyelination increases radial diffusivity in corpus callosum of mouse brain. NeuroImage 26, 132–140.

Simmonds, D.J., Hallquist, M.N., Asato, M., Luna, B., 2014. Developmental stages and sex differences of white matter and behavioral development through adolescence: a longitudinal diffusion tensor imaging (DTI) study. Neuroimage 92, 356–368.

Solso, S., Xu, R., Proudfoot, J., Hagler D.J., Jr., Campbell, K., Venkatraman, V., Carter Barnes, C., Ahrens-Barbeau, C., Pierce, K., Dale, A., Eyler, L., Courchesne, E., 2016. Diffusion tensor imaging provides evidence of possible axonal overconnectivity in frontal lobes in autism spectrum disorder toddlers. Biol. Psychiatry 79 (8), 676–684.

Song, J.W., Gruber, G.M., Patsch, J.M., Seidl, R., Prayer, D., Kasprian, G., 2018. How accurate are prenatal tractography results? A postnatal in vivo follow-up study using diffusion tensor imaging. Pediatr. Radiol. 48, 486–498.

Steele, C.J., Bailey, J.A., Zatorre, R.J., Penhune, V.B., 2013. Early musical training and white-matter plasticity in the corpus callosum: evidence for a sensitive period. J. Neurosci. 33, 1282–1290.

Sullivan, E.V., Adalsteinsson, E., Hedehus, M., Ju, C., Moseley, M., Lim, K.O., Pfefferbaum, A., 2001. Equivalent disruption of regional white matter microstructure in ageing healthy men and women. Neuroreport 12, 99–104.

Sullivan, E.V., Rohlfing, T., Pfefferbaum, A., 2010. Longitudinal study of callosal microstructure in the normal adult aging brain using quantitative DTI fiber tracking. Dev. Neuropsychol. 35, 233–256.

Takahashi, E., Folkerth, R.D., Galaburda, A.M., Grant, P.E., 2012. Emerging cerebral connectivity in the human fetal brain: an MR tractography study. Cereb. Cortex 22, 455–464.

Tamnes, C.K., Roalf, D.R., Goddings, A.L., Lebel, C., 2018. Diffusion MRI of white matter microstructure development in childhood and adolescence: methods, challenges and progress. Dev. Cogn. Neurosci. 33, 161–175.

Tang, Y.Y., Lu, Q., Geng, X., Stein, E.A., Yang, Y., Posner, M.I., 2010. Short-term meditation induces white matter changes in the anterior cingulate. Proc. Natl. Acad. Sci. U. S. A. 107, 15649–15652.

Taubert, M., Roggenhofer, E., Melie-Garcia, L., Muller, S., Lehmann, N., Preisig, M., Vollenweider, P., Marques-Vidal, P., Lutti, A., Kherif, F., Draganski, B., 2020. Converging patterns of aging-associated brain volume loss and tissue microstructure differences. Neurobiol. Aging 88, 108–118.

The SPRINT MIND Investigators for the SPRINT Research Group, Nasrallah, I.M., Pajewski, N.M., Auchus, A.P., Chelune, G., Cheung, A.K., Cleveland, M.L., Coker, L.H., Crowe, M.G., Cushman, W.C., Cutler, J.A., Davatzikos, C., Desiderio, L., Doshi, J., Erus, G., Fine, L.J., Gaussoin, S.A., Harris, D., Johnson, K.C., Kimmel, P.L., Kurella Tamura, M., Launer, L.J., Lerner, A.J., Lewis, C.E., Martindale-Adams, J., Moy, C.S., Nichols, L.O., Oparil, S., Ogrocki, P.K., Rahman, M., Rapp, S.R., Reboussin, D.M., Rocco, M.V., Sachs, B.C., Sink, K.M., Still, C.H., Supiano, M.A., Snyder, J.K., Wadley, V.G., Walker, J., Weiner, D.E., Whelton, P.K., Wilson, V.M., Woolard, N., Wright Jr., J.T., Wright, C.B., Williamson, J.D., Bryan, R.N., 2019. Association of intensive vs standard blood pressure control with cerebral white matter lesions. JAMA 322, 524–534.

Thieba, C., Frayne, A., Walton, M., Mah, A., Benischek, A., Dewey, D., Lebel, C., 2018. Factors associated with successful MRI scanning in unsedated young children. Front. Pediatr. 6, 146.

Toschi, N., Gisbert, R.A., Passamonti, L., Canals, S., De Santis, S., 2020. Multishell diffusion imaging reveals sex-specific trajectories of early white matter degeneration in normal aging. Neurobiol. Aging 86, 191–200.

Tournier, J.D., Mori, S., Leemans, A., 2011. Diffusion tensor imaging and beyond. Magn. Reson. Med. 65, 1532–1556.

Treit, S., Lebel, C., Baugh, L., Rasmussen, C., Andrew, G., Beaulieu, C., 2013. Longitudinal MRI reveals altered trajectory of brain development during childhood and adolescence in fetal alcohol spectrum disorders. J. Neurosci. 33, 10098–10109.

Tsang, T.W., Lucas, B.R., Carmichael Olson, H., Pinto, R.Z., Elliott, E.J., 2016. Prenatal alcohol exposure, FASD, and child behavior: a meta-analysis. Pediatrics 137 (3), e20152542.

Vandermosten, M., Boets, B., Poelmans, H., Sunaert, S., Wouters, J., Ghesquiere, P., 2012a. A tractography study in dyslexia: neuroanatomic correlates of orthographic, phonological and speech processing. Brain 135, 935–948.

Vandermosten, M., Boets, B., Wouters, J., Ghesquiere, P., 2012b. A qualitative and quantitative review of diffusion tensor imaging studies in reading and dyslexia. Neurosci. Biobehav. Rev. 36, 1532–1552.

Vanderwal, T., Kelly, C., Eilbott, J., Mayes, L.C., Castellanos, F.X., 2015. Inscapes: a movie paradigm to improve compliance in functional magnetic resonance imaging. NeuroImage 122, 222–232.

Walton, M., Dewey, D., Lebel, C., 2018. Brain white matter structure and language ability in preschool-aged children. Brain Lang. 176, 19–25.

Wandell, B.A., Yeatman, J.D., 2013. Biological development of reading circuits. Curr. Opin. Neurobiol. 23, 261–268.

Wang, Y., Mauer, M.V., Raney, T., Peysakhovich, B., Becker, B.L.C., Sliva, D.D., Gaab, N., 2017. Development of tract-specific white matter pathways during early reading development in at-risk children and typical controls. Cereb. Cortex 27 (4), 2469–2485.

Wassenaar, T.M., Yaffe, K., van der Werf, Y.D., Sexton, C.E., 2019. Associations between modifiable risk factors and white matter of the aging brain: insights from diffusion tensor imaging studies. Neurobiol. Aging 80, 56–70.

Wedeen, V.J., Hagmann, P., Tseng, W.Y., Reese, T.G., Weisskoff, R.M., 2005. Mapping complex tissue architecture with diffusion spectrum magnetic resonance imaging. Magn. Reson. Med. 54, 1377–1386.

Westlye, L.T., Walhovd, K.B., Dale, A.M., Bjornerud, A., Due-Tonnessen, P., Engvig, A., Grydeland, H., Tamnes, C.K., Ostby, Y., Fjell, A.M., 2010. Life-span changes of the human brain white matter: diffusion tensor imaging (DTI) and volumetry. Cereb. Cortex 20, 2055–2068.

Wickham, H., 2014. Tidy data. J. Stat. Softw. 59.

Wierenga, L.M., van den Heuvel, M.P., Oranje, B., Giedd, J.N., Durston, S., Peper, J.S., Brown, T.T., Crone, E.A., The Pediatric Longitudinal Imaging, Neurocognition, and Genetics Study, 2018. A multisample study of longitudinal changes in brain network architecture in 4-13-year-old children. Hum. Brain Mapp. 39, 157–170.

Wolff, J.J., Gu, H., Gerig, G., Elison, J.T., Styner, M., Gouttard, S., Botteron, K.N., Dager, S.R., Dawson, G., Estes, A.M., Evans, A.C., Hazlett, H.C., Kostopoulos, P., McKinstry, R.C., Paterson, S.J., Schultz, R.T., Zwaigenbaum, L., Piven, J., 2012. Differences in white matter fiber tract development present from 6 to 24 months in infants with autism. Am. J. Psychiatry 169 (6), 589–600.

Wozniak, J.R., Muetzel, R.L., 2011. What does diffusion tensor imaging reveal about the brain and cognition in fetal alcohol spectrum disorders? Neuropsychol. Rev. 21 (2), 133–147.

Yakovlev, P.I., Lecours, A.-R., 1967. The myelogenetic cycles of regional maturation of the brain. In: Minkowski, A. (Ed.), Regional Development of the Brain Early in Life. Blackwell Scientific Publications Inc., Boston, MA, pp. 3–70.

Yap, P.T., Fan, Y., Chen, Y., Gilmore, J.H., Lin, W., Shen, D., 2011. Development trends of white matter connectivity in the first years of life. PLoS One 6, e24678.

Yeatman, J.D., Dougherty, R.F., Rykhlevskaia, E., Sherbondy, A.J., Deutsch, G.K., Wandell, B.A., Ben-Shachar, M., 2011. Anatomical properties of the arcuate fasciculus predict phonological and reading skills in children. J. Cogn. Neurosci. 23, 3304–3317.

Yeatman, J.D., Dougherty, R.F., Ben-Shachar, M., Wandell, B.A., 2012a. Development of white matter and reading skills. Proc. Natl. Acad. Sci. U. S. A. 109, E3045–E3053.

Yeatman, J.D., Dougherty, R.F., Myall, N.J., Wandell, B.A., Feldman, H.M., 2012b. Tract profiles of white matter properties: automating fiber-tract quantification. PLoS One 7, e49790.

Yeatman, J.D., Wandell, B.A., Mezer, A.A., 2014. Lifespan maturation and degeneration of human brain white matter. Nat. Commun. 5, 4932.

Yeatman, J.D., Richie-Halford, A., Smith, J.K., Keshavan, A., Rokem, A., 2018. A browser-based tool for visualization and analysis of diffusion MRI data. Nat. Commun. 9, 940.

Yendiki, A., Panneck, P., Srinivasan, P., Stevens, A., Zollei, L., Augustinack, J., Wang, R., Salat, D., Ehrlich, S., Behrens, T., Jbabdi, S., Gollub, R., Fischl, B., 2011. Automated probabilistic reconstruction of white-matter pathways in health and disease using an atlas of the underlying anatomy. Front. Neuroinform. 5, 23.

Yeo, S.S., Jang, S.H., Son, S.M., 2014. The different maturation of the corticospinal tract and corticoreticular pathway in normal brain development: diffusion tensor imaging study. Front. Hum. Neurosci. 8, 573.

Zanin, E., Ranjeva, J.P., Confort-Gouny, S., Guye, M., Denis, D., Cozzone, P.J., Girard, N., 2011. White matter maturation of normal human fetal brain. An in vivo diffusion tensor tractography study. Brain Behav. 1, 95–108.

Zatorre, R.J., Fields, R.D., Johansen-Berg, H., 2012. Plasticity in gray and white: neuroimaging changes in brain structure during learning. Nat. Neurosci. 15, 528–536.

Zavaliangos-Petropulu, A., Nir, T.M., Thomopoulos, S.I., Reid, R.I., Bernstein, M.A., Borowski, B., Jack Jr., C.R., Weiner, M.W., Jahanshad, N., Thompson, P.M., 2019. Diffusion MRI indices and their relation to cognitive impairment in brain aging: the updated multi-protocol approach in ADNI3. Front. Neuroinform. 13, 2.

Zhang, H., Schneider, T., Wheeler-Kingshott, C.A., Alexander, D.C., 2012. NODDI: practical in vivo neurite orientation dispersion and density imaging of the human brain. NeuroImage 61, 1000–1016.

Ziegler, D.A., Piguet, O., Salat, D.H., Prince, K., Connally, E., Corkin, S., 2010. Cognition in healthy aging is related to regional white matter integrity, but not cortical thickness. Neurobiol. Aging 31, 1912–1926.

Zollei, L., Jaimes, C., Saliba, E., Grant, P.E., Yendiki, A., 2019. TRActs constrained by UnderLying INfant anatomy (TRACULInA): an automated probabilistic tractography tool with anatomical priors for use in the newborn brain. NeuroImage 199, 1–17.

# Chapter 30

# Linking behavior with white matter networks

Sanja Budisavljevic[a], Stephanie Ameis[b], Rok Berlot[c,d], Hanrietta Howells[e], and Marika Urbanski[f,g]

[a]*School of Medicine, University of St Andrews, St Andrews, United Kingdom,* [b]*University of Toronto, Toronto, ON, Canada,* [c]*Department of Neurology, University Medical Centre Ljubljana, Ljubljana, Slovenia,* [d]*Faculty of Medicine, University of Ljubljana, Ljubljana, Slovenia,* [e]*Department of Medical Biotechnology and Translational Medicine, University of Milan, Milan, Italy,* [f]*Department of Neurology, CHU Poitiers, Poitiers, France,* [g]*Labcom I3M, Centre Hospitalier Universitaire de Poitiers, plateforme Ultra-Haut champ 3T-7T, Poitiers University Hospital, Laboratory of Applied Mathematics, CNRS UMR 7348, Poitiers, France*

## 1 Introduction

Science has long been intrigued by how the brain supports complex skills and behaviors. White matter works in concert with gray matter to mediate the extraordinary diversity of human behavior. It can be seen as supporting the transfer of information within distributed neural networks, while gray matter subserves information processing (Filley, 2012). Distributed neural networks are organized to mediate critical aspects of human behavior such as language, cognition, socioemotional, visuospatial, and motor skills, all of which depend on the structural connectivity provided by myelinated tracts. Thus white matter has been conceptualized as an essential human connectome component (Sporns et al., 2005) and the structural basis of human behavior (Catani et al., 2012; Mesulam, 1990). Disrupted white matter could have profound effects, and this will be shown using examples of neurodevelopmental, psychiatric, and neurological disorders.

The advent of diffusion MRI tractography has brought extraordinary momentum to the study of brain-behavior relationships, by enabling for the first time in vivo investigations of white matter pathways and the link to a wide variety of behaviors and functions in health and disease. These developments elevated a new approach, after the traditional clinico-pathological "lesion method" that compares brain lesions examined postmortem to behavioral symptoms and dysfunctions observed during one's life (Damasio et al., 2003).

The objective of this chapter is to provide the reader with an overview of up-to-date knowledge regarding the brain-behavior relationship, focusing on white matter connections explored with diffusion MRI tractography. We review the key white matter tracts and their functional correlates that fall within the five behavioral domains: (i) socioemotional, (ii) cognitive, (iii) language, (iv) motor, and (v) visuospatial (Fig. 1). We limit this review to the following pathways: the superior longitudinal fasciculus (SLF), the arcuate fasciculus, the uncinate fasciculus, cingulum, fornix, and the corticospinal tract (CST); and we discuss the research in the context of two white matter scales, the macrostructure and the microstructure. Macrostructure can be assessed through the gross volume of the tract, for example, through measurements of the number of voxels crossed by the streamlines and fiber density. Microstructure, on the other hand, reflects the underlying tissue properties (myelination, axonal membrane integrity, fiber orientation, etc.) (Beaulieu, 2002; Jones et al., 2013) and includes diffusion-based indices such as fractional anisotropy (FA), mean diffusivity (MD), apparent fiber density (AFD), hindrance modulated orientational anisotropy (HMOA), and others (see Chapters 10, 12, and 23). For simplicity, we structure our discussion in two parts: first explaining how white matter tracts support behavior in a healthy population, and second, their role in disorders that present with functional impairments in one of the five domains.

## 2 Socioemotional functions

### 2.1 White matter tracts and socioemotional processing in a healthy population

Like many behavioral tasks, socioemotional tasks are computationally demanding, requiring real-time integration of multiple processes for successful performance (i.e., face, gesture, language processing, prosody and emotion decoding, mind-reading and reflection, executive functioning, decision-making, and response execution). Performance depends on

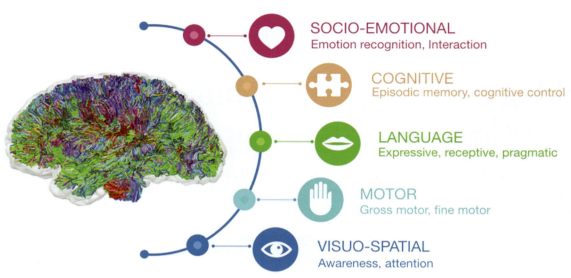

**FIG. 1** The organization of the chapter and five behavioral domains covered.

efficiency and coordination across a number of large-scale distributed neural networks that are structurally integrated via white matter connections (Catani et al., 2013). Although structural connectivity and functional connectivity—that is, associations between hemodynamic time-courses between regions at rest or during task performance, measured using functional MRI (fMRI)—are significantly correlated (Honey et al., 2009), correspondence is imperfect and lower among transmodal as opposed to unimodal regions (Suarez et al., 2020). A number of functional networks that support social processes and are integrated by structural connections have been described.

In terms of the limbic system (Fig. 2), the *temporo-amygdalo-orbitofrontal* network (Mega et al., 1997) connected by the uncinate fasciculus (Catani et al., 2002; Kier et al., 2004) has been linked to recognition of complex emotions in humans (Fujie et al., 2008) and development of joint attention (Elison et al., 2013a,b). The *medial default mode* network (Buckner et al., 2008; Frith and Frith, 2010; Uddin et al., 2007), including nodes connected via dorsal cingulum bundle (Catani and Thiebaut de Schotten, 2008), has been linked to a range of motivational and psychosocial functions (Catani et al., 2002; Schmahmann et al., 2007; Supekar et al., 2010). On the other hand, the *hippocampal-diencephalic retrosplenial* network, which includes the fornix and the ventral cingulum, is thought to be dedicated to memory and spatial attention (Catani et al., 2013).

The inferior longitudinal fasciculus (ILF) and the inferior fronto-occipital fasciculus (IFOF) (Catani et al., 2003; Philippi et al., 2009) have been linked to processing affective facial expressions (Philippi et al., 2009) as part of the face processing system (Adolphs et al., 1999; Pelphrey et al., 2004). The arcuate fasciculus links structures that correspond to the mirror neuron and language processing systems (Catani and Thiebaut de Schotten, 2008) with segments of this white matter tract postulated to be part of a hierarchical anatomical system that supports development of complex language and social communication (Catani and Bambini, 2014).

**FIG. 2** White matter tracts of the limbic system associated with socioemotional processing: Dorsal cingulum *(blue)*, ventral cingulum *(yellow)*, uncinate fasciculus *(green)*, fornix *(dark yellow)*.

Although long-range white matter tracts are visible during fetal development (Huang et al., 2006), their maturation takes on a protracted and variable developmental course (Lebel et al., 2008). Such prolonged environmental input may be required to shape circuits over time to support complex behavior. For example, with respect to development of complex social communication, earlier developing (frontoparietal or "anterior") segments of the arcuate fasciculus are postulated to support processing of earlier developing skills, such as decoding the meaning of others' actions, while later developing segments (frontotemporal or "long," and temporoparietal or "posterior") support more complex features of language processing, such as the attribution of beliefs and mental states to others during conversational (language-based) interactions (Catani and Bambini, 2014). The protracted development of white matter pathways is most prominent within the uncinate fasciculus and cingulum bundle (i.e., typical maturation continues past the second decade of life) (Lebel et al., 2008), two socioemotional processing tracts, suggesting that delayed development in these systems may be beneficial to allow for environmental shaping for development and refinement of complex skills over time.

## 2.2 Frontal and limbic networks and impairments in social-emotional processing

Autism spectrum disorder (ASD), referred to hereafter as autism, is a neurodevelopmental disorder that presents social and communication difficulties among others, affecting how individuals socially interact and process emotions. It has been ~15 years since the global underconnectivity alongside local overconnectivity hypothesis of autism was proposed. This hypothesis postulates that developmental disruption of long-range brain connections may be an important pathogenic factor contributing to the core functional and behavioral impairments present in individuals diagnosed with autism (Belmonte et al., 2004). The narrative review of evidence from 72 diffusion imaging studies published by Ameis and Catani (2015) implicated altered microstructure of white matter pathways supporting social communication in autism including the uncinate and arcuate fasciculi, and cingulum bundle (Cheon et al., 2011; Kumar et al., 2010; Pardini et al., 2009; Poutska et al., 2011). However, white matter alterations were not limited to white matter connecting social processing regions and there was considerable variability in findings reported from study to study. Of note, a relatively large study examining white matter across the brain in autistic adult males, compared to controls, found that while white matter alterations were found in different white matter pathways (including arcuate fasciculus, corpus callosum, uncinate fasciculus, and cingulum bundle), altered numbers of streamlines and reduced FA in the uncinate fasciculus and cingulum bundle correlated with caregiver reported social reciprocity deficits and use of facial expressions (Fig. 3) (Catani et al., 2016). A longitudinal diffusion weighted imaging study in the siblings of children with autism that went on to develop an autism diagnosis as toddlers themselves provides insights into how alterations in neurodevelopment may impact white matter indices and their correlates to behavior in later life. Specifically, in infant siblings that went on to be diagnosed with autism compared to those who did not receive an autism diagnosis in early childhood, white matter development differed in several major white matter pathways across the brain (with no predilection for white matter pathways involved in social processing) (Wolff et al., 2012). Atypical neurodevelopment in the first years of life would likely result in broad downstream effects on brain structure, function, and connectivity that could differ substantially from one affected individual to another (Ameis and Catani, 2015). One might speculate that downstream effects of early developmental deviations to white matter tracts could drive pathological differences beyond early childhood. Such secondary processes could include disruption to the lengthy time window for myelination that is typical for long-range white matter tracts that link socioemotional structures (Lebel et al., 2008), leading to decreased white matter integrity beyond early childhood autism.

A number of methodological issues in diffusion MRI studies have been highlighted that could lead to the variability of findings observed in the literature, including (i) head motion artifacts (that are greater in clinical populations) (Koldewyn et al., 2014; Pardoe et al., 2016), (ii) complex heterogeneity that exists across the current classification of autism (Levy et al., 2011; Szatmari, 2011), (iii) variable data quality and imaging methods used across studies, (iv) small samples ranging widely in age, and (v) diffusion metrics used (e.g., number of streamlines) that may not fully reflect the underlying white matter macro- or microstructure (see also Chapter 23). Thus further studies that comprehensively examine the development of the white matter tracts and their associations with complex social functioning as well as social deficits in larger samples of affected individuals using possibly multimodal methods are needed to clarify their importance in contributing to social communication impairment in autism.

The protracted developmental time course of white matter tracts that are critical to socioemotional processes (e.g., cingulum bundle and uncinate fasciculus) offers the opportunity for therapeutic intervention. Several diffusion imaging studies have now found evidence of practice-induced plasticity within relevant white matter tracts following cognitive or motor training in typically developing children (Bengtsson et al., 2005), and in the context of successful therapeutic interventions in brain-based conditions (Keller and Just, 2009; Trivedi et al., 2008). Therefore future efforts to further elucidate the functional importance of impaired white matter connectivity, and its relevance to social cognitive performance and specific socioemotional behaviors, may uncover new opportunities for biologically targeted interventions for a variety of brain-based conditions where social processes are affected.

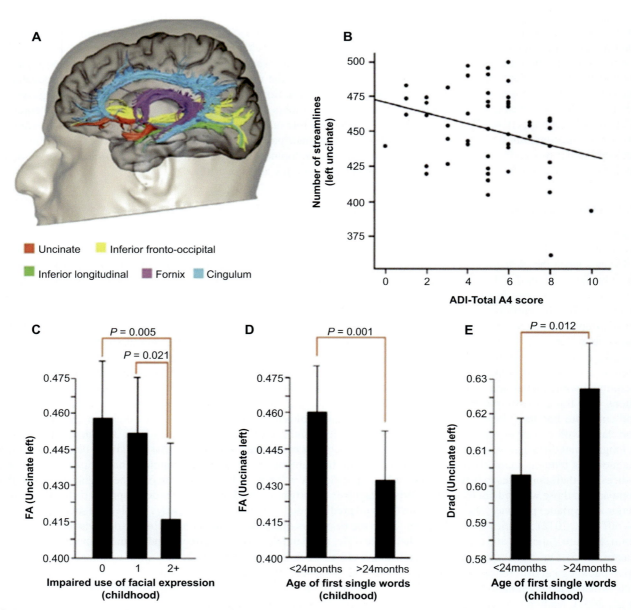

**FIG. 3** The anatomy of the limbic tracts in relation to childhood autism diagnostic interview-revised (ASD) symptoms. (A) Tractography reconstructions of the limbic pathways. (B) Negative correlation between the number of streamlines of the left uncinate fasciculus and the total A4 score for impaired socioemotional reciprocity in the ADI-R (Pearson's correlation = 0.295; $P = .01$) in the ASD group. ASD participants with a (C) significant history of impaired use of facial expression in childhood and (D and E) late use of first single words had a significantly lower fractional anisotropy (FA) and higher radial diffusivity in the left uncinate fasciculus. *(Reproduced with permission from Catani, M, Dell'Acqua, F, Budisavljevic, S, et al., 2016. Frontal networks in adults with autism spectrum disorder. Brain 139 (2), 616–630.)*

## 3 Cognitive functions

### 3.1 Limbic white matter tracts and episodic memory in healthy populations

One part of cognitive research explores the nature of episodic memory, a neurocognitive system that enables conscious recollection of events as they were previously experienced (Wheeler, 2001). These memories typically include information about the time and place of an event, as well as detailed information about the event itself. Correlations between episodic memory and elements of the limbic network have been detected across the spectrum of healthy aging in structural MRI studies, but also in diffusion MRI studies that implicated individual tracts such as the cingulum, uncinate, and fornix (Fig. 2). Consolidation of episodic memory has been most closely associated with the limbic system and the circuit of Papez—a neural loop that goes from the hippocampus, through the mammillary body, the anterior thalamic nucleus, the cingulate and parahippocampal gyrus, and back to the hippocampus (Aggleton and Brown, 2006; Choi et al., 2019).

Forming part of the Papez circuit, the fornix is the principal tract connecting the hippocampus with other structures, including the diencephalon, the basal forebrain, and the prefrontal cortex. Fornix microstructure is associated with episodic memory performance in healthy young and elderly adults (Rudebeck et al., 2009; Metzler-Baddeley et al., 2011). Based on the pattern of its rostral connections, the fornix can be split in two subdivisions separated by the anterior commissure: the precommissural fornix that projects to the basal forebrain and the prefrontal cortex; and the postcommissural fornix projecting to the anterior thalamus and mammillary bodies. Microstructural variations in postcommissural fibers were found to be associated with visual memory performance in healthy elderly individuals (Christiansen et al., 2016).

Other diffusion MRI studies suggest episodic memory is related to structural properties of more distributed white matter networks, in particular tracts linking the medial temporal lobes (Kantarci et al., 2011). In addition, other temporal association tracts have been implicated in episodic memory through rare case studies with focal lesions (Levine et al., 1998; Valenstein et al., 1987). The uncinate fasciculus is a bidirectional pathway that connects the anterior temporal lobe with the medial and orbitofrontal prefrontal cortices (Schmahmann and Pandya, 2006). It plays a role in the integration of cognition, behavior, autonomic, and emotional states (Mesulam, 2000). Microstructure of the uncinate was associated with visual memory performance in healthy aging (Metzler-Baddeley et al., 2011).

The ventral cingulum or hippocampal cingulum is the posterior segment of the cingulum bundle running within the parahippocampal gyrus and retrosplenial cingulate (Catani et al., 2013; Budisavljevic et al., 2015). The cingulum forms an almost complete ring from the orbital cortices, along the dorsal surface of the corpus callosum and toward the temporal lobes. It harbors the connections of the prefrontal cortex with the medial parietal and temporal lobes. FA of the left ventral cingulum was associated with episodic memory in elderly individuals without dementia independent of the hippocampal volume (Ezzati et al., 2016). The exact pattern of the relationship and the extent of white matter network involved differ among studies, partly due to differences of the studied populations, and different sensitivity and specificity due to variations in sample sizes and methodology used. For example, ventral cingulum microstructure was not associated with episodic memory in a different study using tractography (Metzler-Baddeley et al., 2011). Nevertheless, diffusion MRI studies have helped establish a central role of limbic white matter pathways in episodic memory.

## 3.2 Limbic white matter tracts and impairments in episodic memory

Correlations between episodic memory and elements of the limbic network have been detected across the spectrum of age-associated neurodegenerative disorders (Choo et al., 2010; Leube et al., 2008). Diffusion MRI studies have pointed out the role of individual tracts such as the cingulum and fornix in memory decline (Choo et al., 2010; Fellgiebel et al., 2005; Mielke et al., 2009; Zhuang et al., 2012). Thalamic components are also receiving increased attention, as anterior thalamic radiations have been linked to episodic memory in preterm children at risk of memory impairments (Kelly et al., 2021).

In patients with mild cognitive impairment (MCI), the prodromal stage of Alzheimer's dementia (AD), the fornix is structurally compromised (Nowrangi et al., 2013), while the uncinate fasciculus and the ventral cingulum exhibit more subtle microstructural alterations (Berlot et al., 2014; Metzler-Baddeley et al., 2012a). With the fornix compromised, other limbic tracts have a disproportionate effect on memory performance. Left ventral cingulum microstructure is most strongly associated with verbal episodic memory performance, whereas both the ventral cingulum and the uncinate are associated with recognition memory. The recruitment of extrafornical connections suggests a shift in memory strategies from recollection to more familiarity-based processes (Metzler-Baddeley et al., 2012a).

Compared to MCI, microstructural alterations within the fornix are more pronounced in AD (Tang et al., 2017; Shaikh et al., 2022). AD is also characterized by more widespread reductions in white matter integrity (Acosta-Cabronero and Nestor, 2014; Sexton et al., 2011) that include the posterior cingulate and parahippocampal areas, temporoparietal regions, and the splenium. This pattern corresponds with the limbic tracts connecting the circuit of Papez and is highly concordant with the distribution of gray matter atrophy observed in AD (Acosta-Cabronero et al., 2010).

## 3.3 White matter tracts and cognitive control in a healthy population

Cognitive control allows for mobilization of cognitive resources in the face of complex or competing task demands and is essential for what we recognize as intelligent behavior. Studies of patients with brain injuries and functional MRI studies highlight the importance of the medial and dorsal prefrontal cortex as key regions for various aspects of cognitive control, which also depends on a distributed cognitive control network (Cole and Schneider, 2007; Miller, 2000).

In terms of white matter tracts that connect the prefrontal regions implicated in cognitive control, the cingulum has come up as a tract of interest. One of the first insights into the role of cingulum microstructure came from a study using a region of interest approach, which found that variations in the FA both in the anterior and the ventral cingulum correlate with attention and executive function (Kantarci et al., 2011). Further evidence for the role of the cingulum in cognitive control was offered by studies using diffusion MRI tractography. Metzler-Baddeley et al. (2012b) reconstructed the anterior, middle, posterior, and parahippocampal cingulum portions. They observed that aspects of cognitive control are most sensitive to FA variations of the left anterior cingulum bundle, which harbors connections that terminate in the anterior cingulate cortex. In other studies, cingulum microstructure was identified as a predictor of cognitive control performance in healthy elderly participants independently of the volume of white matter hyperintensities and gray matter atrophy (Bettcher et al., 2016; Edde et al., 2020).

Resistance to interference and inhibition of behavioral output also represent crucial components of cognitive control. Successful response inhibition has been associated with a predominantly right lateralized network of cortical and subcortical structures, with converging evidence suggesting a role of the inferior frontal gyrus (IFG) and its connections (Chambers et al., 2009). In a study combining tractography with intraoperative stimulation, the right IFG, its connections to the striatum, and anterior thalamic connections were identified as critical for inhibitory control (Puglisi et al., 2019). Evidence for involvement of a more distributed network in cognitive control also comes from nontractography studies. In a study investigating fighter pilots, a group trained to deal with exceptionally demanding situations, performance in tasks investigating precision choices at speed in the presence of conflicting cues was associated with radial diffusivity of the right dorsomedial frontal region and the right parietal lobe (Roberts et al., 2010).

## 3.4 White matter tracts and impairments in cognitive control

White matter structural degradation is among the key substrates of cognitive aging. While anterior (dorsal) cingulum connections are principally implicated in cognitive control in healthy elderly individuals, more posterior (ventral) connections contribute to cognitive control performance in MCI (Metzler-Baddeley et al., 2012b). Cognitive control is also related to global topological properties of white matter structural networks in both AD and vascular dementia (Reijmer et al., 2013; Lawrence et al., 2014). It was observed that structural network topology mediates the effect of cingulum microstructural alterations on cognitive performance in MCI (Berlot et al., 2016). This suggests that the role of individual tracts is not only to convey information specific to individual cognitive processes, but also to maintain efficient network architecture that allows coordinated function of the connectome (Berlot and O'Sullivan, 2017).

Among other neurodegenerative disorders, cognitive control and behavioral deficits are central features of the behavioral variant of frontotemporal dementia (bvFTD), a common cause of young onset dementia. White matter microstructural alterations are widespread in bvFTD, and structural decline in frontal regions is linked to worsening cognitive control (Yu and Lee, 2019). A study combining voxel-based morphometry and tract-based spatial statistics (TBSS) to investigate both gray and white matter involvement in bvFTD pointed out a network of orbitofrontal, medial frontal, and anterior temporal brain regions together with their connecting white matter tracts that underpins behavioral disinhibition (Hornberger et al., 2011). These results partly conform to the pattern of white matter involvement identified in a study using tractography, where the FA of the left and right anterior cingulate, right posterior cingulate, and left uncinate correlated with executive function in bvFTD independent of gray matter atrophy (Tartaglia et al., 2012).

# 4 Language functions

## 4.1 Arcuate fasciculus and language functions in a healthy population

The capacity for language is a hallmark of the human species. Hence, it is not surprising that the study of language dominated the field of neuroscience for over 150 years, from the early neuroanatomical descriptions of the arcuate fasciculus by Reil and Burdach, the seminal works of Broca and Wernicke, to the recent tractography literature.

The growing consensus is that language processing is supported by distributed, large-scale cortical and subcortical networks and that understanding the connectivity of language can bring crucial insights into its function. The most prominent contemporary model of language neuroanatomy proposes a dorsal stream that maps auditory speech to articulatory (motor) representations and a ventral stream that plays a role in mapping auditory speech sounds to meaning (Hickok and Poeppel, 2000, 2004). Out of the several putative language pathways, the arcuate fasciculus (frontotemporal connection within the dorsal stream) has a central place. The popularity of this tract has been confirmed by a systematic review, showing that the arcuate fasciculus is the most popular tract to be studied with tractography in healthy populations (Forkel et al., 2022). Furthermore, the authors showed that it is also among the most sensitive in relation to behavior.

The classical definition of the arcuate fasciculus describes it as the structural link between Broca's frontal territory for speech production and Wernicke's temporal area for speech understanding. However, the first tractography studies showed that the anatomy of the arcuate fasciculus is more complex than previously thought, and that besides the classical fronto-temporal (so-called long) segment, there is also an anterior segment linking Broca's territory with the inferior parietal lobule (Geschwind's territory) and a posterior segment linking Geschwind's territory with Wernicke's territory (Fig. 4; Catani et al., 2005; Eluvathingal et al., 2007). The arcuate fasciculus is sometimes considered in its entirety and sometimes split into tripartite subdivision, and the differences in opinion have arisen regarding the origins, terminations, and the extent of the arcuate connections (e.g., Catani and Budisavljevic, 2014; Dick and Tremblay, 2012). All of these different approaches also produced differences when it comes to the functional roles that the arcuate fasciculus plays.

When discussing the functions that the arcuate fasciculus may support, we first need to mention the concept of lateralization or asymmetry. Some of the early studies showed that the left functional hemispheric dominance for language is compatible with the early left lateralization of the arcuate fasciculus, already observable in 1–4-month-old infants (Dubois et al., 2009). The left arcuate laterality has since been replicated many times both in terms of macrostructure and microstructure (Thiebaut de Schotten et al., 2011a,b). However, when considering a tripartite subdivision, the posterior (temporoparietal) segment, unlike the long and anterior segments, lateralizes much later in life, during late childhood and adolescence, with changes driven by environmental factors (Budisavljevic et al., 2015). It needs to be noted that the choice of tractography pipeline was found to influence measurements of the arcuate laterality, both in terms of macrostructure and microstructure (Bain et al., 2019), and these methodological considerations should be taken into account when comparing studies.

The functions of the arcuate fasciculus have been studied extensively in healthy populations. Although it seems to be involved in several behavioral domains, language measures remain the most common association with this pathway (Forkel

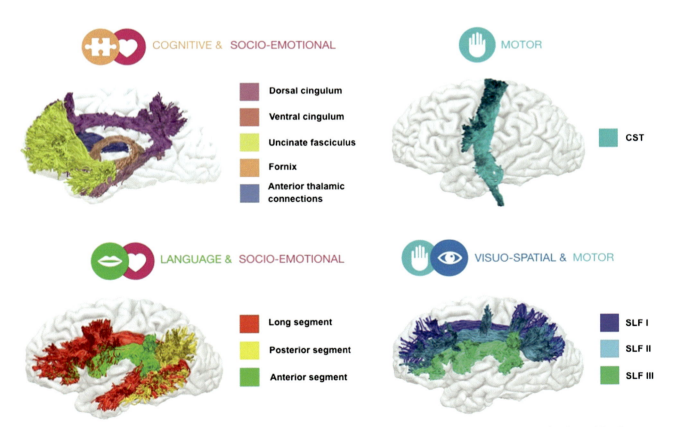

**FIG. 4** An overview of the key white matter tracts discussion in the chapter in relation to five behavioral domains (socioemotional, cognitive, language, motor, and visuospatial). *CST*, corticospinal tract; *SLF*, superior longitudinal fasciculus.

**620 PART | V** Tractography applications

et al., 2022). The interindividual variability in the arcuate fasciculus anatomy has been implicated in learning new words (Lopez-Barroso et al., 2013), verbal recall (Catani et al., 2007), syntax (Berwick et al., 2013), reading skills (Yeatman et al., 2011), and others. The recent systematic review concluded that the classical, long segment could be driving the link with the language domain, while in contrast, the anterior and posterior segments are usually associated with aspects of the memory and attention (Forkel et al., 2022).

## 4.2 Arcuate fasciculus and language impairments

In psychiatric and neurological patients, damage to the arcuate fasciculus is associated with auditory hallucinations in schizophrenia (Catani et al., 2011), aphasia severity in stroke (Forkel et al., 2014), and repetition deficits in primary progressive aphasia patients (Forkel et al., 2020). Some of these studies highlighted tractography as a potential tool to predict patients' recovery and clinical outcomes. For example, a study showed that individual differences in the asymmetry of the arcuate fasciculus detected by tractography could help clinicians to predict language recovery at 6 months in left hemisphere stroke patients, independent of other predictive factors, such as volume of the lesion, initial severity of language deficits, age, and gender of the patients (Forkel et al., 2014). Similarly, FA in the left arcuate fasciculus was found to correlate with the severity of auditory hallucinations in schizophrenia (Psomiades et al., 2016), one of the most salient symptoms where the arcuate anatomy could have predictive value.

Arcuate fasciculus aberrations have been found also in neurodevelopmental disorders like autism (Catani et al., 2016) and developmental language disorder (Verly et al., 2019). In children and adolescents with autism, research found reduced FA in the left ( Jou et al., 2011; Lai et al., 2012) and the right arcuate (Sahyoun et al., 2010; Jou et al., 2011) as well as the loss of typical arcuate lateralization patterns (Fletcher et al., 2010). A tractography study by Catani et al. (2016) in the adult population (61 adult males with autism with normal intelligence and 61 age-matched controls), found significant differences in the microstructure of the left arcuate fasciculus using tractography. Furthermore, a lower number of streamlines of the left anterior segment was associated with the higher severity of stereotyped, repetitive, and idiosyncratic speech symptoms in childhood (however, caution should be taken when using this metric as the number of streamlines might not reliably reflect the underlying white matter). It seems that neuroanatomical differences in the arcuate remain, even when the severity of language deficits declines in adulthood. Future longitudinal studies, conducted in well-characterized large populations and with standardized tractography methods applied to both young and older populations, are necessary to understand age-related changes in structural abnormalities underlying language and communication deficits in autism.

## 5 Motor functions

As action is the product of almost all behaviors, it might be expected that its circuitry would be one of the best studied areas. The major projection pathway, the CST, is routinely shown to be correlated with motor behavior, and this is, like the arcuate fasciculus, one of the most commonly reported associations between white matter anatomy and behavior in the literature (Forkel et al., 2020). Interindividual differences in the volume and microstructural features of the CST have been shown to be directly correlated with motor behavior in healthy adolescents (Angstmann et al., 2016) and adults (Rose et al., 2012) and altered in clinical syndromes, including neurodegenerative disorders (Ciccarelli et al., 2006) and particularly stroke (Puig et al., 2013; Radlinska et al., 2010). It has also been demonstrated that changes in the CST through late childhood are associated with the development of manual dexterity skills (Fuelscher et al., 2021).

One advantage of diffusion tractography is the ability to study whole brain networks, and particularly the relevance of bilateral white matter pathways. This has demonstrated some key insights into anatomofunctional organization of the CST. Postmortem study of the CST has shown it is larger in the left hemisphere than the right, termed a leftward asymmetry (Nathan et al., 1990; Rademacher et al., 2001). It was theorized that this may therefore be linked to the preference to use the right hand over the left, right-handedness; however, systematic studies using tractography in consistently left- and right-handed adults have demonstrated this leftward hemispheric asymmetry exists in both groups (Howells et al., 2020; Westerhausen et al., 2007), and there does not appear to be any link with handedness. As interactions between a hand and an object require the orchestration of finger movements based on the properties of the object and the goal of the action, it is likely that hand preference may be linked to asymmetric pathways "upstream" of connections involved in motor output to the peripheral nervous system, i.e., pathways connecting premotor and sensory association areas. However, these white matter connections are relatively understudied compared with the CST.

The superior longitudinal fasciculi are the major pathways connecting premotor and parietal association areas, running in the anterior to posterior orientation via three branches bilaterally (SLF I: dorsomedial; SLF III: ventrolateral, with the SLF II running between the two; Fig. 4). It has been demonstrated that these pathways are recruited during visuospatial attention tasks, and interindividual variability in their hemispheric asymmetry is linked to performance on visuospatial tests (Thiebaut de Schotten et al., 2011a,b; Quentin et al., 2015). Stimulation of these fibers also directly affects attentional processing (Quentin et al., 2015; Thiebaut de Schotten et al., 2005) (more on the attentional aspects in the following section). However, importantly, these fibers have been well studied in the context of visuomotor control using tracing and intracortical recording in nonhuman primates (Borra and Luppino, 2017). Although the ability to ascertain direct homologs between tracing data and tractography of macaque visuomotor pathways is still in its infancy (Howells et al., 2020), these frontoparietal pathways can be mapped well using tractography in humans (Makris et al., 2005), and correlations with behavior provide a means to indirectly evaluate their function.

The first tractography study to link interindividual differences in the branches of the SLF with visuomotor behavior demonstrated that its hemispheric asymmetry could be correlated with specific kinematic measures recorded during right-hand reaching (Budisavljevic et al., 2017). Greater rightward asymmetry of the middle branch (SLF II) was correlated with higher movement acceleration when comparing 30 right-handed adults. Interestingly, participants who responded faster to stimuli appearing in one hemifield on a visuospatial attention task (the Posner paradigm) also have a larger volume in the contralateral SLF II (Thiebaut de Schotten et al., 2011a,b) and resection of this branch also leads to deficits in selective attention (Howells et al., 2020). This suggests there is a link between this pathway and visuomotor behavior, although a relationship with handedness has not been demonstrated.

Currently, there is evidence to suggest that the dorsomedial branch of the SLF I differs in structural asymmetry between right- and left-handers. While this pathway is left-lateralized in right-handers (Budisavljevic et al., 2017; Howells et al., 2018), it has a more variable pattern of asymmetry in left-handers (Cazzoli and Chechlacz, 2017; Howells et al., 2018). There is also an association between structural asymmetry of the SLF I and behavioral asymmetry in visuomotor performance, measured using a visuomotor pegboard task (Howells et al., 2018). A follow-up to this work examined the effect of resection of this pathway, measuring lateralized hand use during the assembly of a jigsaw puzzle before and after neurosurgical resection in 17 brain tumor patients (Howells et al., 2020). In this work, it was demonstrated that the right hand was selected more than the left to pick up and position puzzle pieces before surgery, but that this dominance in hand use was reduced following surgery and resection of the SLF I, despite no change in general motor ability. This suggests that this pathway could play an important role in lateralized hand selection, which may be a precursor to handedness (i.e., the more established use of one hand over the other).

Other association pathways have more recently been linked to motor behavior, including those running within the occipital lobe connecting the dorsal and ventral streams for visual processing, the vertical occipital fasciculus, and connecting the parietal and temporal lobe, the posterior branch of the arcuate fasciculus (Budisavljevic et al., 2018). Microstructural features detected using tract-specific metrics like hindrance modulated orientational anisotropy (HMOA) are correlated with kinematic metrics during lifting and grasping of objects in both the left and right hemispheres. Kinematic parameters have also been linked to white matter connections running within the frontal lobe. A study of 32 healthy adults demonstrated that interindividual variability in HMOA of the frontal aslant tract, connecting the superior frontal and IFG, is linked to kinematic measures during reaching and reach-to-grasp movements (Budisavljevic et al., 2017). This pathway connects the dorsal and ventral premotor regions, and this study also showed associations with short superficial U-shaped connections between the inferior and middle, and middle and superior, frontal gyri. This finding was particularly novel, as these short cortico-cortical connections have been neglected in the literature until the emergence of tractography has now enabled them to be studied more closely (Catani et al., 2012; Guevara et al., 2020).

The role of superficial white matter fibers in motor behavior is not limited to premotor connectivity. Direct connectivity exists between the precentral and postcentral gyrus via U-shaped connections, which are concentrated particularly in the regions associated with hand function (Catani et al., 2012), rather than the lower limbs or mouth. The concentration of these fibers in this region suggest that they may play a key role in integrating incoming sensory and outgoing motor information for manual ability, and this is supported by tractography studies that have correlated their size with manual dexterity (Rose et al., 2012). One study in particular examined these fibers in 60 healthy adults and 60 individuals with ASD (Thompson et al., 2017). They showed an association between U-shaped fibers in the hand region in the left hemisphere and performance on a peg placing task. In the population with ASD, the behavioral association was with U-shaped fibers in the right hemisphere, potentially consistent with the right shift hypothesis of ASD (Annett, 2002). This may also be linked to the higher incidence of left-handedness in individuals with ASD; however, this has not yet been studied.

These short U-shaped fibers have also been studied using direct electrical stimulation in the intraoperative setting. Before the advent of tractography, causal insights into brain function enabled by electrical stimulation were limited to the cortex. When moving below the cortex, it is not possible to assess which white matter tracts have been affected by subcortical stimulation using the naked eye. Tractography has provided a means to directly target both terminations of white matter tracts on the cortex but also the body of fasciculi subcortically during intraoperative stimulation (Herbert and Duggau, 2020). In this

## 622 PART | V Tractography applications

way, it is possible to assess which circuits are affected by the application of electrical current and how this affects behavior. When assessing motor behavior, neurosurgeons primarily ask patients to perform ongoing flexion and extension of hands and arms, and if this behavior is transiently halted (a "negative motor behavior"), this is an indication that the affected fibers are important for movement. While clinically useful, this may hide variation in motor behaviors reflecting different hierarchical networks, which are of interest for scientific purposes, but may also reflect different potential for reorganization (Viganò et al., 2021). Combining electromyographic (EMG) recording, tractography, and electrical stimulation with more sensitive motor tests, it has been possible to differentiate the role of different populations of U-shaped fibers in the hand region in motor behavior, through characteristic changes in EMG recording and hand grasping behaviors online during surgery (Fornia et al., 2020). Using this approach, it has been possible to show a gradient within the hand region of the precentral gyrus, with more anterior U-shaped fibers connected with premotor regions linked to cognitive-motor aspects of hand movement, and U-shaped fibers connected with somatosensory cortex linked to later stages of motor processing (Viganò et al., 2019). Currently, other motor pathways are less well studied using intraoperative methods with this level of granularity; however, this is likely to change over the coming years.

## 6 Visuospatial functions

### 6.1 Frontoparietal white matter tracts and visuospatial functions in a healthy population

Visuospatial attention in humans is mostly a function under right hemispheric dominance, compatible with the right-lateralization of regions of visual (fusiform gyrus) and parietal (supramarginal and angular gyri) cortices (Caeyenberghs and Leemans, 2014). Those regions are connected with frontal regions through the SLF that subserves visuospatial attention and attentional selection of relevant stimuli, as well as visuomotor processing, as mentioned in the previous section.

As mentioned previously, the SLF is composed of the three parallel longitudinal branches (see Fig. 4). The superior branch (SLF I) is dorsal and runs from the superior parietal lobule and the precuneus (BA7, BA5) to the superior frontal gyrus (BA6, BA8, BA9, up to BA10). The middle branch (SLF II) runs from the angular gyrus (BA39) to the posterior regions of the middle frontal gyrus (BA6, with few projections going further up to BA 46). The inferior branch (SLF III) is ventral and connects the supramarginal gyrus (BA40) to the IFG (BA44, BA45, BA47) (Rojkova et al., 2016; Budisavljevic et al., 2017). The SLF I has been shown to be symmetrically distributed, while the SLF II and SLF III tend to be right-lateralized in a healthy right-handed population (Thiebaut de Schotten et al., 2011a,b). The lateralization of the SLF II and III has been confirmed in vivo by direct electrical stimulation mapping (Sarubbo et al., 2020).

The three branches of the SLF subserve different processes and functional interactions within the attentional networks. It has been suggested that the SLF I supports the dorsal attentional network (DAN), while the SLF III connects the two core regions of the ventral attentional network (VAN), which is also strongly right-lateralized (Corbetta et al., 2008; Vergani et al., 2021; Bernard et al., 2020). On the other hand, the SLF II mediates the communication between DAN (goal-directed attention, represented by the SLF I) and VAN (exogenous attention by identification of salient events by the SLF III) (Carter et al., 2017; Fox et al., 2006; Thiebaut de Schotten et al., 2011a,b). This idea has been supported in a large tractography study (129 healthy subjects) combined with 14 metaanalyses of functional MRI studies (Parlatini et al., 2017). The authors have shown segregation into a dorsal spatial/motor network, which overlaps with the projections of the SLF I and a ventral/nonspatial/motor network, which overlaps with the projections of the SLF III. The projections of the SLF II correspond to a network at the intersection between the dorsal and ventral networks. Similarly, based on Bundesen's theory of visual attention, Chechlacz et al. (2015) showed that higher visual short-term memory capacity correlated with a higher HMOA within the right SLF II and SLF III as well as higher speed of visual information processing. At the macrostructural level, the larger the volume of the right SLF II, the greater was the leftward bias. Chica et al. (2018) showed that the reduced integrity of the left SLF III (rather than the increased integrity of the right SLF III) allowed for more interactions between dorsal and VANs in the left frontal eye-field.

### 6.2 Frontoparietal white matter tracts in unilateral spatial neglect

Left unilateral spatial neglect (USN) is a neurological syndrome resulting frequently from right brain lesion (about 80% in the acute stage of a stroke) (Azouvi et al., 2002) in which patients fail to detect, orient, and/or respond to left-sided objects. These deficits can heavily impact everyday life, through not eating the left portions of meals, bumping into obstacles on the left side, being unable to drive anymore, etc. Thus USN is an important factor adding to the loss of autonomy as well as poor functional recovery (Bartolomeo, 2014). USN can be viewed as a deficit in the exogenous orienting of attention to the left space, combined with nonspatially lateralized deficits (sustained attention and visual working memory) (Chica et al., 2011;

Fabius et al., 2020). USN can result from damage to different functional components depending on the lesion sites (Toba et al., 2018; Vaessen et al., 2016; Verdon et al., 2010).

Functional MRI studies in healthy participants have shown that the orienting of spatial attention relies on the coordinated activity of frontoparietal networks (Nobre, 2001; Corbetta and Shulman, 2002). The first implication of the SLF in USN came from a lesion study by Doricchi and Tomaiuolo (2003) in which they showed that the maximal overlap of chronic neglect patients' lesions was found in the SLF beneath the rostral part of the supramarginal gyrus. This was confirmed by direct electrical stimulation during awake surgery for gliomas in 2 patients who were presented with a line bisection task and showed rightward deviation upon inactivation of the SLF (Thiebaut de Schotten et al., 2005). The involvement of the right SLF in USN was later confirmed in another study in patients with brain tumors (Shinoura et al., 2009). Among the three branches, the right SLF III was found to be most implicated in the USN symptoms in chronic patients (Urbanski et al., 2008, 2011) and in a single case-report of an acute patient (Ciaraffa et al., 2013). However, in healthy populations a stronger predictor for rightwards deviation in line bisection and cued attention tasks was the right SLF II (Thiebaut de Schotten et al., 2011a,b, 2014). However, Ciaraffa et al. (2013) performed the tractography on a patient at an acute stage of the stroke. Urbanski et al. (2011) only found correlations between reduced FA in chronic patients with performances on paper and pencil tasks requiring target/distractor discrimination but not for deviation in a line bisection task.

Using TBSS, Lunven et al. (2015) found a strong association between the severity of USN and decreased FA values in the right SLF II and III but no significant association between FA values and performance on the Bells Test nor on the line bisection task. In a voxel-based study, Pedrazzini and Ptak (2020) showed that damage to the temporoparietal junction and SLF II were significant predictors of the omissions in the Bells Test, whereas damage to the intraparietal sulcus was a significant predictor of the omissions in the letter cancellation test. For the line bisection test, they found damage to the lateral prefrontal cortex/insula and SLF I as significant predictors. Finally, for the reading task, the SLF II was the significant regressor. As for Toba et al. (2018), they found the involvement of the SLF III in the performance for the reading task (but not as a significant regressor), the SLF III as a significant predictor of the deviation in the line bisection task, but not for omissions in the Bells Test regarding FA nor diffusivities.

In summary, differences observed in the studies mentioned here are likely due to the methods used, the type of patients, and the tests performed to explore USN. For example, different measures of white matter disconnections have been employed (FA for Urbanski et al., 2011; lesion as seed for probabilistic diffusion tensor tractography in a healthy brain with BCBtoolkit (Foulon et al., 2018) for Pedrazzini and Ptak, 2020; TBSS regression analysis for Lunven et al., 2015; diffusivities for Toba et al., 2018). Moreover, differences are observed in anatomy between the SLF II and SLF III. For example, the FA-based right SLF III has been shown to project to the angular gyrus and the inferior frontal sulcus according to the Desikan atlas (Desikan et al., 2006), two regions which are associated with the course of the SLF II. The use of diffusion tensor imaging (DTI) tractography to virtually dissect the SLF branches leads to different results than those from the use of spherical deconvolution, which allows for modeling of multiple fiber orientations (Dell'Acqua and Tournier, 2019). Finally, dissociations are often observed in USN between tests used and between line bisection and cancellation tasks (Azouvi et al., 2002; Doricchi et al., 2008). Another difference between studies is the range of lesion onset times of included patients, which implies different dynamics of plasticity (Bartolomeo and Thiebaut de Schotten, 2016). Attention recovery has been shown to be poorer when SLF II/III were damaged (Ramsey et al., 2017), probably due to the fact that when lesioned, the SLF II impairs the lateralized spatial orienting and that the lesioned SLF III impairs the reorienting of attention (Carter et al., 2017). Taken together, all these results converge to the crucial involvement of the SLF (especially the SLF II and III) in both the presence and the severity of neglect signs when disconnected.

# 7 Conclusions

Our current knowledge of white matter tracts and their functional correlates depends upon specific temporal and spatial contexts (stage of development, population characteristics, injury, disorder, neural systems affected, method of inquiry, etc.). This chapter aims to provide an overview of the role of white matter networks in five behavioral domains (socio-emotional, cognitive, language, motor, and visuospatial), with evidence coming from healthy populations and neurodevelopmental, psychiatric, and neurological disorders. It should be emphasized that distinctions between white and gray matter are not absolute and that neurons are the fundamental units of all networks (Filley and Fields, 2016). Thus, as myelocentrism can be as limiting as corticocentrism, a balanced view of the representation of behavior and function is crucial. Furthermore, psychological variables are complex constructs that rely on many different neuronal processes. Nevertheless, combining advanced diffusion tractography more often with functional MRI and other methods will certainly offer productive insights into the brain structure-function relationship in the future.

# References

Acosta-Cabronero, J., Nestor, P.J., 2014. Diffusion tensor imaging in Alzheimer's disease: insights into the limbic-diencephalic network and methodological considerations. Front. Aging Neurosci. 6, 266. https://doi.org/10.3389/fnagi.2014.00266.

Acosta-Cabronero, J., Williams, G.B., Pengas, G., Nestor, P.J., 2010. Absolute diffusivities define the landscape of white matter degeneration in Alzheimer's disease. Brain 133, 529–539. https://doi.org/10.1093/brain/awp257.

Adolphs, R., Tranel, D., Hamann, S., Young, A.W., Calder, A.J., Phelps, E.A., … Damasio, A.R., 1999. Recognition of facial emotion in nine individuals with bilateral amygdala damage. Neuropsychologia 37 (10), 1111–1117 (S0028393299000391 [pii]).

Aggleton, J.P., Brown, M.W., 2006. Interleaving brain systems for episodic and recognition memory. Trends Cogn. Sci. 10, 455–463. https://doi.org/10.1016/j.tics.2006.08.003.

Ameis, S.H., Catani, M., 2015. Altered white matter connectivity as a neural substrate for social impairment in Autism Spectrum Disorder. Cortex 62, 158–181.

Angstmann, S., Skak Madsen, K., Skimminge, A., et al., 2016. Microstructural asymmetry of the corticospinal tracts predicts right-left differences in circle drawing skill in right-handed adolescents. Brain Struct. Funct. 221 (9), 4475–4489.

Annett, M., 2002. Handedness and Brain Asymmetry: The Right Shift Theory. Psychology Press.

Azouvi, P., Samuel, C., Louis-Dreyfus, A., Bernati, T., Bartolomeo, P., Beis, J.M., Chokron, S., Leclercq, M., Marchal, F., Martin, Y., De Montety, G., Olivier, S., Perennou, D., Pradat-Diehl, P., Prairial, C., Rode, G., Siéroff, E., Wiart, L., Rousseaux, M., 2002. Sensitivity of clinical and behavioural tests of spatial neglect after right hemisphere stroke. J. Neurol. Neurosurg. Psychiatry 73 (2), 160–166. https://doi.org/10.1136/jnnp.73.2.160.

Bain, J.S., Yeatman, J.D., Schurr, R., et al., 2019. Evaluating arcuate fasciculus laterality measurements across dataset and tractography pipelines. Hum. Brain Mapp. 40 (13), 3695–3711.

Bartolomeo, P., 2014. Attention disorders after right brain damage. In: Attention Disorders After Right Brain Damage. Springer, London, ISBN: 978-1-4471-5648-2, https://doi.org/10.1007/978-1-4471-5649-9.

Bartolomeo, P., Thiebaut de Schotten, M., 2016. Let thy left brain know what thy right brain doeth: inter-hemispheric compensation of functional deficits after brain damage. Neuropsychologia 93 (Pt B), 407–412. https://doi.org/10.1016/j.neuropsychologia.2016.06.016.

Beaulieu, C., 2002. The basis of anisotropic water diffusion in the nervous system. NMR Biomed 15 (7–8), 435–455.

Belmonte, M.K., Allen, G., Beckel-Mitchener, A., Boulanger, L.M., Carper, R.A., Webb, S.J., 2004. Autism and abnormal development of brain connectivity. J. Neurosci. 24 (42), 9228–9231.

Berlot, R., O'Sullivan, M.J., 2017. What can the topology of white matter structural networks tell us about mild cognitive impairment? Future Neurol. 12 (1). https://doi.org/10.2217/fnl-2016-0022.

Berlot, R., Metzler-Baddeley, C., Jones, D.K., O'Sullivan, M.J., 2014. CSF contamination contributes to apparent microstructural alterations in mild cognitive impairment. NeuroImage 92, 27–35. https://doi.org/10.1016/j.neuroimage.2014.01.031.

Bengtsson, S.L., Nagy, Z., Skare, S., Forsman, L., Forssberg, H., Ullen, F., 2005. Extensive piano practicing has regionally specific effects on white matter development. Nat. Neurosci. 8 (9), 1148–1150. nn1516.

Berlot, R., Metzler-Baddeley, C., Ikram, M.A., Jones, D.K., O'Sullivan, M.J., 2016. Global efficiency of structural networks mediates cognitive control in mild cognitive impairment. Front. Aging Neurosci. 8, 292. https://doi.org/10.3389/fnagi.2016.00292.

Bernard, F., Lemee, J.M., Mazerand, E., Leiber, L.M., Menei, P., Ter Minassian, A., 2020. The ventral attention network: the mirror of the language network in the right brain hemisphere. J. Anat. 237 (4), 632–642. https://doi.org/10.1111/joa.13223.

Berwick, R.C., Friederici, A.D., Chomsky, N., Bolhuis, J.J., 2013. Evolution, brain, and the nature of language. Trends Cogn. Sci. 17 (2), 89–98.

Bettcher, B.M., Mungas, D., Patel, N., Elofson, J., Dutt, S., Wynn, M., Watson, C.L., Stephens, M., Walsh, C.M., Kramer, J.H., 2016. Neuroanatomical substrates of executive functions: beyond prefrontal structures. Neuropsychologia 85, 100–109. https://doi.org/10.1016/j.neuropsychologia.2016.03.001.

Borra, E., Luppino, G., 2017. Functional anatomy of the macaque temporo-parieto-frontal connectivity. Cortex 97, 306–326.

Buckner, R.L., Andrews-Hanna, J.R., Schacter, D.L., 2008. The brain's default network: anatomy, function, and relevance to disease. Ann. N. Y. Acad. Sci. 1124, 1–38.

Budisavljevic, S., Dell'Acqua, F., Rijsdijk, F.V., et al., 2015. Age-related differences and heritability of the perisylvian language networks. J. Neurosci. 35 (37), 12625–12634.

Budisavljevic, S., Dell'Acqua, F., Zanatto, D., Begliomini, C., Miotto, D., Motta, R., Castiello, U., 2017. Asymmetry and structure of the fronto-parietal networks underlie visuomotor processing in humans. Cereb. Cortex 27 (2), 1532–1544. https://doi.org/10.1093/cercor/bhv348.

Budisavljevic, S., Dell'Acqua, F., Castiello, U., 2018. Cross-talk connections underlying dorsal and ventral stream integration during hand actions. Cortex 103, 224–239.

Caeyenberghs, K., Leemans, A., 2014. Hemispheric lateralization of topological organization in structural brain networks. Hum. Brain Mapp. 35 (9), 4944–4957. https://doi.org/10.1002/hbm.22524.

Carter, A.R., McAvoy, M.P., Siegel, J.S., Hong, X., Astafiev, S.V., Rengachary, J., Zinn, K., Metcalf, N.V., Shulman, G.L., Corbetta, M., 2017. Differential white matter involvement associated with distinct visuospatial deficits after right hemisphere stroke. Cortex 88, 81–97. https://doi.org/10.1016/j.cortex.2016.12.009.

Catani, M., Bambini, V., 2014. A model for social communication and language evolution and development (SCALED). Curr. Opin. Neurobiol. 28, 165–171.

Catani, M., Budisavljevic, S., 2014. Contribution of diffusion tractography to the anatomy of language. In: Johansen-Berg, H., Behrens, T.E.J. (Eds.), Diffusion MRI: From Quantitative Measurement to In vivo Neuroanatomy, second ed. Elsevier, pp. 511–529 (Chapter 22).

Catani, M., Thiebaut de Schotten, M., 2008. A diffusion tensor imaging tractography atlas for virtual in vivo dissections. Cortex 44 (8), 1105–1132. https://doi.org/10.1016/j.cortex.2008.05.004 (S0010-9452(08)00123-8 [pii]).

Catani, M., Howard, R.J., Pajevic, S., Jones, D.K., 2002. Virtual in vivo interactive dissection of white matter fasciculi in the human brain. Neuroimage 17 (1), 77–94.

Catani, M., Jones, D.K., Donato, R., Ffytche, D.H., 2003. Occipito-temporal connections in the human brain. Brain 126 (Pt 9), 2093–2107.

Catani, M., Jones, D.K., Ffytche, D.H., 2005. Perisylvian language networks of the human brain. Ann. Neurol. 57, 8–16.

Catani, M., Allin, M.P., Husain, M., Pugliese, L., Mesulam, M.M., Murray, R.M., et al., 2007. Symmetries in human brain language pathways correlate with verbal recall. Proc. Natl. Acad. Sci. U. S. A. 104 (43), 17163–17168.

Catani, M., Craig, M.C., Forkel, S.J., Kanaan, R., Picchioni, M., Toulopoulou, T., et al., 2011. Altered integrity of perisylvian language pathways in schizophrenia: relationship to auditory hallucinations. Biol. Psychiatry 70 (12), 1143–1150.

Catani, M., et al., 2012. Beyond cortical localization in clinico-anatomical correlation. Cortex 48 (10), 1262–1287.

Catani, M., Dell'Acqua, F., Thiebaut de Schotten, M., 2013. A revised limbic system model for memory, emotion and behaviour. Neurosci. Biobehav. Rev. 37 (8), 1724–1737.

Catani, M., Dell'Acqua, F., Budisavljevic, S., et al., 2016. Frontal networks in adults with autism spectrum disorder. Brain 139 (2), 616–630.

Cazzoli, D., Chechlacz, M., 2017. A matter of hand: causal links between hand dominance, structural organization of fronto-parietal attention networks, and variability in behavioural responses to transcranial magnetic stimulation. Cortex 86, 230–246.

Chambers, C.D., Garavan, H., Bellgrove, M.A., 2009. Insights into the neural basis of response inhibition from cognitive and clinical neuroscience. Neurosci. Biobehav. Rev. 33 (5), 631–646. https://doi.org/10.1016/j.neubiorev.2008.08.016.

Chechlacz, M., Gillebert, C.R., Vangkilde, S.A., Petersen, A., Humphreys, G.W., 2015. Structural variability within frontoparietal networks and individual differences in attentional functions: an approach using the theory of visual attention. J. Neurosci. 35 (30), 10647–10658. https://doi.org/10.1523/JNEUROSCI.0210-15.2015.

Cheon, K.A., Kim, Y.S., Oh, S.H., Park, S.Y., Yoon, H.W., Herrington, J., Nair, A., et al., 2011. Involvement of the anterior thalamic radiation in boys with high functioning autism spectrum disorders: a Diffusion Tensor Imaging study. Brain Res. 1417, 77–86.

Chica, A.B., Bartolomeo, P., Valero-Cabre, A., 2011. Dorsal and ventral parietal contributions to spatial orienting in the human brain. J. Neurosci. 31 (22), 8143–8149. https://doi.org/10.1523/JNEUROSCI.5463-10.2010.

Chica, A.B., Thiebaut de Schotten, M., Bartolomeo, P., Paz-Alonso, P.M., 2018. White matter microstructure of attentional networks predicts attention and consciousness functional interactions. Brain Struct. Funct. 223 (2), 653–668. https://doi.org/10.1007/s00429-017-1511-2.

Choi, S.H., Kim, Y.B., Paek, S.H., Cho, Z.H., 2019. Papez circuit observed by in vivo human brain with 7.0T MRI super-resolution track density imaging and track tracing. Front. Neuroanat. 13, 17. https://doi.org/10.3389/fnana.2019.00017.

Choo, I.H., Lee, D.Y., Oh, J.S., Lee, J.S., Lee, D.S., Song, I.C., Youn, J.C., Kim, S.G., Kim, K.W., Jhoo, J.H., Woo, J.I., 2010. Posterior cingulate cortex atrophy and regional cingulum disruption in mild cognitive impairment and Alzheimer's disease. Neurobiol. Aging 31, 772–779. https://doi.org/10.1016/j.neurobiolaging.2008.06.015.

Christiansen, K., Aggleton, J.P., Parker, G.D., O'Sullivan, M.J., Vann, S.D., Metzler-Baddeley, C., 2016. The status of the precommissural and postcommissural fornix in normal ageing and mild cognitive impairment: an MRI tractography study. NeuroImage 130, 35–47. https://doi.org/10.1016/j.neuroimage.2015.12.055.

Ciaraffa, F., Castelli, G., Parati, E.A., Bartolomeo, P., Bizzi, A., 2013. Visual neglect as a disconnection syndrome? A confirmatory case report. Neurocase 19 (4), 351–359. https://doi.org/10.1080/13554794.2012.667130.

Ciccarelli, O., Behrens, T.E., Altmann, D.R., et al., 2006. Probabilistic diffusion tractography: a potential tool to assess the rate of disease progression in amyotrophic lateral sclerosis. Brain 129 (Pt 7), 1859–1871.

Cole, M.W., Schneider, W., 2007. The cognitive control network: integrated cortical regions with dissociable functions. NeuroImage 37, 343–360. https://doi.org/10.1016/j.neuroimage.2007.03.071.

Corbetta, M., Shulman, G.L., 2002. Control of goal-directed and stimulus-driven attention in the brain. Nat. Rev. Neurosci. 3 (3), 201–215. https://doi.org/10.1038/nrn755.

Corbetta, M., Patel, G., Shulman, G.L., 2008. The reorienting system of the human brain: from environment to theory of mind. Neuron 58 (3), 306–324. https://doi.org/10.1016/j.neuron.2008.04.017.

Damasio, A.R., Adolphs, R., Damasio, H., 2003. The contributions of the lesion method to the functional neuroanatomy of emoion. In: Davidson, R.J., Scherer, K.R., Goldsmith, H.H. (Eds.), Handbook of Affective Sciences. Oxford University Press, pp. 66–92.

Dell'Acqua, F., Tournier, J.D., 2019. Modelling white matter with spherical deconvolution: how and why? NMR Biomed. 32 (4), e3945. https://doi.org/10.1002/nbm.3945.

Desikan, R.S., Ségonne, F., Fischl, B., Quinn, B.T., Dickerson, B.C., Blacker, D., Buckner, R.L., Dale, A.M., Maguire, R.P., Hyman, B.T., Albert, M.S., Killiany, R.J., 2006. An automated labeling system for subdividing the human cerebral cortex on MRI scans into gyral based regions of interest. NeuroImage 31 (3), 968–980. https://doi.org/10.1016/j.neuroimage.2006.01.021.

Dick, A.S., Tremblay, P., 2012. Beyond the arcuate fasciculus: consensus and controversy in the connectional anatomy of language. Brain 35 (Pt 12), 3529–3550.

Doricchi, F., Tomaiuolo, F., 2003. The anatomy of neglect without hemianopia: a key role for parietal–frontal disconnection? NeuroReport 14 (17), 2239–2243. https://doi.org/10.1097/00001756-200312020-00021.

Doricchi, F., Thiebaut de Schotten, M., Tomaiuolo, F., Bartolomeo, P., 2008. White matter (dis)connections and gray matter (dys)functions in visual neglect: gaining insights into the brain networks of spatial awareness. Cortex 44 (8), 983–995. https://doi.org/10.1016/j.cortex.2008.03.006.

Dubois, J., Hertz-Pannier, L., Cachia, A., et al., 2009. Structural asymmetries in the infant language and sensori-motor networks. Cereb. Cortex 19 (2), 414–423.

Edde, M., Theaud, G., Rheault, F., Dilharreguy, B., Helmer, C., Dartigues, J.F., Amieva, H., Allard, M., Descoteaux, M., Catheline, G., 2020. Free water: a marker of age-related modifications of the cingulum white matter and its association with cognitive decline. PLoS One 15 (11), e0242696. https://doi.org/10.1371/journal.pone.0242696.

Elison, J.T., Paterson, S.J., Wolff, J.J., Reznick, J.S., Sasson, N.J., Gu, H., … Piven, J., 2013a. White matter microstructure and atypical visual orienting in 7-month-olds at risk for autism. Am. J. Psychiatry 170 (8), 899–908. https://doi.org/10.1176/appi.ajp.2012 (12091150 1669751 [pii]).

Elison, J.T., Wolff, J.J., Heimer, D.C., Paterson, S.J., Gu, H., Hazlett, H.C., Piven, J., 2013b. Frontolimbic neural circuitry at 6 months predicts individual differences in joint attention at 9 months. Dev. Sci. 16 (2), 186–197. https://doi.org/10.1111/desc.12015.

Eluvathingal, T.J., et al., 2007. Quantitative diffusion tensor tractography of association and projection fibers in normally developing children and adolescents. Cereb. Cortex 17 (12), 2760–2768.

Ezzati, A., Katz, M.J., Lipton, M.L., Zimmerman, M.E., Lipton, R.B., 2016. Hippocampal volume and cingulum bundle fractional anisotropy are independently associated with verbal memory in older adults. Brain Imaging Behav. 10, 652–659. https://doi.org/10.1007/s11682-015-9452-y.

Fabius, J., Ten Brink, A.F., der Stigchel, S., Nijboer, T.C.W., 2020. The relationship between visuospatial neglect, spatial working memory and search behavior. J. Clin. Exp. Neuropsychol. 42 (3), 251–262. https://doi.org/10.1080/13803395.2019.1707779.

Fellgiebel, A., Müller, M.J., Wille, P., Dellani, P.R., Scheurich, A., Schmidt, L.G., Stoeter, P., 2005. Color-coded diffusion-tensor-imaging of posterior cingulate fiber tracts in mild cognitive impairment. Neurobiol. Aging 26, 1193–1198. https://doi.org/10.1016/j.neurobiolaging.2004.11.006.

Filley, C.M., 2012. The Behavioural Neurology of White Matter, second ed. Oxford University Press, New York.

Filley, C.M., Fields, R.D., 2016. White matter and cognition: making the connection. J. Neurophysiol. 116 (5), 2093–2104.

Fletcher, P.T., et al., 2010. Microstructural connectivity of the arcuate fasciculus in adolescents with high-functioning autism. NeuroImage 51 (3), 1117–1125.

Forkel, S.J., de Schotten, M.T., Dell'Acqua, F., Kalra, L., Murphy, D.G.M., Williams, S.C.R., Catani, M., 2014. Anatomical predictors of aphasia recovery: a tractography study of bilateral perisylvian language networks. Brain 137 (7), 2027–2039. https://doi.org/10.1093/brain/awu113.

Forkel, S., Rogalski, E., Drossinos Sancho, N., et al., 2020. Anatomical evidence of an indirect pathway for word repetition. Neurology 94 (6), e594–e606.

Forkel, S.J., et al., 2022. White matter variability, cognition, and disorders: a systematic review. Brain Struct. Funct. 227, 529–544.

Fornia, L., Rossi, M., Rabuffetti, M., et al., 2020. Direct electrical stimulation of premotor areas: Different effects on hand muscle activity during object manipulation. Cereb. Cortex 30 (1), 391–405.

Foulon, C., Cerliani, L., Kinkingnéhun, S., Levy, R., Rosso, C., Urbanski, M., Volle, E., de Schotten, M., 2018. Advanced lesion symptom mapping analyses and implementation as *BCBtoolkit*. Gigascience 7 (3), 1–17. https://doi.org/10.1093/gigascience/giy004.

Fox, M.D., Corbetta, M., Snyder, A.Z., Vincent, J.L., Raichle, M.E., 2006. Spontaneous neuronal activity distinguishes human dorsal and ventral attention systems. Proc. Natl. Acad. Sci. U. S. A. 103 (26), 10046–10051. https://doi.org/10.1073/pnas.0604187103.

Frith, C., Frith, U., 2010. Learning from others: introduction to the special review series on social neuroscience. Neuron 65 (6), 739–743. https://doi.org/10.1016/j.neuron.2010.03.015 (S0896-6273(10)00186-8 [pii]).

Fuelscher, I., Hyde, C., Efron, D., et al., 2021. Manual dexterity in late childhood is associated with maturation of the corticospinal tract. NeuroImage 226, 117583.

Fujie, S., Namiki, C., Nishi, H., Yamada, M., Miyata, J., Sakata, D., Sawamoto, N., et al., 2008. The role of the uncinate fasciculus in memory and emotional recognition in amnestic mild cognitive impairment. Dement. Geriatr. Cogn. Disord. 26 (5), 432–439.

Guevara, M., Guevara, P., Román, C., Mangin, J.F., 2020. Superficial white matter: a review on the dMRI analysis methods and applications. Neuroimage 15 (212), 116673.

Herbert, G., Duggau, H., 2020. Revisiting the functional anatomy of the human brain: toward a meta-networking theory of cerebral functions. Physiol. Rev. 100 (3), 1181–1228.

Hickok, G., Poeppel, D., 2000. Towards a functional neuroanatomy of speech perception. Trends Cogn. Sci. 4 (4), 131–138.

Hickok, G., Poeppel, D., 2004. Dorsal and ventral streams: a framework for understanding aspects of the functional anatomy of language. Cognition 92 (1–2), 67–99.

Honey, C.J., et al., 2009. Predicting human resting-state functional connectivity from structural connectivity. Proc. Natl. Acad. Sci. USA 106 (6), 2035–2040.

Hornberger, M., Geng, J., Hodges, J.R., 2011. Convergent grey and white matter evidence of orbitofrontal cortex changes related to disinhibition in behavioural variant frontotemporal dementia. Brain 134, 2502–2512. https://doi.org/10.1093/brain/awr173.

Howells, H., Thiebaut de Schotten, M., Dell'Acqua, F., et al., 2018. Frontoparietal tracts linked to lateralized hand preference and manual specialization. Cereb. Cortex 28 (7), 2482–2494.

Howells, H., Simone, L., Borra, E., et al., 2020. Reproducing macaque lateral grasping and oculomotor networks using resting state functional connectivity and diffusion tractography. Brain Struct. Funct. 225 (8), 2533–2551.

Huang, H., Zhang, J., Wakana, S., Zhang, W., Ren, T., Richards, L.J., Yarowsky, P., et al., 2006. White and gray matter development in human fetal, newborn and pediatric brains. NeuroImage 33 (1), 27–38.

Jones, D.J.K., Knosche, T.R., Turner, R., 2013. White matter integrity, fiber count, and other fallacies: the do's and don'ts of diffusion MRI. Neuroimage 73, 239–254.

Jou, R.J., et al., 2011. Diffusion tensor imaging in autism spectrum disorders: preliminary evidence of abnormal neural connectivity. Aust. N. Z. J. Psychiatry 45 (2), 153–162.

Keller, T.A., Just, M.A., 2009. Altering cortical connectivity: remediation-induced changes in the white matter of poor readers. Neuron 64 (5), 624–631.

Kelly, C.E., Thompson, D.K., Cooper, M., Pham, J., Nguyen, T.D., Yang, J.Y.M., Ball, G., Adamson, C., Murray, A.L., Chen, J., Inder, T.E., Cheong, J.-L.Y., Doyle, L.W., Anderson, P.J., 2021. White matter tracts related to memory and emotion in very preterm children. Pediatr. Res. 89 (6), 1452–1560.

Kier, E.L., Staib, L.H., Davis, L.M., Bronen, R.A., 2004. MR imaging of the temporal stem: anatomic dissection tractography of the uncinate fasciculus, inferior occipitofrontal fasciculus, and Meyer's loop of the optic radiation. AJNR Am. J. Neuroradiol. 25 (5), 677–691.

Koldewyn, K., Yendiki, A., Weigelt, S., Gweon, H., Julian, J., Richardson, H., Malloy, C., et al., 2014. Differences in the right inferior longitudinal fasciculus but no general disruption of white matter tracts in children with autism spectrum disorder. Proc. Natl. Acad. Sci. U. S. A. 111 (5), 1981–1986.

Kumar, A., Sundaram, S.K., Sivaswamy, L., Behen, M.E., Makki, M.I., Ager, J., Janisse, J., et al., 2010. Alterations in frontal lobe tracts and corpus callosum in young children with autism spectrum disorder. Cereb. Cortex 20 (9), 2103–2113.

Lai, G., et al., 2012. Neural systems for speech and song in autism. Brain 135 (Pt 3), 961–975.

Lawrence, A.J., Chung, A.W., Morris, R.G., Markus, H.S., Barrick, T.R., 2014. Structural network efficiency is associated with cognitive impairment in small vessel disease. Neurology 83, 304–311. https://doi.org/10.1212/WNL.0000000000000612.

Lebel, C., Walker, L., Leemans, A., Phillips, L., Beaulieu, C., 2008. Microstructural maturation of the human brain from childhood to adulthood. NeuroImage 40 (3), 1044–1055.

Leube, D.T., Weis, S., Freymann, K., Erb, M., Jessen, F., Heun, R., Grodd, W., Kircher, T.T., 2008. Neural correlates of verbal episodic memory in patients with MCI and Alzheimer's disease—a VBM study. Int. J. Geriatr. Psychiatry 23, 1114–1118. https://doi.org/10.1002/gps.2036.

Levine, B., Black, S.E., Cabeza, R., Sinden, M., Mcintosh, A.R., Toth, J.P., Tulving, E., Stuss, D.T., 1998. Episodic memory and the self in a case of isolated retrograde amnesia. Brain 121, 1951–1973. https://doi.org/10.1093/brain/121.10.1951.

Levy, D., Ronemus, M., Yamrom, B., Lee, Y.H., Leotta, A., Kendall, J., Wigler, M., 2011. Rare de novo and transmitted copy-number variation in autistic spectrum disorders. Neuron 70 (5), 886–897 (S0896-6273(11)00396-5 [pii]).

Lopez-Barroso, D., Catani, M., Ripolles, P., et al., 2013. Word learning is mediated by the left arcuate fasciculus. Proc. Natl. Acad. Sci. USA 110 (32), 13168–13173.

Lunven, M., Thiebaut De Schotten, M., Bourlon, C., Duret, C., Migliaccio, R., Rode, G., Bartolomeo, P., 2015. White matter lesional predictors of chronic visual neglect: a longitudinal study. Brain 138 (3), 746–760. https://doi.org/10.1093/brain/awu389.

Makris, N., Kennedy, D.N., McInerney, S., et al., 2005. Segmentation of subcomponents within the superior longitudinal fascicle in humans: a quantitative, in vivo, DT-MRI study. Cereb. Cortex 15 (6), 854–869.

Mega, M.S., Cummings, J.L., Salloway, S., Malloy, P., 1997. The limbic system: an anatomic, phylogenetic, and clinical perspective. J. Neuropsychiatry Clin. Neurosci. 9 (3), 315–330.

Mesulam, M.M., 1990. Large-scale neurocognitive networks and distributed processing for attention, language and memory. Ann. Neurol. 28 (5), 597–613.

Mesulam, M., 2000. Behavioral neuroanatomy: large-scale networks, association cortex, frontal syndromes, the limbic system, and hemispheric specializations. In: Mesulam, M. (Ed.), Principles of Behavioural and Cognitive Neurology. Oxford University Press, New York, pp. 1–119.

Metzler-Baddeley, C., Jones, D.K., Belaroussi, B., Aggleton, J.P., O'Sullivan, M.J., 2011. Frontotemporal connections in episodic memory and aging: a diffusion MRI tractography study. J. Neurosci. 31, 13236–13245. https://doi.org/10.1523/JNEUROSCI.2317-11.2011.

Metzler-Baddeley, C., Hunt, S., Jones, D.K., Leemans, A., Aggleton, J.P., O'Sullivan, M.J., 2012a. Temporal association tracts and the breakdown of episodic memory in mild cognitive impairment. Neurology 79, 2233–2240. https://doi.org/10.1212/WNL.0b013e31827689e8.

Metzler-Baddeley, C., Jones, D.K., Steventon, J., Westacott, L., Aggleton, J.P., O'Sullivan, M.J., 2012b. Cingulum microstructure predicts cognitive control in older age and mild cognitive impairment. J. Neurosci. 32, 17612–17619. https://doi.org/10.1523/JNEUROSCI.3299-12.2012.

Mielke, M.M., Kozauer, N.A., Chan, K.C.G., George, M., Toroney, J., Zerrate, M., Bandeen-Roche, K., Wang, M.C., vanZijl, P., Pekar, J.J., Mori, S., Lyketsos, C.G., Albert, M., 2009. Regionally-specific diffusion tensor imaging in mild cognitive impairment and Alzheimer's disease. NeuroImage 46, 47–55. https://doi.org/10.1016/j.neuroimage.2009.01.054.

Miller, E.K., 2000. The prefrontal cortex and cognitive control. Nat. Rev. Neurosci. 1, 59–65. https://doi.org/10.1038/35036228.

Nathan, P.W., Smith, M.C., Deacon, P., 1990. The corticospinal tracts in man. Course and location of fibres at different segmental levels. Brain 113 (Pt 2), 303–324.

Nobre, C.A., 2001. Orienting attention to instants in time. Neuropsychologia 39 (12), 1317–1328. https://doi.org/10.1016/S0028-3932(01)00120-8.

Nowrangi, M.A., Lyketsos, C.G., Leoutsakos, J.S., Oishi, K., Albert, M., Mori, S., Mielke, M.M., 2013. Longitudinal, region-specific course of diffusion tensor imaging measures in mild cognitive impairment and Alzheimer's disease. Alzheimers Dement. 9, 519–528. https://doi.org/10.1016/j.alz.2012.05.2186.

Pardini, M., Garaci, F.G., Bonzano, L., Roccatagliata, L., Palmieri, M.G., Pompili, E., Coniglione, F., et al., 2009. White matter reduced streamline coherence in young men with autism and mental retardation. Eur. J. Neurol. 16 (11), 1185–1190.

Pardoe, H.R., Kucharsky Hiess, R., Kuzniecky, R., 2016. Motion and morphometry in clinical and nonclinical populations. NeuroImage 135, 177–185.

Parlatini, V., Radua, J., Dell'Acqua, F., Leslie, A., Simmons, A., Murphy, D.G., Catani, M., Thiebaut de Schotten, M., 2017. Functional segregation and integration within fronto-parietal networks. NeuroImage 146, 367–375. https://doi.org/10.1016/j.neuroimage.2016.08.031.

Pedrazzini, E., Ptak, R., 2020. The neuroanatomy of spatial awareness: a large-scale region-of-interest and voxel-based anatomical study. Brain Imaging Behav. 14 (2), 615–626. https://doi.org/10.1007/s11682-019-00213-5.

Pelphrey, K., Adolphs, R., Morris, J.P., 2004. Neuroanatomical substrates of social cognition dysfunction in autism. Ment. Retard. Dev. Disabil. Res. Rev. 10 (4), 259–271. https://doi.org/10.1002/mrdd.20040.

Philippi, C.L., Mehta, S., Grabowski, T., Adolphs, R., Rudrauf, D., 2009. Damage to association fiber tracts impairs recognition of the facial expression of emotion. J. Neurosci. 29 (48), 15089–15099. https://doi.org/10.1523/JNEUROSCI.0796-09.2009 (29/48/15089 [pii]).

Poutska, L., Jennen-Steinmetz, C., Henze, R., Vomistein, K., Haffner, J., Sieltjes, B., 2011. Fronto-temporal disconnectivity and symptom severity in children with autism spectrum disorder. World J. Biol. Psychiatry. Retrieved from http://informahealthcare.com/doi/full/10.3109/15622975.2011.591824.

Psomiades, M., et al., 2016. Integrity of the arcuate fasciculus in patients with schizophrenia with auditory verbal hallucinations: a DTI-tractography study. NeuroImage Clin. 12, 970–975.

Puglisi, G., Howells, H., Sciortino, T., Leonetti, A., Rossi, M., Conti Nibali, M., Gabriel Gay, L., Fornia, L., Bellacicca, A., Viganò, L., Simone, L., Catani, M., Cerri, G., Bello, L., 2019. Frontal pathways in cognitive control: direct evidence from intraoperative stimulation and diffusion tractography. Brain 142, 2451–2465. https://doi.org/10.1093/brain/awz178.

Puig, J., Blasco, G., Daunis-I-Estadella, J., et al., 2013. Decreased corticospinal tract fractional anisotropy predicts long-term motor outcome after stroke. Stroke 44 (7), 2016–2018.

Quentin, R., Chanes, L., Vernet, M., et al., 2015. Fronto-parietal anatomical connections influence the modulation of conscious visual perception by high-beta frontal oscillatory activity. Cereb. Cortex 25 (8), 2095–2101.

Rademacher, J., Burgel, U., Schormann, T., et al., 2001. Variability and asymmetry in the human precentral motor system. A cytoarchitectonic and myeloarchitectonic brain mapping study. Brain 124 (Pt 11), 2232–2258.

Radlinska, B., Ghinani, S., Leppert, I.R., et al., 2010. Diffusion tensor imaging, permanent pyramidal tract damage, and outcome in subcortical stroke. Neurology 75 (12), 1048–1054.

Ramsey, L.E., et al., 2017. Behavioural clusters and predictors of performance during recovery from stroke. Nat. Hum. Behav. 0038.

Reijmer, Y.D., Leemans, A., Caeyenberghs, K., Heringa, S.M., Koek, H.L., Biessels, G.J., 2013. Disruption of cerebral networks and cognitive impairment in Alzheimer disease. Neurology 80, 1370–1377. https://doi.org/10.1212/WNL.0b013e31828c2ee5.

Roberts, R.E., Anderson, E.J., Husain, M., 2010. Expert cognitive control and individual differences associated with frontal and parietal white matter microstructure. J. Neurosci. 30, 17063–17067. https://doi.org/10.1523/JNEUROSCI.4879-10.2010.

Rojkova, K., Volle, E., Urbanski, M., Humbert, F., Dell'Acqua, F., Thiebaut de Schotten, M., 2016. Atlasing the frontal lobe connections and their variability due to age and education: a spherical deconvolution tractography study. Brain Struct. Funct. 221 (3), 1751–1766. https://doi.org/10.1007/s00429-015-1001-3.

Rose, M., Guzzi, G., Bosco, D., et al., 2012. Motor cortex stimulation in Parkinson's Disease. Neurol. Res. Int. https://doi.org/10.1155/2012/502096.

Rudebeck, S.R., Scholz, J., Millington, R., Rohenkohl, G., Johansen-Berg, H., Lee, A.C.H., 2009. Fornix microstructure correlates with recollection but not familiarity memory. J. Neurosci. 29, 14987–14992. https://doi.org/10.1523/JNEUROSCI.4707-09.2009.

Sahyoun, C.P., et al., 2010. Neuroimaging of the functional and structural networks underlying visuospatial versus linguistic reasoning in high-functioning autism. Neuropsychologia 48 (1), 86–95.

Sarubbo, S., Tate, M., De Benedictis, A., Merler, S., Moritz-Gasser, S., Herbet, G., Duffau, H., 2020. Mapping critical cortical hubs and white matter pathways by direct electrical stimulation: an original functional atlas of the human brain. NeuroImage 205, 116237. https://doi.org/10.1016/j.neuroimage.2019.116237.

Schmahmann, J.D., Pandya, D.N., 2006. Fiber Pathways of the Brain. Oxford University Press, New York.

Schmahmann, J.D., Pandya, D.N., Wang, R., Dai, G., D'Arceuil, H.E., De Crespigny, A.J., Wedeen, V.J., 2007. Association fibre pathways of the brain: parallel observations from diffusion spectrum imaging and autoradiography. Brain 130 (Pt3), 630–653.

Sexton, C.E., Kalu, U.G., Filippini, N., Mackay, C.E., Ebmeier, K.P., 2011. A meta-analysis of diffusion tensor imaging in mild cognitive impairment and Alzheimer's disease. Neurobiol. Aging 32, 2322.e5–2322.e18. https://doi.org/10.1016/j.neurobiolaging.2010.05.019.

Shaikh, I., Beaulieu, C., Gee, M., McCreary, C.R., Beaudin, A.E., Valdés-Cabrera, D., Smith, E.E., Camicioli, R., 2022. Diffusion tensor tractography of the fornix in cerebral amyloid angiopathy, mild cognitive impairment and Alzheimer's disease. NeuroImage Clin. 34, 103002. https://doi.org/10.1016/j.nicl.2022.103002.

Shinoura, N., Suzuki, Y., Yamada, R., Tabei, Y., Saito, K., Yagi, K., 2009. Damage to the right superior longitudinal fasciculus in the inferior parietal lobe plays a role in spatial neglect. Neuropsychologia 47 (12), 2600–2603. https://doi.org/10.1016/j.neuropsychologia.2009.05.010.

Sporns, O., Tononi, G., Kotter, R., 2005. The human connectome: a structural description of the human brain. PLoS Comput. Biol. 1 (4), e42.

Suarez, L.E., Markello, R.D., Betzel, R.F., Misic, B., 2020. Linking structure and function in macroscale brain networks. Trends Cogn. Sci. 24 (4), 302–315.

Supekar, K., Uddin, L.Q., Prater, K., Amin, H., Greicius, M.D., Menon, V., 2010. Development of functional and structural connectivity within the default mode network in young children. NeuroImage 52 (1), 290–301 (S1053-8119(10)00403-9 [pii]).

Szatmari, P., 2011. New recommendations on autism spectrum disorder. BMJ 342, d2456. https://doi.org/10.1136/bmj.d2456.

Tang, S.X., Feng, Q.L., Wang, G.H., Duan, S., Shan, B.C., Dai, J.P., 2017. Diffusion characteristics of the fornix in patients with Alzheimer's disease. Psychiatry Res. Neuroimaging 265, 72–76. https://doi.org/10.1016/j.psychresns.2016.09.012.

Tartaglia, M.C., Zhang, Y., Racine, C., Laluz, V., Neuhaus, J., Chao, L., Kramer, J., Rosen, H., Miller, B., Weiner, M., 2012. Executive dysfunction in frontotemporal dementia is related to abnormalities in frontal white matter tracts. J. Neurol. 259, 1071–1080. https://doi.org/10.1007/s00415-011-6300-x.

Thiebaut de Schotten, M., Urbanski, M., Duffau, H., Volle, E., Lévy, R., Dubois, B., Bartolomeo, P., 2005. Direct evidence for a parietal-frontal pathway subserving spatial awareness in humans. Science 309 (5744), 2226–2228. https://doi.org/10.1126/science.1116251.

Thiebaut de Schotten, M., Dell'Acqua, F., Forkel, S.J., Simmons, A., Vergani, F., Murphy, D.G.M., Catani, M., 2011a. A lateralized brain network for visuospatial attention. Nat. Neurosci. 14 (10), 1245–1246. https://doi.org/10.1038/nn.2905.

Thiebaut de Schotten, M., Ffytche, D.H., Bizzi, A., Dell'Acqua, F., Allin, M., Walshe, M., Murray, R., Williams, S.C., Murphy, D.G.M., Catani, M., 2011b. Atlasing location, asymmetry and inter-subject variability of white matter tracts in the human brain with MR diffusion tractography. Neuro-Image 54 (1), 49–59. https://doi.org/10.1016/j.neuroimage.2010.07.055.

Thiebaut de Schotten, M., Tomaiuolo, F., Aiello, M., Merola, S., Silvetti, M., Lecce, F., Bartolomeo, P., Doricchi, F., 2014. Damage to white matter pathways in subacute and chronic spatial neglect: a group study and 2 single-case studies with complete virtual "in vivo" tractography dissection. Cereb. Cortex 24 (3), 691–706. https://doi.org/10.1093/cercor/bhs351.

Thompson, A., Murphy, D., Dell'Acqua, F., et al., 2017. Impaired communication between the motor and somatosensory homunculus is associated with poor manual dexterity in autism spectrum disorder. Biol. Psychiatry 81 (3), 211–219.

Toba, M.N., Migliaccio, R., Batrancourt, B., Bourlon, C., Duret, C., Pradat-Diehl, P., Dubois, B., Bartolomeo, P., 2018. Common brain networks for distinct deficits in visual neglect. A combined structural and tractography MRI approach. Neuropsychologia 115, 167–178. https://doi.org/10.1016/j.neuropsychologia.2017.10.018.

Trivedi, R., Gupta, R.K., Shah, V., Tripathi, M., Rathore, R.K., Kumar, M., Narayana, … P.A., 2008. Treatment-induced plasticity in cerebral palsy: a diffusion tensor imaging study. Pediatr. Neurol. 39 (5), 341–349 (S0887-8994(08)00344-5 [pii]).

Uddin, L.Q., Iacoboni, M., Lange, C., Keenan, J.P., 2007. The self and social cognition: the role of cortical midline structures and mirror neurons. Trends Cogn. Sci. 11 (4), 153–157.

Urbanski, M., Thiebaut de Schotten, M., Rodrigo, S., Catani, M., Oppenheim, C., Touze, E., Chokron, S., Meder, J.-F., Levy, R., Dubois, B., Bartolomeo, P., 2008. Brain networks of spatial awareness: evidence from diffusion tensor imaging tractography. J. Neurol. Neurosurg. Psychiatry 79 (5), 598–601. https://doi.org/10.1136/jnnp.2007.126276.

Urbanski, M., Thiebaut de Schotten, M., Rodrigo, S., Oppenheim, C., Touzé, E., Méder, J.-F., Moreau, K., Loeper-Jeny, C., Dubois, B., Bartolomeo, P., 2011. DTI-MR tractography of white matter damage in stroke patients with neglect. Exp. Brain Res. 208 (4), 491–505. https://doi.org/10.1007/s00221-010-2496-8.

Vaessen, M.J., Saj, A., Lovblad, K.O., Gschwind, M., Vuilleumier, P., 2016. Structural white-matter connections mediating distinct behavioral components of spatial neglect in right brain-damaged patients. Cortex 77, 54–68. https://doi.org/10.1016/j.cortex.2015.12.008.

Valenstein, E., Bowers, D., Verfaellie, M., Heilman, K.M., Day, A., Watson, R.T., 1987. Retrosplenial amnesia. Brain 110, 1631–1646. https://doi.org/10.1093/brain/110.6.1631.

Verdon, V., Schwartz, S., Lovblad, K.O., Hauert, C.A., Vuilleumier, P., 2010. Neuroanatomy of hemispatial neglect and its functional components: a study using voxel-based lesion-symptom mapping. Brain 133 (3), 880–894. https://doi.org/10.1093/brain/awp305.

Vergani, F., Ghimire, P., Rajashekhar, D., Dell'Acqua, F., Lavrador, J.P., 2021. Superior longitudinal fasciculus (SLF) I and II: an anatomical and functional review. J. Neurosurg. Sci. 65 (6), 560–565. https://doi.org/10.23736/S0390-5616.21.05327-3.

Verly, M., et al., 2019. The mis-wired language network in children with developmental language disorder: insights from DTI tractography. Brain Imaging Behav. 13 (4), 973–984.

Viganò, L., Fornia, L., Rossi, M., et al., 2019. Anatomo-functional characterisation of the human "hand-knob": a direct electrophysiological study. Cortex 113, 239–254.

Viganò, L., Howells, H., Fornia, L., 2021. Negative motor responses to direct electrical stimulation: behavioral assessment hides different effects on muscles. Cortex 137, 194–204.

Westerhausen, R., et al., 2007. Corticospinal tract asymmetries at the level of the internal capsule: is there an association with handedness? Neuroimage 37 (2), 379–386.

Wheeler, M., 2001. Episodic and autobiographical memory: psychological and neural aspects. In: International Encyclopedia of the Social & Behavioral Sciences. Elsevier, pp. 4714–4717, https://doi.org/10.1016/b0-08-043076-7/03514-2.

Wolff, J.J., Gu, H., Gerig, G., Elison, J.T., Styner, M., Gouttard, S., Botteron, K.N., et al., 2012. Differences in white matter fiber tract development present from 6 to 24 months in infants with autism. Am. J. Psychiatry 169 (6), 589–600.

Yeatman, J.D., Dougherty, R.F., Rykhlevskaia, E., Sherbondy, A.J., Deutsch, G.K., Wandell, B.A., Ben-Shachar, M., 2011. Anatomical properties of the arcuate fasciculus predict phonological and reading skills in children. J. Cogn. Neurosci. 23, 3304–3317.

Yu, J., Lee, T.M.C., 2019. The longitudinal decline of white matter microstructural integrity in behavioral variant frontotemporal dementia and its association with executive function. Neurobiol. Aging 76, 62–70. https://doi.org/10.1016/j.neurobiolaging.2018.12.005.

Zhuang, L., Wen, W., Trollor, J.N., Kochan, N.A., Reppermund, S., Brodaty, H., Sachdev, P., 2012. Abnormalities of the fornix in mild cognitive impairment are related to episodic memory loss. J. Alzheimers Dis. 29, 629–639. https://doi.org/10.3233/JAD-2012-111766.

# Chapter 31

# Neurosurgical applications of clinical tractography

Alberto Bizzi[a], Joseph Yuan-Mou Yang[b], Jahard Aliaga-Arias[c], Flavio Dell'Acqua[d,e],
José Pedro Lavrador[c], and Francesco Vergani[c]

[a]*Fondazione IRCCS Istituto Neurologico Carlo Besta, Milan, Italy,* [b]*Neuroscience Advanced Clinical Imaging Service (NACIS), Department of Neurosurgery, the Royal Children's Hospital, Melbourne, VIC, Australia,* [c]*Department of Neurosurgery, King's College Hospital NHS Foundation Trust, London, United Kingdom,* [d]*NATBRAINLAB, Department of Neuroimaging, Institute of Psychiatry, Psychology and Neuroscience, King's College London, London, United Kingdom,* [e]*Department of Forensics and Neurodevelopmental Sciences, Institute of Psychiatry, Psychology and Neuroscience, King's College London, London, United Kingdom*

## 1 Introduction

Surgical approaches aimed at sparing critical brain functions have evolved significantly in the last century. In the early 20th century, Cushing (1902–12) (Pendleton et al., 2012) and Penfield (Penfield and Boldrey, 1937) were among the first to introduce techniques of brain mapping and monitoring when resecting lesions in eloquent brain areas. With the advent of microneurosurgery (Yaşargil, 1999), new surgical approaches and refined surgical techniques have been made available increasing the safety of brain surgery. Today, modern neurosurgery relies on the interaction between advanced imaging and intraoperative monitoring (IOM) techniques for an accurate function-sparing surgery, both at cortical and subcortical level.

Among different imaging techniques, the use of structural connectivity information provided by diffusion tractography offers better understanding of the organization of subcortical white matter and large-scale brain networks at the individual level. As testament and recognition of the clinical relevance of white matter connectivity, tractography has now been applied in the operating theater across different neurosurgical areas, including neuro-oncology (Hart et al., 2017; Sparacia et al., 2020), epilepsy (Busby et al., 2019; Maallo et al., 2020), and neurovascular surgery (Zanello et al., 2017; Kazumata et al., 2017). Moreover, when also combined with functional data from blood oxygen level dependent functional MRI (BOLD fMRI), tractography allows the integration of functional mapping at the cortical level with subcortical connectivity, both preoperatively (Sagar et al., 2019; Duffau, 2020) and intraoperatively (Duffau, 2017). This unique combination is considered a fundamental step forward in improving the outcomes of neurosurgery (Chang et al., 2011; Pallud and Dezamis, 2017).

Diffusion MRI tractography of white matter pathways or, in short, white matter tractography, represents a major advancement in surgical planning. Two major contributions are unique to this technique. First, tractography is the only technique that allows the study of human white matter tracts in vivo (Catani et al., 2002; Catani and Thiebaut de Schotten, 2008). In this, tractography differs from Klingler technique, which has been used for many years to study white matter anatomy, but it is only a postmortem dissection technique (Klingler, 1935). Nevertheless, multiple validation studies have shown that tractography results are consistent with those obtained with the Klingler's method (Fig. 1) (Catani et al., 2012; Lawes et al., 2008; Martino et al., 2011), supporting the use of this imaging technique in clinical application. Second, tractography is the only tool able to provide an individual and personalized assessment of subcortical anatomy, based on patients' own anatomy and MRl data. While other mapping techniques, such as navigated transcranial magnetic stimulation (nTMS), BOLD fMRI, and magnetoencephalography, are limited to the functional assessment of the cortex, tractography is the only preoperative technique allowing a personalized mapping of white matter tracts. It is important to emphasize that tractography, as explained in previous chapters of this book, provides gross anatomical information and that its resolution and application are limited to the major white matter tracts. Also, it does not provide information about the function and the direction of action potential propagation (Duffau, 2014).

The impact of this technique spans across multiple neurosurgical subspecialties. While neuro-oncology leads the way in its clinical application, the clinical benefits of tractography in functional and vascular neurosurgery are increasingly being

---

**Handbook of Diffusion MR Tractography. https://doi.org/10.1016/B978-0-12-818894-1.00038-0**
Copyright © 2025 Elsevier Ltd. All rights are reserved, including those for text and data mining, AI training, and similar technologies.

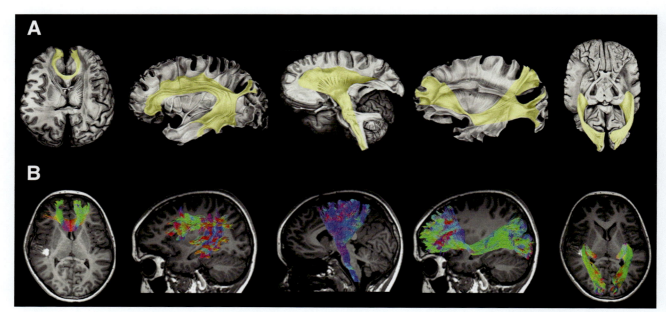

**FIG. 1** (A) Examples of Ludwig and Klingler's classic postmortem fiber dissection of major white matter tracts in humans, highlighted in *yellow*. From left to right: Anterior fibers (forceps minor) of corpus callosum, arcuate fasciculus, corona radiata, corticospinal tract, the inferior fronto-occipital fasciculus, and the optic radiation. (B) Comparable tractography images from an 11-year-old boy with right parietal pediatric high grade glioma. *(Panel (A) reproduced with permission from Yang, J.Y., Yeh, C.H., Poupon, C., Calamante, F., 2021. Diffusion MRI tractography for neurosurgery: the basics, current state, technical reliability and challenges. Phys. Med. Biol., with CC BY 4.0 copyright permission.)*

recognized. This chapter will discuss these advancements and will also explore other less common applications of clinical tractography, such as skull base and pediatric neurosurgery. It is important to note that while the primary role of clinical tractography is for presurgical planning and white matter localization, quantitative analysis of tract-based diffusion metrics, as surrogates for white matter microstructural integrity, have started to emerge in clinical applications. Therefore this chapter will include examples of correlations of diffusion metrics with postoperative functional outcome, as well as examples of tractography applications in neurotrauma.

## 2 Neuro-oncology

Brain tumors have an estimated incidence of about 16 per 100,000 population and represent the first cause of cancer-related death in children and young adults under the age of 40 (UK Office for National Statistics). Gliomas represent the most common form of primary brain tumors in adults; according to their histopathology, gliomas are classified by the WHO into four grades of increasing malignancy (1–4). In recent years, molecular characterization allowed the identification of subgroups of gliomas that are associated with distinct prognosis and predicted treatment response (Stupp and Hegi, 2013). In particular, the diagnostic importance of isocitrate dehydrogenase (IDH) mutational status in diffuse gliomas has been recently recognized (Louis et al., 2016). The incorporation of IDH and other molecular biomarkers into an integrated diagnosis of gliomas provides a more reproducible and clinically meaningful classification of diffuse gliomas in adults (Brat et al., 2020; Eckel-Passow et al., 2015). It is now well recognized that, in contrast to IDH-mutant gliomas, IDH-wildtype astrocytic tumors are distinct clinical and genetic entities with more aggressive clinical behavior (Brat et al., 2020; Louis et al., 2016). IDH-wildtype, grade 4 gliomas, or glioblastoma, are the most aggressive primary brain tumor, with a limited survival. In this context, maximal surgical resection of brain tumors can play a significant role in improving overall and progression-free survival (Jakola et al., 2012; Stummer et al., 2006; de Leeuw and Vogelbaum, 2019; Jackson et al., 2020; Ius et al., 2022). However, one of the factors that frequently limits the extent of resection is the location of tumors in close relationship to eloquent brain areas, where surgery carries an increased risk of inducing permanent neurological deficits. It has been demonstrated that diffuse infiltrative low-grade gliomas have the tendency to affect eloquent regions (Duffau and Capelle, 2004), while a significant percentage of patients (up to 30% of all gliomas in a general neuro-oncology population) will require IOM at surgery (Giamouriadis et al., 2020). The need to mediate between the extent of resection and the risk of inducing neurological deficits is at the heart of modern neuro-oncology and represents the concept of "onco-functional balance" (Duffau and Capelle, 2004). In this context, the development of tractography has played a role in different aspects of neuro-oncological surgery, with reference to surgical planning, risk-stratification of surgical candidates, and patient counseling.

Visualization of the subcortical anatomy for surgical planning is perhaps the major contribution given by tractography to neuro-oncological surgery. The elegant and colorful maps are a very useful tool for risk assessment and the surgical planning of many neurosurgical procedures. The maps are often critical to the neurosurgeon to establish an informed agreement with the patient by balancing the oncological objectives with preservation of function. The reconstructed white matter tracts can be used with available neuronavigation systems for direct surgical planning. In addition, the dissected white matter pathways can be virtually superimposed onto the brain via intraoperative microscope (Lavrador et al., 2020a,b; Kuhnt et al., 2012) (see Fig. 2). The reliability of tractography in delineating the anatomy of major white matter tracts in relation to tumors has been established by multiple groups comparing the preoperative location of motor and language tracts with intraoperative direct electric stimulation (Bello et al., 2010; González-Darder et al., 2010; Henderson et al., 2020). Similar work has been performed comparing intraoperative distances to the corticospinal tract (CST) defined by IOM, preoperative, and postoperative tractography (Ohue et al., 2012). These studies also highlighted that, with the use of this technique, a decreased number of stimulations is generally required to identify relevant subcortical structures, therefore decreasing the risk of induced seizures and shortening the surgical time (Bello et al., 2010). In addition to IOM, which remains the gold standard for subcortical mapping, other groups have reported the use of tractography in association with other preoperative techniques, such as nTMS (Ille et al., 2021; Raffa et al., 2018a,b; Sollmann et al., 2018) and fMRI (D'Andrea et al., 2016) to improve the delineation of specific white matter tracts. For example, nTMS-seeded tractography has been described to increase the specificity of the reconstructed CST (Raffa et al., 2018a,b), as it uses functional information to guide seeding and anatomical reconstruction of the tract. Other authors have described nTMS-seeded tractography for the delineation of the language network (Ille et al., 2021; Sollmann et al., 2018) even though the variable positive predictive values of nTMS for language challenge its application. One of the major limits in the use of tractography in neuro-oncology is the presence of brain edema and the tumoral mass effect, causing distortion of normal anatomy and artifactual reconstructions. However, advanced algorithms and techniques have been designed to overcome the

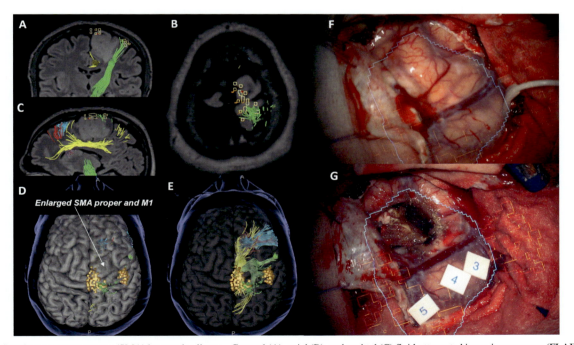

**FIG. 2** Supplementary motor area (SMA) low grade glioma—Coronal (A), axial (B), and sagittal (C) fluid-attenuated inversion recovery (FLAIR) structural MRI images of a WHO Grade 3 oligodendroglioma in the right SMA proper. 3D modeling of the brain and preoperative cortical—navigated transcranial magnetic stimulation (nTMS)—and subcortical mapping—diffusion tensor imaging based tractography (D and E). Intraoperative augmented reality showing overlay of the 3D modeling reconstruction and the brain before (F) and after (G) tumor resection. The subdural strip for continuous motor evoked potential monitoring shown in panel F over the functional area of the upper limb. Correlation between preoperative nTMS and tractography and intraoperative positive stimulation at 7 mA (tags 3, 4, and 5) for the lower limb is documented in panel G. *Blue circular overlay*, tumor model; *blue tract*, frontal aslant tract; *green tract*, corticospinal tract; *red tract*, anterior thalamic radiations; *yellow dots*, nTMS positive responses for upper and lower limbs; *yellow tract*, cingulate.

shortcomings of conventional diffusion tensor imaging (DTI) based tractography in brain tumor surgery, further improving the accuracy of tumor-white matter visualization and surgical planning (Abhinav et al., 2015; Fernandez-Miranda et al., 2012; Yang et al., 2021). Figs. 3 and 4 show a comparison of different tractography methods applied on different patients. The reader can find an exhaustive overview of contemporary diffusion models and algorithms used in tractography in the previous chapters of this book. (See from Chapters 10 to 12 for an overview of different diffusion models and fiber orientation representations; from Chapters 13 to 17 for different tractography algorithms; and finally from Chapters 18 to 22 for more advanced topics and practical consideration of clinical tractography.)

Preoperative patient counseling and prediction of surgical outcome is an extensive area of research (Lavrador et al., 2020a,b; Castellano et al., 2012; Tuncer et al., 2020; Rosenstock et al., 2017a,b; Sollmann et al., 2019, 2020). Understanding the individual relation between the tumor and the surrounding white matter structures is crucial in patient selection (Berntsen et al., 2010) and may change the surgical strategy (Zakaria et al., 2017). Tractography has also been used to predict the extent of resection, since the preoperative infiltration or displacement of some tracts, like the CST and inferior fronto-occipital fasciculus (IFOF), is associated with a lower probability of gross total resection (Castellano et al., 2012). In addition, displacement, or infiltration of specific tracts (such as CST, IFOF, and superior longitudinal fasciculus, SLF) was correlated with an increased risk of a transient neurological deficit after surgery (Bello et al., 2010; Castellano et al., 2012). Finally, tractography-assisted resection has been associated with longer overall survival (Wu et al., 2007) potentially as a surrogate of the previously mentioned impact on neurological deficits and extent of resection.

Either alone or combined with other preoperative cortical mapping techniques, tractography has been used in different risk-stratification scores. Tractography can be helpful in assessing the risk of surgery based on the distance between a specific tract and the lesion that is targeted at surgery (so-called *lesion-to-tract distance*, LTD). For motor function, a score combining the anatomical and functional information provided by nTMS and LTD has recently been published supporting higher preoperative risk for motor deficit in patients with tumors involving the primary motor cortex, abnormal cortical

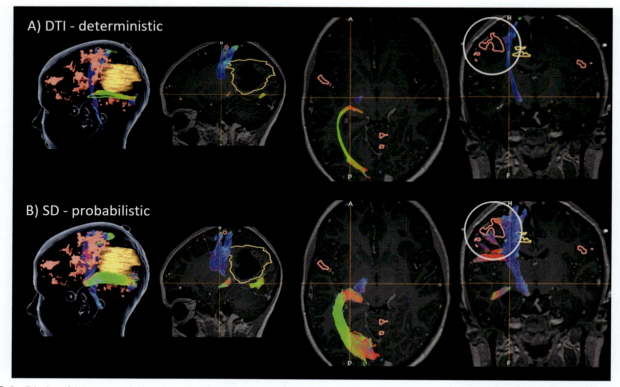

**FIG. 3** Display of the neurosurgical workup using deterministic DTI(A) and probabilistic SD based tractography (B) using the same MRI dataset from an 11-year-old girl with a left parietal high grade glioma. The tumor is rendered in *yellow*, and the finger motor functional MRI activation is in *red*. Larger gaps between the tumor margins and both white matter tracts are visible in Panel A than in B. In Panel A, there is a failure to reconstruct the lateral face-motor projections of the corticospinal tract and lack of spatial overlap between activated finger-motor cortex and the tract, as indicated by the *white circles*. (*Figure modified from Beare, R., Bonnie, A., Aaron, W., Kean, M., et al., 2023. Karawun: a software package for assisting evaluation of advances in multimodal imaging for neurosurgical planning and intraoperative neuronavigation. Int. J. Comput. Assist. Radiol. Surg. 18 (1), 171–179. https://doi.org/10.1007/s11548-022-02736-7.)*

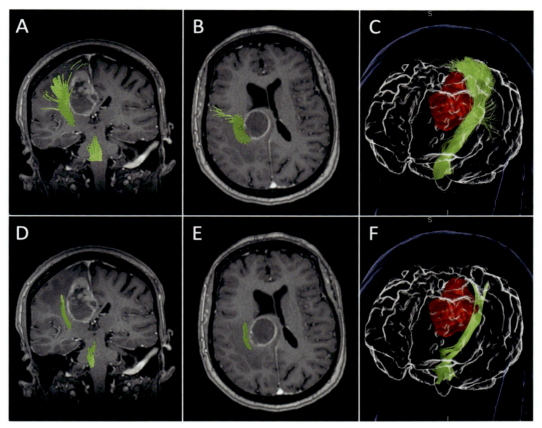

**FIG. 4** Example of the different volumetric extension of spherical deconvolution (SD) and diffusion tensor imaging (DTI) tractography reconstruction of the corticospinal tract (CST) performed for preoperative planning of a case of deep frontal tumor. The panels in the top row show coronal (A), axial (B), and 3D reconstruction (C) of a volumetric T1 postgadolinium sequence with superimposed probabilistic SD tractography of the CST, appearing to be in extended medial contact with the lesion. In the bottom row, on the same anatomical planes and 3D reconstruction (D–F), the CST estimated by deterministic DTI tractography shows a significantly reduced contact with the tumor. Colors legend: *green*, CST; *red*, segmented tumor.

excitability, and distance to CST <8 mm (Rosenstock et al., 2017b) or an LTD threshold ≤12 mm (Sollmann et al., 2020). Prediction of language outcomes is more challenging given the higher complexity of the network involved as well as plasticity potential (Sarubbo et al., 2020; Deverdun et al., 2020; Yogarajah et al., 2010; Southwell et al., 2016). LTD has again been involved in prediction of language outcomes either with region of interest (ROI)-based tractography (Tuncer et al., 2020) or function-based tractography with nTMS (Sollmann et al., 2020). With preoperative infiltration (Tuncer et al., 2020), an LTD ≤16 mm of the arcuate fasciculus (AF) or LTD ≤25 mm for other putative language eloquent tracts—IFOF, uncinate fasciculus (UF), frontal aslant tract (FAT), inferior longitudinal fasciculus (ILF)—have been associated with poor language outcomes (Sollmann et al., 2020). Also, specific anatomical subcortical clusters of injury were related to high-risk persistent aphasic syndromes: the temporo-parieto-occipital junction (AF, IFOF, ILF and the temporo-parietal division of the SLF) and the temporal stem/peri-insular white matter (IFOF, ILF, UF, AF) (Tuncer et al., 2020; Caverzasi et al., 2016). Some safety cut-offs for language-sparing function resection were suggested: LTDs of ≥8 mm for AF and ≥11 mm for SLF, ILF, UF, IFOF (Sollmann et al., 2019).

Evidence of LTD impact on the postoperative risk of motor or language deficits after brain tumor resection has demonstrated overall its potential usefulness in surgical planning and patient counseling. However, substantial variations in estimates of tract dimensions, extension, and proximity to the surgical target are expected when using different tractography algorithms or methods, ranging from the more circumscribed tracts of deterministic DTI to the broadest of probabilistic spherical deconvolution (SD) algorithms (see Figs. 3 and 4). Therefore a main consideration regarding the reliability and efficacy of LTD employment in neurosurgical oncology practice concerns the consistency with the specific tractography methods validated as predictors of the clinical outcome. Figs. 3 and 4 are a clear example of how LTD may change by using different tractography algorithms.

**636** **PART | V** Tractography applications

A complementary approach to prediction of postoperative deficits is based on the evaluation of DTI metrics of potential eloquent tracts. Classically, a tumor can interact with tracts in three different ways: displacement (abnormal morphology and normal fractional anisotropy (FA)), production of vasogenic edema or infiltration (normal morphology with decreased FA), and disruption (inability to dissect the tract in an isotropic area) ( Jellison et al., 2004). FA of the AF has been directly correlated with language outcomes—higher FA in the dominant AF is related to improvement of the postoperative language outcomes as assessed by the Western Aphasia Battery, particularly in the subcategories of naming, reading, and writing (Kinoshita et al., 2014). Similarly, decreased FA (Morita et al., 2011; Li et al., 2013) and increased apparent diffusion coefficient (ADC) were correlated with abnormal motor function in tumors in close relation to the CST at the level of the internal capsule (Morita et al., 2011). Tractography can also provide valuable insight into the interaction between tumors and specific tracts in asymptomatic patients. To this extent, advanced diffusion metrics based on SD methods, such as apparent fiber density (AFD) or hindrance modulated orientational anisotropy (HMOA) (see Chapter 12), have been utilized. In a recent study, changes in HMOA of the CST were found to be correlated with abnormal cortical excitability in asymptomatic patients with a glioma in the motor-eloquent area. Mean and axial diffusivities were also positively correlated with the resting motor threshold as assessed by nTMS (Mirchandani et al., 2020). Furthermore, other studies have similarly demonstrated that both tract-averaged and peritumoral FA reduction and ADC increase, obtained from preoperative nTMS-based CST tractography, can effectively predict postoperative motor deterioration in motor-eloquent adult high-grade gliomas (Rosenstock et al., 2017a). A more recent publication, by the same group, extended the investigation to 65 motor-eloquent high-grade glioma adults. They used AFD extracted along the CST, confirming that advanced diffusion metrics can offer greater specificity in identifying tumor-induced changes when compared to ADC or FA values (Fekonja et al., 2021). In addition to diffusion metrics, tract shape analysis and tract deformations have also shown some potential clinical applications. A recent systematic review and metaanalysis investigated spatial white matter tract alteration patterns reported by tractography studies in glioma patients. It found low-grade gliomas caused tract displacement more often than the high-grade gliomas; while pattern of tract infiltration did not differ between tumor grades (Mahmoodi et al., 2023).

## 3 Intraoperative tractography

The presence of intraoperative brain shift can compromise the accuracy of intraoperative image-guidance using preoperatively reconstructed tractography (Dorward et al., 1998; Nabavi et al., 2001); see Fig. 5. Intraoperative brain shift describes progressive dynamic brain parenchymal deformation resulting from multiple interactive intraoperative factors, such as the resection size, the presence of brain edema, and CSF loss (Romano et al., 2011; Gerard et al., 2017). Acquiring intraoperative MRI (iMRI) data provides updated brain anatomy and addresses the registration inaccuracy caused by brain shift. While other image modalities, such as intraoperative ultrasound, can be used to rapidly assess lesion location and resection status in real time, they do not enable tractography visualization (Sosna et al., 2005; Yeole et al., 2020).

Initial investigations into intraoperative white matter tract shift came primarily from deterministic DTI-based corticospinal tractography studies, using mixed adult and pediatric glioma patients (Nimsky et al., 2006, 2007; Romano et al., 2011). A contemporary pediatric epilepsy and brain tumor surgical series using serial intraoperative high angular resolution diffusion imaging data, combined with a probabilistic SD tracking algorithm (Yang et al., 2017), demonstrated that intraoperative white matter tract shift accounted for approximately half of the ~10 mm registration inaccuracy between IOM validated white matter tract position and preoperative tractography (Berman et al., 2007; Kamada et al., 2005; Mikuni et al., 2007). This 10 mm distance is typically used to define LTD threshold, thus further highlighting the need to consider the different diffusion MRI data types and tracking algorithms, when applying the LTD threshold in an intraoperative setting.

Clinical adoption of intraoperative tractography updates has so far been limited, mostly due to the high costs and logistics of setting up and running an iMRI operating theater. A previous adult mixed low- and high-grade glioma study combining intraoperative corticospinal tractography and IOM, by means of motor evoked potential (MEP), had reported higher gross total resection rates with motor functional preservation (Maesawa et al., 2010). In this study, the distance between MEP responsive sites and intraoperative tractography was significantly correlated with the stimulation intensity but was not with the distance from preoperative tractography.

It is currently not feasible to acquire and process intraoperative advanced diffusion data (i.e., beyond DTI), due to concern about added scanning time prolonging surgical time and related patient's anesthetic and surgical risks. Emerging and ongoing research efforts, such as tractography visualization with real-time feedback (Chamberland et al., 2014; also see real-time tractography mentioned in Chapter 19), semiautomated, supervised intraoperative tractography methods incorporating tumor deformation models (Young et al., 2022), and application of nonlinear registration to preoperatively

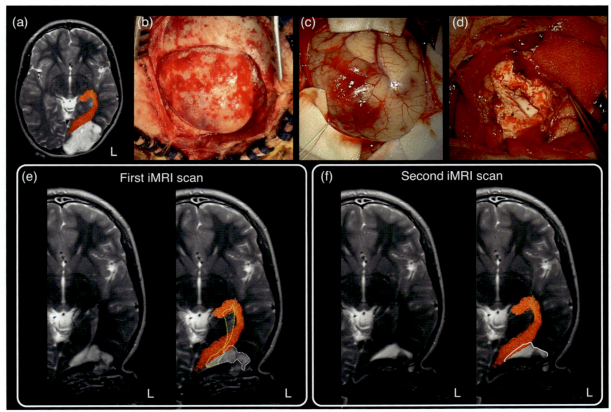

**FIG. 5** A clinical case showing evidence of progressive intraoperative brain shift affecting neuronavigation accuracy using preoperatively reconstructed tractography. This is a 13-year-old girl with drug-resistant epilepsy referable to a left occipital developmental brain tumor (dysembryoplastic neuroepithelial tumor). She had full visual fields on formal perimetry testing before the surgery. (A) Preoperative axial T2-weighted MRI showing the tumor location and with left optic radiation tractography reconstruction *(in orange)*, showing the tract is anteromedially displaced by the tumor. (B–D) Serial intraoperative photos showing dynamic brain shift occurred throughout the surgical course. Different degrees of brain herniations are noted at craniotomy, prior to (B) and following dura opening. (C) The view at the completion of tumor resection, showing the surgical cavity collapses away from the surgeon's view. (D) Note the cavity is lined by both the occipital white matter *(asterisk)* and the resection enters into the occipital horn of the left lateral ventricle *(crosshair)*. Two intraoperative MRI scans were performed, first during (E), and second at completion, confirming gross total tumor resection was achieved (F). The intraoperative MRI is shown with and without the preoperative tractography overlays. In both instances, using this preoperatively prepared tractography would result in inaccurate image-guidance during the surgery. This is evident by the overlaps observed between the tractography, the residual tumor *(dashed white-colored outline)*, and the lateral ventricle *(dashed yellow-colored outline)* in (E); and between the tractography and the surgical cavity *(solid white-color line)* in (F). Abbreviations: *iMRI*, intraoperative MRI; *L*, left. *(Figure adapted and modified from Yang, J.Y., Yeh, C.H., Poupon, C., Calamante, F., 2021. Diffusion MRI tractography for neurosurgery: the basics, current state, technical reliability and challenges. Phys. Med. Biol., with CC BY 4.0 copyright permission.)*

reconstructed tractography images, based on simulated gravity and hydrostatic force dependent brain shift parameters (Negwer et al., 2020) have all shown early promise for intraoperative applications. Advances in MRI hardware and diffusion-weighted imaging (DWI) acquisition schemes, such as simultaneous multislice imaging, also meant advanced diffusion data can now be acquired within comparable time to DTI (Feinberg and Setsompop, 2013).

Similarly, further research is also required to better understand the effects of different diffusion MRI acquisition schemes and data preprocessing pipelines on the tractography accuracy in an intraoperative setting. Intraoperative diffusion data is unique in many ways, such as scans acquired with open cranium, additional imaging artifacts due to the presence of surgical and anesthetic instruments, and brain distortions relating to intraoperative brain shift and surgical resection (see Figs. 5 and 6). These intraoperative factors have potential to complicate the extent and the nature of diffusion echo planar imaging (EPI) distortion in different ways, and thus pose added challenges, compared to using conventional, closed cranium, presurgical MRI data. Recent studies adopting state-of-the-art EPI distortion correction strategies (Yang et al., 2022) and alternative readout segmented EPI sequence with shortened echo train (Elliott et al., 2019) have demonstrated intraoperative diffusion data with less regional EPI susceptibility artifacts and anatomically more faithful image geometries, and its positive impact on tractography accuracy.

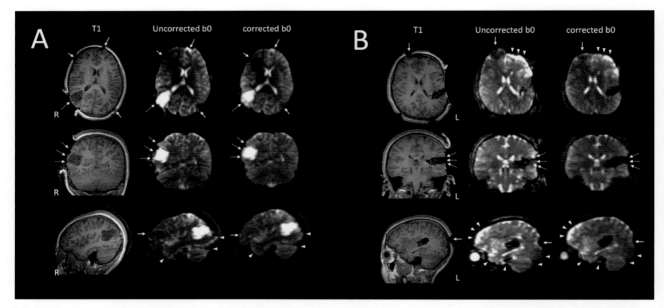

**FIG. 6** Two case examples illustrating the quality of DWI-EPI distortion correction using a state-of-the-art distortion correction technique (Schilling et al., 2019). (A) Intraoperative MRI acquired from a 12-year-old girl with focal drug-resistant epilepsy referable to a right temporoparietal dysembryoplastic neuroepithelial tumor (DNET). The intraoperative MRI was acquired following cranial opening, prior to corticectomy and lesion resection. (B) Intraoperative MRI acquired from a 13-year-old girl with focal drug-resistant epilepsy referable to a left temporoparietal DNET. The intraoperative MRI data was acquired following initial lesion resection. Selected orthogonal T1W images, both distortion corrected and uncorrected b0 images are shown in radiological orientation (*L*, left; *R*, right). Distortion corrections were most notable over the typical frontal, temporal, and occipital regions (*white arrowheads*). Additional distortion corrections are noted at and near the metallic skull pin sites (*solid white arrows*), over the exposed cortical surface at the craniotomy site, at the brain-lesion boundaries, and affected the geometry of the lesions, and the surgical cavities (*dashed white arrows*). (*Figure adapted and modified from Yang, J.Y., Chen, J., Alexander, B., Schilling, K., Kean, M., Wray, A., Seal, M., Maixner, W., Beare, R., 2022. Assessment of intraoperative diffusion EPI distortion and its impact on estimation of supratentorial white matter tract positions in pediatric epilepsy surgery. NeuroImage Clin. 35, 103097, with CC BY 4.0 copyright permission.*)

## 4 Functional neurosurgery

Epilepsy and deep brain stimulation (DBS) surgeries have witnessed significant advances in the last decades due to the introduction of tractography (Szmuda et al., 2016; Piper et al., 2014; Sivakanthan et al., 2016). Significant effort has been made to improve the quality of surgical planning that ultimately will be reflected in better clinical outcomes (See and King, 2017).

Temporal lobe epilepsy (TLE) surgical planning can benefit from tractography for the preoperative identification of Meyer's loop (ML) (Szmuda et al., 2016; Sivakanthan et al., 2016; James et al., 2015). A significant heterogeneity of the anterior extension of the ML has been reported—with some authors suggesting an asymmetry of this tract, with increased anterior extension of the left ML when compared to the right (Winston, 2013). Therefore tractography can assist in a patient-specific planning approach when considering TLE surgery (Yogarajah et al., 2009). The chosen tractography algorithm, diffusion MRI data quality, and image resolution can significantly impact the accuracy of ML reconstruction (Lilja et al., 2014). It has been shown that optic radiation tractography reconstructed for presurgical planning and used for intraoperative neuronavigation (i.e., image guidance) significantly reduces the risk of postoperative visual field deficits, without negatively impacting on the seizure outcomes (Cui et al., 2015). Preservation of the ML has a significant impact on patients' quality-of-life, in particular, the ability to return to driving after surgery, an important outcome determinant in TLE surgery (Winston et al., 2012). Extratemporal epilepsy surgery performed in eloquent brain areas can also benefit from tractography (Radhakrishnan et al., 2011; also see Section 7 for more details). From a prognostication perspective, white matter tract density analysis in TLE surgery has also been suggested as a postsurgical marker of seizure outcome, with an association between residual seizures after surgery and increased tract density in ipsilateral lingual, temporal pole, pars opercularis, inferior parietal, and contralateral frontal pole (Alizadeh et al., 2019).

Recently, tractography has been applied to refine the targeting strategies in DBS for both movement disorders and psychosurgery (Rodrigues et al., 2018; Davidson et al., 2020). Three main approaches summarize the tractography-based approach to surgical planning and clinical research in DBS: (i) tract proximity analysis (correlation of distance between the electrode and the putative functional tracts), (ii) tract activation modeling (tracts generated by seeding in a region of interest within the electrode or the whole electrode), and (iii) direct tract targeting (where the functional tract is the target for electrode implantation (Calabrese, 2016). This methodology stems from the understanding that the clinical effects of the neuromodulation produced by DBS are related not only to the stimulated nuclei but also to the surrounding networks. In essential tremor (ET), targeting the dentato-rubro-thalamic tract (DRTT) was correlated with an increase clinical efficacy derived from electrode implantation in the thalamus (DRTT has terminations in both ventral intermediate—a classical target for ET—and ventral posterior oral nucleus) (Coenen et al., 2011, 2014; King et al., 2017; O'Halloran et al., 2016). Similarly, in Parkinson's disease patients, DBS of white matter structures around the subthalamic nucleus (Zona Incerta and Forel's Fields) has been found to be also clinically effective (Plaha et al., 2006; Vergani et al., 2007). In psychosurgery, the studies are more heterogeneous and include a small number of patients (Schlaepfer et al., 2013; Bewernick et al., 2017; Riva-Posse et al., 2014, 2017; Fenoy et al., 2016). However, tractography-based targeting of the superior and lateral medial forebrain bundle (MFB) (Schlaepfer et al., 2013; Bewernick et al., 2017) and the subgenual area for concomitant modulation of the forceps minor, cingulum, uncinate, MFB, and Brodmann Area 25 (Riva-Posse et al., 2014, 2017; Vergani et al., 2016) have produced promising short-term (Schlaepfer et al., 2013; Riva-Posse et al., 2014, 2017) and long-term (Bewernick et al., 2017; Fenoy et al., 2016) results with an increase in the rate of responders to DBS in treatment-resistant depression. Emergent applications of these tractography-based targeting models in psychosurgery range from pathological aggressiveness (stimulation of the posterior medial hypothalamus with neuromodulation of the MFB (Torres et al., 2020)), posttraumatic stress disorder (white-matter blueprint subcallosal cingulum (SCC) targeting scheme (Hamani et al., 2020)), pain (spinothalamic tract (Hunsche et al., 2013) and gray periaqueductal area (Owen et al., 2008; Boccard et al., 2015)) and obsessive-compulsive disorder (limbic and associative pathways in the anterior limb of the internal capsule (Avecillas-Chasin et al., 2021)). Preoperative tractography has been recently used also to predict surgical response in both obsessive-compulsive disorder (stronger dorsolateral prefrontal-thalamic structural connectivity) and anxiety disorder (stronger dorsal cingulate-thalamic structural connectivity (Zhang et al., 2021)). In summary, there are promising data suggesting that tractography can improve DBS-surgery outcomes, targeting specific white matter tracts.

## 5 Neurovascular

Arteriovenous malformations (AVMs) are vascular lesions that usually present with intracerebral bleeding. Clinical, anatomical, and intrinsic vascular properties of AVMs have been used to determine risk stratification scores and to assist in decision-making regarding their treatment (Grüter et al., 2021). Recent evidence suggests that tractography can add useful information to surgical planning and in predicting the occurrence of focal neurological deficit in motor eloquent AVMs. In particular, the nidus-to-CST distance and the closest distance to CST have been considered as relevant parameters (Jiao et al., 2017b; Li et al., 2019). Multiple cut-offs for lesion-to-eloquent areas have been proposed in the literature, with 5 mm having been suggested as a safety margin (Jiao et al., 2017b, 2018a,b; Li et al., 2019; Lin et al., 2016). A similar approach has been used to predict visual field deficits and speech disturbances in AVMs in temporo-occipital junction and LTD <3.1 mm has been related to significantly higher risk of neurological deficit (Jiao et al., 2016). A randomized control trial assessed the impact of tractography-assisted navigation in surgical planning and resection of AVMs and the results showed significantly better preservation of the visual fields with an LTD (optic radiations) of 5 mm (Tong et al., 2015). From a language outcome perspective, involvement and proximity to the long segment of the AF have been related to worse language outcomes, as well as a location in the Geschwind's area as opposed to classical Broca's and Wernicke's areas (Jiao et al., 2017a,b). Preliminary postoperative evidence supports potential structural and functional plasticity after resection of Geschwind's area AVMs with increased tract volume and number of streamlines in the contralateral AF or ipsilateral IFOF. Also, significant changes in the structural connectivity of patients with brain AVMs located in Broca's area have been identified with decreased connectivity of the right perisylvian regions but enhanced connectivity in the perifocal areas. In this study, weighted brain structural networks were studied using tractography and graph theory (Li et al., 2020). An example of preoperative tractography for surgical planning of AVM resection is illustrated in Fig. 7.

Similar tract-to-function associations were also demonstrated in motor eloquent cavernoma surgery (Januszewski et al., 2016). A recent randomized control trial assessing the impact of tractography in brainstem cavernoma surgery concluded that surgical morbidity, modified Rankin Score (mRS), and motor focal neurological deficits were significantly lower in the group where tractography was used compared to without using tractography (Li et al., 2018). In another study, the lesion-to-CST distance was statistically related to short-term (cut-off <2.55 mm) and long-term (cut-off <2.30 mm) motor deficits, leading to the overall proposal of 3.0 mm as a safety cut-off (Lin et al., 2018); nTMS-based tractography suggested 1 mm cut-off (Zdunczyk et al., 2020) for new motor deficit. Similar conclusions are not universally supported in the literature, as some studies report no relationship between the lesion-to-CST distance, particularly in brainstem (Flores et al., 2015).

**640 PART | V** Tractography applications

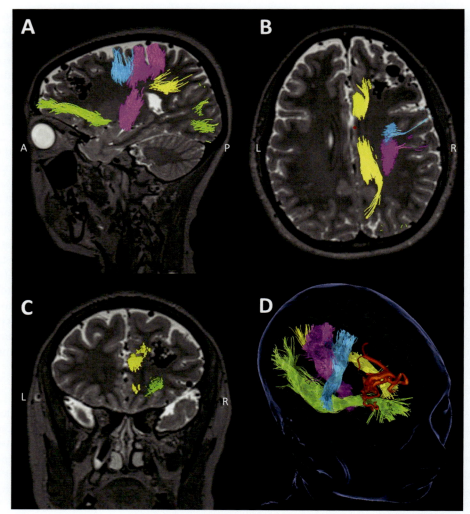

**FIG. 7** Right frontal arteriovenous malformation (AVM) and tractography for surgical planning—sagittal (A), axial (B) , and coronal (C) views of T2-weighted volumetric imaging showing a right frontal AVM with integration with white matter tracts used in surgical planning (*blue*, frontal aslant tract; *green*, inferior fronto-occipital fasciculus; *magenta*, corticospinal tract; *yellow*, cingulate). (D) 3D modeling of the AVM (*dark orange*) with the tracts mentioned for surgical planning and the patient-specific brain surface model. (*Courtesy of Mr. Daniel Walsh, King's College Hospital NHS Foundation Trust, London, United Kingdom.*)

Understanding the impact of the diffusion metrics in the prognostication and preoperative risk stratification in patients with cavernomas is a significant area of research. The literature so far has provided inconsistent results. Some suggested that, even though FA is significantly lower at the level of the lesion, it cannot predict surgical outcome and does not reflect tract-specific changes (not replicated above and below the lesion) (Yao et al., 2015). Other studies suggested consistent rostrocaudal extension of the diffusion metric changes (Faraji et al., 2015), with higher preoperative mean FA values of the CST related to better motor outcomes, as reflected by mRS (Abhinav et al., 2020; Zdunczyk et al., 2020).

## 6 Skull base

Tractography improved the preoperative understanding of the relationship of certain cranial nerves with skull base tumors (Wende et al., 2020). This is particularly true for the facial nerve, where a pooled analysis of vestibular schwannoma series documented that tractography can successfully identify the facial nerve in 96.6% of cases, with an accuracy of 90.6% as demonstrated by IOM. This appears to be particularly relevant in patients with larger lesions (>2.5 cm) (Savardekar et al., 2018). In trigeminal neuralgia, DTI metrics can help in predicting response to treatment as abnormal diffusivity in the trigeminal pontine fibers—increased axial diffusivity, mean diffusivity (MD), and decreased FA in the cisternal component of the trigeminal nerve—is more frequent in nonresponders to surgical treatment (Hung et al., 2017). In addition, attempts have been made to reconstruct the hypothalamic-hypophyseal tract and use it as a surrogate reconstruction for the whole pituitary stalk (Wang et al., 2018). These may be proven useful in the future for surgical planning and for patient counseling with regards to the risk of developing diabetes insipidus.

Reconstructing cranial nerve tractography is challenging and requires higher resolution diffusion data and potentially more sophisticated algorithms to be able to dissect these structures. More research is needed in this area (Behan et al., 2017; Yoshino et al., 2016; Wende et al., 2020).

## 7 Pediatrics

Even though the literature is not as extensive as in adults, the principles of tractography application are similar to what is reported in adults: preoperative planning and risk stratification in brain tumors (Birinyi et al., 2015; Celtikci et al., 2017; Foley and Boop, 2017; Lorenzen et al., 2018), in lesion-based focal drug-resistant epilepsy (Winston et al., 2011), applied tractography alone or combined with other mapping techniques like nTMS (Rosenstock et al., 2020; Schramm et al., 2021) and with BOLD fMRI (Sommer et al., 2013; Yang et al., 2020; Macdonald-Laurs et al., 2021).

Performing tractography in children comes with several challenges. Active brain development and myelination particularly during the first 12 months of life poses a unique challenge of local white matter modeling, which can impact on the tractography reconstruction in this population (Yepes-Calderon et al., 2016). Invasive IOM and brain mapping strategies, including awake craniotomy, are often either not suitable or contraindicated, especially in the young and those with impaired cognition. Similarly, false negative brain mapping may occur due to incomplete myelination and demyelination related to the underlying neuronal migration disorders causing epilepsy (Chitoku et al., 2001; Ng et al., 2009) or due to patient fatigue, as in the case of language mapping in awake surgery (Mandonnet, 2011). The inability to carry out invasive presurgical and surgical evaluation makes tractography integrated multimodal structural and functional imaging more relevant, and the accuracy of these advanced neuroimaging techniques even more paramount in pediatric neurosurgical practice.

There is now emerging evidence coming from selected tertiary pediatric neurosurgical centers reporting favorable seizure and oncological outcomes, with minimal postsurgical functional morbidities in children with focal cortical dysplasia and developmental brain tumors located in the language and motor eloquent areas, including the language dominant operculum and insula (Yang et al., 2020; Macdonald-Laurs et al., 2021). In these series, probabilistic SD-based tractography and BOLD fMRI have been used to inform presurgical planning and integrated with commercial surgical neuronavigation systems to provide intraoperative image-guidance (a.k.a., functional neuronavigation; Beare et al., 2023), without the need for invasive IOM. A case example is shown in Fig. 8.

Another unique aspect of tractography applications in children is that over two-thirds of pediatric brain tumors diagnosed are infratentorial in location (i.e., affecting the cerebellum and/or the brainstem), as opposed to the supratentorial location frequently seen in adult gliomas (Pollack, 1997). Tractography has been used to predict the occurrence of postoperative cerebellar mutism syndrome (POPCMS) after posterior fossa surgery in children, associated with a decrease in FA of the left superior cerebellar peduncle (Vedantam et al., 2019) and increased MD in the inferior olivary nuclei (Yecies et al., 2019). The superior cerebellar peduncle contains proximal efferent cerebellar fibers forming the dentato-thalamo-cortical pathway, the anatomical substrate involved in POPCMS. The inferior olivary nuclei may be involved through its direct connection with the cerebellar dentate nuclei. In children with diffusion pontine gliomas (DPG), along-tract median FA and ADC of the CST and transverse cerebellar-pontine tract were significantly altered compared to healthy controls, and decreased FA was associated with severity of cranial nerve deficits seen in these children (Helton et al., 2006). Another pediatric DPG deterministic DTI-based tractography study demonstrated that serial changes in along-tract FA and MD, combining with tractography evaluation, can be used to assess radiation therapy treatment response and disease progression in these children (Prabhu et al., 2011).

Tractography is also valuable in predicting disease-related functional status and as a preoperative and intraoperative image adjunct for supratentorial glioma surgeries in children (Borja et al., 2013; Moshel et al., 2009). Reduced median FA in optic nerves and optic radiations were significantly associated with poorer visual acuity in children with optic pathway gliomas, with and without neurofibromatosis type 1 (Hales et al., 2018). A retrospective study of optic pathway tractography in children with optic pathway gliomas revealed different patterns of fiber arrangement, based on different tumor locations (i.e., prechiasmatic versus chiasmatic lesions). Such knowledge has potential to further inform surgical neuronavigation in future cases (Lober et al., 2012).

Tractography has also been used in functional neurosurgery planning in children, including standard or tailored temporal lobectomy for TLE (Lacerda et al., 2020) with a particular focus on visual outcomes (Winston et al., 2011). Its use in extratemporal lobe epilepsy surgeries, including resection of focal cortical dysplasia, long-term epilepsy associated tumors, cavernoma, and cerebral gliosis, for motor, language, and visual functional preservation has been reported in both pediatric

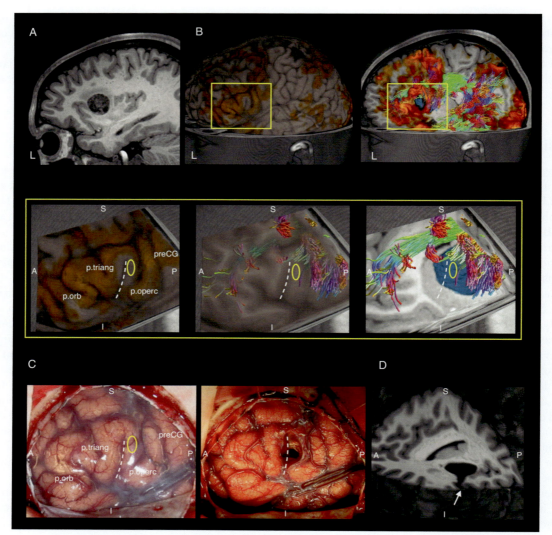

**FIG. 8** This is a 16-year-old boy presenting with ictal expressive aphasia and drug-resistant epilepsy referable to a left (language dominant) insula developmental tumor. (A) Tumor on preoperative sagittal T1-weighted image. (B) Preoperative 3D views of language fMRI activation *(in orange)* and arcuate fasciculus (AF) tractography reconstructed using a SD-based model and a probabilistic tracking algorithm; tumor is in *blue*. The close-up views *(yellow rectangle)* showing the proposed corticotomy site *(yellow ovals)* in the ventral aspect of pars opercularis (a.k.a. the classic "Broca's area"), bordering the vertical ramus of sylvian fissure *(dotted white lines)*, away from the language-activated cortex, and concordant AF terminations. Tractography and fMRI activation blobs were integrated with the surgical neuronavigation system, enabling intraoperative image-guidance using in-house developed software (Beare et al., 2023). (C) Intraoperative craniotomy views, before (left panel) and after (right panel) resection, showing the proposed surgical plan was executed precisely. Gross total tumor resection was achieved through a minimal corticotomy site. (D) Postoperative 3D T1W image render, demonstrating complete resection with minimal overlying cortical disruption *(white arrow)*. Histopathology describes a heavily calcified dysembryoplastic neuroepithelial tumor. This patient had an uneventful course of postoperative recovery, with no evidence of either transient or permanent language deficits. He remained mostly seizure free with occasional aura (Engel 1b) 2 years following the surgery, with good quality of life, and was cleared medically for driving. Abbreviations: *A*, anterior; *I*, inferior; *L*, left; *P*, posterior; *p.operc*, pars opercularis; *p.orb*, pars orbitalis; *p.triang*, pars triangularis; *preCG*, precentral gyrus; *S*, superior.

focused (Yang et al., 2020; Giordano et al., 2017; Macdonald-Laurs et al., 2021) and mixed adult and pediatric epilepsy surgery series (Roessler et al., 2016; Sommer et al., 2013; Sun et al., 2011).

Furthermore, tractography has been used to confirm completeness of callosotomy, and identify candidates with incomplete disconnection, needing further surgeries (Choudhri et al., 2013). Recent case reports have demonstrated the use of tractography to assist tailored presurgical planning, for percutaneous functional hemispherectomy performed through laser interstitial thermal therapy (Mendoza-Elias et al., 2023), and for MR-guided laser ablation surgery of the dominant insula in children (Kaufmann et al., 2021).

Finally, tractography has been applied in traumatic brain injury (TBI) prognostication in pediatric patients (Lindsey et al., 2020; Watson et al., 2019). Despite some specific topics related to pediatric pathology—like the potential impact of hydrocephalus in abnormal global and regional network analysis in the left hemisphere (Yuan et al., 2016)—the principles of clinical tractography applications are similar to those in the adults.

## 8  Traumatic brain injury

While clinical tractography is primarily used for presurgical planning and localization of individual white matter (WM) tracts, the use of quantitative along-tract diffusion metrics has begun to find its way into clinical and clinical research applications.

In particular, diffusion imaging and MRI tractography have been used to provide further insights into white matter damage in TBI patients and to aid prognostication about postinjury mortality and neurocognitive function at injury recovery. Quantitative mapping of diffusion metrics along tract profile provides further information in diffuse axonal injury showing different susceptibilities to injury in different segments of the cingulum ( Jang et al., 2013). DTI metrics derived from the following white matter tracts/systems: cingulum, fornix, hippocampus, dorsolateral prefrontal region, CST, cortico-reticular pathway, ascending reticular activating system, Papez circuit, optic radiation, mammillothalamic tract (Darwazeh et al., 2018; Cho and Jang, 2020; Balasubramanian et al., 2020) have been correlated with cognitive impairment after injury. DTI metrics, such as MD, and streamline metrics, such as average streamline length, number of streamlines, and the general ability to reconstruct tracts, were found to correlate with late mortality after severe TBI (Sener et al., 2016). Higher FA and lower ADC derived from the cingulum were reported to be associated with higher IQ scores in chronic TBI patients (Baek et al., 2013). A recent metaanalysis concluded that higher FA and lower MD/ADC are the most frequently reported DTI metrics associated with better cognition after acute brain injury and that memory and/or attention performances were related to such findings in the corpus callosum, fornix, internal capsule, AF, and UF (Wallace et al., 2018b). The severity of TBI is frequently reported to be associated with a decrease in FA and increase in MD/ADC from the studied white matter tracts (Wallace et al., 2018a).

Tractography may also help provide explanations for patients with decreased consciousness state after acute brain injuries. Balasubramanian et al. reported tractography findings in four patients with spontaneous aneurysmal subarachnoid hemorrhage and bifrontal contusion with acute subdural hematoma. They showed the frontal lobe white matter tracts in these patients had low FA values and reduced tract volumes, in keeping with cognitive impairment reported postoperatively and at injury recovery (Balasubramanian et al., 2020). However, not all studies support the predictive value of the DTI metrics in cognitive impairment after injury, particularly in mild cases ( Jang et al., 2019). For example, a longitudinal study provided preliminary data about recovery of FA in patients with acute injuries but with no direct relation to improvement in disability scores (Edlow et al., 2016). Other studies have investigated changes in DTI metrics in the corpus callosum between multiple time-points after brain injury. In this case, TBI patients continued to show significantly lower FA and axial diffusivity and higher radial diffusivity values when compared with controls (Castaño-Leon et al., 2019). The heterogeneous and multiplicity nature of the injury mechanisms makes summarizing diffusion MRI tractography study findings in TBI patients challenging. More research involving larger patient cohorts with similar injury mechanisms and injury sites are still needed.

## 9  Conclusions

Tractography has emerged as a highly valuable tool for neurosurgery, proving its value in preoperative planning, intraoperative image-guidance, surgical risk evaluation, and postoperative functional and neurocognitive outcome prediction. Over the last decade, tractography has established itself across different neurosurgical disciplines including in neurooncology, vascular, functional, and skull base surgeries and in pediatric neurosurgery. Nevertheless, as with many new technologies, the quality of diffusion MRI data and the sophistication of tractography tools are continuously evolving. Advanced methods and research techniques are either waiting or just starting to be translated into routine clinical practice. A potential emerging area is the combined use of tractography with quantitative diffusion metrics or tract quantification, allowing a better understanding and management of conditions like TBI. As the field progresses, new innovations will further refine our surgical strategies and ultimately enhance patient outcomes.

## References

Abhinav, K., Yeh, F.C., Mansouri, A., Zadeh, G., Fernandez-Miranda, J.C., 2015. High-definition fiber tractography for the evaluation of perilesional white matter tracts in high-grade glioma surgery. Neuro-Oncology 17 (9), 1199–1209. https://doi.org/10.1093/neuonc/nov113. Epub 2015 Jun 27. PMID: 26117712; PMCID.

**644** **PART | V** Tractography applications

Abhinav, K., Nielsen, T.H., Singh, R., Weng, Y., Han, S.S., Iv, M., Steinberg, G.K., 2020. Utility of a quantitative approach using diffusion tensor imaging for prognostication regarding motor and functional outcomes in patients with surgically resected deep intracranial cavernous malformations. Neurosurgery 86 (5), 665–675. https://doi.org/10.1093/neuros/nyz259. PMID: 31360998.

Alizadeh, M., Kozlowski, L., Muller, J., Ashraf, N., Shahrampour, S., Mohamed, F.B., Wu, C., Sharan, A., 2019. Hemispheric regional based analysis of diffusion tensor imaging and diffusion tensor tractography in patients with temporal lobe epilepsy and correlation with patient outcomes. Sci. Rep. 9 (1), 215. https://doi.org/10.1038/s41598-018-36818-x. PMID: 30659215; PMCID: PMC6338779.

Avecillas-Chasin, J.M., Hurwitz, T.A., Bogod, N.M., Honey, C.R., 2021. Tractography-guided anterior capsulotomy for major depression and obsessive-compulsive disorder: targeting the emotion network. Oper. Neurosurg. (Hagerstown). https://doi.org/10.1093/ons/opaa420. opaa420. Epub ahead of print. PMID: 33475697.

Baek, S.O., Kim, O.L., Kim, S.H., Kim, M.S., Son, S.M., Cho, Y.W., Byun, W.M., Jang, S.H., 2013. Relation between cingulum injury and cognition in chronic patients with traumatic brain injury; diffusion tensor tractography study. NeuroRehabilitation 33 (3), 465–471. https://doi.org/10.3233/NRE-130979. PMID: 23949082.

Balasubramanian, S.C., Talluri, S., Kawase, T., Yamada, Y., Murayama, K., Tanaka, R., Miyatani, K., Kojima, D., Kato, Y., 2020. Demystifying white matter injury in the unconscious patients with diffusion tensor imaging. Asian J. Neurosurg. 15 (2), 370–376. https://doi.org/10.4103/ajns. AJNS_55_20. PMID: 32656134; PMCID: PMC7335132.

Beare, R., Bonnie, A., Aaron, W., Kean, M., et al., 2023. Karawun: a software package for assisting evaluation of advances in multimodal imaging for neurosurgical planning and intraoperative neuronavigation. Int. J. Comput. Assist. Radiol. Surg. 18 (1), 171–179. https://doi.org/10.1007/s11548-022-02736-7.

Behan, B., Chen, D.Q., Sammartino, F., DeSouza, D.D., Wharton-Shukster, E., Hodaie, M., 2017. Comparison of diffusion-weighted MRI reconstruction methods for visualization of cranial nerves in posterior fossa surgery. Front. Neurosci. 11, 554. https://doi.org/10.3389/fnins.2017.00554. PMID: 29062268; PMCID: PMC5640769.

Bello, L., Castellano, A., Fava, E., Casaceli, G., Riva, M., Scotti, G., Gaini, S.M., Falini, A., 2010. Intraoperative use of diffusion tensor imaging fiber tractography and subcortical mapping for resection of gliomas: technical considerations. Neurosurg. Focus. 28 (2), E6. https://doi.org/10.3171/2009.12.FOCUS09240. PMID: 20121441.

Berman, J.I., Berger, M.S., Chung, S.W., Nagarajan, S.S., Henry, R.G., 2007. Accuracy of diffusion tensor magnetic resonance imaging tractography assessed using intraoperative subcortical stimulation mapping and magnetic source imaging. J. Neurosurg. 107 (3), 488–494.

Berntsen, E.M., Gulati, S., Solheim, O., Kvistad, K.A., Torp, S.H., Selbekk, T., et al., 2010. Functional magnetic resonance imaging and diffusion tensor tractography incorporated into an intraoperative 3-dimensional ultrasound-based neuronavigation system: impact on therapeutic strategies, extent of resection, and clinical outcome. Neurosurgery 67, 251–264. https://doi.org/10.1227/01.NEU.0000371731.20246.AC.

Bewernick, B.H., Kayser, S., Gippert, S.M., Switala, C., Coenen, V.A., Schlaepfer, T.E., 2017. Deep brain stimulation to the medial forebrain bundle for depression-long-term outcomes and a novel data analysis strategy. Brain Stimul. 10, 664–671.

Birinyi, P.V., Bieser, S., Reis, M., Guzman, M.A., Agarwal, A., Abdel-Baki, M.S., Elbabaa, S.K., 2015. Impact of DTI tractography on surgical planning for resection of a pediatric pre-pontine neurenteric cyst: a case discussion and literature review. Childs Nerv. Syst. 31 (3), 457–463. https://doi.org/10.1007/s00381-014-2587-0. Epub 2014 Nov 19. PMID: 25407831.

Boccard, S.G., Pereira, E.A., Aziz, T.Z., 2015. Deep brain stimulation for chronic pain. J. Clin. Neurosci. 22 (10), 1537–1543. https://doi.org/10.1016/j.jocn.2015.04.005. Epub 2015 Jun 26. PMID: 26122383.

Borja, M.J., Plaza, M.J., Altman, N., Saigal, G., 2013. Conventional and advanced MRI features of pediatric intracranial tumors: supratentorial tumors. AJR Am. J. Roentgenol. 200 (5), W483–W503. https://doi.org/10.2214/AJR.12.9724.

Brat, D.J., Aldape, K., Colman, H., Figrarella-Branger, D., et al., 2020. cIMPACT-NOW update 5: recommended grading criteria and terminologies for IDH-mutant astrocytomas. Acta Neuropathol. 139 (3), 603–608. https://doi.org/10.1007/s00401-020-02127-9.

Busby, N., Halai, A.D., Parker, G.J.M., Coope, D.J., Lambon Ralph, M.A., 2019. Mapping whole brain connectivity changes: the potential impact of different surgical resection approaches for temporal lobe epilepsy. Cortex 113, 1–14. https://doi.org/10.1016/j.cortex.2018.11.003. Epub 2018 Nov 17. PMID: 30557759.

Calabrese, E., 2016. Diffusion tractography in deep brain stimulation surgery: a review. Front. Neuroanat. 10, 45. https://doi.org/10.3389/fnana.2016.00045. PMID: 27199677; PMCID: PMC4852260.

Castaño-Leon, A.M., Cicuendez, M., Navarro, B., Paredes, I., Munarriz, P.M., Cepeda, S., Hilario, A., Ramos, A., Gomez, P.A., Lagares, A., 2019. Longitudinal analysis of corpus callosum diffusion tensor imaging metrics and its association with neurological outcome. J. Neurotrauma 36 (19), 2785–2802. https://doi.org/10.1089/neu.2018.5978. Epub 2019 Jun 7. PMID: 30963801.

Castellano, A., Bello, L., Michelozzi, C., Gallucci, M., Fava, E., Iadanza, A., Riva, M., Casaceli, G., Falini, A., 2012. Role of diffusion tensor magnetic resonance tractography in predicting the extent of resection in glioma surgery. Neuro-Oncology 14 (2), 192–202. https://doi.org/10.1093/neuonc/nor188. Epub 2011 Oct 20. PMID: 22015596; PMCID: PMC3266379.

Catani, M., Dell'acqua, F., Vergani, F., Malik, F., Hodge, H., Roy, P., Valabregue, R., Thiebaut de Schotten, M., 2012. Short frontal lobe connections of the human brain. Cortex 48 (2), 273–291. https://doi.org/10.1016/j.cortex.2011.12.001. Epub 2011 Dec 13. PMID: 22209688.

Catani, M., Thiebaut de Schotten, M., 2008. A diffusion tensor imaging tractography atlas for virtual in vivo dissections. Cortex 44 (8), 1105–1132. https://doi.org/10.1016/j.cortex.2008.05.004.

Catani, M., Howard, R.J., Pajevic, S., Jones, D.K., 2002. Virtual in vivo interactive dissection of white matter fasciculi in the human brain. Neuroimage 17 (1), 77–94. https://doi.org/10.1006/nimg.2002.1136.

Caverzasi, E., Hervey-Jumper, S.L., Jordan, K.M., Lobach, I.V., Li, J., Panara, V., Racine, C.A., Sankaranarayanan, V., Amirbekian, B., Papinutto, N., Berger, M.S., Henry, R.G., 2016. Identifying preoperative language tracts and predicting postoperative functional recovery using HARDI q-ball fiber tractography in patients with gliomas. J. Neurosurg. 125 (1), 33–45. https://doi.org/10.3171/2015.6.JNS142203. Epub 2015 Dec 11. PMID: 26654181.

Celtikci, E., Celtikci, P., Fernandes-Cabral, D.T., Ucar, M., Fernandez-Miranda, J.C., Borcek, A.O., 2017. High-definition fiber tractography in evaluation and surgical planning of thalamopeduncular pilocytic astrocytomas in pediatric population: case series and review of literature. World Neurosurg. 98, 463–469. https://doi.org/10.1016/j.wneu.2016.11.061. Epub 2016 Nov 22. PMID: 27888085.

Chamberland, M., Whittingstall, K., Fortin, D., Mathieu, D., Descoteaux, M., 2014. Real-time multi-peak tractography for instantaneous connectivity display. Front. Neuroinform. 8, 59.

Chang, E.F., Clark, A., Smith, J.S., Polley, M.Y., Chang, S.M., Barbaro, N.M., Parsa, A.T., McDermott, M.W., Berger, M.S., 2011. Functional mapping-guided resection of low-grade gliomas in eloquent areas of the brain: improvement of long-term survival. Clinical article, J. Neurosurg. 114 (3), 566–573. https://doi.org/10.3171/2010.6.JNS091246. Epub 2010 Jul 16. PMID: 20635853; PMCID: PMC3877621.

Chitoku, S., Otsubo, H., Harada, Y., Jay, V., Rutka, J.T., Weiss, S.K., Abdoll, M., Snead 3rd., O.C., 2001. Extraoperative cortical stimulation of motor function in children. Pediatr. Neurol. 24 (5), 344–350.

Cho, M.K., Jang, S.H., 2020. Diffusion tensor imaging studies on spontaneous subarachnoid hemorrhage-related brain injury: a mini-review. Front. Neurol. (11), 283. https://doi.org/10.3389/fneur.2020.00283. PMID: 32411076; PMCID: PMC7198780.

Choudhri, A.F., Whitehead, M.T., McGregor, A.L., Einhaus, S.L., Boop, F.A., Wheless, J.W., 2013. Diffusion tensor imaging to evaluate commissural disconnection after corpus callosotomy. Neuroradiology 55 (11), 1397–1403.

Coenen, V.A., Allert, N., Madler, B., 2011. A role of diffusion tensor imaging fiber tracking in deep brain stimulation surgery: DBS of the dentato-rubro-thalamic tract (DRT) for the treatment of therapy-refractory tremor. Acta Neurochir. 153, 1579–1585. discussion 1585.

Coenen, V.A., Allert, N., Paus, S., Kronenburger, M., Urbach, H., Madler, B., 2014. Modulation of the cerebello-thalamo-cortical network in thalamic deep brain stimulation for tremor: a diffusion tensor imaging study. Neurosurgery 75, 657–669. discussion 669–670.

Cui, Z., Ling, Z., Pan, L., Song, H., Chen, X., Shi, W., Liu, Z., Wang, Q., Zhang, Z., Li, Y., Wang, X., Qing, Y., Xu, X., Mao, Z., Xu, B., Yu, X., Luan, G., 2015. Optic radiation mapping reduces the risk of visual field deficits in anterior temporal lobe resection. Int. J. Clin. Exp. Med. 8 (8), 14283–14295. PMID: 26550412; PMCID: PMC4613097.

D'Andrea, G., Familiari, P., Di Lauro, A., Angelini, A., Sessa, G., 2016. Safe resection of gliomas of the dominant angular gyrus availing of preoperative FMRI and intraoperative DTI: preliminary series and surgical technique. World Neurosurg. 87, 627–639. https://doi.org/10.1016/j.wneu.2015.10.076. Epub 2015 Nov 5. PMID: 26548825.

Darwazeh, R., Wei, M., Zhong, J., Zhang, Z., Lv, F., Darwazeh, M., Guo, Y., Yan, Y., Hoz, S.S., Sun, X., 2018. Significant injury of the mammillothalamic tract without injury of the corticospinal tract after aneurysmal subarachnoid hemorrhage: a retrospective diffusion tensor imaging study. World Neurosurg. 114, e624–e630. https://doi.org/10.1016/j.wneu.2018.03.042. Epub 2018 Mar 13. PMID: 29548966.

Davidson, B., Lipsman, N., Meng, Y., Rabin, J.S., Giacobbe, P., Hamani, C., 2020. The use of tractography-based targeting in deep brain stimulation for psychiatric indications. Front. Hum. Neurosci. 14, 588423. https://doi.org/10.3389/fnhum.2020.588423. PMID: 33304258; PMCID: PMC7701283.

de Leeuw, C.N., Vogelbaum, M.A., 2019. Supratotal resection in glioma: a systematic review. Neuro-Oncology 21 (2), 179–188. https://doi.org/10.1093/neuonc/noy166. PMID: 30321384; PMCID: PMC6374756.

Deverdun, J., van Dokkum, L.E.H., Le Bars, E., Herbet, G., Mura, T., D'agata, B., Picot, M.C., Menjot, N., Molino, F., Duffau, H., Moritz, G.S., 2020. Language reorganization after resection of low-grade gliomas: an fMRI task based connectivity study. Brain Imaging Behav. 14 (5), 1779–1791. https://doi.org/10.1007/s11682-019-00114-7. PMID: 31111301.

Dorward, N.L., Alberti, O., Velani, B., Gerritsen, F.A., Harkness, W.F., Kitchen, N.D., Thomas, D.G., 1998. Postimaging brain distortion: magnitude, correlates, and impact on neuronavigation. J. Neurosurg. 88 (4), 656–662.

Duffau, H., 2014. The dangers of magnetic resonance imaging diffusion tensor tractography in brain surgery. World Neurosurg. 81 (1), 56–58. https://doi.org/10.1016/j.wneu.2013.01.116. Epub 2013 Feb 1. PMID: 23376386.

Duffau, H., 2017. Mapping the connectome in awake surgery for gliomas: an update. J. Neurosurg. Sci. 61 (6), 612–630. https://doi.org/10.23736/S0390-5616.17.04017-6. Epub 2017 Mar 6. PMID: 28263047.

Duffau, H., 2020. Functional mapping before and after low-grade glioma surgery: a new way to decipher various spatiotemporal patterns of individual neuroplastic potential in brain tumor patients. Cancers (Basel) 12 (9), 2611. https://doi.org/10.3390/cancers12092611. PMID: 32933174; PMCID: PMC7565450.

Duffau, H., Capelle, L., 2004. Preferential brain locations of low-grade gliomas. Cancer 100 (12), 2622–2626. https://doi.org/10.1002/cncr.20297. PMID: 15197805.

Eckel-Passow, J.E., Lachance, D.H., Molinaro, A.M., Walsh, K.M., et al., 2015. Glioma groups based on 1p/19q, IDH, and TERT promoter mutations in tumors. N. Engl. J. Med. 372 (26), 2499–2508. https://doi.org/10.1056/NEJMoa1407279.

Edlow, B.L., Copen, W.A., Izzy, S., van der Kouwe, A., Glenn, M.B., Greenberg, S.M., Greer, D.M., Wu, O., 2016. Longitudinal diffusion tensor imaging detects recovery of fractional anisotropy within traumatic axonal injury lesions. Neurocrit. Care. 24 (3), 342–352. https://doi.org/10.1007/s12028-015-0216-8. PMID: 26690938; PMCID: PMC4884487.

Elliott, C.A., Danyluk, H., Aronyk, K.E., Au, K., Wheatley, B.M., Gross, D.W., Sankar, T., Beaulieu, C., 2019. Intraoperative acquisition of DTI in cranial neurosurgery: readout-segmented DTI versus standard single-shot DTI. J. Neurosurg., 1–10.

Faraji, A.H., Abhinav, K., Jarbo, K., Yeh, F.C., Shin, S.S., Pathak, S., Hirsch, B.E., Schneider, W., Fernandez-Miranda, J.C., Friedlander, R.M., 2015. Longitudinal evaluation of corticospinal tract in patients with resected brainstem cavernous malformations using high-definition fiber tractography and diffusion connectometry analysis: preliminary experience. J. Neurosurg. 123 (5), 1133–1144. https://doi.org/10.3171/2014.12.JNS142169. Epub 2015 Jun 5. PMID: 26047420.

Feinberg, D.A., Setsompop, K., 2013. Ultra-fast MRI of the human brain with simultaneous multi-slice imaging. J. Magn. Reson. 229, 90–100.

Fekonja, L.S., Wang, Z., Aydogan, D.B., Roine, T., Engelhardt, M., Dreyer, F.R., Vajkoczy, P., Picht, T., 2021. Detecting corticospinal tract impairment in tumor patients with fiber density and tensor-based metrics. Front. Oncol. 10, 622358.

Fenoy, A.J., Schulz, P., Selvaraj, S., Burrows, C., Spiker, D., Cao, B., Zunta-Soares, G., Gajwani, P., Quevedo, J., Soares, J., 2016. Deep brain stimulation of the medial forebrain bundle: distinctive responses in resistant depression. J. Affect. Disord. 203, 143–151.

Fernandez-Miranda, J.C., Pathak, S., Engh, J., Jarbo, K., Verstynen, T., Yeh, F.C., Wang, Y., Mintz, A., Boada, F., Schneider, W., Friedlander, R., 2012. High-definition fiber tractography of the human brain: neuroanatomical validation and neurosurgical applications. Neurosurgery 71 (2), 430–453. https://doi.org/10.1227/NEU.0b013e3182592faa. PMID: 22513841.

Flores, B.C., Whittemore, A.R., Samson, D.S., Barnett, S.L., 2015. The utility of preoperative diffusion tensor imaging in the surgical management of brainstem cavernous malformations. J. Neurosurg. 122 (3), 653–662. https://doi.org/10.3171/2014.11.JNS13680. Epub 2015 Jan 9. PMID: 25574568.

Foley, R., Boop, F., 2017. Tractography guides the approach for resection of thalamopeduncular tumors. Acta Neurochir. 159 (9), 1597–1601. https://doi.org/10.1007/s00701-017-3257-2. Epub 2017 Jul 3. PMID: 28674731.

Gerard, I.J., Kersten-Oertel, M., Petrecca, K., Sirhan, D., Hall, J.A., Collins, D.L., 2017. Brain shift in neuronavigation of brain tumors: a review. Med. Image Anal. 35, 403–420.

Giamouriadis, A., Lavrador, J.P., Bhangoo, R., Ashkan, K., Vergani, F., 2020. How many patients require brain mapping in an adult neuro-oncology service? Neurosurg. Rev. 43 (2), 729–738. https://doi.org/10.1007/s10143-019-01112-5. Epub 2019 May 19. PMID: 31104183.

Giordano, M., Samii, A., Lawson McLean, A.C., Bertalanffy, H., Fahlbusch, R., Samii, M., Di Rocco, C., 2017. Intraoperative magnetic resonance imaging in pediatric neurosurgery: safety and utility. J. Neurosurg. Pediatr. 19 (1), 77–84.

González-Darder, J.M., González-López, P., Talamantes, F., Quilis, V., Cortés, V., García-March, G., Roldán, P., 2010. Multimodal navigation in the functional microsurgical resection of intrinsic brain tumors located in eloquent motor areas: role of tractography. Neurosurg. Focus. 28 (2), E5. https://doi.org/10.3171/2009.11.FOCUS09234. PMID: 20121440.

Grüter, B.E., Sun, W., Fierstra, J., Regli, L., Germans, M.R., 2021. Systematic review of brain arteriovenous malformation grading systems evaluating microsurgical treatment recommendation. Neurosurg. Rev. https://doi.org/10.1007/s10143-020-01464-3. Epub ahead of print. PMID: 33501562.

Hales, P.W., Smith, V., Dhanoa-Hayre, D., O'Hare, P., Mankad, K., d'Arco, F., Cooper, J., Kaur, R., Phipps, K., Bowman, R., Hargrave, D., Clark, C., 2018. Delineation of the visual pathway in paediatric optic pathway glioma patients using probabilistic tractography, and correlations with visual acuity. NeuroImage Clin. 17, 541–548.

Hamani, C., Davidson, B., Levitt, A., Meng, Y., Corchs, F., Abrahao, A., et al., 2020. Deep Brain Stimulation for treatment resistant post-traumatic stress disorder: a feasibility study. Biol. Psychiatry 1. S0006-3223(20)31624-3. (Epub ahead of print).

Hart, M.G., Price, S.J., Suckling, J., 2017. Functional connectivity networks for preoperative brain mapping in neurosurgery. J. Neurosurg. 126 (6), 1941–1950. https://doi.org/10.3171/2016.6.JNS1662. Epub 2016 Aug 26. PMID: 27564466.

Helton, K.J., Phillips, N.S., Khan, R.B., Boop, F.A., Sanford, R.A., Zou, P., Li, C.S., Langston, J.W., Ogg, R.J., 2006. Diffusion tensor imaging of tract involvement in children with pontine tumors. AJNR Am. J. Neuroradiol. 27 (4), 786–793.

Henderson, F., Abdullah, K.G., Verma, R., Brem, S., 2020. Tractography and the connectome in neurosurgical treatment of gliomas: the premise, the progress, and the potential. Neurosurg. Focus. 48 (2), E6. https://doi.org/10.3171/2019.11.FOCUS19785. PMID: 32006950; PMCID: PMC7831974.

Hung, P.S., Chen, D.Q., Davis, K.D., Zhong, J., Hodaie, M., 2017. Predicting pain relief: use of pre-surgical trigeminal nerve diffusion metrics in trigeminal neuralgia. NeuroImage Clin. (15), 710–718. https://doi.org/10.1016/j.nicl.2017.06.017. PMID: 28702348; PMCID: PMC5491459.

Hunsche, S., Sauner, D., Runge, M.J., Lenartz, D., El Majdoub, F., Treuer, H., Sturm, V., Maarouf, M., 2013. Tractography-guided stimulation of somatosensory fibers for thalamic pain relief. Stereotact. Funct. Neurosurg. 91 (5), 328–334. https://doi.org/10.1159/000350024. Epub 2013 Aug 17. PMID: 23969597.

Ille, S., Schroeder, A., Albers, L., Kelm, A., Droese, D., Meyer, B., Krieg, S.M., 2021. Non-invasive mapping for effective preoperative guidance to approach highly language-eloquent gliomas—a large scale comparative cohort study using a new classification for language eloquence. Cancers (Basel) 13 (2), 207. https://doi.org/10.3390/cancers13020207. PMID: 33430112; PMCID: PMC7827798.

Ius, T., Ng, S., Young, J.S., Tomasino, B., Polano, M., Ben-Israel, D., Kelly, J.J.P., Skrap, M., Duffau, H., Berger, M.S., 2022. The benefit of early surgery on overall survival in incidental low-grade glioma patients: a multicenter study. Neuro-Oncology 24 (4), 624–638. https://doi.org/10.1093/neuonc/noab210. PMID: 34498069; PMCID: PMC8972318.

Jackson, C., Choi, J., Khalafallah, A.M., Price, C., Bettegowda, C., Lim, M., Gallia, G., Weingart, J., Brem, H., Mukherjee, D., 2020. A systematic review and meta-analysis of supratotal versus gross total resection for glioblastoma. J. Neuro-Oncol. 148 (3), 419–431. https://doi.org/10.1007/s11060-020-03556-y. Epub 2020 Jun 19. PMID: 32562247.

Jakola, A.S., Myrmel, K.S., Kloster, R., Torp, S.H., Lindal, S., Unsgård, G., Solheim, O., 2012. Comparison of a strategy favoring early surgical resection vs a strategy favoring watchful waiting in low-grade gliomas. JAMA 308 (18), 1881–1888. https://doi.org/10.1001/jama.2012.12807. PMID: 23099483.

James, J.S., Radhakrishnan, A., Thomas, B., Madhusoodanan, M., Kesavadas, C., Abraham, M., Menon, R., Rathore, C., Vilanilam, G., 2015. Diffusion tensor imaging tractography of Meyer's loop in planning resective surgery for drug-resistant temporal lobe epilepsy. Epilepsy Res. 110, 95–104. https://doi.org/10.1016/j.eplepsyres.2014.11.020. Epub 2014 Nov 27. PMID: 25616461.

Jang, S.H., Kim, S.H., Kim, O.R., Byun, W.M., Kim, M.S., Seo, J.P., Chang, M.C., 2013. Cingulum injury in patients with diffuse axonal injury: a diffusion tensor imaging study. Neurosci. Lett. 543, 47–51. https://doi.org/10.1016/j.neulet.2013.02.058. Epub 2013 Apr 2. PMID: 23562507.

Jang, S.H., Kim, S.H., Lee, H.D., 2019. Traumatic axonal injury of the cingulum in patients with mild traumatic brain injury: a diffusion tensor tractography study. Neural Regen. Res. 14 (9), 1556–1561. https://doi.org/10.4103/1673-5374.255977. PMID: 31089054; PMCID: PMC6557111.

Januszewski, J., Albert, L., Black, K., Dehdashti, A.R., 2016. The usefulness of diffusion tensor imaging and tractography in surgery of brainstem cavernous malformations. World Neurosurg. 93, 377–388. https://doi.org/10.1016/j.wneu.2016.06.019. Epub 2016 Jun 14. PMID: 27312394.

Jellison, B.J., Field, A.S., Medow, J., Lazar, M., Salamat, M.S., Alexander, A.L., 2004. Diffusion tensor imaging of cerebral white matter: a pictorial review of physics, fiber tract anatomy, and tumor imaging patterns. AJNR Am. J. Neuroradiol. 25 (3), 356–369. PMID: 15037456.

Jiao, Y., Lin, F., Wu, J., Li, H., Wang, L., Jin, Z., Wang, S., Cao, Y., 2016. Lesion-to-eloquent fiber distance is a crucial risk factor in presurgical evaluation of arteriovenous malformations in the temporo-occipital junction. World Neurosurg. 93, 355–364. https://doi.org/10.1016/j.wneu.2016.06.059. Epub 2016 Jun 23. PMID: 27345834.

Jiao, Y., Lin, F., Wu, J., Li, H., Chen, X., Li, Z., Ma, J., Cao, Y., Wang, S., Zhao, J., 2017a. Brain arteriovenous malformations located in language area: surgical outcomes and risk factors for postoperative language deficits. World Neurosurg. 105, 478–491. https://doi.org/10.1016/j.wneu.2017.05.159. Epub 2017 Jun 8. PMID: 28602661.

Jiao, Y., Lin, F., Wu, J., Li, H., Chen, X., Li, Z., Ma, J., Cao, Y., Wang, S., Zhao, J., 2017b. Brain arteriovenous malformations located in premotor cortex: surgical outcomes and risk factors for postoperative neurologic deficits. World Neurosurg. 105, 432–440. https://doi.org/10.1016/j.wneu.2017.05.146. Epub 2017 Jun 3. PMID: 28583455.

Jiao, Y., Lin, F., Wu, J., Li, H., Chen, X., Li, Z., Ma, J., Cao, Y., Wang, S., Zhao, J., 2018a. Risk factors for neurological deficits after surgical treatment of brain arteriovenous malformations supplied by deep perforating arteries. Neurosurg. Rev. 41 (1), 255–265. https://doi.org/10.1007/s10143-017-0848-6. Epub 2017 Apr 4. PMID: 28378108.

Jiao, Y., Lin, F., Wu, J., Li, H., Wang, L., Jin, Z., Wang, S., Cao, Y., 2018b. A supplementary grading scale combining lesion-to-eloquence distance for predicting surgical outcomes of patients with brain arteriovenous malformations. J. Neurosurg. 128 (2), 530–540. https://doi.org/10.3171/2016.10. JNS161415. Epub 2017 Mar 31. PMID: 28362235.

Kamada, K., Todo, T., Masutani, Y., Aoki, S., Ino, K., Takano, T., Kirino, T., Kawahara, N., Morita, A., 2005. Combined use of tractography-integrated functional neuronavigation and direct fiber stimulation. J. Neurosurg. 102 (4), 664–672.

Kaufmann, T.J., Lehman, V.T., Wong-Kisiel, L.C., Kerezoudis, P., Miller, K.J., 2021. The utility of diffusion tractography for speech preservation in laser ablation of the dominant insula: illustrative case. J. Neurosurg. Case Lessons 1 (19). CASE21113.

Kazumata, K., Tha, K.K., Uchino, H., Ito, M., Nakayama, N., Abumiya, T., 2017. Mapping altered brain connectivity and its clinical associations in adult moyamoya disease: a resting-state functional MRI study. PLoS One 12 (8), e0182759. https://doi.org/10.1371/journal.pone.0182759. PMID: 28783763; PMCID: PMC5544229.

King, N.K.K., Krishna, V., Sammartino, F., Bari, A., Reddy, G.D., Hodaie, M., Kalia, S.K., Fasano, A., Munhoz, R.P., Lozano, A.M., et al., 2017. Anatomic targeting of the optimal location for thalamic deep brain stimulation in patients with essential tremor. World Neurosurg. 107, 168–174.

Kinoshita, M., Nakada, M., Okita, H., Hamada, J.I., Hayashi, Y., 2014. Predictive value of fractional anisotropy of the arcuate fasciculus for the functional recovery of language after brain tumor resection: a preliminary study. Clin. Neurol. Neurosurg. 117, 45–50. https://doi.org/10.1016/j.clineuro.2013.12.002. Epub 2013 Dec 8. PMID: 24438804.

Klingler, J., 1935. Erleichterung der makroskopischen Präparation des Gehirns durch den Gefrierprozess. Schweiz. Arch. Neurol. Psychiatr. 36, 247–256.

Kuhnt, D., Bauer, M.H., Becker, A., Merhof, D., Zolal, A., Richter, M., Grummich, P., Ganslandt, O., Buchfelder, M., Nimsky, C., 2012. Intraoperative visualization of fiber tracking based reconstruction of language pathways in glioma surgery. Neurosurgery 70 (4), 911–919. discussion 919–920 https://doi.org/10.1227/NEU.0b013e318237a807. PMID: 21946508.

Lacerda, L.M., Clayden, J.D., Handley, S.E., Winston, G.P., Kaden, E., Tisdall, M., Cross, J.H., Liasis, A., Clark, C.A., 2020. Microstructural investigations of the visual pathways in pediatric epilepsy neurosurgery: insights from multi-shell diffusion magnetic resonance imaging. Front. Neurosci. 14, 269. https://doi.org/10.3389/fnins.2020.00269. PMID: 32322185; PMCID: PMC7158873.

Lavrador, J.P., Ghimire, P., Brogna, C., Furlanetti, L., Patel, S., Gullan, R., Ashkan, K., Bhangoo, R., Vergani, F., 2020a. Pre- and intraoperative mapping for tumors in the primary motor cortex: decision-making process in surgical resection. J. Neurol. Surg. A Cent. Eur. Neurosurg. https://doi.org/10.1055/s-0040-1709729. Epub ahead of print. PMID: 32438419.

Lavrador, J.P., Patel, S., Gullan, R., Bhangoo, R., Vergani, F., Ashkan, K., 2020b. Technology in context: a holistic care approach. Clin. Neurophysiol. 131 (2), 577–578. https://doi.org/10.1016/j.clinph.2019.10.026. Epub 2019 Nov 21. PMID: 31791924.

Lawes, I.N., Barrick, T.R., Murugam, V., Spierings, N., Evans, D.R., Song, M., Clark, C.A., 2008. Atlas-based segmentation of white matter tracts of the human brain using diffusion tensor tractography and comparison with classical dissection. NeuroImage 39 (1), 62–79. https://doi.org/10.1016/j.neuroimage.2007.06.041. Epub 2007 Aug 7. PMID: 17919935.

Li, J., Chen, X., Zhang, J., Zheng, G., Lv, X., Li, F., Hu, S., Zhang, T., Xu, B., 2013. Intraoperative diffusion tensor imaging predicts the recovery of motor dysfunction after insular lesions. Neural Regen. Res. 8 (15), 1400–1409. https://doi.org/10.3969/j.issn.1673-5374.2013.15.007. PMID: 25206435; PMCID: PMC4107766.

Li, D., Jiao, Y.M., Wang, L., Lin, F.X., Wu, J., Tong, X.Z., Wang, S., Cao, Y., 2018. Surgical outcome of motor deficits and neurological status in brainstem cavernous malformations based on preoperative diffusion tensor imaging: a prospective randomized clinical trial. J. Neurosurg. 130 (1), 286–301. https://doi.org/10.3171/2017.8.JNS17854. PMID: 29547081.

Li, M., Jiang, P., Guo, R., Liu, Q., Yang, S., Wu, J., Cao, Y., Wang, S., 2019. A tractography-based grading scale of brain arteriovenous malformations close to the corticospinal tract to predict motor outcome after surgery. Front. Neurol. 10, 761. https://doi.org/10.3389/fneur.2019.00761. PMID: 31379715; PMCID: PMC6650564.

Li, M., Jiang, P., Wu, J., Guo, R., Deng, X., Cao, Y., Wang, S., 2020. Altered brain structural networks in patients with brain arteriovenous malformations located in Broca's area. Neural Plast. 2020, 8886803. https://doi.org/10.1155/2020/8886803. PMID: 33163073; PMCID: PMC7604605.

Lilja, Y., Ljungberg, M., Starck, G., Malmgren, K., Rydenhag, B., Nilsson, D.T., 2014. Visualizing Meyer's loop: a comparison of deterministic and probabilistic tractography. Epilepsy Res. 108 (3), 481–490. https://doi.org/10.1016/j.eplepsyres.2014.01.017. Epub 2014 Feb 2. PMID: 24559840.

Lin, F., Zhao, B., Wu, J., Wang, L., Jin, Z., Cao, Y., Wang, S., 2016. Risk factors for worsened muscle strength after the surgical treatment of arteriovenous malformations of the eloquent motor area. J. Neurosurg. 125 (2), 289–298. https://doi.org/10.3171/2015.6.JNS15969. Epub 2015 Dec 4. PMID: 26636384.

**648 PART | V** Tractography applications

Lin, Y., Lin, F., Kang, D., Jiao, Y., Cao, Y., Wang, S., 2018. Supratentorial cavernous malformations adjacent to the corticospinal tract: surgical outcomes and predictive value of diffusion tensor imaging findings. J. Neurosurg. 128 (2), 541–552. https://doi.org/10.3171/2016.10.JNS161179. Epub 2017 Mar 31. PMID: 28362238.

Lindsey, H.M., Lalani, S.J., Mietchen, J., Gale, S.D., Wilde, E.A., Faber, J., MacLeod, M.C., Hunter, J.V., Chu, Z.D., Aitken, M.E., Ewing-Cobbs, L., Levin, H.S., 2020. Acute pediatric traumatic brain injury severity predicts long-term verbal memory performance through suppression by white matter integrity on diffusion tensor imaging. Brain Imaging Behav. 14 (5), 1626–1637. https://doi.org/10.1007/s11682-019-00093-9. PMID: 31134584; PMCID: PMC6879808.

Lober, R.M., Guzman, R., Cheshier, S.H., Fredrick, D.R., Edwards, M.S., Yeom, K.W., 2012. Application of diffusion tensor tractography in pediatric optic pathway glioma. J. Neurosurg. Pediatr. 10 (4), 273–280.

Lorenzen, A., Groeschel, S., Ernemann, U., Wilke, M., Schuhmann, M.U., 2018. Role of presurgical functional MRI and diffusion MR tractography in pediatric low-grade brain tumor surgery: a single-center study. Childs Nerv. Syst. 34 (11), 2241–2248. https://doi.org/10.1007/s00381-018-3828-4. Epub 2018 May 25. PMID: 29802593.

Louis, D.N., Perry, A., Reifenberger, G., von Deimling, A., et al., 2016. The 2016 World Health Organization Classification of Tumors of the Central Nervous System: a summary. Acta Neuropathol. 131, 803–820. https://doi.org/10.1007/s00401-016-1545-1.

Maallo, A.M.S., Granovetter, M.C., Freud, E., Kastner, S., Pinsk, M.A., Patterson, C., Behrmann, M., 2020. Large-scale resculpting of cortical circuits in children after surgical resection. Sci. Rep. 10 (1), 21589. https://doi.org/10.1038/s41598-020-78394-z. PMID: 33299002; PMCID: PMC7725819.

Macdonald-Laurs, E., Maixner, W.J., Bailey, C.A., Barton, S.M., Mandelstam, S.A., Yuan-Mou Yang, J., Warren, A.E.L., Kean, M.J., Francis, P., MacGregor, D., D'Arcy, C., Wrennall, J.A., Davidson, A., Pope, K., Leventer, R.J., Freeman, J.L., Wray, A., Jackson, G.D., Harvey, A.S., 2021. One-stage, limited-resection epilepsy surgery for bottom-of-sulcus dysplasia. Neurology 97 (2), e178–e190.

Maesawa, S., Fujii, M., Nakahara, N., Watanabe, T., Wakabayashi, T., Yoshida, J., 2010. Intraoperative tractography and motor evoked potential (MEP) monitoring in surgery for gliomas around the corticospinal tract. World Neurosurg. 74 (1), 153–161.

Mahmoodi, A.L., Landers, M.J.F., Rutten, G.M., Brouwers, H.B., 2023. Characterization and classification of spatial white matter tract alteration patterns in glioma patients using magnetic resonance tractography: a systematic review and meta-analysis. Cancers (Basel) 15 (14), 3631. https://doi.org/10.3390/cancers15143631.

Mandonnet, E., 2011. Intraoperative electrical mapping: advances, limitations, and perspectives. In: Duffau, H. (Ed.), Brain Mapping From Neural Basis of Cognition to Surgical Applications. Springer, Wien, NY, pp. 101–108.

Martino, J., De Witt Hamer, P.C., Vergani, F., Brogna, C., de Lucas, E.M., Vázquez-Barquero, A., García-Porrero, J.A., Duffau, H., 2011. Cortex-sparing fiber dissection: an improved method for the study of white matter anatomy in the human brain. J. Anat. 219 (4), 531–541. https://doi.org/10.1111/j.1469-7580.2011.01414.x. Epub 2011 Jul 18. PMID: 21767263; PMCID: PMC3196758.

Mendoza-Elias, N., Satzer, D., Henry, J., Nordli Jr., D.R., Warnke, P.C., 2023. Tailored hemispherotomy using tractography-guided laser interstitial thermal therapy. Oper. Neurosurg. (Hagerstown) 24 (6), e407–e413.

Mikuni, N., Okada, T., Enatsu, R., Miki, Y., Hanakawa, T., Urayama, S., Kikuta, K., Takahashi, J.A., Nozaki, K., Fukuyama, H., Hashimoto, N., 2007. Clinical impact of integrated functional neuronavigation and subcortical electrical stimulation to preserve motor function during resection of brain tumors. J. Neurosurg. 106 (4), 593–598.

Mirchandani, A.S., Beyh, A., Lavrador, J.P., Howells, H., Dell'Acqua, F., Vergani, F., 2020. Altered corticospinal microstructure and motor cortex excitability in gliomas: an advanced tractography and transcranial magnetic stimulation study. J. Neurosurg., 1–9. https://doi.org/10.3171/2020.2.JNS192994. Epub ahead of print. PMID: 32357341.

Morita, N., Wang, S., Kadakia, P., Chawla, S., Poptani, H., Melhem, E.R., 2011. Diffusion tensor imaging of the corticospinal tract in patients with brain neoplasms. Magn. Reson. Med. Sci. 10 (4), 239–243. https://doi.org/10.2463/mrms.10.239. PMID: 22214908.

Moshel, Y.A., Elliott, R.E., Monoky, D.J., Wisoff, J.H., 2009. Role of diffusion tensor imaging in resection of thalamic juvenile pilocytic astrocytoma. J. Neurosurg. Pediatr. 4 (6), 495–505.

Nabavi, A., Black, P.M., Gering, D.T., Westin, C.F., Mehta, V., Pergolizzi Jr., R.S., Ferrant, M., Warfield, S.K., Hata, N., Schwartz, R.B., Wells 3rd, W.M., Kikinis, R., Jolesz, F.A., 2001. Serial intraoperative magnetic resonance imaging of brain shift. Neurosurgery 48 (4), 787–797. discussion 797–798.

Negwer, C., Hiepe, P., Meyer, B., Krieg, S.M., 2020. Elastic fusion enables fusion of intraoperative magnetic resonance imaging data with preoperative neuronavigation data. World Neurosurg. 142, e223–e228.

Ng, W.H., Ochi, A., Rutka, J.T., Strantzas, S., Holmes, L., Otsubo, H., 2009. Stimulation threshold potentials of intraoperative cortical motor mapping using monopolar trains of five in pediatric epilepsy surgery. Childs Nerv. Syst. 26 (5), 675–679.

Nimsky, C., Ganslandt, O., Merhof, D., Sorensen, A.G., Fahlbusch, R., 2006. Intraoperative visualization of the pyramidal tract by diffusion-tensor-imaging-based fiber tracking. NeuroImage 30 (4), 1219–1229.

Nimsky, C., Ganslandt, O., Hastreiter, P., Wang, R., Benner, T., Sorensen, A.G., Fahlbusch, R., 2007. Preoperative and intraoperative diffusion tensor imaging-based fiber tracking in glioma surgery. Neurosurgery 61 (1 Suppl), 178–185. discussion 186.

O'Halloran, R.L., Chartrain, A.G., Rasouli, J.J., Ramdhani, R.A., Kopell, B.H., 2016. Case study of image-guided deep brain stimulation: magnetic resonance imaging-based white matter tractography shows differences in responders and nonresponders. World Neurosurg. 96, 613.e9–613.e16.

Ohue, S., Kohno, S., Inoue, A., Yamashita, D., Harada, H., Kumon, Y., Kikuchi, K., Miki, H., Ohnishi, T., 2012. Accuracy of diffusion tensor magnetic resonance imaging-based tractography for surgery of gliomas near the pyramidal tract: a significant correlation between subcortical electrical stimulation and postoperative tractography. Neurosurgery 70 (2), 283–293. discussion 294 https://doi.org/10.1227/NEU.0b013e31823020e6. PMID: 21811189.

Owen, S.L., Heath, J., Kringelbach, M., Green, A.L., Pereira, E.A., Jenkinson, N., Jegan, T., Stein, J.F., Aziz, T.Z., 2008. Pre-operative DTI and probabilistic tractography in four patients with deep brain stimulation for chronic pain. J. Clin. Neurosci. 15 (7), 801–805. https://doi.org/10.1016/j.jocn.2007.06.010. Epub 2008 May 20. PMID: 18495481.

Pallud, J., Dezamis, E., 2017. Functional and oncological outcomes following awake surgical resection using intraoperative cortico-subcortical functional mapping for supratentorial gliomas located in eloquent areas. Neurochirurgie 63 (3), 208–218. https://doi.org/10.1016/j.neuchi.2016.08.003. Epub 2017 Feb 1. PMID: 28161013.

Pendleton, C., Zaidi, H.A., Chaichana, K.L., Raza, S.M., Carson, B.S., Cohen-Gadol, A.A., Quinones-Hinojosa, A., 2012. Harvey Cushing's contributions to motor mapping: 1902-1912. Cortex 48 (1), 7–14. https://doi.org/10.1016/j.cortex.2010.04.006. Epub 2010 Apr 29. PMID: 20510407.

Penfield, W., Boldrey, E., 1937. Somatic motor and sensory representations in the cerebral cortex of man as studied by electrical stimulation. Brain 60, 389–443.

Piper, R.J., Yoong, M.M., Kandasamy, J., Chin, R.F., 2014. Application of diffusion tensor imaging and tractography of the optic radiation in anterior temporal lobe resection for epilepsy: a systematic review. Clin. Neurol. Neurosurg. 124, 59–65. https://doi.org/10.1016/j.clineuro.2014.06.013. Epub 2014 Jun 17. PMID: 25016240.

Plaha, P., Ben-Shlomo, Y., Patel, N.K., Gill, S.S., 2006. Stimulation of the caudal zona incerta is superior to stimulation of the subthalamic nucleus in improving contralateral parkinsonism. Brain 129 (Pt 7), 1732–1747.

Pollack, I.F., 1997. Posterior fossa syndrome. Int. Rev. Neurobiol. 41, 411–432.

Prabhu, S.P., Ng, S., Vajapeyam, S., Kieran, M.W., Pollack, I.F., Geyer, R., Haas-Kogan, D., Boyett, J.M., Kun, L., Poussaint, T.Y., 2011. DTI assessment of the brainstem white matter tracts in pediatric BSG before and after therapy: a report from the Pediatric Brain Tumor Consortium. Childs Nerv. Syst. 27 (1), 11–18.

Radhakrishnan, A., James, J.S., Kesavadas, C., Thomas, B., Bahuleyan, B., Abraham, M., Radhakrishnan, K., 2011. Utility of diffusion tensor imaging tractography in decision making for extratemporal resective epilepsy surgery. Epilepsy Res. 97 (1–2), 52–63. https://doi.org/10.1016/j.eplepsyres.2011.07.003. Epub 2011 Aug 10. PMID: 21835594.

Raffa, G., Conti, A., Scibilia, A., Cardali, S.M., Esposito, F., Angileri, F.F., La Torre, D., Sindorio, C., Abbritti, R.V., Germanò, A., Tomasello, F., 2018a. The impact of diffusion tensor imaging fiber tracking of the corticospinal tract based on navigated transcranial magnetic stimulation on surgery of motor-eloquent brain lesions. Neurosurgery 83 (4), 768–782. https://doi.org/10.1093/neuros/nyx554. PMID: 29211865.

Raffa, G., Quattropani, M.C., Scibilia, A., Conti, A., Angileri, F.F., Esposito, F., Sindorio, C., Cardali, S.M., Germanò, A., Tomasello, F., 2018b. Surgery of language-eloquent tumors in patients not eligible for awake surgery: the impact of a protocol based on navigated transcranial magnetic stimulation on presurgical planning and language outcome, with evidence of tumor-induced intra-hemispheric plasticity. Clin. Neurol. Neurosurg. 168, 127–139. https://doi.org/10.1016/j.clineuro.2018.03.009. Epub 2018 Mar 11. PMID: 29549813.

Riva-Posse, P., Choi, K.S., Holtzheimer, P.E., McIntyre, C.C., Gross, R.E., Chaturvedi, A., Crowell, A.L., Garlow, S.J., Rajendra, J.K., Mayberg, H.S., 2014. Defining critical white matter pathways mediating successful subcallosal cingulate deep brain stimulation for treatment-resistant depression. Biol. Psychiatry 76, 963–969.

Riva-Posse, P., Choi, K.S., Holtzheimer, P.E., Crowell, A.L., Garlow, S.J., Rajendra, J.K., McIntyre, C.C., Gross, R.E., Mayberg, H.S., 2017. A connectomic approach for subcallosal cingulate deep brain stimulation surgery: prospective targeting in treatment-resistant depression. Mol. Psychiatry 23 (4), 843–849. https://doi.org/10.1038/mp.2017.59.

Rodrigues, N.B., Mithani, K., Meng, Y., Lipsman, N., Hamani, C., 2018. The emerging role of tractography in deep brain stimulation: basic principles and current applications. Brain Sci. 8 (2), 23. https://doi.org/10.3390/brainsci8020023. PMID: 29382119; PMCID: PMC5836042.

Roessler, K., Hofmann, A., Sommer, B., Grummich, P., Coras, R., Kasper, B.S., Hamer, H.M., Blumcke, I., Stefan, H., Nimsky, C., Buchfelder, M., 2016. Resective surgery for medically refractory epilepsy using intraoperative MRI and functional neuronavigation: the Erlangen experience of 415 patients. Neurosurg. Focus. 40 (3), E15.

Romano, A., D'Andrea, G., Calabria, L.F., Coppola, V., Espagnet, C.R., Pierallini, A., Ferrante, L., Fantozzi, L., Bozzao, A., 2011. Pre- and intraoperative tractographic evaluation of corticospinal tract shift. Neurosurgery 69 (3), 696–704. discussion 704–695.

Rosenstock, T., Giampiccolo, D., Schneider, H., Runge, S.J., Bahrend, I., Vajkoczy, P., Picht, T., 2017a. Specific DTI seeding and diffusivity-analysis improve the quality and prognostic value of TMS-based deterministic DTI of the pyramidal tract. NeuroImage Clin. 16, 276–285.

Rosenstock, T., Grittner, U., Acker, G., Schwarzer, V., Kulchytska, N., Vajkoczy, P., Picht, T., 2017b. Risk stratification in motor area-related glioma surgery based on navigated transcranial magnetic stimulation data. J. Neurosurg. 126 (4), 1227–1237. https://doi.org/10.3171/2016.4.JNS152896. Epub 2016 Jun 3. PMID: 27257834.

Rosenstock, T., Picht, T., Schneider, H., Vajkoczy, P., Thomale, U.W., 2020. Pediatric navigated transcranial magnetic stimulation motor and language mapping combined with diffusion tensor imaging tractography: clinical experience. J. Neurosurg. Pediatr., 1–11. https://doi.org/10.3171/2020.4. PEDS20174. Epub ahead of print. PMID: 32707554.

Sagar, S., Rick, J., Chandra, A., Yagnik, G., Aghi, M.K., 2019. Functional brain mapping: overview of techniques and their application to neurosurgery. Neurosurg. Rev. 42 (3), 639–647. https://doi.org/10.1007/s10143-018-1007-4. Epub 2018 Jul 13. PMID: 30006663.

Sarubbo, S., Tate, M., De Benedictis, A., Merler, S., Moritz-Gasser, S., Herbet, G., Duffau, H., 2020. Mapping critical cortical hubs and white matter pathways by direct electrical stimulation: an original functional atlas of the human brain. NeuroImage 205, 116237. https://doi.org/10.1016/j.neuroimage.2019.116237. Epub 2019 Oct 15. PMID: 31626897; PMCID: PMC7217287.

Savardekar, A.R., Patra, D.P., Thakur, J.D., Narayan, V., Mohammed, N., Bollam, P., Nanda, A., 2018. Preoperative diffusion tensor imaging-fiber tracking for facial nerve identification in vestibular schwannoma: a systematic review on its evolution and current status with a pooled data analysis of surgical concordance rates. Neurosurg. Focus. 44 (3), E5. https://doi.org/10.3171/2017.12.FOCUS17672. PMID: 29490547.

Schilling, K.G., Blaber, J., Huo, Y., Newton, A., Hansen, C., Nath, V., Shafer, A.T., Williams, O., Resnick, S.M., Rogers, B., Anderson, A.W., Landman, B.A., 2019. Synthesized b0 for diffusion distortion correction (Synb0-DisCo). Magn. Reson. Imaging 64, 62–70.

Schlaepfer, T.E., Bewernick, B.H., Kayser, S., Madler, B., Coenen, V.A., 2013. Rapid effects of deep brain stimulation for treatment-resistant major depression. Biol. Psychiatry 73, 1204–1212.

Schramm, S., Mehta, A., Auguste, K.I., Tarapore, P.E., 2021. Navigated transcranial magnetic stimulation mapping of the motor cortex for preoperative diagnostics in pediatric epilepsy. J. Neurosurg. Pediatr. 28 (3), 287–294.

See, A.A.Q., King, N.K.K., 2017. Improving surgical outcome using diffusion tensor imaging techniques in deep brain stimulation. Front. Surg. (4), 54. https://doi.org/10.3389/fsurg.2017.00054. PMID: 29034243; PMCID: PMC5625016.

Sener, S., Van Hecke, W., Feyen, B.F., Van der Steen, G., Pullens, P., Van de Hauwe, L., Menovsky, T., Parizel, P.M., Jorens, P.G., Maas, A.I., 2016. Diffusion tensor imaging: a possible biomarker in severe traumatic brain injury and aneurysmal subarachnoid hemorrhage? Neurosurgery 79 (6), 786–793. https://doi.org/10.1227/NEU.0000000000001325. PMID: 27352277.

Sivakanthan, S., Neal, E., Murtagh, R., Vale, F.L., 2016. The evolving utility of diffusion tensor tractography in the surgical management of temporal lobe epilepsy: a review. Acta Neurochir. 158 (11), 2185–2193. https://doi.org/10.1007/s00701-016-2910-5. Epub 2016 Aug 26. PMID: 27566714.

Sollmann, N., Kelm, A., Ille, S., Schröder, A., Zimmer, C., Ringel, F., Meyer, B., Krieg, S.M., 2018. Setup presentation and clinical outcome analysis of treating highly language-eloquent gliomas via preoperative navigated transcranial magnetic stimulation and tractography. Neurosurg. Focus. 44 (6), E2. https://doi.org/10.3171/2018.3.FOCUS1838. PMID: 29852769.

Sollmann, N., Fratini, A., Zhang, H., Zimmer, C., Meyer, B., Krieg, S.M., 2019. Associations between clinical outcome and tractography based on navigated transcranial magnetic stimulation in patients with language-eloquent brain lesions. J. Neurosurg. 132 (4), 1033–1042. https://doi.org/10.3171/2018.12.JNS182988. PMID: 30875686.

Sollmann, N., Zhang, H., Fratini, A., Wildschuetz, N., Ille, S., Schröder, A., Zimmer, C., Meyer, B., Krieg, S.M., 2020. Risk assessment by presurgical tractography using navigated TMS maps in patients with highly motor- or language-eloquent brain tumors. Cancers (Basel) 12 (5), 1264. https://doi.org/10.3390/cancers12051264. PMID: 32429502; PMCID: PMC7281396.

Sommer, B., Grummich, P., Coras, R., Kasper, B.S., Blumcke, I., Hamer, H.M., Stefan, H., Buchfelder, M., Roessler, K., 2013. Integration of functional neuronavigation and intraoperative MRI in surgery for drug-resistant extratemporal epilepsy close to eloquent brain areas. Neurosurg. Focus. 34 (4), E4.

Sosna, J., Barth, M.M., Kruskal, J.B., Kane, R.A., 2005. Intraoperative sonography for neurosurgery. J. Ultrasound Med. 24 (12), 1671–1682.

Southwell, D.G., Hervey-Jumper, S.L., Perry, D.W., Berger, M.S., 2016. Intraoperative mapping during repeat awake craniotomy reveals the functional plasticity of adult cortex. J. Neurosurg. 124 (5), 1460–1469. https://doi.org/10.3171/2015.5.JNS142833. Epub 2015 Nov 6. PMID: 26544767.

Sparacia, G., Parla, G., Lo Re, V., Cannella, R., Mamone, G., Carollo, V., Midiri, M., Grasso, G., 2020. Resting-state functional connectome in patients with brain tumors before and after surgical resection. World Neurosurg. 141, e182–e194. https://doi.org/10.1016/j.wneu.2020.05.054. Epub 2020 May 16. PMID: 32428723.

Stummer, W., Pichlmeier, U., Meinel, T., Wiestler, O.D., Zanella, F., Reulen, H.J., ALA-Glioma Study Group, 2006. Fluorescence-guided surgery with 5-aminolevulinic acid for resection of malignant glioma: a randomised controlled multicentre phase III trial. Lancet Oncol. 7 (5), 392–401. https://doi.org/10.1016/S1470-2045(06)70665-9. PMID: 16648043.

Stupp, R., Hegi, M.E., 2013. Brain cancer in 2012: molecular characterization leads the way. Nat. Rev. Clin. Oncol. 10 (2), 69–70. https://doi.org/10.1038/nrclinonc.2012.240.

Sun, G.C., Chen, X.L., Zhao, Y., Wang, F., Song, Z.J., Wang, Y.B., Wang, D., Xu, B.N., 2011. Intraoperative MRI with integrated functional neuronavigation-guided resection of supratentorial cavernous malformations in eloquent brain areas. J. Clin. Neurosci. 18 (10), 1350–1354.

Szmuda, M., Szmuda, T., Springer, J., Rogowska, M., Sabisz, A., Dubaniewicz, M., Mazurkiewicz-Bełdzińska, M., 2016. Diffusion tensor tractography imaging in pediatric epilepsy—a systematic review. Neurol. Neurochir. Pol. 50 (1), 1–6. https://doi.org/10.1016/j.pjnns.2015.10.003. Epub 2015 Oct 29. PMID: 26851683.

Tong, X., Wu, J., Lin, F., Cao, Y., Zhao, Y., Jin, Z., Ning, B., Zhao, B., Li, Y., Wang, L., Zhang, S., Wang, S., Zhao, J., 2015. Visual field preservation in surgery of occipital arteriovenous malformations: a prospective study. World Neurosurg. 84 (5), 1423–1436. https://doi.org/10.1016/j.wneu.2015.06.069. Epub 2015 Jul 3. PMID: 26145824.

Torres, C.V., Blasco, G., Navas García, M., Ezquiaga, E., Pastor, J., Vega-Zelaya, L., Pulido Rivas, P., Pérez Rodrigo, S., Manzanares, R., 2020. Deep brain stimulation for aggressiveness: long-term follow-up and tractography study of the stimulated brain areas. J. Neurosurg., 1–10. https://doi.org/10.3171/2019.11.JNS192608. Epub ahead of print. PMID: 32032944.

Tuncer, M.S., Salvati, L.F., Grittner, U., Hardt, J., Schilling, R., Bährend, I., Silva, L.L., Fekonja, L.S., Faust, K., Vajkoczy, P., Rosenstock, T., Picht, T., 2020. Towards a tractography-based risk stratification model for language area associated gliomas. NeuroImage Clin. 29, 102541. https://doi.org/10.1016/j.nicl.2020.102541. Epub ahead of print. PMID: 33401138; PMCID: PMC7785953.

Vedantam, A., Stormes, K.M., Gadgil, N., Kralik, S.F., Aldave, G., Lam, S.K., 2019. Association between postoperative DTI metrics and neurological deficits after posterior fossa tumor resection in children. J. Neurosurg. Pediatr., 1–7. https://doi.org/10.3171/2019.5.PEDS1912. Epub ahead of print. PMID: 31323626.

Vergani, F., Landi, A., Antonini, A., Parolin, M., Cilia, R., Grimaldi, M., Ferrarese, C., Gaini, S.M., Sganzerla, E.P., 2007. Anatomical identification of active contacts in subthalamic deep brain stimulation. Surg. Neurol. 67 (2), 140–146.

Vergani, F., Martino, J., Morris, C., Attems, J., et al., 2016. Anatomic connections of the subgenual cingulate region. Neurosurgery 79 (3), 465–472. https://doi.org/10.1227/NEU.0000000000001315.

Wallace, E.J., Mathias, J.L., Ward, L., 2018a. Diffusion tensor imaging changes following mild, moderate and severe adult traumatic brain injury: a meta-analysis. Brain Imaging Behav. 12 (6), 1607–1621. https://doi.org/10.1007/s11682-018-9823-2. PMID: 29383621.

Wallace, E.J., Mathias, J.L., Ward, L., 2018b. The relationship between diffusion tensor imaging findings and cognitive outcomes following adult traumatic brain injury: a meta-analysis. Neurosci. Biobehav. Rev. 92, 93–103. https://doi.org/10.1016/j.neubiorev.2018.05.023. Epub 2018 May 24. PMID: 29803527.

Wang, F., Jiang, J., Zhang, J., Wang, Q., 2018. Predicting pituitary stalk position by in vivo visualization of the hypothalamo-hypophyseal tract in craniopharyngioma using diffusion tensor imaging tractography. Neurosurg. Rev. 41 (3), 841–849. https://doi.org/10.1007/s10143-017-0933-x. Epub 2017 Nov 28. PMID: 29185147.

Watson, C.G., DeMaster, D., Ewing-Cobbs, L., 2019. Graph theory analysis of DTI tractography in children with traumatic injury. NeuroImage Clin. 21, 101673. https://doi.org/10.1016/j.nicl.2019.101673. Epub 2019 Jan 10. PMID: 30660661; PMCID: PMC6412099.

Wende, T., Hoffmann, K.T., Meixensberger, J., 2020. Tractography in neurosurgery: a systematic review of current applications. J. Neurol. Surg. A Cent. Eur. Neurosurg. 81 (5), 442–455. https://doi.org/10.1055/s-0039-1691823. Epub 2020 Mar 16. PMID: 32176926.

Winston, G.P., 2013. Epilepsy surgery, vision, and driving: what has surgery taught us and could modern imaging reduce the risk of visual deficits? Epilepsia 54 (11), 1877–1888. https://doi.org/10.1111/epi.12372. Epub 2013 Sep 20. PMID: 24199825; PMCID: PMC4030586.

Winston, G.P., Yogarajah, M., Symms, M.R., McEvoy, A.W., Micallef, C., Duncan, J.S., 2011. Diffusion tensor imaging tractography to visualize the relationship of the optic radiation to epileptogenic lesions prior to neurosurgery. Epilepsia 52 (8), 1430–1438. https://doi.org/10.1111/j.1528-1167.2011.03088.x. Epub 2011 May 13. PMID: 21569018; PMCID: PMC4471629.

Winston, G.P., Daga, P., Stretton, J., Modat, M., Symms, M.R., McEvoy, A.W., Ourselin, S., Duncan, J.S., 2012. Optic radiation tractography and vision in anterior temporal lobe resection. Ann. Neurol. 71 (3), 334–341. https://doi.org/10.1002/ana.22619. PMID: 22451201; PMCID: PMC3698700.

Wu, J.S., Zhou, L.F., Tang, W.J., Mao, Y., Hu, J., Song, Y.Y., et al., 2007. Clinical evaluation and follow-up outcome of diffusion tensor imaging-based functional neuronavigation: a prospective, controlled study in patients with gliomas involving pyramidal tracts. Neurosurgery 61, 935–949.

Yang, J.Y., Beare, R., Seal, M.L., Harvey, A.S., Anderson, V.A., Maixner, W.J., 2017. A systematic evaluation of intraoperative white matter tract shift in pediatric epilepsy surgery using high-field MRI and probabilistic high angular resolution diffusion imaging tractography. J. Neurosurg. Pediatr. 19 (5), 592–605.

Yang, J.Y., Menon, R., Barton, S., Mandelstam, S.A., Kerr, R., Wrennall, J., Bailey, C., Freeman, J., Maixner, W.J., Harvey, A.S., 2020. One-stage, language-dominant, opercular-insular epilepsy surgery with multimodal structural and functional neuroimaging evaluation. In: Proceedings of ISMRM, 234.

Yang, J.Y., Yeh, C.H., Poupon, C., Calamante, F., 2021. Diffusion MRI tractography for neurosurgery: the basics, current state, technical reliability and challenges. Phys. Med. Biol. 66 (15), TR01. https://doi.org/10.1088/1361-6560/ac0d90.

Yang, J.Y., Chen, J., Alexander, B., Schilling, K., Kean, M., Wray, A., Seal, M., Maixner, W., Beare, R., 2022. Assessment of intraoperative diffusion EPI distortion and its impact on estimation of supratentorial white matter tract positions in pediatric epilepsy surgery. NeuroImage Clin. 35, 103097.

Yao, Y., Ulrich, N.H., Guggenberger, R., Alzarhani, Y.A., Bertalanffy, H., Kollias, S.S., 2015. Quantification of corticospinal tracts with diffusion tensor imaging in brainstem surgery: prognostic value in 14 consecutive cases at 3T magnetic resonance imaging. World Neurosurg. 83 (6), 1006–1014. https://doi.org/10.1016/j.wneu.2015.01.045. Epub 2015 Mar 5. PMID: 25749578.

Yaşargil, M.G., 1999. A legacy of microneurosurgery: memoirs, lessons, and axioms. Neurosurgery 45 (5), 1025–1092. https://doi.org/10.1097/00006123-199911000-00014. PMID: 10549924.

Yecies, D., Jabarkheel, R., Han, M., Kim, Y.H., Bruckert, L., Shpanskaya, K., Perez, A., Edwards, M.S.B., Grant, G.A., Yeom, K.W., 2019. Posterior fossa syndrome and increased mean diffusivity in the olivary bodies. J. Neurosurg. Pediatr., 1–6. https://doi.org/10.3171/2019.5.PEDS1964. Epub ahead of print. PMID: 31349230.

Yeole, U., Singh, V., Mishra, A., Shaikh, S., Shetty, P., Moiyadi, A., 2020. Navigated intraoperative ultrasonography for brain tumors: a pictorial essay on the technique, its utility, and its benefits in neuro-oncology. Ultrasonography 39 (4), 394–406.

Yepes-Calderon, F., Lao, Y., Fillard, P., Nelson, M.D., Panigrahy, A., Lepore, N., 2016. Tractography in the clinics: implementing a pipeline to characterize early brain development. NeuroImage Clin. 14, 629–640. https://doi.org/10.1016/j.nicl.2016.12.029. PMID: 28348954; PMCID: PMC5357703.

Yogarajah, M., Focke, N.K., Bonelli, S., Cercignani, M., Acheson, J., Parker, G.J., Alexander, D.C., McEvoy, A.W., Symms, M.R., Koepp, M.J., Duncan, J.S., 2009. Defining Meyer's loop-temporal lobe resections, visual field deficits and diffusion tensor tractography. Brain 132 (Pt 6), 1656–1668. https://doi.org/10.1093/brain/awp114. Epub 2009 May 21. PMID: 19460796; PMCID: PMC2685925.

Yogarajah, M., Focke, N.K., Bonelli, S.B., Thompson, P., Vollmar, C., McEvoy, A.W., Alexander, D.C., Symms, M.R., Koepp, M.J., Duncan, J.S., 2010. The structural plasticity of white matter networks following anterior temporal lobe resection. Brain 133 (Pt 8), 2348–2364. https://doi.org/10.1093/brain/awq175. PMID: 20826432; PMCID: PMC3198261.

Yoshino, M., Abhinav, K., Yeh, F.C., Panesar, S., Fernandes, D., Pathak, S., Gardner, P.A., Fernandez-Miranda, J.C., 2016. Visualization of cranial nerves using high-definition fiber tractography. Neurosurgery 79 (1), 146–165. https://doi.org/10.1227/NEU.0000000000001241. PMID: 27070917.

Young, F., Aquilina, K., Clark, C.A., Clayden, J.D., 2022. Fibre tract segmentation for intraoperative diffusion MRI in neurosurgical patients using tract-specific orientation atlas and tumour deformation modelling. Int. J. Comput. Assist. Radiol. Surg. 17 (9), 1559–1567.

Yuan, W., Meller, A., Shimony, J.S., Nash, T., Jones, B.V., Holland, S.K., Altaye, M., Barnard, H., Phillips, J., Powell, S., McKinstry, R.C., Limbrick, D.-D., Rajagopal, A., Mangano, F.T., 2016. Left hemisphere structural connectivity abnormality in pediatric hydrocephalus patients following surgery. NeuroImage Clin. 12, 631–639. https://doi.org/10.1016/j.nicl.2016.09.003. PMID: 27722087; PMCID: PMC5048110.

**652 PART | V** Tractography applications

Zakaria, H., Haider, S., Lee, I., 2017. Automated whole brain tractography affects preoperative surgical decision making. Cureus 9 (9), e1656. https://doi.org/10.7759/cureus.1656. PMID: 29147631; PMCID: PMC5673476.

Zanello, M., Wager, M., Corns, R., Capelle, L., Mandonnet, E., Fontaine, D., Reyns, N., Dezamis, E., Matsuda, R., Bresson, D., Duffau, H., Pallud, J., 2017. Resection of cavernous angioma located in eloquent areas using functional cortical and subcortical mapping under awake conditions. Outcomes in a 50-case multicentre series. Neurochirurgie 63 (3), 219–226. https://doi.org/10.1016/j.neuchi.2016.08.008. Epub 2017 May 12. PMID: 28502568.

Zdunczyk, A., Roth, F., Picht, T., Vajkoczy, P., 2020. Functional DTI tractography in brainstem cavernoma surgery. J. Neurosurg., 1–10.

Zhang, C., Kim, S.G., Li, J., Zhang, Y., Lv, Q., Zeljic, K., Gong, H., Wei, H., Liu, W., Sun, B., Wang, Z., Voon, V., 2021. Anterior limb of the internal capsule tractography: relationship with capsulotomy outcomes in obsessive-compulsive disorder. J. Neurol. Neurosurg. Psychiatry. https://doi.org/10.1136/jnnp-2020-323062. Epub ahead of print. PMID: 33461976.

# Chapter 32

# Preclinical and ex vivo tractography: Techniques and applications at high field

## Manisha Aggarwal
*Department of Radiology and Radiological Science, Johns Hopkins University School of Medicine, Baltimore, MD, United States*

## 1 Introduction

Compared to clinical applications of tractography, preclinical and ex vivo diffusion MRI (dMRI) studies are often performed on high-field narrow-bore systems that can allow imaging at higher spatial resolutions. In recent years, there have been several emerging applications of dMRI for mesoscale mapping of connectivity in the postmortem human brain and for structural connectomics in animal models. In addition, the hardware capabilities of preclinical scanners can benefit the acquisition of extensive dMRI datasets that are important for comparison and cross-validation of diffusion modeling and tractography approaches.

Compared with clinical MRI field strengths of 1.5–7 T, preclinical MRI systems range in field strengths from 4.7 T up to 21 T. Typical peak gradient amplitudes on preclinical scanners are in the range of 200–800 mT/m, with microimaging systems capable of generating gradients up to 3000 mT/m. While the basic principles of tractography for preclinical and ex vivo dMRI are similar to those for clinical diffusion studies, the specific hardware and tissue properties require some specific technical considerations to achieve optimal image quality. The higher field and gradient strengths, specialized radiofrequency (RF) coils, and potential to scan for longer times can enable dMRI of ex vivo human brain specimens at isotropic resolutions of a few hundred microns, which is nearly an order of magnitude higher than typically achievable resolution limits on clinical MRI hardware (Roebroeck et al., 2019). For preclinical tractography applications, imaging at significantly higher spatial resolutions is necessary in order to resolve anatomically comparable details. For instance, compared to an adult human brain, a mouse brain is ~3000 times smaller. Using tailored acquisition pulse sequences, ex vivo diffusion microimaging of small animal brains can be performed at isotropic resolutions down to 25–50 μm (Aggarwal et al., 2010; Wang et al., 2018). Both deterministic and probabilistic tractography approaches have been used to generate structural connectomes of rodent brains (Calabrese et al., 2015b; Chen et al., 2015; Crater et al., 2022; Ingalhalikar et al., 2014; Sinke et al., 2018), with important emerging applications for mapping of altered connectivity in experimental models of neurological disorders.

Another focus of preclinical or ex vivo dMRI studies is cross-validation of tractography results using other imaging modalities, which can be performed in the same specimens. Diffusion tractography relies on an indirect inference of intravoxel fiber orientation distributions (FODs) based on estimates of signal loss due to water mobility. The reconstruction and tractography processes invariably introduce a number of approximations or biases that can limit the anatomical accuracy of the results (Maier-Hein et al., 2017; Schilling et al., 2019d; Thomas et al., 2014; Yendiki et al., 2022). Therefore validation of tractography in dMRI studies remains an important, yet unresolved, question. Preclinical and ex vivo studies can allow comparisons with conventional axonal tracers, histological staining methods, or emerging microscopy techniques sensitive to the optical anisotropy or birefringence of axonal myelin sheaths, which can serve as a potential "round truth" for validation of fiber orientation estimates derived from different dMRI reconstruction and tractography approaches. While most current techniques are limited to 2D sectioning or relatively small tissue volumes, the ability to integrate data across multiple modalities and spatial scales may lead to improved accuracy of FOD estimates, and is another growing area of interest (Goubran et al., 2019; Howard et al., 2019).

This chapter provides a comprehensive overview of the techniques and applications for dMRI-based tractography on high-field preclinical scanners, focusing on mesoscale mapping of structural connectivity in the ex vivo human brain, diffusion microimaging in small animal models, as well as cross-validation approaches using complementary methods for inferring white matter orientations. The chapter starts with a brief overview of the main technical considerations

---

Handbook of Diffusion MR Tractography. https://doi.org/10.1016/B978-0-12-818894-1.00025-2
Copyright © 2025 Elsevier Ltd. All rights reserved, including those for text and data mining, AI training, and similar technologies.

**654 PART | V** Tractography applications

involved in obtaining high-quality and high-signal-to-noise ratio (SNR) preclinical or ex vivo dMRI data for tractography, in terms of both acquisition pulse sequences and $q$-space sampling. Specific applications for mapping of mesoscopic connectivity in the postmortem human brain, dMRI-based structural connectomics in preclinical models of neurological disorders, and comparison of different diffusion modeling and tractography approaches are then discussed. Finally, the chapter presents an overview of current approaches and potential pitfalls for validation of dMRI-based fiber orientation estimates and tractography results using neuronal tracers, microscopy, and optical imaging techniques.

## 2  Technical considerations

The accuracy and robustness of fiber tractography depends foremost on the acquisition of high-quality dMRI data with minimal artifacts, high SNR, and optimal angular contrast. The major challenges for diffusion acquisition and microimaging at high field are the shorter relaxation times ($T_2/T_2^*$) and increased susceptibility to magnetic field inhomogeneity effects. In addition, for fixed tissues, the inherently lower $T_2$ and diffusivity both affect the acquired signal quality adversely and impose competing constraints on the achievable SNR and diffusion-weighted contrast. The combination of these factors typically precludes the use of single-shot diffusion-weighted echo planar imaging (ss-dwEPI) sequences that are most widely used for clinical diffusion imaging (Chapters 6 and 7). As a result, sequences in popular use for clinical dMRI cannot be directly translated for high-resolution diffusion imaging on preclinical systems. Following, we discuss the main technical considerations for acquiring high-quality preclinical and ex vivo dMRI data in terms of both acquisition pulse sequences (encoding in $k$-space) and diffusion sensitization (sampling in $q$-space).

### 2.1  Acquisition pulse sequences

Two-dimensional ss-dwEPI is the most widely used method of choice for clinical diffusion imaging. At preclinical field strengths, the decrease in $T_2^*$ and increased field inhomogeneity effects can render the use of lengthy echo trains suboptimal for dMRI and tractography; ss-dwEPI is prone to magnetic susceptibility artifacts in the form of $T_2^*$-induced signal decay and off-resonance effects arising from $\Delta B_0$-induced frequency dispersions. Multishot segmented or interleaved EPI (ms-dwEPI) readouts can be used to mitigate the SNR loss and susceptibility artifacts by shortening the echo train length, and they have been used for preclinical dMRI and tractography in both in vivo and ex vivo applications (Aggarwal et al., 2013; Harsan et al., 2010; Schilling et al., 2019b; Thomas et al., 2014). While EPI readouts provide the highest scan efficiency (SNR per unit time), the anatomical fidelity depends on achieving a uniform field across the imaging volume and can be adversely affected by inhomogeneity-related geometric distortions and ghosting. Because of their high efficiency, ms-dwEPI sequences are ideal for high angular resolution dMRI acquisitions. SNR and bandwidth related constraints on the achievable spatial resolution, however, can limit their use for diffusion microimaging applications.

Diffusion-weighted spin-echo (dwSE) or single line readouts have also been used at preclinical field strengths (D'Arceuil et al., 2007; Guilfoyle et al., 2003). These offer high anatomic fidelity and are largely insensitive to $B_0$ inhomogeneity effects. However, due to the low SNR efficiency η (=SNR/$\sqrt{}$scan time), they can require prohibitively long scan times for acquisitions at combined high spatial and angular resolutions. Compressed sensing can alternatively be used to speed up the acquisition, with acceleration factors of up to 8 reported for dMRI of the fixed mouse brain (Wang et al., 2018). Readouts with echo trains consisting of multiple refocusing RF pulses can offer an optimal trade-off between SNR efficiency and image quality. These include 3D rapid acquisition with relaxation enhancement (dwRARE) (Tyszka and Frank, 2009) and 3D gradient-and-spin-echo (dwGRASE) (Aggarwal et al., 2010), which can enable typical acceleration factors of ~4–12 compared to standard dwSE, while being less prone to inhomogeneity-related artifacts than ms-dwEPI.

As the Carr-Purcell-Meiboom-Gill condition of multiple RF-pulse echo trains tends to break down with the application of strong diffusion gradients (Pipe et al., 2002), these techniques require sophisticated methods for $k$-space phase correction. An example of typical phase modulation in $k$-space and its effect on the image quality is illustrated in Fig. 1: eddy currents induced by the strong diffusion-encoding gradients can lead to severe image artifacts even in the absence of bulk motion. If uncorrected for, these can be particularly problematic at higher $b$-values often necessary for tractography. Further, as the eddy current-induced artifacts depend on both the magnitude and orientation of the applied diffusion-encoding gradients, they can particularly confound tractography results. Approaches to map and correct for the $k$-space phase modulation include, but are not limited to, the acquisition of reference scans (Tyszka and Frank, 2009) and dual navigator echoes (Aggarwal et al., 2010; Mori and van Zijl, 1998). With 3D volumetric excitations, the $k$-space trajectory in dwGRASE can also be designed to separate the phase modulation due to eddy currents and off-resonance effects along different phase-encoding axes (Oshio and Feinberg, 1991). This allows the phase differences due to both these

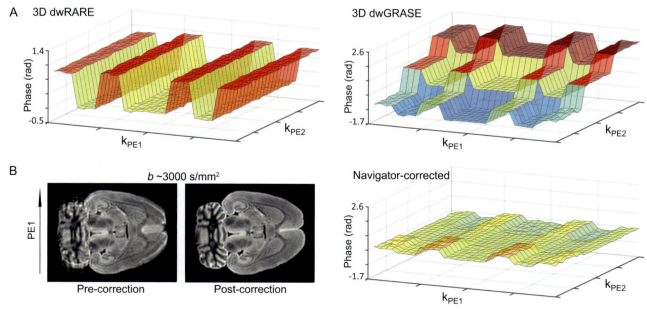

FIG. 1 Correction of phase-related artifacts in 3D multiple RF-pulse sequences for high-resolution preclinical dMRI. (A) Plots illustrating the typical stair-step patterns of phase modulation across k-space due to eddy currents induced by strong diffusion gradients, for dwRARE (left) and dwGRASE (right). For illustration, the phase of a representative point along the readout axis measured in the absence of phase-encoding gradients is plotted. (B) Representative images from a fixed rat brain ($b \sim 3000 s/mm^2$) acquired with 3D dwGRASE at 11.7T demonstrate the resulting ghosting artifacts and their minimization using dual-navigator-echo based phase correction. The plot in (B) shows minimization of the phase modulation errors along both the primary and secondary phase-encoding axes ($k_{PE1}$ and $k_{PE2}$) after navigator-based correction.

effects to be effectively corrected by the use of dual navigator echoes (as illustrated in Fig. 1B). Other sequence modifications to combat the effects of eddy currents include the use of quadratic phase cycling along the RF pulse train and gradient crushing schemes.

*Diffusion preparation:* In addition to imaging readouts, modifications to the diffusion weighting preparation can also be used to reduce artifacts or combat the effects of reduced $T_2$ at high fields in order to optimize the image quality for tractography. Double-refocused or bipolar diffusion-encoding gradients can be used to minimize the effects of eddy currents by reducing the interval between positive and negative gradient ramps (Reese et al., 2003), and have been used for preclinical diffusion microimaging (Aggarwal et al., 2010; Tyszka and Frank, 2009). Diffusion-weighted stimulated echo (dwSTE) acquisitions allow long diffusion times without incurring magnetization loss due to $T_2$ decay during the mixing time, which may be advantageous for tractography applications (Rane et al., 2010). Note that dwSTE acquisitions have an inherent 50% signal loss, but may be suitable when the desired diffusion times are long as compared to $T_2$, and $T_1 \gg T_2$, conditions that are likely to be met at high fields especially for ex vivo dMRI as the tissue $T_2$ is further reduced (Fritz et al., 2019); dwSTE acquisitions have been used for preclinical dMRI (Boretius et al., 2007; Guilfoyle et al., 2011), but these can require correction of the B-matrix to account for nontrivial effects of imaging and crusher gradients as compared to spin-echo preparations (Lundell et al., 2014). Other techniques such as diffusion-weighted steady-state free precession (dwSSFP) can also be advantageous for ex vivo dMRI and tractography at high fields by allowing data acquisition at shorter TEs (Foxley et al., 2014). However, dwSSFP acquisitions introduce a complex voxel-wise signal dependence on flip angle, $T_1$, $T_2$, and b-value, and require additional estimation of $T_1$, $T_2$, and $B_1$ field maps; thus their use for tractography on preclinical systems has been relatively less explored.

## 2.2 q-space sampling schemes

Designing optimal q-space sampling strategies typically involves a trade-off between spatial and angular resolutions, number of shells, diffusion contrast versus SNR, and total acquisition time, as also discussed in other chapters of this book (Chapters 6, 7, 10, and 11). For a standard pulsed-gradient spin echo (PGSE) sequence, the diffusion wave vector **q** as defined by Callaghan (1991) depends on the gradient duration ($\delta$), magnitude, and orientation of the gradient vector **G**, as $\mathbf{q} \propto \gamma \delta \mathbf{G}$, where $\gamma$ is the gyromagnetic ratio, and the b-value is related to the square magnitude of the **q** vector and the diffusion time, as $b = q^2 t$. Under the narrow pulse approximation, the q-space formalism links the diffusion-weighted

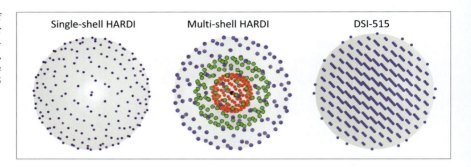

FIG. 2 Schematic representations of common $q$-space sampling schemes used for preclinical and ex vivo tractography applications: single-shell HARDI, multishell HARDI, and diffusion spectrum imaging (DSI). The DSI scheme shown is for sampling of 515 points on a Cartesian lattice in $q$-space.

signal to the ensemble average propagator (EAP) by a 3D Fourier relationship. Here we assume that the effective diffusion time is fixed, although that can also be varied.

Designing acquisition strategies for tractography therefore involves either sufficiently dense sampling of 3D $q$-space on a Cartesian lattice to estimate the EAP using an inverse Fourier transform (Wedeen et al., 2005), or subsampling schemes aimed at estimating the angular information of the propagator in the form of the diffusion orientation distribution function (ODF) or its variants. The latter can be broadly based on model-free approaches or methods using mixture models, most of which require high angular resolution diffusion imaging (HARDI) or sampling of $q$-space over one or more spherical shells (Alexander, 2005; Behrens et al., 2003; De Luca et al., 2020; Dell'Acqua et al., 2013b; Dell'Acqua and Tournier, 2019; Descoteaux et al., 2009; Frank, 2001; Jansons and Alexander, 2003; Jeurissen et al., 2014; Özarslan et al., 2006; Tournier et al., 2004; Tuch, 2004; Yeh et al., 2010). In the simplest case, relatively few orientations over a single shell can be acquired for diffusion tensor imaging (DTI) (Basser et al., 1994), which approximates the EAP by a multivariate zero-mean Gaussian distribution but has known constraints in regions of complex intravoxel fiber orientations (Alexander et al., 2002; Frank, 2002). Fig. 2 shows some examples of common $q$-space sampling schemes that have been used for preclinical and ex vivo dMRI.

While there is no common consensus on the optimal $q$-space sampling strategies, the choice is often driven by the specific application and tractography methods to be used and on the biophysical properties of the tissue. For ex vivo dMRI, in particular, the $b$- or $q$-values need to be significantly higher in order to offset the effect of reduced diffusivity and to obtain diffusion contrast comparable to in vivo applications for tractography. For example, $b$-values 2–3 times as high as those used in vivo are often necessary for optimal contrast (D'Arceuil et al., 2007; Dyrby et al., 2011; Roebroeck et al., 2019). For multishell acquisitions, the higher shells need to be appropriately scaled in order to optimize the trade-off between diffusion-weighted contrast and SNR.

## 3 Applications

In recent years, high-resolution dMRI and tractography methods have been increasingly used to investigate connectivity in both ex vivo human brain and preclinical animal brain studies from mesoscopic to microscopic scales. The advantages of higher field and/or gradient strengths, smaller and more sensitive RF coils, longer scan times, and tailored acquisition sequences can allow pushing the resolution envelope for dMRI to well beyond the current limits on clinical MRI systems. As discussed in the previous section, studies on preclinical field strengths can span a gamut of acquisition schemes and tractography methods, from DTI-based deterministic fiber tracking to model-free approaches such as diffusion spectrum imaging (DSI). Additionally, the acquisition of extensive high-quality and high-resolution dMRI datasets on preclinical scanners is important for comparison and potential validation of different tractography and diffusion signal modeling approaches.

### 3.1 Mesoscale connectivity mapping in the human brain

While in vivo dMRI has been widely used to trace macroscopic white matter pathways in the human brain, the acquisition of high-resolution postmortem datasets in recent years has enabled investigations of 3D structural connectivity in the human brain at mesoscopic scales. Compared to typical in vivo voxel sizes of 1–2.5 mm, ex vivo dMRI can be performed at isotropic resolutions <300 μm, resulting in voxel volumes that are nearly three orders of magnitude smaller. Recent studies have used diffusion tractography based on acquisitions at preclinical field strengths to investigate connectivity in both cortical and subcortical structures.

Dell'Acqua et al. used HARDI with 30 directions and an Euler-like streamline tractography algorithm to reconstruct details of axonal connectivity in the fixed human cerebellum at 7 T (Dell'Acqua et al., 2013a). At 100 μm in-plane resolution, details of cerebellar intracortical connectivity such as parallel fibers and white matter fibers running into individual cerebellar folia could be reconstructed. HARDI of the human brainstem at 7 T or 11.7 T has been used to reconstruct crossing white matter pathways based on both deterministic and probabilistic tractography approaches (Aggarwal et al., 2013; Calabrese et al., 2015e; Henssen et al., 2019b). Fig. 3 shows the level of anatomical detail resolved with high-field ex vivo HARDI of the brainstem at 255 μm isotropic resolution compared to in vivo dMRI at 2.2 mm resolution. As discussed in Section 2, high $b$-values (~4000 s/mm$^2$) were used in the preceding studies to both offset the reduced diffusivity of fixed tissue and to enable sharper delineation of peaks for resolving multiple intravoxel orientations in the reconstructed fiber ODFs. Fig. 4 illustrates the fiber ODFs and color-coded track density image of the pons reconstructed using superresolution constrained spherical deconvolution (CSD) and probabilistic streamline tractography based on single-shell HARDI with 30 directions. As seen in Fig. 4, fine details of white matter pathways including interdigitating fiber fascicles of the corticospinal tract and transverse pontine fibers could be reconstructed. Calabrese et al. (2015e) used probabilistic tractography based on HARDI of the brainstem and thalamus at 7 T to reconstruct the dentatorubrothalamic tract as a reference for deep brain stimulation targeting.

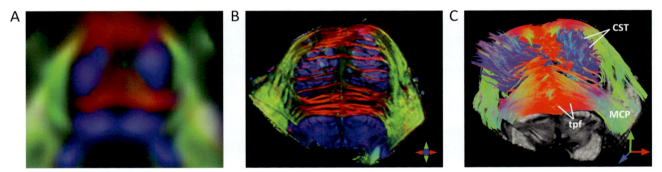

**FIG. 3** Comparison of in vivo (2.2 mm isotropic resolution) and ex vivo (255 μm isotropic resolution) dMRI of the human brainstem and deterministic tractography showing reconstruction of major white matter tracts. (A, B) Axial direction-encoded color map slices at the level of the pons, from in vivo data acquired with $b = 1000$ s/mm$^2$, and ex vivo data acquired with $b = 4000$ s/mm$^2$. (C) Interdigitating fascicles of the corticospinal tract (CST) and transverse pontine fibers (tpf) and the middle cerebellar peduncle (MCP) reconstructed using deterministic tractography. *(Reproduced with permission from Aggarwal, M., Zhang, J., Pletnikova, O., Crain, B., Troncoso, J., Mori, S., 2013. Feasibility of creating a high-resolution 3D diffusion tensor imaging based atlas of the human brainstem: a case study at 11.7 T. NeuroImage 74, 117–127.)*

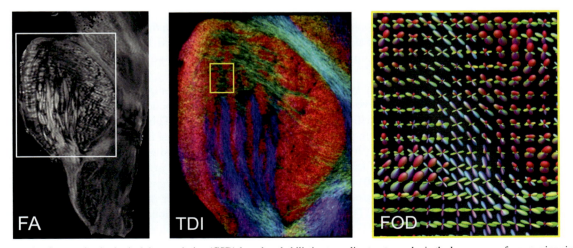

**FIG. 4** Example of constrained spherical deconvolution (CSD)-based probabilistic streamline tractography in the human pons from ex vivo single-shell HARDI data at 11.7 T ($b$-value ~4000 s/mm$^2$). A sagittal slice from the fractional anisotropy (FA) map of the intact brainstem and superresolution trackdensity image (TDI) of the pons based on probabilistic tractography are shown. The TDI slice shows fine interdigitating orientations of the corticospinal tract and transverse pontine fibers that could be resolved with combined high spatio-angular resolution dMRI. A zoomed-in view of the fiber orientation distributions (FODs) in a small region indicated by the *yellow* rectangle is shown in the right panel.

**658 PART | V** Tractography applications

Probabilistic tractography approaches based on high-field ex vivo HARDI have been used to investigate layer-specific connectivity in human cortical areas. Tracking of intracortical fibers is challenging in the human brain in vivo, due to both the relatively low anisotropy in gray matter regions and the limited spatial resolution that is coarser than the thickness of cortical layers. Studies using high spatial-angular resolution acquisitions on preclinical field strengths have demonstrated reconstruction of layer-specific fiber ODFs in different cortical areas (Aggarwal et al., 2015; Kleinnijenhuis et al., 2013; Leuze et al., 2014). CSD-based probabilistic tractography results showed close correspondence with the organization of tangential fibers in the visual cortex revealed by polarized light imaging (PLI) (Leuze et al., 2014) and region-specific myeloarchitecture of distinct cortical areas as seen with silver staining (Aggarwal et al., 2015).

The reduced partial volume effects and higher SNR achievable with ex vivo high-field dMRI have also been used to examine connectivity in other gray matter regions, such as the hippocampus, and their changes with pathologies. Modo et al. (2016) used deterministic tractography based on dMRI of surgically excised hippocampal tissue at 11.7 T to compare aberrant mossy fiber connections in temporal lobe epilepsy. With dMRI of the hippocampus at an isotropic resolution of 300 μm, Beaujoin et al. (2018) used streamline regularized deterministic or probabilistic tractography algorithms based on DTI or Q-ball imaging (QBI) to track intrahippocampal connections. Mollink et al. (2019) combined probabilistic tractography and PLI in the same postmortem hippocampal specimens to investigate degeneration of the perforant path in amyotrophic lateral sclerosis. Other studies have used ex vivo dMRI on high-field systems in combination with histological staining or PLI to trace specific pathways in the amygdala (Mori et al., 2017), medulla (Henssen et al., 2019a), or spinal cord (Henssen et al., 2019d).

The increasing availability of high-quality and high spatio-angular resolution datasets at high fields can open up novel avenues for 3D mapping of mesoscale connectivity in the human brain, and allow bridging the gap in spatial scales between in vivo dMRI and cellular-level microscopic imaging techniques, as discussed in Section 4.

## 3.2 Microimaging and preclinical applications

Tractography applications in small animal imaging require acquisitions at considerably higher resolutions in order to resolve details that are anatomically comparable to clinical human brain studies. Typical resolutions for in vivo tractography studies in small animals range from 100 to 200 μm, often with lower through-plane resolutions, whereas ex vivo studies are typically performed at microimaging resolutions with isotropic voxel dimensions < 100 μm.

In vivo DTI-based deterministic methods have been used to reconstruct major white matter tracts in the mouse brain with 6–12 directions and $b = 1000$ s/mm$^2$ (Boretius et al., 2007; Wu et al., 2013). Based on in vivo single-shell HARDI with 30 directions and $b = 1000$ s/mm$^2$, Harsan et al. (2010) used both deterministic and probabilistic tractography approaches to reconstruct limbic and visual tracts in the mouse brain. In another study, Harsan et al. (2013) used a global fiber tracking algorithm based on single-shell HARDI to compare thalamocortical projections in wild-type and *reeler* mutant mice, and validated the results using axonal tracing. While most in vivo dMRI studies of rodent brains are performed using 2D multi-slice acquisitions with lower through-plane resolutions depending on the SNR and available gradient strength for slice-selective excitation, one caveat is that anisotropic voxel sizes can introduce biases in the estimates of anisotropy measures and tractography (Basser et al., 2000; Neher et al., 2013). A few studies have also used deterministic or probabilistic tractography with isotropic resolutions of 100–200 μm to examine connectivity in the in vivo rat brain (Tu et al., 2014) or mouse hippocampus (Wu and Zhang, 2016).

Ex vivo dMRI studies in small animals are typically performed with higher $b$- or $q$-values to offset the reduced diffusivity, as discussed in Section 2.2. For ex vivo dMRI, gadolinium-based diamagnetic contrast agents are often used to shorten the tissue $T_1$, thereby allowing acquisitions with shorter repetition times. This can be done either by adding the contrast agent during the specimen preparation phase, or by introducing the contrast agent directly during perfusion fixation, in a technique known as active staining ( Johnson et al., 2012). Efficient acquisition sequences such as dwGRASE or active staining methods have allowed 3D diffusion microimaging and tractography in the mouse brain with isotropic resolutions down to 25–55 μm (Aggarwal et al., 2014, 2010; Jiang and Johnson, 2010; Wang et al., 2020). In recent years, high-resolution ex vivo dMRI and tractography-based atlases have been introduced for developing and adult mouse brains (Aggarwal et al., 2009; Jiang and Johnson, 2011), rat brains (Calabrese et al., 2013; Johnson et al., 2012), and the macaque brain (Calabrese et al., 2015a; Feng et al., 2017; Warrington et al., 2020). Diffusion microimaging, atlas-based segmentation of seed regions, and deterministic or probabilistic tractography approaches have also been used to construct tractograms or structural connectomes of the mouse brain, representing brain-wide maps of interconnectivity across multiple seed regions (Chen et al., 2015; Crater et al., 2022; Ingalhalikar et al., 2014; Sinke et al., 2018). Fig. 5 shows a schematic of the steps for probabilistic tractography-based connectome generation using ex vivo diffusion microimaging of the mouse brain and atlas-based segmentation into 148 anatomical structures (Calabrese et al., 2015b).

Preclinical and ex vivo tractography: Techniques and applications at high field **Chapter | 32** 659

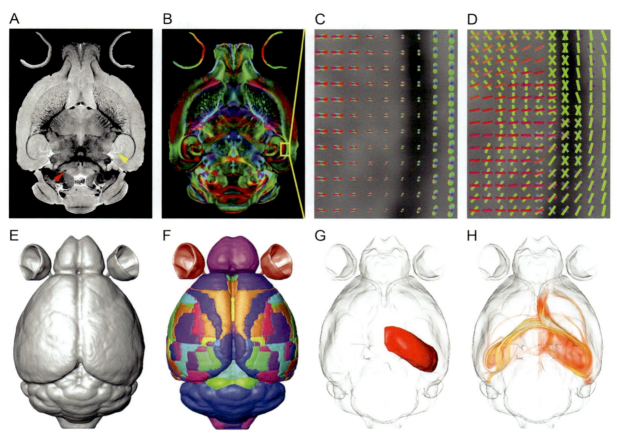

**FIG. 5** Schematic of steps for ex vivo diffusion microimaging-based probabilistic tractography and structural connectome generation for the mouse brain. (A, B) Representative axial slices from anatomic and color-coded fractional anisotropy images of the mouse brain at 43-μm isotropic resolution. (C, D) Diffusion orientation distribution functions and estimated fiber orientations in a region of the hippocampus. (E, F) Whole brain mask and atlas-based segmentation of 148 structures. (G, H) Probabilistic tractography for the right hippocampus as the seed region. Tractography data from all seed regions can be combined to generate a brain-wide connectivity matrix. *(Reproduced with permission from Calabrese, E., Badea, A., Cofer, G., Qi, Y., Johnson, G.A., 2015b. A diffusion MRI tractography connectome of the mouse brain and comparison with neuronal tracer data. Cereb. Cortex (New York, N.Y.: 1991) 25, 4628-4637.)*

Establishing such tractography-based structural connectomes has paved the way for whole mouse brain "connectomics," with emerging applications in mapping of altered brain connectivity in genetic mouse models of neurological disorders. For instance, Badea et al. (2019) used whole-brain connectomics based on ex vivo HARDI to identify different connectivity patterns in mouse models of genetic risk factors for late-onset Alzheimer's disease. Other studies have used diffusion tractography-based structural connectomes to investigate aberrant brain connectivity in the BTBR mouse strain, a genetic model for human autism (Badea et al., 2019; Edwards et al., 2020; Vega-Pons et al., 2017). DTI-based deterministic tractography revealed an absence of interhemispheric connections in the corpus callosum and dorsal hippocampal commissure in the BTBR mouse brain, along with a rostrocaudal reorganization of white matter tracts as compared to control C57BL/6 mouse brains (Vega-Pons et al., 2017). Johnson et al. (2019) compared connectomes generated from whole-brain tractography based on a HARDI/generalized $q$-space sampling approach, and found differences in the connectivity matrices of the two strains reflecting a disruption of callosal and commissural fibers and relative sparseness of connectivity in the BTBR mouse (Fig. 6).

## 3.3 Comparison of diffusion modeling and tractography approaches

The fidelity of fiber tractography in the brain depends on a number of factors (Chapters 26 and 27). Differences across acquisition or sampling schemes, diffusion models, and tractography algorithms and parameters all introduce assumptions and biases affecting the reliability and reproducibility of reconstructed fiber tracts. The higher resolutions and higher SNR

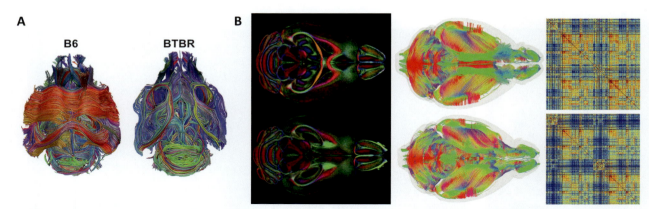

**FIG. 6** Mapping of aberrant connectivity in the BTBR mouse model of autism based on ex vivo whole-brain diffusion tractography and structural connectomics. (A) DTI-based whole-brain deterministic tractography shows a lack of interhemispheric connections in the corpus callosum and dorsal hippocampal commissure in the BTBR mouse brain as compared to normal C57BL/6 mice. (B) Comparison of tractography-based connectomes for the C57BL/6 (top row) and BTBR (bottom row) mouse brains. From left to right: Axial slice from direction-encoded color maps, whole-brain deterministic tractography based on a HARDI/generalized $q$-space sampling approach, and corresponding connectome matrices for the two strains illustrating differences in connectivity across the entire brain. *(Modified with permission from Vega-Pons, S., Olivetti, E., Avesani, P., Dodero, L., Gozzi, A., Bifone, A., 2017. Differential effects of brain disorders on structural and functional connectivity. Front. Neurosci. 10 and Johnson, G.A., Wang, N., Anderson, R.J., Chen, M., Cofer, G.P., Gee, J.C., Pratson, F., Tustison, N., White, L.E., 2019. Whole mouse brain connectomics. J. Comp. Neurol. 527, 2146–2157.)*

achievable on preclinical field strengths, especially with ex vivo acquisitions, are also useful for evaluation and comparison of different diffusion modeling and tractography approaches.

Several studies have used high-resolution ex vivo dMRI to compare different signal models and tractography approaches for reconstructing specific pathways or brain-wide connectomes. For instance, Moldrich et al. (2010) used HARDI of mouse brains at 0.1 mm isotropic resolution and 30 directions to compare various deterministic and probabilistic tractography algorithms across a range of regions of interest. Results from DTI, HARDI/Q-ball reconstruction on the spherical harmonic basis, CSD, and a Bayesian automatic relevance detection model were compared for tractography in both white and gray matter regions. Considering pathways traversing gray matter regions, CSD-based probabilistic tractography was found to most closely reproduce the known topographical organization of thalamocortical projections in the mouse brain. Using ex vivo HARDI of rat brains at 0.15 mm isotropic resolution with 60 directions, Sinke et al. (2018) compared brain-wide structural connectomes reconstructed from DTI-based, CSD-based, and global tractography algorithms against standard neuronal tracer data. For 106 cortical regions considered, CSD-based tractography was found to show the highest sensitivity for both inter- and intrahemispheric connections, but the lowest specificity, whereas global tractography resulted in the lowest sensitivity but highest specificity. Spherical deconvolution informed filtering of tractograms led to increased specificity but reduced sensitivity for both DT-based and CSD-based tractography. These studies also compared the effects of various parameters on tractography results, including streamline number, step size, curvature or angular threshold, fiber ODF threshold, and whole-brain versus region-of-interest (ROI) seeding.

High-resolution ex vivo HARDI of primate brains has also been used for comparison of different deterministic and probabilistic tractography approaches based on DTI, CSD, ball and stick, and Q-ball reconstructions (Schilling et al., 2019a; Thomas et al., 2014). Based on 3D HARDI of the macaque brain at 250 μm isotropic resolution and 121 directions, Thomas et al. (2014) compared tractography results with tracer data for two cortical ROIs from an established atlas. Among the deterministic techniques, CSD was found to show the highest sensitivity across the different ROIs, but also the lowest specificity. In contrast, the sensitivity and specificity of Q-ball and DTI approaches were found to vary widely with the seed locations.

The gradient capabilities of preclinical systems can also be harnessed for comparing tractography results based on gradient-intensive acquisition and sampling schemes such as high-resolution DSI. Using a peak gradient intensity ($|G|_{max}$) of 380 mT/m, Wedeen et al. (2008) acquired DSI data of fixed macaque brains with 515 $q$-space points and $b$-values up to 40,000 s/mm$^2$, and compared results of DSI- and DTI-based tract reconstructions in several brain regions. DSI-based tractography allowed reconstruction of known anatomic fiber decussations in the optic chiasm and crossing fibers in the pons, which could not be reconstructed accurately based on DTI (Fig. 7). In addition to resolving crossing fibers within white matter, DSI-based tractography was found to demonstrate improved ability to resolve fiber fascicles penetrating into gray matter regions such as the cerebral and cerebellar cortices and subcortical nuclei.

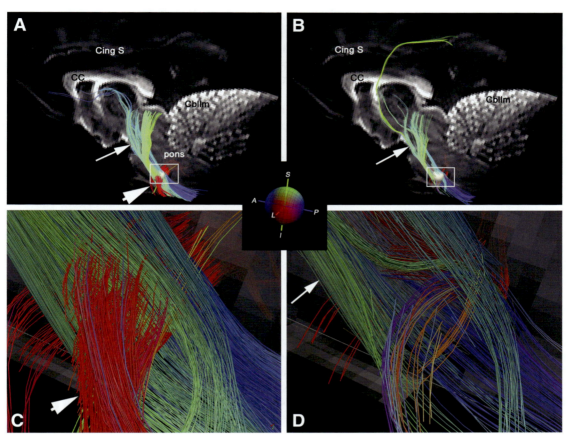

FIG. 7 Comparison of DSI and DTI tractography results in the macaque pons based on ex vivo dMRI at 4.7 T with 515 $q$-space points sampled on a cubic lattice. (A) DSI-based tractography shows reconstruction of descending corticofugal fibers (*green* and *blue*, indicated by *white arrow*) intersecting with transversely oriented pontocerebellar fibers (*red*, indicated by arrowhead). (B) In comparison, DTI-based tractography shows limited ability to resolve intersections of the descending and transverse fibers. Panels (C) and (D) show higher magnification views of regions within the white boxes in (A) and (B), respectively. CC: corpus callosum, Cing S: cingulate sulcus, Cbllm: cerebellum. *(Adapted with permission from Wedeen, V.J., Wang, R.P., Schmahmann, J.D., Benner, T., Tseng, W.Y.I., Dai, G., Pandya, D.N., Hagmann, P., D'Arceuil, H., de Crespigny, A.J., 2008. Diffusion spectrum magnetic resonance imaging (DSI) tractography of crossing fibers. NeuroImage 41, 1267–1277.)*

Other studies have compared the effects of imaging protocols and acquisition parameters, including angular resolution, spatial resolution, and $b$-value, on tractography-based reconstruction of anatomically well-defined pathways or whole-brain connectomics in mouse models (Anderson et al., 2020; Crater et al., 2022). Using ex vivo dMRI of the macaque brain, Calabrese et al. (2014) also demonstrated the effects of trade-offs between angular and spatial resolutions on tractography results based on six different time-matched $q$-space sampling schemes including DTI, QBI, and DSI.

## 4 Towards validation: Comparison with neuronal tracing, microscopy, and optical imaging modalities

Tractography involves an indirect inference of fiber orientations based on signal attenuation due to restricted water diffusion in neural tissues. Inferring fiber orientations from the estimated water diffusion displacement profiles in dMRI is inherently a complex and ill-posed inverse problem, often confounded by the effects of finite number of sampling directions, partial voluming, and limited SNR. Comparison and cross-validation with other techniques that can directly assess structural connectivity or fiber orientational information is therefore an important step towards robust interpretation of tractography results (see also Dyrby, Chapter 26; and Schilling, Chapter 27).

Preclinical and ex vivo studies enable direct comparisons with complementary approaches including neuronal tracing or optical imaging in the same specimens, and have proved critical for understanding the advantages and pitfalls of tractography and different model-based fiber reconstruction methods (Schilling et al., 2019d; Thomas et al., 2014; Yendiki et al.,

2022). Currently, there is no single modality that serves as a universal gold standard for validation of tractography in neural tissues. Different microscopy methods have specific limitations in terms of resolution scale, dimensionality (2D versus 3D), maximum tissue coverage, and specificity to structural orientation or biophysical properties of white matter (e.g., optical birefringence of myelin). This section discusses the techniques used for validation of tractography and FOD estimates in preclinical and ex vivo studies, from conventional axonal tracing to more recent tissue clearing and optical microscopy methods.

*Neuronal tracers and histological staining:* Anatomical tracing via injection of anterograde or bidirectional dyes is one of the most widely used conventional methods for inferring axonal connectivity in animal models, and has been used for comparison of dMRI-based tractography approaches (Dauguet et al., 2007; Donahue et al., 2016; Dyrby et al., 2007; Girard et al., 2020; Grisot et al., 2021). The increasing availability of comprehensive large-scale neuronal tracing databases for the mouse and macaque brains (Oh et al., 2014; Stephan et al., 2001) has opened up avenues for validation of whole-brain structural connectomes based on dMRI (Chen et al., 2015; van den Heuvel et al., 2015).

Estimates of FOD can also be derived from conventional histological myelin-stained or fluorescence microscopy sections, for quantitative comparison with dMRI-based ODFs. Methods used to derive 2D in-plane orientation distribution estimates from histological sections include manual tracing (Leergaard et al., 2010), Fourier domain filtering (Choe et al., 2012), and structure tensor analysis (Budde and Frank, 2012; Seehaus et al., 2015). Assuming registration accuracy, quantitative errors between peak orientations derived from histology and dMRI can be analyzed in regions of varying fiber complexity. The angular errors reported in these studies range from 5 to 6 degrees for ODFs estimated from $q$-space imaging in crossing-fiber regions (Leergaard et al., 2010), and less than 10 degrees for diffusion tensors in parallel-fiber regions (Choe et al., 2012). Block-face histology with optical modalities such as confocal microscopy and optical coherence tomography have also been used for validation of dMRI and fiber orientation estimates (Lefebvre et al., 2017, 2018). Structure tensor analysis can be applied to confocal microscopy images in order to extract 3D FODs (Khan et al., 2015; Schilling et al., 2016, 2019b). Using such "ground-truth" histologically derived FODs, Schilling et al. (2018) compared orientation distribution estimates from various dMRI methods including CSD, QBI, diffusion orientation transform, and persistent angular structure in the ex vivo monkey brain, to evaluate overall agreement in FOD shape, correct assessment of the number of fiber populations, and angular accuracy in orientation. Across the dMRI methods considered, they reported median angular errors of ~10 degrees for the primary fiber direction and ~20 degrees for the secondary fiber, where present.

*3D tissue clearing*: In addition to 2D methods, tissue clearing techniques such as CLARITY (Chung and Deisseroth, 2013) or CUBIC (Susaki et al., 2014) enable optical microscopy of 3D tissue blocks up to the size of a whole mouse brain. Immunolabeling of neurofilament or myelin-related proteins and confocal or two-photon microscopy of cleared brains can allow visualizing 3D neuronal fibers in intact tissue (Chang et al., 2017), and have been used to qualitatively compare DTI-based tractography in mouse brains with connectivity defects (Gimenez et al., 2017). 3D structure tensor analysis has also been applied to generate CLARITY-based FODs and streamlines, allowing quantitative comparisons with ex vivo dMRI-based fiber orientation estimates and tractography results from both deterministic and CSD-based probabilistic approaches (Goubran et al., 2019; Leuze et al., 2021; Stolp et al., 2018).

*Optical microscopy techniques based on myelin birefringence:* PLI is an optical microscopy technique that utilizes the birefringence of myelin in white matter to quantitatively estimate fiber orientations in histological tissue sections at micrometer in-plane resolution (Axer et al., 2011). Qualitative comparisons of dMRI tractography with PLI in the same specimens have been performed in ex vivo high-field studies of the human brainstem (Henssen et al., 2019b) and visual cortex (Leuze et al., 2014). The histogram of PLI vector orientations over a local neighborhood of voxels can also be computed to derive an analytical FOD (Alimi et al., 2020; Axer et al., 2016). Mollink et al. (2017) quantitatively compared the in-plane dispersion of fiber orientations in human white matter as estimated from ex vivo dMRI and structure tensor analysis of myelin-stained histology or PLI sections, and found correlation coefficients of $r = 0.79$ between dMRI and histology, and $r = 0.60$ between dMRI and PLI. Beyond validation, microscopy data may also be integrated with dMRI for modeling of FODs by exploiting the complementary strengths of each modality. Joint modeling of coregistered dMRI and microscopy or PLI data in the same tissue sample has been proposed, for example, to overcome the degeneracy between fiber dispersion and radial diffusion (Howard et al., 2021, 2019).

Polarization-sensitive optical coherence tomography (PSOCT) is another technique that uses polarized light to probe the optical birefringence of myelinated axons (de Boer et al., 1997). Wang et al. (2014) cross-validated PSOCT- and DTI-based in-plane estimates of fiber orientations in the ex vivo human medulla, and found close agreement with a mean orientation mismatch of 5.4 degrees in pixels with FA > 0.15. Using PSOCT and dMRI of ex vivo human brain samples at 9.4 T, Jones et al. (2020) compared the accuracy of dMRI orientation estimates from various grid- and shell-based sampling schemes and reconstruction methods including DTI, ball-and-stick model, DSI, QBI, and generalized $q$-sampling imaging.

For regions with varying fiber complexity, they showed that QBI yielded the best performance for single-shell data, whereas for multishell data the best performance was achieved with the ball-and-stick model.

## 5 Conclusion

This chapter covered the main technical considerations, applications, and validation approaches for preclinical and ex vivo dMRI-based tractography. With spatial resolution ranging from a few hundreds down to tens of microns, ex vivo dMRI is currently the only technique that allows 3D mapping of structural connectivity in the brain at the mesoscopic scale. The increasing availability of high spatio-angular resolution dMRI datasets of both human and animal brains with corresponding optical imaging and microscopy data offers the potential for improved mapping of brain connectivity by integrating information across multiple spatial scales. With ongoing advances in both high-field MRI hardware and acquisition techniques, preclinical and ex vivo tractography applications will be important to help bridge the gap between low-resolution clinical dMRI and cellular-resolution microscopy techniques.

## References

Aggarwal, M., Zhang, J., Miller, M.I., Sidman, R.L., Mori, S., 2009. Magnetic resonance imaging and micro-computed tomography combined atlas of developing and adult mouse brains for stereotaxic surgery. Neuroscience 162, 1339–1350.

Aggarwal, M., Mori, S., Shimogori, T., Blackshaw, S., Zhang, J., 2010. Three-dimensional diffusion tensor microimaging for anatomical characterization of the mouse brain. Magn. Reson. Med. 64, 249–261.

Aggarwal, M., Zhang, J., Pletnikova, O., Crain, B., Troncoso, J., Mori, S., 2013. Feasibility of creating a high-resolution 3D diffusion tensor imaging based atlas of the human brainstem: a case study at 11.7 T. NeuroImage 74, 117–127.

Aggarwal, M., Gobius, I., Richards, L.J., Mori, S., 2014. Diffusion MR microscopy of cortical development in the mouse embryo. Cereb. Cortex 25, 1970–1980.

Aggarwal, M., Nauen, D.W., Troncoso, J.C., Mori, S., 2015. Probing region-specific microstructure of human cortical areas using high angular and spatial resolution diffusion MRI. NeuroImage 105, 198–207.

Alexander, D.C., 2005. Multiple-fiber reconstruction algorithms for diffusion MRI. In: White Matter in Cognitive Neuroscience: Advances in Diffusion Tensor Imaging and Its Applications. vol. 1064. Annals of the New York Academy of Sciences, pp. 113–133.

Alexander, D.C., Barker, G.J., Arridge, S.R., 2002. Detection and modeling of non-Gaussian apparent diffusion coefficient profiles in human brain data. Magn. Reson. Med. 48, 331–340.

Alimi, A., Deslauriers-Gauthier, S., Matuschke, F., Müller, A., Muenzing, S.E.A., Axer, M., Deriche, R., 2020. Analytical and fast fiber orientation distribution reconstruction in 3D-polarized light imaging. Med. Image Anal. 65, 101760.

Anderson, R.J., Long, C.M., Calabrese, E.D., Robertson, S.H., Johnson, G.A., Cofer, G.P., O'Brien, R.J., Badea, A., 2020. Optimizing diffusion imaging protocols for structural connectomics in mouse models of neurological conditions. Front. Phys. 8, 88.

Axer, M., Amunts, K., Grässel, D., Palm, C., Dammers, J., Axer, H., Pietrzyk, U., Zilles, K., 2011. A novel approach to the human connectome: ultra-high resolution mapping of fiber tracts in the brain. NeuroImage 54, 1091–1101.

Axer, M., Strohmer, S., Gräßel, D., Bücker, O., Dohmen, M., Reckfort, J., Zilles, K., Amunts, K., 2016. Estimating fiber orientation distribution functions in 3D-polarized light imaging. Front. Neuroanat. 10.

Badea, A., Wu, W., Shuff, J., Wang, M., Anderson, R.J., Qi, Y., Johnson, G.A., Wilson, J.G., Koudoro, S., Garyfallidis, E., Colton, C.A., Dunson, D.B., 2019. Identifying vulnerable brain networks in mouse models of genetic risk factors for late onset Alzheimer's disease. Front. Neuroinform. 13.

Basser, P.J., Mattiello, J., LeBihan, D., 1994. Estimation of the effective self-diffusion tensor from the NMR spin echo. J. Magn. Reson. B 103, 247–254.

Basser, P.J., Pajevic, S., Pierpaoli, C., Duda, J., Aldroubi, A., 2000. In vivo fiber tractography using DT-MRI data. Magn. Reson. Med. 44, 625–632.

Beaujoin, J., Palomero-Gallagher, N., Boumezbeur, F., Axer, M., Bernard, J., Poupon, F., Schmitz, D., Mangin, J.-F., Poupon, C., 2018. Post-mortem inference of the human hippocampal connectivity and microstructure using ultra-high field diffusion MRI at 11.7 T. Brain Struct. Funct. 223, 2157–2179.

Behrens, T.E.J., Woolrich, M.W., Jenkinson, M., Johansen-Berg, H., Nunes, R.G., Clare, S., Matthews, P.M., Brady, J.M., Smith, S.M., 2003. Characterization and propagation of uncertainty in diffusion-weighted MR imaging. Magn. Reson. Med. 50, 1077–1088.

Boretius, S., Würfel, J., Zipp, F., Frahm, J., Michaelis, T., 2007. High-field diffusion tensor imaging of mouse brain in vivo using single-shot STEAM MRI. J. Neurosci. Methods 161, 112–117.

Budde, M.D., Frank, J.A., 2012. Examining brain microstructure using structure tensor analysis of histological sections. NeuroImage 63, 1–10.

Calabrese, E., Badea, A., Watson, C., Johnson, G.A., 2013. A quantitative magnetic resonance histology atlas of postnatal rat brain development with regional estimates of growth and variability. NeuroImage 71, 196–206.

Calabrese, E., Badea, A., Coe, C.L., Lubach, G.R., Styner, M.A., Johnson, G.A., 2014. Investigating the tradeoffs between spatial resolution and diffusion sampling for brain mapping with diffusion tractography: time well spent? Hum. Brain Mapp. 35, 5667–5685.

Calabrese, E., Badea, A., Coe, C.L., Lubach, G.R., Shi, Y., Styner, M.A., Johnson, G.A., 2015a. A diffusion tensor MRI atlas of the postmortem rhesus macaque brain. NeuroImage 117, 408–416.

Calabrese, E., Badea, A., Cofer, G., Qi, Y., Johnson, G.A., 2015b. A diffusion MRI tractography connectome of the mouse brain and comparison with neuronal tracer data. Cereb. Cortex (New York, NY: 1991) 25, 4628–4637.

Calabrese, E., Hickey, P., Hulette, C., Zhang, J., Parente, B., Lad, S.P., Johnson, G.A., 2015e. Postmortem diffusion MRI of the human brainstem and thalamus for deep brain stimulator electrode localization. Hum. Brain Mapp. 36, 3167–3178.

Callaghan, P.T., 1991. Principles of Nuclear Magnetic Resonance Microscopy. Oxford University Press, Oxford.

Chang, E.H., Argyelan, M., Aggarwal, M., Chandon, T.-S.S., Karlsgodt, K.H., Mori, S., Malhotra, A.K., 2017. The role of myelination in measures of white matter integrity: combination of diffusion tensor imaging and two-photon microscopy of CLARITY intact brains. NeuroImage 147, 253–261.

Chen, H., Liu, T., Zhao, Y., Zhang, T., Li, Y., Li, M., Zhang, H., Kuang, H., Guo, L., Tsien, J.Z., Liu, T., 2015. Optimization of large-scale mouse brain connectome via joint evaluation of DTI and neuron tracing data. NeuroImage 115, 202–213.

Choe, A.S., Stepniewska, I., Colvin, D.C., Ding, Z., Anderson, A.W., 2012. Validation of diffusion tensor MRI in the central nervous system using light microscopy: quantitative comparison of fiber properties. NMR Biomed. 25, 900–908.

Chung, K., Deisseroth, K., 2013. CLARITY for mapping the nervous system. Nat. Methods 10, 508–513.

Crater, S., Maharjan, S., Qi, Y., Zhao, Q., Cofer, G., Cook, J.C., Johnson, G.A., Wang, N., 2022. Resolution and b value dependent structural connectome in ex vivo mouse brain. NeuroImage 255, 119199.

D'Arceuil, H.E., Westmoreland, S., de Crespigny, A.J., 2007. An approach to high resolution diffusion tensor imaging in fixed primate brain. NeuroImage 35, 553–565.

Dauguet, J., Peled, S., Berezovskii, V., Delzescaux, T., Warfield, S.K., Born, R., Westin, C.-F., 2007. Comparison of fiber tracts derived from in-vivo DTI tractography with 3D histological neural tract tracer reconstruction on a macaque brain. NeuroImage 37, 530–538.

de Boer, J.F., Milner, T.E., van Gemert, M.J., Nelson, J.S., 1997. Two-dimensional birefringence imaging in biological tissue by polarization-sensitive optical coherence tomography. Opt. Lett. 22, 934–936.

De Luca, A., Guo, F., Froeling, M., Leemans, A., 2020. Spherical deconvolution with tissue-specific response functions and multi-shell diffusion MRI to estimate multiple fiber orientation distributions (mFODs). NeuroImage 222, 117206.

Dell'Acqua, F., Tournier, J.-D., 2019. Modelling white matter with spherical deconvolution: how and why? NMR Biomed. 32, e3945.

Dell'Acqua, F., Bodi, I., Slater, D., Catani, M., Modo, M., 2013a. MR diffusion histology and micro-tractography reveal mesoscale features of the human cerebellum. Cerebellum (London, England) 12, 923–931.

Dell'Acqua, F., Simmons, A., Williams, S.C., Catani, M., 2013b. Can spherical deconvolution provide more information than fiber orientations? Hindrance modulated orientational anisotropy, a true-tract specific index to characterize white matter diffusion. Hum. Brain Mapp. 34, 2464–2483.

Descoteaux, M., Deriche, R., Knosche, T.R., Anwander, A., 2009. Deterministic and probabilistic tractography based on complex fibre orientation distributions. IEEE Trans. Med. Imaging 28, 269–286.

Donahue, C.J., Sotiropoulos, S.N., Jbabdi, S., Hernandez-Fernandez, M., Behrens, T.E., Dyrby, T.B., Coalson, T., Kennedy, H., Knoblauch, K., Van Essen, D.C., Glasser, M.F., 2016. Using diffusion tractography to predict cortical connection strength and distance: a quantitative comparison with tracers in the monkey. J. Neurosci. 36, 6758.

Dyrby, T.B., Søgaard, L.V., Parker, G.J., Alexander, D.C., Lind, N.M., Baaré, W.F.C., Hay-Schmidt, A., Eriksen, N., Pakkenberg, B., Paulson, O.B., Jelsing, J., 2007. Validation of in vitro probabilistic tractography. NeuroImage 37, 1267–1277.

Dyrby, T.B., Baaré, W.F.C., Alexander, D.C., Jelsing, J., Garde, E., Søgaard, L.V., 2011. An ex vivo imaging pipeline for producing high-quality and high-resolution diffusion-weighted imaging datasets. Hum. Brain Mapp. 32, 544–563.

Edwards, T.J., Fenlon, L.R., Dean, R.J., Bunt, J., Sherr, E.H., Richards, L.J., 2020. Altered structural connectivity networks in a mouse model of complete and partial dysgenesis of the corpus callosum. NeuroImage 217, 116868.

Feng, L., Jeon, T., Yu, Q., Ouyang, M., Peng, Q., Mishra, V., Pletikos, M., Sestan, N., Miller, M.I., Mori, S., Hsiao, S., Liu, S., Huang, H., 2017. Population-averaged macaque brain atlas with high-resolution ex vivo DTI integrated into in vivo space. Brain Struct. Funct. 222, 4131–4147.

Foxley, S., Jbabdi, S., Clare, S., Lam, W., Ansorge, O., Douaud, G., Miller, K., 2014. Improving diffusion-weighted imaging of post-mortem human brains: SSFP at 7 T. NeuroImage 102 (Pt 2), 579–589.

Frank, L.R., 2001. Anisotropy in high angular resolution diffusion-weighted MRI. Magn. Reson. Med. 45, 935–939.

Frank, L.R., 2002. Characterization of anisotropy in high angular resolution diffusion-weighted MRI. Magn. Reson. Med. 47, 1083–1099.

Fritz, F.J., Sengupta, S., Harms, R.L., Tse, D.H., Poser, B.A., Roebroeck, A., 2019. Ultra-high resolution and multi-shell diffusion MRI of intact ex vivo human brains using kT-dSTEAM at 9.4T. NeuroImage 202, 116087.

Gimenez, U., Boulan, B., Mauconduit, F., Taurel, F., Leclercq, M., Denarier, E., Brocard, J., Gory-Fauré, S., Andrieux, A., Lahrech, H., Deloulme, J.C., 2017. 3D imaging of the brain morphology and connectivity defects in a model of psychiatric disorders: MAP6-KO mice. Sci. Rep. 7, 10308.

Girard, G., Caminiti, R., Battaglia-Mayer, A., St-Onge, E., Ambrosen, K.S., Eskildsen, S.F., Krug, K., Dyrby, T.B., Descoteaux, M., Thiran, J.-P., Innocenti, G.M., 2020. On the cortical connectivity in the macaque brain: a comparison of diffusion tractography and histological tracing data. NeuroImage 221, 117201.

Goubran, M., Leuze, C., Hsueh, B., Aswendt, M., Ye, L., Tian, Q., Cheng, M.Y., Crow, A., Steinberg, G.K., McNab, J.A., Deisseroth, K., Zeineh, M., 2019. Multimodal image registration and connectivity analysis for integration of connectomic data from microscopy to MRI. Nat. Commun. 10, 5504.

Grisot, G., Haber, S.N., Yendiki, A., 2021. Diffusion MRI and anatomic tracing in the same brain reveal common failure modes of tractography. NeuroImage 239, 118300.

Guilfoyle, D.N., Helpern, J.A., Lim, K.O., 2003. Diffusion tensor imaging in fixed brain tissue at 7.0 T. NMR Biomed. 16, 77–81.

Guilfoyle, D.N., Gerum, S., Hrabe, J., 2011. Murine diffusion imaging using snapshot interleaved EPI acquisition at 7T. J. Neurosci. Methods 199, 10–14.

Harsan, L.-A., Paul, D., Schnell, S., Kreher, B.W., Hennig, J., Staiger, J.F., von Elverfeldt, D., 2010. In vivo diffusion tensor magnetic resonance imaging and fiber tracking of the mouse brain. NMR Biomed. 23, 884–896.

Harsan, L.-A., Dávid, C., Reisert, M., Schnell, S., Hennig, J., von Elverfeldt, D., Staiger, J.F., 2013. Mapping remodeling of thalamocortical projections in the living *reeler* mouse brain by diffusion tractography. Proc. Natl. Acad. Sci. 110, E1797–E1806.

Henssen, D.J.H.A., Derks, B., van Doorn, M., Verhoogt, N.C., Staats, P., Vissers, K., Van Cappellen van Walsum, A.M., 2019a. Visualizing the trigeminovagal complex in the human medulla by combining ex-vivo ultra-high resolution structural MRI and polarized light imaging microscopy. Sci. Rep. 9, 11305.

Henssen, D.J.H.A., Mollink, J., Kurt, E., van Dongen, R., Bartels, R.H.M.A., Gräßel, D., Kozicz, T., Axer, M., Van Cappellen van Walsum, A.-M., 2019b. Ex vivo visualization of the trigeminal pathways in the human brainstem using 11.7T diffusion MRI combined with microscopy polarized light imaging. Brain Struct. Funct. 224, 159–170.

Henssen, D.J.H.A., Weber, R.C., de Boef, J., Mollink, J., Kozicz, T., Kurt, E., van Cappellen van Walsum, A.-M., 2019d. Post-mortem 11.7 tesla magnetic resonance imaging vs. polarized light imaging microscopy to measure the angle and orientation of dorsal root afferents in the human cervical dorsal root entry zone. Front. Neuroanat. 13.

Howard, A.F.D., Mollink, J., Kleinnijenhuis, M., Pallebage-Gamarallage, M., Bastiani, M., Cottaar, M., Miller, K.L., Jbabdi, S., 2019. Joint modelling of diffusion MRI and microscopy. NeuroImage 201, 116014.

Howard, A., Huszár, I., Mars, R., Mollink, J., Scott, C., Daubney, G., Sibson, N., Smart, A., Sallet, J., Jbabdi, S., Miller, K., 2021. The microscopy connectome: towards 3D PLI tractography in the BigMac dataset. Proc. Int. Soc. Mag. Reson. Med. 29, 0863.

Ingalhalikar, M., Parker, D., Ghanbari, Y., Smith, A., Hua, K., Mori, S., Abel, T., Davatzikos, C., Verma, R., 2014. Connectome and maturation profiles of the developing mouse brain using diffusion tensor imaging. Cereb. Cortex 25, 2696–2706.

Jansons, K.M., Alexander, D.C., 2003. Persistent angular structure: new insights from diffusion magnetic resonance imaging data. Inverse Problems 19, 1031–1046.

Jeurissen, B., Tournier, J.-D., Dhollander, T., Connelly, A., Sijbers, J., 2014. Multi-tissue constrained spherical deconvolution for improved analysis of multi-shell diffusion MRI data. NeuroImage 103, 411–426.

Jiang, Y., Johnson, G.A., 2010. Microscopic diffusion tensor imaging of the mouse brain. NeuroImage 50, 465–471.

Jiang, Y., Johnson, G.A., 2011. Microscopic diffusion tensor atlas of the mouse brain. NeuroImage 56, 1235–1243.

Johnson, G.A., Calabrese, E., Badea, A., Paxinos, G., Watson, C., 2012. A multidimensional magnetic resonance histology atlas of the Wistar rat brain. NeuroImage 62, 1848–1856.

Johnson, G.A., Wang, N., Anderson, R.J., Chen, M., Cofer, G.P., Gee, J.C., Pratson, F., Tustison, N., White, L.E., 2019. Whole mouse brain connectomics. J. Comp. Neurol. 527, 2146–2157.

Jones, R., Grisot, G., Augustinack, J., Magnain, C., Boas, D.A., Fischl, B., Wang, H., Yendiki, A., 2020. Insight into the fundamental trade-offs of diffusion MRI from polarization-sensitive optical coherence tomography in ex vivo human brain. NeuroImage 214, 116704.

Khan, A.R., Cornea, A., Leigland, L.A., Kohama, S.G., Jespersen, S.N., Kroenke, C.D., 2015. 3D structure tensor analysis of light microscopy data for validating diffusion MRI. NeuroImage 111, 192–203.

Kleinnijenhuis, M., Zerbi, V., Küsters, B., Slump, C.H., Barth, M., van Cappellen van Walsum, A.-M., 2013. Layer-specific diffusion weighted imaging in human primary visual cortex in vitro. Cortex 49, 2569–2582.

Leergaard, T.B., White, N.S., de Crespigny, A., Bolstad, I., D'Arceuil, H., Bjaalie, J.G., Dale, A.M., 2010. Quantitative histological validation of diffusion MRI fiber orientation distributions in the rat brain. PLoS ONE 5, e8595.

Lefebvre, J., Castonguay, A., Pouliot, P., Descoteaux, M., Lesage, F., 2017. Whole mouse brain imaging using optical coherence tomography: reconstruction, normalization, segmentation, and comparison with diffusion MRI. Neurophotonics 4, 041501.

Lefebvre, J., Delafontaine-Martel, P., Pouliot, P., Girouard, H., Descoteaux, M., Lesage, F., 2018. Fully automated dual-resolution serial optical coherence tomography aimed at diffusion MRI validation in whole mouse brains. Neurophotonics 5, 045004.

Leuze, C.W.U., Anwander, A., Bazin, P.-L., Dhital, B., Stüber, C., Reimann, K., Geyer, S., Turner, R., 2014. Layer-specific intracortical connectivity revealed with diffusion MRI. Cereb. Cortex (New York, NY: 1991) 24, 328–339.

Leuze, C., Goubran, M., Barakovic, M., Aswendt, M., Tian, Q., Hsueh, B., Crow, A., Weber, E.M.M., Steinberg, G.K., Zeineh, M., Plowey, E.D., Daducci, A., Innocenti, G., Thiran, J.P., Deisseroth, K., McNab, J.A., 2021. Comparison of diffusion MRI and CLARITY fiber orientation estimates in both gray and white matter regions of human and primate brain. NeuroImage 228, 117692.

Lundell, H., Alexander, D.C., Dyrby, T.B., 2014. High angular resolution diffusion imaging with stimulated echoes: compensation and correction in experiment design and analysis. NMR Biomed. 27, 918–925.

Maier-Hein, K.H., Neher, P.F., Houde, J.-C., Côté, M.-A., Garyfallidis, E., Zhong, J., Chamberland, M., Yeh, F.-C., Lin, Y.-C., Ji, Q., Reddick, W.E., Glass, J.O., Chen, D.Q., Feng, Y., Gao, C., Wu, Y., Ma, J., He, R., Li, Q., Westin, C.-F., Deslauriers-Gauthier, S., González, J.O.O., Paquette, M., St-Jean, S., Girard, G., Rheault, F., Sidhu, J., Tax, C.M.W., Guo, F., Mesri, H.Y., Dávid, S., Froeling, M., Heemskerk, A.M., Leemans, A., Boré, A., Pinsard, B., Bedetti, C., Desrosiers, M., Brambati, S., Doyon, J., Sarica, A., Vasta, R., Cerasa, A., Quattrone, A., Yeatman, J., Khan, A.R., Hodges, W., Alexander, S., Romascano, D., Barakovic, M., Auría, A., Esteban, O., Lemkaddem, A., Thiran, J.-P., Cetingul, H.E., Odry, B.L., Mailhe, B., Nadar, M.-S., Pizzagalli, F., Prasad, G., Villalon-Reina, J.E., Galvis, J., Thompson, P.M., Requejo, F.D.S., Laguna, P.L., Lacerda, L.M., Barrett, R., Dell'Acqua, F., Catani, M., Petit, L., Caruyer, E., Daducci, A., Dyrby, T.B., Holland-Letz, T., Hilgetag, C.C., Stieltjes, B., Descoteaux, M., 2017. The challenge of mapping the human connectome based on diffusion tractography. Nat. Commun. 8, 1349.

Modo, M., Hitchens, T.K., Liu, J.R., Richardson, R.M., 2016. Detection of aberrant hippocampal mossy fiber connections: ex vivo mesoscale diffusion MRI and microtractography with histological validation in a patient with uncontrolled temporal lobe epilepsy. Hum. Brain Mapp. 37, 780–795.

Moldrich, R.X., Pannek, K., Hoch, R., Rubenstein, J.L., Kurniawan, N.D., Richards, L.J., 2010. Comparative mouse brain tractography of diffusion magnetic resonance imaging. NeuroImage 51, 1027–1036.

Mollink, J., Kleinnijenhuis, M., van Cappellen van Walsum, A.-M., Sotiropoulos, S.N., Cottaar, M., Mirfin, C., Heinrich, M.P., Jenkinson, M., Pallebage-Gamarallage, M., Ansorge, O., Jbabdi, S., Miller, K.L., 2017. Evaluating fibre orientation dispersion in white matter: comparison of diffusion MRI, histology and polarized light imaging. NeuroImage 157, 561–574.

Mollink, J., Hiemstra, M., Miller, K.L., Huszar, I.N., Jenkinson, M., Raaphorst, J., Wiesmann, M., Ansorge, O., Pallebage-Gamarallage, M., van Cappellen van Walsum, A.M., 2019. White matter changes in the perforant path area in patients with amyotrophic lateral sclerosis. Neuropathol. Appl. Neurobiol. 45, 570–585.

Mori, S., van Zijl, P.C.M., 1998. A motion correction scheme by twin-echo navigation for diffusion-weighted magnetic resonance imaging with multiple RF echo acquisition. Magn. Reson. Med. 40, 511–516.

Mori, S., Kageyama, Y., Hou, Z., Aggarwal, M., Patel, J., Brown, T., Miller, M.I., Wu, D., Troncoso, J.C., 2017. Elucidation of white matter tracts of the human amygdala by detailed comparison between high-resolution postmortem magnetic resonance imaging and histology. Front. Neuroanat. 11.

Neher, P., Stieltjes, B., Wolf, I., Meinzer, H.-P., Maier-Hein, K., 2013. Analysis of tractography biases introduced by anisotropic voxels. In: Proc. of the ISMRM Annual Meeting.

Oh, S.W., Harris, J.A., Ng, L., Winslow, B., Cain, N., Mihalas, S., Wang, Q., Lau, C., Kuan, L., Henry, A.M., Mortrud, M.T., Ouellette, B., Nguyen, T.N., Sorensen, S.A., Slaughterbeck, C.R., Wakeman, W., Li, Y., Feng, D., Ho, A., Nicholas, E., Hirokawa, K.E., Bohn, P., Joines, K.M., Peng, H., Hawrylycz, M.J., Phillips, J.W., Hohmann, J.G., Wohnoutka, P., Gerfen, C.R., Koch, C., Bernard, A., Dang, C., Jones, A.R., Zeng, H., 2014. A mesoscale connectome of the mouse brain. Nature 508, 207–214.

Oshio, K., Feinberg, D.A., 1991. GRASE (gradient- and spin-echo) imaging: a novel fast MRI technique. Magn. Reson. Med. 20, 344–349.

Özarslan, E., Shepherd, T.M., Vemuri, B.C., Blackband, S.J., Mareci, T.H., 2006. Resolution of complex tissue microarchitecture using the diffusion orientation transform (DOT). NeuroImage 31, 1086–1103.

Pipe, J.G., Farthing, V.G., Forbes, K.P., 2002. Multishot diffusion-weighted FSE using PROPELLER MRI. Magn. Reson. Med. 47, 42–52.

Rane, S., Nair, G., Duong, T.Q., 2010. DTI at long diffusion time improves fiber tracking. NMR Biomed. 23, 459–465.

Reese, T.G., Heid, O., Weisskoff, R.M., Wedeen, V.J., 2003. Reduction of eddy-current-induced distortion in diffusion MRI using a twice-refocused spin echo. Magn. Reson. Med. 49, 177–182.

Roebroeck, A., Miller, K.L., Aggarwal, M., 2019. Ex vivo diffusion MRI of the human brain: technical challenges and recent advances. NMR Biomed. 32, e3941.

Schilling, K., Janve, V., Gao, Y., Stepniewska, I., Landman, B.A., Anderson, A.W., 2016. Comparison of 3D orientation distribution functions measured with confocal microscopy and diffusion MRI. NeuroImage 129, 185–197.

Schilling, K.G., Janve, V., Gao, Y., Stepniewska, I., Landman, B.A., Anderson, A.W., 2018. Histological validation of diffusion MRI fiber orientation distributions and dispersion. NeuroImage 165, 200–221.

Schilling, K.G., Gao, Y., Stepniewska, I., Janve, V., Landman, B.A., Anderson, A.W., 2019a. Anatomical accuracy of standard-practice tractography algorithms in the motor system - a histological validation in the squirrel monkey brain. Magn. Reson. Imaging 55, 7–25.

Schilling, K.G., Gao, Y., Stepniewska, I., Janve, V., Landman, B.A., Anderson, A.W., 2019b. Histologically derived fiber response functions for diffusion MRI vary across white matter fibers—an ex vivo validation study in the squirrel monkey brain. NMR Biomed. 32, e4090.

Schilling, K.G., Nath, V., Hansen, C., Parvathaneni, P., Blaber, J., Gao, Y., Neher, P., Aydogan, D.B., Shi, Y., Ocampo-Pineda, M., Schiavi, S., Daducci, A., Girard, G., Barakovic, M., Rafael-Patino, J., Romascano, D., Rensonnet, G., Pizzolato, M., Bates, A., Fischi, E., Thiran, J.-P., Canales-Rodríguez, E.J., Huang, C., Zhu, H., Zhong, L., Cabeen, R., Toga, A.W., Rheault, F., Theaud, G., Houde, J.-C., Sidhu, J., Chamberland, M., Westin, C.-F., Dyrby, T.B., Verma, R., Rathi, Y., Irfanoglu, M.O., Thomas, C., Pierpaoli, C., Descoteaux, M., Anderson, A.W., Landman, B.A., 2019d. Limits to anatomical accuracy of diffusion tractography using modern approaches. NeuroImage 185, 1–11.

Seehaus, A., Roebroeck, A., Bastiani, M., Fonseca, L., Bratzke, H., Lori, N., Vilanova, A., Goebel, R., Galuske, R., 2015. Histological validation of high-resolution DTI in human post mortem tissue. Front. Neuroanat. 9.

Sinke, M.R.T., Otte, W.M., Christiaens, D., Schmitt, O., Leemans, A., van der Toorn, A., Sarabdjitsingh, R.A., Joëls, M., Dijkhuizen, R.M., 2018. Diffusion MRI-based cortical connectome reconstruction: dependency on tractography procedures and neuroanatomical characteristics. Brain Struct. Funct. 223, 2269–2285.

Stephan, K.E., Kamper, L., Bozkurt, A., Burns, G.A., Young, M.P., Kötter, R., 2001. Advanced database methodology for the collation of connectivity data on the macaque brain (CoCoMac). Philos. Trans. R. Soc. Lond. Ser. B Biol. Sci. 356, 1159–1186.

Stolp, H.B., Ball, G., So, P.W., Tournier, J.D., Jones, M., Thornton, C., Edwards, A.D., 2018. Voxel-wise comparisons of cellular microstructure and diffusion-MRI in mouse hippocampus using 3D bridging of optically-clear histology with neuroimaging data (3D-BOND). Sci. Rep. 8, 4011.

Susaki, E.A., Tainaka, K., Perrin, D., Kishino, F., Tawara, T., Watanabe, T.M., Yokoyama, C., Onoe, H., Eguchi, M., Yamaguchi, S., Abe, T., Kiyonari, H., Shimizu, Y., Miyawaki, A., Yokota, H., Ueda, H.R., 2014. Whole-brain imaging with single-cell resolution using chemical cocktails and computational analysis. Cell 157, 726–739.

Thomas, C., Ye, F.Q., Irfanoglu, M.O., Modi, P., Saleem, K.S., Leopold, D.A., Pierpaoli, C., 2014. Anatomical accuracy of brain connections derived from diffusion MRI tractography is inherently limited. Proc. Natl. Acad. Sci. 111, 16574–16579.

Tournier, J.D., Calamante, F., Gadian, D.G., Connelly, A., 2004. Direct estimation of the fiber orientation density function from diffusion-weighted MRI data using spherical deconvolution. NeuroImage 23, 1176–1185.

Tu, T.-W., Turtzo, L.C., Williams, R.A., Lescher, J.D., Dean, D.D., Frank, J.A., 2014. Imaging of spontaneous ventriculomegaly and vascular malformations in Wistar rats: implications for preclinical research. J. Neuropathol. Exp. Neurol. 73, 1152–1165.

Tuch, D.S., 2004. Q-ball imaging. Magn. Reson. Med. 52, 1358–1372.

Tyszka, J.M., Frank, L.R., 2009. High-field diffusion MR histology: image-based correction of eddy-current ghosts in diffusion-weighted rapid acquisition with relaxation enhancement (DW-RARE). Magn. Reson. Med. 61, 728–733.

van den Heuvel, M.P., de Reus, M.A., Feldman Barrett, L., Scholtens, L.H., Coopmans, F.M., Schmidt, R., Preuss, T.M., Rilling, J.K., Li, L., 2015. Comparison of diffusion tractography and tract-tracing measures of connectivity strength in rhesus macaque connectome. Hum. Brain Mapp. 36, 3064–3075.

Vega-Pons, S., Olivetti, E., Avesani, P., Dodero, L., Gozzi, A., Bifone, A., 2017. Differential effects of brain disorders on structural and functional connectivity. Front. Neurosci. 10.

Wang, H., Zhu, J., Reuter, M., Vinke, L.N., Yendiki, A., Boas, D.A., Fischl, B., Akkin, T., 2014. Cross-validation of serial optical coherence scanning and diffusion tensor imaging: a study on neural fiber maps in human medulla oblongata. NeuroImage 100, 395–404.

Wang, N., Anderson, R.J., Badea, A., Cofer, G., Dibb, R., Qi, Y., Johnson, G.A., 2018. Whole mouse brain structural connectomics using magnetic resonance histology. Brain Struct. Funct. 223, 4323–4335.

Wang, N., White, L.E., Qi, Y., Cofer, G., Johnson, G.A., 2020. Cytoarchitecture of the mouse brain by high resolution diffusion magnetic resonance imaging. NeuroImage 216, 116876.

Warrington, S., Bryant, K.L., Khrapitchev, A.A., Sallet, J., Charquero-Ballester, M., Douaud, G., Jbabdi, S., Mars, R.B., Sotiropoulos, S.N., 2020. XTRACT—standardised protocols for automated tractography in the human and macaque brain. NeuroImage 217, 116923.

Wedeen, V.J., Hagmann, P., Tseng, W.-Y.I., Reese, T.G., Weisskoff, R.M., 2005. Mapping complex tissue architecture with diffusion spectrum magnetic resonance imaging. Magn. Reson. Med. 54, 1377–1386.

Wedeen, V.J., Wang, R.P., Schmahmann, J.D., Benner, T., Tseng, W.Y.I., Dai, G., Pandya, D.N., Hagmann, P., D'Arceuil, H., de Crespigny, A.J., 2008. Diffusion spectrum magnetic resonance imaging (DSI) tractography of crossing fibers. NeuroImage 41, 1267–1277.

Wu, D., Zhang, J., 2016. In vivo mapping of macroscopic neuronal projections in the mouse hippocampus using high-resolution diffusion MRI. NeuroImage 125, 84–93.

Wu, D., Xu, J., McMahon, M.T., van Zijl, P.C.M., Mori, S., Northington, F.J., Zhang, J., 2013. In vivo high-resolution diffusion tensor imaging of the mouse brain. NeuroImage 83, 18–26.

Yeh, F.C., Wedeen, V.J., Tseng, W.Y., 2010. Generalized q-sampling imaging. IEEE Trans. Med. Imaging 29, 1626–1635.

Yendiki, A., Aggarwal, M., Axer, M., Howard, A.F.D., van Cappellen van Walsum, A.-M., Haber, S.N., 2022. Post mortem mapping of connectional anatomy for the validation of diffusion MRI. NeuroImage 256, 119146.

# Chapter 33

# Multicenter studies and harmonization: Problems, solutions, and open challenges

**Chantal M.W. Tax[a,b], Suheyla Cetin Karayumak[c], Kurt G. Schilling[d,e], Daniel Moyer[f], Bennett A. Landman[d,e,g], Neda Jahanshad[h], and Yogesh Rathi[c]**

[a]CUBRIC, Cardiff University, Cardiff, United Kingdom, [b]UMC Utrecht, Utrecht University, Utrecht, The Netherlands, [c]Brigham and Women's Hospital, Harvard Medical School, Boston, MA, United States, [d]Department of Radiology and Radiological Sciences, Vanderbilt University Medical Center, Nashville, TN, United States, [e]Vanderbilt University Institute of Imaging Science, Vanderbilt University Medical Center, Nashville, TN, United States, [f]CSAIL, Massachusetts Institute of Technology, Cambridge, MA, United States, [g]Department of Electrical and Computer Engineering, Vanderbilt University, Nashville, TN, United States, [h]Imaging Genetics Center, Mark and Mary Stevens Neuroimaging and Informatics Institute, University of Southern California, Marina del Rey, CA, United States

## 1 Introduction

As outlined in other chapters of this book, diffusion magnetic resonance imaging (dMRI) can map brain connectivity and microstructural measures noninvasively and thus provide critical tools for understanding focal as well as network level abnormalities in the brain. The measures derived from dMRI are quite sensitive to the structure of the underlying tissue and have contributed to a wealth of knowledge about the abnormalities in several neurological and psychiatric disorders. However, results from neuroimaging studies often show inconsistent and sometimes contradictory results due to small sample sizes as well as differences in acquisition parameters and postprocessing methods (Melonakos et al., 2011). To address these challenges, several neuroimaging studies from a multitude of consortia are collecting a huge repository of brain imaging data pertaining to healthy subjects and those with neuropsychiatric disorders, including the Human Connectome Project in disease (Van Essen et al., 2014), the Adolescent Brain Cognitive Development study (Barch et al., 2018), the UK Biobank (Miller et al., 2016), and the Healthy Brain Network biobank (Alexander et al., 2017), which will collect phenotypic and neuroimaging data from over tens of thousands of subjects in total at over 50 sites around the world. These large datasets have huge potential scientific benefit, including the opportunity to understand complex neural systems across the lifespan and across mental disorders. Enabling pooled large-scale analysis of these datasets represents an outstanding opportunity for improving our understanding of the human brain in health and disease, often by reusing datasets for applications beyond their original purpose. Comparing and contrasting the overlap in abnormalities across diagnostic categories becomes feasible when datasets to study disparate conditions are merged. For example, it will be extremely useful to study the common and disparate patterns of abnormalities between Alzheimer's disease (Harrison et al., 2020) and fronto-temporal dementia. However, such research requires large sample sizes, which can only be obtained by appropriate pooling of dMRI data acquired from multiple MRI systems. Oftentimes, the scientific potential of pooling dMRI data is not realized until after data collection, and thus the pooling must be done retrospectively. While more and more datasets are now becoming openly available for data reuse, our ability to jointly analyze them is currently hampered by important technical challenges.

When combining acquisitions—where an acquisition is typically constituted of a collection of dMRIs at a specific time point, from a single subject, in an uninterrupted MR session, with a specific set of MR parameters—one is ultimately interested in variability purely associated with the anatomy, but other sources of variability can obscure the effect of interest. For example, even with the same acquisition parameters, studies investigating variability through region of interest (ROI)-based analysis on traveling subjects show that the interscanner variability in fractional anisotropy (FA) in the white matter could be as much as 5%–10%, and more than 10%–25% in the gray matter regions (Vollmar et al., 2010). On the other hand, differences in FA due to changes in the neurobiology of the tissue in schizophrenia and other psychiatric disorders could be as low as 5% (Del Re et al., 2019). Thus a simple aggregation or "pooling" of data from multiple sites and MRI scanners, even with the same acquisition parameters, can lead to erroneous results. Apart from the previously

---

Handbook of Diffusion MR Tractography. https://doi.org/10.1016/B978-0-12-818894-1.00039-2
Copyright © 2025 Elsevier Ltd. All rights are reserved, including those for text and data mining, AI training, and similar technologies.

**670 PART | V** Tractography applications

mentioned variabilities, the problem of combining datasets from multiple sites becomes even more challenging when one considers that variability can arise at each step of the typically lengthy dMRI acquisition and processing pipeline. Thus harmonization or statistically accounting for such differences along with a consistent processing pipeline is of utmost importance for reproducible neuroscientific research with increased statistical power. This chapter focuses on the sources of variability in multicenter studies, solutions that have been proposed to deal with variability in joint analyses, and open challenges in the era of "big data" for dMRI and tractography.

## 2 The problem

When assessing population differences with respect to a specific trait (e.g., case/control differences in white matter microstructure), statistical approaches most often try to understand how the biological differences explain the variability in the overall distribution of the trait. However, the variability in the population is not purely driven by biological variables and can be confounded with methodological choices and noise. Variability can originate at different steps of the acquisition and processing pipeline. In this section we will discuss how variability is typically quantified and outline the different sources of variability in more detail.

### 2.1 Quantifying variability

Measurement variability or dispersion across different instances of an experimental setting (while keeping other settings fixed) can be quantified in different ways. Measures of statistical variability or dissimilarity are generally zero if all the data are the same and increases as the data become more diverse, and vice versa for measures of correlation or similarity. We will describe here some commonly used strategies to quantify variability of diffusion measures, tractography streamlines, and connectivity matrices.

#### 2.1.1 Scalar diffusion measures

The variability in scalar measures such as FA or mean diffusivity (MD) within an ROI can be quantified with common approaches such as the standard deviation, range, or mean absolute difference, and their robust variants such as median absolute deviation and interquartile range. While these measures have the same units as the quantity being measured, the coefficient of variation (or ratio of standard deviation to the mean) has been adopted as a dimensionless measure, for example, to quantify interscanner variability (Grech-Sollars et al., 2015). The intraclass correlation coefficient (ICC) is a dimensionless measure that can be used when measurements are organized into groups, that is, to describe how strongly measurements in the same group resemble each other. This has been used to assess reproducibility of intrascanner measurements, that is, multiple repeated measurements across different subjects (Koller et al., 2021).

#### 2.1.2 Tractography streamlines

For tractography results, variability can be assessed by computing the volume Dice overlap between binary volumes in which voxels have an intensity of 1 if they are traversed by a streamline. Instead of comparing binary volumes, streamline-density maps can be compared, for example, by calculating a Pearson correlation coefficient (Rheault et al., 2020). Another volume-based measure that has been used is bundle adjacency as a modification of the Hausdorff distance, that is, the average of the nearest distance from nonoverlapping voxels in one bundle to a second bundle and vice versa. If variability in bundle segmentation is assessed from the same set of underlying streamlines, a streamline Dice overlap can be computed that represents the intersection of streamlines divided by the union of all streamlines in a bundle (Schilling et al., 2020). Another potential metric to calculate overlap between fiber bundles is to calculate a probability distribution of the spatial coordinates of the streamlines and compute the Bhattacharyya distance between the probability distributions (Rathi et al., 2013). This also has the advantage (like the Dice overlap) of the measure being restricted to between 0 and 1.

#### 2.1.3 Connectivity matrices

Some studies have also assessed variability in connectivity matrices generated from tractography. For example, the mean difference in normalized connectivity—defined as the fraction of streamlines connecting different regions—can give insight into the overall variability (Girard et al., 2015b). In addition, the Pearson correlation coefficient has been used, as well as more advanced measures of variability using graph matching techniques (Osmanlıoğlu et al., 2020). Finally the variability of derived graph theoretical measures such as modularity or global efficiency can be evaluated with previously discussed strategies such as ICC (Roine et al., 2019).

## 2.2 Sources of variability

Variability in multicenter studies can occur at each stage where researchers can consciously or unconsciously make different choices. The sources discussed in this section cover aspects from the design of the experiment to the acquisition and finally the processing, and it is important to realize that sources of variability can accumulate with each step.

### 2.2.1 Intrinsic variability

Sources of intrinsic variability include those that are often beyond the direct control of the researcher performing the experiment, that is, variability that (1) necessarily or intrinsically occurs when the same person is scanned on the same scanner in two or more consecutive sessions, (2) is related to the biology of the subject, and (3) occurs when the same person is scanned on different scanners and at different sites (Fig. 1).

#### Intrascanner

Intrascanner variability, or test-retest variability, is the intrinsic or necessary variability when the same experiment is performed on the same subject successively. Thermal noise, produced by the stochastic motion of electrons in the coil and inductive eddy currents in the subject, adds a random source of variation to each image that cannot be controlled (Aja-Fernández and Vegas-Sánchez-Ferrero, 2016). The usage and load of the scanner prior to the experiment can affect the measurement; patterns of signal intensity drift are, for example, different on a "cold" scanner than on a scanner that has been used just prior to the experiment (Vos et al., 2017). Furthermore, it is difficult to position the participant in exactly the same way in successive experiments, although strict guidelines and 3D-printed padding can reduce variability, for example, in longitudinal studies. Nonuniformities of the gradients, which become more prominent with stronger gradients (Setsompop et al., 2013; Jones et al., 2018) or gradient head inserts, are spatially varying and thus can amplify variability in geometric distortions and diffusion weighting when the subject is positioned differently. Variability can also be caused by

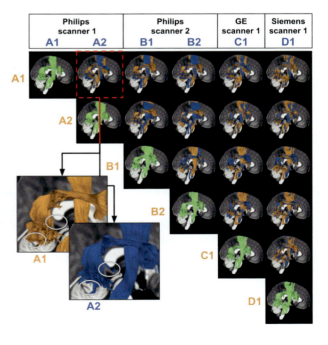

**FIG. 1** Examples of intrinsic variability: variability that occurs on the same scanner, across different scanners, and across different vendors. A single subject was scanned on four scanners (A–D) with repeat scans on scanner A (A1 and A2) and scanner B (B1 and B2). Comparisons of scalar measures (FA) are shown on the *left*, and visualization of tractography differences on the *right*. For FA, the derived value is shown along the diagonal, while off diagonals show the magnitude of differences between sessions (scaled from −0.2 to +0.2). For tractography, streamlines from different sessions are displayed as different colors. Data was collected from part of the MASiVar database (Cai et al., 2021).

## 672 PART | V Tractography applications

**TABLE 1** Overview of studies assessing intra- and interscanner variability of dMRI measures; see also Vollmar et al. (2010) and Zhu et al. (2019).

| Study | Scanner | Acquisition | Subjects | Repeated scans |
|---|---|---|---|---|
| Bisdas et al. (2008) | 3T Philips Intera | 16 Directions, $2 \times 2 \times 3$ mm | 12 | Intrascanner 2 rescan |
| Jansen and Kooi (2007) | 3T Philips Achieva | 15 Directions, $2 \times 2 \times 2$ mm | 10 | Intrascanner 2 rescan |
| Bonekamp and Nagae (2007) | 1.5T GE | 15 Directions $\times 2$, $2.5 \times 2.5 \times 5$ mm | 10 | Intrascanner 2 rescan |
| Heiervang and Behrens (2006) | 1.5T Siemens Sonata | 60 Directions, $2 \times 2 \times 2$ mm | 8 | Intrascanner 2 rescan |
| Ciccarelli and Parker (2003) | 1.5T GE Signa | 60 Directions, $2.5 \times 2.5 \times 2.5$ mm | 10 | Intrascanner rescan |
| Cercignani and Bammer (2003) | 1.5T Philips Gyroscan and Siemens Vision | 6 Directions $\times 10$ or 8 directions $\times 8$, $1.95 \times 1.95 \times 5$ mm | 12 | Intrascanner 2 rescan, intersequence 3 rescan, interscanner 2 rescan |
| Pfefferbaum and Adalsteinsson (2003) | 1.5T GE Echospeed and GE Twinspeed | 6 Directions $\times 6$ | 10 | Intrascanner 2 rescan, interscanner 2 rescan |
| Vollmar et al. (2010) | 3T GE Signa | 32 Directions, $2.4 \times 2.4 \times 2.4$ mm | 9 | Intrascanner 2 rescan, interscanner 2 rescan |
| Grech-Sollars et al. (2015) | 1.5T Siemens Avanto $\times 3$ and 1.5T Siemens Symphony 1.5T and 3T Philips Achieva $\times 4$ | 15–60 Directions, variable resolution | 4–8 and ice water phantom | Intrascanner rescan, interscanner rescan |
| O'Connor et al. (2017) | 1.5T Siemens Avanto | 64 Directions $2 \times 2 \times 2$ mm | 13 | Intrascanner 11 rescan |
| Tong et al. (2019) | 3T Siemens Magnetom Prisma $\times 8$ | 90 Directions $1.5 \times 1.5 \times 1.5$ mm | 3 | Interscanner, intrascanner 2 rescan |
| Koller et al. (2021) | 3T Siemens Connectome | 233 Directions $2 \times 2 \times 2$ mm | 6 | Intrascanner 4 rescan |

variations in calibration of the magnetic field gradients in the scanner. Finally, shimming is commonly performed prior to each experiment to make the magnetic field more homogeneous and can introduce further differences.

Several studies have investigated test-retest reproducibility by rescanning the same subjects on the same scanner (Table 1), and results generally vary with the region studied and analysis approach taken. Vollmar et al. (2010) generated reproducibility maps of repeated scans and found a 5% relative change of FA in the retest scan in WM, but up to 25% in GM. Koller et al. (2021) reported an average CV of up to 1% for FA and MD in different tracts, and up to 2.1% for the restricted signal fraction from CHARMED (Assaf and Basser, 2005).

When the same subject is scanned on the same scanner with the same protocol some time apart, scanner-software updates can introduce differences. It is often difficult to predict how software updates affect the results, and whether this is specific to a certain stage (raw data storage, reconstruction) or sequence. Unless extensive information is available from the vendor and the data can be retrospectively adjusted, software updates are ideally avoided entirely during the course of a study.

### Subject-related

Intrinsic sources of variability can also be biological in nature, and may be undesired if it is not the effect of interest. For example, it is well known that dMRI measures vary as a function of age, which can hamper longitudinal studies not primarily looking at age effects. Additional subject-related sources of variability could include caffeine intake prior to the experiment, amount of sleep, time during the menstrual period, and time of day. For example, Thomas et al. (2018) have shown an increase of mean diffusivity from morning to afternoon scans at the interface of the gray matter/cerebrospinal fluid (CSF), highlighting the time of day as a possible confounding factor in experiment design.

### Interscanner and intersite

Several studies have shown variability in dMRI measures when the same subject is scanned on different scanners with the same acquisition protocol (Table 1). Such variability was observed both across scanners from different vendors and across scanners from the same vendor. While other sources discussed before may have played a role in these studies (e.g., participant positioning, shimming, time of day), scanner-specific sources can further contribute to the observed variation. Tong et al. (2019) investigated reproducibility across eight scanners of the same vendor, type, and software version, with identical multishell dMRI acquisition protocols. They found that intrascanner reproducibility of track density and structural connectomes generated from multishell data was higher than interscanner, and that multishell schemes can produce higher reproducibility and precision across scanners compared to single-shell schemes.

Both software and hardware differences can further contribute to interscanner variability. For example, scanners can vary in field strength (e.g., 1.5, 3, or 7T) and gradient strength (e.g., 40, 80, or 300 mT/m), and the designs of coil and gradient systems can vary accordingly or between different types of scanners with the same field or gradient strength. This yields, for example, differences in gradient nonlinearities, which are more pronounced at ultrastrong gradient strengths (Bammer et al., 2003; Setsompop et al., 2013; Jones et al., 2018). Tax et al. (2018) acquired a database of the same 14 subjects scanned on three different 3T scanners with different gradient strengths to facilitate the development of cross-scanner harmonization strategies. Using this database, Chamberland et al. (2018) investigated the effect of acquisition and gradient strength on the virtual reconstruction of the anterior portion of the optic radiation (Meyer's loop), relevant for temporal lobe epilepsy surgery. Stronger gradients facilitated the acquisition of higher $b$-values, resolution, and increased number of directions, and resulted in increased anatomical fidelity in agreement with ex vivo dissection studies. Grech-Sollars et al. (2015) found an average interscanner CV of less than 4% for DTI measures MD and FA in WM and GM, in agreement with values reported by Vollmar et al. (2010). Finally, head coils can differ in their design and number of channels, and an independent quantification is necessary to assess its effect.

### 2.2.2 Acquisition

A typical MRI acquisition consists of an encoding block to sensitize the signal to different MR phenomena, and a readout block to record the image. Here we will discuss sources of variability at the acquisition stage, that is, (1) the encoding of diffusion, (2) the encoding of other phenomena such as relaxation, and (3) image readout and reconstruction.

### Encoding of diffusion

The diffusion-encoding block consists of the deliberate application of magnetic field gradients to sensitize the signal to the random motion of molecules. The timings and strengths of these gradients along each axis ultimately determine how the diffusion process is encoded in the signal. Taking into account hardware constraints (typically a maximum gradient strength of 40 mT/m and slew rate of 200 mT/m/s for clinical scanners) and sequence constraints (Sjölund et al., 2015), a plethora of options—and thus sources of variability—exist for the design of such gradient waveforms. The most commonly used encoding scheme is Stejskal-Tanner encoding (Stejskal and Tanner, 1965, also called pulsed gradient or linear tensor encoding), which consists of two trapezoidal pulses defined by their duration $\delta$, separation $\Delta$, and gradient strength along each axis $\mathbf{g}$ (with $g = |\mathbf{g}|$). In practice, scanner vendors often ask the user to define a $b$-value $b = \gamma^2 g^2 \delta^2 (\Delta - \delta/3)$ without explicit control over the gradient strengths and timings of the gradients. In addition to differences in $b$-value (Fig. 2a), differences in timings for the same $b$-value can further cause variability between scanners of different vendors or software versions. In addition, it is not always possible to input a customized set of gradient directions, causing additional variation if the available options are not harmonized across platforms. Other commonly used encoding schemes include the use of bipolar gradients on each side of the refocusing pulse to reduce the effect of eddy currents (Alexander et al., 1997).

### Encoding of other phenomena

During the time for diffusion-encoding, other phenomena inherently take place such as longitudinal and transverse relaxation, magnetization transfer, and exchange. The duration of the encoding block thus also influences the degree to which the signal is modulated by these other processes; dMRI is most commonly performed within the context of the spin-echo (SE) experiment, which consists of a 90-degree and a 180-degree radio-frequency pulse. Differences in the timings of these pulses—commonly summarized by the echo time (TE) and repetition time (TR)—can lead to additional variability. For

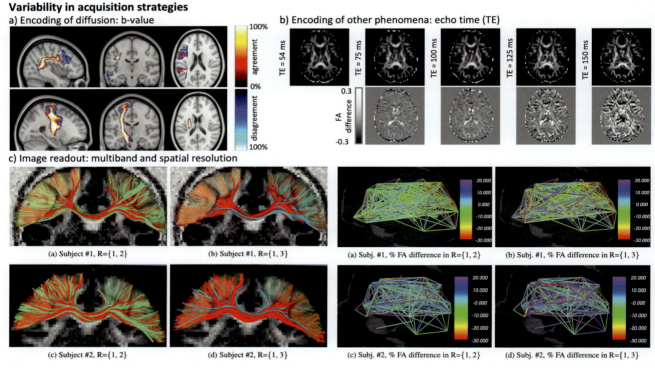

FIG. 2 Examples of variability introduced by differences in processing. (a) Tractography variability of the Arcuate Fasciculus (*top*) and corticospinal tract (*bottom*) between b-value 1200 and 3000 s/mm². Maps show overlap (i.e., agreement) and nonoverlap (i.e., disagreement), averaged across multiple subjects. The percent agreement indicates areas where a pathway is consistently located and is shown using a "hot" colormap, while the percent disagreement indicates areas without consistent overlap and is shown using a "cold" colormap (Schilling et al., 2021). (b) Variability due to differences in the encoding of other phenomena than diffusion: diffusion estimates depend on echo time (Tax et al., 2021). *Top row*: Estimated FA per TE, *bottom row*: difference in FA with the shortest TE. (c) Tractography for different acceleration factors *R*, for a spatial resolution of 2 mm isotropic (Subject 1, *top*) and 2.5 mm isotropic (Subject 2, *bottom*). *Left*: Qualitative overlap of tractography of the interhemispheric precentral fibers. *Right*: Connectivity graph color coded according to the percentage difference in FA, showing an increase in difference for $R = 3$. *(From Rathi, Y., Gagoski, B., Setsompop, K., Grant, P.E., Westin, C.F., 2014. Comparing simultaneous multi-slice diffusion acquisitions. In: Schultz, T., Nedjati-Gilani, G., Venkataraman, A., O'Donnell, L., Panagiotaki, E. (Eds.), Computational Diffusion MRI and Brain Connectivity. Springer International Publishing, Cham, pp. 3–11, with permission.)*

example, the TE is usually set as short as possible to reduce the effect of transverse relaxation and therefore increase the signal-to-noise ratio (SNR), but the minimum TE depends on the desired b-value and maximum gradient strength of the hardware, among others (Jones et al., 2018). In multicenter studies with different gradient hardware, a balanced trade-off therefore has to be made between reducing variability and optimizing SNR. The TR influences the duration of the scan and is therefore ideally kept as short as possible, but if kept too short the SNR will be penalized because of incomplete longitudinal relaxation. Cardiac gating can furthermore influence the TR and cause variability in data quality if not used consistently between acquisitions.

### Image readout and reconstruction

Another source of variability in the data stems from the image readout and reconstruction algorithms used. To reduce acquisition time by exploiting redundancy in the data, only a partial set of measurements are acquired for each slice using an echo-planar imaging (EPI) sequence (a standard method typically used for in vivo dMRI experiments), that is, the so-called partial Fourier readout. In this case, typical choices are six-eighth or seven-eighth partial Fourier acquisition. The rest of the data in k-space can be estimated using different algorithms, including zero-filling or projection onto convex sets (McGibney et al., 1993). In addition, variation exists in the number of receiver coils used (16, 20, 32, or 64 channels), and the coil combination algorithm used during parallel imaging. Parallel imaging involves skipping the $k_y$-lines (from k-space Fourier data) during acquisition and subsequently estimating them using a particular algorithm. For example, one could use SENSE (Pruessmann et al., 1999) or GRAPPA (Griswold et al., 2002) to estimate these missing $k_y$-lines and finally combine the data from each channel to obtain the full k-space data. For example, one could use the sum-of-squares algorithm or adaptive combine to combine data from each of the coils, depending on what is made available by the vendor. The final image-space data will have a different noise distribution depending on the type of algorithm used to combine the coil data. Other factors such as the use of multiband sequences also affect the reconstructed data and hence contribute to bias and variability in the acquired data (Fig. 2c; Rathi et al., 2014).

### 2.2.3 Processing

Once the data is acquired, it needs to be processed for further (statistical) analysis. Here we discuss variability that can occur during (1) preprocessing to correct for image artifacts, (2) model estimation, and (3) tractography.

#### Preprocessing

The dMRI preprocessing pipeline typically consists of multiple steps, such as correction for subject motion and geometrical image deformations due to eddy currents and susceptibility differences (see Chapter 9). However, no consensus exists on which steps to include (an example of variability when omitting drift correction is shown in Fig. 3a), in which exact order the steps are to be performed, and which software to use for each step. It is this lack of consensus that increases variability when datasets are processed in different centers. In addition, some algorithms require specific additional information or images that are not always available. For example, to correct for geometrical EPI image distortions due to susceptibility differences, it has become common practice to acquire data with opposite phase encoding direction to estimate the deformation field. However, legacy data without such acquisitions require an alternative strategy to correct for these distortions (e.g., by registering to a structural $T_1$-weighted image), which creates a necessary variability in the preprocessing when pooled with data for which reverse phase encoding images are available (as, e.g., in Tax et al., 2019a; Ning et al., 2020).

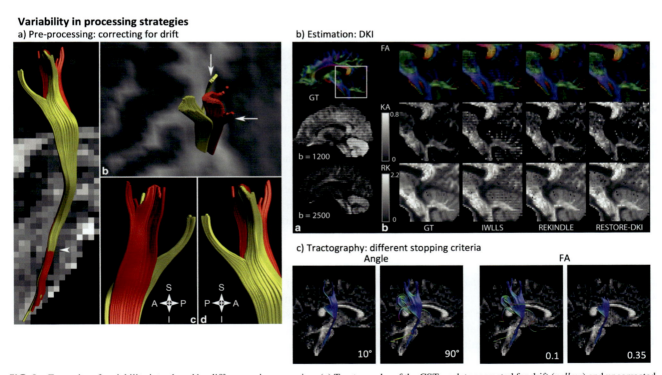

**FIG. 3** Examples of variability introduced by differences in processing. (a) Tractography of the CST on data corrected for drift (*yellow*) and uncorrected (*red*) from the same single-voxel seed location (*left*). Streamlines terminate in different gyri (*right top*). (b) Estimation of the diffusion and kurtosis tensor and derived measures (fractional anisotropy [FA], kurtosis anisotropy [KA], and radial kurtosis [RK]). Outliers were simulated in ground truth (GT) data, and different estimation algorithms can lead to variability in the parameter maps. (c) The effect of different stopping criteria for tractography: angle threshold (*left*) and FA threshold (*right*). ((a) From Vos, S.B., Tax, C.M.W., Luijten, P.R., Ourselin, S., Leemans, A., Froeling, M., 2017. The importance of correcting for signal drift in diffusion MRI. Magn. Reson. Med. 77, 285–299. https://doi.org/10.1002/mrm.26124, with permission; (b) From Tax, C.M.W., Otte, W.M., Viergever, M.A., Dijkhuizen, R.M., Leemans, A., 2015. REKINDLE: robust extraction of kurtosis indices with linear estimation. Magn. Reson. Med. 73 (2), 794–808. https://doi.org/10.1002/MRM.25165, with permission.)

## Estimation

Once the data is preprocessed, a model or representation can be fit to extract meaningful features from the data (see Chapters 10–12). The breadth of approaches developed by the community can in and of itself introduce variability; there are, for example, multiple ways to extract and represent a fiber orientation distribution function (fODF) including spherical harmonics (Tournier et al., 2004), via discrete directions on the unit sphere (Dell'Acqua et al., 2007), or a Watson distribution (Zhang et al., 2012). Such differences in representation may introduce differences in the estimated fiber direction or apparent fiber density (Raffelt et al., 2012; Dell'Acqua et al., 2013) ultimately used for analysis. But even for a given representation, variations in parameter estimation strategies can affect the outcome. For example, to estimate the six parameters of the diffusion tensor (DT, i.e., $D_{xx}, D_{xy}, D_{xz}, D_{yy}, D_{yz}, D_{zz}$), one can choose between a range of fitting methods such as least squares (LS), maximum likelihood, and Bayesian methods. LS approaches have been extensively studied and can be fast, and even within this class of methods different implementations are employed by different software packages such as nonlinear least squares (NLS), linear least squares (LLS), and weighted linear least squares (WLLS). NLS and WLLS are iterative strategies and the outcome may depend on how the first iteration is initiated (Veraart et al., 2013). Robust approaches aim to reduce the effect of erroneous measurements (outliers), which can occur, for example, due to motion. Not employing robust estimation strategies in the presence of outliers can significantly affect estimates and tractography (Fig. 3b). In addition to different estimation methods, different constraints and regularizations can be imposed that affect the result. Common constraints on the DT aim for its eigenvalues to be positive, and spatial regularization can be employed to suppress the influence of noise by using neighborhood information.

## Tractography

In its most straightforward form, the process of fiber tractography pieces together the voxel-wise estimates of fiber orientation to represent global fiber trajectories (Chapters 13 and 14). A number of choices in this process may introduce variability if done differently in different studies. First, intuitively, tractography is highly dependent on the local fiber orientation estimation. The DT, described earlier, is currently the most widely employed representation for characterizing orientation. However, the DT is only capable of estimating a single orientation per voxel, whereas the brain is known to contain a complex mixture of crossing, fanning, and branching fiber systems. A range of methods, based on so-called "high angular resolution diffusion imaging," or HARDI, acquisitions have been developed to mitigate this problem (Chapters 10 and 11). These reconstruction methods each have their own strengths and limitations, but can lead to differences in estimates of fiber orientation and the number of fibers within each voxel. The second challenge lies in linking discrete fiber orientation estimates together into a continuous trajectory (Chapters 13 and 14). Differences in numerical integration procedures, discrete step sizes, and interpolation procedures have influenced resulting streamlines in simulations, phantoms, and in vivo data. The next source of variability lies in choosing where to start, and when to terminate, the tracking process. Typically, tracking can be initiated from seed points throughout the whole brain or at the interface of gray and white matter, and can be stopped when streamlines enter the gray matter, regions of low FA, or regions of high curvature. The choices of seeding and stopping strategies can lead to biases associated with pathway length, volume, locations, and resulting connectivity (Fig. 3c). Rather than using a fixed set of tractography parameters, the optimal settings can be very different between brain regions as the scale and curvature of fiber bundles varies (Chamberland et al., 2014).

In addition to the decisions mentioned here, there are several families of tracking algorithms that differ in how they generate and interpret streamlines. A deterministic algorithm utilizes a unique fiber orientation estimate in each voxel, and will always result in the same streamline if initiated at the same seed point (Chapter 13). On the other hand, probabilistic algorithms utilize a stochastic propagation and result in a distribution of possible streamlines that reflects a possible anatomical distribution of possible streamlines and/or statistical uncertainty in the data (Chapter 14). Algorithms may also be classified as local or global. The local approach is most common, and steps along the local vector fields to create individual streamlines one at a time. A global approach, on the other hand, tries to reconstruct all connections simultaneously by finding the configuration of streamlines that best describes the measured data (Chapter 16). An intermediate, semiglobal, approach is to reconstruct a large number of streamlines throughout the whole brain and filter streamlines that do not match certain criteria or well-explain the data.

Finally, study-dependent variability is introduced when the results of tractography are quantified. Tractography can be used to assess the human connectome (i.e., the mapping of all neural connections of the brain) or to assess individual white matter pathways. As the name implies, connectomics attempts to quantify the connectivity of all neural connections of the brain. Summary statistics depend on how the "strength of connection" from one region to another is defined, where one may use raw streamline counts, or streamlines weighted by volume, surface areas, lengths, or microstructural metrics. Additionally, connectomes are dependent on choices of atlases and brain parcellations to defined network nodes, as well as

Multicenter studies and harmonization: Problems, solutions, and open challenges **Chapter | 33 677**

how streamlines are assigned to graph edges. For bundle segmentation, variability exists in the rules and constraints used to dissect, or isolate, a specific pathway, with a number of automated, semiautomated, and manual methods proposed for a number of bundles of interest (Schilling et al., 2021). Moreover, there is significant intra- and interrater variability even if a specific bundle segmentation protocol is provided (Rheault et al., 2020). Finally, quantifying properties of interest of these bundles takes many forms. Most commonly, a simple measure like the FA can be assigned to a bundle by taking the average FA within the entire volume, by weighting the FA by streamline density, including only the core or middle portion of the bundle, or measuring and quantifying the FA at different positions along the bundle.

## 2.3 Magnitude of variability

The previous sections have highlighted many sources of variability in the interpretation of dMRI data and tractography derivatives. Here, it is attempted to put numbers to these sources, but note that due to the complex relationships between sources of variability, the absolute values described are not directly comparable across studies, but are noted in order to act as a reference for comparisons against expected variations across subjects, or across time. Region-based diffusion measures, most commonly derived from DTI, are often applied in a patient-control comparison without the use of tractography itself; hence, most of our knowledge of precision and variance is derived from voxel-based or atlas-based analysis of traveling phantoms and traveling subjects (i.e., multisite and multiscanner studies).

Generally, FA and MD show relatively little variation within scanners, with average CVs typically ranging between 0.5% and 4% and ICC> 0.9, and slightly more variation across sites, scanners, and manufacturers, with 1%–7% variation and ICC > 0.8 (Magnotta et al., 2012; Vollmar et al., 2010; Grech-Sollars et al., 2015; Fortin et al., 2017; Teipel et al., 2011; Palacios et al., 2017; Koller et al., 2021; Andica et al., 2020). Intrasite and intersite variabilities are nonuniform across white matter, higher in gray matter regions with values ranging from 8% to 25% variation, and are dependent upon spatial resolution and angular resolution (Vollmar et al., 2010; Grech-Sollars et al., 2015; Tax et al., 2019a; Ning et al., 2020; Chamberland et al., 2018; Farrell et al., 2007).

The use of tractography to derive tract-averaged DTI measures leads to slightly increased CV and decreased ICC compared to atlas-based regions. Intrasite FA and MD variations range from 3% to 6%, with intersite 5%–10% across most pathways (Tong et al., 2019; Nath et al., 2020). Going beyond DTI, measures of track density and volume have been shown to be less reproducible than measures of FA and MD, with intra- and intersite track density reported as 5%–30% (ICC > 0.48), and track volume 3%–25% (ICC > 0.53) (Tong et al., 2019; Nath et al., 2020). In both cases, larger variation is observed when using less data, lower data quality, and in smaller pathways. Similarly, measures of Dice overlap of bundles across scanners is consistently 0.6 or greater, with only a small decrease in overlap (0.05 decrease) across scanners (Nath et al., 2020). Again, there is a difference across tracts, with some pathways more reproducible than others.

Measures of region-to-region connectome matrices have also been shown to be highly reproducible within sites. Test retest shows generally high ICC (>0.75), high correlation of matrices (>0.85), and low variation (<10%) of network-based indices of connectomes, with intrasubject differences 30%–60% less than intersubject differences (Bastiani et al., 2012; Roine et al., 2019; Buchanan et al., 2014; Dennis et al., 2012; Zhong et al., 2015; Tsai, 2018; Girard et al., 2015a, b). For both bundle segmentation and connectome analysis, acquisition variance and differences in protocols and workflows may lead to greater variability than scanner and site effects. For example, larger variation in bundle segmentation is observed when using fewer directions and $b$-values, lower data quality, and in smaller pathways of the brain (Bastiani et al., 2012). Acquisition acceleration (e.g., multiband) may result in 5%–10% variation in DTI parameters, and similar variations in connectivity and bundle overlap, with larger effects at higher resolution (Rathi et al., 2014).

For connectome analysis, streamline filtering has been shown to increase reproducibility, as well as increasing connectome thresholds, although with an increase in network sparsity (Smith et al., 2013, 2015; Bastiani et al., 2012). Differences in seeding strategies, connection weights, track propagation (deterministic vs. probabilistic) and reconstruction methods also alter variability (Osmanlıoğlu et al., 2020; Bastiani et al., 2012). For example, DTI reconstruction actually has higher reproducibility despite known false negatives (Prčkovska et al., 2016). For bundle segmentation using manual protocols, variation within the rater performing segmentation and across raters results in moderate to high reproducibility with high ICC of volumes (ICC usually > 0.6), high ICC of derived-FA (ICC > 0.7), high Dice overlaps (0.77 and 0.65 for intra- and interrater), and high density correlation maps (correlation = 0.90) (Wakana et al., 2007; Rheault et al., 2020). Again, variability depends on SNR and resolution, with a wide range of average volume overlaps and ICCs dependent upon pathway, ranging from 0.95 for major white matter tracts to 0.2 for smaller, more complex, pathways (Rheault et al., 2020; Bayrak et al., 2019; Schilling et al., 2021). Finally, bundle segmentation is highly dependent upon the protocol (manual or automated) used to segment bundles. Although most show the same basic shape and location, there is large variation in overall volume and overlap (0.2–0.6 Dice overlap across protocols for the same pathway) (Schilling et al., 2021).

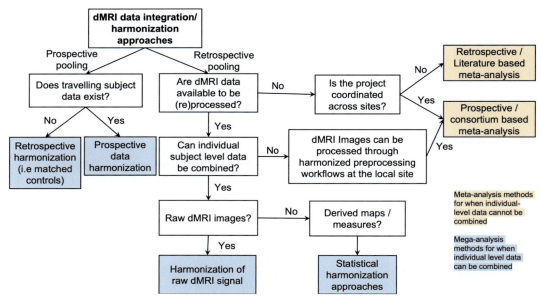

FIG. 4 Overview of solutions to address dMRI scanner differences.

## 3 Proposed solutions

In order to mitigate scanner and protocol effects, various data analysis and harmonization methods have been proposed. We have grouped the proposed solutions into two main categories, depending on the input data and in which stage of the dMRI data processing pipeline the harmonization is applied (Fig. 4): Section 3.1 discusses statistical approaches for harmonization of diffusion MRI derived measures and Section 3.2 focuses on harmonization of DWI data.

### 3.1 Statistical approaches for harmonization of diffusion MRI measures

The approaches under this category use the diffusion maps, such as FA, Mean Kurtosis (MK), etc., as an input from multiple sites or scanners and apply statistical data pooling at the final stage of the processing pipeline to reduce the unwanted intra- and intersite variability across the input diffusion maps. There are several advantages to statistical harmonization approaches. They are fast and easy to employ, very beneficial when only diffusion maps of individual subjects or statistical maps from different sites are available, but DWI data are not necessarily available. On the other hand, one limitation is that they require a sample size that guarantees a statistically representative sample from each scanner included in the study. Additionally, the statistical data pooling should be performed for each diffusion measure separately, thus making the harmonization procedure entirely model-specific (e.g., FA maps collected from multiple studies must be harmonized independently from the MD maps). These approaches allow diffusion maps obtained from different diffusion acquisition schemes (diffusion directions, b-values, repetition time, echo time, etc.) to be harmonized, and subject groups across sites with different age, gender, handedness, and socioeconomic status are not required to be matched. However, it is a known fact that many diffusion measures have a nonlinear relationship with acquisition parameters (e.g., FA and b-value) and demographics (e.g., FA and age) or cognitive variables (e.g., FA and IQ) (Mirzaalian et al., 2018; Cetin-Karayumak et al., 2019, 2020a). Therefore it is not clear how scanner- or site-specific nonlinearities in the signal propagate through the preprocessing techniques and model fitting (Cetin-Karayumak et al., 2019). Finally, recent studies have demonstrated that even the simplest changes in the processing pipelines can add drastic effects to the statistical harmonization approaches (Cetin-Karayumak et al., 2020b, c). Therefore it is highly recommended to apply the same preprocessing steps and software across sites before applying any statistical harmonization method described later. We also note that the majority of the existing statistical harmonization tools are not suitable for machine learning models aimed at clinical translation.

The most well-known statistical approaches are described in following sections: Section 3.1.1 discusses metaanalysis and Section 3.1.2 focuses on statistical covariates.

### 3.1.1 Metaanalysis

Access to data and computational resources are often limiting factors in large-scale statistical analyses. Even if subject-level data is not accessed (whether it be derived measures or scans), information and trends across datasets can still be pooled. In literature-based metaanalysis, summary statistics from published studies are pooled in an attempt to come to a consensus across studies and to determine a more robust effect size. However, methodological variability in the studies after they are published, along with publication biases, may affect results. Recently, large-scale consortia such as ENIGMA have conducted coordinated metaanalyses, where the image processing and the statistical analyses are harmonized across different independent studies and datasets from around the world. Rather than each group publishing individually on their results, statistical results, including effect sizes, are pooled together in a metaanalysis. This form of metaanalysis does not aim to harmonize the imaging-derived metrics, but the analysis plan. To pool the statistical findings, an inverse-variance-based metaanalysis is often performed. Here, the resulting effects from individual datasets are often weighted by the inverse of the variance of that dataset's statistical effect. Thus if a dataset has a very noisy acquisition protocol, it likely will have a large variance around the effect size and not contribute as much to the overall effect. Metaanalytical approaches are limited in that they require all datasets to be able to perform the same analysis. For example, in a case-control test, it is required that all datasets have a sufficient sample size of both cases and controls; studies that only have collected data on cases or controls are not able to contribute. In those cases, individual-level data will need to be pooled to perform analyses. These analyses are often referred to as megaanalyses. The remainder of this section discusses approaches for megaanalyses.

### 3.1.2 Statistical covariates

The approaches using statistical covariates aim to regress out the effects of intersite variability from the diffusion maps while preserving only biological effects.

#### Linear correction factor-based harmonization

Pohl et al. (2016) used information from three traveling subjects to obtain a linear correction factor for scanner-related effects in FA, while a different correction factor for each ROI was estimated. The main advantage of the correction factor is its simple derivation; however, the correction factor might not necessarily be sufficient because of the intrinsically highly nonlinear and also regionally varying scanner-related effects (Cetin-Karayumak et al., 2019; Mirzaalian et al., 2016). Thus using a single linear regressor for large ROIs can lead to erroneous results in the aggregated data (Mirzaalian et al., 2016).

#### ComBat harmonization

Fortin et al. (2017) have proposed a powerful and fast statistical data pooling tool that uses ComBat, a batch-effect correction tool used in genomics. This method can estimate an additive and a multiplicative site-effect coefficient at each voxel, thus accounting for voxel-wise scanner differences (of note, ComBat harmonization can be applied at the ROI level, too). Despite being a powerful harmonization tool, ComBat's optimization procedure assumes that the site-effect parameters follow a particular parametric prior distribution (Gaussian and inverse-gamma), which might not generalize to all scenarios or measures derived from other models (e.g., multicompartment models) (Cetin-Karayumak et al., 2019). Thus a few studies have thoroughly explored the reliability of ComBat (Cetin-Karayumak et al., 2020c; Garcia-Dias et al., 2020). For example, a recent study (Cetin-Karayumak et al., 2020c) demonstrated that ComBat altered the intergroup variability at some sites and in some regions the direction of intersubject biological differences were even flipped (measured by effect sizes) (Cetin-Karayumak et al., 2020c). However, the most recent versions of ComBat such as generalized additive model (GAM)-ComBat (Pomponio et al., 2020) might mitigate some of the problems.

#### FA-based hardware-phantom harmonization

Timmermans et al. (2019) built comprehensive diffusion single-strand phantoms for the estimation of the mean and standard deviation of FA encompassing hardware-related (scanner vendor and head coil), acquisition-related (bandwidth, TE, and TR), and quality-related (SNR and mean residual) effects. Several models with different fixed and random effects (site was included as a random effect) were built, compared, and validated by taking into account effects that are constant across all individuals, and random predictors relate to effects that vary across individuals. In general, hardware phantoms can be very beneficial. They can be transferred to multiple sites easily, and scanned multiple times for a longer time at any time of the day while not being affected by motion artifacts. The main limitation is that current phantoms are not complex enough to represent the architectural configurations of human brain tissues. Thus available phantoms, including Timmermans et al. (2019), provide a "global scaling" of the data to achieve harmonization, which cannot be used for voxel-wise harmonization. Despite the current limitations, future research on hardware phantoms that can mimic the human brain can be very significant for robust harmonization.

## 3.2 Harmonization of DWI data

Statistical data pooling approaches have been widely preferred in the literature for joint analysis of diffusion maps due to their simplicity in neuroimaging applications. However, in addition to the limitations mentioned in Section 3.1, none of the statistical data pooling approaches can reconstruct the harmonized signal, which is essential for consistent microstructural, tractography, and connectivity studies. To overcome these limitations, recent studies have demonstrated the significance of harmonization at the signal level (Cetin-Karayumak et al., 2019). Hence, the second group of harmonization approaches uses diffusion weighted images (DWIs) at the signal level as early in the dMRI processing pipeline as possible. One benefit of this type of harmonization is that it can minimize the effects of preprocessing as well as subsequent data modeling in harmonization. Other advantages are that any model fitting can be applied in later stages of the processing pipeline after harmonization, and any scanner-specific nonlinearities in the diffusion signal can be modeled. On the other hand, they require groups of healthy subjects from multiple sites for learning the scanner differences, and can most straightforwardly be applied to DWI data acquired with similar spatial resolution, number of $b$-shells, $b$-values, or number of diffusion gradient directions.

DWI data harmonization approaches can be divided into two groups based on the availability of traveling heads: Section 3.2.1 discusses retrospective harmonization using linear regression, and Section 3.2.2 focuses on prospective harmonization with traveling subjects. The approaches of Moyer et al. (2020) and St-Jean et al. (2020) are both applicable retrospectively, but have primarily been evaluated in a prospective scenario and will be discussed in the next section.

### 3.2.1 Retrospective DWI data harmonization

Retrospective approaches were designed to learn the scanner differences across retrospectively collected DWI data of groups of subjects from different sites, which are matched by age, gender, IQ, handedness, and other socioeconomic factors across sites. Of note, these subjects from different sites are not necessarily one-to-one matched; the matching is usually achieved at the group level (i.e., to statistically test the matching, unpaired $t$-test can be used). The best-known approach in this category is rotation invariant spherical harmonics (RISH) with linear regression (Linear-RISH) (see details in the "Linear-rotation invariant spherical harmonics" section) (Mirzaalian et al., 2015, 2016, 2018; Cetin-Karayumak et al., 2019). Another approach, the method of moments (Huynh et al., 2019), can be potentially included in this category, but at the time of writing only the theory of the method has been presented, and extensive validation is needed on the retrospectively collected DWI data.

**Linear-rotation invariant spherical harmonics**

Mirzaalian et al. (2015, 2016, 2018) proposed the proof of concept of RISH for dMRI signal harmonization, and this method has been improved by Cetin-Karayumak et al. (2019) to enable the Linear-RISH method to harmonize DWI data with different $b$-values and spatial resolutions. The Linear-RISH method has been extensively validated (Cetin-Karayumak et al., 2019; Ning et al., 2020) and its feasibility has been demonstrated in multiple neuroimaging applications of psychiatric disorders (Cetin-Karayumak et al., 2020a; Seitz et al., 2021; Hegde et al., 2020; Lv et al., 2021).

RISH features can be modeled using spherical harmonics (SH) by computing the energy of the SH coefficients at each SH order. Thus RISH features are independent of the orientation and represent different aspects of the diffusion signal (Fig. 5). Prior to RISH harmonization, one site must be selected as the reference (R) site, and the other sites to be

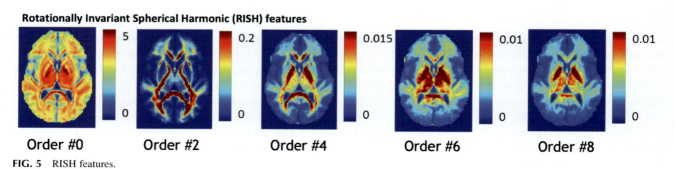

**FIG. 5** RISH features.

mapped/harmonized to the R site are set as target (T) individually. This mapping is achieved using voxel-wise linear regression. The decision on the R site is generally based on multiple criteria: (i) the R site is usually the most comprehensive site based on the demographics of the healthy controls, that is, R is preferred to include a wider age range covering the age range of all target sites, as well as evenly distributed healthy female and male subjects; (ii) the R site generally has ideally higher quality (e.g., higher spatial resolution) and SNR DWI data; (iii) a site with the most common or similar acquisition parameters can be selected as an R site to minimize the effects of angular and spatial resampling.

RISH harmonization involves two steps: (i) Learning the voxel-vise linear mapping from T to R using RISH features of at least 16 matched healthy controls (Cetin-Karayumak et al., 2019) (recall the previously mentioned matching criteria); and (ii) the learned mapping is later applied to the DWI of all subjects of T, including healthy controls and patients. Open-source code for Linear-RISH is available for both single-shell (https://github.com/pnlbwh/dMRIharmonization) as well as multishell (https://github.com/pnlbwh/multi-shell-dMRIharmonization) DWI data harmonization.

### 3.2.2 Prospective DWI data harmonization with traveling subjects

Ideally, the interscanner and/or interprotocol differences can be learned using individual subjects rescanned on different systems (i.e., traveling subjects) in relatively quick succession, such that measurement differences can only be attributed to scanner- or protocol-related differences, and hence not to biological differences. This should allow scanner-specific effects to be captured with fewer subjects. It is, however, difficult to acquire such scans due to the fact that rescanning requires time-dedication of the subjects and can be expensive; therefore the majority of the existing databases include only a limited number of traveling-subject scans. Here, we will discuss harmonization methods that have been designed for and/or evaluated on traveling-subject databases (1) within field strengths and (2) across field strengths.

#### Harmonization within magnetic field strengths

Tax et al. (2018) and Ning et al. (2020) presented a publicly available benchmark database of 15 subjects scanned with multiple standard and state-of-the-art acquisitions on three 3T scanners. Using this benchmark database, two challenges were organized at the Medical Image Computing and Computer Assisted Intervention (MICCAI) 2017 and 2018 conferences, which evaluated 5 different algorithms for harmonization of single-shell and 19 for multishell dMRI data, respectively. Various approaches, including linear regression (i.e., see the "Linear-rotation invariant spherical harmonics" section), nonlinear neural network, dictionary learning, and model-based harmonization algorithms have been evaluated in these challenges. Briefly, the best performing algorithms show comparable performance globally across subjects, but there is variability across regions: Tax et al. (2018) report larger errors in WM in anisotropic measures such as FA, whereas isotropic measures such as MD are more comparable across WM and GM. Ning et al. (2020) reported a higher average absolute percentage error across evaluated measures in GM than in WM, with the highest errors in the superior parietal, the entorhinal, the temporal pole, and the frontal pole cortical regions. They furthermore found a higher error for the prediction of measures derived from higher $b$-values for most algorithms. Notably, the Linear-RISH (see the "Linear-rotation invariant spherical harmonics" section) method was one of the best performing approaches in the MICCAI 2018 challenge (Ning et al., 2020). Here, we summarize a few other approaches that have been evaluated on traveling subject data (see also Tax et al., 2019b and Ning et al., 2020).

***Spherical harmonic residual network (SHResNet)*** Koppers et al. (2018) proposed a deep learning network structure-based SHResNet algorithm on the novel concept of residual structure (He et al., 2016), which introduces a subtraction path from the input to the network output, resulting in very robust performance. The network focuses only on the difference between the input and its corresponding target signal. Such residual structures are efficiently trainable, even for very deep networks. Nonetheless, the harmonization is done per harmonic order of the SH signal, so, the signal from both T and R should have the same SH orders.

***Multistage prediction networks*** Blumberg et al. (2019) expand on the supervised paradigm of the SHResNet and similar methods, ensembling multiple networks trained for different tasks together during a second, refinement ("fine tuning"), training stage. In the harmonization context a practitioner may train image-to-image transfer networks between each pair of scanners, and then refine the networks into an image transfer network for a specific pair of sites that is better than its component parts.

***Invariant representations*** Moyer et al. (2020) proposed the construction of site-conditional autoencoders. The corresponding image or patch encodings can then be regularized to enforce statistical invariance to a site indicator variable;

682 **PART | V** Tractography applications

this results in an image encoder-decoder scheme that transmits images through an invariant code. By the data processing inequality, all of the decoded images will be "uninformed" of their original site.

***Dictionary learning*** St-Jean et al. (2017, 2020) presented a sparse dictionary learning approach, wherein different dictionaries are learned for each scanner/site. This carries the belief that site-wise signals will be encoded separately, with disjoint components for each site-specific signal. Harmonization is undertaken by encoding each scan using the dictionary from the target scanner.

### Harmonization across different magnetic field strengths

Better scanner technologies as well as higher magnetic field strength scanners are becoming more popular, as they provide higher contrast and resolution in DWI. For instance, data from a 7T scanner can reveal anatomical details not visible at 3T. Cetin-Karayumak et al. (2018) proposed harmonization approaches for multishell DWI data from different magnetic strengths for joint analysis. In particular, two methods were proposed: voxel-wise linear mapping (see the "Linear-rotation invariant spherical harmonics" section), and patch-based nonlinear mapping using deep convolutional neural networks (CNNs). In multiple experiments, the authors demonstrated the efficacy of using these techniques to harmonize dMRI data from vastly different scanners in a model-free manner. Validation was done on 30 test subjects by comparing the DTI metrics as well as the tractography performances before and after harmonization.

## 4  Open challenges

Despite the progress made so far on harmonizing dMRI data, several challenges remain to be addressed, both in techniques that use statistical approaches and those that harmonize the dMRI signal. Section 2.2 discussed several sources of variability, which each can cause differences in the results when not harmonized in multicenter studies. However, these sources may interact with each other and propagate through the acquisition and processing pipeline in a nonlinear and nontrivial way. We will discuss open challenges related to the effect of acquisition, preprocessing, and estimation on multicenter variability of microstructural measures in Section 4.1. If tractography is used for the analysis, for example, to define ROIs or to establish connectivity matrices, one should also take into consideration the variability associated with fiber direction estimates and bundles, which will be discussed in Section 4.2. Finally, we will discuss open challenges in the case of patient and longitudinal data and evaluation.

### 4.1  Multicenter variability of microstructural estimates

An important aspect that remains unexplored is the effect of the differences in the $b$-values, the number of gradient directions, the spatial resolution, and other acquisition parameters on the ability of statistics-based methods to perform proper harmonization. All of these variables affect the measure of interest (e.g., FA) in a highly nonlinear fashion, thereby making it difficult for linear covariates to separate scanner effects from differences in acquisition parameters. While some signal-based methods address this issue by accounting for these acquisition differences (Cetin-Karayumak et al., 2018, 2019), it is important for statistical methods to address such challenges.

Along similar lines, it is unclear how design choices for the different data preprocessing pipelines can affect harmonization results. For example, in cases without traveling subjects, image registration plays an important role in most harmonization methods (statistical as well as signal based). If registration fails or is inaccurate, it can lead to inaccuracies in the harmonization results. This is of particular importance when harmonization involves using data with very large variability (e.g., subjects with very large ventricles). Thus care must be taken and ideally manual checks performed to ensure proper registration. Other preprocessing steps include interpolation (in the case of data with different spatial resolutions) and image denoising, but it remains unclear how scanner-related effects propagate through these steps and affect the measures of interest. This is especially true for statistics-based methods. Some preprocessing steps require additional data, for example, reversed phase encoding images for correction of susceptibility-related distortions, which may not be available for all datasets in a multicenter study. A trade-off will have to be made between using consistent methods between centers, or performing different corrections for those datasets that have additional information available but not for others. Some preprocessing pipelines may reduce the variability between scanners more than others.

Regarding estimation, recent work (Cetin-Karayumak et al., 2020b,c) has shown inconsistencies in harmonization when using different software packages to estimate the DT (all other steps being the same). Further, different types of models (e.g., single tensor, kurtosis, spherical deconvolution, diffusion propagator, multicompartment, etc.) are used by research groups to estimate tissue microstructural indices. The impact of scanner biases on the estimation of these models and the derived measures (used in statistical methods) is still an area open to research. Additionally, it is an open problem to determine whether scanner biases are consistently modeled by statistical methods for the various measures (e.g., FA or mean kurtosis) obtained from different dMRI models.

## 4.2 Fiber direction estimates and tractography

Tractography can be used to analyze differences between microstructural measures. First, it is important for approaches that directly harmonize the dMRI signal to ensure that the harmonized signal preserves fiber orientation. While the RISH-based approaches were designed and validated to ensure fiber orientations are preserved (Mirzaalian et al., 2016), methods that do not use rotational invariance should take extra precaution and ensure that fiber tractography is not affected during the harmonization process. However, even when orientational information remains untouched, differences in the tracks being traced in the same individual but on different scanners can still cause variability in, for example, the measured FA along a track. Such differences, for example, occur because tractography methods are sensitive to noise in the data, with varying noise distributions in different regions affecting the estimation of the fiber orientation peaks. Additionally, partial volume effects, b-values, and number of gradient directions all play a significant role in the fiber bundles being traced.

Further, the same tractography settings across datasets can yield different results; differences in the reconstruction methods and the acceleration factors used by each scanner, among others, can lead to significant variability in the estimation of dMRI-derived metrics like FA (see Fig. 2c) often used for seeding and determining stopping criteria. While some tractography algorithms address this issue by using a seeding and stopping mask (obtained from T1-weighted anatomical scans), other algorithms still use dMRI-derived metrics. This can lead to differences in the number and length of tracks being traced on different sites. Additionally, differences in gradient nonlinearities and geometric distortions can lead to significantly different tracks being traced for the same subject but from different scanners, for example, the corticospinal tract (Pierpaoli, 2010). Thus the extent of the tracks being traced can be different due to scanner differences (Chamberland et al., 2018). Note that, in such cases, it may be inappropriate to use track-based harmonization methods such as metaanalysis or statistical covariate-based methods as the tractography itself is affected in a nonlinear way between sites.

## 4.3 Disease

Another area that needs significant focus is harmonization in the presence of tumors or lesions, or without the availability of age- and sex-matched healthy subject data. Addressing this issue will be of significant clinical benefit, making it easier to compare multisite scans acquired at different time points to, for example, determine tumor progression in the same subjects (but different scanners). Finally, harmonization will potentially solve another major hurdle that relates to adoption of dMRI in clinical trials.

## 4.4 Longitudinal analysis

This chapter has mostly focused on harmonizing data from multiple centers, but variability may also occur longitudinally when scanners receive hardware and/or software updates. It remains an open question how to best approach this; scanner vendors potentially play an important role in providing quantitative information about the variability introduced by such updates.

## 4.5 Evaluation of harmonization

Overall, extensive testing and validation of harmonization methods are critical for the field to move forward, especially if they are to be used in a clinical setting as well as for quantitative analysis. Ultimately, the goal is to bring the magnitude of the intersite variability for a single subject scanned on two scanners to the level of intrasubject variability within the same scanner. Traveling-subject and test-retest data are necessary for evaluation, and public databases start to emerge (Cai et al., 2021; Tax et al., 2018; Koller et al., 2021). Acquiring traveling subject data is logistically challenging, and an open question is how many subjects are required for accurate harmonization (traveling or age- and sex-matched), and to what extent this can be compensated for with test-retest scans. Hardware phantoms have been used for harmonization and provide a possible solution as they can be scanned multiple times for long periods of time, they do not suffer from motion artifacts, and they are stable over time. However, phantoms are often simplistic and hence do not support voxel-wise harmonization; future efforts

**684 PART | V** Tractography applications

should focus on increasing the complexity to better match in vivo data. Finally, it is unclear how traveling subject or phantom data can be used by statistical pooling methods for harmonization of nontraveling subject data.

## 5 Conclusion

Combining datasets from multiple scanners and sites provides a unique opportunity to address a wealth of clinical questions. However, variability is introduced at each step in the acquisition and processing pipeline. In addition to effects across sessions, scanners, and sites, care must be taken when comparing and interpreting results using different acquisitions, reconstructions, and processing pipelines. Harmonization of dMRI data has gained popularity as a solution to undesired variability in multicenter studies, with proposed methods either addressing the differences statistically or by harmonizing the raw data. However, with the many sources of variability affecting the data in a nontrivial way, the harmonization problem is complex and requires further development and improved strategies for evaluation.

## References

Aja-Fernández, S., Vegas-Sánchez-Ferrero, G., 2016. Statistical Analysis of Noise in MRI. Springer International Publishing, Cham, https://doi.org/10.1007/978-3-319-39934-8.

Alexander, A.L., Tsuruda, J.S., Parker, D.L., 1997. Elimination of eddy current artifacts in diffusion-weighted echo-planar images: the use of bipolar gradients. Magn. Reson. Med. 38 (6), 1016–1021.

Alexander, L.M., Escalera, J., Ai, L., Andreotti, C., Febre, K., Mangone, A., Vega-Potler, N., Langer, N., Alexander, A., Kovacs, M., et al., 2017. An open resource for transdiagnostic research in pediatric mental health and learning disorders. Sci. Data 4, 170181.

Andica, C., Kamagata, K., Hayashi, T., Hagiwara, A., Uchida, W., Saito, Y., Kamiya, K., Fujita, S., Akashi, T., Wada, A., et al., 2020. Scan-rescan and inter-vendor reproducibility of neurite orientation dispersion and density imaging metrics. Neuroradiology 62 (4), 483–494.

Assaf, Y., Basser, P.J., 2005. Composite hindered and restricted model of diffusion (CHARMED) MR imaging of the human brain. NeuroImage 27 (1), 48–58. https://doi.org/10.1016/j.neuroimage.2005.03.042.

Bammer, R., Markl, M., Barnett, A., Acar, B., Alley, M.T., Pelc, N.J., Glover, G.H., Moseley, M.E., 2003. Analysis and generalized correction of the effect of spatial gradient field distortions in diffusion-weighted imaging. Magn. Reson. Med. 50 (3), 560–569. https://doi.org/10.1002/mrm.10545.

Barch, D.M., Albaugh, M.D., Avenevoli, S., Chang, L., Clark, D.B., Glantz, M.D., Hudziak, J.J., Jernigan, T.L., Tapert, S.F., Yurgelun-Todd, D., et al., 2018. Demographic, physical and mental health assessments in the adolescent brain and cognitive development study: rationale and description. Dev. Cogn. Neurosci. 32, 55–66.

Bastiani, M., Shah, N.J., Goebel, R., Roebroeck, A., 2012. Human cortical connectome reconstruction from diffusion weighted MRI: the effect of tractography algorithm. NeuroImage 62 (3), 1732–1749.

Bayrak, R.G., Schilling, K.G., Greer, J.M., Hansen, C.B., Greer, C.M., Blaber, J.A., Williams, O., Beason-Held, L.L., Resnick, S.M., Rogers, B.P., et al., 2019. TractEM: fast protocols for whole brain deterministic tractography-based white matter atlas. bioRxiv, 651935.

Bisdas, S., et al., 2008. Reproducibility, interrater agreement, and age-related changes of fractional anisotropy measures at 3t in healthy subjects: effect of the applied b-value. Am. J. Neuroradiol. 29 (6), 1128–1133.

Blumberg, S.B., Palombo, M., Khoo, C.S., Tax, C.M.W., Tanno, R., Alexander, D.C., 2019. Multi-stage prediction networks for data harmonization. In: International Conference on Medical Image Computing and Computer-Assisted Intervention, pp. 411–419.

Bonekamp, D., Nagae, L.M., 2007. Diffusion tensor imaging in children and adolescents: reproducibility, hemispheric, and age-related differences. NeuroImage 34 (2), 733–742.

Buchanan, C.R., Pernet, C.R., Gorgolewski, K.J., Storkey, A.J., Bastin, M.E., 2014. Test-retest reliability of structural brain networks from diffusion MRI. NeuroImage 86, 231–243.

Cai, L.Y., Yang, Q., Kanakaraj, P., Nath, V., Newton, A.T., Edmonson, H.A., Luci, J., Conrad, B.N., Price, G.R., Hansen, C.B., et al., 2021. MASiVar: multisite, multiscanner, and multisubject acquisitions for studying variability in diffusion weighted MRI. Magn. Reson. Med. 86, 3304–3320.

Cercignani, M., Bammer, R., 2003. Inter-sequence and inter-imaging unit variability of diffusion tensor MR imaging histogram-derived metrics of the brain in healthy volunteers. AJNR Am. J. Neuroradiol. 24 (4), 638–643.

Cetin-Karayumak, S., Kubicki, M., Rathi, Y., 2018. Harmonizing diffusion MRI data across magnetic field strengths. In: Frangi, A.F., Schnabel, J.A., Davatzikos, C., Alberola-López, C., Fichtinger, G. (Eds.), Medical Image Computing and Computer Assisted Intervention—MICCAI 2018, Springer International Publishing, Cham, pp. 116–124.

Cetin-Karayumak, S., Bouix, S., Ning, L., James, A., Crow, T., Shenton, M., Kubicki, M., Rathi, Y., 2019. Retrospective harmonization of multi-site diffusion MRI data acquired with different acquisition parameters. NeuroImage 184, 180–200.

Cetin-Karayumak, S., Di Biase, M., Chunga-Iturry, N., Reid, B., Somes, N., Lyall, A., Kelly, S., Pasternak, O., Vangel, M., Pearlson, G., Tamminga, C., Sweeney, J., Clementz, B., Schretlen, D., Viher, P., Stegmayer, K., Walther, S., Lee, J., Crow, T., James, A., Voineskos, A., Buchanan, R., Szeszko, P.-R., Malhotra, A., McCarley, R., Matcheri, K., Shenton, M., Rathi, Y., Kubicki, M., 2020a. White matter abnormalities across the lifespan of schizophrenia: a harmonized multi-site diffusion MRI study. Mol. Psychiatry 25, 3208–3219. https://doi.org/10.1038/s41380-019-0509-y.

Cetin-Karayumak, S., O'Sullivan, L., Lyons, M.G., Billah, T., Pasternak, O., Bouix, S., Kubicki, M., Rathi, Y., 2020b. Reproducibility crisis in diffusion MRI: contribution of software processing pipelines. In: ISMRM, p. 4380.

Cetin-Karayumak, S., O'Sullivan, L., Lyons, M.G., Kubicki, M., Rath, Y., 2020c. Exploring the reliability of ComBat for multi-site diffusion MRI harmonization. In: ISMRM, p. 4379.

Chamberland, M., Whittingstall, K., Fortin, D., Mathieu, D., Descoteaux, M., 2014. Real-time multi-peak tractography for instantaneous connectivity display. Front. Neuroinform. 8, 59. https://doi.org/10.3389/fninf.2014.00059.

Chamberland, M., Tax, C.M.W., Jones, D.K., 2018. Meyer's loop tractography for image-guided surgery depends on imaging protocol and hardware. NeuroImage 20, 458–465.

Ciccarelli, O., Parker, G.J., 2003. From diffusion tractography to quantitative white matter tract measures: a reproducibility study. NeuroImage 18 (2), 348–359.

Del Re, E.C., Bouix, S., Fitzsimmons, J., Blokland, G.A.M., Mesholam-Gately, R., Wojcik, J., Kikinis, Z., Kubicki, M., Petryshen, T., Pasternak, O., et al., 2019. Diffusion abnormalities in the corpus callosum in first episode schizophrenia: associated with enlarged lateral ventricles and symptomatology. Psychiatry Res. 277, 45–51.

Dell'Acqua, F., Rizzo, G., Scifo, P., Clarke, R.A., Scotti, G., Fazio, F., 2007. A model-based deconvolution approach to solve fiber crossing in diffusion-weighted MR imaging. IEEE Trans. Biomed. Eng. 54 (3), 462–472. https://doi.org/10.1109/TBME.2006.888830.

Dell'Acqua, F., Simmons, A., Williams, S.C.R., Catani, M., 2013. Can spherical deconvolution provide more information than fiber orientations? Hindrance modulated orientational anisotropy, a true-tract specific index to characterize white matter diffusion. Hum. Brain Mapp. 34 (10), 2464–2483.

Dennis, E.L., Jahanshad, N., Toga, A.W., McMahon, K.L., De Zubicaray, G.I., Martin, N.G., Wright, M.J., Thompson, P.M., 2012. Test-retest reliability of graph theory measures of structural brain connectivity. In: International Conference on Medical Image Computing and Computer-Assisted Intervention, pp. 305–312.

Farrell, J.A.D., Landman, B.A., Jones, C.K., Smith, S.A., Prince, J.L., Van Zijl, P.C.M., Mori, S., 2007. Effects of signal-to-noise ratio on the accuracy and reproducibility of diffusion tensor imaging-derived fractional anisotropy, mean diffusivity, and principal eigenvector measurements at 1.5 T. J. Magn. Reson. Imaging 26 (3), 756–767.

Fortin, J.P., Parker, D., Tunç, B., Watanabe, T., Elliott, M.A., Ruparel, K., Roalf, D.R., Satterthwaite, T.D., Gur, R.C., Gur, R.E., Schultz, R.T., Verma, R., Shinohara, R.T., 2017. Harmonization of multi-site diffusion tensor imaging data. NeuroImage 161, 149–170. https://doi.org/10.1016/J.NEUROIMAGE.2017.08.047.

Garcia-Dias, R., Scarpazza, C., Baecker, L., Vieira, S., Pinaya, W.H.L., Corvin, A., Redolfi, A., Nelson, B., Crespo-Facorro, B., McDonald, C., Tordesillas-Gutiérrez, D., Cannon, D., Mothersill, D., Hernaus, D., Morris, D., Setien-Suero, E., Donohoe, G., Frisoni, G., Tronchin, G., Sato, J., Marcelis, M., Kempton, M., van Haren, N.E.M., Gruber, O., McGorry, P., Amminger, P., McGuire, P., Gong, Q., Kahn, R.S., Ayesa-Arriola, R., van Amelsvoort, T., Ortiz-García de la Foz, V., Calhoun, V., Cahn, W., Mechelli, A., 2020. Neuroharmony: a new tool for harmonizing volumetric MRI data from unseen scanners. NeuroImage 220, 117127. https://doi.org/10.1016/j.neuroimage.2020.117127.

Girard, G., Whittingstall, K., Deriche, R., Descoteaux, M., 2015a. Structural connectivity reproducibility through multiple acquisitions. In: Organization for Human Brain Mapping, Honolulu, USA.

Girard, G., Whittingstall, K., Deriche, R., Descoteaux, M., 2015b. Studying white matter tractography reproducibility through connectivity matrices. In: International Symposium on Magnetic Resonance in Medicine.

Grech-Sollars, M., Hales, P.W., Miyazaki, K., Raschke, F., Rodriguez, D., Wilson, M., Gill, S.K., Banks, T., Saunders, D.E., Clayden, J.D., Gwilliam, M.-N., Barrick, T.R., Morgan, P.S., Davies, N.P., Rossiter, J., Auer, D.P., Grundy, R., Leach, M.O., Howe, F.A., Peet, A.C., Clark, C.A., 2015. Multi-centre reproducibility of diffusion MRI parameters for clinical sequences in the brain. NMR Biomed. 28 (4), 468–485. https://doi.org/10.1002/nbm.3269.

Griswold, M.A., Jakob, P.M., Heidemann, R.M., Nittka, M., Jellus, V., Wang, J., Kiefer, B., Haase, A., 2002. Generalized autocalibrating partially parallel acquisitions (GRAPPA). Magn. Reson. Med. 47 (6), 1202–1210.

Harrison, J.R., Bhatia, S., Tan, Z.X., Mirza-Davies, A., Benkert, H., Tax, C.M.W., Jones, D.K., 2020. Imaging Alzheimer's genetic risk using diffusion MRI: a systematic review. Neuroimage Clin. 27, 102359.

He, K., Zhang, X., Ren, S., Sun, J., 2016. Deep residual learning for image recognition. In: 2016 IEEE Conference on Computer Vision and Pattern Recognition (CVPR), IEEE, pp. 770–778. http://ieeexplore.ieee.org/document/7780459/.

Hegde, R.R., Kelly, S., Lutz, O., Guimond, S., Cetin-Karayumak, S., Mike, L., Mesholam-Gately, R.I., Pasternak, O., Kubicki, M., Eack, S.M., Keshavan, M.S., 2020. Association of white matter microstructure and extracellular free-water with cognitive performance in the early course of schizophrenia. Psychiatry Res. 305, 111159. https://doi.org/10.1016/j.pscychresns.2020.111159.

Heiervang, E., Behrens, T.E., 2006. Between session reproducibility and between subject variability of diffusion MR and tractography measures. NeuroImage 33 (3), 867–877.

Huynh, K.M., Chen, G., Wu, Y., Shen, D., Yap, P., 2019. Multi-site harmonization of diffusion MRI data via method of moments. IEEE Trans. Med. Imaging 38 (7), 1599–1609.

Jansen, J.F., Kooi, M.E., 2007. Reproducibility of quantitative cerebral T2 relaxometry, diffusion tensor imaging, and 1H magnetic resonance spectroscopy at 3.0 Tesla. Invest. Radiol. 42 (6), 327–337.

Jones, D.K., Alexander, D.C., Bowtell, R., Cercignani, M., Dell'Acqua, F., McHugh, D.J., Miller, K.L., Palombo, M., Parker, G.J.M., Rudrapatna, U.S., Tax, C.M.W., 2018. Microstructural imaging of the human brain with a 'super-scanner': 10 key advantages of ultra-strong gradients for diffusion MRI. NeuroImage. https://doi.org/10.1016/J.NEUROIMAGE.2018.05.047. https://www.sciencedirect.com/science/article/pii/S1053811918304610?via%3Dihub.

Koller, K., Rudrapatna, S.U., Chamberland, M., Raven, E.P., Parker, G.D., Tax, C.M.W., Drakesmith, M., Fasan, F., Owen, D., Hughes, G., et al., 2021. MICRA: microstructural image compilation with repeated acquisitions. NeuroImage 225, 117406.

**686 PART | V** Tractography applications

Koppers, S., Bloy, L., Berman, J.I., Tax, C.M.W., Edgar, J.C., Merhof, D., 2018. Spherical harmonic residual network for diffusion signal harmonization. In: Computational Diffusion MRI, Springer.

Lv, J., Biase, M.D., Cash, R.F.H., Cocchi, L., Cropley, V., Klauser, P., Tian, Y., Bayer, J., Schmaal, L., Cetin-Karayumak, S., Rathi, Y., Pasternak, O., Bousman, C., Pantelis, C., Calamante, F., Zalesky, A., 2021. Individual deviations from normative models of brain structure in a large cross-sectional schizophrenia cohort. Mol. Psychiatry 26, 3512–3523.

Magnotta, V.A., Matsui, J.T., Liu, D., Johnson, H.J., Long, J.D., Bolster Jr., B.D., Mueller, B.A., Lim, K., Mori, S., Helmer, K.G., et al., 2012. Multicenter reliability of diffusion tensor imaging. Brain Connect. 2 (6), 345–355.

McGibney, G., Smith, M.R., Nichols, S.T., Crawley, A., 1993. Quantitative evaluation of several partial Fourier reconstruction algorithms used in MRI. Magn. Reson. Med. 30 (1), 51–59.

Melonakos, E.D., Shenton, M.E., Rathi, Y., Terry, D.P., Bouix, S., Kubicki, M., 2011. Voxel-based morphometry (VBM) studies in schizophrenia—can white matter changes be reliably detected with VBM? Psychiatry Res. 193 (2), 65–70.

Miller, K.L., Alfaro-Almagro, F., Bangerter, N.K., Thomas, D.L., Yacoub, E., Xu, J., Bartsch, A.J., Jbabdi, S., Sotiropoulos, S.N., Andersson, J.L.R., et al., 2016. Multimodal population brain imaging in the UK Biobank prospective epidemiological study. Nat. Neurosci. 19 (11), 1523–1536.

Mirzaalian, H., de Pierrefeu, A., Savadjiev, P., Pasternak, O., Bouix, S., Kubicki, M., Westin, C.F., Shenton, M.E., Rathi, Y., 2015. Harmonizing diffusion MRI data across multiple sites and scanners. Med. Image Comput. Comput. Assist. Interv. 9349, 12–19. https://doi.org/10.1007/978-3-319-24553-9_2.

Mirzaalian, H., Ning, L., Savadjiev, P., Pasternak, O., Bouix, S., Michailovich, O., Grant, G., Marx, C.E., Morey, R.A., Flashman, L.A., George, M.S., McAllister, T.W., Andaluz, N., Shutter, L., Coimbra, R., Zafonte, R.D., Coleman, M.J., Kubicki, M., Westin, C.F., Stein, M.B., Shenton, M.E., Rathi, Y., 2016. Inter-site and inter-scanner diffusion MRI data harmonization. NeuroImage 135, 311–323. https://doi.org/10.1016/j.neuroimage.2016.04.041.

Mirzaalian, H., Ning, L., Savadjiev, P., Pasternak, O., Bouix, S., Michailovich, O., Karmacharya, S., Grant, G., Marx, C.E., Morey, R.A., Flashman, L.A., George, M.S., McAllister, T.W., Andaluz, N., Shutter, L., Coimbra, R., Zafonte, R.D., Coleman, M.J., Kubicki, M., Westin, C.F., Stein, M.B., Shenton, M.E., Rathi, Y., 2018. Multi-site harmonization of diffusion MRI data in a registration framework. Brain Imaging Behav. 12, 284–295. https://doi.org/10.1007/s11682-016-9670-y.

Moyer, D., Ver Steeg, G., Tax, C.M.W., Thompson, P.M., 2020. Scanner invariant representations for diffusion MRI harmonization. Magn. Reson. Med. 84, 2174–2189.

Nath, V., Schilling, K.G., Parvathaneni, P., Huo, Y., Blaber, J.A., Hainline, A.E., Barakovic, M., Romascano, D., Rafael-Patino, J., Frigo, M., et al., 2020. Tractography reproducibility challenge with empirical data (TraCED): the 2017 ISMRM diffusion study group challenge. J. Magn. Reson. Imaging 51 (1), 234–249.

Ning, L., Bonet-Carne, E., Grussu, F., Sepehrband, F., Kaden, E., Veraart, J., Blumberg, S.B., Khoo, C.S., Palombo, M., Kokkinos, I., et al., 2020. Cross-scanner and cross-protocol multi-shell diffusion MRI data harmonization: algorithms and results. NeuroImage 221, 117128.

O'Connor, D., Potler, N.V., Kovacs, M., Xu, T., Ai, L., Pellman, J., Vanderwal, T., Parra, L.C., Cohen, S., Ghosh, S., Escalera, J., Grant-Villegas, N., Osman, Y., Bui, A., Craddock, R.C., Milham, M.P., 2017. The Healthy Brain Network Serial Scanning Initiative: a resource for evaluating inter-individual differences and their reliabilities across scan conditions and sessions. GigaScience 6 (2). https://doi.org/10.1093/gigascience/giw011.

Osmanlıoğlu, Y., Alappatt, J.A., Parker, D., Verma, R., 2020. Connectomic consistency: a systematic stability analysis of structural and functional connectivity. J. Neural Eng. 17 (4), 045004. https://doi.org/10.1088/1741-2552/ab947b.

Palacios, E.M., Martin, A.J., Boss, M.A., Ezekiel, F., Chang, Y.S., Yuh, E.L., Vassar, M.J., Schnyer, D.M., MacDonald, C.L., Crawford, K.L., et al., 2017. Toward precision and reproducibility of diffusion tensor imaging: a multicenter diffusion phantom and traveling volunteer study. Am. J. Neuroradiol. 38 (3), 537–545.

Pfefferbaum, A., Adalsteinsson, E., 2003. Replicability of diffusion tensor imaging measurements of fractional anisotropy and trace in brain. J. Magn. Reson. Imaging 18 (4), 427–433.

Pierpaoli, C., 2010. Artifacts in diffusion MRI. In: Diffusion MRI, Oxford University Press, pp. 303–318.

Pohl, K.M., Sullivan, E.V., Rohlfing, T., Chu, W., Kwon, D., Nichols, B.N., Zhang, Y., Brown, S.A., Tapert, S.F., Cummins, K., Thompson, W.K., Brumback, T., Colrain, I.M., Baker, F.C., Prouty, D., De Bellis, M.D., Voyvodic, J.T., Clark, D.B., Schirda, C., Nagel, B.J., Pfefferbaum, A., 2016. Harmonizing DTI measurements across scanners to examine the development of white matter microstructure in 803 adolescents of the NCANDA study. NeuroImage 130, 194–213. https://doi.org/10.1016/J.NEUROIMAGE.2016.01.061.

Pomponio, R., Erus, G., Habes, M., Doshi, J., Srinivasan, D., Mamourian, E., Bashyam, V., Nasrallah, I.M., Satterthwaite, T.D., Fan, Y., Launer, L.J., Masters, C.L., Maruff, P., Zhuo, C., Völzke, H., 2020. Harmonization of large MRI datasets for the analysis of brain imaging patterns throughout the lifespan. NeuroImage 208, 116450. https://doi.org/10.1016/j.neuroimage.2019.116450.

Prčkovska, V., Rodrigues, P., Sanchez, A.P., Ramos, M., Andorra, M., Martinez-Heras, E., Falcon, C., Prats-Galino, A., Villoslada, P., 2016. Reproducibility of the structural connectome reconstruction across diffusion methods. J. Neuroimaging 26 (1), 46–57.

Pruessmann, K.P., Weiger, M., Scheidegger, M.B., Boesiger, P., 1999. SENSE: sensitivity encoding for fast MRI. Magn. Reson. Med. 42 (5), 952–962.

Raffelt, D., Tournier, J.D., Rose, S., Ridgway, G.R., Henderson, R., Crozier, S., Salvado, O., Connelly, A., 2012. Apparent fibre density: a novel measure for the analysis of diffusion-weighted magnetic resonance images. NeuroImage 59 (4), 3976–3994.

Rathi, Y., Gagoski, B., Setsompop, K., Michailovich, O., Grant, P.E., Westin, C.F., 2013. Diffusion propagator estimation from sparse measurements in a tractography framework. In: International Conference on Medical Image Computing and Computer-Assisted Intervention, pp. 510–517.

Rathi, Y., Gagoski, B., Setsompop, K., Grant, P.E., Westin, C.F., 2014. Comparing simultaneous multi-slice diffusion acquisitions. In: Schultz, T., Nedjati-Gilani, G., Venkataraman, A., O'Donnell, L., Panagiotaki, E. (Eds.), Computational Diffusion MRI and Brain Connectivity, Springer International Publishing, Cham, pp. 3–11.

Rheault, F., De Benedictis, A., Daducci, A., Maffei, C., Tax, C.M.W., Romascano, D., Caverzasi, E., Morency, F.C., Corrivetti, F., Pestilli, F., et al., 2020. Tractostorm: the what, why, and how of tractography dissection reproducibility. Hum. Brain Mapp. 41 (7), 1859–1874.

Roine, T., Jeurissen, B., Perrone, D., Aelterman, J., Philips, W., Sijbers, J., Leemans, A., 2019. Reproducibility and intercorrelation of graph theoretical measures in structural brain connectivity networks. Med. Image Anal. 52, 56–67.

Schilling, K.G., Rheault, F., Petit, L., Hansen, C.B., Nath, V., Yeh, F.C., Girard, G., Barakovic, M., Rafael-Patino, J., Yu, T., Fischi-Gomez, E., Pizzolato, M., Ocampo-Pineda, M., Schiavi, S., Canales-Rodríguez, E.J., Daducci, A., Granziera, C., Innocenti, G., Thiran, J.P., Mancini, L., Wastling, S., Cocozza, S., Petracca, M., Pontillo, G., Mancini, M., Vos, S.B., Vakharia, V.N., Duncan, J.S., Melero, H., Manzanedo, L., Sanz-Morales, E., Peña-Melián, Á., Calamante, F., Attyé, A., Cabeen, R.P., Korobova, L., Toga, A.W., Vijayakumari, A.A., Parker, D., Verma, R., Radwan, A., Sunaert, S., Emsell, L., De Luca, A., Leemans, A., Bajada, C.J., Haroon, H., Azadbakht, H., Chamberland, M., Genc, S., Tax, C.M.W., Yeh, P.H., Srikanchana, R., McKnight, C., Yang, J.Y.M., Chen, J., Kelly, C.E., Yeh, C.H., Cochereau, J., Maller, J.J., Welton, T., Almairac, F., Seunarine, K.K., Clark, C.A., Zhang, F., Makris, N., Golby, A., Rathi, Y., O'Donnell, L.J., Xia, Y., Aydogan, D.B., Shi, Y., Fernandes, F.G., Raemaekers, M., Warrington, S., Michielse, S., Ramírez-Manzanares, A., Concha, L., Aranda, R., Meraz, M.R., Lerma-Usabiaga, G., Roitman, L., Fekonja, L.S., Calarco, N., Joseph, M., Nakua, H., Voineskos, A.N., Karan, P., Grenier, G., Legarreta, J.H., Adluru, N., Nair, V.A., Prabhakaran, V., Alexander, A.L., Kamagata, K., Saito, Y., Uchida, W., Andica, C., Masahiro, A., Bayrak, R.G., Wheeler-Kingshott, C.A.M.G., D'Angelo, E., Palesi, F., Savini, G., Rolandi, N., Guevara, P., Houenou, J., López-López, N., Mangin, J.F., Poupon, C., Román, C., Vázquez, A., Maffei, C., Arantes, M., Andrade, J.P., Silva, S.M., Raja, R., Calhoun, V.D., Caverzasi, E., Sacco, S., Lauricella, M., Pestilli, F., Bullock, D., Zhan, Y., Brignoni-Perez, E., Lebel, C., Reynolds, J.E., Nestrasil, I., Labounek, R., Lenglet, C., Paulson, A., Aulicka, S., Heilbronner, S., Heuer, K., Anderson, A.W., Landman, B.A., Descoteaux, M., 2020. Tractography dissection variability: what happens when 42 groups dissect 14 white matter bundles on the same dataset? bioRxiv. https://doi.org/10.1101/2020.10.07.321083.

Schilling, K.G., Tax, C.M.W., Rheault, F., Hansen, C., Yang, Q., Yeh, F.C., Cai, L., Anderson, A.W., Landman, B.A., 2021. Fiber tractography bundle segmentation depends on scanner effects, vendor effects, acquisition resolution, diffusion sampling scheme, diffusion sensitization, and bundle segmentation workflow. NeuroImage 242, 118451. https://doi.org/10.1016/j.neuroimage.2021.118451.

Seitz, J., Cetin-Karayumak, S., Lyall, A., Pasternak, O., Baxi, M., Vangel, M., Pearlson, G., Tamminga, C., Sweeney, J., Clementz, B., Schretlen, D., Viher, P.V., Stegmayer, K., Walther, S., Lee, J., Crow, T., James, A., Voineskos, A., Buchanan, R.W., Szeszko, P.R., Malhotra, A., Keshavan, M., Koerte, I.K., Shenton, M.E., Rathi, Y., Kubicki, M., 2021. Investigating sexual dimorphism of human white matter in a harmonized, multisite diffusion magnetic resonance imaging study. Cereb. Cortex 31, 201–212. https://doi.org/10.1093/cercor/bhaa220.

Setsompop, K., Kimmlingen, R., Eberlein, E., Witzel, T., Cohen-Adad, J., McNab, J.A., Keil, B., Tisdall, M.D., Hoecht, P., Dietz, P., Cauley, S.F., Tountcheva, V., Matschl, V., Lenz, V.H., Heberlein, K., Potthast, A., Thein, H., Van Horn, J., Toga, A., Schmitt, F., Lehne, D., Rosen, B.R., Wedeen, V., Wald, L.L., 2013. Pushing the limits of in vivo diffusion MRI for the human connectome project. NeuroImage 80, 220–233. https://doi.org/10.1016/j.neuroimage.2013.05.078.

Sjölund, J., Szczepankiewicz, F., Nilsson, M., Topgaard, D., Westin, C.F., Knutsson, H., 2015. Constrained optimization of gradient waveforms for generalized diffusion encoding. J. Magn. Reson. 261, 157–168. https://doi.org/10.1016/J.JMR.2015.10.012.

Smith, R.E., Tournier, J.D., Calamante, F., Connelly, A., 2013. SIFT: spherical-deconvolution informed filtering of tractograms. NeuroImage 67, 298–312.

Smith, R.E., Tournier, J.D., Calamante, F., Connelly, A., 2015. The effects of SIFT on the reproducibility and biological accuracy of the structural connectome. NeuroImage 104, 253–265.

St-Jean, S., Viergever, M., Leemans, A., 2017. A unified framework for upsampling and denoising of diffusion MRI data. In: 25th Annual Meeting of ISMRM, p. 3533.

St-Jean, S., Viergever, M.A., Leemans, A., 2020. Harmonization of diffusion MRI data sets with adaptive dictionary learning. Hum. Brain Mapp. 41 (16), 4478–4499.

Stejskal, E.O., Tanner, J.E., 1965. Spin diffusion measurements: spin echoes in the presence of a time-dependent field gradient. J. Chem. Phys. 42 (1), 288–292. https://doi.org/10.1063/1.1695690.

Tax, C.M.W., de Almeida Martins, J.P., Szczepankiewicz, F., Westin, C.F., Chamberland, M., Topgaard, D., Jones, D.K., 2018. From physical chemistry to human brain biology: unconstrained inversion of 5-dimensional diffusion-T2 correlation data. In: ISMRM, p. 1101. http://archive.ismrm.org/2018/1101.html.

Tax, C.M.W., Grussu, F., Kaden, E., Ning, L., Rudrapatna, U., Evans, C.J., St-Jean, S., Leemans, A., Koppers, S., Merhof, D., Ghosh, A., Tanno, R., Alexander, D.C., Zappalà, S., Charron, C., Kusmia, S., Linden, D.E.J., Jones, D.K., Veraart, J., 2019a. Cross-scanner and cross-protocol diffusion MRI data harmonisation: a benchmark database and evaluation of algorithms. NeuroImage 195, 285–299. https://doi.org/10.1016/j.neuroimage.2019.01.077.

Tax, C.M.W., Rudrapatna, U.S., Mueller, L., Jones, D.K., 2019b. Characterizing diffusion of myelin water in the living human brain using ultra-strong gradients and spiral readout. In: ISMRM, p. 1115. https://cds.ismrm.org/protected/19MPresentations/abstracts/1115.html.

Tax, C.M.W., Kleban, E., Chamberland, M., Baraković, M., Rudrapatna, U., Jones, D.K., 2021. Measuring compartmental T2-orientational dependence in human brain white matter using a tiltable RF coil and diffusion-T2 correlation MRI. NeuroImage 236, 117967. https://doi.org/10.1016/j.neuroimage.2021.117967.

Teipel, S.J., Reuter, S., Stieltjes, B., Acosta-Cabronero, J., Ernemann, U., Fellgiebel, A., Filippi, M., Frisoni, G., Hentschel, F., Jessen, F., et al., 2011. Multicenter stability of diffusion tensor imaging measures: a European clinical and physical phantom study. Psychiatry Res. 194 (3), 363–371.

Thomas, C., Sadeghi, N., Nayak, A., Trefler, A., Sarlls, J., Baker, C.I., Pierpaoli, C., 2018. Impact of time-of-day on diffusivity measures of brain tissue derived from diffusion tensor imaging. NeuroImage 173, 25–34. https://doi.org/10.1016/j.neuroimage.2018.02.026.

Timmermans, C., Smeets, D., Verheyden, J., Terzopoulos, V., Anania, V., Parizel, P.M., Maas, A., 2019. Potential of a statistical approach for the standardization of multicenter diffusion tensor data: a phantom study. J. Magn. Reson. Imaging 49 (4), 955–965. https://doi.org/10.1002/jmri.26333.

Tong, Q., He, H., Gong, T., Li, C., Liang, P., Qian, T., Sun, Y., Ding, Q., Li, K., Zhong, J., 2019. Reproducibility of multi-shell diffusion tractography on traveling subjects: a multicenter study prospective. Magn. Reson. Imaging 59, 1–9. https://doi.org/10.1016/j.mri.2019.02.011.

Tournier, J.D., Calamante, F., Gadian, D.G., Connelly, A., 2004. Direct estimation of the fiber orientation density function from diffusion-weighted MRI data using spherical deconvolution. NeuroImage 23 (3), 1176–1185.

Tsai, S.Y., 2018. Reproducibility of structural brain connectivity and network metrics using probabilistic diffusion tractography. Sci. Rep. 8 (1), 1–12.

Van Essen, D.C., Jbabdi, S., Sotiropoulos, S.N., Chen, C., Dikranian, K., Coalson, T., Harwell, J., Glasser, M.F., 2014. Chapter 16—Mapping connections in humans and non-human primates: aspirations and challenges for diffusion imaging. In: Diffusion MRI, pp. 337–358.

Veraart, J., Sijbers, J., Sunaert, S., Leemans, A., Jeurissen, B., 2013. Weighted linear least squares estimation of diffusion MRI parameters: strengths, limitations, and pitfalls. NeuroImage 81, 335–346. https://doi.org/10.1016/J.NEUROIMAGE.2013.05.028.

Vollmar, C., O'Muircheartaigh, J., Barker, G.J., Symms, M.R., Thompson, P., Kumari, V., Duncan, J.S., Richardson, M.P., Koepp, M.J., 2010. Identical, but not the same: intra-site and inter-site reproducibility of fractional anisotropy measures on two 3.0 T scanners. NeuroImage 51 (4), 1384–1394. https://doi.org/10.1016/J.NEUROIMAGE.2010.03.046.

Vos, S.B., Tax, C.M.W., Luijten, P.R., Ourselin, S., Leemans, A., Froeling, M., 2017. The importance of correcting for signal drift in diffusion MRI. Magn. Reson. Med. 77, 285–299. https://doi.org/10.1002/mrm.26124.

Wakana, S., Caprihan, A., Panzenboeck, M.M., Fallon, J.H., Perry, M., Gollub, R.L., Hua, K., Zhang, J., Jiang, H., Dubey, P., et al., 2007. Reproducibility of quantitative tractography methods applied to cerebral white matter. NeuroImage 36 (3), 630–644.

Zhang, H., Schneider, T., Wheeler-Kingshott, C.A., Alexander, D.C., 2012. NODDI: practical in vivo neurite orientation dispersion and density imaging of the human brain. NeuroImage 61 (4), 1000–1016. https://doi.org/10.1016/j.neuroimage.2012.03.072.

Zhong, S., He, Y., Gong, G., 2015. Convergence and divergence across construction methods for human brain white matter networks: an assessment based on individual differences. Hum. Brain Mapp. 36 (5), 1995–2013.

Zhu, A.H., Moyer, D.C., Nir, T.M., Thompson, P.M., Jahanshad, N., 2019. Challenges and opportunities in dMRI data harmonization. In: Bonet-Carne, E., Grussu, F., Ning, L., Sepehrband, F., Tax, C.M.W. (Eds.), Computational Diffusion MRI, Springer International Publishing, Cham, pp. 157–172.

# Part VI

# Appendix

# Appendix A

# Vectors and tensors

**Philippe Karan[a], Gabrielle Grenier[a], and Jon Haitz Legarreta[b]**

[a]*Sherbrooke Connectivity Imaging Laboratory (SCIL), Université de Sherbrooke, Sherbrooke, QC, Canada,* [b]*Department of Radiology, Brigham and Women's Hospital, Mass General Brigham/Harvard Medical School, Boston, MA, United States*

Vectors and tensors are mathematical objects that are essential to compute the white matter tractography maps from diffusion magnetic resonance imaging (dMRI) data. On the one hand, the diffusion signal encodes both dimensional and (diffusion) orientational data, and thus tensors provide the appropriate mathematical framework to deal with such multidimensional values. On the other hand, streamline propagation methods map the locally reconstructed fiber orientations to long-range fiber trajectories steered by the vector orientations. This appendix provides an overview of the foundations of these mathematical objects that support the methods discussed in this book.

## A.1 Vectors

### A.1.1 Mathematics of the vector

#### A.1.1.1 Definitions

An $n$-dimensional vector, denoted $\mathbf{v}$ or $\vec{v}$, is a mathematical entity that is an element in an $n$-dimensional vector space that has magnitude (size) and direction. Moreover, a vector can be represented as an oriented segment going from a point $A$ to another point $B$ and can be written as $\overrightarrow{AB}$.

In an $n$-dimensional Euclidean space ($\mathbb{R}^n$), the vector *scalar components* (or *scalar projections*) are denoted by an $n$-tuple:

$$\mathbf{v} = (v_1, v_2, v_3, \ldots, v_{n-1}, v_n), \quad v_i \in \mathbb{R} \tag{A.1}$$

A vector can also be represented using the standard basis (also called natural basis) vectors of the vector space where the vector is defined. In the $n$-dimensional Euclidean space, the standard basis consists of $n$ distinct vectors such that $\{\mathbf{e}_i : 1 \leq i \leq n\}$, where $\mathbf{e}_i$ denotes the vector with a 1 in the $i$th coordinate and 0s elsewhere. Thus a vector can be written as

$$\mathbf{v} = \mathbf{v}_1 + \mathbf{v}_2 + \mathbf{v}_3 + \cdots + \mathbf{v}_{n-1} + \mathbf{v}_n \tag{A.2}$$

$$= v_1\mathbf{e}_1 + v_2\mathbf{e}_2 + v_3\mathbf{e}_3 + \cdots + v_{n-1}\mathbf{e}_{n-1} + v_n\mathbf{e}_n \tag{A.3}$$

where $\mathbf{v}_1, \mathbf{v}_2, \mathbf{v}_3, \ldots, \mathbf{v}_{n-1}, \mathbf{v}_n$ are the *vector components* (or *vector projections*) of $\mathbf{v}$ on the basis vectors.

In particular, in the three-dimensional Euclidean space ($\mathbb{R}^3$), there are three of these standard basis vectors:

$$\mathbf{e}_1 = (1, 0, 0), \quad \mathbf{e}_2 = (0, 1, 0), \quad \mathbf{e}_3 = (0, 0, 1) \tag{A.4}$$

These have the intuitive interpretation as vectors of unit length pointing up the $X$, $Y$, and $Z$ axes of a Cartesian coordinate system, respectively. Thus, for the three-dimensional Euclidean space ($\mathbb{R}^3$) in particular, vectors are identified with triples of scalar components $\mathbf{v} = (v_1, v_2, v_3)$ or $\mathbf{v} = (v_x, v_y, v_z)$.

In the Euclidean space, the magnitude of a vector is sometimes called its "length" and is computed using the Euclidean norm:

$$\| v \| = \sqrt{\sum_{i=1}^{n} v_i^2} \tag{A.5}$$

**692 PART | VI** Appendix

An arbitrary, $n$-dimensional vector may be converted to a *unit vector* (denoted by $\hat{\mathbf{v}}$) by dividing by its norm:

$$\hat{\mathbf{v}} = \frac{v}{\parallel v \parallel} \tag{A.6}$$

A zero vector, denoted $\mathbf{0}$, is a vector of length 0, and thus has all components equal to zero.

Given a subset $S \subset \mathbb{R}^n$, a vector field is a map $V : S \to \mathbb{R}^n$ that assigns any value in the subset $S$ to a vector in $\mathbb{R}^n$ using the vector-valued function $V$.

### A.1.1.2 Operations

Table A.1 summarizes the set of operations between vectors defined by the vector algebra. The notation assumes operations between two $n$-dimensional vectors $\mathbf{u}$ and $\mathbf{v}$ forming an angle $\theta$ on the plane containing them.

It is interesting to note that:

- The result of the *multiplication of a vector by a scalar* is another vector whose magnitude is stretched by a factor of $r$, and which has the same direction as the original vector $\mathbf{u}$.
- The *dot product* (sometimes called *scalar product*) can also be expressed as (see Section A.1.1.3):

$$\mathbf{u} \cdot \mathbf{v} = [u_1 \ u_2 \ u_3 \cdots u_{n-1} \ u_n] \cdot \begin{bmatrix} v_1 \\ v_2 \\ v_3 \\ \vdots \\ v_{n-1} \\ v_n \end{bmatrix} \tag{A.7}$$

It can also be expressed in geometrical terms as $\mathbf{u} \cdot \mathbf{v} = \parallel \mathbf{u} \parallel \cdot \parallel \mathbf{v} \parallel \cos(\theta)$.

As shown in Fig. A.1, the dot product between adjacent voxel vectors in dMRI tractography can determine which streamline propagation configurations will obey the maximum aperture angle or cone, and hence will map the long-range white matter connections.

- The result of the *cross product* (sometimes called *vector product*) between two vectors is a vector that is orthogonal to both of them, with a direction given by the right-hand rule (denoted $\hat{\mathbf{n}}$) and a magnitude equal to the area of the parallelogram that the vectors span. The cross-product is defined only in three-dimensional space.

### A.1.1.3 Notations

Vectors can be arranged column-wise or row-wise:

- An $n$-dimensional *column vector* contains only one column and $n$ rows. The column vectors with a fixed number of rows form a vector space.
- An $n$-dimensional *row vector* contains only one row and $n$ columns. The row vectors with a fixed number of columns form a vector space.

**TABLE A.1** Vector algebra.

| Operation | Expression | Result entity |
| --- | --- | --- |
| Addition | $\mathbf{u} + \mathbf{v} = \sum_{i=1}^{n} (u_i + v_i)\mathbf{e}_i$ | Vector |
| Subtraction | $\mathbf{u} - \mathbf{v} = \sum_{i=1}^{n} (u_i - v_i)\mathbf{e}_i$ | Vector |
| Multiplication by a scalar | $r\mathbf{u} = \sum_{i=1}^{n} r u_i \mathbf{e}_i$ | Vector |
| Dot product | $\mathbf{u} \cdot \mathbf{v} = \sum_{i=1}^{n} u_i v_i$ | Scalar |
| Cross-product | $\mathbf{u} \times \mathbf{v} = \parallel \mathbf{u} \parallel \cdot \parallel \mathbf{v} \parallel \sin(\theta) \hat{\mathbf{n}}$ | Vector |

Operations between two $n$-dimensional vectors $\mathbf{u}$ and $\mathbf{v}$ forming an angle $\theta$ on the plane containing them.

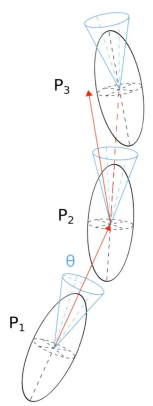

**FIG. A.1** Streamline propagation. The maximum aperture angle (denoted as $\theta$ in the figure) allows propagation of streamlines within the allowed aperture.

In tractography, the common notation is the column-wise vector, which is denoted as

$$\mathbf{v} = \begin{bmatrix} v_1 \\ v_2 \\ v_3 \\ \vdots \\ v_{n-1} \\ v_n \end{bmatrix} = [v_1\, v_2\, v_3 \cdots v_{n-1}\, v_n]^T \tag{A.8}$$

where $T$ denotes the transpose operator.

## A.2 Tensors

### A.2.1 Mathematics of the tensor

#### A.2.1.1 Definitions

Tensors are mathematical objects that live in a vector space, meaning that they can add up and be multiplied by scalars. Moreover, tensors do not depend on the choice of basis, but rather they are represented differently according to the basis. To put it simply, tensors are a generalization of scalars and vectors. These objects can be manipulated with operations, which are presented in Section A.2.1.2.

**694 PART | VI Appendix**

An example of a tensor commonly used in MRI is the diffusion tensor $\mathbf{D}$, which encodes information about the diffusion properties on one point in space. It can be written as

$$\mathbf{D} = \begin{pmatrix} d_{xx} & d_{xy} & d_{xz} \\ d_{yx} & d_{yy} & d_{yz} \\ d_{zx} & d_{zy} & d_{zz} \end{pmatrix} \tag{A.9}$$

where each component describes the diffusion along a direction, given by the choice of basis. In this case, the basis is the common Cartesian basis ($\mathbf{e}_1$, $\mathbf{e}_2$, $\mathbf{e}_3$). One can change the basis by applying the appropriate transformation to the tensor.

The simplest tensor is a scalar, namely a tensor of order 0. A vector, or tensor of order 1, is formed when $n$ scalars are put together into a tuple. Applying the same logic, a tensor of order 2 can be viewed as $m$ vectors, of $n$ scalars each, put together side by side. The diffusion tensor fits in this category, and could also be referred to as a matrix, as any tensor of order 2 can be expressed as a matrix (the opposite is not always true). Following the logic again, a tensor of order 3 can be represented by an $l \times m \times n$ array.

In dMRI, the trace of the matrix can be useful as well as the inverse. The trace (TR) of a matrix is the sum of its diagonal elements. For example, the mean diffusivity of an isotropic (all off-diagonal terms are null) diffusion tensor is calculated as $\mathrm{Tr}(\mathbf{D})/3 = (d_{xx} + d_{yy} + d_{zz})/3$. As for the inverse, it can be interpreted as a kind of division of tensors. Let us introduce the $3 \times 3$ identity matrix:

$$\mathbb{I} = \begin{pmatrix} 1 & 0 & 0 \\ 0 & 1 & 0 \\ 0 & 0 & 1 \end{pmatrix} \tag{A.10}$$

and use it to explain the matrix inversion. The inverse of $\mathbf{D}$ is written $\mathbf{D}^{-1}$ and it follows $\mathbf{D}\mathbf{D}^{-1} = \mathbf{D}^{-1}\mathbf{D} = \mathbb{I}$. It is useful to manipulate matrices around equations to isolate terms or simplify the equation.

### A.2.1.2 Operations

Along with the vector space properties, tensors are equipped with two important operations. First is the tensor product, which allows two tensors to be multiplied in such a way that the resulting tensor has a higher or same order as before. Applied on vectors, this is also called an outer product, while a tensor product on matrices is called a Kronecker product. Second is the contraction, which allows two tensors to be multiplied in order to reduce or maintain the order of the result. The contraction of two vectors is done with the dot product or inner product. For matrices, this operation is said to be a matrix multiplication.

Here is an example of such operations. Let two vectors $\mathbf{a} = [a_1, a_2, a_3]$ and $\mathbf{b} = [b_1, b_2, b_3]$. A tensor product between these two would be written as

$$\mathbf{a}^T \otimes \mathbf{b} = \begin{bmatrix} a_1 \\ a_2 \\ a_3 \end{bmatrix} \otimes [b_1 \; b_2 \; b_3] = \begin{bmatrix} a_1 b_1 & a_1 b_2 & a_1 b_3 \\ a_2 b_1 & a_2 b_2 & a_2 b_3 \\ a_3 b_1 & a_3 b_2 & a_3 b_3 \end{bmatrix} \tag{A.11}$$

while a dot product would look like

$$\mathbf{a} \cdot \mathbf{b}^T = [a_1 \; a_2 \; a_3] \cdot \begin{bmatrix} b_1 \\ b_2 \\ b_3 \end{bmatrix} = a_1 b_1 + a_2 b_2 + a_3 b_3 \tag{A.12}$$

In Eq. (A.11), the order went from 1 to 2, increasing it. The resulting tensor in Eq. (A.12) decreased from 1 to 0, a scalar. Note that for the contraction, the dimensions of the two tensors must fit.

There is another operation on matrices, useful in the field of b-tensor encoding, called the Frobenius inner product and written as follows, for two real matrices $\mathbf{A}$ and $\mathbf{B}$:

$$\langle \mathbf{A}, \mathbf{B} \rangle_F = \sum_{i,j} A_{ij} B_{ij} = \mathrm{Tr}(\mathbf{A}^T \mathbf{B}) \tag{A.13}$$

Vectors and tensors **Appendix | A** **695**

### A.2.1.3 Eigenvalues and eigenvectors

As said before, the representation of a tensor (as a matrix) only depends on the choice of basis. Sometimes, the classical bases like the Cartesian basis lead to a messy representation and a complicated interpretation of it. It can be useful to find a different basis where the tensor appears simpler. Here are presented two similar ways to do so.

In the case where the desired matrix $\mathbf{M}$ is square, *spectral decomposition*, or *eigendecomposition*, can be used to *diagonalize* it. The square matrix is diagonalizable if another matrix $\mathbf{U}$ exists, such as $\mathbf{M} = \mathbf{U}\Lambda\mathbf{U}^{-1}$, where $\Lambda$ is a diagonal matrix. The goal is to find the right matrix $\mathbf{U}$ to diagonalize $\mathbf{M}$ into $\Lambda$, which would be easier to interpret. The columns of $\mathbf{U}$ are then called *eigenvectors*, forming a new basis, and the corresponding values on the diagonal are named *eigenvalues*. Put in another way, for the given matrix $\mathbf{M}$, the following equation must be satisfied: $\mathbf{Mv} = \lambda\mathbf{v}$, where $\mathbf{v}$ is an eigenvector of $\mathbf{M}$ and the scalar $\lambda$ is its eigenvalue. This whole process can be useful to help manipulate matrices more easily, such as the diffusion tensor, which can find itself in an exponential, for example.

In the case of a rectangular matrix, *singular value decomposition*, or SVD, is used for a similar function to eigendecomposition. With this technique, an $m \times n$ real matrix $\mathbf{M}$ can be factorized, as $\mathbf{U}\Lambda\mathbf{V}^{\mathbf{T}}$, where $\mathbf{U}$ is an $m \times m$ real matrix, $\mathbf{V}$ is an $n \times n$ real matrix, and $\mathbf{N}$ is the $m \times n$ diagonal factorized matrix. Each value on the diagonal of $\Lambda$ is nonnegative and real, and those values are called the singular values of the matrix $\mathbf{M}$, hence the name SVD. The columns of $\mathbf{U}$ and $\mathbf{V}$ are called left-singular vectors and right-singular vectors, respectively. All of this is very similar to eigendecomposition, with the difference of SVD working on rectangular matrices, resulting in two new bases $\mathbf{U}$ and $\mathbf{V}$. Now $\mathbf{Mv} = \lambda\mathbf{u}$ and $\mathbf{M}^{\mathbf{T}}\mathbf{u} = \lambda\mathbf{v}$ must be satisfied. Here, $\mathbf{u}$ is called the left-singular vector while $\mathbf{v}$ is the right-singular vector, with $\lambda$ the singular value.

## A.2.2 Applications of tensors in dMRI tractography

### A.2.2.1 Real symmetric diffusion tensor

In diffusion tensor imaging (Chapter 10), often called DTI, the scalar diffusion coefficient is replaced by a diffusion tensor $\mathbf{D}$, a tensor of order 2, which can also be designated as a matrix, as in Eq. (A.9). In particular, it is a real symmetric matrix, that is, $d_{ij} \in \mathbb{R}$ and $d_{ij} = d_{ji}$ for $i, j = x, y, z$. Hence, the tensor can be described by only six elements, giving a minimum of six directions of acquisition to be able to compute the diffusion tensor.

When using software for diffusion tensor reconstruction, DTI metrics calculations, or visualization, a one-dimensional (1D) representation of the tensor $\mathbf{D}$ is employed. However, there are two different conventions for writing this 1D-array of six elements depending on the software. It can either be $[d_{xx}, d_{yy}, d_{zz}, d_{xy}, d_{xz}, d_{yz}]$ or $[d_{xx}, d_{xy}, d_{xz}, d_{yy}, d_{yz}, d_{zz}]$. Using the wrong convention causes major differences in calculations and visualization.

### A.2.2.2 Diagonalization of the diffusion tensor

Once the diffusion tensor is computed, a diagonalization of the resulting square matrix is done. Indeed, the three eigenvectors $(\mathbf{v}_1, \mathbf{v}_2, \mathbf{v}_3)$, and the associated eigenvalues $(\lambda_1, \lambda_2, \lambda_3)$, of the diffusion tensor are key components of the tractography process and the calculation of diffusion metrics. The eigenvectors correspond to the three noncollinear principal axes of the ellipsoid representing the tensor, while the associated eigenvalues indicate the axial apparent diffusivity.

Thus a high diffusion anisotropy in a voxel is synonymous with an elongated ellipsoid that translates to a high $\lambda_1$ and low $\lambda_2$, $\lambda_3$. The tractography algorithm can then consider the eigenvector $\mathbf{v}_1$ as the direction of maximum diffusivity. On the contrary, $\lambda_1 \approx \lambda_2 \approx \lambda_3$ means isotropic diffusion. DTI metrics use the relation between eigenvalues to quantify the anisotropy. For example, fractional anisotropy, equation (Chapter 10), is a derivative of the standard deviation, where 0 means isotropy and 1 anisotropy.

# Appendix B

# Numerical integration

**Jon Haitz Legarreta[a] and Robert A. Dallyn[b]**

[a]*Department of Radiology, Brigham and Women's Hospital, Mass General Brigham/Harvard Medical School, Boston, MA, United States,* [b]*NatBrainLab, Institute of Psychiatry, Psychology and Neuroscience, King's College London, London, United Kingdom*

Within the context of tractography, numerical integration methods are used for propagating locally computed orientation vectors to obtain streamline representations of the fibers. Thus they allow tractograms to be generated from the fiber orientation distribution functions extracted from the diffusion-weighted imaging (DWI) data. In fact, numerical integration tools gave rise to tractography itself when several authors used them in the DWI domain to obtain streamline representations from the diffusion tensor imaging (DTI) fit local orientations at the end of the 1990s (Conturo et al., 1999; Mori et al., 1999; Wedeen et al., 1995). Numerical integration methods have since remained as one of the most stable pieces in the classical tractography pipeline.

## B.1 Introduction

Numerical integration is a field in numerical analysis that focuses on computing (with a given accuracy) the approximate value of a definite integral that cannot be computed analytically. Numerical integration methods usually solve the integral framing the problem as an ordinary differential equation with given initial conditions (i.e., an initial value problem) following

$$
\begin{aligned}
F(x) &= \int_a^x f(s)ds \\
\frac{dF(x)}{dx} &= f(x), \quad F(a) = 0
\end{aligned}
\tag{B.1}
$$

where $F(x)$ is the (primitive) function that is sought to be approximated, and the values of the function $f(s)$ are known at a finite number of points. The interest in rewriting the integral problem as a differential equation lies in the fact that the right-hand side of Eq. (B.1), which is equivalently expressed as $F'(x) = f(x)$, contains only the dependent variable $x$ and not the dependent $F$, which simplifies the development of the theory and resolution methods. Hence, methods developed for ordinary differential equations can be applied to the restated problem and thus be used to evaluate the integral.

## B.2 Methods

Numerical integration methods can be classified according to different criteria:

- *Availability of the data*: Depending on whether all data points for the problem are readily available, *explicit* methods approximate the derivative's value using the past values (i.e., available up to the current point), whereas *implicit* methods find a solution by solving an equation involving both the past values as well as the current value. Thus *explicit* methods are used when data is only available up to the current system state, and *implicit* methods are used when all data is immediately accessible. The two methods are often called *backward difference* and *forward difference*, respectively.
- *Step*: Depending on the number of values used to compute the derivative's next point value, methods can be classified as being *single-step* or *multistep* methods if they use only one or multiple values, respectively. Also, the step size can be fixed or variable (or adaptive) depending on whether the method allows it to change across iterations. Adaptive methods are often used for error control and to ensure stability.

All such algorithms proceed in the following fashion:

- An initial point is taken.

**698 PART | VI** Appendix

- A step forward is taken, and a solution is derived for the new location. In single-step methods only the immediately previous step is used to compute the new result, whereas in multistep methods many previous points may matter. This is not to be confused with the step size—the number of underlying distance units separating each step, which can be nonunitary in either case.
- The process continues in an iterative fashion until the last data point is used.

In the following, the numerical integration methods that are most widely used in the domain of tractography are explained for first-order linear ordinary differential equations (which are the ones encountered in the context of tractography) and single (dependent) variable cases.

## B.2.1 The Euler method

The Euler method is the most basic method for numerical integration of ordinary differential equations with a given initial value. The Euler method is a first-order method, which means that the local truncation error is proportional to the square of the step size ($O(h^2)$), and the global error is proportional to the step size ($O(h)$). The Euler method often serves as the basis to construct more complex methods.

A first-order differential equation is an initial value problem of the form

$$y'(x) = f(x, y(x)), \quad y(x_0) = y_0 \tag{B.2}$$

Starting with the differential equation (B.2), the derivative $y'$ is substituted by the finite difference approximation

$$y'(x) \approx \frac{y(x + h) - y(x)}{h}$$

which when rearranged yields

$$y(x + h) \approx y(x) + hy'(x)$$

and using Eq. (B.2) gives

$$y(x + h) \approx y(x) + hf(x, y(x))$$

The preceding expression is applied recursively to compute estimates of the exact solution at different data points. We denote the value at the $n$th data point by $x_n$, and the computed solution at the $n$th time-step by $y_n$, that is, $y_n \equiv y(x = x_n)$; given a step size $h$ (assumed to be constant for the sake of simplicity, i.e., $h = x_n - x_{n-1}$), the explicit (or forward) Euler method applies the following recursion rule:

$$y_{n+1} = y_n + hf(x_n, y_n) + O(h^2) \tag{B.3}$$

If $y$ is expanded in the neighborhood of $x = x_n$, the expression (B.3) can be assimilated to a Taylor series truncated at order one:

$$y_{n+1} \equiv y(x_n + h) = y(x_n) + h\frac{dy}{dx}\bigg|_{x_n} + O(h^2)$$
$$= y_n + hf(x_n, y_n) + O(h^2) \tag{B.4}$$

The implicit (or backward) Euler method computes the approximations using

$$y_{n+1} = y_n + hf(x_{n+1}, y_{n+1}) + O(h^2) \tag{B.5}$$

This differs from the forward Euler method in that the latter uses $f(x_n, y_n)$ instead of $f(x_{n+1}, y_{n+1})$. The backward Euler method is an implicit method, meaning that to find $y_{n+1}$ an equation must be solved. Fixed-point iteration or (some modification of) the Newton-Raphson method are often used to achieve this.

Note that it is computationally more expensive to solve for implicit methods. The advantage of implicit methods is that they are usually more stable for solving a stiff equation (i.e., equations whose approximation error grows so large that it dominates the calculations), meaning that a larger step size $h$ can be used.

The forward Euler method has a simple computational scheme but requires a very small step size for any *sufficiently* accurate solutions (since the error is proportional to $h$). The modified Euler method (also known as the midpoint method) mitigates this by using more function evaluations: it takes the arithmetic mean of the derivatives at the beginning and end of the interval, yielding an order 2 method:

$$y_{n+1} = y_n + hf\left(x_n + \frac{1}{2}h, y_n + \frac{1}{2}hf(x_n, y_n)\right) + O(h^3) \tag{B.6}$$

Further enhancements to the Euler method lead to the Runge-Kutta and Adams methods. In the context of tractography, only the implicit (or forward difference) form of the Euler method of orders up to two are used.

## B.2.2 The Runge-Kutta method

In general, the accuracy of a numerical integration method can be improved by decreasing the step size and taking more steps at the cost of increased computation. A set of higher-order methods known as Runge-Kutta (RK) optimizes both the accuracy and the computational efficiency.

In the forward Euler method, the information on the derivative of $y$ at the given time step is used to extrapolate the solution to the next time step, resulting in a first-order numerical technique (with $O(h^2)$ local truncation error). RK methods work by approximating the exact solution with the $N$ first terms of the Taylor series. This allows the methods to use the derivative at more than one point to extrapolate the solution to the future time step. Although implicit RK methods do also exist, only the explicit version is covered in this text.

Given the initial value problem of Eq. (B.3), a step size $h$, and the solution $y_n$ at the $n$th time step, the solution $y_{n+1}$ at the next time step $n + 1$ is computed by a weighed sum of two increase estimates, namely $k_1$ and $k_2$:

$$\begin{aligned} k_1 &= hf(x_n, y_n) \\ k_2 &= hf(x_n + \alpha h, y_n + \beta k_1) \\ y_{n+1} &= y_n + ak_1 + bk_2 \end{aligned} \tag{B.7}$$

where the constants $\alpha, \beta, a$, and $b$ have to be evaluated so that the resulting method has a local truncation error of $O(h^3)$. Note that if $k_2 = 0$ and $a = 1$, then Eq. (B.7) reduces to the forward Euler method.

The RK methods take the Euler method's first approximation to compute the first derivative. The aim is to come up with a computational framework to find the values of the parameters $\alpha, \beta, a$, and $b$. This is achieved by writing the Taylor series expansion of $y$ in the neighborhood of $x_n$ up to the $h^2$ term, that is,

$$y(x_{n+1}) = y(x_n) + h\frac{dy}{dx}\bigg|_{x_n} + \frac{h^2}{2}\frac{d^2y}{dx^2}\bigg|_{x_n} + O(h^3) \tag{B.8}$$

However, we know from the initial value problem of Eq. (B.3) that $dy/dx = f(x, y)$ so that

$$\frac{d^2y}{dx^2} = \frac{df(x,y)}{dx} = \frac{\partial f}{\partial x} + \frac{\partial f}{\partial y}\frac{dy}{dx} = \frac{\partial f}{\partial x} + f\frac{\partial f}{\partial y} \tag{B.9}$$

This allows us to rewrite Eq. (B.8) as

$$y_{n+1} = y_n + hf(x_n, y_n) + \frac{h^2}{2}\left[\frac{\partial f}{\partial x} + f\frac{\partial f}{\partial y}\right](x_n, y_n) + O(h^3) \tag{B.10}$$

Substituting the definitions of $k_1$ and $k_2$ in Eq. (B.7),

$$y_{n+1} = y_n + ahf(x_n, y_n) + bhf(x_n + \alpha h, y_n + \beta hf(x_n, y_n)) \tag{B.11}$$

In order to identify the equivalent terms in Eqs. (B.10), (B.11), $f(x, y)$ is developed as a Taylor series in terms of $x_n, y_n$. Keeping only the terms corresponding to the first derivative,

$$f(x_n + \alpha h, y_n + \beta hf(x_n, y_n)) \doteq \left[f + \alpha h\frac{\partial f}{\partial x} + \beta hf\frac{\partial f}{\partial y}\right](x_n, y_n) \tag{B.12}$$

**700 PART | VI** Appendix

Now, substituting for $k_2$ from Eq. (B.12) in Eq. (B.11) and reordering,

$$
\begin{aligned}
y_{n+1} &= y_n + ahf(x_n,y_n) + bh\left[f + \alpha h\frac{\partial f}{\partial x} + \beta hf\frac{\partial f}{\partial y}\right](x_n,y_n) + O(h^3)\\
&= y_n + (a+b)hf(x_n,y_n) + bh^2\left[\alpha\frac{\partial f}{\partial x} + \beta f\frac{\partial f}{\partial y}\right](x_n,y_n) + O(h^3)
\end{aligned}
\tag{B.13}
$$

Note that Eqs. (B.10), (B.13) are identical if:

$$
a + b = 1, \quad \alpha b = \frac{1}{2}, \quad \beta b = \frac{1}{2}
\tag{B.14}
$$

which gives a way to determine the constants. Given that this is an underdetermined system of equations, there are infinitely many choices of $a$, $b$, $\alpha$, and $\beta$ that satisfy Eq. (B.14). Choosing $\alpha = \beta = 1$ and $a = b = 1/2$ gives the classical second-order RK method (RK2):

$$
\begin{aligned}
k_1 &= hf(x_n,y_n)\\
k_2 &= hf(x_n+h,y_n+k_1)\\
y_{n+1} &= y_n + \frac{1}{2}(k_1+k_2)
\end{aligned}
\tag{B.15}
$$

Note that the modified Euler method is a special case of the RK2 method.

In a similar fashion, RK methods of higher order can be developed. The most widely used RK method, including the domain of tractography, is the fourth-order Runge-Kutta (RK4), which has a local truncation error of $O(h^5)$. This leads to 11 equations with 13 unknowns, which are solved by taking arbitrary values for two of them. The most widely used values lead to the following group of equations:

$$
\begin{aligned}
k_1 &= hf(x_n,y_n)\\
k_2 &= hf\left(x_n+\frac{1}{2}h,y_n+\frac{1}{2}k_1\right)\\
k_3 &= hf\left(x_n+\frac{1}{2}h,y_n+\frac{1}{2}k_2\right)\\
k_4 &= hf(x_n+h,y_n+k_3)\\
y_{n+1} &= y_n + \frac{1}{6}(k_1 + 2k_2 + 2k_3 + k_4)
\end{aligned}
\tag{B.16}
$$

Although four evaluations (instead of two) are required for each step, the RK4 method is more efficient than the modified Euler method since the steps can be orders of magnitude larger for the same accuracy.

Both Euler and RK methods are single-step methods. However, RK methods can take some intermediate steps (e.g., a half-step) to obtain a higher-order method, but then discard all previous information before taking a second step.

### B.2.3   The Adams-Bashforth method

As opposed to the Euler and RK methods, which apply implicit variants in tractography, the Adams method used in tractography is the explicit or backward difference variant, called the Adams-Bashforth method.

Adams methods are multistep methods based on the idea of approximating the integrand with a polynomial within the interval $(x_n, x_{n+1})$. Using a $k$th-order polynomial results in a $k + 1$th-order method. Multistep methods use information from the previous $k$ steps to construct a polynomial to approximate the derivative function. The computed values are then used to extrapolate into the next interval. In particular, a linear multistep method uses a linear combination of $y_i$ and $f(x_i, y_i)$ to calculate the value of $y$ for the desired current step. Thus a linear multistep method is a method of the form:

$$
\begin{aligned}
&y_{n+k} + a_{k-1}y_{n+k-1} + a_{k-2}y_{n+k-2} + \cdots + a_0 y_n\\
&= h[b_k f(x_{n+k},y_{n+k}) + b_{k-1}f(x_{n+k-1},y_{n+k-1}) + \cdots + b_0 f(x_n,y_n)]\\
&\Leftrightarrow \sum_{j=0}^{k} a_j y_{n+j} = h\sum_{j=0}^{k} b_j f(x_{n+j},y_{n+j})
\end{aligned}
$$

with $a_k = 1$. The coefficients $a_0, \ldots, a_{k-1}$ and $b_0, \ldots, b_k$ determine the method. If $b_k = 0$, then the method is an explicit one, since the formula can directly compute $y_{n+k}$. If $b_k \neq 0$, then the method belongs to the implicit family, since the value of $y_{n+k}$ depends on the value of $f(x_{n+k}, y_{n+k})$, and the equation must be solved for $y_{n+k}$. Again, iterative methods such as the Newton-Raphson method can be used to solve the implicit equation.

The coefficients $b_j$ can be determined as follows. A polynomial interpolation is used to find the polynomial $p$ of degree $k - 1$ such that

$$p(x_{n+i}) = f(x_{n+i}, y_{n+i}), \quad \text{for } i = 0, \dots, k - 1$$

The Lagrange formula for polynomial interpolation yields

$$p(x) = \sum_{j=0}^{k-1} \frac{(-1)^{k-j-1} f(x_{n+j}, y_{n+j})}{j!(k-j-1)! h^{k-1}} \prod_{\substack{i=0 \\ i \neq j}}^{k-1} (x - x_{n+i})$$

The polynomial $p$ is locally a good approximation of the right-hand side of the differential equation $y' = f(x, y)$ that is to be solved, so the equation $y' = p(x)$ can be considered instead. This equation can be solved exactly; the solution is simply the integral of $p$. The initial value problem can be equivalently written as

$$\frac{dy}{dx} = f(x, y)$$
$$\Leftrightarrow dy = f(x, y)dx$$
$$\Leftrightarrow \int_{x_n}^{x_{n+1}} dy = y_{n+1} - y_n = \int_{x_n}^{x_{n+1}} f(x, y)dx$$
$$\Leftrightarrow y_{n+1} = y_n + \int_{x_n}^{x_{n+1}} f(x, y)dx$$

And hence, for the $k$-step Adams-Bashforth method at hand,

$$y_{n+k} = y_{n+k-1} + \int_{x_{n+k-1}}^{x_{n+k}} p(x)dx \tag{B.17}$$

Note that if the polynomial $p$ is a first-order polynomial, the first-order Adams-Bashforth method simply reduces to the forward Euler method.

The Adams-Bashforth method arises when the formula for $p$ is substituted in Eq. (B.17). The coefficients $b_j$ turn out to be given by

$$b_{k-j-1} = \frac{(-1)^j}{j!(k-j-1)!} \int_0^1 \prod_{\substack{i=0 \\ i \neq j}}^{k-1} (u + i)du, \quad \text{for } j = 0, \dots, k - 1$$

As an illustration, the two-step (i.e., using two previous values, $y_{n+1}$ and $y_n$) Adams-Bashforth method is given by

$$y_{n+2} = y_{n+1} + h \left[ \frac{3}{2} f(x_{n+1}, y_{n+1}) - \frac{1}{2} f(x_n, y_n) \right]$$

## B.3  Aspects of numerical integration

There are a few aspects that should be taken into account when choosing a numerical integration method. The choice will determine, among other things, the achievable accuracy or the stability of the method with respect to the function. The following is a list of those aspects:

- *Order*: The order of the method corresponds to the power of the step size up to which the Taylor series expansion of the approximate solution differs less than a given amount from the true solution. The higher the order of the method, the more accurate the approximation. Usually, this comes at the expense of a higher computational cost since more points need to be evaluated at each step.
- *Truncation error*: The local truncation error of the method is the error accrued in advancing one step of the method; that is, it is the difference between the result given by the method assuming that no error was made in earlier steps, and the exact solution. The global truncation error in turn, is the error sustained in all the steps one needs to reach for a fixed time

**702 PART | VI** Appendix

point in the computation. It follows from the method's order definition that the error is related to the step size: generally, a small increment (step size) will lead to a smaller error. As in any other area of study of numerical approximation methods, error analysis is of cardinal importance when studying numerical integration methods.

- *Stability*: In a stable method early errors (due to the truncation error, an initial value that is slightly incorrect, or round-off errors) are dampened out as the computations proceed, that is, they do not grow without bound. Similarly, there are certain equations whose numerical integration may result in larger errors around some points due to the particular features presented by the underlying true function. Some numerical integration methods may adapt better to such situations. Stability is an important property because it ensures that a small truncation error introduced at each time step does not cause a catastrophic divergence in the solution over time. Stability is also related to the step size: this is so because in explicit methods the solution may become numerically unstable if the selected step is not small enough. Some implicit methods always lead to bounded solutions under some circumstances, and hence explicit methods are said to be *conditionally stable*, whereas implicit methods are (generally) *unconditionally stable*. Thus although explicit methods produce equations that are easier to solve, they require a small step size; conversely, implicit methods allow for a larger step size, but produce equations that are more expensive to solve.

For a further discussion about numerical integration methods and aspects, the reader is referred to specialized texts such as Bradie (2005), Burden et al. (2010), Gerald and Wheatley (2004), or Iserles (2008).

## B.4 Applications and relevance in tractography

Tracking algorithms integrate local estimates of the diffusion direction. Once the local orientation of the fibers at every voxel is estimated using a given reconstruction model (see Chapters 10, 11 and 12), these directions can be joined to reconstruct a complete long-range fiber trajectory and hence approximate the anatomical pathways. Simpler tracking processes essentially assume that each imaging voxel is characterized by a single predominant fiber orientation and piece together these local orientations to infer global fiber trajectories. Where it is believed that a single voxel is likely to contain information from more than one underlying fiber orientation, methods predominately using spherical deconvolution can be applied to deconstruct the voxel signal into a number of discrete fiber bundles. These discrete fiber bundles can then be iterated over separately while tracking using any of the numerical integration methods previously discussed. Mathematically, the set of local fiber orientations can be considered as a three-dimensional (3D) vector field and the streamlines as the global fiber trajectories. Formally, a streamline is any curve that along its trajectory is tangent to the vector field and that can be represented as a 3D space curve $\mathbf{r}(s) = (x(s), y(s), z(s))$, parameterized by its arc length $s$. In order for a streamline to align with the vector field, the tangent at arc length $s$ has to be equal to the vector at the corresponding position:

$$\frac{d\mathbf{r}(s)}{ds} = \mathbf{v}[\mathbf{r}(s)] \tag{B.18}$$

where $\mathbf{r}(s)$ denotes the 3D position along the streamline and $\mathbf{v}$ is the 3D vector field, representing the eigenvector—primary direction—of a (diffusion) tensor model or a different one.

Tractograms are essentially solutions to the first-order differential equation (B.18). Mathematically, the most intuitive way to perform this is by integrating Eq. (B.18) using numerical integration methods: the integral curve $\mathbf{r}(s)$ passing through a given point $r_0$ for $s > 0$ and $s < 0$ (i.e., in both directions, forward and backward) is the solution to:

$$\mathbf{r}(s) = \int_{s_0} \mathbf{r}[(s)]ds \tag{B.19}$$

where $\mathbf{r}(s_0) = r_0$ represents the starting point (initial conditions) of the streamline, which is often referred to as the *seed point*.

The computed local orientations can then be integrated into streamlines in a step-wise (iterative) fashion. As shown in Fig. B.1, starting at a given seed point $r_0$, obtaining the corresponding local fiber orientation $\mathbf{v}(\mathbf{r}_0)$ and then following that direction for a short distance $0 < \Delta \leq 1$, called the (integration) step size, the next point $\mathbf{r}_1 = \mathbf{r}_0 + \Delta\mathbf{v}(\mathbf{r}_0)$ on the streamline is obtained. Thus the entire pathway can be reconstructed by iteratively performing this procedure:

$$\mathbf{r}_{i+1} = \mathbf{r}_i + \Delta\mathbf{v}(\mathbf{r}_i) \tag{B.20}$$

where $0 < \Delta \leq 1$ defines the integration finite step, and giving rise to an entire tractogram.

The described procedure corresponds to the Euler integration method and was proposed in Conturo et al. (1999) to perform the tracking process. Note that Euler integration is a first-order integration method, that is, it uses only the first

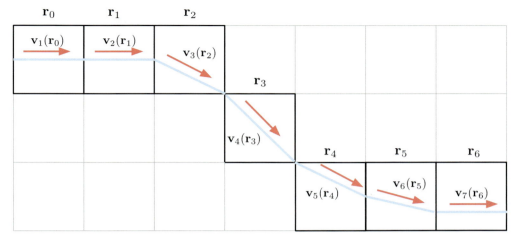

**FIG. B.1** Streamline propagation procedure. A schematic representation of the streamline propagation procedure in tractography: numerical integration methods are used to integrate the local fiber orientation information $\mathbf{v}_{i+1}$ at each point $\mathbf{r}_i$ into a streamline.

two terms of the Taylor expansion. Also, it assumes that the orientation $\mathbf{v}(\mathbf{r}_i)$ is constant at the length scale of step size $\Delta$, which makes the presented approach highly susceptible to overshoots in highly curved regions, especially for larger step sizes. In Basser et al. (2000), authors used a higher-order numerical integration scheme, the fourth-order RK scheme, to take into account the variations of $\mathbf{v}$ between $\mathbf{r}_i$ and $\mathbf{r}_{i+1}$. Such higher-order approximations are much less susceptible to integration errors (Jeurissen et al., 2019). Previously, Lazar and Alexander (2003) had provided an excellent analysis of the accuracy of the numerical integration methods in simulated tractography data. Variants of these methods have been proposed by several authors: see, for example, Tournier et al. (2010), Yeh et al. (2010), and Garyfallidis (2012).

# References

Basser, P.J., Pajevic, S., Pierpaoli, C., Duda, J., Aldroubi, A., 2000. In vivo fiber tractography using DT-MRI data. Magn. Reson. Med. 44 (4), 625–632. https://doi.org/10.1002/1522-2594(200010)44:43.0.CO;2-O.
Bradie, B., 2005. A Friendly Introduction to Numerical Analysis. Pearson.
Burden, R.L., Faires, J.D., Burden, A.M., 2010. Numerical Analysis, ninth ed. BrooksCole.
Conturo, T.E., Lori, N.F., Cull, T.S., Akbudak, E., Snyder, A.Z., Shimony, J.S., McKinstry, R.C., Burton, H., Raichle, M.E., 1999. Tracking neuronal fiber pathways in the living human brain. Proc. Natl Acad. Sci. USA 96 (18), 10422–10427. https://doi.org/10.1073/pnas.96.18.10422.
Garyfallidis, E., 2012. Towards an Accurate Brain Tractography (Ph.D. thesis). Cambridge University.
Gerald, C.F., Wheatley, P.O., 2004. Applied Numerical Analysis, seventh ed. Pearson Education, Inc.
Iserles, A., 2008. A First Course in the Numerical Analysis of Differential Equations, second ed. Cambridge University Press.
Jeurissen, B., Descoteaux, M., Mori, S., Leemans, A., 2019. Diffusion MRI fiber tractography of the brain. NMR Biomed. 32 (4), e3785. https://doi.org/10.1002/nbm.3785.
Lazar, M., Alexander, A.L., 2003. An error analysis of white matter tractography methods: synthetic diffusion tensor field simulations. NeuroImage 20 (2), 1140–1153. https://doi.org/10.1016/S1053-8119(03)00277-5.
Mori, S., Crain, B.J., Chacko, V.P., Van Zijl, P.C.M., 1999. Three-dimensional tracking of axonal projections in the brain by magnetic resonance imaging. Ann. Neurol. 45 (2), 265–269. https://doi.org/10.1002/1531-8249(199902)45:23.0.CO;2-3.
Tournier, J.D., Calamante, F., Connelly, A., 2010. Improved probabilistic streamlines tractography by 2nd order integration over fibre orientation distributions. In: Proceedings of the International Society for Magnetic Resonance in Medicine (ISMRM) 17th Annual Scientific Meeting & Exhibition, International Society for Magnetic Resonance in Medicine, Stockholm, Sweden, p. 1670.
Wedeen, V.J., Davis, T.L., Weisskoff, R.M., Tootell, R.B.H., Rosen, B.R., Belliveau, J.W., 1995. White matter connectivity explored by MRI. In: Proceedings of the First International Conference in Functional Mapping of the Human Brain, Paris, France, vol. 1, p. 69.
Yeh, F.C., Wedeen, V.J., Tseng, W.Y.I., 2010. Generalized q-sampling imaging. IEEE Trans. Med. Imaging 29 (9), 1626–1635. https://doi.org/10.1109/TMI.2010.2045126.

# Appendix C

# Interpolation, splines, and smoothing

**Matteo Battocchio and Simona Schiavi**
*Department of Computer Science, University of Verona, Verona, Italy*

Given an input set of discrete data points, curve fitting is the process of finding the continuous mathematical function that has the best fit to some series of data points. Depending on whether we look for the exact representation or an approximation of our data, the process is defined as *interpolation* or *smoothing*.

## C.1 Data interpolation

### C.1.1 Diffusion image interpolation

Data interpolation is the process of estimating unknown values that lie in between some known data points. Interpolation is widely used in diffusion-weighted magnetic resonance imaging (DWI), where acquisition time constraints for clinical applications heavily undermine image resolution and signal-to-noise ratio (SNR). Low SNR has been proven to bias the estimation of diffusion parameters (Basser et al., 1994), while low spatial resolution increases the severity of partial volume effects, which in turn hamper DWI analysis and the capability to resolve complex fiber configurations in white matter reconstruction (Alexander et al., 2001; Oouchi et al., 2007). On the other hand, high resolution and ultrahigh resolution DWIs can reach high sensitivity by combining strong gradient and long acquisition protocols, but time and resources needed for this type of acquisitions prevent their frequent exploitation.

Data interpolation has been shown to improve resolution and anatomical details of lower quality data, becoming increasingly fundamental in DWI processing. For instance, interpolation of DWIs is commonly used when applying transformations to align images acquired with different gradient directions or when converting DWI datasets acquired with anisotropic voxels to isotropic voxels. Another classic example of DWIs transformation is the registration to a different space, which often implies the resampling of the image to a higher or lower spatial resolution. In this context, superresolution approaches are a class of methods that exploits interpolation kernels to increase voxel-wise anatomical detail (Lehmann et al., 2001). To achieve finer details they interpolate and combine multiple images of the same object at lower resolution, resulting in a single image with increased level of details. Depending on the approach, this usually involves particular acquisition protocols; therefore it cannot be applied to preexisting DWI data (Dyrby et al., 2014).

Formally, the interpolation step represents the convolution of a discrete image sample with a continuous impulse function $h$ of a reconstruction filter $s$ (Lehmann et al., 2001). This class of functions is characterized by a tall narrow spike (impulse) and light tails. Reconstruction filters allow reducing the classic stair-stepped waveforms of a digital input by restoring the frequency, phase, and amplitude of the original signal. For example, in the case of a 2D image, this implies the reconstruction of a continuous signal $s(x, y)$ from one of the corresponding discrete neighbors $s(k, l)$:

$$s(x,y) = \sum_k \sum_l s(k,l) * h(x - k, y - l), \tag{C.1}$$

where $x, y \in \mathbb{R}$ and $k, l \in \mathbb{N}_0$ (Lehmann et al., 2001). An ideal interpolator guarantees that the image is not modified once resampled from the same space grid. An example is the application of an ideal infinite impulse response interpolator 1D filter $h_{\mathrm{1D}}$ in the Fourier domain:

$$h_{\mathrm{1D}}(x) = \frac{\sin(\pi x)}{\pi x} = \mathrm{sinc}(x) \tag{C.2}$$

**706 PART | VI** Appendix

corresponding to a periodic positive kernel defined as follows:

$$\begin{cases} h(0) \equiv 1 \\ h(x) \equiv 0, \quad |x| = 1, 2, \ldots \end{cases} \tag{C.3}$$

Interpolation kernels are often defined over an infinite support domain that needs to be constrained in order to be applied for spatial convolution. This is usually performed via truncation or windowing (Lehmann et al., 2001). The first implies the multiplication of the ideal kernel with a rectangular function in the space domain, which corresponds to a sinc function in the frequency domain. However, discarding higher frequencies in this way leads to ringing effects, which are called Gibb's phenomenon (Aldroubi et al., 1992). Windowing, on the other hand, first truncates the infinite length ideal impulse response, and then makes use of smoother functions to attenuate the side effects on the resulting filter spectrum caused by the crude truncation (Oppenheim et al., 1999). There is a wide range of kernels that approximate the ideal sinc function interpolator with different properties. The most straightforward include nearest neighbor $h_{NN}$ and linear $h_{Lin}$ methods defined as follows:

$$h_{NN} = \begin{cases} 1, & 0 \leq |x| < 0.5 \\ 0, & \text{elsewhere,} \end{cases} \tag{C.4}$$

$$h_{Lin} = \begin{cases} 1 - |x|, & 0 \leq |x| < 1 \\ 0, & \text{elsewhere.} \end{cases} \tag{C.5}$$

These methods are easy to implement but are characterized by aliasing of the data after the cutoff point and high-frequency attenuation. More sophisticated approaches are investigated in Dyrby et al. (2014), where different polynomial interpolation methods are compared and some of them are shown in Fig. C.1. Polynomial interpolation, like cubic and B-spline (Bartels et al., 1987), has many advantages. First of all, it is easy and computationally fast to determine (Lehmann et al., 2001) and, in the case of cubic or higher-order methods, they are twice differentiable continuous (C2-continuous). This in turn allows avoidance of problems of oscillation at the edges, known as Runge's phenomenon (Runge, 1901). Finally, basis splines (B-splines) are piece-wise concatenations of polynomial curves, defined as basis functions, of a certain order $N$. The order of a basis function reflects the extent of its spatial support region, as shown in Fig. C.2. First order, or linear, B-spline interpolator $h_2$ corresponds to the convolution of the rectangular kernel, Eq. (C.4), with itself:

$$h_2(x) = h_1(x) * h_1(x). \tag{C.6}$$

It follows that any order B-splines can be constructed as consecutive convolutions with the rectangular function $h_1(x)$:

$$h_N(x) = h_1^1(x) * h_1^2(x) * h_1^3(x) * \cdots * h_1^{N-1}(x). \tag{C.7}$$

Higher-order B-spline interpolation methods have proven to be more accurate thanks to their ability to exploit neighborhood information but, at the same time, they are more subjective to artifact such as ringing in correspondence with sharp borders, caused by over- or undershoot when transitioning from one side to the other (Dyrby et al., 2014). Notably, this artifact is independent from the sampling factor, that is, the ratio between the original and the interpolated image resolutions, but it is exclusively determined by the grid size of the original image.

Kernels that do not satisfy Eq. (C.3) are defined as approximators or smoothing kernels. Examples of approximators, such as the polynomial kernels, are presented in the next section, which allows uniform approximation of continuous functions over finite intervals.

## C.2 Spline smoothing

In the previous section we saw how interpolation forces a parametric curve or surface to contain the given data. In most cases, however, these are noisy, making interpolation impractical or particularly unstable. In response to that, fitting methods try to capture the general shape of the data leading to more robust estimations at the cost of a lower accuracy. In the case of an unknown data distribution, the estimation becomes nonparametric and it is often also called smoothing (Green and Silverman, 2019). Smoothing can be performed in different ways but all imply the computation of a constraint factor (lambda) driving the accuracy of the estimation. The choice of this parameter is performed empirically looking at the performance of our estimator (Hsieh and Manski, 1984; Jin, 1992).

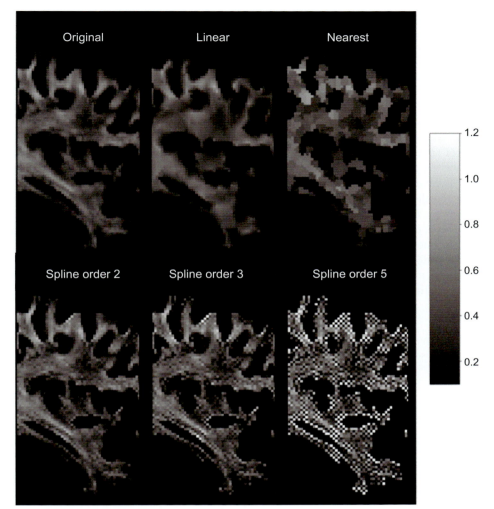

**FIG. C.1** Application of different interpolation approaches on a fractional anisotropy map computed from diffusion tensor imaging reconstruction with voxel resolution of 1.25 isotropic. The linear and nearest neighbor interpolations have been computed from a random selection on 1000 points. For the different order spline interpolations, at the boundaries we adopt a mirror criterion, that is, reflecting the input over the edge of the last voxels.

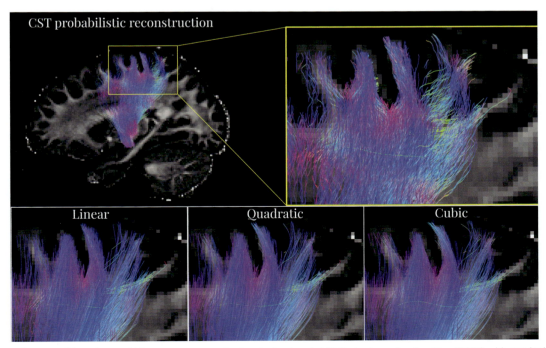

**FIG. C.2** B-spline smoothing of pyramidal tract reconstructed using the local probabilistic tractography algorithm iFOD2 with standard parameters. From left to right are shown the results of linear, quadratic, and cubic spline smoothing using an average number of 11 control points.

Smoothing in DWI is used not only for image reconstruction and registration but has been widely exploited by tractography algorithms with different approaches. Cosine series representation has been widely used to model planar curves (Persoon and Fu, 1977; Staib and Duncan, 1992) and more recently has been applied to encode fiber bundles consisting of multiple 3D curves (Chung et al., 2010). These are represented by so-called Fourier descriptors, in which the coordinates are parameterized as coefficients of cosine series expansion up to a fixed degree.

Another example of spline smoothing can be found in Jbabdi et al. (2007), where B-splines are used to model white matter pathways connecting pairwise regions of the brain. Formally, the approximation B-spline of degree $d$ is a piece-wise concatenation of polynomial curves joining in correspondence of $n + 1$ control points, called knots. The order of the function is determined by the order of the piece-wise polynomial basis functions, where an increasing order results in a larger spatial support region for each basis function. In particular, cubic splines are widely used because (i) the low number of parameters makes their computation fast and easy to perform and (ii) they can include a large variety of paths which makes them suitable for tractography applications.

Given a function $f(x)$ defined on an interval $[a,b]$, and given an ordered set of points $\{(x_0, f(x_0)), (x_1, f(x_1)), ..., (x_n, f(x_n))\}$, we want to fit a curve through the points as an approximation of the function $f(x)$. The cubic spline is a function $S(x)$ on $[a,b]$ defined as

$$S_i(x) = a_i + b_i \cdot (x - x_i) + c_i \cdot (x - x_i)^2 + d_i \cdot (x - x_i)^3, \tag{C.8}$$

characterized by $4n$ unknowns that, in the case of the free boundary case, implies that $a_i = f(x_i)$ for $i = 0, 1, 2, ..., n - 1$. Letting $h_i = x_{x+1} - x_i$, the solution can be found solving the linear system $Ax = b$, where

$$A = \begin{bmatrix} 1 & 0 & 0 & 0 & \cdots & 0 \\ h_0 & 2(h_0 + h_1) & h_1 & 0 & \cdots & 0 \\ 0 & h_1 & 2(h_1 + h_2) & h_2 & \cdots & 0 \\ 0 & 0 & h_2 & 2(h_2 + h_3) & \cdots & 0 \\ 0 & 0 & 0 & h_3 & \cdots & 0 \\ \vdots & \vdots & \vdots & \cdots & \ddots & \vdots \\ 0 & 0 & 0 & \cdots & \cdots & 1 \end{bmatrix} \tag{C.9}$$

and

$$b = \begin{bmatrix} 0 \\ \dfrac{3}{h_1}(a_2 - a_1) - \dfrac{3}{h_0}(a_1 - a_0) \\ \vdots \\ \dfrac{3}{h_{n-1}}(a_n - a_{n-1}) - \dfrac{3}{h_{n-2}}(a_{n-1} - a_{n-2}) \end{bmatrix}. \tag{C.10}$$

Fig. C.3 shows an example of cubic spline smoothing of a 3D dataset. The points approximation is regulated by $\lambda$, called the smoothing parameter, controlling the trade-off between fidelity to the data and smoothness of the function estimate. If $\lambda = 0$ the smoothing spline converges to the interpolating spline (see Fig. C.3, green surface), while for $\lambda \to \infty$ it converges to a linear least squares estimate.

Catmull-Rom (Catmull and Rom, 1974) is a particular class of cubic B-splines showing some useful characteristics (Lehmann et al., 2001; Dyrby et al., 2014): they are cubic, hence can easily adapt to describe different fiber shapes and, differently from other B-splines, their trajectories are forced to cross the knots, including the extremities. The last property is particularly useful to keep the shape of the resulting smoothed fiber under control.

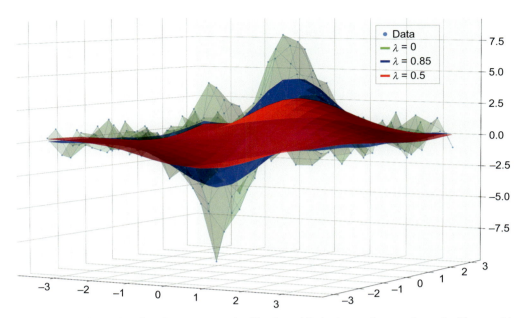

**FIG. C.3** Effects of smoothing using cubic splines. Data are reported as *blue dots*, while the three surfaces are the result of the smoothing using different approximation errors.

# References

Aldroubi, A., Unser, M., Eden, M., 1992. Cardinal spline filters: stability and convergence to the ideal sinc interpolator. Signal Process. 28 (2), 127–138.

Alexander, A.L., Hasan, K.M., Lazar, M., Tsuruda, J.S., Parker, D.L., 2001. Analysis of partial volume effects in diffusion-tensor MRI. Magn. Reson. Med. 45 (5), 770–780.

Bartels, R.H., Beatty, J.C., Barsky, B.A., 1987. An Introduction to Splines for Use in Computer Graphics and Geometric Modeling. Morgan Kaufmann Publishers.

Basser, P.J., Mattiello, J., LeBihan, D., 1994. MR diffusion tensor spectroscopy and imaging. Biophys. J. 66 (1), 259–267.

Catmull, E., Rom, R., 1974. Computer Aided Geometric Design. Academic Press, New York, pp. 317–326.

Chung, M.K., Adluru, N., Lee, J.E., Lazar, M., Lainhart, J.E., Alexander, A.L., 2010. Cosine series representation of 3D curves and its application to white matter fiber bundles in diffusion tensor imaging. Stat. Interface 3 (1), 69.

Dyrby, T.B., Lundell, H., Burke, M.W., Reislev, N.L., Paulson, O.B., Ptito, M., Siebner, H.R., 2014. Interpolation of diffusion weighted imaging datasets. NeuroImage 103, 202–213.

Green, P.J., Silverman, B.W., 2019. Nonparametric Regression and Generalized Linear Models: A Roughness Penalty Approach. Chapman and Hall/CRC.

Hsieh, D.A., Manski, C.F., 1984. Monte-Carlo Evidence on Adaptive Maximum Likelihood Estimation of a Regression. University of Wisconsin-Madison, Social Systems Research Institute. https://ideas.repec.org/p/ags/uwssri/292597.html. SSRI Workshop Series 292597.

Jbabdi, S., Woolrich, M.W., Andersson, J.L.R., Behrens, T.E.J., 2007. A Bayesian framework for global tractography. NeuroImage 37 (1), 116–129.

Jin, K., 1992. Empirical smoothing parameter selection in adaptive estimation. Ann. Stat. 20 (4), 1844–1874.

Lehmann, T.M., Gonner, C., Spitzer, K., 2001. Addendum: B-spline interpolation in medical image processing. IEEE Trans. Med. Imaging 20 (7), 660–665. https://doi.org/10.1109/42.932749.

Oouchi, H., Yamada, K., Sakai, K., Kizu, O., Kubota, T., Ito, H., Nishimura, T., 2007. Diffusion anisotropy measurement of brain white matter is affected by voxel size: underestimation occurs in areas with crossing fibers. Am. J. Neuroradiol. 28 (6), 1102–1106.

Oppenheim, A.V., Schafer, R.W., Buck, J.R., 1999. Discrete-Time Signal Processing, second ed. Prentice-Hall, Inc., Englewood Cliffs, NJ.

Persoon, E., Fu, K.S., 1977. Shape discrimination using Fourier descriptors. IEEE Trans. Syst. Man Cybern. 7 (3), 170–179.

Runge, C., 1901. Über empirische funktionen und die interpolation zwischen äquidistanten ordinaten. Mag. Math. Phys. 46 (224–243), 20.

Staib, L.H., Duncan, J.S., 1992. Boundary finding with parametrically deformable models. IEEE Trans. Pattern Anal. Mach. Intell. 14 (11), 1061–1075.

# Appendix D

# Spherical harmonics

**Gabrielle Grenier[a], Charles Poirier[a], and Jon Haitz Legarreta[b]**

[a]*Department of Computer Sciences, Université de Sherbrooke, Sherbrooke, QC, Canada,* [b]*Department of Radiology, Brigham and Women's Hospital, Mass General Brigham/Harvard Medical School, Boston, MA, United States*

Spherical harmonics (SHs) are a mathematical tool that allows any complex valued spherical function to be represented as a linear combination of SH basis functions. The mathematical definitions in this chapter are adapted from Arfken and Weber (1995). Given the nature of the diffusion encoding gradients used in diffusion-weighted imaging (DWI), SHs form an interesting basis to model the measured signal on a sphere using a series of coefficients. In this regard, SHs parallel the Fourier analysis. Indeed, both allow representing a function in terms of the coefficients of harmonic functions. Since their introduction to the DWI domain, SHs have become an important tool to represent the diffusion signal and have enabled the development of a set of distinctive methods for the computation of the orientation distribution functions.

## D.1 Definition and properties of spherical harmonics

SHs are special functions defined on the sphere. They arise from solving Laplace's equation in spherical coordinates, a second-order partial differential equation (PDE), whose solutions are called "harmonic functions." For $r \in \mathbb{R}^3$, $\theta \in [0, \pi]$, and $\phi \in [0, 2\pi)$, the PDE called the Laplace equation in spherical coordinates is given by

$$\frac{1}{r^2}\frac{\partial}{\partial r}\left(r^2\frac{\partial \psi}{\partial r}\right) + \frac{1}{r^2 \sin^2\theta}\frac{\partial^2 \psi}{\partial \phi^2} + \frac{1}{r^2 \sin\theta}\frac{\partial}{\partial \theta}\left(\sin\theta\frac{\partial \psi}{\partial \theta}\right) = 0. \tag{D.1}$$

With the method of separation of variables, it is possible to solve this PDE. Indeed, the general complex solution for a radius of 1 is

$$\psi(\theta, \phi) = \sum_{\ell=0}^{\infty} \sum_{m=-\ell}^{\ell} c_\ell^m P_\ell^m(\cos\theta)e^{im\phi}, \tag{D.2}$$

where $P_\ell^m$ is the associated Legendre polynomial with $\ell \in \mathbb{N}$ and $-\ell \le m \le \ell$.

**Definition 1.** SHs $Y_\ell^m$ are spherical functions of order $\ell$ and degree $m$ that are particular solutions of the angular part of the Laplace equation in spherical coordinates defined as

$$Y_\ell^m(\theta, \phi) = \sqrt{\frac{2\ell+1}{4\pi}\frac{(\ell-m)!}{(\ell+m)!}}P_\ell^m(\cos\theta)e^{im\phi}. \tag{D.3}$$

Note that the first coefficient of Eq. (D.3) is a normalization factor that gives the orthonormality property as explained later. First, here is a list of SHs for $l = 0, 1, 2$ and nonnegative values of $m$ and an illustration of SHs in Fig. D.1.

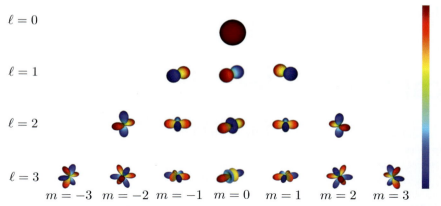

**FIG. D.1** Representation of spherical harmonics up to order 3. The spheres' radii at $(\theta, \phi)$ is scaled by the absolute value of $Re[Y_\ell^m]$. The color map goes from the minimum value of $Re[Y_\ell^m]$ (in *blue*) to the maximum value of $Re[Y_\ell^m]$ (in *red*).

$$\ell = 0, \qquad Y_0^0(\theta, \phi) = \frac{1}{\sqrt{4\pi}},$$

$$\ell = 1, \qquad \begin{cases} Y_1^0(\theta, \phi) = \sqrt{\frac{3}{4\pi}} \cos\theta \\ Y_1^1(\theta, \phi) = \sqrt{\frac{3}{8\pi}} \sin\theta e^{i\phi}, \end{cases}$$

$$\ell = 2, \qquad \begin{cases} Y_2^0(\theta, \phi) = \frac{1}{2}\sqrt{\frac{5}{4\pi}}(3\cos^2\theta - 1) \\ Y_2^1(\theta, \phi) = -\sqrt{\frac{15}{8\pi}} \sin\theta \cos\theta e^{i\phi} \\ Y_2^2(\theta, \phi) = \frac{1}{4}\sqrt{\frac{15}{2\pi}} \sin^2\theta e^{2i\phi}. \end{cases}$$

If we consider the inner product of two complex functions $f$, $g$, on an interval $[a, b]$, given by

$$\langle f, g \rangle = \int_a^b f(t)\overline{g(t)}\, dt, \tag{D.4}$$

with $\overline{g(t)}$ the complex conjugate of $g(t)$, some properties of the SHs can then directly be deduced from the definition:

**Orthogonality:** Let $Y_\ell^m(\theta, \phi)$ and $Y_{\ell'}^{m'}(\theta, \phi)$, $m \neq m'$, $\ell \neq \ell'$, be SHs and $d\Omega$ the integration over the unit sphere; then

$$\langle Y_\ell^m(\theta, \phi), Y_{\ell'}^{m'}(\theta, \phi) \rangle = \int_\Omega Y_\ell^m(\theta, \phi)\overline{Y_{\ell'}^{m'}(\theta, \phi)}\, d\Omega = 0.$$

**Orthonormality:** For $Y_\ell^m(\theta, \phi)$, an SH, $\langle Y_\ell^m(\theta, \phi), Y_\ell^m(\theta, \phi) \rangle = 1$.
**Complex conjugation:.** From the definition of complex numbers, we have $Y_\ell^{-m}(\theta, \phi) = (-1)^m \overline{Y_{\ell'}^{m'}(\theta, \phi)}$.
**Completeness:** Linear combinations of the SHs are dense in $C(S^2)$ and $L^2(S^2)$, respectively, the space of continuous complex functions on $S^2$ and the space of complex squared integrable functions on $S^2$.
**Symmetry:** The transformation $(\theta, \phi) \to (\pi - \theta, \phi + \pi)$ gives this parity:

$$Y_\ell^m(\theta, \phi) \to Y_\ell^m(\pi - \theta, \pi + \phi) = (-1)^\ell Y_\ell^m(\theta, \phi).$$

**Linear independence:** There are no nontrivial linear combinations of SHs equal to zero.

## D.2 Spherical harmonics as a basis

Let $V$ be a vector space. A basis of $V$ is a set of linearly independent (orthogonal) vectors in $V$ that can express any given vector of this vector space as a linear combination of the elements in the basis. The most common example is the canonical basis over $\mathbb{R}^3$, where the three vectors forming the basis are $e_1 = (1, 0, 0)$, $e_2 = (0, 1, 0)$, and $e_3 = (0, 0, 1)$. In formal terms, a basis is defined as follows.

**Definition 2.** Let $V$ be a vector space. A subset of $B$ of $V$ is a basis if:

- $B$ is linearly independent.
- Every vector of $V$ can be expressed as a linear combination of elements of $B$.

In the context of diffusion magnetic resonance imaging (dMRI), signals are often represented as a function of $(\theta, \phi)$ on the sphere. By the properties shown in Section D.1, we see that the set of SH functions forms a basis of the space of complex spherical functions. Thus, like any periodic real-valued integrable function can be written as a Fourier series, any complex spherical function can be expressed as an infinite sum of SH.

Let $\psi$ be a complex function living on the unit sphere. Then, $\psi$ can be expressed as a linear combination of SH functions:

$$\psi(\theta, \phi) = \sum_{\ell=0}^{\infty} \sum_{m=-\ell}^{m=\ell} c_\ell^m Y_\ell^m(\theta, \phi), \qquad (D.5)$$

where $c_\ell^m$ are some coefficients and $Y_\ell^m(\theta, \phi)$ is the SH function of $\ell$th order and $m$th degree.

However, since in the context of dMRI we only consider real and symmetric signals, a basis for representing any complex function on the sphere is not needed. We therefore define a modified SH basis that is real and symmetric. The symmetry property can be enforced by using, in our basis, only even order SH functions and the real property, by alternately considering the real or imaginary part depending on the degree of the harmonic. Various bases, such as the Tournier basis (Tournier et al., 2007) and Descoteaux basis (Descoteaux et al., 2007), have been proposed over the years having such properties and some of them will be discussed in detail in Section D.3.

SHs can also be used to give an approximation of a spherical function $\Psi(\theta, \phi)$ by truncating Eq. (D.5) to a given order $\ell_{max}$:

$$\Psi(\theta, \phi) = \sum_{\ell=0}^{\ell_{max}} \sum_{m=-\ell}^{\ell} c_\ell^m Y_\ell^m(\theta, \phi). \qquad (D.6)$$

One advantage of this approximation is the compression of the signal. Indeed, for a DWI signal acquired over 30 or 100 diffusion encoding gradient directions, the signal can be represented using 15, 28, or 45 SH coefficients per voxel (instead of the 30 or 100 diffusion data values) for an SH basis of order $L = \max(\ell) = 4, 6,$ or $8$, respectively. The effect of truncating the SH series to a maximum order $\ell_{max}$ is shown in Fig. D.2. Higher orders of SH are associated with high frequencies, which corresponds to the details in a signal. The higher $\ell_{max}$ is, the more precise the approximation will be. However, because noise resides in the high frequency spectrum, a truncated SH series has the benefit of reducing the information associated with noise.

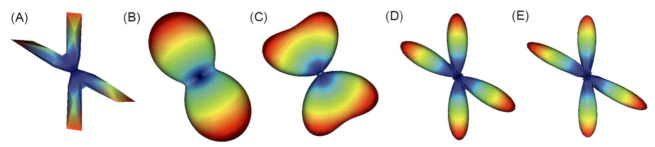

**FIG. D.2** Illustration of the effect of truncation on the reconstruction of a spherical function (A) evaluated for 200 discrete directions. Using (A) as ground truth, (B), (C), (D), and (E) show the best approximation of the signal using a truncated SH series up to a maximum order $\ell_{max}$ of 2, 4, 8, and 16, respectively.

## D.3 Applications and relevance in tractography

The use of SHs as a basis to represent the diffusion signal was proposed by Frank (2002) and Alexander et al. (2002) using two different approaches. The diffusion signal $S$ at each of the gradient encoding directions $(\theta_i, \phi_i)$, $i = 1, ..., N$, can effectively be expressed as an expansion on SH basis functions:

$$S(\theta, \phi) = \sum_{j=1}^{R} c_j Y_j(\theta, \phi) \tag{D.7}$$

where $R$ is the number of SH terms in the series (for $L = \max(\ell)$, the maximum order chosen is $R = (1/2)(L+1)(L+2)$) and $c_j$ the coefficients of the expansion.

Thanks to the SH-based representation, it was shown that analytical solutions are possible to transform the diffusion signal $S$ into the diffusion orientation distribution function (dODF) by means of simple matrix multiplication operations on the SH coefficients (Descoteaux et al., 2007). From the dODF, the principal fiber orientation(s) can be extracted and can integrate the local orientations to perform tractography. Similarly, the fiber orientation distribution function (fODF), which is a sharpened version of the dODF, can also be obtained through the SH coefficients.

Thus, using any given local fiber reconstruction model to recover the fODF, one can represent this fODF using an SH basis. Given that the measured diffusion signal in a single-shell HARDI acquisition is real and antipodally symmetric, most current SH-based reconstruction methods use only even orders to represent the fiber orientation functions, even if the underlying fiber geometry can be asymmetric. A few modified SH bases have been proposed in the literature to represent the diffusion signal expansion:

- Tournier basis (Tournier et al., 2007):

$$Y_j = \begin{cases} Im[Y_\ell^m] & \text{if } m < 0 \\ Re[Y_\ell^m] & \text{if } m \geq 0 \end{cases} \tag{D.8}$$

where $j(\ell, m) = (\ell^2 + \ell)/2 + m$.

- Descoteaux basis (Descoteaux et al., 2007):

$$Y_j = \begin{cases} \sqrt{2} \cdot Re[Y_\ell^m] & \text{if } m < 0 \\ Y_\ell^m & \text{if } m = 0 \\ \sqrt{2} \cdot Im[Y_\ell^m] & \text{if } m > 0 \end{cases} \tag{D.9}$$

where $j(\ell, m) = (\ell^2 + \ell + 2)/2 + m$.

In all these cases, $Y_\ell^m$ is the standard SH function defined in Eq. (D.3); $Re[\cdot]$ and $Im[\cdot]$ represent the real and imaginary part of $Y_\ell^m$, respectively; $-\ell \leq m \leq \ell$ as usual; and $\ell$ even.

Nevertheless, a complete basis including odd orders allows computing of asymmetric orientation distribution functions at a local level, as shown in Fig. D.3. This, in turn, naturally leads to the ability to distinguish voxel-wise fiber patterns such

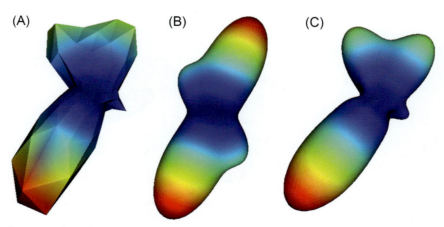

**FIG. D.3** Comparison of reconstructions of an asymmetric spherical function (A) evaluated for 128 sphere directions using a symmetric SH basis consisting exclusively of even order SH functions (B) and a complete SH basis consisting of even and odd order SH functions (C). Both SH bases are truncated at a max order $\ell_{max}$ of 8.

as bending, branching, or fanning. For various strategies to model asymmetric ODFs, see, for example, Delputte et al. (2007), Barmpoutis et al. (2008), Ehricke et al. (2011), Reisert et al. (2012), Bastiani et al. (2017), Karayumak et al. (2018), Wu et al. (2020), and Feng and He (2020).

# References

Alexander, D.C., Barker, G.J., Arridge, S.R., 2002. Detection and modeling of non-Gaussian apparent diffusion coefficient profiles in human brain data. Magn. Reson. Med. 48 (2), 331–340.

Arfken, G.B., Weber, H.J., 1995. Mathematical Methods for Physicists, fourth. Academic Press.

Barmpoutis, A., Vemuri, B.C., Howland, D., Forder, J.R., 2008. Extracting tractosemas from a displacement probability field for tractography in DW-MRI. In: Metaxas, D., Axel, L., Fichtinger, G., Székely, G. (Eds.), Lecture Notes in Computer Science. Medical Image Computing and Computer-Assisted Intervention—MICCAI 2008, Springer, Berlin, Heidelberg, pp. 9–16.

Bastiani, M., Cottaar, M., Dikranian, K., Ghosh, A., Zhang, H., Alexander, D.C., Behrens, T.E., Jbabdi, S., Sotiropoulos, S.N., 2017. Improved tractography using asymmetric fibre orientation distributions. NeuroImage 158, 205–218.

Delputte, S., Dierckx, H., Fieremans, E., D'Asseler, Y., Achten, R., Lemahieu, I., 2007. Postprocessing of brain white matter fiber orientation distribution functions. In: 2007 4th IEEE International Symposium on Biomedical Imaging: From Nano to Macro, pp. 784–787.

Descoteaux, M., Angelino, E., Fitzgibbons, S., Deriche, R., 2007. Regularized, fast, and robust analytical Q-ball imaging. Magn. Reson. Med. 58 (3), 497–510.

Ehricke, H.H., Otto, K.M., Klose, U., 2011. Regularization of bending and crossing white matter fibers in MRI Q-ball fields. Magn. Reson. Imaging 29 (7), 916–926.

Feng, Y., He, J., 2020. Asymmetric fiber trajectory distribution estimation using streamline differential equation. Med. Image Anal. 63, 101686.

Frank, L.R., 2002. Characterization of anisotropy in high angular resolution diffusion-weighted MRI. Magn. Reson. Med. 47 (6), 1083–1099.

Karayumak, S.C., Özarslan, E., Unal, G., 2018. Asymmetric orientation distribution functions (AODFs) revealing intravoxel geometry in diffusion MRI. Magn. Reson. Imaging 49, 145–158.

Reisert, M., Kellner, E., Kiselev, V.G., 2012. About the geometry of asymmetric fiber orientation distributions. IEEE Trans. Med. Imaging 31 (6), 1240–1249.

Tournier, J.D., Calamante, F., Connelly, A., 2007. Robust determination of the fibre orientation distribution in diffusion MRI: non-negativity constrained super-resolved spherical deconvolution. NeuroImage 35 (4), 1459–1472.

Wu, Y., Hong, Y., Feng, Y., Shen, D., Yap, P.T., 2020. Mitigating gyral bias in cortical tractography via asymmetric fiber orientation distributions. Med. Image Anal. 59, 101543.

# Index

Note: Page numbers followed by *f* indicate figures, *t* indicate tables, *b* indicate boxes, and *np* indicate footnotes.

## A

ABrainVis, 435
Absolute fODF threshold, 270
Accumbofrontal pathway, 415–416, 416*f*
Acquisition, 673–675
  encoding of diffusion, 673, 674*f*
  image readout and reconstruction, 674–675
  pulse sequences, 654–655
  schemes, 584
Active staining, 658
Adams-Bashforth method, 700–701
Adolescent Brain and Cognitive Development (ABCD) project, 96, 603
Agglomerative methods, 454
Aging, 590–596
Aleatoric uncertainty, 259–260
Allocortex, 9
Along-streamline profiling analysis, 439–440, 440*f*
  along whole bundle average, 442
  bundle-profiles, 439–440, 442, 445–447
  bundle segmentation, 440
  core streamline definition, 441–442, 441*f*
  functional profile analysis, 445
  multiple comparisons, correcting for, 444–445
  normativemodeling and machine learning, 445–446, 445*f*
  statistical tests, 442–445
  streamline ordering, 440–441
  white matter tissue microstructure, 439, 447, 447*f*
Alzheimer's disease, 373, 439, 586, 597–599, 598–599*f*, 669
Ambient occlusion, 382, 383*t*, 384, 385*f*, 389
AMICOX, 356
Amyotrophic lateral sclerosis (ALS), 439, 445–446
Analysis of variance (ANOVA), 444–445
Anaplastic astrocytoma, 375
Anatomical constraints, 352–354
  ROI-based tractography, 352–353, 354*f*
Anatomically constrained tractography (ACT) algorithms, 363–364, 571
Anatomical model systems
  animal ethics and 3 Rs, 512–513
  fiber orientation, 525–530
  microdissection, 513–517
  neuronal tracers, 518–525
  preclinical ex vivo dMRI, 513, 514*f*
  species differences, 511–512, 512*f*
Anatomical priors, 571, 572*f*

Anatomical validation, limitations in, 577
AND gates, 242–243, 244*f*, 248*f*
Angle threshold, 246–247
Angular correlation coefficient (ACC), 535
Angular error, 535, 662
Angular gyrus (Ag), 406, 407*f*
Angular resolution, 137
Animal electricity, 43
Anisotropic diffusion, 63–64, 177–182
Anisotropic phantoms, for diffusion tractography, 499–500*t*, 504
  biomimetic phantoms, 501–504
  capillary phantoms, 501, 502*f*
  plainfiber phantoms, 498–501, 502–503*f*
  plants, 498, 500*f*
Anisotropic power (AP) map, 398
Anisotropy threshold, 247–248, 248*f*
Anterior commissure (AC), 412, 413*f*
Anterior thalamic radiations, 408, 409*f*
Antialiasing filter, 403
ANTS-ATROPOS, 571
Apparent diffusion coefficient (ADC), 61, 68, 177, 195
Apparent fiber density (AFD), 192, 227–228, 636
Arcuate fasciculus (AF), 372–374*f*, 401, 406, 407*f*, 423, 440–441
  functions of, 619–620
  and language functions, 618–620
  and language impairments, 620
Arteriovenous malformations (AVMs), 639, 640*f*
Aspiny stellate cells, 7, 7*f*
Association fibers, 46–47
Associationist school, 46–47
Association tracts, 9–10, 10*f*
Atlas-based WM masking, 370–371
Augmented reality (AR) technology, 389
Autism spectrum disorder (ASD), 596, 615
Autoencoders (AEs), 338
Automated fiber quantification (AFQ), 435, 584–585
Automatic bundle-segmentation variability, 575, 575*f*
Automatic relevance determination (ARD), 263
Automatic segmentation, 423–424, 424*f*
Autonomic nervous system (ANS), 4–5
AxCaliber models, 68
Axial diffusivity (AD), 583–584
Axon, 485–486
  coarse-graining over, 89–92

orientation of, 487
  as stick, 91–92, 91*f*
Axonal arbor, 25*b*
Axonal branching, 4*f*, 12
Axonal dispersion, 25*b*
Axonal misalignment, 25*b*
Axonal orientation distribution function (aODF), 259, 270
Axonal tracing, 47–48
Axon caliber scale, 89–90
Axon collateral, 25*b*
Axon diameter distributions, 543
Axon diameter index, 595*f*
AxonPacking algorithm, 495–496
Axons, 4, 25*b*, 26*f*, 423, 544–545
  diameter, 32–33
  differential amplification, 28–29
  functions, 27–28
  mapping, 28
  nomenclature of, 27, 27*b*
  synaptic connections, development of, 29
  temporal transformations, 30–32, 31–32*f*
Axoplasm, 27*b*

## B

Baby Connectome Project, 603
Backward Euler method, 698
Ball-and-rackets model, 184, 223
Ball-and-stick model, 183–184, 184*f*, 222–223, 430
Bayesian analysis, 444–445
Bayesian inference, 263
BDA tracers, 551
Behavioral variant of frontotemporal dementia (bvFTD), 618
Benjamini-Hochberg procedure, 444
Betz cells, 7
Biexponential model, 68
Bifrontal contusion, 643
Binary connectomes, 467
Binary tissue masks, 348–349
  limitations of, 349, 349*f*
  seeding mask, 348, 348*f*
Bingham distribution, 229–230*b*, 230*f*
Biomimetic phantoms, 501–504
Bipolar gradient pulse, 104*f*
Black reaction, 45*f*
Blipped-CAIPI SMS-EPI, 131
"Blip-up blip-down" methods, 160
Bloch-Torrey equation, 494
  mesoscopic, 82

717

**718** Index

Block-face images, 528, 530*f*, 662
Blood oxygenation level-dependent (BOLD) effect, 356
BlueMatter method, 306–307, 310
Blurring, 135
BOLD fMRI, 69–70, 641
Bonferroni correction, 444
Boolean operators, 401
Bootstrapping, 262–263
Border-mapping methods, 454
Bottleneck effect, 566, 567*f*
Bottom-up methods, 302
Boutons, 25*b*
Box-counting dimension, 435
Brain
  anatomy, 543
  cellular layering and fiber pattern, 8*f*
  connectome, 70–71
  development, 586–590
  gray and white matter, 4, 5*f*
  maturation, 587
  networks, 4*f*
Brain areas
  connectivity fingerprints, 453*f*, 454, 456
  definition, 452–454
Brain connectivity, 41, 659
  animal models, 48*f*, 49
Brain mapping strategies, 641
Brain network
  anatomy, 485–486
  artificialdMR images, simulation of, 489–494
  connectivity, 470–471, 471*f*
  diffusion model, 490*f*, 491
  MR acquisition model, 490*f*, 491
  numericaldMRI phantoms, 491–494, 492*t*
  tissue microstructure simulations, 494–497
  tissue model, 490, 490*f*
  topology, 472–474, 472*f*
Brain parcellation, 454, 456
  connectivity-basedparcellation, 455
  gradients and soft parcellations, 456
  methods, 453*f*, 454–455
  multimodalparcellation, 455
  somatosensory motor cortex, 456, 457*f*
Brain spacetime curvature, 71, 71*f*
Brain tumor, 374*f*, 375–376, 376*f*, 398–401, 400*f*
Brain volume, 586
Brodmann areas, 8*f*
Brownian motion, 55–56
B-splines, 242
B-spline smoothing, 705–706, 707*f*, 708
*b*-tensors, 215
Bundle adjacency (BA), 432, 434–435
Bundle-based and streamline-based accuracy, 556
Bundle of interest (BOI), 575
Bundle overlap (OL), 323
Bundle overreach (OR), 323
Bundle prediction, 334–337
  multiple bundle labels, 334–335
  multiple bundle labels and one rejection class, 335–336
Bundle-profiles, 439–440, 442, 445–447
Bundle saturation effect, 571–572, 573*f*

Bundle segmentation, 440, 487, 573–575, 677
Bundle-specific tractography (BST), 353–354, 355*f*
Bundle-Wise Deep Tracker, 330*t*
Bundle-wise density, 303*b*
*b*-value, 106
  choice of, 109–110, 110*f*

# C

CAMINO, 494–496
Capillary phantoms, 501, 502*f*
Carr-Purcell-Meiboom-Gill condition of multiple RF-pulse, 654–655
Cell theory, 43
Central limit theorem (CLT), 78–80, 79*f*
Central nervous system (CNS), 4–5
Cerebral cortex
  cortical areas, 19–21
  cortical layers, 21
  gyrification, 22
  input connections, 23–24, 24*f*
  output connections, 22–23
  parcellation of, 8*f*
  receptors, 24–25
  vertical organization, 21
Cerebral gliosis, 375
Cerebrospinal fluid (CSF), 134, 305, 363–364
Cerebrovascular disease, 597–599, 598–599*f*
Chemical synapse, 27*b*
Christoffel symbols, 280–281, 288–289
Cingulum, 404, 405*f*
Cingulum bundle, 440–441, 441*f*
Classic tractography, 315
Classifyber, 337, 433–435, 434*f*
Clinical tractography, 631–632
  functional neurosurgery, 638–639
  intraoperativetractography, 636–637, 637–638*f*
  neuro-oncology, 632–636
  neurovascular, 639–640
  pediatrics, 641–643, 642*f*
  skull base, 640–641
  traumatic brain injury, 643
CNN-based bundle modeling, 332–333, 332*f*
Cognitive associations, 593–596
Cognitive control, white matter tracts and, 617–618
Cognitive development, 601
Cognitive functions, linking behavior with white matter networks, 616–618
  episodic memory, limbic white matter tracts and impairments in, 617
  limbic white matter tracts and episodic memory, 616–617
  white matter tracts and cognitive control, 617–618
Coils
  radiofrequency, 108–109
  shim, 108
Collateral fibers, 6
Collations of Connectivity data of the Macaque (CoCoMac), 545–546
Color-coded FA maps. *See* Directional encoded color (DEC) maps
Column-wise vector, 698–699

ComBat harmonization, 679
Commissural fibers, 46–47
Commissural tracts, 9–10, 10*f*
Commissure, 25*b*
COMMIT$_{AxSize}$, 310–311
COMMIT method. *See* Convex optimization modeling for microstructure informed tractography (COMMIT) method
COMMIT$_{T2}$, 310–312
Complex axonal trajectories, 544–545, 544*f*
Complexity index (CX), 231
Composite hindered and restricted model of diffusion (CHARMED) model, 159, 493, 583–584, 596
Computer graphics, 383*t*, 383*f*
Concomitant gradients, 114
ConFig algorithm, 496
Confusion matrix, 533, 533*f*
Connectivity matrices, 670
Connectome, 577
  diffusion MRI tractography
    brain area, 452–454
    brainparcellation (*see* Brain parcellation)
    deep white matter tracking, 457–459, 458*f*
    distance bias, 461
    edges, 460–461
    filters, 467–468
    graph analysis, 469–475
    gyral bias, 461
    node delineation, 451
    path termination and superficial white matter tracking, 457, 458–460*f*, 459–460
    phantoms, 463–465, 464*f*, 468
    seeding strategy, 458, 458*f*, 461–462
    thresholding, 466–468, 466*f*
    validation, 463, 464*f*, 465, 468
  empirical validations, 531
  measures for validation, 533–534
Connectomics, 381, 573–574, 659
  analysis, 372
  effects on, 576
Consensus thresholding methods, 467
Constrained spherical deconvolution (CSD), 167, 190, 190*f*, 193, 207, 209*f*, 225, 568
Contemporary neuroimaging methods, 49–51, 51*f*
Contrack, 460
Contrast-to-noise ratio (CNR), 109, 111*f*
Controlled aliasing in parallel imaging (CAIPI), 131
Convex optimization modeling for microstructure informed tractography (COMMIT) method, 308–309, 311–312, 460, 468, 494
Convolutional neural networks (CNNs), 435
Corpus callosum (CC), 242*f*, 404, 405*f*, 412, 412*f*, 445*f*
Cortex, mapping and connectivity, 44–46
Cortical areas, 19–21, 20*f*
Cortical layers, 21
Cortical myelogenetic maps, 44*f*
Cortical surfaces, 351–352, 352*f*
  geometricalmodeling, 351–352, 352*f*
  limitations of, 352

Corticobulbar tract (CBT), 545
Cortico-cortical connections, 23, 33
Cortico-descending axons, 27, 32f
Corticospinal pathways, 249, 250f
Corticospinal tract (CST), 374f, 399–401, 400f,
   408, 409f, 426–427, 545, 633–634
Cost function, 156, 301
Crossing fibers, 181–182, 221–222, 225f, 555
Cross product, 697
CSD. See Constrained spherical deconvolution
   (CSD)
CSD-based probabilistic tractography, 658
Cubic spline smoothing, 708, 709f
Cumulants, 77
Cytoarchitectonic and neuronal connectivity, 7–9
Cytoskeleton, 27b

# D

Damped Richardson-Lucy spherical
   deconvolution (dRL-SD) algorithm, 191,
   194, 195f, 225
Data-driven approaches, 207–212
   pros and cons of, 212
Data quality assessment, 152–153, 153f
Davies-Bouldin index, 429
Decussation, 25b
Deep brain stimulation (DBS), 638
Deep2dU-net type networks, 332
Deep learning-based segmentation, 424f
Deep learning (DL), 316
Deep nuclei, 571, 572f
Deep white matter tracking, 457–459, 458f
DeepWMA, 433, 435
Denoising, 164
Density-based thresholding, 466–467
Dentato-rubro-thalamic tract (DRTT), 638–639
Dentatorubrothalamic tract, visualization of,
   552–553
Deterministic algorithm, 676
Deterministic fibertractography, 241
Deterministic-output bundle-specific RNN,
   329–330, 330t, 331f
Deterministic-output RNN, 327, 328f
   using reinforcement learning, 331, 331f
Deterministic tractography, 398, 458, 462–463,
   468
Developmental disorders, 596
Diagonalization, of diffusion tensor, 695
Diaschisis, 44
Dice coefficient, 434
Dice similarity coefficient (DSC), 426–427,
   427t
Dictionary learning, 682
Differential amplification, by axons, 28–29
Differential geometry, 275
Diffusion, 77
   anisotropy, 177–182
   as coarse-graining, 86–89, 87f
   coefficient, 56
   constant, 77–78
   encoding adaptations, 116–117
   encoding directions, 110–111
   encoding of, 673, 674f
   equation, 80

fMRI, 70
   in tissue microstructure, 80–81
Diffusion contrast
   qualitative description of origin, 103–105,
      105f
   quantitative description of origin, 105–107
Diffusion echo planar imaging (EPI) distortion,
   637
Diffusion ellipsoid, 178–179, 178f
Diffusion/IVIM EPI, 61–62, 64f
Diffusion kurtosis imaging (DKI), 68, 165–166,
   201–202, 205–206, 205–206f, 224
Diffusion length, 105–106
Diffusion magnetic resonance imaging (dMRI),
   241, 297, 302, 347, 363, 565, 669
   along-streamline profiling (see Along-
      streamline profiling analysis)
   artifacts, 152f
   augmented/virtual reality (AR/VR) for, 389
   in clinical field, 61–63
   diffusion data analysis, categories of, 439
   fibertractography (FT) algorithm, 241
   first trials, 57–59, 59–60f
   generalized, 60–61
   gradient coils, 108
   in-context visualization of, 387–388, 387f
   non-Gaussian, 68, 69f
   radiofrequency coils, 108–109
   region-of-interest (ROI)-based analyses,
      446–447
   shim coils, 108
   tract-based spatial statistics (TBSS), 439, 446
   voxel-based analysis (VBA), 439, 446
Diffusion measures, 599
Diffusion microimaging, 658
Diffusion modeling, 204–206, 215–216, 490f,
   491, 659–661, 661f
Diffusion MRI fibertractography, 543
Diffusion MRI measures, statistical approaches,
   678–680
   ComBat harmonization, 679
   FA-based hardware-phantom harmonization,
      679–680
   linear correction factor-based harmonization,
      679
   metaanalysis, 679
   statistical covariates, 679–680
Diffusion MRI tractography, 613, 631
   imaging networks with, 12–13, 13f
Diffusion-narrowing regime, 90
Diffusion orientation density function (dODF),
   186, 195, 204–206, 215–216, 223–224,
   224f, 714
   vs.fODF, 225f
Diffusion orientation transform (DOT), 187–188
Diffusion sequences
   multicontrast sensitization of, 135–137,
      136f
   segmented, 129–131
   single-shot, 124–127, 124f, 126–127f
   spatial and angular resolution, 137
Diffusion spectrum imaging (DSI), 182,
   204–205, 205f
Diffusion techniques, 177

Diffusion tensor imaging (DTI), 3, 49, 165–166,
   201, 241–242, 242f, 245, 365–368, 368f,
   398, 398f, 568, 583–584, 633–634, 635f,
   676, 695
   computation, 179
   connecting geometry to, 285–287
   emergence of, 63–67, 65f
   fractional anisotropy (FA) map, 347–348
   fractional anisotropy (FA) mask, 363–364
   imaging, 178–179, 178f
   limitations, 181–182
   metrics, 179–180, 180f
   orientation, 181
   reconstruction, 677
   single-shell, 202f
Diffusion time, 105
Diffusion tractography, 551, 583, 623, 631, 653
Diffusion-weighted images (DWI), 377, 397,
   402, 680, 697, 705, 711
   connecting geometry to, 285–287
Diffusion-weighted magnetic resonance imaging
   (dMRI), 201, 315, 600–601
Diffusion-weighted signal, 82–83
Diffusion-weighted spin-echo (dwSE), 654
Diffusion-weighted steady-state free precession
   (dwSSFP), 655
Diffusion-weighted stimulated echo (dwSTE),
   655
Directional encoded color (DEC) maps, 181
DiSCo, 492t, 494
Disconnection syndromes, 46–47, 47f
Discrete fiber orientation, 221–223, 222f
Discriminative methods, for global tractography,
   302, 302f
   candidate streamlines, 306
   convex optimization modeling for
      microstructure informed tractography
      (COMMIT) method, 308–309
   linear fascicle evaluation (LiFE) algorithm,
      309
   SIFT2 method, 309
   streamline filtering methods, 306–308
Dispersion-based probabilistic tractography, 264
   streamline propagation, 264–265, 265f
   synthetic curve configuration, 267f
Dissimilarity projection, 425
Distance-dependent thresholding methods, 467
Distortions, 135
dMRItractography community, 576–577
   anatomical definitions, reaching consensus
      on, 577
   anatomical validation, limitations in, 577
   brain structural connectivity concept, 577
   group of fibers concept, 576
   nomenclature and terminology, 576–577
   singlefiber concept, 576
DOT. See Diffusion orientation transform
   (DOT)
Dot product, 697
Double average, 85f, 86–87, 87f
DR-BUDDI approach, 161
DSC. See Dice similarity coefficient (DSC)
DTI. See Diffusion tensor imaging (DTI)
DTIPrep, 586

**720** Index

Dual Finsler function, 282–283
Dual Riemann-Finsler metric tensor, 282–283
Dual vector space, 290
DWI-EPI distortion, 638*f*
Dysembryoplasticneuroepithelialtumor
  (DNET), 373–375, 638*f*
Dyslexia, 593

## E

Echo planar imaging (EPI), 61–62, 123, 138,
    674–675
  interleaved, 130
  readout segmented, 130
Echo time (TE), 673–674
Eddy-current-induced distortions, 155*f*, 157
Eddy currents, 112–114, 113*f*, 157–159
Edema correction tractography methods,
    375–376
Educational interventions, 602
Eigendecomposition, 700
Eigenvalue maps, 180
Eigenvalues, 695
Eigenvectors, 181–182, 695
Einstein summation convention, 290
Electrical stimulation, 621–622
Electrical stimulation mapping (ESM), 336
Electrical synapse, 27*b*
Electroencephalography (EEG), 49–50
Electromyographic (EMG) recording, 621–622
Empirical validations
  assessed, 531
  connectomes, 531
  considerations, 533
  definition, 530
  fiber bundles, 531–532
  local reconstruction, 532–533
Encephalomalacia, 375
ENIGMA, 679
*Entrack*, 325, 326*f*
EPI. *See* Echo planar imaging (EPI)
Epilepsy, 638
Episodic memory, 616–617
  limbic white matter tracts and impairments in,
    617
Error matrix, 533
Essential tremor (ET), 638–639
Euclidean distance, 442
Eulaminateisocortex, 8
Euler-Lagrange equation, 281
Euler's method, 243–244, 698–699
European Society for Magnetic Resonance in
    Medicine and Biology (ESMRMB), 59
European Society of Radiology (ECR), 59
ExploreDTI, 402
Extraaxonal signal, 92
Ex vivo diffusion imaging, 143–144

## F

FA-based hardware-phantom harmonization,
    679–680
False negative brain mapping, 641
Family-wise error rate (FWER), 444
Fascicles, 92

Fast marching (FM) method, 248–249
FAT. *See* Frontal aslant tract (FAT)
Feature-based approach, 424
Feedback connection, 23, 25*b*
Feedforward connection, 23, 25*b*
Fetal alcohol spectrum disorder (FASD), 596
Fetal brain tractography, 14
Fetal diffusion imaging, 144–145
Feynman's path integral, 83*np*
FFClust, 429
Fiber assignment by continuous tracking
    (FACT) algorithm, 365
Fiber bundle, 244*f*, 246, 248–249, 292, 531–532
Fiber bundle capacity (FBC) measure, 303
*FiberCup*, 279*f*, 285*f*, 317, 549
  phantom, 463–465
Fiber direction estimates, and tractography, 683
Fiber dispersion and structural complexity,
    228–231
Fiberfox, 465, 468, 493–494
*Fiberfox* software, 319
*FiberMap*, 335–336
Fibernavigator, 435
*FiberNet*, 334
*FiberNet2.0*, 337
Fiber orientation, 221, 525. *See also* Orientation
    distribution function (ODF)
  challenges and opportunities of histology,
    528–530, 530*f*
  direct measures, 525, 525*f*, 527–528, 529*f*
  discrete, 221–223, 222*f*
  indirect measures of, 525–528, 525–526*f*
  local, 227
  measures for validation, 534–535
Fiber orientation distribution function (fODF),
    137, 186, 188–191, 193–194, 196*f*, 207,
    215–216, 225–226, 259, 364*f*, 365, 366*f*,
    368–369, 369*f*, 568, 714
  *vs.*dODF, 225*f*
  withmultishell constrained spherical
    deconvolution, 214*f*
  fromnondiffusion methods, 234
  reconstruction techniques, 546–547, 547*f*
  3D, 234*f*
Fiber orientation distributions (FODs), 653
Fiber response function, 191–192, 192*f*
FiberStars, 435
Fiber tracking. *See* Fibertractography (FT)
Fibertractography (FT), 381, 386–387
  determinisitictractography, 241
  front evolution approaches, 248–249
  history, 241
  ingredients of, 241–249
  methodological considerations, 249–250, 250*f*
  probabilistictractography, 241
  stopping criteria, 242
  streamline FT, 241, 243–246
  tract pathways, 242
  user-defined FT settings, 246–248
  whole-brainfibertractography, 242–243, 243*f*
Fibrous tissue, models of, 207–216
Fick's law, 80
Filtering tractograms, with microstructure
    information, 355

Finite element/difference (FEM/FDM) methods,
    494
Finsler function/Finsler norm, 280
Finsler geodesic tractography, 287
Finsler geometry, 289
Fisher-von-Mises (FvM) distribution, 325
FITs. *See* Frontal insular tracts (FITs)
Fixel, 228, 229*f*
Fixel-based analyses, 192
FMT. *See* Frontomarginal tract (FMT)
Focal cortical dysplasia (FCD), 371
fODF. *See* Fiber orientation distribution function
    (fODF)
Fornix (Fx), 10, 414, 414*f*
Fourier tracts (FouTS), 306
Fourth-order Runge-Kutta (RK4) method,
    244–245, 700
Fractal dimension (FD), 434–435
Fractional anisotropy (FA), 179–181, 180*f*, 221,
    247–248, 316, 442, 583–584, 597*f*, 636,
    669–670
  age-associated decline of, 594–596
*Frankenstein*, 43
Free diffusion, 78*f*
Freesurfer, 371
Freezing/thawing, 514–515
Frénet equation, 243
Frenet-Serret approach, 265
Frog experiment, Galvani's, 43, 43*f*
Frontal aslant tract (FAT), 406, 407*f*
Frontal insular tracts (FITs), 406–407, 408*f*
Frontal networks, social-emotional
    processing, 615
Frontal-temporal connections, 589
Front-evolution-based FT approach,
    248–249
Frontomarginal tract (FMT), 412–413, 413*f*
Fronto-orbito-polar tract (FOP), 412–413,
    413*f*
Frontoparietal white matter tracts
  in unilateral spatial neglect, 622–623
  and visuospatial functions, 622
Frontostriatal projections, 408–409, 409*f*
Fronto-temporal dementia, 669
FSL-FAST, 571
*FS2NET*, 338
Fully convolutional neural network (FCNN),
    432
Functional connectivity, 357
Functional local orientation, 357–358
Functional maps, tractography with,
    356–358
  functional connectivity, 357
  functional local orientation, 357–358
  functional regions, 356–357
Functional MRI (fMRI), 50, 51*f*
  diffusion, 70
  IVIM, 69–70
Functional neurosurgery, 638–639
Functional profile analysis, 445
Functional regions, 356–357
Functional resting state analysis, 50
Funk-Radon transform (FRT), 185

# G

Gadolinium-based diamagnetic contrast agents, 658
Galvani's frog experiment, 43, 43f
Ganglioglioma, 373–375
Gaussian diffusion, 178, 181–182
Gaussian diffusion propagator, 80
Generalized autocalibrated partially parallel acquisitions (GRAPPA), 163
Generalized fractional anisotropy (GFA), 223–224
Generalized Richardson-Lucy (GRL), 191, 194
Generative global tractography methods, 302–303, 302f
  Gibbs tracker, 304–305
  multitissue and multicompartment models, 305
  spin glass models, 304
  streamline-wise approaches, 305–306
Geniculocalcarine fasciculus, 410
Geodesic methods, 300
Geodesic tractography, 275, 289
  data variability effect on geometry, 287–289
  DTI and DWI, connecting geometry to, 285–287
  Riemann-DTI paradigm, 276–289
Geodesic tube, 288f, 289
Geometrical modeling, 351–352, 352f
Geometric distortions, 157–161
  eddy currents, 157–159
  gradient nonlinearities, 161
  susceptibility gradients, 159–161
Geschwind's area AVMs, 639
Geschwind's territory, with Wernicke's territory, 619
Ghosting, 132–133, 133f
Gibbs ringing, 134, 154–155, 154f
Gibbs tracker methods, 304–305
Glia, 19
Glioblastoma multiforme, 375
Global modeling, 331–333
Global tractography, 298–301, 301b, 312
  alternative formulation, 301
  cost function, 301
  definition, 299
  discriminative methods, 302, 302f, 306–309
  forward model, 301
  generative methods, 302–306, 302f
  vs. local tractography, 299, 300f
  multimodal data, 310–311
  optimization, 301
  quantitative metric of connectivity, 303b
  reconstruction paradigm, 301–303
  regularization techniques, 302
  tissue microstructure, advanced models of, 309–310, 310f
  WM organization, anatomical information about, 311–312, 311f
Global tractography algorithms, 569
Glucose, 50–51
Glyphs visualization, 381
Gradient, 138
Gradient coils, 108
Gradient nonlinearity, 114

Gram matrix, 276, 291
Graph analysis, of connectome, 475
  brain network connectivity, 470–471, 471f
  brain network topology, 472–474, 472f
  graph model, 469–470, 470f
  null model, 475
Graph theory analysis, 50, 372, 381, 590, 591f
g-ratio, 27b
Gray matter (GM), 5f, 19–25, 363–364
  development of, 23–24
  morphometry, 594–596
Green's ratio, 304
Ground truth, 317, 577
Group of fibers concept, 576
Growth cone, 25b
gSlider method, 132
gSlider-SMSb, 132, 133f
Gudden method, 44
Gyral bias, 459, 459f, 461
Gyrencephalic brains, 512
Gyrification, 22

# H

Hamiltonian formalism, 281–283
Hamilton-Jacobi theory, 279, 283–284
*HAMLET* (Hierarchical hArMonic filters for LEarning Tracts), 333, 334f
HARDI. *See* High angular resolution diffusion imaging (HARDI)
Harmonic functions, 711
Harmonization
  of DWI data, 680–682
    prospective DWI data harmonization with traveling subjects, 681–682
    retrospective DWI data harmonization, 680–681
  evaluation of, 683–684
  within magnetic field strengths, 681–682
Heaviside function, 112
Hermitian symmetric $k$-space, 128, 129f
Heteroscedasticity, 165
Heterotopic commissural fibers, 10
High angular resolution diffusion imaging (HARDI), 109, 117, 177, 182–188, 195, 201, 202f, 656, 660, 676
  data, 369
  diffusion orientation transform (DOT), 187–188
  HARDI 2013, 317–318
  multitensor and multicompartmental approaches, 183–185, 184f
  nonparametric approaches, 185
  persistent angular structure (PAS), 187–188
  Q-ball imaging (QBI), 185–187, 187f
  spherical harmonics (SH), 182
High-field ex vivo HARDI, 658
High-resolution ex vivo HARDI, 660
Hindered diffusion, 81
Hindrance modulated orientation anisotropy (HMOA), 192, 228, 621, 636
Hippocampal cingulum, 617
Hippocampal-diencephalic retrosplenial network, 614
Histological staining, 662

Homoscedasticity, 165
Homotopic commissural fibers, 10
Hopf-Rinow theorem, 277–278, 289
Horizon, 434–435
Horton-Strahler ordering, 25b
Human behavior, 613
Human brain, mesoscale connectivity mapping in, 656–658, 657f
Human Connectome Project (HCP), 96, 125, 143–144, 603
  minor bundle dataset, 321, 321f
Hybrid-tractography, 557
Hyperstreamline visualization, 382–383, 382f

# I

iFOD2 algorithm, 264–265, 265f, 267f
IFOF. *See* Inferior fronto-occipital fasciculus (IFOF)
ILF. *See* Inferior longitudinal fasciculus (ILF)
Illustrative visualization techniques, 382
Image readout, 674–675
Imaging-related artifact reduction
  cerebrospinal fluid effects, 134
  fat, 134, 134f
  ghosting, 132–133, 133f
  motion, 135
Immersive virtuality, 389
Incoherent motion, 115
In-context visualization, of tractography, 387–388, 387f
Inferior fronto-occipital fasciculus (IFOF), 156, 401, 405, 406f, 446, 515, 545, 571, 634
Inferior longitudinal fasciculus (ILF), 372f, 401, 405, 406f, 589, 614
Informed RESTORE/iRESTORE, 164–165
Inherent sensitivity, 549–550, 550f
Interactive tractography visualizations, 384–388
Interactive visualization, 371
Interhemispheric tract asymmetry, 13–15, 15f
Interleaved EPI, 130
International Federation of Associations of Anatomists (IFAA), 3
International Society of Magnetic Resonance in Medicine (ISMRM) workshop, 56, 63f
Interscanner, 673
Intersite, 673
Intra-axonal components, 4f, 12–13
Intraaxonal signal, 91f, 92
Intraoperative monitoring (IOM), 631
Intraoperative tractography, 636–637, 637–638f
Intrascanner variability, 671–672, 672t
Intravolume movement, 156
Intravoxelfiber orientation dispersion, 544–545
Intravoxel incoherent motion (IVIM), 60–61, 62f, 194
  in clinical field, 61–63
  fMRI, 69–70
  MRI, 61
Intrinsic variability, 671–673, 671f, 672t
In utero development, of white matter tracts, 587–589, 588f
Invalid bundles (IB), 323
Invariant representations, 681–682
Invasive IOM, 641

**722** Index

Inverse fast Fourier transform (IFFT), 204
Inverse Gram matrix, 291
Inverse problem, 297
Inversion recovery (IR), 136
In vivo imaging, of nonhuman primates, 143
Ionotropic receptors, 5
ISMRM 2015 Tractography Challenge dataset, 340
Isocitrate dehydrogenase (IDH), 632–633
Isocortex, 8
Isotropic liquids, 497–498
Isotropic phantoms, 497
Iterative optimization algorithms, 179
Iterative weighted linear least squares (IWLLS), 165
IVIM. *See* Intravoxel incoherent motion (IVIM)

**J**

Jensen-Shannon divergence (JSD), 535

**K**

Klingler's dissection technique, 398, 398*f*, 402, 514–515, 631, 632*f*
Koniocortex, 9

**L**

Label transcription approaches, 338, 338*f*
Lagrange polynomials, 242
Lagrangian formalism, 279–281
Langevin dynamics, 77–78
Language impairments, arcuate fasciculus and, 620
LAP. *See* Linear assignment problem (LAP)
Laplace-Beltrami operator, 186
Laplace-Beltrami regularization, 168
Laplace equation, 711
Laplacian regularization, 168
Larmor frequency, 82
Lateralization, of cognitive functions, 13–14
Least squares (LS), 165
  minimization, 305
Left-right oriented pontocerebellar pathways, 249, 250*f*
Left unilateral spatial neglect (USN), 622–623
Left ventral cingulum microstructure, 617
Length threshold, 402–403
Lesion-to-CST distance, 639–640
Lesion-to-eloquent areas, 639
Lesion-to-tract distance (LTD), 634–635
Leucotomy, 48–49
Levenberg-Marquart algorithm, 187–188
Levi-Civita connection symbols, 292
LiFE, 460, 468
Light imaging techniques, 527
Limbic circuit, 48–49
Limbic networks, social-emotional processing, 615
Limbic white matter tracts, 616–617
  episodic memory, 617
Linear assignment problem (LAP), 433
Linear correction factor-based harmonization, 679
Linear fascicle evaluation (LiFE) algorithm, 309

Linear formulation, 309
Linear independence, 712
Linear least squares (LLS), 165, 179, 676
  optimization, 308–309
Linear operator, 290
Linear-rotation invariant spherical harmonics (Linear-RISH), 680–681, 680*f*
Linear space, 289–290
Linking behavior with white matter networks, 613
  behavioral domains, 613, 614*f*
  cognitive functions, 616–618
    episodic memory, limbic white matter tracts and impairments in, 617
    limbic white matter tracts and episodic memory, 616–617
    white matter tracts and cognitive control, 617–618
  language functions, 618–620
    arcuate fasciculus and, 618–620
    arcuate fasciculus and language impairments, 620
  motor functions, 620–622
  socioemotional functions, 613–615
    social-emotional processing, frontal and limbic networks and impairments in, 615, 616*f*
    white matter tracts and socioemotional processing, 613–615, 614*f*
  visuospatial functions, 622–623
    frontoparietal white matter tracts and, 622
    unilateral spatial neglect, frontoparietal white matter tracts in, 622–623
Lissencephalic brains, 512
Local fiber orientation, 227
Local tractography, 299, 300*f*
Log-Euclidean metrics, 179
Longitudinal analysis, 683
Low-grade gliomas, 373
Lumbosacral nerves, reconstruction of, 146–147, 146*f*

**M**

Machine-learning-based image segmentation techniques, 370
Machine learning (ML), 316, 423–424, 424*f*, 445*f*
  approaches, 316, 323
  intractography, 315
    considerations on, 338–339
    discrete imaging information, 316*f*
    spectrum of, 315
    suitable datasets for, 341
    validation, 339–340
Macrostructure, 613
Magnetic field strengths, harmonization within, 681–682
Magnetic resonance imaging (MRI), 397
  spin dephasing in, 103, 104*f*
  voxels, 487
Magnetic susceptibility, 159
MAGNEticTractography (MAGNET), 353–354
Magnetization transfer, 136–137
Magnetoencephalography (MEG), 49–50

Magniscan, 58
Magnitude of variability, 677
Mann-Whitney U test, 444
Mapping, by axons, 28
Markov Chain Monte Carlo (MCMC) sampling algorithms, 263
Markov Kennedy database, 551
Maslov-Sneppen (MS) rewiring procedure, 475
*b*-Matrix/*b*-tensor, 138
Matrix-variate gamma distributions, 184
Maximum entropy spherical deconvolution (MESD), 188
Maximum likelihood (ML), 165–166
Maxwell terms. *See* Concomitant gradients
Mean diffusivity (MD), 179, 180*f*, 583–584
Mean distance flip, 440–441
Measurement variability, 25
Medial default mode network, 614
Medial forebrain bundle (MFB), 638–639
Medial occipital longitudinal tract (MOLT), 410–412, 411*f*
Medical Image Computing and Computer Assisted Intervention (MICCAI) conference, 317
MEDUSA, 496
Mesh and tessellation-based approaches, 226
Mesoscale connectivity mapping, 656–658, 657*f*
Mesoscopic Bloch-Torrey equation, 82
Mesoscopic fiber tracking (MesoFT), 96, 305
Mesoscopic scale, in brain dMRI, 81–82, 81*f*
Metaanalytical approaches, 679
"Metabolic" functional connectivity, 50–51
Metabotropic receptors, 5
Meyer's loop, 673, 680
Meynert's rule, 11*b*
MI cost-function, 156–157
Microdissection
  freezing/thawing, 514–515
  limitations, 517
  procedure, 515, 516*f*
  3D microdissection, 515–517, 517*f*
Microfilaments, 27*b*
Microglia, 19
Microimaging, 656–658, 659*f*
Micrometer image resolution, 529–530
Microscopic diffusion process, 494–495
Microscopy, 659–661, 661*f*
Microstructural alterations, 617
Microstructural estimates, multicenter variability of, 682–683
Microstructure, 584–585, 613
Microstructure information, tractography with, 355–356
  filteringtractograms with, 355
  guiding trajectories using, 356
Microstructure informed tractography, 309–310
Microstructure modeling, 213–214
Microstructure scale, 82
Microstructure validation, measures for, 535
MicroTrack, 310
Microtubules, 27*b*
Middle temporal gyrus (MTg), 406, 407*f*
Mild cognitive impairment (MCI), 617–618

Minimum direct-flip (MDF) distance, 426–428, 431–432, 434
Min-max normalization, 223–224
MITK Diffusion, 319–320, 325
Model-based fiber responses, 191
Model-driven approaches, 213–215
  microstructuremodeling, 213–214
  pros and cons of, 214–215
Modern connectivity, 47–49
Modern neurosurgery, 631
Modifiable risk factors, 599–600, 600f
MOLT. *See* Medial occipital longitudinal tract (MOLT)
Monoexponential model, 68
Monte Carlo (MC) simulations, 261, 494–495
Motor evoked potential (MEP), 636
MR acquisition model, 490f, 491
Multiband (MB). *See* Simultaneous multislice (SMS)
Multicenter studies and harmonization, 669–670
  challenges, 682–684
    disease, 683
    fiber direction estimates and tractography, 683
    harmonization, evaluation of, 683–684
    longitudinal analysis, 683
    microstructural estimates, multicenter variability of, 682–683
  diffusion MRI measures, statistical approaches for harmonization, 678–680
  dMRI scanner differences, 678, 678f
  harmonization of DWI data, 680–682
    prospective DWI data harmonization with traveling subjects, 681–682
    retrospective DWI data harmonization, 680–681
  magnitude of variability, 677
  quantifying variability, 670
    connectivity matrices, 670
    scalar diffusion measures, 670
    tractography streamlines, 670
  variability, sources of, 671–677
    acquisition, 673–675
    intrinsic variability, 671–673, 672t
    processing, 675–677
    subject-related, 672–673
Multicenter variability, of microstructural estimates, 682–683
Multicompartment models, 167–168
Multicompartment spherical deconvolution, 208–210, 208f, 215
Multilayer network analysis, 475
Multilayer perceptron (MLP), 325
Multiple anisotropic compartments, 210–211, 210f
Multiple bundle labels, 334–336
Multiple RF-pulse, Carr-Purcell-Meiboom-Gill condition of, 654–655
Multiple sclerosis (MS), 364–365, 364f, 377, 439
Multishell acquisition schemes, 112
Multishell CHARMED procedure, 596
Multishell CSD approach, 332
Multishell data, 202–203, 203f
  forfiber tracking, 203–215

Multishell HARDI, 112, 202f
Multishell imaging and modeling, 215, 217f
Multishellmultitissue CSD (MSMT-CSD) algorithm, 191
Multishellmultitissue model, 305
Multistage prediction networks, 681
Multitensors, 183–185, 184f, 365
Multitissue or multicompartment (MT-CSD), 208, 209f
  with multiple anisotropic compartments, 210f
Multitissue spherical convolution
  with multiple $b$-tensor shapes, 216f
  with multiple $b$-values, 208f
Multivariate autoregressive (MAR) model, 357
Myelin, 27b
Myelinated axonal bundles, 9
Myelination, 33
Myelin birefringence, optical microscopy techniques on, 662
Myelin staining, 43–44
Myelin streamline decomposition (MySD), 311
Myelocentrism, 623

# N

NBS. *See* Network-based statistic (NBS)
Nearest neighbor (NN) segmentation, 426–427
Neocortex, 8
Neonatal diffusion imaging, 144–145
Nervous system, 4–5, 5f
Network-based statistic (NBS), 471, 471f
Network control theory, 475
Network neuroscience, 466–467
Neural network based local modeling, 325–326
Neural network-based local track direction predictions, 326–327
Neurite density index (NDI), 584, 587
Neurite orientation density and distribution imaging (NODDI), 228–230
Neurite orientation dispersion and density imaging, 584
Neuroanatomical dissections
  accumbofrontal pathway, 415–416, 416f
  anterior commissure (AC), 412, 413f
  anterior thalamic radiations, 408, 409f
  arcuate fasciculus (AF), 401, 406, 407f
  cingulum, 404, 405f
  corpus callosum (CC), 404, 405f, 412, 412f
  corticospinal tract (CST), 408, 409f
  fornix (Fx), 414, 414f
  frontal aslant tract (FAT), 406, 407f
  frontal insular tracts (FITs), 406–407, 408f
  frontomarginal and fronto-orbito-polar tracts, 412, 413f
  frontostriatal projections, 408–409, 409f
  geniculocalcarine fasciculus/optic radiation, 410, 411f
  inferiorfronto-occipital fasciculus (IFOF), 401, 405, 406f
  inferior longitudinal fasciculus (ILF), 401, 405, 406f
  medial occipital longitudinal tract (MOLT), 410–412, 411f
  pre- and postcentral U-shaped fibers, 410, 410f

superior longitudinal fasciculus (SLFs), 404, 404f
  uncinate fasciculus (UF), 401, 405, 406f
  vertical occipital fasciculus (VOF), 414–415, 415f
Neurochemical visualization, 521, 522f
Neurodevelopment, 615
Neurofilaments, 27b
Neuronal tracers, 662
  anterograde tracers, 518, 518f
  axonal branching/morphology, 519, 520f
  connectional weights and "numbers", 518–519
  heterogeneity of axon diameters, 521
  network analysis, 518
  neurochemical visualization (immunocytochemistry), 521, 522f
  reciprocity, 521
  resources, 521–523, 522–523t
  retrograde tracers, 518
  topography, 519–520
  tracer injections, 523–525, 524f
  weak (auxiliary) connections, 521
Neuronal tracing, 659–661, 661f
Neurons, 4
Neuro-oncology, 631–636
Neuroscience
  history of, 41
  tractography in, 565, 566f
Neurotransmitters, 5
Neurovascular, 639–640
Newton-Raphson method, 698, 700–701
NHPs. *See* Nonhuman primates (NHPs)
Nibabel, 435
Noise, 161–164, 162f
Noise maps, 163
Non-Gaussian diffusion MRI, 68, 69f
Nonhuman primates (NHPs), 143, 402, 511–512
Nonlinear least squares (NLS) approach, 179, 676
Normative modeling and machine learning, 445f
  voxel-based methods, 445
NOT gates, 242–243, 244f, 248f
nTMS-seededtractography, 633–634
Nuclear magnetic resonance (NMR), 56, 58f
  diffusion measurement, 81–86
Number of fiber directions (NuFiD), 232–234
Number of fiber orientations (NuFO), 231, 231f
Numerical dMRI phantoms, 491–494, 492t
Numerical Fiber Generator, 492
Numerical integration
  Adams-Bashforth method, 700–701
  definition, 697
  Euler method, 698–699
  explicit methods, 697
  implicit methods, 697
  order, 701
  Runge-Kutta (RK) method, 699–700
  single-step/multistep methods, 697
  stability, 702
  intractography, 702–703
  truncation error, 701
Numerical simulations, 489–497
Nyquist N/2 ghost, 132, 133f
Nystrom approximation, 429

# O

ODF. *See* Orientation distribution function (ODF)
Offset-Gaussian function, 162
Oligodendrocytes, 4, 33
Onco-functional balance, 632–633
One rejection class, 335–336
OpenIGLink, 370
Optical coherence tomography (OCT), 577
Optical imaging modalities, 659–661, 661f
Optical microscopy techniques, on myelin birefringence, 662
Optic radiation (OR), 410, 411f, 423
Orientation dispersion, 259–261, 260f
Orientation dispersion index (ODI), 584
Orientation distribution function (ODF), 92, 112, 185–186, 204f, 223, 382
   advanced processing of, 232–234, 233f
   continuous, 221
   diffusion, 223–224, 224f
   fiber, 225–226
   fiber dispersion and structural complexity, 228–231
   fiber-specific information and fixels, 227–228, 229f
   limitation of, 230, 231f
   localfiber orientation, 227
   representation, 226–227, 226f
   track, 232
   uncertainty, 232
Oriented receptor field, 47–48
Orthogonality, 712
Orthonormality, 712
Oscillating gradient spin echo (OGSE), 116
Outliers, 164–165
Overlapping sensory maps, 456
Overlapping tracks, differential maturation in, 589–590, 589f
Oxygen extraction fraction (OEF), 50

# P

Papez circuit, 617
Parahippocampal/hippocampal-cingulum white matter, 599
Parallel imaging, 553, 674–675
   EPI-based, 127–128, 128f
Parallel/longitudinal diffusivity, 583–584
Parallel transport tractography (PTT) algorithm, 265, 265f, 267f
Parameter estimation degeneracies, 95
Parent axon, 25b
Parietal-occipital sulcus (POS), 414–415
Parkinson's disease, 439
Partial differential equation (PDE), 711
Partial Fourier acquisition, 128–129, 129f
Path generation, 569–571, 570f
Pediatric diffusion imaging, 144–145
Pediatric neurosurgery, clinical tractography, 641–643, 642f
Percentage overlay maps, 403, 403f
Peripheral nervous system (PNS), 4–5
Peritumoraledematractography, 376f
Perpendicular/transverse diffusivity, 583–584

Persistent angular structure-MRI (PASMRI), 177, 187–188
Persistent angular structure (PAS), 187–188
Phantomas, 493, 504
   physical phantoms, 497–504
Phantom pathways, 249, 250f
Phosphate buffered saline (PBS), 144
Photometry, 439
Photorealistic rendering, 389, 389f
Physically implausible signals (PIS), 154, 154f
Physical phantoms
   anisotropic phantoms, for diffusion tractography, 498–504, 499–500t
   definition, 497
   isotropic liquids, 497–498
   isotropic phantoms, 497
Physics-Oriented Simulated Scanner for Understanding MRI (POSSUM), 493
Plain fiber phantoms, 498–501, 502–503f
Plasticity, 600–602
Point density model (PDM), 426
Polarization-sensitive optical coherence tomography (PSOCT), 662–663
Polarized light imaging (PLI), 25–26, 241, 527–528, 529f, 577
Polycaprolactone (PCL) microfibers, 501–504
Pontocerebellar pathways, 249, 250f
Positron emission tomography (PET), 50, 51f
POSSUM. *See* Physics-Oriented Simulated Scanner for Understanding MRI (POSSUM)
Postoperative cerebellar mutism syndrome (POPCMS), 641
Preclinical and ex vivo tractography, 653
   applications, 656–661
      diffusionmodeling and tractography approaches, comparison of, 659–661, 661f
      human brain, mesoscale connectivity mapping in, 656–658
      microimaging and preclinical applications, 656–658, 659f
      neuronal tracing, microscopy, and optical imaging modalities, 659–661, 661f
   technical considerations, 654–656
      acquisition pulse sequences, 654–655
      q-space sampling schemes, 655–656
Preclinical systems, 143
Pre-emphasis, Eddy current and, 112, 113f
Principal component analysis (PCA), 445–446
Probabilistic propagation and tracking, 261–262, 261f
Probabilistic SD-based tractography, 634f, 641
Probabilistic tissue maps, 349–351, 350f
Probabilistic tractography, 241, 257, 258f, 271, 458, 462, 468
   approaches, 658
   comparison of, 266–269
   curve bundle configuration, 266f
   dispersion-based, 264–265
   interpretation, 269–270
   synthetic data, 266–267
   uncertainty-based, 262–264
   in vivo data, 267–269, 268f

Probability density function (PDF), 204, 204f
Probability mass function sampling, 264
Probtrackx2, 266–267
Progressive intraoperative brain shift, 637f
Projection fibers, 46–47
Projection tracts, 9, 10f
Propagation, 241, 243–249
   error, 244–245, 249
   fiber pathway, 246
   orientation, 245
   tensorline, 245
Proportionality coefficient μ, 307
Pseudo-diffusion, 60–61
Pseudonormalization, 398–399
Pulsed field gradient (PFG) technique, 103, 117
Pulsed gradients, 83–86
Pulsed-gradient spin-echo (PGSE), 83–84, 84f, 104–105, 104f, 117
Pyramidal cells, 7, 7f
Pyramidal neurons, 22
Pyramidal tract (PyT), 545

# Q

q-BALL, 177
Q-ball imaging (QBI), 168, 185–187, 187f, 658
QBI with constant solid angle (QBI-CSA), 186
Q-space imaging, 105, 116, 204, 204f, 655–656
Quantitative anisotropy (QA), 223–224
QuickBundles (QB), 425f, 428, 428f, 493
QuickBundlesX (QBX), 428

# R

Radial diffusivity (RD), 583–584
Radiofrequency coils, 108–109
Radio frequency (RF), 138, 653
Radiological Society of North American (RSNA), 59
Ranvier nodes, 27b
Readout segmented EPI, 130
ReAl-LiFE, 468
Real-time tractography (RTT), 365, 367f, 386–387
Receptor-enriched analysis of functional connectivity by targets (REACT), 51
RecoBundles (RB), 431–432, 431f
Reduced sampling schemes
   parallel imaging, 127–128, 128f
   partial Fourier acquisition, 128–129, 129f
Region-based and bundle-specific tractography, 354
Regions of interest (ROIs), 242–243, 244f, 365, 371, 374f, 375–376, 384–385, 386f, 398, 424, 424f, 446–447
   accumbofrontal pathway, 415–416, 416f
   anatomical placement of, 401, 401f
   analysis, 669–670
   anterior commissure (AC), 412, 413f
   anterior thalamic radiations, 408, 409f
   arcuate fasciculus (AF), 401, 406, 407f
   cingulum, 404, 405f
   corpus callosum (CC), 404, 405f, 412, 412f
   corticospinal tract (CST), 408, 409f

delineation, optimal diffusion maps for, 398–399, 399f
fornix (Fx), 414, 414f
frontal aslant tract (FAT), 406, 407f
frontal insular tracts (FITs), 406–407, 408f
frontomarginal and fronto-orbito-polar tracts, 412, 413f
frontostriatal projections, 408–409, 409f
geniculocalcarine fasciculus/optic radiation, 410, 411f
inferiorfronto-occipital fasciculus (IFOF), 401, 405, 406f
inferior longitudinal fasciculus (ILF), 401, 405, 406f
medial occipital longitudinal tract (MOLT), 410–412, 411f
nonhuman primates (NHPs), anatomical delineation in, 402
pre- and postcentral U-shaped fibers, 410, 410f
seeding ROI, 242f, 246, 247f
superior longitudinal fasciculus (SLFs), 404, 404f
uncinate fasciculus (UF), 401, 405, 406f
vertical occipital fasciculus (VOF), 414–415, 415f
Region-to-region connectome matrices, 677
Reinforcement learning, deterministic-output RNN using, 331, 331f
Rejection sampling methods, 264
Relative fODF threshold, 270
Repetition time (TR), 673–674
Replacement, reduction, and refinement (3 Rs), 512–513
Representative streamline, 441–442
Response function estimation and factorization, 211–212
Restricted diffusion, 81
Restricted signal fraction, 583–584
Retrospective DWI data harmonization, 680–681
Reversible Jump Markov Chain Monte Carlo (RJMCMC), 304–306
Richardson-Lucy spherical deconvolution, 167, 193–194
Rician distribution, 162
Rician noise, 179, 491
Riemann-DTI paradigm, 289
  Hamiltonian formalism, 281–283
  Hamilton-Jacobi theory, 283–284
  heuristics, 276–279
  Lagrangian formalism, 279–281
  operationalization, 284–285, 285f
Riemann-Finsler-DWI paradigm, 281–282, 286–287
Riemannian metric, 276
Robust Estimation of Tensors by Outlier Rejection (RESTORE) algorithm, 164–165
Robust Extraction of Kurtosis INDices with Linear Estimation (REKINDLE) approach, 164–165
ROIs. *See* Regions of interest (ROIs)
Rotational invariance, 94–96

Rotation invariant spherical harmonics (RISH), 680
Row vector, 698
RTT. *See* Real-time tractography (RTT)
Runge-Kutta (RK) method, 244–245, 264–265, 699–700
Runge's phenomenon, 705–706

## S

Scattering light imaging (SLI), 528, 529f
Schizophrenia, 439
Second-order Runge-Kutta (RK2) integration scheme, 244–245
Seeding density, 246, 247f
Seeding mask, 348, 348f
Seeding ROI, 242f, 246, 247f
Seeding strategy, 458, 458f, 461–462, 573
Seeding technique, 440–441
Segmented diffusion sequences, 129–131, 130–131f
SENSE, 674–675
Shading techniques, 382, 383t
Shim, 138
Shim coils, 108
Short U-shaped fibers, 621–622
SHs. *See* Spherical harmonics (SHs)
SIFT2 method, 309
Signal drift, 153–154
Signal oscillations, 154, 154f
Signal-to-noise ratio (SNR), 179, 653–654, 705
Simulated annealing, 304–306
Simulated *FiberCup*, 317, 318f
Simultaneous multislab EPI (SMSb) approaches, 132
Simultaneous multislice (SMS), 131–132, 132f
Sindex (Signature) approach, 69f
Single-shell CSD, 208, 209f
Single-shell imaging, 215, 217f
Single-shell models, 195–196, 215, 217f
Single-shot diffusion sequences, 124–127, 124f, 126–127f
Singular value decomposition (SVD), 700
Skip filter, 402
Skull base, 640–641
Slew rate, 138
SLI. *See* Scattering light imaging (SLI)
*SlicerDMRI*, 336
Slicing techniques, 385–386, 386f
Small-angle X-ray scattering tensor tomography, 241
SNR. *See* Signal-to-noise ratio (SNR)
Social-emotional processing, 615, 616f
Socioemotional functions, linking behavior, 613–615
  social-emotional processing, frontal and limbic networks and impairments in, 615, 616f
  white matter tracts and socioemotional processing, 613–615, 614f
Socioemotional processing, white matter tracts and, 613–615, 614f
Soft parcellation approach, 456
Somatosensory motor cortex, 456, 457f
*k*-Space, 138

Spatial resolution, 137
Spatial resolution diffusion imaging, 603
Specificity trade-off, 549–550, 550f
Spectral clustering, 429
Spectral decomposition, 700
Spectral embedding, 424f
Spherical *b*-tensor encoding sequence diagram, 116, 116f
Spherical convolution, 207, 207f
Spherical-deconvolution informed filtering of tractograms (SIFT) method, 307–308, 468
Spherical deconvolution (SD) approaches, 166–167, 177, 188–194, 189f, 207–208, 241, 635, 635f
  constrained spherical deconvolution (CSD), 193
  fiber response function, 191–192, 192f
  multicompartment, 208–210, 208f, 215
  Richardson-Lucy spherical deconvolution, 193–194, 195f
  solving deconvolution problem, 189–191
Spherical harmonic residual network (SHResNet), 681
Spherical harmonics (SHs), 94, 182, 195–196, 224, 226–227, 680–681, 711
  as basis, 713
  definition and properties, 711–712, 712f
  intractography, 714–715, 714f
Spherical Radon transform. *See* Funk-Radon transform (FRT)
Spin dephasing, in MRI, 103, 104f
Spin-echo (SE) experiment, 673–674
Spin glass models, 304
Spin-history effects, 156
Spiny stellate cells, 7, 7f
Spline smoothing, 706–708, 709f
Spontaneous aneurysmal subarachnoid hemorrhage, 643
Steady-state free precession (SSFP), 144
Stejskal-Tanner technique, 56, 103, 673
Stimulated echo acquisition mode (STEAM) sequence, 116–117, 144, 145f, 147
Streamline-based linear registration (SLR), 431–432, 431f
Streamline-based segmentation methods, 435
Streamline classification, 333–338
Streamline counts, 461–463, 465, 467
Streamline filtering, 677
Streamline FT, 241, 243–246
  Euler's method, 243–244
  Runge-Kutta integration, 244–245
  tensor deflection (TEND), 245–246
Streamline propagation, 703f
  procedure, 702, 703f
Streamlines, 299, 424–425, 425f
  bundle atlas, 435
  distance functions, 425–427
  filtering methods, 306–308
  neuronal pathways, 300
  tractography, 298
  visualization, 381–384
Stroke, 375, 399–401, 400f
Structural connectivity, 12, 357

**726** Index

Structure tensor analysis (STA), 234, 526–527, 526*f*, 662
Student's t-test, 444
Study-dependent variability, 676–677
Subject motion
  correction, 155*f*, 156–157
  effects of, 155–156
Subject-related variability, 672–673
Superficial white matter fibers, 621
Superficial white matter tracking, 457, 458–459*f*, 459–460
Superior cerebellar peduncle, 641
Superior longitudinal fasciculus (SLFs), 373*f*, 404, 404*f*, 515, 545, 620–621
Superior temporal gyrus (STg), 406, 407*f*
Supervised methods, 429–434
  Classifyber, 433–434, 434*f*
  DeepWMA, 433
  linear assignment problem (LAP), 433
  RecoBundles (RB), 431–432, 431*f*
  TractQuerier, 430–431
  TractSeg, 432
  Tracula, 430
  XTRACT, 430
Supplementary motor area low grade glioma, 633–634, 633*f*
Supramarginal gyrus (SMg), 406, 407*f*
Surface-enhanced tractography (SET), 278, 322
Surface rendering, 387
Surgical planning, visualization, 633–634
Susceptibility-by-movement interaction, 156
Susceptibility distortions, 159
Susceptibility gradients, 159–161
Susceptibility-induced off-resonance field, 159
SVD. *See* Singular value decomposition (SVD)
Synapse, 5, 6*f*
System drift, 114–115

## T

Tangent bundle, 277*f*
Tangent space, 277*f*, 292
Task functional localizers, 453
TBSS. *See* Tract-based spatial statistics (TBSS)
Temporal lobe epilepsy (TLE), 638
Temporal transformations, 30–32, 31–32*f*
Temporo-amygdalo-orbitofrontal network, 614
Tensor deflection (TEND), 245–246
Tensorline propagation. *See* Tensor deflection (TEND)
Tensors
  definitions, 693–694
  dMRItractography, applications in, 695
  eigenvalues and eigenvectors, 695
  formalism, 181
  operations, 694
Tensor/scalar diffusion coefficient, 106
Tensor-valued diffusion encoding, 215
Termination thresholds, 246
  angle threshold, 246–247
  anisotropy threshold, 247–248, 248*f*
Texture-based tractography visualization, 383*t*, 384
Tikhonov regularization, 309

Time-based prospective memory reduction, 594–596
Tissue decomposition, 211, 211*f*
Tissue maps, tractography with, 347–352, 348*f*
Tissue model, 490, 490*f*
Tissue specific brain age approach, 596
Tissue volume fractions, 511–512
Top-down methods, 302
TOPUP approach, 160–161
Tracer injections, 523–525, 524*f*
Trace (TR), 179
Track density imaging (TDI), 384
Track orientation distribution (TOD), 232
Trackvis, 402, 434
Tract-based spatial statistics (TBSS), 439, 446, 618, 623
Tractogram, 423, 577
Tractography, 49
  aging, 590–596, 594*f*
    cognitive associations, 593–596
    contributions of, 593, 595*f*
  algorithm, 297–298, 423, 439–440
  applications of, 323–338, 323*f*
  approaches, 659–661, 661*f*
  atypical populations, 596–600
    cerebrovascular and Alzheimer's disease, 597–599, 598–599*f*
    developmental disorders, 596
    modifiable risk factors, 599–600, 600*f*
  brain development, 586–590
    contributions of, 587
    graph theory analysis, 590, 591*f*
    overlapping tracks, differential maturation in, 589–590, 589*f*
    white matter tracts, in utero development of, 587–589, 588*f*
  confounds, 586
  densities, 298
  dMRItractography community, challenges and opportunities for the, 576–577
  emergence of, 63–67
  with functional maps, 356–358
    functional connectivity, 357
    functional local orientation, 357–358
    functional regions, 356–357
  globalmodeling, 331–333
    CNN-based bundle modeling, 332–333, 332*f*
    rotation covariant tract estimation, 333
  hardware phantom, simulated, and ex vivo data, 317–320
    3D VoTEM phantom, 318, 318*f*
    *FiberCup*, 317
    HARDI 2013, 317–318
    ISMRM phantom, 319, 319*f*
    random-configurationFiberfox MRI simulations, 319, 320*f*
    99 simulated brains dataset, 319–320, 320*f*
    Simulated *FiberCup*, 317, 318*f*
  ill-posed inverse problem, 297, 297*f*
  inverse problem, 297
  localmodeling, 324–327
    neural network based local modeling, 325–326

    neural network-based local track direction predictions, 326–327
    random forest-based local modeling, 324–325, 324*f*, 325*t*
  measures, 583–584, 584*f*
  with microstructure information, 355–356
    filteringtractograms with, 355
    guiding trajectories using, 356
  ML-based tractography
    considerations on, 338–339
    suitable datasets for, 341
    validation, state of the art in, 339–340
  in neuroscience, 565, 566*f*
  within pathological brains, 363–371
  pathways and connectomes, 574–576
    bundle segmentation, 574–575
    effects on connectomics, 576
  plasticity, 600–602
    cognitive development, 601
    environmental factors, causal influence of, 602, 603*f*
    white matter development, 601
  plausibility and biological interpretability, 298, 298*f*
  robust to crossing fibers, 368–370
  sequence-basedmodeling, 327–331
    deterministic-output bundle-specific RNN, 329–330, 330*t*, 331*f*
    deterministic-output RNN, 327, 328*f*
    deterministic-output RNN using reinforcement learning, 331, 331*f*
    probabilistic-output RNN, 328–329, 329*f*
  sex differences, 596
  slicing, 385–386, 386*f*
  streamline, 670
  streamline classification, 333–338, 335*f*
    bundle prediction, 334–337
    label transcription approaches, 338, 338*f*
  study cohorts and design, 586
  with tissue maps, 347–352, 348*f*
    binary tissue masks, 348–349
    cortical surfaces, 351–352, 352*f*
    probabilistic tissue maps, 349–351, 350*f*
  training and validation data, 317–322
  trajectories, 298
  types of analysis, 584–586
  virtual reconstruction *versus* underlying anatomy, 565–574
    anatomical priors, 571, 572*f*
    bottleneck effect, 566, 567*f*
    deep nuclei, 571, 572*f*
    local signal challenges, 566–569
    narrow intersection effect, 568–569, 569*f*
    origin and termination, 571
    path generation challenge, 569–571, 570*f*
    and spatial coverage, 571–574
    wall effect, 567–568, 568*f*
  in vivo human datasets, 321–322
    HCP-minor bundle dataset, 321, 321*f*
    TractoInferno, 321–322
    *TractSeg*, 321, 322*f*

Tractography validation, 485
  acquisition, reconstruction, or tractography parameters, no "optimal" combination of, 554
  anatomical length scales, 485–486, 486f
  anatomical models for (see Anatomical model systems)
  anatomy, 543–546
  bridging anatomy and neuroimaging, 487–489, 488f
  bridging anatomy and tractography, 556–557
  empirical validations, 530–533
  inherent sensitivity/specificity trade-off, 549–550, 550f
  as iterative process, 557
  measures for, 533–535, 533f
  methods, 551–553, 553f
  neuroimaging length scales, 486f, 487
  numerical simulations, 489–497
  obstacles, biases and challenges, 554–556
  physical phantoms, 497–504
  reconstruction techniques, fiber orientation distribution, 546–547, 547f
  steps in process and ambiguity, 487, 488f
  WM bundles, valid path, shape, and position of, 547–549, 548f
Tractography visualization
  along-tract mapping and profiling, 388
  computer graphics, employed in, 383t, 383f
  immersivevirtuality, 389
  interactivetractography visualizations, 384–388
  machine learning approaches, 390
  photorealistic rendering, 389, 389f
  scientific visualization, 381–384
  types of, 382, 383f
  uncertainty visualization, 388–389
  web-based visualization, 388
  with and without ambient occlusion, 384, 385f
TractoInferno, 321–322
Tractometry, 138, 388, 585–586
Tract orientation maps (TOMs), 332
TractQuerier, 430–431, 493
TRActs Constrained by UnderLying Anatomy (TRACULA), 584–585
TractSeg, 319–321, 322f, 432, 435, 493
Tracula, 430
TRAFIC, 335
Transformation model, 156
Transparency, 382
Traumatic brain injury (TBI), 375, 439, 643
Twice-refocused spin echo (TRSE), 107, 112–114
Two-dimensional light microscopy, 546–547

## U

UK Biobank, 96
Uncertainty-based probabilistic tractography, 262
  Bayesian inference, 263
  bootstrapping, 262–263
  confidence intervals, 269–270, 269f
  synthetic curve configuration, 267f
Uncertainty orientation distribution function (uODF), 222–223, 232, 259, 259f
Uncertainty visualization, 388–389
Uncinate fasciculus (UF), 401, 405, 406f, 431–432, 440–441, 446, 571, 589, 617
Undulation scale, 90–91
Uniform color scheme, 388
Unilateral spatial neglect (USN), 622–623
  frontoparietal white matter tracts in, 622–623
Up-down oriented corticopontinefiber pathways, 249, 250f
U-shaped fibers, 10, 11b, 26, 410, 410f

## V

Valid bundles (VB), 323
Valid connections (VC), 323
Variability, magnitude of, 677
Varifolds distance, 426
Vector, 425
  definitions, 698–699
  notations, 700–701
  operations, 692t, 699–700, 703f
Vector space, 289–290
Ventral cingulum, 617
Ventricular system, 41, 42f
Ventricular theory, 41
Vertical occipital fasciculus (VOF), 414–415, 415f
Vibration artifacts, 115, 115f
Virtual dissections, 435–436
Virtual MR Elastography, 68–69, 69f
Virtual reality (VR), 389
Visuospatial functions, 622–623
  frontoparietal white matter tracts and, 622
Voxel-based analysis (VBA), 439, 446
Voxel-based morphometry (VBM), 439, 618
Voxel-wise density, 303b

## W

Wall effect, 567–568, 568f
Watson distribution, 229–230b
Web-based tractography visualization, 388
Weight-based thresholding, 466–467
Weighted-Dice coefficient, 434
Weighted linear least squares (WLLS), 165, 179, 676

Welch's t-test, 444
Wernicke's temporal area, 619
Wernicke's territory, Geschwind's territory with, 619
White Matter Query Language (WMQL), 430
White matter (WM), 5f, 19, 25–26, 241, 423, 511–512, 515
  anatomically guided manual dissections, 398–401
  anatomy, 9–11, 10f
  anisotropic diffusion in, 177–178
  architecture, 347
  atlas-based WM masking, 370–371
  bundles, valid path, shape, and position of, 547–549, 548f
  concentric zones, 11b
  cylinder-based substrates, generative algorithms to, 495–496, 495f
  development, 601
  dMRI signal, 92–93
  edema, 375–376, 375f
  fiber pathway, 243
  Fourier tracts (FouTS), 306
  fractal dimension (FD), 435
  lesions, 373
  microstructure, 594–596
  modeling and tracking, 363
  neuroanatomical dissections, 404–416
  nonhuman primates, anatomical delineation in, 402
  organization, anatomical information about, 311–312, 311f
  organization of, 11f
  realistic synthetic substrates
    3D microscopy data of, 497
    generative algorithms to, 495f, 496
  segmentation, 12
  standard model of diffusion in, 93–94, 93f
  statistical indices, 402–403, 403f
  streamlines, 424–427, 425f
  TractQuerier, 430–431
  virtual white matter bundle dissection, 371
White matter tracts
  and cognitive control, 617–618
  protracted developmental time course of, 615
  and socioemotional processing, 613–615, 614f
  in utero development of, 587–589, 588f
Whole-brain tractography, 241, 329
Wiener measure, 83np
Wilcoxon rank test, 444
WM-GM interface, 349
  seeding strategies, 573

## X

XTRACT, 430

Printed in the United States
by Baker & Taylor Publisher Services